HEINZ MEHLHORN (ED.)

Encyclopedic Reference of Parasitology

Second Edition

Diseases, Treatment, Therapy

With contributions by
P.M. ARMSTRONG H. ASPÖCK R.P. BAUGHMAN C. BEHR C. COMBES
J. DE BONT J.F. DUBREMETZ J. FREEMAN J.K. FRENKEL A. GESSNER
M. GUSTAFSSON W. HAAS D.W. HALTON H. HÄNEL O. HANSEN
A. HARDER E.S. KANESHIRO P. KÖHLER M. LONDERSHAUSEN
U. MACKENSTEDT H. MEHLHORN L.H. PEREIRA DA SILVA W. RAETHER
I. REITER-OWONA D. RICHTER M. RÖLLINGHOFF G. SCHAUB T. SCHNIEDER
H.M. SEITZ A. SPIELMAN K.D. SPINDLER H. TARASCHEWSKI A.G.M. TIELENS
A. TURBERG J. VERCRUYSSE W.P. VOIGT V. WALLDORF W.H. WERNSDORFER

With 126 Figures and 107 Tables

Springer

Editor:
Professor Dr. Heinz Mehlhorn
Heinrich-Heine-Universität
Institut für Zoomorphologie, Zellbiologie
und Parasitologie
Universitätsstraße 1
40225 Düsseldorf

ISBN 3-540-66829-2 Springer-Verlag Berlin Heidelberg New York

Die Deutsche Bibliothek – CIP-Einheitsaufnahme
Encyclopedic reference of parasitology [Medienkombination] / Heinz Mehlhorn (ed.). – Berlin ; Heidelberg : Springer
ISBN 3-540-66239-1

Buch.
Biology, structure, function. – 2. ed. – 2001
ISBN 3-540-66819-5

0101 deutsche buecherei
Encyclopedic reference of parasitology [Medienkombination] / Heinz Mehlhorn (ed.). – Berlin ; Heidelberg : Springer
ISBN 3-540-66239-1

Buch.
Diseases, treatment, therapy. – 2. ed. – 2001
ISBN 3-540-66829-2

0101 deutsche buecherei
Encyclopedic reference of parasitology [Medienkombination] / Heinz Mehlhorn (ed.). – Berlin ; Heidelberg : Springer
ISBN 3-540-66239-1

CD-ROM. Encyclopedic reference of parasitology. – 2. ed. – 2001
ISBN 3-540-14825-6

http://www.springer.de

Springer-Verlag Berlin Heidelberg 2001
Printed in Germany

Typesetting: medio Technologies AG, Berlin
Cover design: design & production GmbH, Heidelberg

Printed on acid-free paper SPIN: 10750673 14/3130/ag 5 4 3 2 1 0

Encyclopedic Reference of Parasitology

Springer
Berlin
Heidelberg
New York
Barcelona
Hong Kong
London
Milan
Paris
Singapore
Tokyo

Preface to the Second Edition

During the last decade since the appearance of the first edition, parasitology has made considerable advances and many new diseases have cast their shadows on life, on the welfare of man and his domestic animals. Although worldwide billions of dollars have been spent, in many fields the available control measurements are far from being satisfactory and in others they are completely lacking. The present status is characterized by mild optimism with regard to vaccines, an increasing resistance of parasites towards drugs and by the absence of measurements against several parasitoses on the one hand and decreasing funds for research on the other. Nevertheless parasitological research in total was very successful during the last decade, and new methods revealed new, hopefully promising, insights into the fight "*parasite against host*". Thus this second edition intends to give a comprehensive review of the facts and trends in veterinarian and human parasitology, through contributions from distinguished specialists. In addition to the authors of the first edition, several internationally renowned colleagues joined our crew to help to reach a broad readership. The layout of the book has changed to an encyclopedic arrangement of rather comprehensive key words (chapters) along the subject index being included in the flow of the text. This will help to speed up the search for information when the book is used either in its present shape, as a compact disc or in a future electronic on-line version. The printed version is split into two volumes covering different head topics: Volume 1 – All biological aspects; Volume 2 – all clinical, pathological and therapeutical aspects.

We hope that this second edition will be as well accepted as the first one and will be able to contribute to our common goals in the struggle for a better life.

Düsseldorf, September 2000

For the authors
Prof. Dr. Heinz Mehlhorn (Editor)
Heinrich-Heine-Universität
Düsseldorf, Germany

Preface to the First Edition

Although in recent decades many methods have been developed to control parasitic diseases of humans and animals, chemoresistance and reduction of budgets for control have caused the problems to incease worldwide. Efforts in the "*struggle against parasites*" must be redoubled if we are not to become overwhelmed by human health problems and problems of food production. This absolute need has led to the application of various new methods to classical parasitology. Thus the different fields of parasitological research are at present expanding so rapidly that it is impossible for an individual to follow the main problems and to evaluate and recognize recent progress.

The purpose of this book is to give a comprehensive review of the facts and trends in veterinary and human parasitology, through contributions from distinguished specialists in different fields. The authors have focused their contributions on the most important and promising results, in a way which it is hoped will inform students, teachers, and researchers (zoologists, veterinarians, physicians) about those topics, which may be far from their own working fields, but knowledge of which may be necessary to develop new ideas. Thus, all chapters, the length of which will surely change in future editions, are provided with references opening the literary entrance to each field of research.

We hope that the book will be fruitful and lead to the establishment of new ideas, trends, and techniques in the struggle against parasites.

Bochum, January 1988

For the authors
PROF. DR. H. MEHLHORN (EDITOR)
Ruhr-Universität Bochum, FRG

Acknowledgements

No one could write a book such as this without the help of many people, including our close coworkers. Their material and comments were helpful while selecting and preparing the contributions to this book. We are especially grateful to those colleagues who contributed one or several micrographs:

- Prof. Dr. G. Brugerolle, Clermont-Ferrand
- Prof. Dr. J.F. DeJonckheere, Brussels
- Prof. Dr. I. Desportes, Paris
- Dr. W. Franz, Münster
- Prof. Dr. J. Grüntzig, Düsseldorf
- Prof. Dr. I. Ishii, Japan
- Prof. Dr. K. Hausmann, Berlin
- Prof. Dr. M. Køie, Kopenhagen
- B. Mehlhorn, Neuss
- Prof. Dr. W. Peters, Düsseldorf
- PD Dr. G. Schmahl, Düsseldorf
- Dr. J. Schmidt, Düsseldorf
- Prof. Dr. J. Schrével, Paris
- Prof. Dr. Y. Yoshida, Kyoto

All other micrographs are either from the authors of the particular chapter or from the editor.

The editor and the authors would like to thank Mrs. K. Aldenhoven and Miss Y. Müller for carefully typing large parts of the manuscript, Mrs. H. Horn and Mr. S. Köhler for their excellent preparation of the micrographs, and Mrs. B. Mehlhorn for correcting the proofs. We are in debt to Prof. Dr. Lehmacher (Köln) for proof reading of the chapter of the late Prof. J. Freeman (Harvard University). The beautiful hand drawings were produced by the late Fried Theissen (Essen) and Dr. Volker Walldorf (Düsseldorf).

Furthermore we would like to thank the publishers, especially Dr. R. Lange, Mrs. H. Wilbertz, Mr. M. Schirmer, Dr. W. Reuss and Mr. A. Gösling (at the Publisher, Springer-Verlag Heidelberg), for their cooperation and generous support of our efforts to produce an optimum outline of parasitology.

The Authors

Contents

Biology, Structure, Function

The 1592 entries of this volume mainly cover the following topics:

- Arboviruses
- Behaviour
- Classification
- Ecological aspects
- Habitat selection
- History
- Hormones
- Host finding
- Host-parasite interface
- Immune diagnostic methods
- Invasion procedures
- Live cycles
- Metabolism
- Molecular systematics
- Morphology
- Motility
- Nerve functions
- Nutrition
- Penetration
- Phylogenic aspects
- Reproduction
- Serology
- Systematics
- Transmitted pathogens
- Ultrastructure
- Vector biology

Diseases, Treatment, Therapy

The 565 entries mainly covering the following topics:

- Acaricides
- Chemotherapy
- Clinical symptoms
- Control measurements
- Diagnostics
- Diseases of animals
- Diseases of man
- Drug actions
- Drug resistance
- Drug availability
- Epidemiology
- Host reactions
- Immune reactions
- Insecticides
- Opportunistic agents
- Pathology
- Prophylaxis
- Serology
- Strategies
- Tick transmitted diseases
- Vaccination trials
- Vaccines
- Vector transmitted diseases
- Zoonoses

Main Topics and Contributors

- Acanthocephala (Taraschewski)
- Antibodies (Seitz and Reiter-Owona)
- Arboviruses (Aspöck)
- Behaviour (Taraschewski)
- Cell Penetration (Dubremetz)
- Chemotherapy against helminthoses (Raether and Harder)
- Chemotherapy against protozoan diseases (Raether and Hänel)
- Classification (Mehlhorn)
- Clinical and pathological signs of parasitic infections in domestic animals (Vercruysse and de Bont)
- Clinical and pathological signs of parasitic infections in man (Frenkel, Mehlhorn)
- Connecting entries (Mehlhorn)
- Drug action in ectoparasites (Turberg and Londershausen)
- Drug action in Protozoa and helminths (Harder)
- Drug Tables (Raether)
- Ecological aspects (Combes)
- Ectoparasitizides (Londershausen and Hansen)
- Environmental aspects (Combes)
- Epidemiological aspects (Wernsdorfer)
- Eye parasites (Mehlhorn)
- Fine structure of parasites (Mehlhorn)
- Hormones (Spindler)
- Host finding mechanisms (Haas)
- Host parasite interface (Dubremetz, Mehlhorn)
- Immunodiagnostic methods (Seitz and Reiter-Owona)
- Immunological responses of the host (Gessner and Röllinghoff)
- Insects as vectors (Schaub)
- Life cycles (Mehlhorn and Walldorf)
- Lyme disease (Spielman and Armstrong)
- Mathematical Models (Freeman)
- Metabolism (Köhler and Tielens)
- Molecular systematics (Mackenstedt)
- Morphology (Mehlhorn)
- Motility (Dubremetz, Mehlhorn)
- Nerves-structures and functions (Gustafsson and Halton)
- Nutrition (Köhler and Tielens)
- Opportunistic agents (except *Pneumocystis*, Mehlhorn)
- Pathologic effects in animals (Vercruysse and de Bont)
- Pathologic effects in humans (Frenkel)
- Pathology (Frenkel)
- Pentastomida (Walldorf)

- Phylogeny (Mackenstedt)
- Physiological aspects (Köhler and Tielens)
- Planning of control (Wernsdorfer)
- *Pneumocystis carinii* (Baughman and Kaneshiro)
- Reproduction (Mehlhorn)
- Resistance against drugs (Harder)
- Serology (Seitz and Reiter-Owona)
- Strategy of control measurements (Wernsdorfer)
- Ticks as vectors in animals (Voigt from 1st ed., Mehlhorn)
- Ticks as vectors in humans (Spielman and Armstrong)
- Ultrastructure (Mehlhorn)
- Vaccination
  - Protozoa (Behr and Pereira da Silva)
  - Plathelminthes (Richter)
  - Nemathelminthes (Schnieder)
- Vector biology
  - Insects (Schaub, Mehlhorn)
  - Ticks (Spielman, Mehlhorn, Voigt)

All these topics are presented in either a single, long entry, in several smaller, separate entries and/or as inserts in other longer entries. This cooperation of specialists contributes to a better understanding of the recent complex problems in parasitology.

List of Contributors

ARMSTRONG, Philip M, Dr.
Harvard School of Public Health. Now: Southwest Foundation for Biomedical Research, Department of Virology and Immunology, P.O. Box 760549, San Antonio, TX 78245-0549, USA

ASPÖCK, HORST, Prof. Dr.
Institut für medizinische Parasitologie, Universität Wien, Kinderspitalgasse 15, 1095 Wien, Austria

BAUGHMAN, Robert P., Prof. Dr.
Department of Internal Medicine, University of Cincinnati, College of Medicine, P.O.Box 670564, Cincinnati, OH 45220-5220, USA

BEHR, Charlotte, Dr.
Unité d'Immunologie Moléculaire des Parasites, Institut Pasteur, 25 Rue du Dr. Roux, 75724 Paris Cedex 15, France

COMBES, Claude, Prof. Dr.
Centre de Biologie et d'Écologie Tropicale et Méditerranéenne, Université de Perpignan, 66860 Perpignan Cedex, France

DE BONT, Jan , Prof. Dr.
Department of Virology - Parasitology - Immunology, Faculty of Veterinary Medicine, Laboratory of Veterinary Parasitology, University of Gent, Salisburylaan 133, 9820 Merelbeke, Belgium

DUBREMETZ, Jean François, Dr.
Institut de Biologie, Institut Pasteur de Lille, 1 Rue Calmette, 59019 Lille Cedex, France

FREEMAN, Jonathan, Prof. Dr. (died in 2000)
Department Tropical Public Health, Harvard University, School of Public Health, 665, Huntington Avenue, Boston, MA 02115, USA

FRENKEL, Jack K., Prof. Dr. (retired from University of Kansas City, Kansas)
1252 Vallecit A Drive, Santa Fe, NM 87501-8803, USA

GESSNER, André, Priv. Doz. Dr. Dr.
Institut für Klinische Mikrobiologie und Immunologie, Universität Erlangen-Nürnberg, Wasserturmstraße 3, 91054 Erlangen, Germany

Gustafsson, Margaretha, Prof. Dr.
Åbo Akademie University, Deparment of Biology, Biocity, Artillerigatan 6,
20520 Åbo, Finnland

Haas, Wilfried, Prof. Dr.
Institut für Zoologie I, Universität Erlangen, Staudtstrasse 5, D-91058 Erlangen,
Germany

Halton, David W., Prof. Dr.
School of Biology and Biochemistry, Medical Biology Centre, The Queen's University
of Belfast, Belfast BT7 1NN, Northern Ireland, UK

Hänel, Heinz, Dr.
Aventis AG, Global Purchasing, Poseidon House, 65926 Frankfurt, Germany

Hansen, Olaf, Dr.
Bayer AG, GB Tiergesundheit/Forschung, Institut für Parasitologie, Gebäude 6700,
51368 Leverkusen, Germany

Harder, Achim, Priv. Doz. Dr.
Institut für Parasitologie, Bayer AG, 51368 Leverkusen, Germany

Kaneshiro, Edna S., Prof. Dr.
Department of Biological Sciences, University of Cincinnati, P.O.Box 210006,
Cincinnati, OH 45221-0006, USA

Köhler, Peter, Prof. Dr.
Institut für Parasitologie, Universität Zürich, Winterthurerstrasse 266a, 8057 Zürich,
Switzerland

Londershausen, Michael, Prof. Dr.
Institut für Parasitologie, Bayer AG, 51368 Leverkusen, Germany

Mackenstedt, Ute, Prof. Dr.
Institut für Parasitologie, Universität Hohenheim, Emil-Wolff-Strasse 34,
70593 Stuttgart, Germany

Mehlhorn, Heinz, Prof. Dr.
Institut für Zoomorphologie, Zellbiologie und Parasitologie, Heinrich-Heine-Universität, Universitätsstrasse 1, 40225 Düsseldorf, Germany

Pereira da Silva, Luiz Hildebrando, Prof. Dr.
Parasitologie expérimentale, Institut Pasteur Paris, 25 Rue du Docteur Roux,
75724 Paris Cedex 15, France and Centro de Pesquisas em Medicina Tropical, Secretaria de Saúde do Estado de Rondônia, Rodovia BR 364, Km 4.5, 7870-900 Ronônia,
Brasil

Raether, Wolfgang, Prof. Dr. (retired from Hoechst AG Chemotherapy Department)
Freigasse 3, 63303 Dreieich, Germany

Reiter-Owona, Ingrid, Dr.
Institut für medizinische Parasitologie, Universität Bonn, Sigmund-Freud-Strasse 25, 53008 Bonn, Germany

Richter, Dania, Dr.
Department of Tropical Public Health, Harvard School of Public Health, 655, Huntington Avenue, Boston, MA 02115, USA; now: Institut für Parasitologie der Humboldt Universität, Malteserstr. 74–100, 12249 Berlin, Germany

Röllinghoff, Martin, Prof. Dr.
Institut für Klinische Mikrobiologie und Immunologie, Universität Erlangen-Nürnberg, Wasserturmstraße 3, 91054 Erlangen, Germany

Schaub, Günter, Prof. Dr.
Institut für Spezielle Zoologie, Ruhr Universität Bochum, Unversitätsstrasse 150, 44780 Bochum, Germany

Schnieder, Thomas, Prof. Dr.
Institut für Parasitologie, Tierärztliche Hochschule Hannover, Bünteweg 17, 30559 Hannover, Germany

Seitz, Hans Martin, Prof. Dr.
Institut für medizinische Parasitologie, Universität Bonn, Sigmund-Freud-Strasse 25, 53008 Bonn, Germany

Spielman, Andrew, Prof. Dr.
Department of Tropical Health, Harvard School of Public Health, 665, Huntington Avenue, Boston, MA 02115, USA

Spindler, Klaus Dieter, Prof. Dr.
Institut für Allgemeine Zoologie, Universität Ulm, Albert-Einstein-Allee 1, 89069 Ulm, Germany

Taraschewski, Horst, Prof. Dr.
Zoologisches Institut, TH Karlsruhe, Kaiserstrasse 12, 76128 Karlsruhe, Germany

Tielens, A.G.M., Prof. Dr.
Department of Basic Sciences, Division of Biochemistry, Faculty of Veterinary Medicine, University of Utrecht, P.O. Box 80176, 3508 TD Utrecht, Netherlands

Turberg, Andreas, Dr.
Bayer AG, GB Tiergesundheit/Forschung, Institut für Parasitologie, Gebäude 6700, 51368 Leverkusen, Germany

Vercruysse, Joseph, Prof. Dr.
Department of Virology – Parasitology – Immunology, Faculty of Veterinary Medicine, Laboratory of Veterinary Parasitology, University of Gent, Salisburylaan 133, 9820 Merelbeke, Belgium

Voigt, Wolf P., Prof. Dr.
BGVV, Diedersdorfer Weg 1, 12277 Berlin, Germany

WALLDORF, Volker, Dr.
Institut für Zoomorphologie, Zellbiologie und Parasitologie, Heinrich-Heine-Universität Düsseldorf, Universitätsstrasse 1, 40225 Düsseldorf, Germany

WERNSDORFER, Walter H., Prof. Dr.
Institute of Specific Prophylaxis and Tropical Medicine, University of Wien, Kinderspitalgasse 15, 1095 Wien, Austria; former member of WHO

Introduction

Starting from the early beginnings of human culture, man became aware of parasites. In animals, which developed social contacts via coat-lousing, humans noted first the crucial activities of large amounts of ectoparasites such as ticks, lice, fleas, mosquitoes, flies, etc., as is shown in the earliest written reports of mankind. Furthermore, those endoparasitic worms that occurred in feces in larger numbers and were big enough to be seen with the naked eye were known. Thus the physicians of the Egyptians ($\sim$2000 BC), the Greek physician Hippocrates (460 – 370 BC) and the natural scientist Aristoteles (384 – 322 BC) knew very well ascarids, oxyurids, and of course tapeworms. Their knowledge was passed on to the Romans, who called the round worms *lumbrici teretes* and the plathyhelminths *lumbrici lati*, and from there it became transmitted to later human societies especially by propagation of manuscripts in Christian cloisters or by translations of Greek books that were being used and preserved by physicians in the Near East.

However, only a few remedies were available apart from combing (Fig. 1), catching of parasites (Fig. 2), bathing in water and/or hot sand or eating special plants or spicy

Fig. 1. Redrawn reproduction of a medieval engraving showing a housewife delousing her husband with a comb-like instrument.

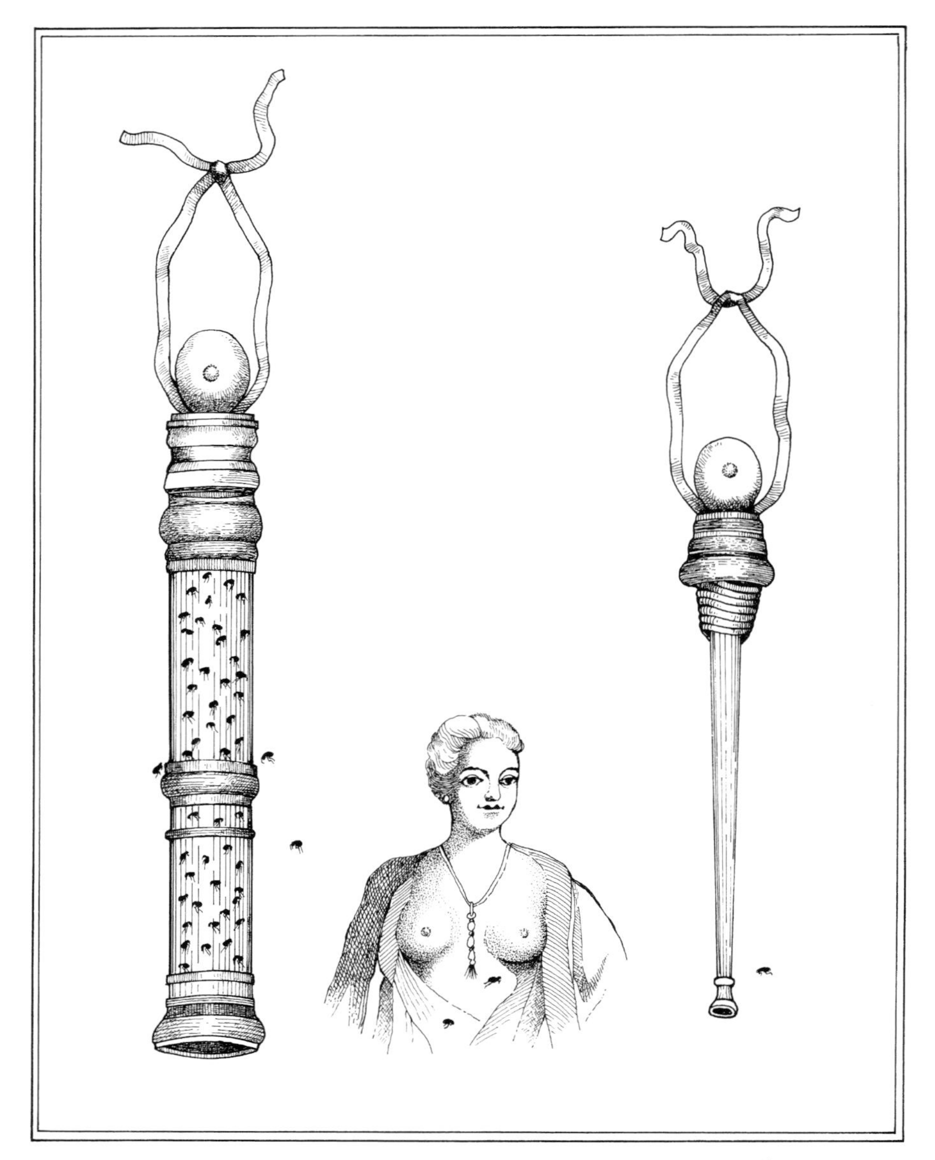

Fig. 2. Redrawn reproduction of a figure from a German book of the 18th century showing two types of lady's necklace used as glooming flea-catcher.

food, which were felt to decrease intestinal worm populations, as, for example, pepper does (Fig. 3). Thus the highly sophisticated physicians of the ancient Egyptian kingdoms surely did know the fatal symptoms of the schistosome-derived diseases, but the transmission pathways and methods of treatment were as nebulous as they were 3000 years later when the Holy Hildegard of Bingen (1098–1179) recommended that worms be treated with, for example, extracts of stinging nettles, dandelions and walnut-tree leaves, as described in her book "*Physica*" (1150–1160) – chapter "*De causis et curis morborum*" (i.e. "On the causes and cures of diseases"). The treatment of dracunculosis by removal of the whole worm from human skin was, however, much more successful. The use of a wooden splinter, onto which this so-called Medina-worm was wound by physicians in the Near East, probably gave rise to the Aesculap-stick of our days – the symbol of an increasingly successful caste – although it is

Fig. 3. Redrawn reproduction of an ambulant Renaissance pharmacist equipped with his main helper plants and therapeutical animals, including snakes and leeches.

not long ago that cupping and/or the use of leeches were universal remedies (Fig. 4, Fig. 5). At the end of the Middle Age, a new interest arose among educated people to study the natural world, and this newly awakened curiosity led people to make detailed investigations of plants and animals. Even human beings were a subject of investigation, provided religion did not prevent this (e.g. dissections of humans – even of executed and thus lawless people – were forbidden for centuries in Christian and Moslem countries. Thus at first, descriptions of the outer morphology of plants, animals and humans became available and later – after the development of microscopical techniques – structural ground-plans and histological insights into organisms were obtained. However, it was not until the middle of the 19th century that the theory of "de-novo creation", (latin: *generatio aequivoca et spontanea*), the creation of organisms from dead or anorganic material (e.g. worms develop from intestinal slime) became replaced by the idea of cellular organisation and the self-reproduction of organisms as postulated in Virchows thesis (1858): "*omnis cellula e cellula*" – each cell derives from a cell. This growing spirit of investigation led to the discovery of numerous species of plants and animals and to the differentiation into pro- and eukaryotic organization of organisms. The knowledge derived from the cell-dependent life of viruses or prions is a fruit of our century. According to their morphology and life cycles – the study of which is not completed even today – species of bacteria, fungi, plants and animals were characterized and ***systematical classifications*** and ***phylogenetic trees*** were established. Such investigations provided a basis for the establishment of phylogenetical theories such as those of Lamarck or Darwin.

Moreover, most of the species of parasites still valid today were described in those times (cf. ***Historical Landmarks***) and the term ***parasite*** (greek: *parasitos* = eaters at the court = meal taster) became fixed as the word to describe those organisms that

Fig. 4. Redrawn reproduction of a Baroque noble using cupping-glasses in order to be bled.

Fig. 5. Redrawn reproduction of a medieval engraving demonstrating the therapeutic use of the leech, *Hirudo medicinalis*, even in middle-class-households.

live on other animals or humans. According to the different life-cycle adaptations the latter may become:

- ***final (definite) hosts*** lodging the sexual stages of the parasite,
- ***intermediate hosts*** lodging asexually reproducing stages of the parasite,
- ***transitory/accidental/paratenic hosts*** lodging parasitic stages without further reproduction or
- ***vectors*** representing blood sucking parasites such as arthropods or leeches which transmit other pathogens and/or parasites during their blood meal.

The constant refinement of microscopical techniques (including the establishment of electron microscopy) and the development of a broad spectrum of molecular biological methods led (especially in the last 30 years) to an explosion of the knowledge on the organization of the parasites, on the parasite-host interface and on host immune reactions, which altogether were used to establish control measurements and to develop prophylactic strategies, drugs and/or vaccines. Thus the two volumes of the book presented here are based on the following pillars:

- life cycles (inclusive behavior and epidemiology),
- morphology (up to molecular insights),
- mechanisms of reproduction,
- metabolism and nutrition,
- host-parasite interactions,
- diseases and pathological effects,
- immune reactions,
- control measurements (inclusive drugs, vaccines, prophylactic strategies).

The key words of both volumes also intend to outline interactions with many other fields of interest and importance (see above: Contents and Topics).

A

Abdominal Pain

Clinical symptom in animals due to parasitic infections (→ Alimentary System Diseases, Animals, → Clinical Pathology, Animals)

Abortion

Premature expulsion of an embryo or a nonviable fetus caused by parasitic infections e.g. with *Toxoplasma gondii*, *Neospora caninum* (syn. *Hammondia heydorni*), *Trypanosoma* species, *Tritrichomonas foetus*.

Abscess

→ Pathology.

Acanthamoebiasis

→ *Acanthamoeba* (Vol. 1) spp. have been found in the throat; mouth pipetting of fluids into cell culture has given rise to contaminated cell cultures in numerous instances. In the throat the → amoebae (Vol. 1) appear to be nonpathogenic. However, in patients with longstanding immunosuppression tissue invasion does occur, usually leading to encephalitis which is fatal and occasionally to focal lesions elsewhere. The inflammation is mononuclear; partially in response to necrosis of brain tissue and is sometimes stated to be granulomatous; hemorrhage may be marked. Large amoebic trophozoites and smaller cysts with an irregular "corrugated" wall are found in the lesions. Often the amoebae are difficult to distinguish from macrophages; the latter have intensely staining nuclei, whereas the amoebae have vesicular nuclei and a "foamy" cytoplasm (→ Pathology/Fig. 4F). The inflammatory reaction is of course variable because of immunosuppression of the patients. The amoebae are not found in the spinal fluid. Patients with the Acquired Immunodeficiency Syndrome (AIDS) showed invasion of the nasopharynx with *Acanthamoeba* spp. Other sites of involvement by *Acanthamoeba* spp. are the cornea, skin, and lung; especially in the eyes *Acanthamoeba* stages are rather common in persons using plastic lenses. Thus these species are not only → opportunistic agents (Vol. 1).

Main clinical symptoms: Chronic brain disturbances, possibly granulomatous encephalitis (GAE); eye: conjunctivitis, keratitis, uveitis
Incubation period: 1 day 2 weeks
Prepatent period: 1 day 2 weeks
Patent period: Weeks to months in chronic cases
Diagnosis: Culture techniques
Prophylaxis: Do not swim in eutrophic lakes; change fluids for contact lenses often
Therapy: see → Treatment of Opportunistic Agents

Acanthocephalacidal Drugs

General Information

The large to medium sized → acanthocephalans (Vol. 1) are thorny (spiny)-headed worms with an elongated proboscis armed with recurved hooks parasitizing the digestive tract of a wide range of vertebrate animals throughout the world, and occasionally are found in humans. They have been placed in their own phylum since their affinities to other parasites are not well defined. The sexes are separate, males being much smaller than females. The life cycle of acanthocephalans infecting mammals involves intermediate hosts. There are a number of genera in the dung beetle family Scara-

Table 1. Control and treatment of acanthocephala infections in humans and animals.

HOST (other information)	PARASITE (other information)	CONTROL AND TREATMENT (nonproprietary name) (miscellaneous comments)
HUMANS (acquired the parasite by ingesting beetles as food)	*Macracanthorhynchus hirudinaceus* (pig), *Moniliformis moniliformis* (rodents)	prevention can be effected by rodent control and keeping of food intended to be eaten cold in beetle-proof containers may help to prevent accidental infection; treatment is not well established; **niclosamide** (→ Cestodocidal Drugs) has been successfully used in Nigeria, loperamid hydrochloride, an antidiarrheal agent, proved very active in *M. hirudinaceus* infected pigs (see below); in China, surgery is often practiced to remove adult worms from heavily infected patients;
PIG, WILD BOAR (other species occurring in carnivores like wolf, domestic dog badger, fox, others are *M. catalinum*, and *M. ingens* and *Oncicola canis*	*M. hirudinaceus* **egg** containing the acanthor larva with rostellar hooks is large, ovoid and has a thick , dark brown, textured shell **egg** of *O. canis* is relatively small, ovoid, brownish, and has a smooth, thick shell	where pigs are kept in small sties or runs regular removal and suitable disposal of feces containing eggs will help in reducing the infection; older drugs such as carbon tetrachloride, tetrachlorethylene and nicotine sulphate have been used; the drug of choice appears to be **loperamid** hydrochloride which at 1.5 mg/kg twice daily x 3 days kills 100% adult and pre-adult worms without showing side effects; **fenbendazole** (→ Nematocidal Drugs, Animals) at 20 mg/kg x 5 days and levamisole may also be effective; a single intramuscular dose of 0.3 mg/kg **doramectin** reduced *M. hirudinaceus* worm burden in naturally infected pigs by 62 %
MONKEYS infection may be common in zoos	*Prosthenorchis elegans* *P. spicula* **eggs** in feces are smaller than those of *M. moniliformis*	insecticides and good sanitation will control the intermediate hosts (cockroaches, *Blattella germanica*); dithiazinine iodide has been effective; other drugs (see pig) may also affect adult worms and may be used by way of trial
HEDGEHOG adult worms of different species measure 0.5– 12 cm in length	*Prosthenorchis rosai* *Nephridiorhynchus major* and others **eggs** in feces have a thick shell	pathogenic effects of adult worms may be ulcerative enteritis, GI disturbance and peritonitis caused by perforation of intestinal wall; intermediate hosts are insects; prevention is not possible; treatment with fenbendazole (20–50 mg/kg x 5 days in feed), other benzimidazole carbamates (→ Nematocidal Drugs, Animals), loperamid or levamisole may be used by way of trial
RODENTS (mice, rats) parasite has a worldwide distribution	*M. moniliformis* **egg** is larger than that of *M. hirudinaceus*, elongated oval, has a thick, smooth, clear shell	strategic use of insecticides may control intermediate host (cockroaches) in laboratories; treatment of rodents is similar to that used with *M. hirudinaceus* (see hedgehog, above); infections have been reported on rare occasions in humans and are acquired by accidentally ingesting of beetles with food; adult worm has a pseudosegmentation of the body; adult female worm can reach > 20 cm in length
AQUATIC BIRDS parasite has minor veterinary importance	*Filicollis anatis* *Polymorphus minutus* (syn. *P. boschadis*) **eggs** in feces are relatively small and spindle shaped	worldwide distribution; prevention is impossible because of the ubiquitously occurring freshwater isopods and other crustacean intermediate hosts; treatment is unknown but fenbendazole (20–50 mg/kg x 5 days in feed), other benzimidazole carbamates (→ Nematocidal Drugs, Animals), and loperamid may be used by way of trial; acanthocephalan infections (*Centrorhynchus lancea, Mediorhynchus taeniatus*) have been reported in free-living houbara bustards in the United Arab Emirates

Table 1. (Continued) Control and treatment of acanthocephala infections in humans and animals.

HOST (other information)	PARASITE (other information)	CONTROL AND TREATMENT (nonproprietary name) (miscellaneous comments)
FINFISH: [1]non-predacious [2]predatory [3]freshwater [4]saltwater [5]brackish water [6]perch, eel [7]salmonoid fishes [8]cod	[1,2,4,5] *Pomphorhynchus laevis* [3,5]*Neoechinorhynchus rutili* [6]*Acanthocephalus lucii* [7]*A. anguillae* [4,8]*Echinorhynchus gadi* [7]*E. truttae*, others	these acanthocephalans may cause heavy infections accompanied by high mortality in aquaculture – the selective breeding and raising of fish in 'fish farms'; commercial fresh water or marine intensive fish farming in certain locations in Europe and particular in India and China may not only suffer disastrous economic losses caused by parasites but also considerable pollution of water by food and chemicals (see Anonymous, Nature 386: 105–110, 1997); in trout, **loperamid** in-feed proved effective at 50 mg/kg/ day x 3 days killing 100% adult and preadult worms without obvious adverse effects; the drug is not licensed for food fish

Data given in this Table have no claim to full information

baeidae and cockroaches containing the infective stage of worm (→ cystacanth (Vol. 1), which is really a young adult) or vertebrates, which act as paratenic hosts (e.g. mice, and frogs) harboring re-encysted cystacanths. No acanthocephalans are primarily human parasites.

Two species of these worms may infrequently infect humans more often than others. One species is *Moniliformis moniliformis*, which commonly parasitized rats and other **rodent hosts**, the other *Macracanthorhynchus hirudinaceus* (adults resemble *Ascaris suum*), a common parasite of pigs and wild boars, ubiquitous in areas where **pigs** are kept free. The final host (swine, occasionally man) becomes infected by ingesting either the infected grubs or the adult beetles and rarely the infected vertebrate. Infections may frequently occur in China and Indonesia and other parts of Southeast Asia, but also in most other countries of the world though *M. hirudinaceus* is absent from Western Europe. In **humans**, the adult worms which are attached to the wall of the small intestine cause diarrhea, GI disturbances, and vomiting but also serious complications, such as severe ulcerative enteritis or perforation of the bowl resulting in peritonitis. The pathogenic significance of *M. hirudinaceus* is similar for pigs causing granuloma formation at the site of attachment in the small intestine, weight loss and, rarely in heavy infections, penetration of the intestinal wall resulting in fatal peritonitis.

Very important parasites of Central and South American **monkeys** are *Prosthenorchis elegans* (very common), and *P. spicula* (less common). These acanthocephalans are now found throughout the world where primates are kept in zoos or elsewhere in captivity and where they have introduced the parasites. In heavy infections there is diarrhea, anorexia and debilitation often associated with death caused by perforation of the intestinal wall by adult worms.

There are two genera which may cause enteritis in **aquatic birds**, e.g. *Filicollis* (*F. anatis*) and *Polymorphus* (*P. minutus*). The intermediate hosts in both cases are crustaceans. Pathogenic effects produced by adult worms attached to the intestinal wall resemble those that are seen in mammals.

Pathogenicity of numerous **fish** acanthocephalans is species specific and varies considerably. There are species which penetrate through the intestinal wall, thereby entering the body cavity of fish. Other species remain in the lumen of intestine, showing frequent or less frequent change of attachment site. Intermediate hosts are crustaceans, and various fishes may serve as paratenic hosts. In the fish industry disastrous economic losses may be due to acanthocephalans, especially when fish farming is practised with overcrowded fish populations.

Prevention and Treatment

Prevention and treatment of Acanthocephala infections can be seen in Table 1, which are in general problematic. In man, usually Acanthocephala eggs are not passed with the feces since worms may not mature to adults. Diagnosis is made by x-ray examination or endoscopy. Serological tests are not available.

Acanthocephalan Infections

Attachment

Generally, acanthocephalans that have a short neck do not deeply penetrate into the host's intestinal wall with their praesoma, i.e. they do not create lesions reaching as deep as the muscular layers of the intestinal wall (see → Acanthocephala/Fig. 4 (Vol. 1), → Acanthocephala/Fig. 5 (Vol. 1), → Acanthocephala/Fig. 6 (Vol. 1)). In contrast, many Acanthocephalans possess a long neck which may comprise a bulbus as an inflated part of the neck (Fig. 1). The bulbus functions as a dowel enabling the worm to occupy a permanent point of attachment at one site. The perforation of the intestinal wall may be supported by proteolytic enzymes or occur in thin walls (Fig. 2).

Cellular Host Responses

Successive Cell Assemblages The specific composition of host cells accumulating at the worm's praesoma follows a certain sequence which is related to the worm's mode of attachment. During the first days p. i. only necrotic tissue surrounding the praesoma can be found. After about 3–5 days p. i. a belt of inflammatory tissue with haemorrhagic involvement starts forming (Fig. 7). From about 10 days p. i. onwards the further succession of cell assemblages depends on the type of attachment. At the praesoma of perforating species the inflammatory tissue becomes dominated by macrophages maturing into epitheloid cells and another belt of connective tissue which attains a blue colour in Azan-stained paraffin sections, starts forming (Fig. 2). Later on, this belt may become considerably reinforced and interspersed with fibroblasts and collagen fibres and an outer belt of connective tissue (Fig. 3B,C) sometimes consisting of plane collagene fibres (Fig. 3C) is built up. The attachment site of the eoacanthocephalan *Neoechinorhynchus rutili*, although non perforating, is characterised by a pronounced accumulation of collagen fibres (Fig. 3A). In contrast, species with shallow attachment seem to change their point of attachment prior to the formation of connective tissue at the site of the temporary anchorage, and a concentric zonation of neoplasic tissue around the parasite's praesoma does not emerge. Haemorrhagic spots (Fig. 7C) may occur in all the tissue belts mentioned here.

Longitudinal Zonation of Defence Cells Among perforating species like *Pomphorhynchus laevis* the chronic stage of infection not only reveals concentrically arranged belts of necrotic, inflammatory and connective tissue (Fig. 1) but also a longitudinal zonation can be figured out. Near or inside the peritoneal cavity the necrotic belt is very conspicuous (Fig. 1). It is very rich in lipids that seem to become absorbed by the worm's tegument. In contrast, at the lumen side of the intestinal tube the necrotic belt almost does not exist and defence cells form a closely fitting belt of solid compensatory tissue. The question arises whether this zonation is host or parasite induced. Taking also the histopathology of *N. rutili* into consideration (Fig. 3A) it appears that the described zonation of the host's tissue is largely brought about by the parasite.

Proboscis Hooks Irrespectively of the type of construction, i. e. whether a tegumental outer confinement exists the hooks always provoke the most pronounced accumulation of granulocytes. In *N. rutili* the small roundish proboscis with its large hooks, the attracting of granulocytes towards the hooks and the surrounding tegument is very severe in salmonids and other fish during the acute initial phase of the infection and it has been suspected already that this is induced by the parasite benefiting from it in aspects related to nutrition or attachment (Figs. 3A, 4B). In comparison, the trout-specific species *Echinorhynchus truttae* only provokes a rather mild granulocyte response.

Infections in fish, birds and mammals

In contrast to the paucity of information available on many aspects of host-parasite-interactions of acanthocephalans, these worms are very common in many wild living vertebrates as well as cultured fish. Species like *Moniliformis moniliformis* can be easily kept in a laboratory in cockroaches and rats. It offers huge quantities of tissue and surface coat useful for physiological and molecular studies.

In fish hosts the inflammatory response is dominated by granulocytes (Fig. 7) and marcrophages/epitheloid cells depending on the conditions described above. Usually eosinophils are the most abundant granulocytes (Fig. 7A) but in eels for instance eosinophilic granulocytes rarely occur and thus heterophilic granulocytes dominate in the tissue near the praesoma of *Paratenuisentis ambiguus* or other parasites of eel. Fish-specific melano-macrophages, also may occur near the acanthocephalan praesoma. Plasma cells, however, (documented from fish) also may occur but do not seem to play a role as for instance in mammals. In fish immunoglobulins develop relatively slowly during the course of an infection, precipitms are rare, and the only immunoglobulin class produced is IgM being better at agglutination and complement activation than precipitation. Very little is known about the use of antibodies of fish against acanthocephalans.

In birds (ducks infected with *Filicollis anatis*, Fig. 6) heterophils represent the major granulocyte fraction (Fig. 7B). Macrophage giant cells frequently occur in the tissue near the parasite's praesoma, and also plasma cells are contained in it depending on the stage of infection.

In mammals a progressed stage of infection is accompanied by the abundance of plasma cells near the worm surface (Fig. 5B) after having passed through an acute phase of infection associated with a mass occurrence of eosinophilic and heterophilic granulocytes (Fig. 5A,C). In rats infected with *Moniliformis moniliformis* the occurrence of plasma cells up from about 10 d. p. i. (Fig. 5B) corresponds with the increase of worm-specific IgE-antibodies as described elsewhere.

Several fish researchers have often been amazed by the high intensities, often exceeding 100 specimens per gut, with acanthocephalans like the perforating species *Pomphorhynchus laevis*, being tolerated without showing external signs of disease.

In birds and mammals, however, different species of perforating acanthocephalans have been involved in morbidity and mortality, especially under elevated worm intensities as shown by a domestic duck infected with *Filicollis anatis* (Fig. 6). This polymorphid palaeacanthocephalan also causes weight loss, anaemia, debility, apathy and somnolence among infected ducks.

In mammals and birds the involvement of secondary infections and the infiltration of bacteria from the deep lesions into the peritoneal cavity seem to contribute to this pronounced pathology. In swine the sites of attachment of the perforating archiacanthocephalan *Macracanthorhynchus hirudinaceus* are marked on the outer surface by a caseous nodule with a reddened annulation around it. It frequently abscesses with bacterial involvement, which may lead to perforation of the gut wall. Human patients in China and Southeast Asia have suffered unbearable abdominal pain during the passage of the entire worm or parts of it into the peritoneal cavity which is accompanied by infiltration of eosinophils and neutrophils, massive oedema and large quantities of serosanguineous exudate in the body cavity near the site of perforation. Prior to 1980 perforating acanthocephalans (mainly *Prosthenorchis elegans* and occasionally *Oncicola spirula*) infecting primates held in zoos were the cause of numerous fatal cases among these hosts. But since then no further cases have been published, probably due to the resulting growing awareness of this threat grown in zoological gardens since then. The perforation canal through the intestinal wall harbouring the parasite was filled with inflammatory exudate enriched with scattered masses of bacteria and acute peritonitis was diagnosed to be the cause of death in many cases reported by several authors. Less than 15 specimens of *P. elegans* are considered sufficient to cause mortality in squirrel monkeys.

In contrast to these descriptions from three perforating acanthocephalans, *Moniliformis moniliformis* with its superficial attachment (Fig. 8) does not create mortality among rats, and in these hosts as well as in accidental human cases where a worm had been measured to be 26,5 cm in length, the symptoms of morbidity are much less pronounced. All these findings suggest that the depth of penetration of an acanthocephalan species is an important criterion influencing the course of pathology at least in host-parasite associations involving mammals or birds as final hosts.

Praesoma Morphology Influencing the Host-parasite Interface The praesoma of eo- and palaeacanthocephalans is invested by a lipoid surface cover (→ Acanthocephala/Integument (Vol. 1)). It probably fulfills defence functions but obviously also incorporates lipids from necrotic host tissue which then becomes absorbed by the parasite (→ Acanthocephala/Fig. 7 (Vol. 1), → Acanthocephala/Fig. 8 (Vol. 1)). Within this coat, osmiophilic films seem to become shed from the parasite's surface. Whether these are loaded with peptides or substances deriving from defence mechanisms of the host will have to be proven in experiments.

The phenomenon of shedding (capping) of surface coat can be commonly observed from *M. moniliformis* in rats (Fig. 5A,C). Large patches of the fuzzy surface coat detach from the praesomal worm surface into the surrounding necrotic inflammatory tissue/exudate. It happens during the acute phase of infection associated with an granulocyte response as well as later on when plasma cells accumulate at the praesoma. One may conclude that host's enzymes and/or antibodies bind to this coat until it becomes shed by the parasites. In transmission electron micrographs patches of detached coat attain a more coarsely structured and more electron lucent appearance compared to the glycocalyx still lining the surface (Fig. 5A,C).

Leakage of Liquids, Cells and Debris from the Lesion The phenomenon of leakage from the lesion created by a worm into the intestinal lumen has been documented in many acanthocephalans (Fig. 4A, Fig. 8). It seems to be most pronounced at the climax of the acute phase of infection at the point of attachment of perforating species (Fig. 4A). Having reached a progressed chronic stage, the opening of the lesion heading towards the intestinal lumen has been almost tightened up by the host so that the leakage has decreased (Fig. 1). Most non-perforating species, on the other hand, seem to switch their site of attachment prior to a severe inflammatory reaction with an accompanying leakage as shown in Fig. 4A.

Due to controverse findings from different host-parasite associations it appears that the loss of plasma proteins through the intestinal wall of fishes, infected by acanthocephalans, into the intestinal lumen depends on different conditions.

Abrasion, Erosion, Compression Usually the mucosal surface within the range of an acanthocephalan metasoma, also becomes mechanically affected by movements and the body pressure of the worms, especially close to the lesion. Also, many acanthocephalans possess body spines which support a burr-like affiliation of parts of the metasoma with the mucosal surface which should have some scratching effect.

Increased Diameter of the Intestinal Wall A local increase in thickness of the intestinal wall at the site of anchorage is well described in acanthocephalans. In addition also other parts of the host's gut which are not in contact with the worms may be enlarged in diameter.

Distention of the Intestine The rat intestine in its length may be extended due to a single specimen of *M. moniliformis*. Also certain appendages like pyloric caeca were found to become almost doubled in diameter due to the acanthocephalans (*Leptorhynchoides thecatus*) inhabiting the caeca. Interestingly intestinal caeca seem to be a preferred environment of several intestinal helminths, and one acanthocephalan has been found to create its own caecal microhabitat. This species *Neoechinorhynchus carpiodi* also exhibits that social clustering exists among acanthocephalans. Up to 20 or more worms can be found together inside deep caverns surrounded by collagen capsules seen as expansions at the outer intestinal wall. The potential involvement of parasite enzymes supporting the mechanical activity of the hooks in the formation of the "caeca" has not yet been investigated.

Occlusion of the Intestine A few acanthocephalans such as *Macracanthorhynchus hirudinaceus* reach considerable size i. e. more than half a meter in length. High worm burdens have been mentioned to have occluded the intestinal tube leading to mortality among piglets.

Negative Influence on Metabolic Parameters As in other helminths which are better investigated than acanthocephalans one should expect that various physiological parameters of the hosts become influenced by an acanthocephalan infection. Indeed, growth, weight gain and blood sugar concentrations in rats infected with *M. moniliformis* were significantly affected by diets containing growth-limiting amounts of carbohydrates and the energy metabolism of starlings experimentally infected with *Plagiorhynchus cylindraceus* was found to be altered under distinct laboratory conditions. Most other observations contributing to this subject remain preliminary or general.

Fever Elevated body temperatures due to acanthocephalan infections have been reported in domestic ducks as well as in primates and humans.

Goblet Cell Hyperplasia Several authors have reported inflated goblet cells along the intestinal mucosa in hosts (fish, rats) infected with acanthocephalans, also their density was found to increase due to acanthocephalan infections (Fig. 4A, Fig. 8). It is rather unlikely that adult non-perforating acanthocephalans would be negatively affected by this measure (even less likely for perforating species), but in secondary infections young worms just entering the gut trying to establish themselves, might be affected by an excess of mucins on the mucosal surface.

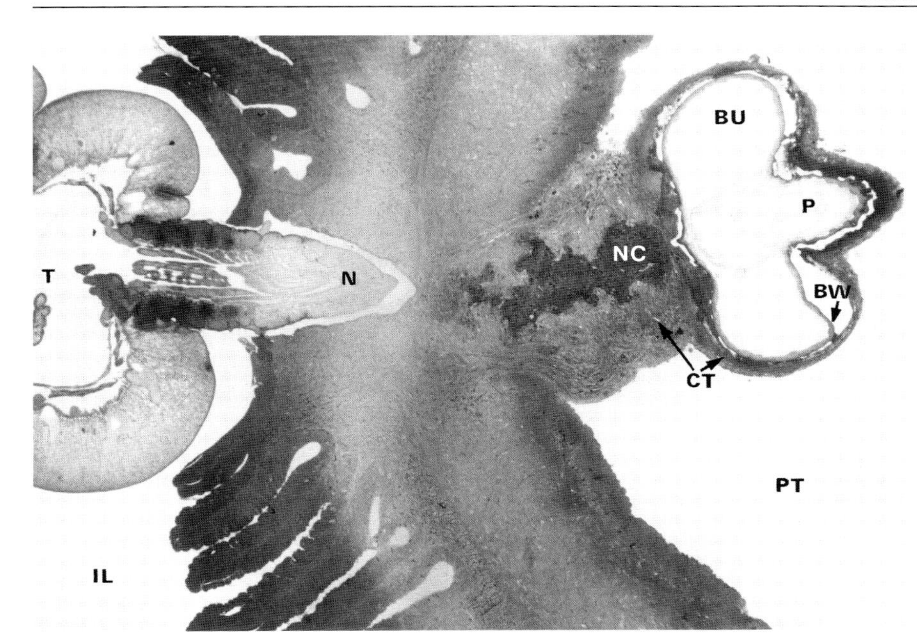

Fig. 1. Micrograph of a semithin section of *Pomphorhynchus laevis* (penetrating species) in a naturally infected adult chub (*Leuciscus cephalus*). The upper presoma (e.g., proboscis; P, bulbus; BU, upper neck, not visible) has penetrated the host's peritoneum (PT). The presoma has become encapsulated by connective tissue (CT) which degenerates in close contact to the worm. x 40. *BW*, bulbus wall; *IL*, intestinal lumen; *N*, neck; *NC*, necrosis; *T*, trunk (metasoma)

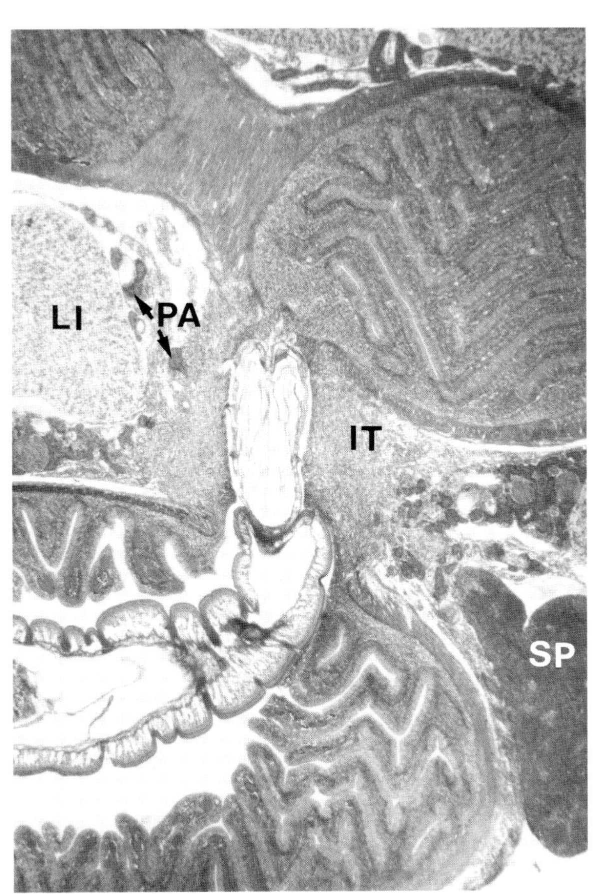

Fig. 2. Light micrograph of a paraffin section showing a longitudinally sectioned male *Acanthocephalus anguillae* in an experimentally infected goldfish fingerling, 18 d. p. i. The worm has perforated the intestinal wall of one loop with its praesoma and the anterior portion of its metasoma and now has ruptured the outer part of another loop's wall. The extraintestinal part of the worm is enclosed by inflammatory tissue (IT). x 80. *LI*, liver; *SP*, spleen, *PA*, pancreas

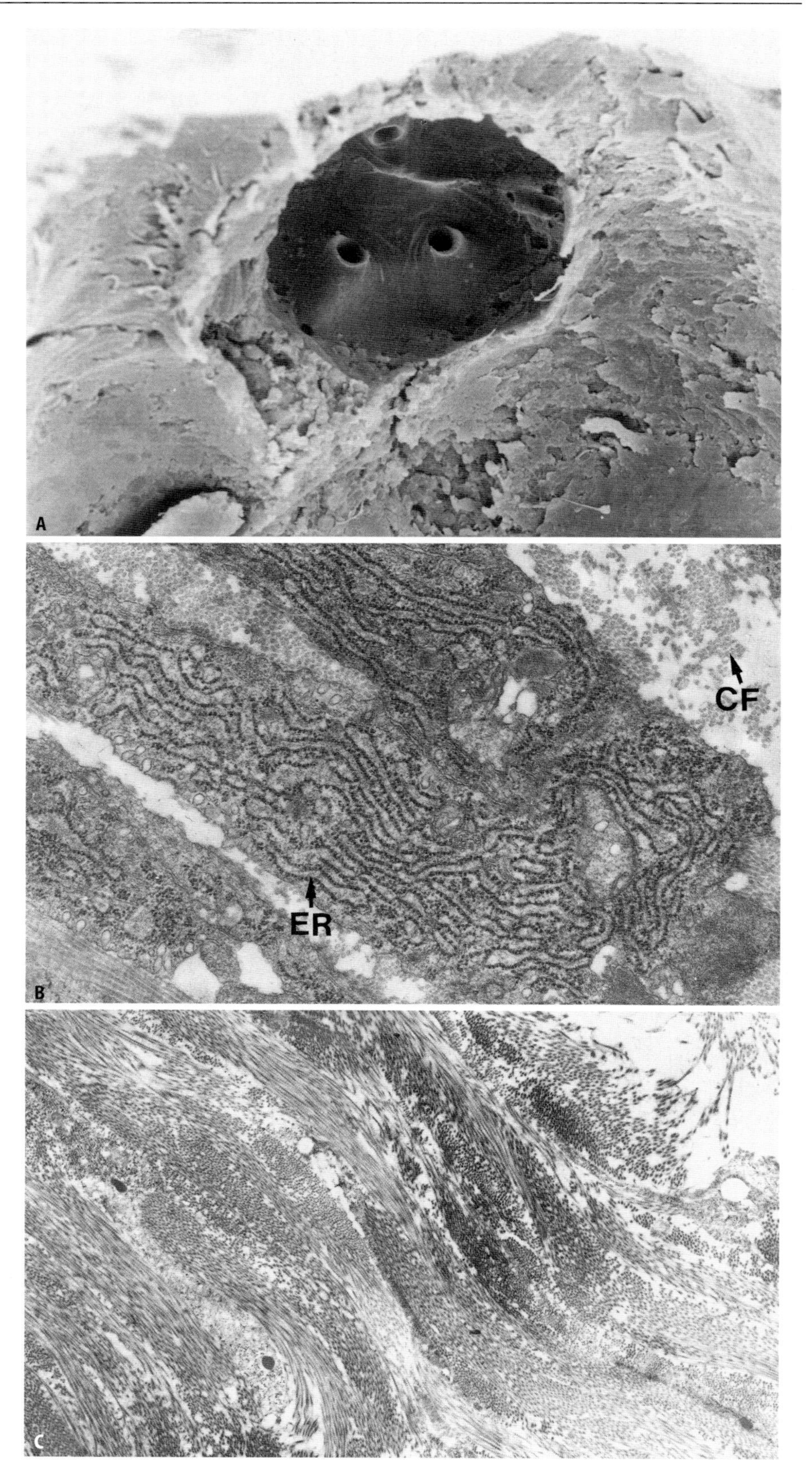

Fig. 3 A-C. Micrographs showing connective tissue in the intestinal wall of fishes infected with acanthocephalans. **A** Scanning electron micrograph showing the former point of attachment of a proboscis of *Neoechinorhynchus rutili* in the intestinal wall of a naturally infected juvenile rainbow trout. Collagen fibres have formed a firm capsule appearing as a "print" of the proboscis. **B** Transmission electron micrograph of fibroblasts (ER: rough endoplasmatic reticulum) and excreted collagen fibres (CF) in the outer connective tissue belt encapsulating the proboscis of *Acanthocephalus anguillae* in an experimentally infected rainbow trout 30 d. p. i. **C** Transmission electron micrograph of firm connective tissue consisting of plane collagen fibres near the outer margin of the neoplasic tissue encircling the bulbus of a *Pomphorhychus laevis* in a natural infection in a chub (*Leuciscus cephalus*).

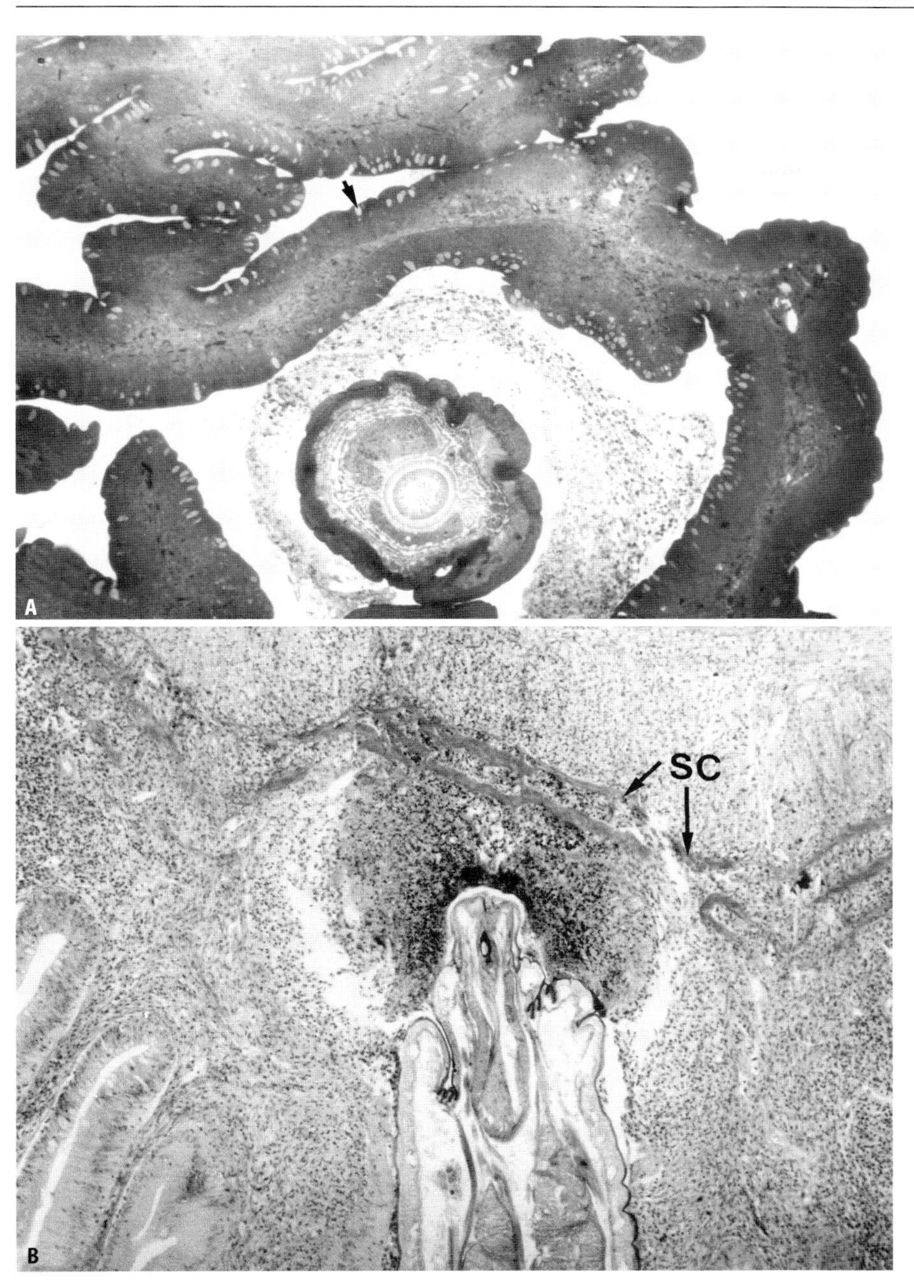

Fig. 4 A-B. Light microscopical micrographs of sections of acanthocephalans and surrounding tissue of the fish hosts' intestinal wall. **A** Semithin cross section through the anterior metasoma of a young adult *Acanthocephalus anguillae* and surrounding intestinal plicae. The worm is surrounded by a "cloud" of cells and liquids leaking from the intestinal wall into the gut lumen. Also note the densely set, slightly hyperplastic goblet cells in the mucosa (arrow).
B Paraffin section through the anterior third of a *Neoechinorhynchus rutili* in a naturally infected grayling (*Thymallus thymallus*). Note the huge inflammatory reaction around the worm's praesoma and the collagenous stratum compactum (SC) in the gut wall of this host functioning as a perforation obstacle for other acanthocephalans like *A. anguillae*.

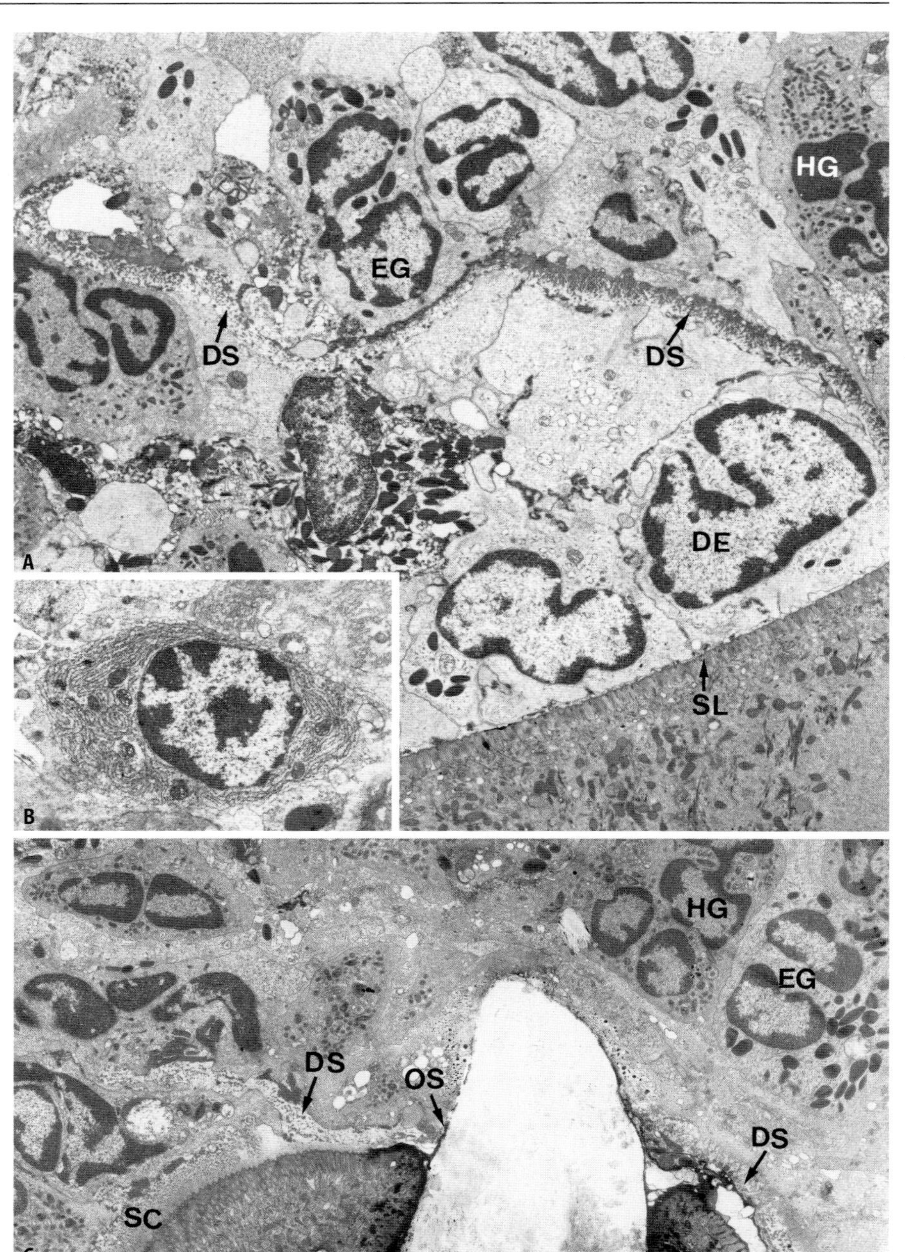

Fig. 5 A-C. Transmission electron micrographs of the proboscis surface and of surrounding host tissue of *Moniliformis moniliformis* in rats. *DS*, detached surface coat; DE, degranulated eosinophilic granulocyte; *EG*, eosinophilic granulocyte; *HG*, heterophilic granulocyte; *SC*, surface coat; *SL*, striped layer of the tegument; *OS*, osmiophilic substance deriving from the annular cleft around the "naked" hook. **A** Acute phase of infection, 10 d. p. i.. Eosinophilic and heterophilic granulocytes form dense populations near the praesoma. Large patches of the surface coat have detached from the worm. **B** Plasma cell out of a dense association of such cells near the surface of a worm, 60 d. p. i. **C** 10 d. p. i. infection. Note the fine fuzzy structure of the surface coat still adhering to the worm's praesomal surface while other portions of it have detached.

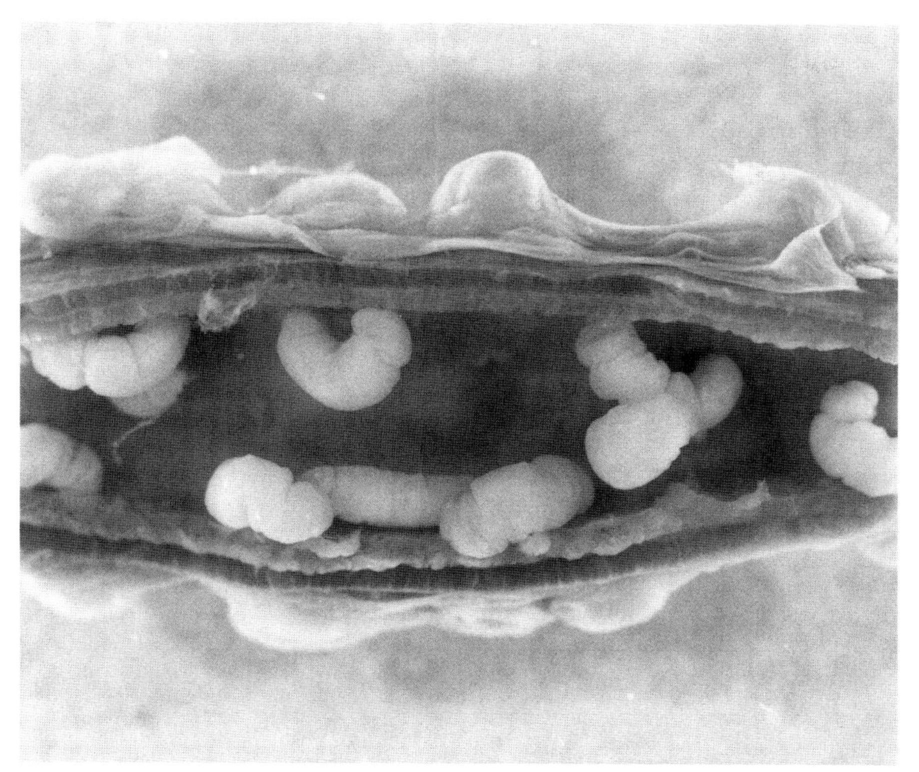

Fig. 6. Light micrograph of an opened intestine of a domestic duck naturally infected with *Filicollis anatis*. The trunks (metasoma) of the worm can be figured out inside the gut lumen whereas the praesomal bulb encapsulated by neoplasic tissue is seen at the outer side of the intestinal wall; The worm density of *F. anatis* shown here potentially crates mortality among ducks.

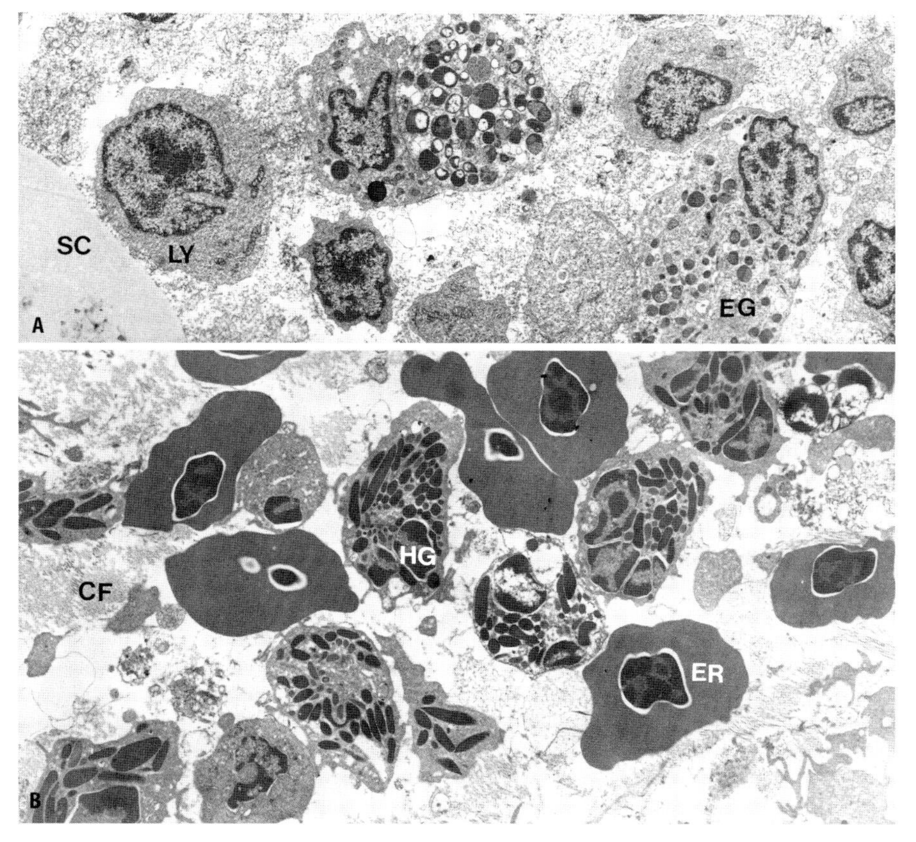

Fig. 7 A,B. Transmission electron micrographs of granulocytes and erythrocytes near the praesoma of acanthocephalan species. **A** Acute infection of *Acanthocephalus anguillae* in a carp fingerling, 14 d. p. i.. The eosinophilic granulocytes (EG) have attained different stages of degranulation. *LY*, lymphocyte; *SC*, surface coat of the worm's praesoma. **B** Inflammed and haemorrhagic loose neoplasic tissue of a naturally infected duck near the anterior bulbus/proboscis of a *Filicollis anatis*. *ER*, erythrocyte; *HG*, heterophilic granulocyte; *CF*, collagen fibres

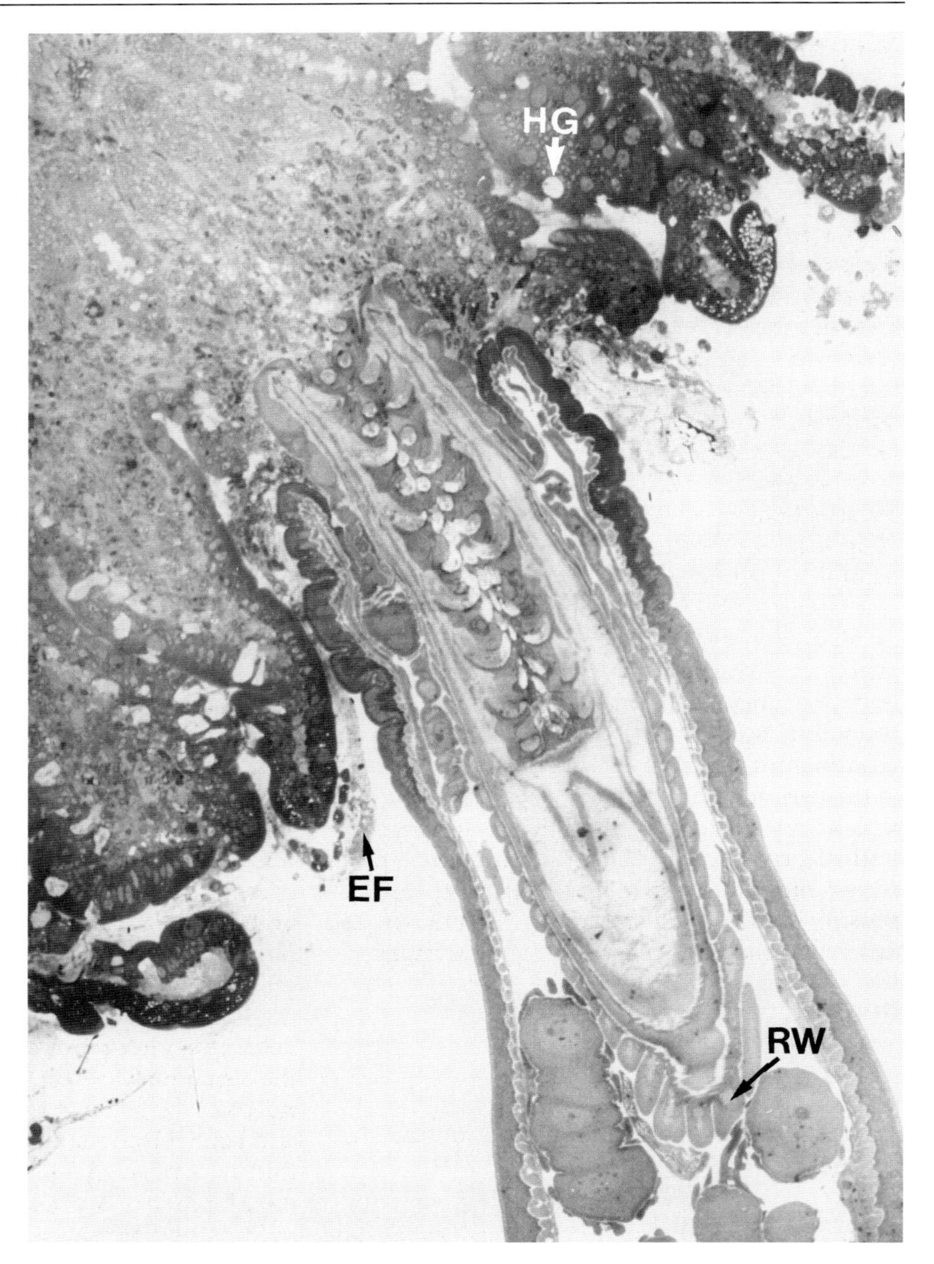

Fig. 8. Semithin longitudinal section of the praesoma and anterior trunk of a 20-d. p. i.-specimen of *Moniliformis moniliformis* in a rat. Note the superficial attachment, the deep proboscis cavity, the hyperplasic goblet cells (HG) and the efflux (EF) from the lesion into the intestinal lumen. *RW*, spirally arranged muscle cords of the outer receptacle wall

Skeletal Deformations Notes exist, that describe brown trout, heavily infected with different acanthocephalan species, to show deformed backbones and/or shortened gill operculae or fins. This may have to do with the recently detected very high absorptive capacity of acanthocephalans for calcium and other minerals.

Infections in Humans

Two species of archiacanthocephalans and three species of palaeacanthocephalans have been found in humans (→ Acanthocephala (Vol. 1)). *Macracanthorhynchus hirudinaceus* has been reported in China, Thailand, regions of the former USSR and CSSR, Madagascar and Brazil, and *Moniliformis moniliformis* in Italy, Israel, Sudan, Iraq, Iran, Bangladesh, USA, and Belize. The latter are located inside the intestine, where they reach sexual maturity. *Acanthocephalus rauschii*, *A. bufonis*, and *Corynosoma strumosum* (parasites of fish, amphibians, and seals, respectively) have been obtained from human patients in Alaska *(A. rauschii* and *C. strumosum*) and Indonesia (*A. bufonis*). *A. rauschii* was located in the peritoneum, and the other two palaeacanthocephalans in the gut.

Infections with *M. hirudinaceus* and *M. moniliformis* usually occur among small children who willingly or accidentally ingest insects. In parts of Asia, however, where raw or undercooked insects are customarily eaten, adult humans also become infected; symptoms of infection such as weight loss, intermittent fever, bulging abdomen, diarrhea, and severe pain are well described. In a hospital in China 115 cases of acute abdominal colic due to *M. hirudinaceus* were reported over a period of only three years. Often *Macracanthorhynchus hirudinaceus* in humans occupies extraintestinal positions. The migration of this perforating acanthocephalan through the gut wall is very painful. An adult volunteer who swallowed infective larvae of *M. moniliformis* suffered from abdominal pain beginning from the 20th day after infection. It seems that *M. moniliformis* does not lead to a great inflammatory reaction; however, abdominal surgery on patients infected with *M. hirudinaceus* revealed a serosanguinous exudate in the peritoneum, inflamed parts of the intestine with nodules of up to 3 cm in diameter, and/or intestinal perforations. Thus, both species seem to accept humans as suitable hosts, showing a pathogenicity similar to that in their major hosts, i.e., rats and swine (*M. hirudinaceus*) respectively.

Therapy

→ Acanthocephalacidal Drugs.

Acariosis, Animals

Several mites infest animals and cause significant dermatologic diseases. These may be occasional parasites e.g. harvest mites (*Trombicula*) or obligatory parasites like *Sarcoptes* and *Demodex* genera. Mites may be free-living on the surface of the skin (*Cheyletiella*, *Chorioptes*), superficial burrowers (→ *Sarcoptes* (Vol. 1)), or may penetrate more deeply (*Demodex*). The parasitic mites of the families Sarcoptidae and Psoroptidae, known as "mange mites", generally give rise to well defined dermatoses. The lesions are the result of mechanical damage to the skin and probably also of → hypersensitivity reactions to toxic secretions (→ Pathology/Fig. 30).

Sarcoptes scabiei (**Sarcoptic mange**) occurs commonly in pigs, dogs and cattle; and more rarely in horses, sheep, goats and cats. The so-called feline scabies is caused by *Notoedres cati* . The several varieties of → *Sarcoptes scabiei* (Vol. 1) may represent strains of the same mite which have become more adapted to particular hosts. The mites burrow into the skin. It has long been suspected that antigens from the mites themselves, their faeces, or their moulting and hatching fluids are responsible for the allergic reactions observed. Clinical signs are similar in all species and consist of a papillar eruption accompanied by severe pruritus. The intense scratching caused by the pruritus may lead to alopecia, secondary bacterial infections, lichens, and hyperpigmentation. The skin becomes thickened in severe cases. Clinically affected animals may become debilitated and lose weight, or fail to properly gain weight. Feed efficiency is reduced and the hide may suffer considerable damages. Severely affected pigs may also become anaemic. The distribution of the lesions is characteristic in the various hosts.

Psoroptic mange is a serious disease in cattle and sheep, less so in horse and goats. The causative mites are species of *Psoroptes* and are host specific. Mites penetrate the epidermis to suck body fluids, and cause a local reaction with formation of vesicles. The exudate from the vesicles coagulates and dries on the skin surface, resulting in the formation of a crust or scab of varying thickness. The mites move to the edge of the scab, and the lesion increases in size. There is marked pruritus, and scratching results in alopecia, erosions and lichenification. Lesions usually begin in areas thickly covered by hair or wool. Debilitation, reduced productivity and occasionally death may follow severe infestations. In goats *Psoroptes cuniculi* is known as the "earcanker" mite because of its predilection for the ear (causing otitis).

Chorioptic mange occurs commonly in cattle, sheep and horses, and more rarely in goats. The *Chorioptes* mites are host specific and live on the surface of the skin. Generally, it is a less severe condition than psoroptic or sarcoptic mange. Lesions consist of alopecia, erythrema, excoriations and (small) crusts associated with pruritus. The mites have a predilection for the perineum, udder, caudal areas of thigh, rump and feet. Lesions caused by *Chorioptes equi* start as a pruritic dermatitis affecting the distal limbs around the foot and fetlock. A moist dermatitis of the fetlocks develops in chronic cases.

Otodectes cynotis is an obligate parasite of the external skin surface, mainly the external ear canal of cats and sometimes dogs. The major lesion is thus otitis externa.

Demodex spp. lives in the hair follicles and sebaceous glands of dogs, cats cattle and goats. Demodicosis is of clinical importance in dogs. It is less common in cats and other animals. Mites are often present on healthy animals without causing obvious lesions, but heavier infestations produce mechanical damage to the skin. As the mites multiply in hair follicles and sebaceous glands, the hairs fall out. Enlargement and rupture of adjacent follicles and glands leads to cyst formation. Secondary pyoderma associated with staphylococcal infection is a common complication in the dog. Skin lesions in this animal may range from small localised patches of alopecia, in which the skin may appear normal, to more generalised dermatitis with loss of hair, thickening and discoloration of the skin. In the pustular form of the condition, small pustules are formed in the hair follicles. Dogs with generalised demodicosis may have a concurrent pododemodicosis, which is characterized by interdigital erythema and alopecia or interdigital furunculosis with associated oedema and pain. Pododemodicosis may be the only manifestation of the disease. In cattle and goats *Demodex* lesions consist of small elevated nodules of varying size. Most lesions appear on the shoulder, head and neck region. Nodular demodicosis is characterised by the permanent formation of new nodules when the older ones disappear. Nodules arise when granulomatous inflammation is complicated by secondary bacterial infection.

Other ectoparasitic mites include *Dermanyssus gallinae*, *Lynxacarus radovsky*, *Trombiculidae* (chiggers) and *Cheyletiella yasguri*, *C. blakei* in dogs and cats, and *Psorogates ovis* in sheep. Clinical signs include erythrema and pruritic papulocrustous eruptions. Cheyletiellosis in dogs and cats is typically more severe in young animals, with the primary lesion being scaling over the dorsal medline. Pruritus is variable.

Therapy

→ Acarizides (Vol. 1), → Nematocidal Drugs, Animals, see → Arthropodicidal Drugs.

Acceptable Daily Intake

Synonym

ADI

Definition

Dose of a drug residue in edible tissues, such as meat, various organs, fat, etc., which during the entire lifetime of a person seems to be without obvious risk to health based on all toxicological data known at the time.

General Information

The ADI for humans may be determined by applying a safety factor of 1:100, or a safety factor of at least 1:1000 in case of a teratogenic drug. Therapeutic claims made by the manufacturer must coincide with safety and tissue residue data for the drug approved by government regulatory agencies (→ Chemotherapy).

Acetylcholine-Neurotransmission-Affecting Drugs

Mode of Action

Fig. 1

Structures

Fig. 2

Organophosphates

Important Compounds Dichlorvos, Diuredosan, Frento, Metrifonate, Coumaphos, Haloxon, Naphthalophos, Vapona.

Synonyms Dichlorvos: Atgard, Dichlorman, DDVP, Equigard, Equigel, Task

Diuredosan: Uredofos, Sansalid

Metrifonate: Trichlorphon, Anthon, Bilarcil, Combot, Dipterex, Difrifon, Dylox, Dyrex, Mastotem, Neguvon, Tugon; in: Bubulin, Combotel, Dyrex T.F., Equizole, Neguvon A, Telmin B

Coumaphos: Asuntol, Baymix, Co-Ral, Meldane, Muskatox

Haloxon: Eustidil, Halox, Loxon; in: Haloxil

Naphthalophos: Amdax, Maretin

Clinical Relevance Metrifonate was introduced as insecticide in 1955. It exerts activity in *Taenia solium* neurocysticercosis. Diuredosan is active against *Taenia spp.*, *Dipylidium caninum*, *Mesocestoides corti* and only slightly against *Echinococcus*

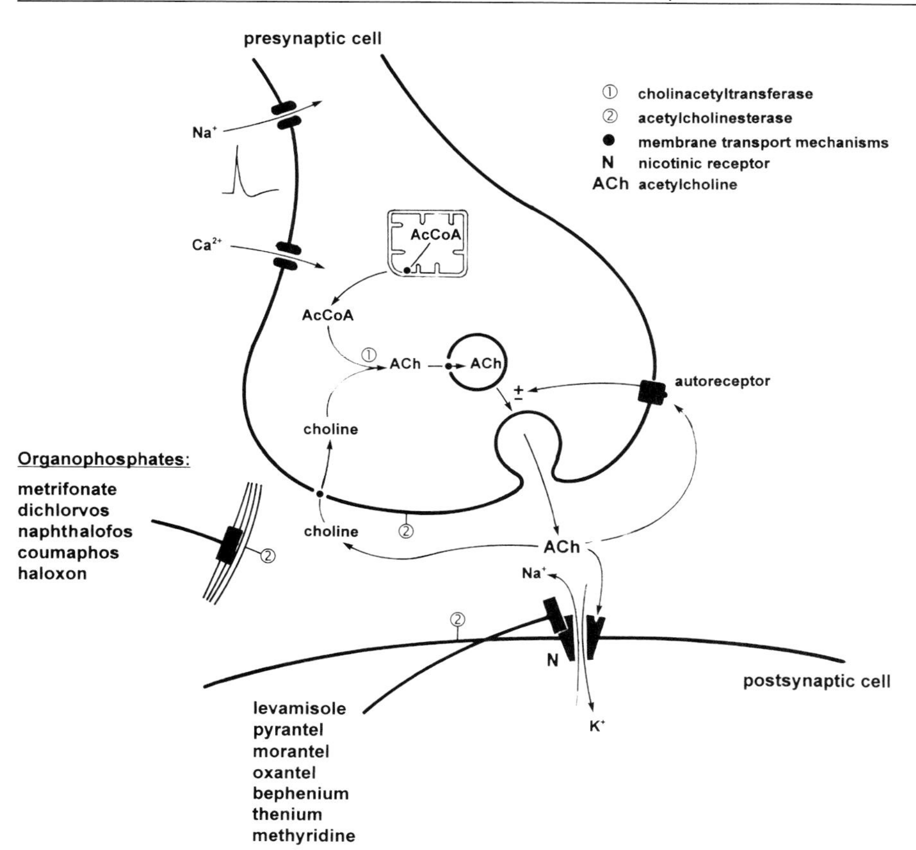

Fig. 1. Model of the action of drugs interfering with acetylcholine mediated neurotransmission.

granulosus. Fospirate is active against *T. hydatigena*.

The antitrematodal activity of metrifonate is directed against *Schistosoma haematobium*, it is only slightly active against *S. mansoni* and *S. japonicum*. Metrifonate is one of the drugs recommended by the WHO for the treatment of urinary schistosomiasis and present in the current Model List of Essential Drugs. For use in ruminants organophosphates had been available only in a limited number of countries. They are generally not as effective as the broad-spectrum anthelmintics and also have a lower therapeutic index. Haloxon, trichlorphon, coumaphos, naphthalophos and crufomate had been used in cattle, haloxon in sheep, dichlorvos in pigs, dichlorvos and trichlorphon in horses, and coumaphos in poultry.

Molecular Interactions Metrifonate is unstable in aqueous solutions and a spontaneous, nonenzymatic transformation into various compounds takes place. One of the degradation products is dichlorvos which exerts high biological activity. Acetylcholinesterase (AChesterase) from helminths as target for organophosphates was first explored in the late 1950's. Metrifonate has in-vitro activity against *Ascaris lumbricoides* by the inhibition of acetylcholinesterases and cholinesterases resulting in an impairment of the action of the neurotransmitter acetylcholine (Fig. 1). AChesterase in nematodes is inhibited at very low concentrations of 10^{-13} M. This leads to an indirect permanent stimulation of excitatory neuromuscular transmission mediated by acetylcholine, followed by a continuous depolarization of the postsynaptic junction resulting in a spastic paralysis. In general, a complete paralysis of the oral sucker is induced at lower metrifonate concentrations than those required to produce complete paralysis of the body musculature. The prolonged paralysis of the intestinal musculature leads to an interruption of peristaltic movements and a starvation of the parasite. The differences in metrifonate susceptibility among schistosome species are explained by differences in the amount of AChesterase located on the surface of adult schistosomes. Thus, *S. haematobium* teguments contain up to 20 times, and *S. bovis* teguments up to 6.9

Fig. 2. Structures of drugs affecting nicotinergic neurotransmission.

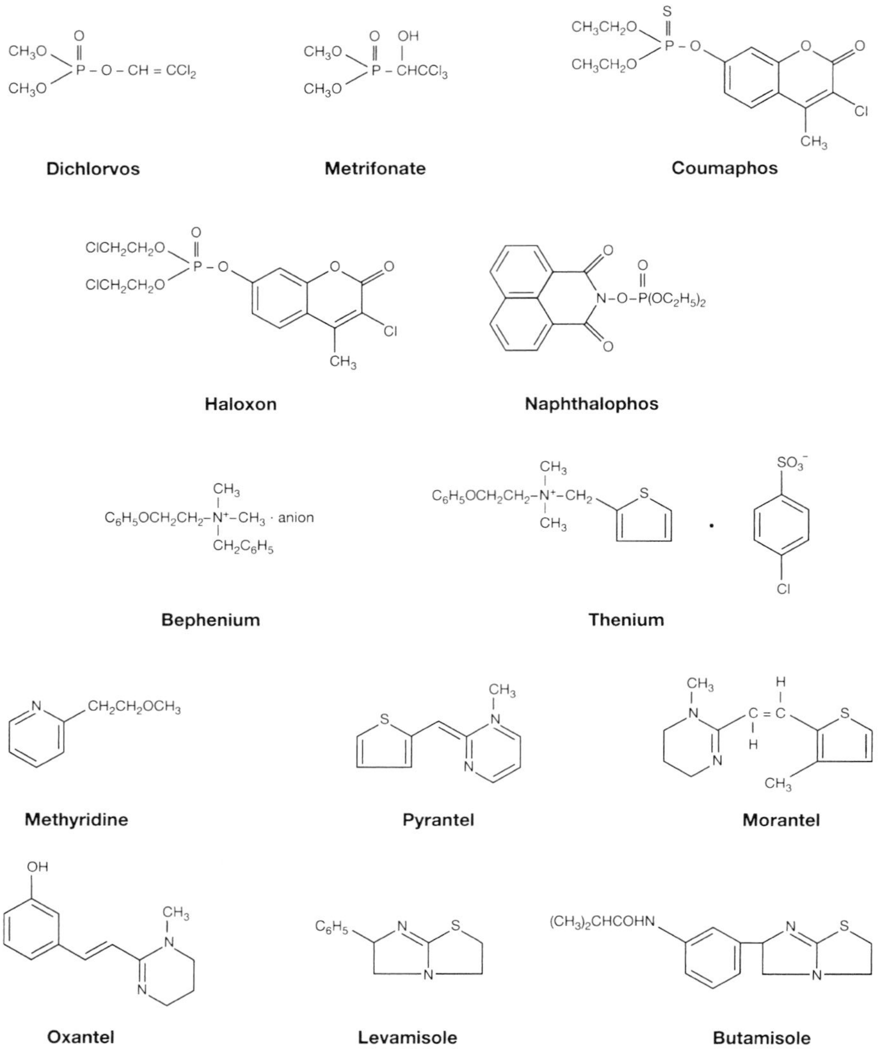

Dichlorvos

Metrifonate

Coumaphos

Haloxon

Naphthalophos

Bephenium

Thenium

Methyridine

Pyrantel

Morantel

Oxantel

Levamisole

Butamisole

times higher AChesterase activity than *S. mansoni* teguments. These quantitative differences correlate well with the relative sensitivities of these species to metrifonate. There is presumably an association of the tegumental AChesterase and nACh receptors located on the dorsal surface of the adult males which may be responsible for the glucose import into schistosomes. Thus, the surface and not the muscle AChesterase functions as the primary target of the metrifonate action.

The antinematodal activity of metrifonate is low. It has little activity against *Trichuris vulpis* and some activity against roundworms and hookworms.

Metrifonate also has antifilarial activity which is directed against microfilariae (→ Inhibitory-Neurotransmission-Affecting Drugs/Table 1) and only to a minor degree against adult worms. Some organophosphates exert strong activity against *Litomosoides carinii* microfilariae. Metrifonate and fenthion are effective against microfilariae of *Dirofilaria immitis*. There is no effect on developing stages of *D. immitis*, *L. carinii* and *Acanthocheilonema viteae* by organophosphates. In cats metrifonate has adulticidal effects.

The action of organophosphates against *L. carinii* microfilariae is complicated. The inhibition of AChesterase which is important for effect against gastrointestinal nematodes and *S. haematobium* is presumably not the mode of action of these compounds against microfilariae. Microfilariae do not become primarily immobilized in-vivo by haloxon or metrifonate. Instead, there is an induction of organophosphate-mediated adherence of phagocytic cells to the microfilariae observable resulting in a final killing of larvae. This effect looks similar but not identical to that of DEC.

Furthermore metrifonate possesses insecticidal activity. In this indication metrifonate is known as trichlorphon (= Dipterex or Dylox).

In the meantime metrifonate is in clinical development in Alzheimer's disease. The symptoms of Alzheimer's dementia are significantly improved. The mechanism of action of metrifonate in this indication relies on the inhibitory effect against brain AChesterase resulting in an increase of acetylcholine in the brain. Thereby, a 70% inhibition of this enzyme is achieved. This leads to an improved function of nerve cells, which is impaired in this disease. Despite this effect, the tolerability of metrifonate is good and side effects are rare.

Resistance Until now resistance against metrifonate is not known. However, against coumaphos and naphthalophos there are resistant *Haemonchus contortus* strains in sheep and goats which led to the ineffectivity of these organophosphates.

Ethanolamines

Important Compounds Bephenium, thenium, methyridin.

Synonyms Bephenium: Alcopar, Francin

Thenium: Bancaris, Canopar; in: Ancaris, Thenatol

Methyridin: Dekelmin, Mintic, Promintic

Clinical Relevance The antinematodal activity of bephenium is directed against *Ascaris lumbricoides*, *Ancylostoma duodenale*, *Trichostrongylus*.

Molecular Interactions Several drugs belong to the class of ethanolamines such as bephenium, thenium and methyridine. They have structural similarity to acetylcholine or nicotine and act as agonists of the acetylcholine receptor (Fig. 1) by binding to the acetylcholine receptors in the nerve cords of the nematodes. The compounds are not inactivated by AChesterase. The result of the agonistic activity is the induction of depolarization and a permanent muscle contraction or spastic paralysis in the worms. The specific toxicity of these compounds is explained by the greater affinity to the parasite's receptors compared to the hosts receptors. At high concentrations they are also toxic for the host.

Pyrantel, Morantel, Oxantel, Levamisole, Tetramisole

Synonyms pyrantel pamoate: in Antiminth, Banminth, Cobantril, Combantrin, Felex, Helmex, Imathal, Nemex, Piranver, Pyraminth, Strongid T; in: Dosalid, Trivexan, Welpan

pyrantel tartrate: in Banminth, Exhelm, Nemex, Pyreguan, Strongid

morantel: in Banminth II, Ibantic, Paratec; rumatel; in: Banminth D, Equiban

oxantel pamoate in: Quantrel

levamisole: Anthelpor, Aviverm, Bionem, Cevasol, Chronomintic, Citarin-L, Cyverm, Dilarvon, Duphamisole, Ketrax, Levadin, Levamisol "Virbac" 10%, Levipor, Levacide, Levasole, Narpenol 5, Nemacide, Nilvern, Ripercol L, Solaskil; in: Ambex, Nilvax, Nilzan, Spectril

tetramisole: Anthelvet, Ascaridil, Citarin, Nemicide, Ripercol, Spartakon

Clinical Relevance Pyrantel has some anticestodal activity. It was effective against tapeworms in horses in field trials. The efficacy against *Anoplocephala perfoliata* is uncertain. However, pyrantel and levamisole are mainly used in the treatment of nematode infections in human and veterinary medicine (→ Microtubule-Function-Affecting Drugs/Table 1). In addition, levamisole has microfilaricidal activity as shown in *Litomosoides carinii* infected *Mastomys coucha*. There is also microfilaricidal efficacy against *Wuchereria bancrofti* and *Brugia malayi* in man (→ Inhibitory-Neurotransmission-Affecting Drugs/Table 1). Of interest is the topical application (eye drops) of levamisole. Levamisole has no effect against microfilariae of *Onchocerca volvulus*. The side effects are similar to that of DEC treatment. Embryotoxic activity is observed in different filarial infections.

Molecular Interactions The mode of action of these drugs relies on their anticholinergic activity. Levamisole is chemically an imidazothiazole, while pyrantel, morantel and oxantel are tetrahydropyrimidine derivatives. They are simultaneously agonists and antagonists of the nematode ACh-receptors. Pyrantel acts as a nicotinic agonist on the acetylcholine receptor in the nematode *Ascaris suum* (Fig. 1). It induces a depolarization and an increase in input conductance of the muscle membrane to sodium and potassium. Pyrantel leads to a spastic paralysis of *Angiostrongylus cantonensis*. Levamisole is taken up by the nematodes via the cuticula, whereas the uptake of pyrantel

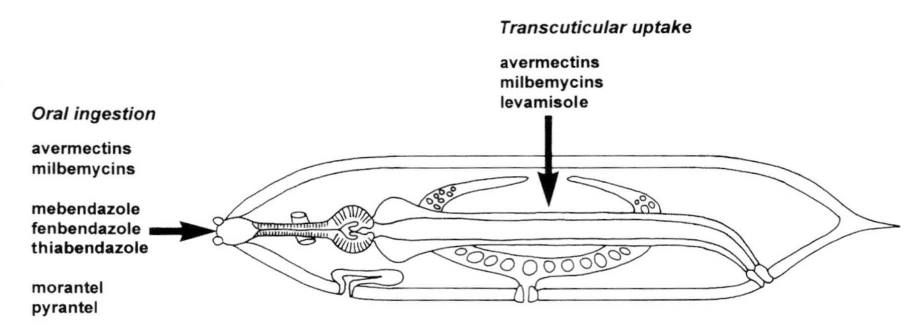

Fig. 3. Route of Uptake of Anthelmintics by Nematodes (representation of the morphology of a sexually mature female worm according to Cox (1996) In: Cox FEG (ed) Modern Parasitology, Blackwell Science, Second edition).

and morantel occurs via the oral route (Fig. 3). In patch-clamp studies these nicotinic anthelmintics open non-selectively cation-channels. Each channel has a characteristic, particular conductance. There are variations of conductance between channels recorded from different patches in the range 19–60 pS. The mean open-time of the channels varies with the anthelmintic in the range of 0.5 ms to 2.5 ms. These nicotinic anthelmintics act on the receptor with properties similar to but not identical with the nicotinic receptors in mammalian and vertebrate hosts because of pharmacological differences. Indeed, there is no significant action of levamisole on host muscle.

The antagonistic effects of levamisole, pyrantel, morantel or oxantel may be explained by the fact that these molecules are too large to pass through the nicotinic acetylcholine receptor channel but they pass through the extracellular opening of the channel and lead to a blockage at the so-called middle ring of the channels. Thus, the middle ring of ACh-channel acts as a bottle neck for passing molecules. Thus levamisole, pyrantel, oxantel and morantel induce a voltage-sensitive channel block at the nicotinic receptors in *A. suum* at higher concentrations.

The antifilaricidal action of levamisole relies presumably on an immunomodulatory effect. Thereby, cell mediated immunity is stimulated resulting in a generally depressed cellular response in microfilariaemic hosts. The levamisole-dependent immunostimulation is observed at low doses. So far there are no experimental data on enhanced immune reactions after levamisol treatment. There is only one report on enhanced filarial antigen-induced inhibition of the migration of macrophages in *B. malayi* infected *Mastomys coucha*.

Besides the antiparasitic activities levamisole exerts antitumor activity.

Resistance Against Levamisole Resistance against levamisole occurs in a variety of nematodes in sheep, goats, cattle and swine. The mode of resistance against levamisole is yet unclear. The heterogeneity in the types of nicotinic receptors are thought to facilitate the development of anthelmintic resistance against nicotinic drugs and perhaps other anthelmintics that act on membrane ion channels. Thus, in *Oesophagostomum dendatum* up to four nicotinic receptor types are present. The α-chains of the nicotinic ACh-receptor from susceptible and resistant *Trichostrongylus colubriformis* and *Haemonchus contortus* show no difference in amino acid sequence. There are different effects on the hatching of eggs between levamisole/morantel susceptible, levamisole/morantel resistant and morantel resistant strains of *H. contortus* and *T. colubriformis* in the presence of levamisole, morantel or pyrantel. Compared to adult susceptible *Haemonchus contortus*, five- to six-fold higher concentrations of ACh and other nicotinic agonists are necessary to cause equivalent contractions in levamosole/morantel resistant worms. It is assumed that levamisole/morantel resistance is due to a reduced number or sensitivity of cholinergic receptors in resistant *H. contortus*. Levamisole resistant mutants of *Caenorhabditis elegans* lack ACh-receptors and there is no response to the nicotinic agonists acetylcholine, nicotine, carbamyl chloride or levamisole which in susceptible *C. elegans* cause contractions. The binding of cholinergic agonists varies between levamisole susceptible and resistant *C. elegans*.

Acquired Immunity

→ Immune Responses.

Acrodermatitis

A. chronica atrophicans is a symptom of phase three of the tick-transmitted → Lyme disease.

ADCC

Antibody-dependent cytotoxicity (→ Immune Responses).

ADI

Synonym
→ Acceptable Daily Intake.

Adjuvant

To increase the immunogenicity of an antigen, various adjuvants are available. Freund's adjuvant, an oil emulsion of heat-killed *Mycobacterium tuberculosis*, saponins and various other formulations have commonly been used in experimental animal models. Because of possibly severe local and systemic reactions, they are considered unsafe for use in people and their use in animals is now restricted. Solely aluminium hydroxides and aluminium phosphate are registered for use in people.

AFC

Antibody forming cell (e.g. plasma cell, B-lymphocyte)

African Swine Fever

Caused by the → ASF virus (Vol. 1) (ungrouped), African swine fever is an acute, contagious disease leading to high mortality in domestic pigs. It is maintained in wild pigs, particularly warthogs (*Phacochoerus* spp.), and in the tick *Ornithodorus porcinus*. It has been reported that infected male *O. porcinus* can infect uninfected females during mating. Warthogs are usually not severely affected, but it can be considered one of the most serious diseases of domestic swine, where it can be transmitted by contact alone and has caused severe economic losses after introduction into European countries.

African Trypanosomiasis

Synonym
→ Sleeping sickness of man, → Nagana of animals

See → Trypanosomiasis, Man, → Serology (Vol. 1).

Agents of Disease

Prions, viruses, bacteria, fungi, parasites

Algid Malaria

Patent malaria also possible in *Plasmodium falciparum* with progressing multiplication of parasites but without fever. Thus this form of disease is often misdiagnosed.

Alimentary System Diseases, Animals

Gastrointestinal parasites affect their hosts, both directly and indirectly, through a wide variety of mechanisms. The diseases are associated mainly with anorexia, loss of productivity, and diarrhoea (→ Clinical Pathology, Animals). The characteristic changes in blood constituents are hypoalbuminaemia and anaemia – for parasites inducing loss of blood. Abomasal parasitism is associated with characteristic increases in the concentration of plasma pepsinogen and liver parasitism with increases in the levels of liver enzymes.

The lesions caused by important parasites of domestic animals have been described many times and will not be dealt with in detail. Briefly, lesions may be almost negligible, as for instance in *Moniezia* infections in ruminants. They may also be confined to the point of attachment of the parasite, as with the tapeworm *Anoplocephala* in horses or the acantocaphalan *Macrocanthorynchus hirudinaceus* in the pig. At the other extreme, lesions caused by some parasites may cover entire parts of the gastrointestinal tract such as during infections with *Ostertagia* in the abomasum, *Strongyloides* in the small intestine, or *Trichuris* in the large bowel. Finally, the liver fluke *Fasciola hepatica* causes hepatic necrosis and haemorrhages during migration, and erosions of the biliary mucosa when it reaches the bile duct.

It is not always possible to establish a link between the severity of the lesions and either the impaired physiology of an organ, or the secondary manifestations of disease, such as anorexia, productivity loss, diarrhoea or haematological changes. For example lambs infected with *Haemonchus contortus* and given a low protein diet show more severe clinical signs of weight loss, anaemia and loss of appetite than others given a normal diet, despite similar levels of blood loss. Dietary protein supplementation may not, however, be successful in animals with concurrent abomasal and intestinal infections in which compensatory digestion and an increase in intestinal absorption of nutrients depend upon the integrity of the latter organ. On the other hand, larval challenge of immune animals may cause impaired production despite low parasite burdens. It is important to consider these nutritional and immunological factors during any field observations. They may well account for some of the conflicting reports on the effect of parasitism. A genetically determined variation in the susceptibility of cattle and sheep to certain helminth infections has also been demonstrated. Inherited resistance, as assessed by faecal egg output, worm burdens or clinical symptoms of disease, has been described e.g. in sheep or cattle infected with *Ostertagia* spp, *Haemonchus contortus*, *Cooperia* spp and *Trichostrongylus* spp. Finally, the interaction between parasitism and the physiological condition of the ewe is well documented: the postparturient rise in worm egg output is attributed to a loss of resistance associated with late pregnancy and lactation. The common clinical signs and pathology of the parasitic diseases of the alimentary tract are summarized in: → Alimentary System Diseases, Ruminants, → Alimentary System Diseases, Horses, → Alimentary System Diseases, Swine, → Alimentary System Diseases, Carnivores.

Therapy and other measurements are presented in the chapters on control and disease control and under → Drugs.

Alimentary System Diseases, Carnivores

The common clinical signs and pathology of the parasitic diseases of the gastrointestinal tract of carnivores are summarized in Table 1; note:

[a] Clinical signs and pathology: 1, ***Weight loss***; 2, ***Anorexia***; 3, ***Vomiting***; 4, ***Diarrhoea***; 5, ***Dehydration***; 6, ***Untriftiness***; 7, ***Anaemia***; 8, ***Abdominal pain***; 9, ***Hypoalbuminaemia***; 10, ***Others***

Occurrence of signs: ±, rare; +, common; ++, very common

* Probably identical with *Neospora caninum*

Protozoal Enteritis

Coccidiosis (***Coccidiosis, animals***) and Cryptosporidiosis have been reported in dogs and cats. *Giardia* spp occurs in dogs and cats, and giardiosis may represent an important problem in these hosts. It causes anorexia, depression, and a mild recurring diarrhoea consisting of soft, light-coloured stools with a characteristic "oatmeal" texture, frequently containing mucus. Growth retardation and cachexia may occur. The mechanism by which giardial malabsorption and diarrhoea occurs is unclear. Epithelial damage, increased turnover of epithelial cells, villous shortening and disaccharidase deficiency have all been reported as manifestations of giardiosis. *Entamoeba histolytica* is the cause of entamoebiasis in humans and among domestic animals. The disease which occurs rarely in dogs is characterized by diarrhoea.

Gastrointestinal Helminthosis

Oesophagus *Spirocerca lupi*, the oesophageal worm, is a parasite usually associated with the formation of nodules in the oesophageal wall. Occasionally, however, it may be found in gastric nodules in cats. Most infected dogs are not clinically affected and the infection is only detected at necropsy. Signs of oesophagal involvement include anorexia, vomiting, dysphagia and bloodstained regurgitus. There is a strong evidence of a causal relationship between *Spirocerca* and oesophageal fibrosarcomes or osteosarcomas. Ventral spondylitis of caudal thoracic vertebrae is also observed, the cause might be migrating worms inducing periosteal irritation.

Stomach The presence of parasites in the stomach of dogs and cats does not commonly produce signs, except in severe infestations. The over-

Table 1. Gastrointestinal parasitic diseases of carnivores (according to Vercruysse and De Bont)

Parasite	Host	Clinical signs and pathology[a]									
		1	2	3	4	5	6	7	8	9	10
Oesophagus											
Spirocerca lupi	Dog	+	+	+				±			Dysphagia haematomesis
Stomach											
Gnathostoma spinigerum	Cat, dog	+		±					+		Anophagia
Ollulanus tricuspis	Cat			+							
Physaloptera spp.	Dog, cat			+							Regurgitation
Small intestine											
Protozoa											
Cryptosporidium	Dog, cat				+						
Cystoisospora spp.	Dog, cat		+		+	+			±		
*Hammondia heydorni**	Dog				±						Abortion in *N. caninum*
Toxoplasma gondii	Cat				+						
Entamoeba histolytica	Dog				+						
Giardia sp.	Dog, cat				+						
Cestoda											
Taenia spp.,	Dog, cat										
Echinococcus spp.											
Mesocestoïdes spp.											
Dipylidium caninum	Dog, cat										Perianal itching (dog)
Nematoda											
Ancylostoma spp.	Dog	+	+		+			++		+	Epistaxis
Ancylostoma tubaeforme	Cat							+			
Uncinaria stenocephala	Dog	+					+	±			
Strongyloïdes stercoralis	Dog	+			+	+		±			
Toxocara canis	Dog	+	+	+	+			±		+	Pot-bellied
Toxocara cati	Cat	+	+	+	+					+	
Toxascaris leonina	Dog, cat	+	+	+							
Large intestine											
Trichuris vulpis	Dog	+	+		+	+		+			

Explanations see page 20

all incidence of these parasites is low. Several nematodes have been identified in the stomachs of dogs, cats and wild carnivora: *Cylicospirura*, *Cyathospirura*, *Spirura*, *Physaloptera* spp., *Gnathostoma spinigerum* and *Ollulanus tricuspis*. Only the latter three parasites are likely to cause signs of gastric parasitism. *Physaloptera* spp. inhabit the stomach and proximal duodenum of many carnivores and may be a cause of vomiting and chronic gastritis. Regurgitation may occur, presumably because of the oesophagitis induced by vomiting. Adults of *G. spinigerum* are found in groups up to 10 in cysts within the gastric mucosa of cats and dogs. The cysts may be 2 cm in diameter, and communicate with the gastric lumen by small openings. Gnathostomosis has been associated with illness and death. Symptoms include anophagia, occasional vomiting, abdominal pain and loss of weight.

Ollulanus tricuspis inhabits the stomach of cats. Although most infections are asymptomatic, *O. tricuspis* should be included in the differential diagnosis for vomiting in cats, especially when vomiting is postprandial.

Small Intestine Several trematodes cause infections in carnivores. *Schistosoma japonicum* live within the mesenteric and hepatic portal veins of dogs (***Schistosomiasis, Animals*** / Ruminants).

Many genera of trematodes (*Apophallus*, *Heterophyes*, *Alaria*, *Nanophyetus* and others) are parasitic in dogs and cats throughout the world, but they are only rarely of any clinical consequence. *Nanophyetus salmincola* is important as a vector of the highly pathogenic *Neorickettsia helminthaeca*, the cause of "salmon poisoning" in dogs, a disease with a high mortality rate.

Dogs may be parasitized by a great number of cestodes. Some common species are *Taenia hydatigena*, *T. pisiformis*, *T. ovis*, *T. multiceps*, *T. serialis*, *Dipylidium caninum*, *Echinococcus* spp., *Mesocestoides* spp. and *Spirometra* spp. Two species occur frequently in cats: *Taenia taeniaeformis* and *Dipylidium caninum*.

In most cases there are few, if any, clinical symptoms. Infections have been associated with diarrhoea and failure to thrive, but concomitant gastrointestinal nematode parasitism may often be of greater significance. The passage through the rectum and anus of the gravid segments of *Dipylidium caninum* produces irritation in dogs, but not in cats. The proglottides passing through the rectum, or lodging in the anal glands, frequently cause the dog to scoot.

Strongyloidosis in dogs *is caused by Strongyloides stercoralis*. Though not common, infection in young animals may have severe consequences. There is enteritis with erosion of the mucosa of the small intestine, and haemorrhages. Bloody diarrhoea occurs in heavy infections. Dehydration develops rapidly, and death may occur.

There are four hookworm species that commonly affect dogs and cats. *Ancylostoma caninum* and *A. tubaeforme* are the most pathogenic species of dogs and cats, respectively. *A. braziliense* is another hookworm parasite of dogs and cats and is not very pathogenic. *Uncinaria stenocephala* is mainly a hookworm of dogs, in cats it is much less common. *U. stenocephala* is the least pathogenic of the hookworms that affect dogs and cats. With *A. caninum*, disease results from blood loss into the bowel which leads to iron deficiency, anaemia (hypochromic, microcytic). In chronic infections there is emaciation, poor appetite, and pica and the animals are frail. There is some diarrhoea and the faeces are dark and sometimes blood-streaked.

With respect to ascarid infections *Toxocara canis* is by far the commonest roundworm of dogs. It is not highly pathogenic and the effects of infection are often overstated. The pulmonary phase may be lethal in pups infected heavily before birth; death occurring within a week after birth. This is the most severe effect of *T.canis* infection. Once the worms have become adults in the small intestine the only clincal signs are those of pot-bellied abdomen, light diarrhoea, and failure to thrive. Adult worms are often passed in the faeces or vomited. The other ascarid of the dog, and rarely the cat, *Toxascaris leonina*, is non-migratory and of little pathological importance.

Toxocara cati, the ascarid of cats, is endemic in all countries. As for infections with *T. canis*, many cases remain asymptomatic.

The Acanthocephalan *Onicola canis* occurs in the small intestine of wild carnivores and occasionally of dogs and cats. The lesions are the same as those described for *Macrocanthorynchus hirudinaceus* in swine (***Alimentary System Diseases, Swine***), although clinical manifestations are rare.

Caecum, Colon, Rectum, Anus The most important species are *T. vulpis* in dogs, and *T. campanula* and *T. serrata* in cats. *Trichuris* is highly prevalent in all parts of the world but rarely causes clinical signs. Heavy infections associated with severe and often haemorrhagic typhlitis or typhlocolitis has been reported in dogs. Clinical manifestations include failure to thrive, weight loss, abdominal pain, diarrhoea, dehydration, and terminal anaemia. In severe cases the faeces may be speckled with fresh blood. The lesions are caused by the adult worms boring tunnels into the mucosa of the large intestine.

Liver

The commonest parasite of the liver of cats, and less frequently in dogs, is *Opisthorchis tenuicollis*. The signs of infection are emaciation, jaundice, and ascites.

Capillaria hepatica occurs only rarely in the liver of cats and dogs, and infection remains inapparent.

Therapy

Chemotherapy, Drugs.

Alimentary System Diseases, Horses

The common clinical signs and pathology of the parasitic diseases of the gastrointestinal tract of horses are summarized in Table 1; note:

[a] Clinical signs and pathology: 1, Loss in performance, → weight loss; 2, → Anorexia; 3, → Diarrhoea; 4, → Constipation; 5, → Abdominal pain; 6, → Hypoalbuminaemia; 7, Others

Occurrence of signs: ±, rare; +, common; ++, very common

Protozoal Enteritis

The only *Eimeria* species occurring in horses, *Eimeria leuckarti*, is not known to cause Coccidiosis (→ Coccidiosis, animals). *Cryptosporidium* has been reported in all domestic animals (→ Cryptosporidiosis, animals) including horses. In horses, symptoms have been particularly observed in animals with inherited or acquired immuno-deficiency. *Giardia* spp. are usually non-pathogenic inhabitants of the small intestine of horses. However, they may cause diseases under certain circumstances, such as in animals which are immunocompromised, malnourished, or very young.

Gastrointestinal Helminthosis

Stomach *Trichostrongylus axei* occurs in the stomach of horses. The worm is rarely a pathogen on its own, most infections are chronic and mild. However, *T. axei* induces typical lesions in horses. The condition has been described as a *gastritis chronica hyperplastica et erosiva circumscripta* for the main lesion is a pad- or cushion-like thickening in the glandular part of the stomach.

Table 1. Gastrointestinal parasitic diseases of horses (according to Vercruysse and De Bont)

Parasite	Clinical signs and pathology[a]						
	1	2	3	4	5	6	7
Stomach							
Protozoa							
Cryptosporidium spp.	+		+				
Nematoda							
Draschia megastoma	+						
Habronema spp.							
Trichostrongylus axei	+						
Arthropoda							
Gasterophilus spp.	+						Yawn
Small intestine							
Cestoda							
Anoplocephala magna	+		±				
Nematoda							
Strongyloides westeri	+		++				
Parascaris equorum	+	+	+	±			Pot-bellied,respiratory signs
Large intestine							
Trematoda							
Gastrodiscus aegyptiacus	+		±				Anaemia
Cestoda							
Anoplocephala perfoliata	+		±		±		Spasmodic colic, ileal impaction colic
Nematoda							
Strongylus spp. *(adults)*	+	+	+		±		
Triodontophorus spp.	+	+	+				
Cyathostominae	+	+	++	+	+	++	
Oxyuris equi	+						Anal pruritis

Explanations see above

The most common parasites of the equine stomach are larvae of botflies of the genus *Gasterophilus*. The most common species which pass to the stomach and settle on the gastric mucosa are *G. intestinalis* followed by *G. haemorrhoïdalis*. Despite the dramatic appearance of a heavy bot infestation, its true veterinary significance is unclear. The parasites may produce significant gastric lesions, but primarily in the non-glandular part of the stomach which plays little role in digestion. This is possibly the reason why the vast majority of infections remain asymptomatic. Yawning and unsatisfactory performance are described as common signs. Erratic movements of the larvae into the abdomen, (with or without peritonitis) are described.

The spiruroid nematodes *Habronema majus*, *H. muscae* and *Draschia megastoma* are also parasitic in the stomach of equidae. *Habronema* species lie on the mucosal surface while *Draschia* burrows in the submucosa to produce large, tumor-like nodules. *D. megastoma* is therefore the most pathogenic of the stomach worms, although clinical signs remain inapparent in most cases. The parasite is often incorrectly blamed for the disease and death of the host because of the very impressive character of the lesions it produces.

Small Intestine The cestodes found in the small intestine are *Anoplocephala magna* and *Paranoplocephala mamillana*. In most cases they induce few, if any, clinical symptoms. The only *Strongyloides* species in horses is *Strongyloides westeri* (→ Strongyloidosis, Animals/Ruminants). Clinical outbreaks principally affect young suckling foals. Signs include anorexia, loss of weight, coughing, diarrhoea (rarely haemorrhagic), dehydration, slight to moderate anaemia. Severe infections may be fatal. The ascarid *Parascaris equorum* is a common parasite in young horses. Both the parasitic migration through the lungs and the intestinal phase of the life cycle may be associated with clinical signs. Symptoms include lethargy, loss of appetite, coughing, nasal discharge and decreased weight gain. In the rare severe infections the intestinal phase may cause impaction, rupture, peritonitis, intussusception, and formation of abscesses.

Caecum, Colon, Rectum, Anus Members of the Trematode genus *Gastrodiscus* may be found in the colon of horses and swine, but they are of little clinical significance. However, the immature trematode has been reported to cause a severe and hyperacute, possibly fatal colitis in horses.

The cestode *Anoplocephala perfoliata* is commonly found in the distal small intestine, caecum and proximal large intestine. A significant number of the parasites often accumulates at the ileocaecal junction and may produce marked lesions there. *A. perfoliata* is a significant risk factor for spasmodic colic and ileal impaction colic in the horse. The risk of spasmodic colic increases with infection intensity.

Members of the Strongylidae are abundant and common nematode parasites of the caecum and colon in equidae. There are 55 species of strongyles, with fewer than 20 commonly found in horses, usually as mixed infections. Therefore, the clinical signs can be considered to be caused by all the worm species collectively. Specific clinical signs may arise due to the larval stages of the *Strongylus* spp. These are described in the section on the cardio-vascular system. Adult large strongyles are inhabitants of the large intestine. They feed by attaching to the glandular epithelium and drawing a plug of mucosa into the buccal capsule. Incidental damage to blood vessels in the plug results in bleeding. Some, as seen with clusters of *Triodontophorus* spp, cause deep ulcers. The damage thus caused results in the formation of crater-like ulcers which may extend deep into the gut wall. These lesions are believed to be the cause of anaemia, failure to thrive and poor performance. The colitis and typhlitis do not often lead to diarrhoea. The adults of the Cyathostominae feed mainly on intestinal contents and are of little pathogenic significance. *Cyathostoma*-associated clinical disease is attributed to the synchronous emergence of large numbers of previously inhibited third- and fourth-stage larvae from the large intestinal wall, resulting in physical disruption of the mucosa and typhlitis/colitis. Animals sometimes develop a fatal syndrome (winter cyathostominosis) characterized by sudden onset of diarrhoea, mild colic signs, failure to thrive or cachexia, and hypoalbuminaemia. Emaciation possibly results from the anorexia, the reduction of absorptive function and the loss of protein through the disrupted intestine. A reduced level of ileo-caeco-colic motility was recorded in ponies experimentally infected with a mixture of strongyles, predominantly cyathostomes. It was argued that this would reduce propulsion of ingesta, and cause the loss of appetite and weight. Additional mechanisms which may contribute to the occurrence of cyathostome-associated colic include intestinal mucosal oedema and/or vasoconstriction induced by the local production of vaso-active substances in response to the presence of cyathostome mucosal stages.

The only oxyurid of importance is *Oxyuris equi*, the large pinworm of horses. It is never regarded as a serious pathogen. The chief feature of oxyuriosis in equines is the anal pruritus produced by the egg-laying females. The irritation caused by the anal pruritus produces restlessness and improper feeding. The animal rubs the base of its tail against any suitable object, causing the hairs to break off and the tail to acquire an ungroomed "rat-tailed" appearance.

Liver

A variety of cestodes, nematodes and trematodes produce inflammation of the liver and bile ducts, but they are of minor importance in horses. *Parascaris equorum*, *Strongylus equinus* and *S. edentatus* migrate through the liver, but only *S. edentatus* has been associated with transient colic. *Echinococcus granulosus* is commonly found in the liver. Although they may involve a large amount of tissue, hydatid infections appear to be well tolerated. *Fasciola hepatica* is occasionally found in the equine liver. Heavy infections are rare and are usually only discovered during post-mortem examination.

Abdominal Cavity

Most of the parasites found in the peritoneal cavity have their final habitat elsewhere and passage through the peritoneum occurs in the normal course of migration, or by accident. Parasites normally migrating through the peritoneal cavity are the immature *Fasciola hepatica*, *Strongylus edentatus*, and *Strongylus equinus*. They can cause acute and chronic peritonitis.

Examples of accidental passages are those of *Parascaris equorum*, and *Gasterophilus*, which enter the cavity through intestinal or gastric perforations.

Therapy

→ Chemotherapy, → Drugs.

Alimentary System Diseases, Ruminants

The common clinical signs and pathology of the parasitic diseases of the gastrointestinal tract of ruminants are summarized in Table 1; note:

[a] Host: C, cattle; S, sheep; G, goats

[b] Clinical signs and pathology: 1, → Anorexia; 2, → Diarrhoea; 3, → Oedema; 4, → Dehydration; 5, → Pale mucosa; 6, → Productivity loss; 7, → Death; 8, → Anaemia; 9, → Hypoalbuminaemia; 10, → Pepsinogen increase

Occurrence of signs: ±, rare; +, common; ++, very common

Protozoal Enteritis

Coccidiosis (→ Coccidiosis, Animals) and Cryptosporidiosis (→ Cryptosporidiosis, Animals) have been reported in ruminants. *Giardia* spp. are usually non-pathogenic inhabitants of the intestine of cattle, sheep and goats. However, they may cause diseases under certain circumstances, such as in animals which are immunocompromised, malnourished, or very young.

Gastrointestinal Helminthosis

Abomasum Various infections of the abomasum have been reported of which → Ostertagiosis is probably the most important disease in grazing sheep and cattle in temperate climatic zones throughout the world. It causes subclinical losses of production and disease. The clinical disease is characterized by diarrhoea, weight loss, decreased production, rough hair coats, partial anorexia, mild anaemia, hypoalbuminaemia, dehydration and in some cases death. *Ostertagia ostertagi* in cattle and *O. (Teledorsagia) circumcincta* in sheep and goats are the most important species.

→ Haemonchosis is a common and severe disease of the ruminant abomasum in many parts of the world. *Haemonchus contortus* infects mainly sheep and goats, while *H. placei* occurs mainly in cattle. The pathogenesis of *Haemonchus* infection is the results of anaemia and hypoproteinaemia caused by the bloodsucking activity of the parasite.

Trichostrongylus axei lives in the abomasum of cattle, sheep and goats. In ruminants, *T. axei* infections are usually part of a mixed abomasal helminthosis and its effects cannot be dissociated from those of other worm species. The worm is rarely a pathogen on its own, as most infections

Table 1. Gastrointestinal parasitic diseases of ruminants (according to Vercruysse and De Bont)

Parasite	Host[a]	Clinical signs and pathology[b]									
		1	2	3	4	5	6	7	8	9	10
Abomasum											
Nematoda											
Haemonchus		±		++		++	++	+	++	+	±
H. placei	C										
H. contortus	S, G										
Ostertagia		++	++	+	+		++	+	±	++	++
O. ostertagi,	C										
O. circumcincta,	S, G										
O. trifurcata											
Trichostrongylus axei	C, S, G	+	+				+			±	±
Mecistocirrus digitatus	C	±		±		+	++	±	++	+	±
Small intestine											
Protozoa											
Cryptosporidium	C, S, G	+	++		±		++				
Eimeria											
E. bovis, E. zuernii	C	+	++		±		++	+	±	+	
E. ovinoidalis	S	+	++		+		++	+		+	
E. ninakohlyakimovae	G	+	++		+		++	+	±	+	
Trematoda											
Calicophoron, Cotylophoron, Fishoederius, Gastrothylax, Paramphistomum, Schistosoma	*C, S, G*	++	++	++			++	+		++	
Cestoda											
Avitellina, Moniezia, Thysaniezia, Thysanosoma	C, S, G						+				
Stilesia globipunctata	S, G	+	++				+				
Nematoda											
Cooperia		+	+				+			+	
C. curticei	S, G										
C. oncophora	C										
C. pectinata, C. punctata	C, S, G										
Hookworms		+	+	+		+	++		++	+	
Bunostomum phlebotomum,	C										
B. trigonocephalum,											
Gaigeria pachyscelis	S, G										
Nematodirus		+	++		+		++	+(S)		+	
N. battus	S										
N. filicollis, N. spathiger	S, G, C										
N. helvetianus	C										
Strongyloides papillosus	S, G, C	+	++				++	+	+	+	
Toxocara vitulorum	C	+	++				++	±		+	
Trichostrongylus		++	++	+	+		++	+	±	++	
T. colubriformis	S, G, C										
T. vitrinus	S, G										
Large intestine											
Nematoda											
Chabertia ovina	S, G	+	+				+			+	
Oesophagostomum		+	+	+			++	±	+	++	
O. columbianum	S, G										
O. radiatum	C										
O. venulosum	C, S										
Trichuris		+	+				+			+	
T. discolor, T. globulosa	C										
T. ovis	S, G										

Explanations see top of page 25

are mild. Animals experimentally infected with large numbers of *T. axei* show a decrease of blood albumin, haemoconcentration and a rise in serum pepsinogen. The clinical signs include diarrhoea, anorexia, progressive emaciation, listlessness and weakness.

The pathogenesis of *M. digitatus* resembles that of *H. placei*. A drop in haematocrit to less than 20% has been reported in cattle infected with more than 1000 worms.

Small Intestine Paramphistome infections may cause significant intestinal problems in ruminants (→ Paramphistomosis). The adult worms live in the rumen, but the pathological effects of infection are caused by the immature stages within the small intestine. The most pathogenic species are thought to be *Paramphistomum microbothrium*, *P. ichikawai*, *P. cervi*, *Cotylophoron cotylophoron* and various species of *Gastrothylax*, *Fishoederius* and *Calicophoron* (→ Digenea (Vol. 1)).

Schistosomes live within the mesenteric and hepatic portal veins of ruminants. Although *Schistosoma* infections in ruminants are highly prevalent in certain regions, the general level of infestation is often too low to cause clinical disease or losses in productivity. Levels sufficiently high to cause outbreaks of clinical schistosomosis do occur occasionally and infestation becomes manifest either as an intestinal syndrome which is usually self limiting, or as a chronic hepatic syndrome, which is usually progressive (Schistosomiasis, animals). The intestinal syndrome is caused by the deposition of large numbers of eggs in the intestinal wall and usually follows a heavy infestation in a susceptible animal, i.e. an animal in which the capacity of the host to suppress the egg laying of the parasite has not been stimulated by previous infestations. This has been reported among cattle, sheep and goats infected with either *S. bovis* or by *S. mattheei*. As the faecal egg counts rise sharply with the onset of egg production the animal develops a mucoid and then haemorrhagic diarrhoea, accompanied by anorexia, loss of condition, general weakness and dullness, roughness of coat, hypoalbuminaemia and paleness of mucous membranes. Death may occur a month or two after the onset of clinical signs. In most cases, the animal makes a spontaneous but slow recovery. The primary cause of the diarrhoea is the passage of large numbers of eggs through the wall of the intestine. The anaemia is usually due to an increased rate of red cell removal from the circulation; while haemodilution and the inability to mount a sufficiently effective erythropoietic response are of secondary importance. The underlying cause of the hypoalbuminaemia is hypercatabolism of albumin due to substantial loss of protein in the gastro-intestinal tract.

In ruminants the more common and widely distributed intestinal → cestodes (Vol. 1) are *Moniezia expansa*, *M. benedeni*, *Thyzaniezia* (*Helicometra*) *giardi*, *Stilesia globipunctata* and *Avitellina* spp. These tapeworms live in the middle third of the small intestine. They are generally of minor pathological importance and only produce harmful effects on the host in rare circumstances. The host-parasite relationship is so well adjusted that there is a surprising absence of apparent pathogenicity, even in such heavy infestations that the passage of intestinal contents may be impeded by the worms and, in most cases there are few, if any, clinical symptoms. Infections have been associated with diarrhoea and ill thriving, but concomitant gastrointestinal nematode parasitism may often be of greater significance. In sheep *S. globipunctata* infections may cause signs of enteritis. The pathogenesis of this parasite is associated with the immature stages, which enter the mucosa of the intestine and induce the formation of nodules.

Several nematode species cause severe gastrointestinal diseases such as → Strongyloidosis, → Trichostrongylosis, → Cooperiosis, and → Nematodirosis. Hookworm infections are caused by the two ancylostomatids occurring in cattle *Bunostomum phlebotomum*, and the little known *Agriostomum vryburgi*. Two species occur in sheep and goats: *Bunostomum trigonocephalum* and *Gaigeria pachyscelis*. Apart from the little damage to the mucosa of the small intestine caused by the bites, the entire pathogenesis of these worms is attributable to bloodsucking, which begins when larvae enter the adult stage. In addition to the withdrawal of great quantities of blood by the parasites, the feeding sites continue to bleed for several minutes after the worms have moved to another area. The signs are those of a rapid loss of blood: pale mucosae, hydraemia, sometimes oedema, lassitude, and loss of weight. The faeces are often diarrhoeic and contain bloody mucus, or may be tarry in character.

The anaemia that develops is first normocytic and normochromic, but later becomes microcytic

Table 2. Parasitic diseases of the liver (ruminants) (according to Vercruysse and De Bont)

Parasite	Clinical signs
Fasciola hepatica, F. gigantica	Acute (sheep): anorexia, distended abdomen pain, death Mild hypoalbuminaemia, hyperglobulinaemia eosinophilia, light anaemia Chronic: loss of appetite, pale mucosae, oedema, productivity loss Hypoalbumiunaemia, anaemia, SGOT and -GT rise
Fascioloides magna	Sheep: weakness and death
Dicrocoelium spp.	Loss of condition
Gigantocotyle explanatum	Decreased production, enlargement of bile ducts and inflammatory reactions
Schistosoma spp.	Loss of condition, diarrhoea

hypochromic as the animal becomes iron-deficient. Other pathological changes recorded are hypoproteinaemia, hypocalcaemia, hyperglycemia and eosinophilia. The lowered plasma protein levels may be caused by both a compensatory replacement of haemoglobin at the expense of circulating plasma proteins and the loss of blood taken by the parasites. The non-specific loss of body fluid by leakage would account for the lowered plasma calcium values; the calcium associated with plasma albumin being lost with the protein.

Toxocara vitulorum, the pathogenic ascarid of large ruminants, is most commonly found in buffalo and cattle calves of less than 4 months of age. Infection occurs through ingestion of milk from the mother. The clinical signs in calves relate primarily to the bulk of parasites in the small intestine, which impedes the passage of ingesta and impairs the assimilation of food. The clinical signs include anorexia, diarrhoea or constipation, dehydration, steatorrhoea, abdominal pain and a butyric odour of the breath.

Caecum, Colon, Rectum, Anus *Trichuris* spp., the whipworms, inhabit the caecum and occasionally the colon of ruminants (→ Trichuriasis, animals). Members of the genus *Oesophagostomum* infect cattle, sheep and goats (→ Oesophagostomosis).

Chabertia ovina is found in the colon of sheep, cattle, goats and deer throughout the world. Infections are usually light in intensity, and outbreaks of clinical disease are sporadic. Disease in sheep is associated with the presence of fifth stage and adult worms in the colon. Affected animals have a severe diarrhoea, sometimes blackened by blood. Ill thriving has been reported. The adults penetrate the muscularis mucosae and take a plug of mucosa into the buccal capsule; minor haemorrhage may be related to physical trauma to the mucosa but this blood loss is insufficient to induce anaemia in natural infections. More significant is the loss of plasma protein from the mucosa, which may cause hypoalbuminaemia and weight loss.

Liver

The common clinical signs and pathology of the parasitic diseases of the liver of ruminants are summarized in Table 2.

A variety of cestodes, nematodes and trematodes produce inflammation of the liver and bile ducts. Some of the parasites produce hepatic lesions in the course of their natural or accidental migration to their final habitat in the guts. The traumatic lesions produced by such larvae rarely cause disease, and are mostly found incidentally during post-mortem examination. Very heavy infections in sheep with the oncospheres of *Cysticercus tenuicollis* may induce severe haemorrhagic migration tracts in the liver. Animals are weak, show abdominal pain, and have an enlarged, often palpable liver. Death may occur.

The most important parasites of the liver are those which have their final habitat in this organ.

Trematodes A variety of trematodes are very important parasites of the liver of animals. They belong to the families Fasciolidae (*Fasciola hepatica*, *F. gigantica* and *Fascioloides magna*), Dicrocoeliidae (*Dicrocoelium dendriticum*, *D. hospes*), and Paramphistomatidae (*Gigantocotyle explanatum*).

Fasciola hepatica, the common liver fluke, is the most widespread and important of the group. *F. gigantica* occurs in the tropics. It occurs mainly in sheep and cattle but a patent infection can develop in horses, pigs, wild animals, and in humans. The pathogenesis of fascioliasis is attributable in part to the invasive stages in the liver and in part to the blood feeding by the adults in the bile ducts. The process in all hosts shows close similarities, but considerable variation in severity occurs (→ Fasciolosis, Animals).

The host range of *Fascioloides magna* includes cattle, bison, sheep, goat and horse but it is only recognized as a pathogen of sheep. In sheep this parasite wanders continuously in the liver and causes extensive parenchymal destruction. Even a few flukes may kill a sheep. *Dicrocoelium*, as with many liver flukes, has a wide host range including both sheep and cattle. The pathological changes in the liver are less severe than those seen in *F. hepatica* infection (→ Fasciolosis, Animals) and even in heavy infections there may be no clinical signs.

A chronic hepatic syndrome has been described in cattle infected with *Schistosoma mattheei*. This syndrome is less common than the intestinal syndrome (see above). It is characterized by progressive hepatic fibrosis and may be manifested clinically as chronic hepatic insufficiency, with loss of condition, or as acute terminal hepatic failure, which often provokes nervous signs as ataxia or mania. The established condition is invariably fatal. The syndrome is of immunological origin and apparently depends on the ability of the ox to limit the infestations by destruction of adult parasites. It is probably initiated by an immunological reaction in the hepatic portal veins to antigen released by dead or dying worms in the portal system. The lesion is analogous to Symmer's pipe-stem fibrosis in man.

Gygantocotyle explanatum occurs in the bile ducts of ruminants in Asia. Heavy infections induce inflammatory reactions, with enlargement of the bile ducts. Decreased production has been reported.

Cestodes *Thysanosoma actinoides* and *Stilesia hepatica* are parasites of the bile ducts of ruminants. They are both practically non-pathogenic.

Abdominal Cavity

Most of the parasites found in the peritoneal cavity have their final habitat elsewhere and passage through the peritoneum occurs in the normal course of migration, or by accident. The only parasite of importance is *Fasciola hepatica* as it can cause acute and chronic peritonitis.

Other parasites, such as the commonly seen *Setaria*, use the peritoneal cavity as their final habitat. All members of this genus live as well-adjusted symbiants in their normal host and do not cause peritoneal lesions.

Therapy

→ Chemotherapy, → Drugs.

Alimentary System Diseases, Swine

The common clinical signs and pathology of the parasitic diseases of the gastrointestinal tract of swine are summarized in Table 1; note:

[a] Clinical signs and pathology: 1, → Loss in performance, → weight loss; 2, → Anorexia; 3, → Diarrhoea; 4, → Constipation; 5, → Abdominal pain; 6, → Hypoalbuminaemia; 7, Others

Occurrence of signs: ±, rare; +, common; ++, very common

Protozoal Enteritis

Coccidiosis (→ Coccidiosis, animals) and Cryptosporidiosis (→ Cryptosporidiosis, Animals) have been reported in pigs. *Balantidium coli* occurs in the large bowel in pigs and sometimes other mammals. It is normally present as a commensal in the lumen but is capable of invasion of tissues injured by other diseases (e.g., *Trichuris* infection, see below).

Gastrointestinal Helminthosis

Stomach There are several helminths that possibly live in the stomach of swine. *In contrast to ruminants, Trichostrongylus axei* occurs rarely in pigs (→ Trichostrongylosis). *Hyostrongylus rubidus* is the most important parasite of the stomach of swine. This trichostrongylid nematode produces lesions resembling those caused by *Ostertagia* in ruminants, but with less severe consequences. Most experimental infections of moderate intensity do not produce obvious clinical signs or loss of production. After heavy infections, plasma protein loss and an increase in plasma pepsinogen values have been reported. Listlessness, thirst, anorexia, anaemia, diarrhoea and reduced weight gains also occurred. Under field conditions hyostrongylosis is associated with the "thin-sow syndrome". However, mixed infections of *H. rubidus* and *Oesophagostomum* spp. commonly occur in the field, which complicates the diagnosis of hyostrongylosis based on clinical signs. Parasite or host-strain differences may account for some

Table 1. Gastrointestinal parasitic diseases of swine (according to Vercruysse and De Bont); explanations see page 29.

Parasite	Clinical signs and pathology[a]						
	1	2	3	4	5	6	7
Stomach							
Nematoda							
Hyostrongylus rubidus	+	+				+	+
Spirurid infections	+	+					
Small intestine							
Protozoa							
Isospora suis	+	+	++	+	+		
Cryptosporidium spp.	+	+	++	+			
Nematoda							
Strongyloïdes ransomi	+	+	+			+	+
Ascaris suum	+	+	+	+			+
Large intestine							
Protozoa							
Balantidium coli			+				
Nematoda							
Oesophagostomum dentatum, O. quadrispinulatum	+	+	+			+	+
Trichuris suis	+	+	+		+	+	+

of the variations recorded in the literature regarding the effects of experimental *H. rubidus* infections on swine health and performance.

Gnathostoma hispidum live deep in the gastric mucosa where they may cause severe ulceration. Migrating larvae may be found in the liver where they leave necrotic tracks. *Physocephalus sexalatus* and *Ascarops strongylina* may be found free in the lumen or partly embedded in the mucosa. *Simondsia paradoxa* female worms form palpable nodules in mucosal crypts. Clinical signs of acute or chronic gastritis, and occasional ulceration have only been observed in the rare heavy infections.

Small Intestine The trematode *Schistosoma japonicum* lives within the mesenteric and hepatic portal veins of pigs (→ Schistosomiasis, Animals).

Strongyloidosis caused by *Strongyloides ransomi* occurs in swine. Clinical outbreaks principally affect piglets. Signs include anorexia, loss of weight, diarrhoea (rarely haemorrhagic), dehydration, slight to moderate anaemia. Severe infections may be fatal.

Hookworms of the genus *Globocephalus* appear to be of little significance in swine (→ Alimentary System Diseases, Ruminants).

Ascaris suum is a very common parasite of pigs. Its importance is related both to the sometimes very significant lesions caused by larvae during migration through the liver and lungs, and to the effects of adult worms in the small intestine. In the liver, the travelling larva causes intralobular necrosis and a granulation reaction known as 'milk spots'. However, these lesions are not associated with symptoms. In the ordinary moderate infections, adults have no defined pathogenesis. In heavy infection, there is experimental evidence of reduction in growth rate, diarrhoea and ill thriving. Intestinal obstruction is rare.

The Acanthocephalan *Macrocanthorynchus hirudinaceus* is the thorny-headed worm of swine, which infects the small intestine. The worm causes trauma at the site of attachment with its thorny proboscis. The proboscis may penetrate as far as the serous coat, and nodules of up to a centimetre in diameter may be visible on the serosal surface of the gut. They occasionally perforate, causing peritonitis. Severely infected pigs may suffer ill thriving and anaemia. The latter is probably related to the loss of plasma protein and to the haemorrhages from numerous ulcerative lesions. The pigs often show signs of acute abdominal pain during these infections.

Caecum, Colon, Rectum, Anus The whipworm *Trichuris suis* inhabits the caecum of pigs. Heavy infections associated with severe and often haemorrhagic typhlitis or typhlocolitis has been reported. Clinical manifestations include anorexia, dysentery, dehydration, weight loss and terminal anaemia. In severe cases the faeces may be markedly haemorrhagic or even all blood. The lesions are caused by the adult worms boring tunnels into the mucosa of the large intestine. Penetration of the mucosa by the parasites produce nodules in the intestinal wall (→ Alimentary System Diseases, Carnivores). It has been shown that concurrent infections with *T. suis* enhances the ability of opportunistic bacteria to multiply and cause disease and pathology.

→ Oesophagostomosis has been reported in pigs. The two common species found in pigs are *O. quadrispinulatum* and *O. dentatum*. Though the parasites themselves are generally highly prevalent, clinical oesophagostomosis is not common in pigs.

Liver

The most important parasitic condition affecting the liver of pigs is the "milk spots" caused by the passage of ascarid larvae (see supra). The pathogenesis of *Stephanurus dentatus* infections (→ Urinary System Diseases, Animals) is primarily related to the damage caused by the larvae during their migration through the liver. Hepatitis cysticercosa, resulting from the migration of *Cysticercus tenuicollis*, may be rarely observed in pigs (→ Alimentary System Diseases, Ruminants).

Fasciola hepatica has been found in pigs, but infections are very rare. Hepatic schistosomosis occur in swine infected with *S. japonicum*. The lesions resemble those observed in cattle (→ Alimentary System Diseases, Ruminants).

The opisthorchid flukes (*Opisthorchis tenuicollis*, *Clonorchis sinensis*) are normally parasitic in the bile ducts of carnivores. However, they may also occur in swine and humans. The signs of heavy infection are emaciation, jaundice and ascites.

Abdominal Cavity

Most of the parasites found in the peritoneal cavity have their final habitat elsewhere and passage through the peritoneum occurs in the normal course of migration, or by accident. Parasites normally migrating through the peritoneal cavity are the immature *Fasciola hepatica* and *Stephanurus dentatus*. Both can cause acute and chronic peritonitis.

Therapy

→ Chemotherapy, → Drugs.

Allergen

Substance (e.g. house dust) or portions of an animal (e.g. mite), that cause an increased immune-reaction in humans.

Allergy

Increased immune reactivity of humans on repeated contact with allergenic substances.

Alopecia

Clinical and pathological symptoms (e.g. loss of hair) of infections with skin parasites (→ Skin Diseases, Animals, → Demodicosis, Man, → Acariosis, Animals, → Lice (Vol. 1)).

Aluminium Hydroxide

The only adjuvant for vaccines licensed for use in humans.

Alveococcosis

→ Echinococcosis due to infection with → *Echinococcus multilocularis* (Vol. 1).

American Trypanosomiasis

→ Chagas' Disease, Man.

Amoebiasis

Synonyms
Amoebic dysentery; *Entamoeba histolytica* infection.

Distribution
Fig. 1

Pathology
Infectious disease caused by the amoeba → *Entamoeba histolytica* (Vol. 1) which may be commensal in the gut for long periods of time, or it may invade the mucosa soon after infection. The dose of infection, the strain of amoebae, the nutritional state of the patient and the nature of the intestinal flora are likely to be determinants of the development of disease. Axenic animals with an intestine free of bacteria do not appear to develop invasive amoebic infection. Some strains of amoebae are more pathogenic and appear to share isoenzyme or genomic patterns, described as zymodemes or schizodemes. Although multiple pathogenic mechanisms of *E. histolytica* have been described, none have been definitively linked with pathogenicity. The amoebic trophozoites are best demonstrated with the periodic acid Schiff (PAS) technique (→ Pathology/Fig. 4C,D) because most trophozoites contain glycogen. Also, their cytoplasm stains more distinctly with Giemsa or eosin than the smaller macrophages. They adhere to the intestinal epithelium and generally invade in areas where the intestinal mucus appears depleted. The amoebae fan out in the lamina propria and submucosa, giving rise to "flask-shaped" ulcers (→ Pathology/Fig. 4A). The histiolytic nature of the amoebae is suggested by the lysis of the surrounding cells, seen light microscopically by lysis of the nuclei and by ultrastructural changes in the cytoplasm. However, the clear space around individual amoebae is largely a fixation artifact. Neutrophils are attracted by the amoebae and are degranulated and lysed (→ Pathology/Fig. 4B). This release of neutrophil granules may contribute to tissue destruction. The absence of neutrophils around the ulcers is notable. The absence of fecal leukocytes with positive results of a guaiac test for blood is a useful diagnostic finding. However, eosinophilic leukocytes are often seen histologically and → Charcot-Leyden crystals may be found in the stool. The magna-stages phagocytose red blood cells and cell debris which distinguishes *E. histolytica* from nonpathogenic amoebae. Intestinal ulcers may extend through the muscularis (→ Pathology/Fig. 4C) and lead to intestinal perforations. This complication is made more likely by the administration of anti-inflammatory corticosteroids which can occur if an erroneous diagnosis of inflammatory bowel disease has been made. However AIDS patients do not appear to be especially susceptible to recrudescenses.

The amoebae often invade the veins in the submucosa of the gut (→ Pathology/Fig. 21B) and are transported to the liver and rarely other organs (lung, brain, skin), where they may set up foci of infection. Liver abscesses may reach a size of several centimeters. They usually develop in the right side according to the laminar flow of the portal vein drainage from the colon. The amoebae colonize, lyse,, and digest the liver cells, giving rise first to amoebic hepatitis (→ Pathology/Fig. 4D). Only when the focus of destroyed liver parenchyma is too large for the lysed debris to be absorbed into the lymphatic and venous circulation will an abscess result. The center of the abscess is formed by brownish, semiliquid fluid which is said to re-

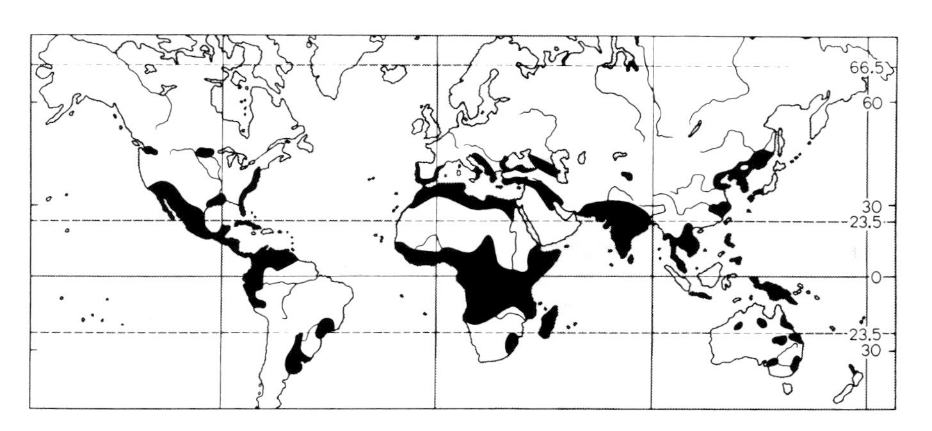

Fig. 1. Distribution map of amoebic dysentery.

semble "anchovy paste". A liver abscess may extend through the diaphragm into pleura and lung with the abscess material being coughed up. Older abscesses are surrounded by fibrinous chronic inflammatory reaction products or by fibrosis. The amoebic trophozoites are found in the periphery of the abscess between the liver cells, best identified in PAS-stained sections (→ Pathology/Fig. 4D). The delicate nuclei of the trophozoites are shown with hematoxylin and eosin, but they stain less intensely than the nuclei of macrophages from which they need to be distinguished. Amoebic cysts are found only in the stools and not in the tissues. Invasive amoebiasis is enhanced by immunosuppression of whatever cause, e.g., by corticosteroid administration, as mentioned above, and in patients with AIDS. *E. histolytica* infections also occur in extraintestinal sites, penile infections following anal intercourse, infections of the cervix uteri and the buccal mucosa.

Immune Responses

Intestinal amoebiasis is characterized by fulminant diarrhea and intestinal hemorrhage. The associated disruption of the intestinal epithelium allows haematogenous dissemination of the parasite leading to granuloma formation most commonly found in the liver. Survival of amoebic trophozoites may be favored by a transient immunosuppression associated with hepatic infections. The downregulation of the host's immune response involves macrophage effector and accessory cell function as well as T cell functions (→ Immune Responses).

Innate Immunity SCID mice infected with *E. histolytica* were used as a model of amoebic liver abscess formation and to study the functional role of neutrophils in vivo. Neutrophil-depleted animals developed significantly larger liver abscesses at early stages of infection, which lacked the prominent inflammatory cell ring observed in control SCID mice. These findings suggest a protective role of neutrophils in the early host response to amoebic infection in the liver (→ Innate Immunity).

B Cells and Antibodies A majority of *E. histolytica* isolates were found to be lysed by the alternative complement pathway in vitro in the absence of specific antibodies. A protective role of specific antibodies is suggested by experiments using passive immunization protocols of SCID mice. It could be shown that serum or purified antibodies from patients with amoebic liver abscesses were able to significantly reduce the mean abscess size in the experimental animals when applied 24 hrs before intrahepatic infection with *E. histolytica.* In addition, specific antibodies against the serine-rich *E. histolytica* protein (SREHP) prevented amoebic liver abscesses in SCID mice. Furthermore, → vaccination with the immunodominant galactose/N-acetylgalactosamine-inhibitable lectin of *E. histolytica* induced protective immunity to amoebic liver abscesses at least in some animals, while in others of the same species exacerbation of disease was observed after vaccination. Lotter et al. recently showed that protective immunity is due to the development of an antibody response to a region of 25 amino acid residues of the lectin, while exacerbation of the disease is caused by antibodies against the NH2-terminal region of the lectin. These findings might be of clinical relevance, since individuals who are colonized with *E. histolytica* but resistant to invasive disease have a high prevalence of antibodies to the protective epitope.

T Cells There is good evidence suggesting that cell mediated immunity is centrally involved in the defense against *E. histolytica.* Human and mouse macrophages activated with IFN-γ are able to kill trophozoites in a contact-dependent manner. This killing involves both oxidative and nonoxidative mechanisms. With mouse macrophages it has been demonstrated that one of the principal effector molecules is nitric oxide (NO). The responsible induction of iNOS is enhanced by → TNF (Vol. 1) produced in an autocrine fashion by stimulated macrophages. The requirement for macrophage activation suggests that a Th1-type immune response might be necessary for an efficient immune response against amoebae. In line with this, T cells from patients with treated amoebic liver abscesses secreted macrophage activating cytokines leading to amoebic killing. IL-2 and IFN-γ production of T cells can be induced by galaxies/N-acetylgalactosamine-inhibitable lectin, an immunodominant molecule of *E. histolytica.* The analysis of cytokine production patterns in gerbils infected with *E. histolytica* showed that resistance to reinfection correlated with a Th1-like response.

Evasion Mechanisms Patients infected with *E. histolytica* generate specific IgGs that often do not prevent invasive or recurrent infection. One mechanism by which the parasite might avoid binding of the specific antibodies is the production of extracellular cysteine proteases. It has been shown that these proteases cleave IgGs near their hinge region resulting in fragments binding less efficiently to *E. histolytica* trophozoites (→ Evasion Mechanisms).

The invasion and disease associated with *E. histolytica* has long been connected with suppression of host cellular immunity. Several studies have begun to clarify, at the cellular level, the dampening effects which *E. histolytica* exerts on immune cell and effector cell functions. Although, as discussed above, macrophages are potent cells for amoebicidal activity, the cytotoxic activity of abscess-derived macrophages is reduced during acute hepatic amoebiasis, while macrophages distal from the site of infection are in a heightened state of activity. Obviously, macrophage suppression is a local event, most likely mediated by direct exposure to the parasite or its products. It has been shown that soluble amoebic proteins (SAP) decreased the IFN-γ-induced upregulation of MHC class II molecules on murine macrophages. This inhibitory effect, which is mediated at the transcriptional level, involves the production of prostaglandin E2 (PGE2). The fact that inhibition of PGE2 synthesis by indomethacin reversed SAP-mediated suppression of MHC class II expression by 60 %, establishes that PGE2 induction is the main but not the only mechanism involved. Thus, PGE2 and related products, which may also be produced by *E. histolytica* trophozoites themselves, could indirectly inhibit T cell receptor recognition of parasite antigens. In addition, pretreatment of macrophages with SAP results in suppression of TNF and nitric oxide synthesis. While addition of indomethacin restored TNF production, it had no effect on iNOS mRNA levels or NO production. Thus, *E. histolytica* appears to have the ability to inhibit different macrophage functions by separate pathways.

To date, T cell deficiencies in amoebiasis are less well characterized. However, the reduced delayed-type hypersensitivity reaction during acute amoebiasis seen in patients as well as the reduced mitogenic response of purified T cells in vitro argue for a relative paucity of T cells mediating defensive responses against amoebae. A major phosphorylated, lipid-containing glycoconjugate surface molecule of *E. histolytica* may function similarly to lipophosphoglycan (LPG) of *Leishmania* by inhibiting PKC signal transduction. Other amoebic molecules with potential suppressive roles include a 220 kDa lectin, which induces the production of IL-4 and IL-10.

Main clinical symptoms: Abdominal pain, bloody-slimy diarrhoea, liver dysfunction in case of liver abscess.
Incubation period: 2–21 days
Prepatent period: 2–7 days
Patent period: Years
Diagnosis: Microscopical determination of cysts in fecal samples, → serology (Vol. 1)
Prophylaxis: Avoidance of uncooked food/water in endemic regions
Therapy: Treatment see → Antidiarrhoeal and Antitrichomoniasis Drugs

Amoebic Infections

Several species of → amoebae (Vol. 1) (e.g. → *Entamoeba histolytica* (Vol. 1) causing → Amoebiasis) give rise to infections of the intestinal lumen (→ Protozoan Infections, Man/Table 1). Most are commensal, feeding on bacteria and producing neither lesions nor functional abnormalities. However, *Dientamoeba fragilis* does give rise to diarrhea often accompanied by → eosinophilia in both the stool and blood but without apparent tissue invasion. Its transmission in the absence of a cyst has been linked circumstantially to the nematode → *Enterobius vermicularis* (Vol. 1).

Therapy

→ Antidiarrhoeal and Antitrichomoniasis Drugs.

Anaemia

Clinical symptom due to infections with e.g. → hookworms (Vol. 1), → *Babesia* (Vol. 1) species, → *Gastrodiscus aegyptiacus* (Vol. 1) (trematodes in horses), → Fascioliasis, Man, → Fasciolosis, Animals, → Schistosomiasis, Man, → Schistosomiasis, Animals, blood sucking of → fleas (Vol. 1), → mites (Vol. 1) (e.g. *Dermanyssus*).

Anal Pruritus

Skin reaction in humans and horses due to infections with pinworms (*Enterobius vermicularis*, *Oxyuris equi*) (→ Alimentary System Diseases, Horses).

Anaphylactic Shock

Hypersensitivity-shock-reaction due to repeated setting free of masses of parasitic antigens inside a host (→ Echinococcosis), fly larvae (→ Myiasis, Animals, → Myiasis, Man).

Anaplasmosis

Disease in bovines due to tick transmitted *Anaplasma marginale*-stages (→ Tick Bites: Effects in Animals).

Ancylostomiasis

Synonym

→ Hookworm (Vol. 1) infection (→ Hookworm (Vol. 1) Disease)

Soil-transmitted helminthic infection (→ Nematocidal Drugs, Man / Table 1).

Anderson and May Model

→ Mathematical Models of Vector-Borne Diseases.

Angiostrongylosis

Abdominal Angiostrongylosis is an accidental parasitosis of man by *Angiostrongylus costaricensis* – a species usually found in rats, acquired by eating third stage larvae from slugs or their mucus left on unwashed vegetables (see volume 1). Most of the cases have been found in the Americas, and only a few in Africa. The larvae enter the intestinal wall and migrate to the arteries of the ileocolic region where they become adult (→ Pathology/Fig. 23B). The eggs are deposited in the vessel; and are propelled into the intestinal mucosa where they give rise to an intense granulomatous inflammation with eosinophils and fibrosis. Most of the eggs degenerate; although larvae are occasionally seen in the tissues, they have not been found in the stool where they are shed by the natural host (→ Pathology/Fig. 29A-D). Many of the arterioles containing adults become thrombosed after the worms die. The lesions may simulate appendicitis or give rise to large inflammatory ileocolic masses which can cause intestinal obstruction and must be resected surgically. **Cerebral angiostrongylosis** is due to infection with *Angiostrongylus cantonensis* (see Vol. 1) showing the following-signs:

Main clinical symptoms: Eosinophilic meningoencephalitis, paralysis
Incubation period: 2–3 weeks
Prepatent period: Larvae are found within 2 days in the brain
Patent period: Months until the larvae die.
Diagnosis: Serodiagnostic methods
Prophylaxis: Avoid eating raw food (meat, snails, crabs)
Therapy: Treatment see → Nematocidal Drugs, Man

Animal Reservoirs

Animals containing identical stages of parasites as found in humans, but symptoms of disease are mostly less strong, so that these animals are often the source for human infections.

Anisakiasis

Anisakiasis comprises accidental human infection; with a variety of nematodes normally infecting marine fish (salmon, herring), which was consumed raw, marinated, or undercooked (→ *Anisakis* (Vol. 1)). The worms are often expelled or vomited. Occasionally they penetrate the gastric or intestinal wall where they give rise to an eosinophilic abscess, or in the omentum, or peritoneal lining associated with an adult nematode that may still be alive or dead (→ Pathology/Fig. 28D). The specimen is often recovered surgically after an abrupt onset of severe abdominal discomfort. The structures of the worm may be well preserved in acute cases or may have degenerated after an illness of a few days.

Main clinical symptoms: Abdominal pain, intermittent diarrhoea, loss of weight
Incubation period: 1 hour – 1 week
Prepatent period: Larvae are present from the moment of oral uptake
Patent period: Weeks until the worms have been killed within granulomas
Diagnosis: Gastroscopical methods
Prophylaxis: Avoid eating raw fish (saltwater)
Therapy: Treatment see → Nematocidal Drugs, Man.

Anopluridosis

Disease due to infestation with lice; see Table 1 (pages 37–38)

Anorexia

Clinical symptom (= reduction in voluntary food intake) in animals due to parasitic infections (→ Alimentary System Diseases, → Clinical Pathology, Animals).

Anthroponoses

Diseases affecting only humans. The major representatives of parasitic anthroponoses are listed in Table 1 (page 39).

Therapy
→ Chemotherapy, → Drugs.

Anthropozoonoses

Synonym
Zooanthroponoses

General Information
Diseases affecting both man and other animals. The major representatives of parasitic zoonoses are listed in Table 1 (pages 40–42).

Therapy
→ Chemotherapy, → Drugs.

Antibiotica

General term to describe drugs acting in various ways on bacteria (e.g., they may have bacteriostatic (tetracyclines) or bacteriocidic (e.g. penicillines) activities. See → Lyme disease, → Streptothricosis, → Borreliosis, → Rickettsiae (Vol. 1).

Antibody

Synonym
Immunoglobulin (Ig)

General Information
The function of antibodies is to remove antigen from the system and to support an effective immune response. Circulating antibodies which are detected in serodiagnostic systems are secreted by antibody forming cells (plasma cells) after contact between B cells and antigen. They circulate freely throughout the blood and lymph. Antibodies are bifunctional molecules with antigen-binding sites (Fab fragment) and a region which is involved in other aspects of the immune regulation (Fc fragment). In most higher mammals and man four of the five immunoglobulin classes of molecules are involved in the humoral immune response to parasites: IgG, IgM, IgA, and IgE.

Classes
Each immunoglobulin class differs in the heavy-chain polypeptides which determine its function at particular stages of the maturation of the immune response. IgM class antibodies are predominately antibodies of the early immune response and they are distributed to a high degree intravascularly. A specific IgM class antibody response is indicative for an acute infection mainly with protozoan parasites. IgM serodiagnostic tests are used to discriminate between acute/recent and latent infections with *Toxoplasma*, *Trypanosoma cruzi*, and *Babesia*. In contrast, during the early phase of West African trypanosomiasis specific IgM antibodies may be undetectable in spite of a rising total serum IgM concentration. In helminthic infections diagnostic IgM class antibodies do not necessarily proceed an IgG response. This may be due either to a "diagnostic window" as seen 4–6 weeks after *Trichinella*-infection or, for example, to the late onset and long-term persis-

Table 1. Sucking animal lice and control measurements (according to Hansen and Londershausen)

Parasite	Host	Vector for	Symptoms	Country	Therapy		
					Products	Application	Compounds
Linognathus setosus	Dog	–	Blood loss, itching, sec. inf., urticaria	Worldwide	Advantage™ (Bayer)	Spot on	Imidacloprid
					Bolfo™ Flohschutz-Puder (Bayer)	Dermal powder	Propoxur
					Mycodex™ Pet Shampoo, Carbaryl (Pfizer)	Shampoo	Carbaryl
Haematopinus eurysternus	Cattle	–	Blood loss, irritation	Worldwide	Rabon™ 3% Dust (Agri Labs)	Self Treating Dust Bags	Tetrachlorvinphos
					Warbex Famphur Pour-on For Cattle™ (Mallinckrodt)	Pour-on	Famphur
					Tiguvon™ Cattle Insecticide Pour on (Bayer Corp.)	Pour on	Fenthion
					Ivomec™ 1% Injection For Cattle (Merial)	Injection	Ivermectin
					Dectomax™ (Pfizer Animal Health)	Injection	Doramectin
					Asuntol™-Puder 1% (Bayer)	Dermal powder	Coumaphos
					Cydectin™ (Bayer)	Injection	Moxidectin
Linognathus vituli	Cattle	–	Blood loss, irritation	Worldwide	Rabon™ 3% Dust (Agri Labs)	Self Treating Dust Bags	Tetrachlorvinphos
					Warbex Famphur Pour-on For Cattle™ (Mallinckrodt)	Pour-on	Famphur
					Tiguvon™ Cattle Insecticide Pour on (Bayer Corp.)	Pour on	Fenthion
					Ivomec™ 1% Injection For Cattle (Merial)	Injection	Ivermectin
					Dectomax™ (Pfizer Animal Health)	Injection	Doramectin
					Asuntol™-Puder 1% (Bayer)	Dermal powder	Coumaphos
					Cydectin™ (Bayer)	Injection	Moxidectin
Solenopotes capillatus	Cattle	–	Smallest cattle sucking lice, Blood loss, irritation	Worldwide	Rabon™ 3% Dust (Agri Labs)	Self Treating Dust Bags	Tetrachlorvinphos
					Warbex Famphur Pour-on For Cattle™ (Mallinckrodt)	Pour-on	Famphur

Table 1. (continued)

Parasite	Host	Vector for	Symptoms	Country	Therapy		
					Products	Application	Compounds
Solenopotes (continued)		–			Tiguvon™ Ca ttle Insecticide Pour on (Bayer Corp.)	Pour on	Fenthion
					Ivomec™ 1% Injection For Cattle (Merial)	Injection	Ivermectin
					Dectomax™ (Pfizer)	Injection	Doramectin
					Cydectin™ (Bayer)	Injection	Moxidectin
Linognathus ovillus	Sheep	–	Blood loss, irritation	Scotland			
Linognathus oviformis	Sheep, Goat	–		Outside Europe wide spread			
Linognathus pedalis	Sheep, Goat	–		Worldwide, but outside Europe			
Linognathus stenopsis	Goat	–		Worldwide			
Haematopinus asini macrocephalus	Horse	–	Strong concern, itching	Worldwide			
Haematopinus suis	Pig	Classical swine fever (hog cholera), Swine pox virus	Blood loss, host specific, all ages, severe itching, irritation, concern, loss of appetite	Worldwide	Rabon™ 3% Dust (Agri Labs)	Self Treating Dust Bags	Tetrachlorvinphos
					Neguvon™ (Bayer): No treatment during migration	Wash or Spray	Trichlorfon/ Metrifonate
					Ivomec™ 0.27% Sterile Solution (Merial)	Injection	Ivermectin
					Point-Guard™ Miticide/Insecticide	Pour-on	Amitraz
					Asuntol™-Puder 1% (Bayer)	Dermal powder	Coumaphos

Table 1. Important parasitoses which are usually restricted to humans as vertebrate hosts (according to Wernsdorfer)

Parasite species	Infective Stage	Mode of Infection	other obligatory hosts	Disease
Protozoa				
Trypanosoma gambiense	Metacyclic trypanosome	Bite of *Glossina* spp.	*Glossina* spp.	Sleeping sickness (Gambian)
Trichomonas vaginalis	Trichomonad	Mainly sexual intercourse	None	Human trichomoniasis
Hartmanella spp. *Acanthamoeba spp.* *Naegleria spp.*	Vegetative forms (cysts?)	Entry through nasal mucosa	None	Amoebic meningoencephalitis
Plasmodium falciparum	Sporozoite	Bite of *Anopheles* spp.	*Anopheles* spp.	Falciparum malaria
Plasmodium ovale	Sporozoite	Bite of *Anopheles* spp.	*Anopheles* spp.	Ovale malaria
Plasmodium vivax	Sporozoite	Bite of *Anopheles* spp.	*Anopheles* spp.	Vivax malaria
Trematodes				
Schistosoma haematobium	Cercaria	Transdermal invasion	*Bulinus* spp.	Urinary schistosomiasis (bilharzia)
Schistosoma intercalatum	Cercaria	Transdermal invasion	*Bulinus* spp.	Schistosomiasis intercalatum
Schistosoma mansoni	Cercaria	Transdermal invasion	*Biomphalaria* spp.	Mansonian schistosomiasis
Nematodes				
Ascaris lumbricoides	Eggs containing 2nd-stage larva	Ingestion	None	Human ascaridiasis
Enterobius vermicularis	Eggs	Ingestion	None	Human pinworm infection
Trichuris trichiura	Embryonated egg	Ingestion	None	Human trichuriasis
Ancylostoma duodenale *Necator americanus*	Strongyliform larva	Transdermal invasion	None	Human ancylostomiasis
Wuchereria bancrofti	Filiariform 3rd-stage larva	Mosquito bite	Mosquitos (*Culex*, *Aedes*, *Anopheles*, *Mansonia* spp.)	Lymphatic (bancroftian) filariasis
Onchocerca volvulus	Filiariform 3rd-stage larva	Blackfly bite	*Simulium* spp.	Onchocerciasis

tence of gut-associated IgM antibodies after *Schistosoma*-infection. The human IgG, which can be subdivided into the four subclasses IgG1-IgG4, is the major antibody of the primary and secondary immune response and the major serum immunoglobulin. In chronic helminth infections elevated IgG and IgE specific responses develop. The prominent IgG subclasses recognizing parasite specific antigens are IgG4> IgG1> IgG3, rarely IgG2. A high IgG4 response may indicate a successful parasite infection. The detection of parasite specific IgG4 is essential for a species specific diagnosis of *Onchocerca volvulus*, *Wuchereria bancrofti*, *Strongyloides*, *Ascaris*, and *Echinococcus*. IgG molecules of all subclasses can cross the placenta. Only the detection of specific IgM and/or IgA in the serum of the fetus or newborn is of relevance for the serological diagnosis of congenital infection with *Toxoplasma*, *Trypanosoma cruzi*, or *Leishmania*. IgM and IgG1-3 antibody classes are able to activate the complement cascade, but not IgG4, IgA, and IgE. IgA is the predominant immunoglobulin in seromucous secretion, IgA1 being the predominant subclass in the serum. IgA antibodies are of additional value as markers for an acute infection as documented for toxoplasmosis.

Table 1. Important zooanthroponotic and anthropozoonotic parasitoses (according to Wernsdorfer)

Parasite species	Principal hosts (besides humans)	Infective stage	Mode of infection	Other obligatory hosts	Disease
Protozoa					
Leishmania donovani	Various domestic animals	Promastigote	Sand fly bite	*Phlebotomus* spp.	Visceral leishmaniasis (Kala-azar)
Leishmania tropica	Dogs	Promastigote	Sand fly bite	*Phlebotomus* spp.	Oriental sore
Leishmania brasiliensis	Dogs	Promastigote	Sand fly bite	*Phlebotomus* spp.	Mucocutaneous leishmaniasis (Espundia)
Trypanosoma rhodesiense	Wild and domestic mammalians	Metacyclic trypanosome	*Glossina* bite	*Glossina* spp.	Sleeping sickness (Rhodesian)
Trypanosoma cruzi	Wild and domestic mammalians	Metacyclic trypanosome	Invasion through bite wound from reduviid feces	Reduviid bugs (*Triatoma, Panstrongylus, Rhodnius* spp.)	Chagas disease
Giardia duodenalis	Wild and domestic mammalians	Cyst	Ingestion	None	Lambliasis (Giardiasis)
Balantidium coli	Pigs	Cyst	Ingestion	None	Balantidiosis (dysentery)
Entamoeba histolytica	Various mammalians	Cyst	Ingestion	None	Amoebiasis (dysentery and amoebic abscesses)
Toxoplasma gondii	Domestic mammalians	Oocyst, bradyzoite	Ingestion and transplacental	None	Toxoplasmosis
Babesia bigemina	Cattle	Sporozoite	Tick bite	Ticks (especially *Boophilus* spp.)	Texas cattle fever, babesiosis
Babesia canis	Dogs	Sporozoite	Tick bite	Ticks (various spp.)	Canine babesiosis
Plasmodium malariae	*Pan troglodytes*	Sporozoite	Bite of *Anopheles* spp.	*Anopheles* spp.	Quartan malaria
Trematodes					
Fasciola hepatica	Sheep, cattle	Metacercaria	Ingestion	Aquatic snails (*Lymnaea, Succinea, Fossaria, Practicolella* spp.)	Fascioliasis (liver rot of sheep and cattle)
Fasciolopsis buski	Pigs	Metacercaria	Ingestion	Aquatic snails (*Planorbis* and *Segmentina* spp.)	Fasciolopsiasis
Dicrocoelium dendriticum	Various domestic mammalians	Xiphidiocercaria	Ingestion with ants	Land snails (mainly *Cionella* spp.) followed by ants (mainly *Formica fusca*)	Dicrocoeliasis

Table 1. (continued)

Parasite species	Principal hosts (besides humans)	Infective stage	Mode of infection	Other obligatory hosts	Disease
Trematodes					
Opisthorchis felineus	Felines	Cercaria	Ingestion	Aquatic snails (*Bithynia* spp.) followed by cyprinid fish	Opisthorchiasis
Clonorchis sinensis	Canines, felines	Cercaria	Ingestion	Various aquatic snail species, followed by cyprinid fish	Clonorchiasis
Paragonimus westermanni	Cats	Metacercaria	Ingestion	Molluscs (*Semisulcospira, Tarebia, Brotia* spp.), followed by crustaceans (mainly freshwater crabs)	Paragonimiasis
Paragonimus kellikotti	Canines, felines and other mammalians	Metacercaria	Ingestion	Aquatic snails followed by crayfish	Paragonimiasis
Schistosoma japonicum	Wild and domestic mammalians	Cercaria	Transdermal invasion	*Oncomelania* spp.	Japanese schistosomiasis
Cestodes					
Diphyllobothrium spp.	Canines, felines and other fish eaters	Procercoid	Ingestion	Copepods (e.g., *Diaptomus* spp.), followed by fish	Fish tapeworm infection
Taenia solium	Pigs (adult worm), cysticercus also in other animals	Egg and cysticercus	Ingestion	No obligatory host species alternation	Cysticercosis cellulosae and pig tapeworm infection
Taenia saginata	Cattle (cysticercus), humans (adult worm)	Cysticercus	Ingestion	Humans (adult worm), cattle (cysticercus)	Taeniasis saginata (cattle tapeworm infection)
Echinococcus granulosus	Canines (adult worm), various mammalians (hydatid cysts)	Egg and hydatid larva	Ingestion	Canines (adult worm), various mammalians (hydatid cysts)	Hydatid disease
E. multilocularis	Fox, mice	Egg for man, alveolar cyst for fox	Ingestion	Dogs, cats, mice, many animals	Alveolar disease
Hymenolepis nana	Rodents	Egg	Ingestion	None	Hymenolepiasis nana
Hymenolepis diminuta	Rodents	Cysticercoid larva	Ingestion with infected roach	Roaches (*Tembrio, Pyralis, Anisolobis* spp., etc.)	Hymenolepiasis diminuta
Nematodes					
Angiostrongylus cantonensis	Rats	3rd-stage ensheathed larva	Ingestion with infected slug or snail	Slugs and aquatic snails	Human angylostrongyliasis (eosinophil meningoencephalitis)
Ascaris lumbricoides	Various vertebrates	Egg with 2nd-stage larva	Ingestion	None	Ascariasis

Table 1. (continued)

Parasite species	Principal hosts (besides humans)	Infective stage	Mode of infection	Other obligatory hosts	Disease
Nematodes					
Toxocara canis	Canines	Embryonated egg or 2nd-stage larva	Ingestion	None	Human visceral larva migrans
Toxocara cati	Felines				
Anisakis spp.	Marine mammalians and birds	3rd-stage larva	Ingestion	Insufficiently known	Human anisakis infection (due to 3rd-stage larvae)
Trichinella spiralis	Various mammalians	Infective larva	Ingestion	None	Trichinellosis
Gnathostoma spinigerum	Canines and felines	3rd-stage larva	Ingestion	*Cyclops* spp., followed by fish, frogs or snakes	Gnathostomiasis
Capillaria hepatica	Various mammalians	Embryonated egg, after passage through "transport" host	Ingestion	Only "transport" host	Capillariosis, human visceral larva migrans
Strongyloides stercoralis	Various mammalians	Filariform larva	Transdermal penetration	None	Strongyloidosis of humans and sheep
Dracunculus medinensis	Canines, felines, equines, monkeys	3rd-stage larva	Ingestion (with *Cyclops*)	*Cyclops* spp.	Guinea worm disease
Brugia malayi	Felines and non-human primates	Filiariform 3rd-stage larva	Mosquito bite	Mosquitos (usually *Anopheles* and *Mansonia* spp.)	Lymphatic (Brugian) filariasis
Loa loa	Baboons	3rd-stage larva	Tabanid bite	Tabanids (*Chrysops* spp.)	Loiasis

An elevation of total serum IgE, but not necessarily specific IgE class antibodies, is often related to infections with helminth parasites and may indicate an active disease. In contrast, protozoan parasites do not induce a significant IgE class antibody response. Specific IgE is not a reliable diagnostic marker for acute *Toxoplasma*-infection due to many non- or low-responders within the population.

Molecular Weights

IgM app. 970 000 daltons; IgG app. 146 000 daltons, IgA app. 160 000 daltons, IgE app. 188 000 daltons, IgD app. 184 000 daltons

Binding Sites and Affinity

Antibodies are highly specialized in recognizing small regions of antigens but occasionally recognize similar epitopes on other related or unrelated molecules (cross-reaction). The binding between the antibody and the epitope of the antigen is non-covalent. The strength of bond between an antibody-combining site with an antigen is characterized as antibody affinity, the strength with which a multivalent antibody binds a multivalent antigen as antibody avidity. The antibody affinity/avidity to many antigens increases during an immune response and determines the biological effectiveness of the antibody. High- affinity antibodies are superior to low-affinity antibodies in respect to many biological reactions (e.g. haemagglutination, complementfixation) and in achieving antigen-binding more effective even at low antigen concentration. Today, this immunological mechanism is utilized for the differential diagnosis of a recent and latent *Toxoplasma*-infection in man.

Half Lives

The half-life of antibodies differs according to the antibody classes. It is high for IgG (21 days), medium for IgM (10 days) and low for IgA, and IgE (6–2 days). The complexed antibody is catabolized before clearance.

Biological Activity

In general, a primary antibody response can be divided into four phases: a first period when no antibodies are detectable, followed by an immune response of primarily IgM antibodies with a progressive change to IgG antibodies and an increasing antibody titer. The plateau phase with stable antibody titers is followed by an antibody decline. During chronic infections, when the parasite is established at its permanent habitat, antibody production is mainly stimulated by circulating (excretory-secretory) antigens. The antibody secretion by B cells continues as long as the antigen(s) persists. Persisting antigen may increase the strength of the immune response. The primary antibody response is not only influenced by the nature of the antigen, its dose and infection route but also by the genetic background of a host. The time of antibody maturation and the final antibody concentration can vary considerably in individuals. Especially low responders or people with a recently contracted infection or patients from low endemic areas may not be detected in all serological test systems. Seroepidemiological studies in *L. infantum*-infected dogs indicate that a high sensitivity of the IFAT is reached only after an incubation period of 8–9 months when first clinical symptoms occur. A complete antibody clearance is only possible after a complete elimination of the parasite (blood parasites after therapy) or its antigenic epitopes. Also, antigenic epitopes may be hidden by "walling off" in the host tissue (*Echinococcus*, *Cysticercus*). However, after elimination of the antigen specific lymphocytes remain in circulation and may respond to a subsequent challenge (immunological memory). A challenge by antigens released after surgical or therapeutical measures (*Echinococcus*, *Cysticercus*) or reinfection induces a secondary antibody production with a steep increase of mainly IgG class antibodies.

Clinical Relevance

Generally, the presence of circulating antibodies indicates the recognition of a foreign antigen. In parasitic infections antibody detection does not correlate with disease, protective immunity or the parasite load. The serum antibodies are the most important defence mechanism of the host against extracellular parasitic forms in blood and body fluids like *Trypanosoma*, *Toxoplasma*, *Plasmodium*, or *Babesia* during acute infection. Symptomatic individuals normally present high antibody levels whereas asymptomatic carriers show a low antibody concentration. Antibodies in individuals without clinical symptoms may indicate an abortive infection as described for *E. multilocularis* or an infection with a long latency/incubation period. A correlation between cyst burden and antibody detection is known for porcine and human cysticercosis (*Taenia solium*). → Cysticercosis or echinococcosis in asymptomatic or symtomatic patients is not excluded by negative test results. An assessment of the parasite load is not possible by conventional serological test systems. Recent developments indicate that specific antibody assays are useful for the determination of the egg-load after *Schistosoma*-infection. However, a better quantitation is by direct and indirect immunoassays which measure circulating parasitic antigens in body fluids. The quality of antibodies induced after natural infection or vaccination may differ in many aspects. It has to be considered that post-vaccination antibodies may not be detectable in conventional test systems. An evaluation of parasitic treatment is difficult by serology. Only a complete parasitological cure is confirmed by negative serological results. Antibody disappearance is described for hepatic fascioliasis in humans 6–12 months after specific therapy and goats 1–5 months after therapy. In patients with Chagas' disease humoral response may persist for years after a negative result in direct parasite examination. This is also true for schistosomiasis, echinococcosis , cysticercosis, and filariosis. However, there are promising results that a follow-up of the antibody response to selected antigen(s) may be efficient in post-therapeutic monitoring.

In parasitic infections the antibody isotype profiles may vary with time and the clinical manifestation. Differences in the IgG isotype pattern of the paired sera from mother and newborn add to the diagnosis of congenital toxoplasmosis. Congenital toxoplasmosis or congenital infection with *Trypanosoma cruzi* is proven by the demonstration of specific IgM and/or IgA antibodies in the newborn serum.

Antidiarrhoeal and Antitrichomoniasis Drugs

Drugs Acting on Giardiasis

Giardia lamblia (syn. *G. intestinalis*, *G. duodenalis*) is distributed worldwide and has been identified in humans and domestic livestock, particularly in young animals such as calves, lambs, piglets and foals and birds. The flagellate (with four pairs of flagella) living in the small intestine may produce acute or chronic enteritis with profuse and heavy diarrhea and growth rate reduction in mammals and birds. *Giardia* spp. infections in animals may pose a serious zoonotic threat to humans. Humans may acquire infection either through waterborne transmission by *Giardia* cysts occurring in the drinking water contaminated with infectious feces or by direct contact with contaminated feces. Although quinacrine and furazolidone resistance have been induced in *G. duodenalis* trophozoites, the substituted acridine dye derivative quinacrine can be of value in humans suffering from giardiasis, mainly in patients showing reduced response to 5-nitroimidazoles. However, quinacrine may cause exacerbation of psoriasis, ocular toxicity, and toxic psychoses or distinctly enhances toxicity of 8-aminoqunolines, e.g. primaquine (→ Malariacidal Drugs).

Benzimidazole carbamates such as mebendazole, fenbendazole or albendazole may exhibit the most pronounced activity against *Giardia* infections in man, farm animals and dogs; however, repeat treatments may be necessary to eliminate parasites in reinfected farm animals. Currently used drugs in *Giardia* infected humans are summarized in Table 1. **The control** and **prevention** of *Giardia* should be directed at young animals. They are highly susceptible to the parasite and show a markedly higher output of cysts than adult animals. Control measures for humans should be centered on environmental and personal hygiene issues such as routine hand-washing, control of insects to prevent their contact with infected stools. **Iodine** seems to be an effective disinfectant for drinking water; filtration systems are also recommended (drawbacks may be clogging and safe removal of contaminated filters); killing of *Giardia* cysts by boiling the water is most effective but will cost energy. The chances of a protective vaccine against giardiasis are poor.

Drugs Acting on Trichomoniasis in Humans and Cattle

Trichomonas vaginalis (motile trophozoite) is a common pathogen of the urogenital system in men and women but is uncommon in preadolescent girls. A frequent transfer of this flagellate is due to sexual intercourse, which explains the need for a therapy of both partners. In childbearing women, there may be an overall prevalence of 30%, which is significantly higher than that in men (5% or more). Very frequently *T. vaginalis* infection is asymptomatic, especially so in the male. Clinical signs may develop if parasites cause degeneration and desquamation of the vaginal epithelium followed by inflammation of the vagi-

Table 1. Drugs acting on *Giardia*, *Trichomonas*, Amoebae, and *Balantidium* in humans.

DISEASE **nonproprietary name** (chemical group)	**Brand name** other information	**Adult dosage/*pediatric dosage** (mg/kg body weight, or total dose/ individual, oral route), miscellaneous comments
GIARDIASIS		
Giardia lamblia may contaminate drinking water, which is a common source of infection for humans and mammals; flagellated trophozoites attach by their 'suckers' to the surface of mucosa of small intestine, producing partial villous atrophy of the duodenum or jejunum; the ovoid cyst is passed in feces and has a very distinctive form; although *G. lamblia* is commensal in many individuals, it may be particularly pathogenic in debilitated and immunosuppressed patients, and is a common cause of diarrhea and a malabsorption syndrome characterized by excessive amounts of fat in the stool (steatorrhea) in travellers		
metronidazole (5- nitroimidazole)	Flagyl, Clont, others	250 mg tid x5d; *15 mg/kg/d in 3 doses x 5d (not licensed in the USA but considered investigational for this condition by the FDA)
tinidazole (5- nitroimidazole)	Fasigyn, others	2 grams once; *50mg/kg once (max. 2 g) treatment should be followed by administration of iodoquinol or paromomycin in doses as used to treat asymptomatic amoebiasis (see below)
furazolidone (nitrofuran)	Furoxone, others	100 mg qid x 7–10d; *6 mg/kg/d in 4 doses x 7–10d toxic effects may be serious and common, hypersensitivity reactions (urticaria)
paromomycin (aminoglycoside)	Humatin	25-35 mg/kg/d in 3 doses x 7d. (use of drug may be suitable in pregnancy as it is not absorbed) (not licensed in the USA but considered investigational for this condition by the FDA)
TRICHOMONIASIS		
motile trophozoites of *Trichomonas vaginalis* can be found in the foamy vaginal discharge of trichomonal vaginitis; the creamy discharge is often secondarily infected with the yeast *Candida albicans*; infection is transmitted by sexual intercourse; asymptomatic infection in the male (and therefore not treated) may be often a source of trichomonal infection in the female partner; for this reason sexual partners should be treated simultaneously; metronidazole-resistant strains have been reported, and enhanced doses of the drug for longer periods or use of tinidazole may be effective against these strains. In cattle the *Tritrichomonas foetus* infection of the cow often leads to an abortion and sterility; infection is transmitted by coitus; in birds, particularly in pigeons, heavy *Trichomonas gallinae* infections may lead to high mortality caused by greet necrotic lesions in the mouth, crop, and the esophagus with extension to the bones of the skull, the liver and elsewhere		
metronidazole (5- nitroimidazole)	Flagyl, Clont, others	2 grams once; or 250 tid or 375 mg bid x 7d, *15 mg/kg /d in 3 doses x 7d
tinidazole (5- nitroimidazole)	Fasigyn, others	2 grams once; *50 mg/kg once (max. 2 g) (the drug seems to be better tolerated than metronidazole, it is not marketed in USA)
AMOEBIASIS		
Entamoeba histolytica infection occurs by oral uptake of cysts, which are quite refractory to environment; the cysts are ingested with drinking water or food, which is not readily prepared or thoroughly cooked; as the disease is not characterized by a certain incubation time it is difficult to determine the onset of the amoebiasis; fever and feces containing blood and foamy fluid may be indications for acute amoebiasis; treatment should be started immediately after diagnosis has been made to avoid colonization of the liver by so-called "magna-forms" (tissue stages causing necrosis of liver parenchyma followed by formation of an abscess); large liver abscesses should be aspirated prior to treatment; there is no drug resistance in amoebiasis		
ASYMPTOMATIC CASES		
asymptomatic carriers extruding cysts (quadrinucleate form) in the feces are an important source of infection; there are no therapeutic drugs, which may affect cysts; **luminal drugs** act directly (by contact) on (uninucleated) trophozoites of *E. histolytica* living in the lumen of large intestine; paromomycin may also act by modifying intestinal bacterial flora		
iodoquinol (8-hydroxyquinoline)	Yodoxin, others (drug of choice)	600 mg tid x 20 d; *30–40 mg/kg/d (max. 2 g) in 3 doses x 20 d; toxic effect may be SMNO (= subacute myelooptic neuropathy) after high doses and for long periods

Table 1. Continued

DISEASE **nonproprietary name** (chemical group)	**Brand name** other information	**Adult dosage/*pediatric dosage** (mg/kg body weight, or total dose/ individual, oral route), miscellaneous comments
paromomycin (aminoglycoside)	Humatin (drug of choice)	25–35 mg/kg/d in 3 doses x 7d (may be useful in pregnancy as it is not absorbed) *25–35 mg/kg/d in 3 doses
diloxanide furoate (dichloroacetamide)	Furamide (alternative drug)	500 mg tid x 10d; *20mg/kg/d in 3 doses x 10d remarkably safe drug for treatment of carriers
MILD TO MODERATE INTESTINAL DISEASE		
uninucleated trophozoites of *E. histolytica* invade the intestinal epithelium, principally in the cecum and the ascending colon causing lytic necrosis of tissues; 5-nitroimidazoles are well absorbed from the intestine and exhibit a marked systemic effect on extraintestinal amoebiasis		
metronidazole (5- nitroimidazole)	Flagyl, Clont, others (drug of choice)	500-750 mg tid x 10d; *35–50 mg/kg/d in 3 doses x 10d
tinidazole (5- nitroimidazole)	Fasigyn (drug of choice)	2 grams/d x3d; *50 mg/kg (max. 2 g) qd x 3d (drug may be better tolerated than metronidazole; it is not marketed in the USA
ornidazole	Tiberal	2 g once a day x 3d; amoebicidal activity of ornidazole is similar to that other 5-nitroimidazole, not licensed in the USA.
SEVERE INTESTINAL DISEASE, EXTRAINTESTINAL AMOEBIASIS (HEPATIC ABSCESS)		
in fulminating amoebic dysentery with loose feces, containing mucus and blood, the ameba penetrate more deeply into the intestinal wall thereby damaging all layers of the intestinal wall resulting in confluent ulceration; ameba may pass into the lymphatics or mesenteric venules and invade other tissues of the body, especially the liver, but also skin or genital organs; the most common form is a large single abscess in the right lobe of liver		
metronidazole	Flagyl (drug of choice)	750 mg tid x 10d; *35–50 mg/kg/d in 3 doses x 10d
tinidazole	Fasigyn (drug of choice)	600mg bid or 800 tid mg grams/d x5d; *50 mg/kg or 60 mg/kg (max. 2 g) qd x5d (drug may be better tolerated than metronidazole; it is not licensed in the USA)
ornidazole (5- nitroimidazole)	Tiberal	2 g once a day x 3d; amebicidal activity of ornidazole is similar to that other 5-nitroimidazole; (not licensed in the USA)
AMOEBIC INFECTIONS OF CENTRAL NERVOUS SYSTEM (CNS) AND THE EYE		
free-living aquatic amoeba such as *Naegleria* spp. *Acanthamoeba* spp., and *Balamuthia mandrillaris* may cause various CN disorders, which may be fatal; *N. fowleri* causes a rapidly fatal infection known as 'primary amoebic meningoencephalitis' (PAM); most patients with naeglerial infection have had a history of recent swimming in fresh water during hot summer weather; a number of *Acanthamoeba* species may produce a chronic CNS infection called 'granulomatous amoebic encephalitis' (GAE) or an eye infection characterized by a chronic progressive ulcerative keratitis; *B. mandrillaris* also causes GAE; unlike naeglerial infection, GAE does not appear to be associated with swimming; infections are often seen in individuals debilitated or immunocompromized including patients with AIDS; diagnosis is made by microscopic identification of living or stained amoeba trophozoites in CSF (Giemsa stained spinal fluid smears) or corneal scrapings, and motile amoeba (trophozoites) can be readily seen in wet-mount preparations; **chemotherapy** of PAM and GAE is problematic; almost all cases of PAM and GAE have been fatal because of the foudroyant course of PAM and lack of an effective causative treatment; considerable toxic amphotericin B seems to be the only drug with clinical efficacy.		
amphotericin B (polyene antibiotic)	Amphotericin B powder	there are some known survivors of **PAM**, children from Australia, the UK, India and the USA treated with amphotericin B: 1–(15) mg/kg/d x 3d i.v. followed by 1mg/kg/d x 6d i.v. or longer (additional amphotericin B given intrathecally plus miconazole or rifampicim i.v.)
miconazole	Daktar, solution for injection	**GAE** caused by *Acanthamoeba* has been successfully treated by total excision granulomatous brain tumor and administration of ketoconazole, *Acanthamoeba* meningitis with penicillin and chloramphenicol; *Acanthamoeba* and *Balamuthia* form cys issues; thus, a potential effective drug for GAE must be capable of damaging cysts and trophozoites to prevent relapse after course of treatment;

Table 1. Continued

DISEASE **nonproprietary name** (chemical group)	**Brand name** other information	**Adult dosage/*pediatric dosage** (mg/kg body weight, or total dose/ individual, oral route), miscellaneous comments
ketoconazole (azoles)	Nizoral	strains of *Acanthamoeba* isolated from fatal GAE cases were susceptible to pentamidine, ketoconazole, flucytosine *in vitro*;
flucytosine (pyrimidone analog)	Ancobon	today, *Acanthamoeba* **keratitis** can be managed by medical treatment alone, if diagnosis has been made soon enough; successful regimens may be topical propamidine, miconazole and neosporin with epithelial debridement (removal of infected tissue from the lesion to expose intact tissue) or ciotrimazole, systemic ketoconazole or itraconazole with topical miconazole and surgical debridement
various topical formulations of azoles and other drugs for *Acanthamoeba* keratitis		
BALANTIDIASIS		
Balantidium coli (a ciliate) is a common commensal of the large intestine of wild and domestic pigs, but it can be pathogenic to humans and primates in which it produces various clinical forms (asymptomatic carrier condition, acute cases with fulminant diarrhea or chronic cases: diarrhea changes with constipation); however human infections are infrequent and pigs act as the main reservoir for human infections; motile flagellated trophozoites can invade the submucosa of large intestine causing an ulcerative enteritis in severe cases; trophozoites and cysts are passed in the feces and can readily be found in fresh wet-mount preparations; the main endemic areas are in the tropical and subtropical areas and where there is close contact between humans and pigs; humans become infected by ingestion of cyst-contaminated food and water		
tetracycline (tetracyclines)	various	500 mg qid x 10d; *40mg/kg /d (max. 2 g) in 4 doses x 10d; use of tetracyclines is contraindicated in pregnancy and children <8 years of age; they discolor teeth in growing children; (not licensed in the USA but considered investigational for this condition by the FDA)
metronidazole (5-nitroimidazole)	Flagyl (drug of choice)	750 mg tid x 5d; *35–50 mg/kg/d in 3 doses x 5d (not licensed in the USA but considered investigational for this condition by the FDA)
iodoquinol (8-hydroxyquinoline)	Yodoxin alternative drug	650 mg tid x 20d; *40 mg/kg/d in 3doses x 20d (not licensed in the USA but considered investigational for this condition by the FDA)

Abbreviations: the letter stands for day (days); qd = daily (quaque die); bid = twice daily; tid = three times per day; qid = four times per day (quarter in die); p.c. (post cibum) = after meals
Dosages listed in the table refer to information from manufacturer or literature (e.g. Medical Letter).
Data given in this Table have no claim to full information

na and vulva and a leucocytic discharge may be evident. Topical treatment can be tried with clotrimazole or pimaricin but appears to be of little value because these drugs lack systemic action, which is essential in cryptic infection of men. Current **therapeutic drugs** with good systemic activity against *T. vaginalis* infections, are the well tolerated **5-nitroimidazoles** metronidazole, ornidazole, tinidazole and others showing high cure rates even after a single dosing (Table 1). These drugs may differ somewhat in their pharmacologic properties but not so in efficacy or toxicity. They have a wide spectrum of activity, including *Entamoeba histolytica*, *Giardia lamblia*, *Bacterium coli* and anaerobic bacterial infections. Drug resistance of *T. vaginalis* strains has been reported infrequently. This step forward in the 5-nitroimidazole-therapy becomes apparent when comparing the list of more than 170 different treatment regimens with exotic compounds like picric acid and mercuric chloride cited in a book on trichomoniasis published about 50 years ago.

Also other flagellates may occur in the digestive tract or in the reproductive system being potential pathogens for diseases in mammals (monkeys, dogs, cats, zoo animals or ruminants), and birds. *Tritrichomonas foetus* with three anterior flagella may be of significance in the veterinary medicine causing an abortion in cows followed by prolonged, and at times permanent, sterility or a closed pyometra (purulent endometritis). In the bull, the principal infection site is the preputial cavity. A *T. foetus* infection is transmitted during coitus or by artificial insemination, and by gynecological examination of cows. The major control measure for trichomoniasis in cattle is the use of artificial insemination. The administration of **5-nitroimidazoles** like metronidazole or tinidazole is probably the most effective and expensive therapy for most of these protozoan infections. However, in some countries the use of 5-nitroimidazols in food animals is now subject to regulatory actions of government agencies. The aim is to remove drugs of this chemical group from the market because of potentially cancerogenic action in rodents. Benign epithelial tumors (adenoma) have been observed in mice after being treated with high daily doses of metronidazole and others 5-nitroimidazoles administered for prolonged periods such as 6 months or so.

Drugs Acting on Histomoniasis (Blackhead Disease) of Birds

Histomanas meleagridis is found in the ceca and liver of various birds (turkey, chicken, peafowl, guinea fowl, pheasant, partridge). The organism is transmitted by ingestion of embryonated eggs of the cecal worm *Heterakis gallinarum*. The amoeboid stage of the flagellate may cause enterohepatitis in turkeys (or chickens) under range or yard management and thus high economic losses in the turkey industry. Control of enterohepatitis consists of good husbandry and preventive medication; contact between turkeys and chickens must be avoided, and regular treatment of all birds with anthelmintics should be performed to reduce the incidence of *H. gallinarum* (passed eggs are still infective for some time after treatment). Drugs used as additives in-feed and/or therapeutic drugs in-water belong to different chemical groups. Various **5-nitroimidazoles** such as dimetridazole, ronidazole, carnidazole or ipronidazole may be used in-feed (the latter drug at dose levels between 50 and 85 ppm, withdrawal time 6 days, EC directives, 1999). Nifursol (use level in-feed, 50–75 ppm, withdrawal time 5 days, EC directives, 1999) or furazolidone, and others belong to the group of **5-nitrofuranes.** However, in various countries there are increasingly legislative restrictions concerning the use of additives in-feed for the prevention of histomoniasis because of potential cancerogenicity of these drugs in rodents.

Drugs Acting on Intestinal and Extraintestinal Amoebiasis of Humans

Entamoeba histolytica in humans, monkeys, dogs, cats, and zoo-animals may produce acute or chronic enteritis associated with profuse and heavy diarrhea. Acute amoebiasis must be differentiated from the chronic form. The use of amoebicidal drugs in the treatment of patients with various clinical forms of amoebiasis (Table 1) has contributed to reduction of morbidity and mortality of the disease. Luminal drugs are used to treat asymptomatic carriers passing cysts of *E. histolytica* in stool being an important source of infection. Various drugs are available for patients with invasive intestinal amebiasis associated with ulceration of deep layers of the intestinal wall of cecum and colon (Table 1). The same agents (e.g. 5-nitroimidazoles) can be used in treating invasive extraintestinal amoebiasis, involving liver, skin and other organs. Systemic amoebicides currently seldom used because of their high toxicity or only limited action (e.g. chloroquine) are emitine and dehydroemitine (alkaloids), which act in the liver, and intestinal wall (other tissues), and chloroquine, which acts in the liver only. Emitine and dehydroemitine administered intramuscularly are fairly toxic drugs producing cardiac arrhythmia, and asystole, but also adverse effects of the central nervous system.

Drugs Acting on Amoebic CNS Diseases and Amoebic Keratitis

A number of free living opportunistic amoebae are known to be potential pathogens for humans. *Acanthamoeba* spp., *Balamuthia mandrillaris* and *Naegleria fowleri* can cause lethal infections particularly in children having close contact with water. Their distribution is worldwide in fresh water and soil. The protozoans live preferably in aquatic habitats where they feed on bacteria (e.g. in warm lakes or swimming pools). Infection by *N. fowleri* is acquired by intranasal absorption of amoebae (flagellates or cysts); trophozoites invade the nasal mucosa, cribriform plate and olfactory bulbs of the brain (cysts are absent in infected tissues). *Acanthamoeba* spp., and *Balamuthia* may infect man via nasal mucosa causing fatal CNS disease or by direct invasion of the cornea through lesions of the eye or the wearing of contaminated contact lenses of healthy persons producing *Acanthamoeba* keratitis. Primary amoebic meningoencephalitis (PAM) due to *N. fowleri* is usually diagnosed after death of patient and is characterized by an acute, hemorrhagic, necrosing meningoencephalitis. *Acanthamoeba* spp. and *Balamuthia* cause focal granulomatous amoebic encephalitis (GAE) and trophozoites and cysts occur in most infected tissues. **Chemotherapy** of PAM and GAE is problematic and only a few cases have been treated successfully with the highly toxic polyene antibiotic amphotericin B (Table 1). *Acanthamoeba* keratitis, which increasingly occurs in immunocompromised patients, can be treated with success if it is diagnosed at an early stage. Various drugs (dibromopropamidine, neomycin azoles, alone or in combination) have been administered topically as ointments and drops (see Table 1).

Drugs Acting on Balantidiasis

Balantidium coli, which possesses cilia for locomotion, is a natural inhabitant of the digestive tract of mammals and widespread in swine; the vegetative forms of *Balantidium* (man, monkeys and pigs) may produce acute or chronic enteritis with profuse and heavy diarrhea. Good hygiene and sanitation may prevent infections, which are associated with close contact to pigs. Acute *Balantidium coli* infections may be treated with **tetracycline**. Alternative drugs may be the 5-nitroimidazole **metronidazole** (Table 1). In many countries its use in food animals is now subject to regulatory actions of government agencies; the aim is to remove drugs of this chemical group from the market because of their potentially cancerogenic action in rodents after prolonged administration.

Antigen Presentation

→ Immune Responses.

Antigens

Synonym

Immunogens

Introduction

Antigens are molecules – proteins, peptides, carbohydrates, nucleic acids, lipids or any other compound – that induce the production of antibodies and bind to antibodies. Parasites may present a great quantity and variety of antigens to the host that change with time as a consequence of maturation through different life cycle stages. Parasitic antigens may be divided into several categories: diagnostic- protective – pathologic. An antigen acts as immunogen when it is able to induce an immune response (immunogenicity). In general, different antibodies are produced to an antigen which bind to different epitopes. The hosts immune response depends on the presentation, quantity and kind of circulating antigen. Cross-reacting epitopes exist between the different stages of the same species, and between different species of the same genus. In helminth infections there is a high degree of cross-reactivity between species and genus. The antigens of helminths commonly exposed to the immune system are excretory and/or secretory (E/S) antigens or surface molecules (somatic antigen). Antigens of Protozoa have a lower degree of variability between species and their immunogens are mainly surface antigens. For each parasite the diagnostic potency of surface, E/S and somatic antigens has to be evaluated in regard to their effectiveness in a specific test system.

Antigen Processing

Due to the often complicated life cycle of parasites, including vector transmission, it may be difficult to maintain helminth parasites in laboratory animals or in vitro. In practice it is impossible to recover large quantities of antigen from each pathogen species. It is a common laboratory practice to select one single, highly cross-reacting species for antibody screening within a whole genus (*Brugia malayi* for filariasis, *Schistosoma mansoni* for schistosomiasis, *Leishmania donovani/infantum* for leishmaniasis, *Plasmodium falciparum* for malaria). Until recently the primary source of antigen for most of the serological assays was parasite maintained in vivo (helminths) or in continuos in vitro culture systems (Protozoa). Complete cells, fragments or cryosections of parasites (cellular antigens) are used as antigen in IFAT, DT, and agglutination test. Crude or purified soluble extract antigens or culture derived E/S antigens are utilized in other test systems like ELISA, IHAT, CFT, CIEP, and gel precipitation. Today, new technologies which allow the construction of synthetic peptides, recombinant proteins or DNA-based immunization have opened up the possibility for large scale production of defined antigens or specific antibodies. However, for each preparation the diagnostic potency must be proven. It could be shown that a single antigen or a monoclonal antibody may be satisfactory for individual diagnosis but not for screening purposes. A combination of several defined antigens may improve the sensitivity, a polyclonal antibody the binding capacity of a test system.

Antigen Assays

→ Immunoassays for antigen detection are increasingly developed. Urinary antigen detection is of diagnostic interest for parasites with developing stages associated to the genito-urinary tract (*Schistosoma haematobium*, *Trichomonas vaginalis*) or parasitic antigens excreted during the acute or chronic infection stage in urine (*Wuchereria bancrofti*, *Trypanosoma cruzi*). Circulating antigen detection is useful for a quantification of blood parasites (*Trypanosoma* spp., microfilariae) or detection of E/S antigens (*Schistosoma* spp.). It may provide information on the presence and/or activity of the parasite. Antigen detection in stool specimens enables a species-specific diagnosis which is impossible after microscopic examination (*Entameba histolytica/dispar*, *Echinococcus* spp.).

Clinical Relevance

Many stage specific antigens of different parasites have been characterized so far and some of them are already applied in serodiagnostic test systems. A tachyzoite-specific, recombinant surface antigen (P30) is of proven diagnostic value for the detection of *Toxoplasma* antibodies. Species specific antigens of *Echinococcus multilocularis* like Em18 or the recombinant antigen Em2 are in use for the serological differentiation of infection with the alveolar or cystic form. ELISAs with improved specificity by use of recombinant antigens were developed for parasitic infections like Chagas' disease, leishmaniasis, *Strongyloides* and many others. Defined glycoproteins of the cyst fluid from *Taenia solium* are highly specific and allow the identification of patients with cysticercosis by use of the WB. The seroreactivity to a specific antigen as indication for active infection was described for visceral leishmaniasis and amebiasis. The detection of one or both genus specific *Schistosoma*-antigens (circulating anodic -CAA- and cathodic -CCA- antigens) indicate an active schistosomiasis. A positive correlation between *Schistosoma haematobium* egg excretion and levels of CAAs was reported. The level of complexed antigen in serum is a good indication for the microfilarial load in Bancroftian filariasis.

Not only a quantification of the parasite load but also a discrimination between recent and past infection is possible by urinary antigen detection. A better discrimination between acute and congenital infected patients was reported for Chagas' disease. The immunodiagnosis of intestinal parasitic infection has advanced significantly through the development of coproantigen detection methods. Commercial ELISA systems which use monoclonal antibodies for the detection of faecal parasitic antigens like *E. histolytica/dispar*, *Giardia lamblia/Giardia intestinalis* and/or *Cryptosporidium* spp. are now available and are going to replace microscopic examination in many laboratories. However, the specificity, sensitivity and clinical value of the available tests are still under investigation. The detection of *Echinococcus*-antigen in faecal samples of the dog (*E. granulosus*) or fox (*E. multilocularis*) by use of polyclonal antibodies may become an important diagnostic tool in veterinary medicine.

Aphanipteridosis

Disease due to → flea (Vol. 1) bites; see page 550.

Aphanipteriosis

→ Fleas (Vol. 1), → Siphonapteridosis; see page 550.

Apparent Infection

Infection followed by symptoms of disease

Argasidiosis

Disease due to Argasid ticks (see Table 1, page 51).

Arteritis

→ Cardiovascular System Diseases, Animals.

Table 1. Argasid (soft) ticks[1] and Control Measurements (according to Hansen and Londershausen).

Parasite	Host	Vector for	Symptoms	Country	Therapy		
					Products	Application	Compounds
Otobius megnini (Spinose ear tick)	Cattle, Dog, Horse, Sheep, Man	Tularemia (*Francisella tularensis*), Q-fever (*Coxiella burnetii*), *Rickettsia rickettsii* (Rocky Mountain spotted fever), Colorado tick fever virus	Ear inflammation, tick paralysis	America, Southern Africa, India	Taktic™ E.C. (Intervet)	Spray or Dip	Amitraz
Ornithodoros moubata	Ruminants, Pig, Man	Q-fever *(Coxiella burnetii)*, Spirochaetosis, Relapsing or African swine fever (virus)	Blood loss, dermatitis	Africa			

[1] Genus *Argas*: see Vol. 1; room treatment: Blattanex™ (Bayer).

Arthropodicidal Drugs

General Information

Much of the increase in animal productivity as well as prevention and treatment of parasitic diseases over the past half century has been due to more efficacious and economical control of arthropods through the use of synthetical chemical compounds.

Chemical control of parasitic arthropods is a critical resource for world wide food production as well as human or animal health. There is little doubt that it will remain so for the foreseeable future, but it is also important that progress be made to sustain and enhance its contribution. Achieving this progress depends to a great extent on the synthesis of new parasiticides, a field in research orientated life science companies that is still highly dynamic.

In the livestock- and companion animal field the demand for chemical control measures has grown with increasing animal numbers and the world wide increase in consumption of meat (Fig. 1).

Arthropod attack causes a variety of damaging effects including reduction in feed efficiency, growth rate, milk and wool production. In addition to lower quantities of finished products, their quality is also impaired and parasite infestation contributes to a loss of general condition and an increase in susceptibility to secondary infections or further parasite attack.

While the level of parasite infestation is generally less severe in pets than in livestock species, the main health concern with dogs and cats is the danger of transmission of fleas, worms and tick-born diseases to human beings.

The importance of drugs to control arthropods is reflected by the market development for insecticides, acaricides and endectocidals displaying activity against some arthropods and various helminths (avermectinoid drugs).

Arthropodicidal products currently account for around 14 % the world animal health market, which has a value of about 17 billion US$ (including feed additives). The performance of arthropodicidal drugs has been notable in recent years, while others have been static or declining. This sector has continually expanded world wide. An amount of 1.600 Mill. US$ was spent in 1993 for livestock- and companion animal drugs useful for the treatment of parasitic arthropods (insecticides, acaricides and endectocides without environmental health insecticides). In 1998 the corresponding value increased by about 40 % to 2.300 Mill. US$. This trend looks yet to continue and it is expected that the market may increase by 12 % until 2002.

In terms of volume the growth rate for livestock arthropodicides is expected to be low. In context, however, with the introduction of new highly innovative drugs against ectoparasiticides of dogs and cats such as advantage (imidacloprid), program (lufenuron) and frontline (fipronil) a significant expansion is expected in the field of pets.

Resistance

The spectrum of parasite resistance is one of the major problems affecting the usage of arthropodicidal drugs.

In some regions the development of this trait has rendered compounds practically useless. One

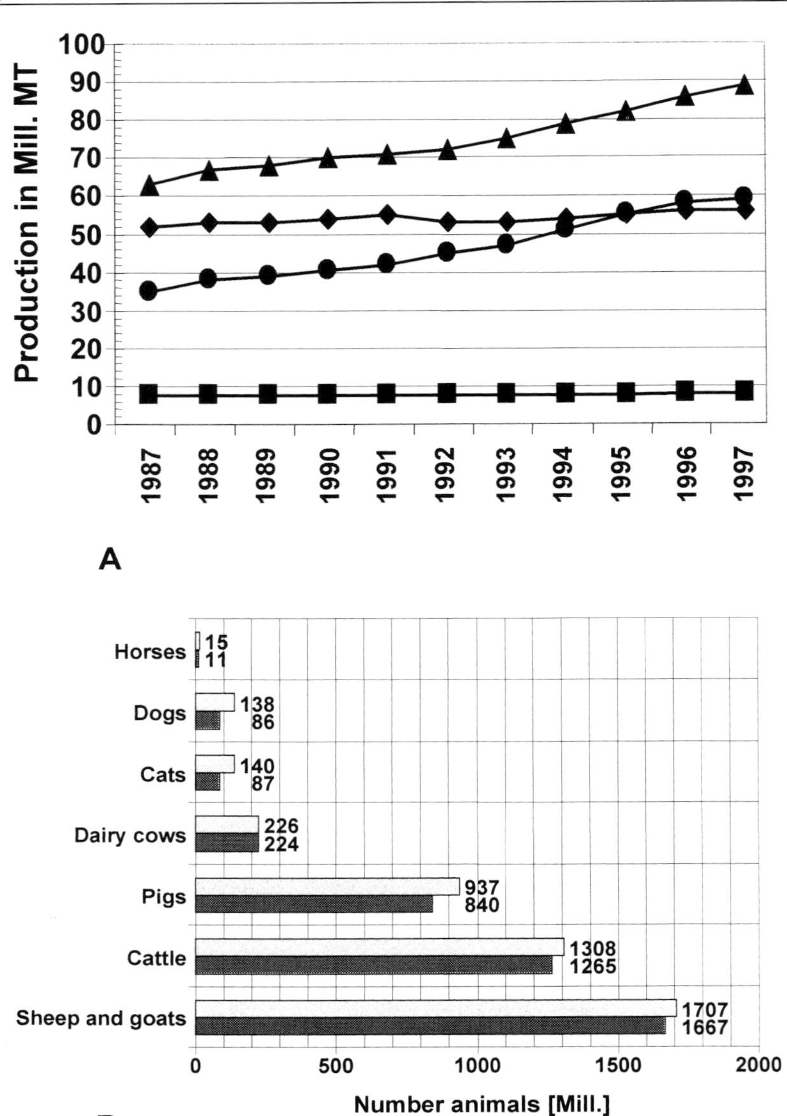

Fig. 1. A Development of worldwide production capacities of meat between 1986 and 1997. (◇) Beef and veal; (■) mutton and lamb; (▲) pork; (●) poultry. B Standing animal populations between 1987 and 1997.

solution is to put more effort into careful promotion of the proper use and rotation or combination of compound used in parasite control in order to prolong the life cycle of the products on the market.

Another possibility, and probably the best long term strategy is to increase the search for drugs displaying an alternative mode of action not interfering with the resistance mechanisms already developed in relevant arthropod pests. Good examples in the area of pets are again imidacloprid, lufenuron and fipronil which act on the nicotinic acetycholin receptor, chitinbiosynthesis and y-aminobutyric acid receptor, respectively.

Improvement of Chemical Product Quality

Particularly in livestock the world trend has now moved firmly away from purely quantitative considerations towards quality of products. Concerns about critical side effects of arthropodicidal compounds which affect the classical targets (acylcholinesterase: organophosphates, carbamates; sodium channel: DDT, pyrethroids; GABA-aminobutyric acid gated chloride channel: chlorinated hydrocarbons) have led the regulatory agencies to demand additionally information and studies in order to define more clearly product efficacy, safety, residue profile and quality. Accompanied by the increasing levels of funding used for regu-

Table 1. Combination products with arthropodicidal drugs (according to Hansen and Londershausen).

Example of combination	Main indications	Formulation/Application
Large animal use		
Abamectin/Praziquantel	Control of round- and tapeworms as wells as bots in horses	Paste
Amitraz/Diazinon	Sheep ectoparasites: flies, keds, lice, ticks	Dip
Cypermethrin/Chlorfenvinphos	Broad tick activity for various host animals, catt lice and flies	Dip/Spray
Cypermethrin/Rotenone	Longterm protection for sheep against itchmite and lice	Dip
Small animal use		
Lufenuron/Milbemycin-oxime	Canine helminths and immature flea stages	Tablet
Permethrin/Pyriproxyfen	Larval development stages and adult fleas	Spray
Propoxur/Flumethrin	Control of adult fleas and ticks	Collar
Propoxur/Methoprene	Environmental control of flea development stages	Spray
Pyrethrin/S-Methoprene	Control of fleas and ticks	Spray

latory compliance the improvement and innovation in delivery formulations of existing chemical classes was and will be a preferable method for expanding product ranges. Good examples are the development of easy to apply pyrethroid pour on's and spot on's (application of an arthropodicidal drug along the backbone of an animal, or as a spot on in one place) which have considerably infringed the share of environmentally less favourable dips and sprays.

Increasing expenditure on development have led to a tendency to exploit drug properties also by combining different compounds in one product, thereby providing a broader range of activity within one treatment. In particular the combination of anthelmintic with arthropodicidal drugs such as levamisole and famphur to "produce" activity against stomach worms, lung worms, cattle grubs and lice or the combination of insect growth regulators with avermectinoid drugs against intestinal worms and flea larvae are of significant interest in this context (Table 1).

Formulation and Delivery Systems

The ease of delivering an arthropodicidal drug has become almost as important in terms of it's consumer acceptance as is it's activity against parasites. With the increasing levels of funding needed to develop new drugs, the improvement in formulation of existing drugs is fast becoming an important method to expand product ranges.

The future trend of innovative delivery systems will probably focus on stress reduction and improvement of convenience. Losses in weight gain and growth performance as well as costs for handling of livestock species can almost reach the spending on medication. Therefore systems which extend the delivery time such as tablets, boluses or other controlled release devices, will reduce handling frequency and may open up possibilities for new administration technologies (Table 2).

As miniaturisation of electronic devices continues to advance one innovative example might be the development of new bolus types that will release ectoparasiticides, for example insect growth regulators, self- or externally triggered over a long period of time. By eliminating larval development in the pasture and reducing re-infestation such products might play an important role in the strategic management of fly control programs.

Another step forward could be the development of more advanced topical applications such as pour ons or spot ons involving for example micro-encapsulation technology for livestock and pets.

Wash-off resistance and decreased dermal penetration of adulticidal drugs such as pyrethroids or others will further stimulate the usage of this easy and stress minimising application.

Classes of Chemical Compounds

A wide variety of ectoparasites are relevant for livestock and pets. These resolve into a relatively

Table 2. Delivery systems of arthropodicidal drugs (according to Turberg and Londershausen).

Method of application	Typical examples of actives or chemical classes used
Major topical formulations	
Collar	Propoxur, Amitraz, Permethrin, Flumethrin, Diazinon, Deltamethrin,
Dips	Methoprene
Pour-ons	Organophosphates, Pyrethroids
Spot-ons	Avermectins, Milbemycins, Pyrethroids
Sprays/Jetting fluids/Mist sprays	Fenthion, Fipronil, Imidacloprid, Permethrin Amitraz, Carbamates, Organophosphates, Pyrethroids
Other topical formulations	
Aerosols	Organophosphates, Pyrethroids, Carbamates
Dust bags/Back-rubbers / Dusts	
Eartags/Strips	
Foams/Shampoos	
Ointments	
Powders	
Washes	
Oral formulations	
Boluses	Methoprene, Avermectins, Milbemycins
Tablets	
Injection	Avermectins, Milbemycins

small number of major problems with a wide variety of products, ingredients and formulation for their control. For information on → repellents (Vol. 1) and → synergists please refer to the respective entries.

Arthropodicidal drugs can be divided into different groups according to their structure and mode of action (Table 3) assuming that similar chemical structure leads to the same basic mode of action. This certainly is a much too simplified view which doesn't reflect realistically an overall biological activity. Biological effects result from the superimposement of primary mode of action, pharmakocinetic behaviour, degradation and excretion driven by the physiochemical parameters of drugs. Chemical compounds may display significantly different properties in this respect, even if they basically belong to the same chemical class.

Organochlorides

Important Compounds Bromocyclen, chlordane, DDT, lindane (gamma benzene hexachloride), methoxychlor.

General Information Lindane and several cyclodienes were demonstrated to stimulate the central nervous system by influencing synaptic transmission and causing hyperexcitation. In the 1980s it was shown that these compounds block the (GABA)-influenced Cl^-current in neurons by binding to the GABA receptor chloride channel. In a set of elegant experiments this theory has been confirmed by showing that target site insensitivity to cyclodienes was associated with a resistance gene (Rdl = resistance to dieldrin) which codes for a GABA receptor subunit in Drosophila melanogaster. The responsible point mutation in the membrane spanning region M2 could be identified by cloning of the M2 region with degenerated PCR primers amongst others in a variety of resistant dipteran flies.

Chlorinated hydrocarbons applicated in sprays or dips once were common products for example in the treatment of myiasis and lice infestation in cattle and other species. Nowadays organochlorides such as chlordane, lindane are no longer approved for veterinary use in many countries, particularly because of their long persistence in the environment. Occasionally compounds like bromocyclen or methoxychlor are used against mange mites biting and sucking lice and flies such as Melophagus or Hippobosca species.

Table 3. Major arthropodicidal chemical classes used in veterinary medicine (according to Turberg and Londershausen).

Chemical class/examples	Mode of action	Chronology of discovery for veterinary use*
Chlorinated hydrocarbons		
DDT	Voltage dependent sodium channel	1943
Cyclodiens, Lindane	GABA-gated chloride channel	~ 1945–1959
Organophosphates	Acetylcholinesterase	~ 1950–1965
Carbamates	Acetylcholinesterase	~ 1960–1970
Pyrethroids	Voltage dependent sodium channel	~ 1970–1985
Amidines	Sites responsive to biogenic amines (Octopamine receptors, monoamine oxidases)	~ 1975–1980
Avermectins / Milbemycins	Glu-gated chloride channel	~ 1981–1998**
Insect growth regulators		
Cyromazine	Interference with cuticle sclerotization	1979
Benzoylphenylureas	Inhibit cuticle deposition/ chitin biosynthesis	~ 1985–1995
Juvenoids	Mimic juvenile hormone effects	~ 1988–1996
Arylpyrazole (Fipronil)	GABA-gated chloride channel	1994**
Chloronicotinyles (Imidacloprid)	Nicotinic acetylcholin receptor	1996**

* Area of entry date based on the introduction of important actives of the respective chemical class into veterinary market
** The discovery and development of further products can be expected due to intensive research efforts in various veterinary companies

Organophosphates/Carbamates

Important Compounds Azamethiphos, bendiocarb, bromopropylate, carbaryl, chlorfenvinphos, chlorpyrifos, coumaphos, cythioate, diazinon, dichlorvos, dicrotophos, dioxathion, ethion, famphur, fenchlorphos, fenitrothion, fenthion, heptenophos, iodofenphos, malathion, methomyl, phosmet, phoxim, pirimiphos, promacyl, propetamphos, propoxur, temephos, tetrachlorvinphos, trichlorfon.

General Information The concept that toxicity to insects and ticks by organophosphates and carbamates results from inhibition of acetylcholinesterase (AChE) was proven to be useful in correlating structure with activity for various compounds from both chemical classes, although some disagreement exists as to the exact mode by which AChE-inhibitors kill ectoparasites.

Organophosphates and carbamates have a wide range of activities against various arthropods, involving blowfly larvae, keds, lice, fleas, mites and ticks in companion and livestock animals. Products with almost any application type have been developed ranging from dips, sprays and pour on's to ear tags and collars. Organophosphates in particular have been popular due to their rapid onset of activity, but in recent years concerns have been raised regarding environmental effects and some adverse reactions. In contrast to organophosphates the carbamylation of AChE by carbaryl and propoxur results in a reversible inhibition of AChE whereas organophosphates inhibit AChE almost irreversibly. The somewhat lower safety level compared to other insecticides has limited their use on older animals (> 3 months) and the development of resistance nowadays restricts their applicability significantly. This problem is especially important in such sectors as the ear tag application.

Pyrethroids and DDT

Important Compounds Alphamethrin, bioallethrin, cyfluthrin, (beta-)cyfluthrin, cyhalothrin, cypermethrin, deltamethrin, fenvalerate, flucythrinate, flumethrin, (tau-)fluvalinate, permethrin, phenothrin, resmethrin, tetramethrin.

General Information The symptoms of pyrethroid and DDT poisoning in insects are characterised by different forms of hyper-excitation caused by repetitive discharge of the nervous systems. In patch clamp investigations of single sodium channel currents it has been demonstrated, that under the influence of pyrethroids and DDT individual channels are modified to remain open for an unphysiological long period of time. The symptoms of intoxication differ somewhat between type II and type I pyrethroids based on the presence or absence of a cyano group at the á-position. Taking into account this structural as well as pharmacokinetic differences between various pyrethroids, the basic mechanism of action at the channel and cellular level is the modulation of the gating kinetics of individual sodium channels.

Since natural pyrethroids are not stable enough under atmospheric influences, the use of synthetic pyrethroids in animal health has become an important method to control arthropod pests such as flies and ticks. Various pyrethroids have been developed for animal health indications expressing more or less pronounced broad spectrum activity against ectoparasites. In this context the relation between dosage and occurrence of adverse drug reactions of the host animal became relevant in differentiating the range of optimal use for these pyrethroids.

Amidines

Important Compounds Amitraz, cymiazole.

General Information The formamidines are structurally related to octopamine. Since they mimic the action of this phenolamine in a number of insect preparations it was concluded that these compounds act via effects on octopamine sensitive adenylate cyclase (octopamine-receptors). In addition formamidines are reasonable potent inhibitors of monoamine oxidase (MAO) in cattle ticks and mammals. However MAO activity in insect nervous tissue is low and no consistent similarities or structure activity relationship have been observed between the biological effects of known MAO inhibitors and those caused by amidines in insects and ticks. It still remains to be established whether this action is causative of any of the various responses of veterinary or agricultural important insects and acarines.

Numerous biochemical targets have been suggested for formamidines, however on the basis of current knowledge the action on sites responsive to biogenic amines seem to be most likely. In animal health formamidines, in particular amitraz, displays a high acaricidal activity and offers an alternative to control organophosphorus and/or organochlorine and/or pyrethroid resistant ticks and mange mites of cattle, sheep and pigs.

Amitraz causes ticks to withdraw their mouthparts rapidly and fall off the host animal, which from the viewpoint of practical control of tick-born diseases is highly desirable.

Avermectins/Milbemycins

Important Compounds Abamectin, doramectin, eprinomectin, ivermectin, selamectin, moxidectin, milbemycin-oxime.

General Information Avermectins, milbemycins and their active derivatives inactivate nematodes, arachnids and some insects by disrupting their nerve transmission. Most of the studies have been performed with ivermectin.

In GABA sensitive fibres from insect skeletal muscles ivermectin induced increased chloride conductance which was reversible at low concentrations (nM range) but irreversible in μM range. GABA insensitive muscle fibres produced only irreversible responses to ivermectin, particularly the drug inhibited chloride conductance gated by glutamate receptors in locust muscle fibres. These and other investigations led to the conclusion that this class of compounds act through their effects on glutamate gated chloride channels.

The activity against arthropods varies to some extent depending on the milbemycin or avermectin type of drug and the delivery method.

Insect Growth Regulators

Insect or acarid growth regulators may provide a complementary method to parasite control relying on adulticide drugs. Growth regulators often have a favourable safety margin which enables them to be used also in systemic applications either as in-feed products, boluses or tablets.

After uptake their presence in manure, body fluid or skin interrupts the life cycle of various ectoparasites. Growth regulators display their mode of action either by preventing development of eggs, larvae or nymphs (e.g. tick Boophilus microplus) directly on contact or by interfering with egg development and hatching after being ingested during feeding of adult females (e.g. flea: Ctenocephalides felis).

Since growth regulators only operate on a specific part of the ectoparasite life cycle they usually have little direct effect on the adult parasite population and achieve control after a longer period of time when compared to fast acting neuroactive compounds.

Growth regulators provide a valuable tool for long term control when used in conjunction with other arthropodicides which clear the host from adult parasite and final larval stages.

Triazine The drug Cyromazine is an insect growth regulator which is still used to control blow fly strike and fly development in poultry manure. Cyromazine interferes with moulting and pupation and causes reduced growth of treated larvae. A direct effect on chitin biosynthesis has not been confirmed, but evidence has been presented that interference with cuticle sclerotisation and elasticity might be responsible for the antiparasitic effect of this substance.

Benzoylphenylureas Important drugs are diflubenzuron, fluazuron, lufenuron, and triflumuron.

The exoskeleton of arthropods consists of different layers, of which typically the procuticle contains the amino sugar polysaccharide chitin, a major component of insect cuticle.

Nucleoside peptide antibiotics and benzoylphenylureas (BPUs) are the major groups of compounds acting primarily at chitin synthesis as target site.

For animal health indications only BPUs have been developed and commercialised internationally. Original studies of the mode of action of BPUs have implicated that synthesis and deposition of chitin are disturbed after application of BPUs.

Since these findings several hypotheses for their mode of action have been proposed, but the underlying biochemical mechanism still remains unclear. So far inhibition of precursor transport, proteolytic activation of chitin synthase zymogen, direct inhibition of chitin synthase, indirect hormonal effects and effects on cell membranes in combination with vesicle transport have been discussed as the primary mode of action.

The inhibition of cuticle deposition after BPU treatment results in death of larvae, pupae and non viable adults. In addition egg development and hatching are interrupted if particular systemically active BPUs are ingested during feeding of female ticks (fluazuron) or fleas (lufenuron). Even if growth regulators require more time to reduce ectoparasite populations they are useful tools, particularly in combination with compounds displaying an alternative mode of action, to control different life stages of ectoparasites.

Juvenoids Important drugs are fenoxycarb, hydroprene, methoprene, and pyriproxyfen.

The development of arthropod parasites comprises an orderly series of stages in the course of which they become transformed from larval stages into an adult. This process involves a series of moults and a metamorphosis which can be disturbed with drugs which act similarly to insect juvenile hormones.

The main effects contributing to insect control often depend on the target pest and the timing of application.

In the veterinary field juvenoids are particularly useful in controlling flea development either used in topical on-animal applications or in sprays to prevent hatching of fleas from companion animal surroundings.

Hygiene pests such as various species of mosquitoes, house flies and to a lesser extent cockroaches can be controlled with juvenoids, too. Due to their improved stability particularly fenoxycarb and pyriproxyfen are useful for these indications.

The exact mode of action of juvenoids has not been resolved so far. A number of factors other than receptor binding may influence the activity of a juvenoid significantly when applied at a sensitive stage in growth. Penetration through the cuticle and transport to the target tissue, rates of degradation and excretion, inhibition of the juvenile hormone (JH) degradation system as well as competitive binding to JH carrier proteins are still under evaluation.

Arylpyrazoles

Important Compound Fipronil

General Information γ-Aminobutyric acid (GABA) is a major inhibitory transmitter at the neuromuscular junction not only in nematodes but also in insects and most likely in acarines, too. Upon binding of GABA to its receptor a rapid influx of chloride into the cell is induced which results in a hyperpolarisation of the membrane potential. In most cases an inhibitory effect on the respective cell is the consequence of this GABA activity. In vertebrates two subtypes have been identified. $GABA_A$ receptors which form a ligand gated chloride channel as well as contain modulatory sites for various drugs; and $GABA_B$ receptors, G-proteins which display various effects on calcium and potassium channels. Most GABA receptors in insects resemble $GABA_A$ subtype, but they are clearly distinct from the vertebrate $GABA_A$ receptors. The GABA receptor is an important target for some older arthropodicides such as lindane cyclodiens but is also affected by fipronil, a new insecticide and acaricide which belongs to the chemical class of arylpyrazoles. This has been recently confirmed by cloning of a Drosophila GABA-gated chloride channel subunit using genetic mapping of a mutation which causes resistance against dieldrin (RdI). Fipronil was developed as an agrochemical and veterinary arthropodicide. In the veterinary field the compound is used for the treatment of fleas in dog and cats and ticks in dogs.

Chloronicotinyles

Important Compound Imidacloprid

General Information Cholinergic synapses play a critical role in transmission in arthropods and nematodes. Recently a new drug, imidacloprid, was discovered displaying a new mode of action. Insecticides of this type act as agonists at the nicotinic acetylcholin receptor (nAChR) as was demonstrated by electrophysiological techniques and biochemical competition assays. The compound is highly effective against target insects, such as fleas or cockroaches but has virtually no effect on the mammalian nervous system. This might be explained by the fact that imidacloprid is not absorbed into the host animal bloodstream or internal organs and nAChRs from mammalian sources are much less or not at all sensitive to the agonistic action of insecticidal chloronicotinyles. In this context the fast killing of adult fleas on dogs and cats on contact not only prevents re-infestation for a long period of time (at least 4 weeks) but also significantly reduces the occurrence of flea allergy dermatitis caused by flea saliva allergens.

Rotenone

Rotenone is a natural product derived from plant roots of Derris and Lonchocarpus. This drug is a highly effective inhibitor of complex I of the mitochondrial respiratory chain and inhibits NADH cytochrome c-reductase in nmole per mg protein range. The drug has largely been replaced by modern compounds, but is still used in combination products against ear mites and demodectic mange.

Benzylbenzoate

Benzylbenzoate is an old acaricidal drug which is useful for the treatment of sarcoptic mites in dogs.

Sulphur

Due to the availability of new and safer products nowadays the use of sulphur for mite treatment has been largely banned not only because of toxicological but also pollution concerns.

Arthur's Phenomenon

→ Immune Complex.

Arthus Reaction

Allergic immediate reaction with severe tissue inflammation, oedema, hemorrhages and necrosis named after the Swiss microbiologist M. Arthus (1862–1935).

Ascariasis, Man

Pathology

Ascariasis is an infection with *Ascaris lumbricoides*, a large, lumen-dwelling nematode contracted by the ingestion of its eggs (→ Ascaris/Fig. 1 (Vol. 1)). The larvae penetrate the small intestine wall and migrate through the lymphatics and blood stream to the liver, and then to the lungs where they enter the alveoli. There they pause for at least 2–3 weeks and molt, giving rise to allergic bronchopneumonia in previously infected and sensitized individuals. Later, they wander up the bronchi and trachea, giving rise to bronchitis with bronchospasm, urticaria, and occasionally, larvae in the sputum. Most larvae are swallowed and grow to adulthood in the small intestine. The adult worms are up to 30 cm long and 4 mm wide, and give rise to mechanical problems because of their size and, especially in children to a severe nutritional drain, because of the worm number and mass. A temperature elevation to 39°C, certain drugs, and some unknown influences may cause the worms to congregate, sometimes resulting in intestinal obstruction and migration out of the gut into the bile duct, esophagus, mouth, pancreatic duct or appendix, and occasionally the liver.

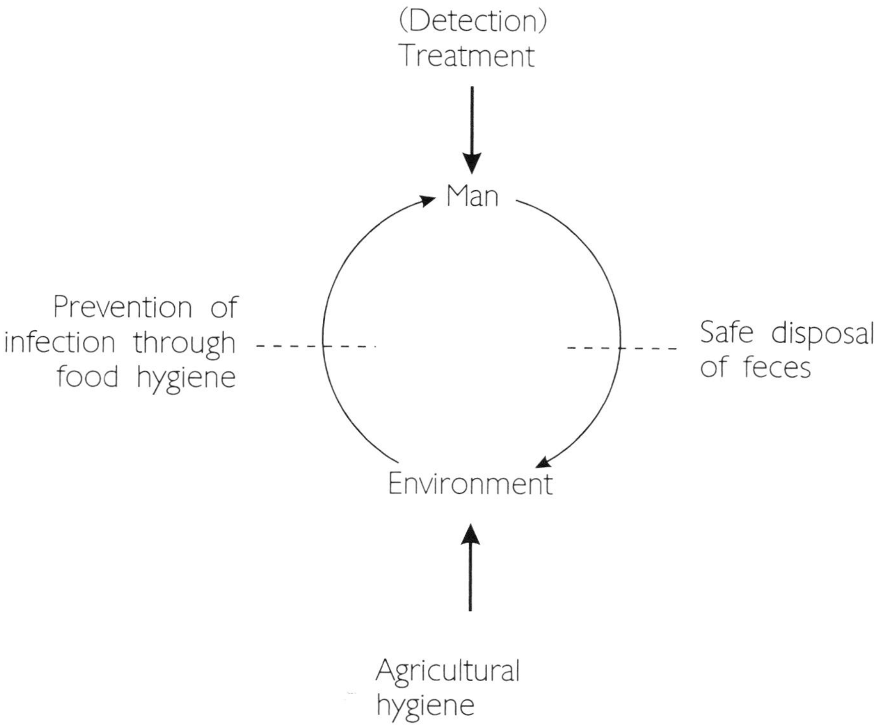

Fig. 1 Targets and approaches for the control of ascariasis

The migration leaves necrotic tracts in the liver with hypersensitive inflammation produced by adults and eggs (→ Pathology/Fig. 27A). Adult worms may perforate the intestine and pass out of the gut, leading to peritonitis.

Targets for Intervention

Eggs of *Ascaris lumbricoides* are not immediately infective after leaving the infected host. They require a holding period in a suitable environment and become infective once second-stage larvae have developed in the eggs. Fig. 1 shows the carrier and the infection cycle as targets of intervention. Control may be achieved by the detection and treatment of infected persons, safe disposal of feces, and improving agricultural hygiene and food hygiene.

Main clinical symptoms: Eosinophilia, abdominal pain, vomiting, enteritis, ileus verminosus
Incubation period: Lung: 7 days, intestine: 3 weeks (however 85% of the patients remain symptomless)
Prepatent period: 2 months
Patent period: 9–15 months
Diagnosis: Microscopic determination of eggs in fecal samples
Prophylaxis: Avoid eating uncooked vegetables and avoid human feces.
Therapy: Treatment see → Nematocidal Drugs, Man

Atopic Reactions

Clinical symptoms of allergic hypersensitivity reactions of type 1 (e.g. rhinitis, asthmatic symptoms, eczema) due to contacts e.g. with dust mites.

Autoimmunity

→ Chagas' Disease, Man/Immune Responses.

Avian Spirochaetosis

Disease due to spirochaetan bacteria transmitted by the bite of *Argas* ticks (→ Tick Bites: Effects in Animals)

B

B-Cells

→ Immune Responses.

Babesiacidal Drugs

See Table 1

Economic Importance

Members of the genus *Babesia* occur throughout the world and may cause a wide range of clinical syndromes in most domestic animals and humans due to differences in virulence within each *Babesia* species. They are transmitted by hard → ticks (Vol. 1) (→ Ixodidae (Vol. 1)) during blood meals, and may produce diseases in their hosts, which are characterized by an acute febrile reaction, jaundice, hemolytic anemia, hemoglobinuria and variable mortality. The babesiosis is of major economic importance in cattle because the majority of about 1.2 billion cattle in the world are potentially exposed to *Babesia* spp. because of the extensive husbandry methods employed in raising these animals. *Babesia* spp. infections may also occur by the frequent introduction of *Babesia*-free animals into *Babesia*-enzootic areas, and by introduction of tick vectors and *Babesia* spp. into clean zones. The greatest economic loss is due to *Babesia* spp. infections in cattle, particularly in the USA, Australia (State of Queensland), South Africa, and South America. Indigenous animals are normally protected from babesiosis in early life by premunition resulting from continuous reinfection. In regions with enzootic stability babesiasis is not a disease problem, and control measures are not necessary. In these zones calves are still protected by maternal antibodies which they had received through the colostrum. Subsequently the animals remain carriers by repeated natural infections.

Epizootiology and Enzootic Stability of Bovine Babesiasis

The idea of maintaining enzootic stability is to allow limited challenge of the parasite and the tick to cattle without producing disease or loss of production. The immune status of a herd can be monitored serologically, and in cases where more than 50% of cattle show no antibabesial serum titres the whole herd should be vaccinated. Thus, the efficient application of vaccination procedures depends considerably on the epizootiology of babesiasis, and involves knowledge of the complex interaction of the host, *Babesia* parasite, and vector. The aim of premunization (Premunization and Use of Antibabesial Drugs (Chemoimmunization)) or immunization with live blood-derived vaccines of attenuated strains of *B. bovis* and *B. bigemina* is to maintain herd immunity, and thus the balance between host and vector. In various countries the idea of reestablishing areas of enzootic stability has strongly been supported, particularly after the introduction of so-called "tick-resistant" cattle, and the occurrence of increasing failures of chemical control measures.

Strategic Tick Control in Areas with Enzootic Instability of Babesiasis

Attempts at controlling babesiasis in districts with enzootic instability (i.e., where less than 90% of cattle show premunition) can be directed against the tick vector (for more details see control of ticks, and application techniques). There are one-host ticks (e.g., *Boophilus* spp.), two-host ticks, and three-host ticks (e.g., *Rhipicephalus* spp., *Haemaphysalis* spp.; *Ixodes ricinus*) which can be controlled by aggressive or strategic dipping (this latter method may involve three dippings in the dry season and is used for example in Australia) plus pasture improvement and/or pasture spelling at intervals of 8–10 weeks in summer or early autumn. Aggressive dipping is still imperative in

Table 1. Babesiacidal drugs used in domestic animals.

Chemical group (approx. dose[a], mg/kg body weight, route, animal)	**Nonproprietary name,** *brand name (company, manufacturer)	**Characteristics**[b] and miscellaneous comments
AZO-NAPHTHALENE DYES (cattle, 2–3, i.v.; dog, 10, 1% solution, i.v.) [c] 1909	**Trypan blue** (syn. Congo blue, Niagara blue) *Trypan Blue SS (Centauer Lab., S.-Africa)	first specific drug with activity against *B. bigemina* and other large species; i.v. application leads to blue staining of meat, body secretions and milk (today, largely replaced by diamidines); discoloration of mucous membranes and plasma interferes with clinical and laboratory parameters; there are severe local reactions after extravascular injection; drug can be used in treating *B. canis* infection in dog but relapse may occur; diminazene may sterilize infection
ACRIDINE DERIVATIVES		
(cattle, 4.4 ml/100 kg, 5 % solution, i.v.) [c] 1919	**Acriflavine** hydrochloride (euflavine, trypaflavine) * Gonacrine (May and Baker)	is active against *B. bigemina* and other large *Babesia* spp.; like trypan blue, it proves less effective against *B. bovis*; i.v. administration limits its use under field conditions; drug is relatively well tolerated at recommended dose; discoloration of tissues is less pronounced than that caused by trypan blue
Urea derivatives (quinolone derivative)(cattle, 1–2 ml/100 kg, 5 % Acaprin solution, 1–2 injections, 24 h interval, s.c.; dogs, 0.25 ml/5 kg, 0.5 % Acaprin solution, s.c.) [c] 1935	**Quinuronium** sulfate * Acaprin * Ludobal (Bayer) * Babesan (ICI) * Akiron, Pirevan, Piroplasmin, and others	for many years the drug of choice in treating bovine babesiosis (*B. bigemina, B. bovis, B. divergens*); it is active against large *Babesia* spp. of swine, horse, and dog; drug has a low therapeutic index and may stimulate parasympathetic nervous system (excessive salivation, frequent urination, or dyspnea caused by anticholinesterase activity; antidotes are atropine or epinephrine; in fatal cases volvulus of jejunum has been observed which may result from hyperperistalsis); its mode of action is uncertain; action of drug on *B. felis* see text
DIAMIDINES		
Carbanilides(cattle, 5–10, i.m.; horses 8, i.m.) [c] 1964	**Amicarbalide** diisethionate * Diampron(May and Baker)	is effective against *B. bigemina, B. caballi, B. motasi* but there are differences in efficacy against large *Babesia* spp.; its effect on small *Babesia* spp. is less pronounced (clinical remission only);
(cattle, 1.2–2.4, s.c.; horse, 2–2.4, i.m.; 2 injections, 24 h interval; dog 5–6, i.m. 1–2 injections, 24 h interval) [c] 1969	**Imidocarb** dipropprionate * Imizol (Burroughs Wellcome) *Carbesia (F) *Forray-65 (Hoechst)	is effective in preventing and treating bovine babesiosis without interfering with development of immunity; protection period may last for several weeks depending on dose used; drug shows efficacy against *B. canis* infections, and equine babesiosis. Fine-structural alterations of *B. herpailuri* in cats treated with the drug were less pronounced and resulted in widening of subpellicular endoplasmic reticulum and perinuclear; drug is very slowly eliminated; bile is an important route of excretion; there is a long preslaughter withdrawal period (cattle; sheep) for edible tissues and milk of approx. 90 days and 21 days, respectively; data on toxicity are somewhat erratic; there is a wide range of individual animal tolerance to the drug; side effects are due to anticholinesterase activity, and fatal toxicosis is usually associated with renal disorder (necrosis), edema, hydrothorax, hydroperitoneum, and pulmonary congestion; mortality in equines (which can occur at 4 mg/kg or higher doses) is attributed to acute tubular renal necrosis; mechanism of drug action is known (possibly interferes with nucleic acid synthesis, page 63, diminazene); excess of polyamines (e.g., spermidine) can nullify trypanocidal activity of imidocarb and amicarbalide; the diamidine imidocarb also proved active against *Trypanosoma brucei* mouse infection, and *Anaplasma marginale* infection in cattle at 3mg/kg b.w. (elimination of parasites is not achieved).

Table 1. (Continued) Babesiacidal drugs used in domestic animals.

Chemical group (approx. dose[a], mg/kg body weight, route, animal)	**Nonproprietary name,** *brand name (company, manufacturer)	**Characteristics**[b] and miscellaneous comments
AROMATIC DIAMIDINES (dog, 5–10, i.m., 15, s.c. 1–2 injections, 24h (48h: s.c.) interval; horse, 8, i.m., 1–2 injections, 24 h interval) [c] 1939	**Phenamidine** diisethionate * Lomadine (May and Baker) * Oxopirvedine (Merieux)	is used mainly in treating canine and equine babesiosis; it has also been used successfully in *B. bigemina* infections; frequent relapses may occur in *B. gibsoni* infections in dogs; drug can be used for eradicating carrier infections in horses; side effects of drug are similar to those noticed with pentamidine; it is well tolerated at recommended dose although s.c. injections may lead to moderate swelling (necrosis, mainly in horses); drug may cause transient immunosuppression in dogs; mechanism of drug action is uncertain but may be similar to that of pentamidine and diminazene (→ Trypanocidal Drugs, Animals)
(experimental drug, approved for use in humans only; cattle, 1–5, s.c. [c] 1939	**Pentamidine** diisethionate * Lomidine (May and Baker)	is reported to be effective against *B. bigemina* allowing carrier state to persist in chemoimmunization (see text); drug is active against *B. canis* and *B. gibsoni* infections in dogs; relapses occurred with *B. gibsoni* at 16.5 mg/kg i.m.; common side effects are vomiting (dogs), nausea, hypotension, tachycardia, and pain at injection site; it may affect *B. microti* infections in man (see text); pentamidine is highly active in treating early stage of *T. b. gambiense* infection in man (→ Trypanocidal Drugs, Man)
[3] (3.5, i.m., cattle, horse) [c] 1955	**Diminazene** aceturate [3c]* Berenil (Intervet), not recommended for use in dog * Ganaseg (Squibb)	is highly active against bovine, ovine, porcine, equine, and canine babesiosis; small *Babesia* spp. are generally more refractory to treatment than large ones; it can be used in chemoimmunization programs, moderating clinical signs and allowing development of premunition (see text); there are various treatment regimens for eliminating babesiosis in cattle, horses, and dogs: in most cases recommended dose has been varied, e.g., 5 mg/kg, twice in 24 h interval to eradicate *Babesia* spp. infections in horses, or 1.75 mg/kg twice in 24 h interval to reduce or avoid neurotoxic side effects in dogs (e.g., ataxia, opisthotonus, nystagmus, extensor rigidity, coma, and even death); however, there seems to be a wide range of individual animal tolerance to the drug; it is well tolerated in equines at recommended dose, but higher doses may cause severe side effects (page 621); in camels there may be mortality at recommended dose but ruminants tolerate the drug at higher doses (7–10 mg/kg); preslaughter withdrawal period (cattle; sheep) for edible tissues and milk is at least 21 days and 3 days, respectively; diminazene was shown to be active against *B. microti* infections in humans (see text); for its activity against African trypanosomes in animals and humans see → Trypanocidal Drugs, Animals and → Trypanocidal Drugs, Man, respectively
8-AMINO-QUINOLINES (experimental drug approved for use in humans only)	**Primaquine** diphosphate * Primaquine (Bayer) * Neo-Quipenyl * Neo-Plasmochin	is active against *B. felis* (base 0.5 mg/kg, i.m., or p.o. repeated doses well tolerated); maximum tolerated dose of base was 1 mg/kg, higher doses caused mortality; standard antibabesial drugs proved ineffective against *B. felis*; for its action on exoerythrocytic stages of *P. vivax* see → Malariacidal Drugs (Malaria antirelapse drug)
TETRACYCLINES		
(long-acting preparation)	**Oxytetracycline** * Terramycin, LA (Pfizer)	*B. divergens* field infections in cattle could be controlled by continuous administration of 20 mg/kg every 4 days during natural exposure on grazing heavily infested with *Ixodes ricinus*; no parasites were seen in treated cattle in comparison with untreated animals which developed a patent parasitemia; the antibiotic seems to be useful in chemoimmunization (see text) against bovine babesiosis allowing development of premunition

Table 1. (Continued) Babesiacidal drugs used in domestic animals.

Chemical group (approx. dose[a], mg/kg body weight, route, animal)	**Nonproprietary name**, *brand name (company, manufacturer)	**Characteristics**[b] and miscellaneous comments
TETRACYCLINES	**Chlortetracycline** *Aureomycin (Lederle) and others	was shown to control parasitemia in the early course of *B. equi* infection; antibabesial activity was demonstrated by six daily injections of 0.5, 2.5, and 2.6 mg/kg; the antibiotic proved active against *T. parva* (→ Theileriacidal Drugs)
MACROLIDE ANTIBIOTICS	**Clindamycin** (Upjohn) in combination with **vancomycin** HCl (glycopolypeptide) (Lilly)	single drug, and clindamycin + quinine were shown to be effective in treating experimental *B. microti* infection in hamsters; clindamycin/ vancomycin combination has successfully been used in eradicating parasitemia and bringing about remission of *B. microti* infections in humans; for more details on standard therapy of human babesiosis see text; side effects caused by clindamycin may be allergic reactions, diarrhea (enterocolitis), hepatotoxicity, occasionally hypotension, ECG changes; vancomycin may reduce antibiotic-associated colitis
HYDROXYNAPHTHOQUINONES		
	Parvaquone *Clexon (Wellcome) buparvaquone *Butalex (Wellcome	antitheilerial drugs (→ Theileriacidal Drugs) have shown some efficacy against *B. equi* (belongs to small *Babesia* spp.) in horses; for more details on *Theileria*-like characteristics of small babesias see text. In splenectomised horses given at 5 mg/kg body weight 4 times at 48 h intervals the parasitemia was eliminated.

[a]**Doses** listed refer to recommended dose of the manufacturer and/or to literature on the subject
[b] For more details see → Drugs
[c] First practical (commercial) application (approx. year)
Data given in the table have no claim to full information
The primary or original manufacturer is indicated

areas where, in addition to babesiosis, East Coast fever, heart water due to *Cowdria ruminantium* infection and anaplasmosis (*Anaplasma centrale* and *A. marginale*) occur. This method involves acaricide administration at regular intervals, either throughout the year or during tick seasons, with the aim of eliminating the tick vector. The application of acaricide can be done in any of three ways. These may be (1) a plunge bath, communal dip, or dip on large farms, (2) a spray-race (system of pipes fitted with spraying nozzles) on large farms, or (3) a hand-spraying system consisting of a large water container and a hand-pump or small motor-pump. The strategic dipping system, i.e., administration of an acaricide at certain time intervals, may lead to temporarily unstable situations since transmission of *Babesia* spp. and *Anaplasma* spp. is not completely interrupted; in this case an additional vaccination against both organisms is advisable.

Tick Control in Livestock to Prevent Babesiosis

Today, the only conventional control measure against argasid and ixodid ticks is the use of acaricides for animals at risk. Pesticides belong to different chemical groups such as chlorinated hydrocarbons, organophosphates, carbamates, diamidines, synthetic pyrethroids, and avermectins. Drugs of these groups have different mode and site of actions. For instance the growth regulator **fluazuron** inhibits the chitin formation in ticks. Its effect results in reduced production of viable eggs in engorged females and may reduce pasture contamination.

Acaricidal control of vectors to protect animals from disease has been applied in many countries although with only limited success, as ticks rapidly developed resistance to pesticides and acaricides (→ Ectoparasitocidal Drugs, → Arthropodicidal Drugs). However, acaracide treatment continues to be necessary although drug tolerance of ticks to most agents, except the macrolytic lactones, is common. In the field, there may occur tremendous economic losses due to breakdown of dipping systems in various regions, the increasing cost of acaricides often beyond the means of many farmers, and the lack of veterinary infrastructure in many countries have made the eradication of tick vectors impossible.

Closantel used for the control of helminthic infections (→ Nematocidal Drugs, Animals, → Nematocidal Drugs, Man, → Trematodocidal Drugs) of sheep and cattle, and larval stages of nasal bot fly (*Oestrus ovis*) of sheep has been shown to have a marked effect also on fecundity and egg viability of ticks. The strongly plasma bound drug is ingested by ticks while feeding and probably affects their mitochondrial energy production by inhibiting oxidative phosphorylisation and thereby adenosin triphosphate synthesis. Thus closantel may play a role in tick control strategies, due to its effect on tick reproductive cycle being greatest at time of dosing. Current alternative control measures and their effective use in the field are still of minor importance. Biological control of ticks with living antagonists distributed by man to lower pest (parasite) populations may lead to their reduction and so to acceptable subclinical densities. In the field, however, the use of biological agents as formulations (products) of viruses, bacteria, fungus, protozoans (e.g., microsporidia), nematodes, and insects, or pheromones, are very much limited and less promising in their effects today. A more successful control measure against widely distributed *Boophilus* spp. and other tick species seems to be the immunisation of cattle with recently developed anti-tick vaccines. The antigens of interest against ticks are native or recombinant proteins located in the gut cells' plasmatic membrane. They are called " novel, concealed, or occult antigens" causing lesions in the tick's intestine thereby reducing the fecundity of ticks. Other putative antigens seem to be located in cells of the salivary glands preventing feeding of the ticks and thus depressing their fertility.

Control of Babesiosis by Various Application Techniques of Acaricides

Acaricides can be applied to animals by different application techniques. When treating animals for external parasites it is important that agents not absorbed through the skin or from the digestive tract or parenteral injection be so applied that contact with the parasite will occur. Dips and sprays generally are suited for treating most animals (especially herds) except when temperatures are below freezing or extremely thirsty animals are to be treated. Systemic drugs, usually the organophosphates and macrocyclic lactones (ivermectin, moxidectin, doramectin, and eprinomectin) are applied as pour-ons, spot-ons, injectables, sprays and feed additives, or via dipping vats (tanks). These agents gain access to the host circulatory system and are then distributed throughout the body. **Dipping** probably offers the best means and most cost effective method of tick control for cattle in tropical areas, and for dogs (is less frequently used for cats). Dipping with acaricides has the advantage of thorough coverage of the skin, coat, and head of cattle (if deep tanks are used). When using such agents caution must be exercised to prevent contaminating humans and their food supply and environment.

Premunization and Use of Antibabesial Drugs (Chemoimmunization)

Bovines of an enzootic area commonly acquire a so-called infection-immunity or **premunition** against babesiosis in the first six months of life. As a result, most of the cattle of enzootic areas are carriers of a few parasites and will therefore develop a certain degree of protective immunity against local *Babesia* strains without showing signs of disease. In contrast, elimination of the infection by curative agents is soon followed by loss of resistance to the parasite in most hosts. Cattle introduced from areas free of babesia or with parasites of different antigenic strains, may acquire babesiosis and die. **Premunization** (artificial induction of premunition) has allowed the introduction of quality cattle into enzootic areas of Australia, USA, Latin America, and elsewhere. Animals introduced are injected with blood from babesia carriers and monitored for the presence of fever and parasitemia. Soon after clinical signs are apparent, animals are treated with subcurative doses of diminazene aceturate or imidocarb dipropionate (Table 1), thereby killing enough parasites to prevent an outbreak of disease but allowing some surviving parasites to induce protective immunity to natural challenges in hosts. Animals can be injected simultaneously with a standardized dose (several million organisms of each species per animal) of *B. bigemina*, *B. bovis*, and *Anaplasma marginale* (*Rickettsia* species) derived from blood of donor bovines often splenectomized. They are then treated well timed with Imidocarb or with long-acting tetracyclines, which are active against both Babesia and Anaplasma. Premunization has several drawbacks. The main obstacles, like the occurrence of hemolytic disease in newborn calves (antibodies against erythrocyte isoantigens), transmission of other blood-born pathogens (e.g. leukosis virus), or storage and transportation problems of the '**vaccines**', have largely been overcome but some problems remain. Thus premunization prevents eradication of the parasites, may cause economic losses (outbreak of disease or occurrence of mortality in herds), may induce a variable resistance status in cattle, and is expensive. On the other hand there may be some advantages concerning premunization. Thus its application is still needed in large enzootic areas because refinements to the system of immunization with attenuated parasites are not yet satisfactory with regard to tolerability of vaccines. Occasionally there are adverse effects such as abortion, hemolytic neonatal disease, or induction of severe babesiosis. The safety of blood derived vaccines is not always guaranteed and transmission of other hematogenous infections cannot be absolutely excluded. This may also be true for the stability of vaccines. They may fail to induce strong herd immunity because of instability of vaccinal parasites or changes in selection of field strains, or prolongation of vaccine shelf life beyond expiration date. Intolerability of vaccines may be shock and disturbances of the blood clotting mechanisms, and calcium balance.

Chemotherapy of Babesiosis in Animals

The elimination of babesias (Table 1) in cattle or horses may play an important role for those animals which have a low-grade infection premunition. In these cases it must be guaranteed that carrier animals are free from infection before being imported into *Babesia*-free areas. However, when drugs are used therapeutically in endemic regions the aim is to promote clinical recovery only, and to allow some parasites to survive, reestablishing premunition. Thus, instead of the chemoimmunization programs (Premunization and Use of Antibabesial Drugs) that are used in districts with enzootic stability or instability, in countries where babesias are rare the so-called diagnosis-treatment method is employed. There are several drugs in use, all of which are "old-timers" and are more or less afflicted with adverse effects involving long withdrawal periods for meat and other edible tissues. Drugs commonly used in treating acute ovine or porcine babesiosis are quinuronium sulfate, imidocarb, and diminazene, which have sufficient efficacy against clinical attacks. Infections with *Babesia* spp. in sheep, goat, and swine must be treated with somewhat higher doses than those normally recommended in cattle. Repeated administration of drugs may be necessary to cure *B. ovis* infections (for details regarding activity and toxicity of antibabesial drugs see Table 1, → Trypanocidal Drugs, Animals; for pharmacokinetics of diminazene diaceturate and ethidium bromide see → Trypanocidal Drugs, Animals/Pharmacokinetics of Trypanocides and Chemical Residues in Edible Tissues and Milk). The effect of an antibabesial drug may vary and can be modified by the severity of the disease, the dosage used, the timing of treatment in the course of infection, and the length of time that the drug is present to affect the parasite. As a rule, large *Babesia* spp. are distinctly more susceptible to chemotherapeutic agents than are the small ones. In general, the latter respond variably to antibabesial drugs. **Sterilization of infection**, i.e., complete elimination of parasites, is usually not achieved with small *Babesia* spp. possibly due to adherence of parasites to capillary walls and consequent obstruction of blood flow. There may be differences in relation and metabolism between small and large *Babesia* spp., and as a result the target and biochemical mode of action of drugs differ. Recovery of ill animals can be achieved if specific and effective treatment is given prior to the onset of severe anemia or disorders of the nervous system, i.e., in the early course of infection. Prognosis is poor for those animals already showing cerebral signs; these are caused by clumps of parasitized erythrocytes blocking capillary blood vessels of cerebral cortex. In severe cases the aim of supportive treatment (e.g. blood transfusion, fluid therapy) is to reduce the occurrence of shock and disturbances of the blood clotting mechanisms and calcium balance.

Elimination of Babesia in Horses

Equine babesiasis is widespread; severe clinical disease and mortality may occur occasionally. Therefore, the elimination of carrier infection in horses being shipped from endemic zones to *Babesia*-free areas gains increasing importance. Various drugs may be used for clearing *Babesia caballi* infections. Amicarbalide (8.8 mg/kg body weight, x2, 24 h interval), imidocarb (1–2 mg/kg body weight, x2, 24 h interval), diminazene (5 mg/kg body weight, x2, 24 h interval), and phenamidine (8.8 mg/kg body weight, x2, 24 h interval) may show sufficient action. *B.equi* (small species) has recently been transferred to the genus *Theileria*.

Elimination of Babesia in Dogs and Cats

Canine babesiosis is becoming increasingly widespread in the USA, Europe, Africa (*B. canis*) and

Asia (*B. gibsoni*). The disease can be treated with a few antibabesial drugs causing more or less toxic side effects, such as diminazene (which is not recommended for use in dogs by the manufacturer, Table 1), imidocarb, amicarbalide, phenamidine and trypan blue. *B. canis* may cause uncomplicated infections (fever, depression, acute hemolysis with a mild to severe anemia, pale mucous membranes, and splenomegaly), or complicated ones (coagulopathy, hepatopathy, immune-mediated hemolytic anemia, renal failure, cerebral signs, pulmonary oedema, and shock). The specific therapy (if started too late, or in case of complicated infection per se) must be combined with supportive treatment (fluid infusion followed by blood transfusion, liver protectants, diuretics, vitamin B complexes, prednisolone). In general, drugs are distinctly less active against *B. gibsoni* than against *B. canis*. Relapses in *B. gibsoni* infections are common, and may also occur after administering markedly higher doses than those recommended. Thus, diminazene in doses of 7–10 mg/kg body weight only suppresses parasitemia, but these doses and even lower ones may cause severe side effects and occasionally mortality in dogs (Raether unpublished). *B. felis* infection, which may cause anemia and icterus in the domestic cat, has been reported to respond to trypan blue and quinuronium. These results are inconsistent with those of Potgieter who found that all known antibabesial drugs failed to affect *B. felis*; successful treatment was achieved in using primaquine diphosphate. Concerning differential diagnosis of feline babesiosis *Cytauxzoon felis* infection (*Theileria* species) should be considered; it may occur in North America and parasitizes lymphocytes and erythrocytes of cats.

Chemoprophylaxis of Babesiasis in Animals

The aim of chemoprophylaxis is to protect susceptible animals from clinical signs of babesiosis caused by natural tick infection, or to moderate the clinical course of infection in immunization programs (Premunization and Use of Antibabesial Drugs). The administration of drugs like **diminazene**, **imidocarb,** or **oxytetracycline** (Table 1) should allow the development of premunition. Imidocarb, which exhibits a fairly long effect on *Babesia* spp. in cattle (4–12 weeks at 2 mg/kg body weight, depending on *Babesia* species, and infection pressure), can also be used for short-term protection of susceptible animals after their introduction into *Babesia*-infected areas.

A single subcutaneous dose of 2.4 mg/kg body weight **imidocarb** may protect dogs from *B. canis* infections for about 4 weeks. However, controversy exists concerning the duration of its prophylactic efficacy. So in Beagle dogs experimentally infected with *B. canis* (merozoites), a single dose of 6mg/kg body weight resulted in a 2-week protection period only. In enzootic *Babesia* areas, dogs can also be protected by long acting acaricides applied in 4-day-intervals, or by application of an inactivated vaccine (Pirodog®, available in France, Switzerland). Although this vaccine allows *Babesia* infection under high infection conditions it may prevent infected dogs from mortality.

Resistance

Although malaria and babesiosis have many similarities, and their causative agents are related, resistance of Babesia and malaria parasites to drugs differs markedly. While resistance of *P. falciparum* to drugs has probably become the most important threat to effective control of malaria, drug resistance in large *Babesia* spp. seems to be a minor problem in the chemotherapy of babesiosis. Resistance to antibabesial drugs can be induced experimentally. Thus in vitro micro-titres tests (96-well flat-bottom plates) may be used to assess drug responsiveness of *B. bovis* or *B. bigemina* to various antibabesial compounds, thereby selecting drug-adapted lines by the presence of sub-inhibitory drug concentrations. Under field conditions drug resistance in *Babesia* spp. may emerge if drugs are used prophylactically or in chemoimmunization programs. Using such dose regimes it is likely that subtherapeutic, low concentrations of the drug are in temporary or permanent contact with the parasites, thereby causing selection of drug-resistant organisms. However, it must be emphasized that the innate poor response of small *Babesia* spp. (e.g., *B. bovis*, *B. ovis*, *B. gibsoni*, *B. felis*, *B. microti*) to antibabesial drugs should be distinguished from an acquired drug resistance occasionally occurring in large babesias. Compared to the greater drug sensitivity of large *Babesia* spp., the varied action of drugs against small forms ("natural resistance") may well be connected with differences in their relation. Small species undergoing schizogony in lymphocytes (e.g. *B. microti*) appear in their fine structure very similar to *Theileria* spp. Molecular analyses of the small subunit ribosomal RNA genes (rDNA) suggest that small *Babesia* spp. may have a close relation to *Theileria* spp. There were a number of reasons why *B. equi* was transferred to the genus *Theileria*.

Chemotherapy of Human Babesiosis

Since the first description of human babesiosis in 1957 in a splenectomized patient in Yugoslavia there have been several case reports from the USA, Europe, and other countries on babesiosis in man. The acquisition of human babesiosis may depend on contact with subadult stages of certain *Ixodes* ticks, e.g. *I. dammini* (its main host is white-footed mouse, *Peromyscus leucopus)*, and *I. ricinus* possibly transmitting *Babesia microti* (a rodent piroplasm) and *B. divergens* (a parasite of cattle), respectively. *B. gibsoni* is a parasite of dogs, and WA-1, a *B. gibsoni*-like piroplasm, has been documented in residents along the Pacific Coast of the USA, and may infect humans in Taiwan and South Africa too. The host (vector/tick) of WA-1 and its reservoir are still unknown. Although closely related to *B. gibsoni*, WA-1 should not infect dogs. However, it is highly pathogenic for most rodents. In humans WA-1 and *B. microti* may cause a similar course of infection and symptoms. Apparent **clinical signs** are parasitemia (intraerythrocytic parasites show characteristic tetrad forms), fever, rigors, cough, headache, vomiting, anorexia, and dark-colored urine. Spleen-intact humans infected with *B. microti* or WA-1 may well respond to a combination of **clindamycin** (macrolide antibiotic) and **quinine** (chinchona alkaloid, Table 1, → Malariacidal Drugs/Malaria of Human) although quinine proved to be totally ineffective against *B. microti*. In many areas of Europe, the enzootic cycle of *B. microti* may obviously depend on uniquely mouse specific tick, *I. trianguliceps*. Possibly for that reason this species is not transmitted to humans, or its European strains are not pathogenic to humans. Reported cases of human babesiosis in Europe are due to *B. divergens* and chiefly occur in farmers and other persons frequently in contact with cattle. As a rule, *B. divergens* infections in man show rapid increasing parasitemia, and damage of large numbers of infected red cells causing massive hemoglobinuria, intravascular hemolysis and renal failure. Therefore an early start of specific treatment is important for any patients infected with *B. divergens*. The treatment of choice seems to be the immediate administration of clindamycin (adult dosage: 1.2 grams 2x/d IV or 600mg 3x/d PO x 7d) plus quinine (650mg 3x/d PO x 7d; all doses cited are from Medical Letter, 1995; pediatric dosage for both drugs see there). Chemotherapy is followed by a massive exchange transfusion to reduce parasitemia, i.e., to remove physically large numbers of infected red cells and prevent extensive hemolysis and renal failure. Exchange transfusion may be used as an alternative or in addition to chemotherapy relying on relatively toxic drugs. However, this therapy with its attendant risks should be reserved for heavily parasitized and seriously ill patients only. Symptomatic and other supportive care should be associated with specific treatment. Clindamycin plus quinine treatment proved to be insufficient in immunosuppressed patients or in HIV infected ones. Therefore, in foudroyant infection courses massive blood exchange may be life saving or application of **atovaquone** as was recently shown.

Pentamidine isethionate (a trypanocide, see → Trypanocidal Drugs, Man) has shown good activity against *B. microti* but failed to eliminate parasites from blood completely. **Diminazene aceturate** (Table 1 and → Trypanocidal Drugs, Animals), an aromatic diamidine like pentamidine and approved for use in animals only, has also been tested in humans. The drug was successful in treating a *B. microti* infection, which did not respond to therapy with oral chloroquine phosphate. The infection was eliminated, but the patient developed acute iodiopathic polyneuritis, probably related to the diminazene therapy.

Babesiosis, Animals

Synonym

Piroplasmosis, Texas Fever, Meadow Red, Redwater Disease.

General Information

Babesiosis is caused by infection with species of tick-borne, intra-erythrocytic and generally host-specific protozoan parasites of the genus → *Babesia* (Vol. 1). It occurs in a wide variety of vertebrate hosts and has a very wide distribution around the world. The two major factors involved in the pathogenesis of babesiosis are the release of pharmacologically-active agents and intravascular haemolysis. The relative importance of each varies with the species of *Babesia*, e.g. the pathogenesis of *B. bigemina* is almost entirely related to haemolysis while with *B. bovis* the release of proteolytic enzymes which directly or indirectly affect the microcirculation and the viscosity of blood is the most important. Effects on the microcirculation include vasodilatation and increased vascular permeability, leading to hypotension and oedema. Higher viscosity and coagulability of blood and increased stickiness of the membrane of blood cells cause aggregation of cells in capillaries and obstruction of the blood flow. This may lead to congestion and degenerative changes in the spleen, lymph nodes, kidneys and brain. Whatever the species involved, the disease may follow acute, subacute or chronic courses, and there may be great variation in the clinical manifestations of infection.

Pathology

Cattle Various species of *Babesia (B. bovis, B. bigemina, B. divergens* and *B. yakimovi)* are transmitted by ticks into wild and domesticated cattle. Large numbers of ruminants are killed by this parasite in tropical and subtropical areas of Australia, South America, South USA, and Africa where the parasites are endemic. The fight against babesiosis is a major economic factor in these areas. *Babesia bovis* infection is probably the most important cause of "tick fevers" of cattle. The fact that some breeds of *Bos indicus* are relatively resistant to the effects of this parasite has led to the suggestion that the organism evolved in these breeds. Young cattle have pronounced resistance to severe infection. The first clinical signs are fever (>40 °C), loss of appetite and listlessness. Anaemia and haemoglobinuria (Redwater) follow, and signs of jaundice develop. Diarrhoea is common and pregnant cows may abort. A form of "cerebral babesiosis" develops in some animals with such clinical signs as hyperaesthesia, nystagmus, grinding of teeth, circling, head pressing, mania, ataxia and convulsion. In terminal cases muscle tremors and wasting may appear, followed by coma and death. At postmortem examination the blood appears to be thin and watery, and there is haemoglobinuria. All connective and fat tissues are oedematous and show evidence of icterus. The spleen is characteristically enlarged. The liver is swollen and the gallbladder distended. In cases of cerebral babesiosis the grey matter of the brain is congested and shows a typical red discoloration. Infection with *B. bigemina* causes similar but usually less severe clinical signs than infections with *B. bovis*, though acute anaemia and death also occur. The pathogenesis is here almost entirely related to a rapid haemolysis, with signs of haemoglobinuria and jaundice. There is no cerebral involvement. *Babesia divergens* and *B. major* are considered less pathogenic than other species in the tropics. In temperate regions, *B. divergens* causes the most severe signs and may be responsible for heavy economical losses. Clinical signs include fever, anaemia, bilirubinuria and haemoglobinuria. Diarrhoea, icterus, nervous signs and abortion may also occur in severe cases.

Horses The horse parasite *Babesia equi* has recently been redescribed as *Theileria equi.* The only *Babesia* spp. which occurs in horses is *B. caballi.* The pathogenesis and clinical signs of infection are similar to those described above for other *Babesia* spp., i.e. that of an haemolytic anaemia which is compounded by phagocytosis of erythrocytes by macrophages. However, most infected animals are not clinically affected. Equine babesiosis is an important reason for the restriction of movements of horses between some countries.

Dogs *Babesia canis* infection is a common cause of death in dogs. The pathogenesis resembles that of *B. bovis* in that mechanisms other than a haemolytic anaemia appear to be involved. However, in contrast with other animals, infection is more frequent in young dogs than in older animals. The clinical picture of dogs suffering from babesiosis is diverse and may follow a hyperacute, acute, or chronic course. Experimental infection of dogs with *B. canis* isolates from geographically different areas reveals different pathology which suggest that the aetiology of the disease caused by these isolates is different. Mildly affected animals develop anaemia and fever, are lethargic and have poor appetite. They show no visible signs of jaundice or haemoglobinuria and recover after a few days. More severe cases show a wide variety of signs, including severe depression, salivation, vomiting, diarrhoea, jaundice and haemoglobinuria or haematuria. Pulmonary involvement occurs frequently with quickened respirations and frothy blood-stained spittle. Anorexia and weight loss may be persistent. Nervous signs may occur. Infections with *B. gibsoni* are less severe and often resemble an uncomplicated haemolytic anemia.

Other Species In most cases, babesial infections of other vertebrate host species are mild or clinically inapparent. However, severe reactions have been described, e.g. during *B. perroncitoi* infections in pigs and *B. felis* infections in cats.

Vaccination

The ticks responsible for transmission of cattle babesiosis are, particularly in Australia, resistant against most of the commercial akarizides, so that vaccination against the parasite is the only mean of fighting the disease. It has been known for a long time that cattle can develop a prominent and long-lasting premunition against the parasite after recovery from babesia infection. From 1897 until the mid-1960s blood from recovered cattle was used as a simple blood vaccine. More recently this live vaccine was refined by making the *B. bovis* parasite less virulent, passing it through splenectomised calves, and diluting the erythrocytes in a cell-free, plasma like medium. This live vaccine was mainly produced and used in Australia but was also and still is used in Africa. About 10^7 parasites are administered per vaccination. The vaccine has a short half life about 6 days and, like most live vaccines, its quality can vary batch to batch. Lives vaccines prepared in splenectomised calves are used with success in Israel against *Babesia bovis, B. bigemina* and *Anaplasma centrale.* They have been proven to be better than vaccines prepared from tissue cultures. The vaccines are stored and dispatched to the field in a concentrated frozen state. The extensive use of the same type of vaccine in Australia from 1959 to 1996, with 27 million doses has been recently reviewed by Callow et al in 1997 with very favorable results. However, this dependence of cold chain facilities is an insurmountable difficulty for poor and tropical areas of Africa and South America in the use of attenuated live vaccines. The search for alternative vaccines has been developed, such as attempts to use irradiated parasites with ^{60}Cobalt with satisfactory results. Efforts to produce parasite extracts or fractions to be used as vaccines have also failed. A breakthrough in the search for a defined vaccine came when it became possible to produce *B. bovis* in a Trager-Jensen-like culture system used for growing *Plasmodium falciparum.* Parasites and culture supernatant, which contains soluble parasite exo-antigens, became available in large quantities and both produced protective immunity in cattle. Lately purified *Babesia bovis* antigens have been successfully used in animal vaccination trials with reference to the ability to induce protection against heterologous strains challenge. These types of vaccines have been used also in tropical areas of India and Brazil with success. However, antigen polymorphism of exo-antigens is a limiting factor for generalization of commercially available vaccines of this type.

There is only a vaccine against *Babesia canis* commercially available (Pirodog®) based on culture-derived antigens. Vaccination, however, has to be repeated regularly and induces poor protection against heterologous *B. canis* parasites.

A second generation of controlled babesiosis vaccines by gene technology methods is being intensively pursued in various laboratories. A large number of antigens to be used in sub-unit vaccines have been identified and cloned. However, again, antigenic polymorphism and lack of knowledge on the immune effector mechanisms responsible for protection are limiting factors for preparing practically useful recombinant or synthetic vaccines which can be used in mass vaccinations.

A 37 kDa glycoprotein of *B. bigemina* has been recently identified as the major component of a protective fraction obtained from cultures of these parasites.

Therapy

→ Babesiacidal Drugs.

Babesiosis, Man

Synonym

Piroplasmosis

General Information

Babesiosis is usually contracted accidentally from the bite of an ixodid tick which transmits → *Babesia* (Vol. 1) spp., an unpigmented protozoan multiplying in the red blood cell, usually in an asplenic individual. Several species of *Babesia* give rise to natural infections in cattle, other domestic animals, and field mice, on which one stage of the ticks normally feeds (→ Babesiosis, Animals). Transmission to man occurs in the next stage. *Ixodes dammini* and *I. ricinus* are the principal vector of *Babesia microti* which is zoonotic in New York State and Massachusetts. *Ixodes ricinus* transmits *B. divergens.* Most European cases of human babesiosis are caused by → *Babesia divergens* (Vol. 1). In humans the *Babesia* spp. undergo binary fission and often are located at the periphery of mature red blood cells, which are destroyed without production of insoluble pigment.

Pathology

Bilirubinemia, jaundice, hemoglobinuria, fever, and hematuria often accompany heavy infections of humans with *B. divergens* of cattle or *B. microti* of rodents. However, many infections appear to pass asymptomatically. The infection is lethal in splenectomised individuals but it is very rare and is restricted to small isolated endemic areas.

Main clinical symptoms: Abdominal symptoms, diarrhoea, continuous fevers of 40–41 °C, anaemia, death
Incubation period: 1–4 weeks
Prepatent period: 1 week
Patent period: 1 year
Diagnosis: Microscopical determination of blood stages in Giemsa stained smear preparations (pigment does never occur), → serology (Vol. 1)
Prophylaxis: Avoid the bite of ticks
Therapy: Treatment see → Babesiacidal Drugs.

Bacillus thuringiensis

→ Disease Control, Methods.

Baker's Itch

→ Mites (Vol. 1).

Balantidiasis, Man

Disease due to infections with the porcine ciliate → *Balantidium coli* (Vol. 1) via oral uptake of cysts from feces.

Main clinical symptoms: Diarrhoea, nausea, obstipation
Incubation period: Days to weeks
Prepatent period: 4 days – weeks
Patent period: Years
Diagnosis: Microscopic determination of cysts and trophozoites in fecal smears.
Prophylaxis: Avoid contact with human or pork feces
Therapy: Treatment see → Antidiarrhoeal and Antitrichomoniasis Drugs.

Balantidosis, Animals

→ *Balantidium coli* (Vol. 1), → Alimentary System Diseases, Swine.

BALF

Abbreviation for bronchoalveolar fluid obtained by lavage in order to detect *Pneumocystis* stages.

Bee Dysentery

Infectious disease of bees caused by the microsporidian species → *Nosema apis* (Vol. 1), see also → Nosematosis.

Benign Theileriosis

Disease in cattle due to infection with *Theileria mutans* (→ Theileriosis).

Bernoulli Trial

→ Mathematical Models of Vector-Borne Diseases.

Besnoitiosis

Disease in cattle and goats due to *Besnoitia besnoiti* with formation of tissue cysts inside skin (and eye) leading to the so-called elephant-skin. Transmission by body contact; life cycle unclear (→ Skin Diseases, Animals/Protozoa, → Coccidia (Vol. 1)).

Bilharz, Theodor

German physician (1825–1862); in 1852 he discovered the human schistosomes, which originally were described as members of the genera *Diplostomum* or *Bilharzia*.

Bilharziomas

Granulomatous and fibrotic lesions that develop around egg masses of schistosomes in organs (e.g. liver) away from the mucosa.

Bilharziosis

Synonym

→ Schistosomiasis, Man.

The name was given in honour of Theodor Bilharz (1825–1862), a German physician who discovered the *Schistosoma* worms in Egypt in 1851–1852 and described them originally as *Diplostomum*.

Bilirubinuria

Clinical symptom in hosts infected e.g. with *Babesia* species or *Theileria equi* (Piroplasmea).

Biological Methods

→ Disease Control, Methods.

Biological Systems

→ Disease Control, Strategies.

Black Disease

Disease due to combined infection with the trematode *Fasciola hepatica* and the bacterium *Clostridium novji* type B.

Black Sickness

→ Visceral Leishmaniasis.

Blackhead Disease

→ *Histomonas meleagridis* (Vol. 1). The 5–30 µm sized, uniflagellated (rarely 2–4 flagella) or amoebic stages of *H. meleagridis* live in the caecum of many species of turkeys, chicken or other related birds. Amoeboid stages may enter the intestinal wall, reach the liver and introduce there necrosis due to repeated binary divisions. Intestinal trophozoites are often included in *Heterakis*-worm eggs keeping their infectivity for up to 4 years thus representing an important transmission pathway.

Main clinical symptoms: Diarrhoea, anaemia, black crest, general weakening, death.
Incubation period: 2 days
Patent period: Several months up to life long
Diagnosis: Microscopic determination of trophozoites in fresh feces
Prophylaxis: Separation of young from older animals
Therapy: Oral application of different nitro-imidazoles (→ Antidiarrhoeal and Antitrichomoniasis Drugs)

Blastocystosis, Man

Disease due to infection with → *Blastocystis hominis* (Vol. 1) cysts from human or animal feces.

Main clinical symptoms: Nausea, diarrhoea, abdominal pain
Incubation period: Days to weeks
Prepatent period: 2–3 days, weeks
Patent period: 2–3 weeks
Diagnosis: Microscopic determination of cysts in fecal samples
Prophylaxis: Avoid contact with human feces
Therapy: Curative treatment unknown; see → Treatment of Opportunistic Agents

Blocker / Effectors of Aminergic Transmission / Ectoparasiticides

Structures Amidines

Important Compounds Amitraz, Cymiazole.

General Information Formamidines are agonists of the octopamine receptors in the arthropod nervous system, causing an increase in nervous activity, reduction of feeding and disruption of reproductive behaviour. Octopamine can act as a neu-

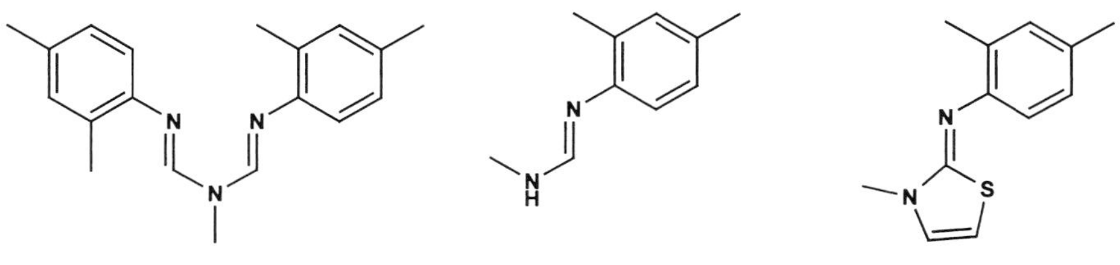

Fig. 1. Structures of amidines.

rotransmitter, neuromodulator or even as a circulating neurohormone in insects. Octopamine is involved in energy mobilisation and stress responses. It has a function as a modulator of muscle contraction and controls the release of adipokinetic hormone. Formamidines have been shown to exert ovicidal effects in insects and acari. Additionally, formamidines of the chlordimeform type have been shown to efficiently inhibit monoamine oxidase from rat liver. Chlordimeform was withdrawn from the market in the late 1980s. General side effects of formamidines in mammals are possible alterations in the animals ability to maintain homeostasis for at least 24 hours after treatment. A symptom often observed with formamidine treated mammals is a reversible sedative effect.

Amitraz is a non-systemic acaricide and insecticide with contact and respiratory action. It has been shown that amitraz is rapidly metabolised in arthropods and that one metabolite (BTS-27271) shows a biological activity superior to that of amitraz itself. Therefore, it is thought that amitraz acts as a pro-insecticide and pro-acaricide. The tickicidal effect comes with an expelling action causing ticks to withdraw mouth parts rapidly and fall off the host animal. The compound is used as an animal ectoparasiticide for the control of ticks, mites and lice on cattle, dogs, goats pigs and sheep. The acute oral LD_{50} for rats is 650 mg/kg (acute percutaneous LD_{50} for rats is >1600 mg/kg). The compound is toxic to fish and other aquatic organisms but rapid hydrolysis makes it unlikely that toxicity will be observed in natural aquatic systems. Amitraz shows low toxicity to bees and predatory insects. Withdrawal period for the compound in meat is 24 hours (7 days for sheep). In mammals rapid breakdown occurs and 4-amino-3-methylbenzoic acid and to a lesser extent N-(2,4-dimethylphenyl)-N'-methylformamidine are excreted as conjugates. **Cymiazole** (CGA 50439) is a non-systemic ectoparasiticide with contact and respiratory action. It shows a good killing-effect as well as inhibition of viable egg production and has a pronounced detaching effect on ticks. Cymiazole is also used as a systemic compound against varroa mites feeding on bees carrying the compound in their body fluid. The compound has a three-day withdrawal period for meat and can be used in milk producing cattle without restrictions. Cymiazole has an acute oral LD_{50} for rats of 725 mg/kg (acute percutaneous LD_{50} >3100 mg/kg). The compound shows only weak toxicity to fish and shows no significant toxicity to bees.

Resistance Resistance against amitraz has been observed in several strains of the southern cattle tick in Australia. No resistance against amidine acaricides was found in multi-host ticks so far.

Blocker of Arthropod Development / Ectoparasiticides

Mode of Action Chitin synthesis inhibitors (Fig. 1, page 74)

Structures Fig. 2, page 75.

Important Compounds Diflubenzuron, Fluazuron, Lufenuron, Triflumuron.

General Information The chemical class of benzoylphenylurea (BPU) compounds is widely used as an insecticide also in animal health applications. The BPU compounds are inhibitors of chitin synthesis thus interfering with the formation of the insect cuticle. There is no inhibition of other poly-sugar synthesis pathways (e.g. hyaluronic acid synthesis) by BPU compounds indicating high specificity. The biochemical target within the chitin biosynthesis pathway has not yet been clearly identified. Recent findings showed interference of BPUs with GTP-mediated Ca-transport in intracellular vesicles in chitin depositing integument cells from American cockroaches. Chitin

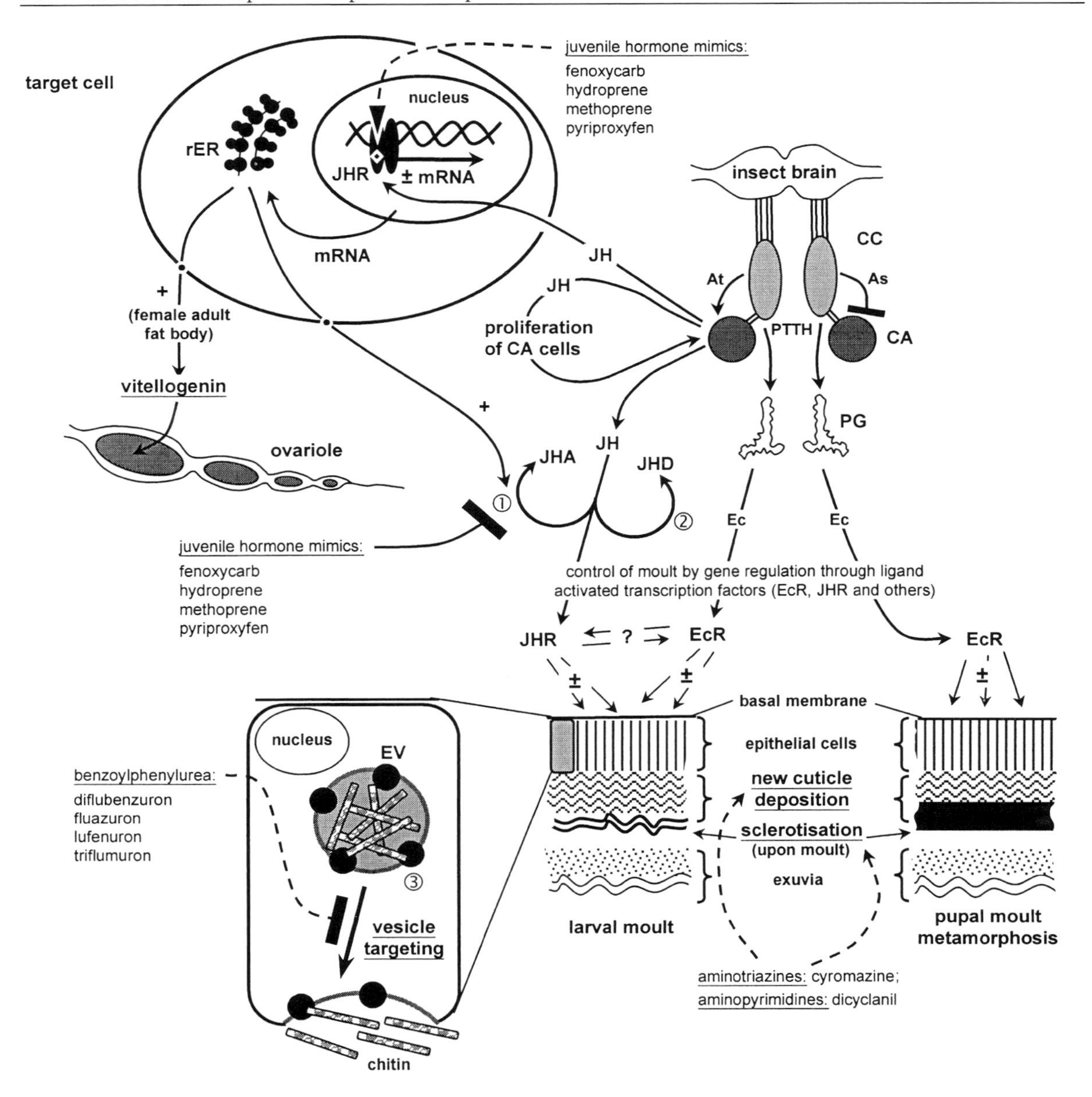

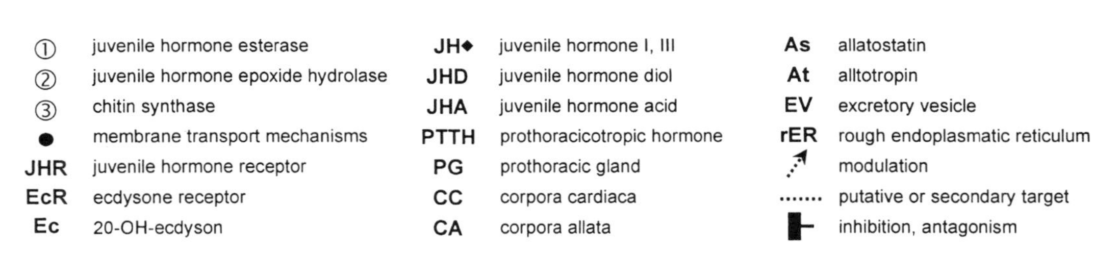

Fig. 1. Model of drug interaction with arthropod development.

Benzoylphenylurea (BPU)

diflubenzuron

fluazuron

lufenuron

triflumuron

Juvenile hormes mimics

methoprene

hydroprene

fenoxycarb

pyriproxyfen

Aminotriazine, aminopyrimidine

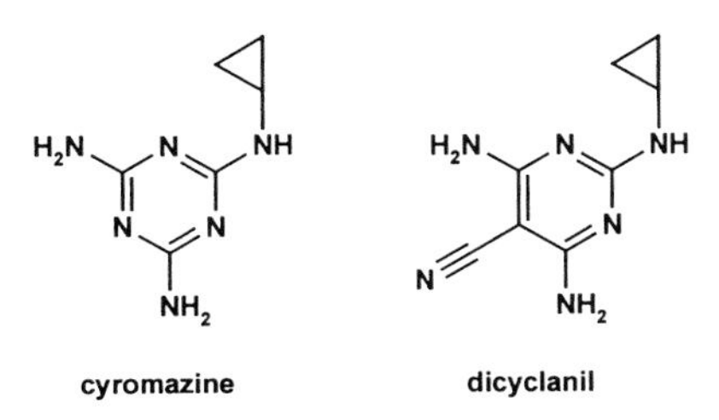

Fig. 2. Structures of ectoparasiticidal drugs interfering with arthropod development.

synthase itself is not inhibited by BPU compounds. Inhibition of chitin synthesis by BPU compounds depends on intact cells.

Diflubenzuron is a non-systemic insect growth regulator with contact and stomach action. It shows activity against moulting larvae or hatching eggs. The acute oral LD_{50} for rats is >4640 mg/kg (acute percutaneous LD_{50} for rats >10000 mg/kg). The compound shows very low toxicity for fish and is not toxic to bees and predatory insects. In mammals diflubenzuron is partly eliminated as the parent compound with the faeces following oral administration. The other part is excreted mainly as hydroxylated metabolites. **Fluazuron** is a non-systemic ixodid growth regulator with contact and stomach action. The compound was recently launched for strategic tick control in Australia. It shows activity against all developmental stages of the cattle tick *Boophilus microplus* including all resistant strains. The acute oral LD_{50}

for rats is >5000 mg/kg (acute percutaneous LD_{50} for rats >2000 mg/kg). Acatak has a withdrawal period of 42 days and treatment of dairy cows and sucking cattle is not allowed. The compound is harmful for fish but not toxic to bees. In mammals the compound is virtually not metabolised following oral administration. **Lufenuron** is an insect growth regulator that mostly acts by ingestion. The compound has been introduced in the pet market as a systemic flea growth regulator. Adult fleas feeding from blood of systemically treated animals lay non-fertile eggs. Larvae feeding from faeces produced by treated adult fleas will be unable to moult and also cease feeding. The acute oral LD_{50} for rats is >2000 mg/kg (acute percutaneous LD_{50} for rats >2000 mg/kg). The compound shows very low toxicity for fish and is only slightly toxic to adult bees. In mammals lufenuron is mainly eliminated as the parent compound with the faeces following oral administration. **Triflumuron** is a non-systemic insect growth regulator with stomach action. It shows activity against moulting larvae and causes infertility of eggs. The compound is used against blowfly, fly larvae in animal houses, cockroaches and flea larvae. The acute oral LD_{50} for rats is >5000 mg/kg (acute percutaneous LD_{50} for rats >5000 mg/kg). Triflumuron shows very low toxicity for fish and is not toxic to predatory insects. In mammals the compound is metabolised by hydrolytic cleavage forming conjugated or partly hydroxylated metabolites containing the 2-chlorophenyl ring and correspondingly the 4-trifluormethyl-methoxyphenyl ring.

Resistance In animal health resistance of ectoparasites against BPU is still rare. This might change with the growing market share of BPU as a sheep ectoparasiticide. In diflubenzuron resistant strains of the housefly *Musca domestica* oxidation seems to be the predominant route of detoxification. Another mechanism of detoxification of BPUs are hydrolases as could be demonstrated by the synergising effect of esterase inhibitors. Metabolic resistance against BPU has been demonstrated for a multi-resistant housefly population. Glutathion-S-transferase and mixed function oxidase enzyme activities were determined and showed elevated levels. A multi-resistant flea strain collected from Florida showed resistance to some BPU compounds in laboratory in-vivo trials.

Juvenile Hormone Mimics

Important Compounds Methoprene, Hydroprene, Fenoxycarb, Pyriproxyfen.

General Information The class of insecticides comprises different chemical classes causing similar phenotypic damage to treated insects. They are mimics of the endogenous juvenile hormone of insects, preventing metamorphosis to viable adults when applied to larval stages. Juvenile hormone mimics also exert ovicidal effects when applied to adults. Up to now, two primary targets of juvenoids have been identified. The compounds fulfil a dual function by inhibiting the juvenile hormone esterase from degrading endogenous juvenile hormone as well as by their weak agonistic effect on juvenile hormone receptors. This adds to the endogenous juvenile hormone effects thus compensating for the naturally occurring degradation of the juvenile hormone producing corpora allata glands. In adult insects juvenile hormones are involved in regulation of vitellogenesis of the eggs. Altering homeostasis in this developmental stage could cause infertile eggs. The complete cascade of effects remains to be established. Several candidates for the endogenous juvenile hormone receptor – in all probability a member of the ligand activated nuclear transcription factor family – are currently under discussion in *Drosophil*a e.g. the gene products of *ultraspiracle* and *methoprene tolerant*. Juvenile hormones and juvenile hormone mimics act as suppressers and stimulators of gene expression depending on the developmental stage and type of regulated protein. Several genes under the control of juvenoids have been identified. This explains the variety of effects observed with juvenoid treated insects.

Fenoxycarb is an insect growth regulator with contact and stomach action. The compound exhibits a strong juvenile hormone activity, inhibiting metamorphosis to the adults stage and interfering with the moulting of early instar larvae. The acute oral LD_{50} for rats is >10000 mg/kg (acute percutaneous LD_{50} for rats >2000 mg/kg). The compound shows low toxicity for fish and is non-toxic to adult bees. In mammals, the major metabolic path for fenoxycarb is ring hydroxylation to form ethyl-[2-[p-(p-hydroxyphenoxy)phenoxy]ethyl]-carbamate. **Hydroprene** is an insect growth regulator closely related to methoprene, predominantly used against hygiene pests. **Methoprene** is an insect growth regulator (juvenile hormone mimic) preventing metamorphosis to viable adults when applied to larval stages. The acute oral LD_{50} for rats is >34600 mg/kg. The compound shows very low toxicity for fish and is non-toxic to adult bees. In mammals, methoprene is metabolised to simple acetates and also cholesterol has been identified as a secondary metabolite. The metabolites were present in milk and blood but have not been detected in tissues. Upon oral administration methoprene is not metabolised and excreted via the faeces and the urine. **Pyriproxyfen** is an insect growth regulator acting as a suppresser of embryogenesis and adult formation (juvenile hormone mimic). The compound has been introduced in the pet market as a potent flea growth regulator and is used for the control of public health insect pests. The acute oral LD_{50} for rats is >5000 mg/kg (acute percutaneous LD_{50} for rats >2000 mg/kg).

Resistance Field resistance of ectoparasites against juvenile hormone mimics has not yet been observed. For multi-resistant houseflies a 100-fold resistance against juvenoid compounds has been observed. For the fruit fly a mild methoprene resistance has been observed in the field. High methoprene resistance was artificially produced in *Drosophila melanogaster* and linked to a gene locus called *methoprene-tolerant* which has been speculated to be the insensitive target of juvenile hormone. Altered binding properties of a cytosolic protein have been identified in juvenile hormone resistant fruit flies. Despite several reports about the isolation of a candidate juvenile hormone receptor gene or protein, the identity of the receptor remains still uncertain (see above).

Aminotriazines/Aminopyrimidines

Important Compounds Cyromazine, Dicylanil.

General Information Currently, two structurally closely related drugs are marketed as insect growth regulators against larvae causing myiasis in animals and developing fly larvae in manure. The first of them, **cyromazine**, entered the market in 1979. The biochemical mode of action remains unclear. Cyromazine has been tested in dihydrofolate reductase and tyrosinase assays but showed no inhibition of these enzymes while another study demonstrated an inhibition of dihydrofolate reductase. The compound is definitely not involved in inhibition of chitin synthase. There are strong hints on involvement of cyromazine in **sclerotisation** of the cuticle. Cyromazine has its highest efficacy against first instar larvae and leads to changes in the elasticity of the cuticle which might cause physical instability and lesions in the cuticle, finally preventing further development. Cyromazine treated *Lucilia cuprina* larvae do not show signs of cuticle apolysis. There was evidence of an abnormal continuous deposition of cuticle material by the epidermal cells. This also holds true for cuticle deposition in the foregut. The sum of observations is indicative of a fundamental interference with insect moulting at the hormonal level.

Cyromazine is an insect growth regulator with contact action interfering with moult and pupation. The topically applied compound has a pronounced residual effect and protects sheep against blowfly strike for eight weeks. There is a withdrawal period of 7 days for meat. Acute oral LD_{50} for rats is 3387 mg/kg (acute percutaneous LD_{50} for rats is >3100 mg/kg). Cyromazine is non-toxic to fish and adult honey bees. The compound is efficiently excreted in mammals, mainly as the parent compound. **Dicyclanil** is an insect growth regulator recently introduced into the market. The compound prevents development of larvae into pupae or adults when incorporated into the insects breeding substrate. Dicyclanil shows a high specificity against developing flies and fleas. Its biological efficacy is higher compared to that of cyromazine.

Resistance against Cyromazine Although 20 years of cyromazine treatment has passed, no proven resistance has been reported from the Australian blowfly. From all field strains assayed only three strains had a few survivors at the discriminating dose, but these flies were unable to reproduce successfully. Cyromazine was able to control even a high level multi-resistant housefly strain with a slightly higher LC_{50} (tolerance factor 1.7). There seems to be no cross-resistance to other insecticides.

Blocker of Chloride Channels / Ectoparasiticides

Mode of Action

Fig. 1

Structures

Fig. 2

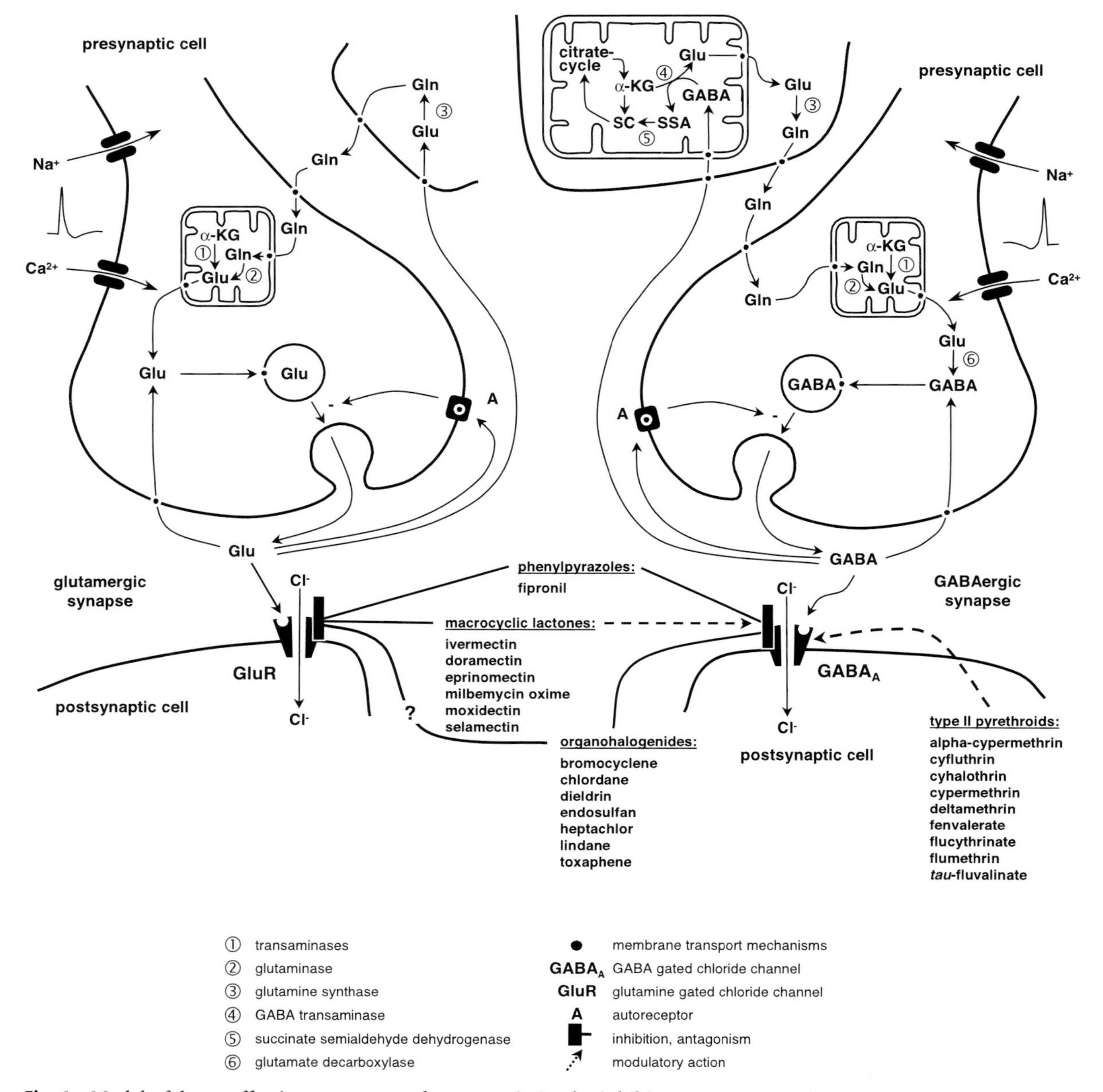

Fig. 1. Model of drugs affecting neuromuscular transmission by inhibitory neurotransmitters.

Organohalogenides

Important Compounds Chlordane, Bromocyclene, Endosulfan, Lindane, Heptachlor, Toxaphene.

General Information **Lindane** was isolated from the mixture of isomers in 1912. In 1943 lindane was identified as the insecticidal principle of hexachlorhexane. Insecticidal properties of chlordane, a cyclodiene, were at first identified in the 1940s. The action of lindane and cyclodienes to stimulate synaptic transmission was demonstrated 40 years ago. However, it was not until the early 1980s that the GABA-gated chloride channel complex was suggested to be their target site. In nerve and muscle preparations lindane and cyclodiene insecticides antagonised GABA-stimulated ^{36}Cl uptake and competed with the TBPS (*t*-butylbicyclophosphorothioate) binding site at the GABA gated chloride channel. Recent studies with cockroach neurones revealed that lindane and cyclodienes decrease the frequency of the GABA-gated chloride channel opening without changing

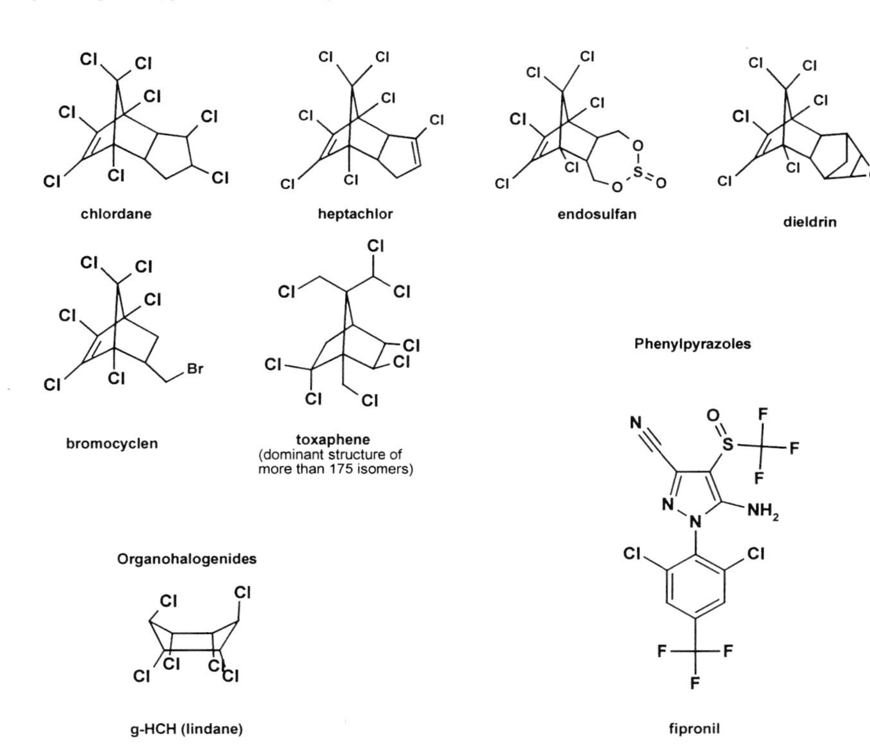

Fig. 2. Structures of antiparasitic drugs affecting GABA- or glutamate-gated chloride channels (for structures of avermectins see → Nematocidal Drugs, Animals, → Nematocidal Drugs, Man).

the mean opening time. Dieldrin suppressed the GABA-induced current in a non-competitive manner. Since picrotoxin attenuates the inhibitory effect of dieldrin, it is speculated that cyclodienes bind to the picrotoxin/TBPS binding site. It follows that lindane and cyclodienes are antagonists at GABA-gated chloride channels directly responsible for excitatory symptoms in poisoning of mammals and insects.

Chlordane (cyclodiene compound) is a non-systemic insecticide with contact, stomach and respiratory action and long residual activity. It controls household insects as well as pests of domestic animals and man. The compound has impurities of different stereo-isomers and heptachlor. Acute oral LD_{50} for rats is 133–649 mg/kg (acute percutaneous $LD_{50}$for rats 217 mg/kg). The compound accumulates in body fat and lipid-containing organs. It shows serious chronic and cumulative toxicity. The compound is toxic to fish and bees. In mammals chlordane is metabolised mainly to hydroxylated products. **Bromocyclene** is a non-systemic insecticide with contact and stomach action. It is used against ectoparasites in sheep and against fleas on pets. **Endosulfan** (cyclodiene compound) is a non-systemic insecticide with contact and stomach action. In animal and public health applications it is used for the control of tsetse flies. Acute oral LD_{50} for rats is 70–240 mg/kg depending on isomer and formulation (acute percutaneous LD_{50} for rats is >4000 mg/kg). Endosulfan is toxic to fish in-vitro but no toxicity has been observed under field conditions. The compound shows no significant toxicity to bees. In mammals endosulfan is eliminated via the faeces within 48 hours. Residues accumulate in the kidney rather than in fat but are eliminated with a half life of 7 days. The compound is rapidly metabolised to less-toxic metabolites and to polar conjugates. **Lindane (γ-HCH)** is one (>99%) isomer of the technically synthesised mixture of hexachlorhexane isomers. It is an insecticide with contact, stomach, and respiratory action and controls a broad spectrum of insects in public health and animal ectoparasites e.g. mites, sucking and biting lice and ticks. Acute oral LD_{50} for rats is 88–270 mg/kg depending on the carrier (acute percutaneous LD_{50} for rats is 900–1000 mg/kg). The compound shows toxicity to fish and to bees. In mammals lindane is found in the milk, body fat, and kidney after oral administration but rapid elimination occurs. Metabolites formed are less chlorinated compounds which are excreted as glucuronic acid conjugates.

Resistance Most of the organohalogenide compounds have been withdrawn from the market for toxicological and environmental reasons. From 1948 to 1954 chlorinated hydrocarbons like DDT (→ Blocker/effectors of Sodium Channels/Organohalogenides) and dieldrin were extensively used for flystrike control. 1954 field monitoring detected no resistant blowfly populations whereas in 1958, the year of the withdrawal of aldrin/dieldrin insecticides for sheep treatment, 70% of the fly strains tested were resistant to dieldrin. The fast development of resistance against dieldrin was speculated to be pre-selected by the previous use of lindane, increasing the frequency of the rdl-gene (encoding for a dieldrin insensitive insect GABA receptor, see below). Resistance against this class of compounds is still present in the ectoparasite populations. For dieldrin resistance e.g. in blowfly populations 2–3% of the individuals still show the resistant phenotype. Therefore, application of selection pressure with new insecticides cross-resistant with this phenotype will favour the resistant individuals leading again to resistant populations in a few generations. Resistance against lindane was reported for Australian strains of the cattle tick *Boophilus microplus*.

Molecular biology techniques enabled the identification of the cyclodiene resistance mediating gene (rdl) that was identified as a member of the ligand-gated chloride channel gene family sensitive to GABA.

Macrocyclic Lactones

Important Compounds Doramectin, Eprinomectin, Ivermectin, Milbemycin Oxime, Moxidectin, Selamectin.

General Information The class of 16-membered macrocyclic lactones comprises two families the avermectins and the milbemycins. The avermectines have been discovered by Omura at the Kitasato Institute in Japan. The compounds are active against ecto- and endoparasites and thus defined as endectocides. The biochemical mode of action is multifunctional. Recent reports established that the major target of the avermectins is the glutamate gated chloride channel. The compounds also exert modulatory agonistic activity at the GABA gated chloride channel, thus causing paralysis. Detailed discussion of the mode of action of avermectins and target site identification has been described in → Nematocidal Drugs, Animals and → Nematocidal Drugs, Man. The following compilation concentrates on the ectoparasiticidal activities of the macrocyclic lactones.

Doramectin is a broad spectrum insecticide and acaricide with contact and stomach action. The 25-cyclohexyl-avermectin B_1 compound was the fourth avermectin derivative introduced into the animal health market. It is a fermentation product of a mutant *Streptomyces avermitilis* strain. The lipophilic cyclohexyl-moiety seems to be responsible for the greater tissue half-life of doramectin. It is used as a systemic endectocide with ectoparasiticidal activity against warble fly, screw worm, lice, mite and ticks including multi-resistant strains. **Eprinomectin** is an ivermectin derivative with an amino-modification of the bisoleandrosyl moiety resulting in enhanced insecticidal efficacy and favourable pharmacokinetics with no withdrawal periods for meat and milk. The compound retains the insecticidal and acaricidal properties of ivermectin. Eprinomectin was introduced into the market as an endectocide in 1997. **Ivermectin** is a potent insecticide and acaricide with stomach and contact action. The product is a semi-synthetic derivative of avermectin analogue of *Streptomyces avermitilis* and consists of 22,23-dihydroavermectin B_{1a} and 22,23-dihydroavermectin B_{1b} (4:1 mixture). The compound is used in different formulations (injectable, pour-on, bolus) against ticks, sucking and biting lice, cattle grubs, mites, horn flies and bot flies. Withdrawal period for meat is 28 days. Ivermectin is not allowed for use with cattle producing milk for human consumption. Withdrawal period for dairy cows is 28 days prior to calving. Acute oral LD_{50} for rats is 10–50 mg/kg. The compound is toxic to fish and other aquatic organisms and toxic to honey bees. Excreted with faeces ivermectin is toxic to coprophagous insects. **Milbemycin-Oxime** is an insecticide, acaricide and nematicide with contact and stomach action. The compound is a mixture of milbemycin A3 and milbemycin A4 (3:7) produced by *Streptomyces hygroscopicus*. Subsequent derivatisation of the hydroxyl group at position 5 to a ketoxime resulted in a less potent compound with a favourable toxicological profile. The compound is predominantly active against nematodes and shows only weak ectoparasiticidal activities at the concentrations used in animal treatment. **Moxidectin**, the third macrocyclic lactone introduced into the market of endectocides

Table 1. Ectoparasitic control by ivermectin, doramectin and moxidectin in cattle at 200 μg kg^{-1}.

Ivermectin	Doramectin	Moxidectin
Grubs/Myiasis		
Dermatobia hominis larvae	*Dermatobia hominis* larvae	–
Cochliomyia hominivorax+	*Cochliomyia hominivorax*	–
Hypoderma bovis	na	na
Hypoderma lineatum	na	na
Lice		
Linognathus vituli	*Linognathus vituli*	*Linognathus vituli*
Haematopinus eurysternus	*Haematopinus eurysternus*	na
Solenopotes capillatus	*Solenopotes capillatus*	*Solenopotes capillatus*
Damalina bovis++	*Damalina bovis*++	na
Mites++		
Psoroptes ovis	*Psoroptes ovis*	*Psoroptes ovis*
Sarcoptes scabiei var. bovis	na	na
Chorioptes bovis	na	na
Ticks ++		
Boophilus microplus	*Boophilus microplus*	*Boophilus microplus*
Boophilus decoloratus	na	na
Ornithodorus savignyi	na	na
Flies		
–	*Haematobia irritans*	–

na, not available; –, not effective at 200 μg kg^{-1}; +, prophylactic (injection) or curative (topical)treatment; ++, parasitic control

is produced by a combination of fermentation and chemical synthesis by 23-methoxime derivatisation of nemadectin, a milbemycin produced by *Streptomyces cyanogriseus noncyanogenus*. The efficacy against endo- and ectoparasite is comparable to that of ivermectin. Withdrawal periods are 28–49 days for meat depending on product. The compound is less toxic to coprophagous insects when compared to ivermectin. **Selamectin** is a semisynthetic doramectin analogue which has been introduced into the market of pet endectocides in 1999. The compound is active against fleas, some ticks, intestinal hookworms, ascarids, and immature heartworms. The favourable toxicology profile was achieved through introduction of a 5-ketoxime group and cleavage of one sugar moiety.

Resistance Broad resistance of parasitic arthropods against avermectins or milbemycins has not yet been reported. No cross-resistance has been found to other compounds. Reports of a milbemycin derivative breaking avermectin resistance of nematodes are under discussion. It is speculated that moxidectin has higher efficacy against specific nematodes compared to ivermectin leading to the control of ivermectin resistant strains of these nematodes. A laboratory selection of sheep blowfly larvae with ivermectin was recently published. A pooled field strain was subjected to ivermectin treatment at a concentration producing more than 70% mortality. The larvae were selected over 60 generations, giving a 2-fold increase in the LC_{50} after the first selection and finally resulting in an 8-fold resistant strain. After relaxation of the selection pressure the selected strain reverted towards susceptibility fairly rapidly. Within eight generations the LC_{50} values dropped from 7-fold to 2-fold compared to the parental strain.

Phenylpyrazoles

Important Compounds Fipronil

General Information Phenylpyrazole intoxication of blowfly causes hyperexcitability and elevates nerve discharge. Inhibitory effects of GABA on *D. melanogaster* motor neurone discharge are reverted in a manner similar to that of picrotoxin and cyclodienes. Phenylpyrazoles are antagonists of the GABA-gated chloride channel and there is evidence that these compounds share a common binding site with cyclodienes, TBPS, and picrotoxin on the GABA receptor.

Fipronil is a broad-spectrum non-systemic insecticide and acaricide with contact and stomach action and good residual activity. It is used for the control of ectoparasites of pets and livestock as well as an insecticide in public health. Acute oral LD_{50} for rats is 97 mg/kg (acute percutaneous LD_{50} for rats is >2000 mg/kg). Fipronil is harmful to some species of birds and fish and highly toxic to bees (direct contact and ingestion). In mammals fipronil is rapidly distributed and metabolised upon adsorption. Fipronil and its sulfone are eliminated mainly via the faeces. Urinary metabolites have been identified as conjugates of ring-opened pyrazole products.

Resistance Analysis of the *rdl* (resistant to dieldrin) gene isolated from cyclodiene resistant mosquitoes and flies showed a specific point mutation (A302S) responsible for the altered insecticide-susceptibility of the target. Cyclodiene resistant fruit fly strains showed a more than 25-fold resistance against phenylpyrazoles. Phenylpyrazole topical treatment of cyclodiene resistant and susceptible *Blattella germanica* resulted in LD_{50} values of 40 µg/insect and 0.07 µg/insect respectively, indicating a 570-fold cross-resistance. A cyclodiene resistant housefly strain showing nearly 2900-fold resistance to dieldrin had an LC_{50} value of 36 ppm for fipronil whereas LC_{50} of a susceptible reference strain was 0.4 ppm revealing a 90-fold cross-resistance. Perhaps more interestingly, despite the fact that the A302S mutation greatly reduces the rate of GABA receptor desensitisation, the fitness of resistant flies does not seem to be decreased.

Blocker / Effectors of Cholinergic Neurotransmission / Ectoparasiticides

Mode of Action Fig. 1

Structures Organohalogenides (Fig. 2)

Important Compounds Bromopropylate

General Information Bromopropylate (phenisobromolate) [isopropyl 4,4'-dibromobenzilate] acts similar to organophosphorous or carbamate compounds and is a weak inhibitor of the insect acetylcholine esterase. Additionally, the compound has a measurable effect on voltage sensitive sodium channels of insect neurones. Despite its

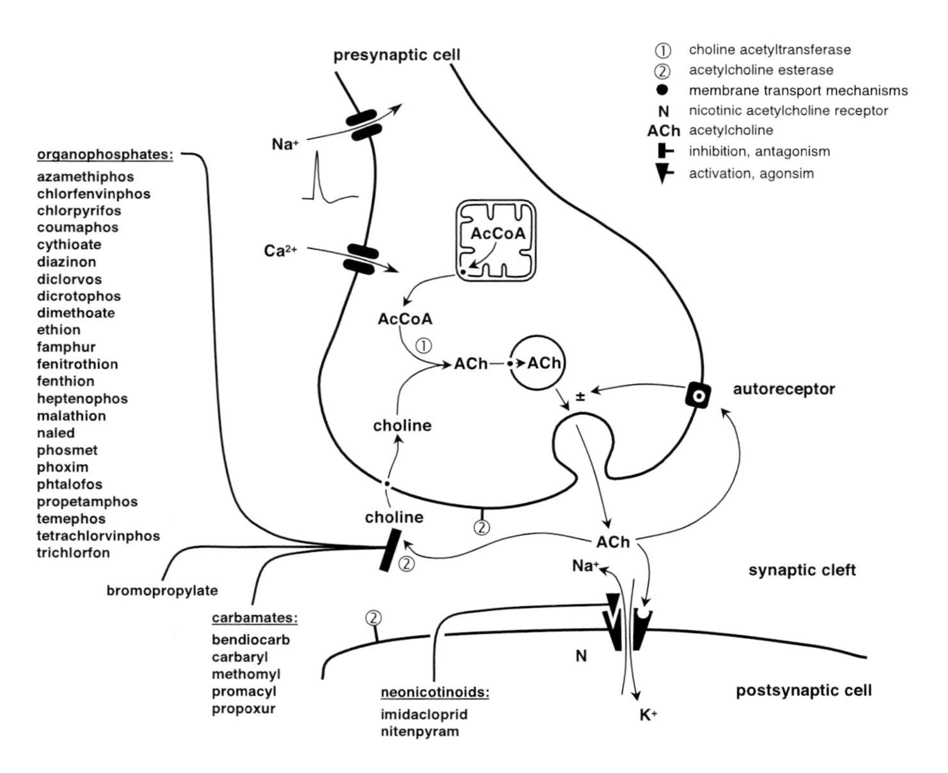

Fig. 1. Model of the action of drugs interfering with acetylcholine mediated neurotransmission.

Monothiophosphates

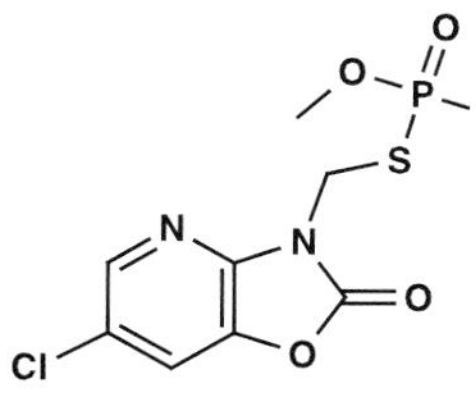

azamethiphos

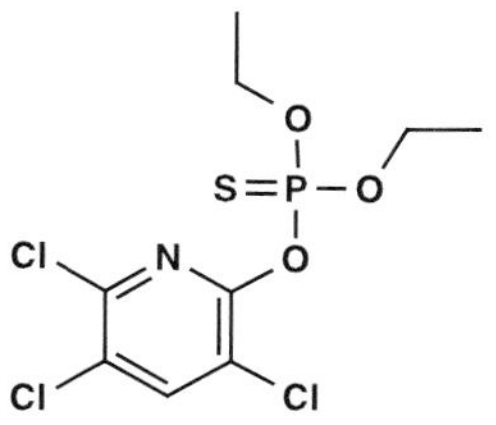

chlorpyriphos

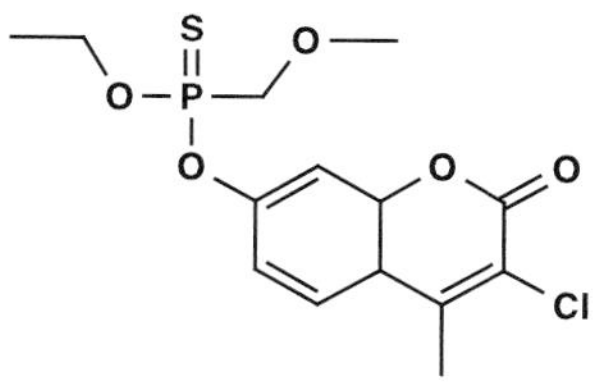

coumaphos

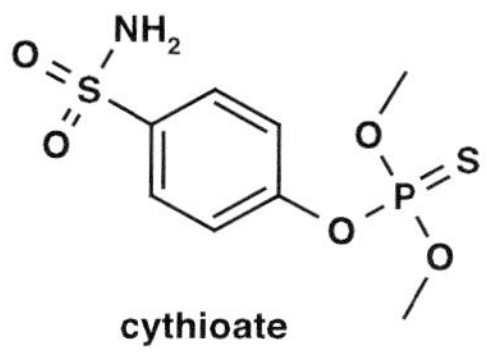

cythioate

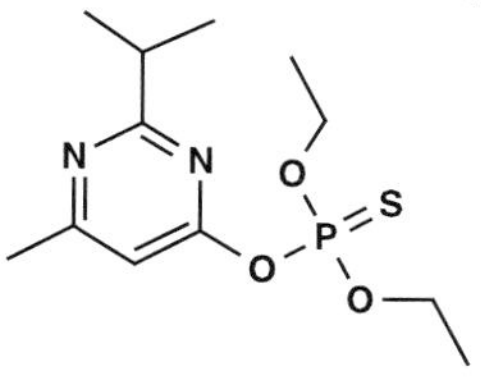

diazinon (dimpylate)

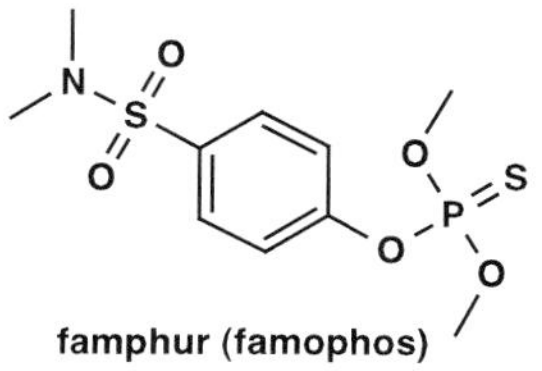

famphur (famophos)

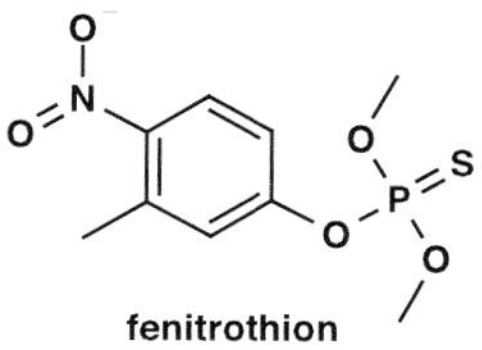

fenitrothion

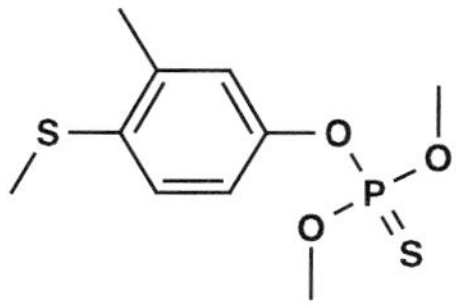

fenthion/MPP

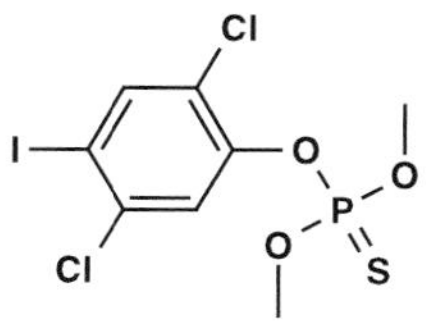

iodofenphos (jodfenphos)

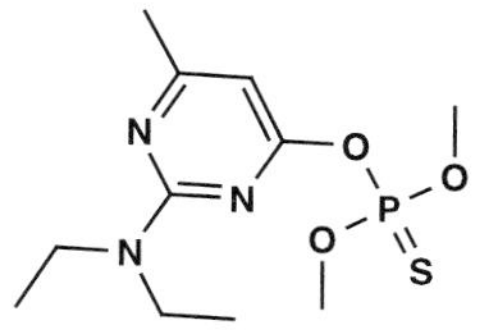

pirimphos-methyl

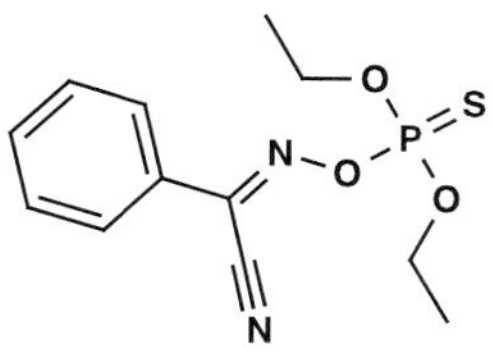

phoxim

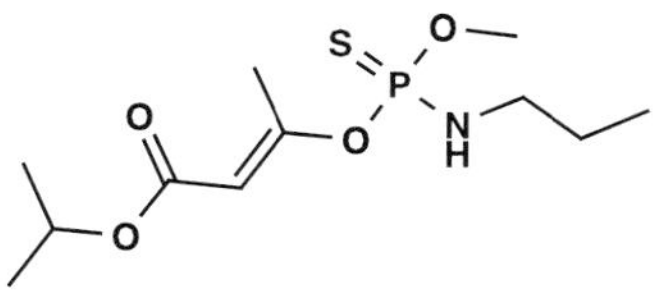

propetamphos

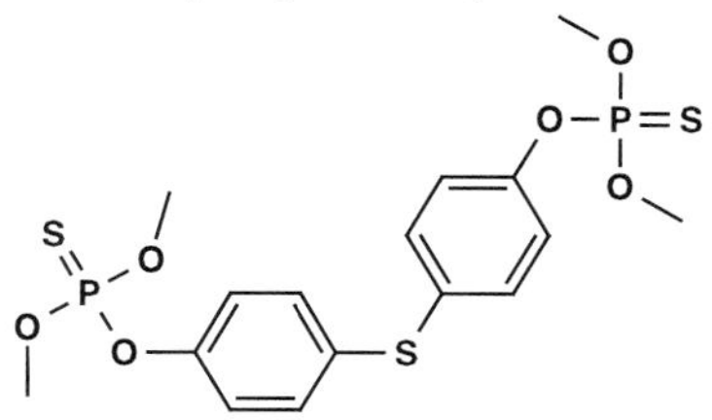

temephos

Dithiophosphates

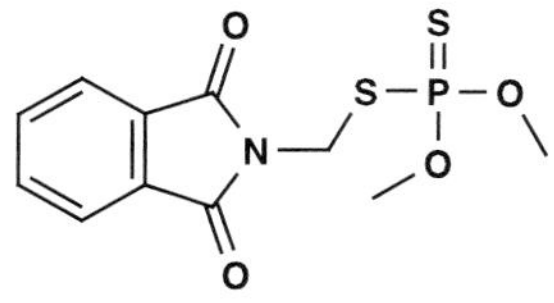

phosmet (PMP)

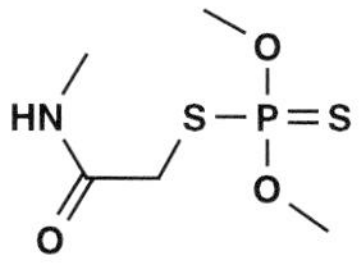

dimethoate

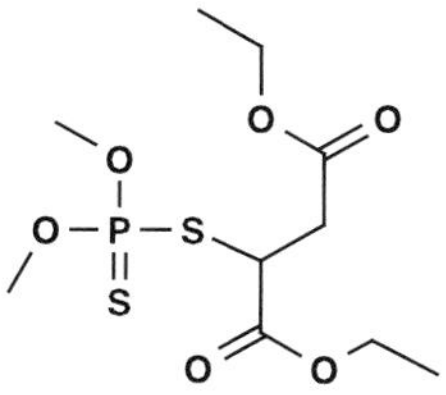

malathion

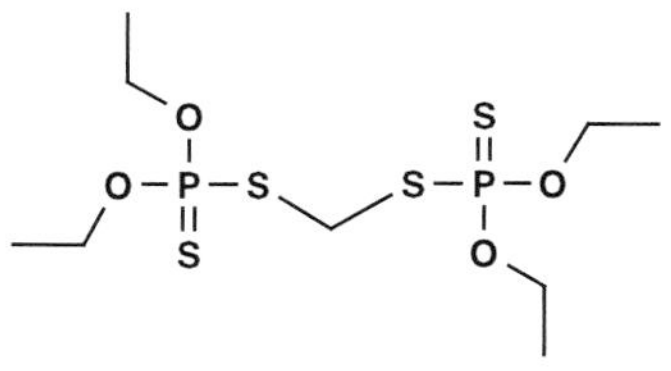

ethion (diethion)

Fig. 2. Structures of drugs affecting cholinergic neurotransmission.

Fig. 2. (continued)

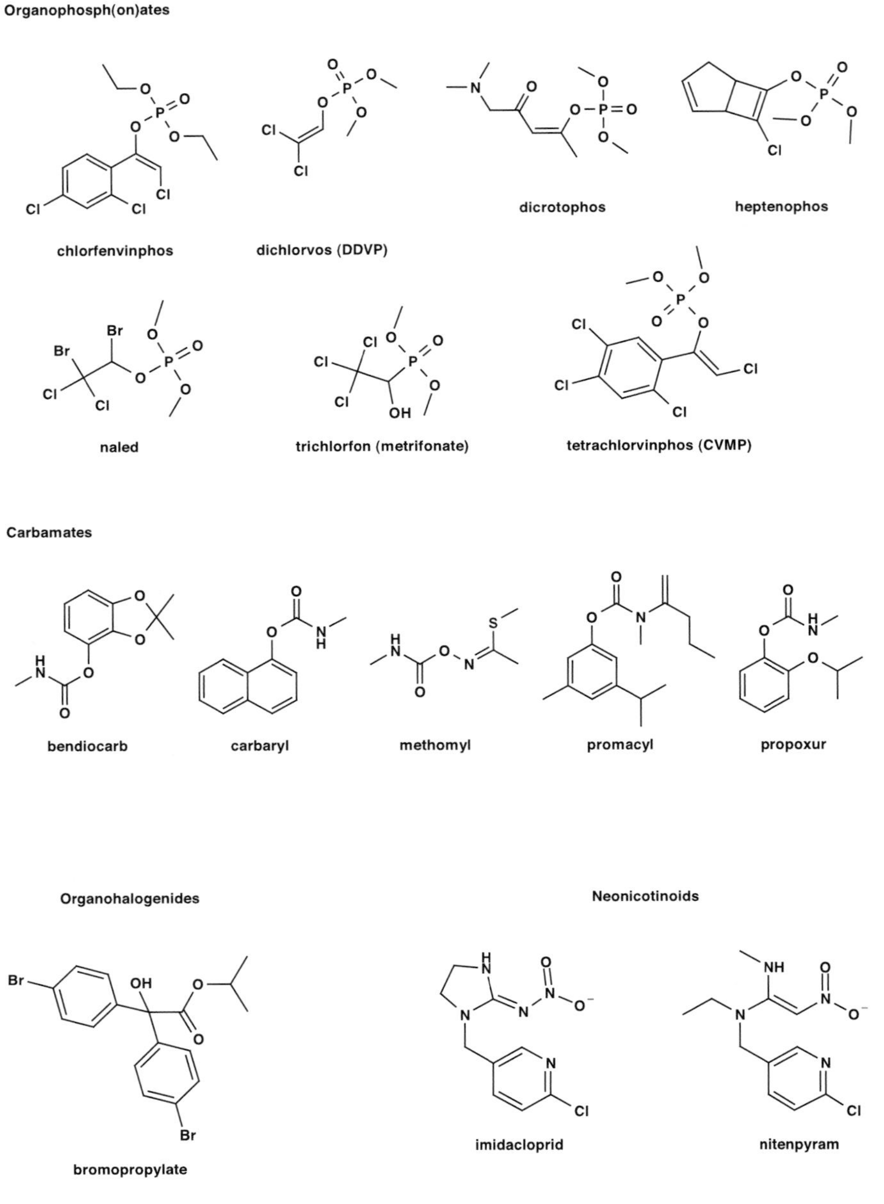

Organophosph(on)ates

chlorfenvinphos

dichlorvos (DDVP)

dicrotophos

heptenophos

naled

trichlorfon (metrifonate)

tetrachlorvinphos (CVMP)

Carbamates

bendiocarb

carbaryl

methomyl

promacyl

propoxur

Organohalogenides

bromopropylate

Neonicotinoids

imidacloprid

nitenpyram

structural similarity with DDT or methoxychlor this effect is not comparable to that of DDT.

Bromopropylate is a non-systemic acaricide with contact action and long residual activity. The compound is used against ectoparasites for diagnosis and control of mite infestations in bees. LD_{50} acute oral toxicity for rats is >5000 mg/kg (acute percutaneous LD_{50} for rats is >4000 mg/kg). The compound is toxic to fish. In mammals bromopropylate is rapidly and efficiently eliminated. Metabolism occurred mainly by cleavage of the isopropyl ester and to a minor extent by oxidation. Metabolites of the oxidation products were 3-hydroxybenzilate and conjugates.

Resistance Resistance of ectoparasites against Bromopropylate is based on sequestration by enhanced hydrolytic activities (carboxylesterases) as well as on oxidation by mono-oxygenases. Bromopropylate resistant varroa mites have not yet been reported.

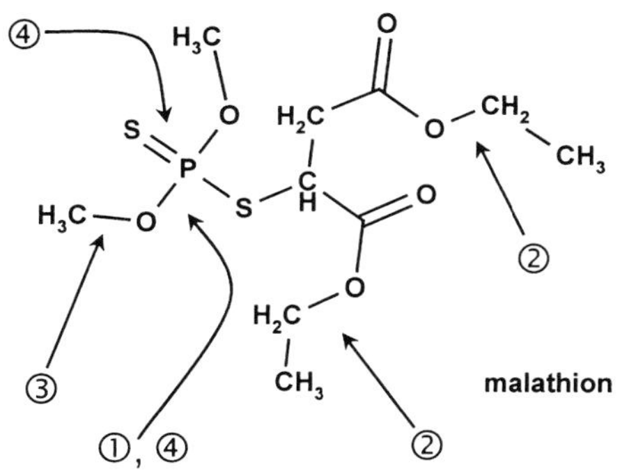

malathion

① phosphodiesterhydrolase
② carboxylesterase
③ glutathion-S-transferase
④ mixed function oxidase

Fig. 3. Enzymes involved in detoxification of malathion.

Organophosphosphorous Compounds

Important Compounds Organophosphates, Organophosphonates, Monothiophosphates, Dithiophosphates.

General Information Insecticidal organophosphorous compounds were synthesised for the first time by G. Schrader nearly 60 years ago. Organophosphates are inhibitors of hydrolases, e.g. carboxyl esterases (ali-esterase), acetylcholine esterases. Several different isoforms of acetylcholine esterases have been identified in the insect central nervous systems which are differentially inhibited by organophosphates. The reaction mechanism of organophosphates with acetylcholine esterase resembles that of the endogenous substrate acetylcholine. The hydroxyl group of a specific serin residue of the acetylcholine esterase is acetylated by acetylcholine, phosphorylated by organophosphates and carbamylated by carbamates. This indicates a competitive mode of action. Since the inhibition constant of organophosphate compounds is small the compounds are occupying the active site thus reducing hydrolysis of acetylcholine in the synaptic cleft. Neuronal transmission is influenced by stimulation of cholinergic synapses followed by depression and paralysis.

Resistance Tolerance or resistance occurs against all of the organophosphates currently on the market. However, resistance is still restricted to limited areas and specific ectoparasites. Reported resistance factors for ectoparasites like cattle tick or cat flea are in the range of 1.5-fold (which in fact should be called biological variability or at the most tolerance) to >100-fold. Organophosphate resistant tick strains of the southern cattle tick *Boophilus microplus* have been identified in Australia, South Africa, Argentina, Brazil, Colombia, Ecuador, Costa Rica and Uruguay. Laboratory tests demonstrated different degrees of resistance from tolerance (1–4-fold reduced susceptibility) via slightly resistant (5–20-fold) to resistant strains (>20-fold). Field testing revealed that organophosphate resistance was widespread among Australian strains of *Boophilus microplus*. Several organophosphate and carbamate insecticides have been tested against a resistant flea strain (*Ctenocephalides felis*) from Florida and compared to a susceptible strain. Malathion (Dithiophosphates) and carbaryl (Carbamates) have been used in flea control at the Florida strain origin for about 30 years. The results indicated a <10-fold tolerance for organophosphates other than malathion which had a 25-fold higher LC_{50} than with susceptible control fleas. Under laboratory conditions a 12-fold resistance against malathion has been observed upon continuous selection of adult fleas for eight generations. In a organophosphate resistant field strain of the cat flea *Ctenocephalides felis* two possible resistance mechanisms have been identified: detoxification by glutathion-S-transferases and target site insensitivity of acetylcholine esterase. Hence, it is speculated that rotation with carbamate insecticides would not control this flea strain.

Resistant strains of the housefly showed reduced cuticle adsorption of organophosphates, reduced formation of biologically highly active oxon derivatives, accelerated formation of phosphates from thiophosphate derivatives, and reduced sensitivity of choline esterases.

For decades, organophosphates have been the only products for sheep treatment against blowfly strike and lice in Australia. Resistance against e.g. diazinon was detectable in 1965 in 20% of the adult blowflies at a low level. In 1970 about 95% of the flies were diazinone resistant. In 1995 this value had stabilised at about 97%. In case of blowfly control the occurrence of resistance did not mean a total control failure. For diazinon the protection period was reduced from about 12 weeks to 4–6 weeks. The actual resistance factors for more than hundred field populations of blowfly are in the range between 2-fold (tolerance) and 42-fold in laboratory trials predominantly concentrating at the higher range.

For diazinon (Monothiophosphates) and fenthion a delayed cuticular adsorption was observed with resistant housefly strains. For most ectoparasites resistant or tolerant against organophosphates one or more of the following resistance mechanisms have been identified: hydrolysis by carboxylesterases (ali-esterases), thus quenching the inhibitor, oxidation by microsomal enzymes (e.g. cytochrome P450 mono-oxygenases), dealkylation by glutathion-S-transferases and target enzyme insensitivity towards organophosphates. By molecular biology techniques additional resistance mechanisms have been demonstrated e.g. multiple alleles of choline esterases with altered specificity, enhanced enzyme expression levels.

Organophosphorous compounds play an ongoing important role in the treatment of ectoparasitic diseases. Resistance has been encountered with synergistic additives inhibiting metabolic enzyme activities e.g. piperonyl butoxide (PBO) as mixed function oxidase inhibitor, S,S,S-tributyl phosphorotrithioate (TBPT) and triphenyl phosphate (TPP) as inhibitors of detoxifying esterases, and diethyl maleate (DEM) as an inhibitor of glutathione-S-transferase mediated detoxification.

Organophosphates and Organophosphonates

Important Compounds Chlorfenvinphos, dichlorvos (DDVP), dicrotophos, heptenophos, naled, tetrachlorvinphos (CVMP), trichlorfon/metrifonate.

General Information **Chlorfenvinphos** is a broad spectrum insecticide and acaricide with contact and stomach action. It has a long residual activity and requires withdrawal periods of three to 21 days (depending on the respective formulation) for meat producing animals treated with the compound. It is used against blowfly strike and also for control of ticks, lice, keds and screw-worm. Chlorfenvinphos shows fish toxicity and only slight bee toxicity. Among the organophosphorous compounds it shows a relatively high acute oral toxicity against rats (9.6 to 39 mg/kg LD_{50}). On the other hand the compound is metabolised quickly in mammals by de-esterification to give 2-chloro-1-(2,4-dichlrophenylvinyl ethyl hydrogen phosphate. Excretion products are glucuronic acid conjugate and mono-hydrolysed metabolites. **Dichlorvos (DDVP)** is a rapid knockdown insecticide and acaricide with respiratory, contact and stomach action which rapidly decomposes in soil. Besides applications on pets and cattle dichlorvos has been used also against ectoparasites on salmon and trout. Because of its ability to act systemically dichlorvos is also used in animal treatment against nematodes, trematodes or cestodes. It is metabolised in mammals by hydrolysis and O-methylation. Acute oral toxicity for rats is IC_{50} mg/kg (LD_{50}, acute percutaneous 90 mg/kg). High toxicity to birds and bees has been observed. The metabolism of DDVP in mammals occurs rapidly with a half life of 25 min by hydrolysis and O-methylation in the liver. It is non-persistent in the environment. **Dicrotophos** is a systemic insecticide and acaricide with contact and stomach action and moderate persistence. Acute oral toxicity for rats is 17–22 mg/kg (LD_{50}, acute percutaneous 110–180 mg/kg). Very toxic to honey bees but due to rapid decline of compound on surfaces with little effect in practice. Dicrotophos is completely metabolised and eliminated in rats and dogs a few days after oral administration. **Heptenophos** represents a systemic insecticide with contact, stomach and respiratory action with fast initial activity and short residual effect. Acute oral toxicity for rats is 96–121 mg/kg (LD_{50}). In mammals the compound is excreted to 96% within 6 days in metabolised form in urine and faeces. **Naled** is a fast-acting non-systemic insecticide and acaricide with contact and stomach and some respiratory action. The biological activity may be due to in-vivo debromination forming dichlorvos (q.v.). The compound is used against hygiene pests and in animal houses and against ticks and fleas on pets. Acute oral toxicity for rats is 430 mg/kg (LD_{50}). The compound causes skin irritation and eye burns on rabbits at an acute percutaneous LD_{50} of 1100 mg/kg. **Tetrachlorvinphos (CVMP)** is a non-systemic insecticide and acaricide with contact and stomach action. The compound is active against, flies, fleas ticks and lice on pets and cattle, swine and poultry and in animal houses. LD_{50} acute oral toxicity for rats is 4000–5000 mg/kg. Low toxicity for birds and high toxicity for bees have been observed. In mammals tetrachlorvinphos is metabolised and eliminated quickly. Major metabolites found in urine of rats and dogs were glucuronic acid derivatives, mainly 2,4,5-trichlorophenyllethandiol glucuronide, 2,3,5-trichloromandelic acid, and 2-chloro-1-(2,4,5-trichlorophenyl)-vinylmethyl hydrogen phosphate. **Trichlorfon** is an insecticide with contact and stomach action. It is used to control fleas flies, keds, lice, warble flies and mange mites on farm animals. It is also active against a variety of fish ectoparasites. Acute oral toxicity for rats is 560–630 mg/kg (LD_{50}; acute percutaneous > 2000 mg/kg). Toxicity to fish and bees is moderate. In mammals the compound is rapidly degraded in the blood. Trichlorfon excretion in the urine is complete within 6 hours. Major metabolites are dimethylphosphoric acid, monomethylphosphoric acid and dichloroacetic acid. As metrifonate the compound is used as an anthelmintic against nematodes, trematodes or cestodes.

Monothiophosphates

Important Compounds Azamethiphos, chlorpyrifos, coumaphos, cythioate, diazinon (dimpylate), famphur (famophos), fenitrothion, fenthion (MPP), iodofenphos, phoxim, propetamphos, temephos.

General Information **Azamethiphos** is an insecticide and acaricide with predominantly contact action that shows quick knockdown and good residual activity. The acute oral toxicity for rats is low (LD_{50} 1180 mg/kg; percutaneous >2150 mg/kg) but bee toxicity has been observed and it was classified as highly toxic to fish (rainbow trout). However, the compound has been tested against fish lice on salmon with positive results (efficacy at 0.01 ppm) and good tolerance by salmon. Major metabolite in mammals is the glucuronic acid conjugate of 1-amino-3hydroxy-5-chloro-pyridine. Azamethiphos is mainly used for control of public hygiene pests and insect pests in animal houses. **Chlorpyriphos** is a non-systemic compound with contact, stomach and respiratory action and efficacy against fleas, ticks and sarcoptic mite. Acute oral toxicity for rats is 135–163 mg/kg (LD_{50}; >2000 mg/kg acute percutaneous) but it shows bee toxicity and high fish toxicity (0.003 mg/l LC_{50} for rainbow trout). Its slow degradation in soil to 3,5,6-trichloro-pyridin-2-ol with a half-life of 80–100 days is responsible for the long persistence in environment. Metabolism in mammals following oral administration is rapid and leads to the same major metabolite which is excreted via urine. **Coumaphos** is a broad spectrum insecticide and acaricide with predominantly contact action. Coumaphos (as 25% WP) has been approved APHIS-USDA-permitted pesticide for treatment of screwworms, scabies and ticks in federal eradication programs. Systemic action in the host animal is directed against warble flies. Withdrawal periods for meat are 15 days to 3 weeks depending on formulation. LD_{50} acute oral toxicity in rats is 16–41mg/kg (LD_{50} percutaneous for rats 860 mg/kg) Because of its relatively low toxicity for bees it is also used as a miticide against *Varroa* mites in bee hives. Because of its ability to break existing resistance interest in the varroatosis treatment with coumaphos has recently grown especially in the southern part of the United States. Coumaphos is also used in animal treatment against nematodes, trematodes or cestodes. **Cythioate** is rapidly absorbed from the gastro-intestinal tract after oral dosing with maximum drug effect for up to eight hours after administration. Fleas and other ectoparasites like ticks and mange mites are killed when they ingest the body fluids of the host. LD_{50} acute oral toxicity for rats is 107–246 mg/kg. The compound is rapidly eliminated from the body. **Diazinon (dimpylate)** is a broad spectrum insecticide and acaricide with contact, stomach and respiratory action. Withdrawal times for meat range from 3 to 14 days depending on formulation and host species. Acute oral toxicity for rats has LD_{50} from 240–480 mg/kg body weight, The compound shows a pronounced bee toxicity. The bird toxicity is impaired with bird repelling properties. Major metabolites in mammals are diethyl thiophosphate and diethyl phosphate. **Famphur** is used as ectoparasiticide against horn flies, grubs and lice in cattle and reindeer. LD_{50} acute oral toxicity for rats is 27–62 mg/kg for rats. **Fenitrothion** is a non-systemic insecticide with contact and stomach action and ectoparasiticidal activity against fleas. It is used as a public health insecticide and for control of flies in animal houses. LD_{50} acute oral toxicity for rats is between 250 and 800 mg/kg (acute percutaneous LD_{50} for rats 890 mg/kg). Fenitrothion is rapidly excreted in the urine and faeces (90% after three days in rats). Major metabolites are dimethylfenitrooxon and 3-methyl-4-nitrophenol. **Fenthion** is a systemic broad spectrum insecticide with contact, stomach and respiratory action against a variety of ectoparasitic insects and hygiene pests. Acute oral toxicity for rats is 250 mg(kg (LD_{50}, acute percutaneous 700 mg/kg). After oral administration the compound is eliminated in mammals mainly in the form of hydrolysis products in the urine. The major metabolites are fenthion sulfoxide, fenthion sulfone and their oxygen analogues. These metabolites are further hydrolysed to the corresponding phenols. **Iodofenphos** is a non-systemic insecticide and acaricide with contact and stomach action used as a premise ectoparasiticide in poultry houses. It shows low toxicity to mammals and is non-toxic for birds. **Phoxim** is a broad spectrum insecticide and acaricide with contact and stomach action and a short-term activity. It controls mange mites, lice, keds, flies, fleas and fly larvae. Acute oral toxicity for rats is 1976–2170 mg/kg (LD_{50}, acute percutaneous >1000 mg/kg). The compound is toxic for fish and bees (contact and respiratory action). In mammals it is rapidly metabolised to diethylphosphoric acid and desethylphoxim. The nitrile is metabolised to phoxim carboxylic acid and metabolism of the oxon is also unusually fast. 97% is secreted within 24 hours in the urine and faeces. **Propetamphos** is an insecticide and acaricide with contact and stomach action and long residual activity. Acute oral toxicity for rats is 60–119 mg/kg (LD_{50}, acute percutaneous 2825 mg/kg for male rats). In mammals, propetamphos is completely metabolised and rapidly excreted mainly via urine and exhaled air. It is detoxified through hydrolytic reactions involving the phosphorus and carboxylic ester bonds followed by conjugation and through oxidation processes leading ultimately to CO_2. **Temephos**, a non-systemic insecticide is used in mosquito larvae control and for treatment of pets against fleas and treatment of humans against lice. The compound shows a low toxicity against mammals (LD_{50} acute oral toxicity for rats 4204–>10000 mg/kg; LD_{50} acute percutaneous toxicity for rats >4000 mg/kg) , birds and fish but is highly toxic to bees. In mammals, temephos is mainly eliminated unchanged in the urine and faeces.

Dithiophosphates

Important Compounds Dimethoate, ethion (diethion), malathion, phosmet (PMP, phtalofos).

General Information **Dimethoate** is a fast acting insecticide and acaricide which quickly penetrates the insect cuticle. High initial penetration rate and slow detoxification rate have been observed with the housefly *Musca*. The oxon compound shows highest insecticidal activity but is more toxic to mammals. Acute oral LD_{50} of dimethoate for rats is 250 mg/kg, acute oral LD_{50} of the dimethoxon is 30 mg/kg. Dimethoate is toxic to bees, fish and arthropod aquatic organisms. **Ethion** is a non-systemic acaricide with predominantly contact action. Only combinations with pyrethroids (deltamethrin, permethrin) or other organophosphorous compounds (dichlorvos) are marketed. Acute oral toxicity for rats is 208 mg/kg (LD_{50}). The compound is toxic to fish and bees. **Malathion** is a non-systemic pro-insecticide and acaricide with contact stomach and respiratory action. The compound is activated by metabolic desulfuration to the corresponding oxon. It is extensively used for vector control in public health and against ectoparasites of cattle, poultry, dogs, and cats. Malathion is also active against human head and body lice. The lice and their eggs are killed quickly upon treatment with 0.003% and 0.06% malathion in acetone. Acute oral toxicity for rats is 1375–2800 mg/kg. Malathion shows low toxicity to birds and is toxic for fish and bees. In mammals the major part of the dose is excreted in the urine and faeces 24 hours after oral administration. Microsomal liver enzymes start detoxification by formation of malaoxon that is subsequently hydrolysed by carboxylesterases. **Phosmet** is a non-systemic insecticide and acaricide with predominantly contact action. Phosmet controls lice, horn flies, mange mites and ticks. The compound has also been formulated to act systemically against warble fly larvae and mange mites. Acute oral toxicity for rats is 113–160 mg/kg (LD_{50}). It is toxic to fish and bees. In mammals it is metabolised rapidly to phthalamic acid and phthalic acid (and derivatives) which are excreted in the urine.

Resistance Malathion: Organophosphorous compounds

Carbamates

Important Compounds Bendiocarb; carbaryl; methomyl; promacyl, propoxur.

General Information Insecticidal **carbamate** compounds are inhibitors of acetylcholine esterases and have been synthesised in 1954 at the first time. The kinetic of acetylcholine esterase inhibition with carbamates compounds has a half life of about

20 minutes for the carbamylated enzyme complex which is more reversible than with organophosphates. The more potent carbamates are structurally closely related to acetylcholine. Their weaker reactivity at the active site is compensated by more pronounced enzyme binding abilities. The signs of intoxication are similar to those of organophosphates. Stimulation of cholinergic synapses is followed finally by paralysis of the ectoparasite.

Bendiocarb is an insecticide with contact and stomach action that gives rapid knock-down and has good residual activity. LD_{50} of acute oral toxicity in rats is 40–156 mg/kg (LD_{50} acute percutaneous toxicity in rats 566–800 mg/kg). The compound is toxic to bees and fish. In mammals bendiocarb is rapidly absorbed after oral administration or inhalation. It is rapidly detoxified and eliminated almost completely after 24 hours as sulphate or glucuronide conjugate of 2,2-dimethyl-1,3-benzodioxol-4-ol its major metabolite. **Carbaryl** is a widely used insecticide with slightly systemic properties and contact and stomach action. The compound was introduced onto the market in 1956. Due to its relatively low toxicity to mammals it is used against ectoparasites. It shows weak inhibition of cholinesterase. Acute oral toxicity for rats (LD_{50} 500–850 mg/kg; acute percutaneous LD_{50} for rats >4000 mg/kg). The compound is toxic to adult bees but carbaryl is not transferred to the breed. The compound shows moderate toxicity to fish if applied as aqueous solution. Carbaryl has very low toxicity for birds (>2000 mg/kg for young pheasants) and has been developed for treatment of ectoparasitic diseases on poultry and other cage birds. Carbaryl does not accumulate in mammals body tissues and is rapidly metabolised to non-toxic substances, particularly 1-naphthol. This metabolite and its glucuronic acid conjugate is eliminated in the urine and faeces. **Methomyl** is a mixture of (Z) and (E) isomers, the former predominating. It is a systemic insecticide and acaricide with contact and stomach action. It is used for control of flies in animal houses. Today, direct application is being replaced more and more by bait formulations, reducing the risk of mammalian intoxication. The acute oral toxicity for rats is 17–24 mg/kg (LD_{50}, acute percutaneous toxicity for rabbits >5000 mg/kg). **Promacyl** was used until recently as a special tickicide with contact and stomach action against a variety of tick species on cattle in Australia. Since 1997 the compound has not been produced any more. It has favourable withdrawal periods of 24 hours (meat) and no withdrawal period for milk. Acute oral toxicity for rats is 1220 mg/kg (LD_{50}, acute percutaneous toxicity for rats >4000 mg/kg). **Propoxur** is a non-systemic insecticide with contact and stomach action. It gives rapid knock-down and long residual activity. The compound is active against fleas ticks, lice and biting lice in dogs and cats. LD_{50} acute oral toxicity for rats is 95–104 mg/kg (acute percutaneous LD_{50} for male rats is 800–1000 mg/kg). The compound is highly toxic to adult bees but shows low toxicity for birds. In rats main metabolites are 2-hydroxyphenyl-N-methylcarbamate and 2-isopropoxyphenol that are rapidly excreted in the urine. The carbamic acid residue is decomposed and carbon dioxide is exhaled.

Resistance Several different resistance mechanisms have been demonstrated in resistant hygiene pests and ectoparasites. Besides observations of reduced cuticle permeability of carbamates in resistant housefly strains detoxification through metabolising enzymes is the dominating mechanism of resistance against carbamates. Other than for organophosphates carbamate resistance depends more on oxidative metabolism. O-dealkylation, N-dealkylation, and N-methylhydroxylation and to a lesser extent hydroxylation are responsible for detoxification of carbamates like propoxur or carbaryl. In addition, turnover of these reactions is elevated by factors two to three in resistant strains compared to susceptible populations. Furthermore, for carbamate resistant strains of cattle tick (*Boophilus microplus*), sheep blowfly (*Lucilia cuprina*), and housefly (*Musca domestica*) choline esterases insensitive to carbamates have been identified. Carbamate insecticides have been extensively used in flea treatments in Florida. Several flea strains have been isolated that showed tolerance or resistance to carbamates and to organophosphates. Resistance factors for propoxur, carbaryl, and bendiocarb are 4-, 20- and 28-fold, respectively. Other authors reported strains of *Boophilus microplus* being resistant against carbamate ectoparasiticides.

Neonicotinoids

Important Compounds Imidacloprid, nitenpyram.

General Information **Nicotine** is not registered anymore for use against ectoparasites in countries with high registration standards for reasons of its intrinsic high mammalian toxicity. However, the natural product nicotine guided the discovery of a new class of insecticides. About 2000 derivatives have been synthesised on the way to a group of ten candidates with highest insecticidal activity finally leading to the synthesis of **imidacloprid**. In comparison to nicotine, the compound impairs a 9-fold lower acute mammalian toxicity and an average >900-fold higher insecticidal activity.

Available data on the mode of action of neonicotinoids predominantly have been generated with imidacloprid. In general, other neonicotinoids behave similarly. From several studies it can be concluded that potent insecticidal neonicotinoids are (partial) agonists of the postsynaptic acetylcholine receptors of motoneurones in insects. Imidacloprid induces slow depolarisation in cell bodies of motor-neurone from cockroach nerve cord preparations which is sensitive to nicotinic antagonists like dihydro-β-erythroidine. Imidacloprid is a more potent agonist than nicotine, however maximum depolarisation is slightly lower than with nicotine. Electrophysiological studies with mammalian nicotinic acetylcholine receptors from rat muscle expressed in *Xenopus* oocytes revealed that imidacloprid acts as a 1000-fold less potent agonist to mammalian nicotinic receptors compared to the naturally occurring agonist acetylcholine. Additionally, the open state phases are significantly reduced compared to acetylcholine. Binding studies with insect and mammalian brain tissue incubated with tritiated imidacloprid revealed high affinity binding sites in insect neuronal tissues but not in the mammalian brain. Hence, the favourable toxicity profile of neonicotinoids is also based on target specificity for the different subtypes of nicotinic acetylcholine receptors present in mammals and insects. Additional insecticidal benefit results from the reduced positive charge under physiological conditions at the receptor-binding nitrogen of neonicotinoid compared to nicotine alkaloids. The positively charged nicotine alkaloids are barely able to cross the lipophilic cuticle of insects, leading to poor contact activity of nicotine. Neonicotinoids show good and long-term systemic action in plants. They are also quickly distributed in mammals upon oral ingestion. However, this efficacy is a short term response due to rapid elimination with dramatic loss of efficacy after less than four days. Residual activity on animals for at least four weeks is achieved only through topical application (e.g. imidacloprid).

Imidacloprid is an insecticide with contact and stomach action introduced in 1992 as a new chemical entity insecticide in crop protection. Four years later imidacloprid was introduced as the active ingredient of a new and highly effective ectoparasiticide against adult fleas. In topical formulations imidacloprid has a long residual activity. Acute oral LD_{50} for rats is 450 mg/kg (acute percutaneous LD_{50} for rats is >5000 mg/kg). Very low toxicity for birds and fish was measured. In rats imidacloprid is quickly absorbed from the gastrointestinal tract and eliminated to 96% within 48 hours mainly via the urine. 15% was eliminated unchanged. The most important metabolic steps were hydroxylation at the imidazoline ring, hydrolysis to 6-chloronicotinic acid, loss of the nitro group with formation of the guanidine and conjugation of 6-chloronicotinic acid with glycine. **Nitenpyram** is an insecticide with contact and stomach action similar to imidacloprid. It has been very recently introduced in selected markets as an oral formulation for acute treatment of fleas on pets. Upon oral administration ectoparasiticidal activity lasts not more than three days due to the quick elimination and excretion of the compound via urine. Nitenpyram is marketed only in combination with longer acting compounds. Acute oral LD_{50} for rats is 1600 mg/kg (acute percutaneous LD_{50} is >2000 mg/kg).

Resistance In the four years after introduction of imidacloprid as a pet flea product, field resistance of ectoparasites against imidacloprid has not been reported. Laboratory trials with the multiresistant (organophosphates, carbamates, pyrethroids, benzoylphenylureas, cyclodienes) field strain "cottontail" from Florida revealed no cross resistance for imidacloprid to other flea insecticides. It can be speculated from data obtained with insects relevant to crop protection (aphids, whiteflies) that mechanisms of detoxification of neonicotinoids in fleas will be based on similar mechanisms. For whitefly and aphids the dominant route of detoxification relies predominantly on oxidative processes, however some of the first step oxidation products retain insecticidal activity. To date, no target insensitivity was observed in insect ectoparasites collected from the field. Susceptibility to imidacloprid was not significantly different in housefly strains resistant against organophosphates or pyrethroids respectively compared to fully susceptible flies. Generally, the housefly is less susceptible to neonicotinoids. This could be due to fast detoxification by mixed function oxidases as can be concluded from experiments enhancing the efficacy of neonicotinoids against houseflies through synergists like piperonyl butoxide. Additionally, slow penetration through the cuticle of the housefly is also discussed.

Blocker / Effectors of Sodium Channels / Ectoparasiticides

Mode of Action Fig. 1

Structures Organohalogenides, Fig. 2

Important Compounds DDT, Methoxychlor

General Information **DDT** is a broad spectrum insecticide introduced onto the market in 1943. Nowadays, the compound has been replaced by less persistent insecticides in nearly all countries. DDT is a persistent non-systemic insecticide with contact and stomach action. DDT acts on the voltage gated sodium channel causing slow open and closing characteristics of the ion channel. This results in an increased negative afterpotential, prolonged action potentials repetitive firing after single stimulus, and spontaneous trains of action potentials. Temperature has a profound effect on the insecticidal activity of DDT. Its potency to induce repetitive discharges from cockroach sensory neurones increases with lowering of the temperature with a Q_{10} of 0.2. While in most countries the use of DDT is prohibited by law, in some countries DDT is still used in mosquito eradication programs. Acute oral LD_{50} for rats is 113–118 mg/kg (acute percutaneous LD_{50} for rats is 2510 mg/kg). DDT is toxic to fish and aquatic life. A side effect of DDT is the inhibition of a Ca-ATPase that is necessary for the calcification of egg shells. The compound accumulates in fatty tissues of mammals and is excreted in milk. The high potential for bioaccumulation and its long persistence has been discussed to be a major hazard to the environment. **Methoxychlor** is an insecticide closely related to DDT with contact and stomach action. It is used for the control of insect pests in animal houses, dairies and household. Acute oral LD_{50} for rats is >6000 mg/kg The compound is toxic to

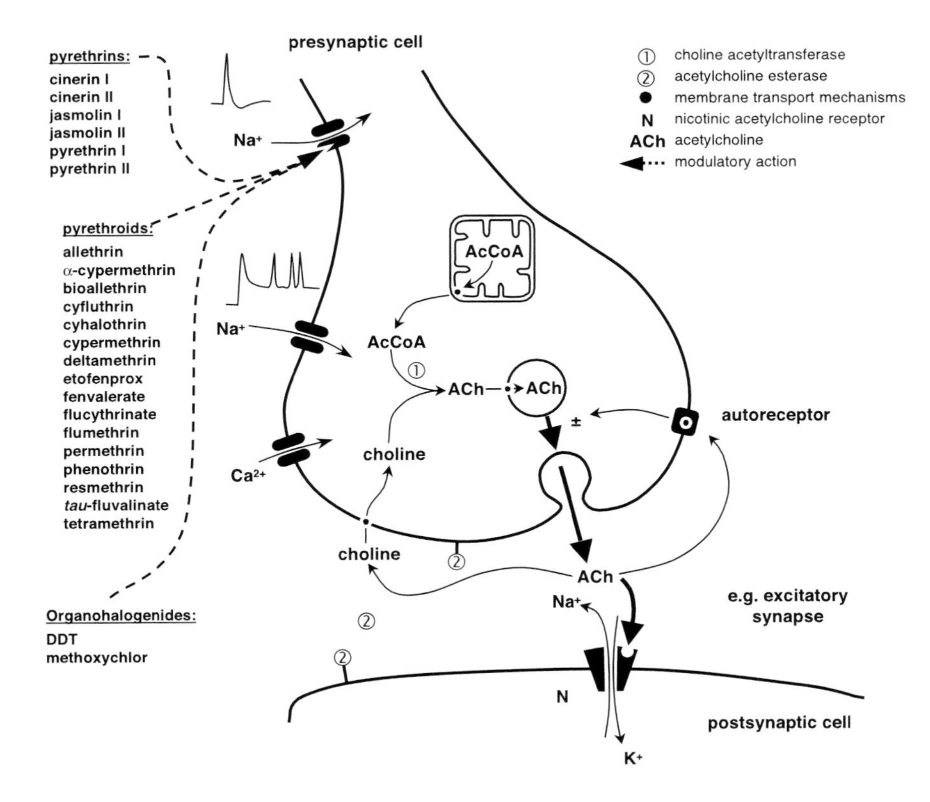

Fig. 1. Model of drugs affecting neural transmission at voltage sensitive sodium channels.

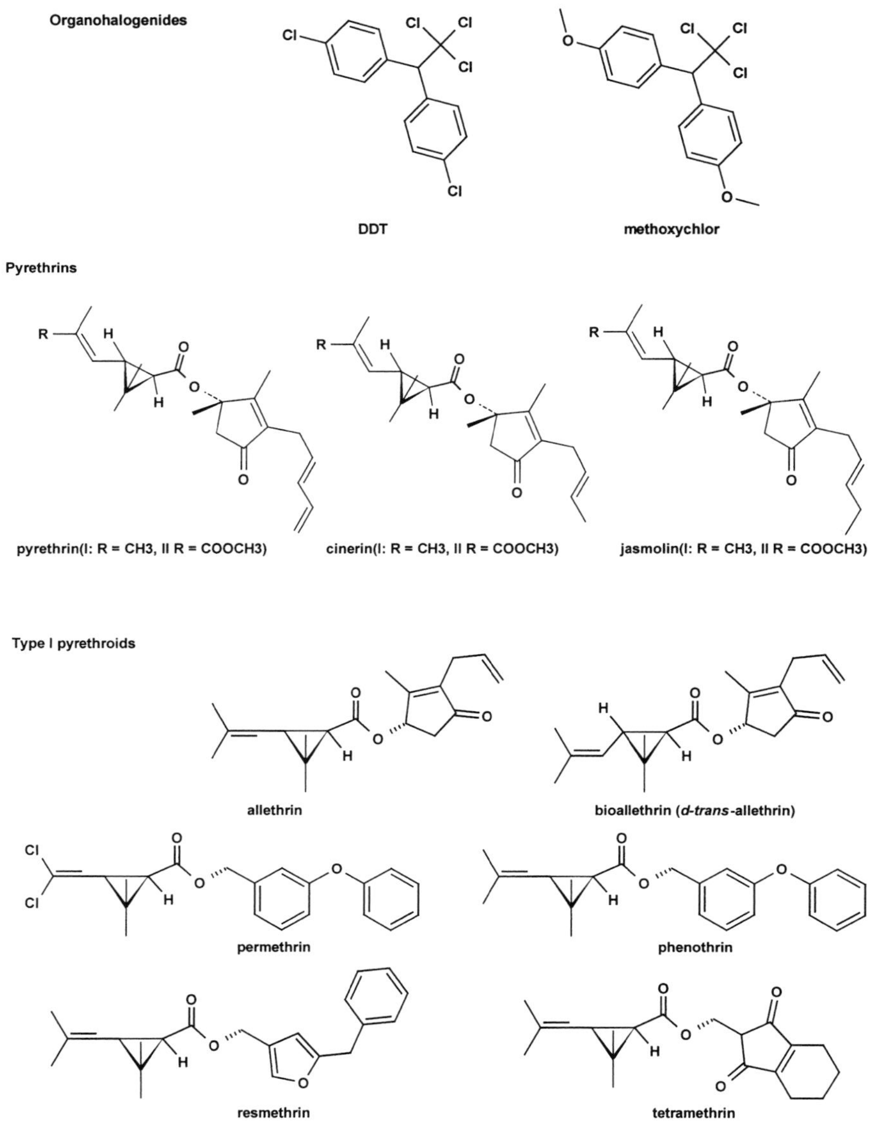

Fig. 2. Structures of antiparasitic drugs affecting voltage sensitive sodium channels.

Organohalogenides

DDT

methoxychlor

Pyrethrins

pyrethrin(I: R = CH3, II R = COOCH3)

cinerin(I: R = CH3, II R = COOCH3)

jasmolin(I: R = CH3, II R = COOCH3)

Type I pyrethroids

allethrin

bioallethrin (*d-trans*-allethrin)

permethrin

phenothrin

resmethrin

tetramethrin

aquatic life and fish. In mammals the compound is degraded by O-dealkylation to the corresponding phenol and diphenol, and by dehydrochlorination to 4,4'-dihydroxybenzophenone.

Resistance Resistance of ectoparasites against DDT and methoxychlor is mainly of metabolic origin. Primarily, elevated levels of dehydrochlorinase activity together with mixed function oxidases are involved in detoxifying DDT. Reduced penetration of DDT through the cuticular layers is another resistance mechanism e.g. in DDT resistant flies. Furthermore, DDT susceptibility is reduced in target site mutant kdr and super-kdr housefly strains. Housefly strains resistant to DDT show reduced susceptibility for methoxychlor. However, methoxychlor resistant flies are fully susceptible to DDT indicating a different mechanism for methoxychlor detoxification. A possible pathway could be demethylation and subsequent conjugation.

Pyrethrins

Important Compounds Pyrethrin I, Pyrethrin II, Cinerin I, Cinerin II, Jasmolin I, Jasmolin II.

General Information Pyrethrum comprises a mixture of naturally occurring pyrethrins (pyrethrin

Type II pyrethroids

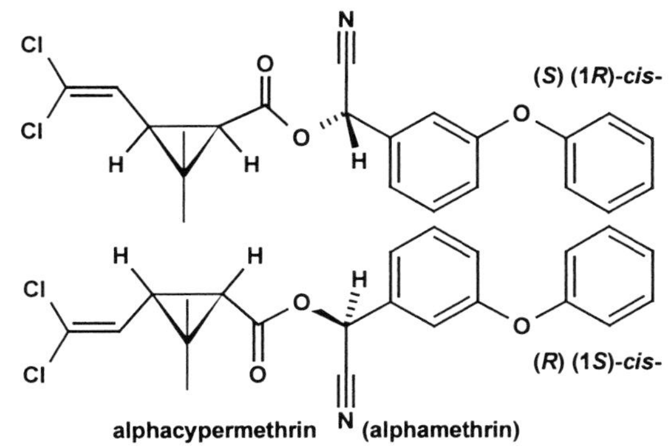

(*S*) (1*R*)-*cis*-

(*R*) (1*S*)-*cis*-

alphacypermethrin (alphamethrin)

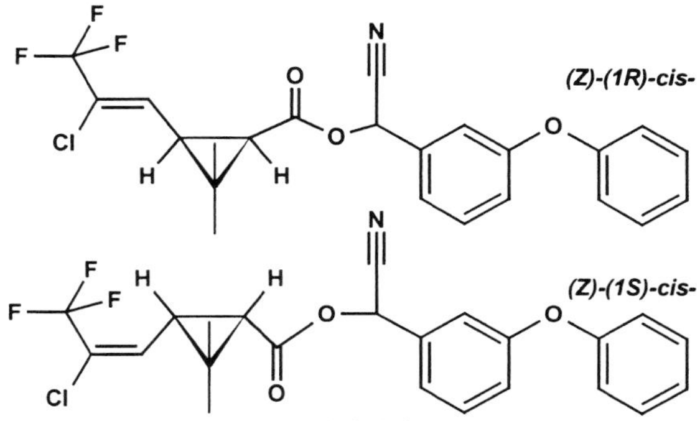

(*Z*)-(1*R*)-*cis*-

(*Z*)-(1*S*)-*cis*-

cyhalothrin

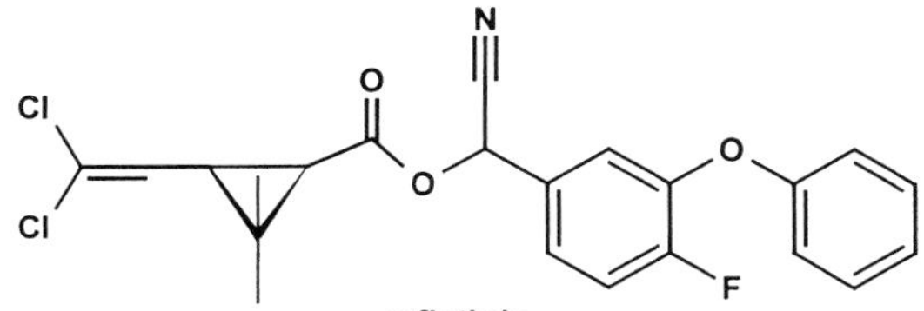

cyfluthrin

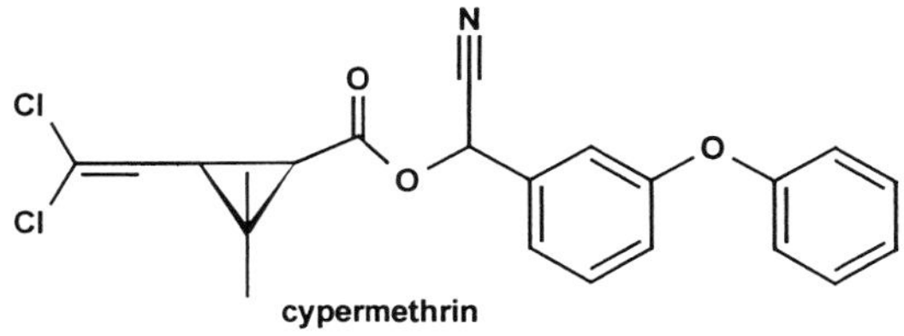

cypermethrin

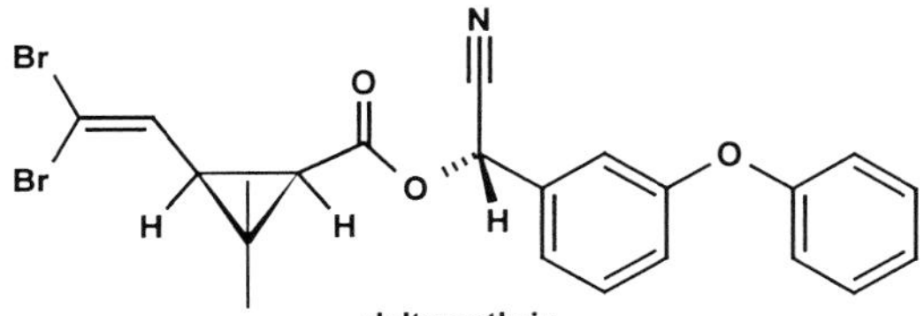

deltamethrin

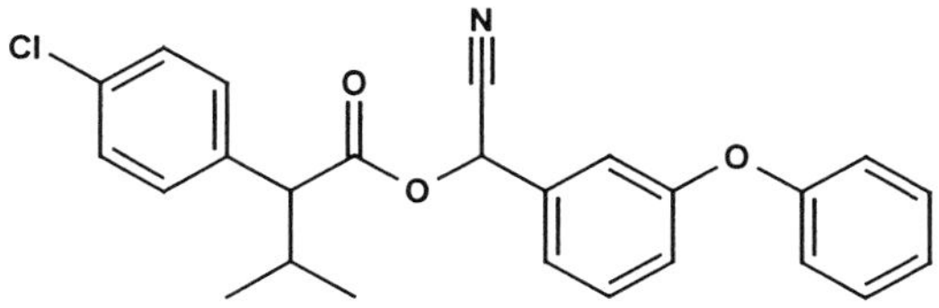

fenvalerate

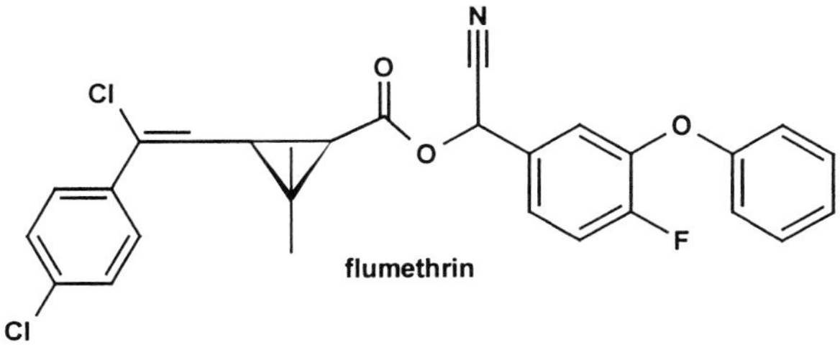

flumethrin

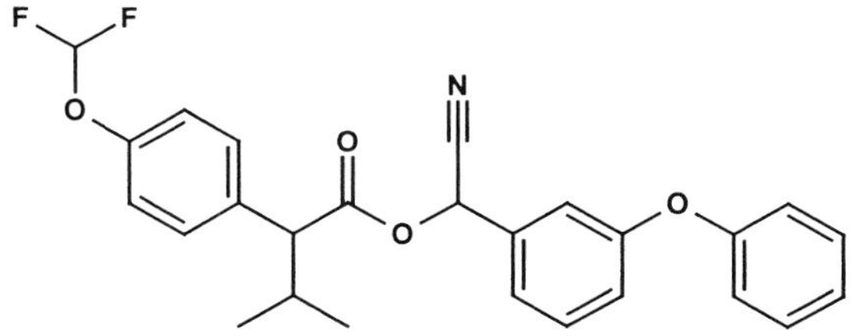

flucythrinate

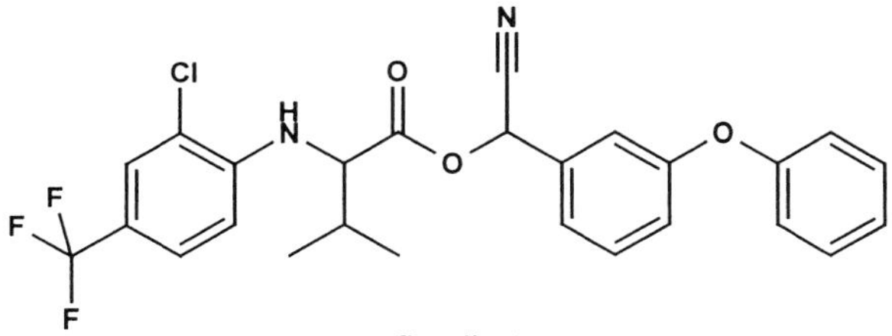

tau-fluvalinate

Non-ester pyrethroids

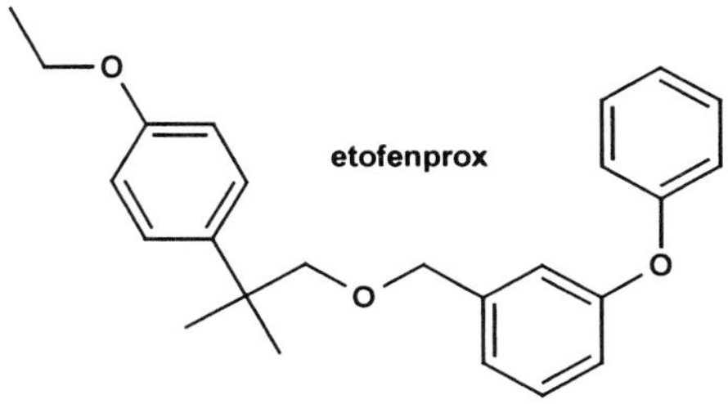

etofenprox

Fig. 2. (continued)

I, pyrethrin II, jasmolin I, jasmolin II, cinerin I, cinerin II) CNA; (Z)-(S)-2-methyl-4-oxo-3-(penta-2,4-dienyl)cyclopent-2-enyl (+)-trans-chrysantemate (cinerin I); (Z)-(S)-2-methyl-4-oxo-3-(penta-2,4-dienyl)cyclopent-2-enyl (+)-trans-chrysantemate (pyrethrin I); (Z)-(S)-3-(but-2-enyl)-2-methyl-4-oxocyclopent-2-enyl (+)-trans-chrysantemate (cinerin I); (Z)-(S)-2-methyl-4-oxo-3(pent-2-enyl)cyclopent-2-enyl (+)-trans-chrysantemate (jasmolin I); (Z)-(S)-2-methyl-4-oxo-3-(penta-2,4-dienyl)cyclopent-2-enyl pyrethrate (pyrethrin II); (Z)-(S)-3-(but-2-enyl)-2-methyl-4-oxocyclopent-2-enyl pyrethrate (cinerin II); (Z)-(S)-2-methyl-4-oxo-3(pent-2-enyl)cyclopent-2-enyl pyrethrate (jasmolin II)) isolated from *Tanacetum* (= *Chrysantemum* = *Pyrethrum*) *cinerariaefolium*. Pyrethrum is an ancient insecticide which was identified in China and spread to Dalmatia, France, USA, Japan in the 19th century (dried powdered flower heads were called Persian insect powder). Pyrethrum is a non-systemic insecticide with some acaricidal activity. It causes immediate paralysis with death occurring later and shows quick knock-down and short acting activities. The primary molecular target of pyrethrins is the neuronal pre-synaptic voltage sensitive sodium channel. Pyrethrum is used in treatment of ectoparasites on companion and farm animals. The biological efficacy of pyrethrins and pyrethroids seems to be less pronounced in the larval stage of most insect species which might be due to the higher activity of the adult stages compared to larvae. Generally, it is combined with synergists like piperonyl butoxide to encounter quick detoxification and to enhance potency. In mammals pyrethrins are rapidly degraded in the stomach by hydrolysis of the ester bond to non-toxic metabolites. The LD_{50} for acute oral toxicity in rats is 584–900 mg/kg (LD_{50} acute percutaneous toxicity >1500 mg/kg in rats). Synergists seem not to enhance the toxicity of pyrethrins to mammals. Pyrethrins are highly toxic to fish and toxic to bees but show a repelling effect.

Pyrethroids

General Information Pyrethrins served as a lead structure for the synthesis of the pyrethroids. The chemical class of pyrethroids is divided into structurally related sub-classes. The type I pyrethroids are ester bond pyrethroids without α-cyano-residue, the type II pyrethroids include all ester bond pyrethroids containing a cyano-group at the α-carbon atom. An example for the class of non-ester bond pyrethroids is also given. A variety of structures were introduced onto the market during the 1970s and 1980s.

The insecticidal symptoms of type I pyrethroids are characterised by hyperexcitation, ataxia, convulsions and paralysis. Type II pyrethroids cause hypersensitivity, tremors and paralysis. The primary target of pyrethrins and all pyrethroids is the voltage sensitive sodium channel on the pre-synaptic side of insect neuronal synapses.

The pyrethroids slow kinetics of both opening and closing of individual sodium channels resulting in delayed and prolonged openings. Pyrethroids also cause a shift of the activation voltage in the direction of hyperpolarisation. There then follows a membrane depolarisation and an increase in depolarising after-potential. The latter reaches the threshold for excitation causing repetitive after-discharges. The membrane potential of sensory neurones increases discharge frequency, and that of nerve terminals increases the release of neurotransmitter and the frequency of spontaneous miniature postsynaptic potentials. The corresponding symptoms of mammalian or arthropod intoxication are hyperexcitation, hypersensitivity, convulsions and tremors. As described for DDT, pyrethroids also show a negative temperature coefficient for their activity on voltage sensitive sodium channels revealing a Q_{10} value of 0.18 for tetramethrin between 25°C and 35°C. Type II pyrethroids also show some modulatory activity at a secondary target site identified as GABA-gated chloride channel.

Type I Pyrethroids Important type I pyrethroids are allethrin, bioallethrin, permethrin, phenothrin, resmethrin, and tetramethrin.

Allethrin [(1R)-isomers] is a non-systemic insecticide with contact, stomach and respiratory action. It gives rapid knock-down and paralysis before killing. It is used for insect control in animal houses and as an animal ectoparasiticide. LD_{50} of acute oral toxicity for rats is 900–2150 mg/kg. In mammals the compound is detoxified after oral administration in the liver by oxidation of one terminal methyl group of the chrysantemic acid moiety to a carboxyl group via an alcohol group. The compound is eliminated via urine and faeces within 2–3 days after treatment. **Bioallethrin**; d-trans-allethrin [$\geq$ 93% (1R)-, $\geq$ 90% trans, $\leq$ 3% cis-isomer) is a potent contact non-systemic, non-residual insecticide producing rapid knock-down. It is used mainly in household and public health and in some countries is marketed for ectoparasite treatment of companion animals. Mammalian toxicity is slightly higher than with allethrin. Detoxification and elimination occurs as described for allethrin. Cis/trans isomerisation has not been observed in soil. **Permethrin** is a non-systemic insecticide with contact and stomach action and slight repellent effect. It is used for repellence and control of biting flies and is also active against biting and sucking lice on cattle. Acute oral LD_{50} for rats is 4000 and 6000 mg/kg for a cis:trans isomer mixture of 40:60 and 20:80 respectively (acute percutaneous LD_{50} for rats is >4000 mg/kg). Permethrin is toxic to fish and bees. In mammals hydrolysis of the ester bond occurs and the compound is eliminated as the glycoside conjugate. **Phenothrin** [(1R)-isomers]is a non-systemic insecticide with contact and stomach action that gives rapid knock-down. It is used in public health against a variety of injurious and nuisance insects and as a combination product with allethrin or tetramethrin for control of fleas and ticks on dogs and cats. Acute oral LD_{50} for rats is >10000 mg/kg (acute percutaneous LD_{50} for rats is >10000 mg/kg). The compound is toxic to fish and bees. **Resmethrin** is a non-systemic insecticide with contact action, acting in a similar manner to the natural pyrethrins but is not synergised by pyrethrum synergists. It is often used in combination with more persistent insecticides. Acute oral LD_{50} for rats is >2500 mg/kg (acute percutaneous LD_{50} for rats is >3000 mg/kg). The compound is toxic to fish and to bees. Metabolism in hens was principally by ester hydrolysis and oxidation, followed by conjugation. **Tetramethrin** is a non-systemic insecticide with contact action that gives rapid knock-down. It is used as flea insecticide for pets. Tetramethrin is often combined with synergists or other pyrethroid insecticides. Acute oral LD_{50} for rats is >4640 mg/kg (acute percutaneous LD_{50} for rats is >5000 mg/kg). The compound is toxic to fish and to bees. Tetramethrin seems to be metabolised in a similar way to the natural pyrethrins. In mammals, following oral administration, about 95% of the metabolised tetramethrin is eliminated in the urine and faeces within 5 days. The principal metabolite is 3-hydroxycyclohexane-1,2-dicarboximide.

Type II Pyrethroids Important type II pyrethroids are alpha-cypermethrin, cyfluthrin, cyhalothrin, cypermethrin, deltamethrin, fenvalerate, flucythrinate, flumethrin, and *tau*-fluvalinate.

Alpha-cypermethrin (formerly also known as alphamethrin) is a non-systemic α-cyano-pyrethroid with contact and stomach action. It is used mainly to control body lice and blowfly strike on sheep with no withdrawal period required. Acute oral LD_{50} for rats is 474 mg/kg (acute percutaneous LD_{50} for rats is >2000 mg/kg; tech. grade). The compound is toxic to fish and bees under experimental conditions but no toxic effects could be observed under field conditions. **Cyhalothrin** is a non-systemic insecticide and acaricide with contact and stomach action and repellent properties. It is mainly used for control of animal ectoparasites on sheep and cattle. Acute oral LD_{50} for rats is 114–166 mg/kg (acute percutaneous LD_{50} for rats is 200–2500 mg/kg). The compound is toxic to fish and other aquatic organisms. In mammals the orally administered compound is hydrolysed at its ester bond and polar conjugates are formed from both moieties. Cyhalothrin is rapidly eliminated via urine and faeces. **Cyfluthrin** is a non-systemic insecticide with contact and stomach action with rapid knock-down efficacy and long residual activity. The compound is used in public health, against stored product pests and against flies in animal health. Acute oral LD_{50} for rats is 590 mg/kg (acute percutaneous LD_{50} for rats is >5000 mg/kg). Toxicity to bees and fish has been observed. Cyfluthrin was largely and quickly eliminated in mammals. 98% of the administered amount was eliminated within 48 h via urine and the faeces. **Cypermethrin** is a non-systemic insecticide and acaricide with contact and stomach action. It also deters ovipositioning blowflys on treated sheep. It has a broad activity against ectoparasites on farm animals. Withdrawal periods are 7 days for meat and at least 6 h for milk. Acute oral LD_{50} for rats is 200–800 mg/kg (acute percutaneous LD_{50} for rats is >1600 mg/kg). The compound is toxic to fish and bees. **Deltamethrin** is a non-systemic fast acting insecticide and acaricide with contact and stomach action. It is used against a variety of ectoparasitic species on livestock. Acute oral LD_{50} for rats is 128–139 mg/kg (acute percutaneous LD_{50} for rats is >5000 mg/kg in aqueous solution). The compound is toxic to fish and bees under experimental conditions, but exhibits a repellent effect. **Fenvalerate** is a non-systemic insecticide and acaricide with contact and stomach action. It controls a broad spectrum of parasitic arthropods and exhibits repellent activity. Acute oral LD_{50} for rats is 451 mg/kg (acute percutaneous LD_{50} for rats is >2500 mg/kg). The compound is toxic to fish and bees. In mammals fenvalerate is rapidly metabolised. Up to 96% of the compound administered orally is excreted in the faeces within 6–14 days. **Flucythrinate** is a non-systemic insecticide with contact and stomach action. It is registered for the control, of flies, fleas and other insects. Acute oral LD_{50} for rats is 67–81 mg/kg. The compound is moderately toxic to fish and toxic to bees but shows a repellent effect. In mammals flucythrinate is eliminated within 24 h (60–70%) to 8 days (>95%) in the faeces and urine. Major metabolic pathways are hydrolysis followed by hydroxylation of the hydrolysis products. **Flumethrin** is a non-systemic insecticide and acaricide with contact and stomach action. It is used for the control of ticks, biting and sucking lice, mites and for diagnosis and control of varroatosis in beehives. At sub-lethal doses a sterilising effect of the pour-on formulation has been demonstrated for *Hyalomma* ticks. Specific formulations have been granted nil withdrawal periods for meat and milk. Acute oral LD_{50} for rats is mg/kg (acute percutaneous LD_{50} for rats is mg/kg). The compound is moderately toxic to fish and shows low toxicity to bees which enables selective treatment against *Varroa* mites. *tau*-**fluvalinate** is a non-systemic broad range insecticide and acaricide. In animal health applications it is marketed as for the control of the *Varroa* mite in beehives. Acute oral LD_{50} for rats is >3000 mg/kg (acute percutaneous LD_{50} for rats is >20000 mg/kg). *tau*-Fluvalinate is toxic to fish and other aquatic organisms.

Non-ester Pyrethroids **Etofenprox** is a non-ester pyrethroid insecticide with contact and stomach action. The compound is mainly used in crop protection but is also used to control public health pests and on livestock against insects. Acute oral LD_{50} for rats is >42880 mg/kg (acute percutaneous LD_{50} for rats is >2140 mg/kg). The compound is slightly toxic to fish.

Resistance Pyrethroid resistant strains of the cattle tick *Boophilus microplus* have been isolated in countries in Central and Southern America, Southern Africa and Australia. Resistance of Australian strains of *Boophilus microplus* against pyrethroids has been reviewed by J. Nolan. Monitoring of horn fly control measures with pyrethroid dips, sprays or pour-ons revealed significantly reduced susceptibility. The period of spray efficacy fell from 30 to 20 and even 5 days in some areas of Argentina. Dips regularly able to control horn flies for 15 days protected animals ranging from 0 to 6 days. Pour-ons of deltamethrin, cypermethrin, cyhalothrin or cyfluthrin experienced a decline in efficacy from 45–60 days down to less than four weeks during the study. Knock-down resistant (*kdr*) and *super kdr* phenotypes have been described in several housefly strains of different origin. Recently, point mutations responsible for pyrethroid and DDT resistance have been identified in the coding region of the *para* sodium channel gene (the insect analogue of the vertebrate voltage sensitive sodium channel) from *kdr* and *super kdr* strains of the housefly as well as kdr-strains of the German cockroach.

Since 1981 pyrethroids have been used as ectoparasiticides against sheep lice as pour-on or wet-dip formulations. By mid 1985 reports of failures of the pour-on pyrethroids were becoming more frequent. About seven years later a lice strain was isolated in New South Wales (Australia) that was found to be 642x resistant to cypermethrin and able to survive pour-on and full immersion dips. Addition of piperonyl butoxide to the formulation resulted in 81% reduction of lice number again.

Blood Diseases, Animals

Synonym

Parasitaemias

General Information

Haemoprotozoal diseases (***Babesiosis, Animals***, ***Trypanosomiasis, Animals***, ***Theileriosis*** and ***Leishmaniasis, Animals***) are caused by infections which localize either in the blood alone or in the blood and solid tissues. Blood infections affect primarily the microcirculation. However, more important are the lesions caused in solid tissues and organs. They are attributable either to systematic malfunction of the circulation or to disturbances in the microcirculation of a specific organ. An important common feature in all the pathogenesis is the release of large amounts of antigens into the plasma. The interactions of these antigens with the host 's immunological responses result also in mechanisms of cellular injury.

Table 1. Parasites of the haematopoietic system (according to Vercruysse and De Bont)

Parasite	Host	Location	Clinical presentation	Principal lesions
Babesia				
B. bigemina	Cattle	Red blood cells	Fever, malaise, loss of appetite, listlessness, anaemia haemoglobinuria, icterus	Findings of acute intravascular haemolytic crisis: capillary congestion of most organs, splenomegaly
B. bovis	Cattle		*B. bovis, B. canis*: nervous symptoms	*B. bovis, B. canis*: congestion grey matter throughout the brain
B. divergens	Cattle			
B. major	Cattle			
B. caballi	Horse			
B. canis	Dog			
B. gibsoni	Dog			
B. felis	Cat			
Leishmania				
L. infantum	Dog	Monocyte-macrophage system (visceral and cutaneous)	Lymphadenomegaly, anaemia, splenomegaly, dry exfoliative dermatitis, ulcerations, weight loss, onycogryphosis, ocular signs, anorexia	Haemo-lymphatic hypertrophy with macrophage proliferation and focal granulomas
Theileria				
T. parva	Cattle	Lymphocytes (lymphoblasts), erythrocytes	E C F: fever, dullness, listless, anorexia, hyperplasia lymph nodes, salivation, lacrimation diarrhoea, watery cough and dyspnea	Lymphoid hyperplasia, splenomegalia petechial haemorrhages over most of the serous and mucous surfaces, interlobular oedema, emphysema and hyperaemia of the lungs
T. annulata	Cattle			
T. lawrencei	Cattle			
T. mutans	Cattle			
T. hirci	Sheep, goats			
T. equi	Horse			
Trypanosoma				
T. congolense	Ruminants	Bloodplasma, perivascular tissues	Acute: severe symptoms and dead after 3 weeks Chronic: remittent fever, anaemia and progressive emaciation. Less common: watery diarrhoea, abortion, stillbirths, corneal opacity, bottle jaw	Generalizes lymphoadenopathia, heart is flabby, liver and spleen are enlarged
T. vivax	Ruminants, Horses			
T. evansi	Horses, Dogs			
T. brucei	Equines, Ruminants, Dogs, cats			
T. simiae	Pigs			

ECF = East Coast Fever

The clinical signs are non-specific and consist of fever, lethargy, weakness and emaciation. Parasitaemias are often associated with anaemia, leucopenia, oedema and hemorrhages (Table 1, page 97).

Therapy
See → Chemotherapy, → Drugs.

Borrelia

Genus of spirochaete-like bacteria transmitted by ticks (e.g. *R. burgdorferi* → Lyme Disease; *Borrelia recurrentis* → Relapsing Fever, *B. duttoni*, Tick Relapsing Fever).

Borreliosis

→ *Ixodes* species (Vol. 1), → Lyme Disease, → Ticks as Vectors of Agents of Diseases, Man.

Boutonneuse Fever

Boutonneuse fever is caused by *Rickettsia conori*, which is widespread in Africa, the Mediterranean region, and parts of Southeast Asia. The tick bite lesion takes on a black, buttonlike appearance (hence the name), with a central dark necrotic area. A large number of tick species appear able to transmit the disease, which can also be acquired by contact with tick tissues when the tick is crushed, for instance when *Rhipicephalus sanguineus* is removed from dogs.

Therapy
Application of antibiotics

Related Entries
→ Rocky Mountain Spotted Fever, → Tick Typhus.

Bronchoalveolar Lavage

Method to obtain fluid from lung in order to diagnose → *Pneumocystis carinii* (Vol. 1) stages.

Bronchoalveolar Lavage Fluid

→ BALF, → Pneumocystosis.

Brown-Brenn Stain

→ Microsporidiosis.

Brugiasis, Man

Disease due to infection with the filarial worm → *Brugia malayi* (Vol. 1).

Main clinical symptoms: see → Elephantiasis tropical, → Filariasis Lymphatic Tropical
Incubation period: 30–60 days
Prepatent period: 50–90 days
Patent period: 8–10 years
Diagnosis: see → Filariasis Lymphatic Tropical
Prophylaxis: Avoidance of mosquito bites
Therapy: Treatment see → Nematocidal Drugs, Man

BSE

Bovine spongious encephalopathy in cattle due to → prions, which apparently are transmitted to many hosts via oral uptake of infected brain (nerves) or contaminated fly larvae and other destruents (?). Transmission to man is possible introducing the so-called new Creutzfeldt-Jacob-Disease (CJD).

Therapy
Unknown

C

Calabar Swelling

Oedema due to infection with wandering *Loa loa* stages in human skin.

Capillariasis

Capillariasis is diagnosed by finding the distinctive *Trichuris*-like eggs either in the liver (→ *Capillaria (Vol. 1) hepatica)* or in the feces *(C. philippinensis)*. In *C. hepatica* infection the embryonated eggs are ingested with soil, and the larvae mature in the liver where adults lay their eggs. They have been found in a number of humans, surrounded by fibrosis. Because *C. hepatica* is normally found in rats, the adults die in humans. Infection with *C. philippinensis* has been described in Thailand and the Philippines. Large numbers of viviparous adults, larvae, and eggs have been found in the lumen and mucosa of the small intestine of people who are believed to have been infected by eating raw freshwater fish. Autoinfection with larvae appears to explain the intensity of the infection. Severe protein-wasting enteropathy with edematous jejunal walls but with little inflammation and an inconstant presence of eosinophils are considered characteristic. → Malabsorption with voluminous diarrhea, followed by emaciation and hypoproteinemic edema, often precedes a fatal outcome after an illness of 2 weeks to 2 months.

Therapy

→ Nematocidal Drugs, Man.

Capillariosis

→ *Capillaria* Species (Vol. 1), syn. Capillariasis.

Cardiovascular System Diseases, Animals

Myocarditis

Most parasites which form cysts in striated muscles may also invade the myocardium. Such cysts include the tiny tubular sarcocysts formed by the protozoan *Sarcocystis*, the metacestodes of the *Taeniidae* (cysticerci and hydatid cysts), or the encysted larvae of the nematode *Trichinella spiralis*. These infections generally produce few or no symptoms and in most cases the cysts are found incidentally on post-mortem examination of the cardiac muscle. However, in the unusual event of a massive infection death may occur after a febrile course.

The pathognomonic lesion in *Gedoelstia*-fly larva-infection (uitpeuloog) is a thrombo-endophlebitis which varies in intensity, location and distribution and is frequently accompanied by thromboendoarteritis. In sheep coronary thrombosis causes myocarditis or myomalacia cordis with sudden death as the sequel.

Trypanosoma cruzi is the cause of human trypanosomiasis in South America (→ Chagas' Disease, Man). It also develops and causes disease in cats, dogs and pigs which, together with numerous wild animals act as reservoir hosts for the parasite. *T. cruzi* multiply within the cytoplasm of the cells of their mammalian host, particularly those in the skeletal and cardiac muscles (amastigote form). The heart is particularly affected, with development of multifocal to diffuse severe granulomatous myocarditis. In dogs non-specific clinical signs such as weight loss, lymphodenomegaly, diarrhoea and anorexia are generally accompanied by signs referable to cardiac dysfunction, such as tachycardia, ascites, weak pulse, lethargy and hepatomegaly.

Vasculitis

Arteries Parasitic arteritis is a very common form of arterial disease in animals (Table 1). The lesions are characterized by inflammatory reaction and thickening of the arterial wall, often accompanied by endothelial damage and thrombosis. Thrombi may partially or completely occlude the artery and emboli arising from pieces of thrombi that have broken loose may occlude vessels distal to the site where parasites have lodged. Rupture of arteries as a result of parasitic lesions is rare. The two parasite species which cause by far the most severe damage to arteries are *Strongylus vulgaris* in horses and *Dirofilaria immitis* in dogs. Other species eliciting varying degrees of arteritis are *Onchocerca armillata* in cattle, *Elaeophora schneideri* in sheep, *Angiostrongylus vasorum* and *Spirocerca lupi* in dogs, and *Aelurostrongylus abstrusus* in cats.

Strongylus vulgaris is often referred to as the most important parasite of horses. It is widely distributed and infects horses of all ages, fatal natural infection having been observed in foals as early as 21 days after birth. The pathogenesis of *S. vulgaris* infection is associated with the larval migration of the parasite through the tissues of the host. Third-stage infective larvae ingested by the grazing animal penetrate the wall of the intestine, molt in the submucosa to become fourth-stage larvae, and invade terminal branches of intestinal arteries to begin their migration. This penetration and early migration of larvae results in inflammatory reactions and small haemorrhages throughout the intestinal wall. This coincides with the rise in body temperature detected 5–7 days after heavy infections. The temperature reaction generally subsides as the fourth-stage larvae migrate in the intima of the mesenteric arteries causing inflammatory reactions and mural thrombi. By 2–3 weeks post-infection the larvae reach the cranial mesenteric artery where they remain for several months. They molt to become fifth-stage larvae and eventually return to the lumen of the large intestine where they complete their maturation and start producing eggs 6–7 months after infection. During their long stay in the cranial mesenteric artery, the larvae induce a severe fibrinous inflammatory reaction which can involve all layers of the arterial wall (verminous arteritis). In chronic infections, the wall of the artery becomes thickened and the lumen partly occluded by thrombi, cellular debris, and larvae which remain firmly attached to the intima. Clinical signs in animals include dullness, progressive weight loss and varying degrees of pyrexia, often with intermittent abdominal discomfort. At post mortem, aneurysmal dilatation of the artery may sometimes be observed. Larval migration of *S. vulgaris* was generally recognized as a major etiological factor of equine colic, although its incidence is decreasing due to the availability of efficient anthelmintics. Acute colic due to thromboembolic infarction of the caecum or large intestine is a well-documented complication of verminous enteritis. Colic signs have also been observed in early massive infections, before verminous lesions develop in the cranial mesenteric artery. Possible mechanisms for colic include damage to and impairment of nervous innervation to the intestine by migrating larvae, release of toxins by degenerating larvae, and hypersensitivity or allergic reactions to *S. vulgaris*. A diarrhoeic syndrome in field cases of verminous arteritis has also been described. The pathophysiology is poorly understood but it may be a response to altered intestinal circulation, local irritation and/or severe ulceration of the mucosa of the caecum and colon caused by thromboembolism. Clinical signs and lesions have been associated with aberrant larval migration in the aorta, coronary, iliacic, spermatic and renal arteries, and in the heart, kidney, brain and spinal cord.

The filarial worm *Dirofilaria immitis*, or "heartworm", is an important parasite of the dog. The adult worms are 12–30 cm long and live in the right ventricle of the heart and in the pulmonary arteries. Most cases with light infection remain asymptomatic. In heavier infestations, the presence of large numbers of heartworms does both mechanically interfere with the circulation through the right heart and lungs, and induces endarteritis with severe intravascular changes. Such heavy infection invariably leads to circulatory and respiratory distress. The mechanical interference with the circulation through the right heart generally causes a compensatory hypertrophy of the right ventricle. At the same time, adult worms and juveniles spread within the pulmonary arterial system and cause inflammatory reactions. They also induce the development of typical myointimal proliferative lesions which gradually reduce the lumen of small pulmonary arteries, leading to pulmonary hypertension. Endothelial damage, with thrombosis and thromboembolism has sometimes been reported. Chronic infections eventually lead to insufficiency of the right heart, which results in passive congestion of the liver, spleen and lungs, ascites and peripheral edema. The onset of clinical signs of dirofilariasis is insidious and may go unrecognized until substantial vascular damage has occurred. A deep, usually soft chest cough is a common early sign. Coughing may be aggravated by exercise, severe coughing being sometimes accompanied by hemoptysis, a sign highly suggestive of heartworm infection. In the late stages some dogs display characteristic respiratory movements: the rib cage remains expanded, and there is an extra inspiratory effort. The very poor exercise tolerance is a consistent clinical finding in advanced cases. Dogs which are suddenly forced to exercise may rapidly become ataxic or experience syncope. Emaciation develops gradually. The final stage of congestive failure of the right heart is only seen in a small proportion of affected dogs. A relative infrequent complication of *D. immitis* heavy infestations is the occurrence of pulmonary embolism by adult worms – or worms killed by chemotherapy – which occlude small pulmonary arteries and cause infarction. Also described in very heavy infections is the "liver failure syndrome" caused by worms which have invaded the vena cava and hepatic veins. The cause of the distribution is not known but it may be related to overcrowding in the normal pulmonary arterial habitat. The syndrome is characterized by sudden onset of weakness, anorexia, bilirubinuria, haemoglobinuria, haemoglobinaemia and anaemia. The anaemia develops due to disseminated intravascular coagulation and fragmentation of red cells. Azotaemia develops, and death usually occurs in 1 to 3 days. Another possible complication of *Dirofilaria* infection is immune complex glomerulonephritis. Microfilaria are usually of little consequence for the dog. If microfilariae block the arterioles of the skin there is erythema, pruritus, and loss of hair. This effect may however be due to concurrent dipetalonemiasis. Infections with *D. immitis* have also been reported in the cat, wild Canidae and Felidae, and rarely in humans.

Onchocerca armillata and *Elaeophora schneideri* are another two filarial parasite species which live in large blood vessels, produce microfilariae, and have biting insects as obligate intermediate hosts. *O. armillata* is focally highly prevalent in ruminants in Africa and Asia. In cattle it produces striking lesions in the intima and media of the thoracic aorta: tunnels, nodules, roughening and calcifications. However, the only cardio-vascular disturbances that have been reported appear to be slight aneurysmal changes, despite the presence in the aortic wall of worms which may be up to 70 cm long. Other symptoms, such as repetitive episodes of collapse and tetanic convulsions, periodic ophthalmic and even blindness sometimes observed in heavily infected animals have been attributed to the microfilariae. *E. schneideri* is common in wild ruminants and sheep in northern America. The adult worms live in the cephalic arteries of the host, while microfilariae are found in the skin, mainly in the head region. In elk and moose, the infection often causes inflammatory reactions and thrombosis of the cephalic arteries, which may lead to obstruction and ischaemic necrosis. The disease, if not fatal, may induce various neurologic signs such as blindness, and necrotizing lesions in the skin of the head. In sheep, arterial lesions are minimal or absent, but hypersensitivity to microfilariae often results in severe exudative dermatitis, primarily over the head.

Angiostrongylus vasorum and *Aelurostrongylus abstrusus* are small metastrongyle worms. The adult *A. vasorum* lives in the pulmonary artery and right ventricle of the dog where it causes a proliferative endarteritis and trombosis in pulmonary arteries comparable to that induced by *D. immitis*. *A. abstrusus* is a lungworm of cats. In addition to causing pulmonary nodules, it is believed to incite the smooth muscle hypertrophy and medial hyperplasia of the pulmonary arteries which may focally be observed in up to 70% of cats. The respiratory clinical signs of these two infections are described in Respiratory system diseases, horses, swine, carnivores.

The spiruroid worm *Spirocerca lupi* is primarily a parasite of dogs. The major clinical signs of spirocercosis are associated with the presence of adult worms in the oesophagus. They are described in → Alimentary System Diseases, Carnivores. However, significant lesions are also caused by the migrating larvae which spend about 3 months in the adventitia and media of the aorta. Aortic lesions include the formation of small nodules containing larvae, thickening of the intima and media, and deposition of atheromatous plaques. Weakness of the aortic wall and partial rupture of the layers sometimes leads to the development of shallow aneurysmal pouches, with possible rupture and fatal hemorrhage. In most cases, aortic lesions do not produce any clinical signs.

Veins Schistosomes are important parasites of the venous vascular system of vertebrates (Table 1). In Africa, *Schistosoma bovis*, *S. mattheei* and *S. curassoni* occur commonly in ruminants, and *S. rhodhaini* has been reported to infect dogs. In Asia, *S. indicum*, *S. spindale* and *S. nasale* occur in a variety of domesticated animals including cattle, water buffalo, sheep, goat, and horse. *S. incognitum* has been reported in pigs, dogs, sheep, goats, and occasionally cattle. Finally, *S. japonicum* is an important zoonosis in southeast Asia. In China, dogs, goats, rabbits, cattle, sheep, pigs, horses, and water buffaloes have all been found to harbour the infection (→ Schistosomiasis, Animals). All these parasites live in the mesenteric and hepatic veins of the host, except for *S. nasale* which is found in the veins of the nasal tissue. The presence of adult worms in the veins does not induce reactions from the host, except when dead worms are occasionally blocked into small veins of the intestine or swept into the liver where they give rise to focal inflammatory reactions (phlebitis) and lymphoid nodules. Such lesions are only of clinical consequence after chemotherapy. Schistosomes are relatively large worms (>10 mm) and the sudden accumulation of considerable numbers of them (up to several tens of thousands) in the portal veins may cause diffuse

Table 1. Parasites inducing myocarditis and vasculitis (according to Vercruysse and De Bont)

Parasite	Host	Location	Clinical presentation	Principal lesions
Protozoa				
Trypanosoma cruzi	Man, wild mammals, dog, cat, pig	Leishmanial forms preference for skeletal and cardiac muscle	Systemic disease, cardial dysfunction as tachycardia ascitis, weak pulse, lethargy	Heart: mild multifocal to diffuse severe granulomatous myocarditis
Trematoda				
Schistosoma curassoni, *S. bovis*, *S. mattheei*	Ruminants, horses	Mainly mesenteric and portal veins	Mucoid and haemorrhagic diarrhoea, anorexia, loss of condition, general weakness, anaemia, hypoalbuminaemia	Granulomatous lesions in intestine and liver
Nematoda				
Angiostrongylus vasorum	Dog, fox	Pulmonary artery	Dyspnea	Granulomatous interstitial pneumonia, endoarteritis
Dirofilaria immitis	Dogs and rarely other animals	Pulmonary aretries and chambers of the right heart	Deep soft chest cough, loss of exercise tolerance, emaciation	Intense endovascular reaction of the pulmonary arteries sometimes in hepatic veins
Elaeophora schneideri	Sheep, deer, elk	Cephalic arteries	Hyperplasia and occlusion of cephalic and other arteries, microfilariae cause dermal lesions	Mild intimal sclerosis to prominent fibrosous thickening of the vessel wall
Onchocerca armillata	Cattle	Thoracic aorta	Microfilariae cause epileptiform signs and ophthalmia	Nodules, roughening and calcifications in the aortic walls
Spirocerca lupi	Dog	Aorta	Usually no clinical signs during the prepatent phase	Local thickening of the intima and media of the aorta
Strongylus vulgaris	Equines	Anterior mesenteric artery and the adjacent aorta and other trunks	Pyrexia, anorexia, weight loss, dullness, colic and death in severe infections, otherwise varying pyrexia, colic, diarrhoea	Verminous arteritis, thrombus formation, infarction of colon and caecum

phlebitis and extensive thrombosis. The pathogenesis of schistosome infection is mainly caused by the migration of millions of eggs through the intestinal wall, and their accumulation in the liver. These clinical syndromes are discussed under → Alimentary System Diseases, Ruminants.

Lymphatic Vessels Filariid worms of the genus *Brugia* parasitize the lymphatic system of dogs and cats in tropical areas. They do not usually cause lymphoedema and elephantiasis as in humans. *Dracunculus insignis* may occasionally infect dogs in North America, causing obstruction and inflammation of the lymphatic vessels, leading to lymphoedema.

Therapy

→ Chemotherapy, → Drugs.

Ceratopogonidosis

Disease due to bite of midges (Table 1).

Cercarial Dermatitis

Synonym

Swimmers' Itch.

Skin reaction in humans due to penetration of cercariae of the schistosomal flukes *Trichobilharzia* spp., *Bilharziella* and relatives, which use birds as final hosts. *Trichobilharzia* cercariae die in the skin leaving inflamed and itching pustules, which disappear after some days.

Cerebral Malaria

→ Insulin (Vol. 1).

Cestode Infections

General Information

→ Cestodes (Vol. 1) live as intestinal parasites firmly attached to the mucosa of the gut in their definitive hosts where they can live for years (e.g., → Taeniasis, Animals, → Taeniasis, Man). Gravid proglottids or eggs are released in the feces and ingestion of eggs by susceptible intermediate hosts results in the release of larvae, also called oncospheres within the gastrointestinal tract. The oncospheres penetrate the intestinal mucosa and migrate into tissues of the host where they can develop into mature metacestodes (e.g., → Cysticercosis, → Echinococcosis). Important cestode infections are listed in → Platyhelminthic Infections, Man, Pathology.

Immune Responses

While relatively little is known on immunity against the adult tapeworms in the gut, mechanisms of host-parasite interactions and the immune response have been extensively studied for the tissue stages. In larval cestode infections the intermediate mammalian host harboring egg-derived metacestodes in the tissues becomes completely immune to reinfection with eggs. In contrast, autoinfections with eggs occur in the case of → *Taenia solium* (Vol. 1) when the host is initially infected with metacestode-derived adult tapeworms in the gut lumen.

Therapy

→ Cestodocidal Drugs.

Cestodocidal Drugs

Table 1, Table 2; page 104 ff

Table 1. Midges and Control Measurements

Parasite	Host	Vector for	Symptoms	Country	Therapy		
					Products	Application	Compounds
Culicoides spp. (biting midges, no-see-ums or punkies)	Ruminants, Horse	Blue tongue virus (Cattle, Sheep), African Horse sickness virus and *Onchocerca cervicalis* (Horse)	Edema, allergic reactions	World-wide	1% Vapona insecticide™ (Durvet)	Spray	Diclorvos

Table 1. Drugs used against cestode infections of domestic animals.

Chemical group, Nonproprietary name (approx. dose, mg/kg body weight, oral route) other information	***Brand name** (manufacturer, company); other information	**Characteristics** (chemotherapeutic effects, adverse effects, miscellaneous comments)
ALKALOIDS		
arecoline hydrobromide (1–1.5; repeated administration may be necessary) in veterinary use since 1921	*Hydarex (Parke Davis) bitter-tasting powder; addition of sucrose (15%) is recommended; may be given as 1.5% solution	"oldtimer" that remains useful for the diagnosis of cestode infections in dogs (*Taenia*, *Echinococcus* spp.); it has strong parasympathomimetic actions, causing purging which removes paralyzed worms from the intestine (should not be used in cats, pregnant bitches, or in pups less than 6 months of age); its safety margin is narrow; atropine sulfate (0.04 mg/kg) may be used as antidote and does not interfere with the cestocidal action of arecoline; its action on *E. granulosus* is variable; there may be high efficacy against *T. ovis*, *T. pisiformis*, and other *Taenia* spp., and/or *Dipylidium caninum* if purging occurs sufficiently; purging is a local effect resulting from cholinergic action on gastrointestinal muscle and glands
arecoline acetarsol (5)	*Cestarsol (May and Baker) *Tenoban (Wellcome) *Nemural ((Winthrop) tasteless powder	the powder is relatively well tolerated in cats and dogs (is not recommended for pregnant animals, pups less than 3 months of age and cats less than 6 months of age); its action is similar to that of the hydrobromide and it should be given to dogs after a light meal; the complex is hydrolyzed in stomach, releasing arecoline which causes paralysis of worm and purging
INORGANIC COMPOUNDS		
di-*n*-butyl *tin* dilaureate (100–125, given in feed) tin compounds have been used in **humans** over some centuries; today the regimen (daily dose over several days) limits their use in man lead arsenate	*Davainex *Tinostat and others crystals or liquid	tin compounds are 'oldtimers' with moderate cestocidal activity (70–90%) against tapeworms in humans; their action may depend on coating tapeworm's cuticle with a thin layer of tin particles which renders strobila susceptible to digestion; di-*n*-butyl tin laureate is still used occasionally in **poultry** flocks and in cage birds; today the compound has been largely replaced by niclosamide; it is highly effective against *Raillietina* spp.; its activity against *Choanotaenia* spp. and *Davainea proglottina* is somewhat erratic; drug is well tolerated at recommended dose but may cause a temporary drop in egg production; **lead arsenate**, first used as an insecticidal drug, has been found by chance to be effective also against tapeworms of sheep; for a long time it has been used world-wide as a inexpensive anticestodal drug with almost 100% effectivity against *Monezia* infections in lambs, kids and calves (single dose of 0.5 and 1g/ head); its safety margin is low, and 2g/head daily for 2 days may cause mortality in sheep; the drug is toxic in poultry

Table 1. (Continued) Drugs used against cestode infections of domestic animals.

Chemical group, Nonproprietary name (approx. dose, mg/kg body weight, oral route) other information	***Brand name** (manufacturer, company); other information	**Characteristics** (chemotherapeutic effects, adverse effects, miscellaneous comments)
PHENOL DERIVATIVES		
dichlorophen(e) (200 sheep; 220 dog, cats) 12h fast before treatment; introduced in 1946 + **toluene**; **bithionol** (~200x1 oral, cats, dogs, sheep, goats, quail; 2x200, 4 days interval, chicken; 600x1 geese) **biothionol** sulfoxide (sulfene or sulphene)	*Dicestal (May and Baker) *Taeniathane (Pitman Moore) *Diphenthane-70 *Vermiplex, *Tri-plex, others *Bithin, others; chicken: in-feed other animals capsules, tablets, boluses,	gave an alternative to arecoline which was unpleasant in use; similar to niclosamide (both are phenol derivatives) it has bactericidal, fungicidal and anticestodal activity; exhibits variable activity (only destrobilating action) against *Taenia* spp. and *Dipylidium caninum* in dogs and proved 'ineffective' against *E. granulosus*; it has limited efficacy against *Moniezia* spp. in sheep; in small animal practice, **dichlorophen +toluene** (antinematodal drug) may be used in dogs for removal of ascarids (*Toxocara canis, Toxascaris leonina*), and hookworms (*Ancylostoma caninum, Uncinaria stenocephala, Dipylidium* and *Taenia*: sum of efficacies ~84%); dichlorophen shows low toxicity (LD50 in rats 2.6 g/kg); this is due to poor absorption of the drug from the alimentary tract; however, drug may be toxic in cats; adverse effects are vomiting and colic diarrhea; **bithionol** is a phenolic compound which may be used for treatment of common cestodes in dogs, cats (little effective against *D. caninum*), poultry (*Raillietina* spp. *Choanotaenia* in chickens, geese, quail), and *Monezia, Thysanosoma* and *Paramphistomum* (~85% efficacy) in ruminants (cf. → Trematodocidal Drugs); it is well tolerated in these animals but may stimulate purgation in dogs and cats; in contrast to bithionol, **biothionol sulfoxide** has the advantage of a lower therapeutic dose (60 mg/kg) against adult cestodes (adults are expelled intact); its antitrematodal efficacy against liver flukes in sheep and cattle is superior to that of the parent compound
HEXYLOXYNAPHTHAMIDINES		
bunamidine chloride (25–50) has been used worldwide after its introduction in 1965 against tapeworm infections of dogs and cats; it quickly replaced arecoline and dichlorophen for routine treatment	*Scolaban (Mallinckrodt) coated tablets tablets should not be crushed or dissolved in water or liquid prior to administration; there may be irritation of oral mucosa and as a result enhanced drug levels in blood plasma and thus unexpected toxic effects; detoxification of drug likely occurs in the liver	active against *Taenia* spp., *D. caninum* (effect varies), *Spirometra* spp., and *E. granulosus*, in cats and dogs; for some years it was the drug of choice against *E. granulosus* (50 mg/kg, repeated after 48 h, empty stomach) although its effect allows some worms to survive; it was replaced in hydatid control schemes by praziquantel about 10 years later; it exhibits variable activity against *Monezia expansa* in sheep, and ascarids in dogs and cats; drug affects tegument of *H. nana* causing disruption of the outer layer which leads to decrease in the rate of glucose uptake and an increase in the rate of glucose efflux; damaged outer layer allows digestion of the worm in the host; it is relatively well tolerated at recommended dose; side effects are transient diarrhea and occasional vomiting; cases of sudden death have occasionally been seen in dogs without evidence of hepatic dysfunction; latter may lead to higher levels of bunamidine in the circulation; in excited dogs high levels of epinephrine may then cause ventricular fibrillation in heart sensitized by bunamidine to endogenous catecholamines; excitement and exertion should be avoided after treatment; reduced spermatogenesis was found in dogs but not in cats; the drug is well tolerated in bitches (all stages of pregnancy)

Table 1. (Continued) Drugs used against cestode infections of domestic animals.

Chemical group, Nonproprietary name (approx. dose, mg/kg body weight, oral route) other information	***Brand name** (manufacturer, company); other information	**Characteristics** (chemotherapeutic effects, adverse effects, miscellaneous comments)
bunamidine hydroxynaphthoate (25–50) has been used in the UK as a drench in ruminants	*Buban (Mallinckrodt) is less irritant to mucous membranes than the hydrochloride	exhibits marked efficacy against *Moniezia expansa* and *M. benedeni* in naturally infected sheep and goats; metaphylactic treatment (single dose) is during spring/summer and in autumn if reinfection occurs; the drug is tolerated in sheep and goats at twice the recommended dose; given in feed, drug shows variable activity against *Taenia* spp. in dogs; it was tested in *poultry* with natural infection of *Raillietina* spp. and *Amoebotaenia sphenoides* and was found to be highly active at 400 mg base/kg
SALICYLANILIDES		
niclosamide (monohydrate) (100–150) **derivatives** have been prepared mainly by Russians as various salts and esters; among them the piperazine salt **phenolsulfonphthalein** (PSP) is ~ 2 times more effective against *M. expansa* in lambs than the parent drug ; in countries of the earlier USSR it is still widely used in animals and humans	*Yomesan (Bayer) was marketed for use in cats and dogs, tablets *Mansonil (Bayer) for use in other domestic animals *Various others good tolerability, a wide safety margin and excellent efficacy against *Taenia* in humans and animals have accounted for its widespread use as a taeniacide in human and veterinary medicine	announced in 1958 and introduced into the market in 1960; since then it has been used world-wide against tapeworm infections of animals and humans, and as a molluscicide (WHO has recommended the drug for control of fresh water snails infected with *Schistosoma* spp. cf. 'Control of trematodes'); highly active against tapeworm infections in dogs, cats, ruminants, and poultry; shows marked activity against *Taenia* spp. but erratic activity against *D. caninum, Mesocestoides corti,* and poor efficacy against *M. lineatus, E. granulosus* in dogs; immature stages *E. granulosus* may be susceptible to 50mg/kg given on 2 consecutive days; the drug exhibits excellent activity against *Moniezia* spp., *Thysanosoma actinoides* (fringed tapeworm), *Thysaniezia giardi* and *Avitellina* spp. in ruminants and is widely used against infections of anoplocephalids in horses (80–100 mg/kg) and those of *Hymenolepis* spp., and *Raillietina* spp. in birds (100 mg/kg) as well as cestode and skin fluke (*Gyrodactylus*, 0.1mg/l, 60 min) infections of fish (*Bothriocephalus*: 40mg/kg daily over 3 days, or a 0.5% in-feed concentration); it is a very safe drug (poor gastrointestinal absorption) also during pregnancy and lactation in cattle and sheep; dogs and cats appear to be more sensitive to the drug although twice the normal dose is well tolerated; the drug should not be used in combination with organophosphate compounds (toxic effects); mode of action: niclosamide inhibits the formation of mitochondrial energy, i.e., oxidative phosphorylation; in susceptible adults uptake of oxygen and glucose is blocked; its major action takes place in the scolex and proximal segments; drug produces spastic and/or paralytic action on various preparations of helminths, e.g., *D. caninum, F. hepatica*
niclosamide: +levamisole; +oxibendazole;	*Stromiten, others *Vitaminthe (Virbac) and others	niclosamide may be combined with nematocidal drugs to have an advantage in that both nematodes and cestodes can be treated in dogs and cats (for further information on dosages, brand names, companies, and characteristics of these drug combinations see → Nematocidal Drugs, Animals).
4'-BROMO-γ-RESORCYLANILIDE		
Resorantel (65, drench)	*Terenol (Hoechst, discontinued) powder; commercially available in parts of Europe and Russia	is highly effective (95–100%) against various cestodes (e.g., *Moniezia* spp. infections in sheep and cattle, *Thysaniezia giardi* and *Avitellina* spp. infections in sheep); the drug is also highly effective (90% against mature and immature rumen flukes (*Paramphistomum cervi*) in cattle, sheep, and goats (cf. Trematodocidal Drugs); there may be slight, transient diarrhea following treatment at recommended dose; the drug is excreted quite rapidly (total residues 3 days after treatment is about 0.1% of total)

Table 1. (Continued) Drugs used against cestode infections of domestic animals.

Chemical group, Nonproprietary name (approx. dose, mg/kg body weight, oral route) other information	***Brand name** (manufacturer, company); other information	**Characteristics** (chemotherapeutic effects, adverse effects, miscellaneous comments)
SUBSTITUTED DIPHENYLETHER		
nitroscanate (50, micronized) mode of action in cestodes appears to be an uncoupler of oxidative phosphorylation	*Lopatol *Cantrodifene (Ciba Geigy) tablets (micronized particles) the drug inhibits ATP synthesis in *Fasciola hepatica*	broad-spectrum compound introduced for use in dogs in 1973; it is active against **roundworms** (*Toxocara canis*, *Toxascaris leonina*), **hookworms** (*Ancylostoma caninum*, *Uncinaria stenocephala*), and tapeworms (*Taenia* spp., and *D. caninum*) of dogs (cf. also → Nematocidal Drugs, Animals); its action on whipworms is poor and that on *E. granulosus* is somewhat erratic at recommended dose; even at repeated doses of 200 mg/kg, total elimination of adult *E. granulosus* is not always achieved; therefore it is not recommended against these worms; the drug is poorly absorbed from gastrointestinal tract but irritates gut's mucosa resulting in relatively high incidence of vomiting (10%–20% of treated dogs) within 3–5 h after treatment; fasting, 12–24 h prior to treatment followed by a small quantity of food thereafter will markedly reduce vomiting; nitroscanate should not be used in cats, as it frequently provokes adverse side effects at therapeutic dose;
PYRAZINOISOQUINOLINES		
praziquantel (5 p.o., s.c., i.m. dog and cat.; *Spirometra* spp., *Diphyllobothrium. latum*: 25 on 2 consecutive days,) (ruminants, 2.5–10) praziquantel has been introduced in 1975 for treatment of cestode infections in cats, and dogs, (humans) drug of choice for treatment of schistosomiasis in humans (see → Trematodocidal Drugs); it has no action on nematodes	*Droncit (Bayer) tablets, pellets, solution for injection, spot on *Fluxacur (Intervet) *Parkevermin (Pharmacia & Upjohn) *Biltricide (Bayer) coated tablets for use in humans only *Cesol; Cestox; Cisticid (Merck) coated tablets for use in humans only *Cestocur (Bayer) for sheep	there is almost complete absorption from the alimentary tract following oral administration; the drug is conveyed throughout the body and reaches high plasma levels in tissues of almost all organs; this puts it in a position to be in close contact with larval and adult stages of cestodes that have highly varied locations in the host; its major pathway of biotransformation is the liver; inactive drug metabolites are excreted mainly by the liver into bile; the drug is extremely active in a single oral, subcutaneous(being less effective), or intramuscular dose (dose depends on cestode infection) against juvenile and adult tapeworms of carnivores as *Taenia* spp., *D. caninum*, *D. latum*, *Spirometra* spp. (*S. mansonoides*, *S. erinacei*), *M. corti*, *E. granulosus*, *E. multilocularis*, bile duct cestodes (e.g., *Stilesia hepatica*, *Thysanosoma actinoides* in ruminants), and a range of cestodes in ruminants, birds, snakes and fish (skin fluke *Gyrodactylus*); *Taenia* and *Diphyllobothrium* infections in *humans* are eliminated by oral doses of 10 and 25 mg/kg, respectively; there is some ovicidal action on eggs of *E. granulosus* released from the proglottid but not against those contained in the proglottid (this effect has therefore little epidemiological value); its activity against larval forms (hydatid cysts) of *E. granulosus* is variable and benzimidazole carbamates (mebendazole, albendazole) prove to be more active against hydatidosis in man; in cattle and sheep its action on cysticercus of most *Taenia* spp. is about 100%; this is also true for intermediate stages of the human tapeworm (*T. saginata*) in cattle; it is a safe drug when given orally or parenterally (no embryotoxic or teratogenic effects in rats or rabbits); the drug is rapidly taken up by cestodes and trematodes; however, uptake of the drug is no guarantee of therapeutic activity (e.g., *F. hepatica* is unaffected by the drug); action of praziquantel results in a rapid vacuolization of tegumental layer in the growth zone of the neck region of cestodes; vacuolization leads to disruption of the apical tegument layer (molecular mechanism leading to these alterations is till now not well understood); contraction of parasite musculature (spastic and/or paralytic) depends on drug concentrations used in isolated host tissue preparations; contraction of *H. diminuta* muscle depends on endogenous $Ca2+$ (as in vertebrate skeletal muscle); contraction of *S. mansoni* muscle depends on uptake of external $Ca2+$ (as in vertebrate smooth muscle)

Table 1. (Continued) Drugs used against cestode infections of domestic animals.

Chemical group, Nonproprietary name (approx. dose, mg/kg body weight, oral route) other information	***Brand name** (manufacturer, company); other information	**Characteristics** (chemotherapeutic effects, adverse effects, miscellaneous comments)
praziquantel + nematocidal drugs:		
+pyrantel	*Drontal cat (Bayer)	antinematodal drugs may be active against certain cestodes and flukes (cf. → Trematodocidal Drugs); some cestocidal benzimidazole carbamates have good activity against *Taenia* spp. in cats and dogs; however , none of these drugs are sufficiently effective against *Dipylidium caninum* in small animals (for more details see this table under benzimidazole carbamates); pyrantel pamoate is widely used to treat nematode infections (cf. → Nematocidal Drugs, Animals, aninematodal drugs in horses); at 13,2 mg/kg it proved to be active against common equine tapeworms although its efficacy against common tapeworms in small animals is lacking; commercially available combinations of praziquantel with these compounds make it possible to deworm animals for tapeworms and certain nematodes simultaneously; the majority of these all-wormers are for use in dogs and cats; they are highly effective against tapeworms and have different anthelmintic spectra with varying activities against nematodes like roundworms, hookworms, whipworms, and heartworms; the characteristics of these drug combinations, and other information on drugs are given in more details in → Nematocidal Drugs, Animals (antinematodal drugs in dogs)
+pyrantel +febantel	*Drontal Plus for dogs	
+pyrantel +oxantel	*Canex cube (Pfizer)	
+fenbendazole	*Caniquantel Plus (Animedica)	
+mebendazole		
(+ ivermectin;)	under development for heartworm infections in dogs	
+abamectin;	*Equimax (Virbac) for use in horses	
HYDROPYRAZINOBENZAZEPINE		
epsiprantel (cat > 7week old 2.25) (dog > 7week old 5.5) (*Echinococcus granulosus*: 7.5) chemically related to praziquantel; since 1989 on the market (first USA, early 90's Canada and Taiwan)	*Cestex (Pfizer Animal Health), tablets	unlike praziquantel, it is poorly absorbed from the gastrointestinal tract and there seem to be no detectable metabolites in the urine of dogs; this puts the compound in a position to be in contact with tapeworms of the alimentary tract for a longer time; drug is used for removal of common cestodes of dogs (*Dipylidium caninum*, *Taenia pisiformis*), and cats (*D. caninum*, *T. taeniaeformis*); unless exposure to infected intermediate hosts is controlled, reinfection will be likely and consequently a retreatment; in cases of *D. caninum* infections an additional flea control program should be applied therefore; immature/adult stages of *Echinococcus granulosus* are eliminated by the drug (5 mg/kg : 94/99%; 7.5 mg/kg: 100%); there is no other information on efficacy against common tapeworms in livestock; the drug is claimed to have a wide safety margin in dogs and cats: at 5 times the normal dose (once daily for 3 days) no adverse effects have been observed; at 40 times the recommended dose (once daily for 4 days) only slight clinical signs were seen; animals under seven weeks of age should not be treated with the drug; mode of action is not yet known on a molecular level but assumed to be similar to that of praziquantel; the tegument of tapeworms becomes damaged, rendering it susceptible to lysis and digestion by the host
epsiprantel + pyrantel (cat, dog)	*Dorsalid (Pfizer)	

Table 1. (Continued) Drugs used against cestode infections of domestic animals.

Chemical group, Nonproprietary name (approx. dose, mg/kg body weight, oral route) other information	***Brand name** (manufacturer, company); other information	**Characteristics** (chemotherapeutic effects, adverse effects, miscellaneous comments)
BENZIMIDAZOLE CARBAMATES		
fenbendazole (dog 50 for 3 days) (sheep 5–10) (horse 10 for 3–5 days) **luxabendazole** (sheep 7.5–10) **mebendazole** (dogs 22 for 5 days) horse 20 for 5 days) **albendazole** (sheep, cattle, 7.5) **oxfendazole** (sheep, cattle, 5) (for formulations, and other drugs see → Nematocidal Drugs, Animals)	*Panacur (Intervet) (Hoechst, discontinued) *Telmin KH (Janssen) tablet, 100 mg *Vermox (Russia) *Telmintic (Pitman Moore) *Valbazen (Pfizer) *Synanthic (Norden) *Systamex (Mallinckrodt, Essex)	benzimidazoles carbamates are highly active antinematodal drugs (cf. → Nematocidal Drugs, Animals), that were developed after tiabendazole, had the 5-position blocked by various groups to slow down the rate of metabolism and excretion; the replacement of the thiazole ring by methylcarbamate may also lead to longer half-lives of compounds; **mebendazole, fenbendazole, oxfendazole** and **albendazole** have this pharmacokinetic property; this allows the compounds to be in contact with the parasites for longer periods, thus exhibiting a broader spectrum of activity; their marked effect against cestodes is less following a single oral dose in dogs and cats than in ruminants and horses; thus, a divided dose generally leads to increased activity as a result of longer persisting drug levels of the poorly absorbed benzimidazoles; **oxfendazole** and **fenbendazole** have a high activity against *Moniezia* spp. and *Thysanosoma actinoides* in sheep and calves; **albendazole** has the broadest anthelmintic spectrum with activity against nematodes, trematodes (liver fluke), and cestodes in sheep and cattle, particularly against intestinal cestodes; it is also effective against larval *T. saginata* in cattle at 50 mg/kg; **mebendazole** exhibits activity against larval hydatids in humans and against larval cestode infections in sheep, cattle, pigs (*Taenia hydatigena*), rabbits (*T. pisiformis*) and mice (tetrahyridia of *Mesocestoides corti*) ; **fenbendazole, albendazole** and **mebendazol** prove active against larval *Echinococcus* spp., *T. ovis*, and *T. hydatigena*; however, a long course of treatment with **mebendazole** or **albendazole** is usually necessary to achieve improvement of hydatid disease in humans; the cost of such treatment precludes their use in animals; mentioned benzimidazoles have a wide margin of safety; **oxfendazole** and **fenbendazole** are not compatible with **bromsalans** (cf. → Trematodocidal Drugs) in cattle (death may occur within 24–48 h); combinations of benzimidazole with praziquantel see above
ANTIBIOTICS		
paromomycin sulfate (identical to aminosidin sulfate and others) isolated in 1955 from *Streptomyces rimosus* var. *paromomycinus*; first use as cestocide in 1963	*Humatin (Parke-Davis)	**antimicrobial substance** with activity against gram-positive and gram-negative bacteria as well as intestinal protozoa (e.g., *Entamoeba histolytica*: → Antidiarrhoeal and Antitrichomoniasis Drug); in humans it proved to be effective against *Taenia saginata* and *T. solium* at doses of 40 mg/kg for 5 days or at a single dose of 75 mg/kg; *Hymenolepis nana* infections in man has been eliminated 100% at doses of 40 mg/kg for seven days; its further use became limited because of the availability of modern drugs with simple regimens for patients; in animals the antibiotic was shown to exhibit a promising effect against *T. taeniaeformis* (cat) but a less pronounced one against *Hymenolepis* spp. (rat, mouse); it is poorly absorbed from alimentary tract and may cause slight or moderate diarrhea; its anticestodal action (*T. saginata*) may be based on fine-structural changes of the tegumental membrane making the parasite susceptible to digestive system of host

Doses listed refer to information from manufacturer literature.
Data given in this Table have no claim to full information
The primary or original manufacturer is indicated if there is lack of information on the current one(s).

Table 2. Drugs used against cestode infections in humans.

Parasite DISEASE distribution, pathology	**Stages** affected (location), morphology of **eggs**	**Chemical class** other information	**Nonproprietary name** adult dosage (oral route), comments	**Miscellaneous comments**
INTESTINAL TAPEWORMS: adult worms consist of a head (scolex). a neck (growth region), and a strobila (series of segments = proglottids)				
DIPHYLLOBOTHRIASIS: first intermediate host is a copepod, second intermediate host are various fresh water fishes due to lack of specificity of the plerocercoid (sparganum lacking a bladder, see sparganosis); man becomes infected by ingestion of raw or undercooked fish-intermediate host containing plerocercoids; adult tapeworm may live for 25 years or longer in humans and may cause rarely vitamin B12 deficiency (due to absorption of B12 by the worm) and megaloblastic anemia; *D. latum* eggs may be confused with those of *Paragonimus westermani* (cf. → Trematodocidal Drugs)				
Diphyllobothrium latum (fish tapeworm), common where cold, clear lakes are abundant; occurs in N Europe, N America, Japan in fish-eating mammals, and man	**adults** (5–15 m long) (upper half of the intestine) **eggs** (unembryonated, ovoid, operculate, small knob at end opposite operculum) pass in feces into water	**pyrazinoisoquinolines**	**praziquantel** (5–10 mg/kg once); contraindicated in ocular cysticercosis	safe and well tolerated, occasionally skin rashes
		halogenated salicylanilides	**niclosamide** (2 g once)	safe drug, also in pregnant or debilitated patient; occasionally abdominal pain, pruritus
Dipylidium caninum: (dog tapeworm)	humans become infected by swallowing fleas containing infective larval stages (cysticercoids)			
distribution worldwide; infrequent in humans, pathogenic effects are unknown; infection is due to close contact with infected dogs and their fleas	**adults** (approx. 50 cm long) (small intestine) **eggs** (thin-shelled, small, spherical; contain six-hooked onchosphere) pass in feces (typical egg packets each containing up to 15 eggs)	**salicylanilides**	**niclosamide** (2 g once)	definite hosts (dogs, cats) should be regularly treated to avoid infection in humans (especially in children); simultaneous eradication of fleas (pets carpets, rugs, and other sites) is needed
		pyrazinoisoquinolines	**praziquantel** (5–10 mg/kg once)	
HYMENOLEPIASIS: humans become usually infected by ingestion of embryonated eggs (contaminated food and water); human infection also is possible by ingestion of arthropods (e.g., beetles) containing the infective larval stage (cysticercoid)				

Table 2. (Continued) Drugs used against cestode infections in humans.

Hymenolepis nana (dwarf tapeworm) occurs worldwide; infects simian primates, rodents, man; onchosphere can hatch also from embryonated *egg* within intestine causing **autoinfection**; this renders eradication difficult; amounts of cysticercoids may damage intestinal mucosa and produce diarrhea	**adults** (approx. 2–4 cm long) (lumen of small intestine; onchosphere undergoes development in lamina propria of a villus to a pre-adult cysticercoid) **eggs** (small, thin-shelled, spherical to sub-spherical, six-hooked larva surrounded by a membrane with two polar 'knobs' from which 4–8 filaments arise) pass in feces	**pyrazinoisoquinolines**	**praziquantel** Bittricide® (25 mg/kg once)	is active against both juvenile and adult stages;
		salicylanilides	**niclosamide** Yomesan® (2 g x 1d; 1g x 6d)	acts only against adults and mature cysticercoids, and thus treatment is necessary for 5–7 days;
		aminoglycosides	**paromomycin** (cf. → Cestodocidal Drugs) side effects are minimal; it is poorly absorbed from gut (<1%) and excreted unchanged in urine	treatment schedule of the antibiotic is similar to that of niclosamide: 40 mg/kg /d for 7d may cause 100% elimination of worm burden
H. diminuta (rat tapeworm) occurs worldwide; common parasite of rats and occasionally a parasite of man) Diagnostic problems: sometimes confusion with eggs of *H. nana*	**adult** (20–60 cm long) (small intestine) **eggs** (large, thick shelled. spherical, six hooked onchosphere, surrounded by a membrane without polar filaments and separated considerably from outer shell) pass in feces	**pyrazinoisoquinolines**	**praziquantel** (25 mg/kg once)	clinical manifestations are inconspicuous (unknown); prevention is rodent control and protection of food from insect intermediate hosts (infection may be acquired by accidental ingestion of infected beetles present in various grain products
		salicylanilides	**niclosamide** (2 g x 1d; 1g x 5–7d)	
TAENIASIS: humans become infected by ingestion of raw or undercooked muscles containing infective cysticercus; diagnostic problems: eggs of human and animal taeniid species (*Taenia* and *Echinococcus* spp) are all indistinguishable from each other; *T. saginata* and *T. solium* diagnostic is usually done by examination of gravid proglottids injected with India ink or stained by permanent stains to visualize characteristic number of lateral uterine branches (*T. saginata*: 15–30, and *T. solium*: 7–13); some pollen grains found in feces may closely resemble eggs of *Taenia* spp.; taeniid *eggs* contain a characteristic six-hooked embryo (onchosphere); caution should be used in handling unidentified gravid (mature) taeniid proglottids since eggs of *T. solium* and *Echinococcus* spp. are infective to humans and can cause larval tapeworm infections (see larval tapeworm infections)				
Taenia saginata (beef tapeworm), worldwide distribution (parallels stock keeping); adult worm causes no distinct lesions; cattle act as intermediate host	**adults** (up to 8 m long) (upper half of small intestine) **eggs** (embryonated spherical, yellow-brown, thick shell striated radially, 30–44μ ∅) pass in feces	**salicylanilides**	**niclosamide** (2 g once)	in most cases there are no clinical signs; on an individual level, infection is fully preventable by thorough cooking of meat
		pyrazinoisoquinolines	**praziquantel** (5–10 mg/kg once)	
		benzimidazole carbamates	**mebendazole** Vermox® (200 mg bid x3d)	

Table 2. (Continued) Drugs used against cestode infections in humans.

T. solium (pork tapeworm), worldwide distribution (parallels pig keeping); pigs and humans act as intermediate host; they become infected with embryonated eggs by ingestion of contaminated food or by autoinfection	**adult** (up to 6 m long) (small intestine) **eggs** (morphology see 'Taeniasis', and *T. saginata*) pass in feces; in contrast to those of and *T. saginata* they are infective to humans and may cause **cyticercosis**	**pyrazinoisoquinolines**	**praziquantel** (5–10 mg/kg once) (has also a moderate effect on larval stages)	niclosamide has no action on embryonated eggs; thus larvae liberated from dead (gravid) segments within upper intestine can cause cysticercosis; no purgatives should be used because purging increases risk of cysticercosis
		salicylanilides	**niclosamide** (alternative drug, 2 g once)	
		benzimidazole carbamates	**mebendazole** (200 mg bid x3d)	

LARVAL TAPEWORMS:
there are two types invading human tissues, (1) larva with a bladder, and (2) larva lacking a bladder (sparganum)

LARVAL CESTODE DISEASES:
man becomes infected by ingesting the onchosphere-containing (embryonated) eggs, or by ingesting the larval stages of tapeworms belonging to the genus *Spirometra* (for more details cf. sparganosis, page 113)

CYSTICERCOSIS, NEUROCYSTICERCOSIS
humans become infected by ingesting contaminated food containing *T. solium* eggs; *autoinfection* is possible; cyticerci with a fibrous tissue capsule may undergo calcification, thereby releasing antigens that cause inflammatory reactions)

T. solium (**larva type**: a single scolex invaginated into a bladder) cerebral cysticercosis may occur where pig keeping is done; prevention of the disease see 'Taeniasis' in this table)	**cysticercus cellulosae** (almost any tissue, often brain: tend to grow and cause space-occupying lesions; within a ventricle cysticercus may lead to hydrocephalus);	**benzimidazole carbamates**	**albendazole** Eskazole® (drug of choice, 400 mg bid x8–30d, repeated as necessary)	**surgical removal** of accessible lesions is the only directly effective therapy, and must include symptomatic therapy with anticonvulsant drugs; chemotherapy of neurocysticercosis must be further evaluated to define guidelines for use of such therapy
		pyrazinoisoquinolines (for more information on treatment, and contraindication of chemotherapy)	**praziquantel** (50 mg/kg/d in 3 doses x15d) (**mebendazole** metrifonate flubendazole experimental drug)	

COENURIASIS (infrequent):
adult worm occurs in dog, fox, coyote (definitive hosts are canidae); intermediate hosts are herbivores (cattle, horse, sheep, wild animals, rarely man; latter become infected by ingestion of contaminated food containing embryonated eggs; human coenurus disease mainly occur in Africa; space-occupying larva usually invades brain

Taenia multiceps (= *Multiceps multiceps*) (**larva type**: multiple invaginated scoleces into a bladder), infrequent	**coenurus cerebralis**, cyst contains several hundred protoscoleces (central nervous system, usually brain)	Prevention is not possible (reservoir hosts include numerous species of wild animals)	**no chemotherapy** is available against space-occupying lesions	treatment involves **surgery** if cyst is accessible

HYDATID DISEASE (hydatidosis): **larva-type**: unilocular hydatid cyst
infection of intermediate host (usually sheep and other herbivores such as cattle, horse, occasionally man) is due to close contact with dogs and other canids (definitive hosts) infected with large numbers of adult worms (3–6 mm long, single, gravid proglottid that is longer than wide, and contains typical *Taenia*-like eggs); feces of dogs contain embryonated eggs infective to intermediate hosts (occasionally man); cestode larval stage in human tissues is characterized by multiple daughter bladders or '**brood capsules**' with multiple invaginated protoscoleces budding from their walls (inner layer of germinal epithelium of cystic cavity); material consisting of death and degenerated scoleces in milky fluid of parent cyst is referred to as **hydatid sand**

Table 2. (Continued) Drugs used against cestode infections in humans.

Echinococcus granulosus in sheep and cattle-keeping countries; prevention of hydatidosis may be largely achieved if abattoir and disposal of infected offal are strictly controlled	**hydatid cyst, unilocular** (rarely multilocular) particularly in *liver*, and lung but also in other tissues depending on intermediate host; rupture of a cyst into the tissue results in dissemination and further growth of scoleces in new forming cysts	**benzimidazole carbamates** **pyrazinoisoquinolines**	**albendazole** (drug of choice, 400 mg bid x28d, repeated as necessary **praziquantel** (regimen see 'Cysticercosis') may be useful preoperatively or in case of fluid spill during surgery	some patients may require surgical resection of hydatid cysts (for more information on criteria for need of surgery, treatment, contraindication of chemotherapy and 'Percutaneous aspiration-injection and re-aspiration' = PAIR and possible complications see Drugs Acting on Adult and Larval Tapeworm Infections of Humans)
ALVEOLAR HYDATID DISEASE: **larva-type:** multilocular or alveolar hydatid cyst large numbers of adult worms (3–5 mm long) primarily occurs in fox (definitive host, occasionally cats, dogs, and wolves); several microtine rodents serve commonly as intermediate host (sylvatic cycle); major source of human infection (which are rare) is through fruits and vegetables contaminated by feces containing embryonated eggs of the taeniid type; in humans **metastasis** of the laminated membrane by the alveolar hydatid resembles lesion of a neoplasm; lesions of liver usually are only membranous; in contrast to hydatid of *E. granulosus*, neither protoscoleces nor hydatid sand (calcareous corpuscles) are identifiable in such cysts in humans				
Echinococcus multilocularis; *E. vogeli* (infrequent) causing **polycystic** echinococcosis	growth of **alveolar (multilocular) cyst** is peripheral and invasive; metastases are frequent (liver, adjacent tissue)	**benzimidazole carbamates** experimental drugs: **flubendazole, fenbendazole**	long-term treatment (3–24 months: **albendazole** (10 mg/kg), **mebendazole** (40–50 mg/kg); high doses are necessary to affect larva (their use has been suggested (WHO Group Bull WHO, 74:231, 1996)	no therapy is fully effective against tumor-like growth of **alveolar typ***e* of cyst; if early diagnosed, surgical excision of lesions is a reliable means of treatment;
SPARGANOSIS: **larva type**: solid-bodied larva lacking a bladder adult worm (pseudophyllidean tapeworm) occurs in cats, dogs, wild canids or felids and is of little significance to definitive hosts; first intermediate host is a copepod, second one any vertebrate due to lack of specificity of the plerocercoid; humans become infected (1) by ingestion of a copepod (crustacea) containing **procercoids** (first larval stage), (2) by ingestion of raw or undercooked flesh or organs of any vertebrate (amphibians, reptiles, mammals, particularly feral pigs raised for human consumption) containing **plerocercoids** (=**sparganum**, second larval stage, easily mistaken for nerves), or (3) by local application of flesh (poultice to wounds or to the eye) containing plerocercoids sparganum				
Spirometra spp. (infrequent), life cycle is similar to that of *D. latum* (fish tapeworm of humans, see this Table)	**procercoid** (migrate into subcutaneous tissue and musculature) **plerocercoid** (migrate into connective tissue of muscles, in abdomen, in hind legs; under peritoneum, pleura)	there is no reliable chemotherapy (**praziquantel** may show some larvicidal effects)	treatment **is surgical** prevention is difficult because of entrenched eating habits and other customs	larval worms cause during subcutaneous migration painful edema, urticaria, inflammation, fibrosis; spargana grow into irregular nodules of subcutaneous tissues

The letter stands for day (days)
Dosages listed in the table refer to information from manufacturer or literature (e.g. from Medical Letter (1998) 'Drugs for parasitic infections'. The Medical Letter (publisher), vol 40 (issue 1017): 1–12. New Rochelle New York).
Data given in this Table have no claim to full information .

Economic Importance and Epizootiology

→ Tapeworms (Vol. 1), which belong to the phylum Plathyhelminthes, are hermaphroditic, endoparasitic, elongate, flat worms without a body cavity or alimentary tract, a few millimeters to several meters in length. As a rule economic loss, resulting from cestode infections is less severe than that due to trematode or nematode infections. Adult stages of tapeworms living in the alimentary tract of the final host are remarkably benign although adults may be up to 8 or 15 m in length (e.g. *Taenia saginata*, *Diphyllobothrium latum*). The scolex of the strobilate stage in eucestodes is provided with holdfast organs (suckers = acetabula) which may be armed with hooks and a rostellum with two rows of hooks. Transmission of many important cestodes in livestock, such as *Taenia* spp. and *Echinococcus* spp., usually involves 'predator -prey' relationships between carnivores or omnivores (e.g., man) acting as final hosts and herbivores (food animals, occasionally man) serving as intermediate hosts. Food animals like ruminants and horses may also acquire adult tapeworms (*Moniezia* spp. or *Anoplocephala* spp.) by ingestion of arthropod intermediate hosts (mites of the family Oribatidae) with herbage. Large numbers of *A. perfoliata* and *A. magna* may cause clinical signs in horses and donkeys, e.g., catarrhal or hemorrhagic enteritis, ulcerative lesions, and occasionally perforation of the intestine. Although lambs, kids, and calves under 6 months of age may be substantially infected with *Moniezia* spp.; pathogenic effects of these tapeworm infections appear to vary considerably. Major economic impact may result from **zoonotic infections** caused by members of the family of Taeniidae (*Taenia* spp., *Echinococcus* spp.). Close contact of humans, dogs, and foxes (final hosts) with feedlot cattle and other ruminants, rodents and by chance, humans, acting as intermediate hosts can lead to larval tapeworm infections. Thus, humans infected with *T. saginata* may pass thousands of eggs daily, which may be transmitted to cattle directly in feed or water or via pasture. Pigs running loose scavenging for food and with easy access to human feces may become infected by ingestion of gravid proglottids. Invertebrates such as blowflies, beetles, or earthworms may disperse *Taenia* spp. eggs, and they will remain viable for about 6 months. Cysticerci such as *Cysticercus bovis* (infectious stage of *T. saginata*) or *C. cellulosae* (infectious stage of *T. solium*) develop primarily in skeletal and cardiac muscle. **Neurocysticercosis** of humans may occur by ingestion of *T. solium* eggs with contaminated food or by **autoinfection.** Autoinfection is possible if eggs are released in the upper intestine, and regurgitated into the stomach. Oncospheres released from digested eggs then reenter the intestine where they initiate the life cycle.

Economic loss in livestock may result also from condemnation of carcasses or offal as unsuitable for human consumption at the abattoirs. For instance, *C. bovis* in the musculature of cattle becomes infective about 10 weeks after infections and remains viable for up to 9 months. Humans become infected by the ingestion of raw or undercooked infected beef (Table 2). Despite the availability of more sensitive and specific immunodiagnostic tests (recombinant antigens, PCR techniques) their role for routine diagnosis of cyticercosis in cattle will be very limited. Such tests could be used, however, on a herd basis to find out whether a herd is free of cyticerci or infected. This would be a valuable approach in epizootiology of cysticercosis. Today there are no effective drugs, which may be used economically in the treatment of metacestode infections in livestock. **Prophylactic measures** should therefore always involve treatment of infected persons and management adjustments such as proper meat inspection. However, the latter measure is estimated to detect only about 50% of the infected carcasses. To kill cysticerci, beef carcasses must be frozen, e.g., for six days at –20 °C, or carcasses should be cooked. There are several reports that cattle and sheep (goats) can be successfully vaccinated against *T. ovis* and *T. saginata* (beef tapeworm) with recombinant antigens inducing high levels of protective immunity. This may give some hope for the commercial use of such vaccines against infection in the near future.

Through chemotherapy, complete elimination of *Echinococcus* spp. infections in dogs must be achieved because of the major agricultural and public health problems due to **hydatidosis** in intermediate hosts. Primarily sheep but also other herbivores such as goats, cattle, horses, camels (pigs), and by chance humans serve as intermediate hosts. There are phylogenetic variations in *Echinococcus*, i.e., several strains of *E. granulosus* differing in their morphology and their isoenzyme patterns. Some of these strains appear to be specific for a particular intermediate host and an endemic area. The horse strain does not appear to infect humans although the sheep strain does. Human hydatidosis (infectious stage of *E. granulosus*, cf. Table 2) is often associated with severe clinical signs, particularly if the brain (neurocysticercosis with calcifications) or heart is involved. This is also true in human alveolar → hydatidosis caused by *E. multilocularis* and *E. vogeli* (Table 2). Hydatidosis in domestic animals rarely produces clinical signs despite heavy infections. Adult *E. granulosus*, only 3.5–5.6 mm in length, is principally harmless to the dog although large numbers of adults may cause enteritis. There has been substantial progress towards successful **vaccination** against *E. granulosus*; recombinant antigens (e.g., 95 kDa-GST fusion) from *E. granulosus* generated more than 95% protection against challenge in vaccinated sheep.

Old Remedies and Modern Compounds with Cestocidal Activity

There are many old remedies showing more or less activity against adult tapeworms. Thus pumpkinseeds, powdered areca (fruits of betel palm, *Areca catechu*), kousso (flowers of an Abyssinian tree, *Hagenia abyssinica*), turpentine (oily mixture of exsudates from coniferous trees, especially longleaf pine), pomegranate root bark (tropical Asian and African tree, *Punica granatum*), and male fern (*Dryopteris filix-mas*) were used as anticestodal remedies. These and other plant products had been gradually replaced by arecoline (alkaloid obtained from seeds of betel palm), organic tin compounds, lead arsenate or dichlorophen(e) during the first half of this century. Since 1921, **arecoline** has been used in veterinary medicine for many years against *Echinococcus granulosus* and *Taenia spp.* in dogs. Because of its relative low efficacy and its severe side effects, it is no longer recommended as a therapeutic drug in dogs and cats. However, its strong parasympathomimetic action causes purging and thus partial removal of paralyzed worms from the intestine. This action makes arecoline a useful diagnostic agent, which may give valuable information on whether a group of dogs on a farm is infected with *Taenia spp.* or *Echinococcus spp.* or not.

Several 'modern' **synthetic compounds** are in current use for the control of tapeworm infections in livestock and pets (Table 1). Such drugs are for instance dichlorophen (mainly pets), niclosamide, resorantel, bunamidine, benzimidazole carbamates, nitroscanate (pets), pyrantel (e.g. horses), praziquantel and epsiprantel (preferably pets). They may exhibit high activity against immature and adult stages of intestinal tapeworms while praziquantel (Table 2) and benzimidazole carbamates (mebendazole, albendozole) show action against larval stages of certain cestodes. Thus, mebendazole and albendozole may improve clinical illness of hydatid disease (*Echinococcus* spp.) in humans following long-term treatment.

Drugs Acting on Adult Tapeworm Infections in Dogs and Cats

Echinococcus spp., *Dipylidium caninum* (highly active proglottids may cause anal irritation and thus scooting), *Diphyllobothrium latum*, and *Spirometra* spp. have public health implications (Table 2). Other cestode species occurring in dogs and cats are *Taenia* and *Mesocestoides*. Routine treatment with effective cestocidal drugs at intervals less than the prepatent periods of the parasites should be used in working dogs to reduce the incidence of *Echinococcus* spp. infections, particularly in rural areas. Other control measures should also be considered to prevent hydatidosis in livestock and man, such as control of home slaughtering of small ruminants and consequent condemnation of infected viscera of infected animals, and quarantine regulations for dogs.

The majority of compounds recommended for treatment in dogs and cats (Table 1), such as bunamidine HCl, some benzimidazole carbamates (e.g., mebendazole, fenbendazole), niclosamide, nitroscanate, praziquantel and epsiprantel, prove highly effective against adult stages of *Taenia* spp. and *D. caninum*. The action of **niclosamide** is, however, somewhat erratic against *D. caninum*. **Bunamidine** appears to be the only drug that shows efficacy against *Spirometra* spp., particularly against *S. erinacei* in dogs and cats. Like praziquantel, it also exhibits activity against *Mesocestoides* spp. but requires a ten times higher dosage than that drug. **Nitroscanate**, introduced in 1973 as a broad-spectrum agent for use in dogs, has marked activity against roundworms (*Toxocara canis*, *Toxascaris leonina*), hookworms (*Ancylostoma caninum*, *Uncinaria stenocephala*) and tapeworms (*Taenia* spp. and *D. caninum*). However, the drug is not recommended for use against *E. granulosus* infection since 100% elimination of adults and juvenile worms is not achievable. **Praziquantel** (Table 1, Table 2) is a safe and highly active drug against a broad range of mature and immature tapeworms (including *Echinococcus* spp.), and most trematodes (particularly *Schistosoma* spp. cf. → Trematodocidal Drugs). **Epsiprantel** one of the latest cestocidal products on the market is chemically related to praziquantel. It exhibits good activity against common tapeworm infections in cats (*D. caninum* and *Taenia taeniaeformis*), and dogs (*Taenia pisiformis* and *D. caninum*), and shows an excellent action on adult *E. granulosus*. The drug is poorly absorbed from the gastrointestinal tract allowing the compound to be in contact with the parasites for a prolonged period.

There are several combinations of praziquantel with antinematodal compounds. **Febantel + pyrantel + praziquantel** (Drontal Plus for dogs) is an all wormer against intestinal nematodes and cestodes. The fixed combination of **praziquantel + pyrantel** (Drontal® for cats) has a high activity against *Ancylostoma caninum* and cestodes but only moderate activity against *Toxocara canis* and *Trichuris vulpis*. **Albendazole + praziquantel** may control various tapeworm infections (*Taenia* spp., *D. caninum*, *Echinococcus* spp.) and nematode infections (roundworms such as *Toxocara* spp., hookworms as *Ancylostoma* spp., the whipworm *Trichuris vulpis*, and *Strongyloides* spp.). **Ivermectin + praziquantel** currently under development is designated for heartworm prophylaxis (*Dirofilaria immitis*) and treatment of tapeworm infections in dogs.

Drugs Acting on Adult and Larval Tapeworm Infection of Sheep and Cattle

There are several species (e.g., *Moniezia expansa*, *M. benedeni*, *Avitellina* spp., and *Thysaniezia giardi*) which occur in the intestine of sheep, goats, cattle, and other ruminants in most parts of the world. *Stilesia hepatica* and *Thysanosoma actinioides* (the fringed tapeworm) are found in the bile ducts of cattle, sheep, and wild ruminants (e.g., deer) in Africa and Asia, or in North and South America (*Thysanosoma actinioides*). None of these relatively large species appears to be very pathogenic, and therefore there may only be economic return from treatment of lambs and calves if the administered compound is effective against both cestodes and nematodes. Such drugs include the **benzimidazole carbamates** (e.g., albendazole, oxfendazole); they exhibit a high efficacy against various intestinal nematodes (→ Nematocidal Drugs, Animals) and cestodes (Table 1) in food animals. **Niclosamide** exerts good activity against *Moniezia* spp. and *Avitellina* spp. in ruminants. **Albendazole, fenbendazole** and **praziquantel** are also effective against cestodes occurring in the bile ducts. However, enhanced doses (double and four times the recommended dose) of these drugs are necessary to eliminate the tapeworms.

High doses of praziquantel (20–40 times the recommended dose) are required to affect *T. saginata* or *T. hydatigena* **larval stages** in cattle and sheep; cost for treatment precludes, however, the use of praziquantel in the field. The same is true for albendazole showing action on larval *T. saginata* in cattle at ten times the recommended dose. The mode of action of the **benzimidazole carbamates** is generally thought to be primarily due to alterations in microtubule polymerization via a direct binding of these compounds to parasite tubulin. Thus, the co-administration of **fenbendazole** and the phenylguanidine **netobimin** (a prodrug of albendazole, cf. → Nematocidal Drugs, Animals, → Nematocidal Drugs, Man) has been shown to inhibit solidly the regenerative capacity of hydadit material in gerbils. Recurrence of hydatidcyst never took place when using co-administration of both drugs. Different effects have been observed after the administration of **praziquantel** in laboratory animals against larval stages of *Echinococcus*. Thus it affects protoscoleces of *E. multilocularis* but does not inhibit growth of cysts of both *E. multilocularis* and *E. granulosus*.

Drugs Acting on Adult Tapeworm Infections in Horses and Donkeys

Clinical signs associated with tapeworms in equines are uncommon. Occasionally *Anoplocephala magna* and *Paranoplocephala mamillana* occurring in the small intestine and rarely stomach may cause catarrhal (rarely hemorrhagic) enteritis. *A. perfoliata* found in the large and small intestine is often localized near the ileocecal orifice and ulcerative lesions and edema may be produced where the scoleces are attached to the cecal wall. Clinical signs (e.g., diarrhea) are usually seen in autumn and spring when infected oribatid mites may be present in great numbers. Because of the low pathogenicity of these tapeworms, treatment should be considered carefully since use of a compound that acts only specifically is unlikely to be economic. It may be of value to use a compound that combines activity against both cestodes and nematodes. Thus, **pyrantel embonate** (an antinematodal agent, cf. → Nematocidal Drugs, Animals) has been found to have a distinct activity against *A. perfoliata* at twice the nematocidal dose. It is only variably active against *A. mamillana* (Strongid-P paste, 38 mg/kg). However, in confirmed cases of equine cestodiasis **praziquantel** (Droncit Injectable given by stomach tube, 1 mg/kg: *A. mamillana*; 0.5 mg/kg: *A. perfoliata*,) exhibited a marked effect against these tapeworms. **Fenbendazole** (3x 10–20 mg/kg, cf. → Nematocidal Drugs, Animals) and other related benzimidazole carbamates also show action on equine tapeworms. However, widespread benzimidazole resistance of equine nematodes may limit the economic value of such a therapy. **Niclosamide** (80–100 mg/kg body weight) has good activity against anoplocephalids of all ages (Table 1). A combination of **praziquantel + abamectin** (Equimax paste, 1x 5.4g paste/100 kg) licensed for use in horses in Australia (cf. → Nematocidal Drugs, Animals) may be used against cestodes, nematodes and bots. Further combinations of praziquantel with 'endectocide' (milbemycins and avermectins) seems to be in the pipeline and will probably be launched soon (patented **ivermectin + praziquantel** combination).

Drugs Acting on Adult Tapeworm Infections in Birds

In the poultry industry the meat production has increased continuously worldwide (cf. → Coccidiocidal Drugs/Economic Importance). In 1994, the annual meat production amounted to ~13 million tons in the USA, ~6.1 million tons in China, and 628.000 tons in Germany. In birds, losses due to moderate and seldom heavier tapeworm infections may be diarrhea (enteritis), weight depression, emaciation, and rarely mortality. Free-range birds, e.g., those in backyard flocks, and cage birds in aviaries with earthen floors are often hosts to many species of tapeworms (e.g., *Davainea proglottina*, *Raillietina* spp., *Cotugnia* spp., *Amoebotaenia cuneata*, *Choanotaenia infundibulum*, *Hymenolepis* spp., *Fimbriaria* spp., and other cestodes). Their life cycles require the development of cysticercoids (larval stages) in a large number of intermediate hosts such as various copepods, snails and insects. Modern husbandry methods may largely prevent access to the various intermediate hosts, thus preventing tapeworm infections in commercial poultry. Consequently, only little information is available about adequately tested compounds in birds. **Niclosamide** (250 mg/kg, per os) seems to be effective and safe in most cases (can be toxic to geese). **Praziquantel** (5–10 mg/kg) is effective against a wide range of immature and mature cestodes in poultry, waterbirds (ducks, geese) and game birds (pheasant, partridge). However, an economic return to its use seems to be questionable. **Benzimidazole carbamates** (fenbendazole, mebendazole, other compounds, → Nematocidal Drugs, Animals), exhibit variable actions on avian cestodes. Fenbendazole is somewhat erratically effective against *Davainea proglottina*, and the activity of mebendazole is limited to *Raillietina* and *Hymenolepis*.

Drugs Acting on Adult and Larval Tapeworm Infections of Humans

Among the adult (intestinal) tapeworms in man, *T. solium* (pork tapeworm, common in Latin America) and *D. latum* (broad tapeworm or fish tapeworm, common in regions where freshwater fishes occur) are particularly important to human health. *T. solium* may cause cysticercosis (Table 2) and *D. latum* pernicious anemia due to vitamin B_{12} deficiency. Other adult tapeworm species affecting humans are *T. saginata* (beef tapeworm, worldwide distribution, approx. 60 million people currently infected), and *Hymenolepis nana* (dwarf tapeworm, infects not only man but also mice and rats). *T. saginata* may cause economic loss due to taeniasis or cysticercosis in cattle and *H. nana* play a role as zoonosis since cross-infections between humans and rodents are possible. Mainly two drugs, the older **niclosamide** (a nitrosalicylanilide) and the newer **praziquantel** (pyrazinoisoquinoline) are used for elimination of these tapeworms from the intestinal tract (dosage, cf. Table 2). Old remedies (certain seeds from locally grown plants) or other drugs may still be in use (Table 1, Table 2).

Larval cestodes of certain tapeworms lodge preferably in the central nervous system, or in the liver (lung or other organs). Thus the cysticercus of *T. solium* causes neurocysticercosis. The hydatid cyst of *E. granulosus* produces cystic echinococcosis, and that of *E. multilocularis* alveolar echinococcosis. *E. vogeli* causes polycystic echinococcosis in humans, and characteristics of *E. vogeli* metacestode are considered intermediate to those of *E. granulosus* and *E. multilocularis*. Because of the tumor-like growth and proliferation of larval stages, clinical manifestations and pathology are characterized by their great diversity of moderate to severe symptoms. **Treatment** of diseases caused by larval stages is based mainly on three measures focusing on symptomatic and causal therapy. (1) Palliative drugs (e.g. antiepileptics and corticosteroids) are used to control seizures and inflammation associated with tissue reactions caused by cestode larva. (2) Measures of palliative surgery are performed to drain CSF by placement of ventricular shunts, and those of causative surgery to remove solitary brain cysticerci or hydadit cyst from liver or lung. (3) Long-term chemotherapy (indicated if surgical resection of cyst is not possible) may result in complete and permanent disappearance of cyst if it is surrounded by minimal adventitial reactions. Patients with complicated cysts (multiple compartments and numerous daughter cysts) and surrounded by solid fibrous reactions appear to be considerably refractory to treatment (drugs used see Table 2). Children suffering from larval cestode infection usually respond more sensitively to chemotherapy than adults. **Albendazole** and **praziquantel** may be useful for affecting and eliminating living cysticerci in brain parenchyma and in the subarachnoidal space. **Antiepileptic drugs** are the treatment of choice for patients suffering from epileptic attacks caused by neurocyticercosis with calcifications. The administration of such drugs is indicated when seizures are the only manifestation of the disease and there is no imaging and immunological evidence of living parasites. **Corticosteroids** should be used simultaneously with chemotherapy (for two or three days prior to and during drug treatment) to reduce exacerbation of neurological symptoms. Any cysticercocidal drug may cause irreparable damage when used to treat ocular or spinal cysts, even when steroids are used. An ophthalmic examination should be carried out before treatment. **Surgical removal** of cystic echinococcosis (*E. granulosus*) appears to be the treatment of choice if cysts are large (>10 cm ∅). When surgical excision cannot be performed because of general condition of the patient and the extent and location of the cyst, **long-term chemotherapy** should be performed either with **albendazole** (10 mg/kg, b.w.) or **mebendazole** (40–50 mg/kg, b.w.) for at least 3 months. These benzimidazole carbamates inhibit the growth of larval *E. multilocularis*, reduce metatases, and may enhance both the quality and length of patient's survival. Sometimes they may be larvicidal after prolonged therapy but often recurrence may occur.

Alveolar hydatid disease (*E. multilocularis*) is often diagnosed too late so that lesion becomes inoperable. This may also be true for surgical resection of extensive lesions caused by polycystic echinococcosis (*E. vogeli*). Liver transplantation has been successfully performed on otherwise terminal cases. Often combination of surgery with chemotherapy (albendazole) is more likely to be successful in cases in which resection is difficult and usually incomplete. Postoperative chemotherapy should be done routinely for at least 2years after radical surgery. Also prolonged follow up of patients for at least 10 years with ultrasound or other imaging procedures is necessary to track down possible recurrence of cysts. Combined methods of percutaneous cyst puncture and drainage under ultrasound guidance can be carried out with or without injection of a protoscolicidal compound (95% ethanol, or 20% sodium chloride solution). Thus liquid cyst content is aspirated (or drained), the chemical installed and then re-aspirated (or drained). Secondary echinococcosis, which may be caused by accidental spill of cyst fluid during this procedure, can be minimized with simultaneous administration of albendazole or preoperative use of praziquantel. Ivermectin directly injected into hydatid cyst of *E. granulosus* has been found to damage viable protoscoleces in gerbils 6 to 8 weeks after injection.'Percutaneous drainage, i.e., aspiration-injection and re-aspiration' (= PAIR) should be performed only if cysts have no biliary communication. Otherwise, sclerosing cholangitis may occur when the chemical is installed. Therefore prior to PAIR, cyst fluid has to be investigated for the presence of bilirubin.

Chagas' Disease, Animals

→ Cardiovascular System Diseases, Animals.

Chagas' Disease, Man

Pathology

→ Trypanosoma cruzi (Vol. 1) infection occurs in Central and South America and is transmitted by triatomid → bugs (Vol. 1) (→ kissing bugs (Vol. 1)). These hematophagous hemiptera evacuate their intestinal contents containing the infections metacyclic forms, while feeding, in order to ingest more blood. Metacyclic forms enter either the skin through the site of the bite, are inoculated by scratching the itching bite wound, or penetrate the mucosa of the conjunctiva or mouth. Amastigotes multiply in adjacent muscle and fat cells in the dermis, producing an inflammatory nodule, the chagoma, which is composed of parasitized cells, histiocytes, and a periphery of neutrophilic granulocytes. This chagoma may persist for several weeks. Trypanosomes are disseminated via the lymphatics, often enlarging the regional lymph nodes. Hematogenous dissemination follows and amastigotes parasitize many tissues, especially histiocytes, adipocytes, myocardial fibers, and autonomic ganglia in the gastrointestinal tract; this is accompanied by fever. The parasitized cells are usually destroyed (→ Pathology/Fig. 13D). When the infection is transmitted by blood transfusion it is particularly severe because it starts with the hematogenous phase of dissemination without any preceding immunization during the chagoma stage. However, the majority of patients with naturally acquired *T. cruzi* infection experience a benign or asymptomatic infection.

The most important lesions are in the heart, the esophagus, and the colon. The myocardium contains large pseudocysts of amastigotes without surrounding inflammation. Destroyed myocardial cells are also found, with lymphocyte, plasma cell and macrophage infiltration often forming "microabscesses" that later heal by fibrosis (→ Pathology/Fig. 13D). During chronic Chagas' disease, which may last for years, the heart usually undergoes marked hypertrophy (400–800 g) and dilatation. Mural thrombi are commonly found. Next to the apex of the left ventricle myocardial inflammation and fibrosis are often most advanced, and the ventricular wall may become so thin as to be transilluminable. After the Purkinje fibers are destroyed, conduction defects, such as arrhythmias, hypotension, tachycardia, right bundle branch block, and later bradycardia make their appearance. Myocardial failure responds poorly to digitalis and results in death, as does ventricular fibrillation. There are usually few amastigotes in the lesions of chronic Chagas' myocarditis, but the myocarditis with diffuse fibrosis still progresses. This has led to various hypotheses suggesting that delayed hypersensitivity or autoimmunity may participate in the pathogenesis. However, reviewing these, here is enough pathologic evidence to support a direct effect of *T. cruzi* on infected cells.

The autonomic ganglia of the esophagus and colon may be destroyed either by parasitization or by an undetermined process. The destruction leads to an interruption of the peristaltic wave and dilatation of the viscus proximal to the destroyed ganglia, often resulting in megaesophagus and megacolon.

Chagas' encephalitis is seen especially in young children, with parasitization of neurons by amastigotes and destruction of even unparasitized cells. There is marked focal neuronal damage accompanied by lymphocytic infiltration and extending into the meninges. Death may occur after a disease course of only 1–2 months.

There are many cases known of spontaneous cure and of chronic asymptomatic parasitemia lasting for 20–40 years without apparent progressive organ lesions. Minor lesions include the lipochagoma resulting from the destruction of adipocytes by trypanosomes, and a painful lipogranuloma which, if it occurs in the cheeks, interferes with eating.

Placental infection may occur and lead to abortion. Sometimes the fetus becomes infected, resulting in encephalitis and death a few days or weeks after delivery.

Immune Responses

T. cruzi is able to infect virtually any nucleated cells of mammals. In infected mammals *T. cruzi* can be found either as flagellated trypomastigote or as amastigote replicating inside of cells. Amastigotes can be found both within phagocytic vacuoles as well as free in the cytoplasm. After rupture of the host cell, amastigotes transform into trypomastigotes, which are the most infective forms. The intracellular habit may be the reason why *T. cruzi* did not evolve mechanisms of adaptive genetic variation to alter expression of surface proteins such as occurring in African trypanosomes. In humans, the majority of infections with *T. cruzi* are thought to be asymptomatic. Chagas' disease typically represents a chronic disease, with a long term interaction of the host´s immune system and the parasite. Sterile immunity is seldom or never achieved and human *T. cruzi* infections are often dormant for decades. As in the case of *T. gondi*, reactivation has been described to occur during intentional immunosuppression for organ transplants or in AIDS patients. Mouse models of *T. cruzi* infection have been used by many investigators to study both acute and chronic infections to elucidate immune response mechanisms leading to the control of the parasite or to autoimmune pathology. Differences in *T. cruzi* strains can greatly influence the immunological characteristics and disease outcome. In addition, inbred mouse strains have been classified as being susceptible or resistant to T. cruzi infection, but this responsiveness clearly depends on the parasite strain as well.

A typical feature of *T. cruzi* infections in mice is a profound suppression in multiple components of the immune system, which for example can result in exacerbated viral infections (e.g. Murine leukemia virus). Likewise, virtually all immune defense mechanisms appear to be involved in the control of *T. cruzi* infection and disease development.

Innate Immunity During the very early phase of infection NK cells are activated via IL-12 produced by macrophages in response to live, infective trypomastigotes. IFN-γ, produced very early by activated NK cells has a protective effect on experimental *T. cruzi* infection in vivo. Both, neutralization of IFN-γ or depletion of NK cells by mAb treatment rendered relative resistant mice more susceptible to the infection and increased parasitemia. However, IFN-γ produced in the later phases of acute infection may not be relevant for protection.

As the acute infection progresses, polyclonal activation of essentially all peripheral lymphoid cell types is induced. One minor subset intensely expanded is the $\gamma\delta$ T cells. Although the relevance of this subset for host protection is still uncertain, one study showed that $\gamma\delta$ T cells are able to modulate the intensity of parasitemia, mortality and tissue inflammation.

B Cells and Antibodies Polyclonal activation of B cells during the acute phase of infection leads to marked Ig production of all isotypes, but only a minor component of the Ig's appears to be parasite-specific. However, transfer of serum containing *T. cruzi*-specific antibodies mediated significant protection against the infection. In this regard, *T. cruzi* differs from other intracellular pathogens, in that humoral responses do play a protective role against circulating trypomastigotes. $CD5^+$ B1 cells may play an important regulatory role by their capacity to produce IL-10. It has been reported, that mice carrying the xid defect and therefore lacking B1 cells, showed an increased IFN-γ production and were more resistant than the appropriate control mice.

T Cells A number of studies demonstrated that both $CD4^+$ and $CD8^+$ T cells were needed for parasitemia control and survival in acute infection. Elimination of $CD4^+$ T cells by antibody treatment or in MHC class II deficient mice resulted in markedly increased susceptibility to *T. cruzi* infection while athymic mice receiving syngeneic $CD4^+$ T cells were better able to contain parasitemia. The need for $CD4^+$ T cells has been attributed to the production of cytokines stimulating macrophages to kill intracellular parasites and/ or to help in the production of protective parasite-lytic antibodies. Elimination of $CD8^+$ T cells by preventing MHC class I expression in gene deficient mice or by treatment with mAb also resulted in markedly increased susceptibility to *T. cruzi* infection. Mice made deficient in both $CD4^+$ and $CD8^+$ T cells are even more susceptible to increased parasitemia and death, which is consistent with a crucial role for IFN-γ produced by both $CD4^+$ and $CD8^+$ T cells.

In contrast to other experimental parasitic infections, infection with *T. cruzi* does not result in a clear polarization of Th1 or Th2 cytokine responses, as both IFN-γ and IL-10 were found to be produced simultaneously. Instead, separate Th1 and Th2 waves may occur at different stages of infection within the same resistant host and a higher amplitude and more rapid onset of a Th1-type response correlated with resistance.

At least in the Brazilian strain model of infection, most of the IFN-γ appears to be produced from a very unusual subset of ab TCR^+, $NK1.1^-$, $CD4^-$ and $CD8^-$ (DN) T cells. However, for reasons unclear so far this cell population does not seem to play any major protective or inflammatory role in *T. cruzi* infection.

Immunosuppression The observation that effector mechanisms were impaired in mice infected with *T. cruzi* led to the search for soluble mediators of this deficiency. TGF-β, which exerts suppressive effects on macrophages and T cells, was found to be released in a biologically active, processed form by cultured spleen cells from *T. cruzi*-infected mice already on day 8 postinfection. Since in most cases TGF-β is produced in an inactive latent form, it was analyzed, how activation occurs in *T. cruzi* infection. Recent findings suggest, that *T. cruzi* trypomastigotes themselves have a potent enzyme capable of activating TGF-β. Since administration of recombinant TGF-β to resistant

B6D2 F1 mice significantly increased both morbidity and mortality in these animals, TGF-β induced and activated by *T. cruzi* may represent a novel mechanism for pathogenesis.

Another candidate for a mediator of macrophage inhibition is IL-10. Like the "host protective" IL-12, IL-10 can also be produced by macrophages in response to live infective trypomastigotes. Since in vitro IL-10 is able to inhibit the ability of macrophages to kill intracellular *T. cruzi*, it has been considered as a "parasite protective" cytokine. In line with this, administration of neutralizing anti-IL-10 mAb could block the development of acute disease in susceptible C57BL/6 mice. In addition, it has been established that infections with *T. cruzi* (strains CL or Tulahuen) trigger a significantly stronger IL-10 production in susceptible than in resistant mice. On the other hand, IL-10 may also possess protective effects, since IL-10 knockout mice infected with the Tulahuen strain showed accelerated mortality, as compared to wild type mice despite a reduced parasite load. In these IL-10 deficient mice, not only rIL-10 but also anti IL-12 mAb reversed the susceptibility, leading to the conclusion, that IL-10 may prevent toxic immune responses characterized by overproduction of IL-12 and IFN-γ. A second mechanism of immunosuppression beside soluble mediators involves the induction of activation-induced programmed cell death, apoptosis. Selective triggering of $CD4^+$ T cell apoptosis has been described in experimental *T. cruzi* infection, which is mediated via the TCR-CD3 pathway and not via CD69 or Ly-6.

Autoimmunity There is strong evidence that much of the pathology of chronic Chagas disease, such as cardiac manifestations or peripheral neuropathy, is caused by autoimmune mechanisms. One antigen of *T. cruzi* (FL-160) has been implicated in molecular mimicry of nervous tissues. About 40 % of persons with chronic *T. cruzi* infections had antibodies to the FL-160 antigen, and antibodies to two epitopes in this protein have been shown to bind to human siatic nerve. Ribosomal P proteins of T. cruzi are also able to trigger antibody responses against host P proteins, which, interestingly, have previously been implicated as autoimmune targets in SLE patients. Furthermore, human T cells and antibodies reacting with myosin and most likely involved in the pathogenesis of Chagas cardiomyopathy may be triggered by a cross-reactive antigen of *T. cruzi*, which was identified recently.

Vaccination

The approximately 12–20 million individuals in Central and South America infected with *Trypanosoma cruzi* represent an important medico social problem. Various chemotherapeutic drugs have become available but no vaccines have been developed. An important effort of Public Health Organization in Brazil, Uruguay and Argentina has successfully controlled the domestic hemiptera vector *Triatoma infestans*, and transmission of Chagas' disease in these countries, is practically reduced to blood transfusion accidents. In other South and Central America countries anti-vector campaigns are successfully in progress. This success in transmission control considerably decreases the interest in vaccination against Chagas' disease. However, a large number of animal species from different orders of mammals are *T. cruzi* reservoirs in nature, including domestic animals like cats and dogs, which represent potential sources of infection for man. Thus, if available, vaccination would have a double interest: (1) Control of animal's sources of infection and (2) Protection of humans migrating in new eventual endemic areas will be possible in future in the Amazon Region. No evidence of antigenic variation has been found in *T. cruzi* as described for the African trypanosomes. However, two main lineages of parasites have been recently identified by molecular markers, one more frequent in human infections and the other in animal reservoirs. In both lineages, the parasite from the vector insect stage (Triatomidae) can be grown easily in liquid culture with semi-synthetic or defined medium while the vertebrate form can also easily grow in tissue culture using a variety of cells from mammals or birds. Thus, the preconditions for developing vaccines directly from parasites' materials are there, either for development of attenuated lines or chemical purification of antigen components.

About 90% of infected persons recover from the acute stage, but continue to have low parasitemias for an indefinite period of time without clinical signs. A small proportion of them later develop the chronic form of Chagas' disease, with chronic myocarditis or mega- esophagus or mega-column syndromes. In the asymptomatic period, carriers have antibodies to *T. cruzi* and this immune response has been considered responsible for resistance to challenge infections in experimental models and man.

Attempts have been made to develop live vaccines using parasites attenuated by chemical modification, by culture passage, and by X ray irradiation. Most of these live vaccines have induced partial immunity but have also shown pathogenic side effects. Some authors have reported protection of laboratory animals immunized with subcellular fractions of *T. cruzi*. Others have described a 90-kilodalton glycoprotein which exists on the parasite's cell surface throughout the life cycle and which was partially protective in mice and in marmosets. An important objection to the use of natural sources of antigens from *T. cruzi* for vaccination is the description by different authors of molecular mimicry between *T. cruzi* antigens and different cell proteins from the human host with description of parasite-antigen driven human T-cell clones that react with cardiac myosin. Antibodies to ribosomal P proteins of *T. cruzi* have been shown to lead to cross-rection with heart tissue and eventually be implicated in the induction of myocarditis. Because of these potential cross-reactions and consequent autoimmune pathologies, some authors consider that the development of a classical vaccination against Chagas' disease is not suitable. There are apparently two types of antibodies with different functional activities. Protective or complement-dependent "lytic" antibodies (LA) are associated with host resistance, whereas "conventional serology" antibodies (CSA) are not. Chronically infected patients or animals have both LAs and CSAs, whereas experimentally immunized animals only have CSAs. However, it is not yet possible to decide whether a vaccine induces protection without knowing first if some of its components will not provoke autoimmune pathology with later manifestations as in natural infection.

Different groups have recently shown that $CD8^+$ T cells, IFN-gamma and macrophages play an essential role in the control of parasite multiplication in the acute phase. They have identified cytotoxic epitope in the trypanosome surface antigen 1 (TSA-1) and in trypanosome surface proteins members of the trans-sialidase superfamily. Effective protection against the acute infection was obtained with immunisation by plasmid DNA containing genes corresponding to these antigens.

Main clinical symptoms: Chagom at bite site, lymphadenitis, oedema, fever, hepatosplenomegaly, cardiomegaly, aneurisms
Incubation period: 5–20 days
Prepatent period: 1–2 months
Patent period: 20 years
Diagnosis: Serologic tests, microscopic examination of blood smears, xenodiagnosis using uninfected bugs, → serology (Vol. 1)
Prophylaxis: Avoid bites of triatomid bugs
Therapy: Treatment see → Trypanocidal Drugs, Man.

Chancre

Papular and later ulcerating lesion caused by African trypanosomes which are present extracellularly in the subcutaneous tissue at the site of the bite of the → tsetse fly (Vol. 1) (→ Sleeping Sickness).

Charcot-Leyden Crystals

→ Pathology, → Schistosomiasis, Animals, → Schistosomiasis, Man, → Paragonimiasis, Man, → Sparganosis, Man; also found in Amoebiasis.

Chemoprophylaxis

Drug application prior to appearance of clinical signs may prevent outbreak of disease (→ Chemotherapy).

Chemotherapy

Definition

Drug application after appearance of clinical signs may eliminate or damage pathogens/parasite and lead to recovery of health in the patient.

General Information

Chemotherapy (CT) and → chemoprophylaxis (CP) are experimental sciences dealing with drugs that selectively inhibit or destroy parasites or other pathogens. Thus, attention to selective toxicity of antiparasitic drugs is central to CP and CT, and useful chemotherapeutic agents should affect the parasite more adversely than the host. Until now, there is no completely safe drug. Knowledge of sources of drugs (pharmacognosy), action and fate in the body (pharmacodynamics), use in the treatment of disease (therapeutics), adverse effects (toxicology) and contraindications is a must to evaluate potential risks as well as possible benefits of drugs to the patient's well-being.

Withdrawal Time of Drugs in Target Animals
Claims made by the drug manufacturer for a drug in treatment of target animals consumed by humans (poultry, bovines, equines) must have been investigated and evaluated in the target animal, besides toxicological studies in rodents (rat, mouse) and other species. The latter are required to determine the **ADI.** The acceptable daily intake is the dose of a drug residue in edible tissues, like meat, various organs, fat, etc., that during the entire lifetime of a person seems to be without obvious risk to health on the basis of all toxicological data known at the time. The **ADI for humans** may be determined by applying a safety factor of 1:100, or a safety factor of at least 1:1000 in case of a teratogenic drug. Therapeutic claims made by the manufacturer must coincide with safety and tissue residue data for the drug approved by government regulatory agencies. The preslaughter withdrawal time refers to the interval from the time an animal is removed from medication until the permitted time of slaughter. During this interval the residue of toxicological concern will reach a safe concentration as defined by the **MRL** (maximum residue limit). Recent workshops give guidance to the analytical approaches. There may be three **types of tolerances** in defining allowable concentrations of drug residues at the time of slaughter up to the time of consumption by humans. (1) The finite tolerance (measurable amount of drug residue permitted in food), (2) the negligible tolerance (insignificant amounts of residue, i.e., a small fraction of maximum ADI), or (3) the zero tolerance (no residue is permitted in feed or food because of extreme toxicity or because the drug is carcinogenic).

Drugs

General Information Drug can be broadly defined as any chemical compound that affects living processes and is used in the diagnosis, prevention, treatment (cure) of disease(s), or for controlling or improving any physiological or pathological disorder or for relief of pain in animals and humans. A drug may have various names: one or more **chemical names** depending on rules of chemical nomenclature used: **(1)** Usually one international nonproprietary name **(INN)**, **(2)** one or more nonproprietary names, e.g. in different countries, and **(3)** usually several proprietary names or brand names. **Products** are galenic, pharmaceutical, and/or medicinal preparations of drugs, e.g. various dosage forms. These may be tablets, pills, capsules, sustained-release boli, or several liquid preparations, e.g. mixtures, or emulsions for oral administration. Dosage forms for injections are ampules, multi-dose vials, large-volume capped bottles, and implants that may be hard, sterile pellets inserted under skin. The parenteral administration of such preparations may be done either by subcutaneous (s.c.), intramuscular (i.m.), intravenous (i.v.), intraperitoneal (i.p.), or intrathecal (i.e., into the subarachnoid space) injection. For application to skin surface several external preparations of drugs may be used such as liniments, lotions, ointments, creams, dusting powders, or aerosols, e.g. topical insecticides.

Critical Use of Drugs in Concert with Other Control Measures The administration of chemotherapeutic or chemoprophylactic agents can only be regarded as a suitable measure when individual cases or parts of the population or herds are treated under controlled conditions after a diagnosis has been made. Parasitism in the field is often a multifactorial problem. Thus, methods of control must include epizootiological/epidemiological and therefore ecological and economic aspects as well as consideration of the development of resistance to drugs by the parasites. Drug application is important but not the only control measure in a large-scale, complex **control strategy**. The aim of each of the measures that can be taken is to reduce or even completely eliminate the parasite population in its environment. However, the tasks, solutions, and objectives of a certain strategy require profound knowledge of the biology of the parasites and the → epizootiology or → epidemiology of parasitism. Thus, when planning and implementing programs of this type, account must be taken of the many different interactions between host and parasite on the one hand (mode of transmission and susceptibility of the population) and those between parasite and/or host and biotype on the other (e.g., pathogen reservoir). Knowledge of the **environmental factors** such as temperature, humidity, soil structure, and prevailing weather conditions which are critical within the infection chain are of practical importance for forecasting parasitic risk in certain areas.

In order for the chemotherapeutic agent administered to have its optimum effect against the parasites, the **time of treatment** must be adapted to the development cycle of the parasite. An example of this is the → metaphylaxis often practiced in veterinary medicine when the aim is to prevent the outbreak of disease in animals already infected, i.e., to kill off certain parasitic developmental stages in the host before serious damage occurs. The correct timing of the treatment in relation to the course of the infection is therefore critical for the success or failure of drug metaphylaxis.

As is well known, the acquired and often species- or even life-cycle-stage-specific **immunity** (→ immune responses) of the host plays an important role in the elimination of parasites. It would be ideal if CT and immunoprophylaxis had a synergistic effect. However, drugs can suppress the development of immunity and the immune response of the host may not materialize. Thus, the lack of an immune response may result from starting treatment at the incorrect time in relation to the course of the infection. This may be true for instance when treatment and infection take place at the same time, or the treatment starts shortly after the infection, as a result of which the parasites are immediately eliminated (without antigenic effect).

In modern livestock farming in which large numbers of animals are reared and fattened in a confined space, continuous drug application must often be performed via the feed. This is because other preventive measures, such as methods for maintaining good hygiene and disinfection, have either not yet been developed or are as yet inadequate (such as reliable immunoprophylaxis) against the high risk of parasitism that exists with this form of farming. Thus, without prophylactic medication, the rapidly increasing pressure of infection would lead to an outbreak of disease in a herd. Consequently, in the fattening of poultry feed additives such as anticoccidiostats can control outbreaks of coccidiosis and therefore serious financial losses can be reduced to a minimum during the fattening period.

A few examples clearly show that prophylactic or therapeutic agents must not be used in a stereotyped fashion. Often their use must be adapted to varying relationships between the parasite and host populations and their environment; only then will a treatment campaign have the desired success. Flexible handling of the chemotherapeutics also considerably slows down the development and therefore the spread of drug-resistant parasites. Before the start of every course of treatment adverse effects of the chemotherapeutic agent must be taken into account in the treatment plan. They should be reduced to a minimum by varying the dosage regimen and the treatment intervals. Anaphylactic reaction to the drug used or the intrinsic toxicity of the drug including its toxic metabolites may induce a severe and life-threatening risk to a patient. This may be applied also to severe allergic reactions caused through toxic metabolic products of endoparasites released after the drug has killed parasites.

Related Entries

→ Acanthocephalacidal Drugs, → Antidiarrhoeal and Antitrichomoniasis Drugs, → Babesiacidal Drugs, → Cestodocidal Drugs, → Coccidiocidal Drugs, → DNA-Synthesis-Affecting Drugs I, → DNA-Synthesis-Affecting Drugs II, → DNA-Synthesis-Affecting Drugs III, → DNA-Synthesis-Affecting Drugs IV, DNA-Synthesis-Affecting Drugs V, → Drugs against Sarcocystosis, → Drugs against Microsporidiosis, → Ectoparasitocidal Drugs, → Energy-Metabolism-Disturbing Drugs, → Hem(oglobin) Interaction, → Inhibitory-Neurotransmission-Affecting Drugs, → Leishmaniacidal Drugs, → Malariacidal Drugs, → Microtubule-Function-Affecting Drugs, → Myxosporidiacidal Drugs, → Nematocidal Drugs, Animals, → Nematocidal Drugs, Man, → Opportunistic Agents (Vol. 1), → Treatment of Opportunistic Agents, → Theileriacidal Drugs, → Trematodocidal Drugs, → Trypanocidal Drugs, Animals, → Trypanocidal Drugs, Man, → Vaccination Against Protozoa

Cheyletiellosis

→ Mange, Animals/Cheyletiellosis.

Chorioptic Mange

→ Mange, Animals/Chorioptic Mange.

Chronic Infections

Long persistent infections mostly occurring in natural hosts and characterized by regular fluctuations in the level of parasitaemia (e.g. *Plasmodium* species, trypanosomiasis) or in the quantity of the parasitic load (e.g. filariae).

CIC

Circulating immune complexes (→ Immune Responses).

Circulating Antigen

→ Antigen (Vol. 1).

Circumsporozoite Protein

→ Malaria, → CSP (Vol. 1), → Vaccination.

CL

→ Cutaneous Leishmaniasis.

Clinical Pathology, Animals

General Information

It is difficult to give an estimate of the (economic) importance of parasitic diseases at the world level. Firstly, because the effect of a particular parasitic disease may vary completely from one region to another. Secondly, our knowledge of some parasitic infections is based primarily on experimental infections, which often differ from natural conditions. Finally, appropriate data are missing for a number of diseases. Some parasites are known to induce severe, sometimes highly fatal clinical syndromes, and have a recognized economical importance. Most of the others only produce subclinical, or even asymptomatic infections. However, it has now been established that such subclinical infections cause significant losses due to long-term effects on animal growth and productivity and increased susceptibility to other parasitic or bacterial/viral diseases.

Parasitism causes a very wide variety of clinical and pathological changes. This is easy to understand as parasites occur in all organs of the host (Fig. 1). However, most signs are non-specific and a clinical examination alone is often insufficient to

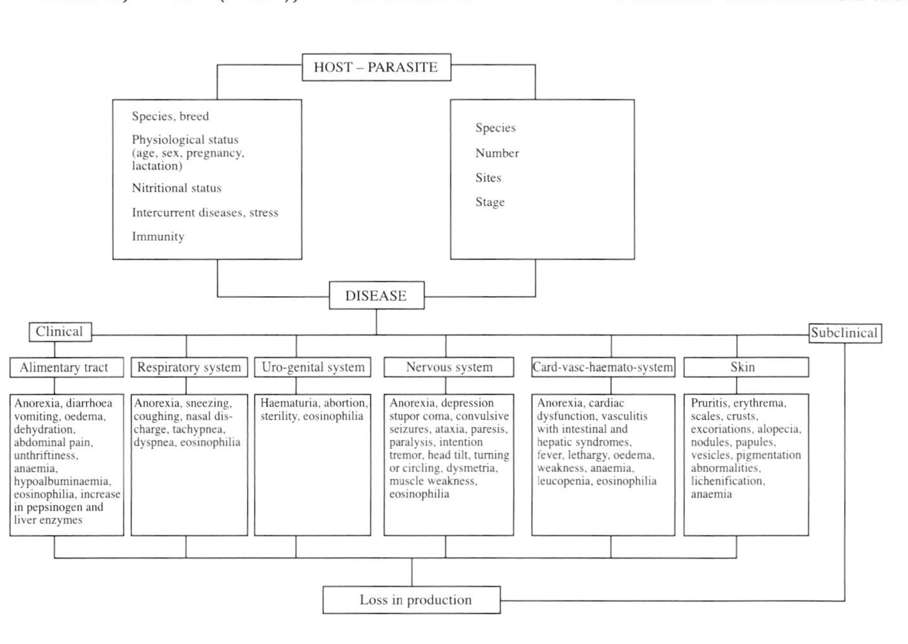

Fig. 1. General outline of host-parasite relationship

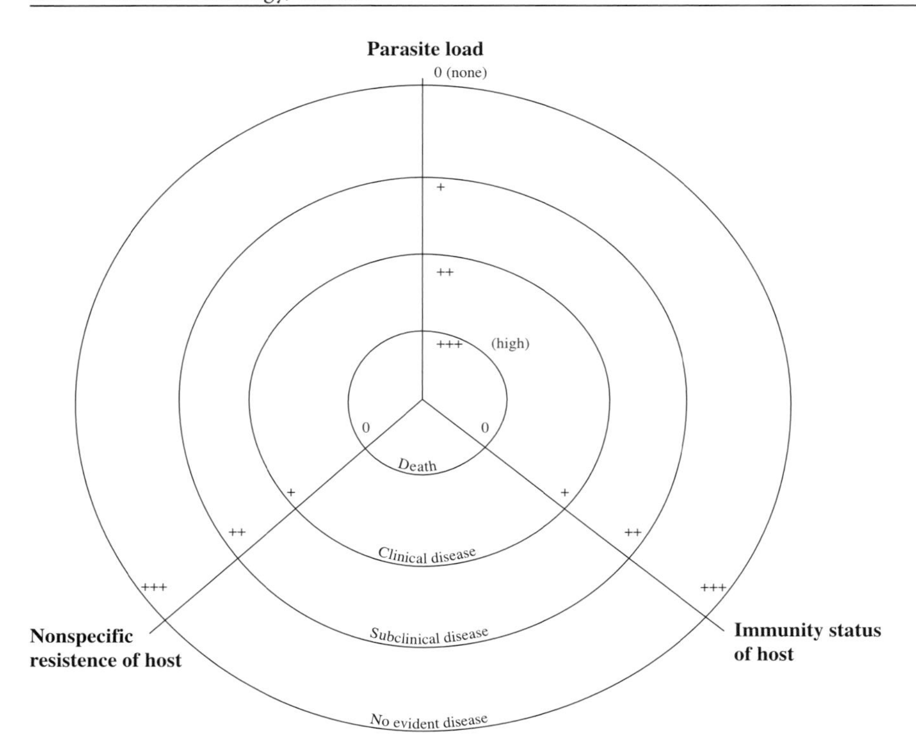

Fig. 2. Relationship between parasite load, immunity and nonspecific resistance of host, and severity of disease

establish an aetiological diagnosis. The clinician is therefore dependent on other diagnostic methods such as laboratory examination of faeces, blood and urine.

The pathogenicity of a particular parasite population will depend on the "parasite load", and on the major site of infection. The heavier the infection the more dramatic is the effect, although there is both a lower and upper threshold. However, in the field many different factors contribute to determine the clinical outcome of a parasitic infection. Age and previous parasite experience increase resistance to infection with most parasites. It is also generally accepted that the plane of nutrition of the animal affects its overall response to the infection, either by affecting the rate of establishment of the parasite or modifying the effects of the infection. Finally, resistance to infection is often reduced by stressful conditions, such as during intercurrent diseases and around parturition (Fig. 2). Caution is needed when applying the results of experimental infections with individual parasites to the field situation in which animals normally acquire several parasite species concurrently. The combined pathophysiological effects of several parasite species may be greater than could be anticipated from the effects of monospecific infections.

Anorexia

Anorexia, or reduction in voluntary food intake, is a feature of most parasitic infections. The degree of appetite loss can vary considerably. In moderate to heavy infections, food intake can be reduced by 20% or more. It may be less obvious in lighter infections. The degree of appetite loss has been shown to vary not only with the level and duration of parasitism but also with the level of protein nutrition. Despite the obvious importance of loss of appetite in parasitized animals, it is still not known why it occurs. The subject has been reviewed by several authors who concluded that anorexia in parasitized animals is certainly complex and that a complete understanding may be difficult to obtain. Possible factors in gastrointestinal parasitism of ruminants are abdominal pain, anatomical or chemical responses to infection, changes in protein digestion and in the availability of amino acids for absorption, changes in plasma levels of hormones (gastrin, cholecystokinin), alterations in the rate of passage of ingesta or direct effects on the central nervous system.

Production Losses

Parasites can have a range of effects on animal production. A reduction in growth rate, and even losses of weight, are probably the most widely described effects of parasitism. It has been recorded for infections with protozoa, nematodes, cestodes, trematodes and arthropods and it is best studied in cattle, sheep, and pigs. The effect on body weight is independent of the site of infection and the heavier the infection the more dramatic the effect. However, there is both a lower and upper threshold in intensity of infection where weight ceases to be affected. In the latter case, this is because of the early death of the animal. Finally, at equal worm burdens young animals or animals on a poor plane of nutrition are more severely affected than those maintained on a better diet, or adults. In addition to changes in body weight, alterations in body composition also occur. These can be of considerable importance when assessing the economic impact of parasitism. Most authors have found that infected animals had a higher percentage of water and a lower deposition of fat, protein, and skeletal calcium and phosphorus, compared with parasite-free controls. The reasons for impaired growth or loss of weight are complex. They have been studied more particularly in intestinal trichostrongylid infections. Because of the loss of appetite, the enteric losses of proteins and the increased rates of intestinal metabolism, there is a net movement of amino acid nitrogen from the muscles, and possibly the skin, to the liver and intestines. This decreases the availability of nitrogen for growth, milk and wool production. The drop in appetite was responsible for nearly 73 % of the difference in live weight gain between calves infected with *Ostertagia ostertagi* and control animals. Table 1 (page 128) presents a list of gastrointestinal parasites with their effect upon body weight.

There has been considerable controversy regarding the effect of parasites on the milk production and apparently conflicting results have been reported. Some studies have shown that anthelmintic treatment in early lactation is associated with an increased milk production of approximately 5% over the whole lactation, while other studies have not shown such an effect. Evaluation of such experiments is made difficult because of the impossibility of identifying those situations where subclinical infections of helminths are likely to cause depression in milk production. Very few experiments have been conducted to evaluate the direct effect of nematode parasitism on milk production. The administration to cows of 130,000 to 200,000 infective larvae of mixed trichostrolongylid species, given once or over several weeks in early lactation, resulted in a reduced milk production of 1 to 3 kg/day. Ewes infected weekly (for 12 weeks, starting last 6 weeks of pregnancy) with 2500 L3 larvae of *Haemonchus contortus* produced 23% less milk than control animals, despite a similar feed intake. It has also been shown that arthropod parasites such as horn flies, stable flies, cattle grubs and mites can reduce milk production by up to 40% – 60 %.

Many helminths infections adversely affect the quantity and quality of wool. Wool growth can be reduced by up to 66 % in some parasitic infections. The reduction in growth rate of wool is accompanied by decreases in staple length and fibre diameter. There are indications that these effects occur at levels of infection below the threshold normally considered to affect body weight.

Reproductive Performance

The impact of parasitism on reproductive performance is of major economic importance. Parasites such as *Toxoplasma gondii* in sheep and goats and *Tritrichomonas foetus* and *Neospora caninum* (syn. *Hammondia heydorni*?) in cattle often induce abortion. Preventive treatment of gastrointestinal trichostrongylosis or fasciolosis in ruminants may result in earlier breeding, higher pregnancy rates and better birth weights.

Clonorchiasis

Synonym

→ Opisthorchiasis, Man.

Coccidiasis, Man

Infectious diseases of the human small intestine caused by various → coccidians (Vol. 1) (→ Isosporosis, Man, → Cyclosporiasis, → Cryptosporidiosis, Man, → Sarcosporidiosis, Man).

Main clinical symptoms: Diarrhoea, vomiting, loss of weight
Incubation period: 2–13 days

Table 1. Effects of some parasites on body weight (according to Vercruysse and De Bont)

Parasite	Host, localization	Experimental design	Weight changes imputed to parasites (comparison with controls)
Protozoa			
Isospora suis	Pig, small intestine	Pigs of 3 days old were infected with 300,000 oocysts	42% reduced weight gain at 21 day old litter weight
Trematoda			
Fasciola hepatica	Calf, liver	Calves, 8-9 months old, received a subclinical (600 cercariae) and clinical (1000 cercariae) infection	Reduced weight gain during the first 6 months subclinical 8%, clinical 28%
Schistosoma bovis	Goat, mesenterial veins	Goats, 6 months old, were infected with 4000 cercariae	20 weeks p.i. infected goats showed a weight increase of only 20%, controls 64%
Nematoda			
Cooperia punctata	Calf, small intestine	Calves, 8 to 13 weeks old, were given a single oral dose of 250,000 L3 larvae	The average weekly weight gain during an 11 week period was reduced by 38%
Haemonchus contortus	Sheep, abomasum	Ewes were weekly infected with 2500 larvae over the last 6 weeks of pregnancy and first 6 weeks of lactation	Infected groups lost 13 kg during the 12 weeks, compared with a loss of 0.5 kg by controls
Hyostrongylus rubidus	Pig, stomach	Pigs 2 months old, were infected with 550 larvae/kg bodyweight, 15 days later with 220 larvae/kg body weight	Average daily gain was reduced by 18%
Oesophagostomum spp.	Pig, large intestine	Pigs weighing average 24.4 kg were given 1500, 3000 and 4500 L3 larvae/kg body weight	At day 21 p.i. the average daily weight gain was reduced by 5, 12 and 23% in the 1500, 3000 and 4500 group, respectively
Ostertagia/ *Cooperia* spp.	Calf, abomasum and small intestine	One group of calves was given a subclinical infection (60,000 L3 larvae), the other a clinical (600,000) infection	Weight gains after 5 weeks were reduced by 16% in the subclinical group and 74% in the clinical group
Ostertagia/ *Cooperia* spp.	Calf, abomasum and small intestine	Calves were exposed to 9 different levels of infection during 147 days	A significant negative linear relationship between the level of exposure and growth performance was found
Strongyloides ransomi	Pig, small intestine	Pigs, weighing average 21.8 kg were given 5000, 10,000 or 20,000 L3 larvae per kg bodyweight	Average daily gain was reduced by 9, 10 and 28% in the 5,000, 10,000 and 20,000 group, respectively

Prepatent period: 7–9 days
Patent period: 2 weeks until 1–2 months (in case of AIDS patients)
Diagnosis: Microscopical determination of oocysts in fecal samples
Prophylaxis: Avoid contact with human feces
Therapy: Treatment see → Coccidiocidal Drugs.

Coccidiocidal Drugs

Economic Importance

Most of the anticoccidials used for the control of the coccidia proper in livestock and poultry are approved by government agencies for the prevention of coccidiosis in chickens (Table 1, page 141 ff). Despite the use of anticoccidials in continuous medication programs global losses chiefly due to subclinical coccidiosis in broiler poultry is estimated at up to US$ 750 million. The enormous expansion of broiler production over the last 30 years has been reflected in the world market for anticoccidial drugs. In 1991 the total turnover per year was about US$ 450 million. This figure includes in-feed anticoccidials (~ 80% ionophores) at about US$ 370 million, in-water anticoccidials at US$ 30–40 million, and biologicals at US$ 25–30; the broiler market with about 80% of the total has been large enough to stimulate major screening and development programs in the American and European pharmaceutical industries. However, the rapid emergence of drug resistance can result in a short market life for some drugs (less than 6 months for buquinolate). However, high cost of obtaining government clearance (registration costs continue to rise and may come to US$ more than 20 million per drug), and the delays involved, have all discouraged pharmaceutical companies. Discovering, characterizing, developing, and registering a new drug may take 8–10 years. The risk of drug resistance may also jeopardize any hope of benefit from the high capital expenditure on registering a new anticoccidial. Several pharmaceutical companies have consequently taken a new approach and abandoned anticoccidial-screening programs although the use of novel high-throughput screening methods would permit large numbers of compounds to be investigated within short periods. In future the most promising alternative to controlling coccidia infections with chemotherapy may be immune prophylaxis. As a result of public pressure against continuous medication, drug resistance problems, and the high cost of drug clearance vaccines, which are able to induce sufficient immunity (both humoral and cellular response) and thus provide flock protection against morbidity and mortality, would be an appropriate alternative.

Epizootiology and Control Measures

The application of any control methods, whether hygienic, chemotherapeutic or immunological, needs profound knowledge of the epizootiology of various coccidioses. The term coccidiasis is used to describe relatively nonpathogenic and usually mixed infections, whereas coccidiosis is a severe disease in the host. In papers related to human medicine the ending *'iasis'*, however, is used to indicate an acute phase of disease with severe clinical symptoms. Thus the parasitological literature may lead to confusion on the severity of the disease. Mixed infections, i.e., those where animals are infected with more than one species of coccidia, are very common and only some species are highly pathogenic. The severity of the disease is the result of the combined actions of the particular mixture of coccidia, the number of sporulated oocysts ingested with feed or water. However, it may be influenced also by the nutritional condition of the host, environmental and climatic factors (such as temperature, moisture, oxygen tension, and sunlight) and the management practices used. Crowding of animals due to intensive rearing, for example in the broiler industry where large numbers of chickens are kept in enormous houses, creates conditions favorable to sudden outbreaks of severe coccidiosis. Fecal debris may concentrate large numbers of oocysts, which can rapidly sporulate and become infective under warm and moist conditions. Severe coccidiosis is, therefore, mostly a "man-made" problem with domestic animals and is not a general problem in the wild or on pastures. When coccidiosis is suspected and oocysts are found, the species present should be identified. A periodic examination of feces for oocysts and a post mortem examination in a few animals will give valuable information on the status of infection.

Coccidiosis cannot be recognized clinically until tissue damage associated with second- or third-generation schizogony occurs. When the disease is present, the **clinical signs** are catarrhalic or hemorrhagic enteritis. The severity of disease is often complicated by the presence of secondary bacterial infections. Moderately affected animals show poor weight gain or weight loss, weakness, and emaciation, severely affected animals (chiefly young ones) may die very soon after the first clinical signs are seen. Since coccidia have a self-limiting life cycle, the acute phase of infection may have already passed before therapeutic treatment can be started. Thus, **treatment** is usually too late to prevent economic loss. All classes of domestic animals can be affected by coccidia, and losses due to coccidiosis in mammals (particularly in bovines: cattle, sheep, goats) are difficult to determine. The expenses of management practices and the cost of preventive or therapeutic drugs are the main points of consideration. Using suitable control measures can minimize symptoms of the disease. There should be **strict sanitation** to prevent feed and water being contaminated by feces, and feedlots, pens, cages, or hutches should be kept dry and well drained and be cleaned out regularly (preferably every day in rabbitries). When outbreaks occur in pasture, water holes and ditches should be fenced off. Crowding of young animals should be avoided.

Negative effects of intensive animal production are numerous. Thus manure/mineral (phosphate/nitrate) accumulation (used as fertilizer), ammonia emission, (one of main causes for acid rain), dead animal disposal, flies nuisance in densely populated areas, and availability of sufficient water of feed quality may increasingly lead to legislative restrictions and costs involved in preserving protection.

Development of Acquired Immunity

Anticoccidials (Table 1) used in poultry may affect stages of the parasite inducing immunity. Pullets intended as commercial layers must develop immunity to coccidiosis when they receive preventive drugs. However, **anticoccidials**, which are effective against second-generation schizonts, can seriously delay the development of immunity if they are given at the dose levels recommended for broilers. Therefore, they should be fed at the lowest possible dose level and for the shortest practical period that give sufficient anticoccidial protection and allow progressive development of immunity in replacement pullets. One of the effects of immunity is to reduce the biotic potential of the coccidia; each oocyst of the common chicken coccidium *E. tenella* is theoretically able to produce about 2.5 million second-generation merozoites although this maximum number may only occur in the case of initial infection. In partially immune animals, however, only some of the parasites complete their life cycle and produce variable oocysts. In completely immune animals a few or no oocysts are produced for prolonged periods of time; some of the parasites may persist in an asexual stage within the host and thus fail to get further than the initial stage of trophozoite or first-generation meront. As a result, immunity to a challenge inoculum usually leads to a reduction in the clinical signs and in parasite multiplication. The specificity of immunity to *Eimeria* spp. is well known, although there may be considerable strain variations in immunity in some species of coccidia in commercial poultry houses as has been shown with *E. acervulina* and the very immunogenic *E. maxima*. The duration of protective immunity is uncertain and depends on several factors like mode of immunization, inoculum dose, age of the host, as well as on *Eimeria* species and strain variation. It is not yet clear whether immunity is of the sterile or premunition type. It seems likely that immunological control of poultry coccidiosis is achievable. In the not too distant future it will replace chemoprophylaxis coming increasingly under public pressure because of drug residues in edible tissues and occasionally in eggs (→ Chemotherapy/Withdrawal Time of Drugs in Target Animals). **Immunoprophylaxis** with attenuated or precocious strains of *Eimeria* spp. will be therefore an attractive alternative for parasite control since it lacks residual problems. In general, anticoccidials do not adversely affect build up of immunity after vaccination programs against bacterial and viral infections.

Coccidiosis in the Domestic Fowl

The disease is responsible for considerable losses in the poultry industry. Rearing of thousands of birds on litter-covered floors in enormous houses may result in a tremendous and dangerous build-up of the oocyst population (→ *Eimeria* (Vol. 1)). A change from litter-covered floors to wire-floored pens greatly reduces the exposure to coccidia. Thus, outbreaks of coccidiosis in laying hens maintained in cages rarely occur. In general, the prophylactic use of anticoccidial drugs is not required if the cages are kept clean and the feces do not contaminate watering and feeding systems. However, discouraging results have been obtained from experiments to convert broiler and breeder flocks entirely to cage operations (the most obvious problems being high equipment and maintenance costs, breast blisters, leg problems, removal of droppings and dead birds, and housefly control). Today most poultrymen rely on **floor rearing methods** for broiler production or breeder flocks and use continuous medication programs. Poultry producers also attempt to control coccidiosis by employing good sanitary programs. Litters should be kept dry so that oocysts cannot sporulate – many outbreaks occur after leaks in roofs or waterers. Wet litter must be cleaned out and replaced with dry litter. When broiler houses are emptied for a new batch of chickens the litter should be piled up for about 24 h so that the heat generated can destroy the majority of oocysts. **Disinfection** is usually impractical since oocysts are resistant to disinfectants used against bacteria, viruses, or fungi. In parasitology laboratories where disinfection is needed, heat (30 min at 60°C) or "effective" fumigants such as ammonia and methyl bromide may be used.

Despite the use of continuous medication programs (see Drugs Acting on Coccidiosis of Domestic Fowl, Table 1), coccidiosis is still the most important parasitic disease in chickens. As a rule young birds are prone to mixed infections and older birds are carriers. In many outbreaks, however, clinical signs can be ascribed to one species or a combination of two or rarely three; signs of coccidiosis become apparent about 3 days after infection; chickens cease feeding and huddle together for warmth; on the 4th day of infection blood may appear in the feces.

The *Eimeria* species of the domestic fowl show marked differences in their **pathogenicity**. *E. tenella* is the most pathogenic and important species occurring in the epithelial cells and submucosa of the ceca; it may produce severe hemorrhagic enteritis, which leads to high mortality in young birds. *E. necatrix* is also a common and highly pathogenic species, which occurs in the small intestine (first- and second-generation schizonts) and in the ceca (third-generation schizont and gamonts). It tends to cause predominantly chronic enteritis in older birds but in acute cases severe submucosal hemorrhaging and even death may occur. *E. acervulina* (probably the most common species seen) occurs in the small intestine. The pathogenicity of *E. acervulina* strains may vary, and the clinical signs consist of weight loss and watery, whitish diarrhea. *E. maxima* is also common; it occurs in the small intestine and is moderately pathogenic causing numerous petechial hemorrhages and a marked production of mucus. *E. brunetti* is markedly pathogenic but relatively uncommon. In heavy infections, characteristic necrotic (hemorrhagic) enteritis is evident in the lower small intestine, colon, and tubular part of the ceca. *E. mitis* is common worldwide and occurs throughout the small intestine, even in the tubular part of the ceca. This species is slightly to moderately pathogenic; in severe infections numerous petechial hemorrhages may be present. *E. mivati* is fairly common in the USA and Canada (presumably worldwide) and seems to be more pathogenic than *E. acervulina*. It primarily occurs in the upper small intestine and has been included by several authors as a variant of *E. acervulina*. The pathogenicity of *Eimeria* spp. in particular that of *E. acervulina* and *E. tenella* seems to be reduced by intestinal digesta viscosity reducing enzymes as xylanases, glucanases and pectinases. These enzymes may improve the nutritional value of wheat- or maize-based diets but mechanisms involved in reducing growth depression in coccidial infections are not yet known and need to be elucidated.

Economic losses (see economic Importance) due to subclinical infection of avian coccidia are enormous and much more common than coccidiosis, and coccidiasis should be regarded as ubiquitous in commercially reared poultry. Because some species of coccidia are highly pathogenic while others are only slightly or moderately pathogenic the species present must be identified. Scrapings from the mucosal surface of the intestine of affected birds must be examined microscopically for oocysts and endogenous stages of the parasite; the location and type of lesions must be determined to establish a definite diagnosis.

Coccidiosis of Turkeys The disease may cause heavy economic loss in the turkey industry; it is primarily a disease of young turkey poults (aged 3 to 10 weeks); older birds are carriers. **Clinical signs** are enteritis, watery or mucoid diarrhea, and anorexia. The most pathogenic and important coccidia are *E. meleagrimitis* (located in the jejunum) and *E. adenoeides* (located in the lower ileum, ceca, and rectum); several species (in particular *E. innocua*, *E. subrotunda* and *E. meleagridis*) are nearly nonpathogenic. Therefore, oocysts of the latter species must be differentiated from the pathogenic ones as well as those of *E. gallopavonis* (lower small intestine, ceca, and rectum) and *E. dispersa* (duodenum, jejunum, ileum) which are not so common, and only slightly or moderately pathogenic. The presence of oocysts (even large numbers) in the feces can only be a tentative diagnosis for coccidiosis; satisfactory diagnosis can only be made at post mortem.

Coccidiosis of Geese and Ducks The disease seems to be of relatively little importance although several homoxenous coccidia are known. Some of these may be associated with severe disease and even death. *E. truncata* is highly pathogenic to goslings, and occasionally causes up to 100% mortality within a few days; older birds are carriers. This species occurs in the kidney tubules; its endogenous stages destroy the epithelial cells, causing enlarged and light-colored kidneys. Other species (goose: *E. anseris*, *E. nocens*, *E. truncata*, *E. kotlani*, duck: *E. kotlani*, *E. danailovi*, *Tyzzeria perniciosa*) appear to be important in areas where crowding and poor sanitation are present. Thus, large numbers of sporulated oocysts in an unhygienic environment often lead to sporadic outbreaks of intestinal coccidiosis.

Action of Drugs on Developmental Stages of Eimeria spp.

The terms coccidiostatics and coccidiocidals (coccidiocides) for drugs, which characterize actions on coccidia, are often used confusingly. Drugs with a **coccidiostatic action**, such as clopidol, decoquinate, buquinolate, methylbenzoquate (Table 1), arrest the development of certain parasite stages in a reversible way; thus withdrawal of the drugs leads to completion of the life cycle and possibly the appearance of clinical signs several days after medication is discontinued. Drugs with **coccidiocidal action**, such as arprinocid, halofuginone, polyether antibiotics, toltrazuril, diclazuril and clazuril (Table 1), kill or irreversibly damage most of certain parasite stages, and there is no evidence of clinical relapse after drug withdrawal. Some drugs (e.g., nicarbazin, amprolium) may have both coccidiostatic and coccidiocidal activity depending on how long the drugs and in which concentration they have been given in the feed or water. Continuous medication of these drugs for more than 48 h usually results in the majority of parasitic stages being damaged. Polyether antibiotics mainly affect asexual *Eimeria* stages as sporozoites, schizonts of the first and second generation. Flow cytometric analysis has been shown to be a reliable and sensitive technique for characterizing coccidiocidal effects on sporozoites exposed to various ionophores in vitro. Diclazuril acts against most stages of *Eimeria* spp. but this may vary between species (e.g. in *E. maxima* gamonts only); this is also true for clazuril. Prophylactic drugs may preferably act on early and/or late asexual stages in the life cycle, like polyether antibiotics, zoalen, decoquinate, clopidol, robenidin, nicarbazin, arprinocid, halofuginone, diclazuril, amprolium+ethopabate, methylbenzoquate+clopidol, and other combinations. A few **therapeutic drugs** used for treating outbreaks of coccidiosis may affect stages of second- and third-generation schizogony and to a certain degree gamonts (sexual stages) as well; drugs reducing clinical signs of coccidiosis are amprolium, sulfonamides, clazuril, toltrazuril and combinations of sulfonamides+dehydrofolate reductase inhibitors (Table 1). Toltrazuril administered in drinking water has brought substantial progress in treatment of coccidiosis in various animals. It has a broad spectrum of activity against various parasitic protozoa, and unlike other anticoccidials, it acts against schizonts and gamonts of various *Eimeria* and *Isospora* spp.

Drugs Acting on Coccidiosis of Domestic Fowl

Today, chemoprophylaxis in continuous medication programs appears to be the only effective tool for controlling coccidiosis in floor-reared poultry although drug resistance in *Eimeria* spp. populations is a widespread problem in the broiler industry today. Strategic use of anticoccidials may thwart the development of resistance processes (see Measures Against Drug Tolerance in the Broiler Industry). However, neither improved sanitation nor vaccination of birds with pathogen species of live precocious oocysts (**Paracox®, Livacox®**) via drinking water or feed can sufficiently substitute continuous medication programs today. The prophylactic use of anticoccidials is based on the application of additives in-feed, i.e., drugs are added directly to bird feed in small quantities, usually in concentrations of a few ppm (part per million). Such anticoccidials are approved by governmental agencies for the use in growing birds (e.g. chickens, turkey, others). Thus, prescribed concentrations rations of these drugs can be fed to licensed table birds ad libitum from the beginning of their life (fattening/growing period) to a prescribed pre-slaughter withdrawal time, or age at first egg, or start of cage or battery keeping of pullets. Additives in-feed used worldwide for the prevention of poultry coccidiosis and therapeutic anticoccidials are listed in Table 1. There are (1) synthetic compounds (amprolium, clopidol, decoquinate, diclazuril, halofuginone, nicarbazin, robenidine, zoalene), (2) polyether antibiotics or ionophores (lasalocid, maduramicin, monensin, narasin, salinomycin, semduramicin) and (3) drug combinations (amprolium+ethopabate, clopidol+methylbenzoquate, narasin+nicarbazin, maduramicin+-nicarbazin). Most **continuous medication** programs may last 35 to 40 days, i.e., from the beginning of the fattening period to a fixed time prior to slaughter of birds. **Withdrawal times** (→ Chemotherapy/Withdrawal Time of Drugs in Target Animals) for most prophylactically used anticoccidials may last 0 to 5 days and longer (e.g. nicarbazin, 9 days in Europe), or may exceed 5 days for other reasons, e.g. to minimize costs, or to promote "compensatory growth". A withdrawal time longer than 5 days may lead to enhanced risk of coccidiosis at the end of fattening period, particularly if environmental hygiene is poor. The prophylactic use of anticoccidials in-feed in **egg-laying birds** is not permitted, although the occurrence of residues of anticoccidials in eggs has been widely proven; carry-over of medicated feed in the feedmill or elsewhere must be the cause. Sensitive and specific analytical assays, as spectophotofluorometry, liquid chromatography, high performance liquid chromatography (HPLC), the qualitative vanillin test and others, are available to detect the parent compound or their residues in feed, edible tissues, and eggs of target animals or non-target animals. In case of outbreaks of coccidiosis **therapeutic drugs** (amprolium, sulfonamides, dehyrofolate reductase inhibitors plus sulfonamides, toltrazuril) are preferably administered via drinking water for a 3- or 5-day treatment course; withdrawal periods between 14 and 28 days may be necessary. In practice, this means that medication towards the end of the growing life of birds is not economically possible, and this is also true for drugs used outside their product license; the latter may also require withdrawal periods up to 28-days or even longer. So often the medicines regulations how they relate to the withdrawal periods stipulated for therapeutic drugs and given to food animals withhold birds (meat chickens, ducks, turkeys) from the possibility of treatment for a significant part of their life.

Drugs Acting on Coccidiosis of Turkeys, Geese, Ducks and Gamebirds

Approved additives in feed, and maxima concentrations (EC directives) for prevention of coccidiosis in turkeys are shown in Table 1. Drugs may be Amprolium + ethopabate (133 ppm/withdrawal time 3 days); metichlorpindol + methylbenzoquate (110 ppm/5 days); (diclazuril, 1 ppm/5 days) halofuginone (3 ppm/5 days), lasalocid (125 ppm/5 days), monensin (100 ppm/3 days), and robenidine (36 ppm/5 days). The turkey industry (mainly found in the USA) has till now been rather small compared to that of chicken broilers; the total market for anticoccidials may come to ~US$ 30 million. The most suitable drug program is continuous medication but anticoccidial treatment covers only a part of the growing period of birds (till 12–16 weeks of age or 8–10 weeks of age if birds are moved to outside pastures or larger facilities). At this time acquired immunity may be established in most of the birds; however, some birds may still be susceptible to damaging infections 12 weeks after the growout start. Intermittent medication is used relatively often since continuous anticoccidial programs are not universally practiced with turkeys up to 8 weeks of age because of the high resulting costs. Concentrations in the feed are generally in the same range as those administered in chicken broiler rations. There is evidence from field experience that drug resistance in coccidia of turkeys may exist. Generally it seems not to be a major problem although monensin resistance has been demonstrated in *E. meleagrimitis* field isolates. Under experimental conditions a drug sensitive laboratory strain became resistant to the drug after 10 generations of selection. (see Therapeutic drugs used against outbreaks of coccidiosis: Drugs Acting on Coccidiosis of Domestic Fowl)

Drugs Acting on Coccidiosis (Eimeriosis) of Ruminants and Horses

Bovine eimeriosis (the most pathogenic species are *E. zuernii* and *E. bovis,* Table 1) is primarily a disease in calves between the ages of 3 weeks and 6 months; older calves and adult animals are usually symptomless carriers. Crowding and lack of sanitation greatly increase outbreaks of disease. Coccidiosis in sheep and goats is often a disease of feedlot lambs, and often occurs in breeding flocks. Mixed infections build up to a peak that may last 1–4 weeks and then decline. *E. ovina* (sheep), *E. ashata* (sheep), *E. ovinoidalis* (sheep), *E. arloingi* (goat) and *E. christenseni* (goat) are of clinical importance. Batch rearing of lambs or calves in groups of similar ages may limit the build up and spread of oocycts to younger animals thereby targeting potential treatment measures. Little information is available on the eimeriosis of **equines**, mainly caused by *E. leuckarti* in the small intestine or *Klossiella equi* in the kidneys of foals. Coccidiosis in horses (including asses and mules) seems to be very rare. Beware of polyether antibiotics in-feed in environment of horse facilities; they are highly toxic to equines.

Most drugs used in **cattle** are approved for the prevention of coccidiosis in poultry. The demand for grain-fed beef has led to cattle rearing techniques (large feedlot complexes) that may encourage damaging infection pressure. Only a few drugs have been approved for treatment in cattle by some governments. **Decoquinate** is licensed in several countries for the treatment and prophylaxis of bovine coccidiosis. The tissues of treated calves that had been on continuous medication with decoquinate were free of schizonts, gamonts and oocysts. The drug apparently kills *E. bovis* sporozoites or arrests their further development if administered at 1.5 mg/kg body weight in the feed.

Coccidiosis in bovines is often wrongly considered to be a sporadic disease with the result that drugs have rarely been used for prophylaxis. Outbreaks of the disease are still handled by spot treatment using therapeutic drugs as sulfonamides, e.g., sulfaguanidine, sulfamethazine, and sulfadimidine or sulfaquinoxaline. Combinations of sulfonamides with trimethoprim are used in manifest coccidiosis in cattle, sheep and goats. Nitrofurans, e.g., nitrofurazone, or amprolium are moderately effective against bovine coccidia. Amprolium has been widely used in the USA for treatment of clinically ill calves and lambs. Toltrazuril in-water and diclazuril in-feed have been proved to be very effective against bovine coccidiosis at a single oral dosage of 20 mg/kg body weight in bovines. **Preventive drugs** as ionophorous antibiotics such as monensin, lasalocid, and salinomycin licensed for prevention of coccidiosis in poultry exhibit distinct activity against coccidiosis in ruminants. The doses of monensin, lasalocid, and salinomycin used for improving feed efficiency in feedlot and pasture cattle correspond to the doses reported to prevent coccidiosis; it is assumed that the anticoccidial "side effect" of these "growth promotors" may considerably reduce coccidiosis problems in feedlot cattle. Calves artificially infected with *Eimeria bovis* (88%) and *E. zuernii* were given lasalocid in-feed at 0.50, 0.75, or 1.0 mg/kg body weight, daily for 45 days. There were no dose-dependent effects but equal reduction rates in oocyst output and preventing clinical coccidiosis. Calves given lasalocid, decoquinate, or monensin in-feed at 33 ppm for 46 days had significantly fewer oocysts in feces and fewer clinical signs of coccidiosis than those given nonmedicated rations. Mixing lasalocid in milk replacer or fresh milk (1mg/kg body weight/day) is an effective method of protecting young calves against early infection with coccidia. Decoquinate used as a creep feed additive is licensed (e.g. in the UK) for the prevention of coccidiosis in **lambs**, and has also been evaluated in the USA as an anticoccidial against coccidiosis in **goats**. Ionophores in-feed such as monensin, lasalocid or salinomycin prove effective also in preventing coccidiosis in lambs, and goats; against caprine coccidia they are, however, only moderately active. Treatment of coccidiosis in lambs and kids is done with drugs used in cattle. Often animals are clinically ill when the disease is diagnosed. At this time the intestinal mucosa is already extensively damaged, and consequently treatment cannot lead to a radical cure. As a rule, all lambs and kids in a flock should be treated, as even those showing no symptoms are likely to be infected. Fluid therapy using either oral rehydration solutions or parenteral solutions, and appropriate anthelminthic treatment are indicated in severely affected animals.

Drugs Acting on Coccidiosis (Eimeriosis) of Rabbits

Coccidiosis in rabbits is essentially restricted to the young (adults are carriers) and occurs particularly in breeding and rearing establishments (rabbitries) although outbreaks of coccidiosis in warrens or similar types of habitat are not uncommon. The most important species seems to be *Eimeria stiedai*, which is common, worldwide and occurs in the walls of the bile ducts in the liver causing hepatic coccidiosis. Other important and pathogenic species, which may occur in the intestine, are *E. irresidua*, *E. magna*, *E. intestinalis*, *E. media*, and *E. perforans*; mixed infections are the rule. The presence of parasites in a case of enteritis does not necessarily indicate the cause. Coccidia may be present in large numbers without any serious clinical signs. Therefore, the most satisfactory diagnosis is made at post mortem; only the presence of characteristic lesions are evident in the liver/or the intestine.

There are only a few additives in-feed approved for the prevention of rabbit coccidiosis in Europe or elsewhere. These are metichlorpindol or clopidol (125–200 ppm), meticlorpindol +methyl benzoquate (220 ppm), robenidine (50–66 ppm), salinomycin (20–25 ppm) and possibly diclazuril (1ppm). Preslaughter withdrawal times of these drugs may range between 5 and 7 days. Robenidine at 33 ppm and meticlorpindol at 200 ppm are somewhat erratic in their anticoccidial activity and prove only partially effective in controlling coccidiosis. Meticlorpindol has been found to be superior to a mixture of sulfaquinoxaline plus pyrimethamine, or sulfadimidine plus robenidine in suppressing oocyst output. Preventive administration of monensin and that of salinomycin in pelleted feed at 50 and 25 ppm may not only reduce markedly the oocyst production but lead to almost total inhibition of hepatic and intestinal lesions as well. This is also true for the administration of toltrazuril in the drinking water at 10–15 ppm. Several **sulfonamides** alone, or in combination with dehydrofolate reductase inhibitors (Table 1) can be used for **therapy** or prevention of coccidiosis. They may be given in various routes and dosage forms (in-feed, in-water, parenterally, and as injection). Their preslaughter withdrawal times range between 8 and 15 days. **Sulfaquinoxaline** (relatively cheap and water soluble) seems to be the most widely used sulfonamide for treating rabbit coccidiosis, and it has been found to be just as helpful in preventing coccidiosis when given in continuous medication programs in feed or water. Treatment with amprolium proved to be less effective than that with sulfonamides in reducing mortality in rabbits suffering from coccidiosis; there was no protection against hepatic coccidiosis at 0.02% amprolium in pelleted feed. The drug of choice for treatment of intestinal and hepatic coccidiosis appears to be toltrazuril (not licensed for rabbits). However, a proper management and good hygiene in rabbitries can markedly reduce long-term medication and treatment with anticoccidials.

Drug Acting on Coccidiosis of Swine, Dogs and Cats

Coccidiosis in **swine** is mainly restricted to young pigs; older pigs are carriers. *E. debliecki* and *E. scabra* are probably the most pathogenic species. *Isospora suis* has also been found to cause severe enteritis; however, intestinal disorders are so common in baby and young pigs and are caused by so many different pathogens that a diagnosis based only on fecal examination is not definite. Only a very large number of oocysts can indicate that coccidiosis is present. *Isospora suis* is an important causative agent of porcine neonatal diarrhea worldwide; it may occur on any farm with any type of management system and at any time of the year although feces samples of sows always proved negative. Most canine and feline coccidia are usually nonpathogenic or only moderately pathogenic; however, coccidiosis in **dogs** and **cats** may be an important cause of the diarrheic syndrome often associated with secondary infections in puppies and kittens. Affected animals are frequently seen in breeding kennels and runways where sanitation is poor. Diagnosis is not definite if it is only based on clinical signs (e.g., diarrhea with blood in the feces), or on the presence of large numbers of *Isospora* spp. oocysts in the feces; post mortem examination is the most adequate diagnosis method. Coccidial infections in dogs and cats may also be caused by heteroxenous genera (*Sarcocystis* spp. in dogs, and *Sarcocystis* spp. and *Toxoplasma gondii* in cats, see respective entries).

The most commonly used compounds for treatment of coccidiosis in dogs, cats (and piglets) are sulfonamides, toltrazuril (and amprolium) affecting mainly later asexual stages of the schizogony cycle and to a certain degree sexual stages (gamonts). There are only a few controlled studies with experimental infections and no reliable evidence of practical problems associated with *Eimeria* spp. in swine. In contrast, *Isospora suis* infections appear to have become an economically important diarrheal disease in young piglets during the last two decades, and modern production systems seem to encourage the disease. Mortality due to *I. suis* infections may reach 10%–20% in nursing pigs, and a similar percentage may be severely stunted. *I. suis* infections causing severe enteritis or even death in piglets have been treated successfully with amprolium; medication of sows with amprolium for 1–2 weeks before and after farrowing and of neonatal piglets may be helpful in reducing morbidity and mortality. There are no anticoccidials approved for the prevention of coccidiosis in swine. Prophylactic administration of toltrazuril in piglets (a single 1.0 ml dose: 20 to 30 mg/kg, orally between 3 and 6 days of age) has been shown to reduce the morbidity in piggeries. There was a significant reduction in the number of antibacterial treatments given to piglets, fewer piglets developing diarrhea, and a significant improvement in growth rate of piglets. *I. suis* was detected in 38% to 50% of fecal samples from several piggeries and in 93% of those from the experimental piggery. Supportive treatment with antibiotics (e.g., chlortetracycline or oxytetracyclines) may reduce secondary bacterial infections in the intestine of piglets with severe enteritis. **Good sanitation** can effectively reduce diarrhea caused by *I. suis* infection in neonatal piglets in large farrowing facilities.

Drug Tolerance Problems in the Broiler Industry

A new successful drug should have competitive advantages over other available drugs, for example high potency and broad-spectrum activity. Furthermore, at the recommended doses the coccidia must not be allowed to develop resistance or to survive for a longer period. The reduction in coccidial sensitivity to any drug (partial drug resistance) encourages the development of subclinical or subacute coccidiosis. Low levels of infection usually cause a moderate drop in feed conversion and thus lead to considerable economic losses in the poultry industry.

The development of drug resistance may be evident if a change in the parasite can be demonstrated by comparing sensitivity before and after exposure to the anticoccidial. Possible causes for **drug failure** in the field may result from selection of resistant organisms, which rapidly become the dominant phenotype in broiler houses. Resistance is commonly believed to arise initially in the presence of "subtherapeutic" or lower than recommended drug concentrations. Contrary to that suggestion, it has been argued that resistance may occur more rapidly in the presence of higher drug concentrations as a result of a more rapid change in gene frequency caused by increasing the intensity of selection. Studies on the incidence of drug resistance in the field and in experimental investigations have shown that resistance among strains of coccidia to **synthetic drugs** such as 4-hydroxyquinolines (e.g., decoquinate, methylbenzoquate), arprinocid, meticlorpindol, and halofuginone is relatively high. The resistance may result from the selection of preexisting mutants in the parent population. Thus drug resistance appears to be a genetic trait and tends to remain in a coccidia population for some years. Recent investigations on intraspecific polymorphisms of *Eimeria* spp. due to drug resistance indicate that differences in drug sensitivity correlate with genetic differences and polymorphisms detected by random amplified polymorphic DNA (RAPD) might facilitate the selection of molecular markers for resistance genotyping.

In vitro studies on uptake of [^{14}C] monensin by *E. tenella* sporozoites in primary chicken kidney cell cultures infected with ionophore-sensitive (IS) or ionophore-resistant (IR) isolates showed significant differences in [^{14}C] monensin accumulation between IS and IR isolates. The latter isolates had decreased uptake of monensin and the amount of the drug required to inhibit development of *E. tenella* by 50% was 20–40 times higher for IR isolates of *E. tenella* which might reflect differences in membrane chemistry. Studies on an *E. tenella* field isolate that was resistant to monensin, salinomycin and lasalocid at double use level and resistant to narasin and maduramicin at normal use level showed good agreement between in vitro and in vivo results. Flowcytometric analysis of fluorescence after simultaneous exposure to fluorescein diacetate (FDA) and propidium iodide is a suitable indicator of cellular viability and proved to be a valuable technique in the study of sporozoite response to anticoccidials. Thus, various improved in vitro techniques in research programs have become increasingly important in recent years. They may be a supplement to and, occasionally, a substitute for in vivo experiments in certain fields of the basic research, as in the field of biotechnology of poultry and farm animals. A first practical approach to countering **drug resistance** in the field may be an increase in the drug concentration. Increasing the drug concentration to compensate drug tolerance of coccidia is not only uneconomical but can also lead to toxic problems since several anticoccidials (e.g., nicarbazine, halofuginone, and various ionophorous antibiotics) are used at doses which are close to toxicity levels in birds. With the exception of some ionophorous antibiotics, such as monensin, salinomycin and lasalocid, none of the synthetic drugs introduced have enjoyed prolonged marketability. The continuing success of the ionophorous antibiotics (monensin was introduced in the USA in 1971) is a result of their broad-spectrum activity, and possibly because resistance in coccidia may not be able to develop by the mechanisms known to occur for synthetic drugs. There have also been early reports that monensin "resistant" *E. maxima* strains may be present in the field. Later selected *Eimeria* spp. from broiler farms in USA and Europe revealed that control of some isolates to *ionophorous drugs* was poor although none of the isolates judged in sensitivity tests was found to be completely resistant to the ionophores tested. Thus, prolonged use of ionophores in poultry units for nearly 10 years led to a decrease in their anticoccidial activity. Salinomycin provided the best overall control, followed by lasalocid and monensin. The latter findings are in agreement with results obtained from field studies on the comparative efficacy of salinomycin and monensin in 17 controlled field trials. They included more than 2 million broilers and were carried out in several European countries, and Salinomycin at 60 ppm showed performance advantages over monensin at 100 ppm. Maduramicin proved to be more effective against ionophore-tolerant field isolates of broilers than monensin and narasin, but showed similar activity to salinomycin in reducing lesions and mortality and in protecting performance. Although **side-resistance** (co-lateral resistance) among related ionophores such as monensin salinomycin, and narasin may occur 'cross-resistance' between maduramicin and not related ionophores has also been reported; it is believed by some investigators that maduramicin is effective against strains resistant to the other ionophores. Results from drug-sensitivity tests with isolates of coccidia from broiler chickens in the USA, Brazil, Argentina, and Europe suggested in many cases incomplete side resistance of coccidia to polyether ionophorous drugs. Some field isolates revealed even complete multiple **cross-resistance** to synthetic and polyether antibiotics including maduramicin, and semduramicin (latest ionophore introduced into the market). An *E. acervulina* laboratory strain, which was resistant to monensin by passaging the strain 14 times in the presence of the drug (100 ppm), was shown to be sensitive to lasalocid but not to maduramicin, narasin, and salinomycin. Cross-resistance between the latter ionophores may suggest a similar mode of action in coccidia. Thus, monensin, maduramicin, narasin, and salinomycin, preferentially form complexes with monovalent rather than divalent alkali metals (e.g., lasalocid), e.g., Rb^+, Na^+, K^+, Cs^+, which may mediate electrically neutral exchange diffusion cation transport across membranes. Contrarily, a monensin-resistant *E. meleagrimitis* strain exhibited cross-resistance to narasin and lasalocid and a field isolate of *E.tenella* showed cross-resistance between lasalocid and several ionophores preferentially forming complexes with monovalent cations. To slow down development of drug resistance in coccidia alternate use of anticoccidials is widely practiced in the poultry industry (Measures Against Drug Tolerance in the Broiler Industry).

Measures Against Drug Tolerance in the Broiler Industry

As shown by epizootiological investigations, coccidial drug resistance poses a serious economic problem to intensive poultry farming both in the USA and in Europe. Today, there is therefore an increasing use of **shuttle** or **dual programs** (rapid rotations), i.e., the switching of anticoccidials during broiler growout. The drug(s) to be switched should belong to different chemical classes (e.g. ionophore/synthetic drug or vice versa during starter/final phase of production) and should be effective at different stages of the parasites' life cycles. Any resistant coccidia, which appear during the use of the first drug, should be affected by the second. In **straight** or **slow rotation** programs a single and often the same drug may be used for several broiler growouts (e.g. about 6 months; each fattening period may last 35–40 days or longer); it is then replaced by an alternative drug. Shuttle programs are still being discussed with regard to their value in delaying drug resistance. Although alternation of the drugs between crops may delay the appearance of resistance, it is likely that the outcome will be the acquisition of multiple resistance. There may also be the risk of underdosing or insufficient activity of drug mixtures resulting from switching the drugs. This might allow resistance to be developed faster. However, by switching the drug underdosing can be excluded if new and exactly medicated feed is poured onto remaining feed; then the two drug fractions at the base of the silo funnel must add up to 100% of any resulting mixture. Mixtures of ionophorous antibiotics were as effective in controlling drug-sensitive *E. tenella* and *E. acervulina* infections as a single antibiotic mixed in the feed at recommended prophylactic concentrations. However, such mixtures (e.g., monensin + narasin or salinomycin) may have limited additive action if there is already a partial resistance to one of the partners. In these experiments it was also shown that combinations of ionophorous antibiotics and synthetic drugs are complementary, even against field isolates with unknown drug response.

Drug combinations (Table 1) have commonly been used for synergism in order to minimize the occurrence of drug resistance (e.g., Lerbek® → methyl benzoquate +meticlorpindol), or to extend the species spectrum to all six pathogenic *Eimeria* spp. in chickens. Amprolium and dinitolamide which show activity against *E. tenella* and *E. necatrix* have been combined with ethopabate and organic arsenicals respectively, in order to expand their spectrum to include the upper intestinal species *E. maxima*, *E. acervulina*, and *E. mitis*. Joyner and Norton found the only experimental evidence of retardation of drug resistance by synergistic effects of drugs. They observed complete resistance to methyl benzoquate 10 ppm and meticlorpindol 125 ppm, respectively, after three passages of *E. maxima*, but not with a 8.35 and 100 ppm combination of the two drugs.

To keep up their competitive position with other anticoccidials, narasin and maduramicin have been combined with nicarbazin, an oldtimer among anticoccidials; these products, however, reveal no performance advantages over related mono-drugs.

Drugs Acting on Cryptosporidiosis in Birds

Species of the genus → Cryptosporidium (Vol. 1) are coccidian parasites that infect epithelial cells (extracytoplasmic) of the intestinal and respiratory tract of vertebrates (see Opportunistic Infections). Although immunocompetent hosts show no or only mild clinical signs after *Cryptosporidium* infections particularly young birds under stress may suffer from life-threatening watery diarrhea, or severe respiratory symptoms. Cryptosporidiosis in chickens, turkeys, quail, and pheasants is usually manifest as respiratory disease caused predominantly by *C. baileyi* or as enteritic disease (small intestine) caused by *C. baileyi* and *C. meleagridis*. The severity of infection depends on the immunocompetence of the host. Infections are due to aerosol transmission of infective oocysts coughed up by carrier (seeder) birds, or may be transmitted by feed or water supplies containing sporulated oocysts derived from feces of infected birds. Clinical signs in birds are coughing, mucoid discharge, dyspnoe, diarrhea, dehydration, weakness and weight loss.

Causal therapy and chemoprophylaxis of chicken cryptosporidiosis with ionophorous antibiotics is problematic. Many approaches to anticryptosporidial efficacy of commercial drugs have failed to improve symptoms in birds suffering from *Cryptosporidium* infections. Several other anticoccidials as sulfonamides, lasalocid sodium, halofuginone, and decoquinate, or other antibiotics (e.g., paromomycin) available as additives in-feed (Table 1), or as other dosage forms for oral administration have proved to be insufficiently effective in controlling or even eradicating *Cryptosporidium* infection in birds. The drugs may exhibit positive short-term effects such as improvement of watery diarrhea and reduction of oocyst output in feces due to their 'static' rather than 'cidal' action on cryptosporidia.

Coccidia of Humans

Coccidia that may cause clinical signs in humans are rare. *Isospora belli*, a single host species in the small intestine, may be responsible for mild intestinal symptoms in some cases. Infections due to heteroxenous species like → Sarcocystis (Vol. 1) spp. (→ Sarcocystosis) or → Toxoplasma gondii (Vol. 1), and related protozoans with a two-host life cycle are seen more often, especially in areas where raw meat is eaten and human beings have contact with cats or other felines. Moreover, *Isospora belli*, *Cryptosporidium parvum*, *Toxoplasma gondii*, and *Cyclospora* spp. belong to the opportunistic protozoa associated with immunosuppression caused by HIV (AIDS), or other pathogens and pathogenic processes (compare page 128).

Coccidiosis, Animals

General Information

The coccidia are members of the suborder Eimeriina. They are typically highly host-, organ-, and tissue-specific. Under natural conditions, most mammals pass small numbers of coccidial oocysts in their faeces, without apparent clinical effect. Coccidiosis becomes important as a disease when animals are reared under conditions that permit the build-up of high numbers of infective oocysts in the environment. This is because the degree of damage caused by coccidia depends upon the numbers of parasites able to replicate in any given site, which depends firstly upon the numbers of infective stages (oocysts) ingested. This is different from other protozoa which may reproduce indefinitely by binary fission, until halted by host immunity or death. *Eimeria* infections are self limiting because the parasites only pass through a limited number of asexual multiplications. Coccidiosis involves (extensive) destruction of the intestinal epithelia. The effects of intestinal coccidiosis in mammals vary with the host-parasite system. They are mainly related to malabsorption induced by villous atrophy and reduction of brush border enzymes, or to anaemia, hypoproteinaemia and dehydration due to exudative enteritis and colitis caused by epithelial erosion and ulceration. High mortality may occur in ruminants infected by the most pathogenic species (see also → Nervous System Diseases, Ruminants).

Pathology

Cattle About 15 species of *Eimeria* parasitize cattle; of these *E. zuernii* and *E. bovis* are potentially highly pathogenic. Other species, such as *E. auburnensis* may at times contribute to the general clinical picture. In general, the infection occurs in calves or weaned feeder cattle under one year of age, but clinical disease occasionally occurs in adults, especially if massive infections are acquired during stressful situations. The diseases caused by *E. zuernii* and *E. bovis* are very similar; they are characterized by a haemorrhagic diarrhoea which may become so severe that pure blood is passed instead of faeces. Tenesmus is marked, and there is anaemia, weakness, anorexia and emaciation. In severe infections death may occur. The first clinical signs appear just before the peak in oocysts output (day 18–19). At that moment there is maximal loss of epithelium in the large intestine due to destruction of cells by second generation schizogony and gamogony. This causes the exposure of the lamina propria and the formation of diphtheritic membranes. The destruction of the epithelium leads to reduction in the reabsorption of water, Na+ and Cl- from the intestinal contents. The abrupt loss of weight and the reduction of the plasma concentration of these two ions support this contention. Exposed capillaries of the large intestinal lamina propria may rupture, leading to loss of erythrocytes and plasma.

Sheep and Goats Coccidial infection is virtually universal in sheep and goats, and large numbers of oocysts may be found in the faeces of clinically normal animals. Coccidiosis in small ruminants is chiefly restricted to young animals up to 4–10 weeks of age. Close morphological similarity between the oocysts of *Eimeria* species from sheep and goats has caused some confusion in the literature. Relatively little is known about the pathogenicity of the different species, but it has been established that *E. ovinoidalis* in sheep and its analogue in goats, *E. ninakohlyakimovae* can be very pathogen. Other species such as *E. bakuensis* (*E. ovina*) and *E. crandallis* in sheep, and *E. arloingi* and *E. christenseni* in goats may exacerbate the symptoms of the former two species. Outbreaks of coccidiosis are usually acute and characterized by moderate morbidity and low mortality. There is a green or yellow watery diarrhoea with a fetid odour, occasionally with blood. Abdominal pain, some anaemia (macrocytic, hypochromic), loss of appetite, dehydration, tenesmus, weakness and loss of weight occur. Depression, inactivity, and recumbency are prominent. Pathological changes include thickening of the caecum and colon mucosa, oedema, haemorrhage and hyperaemia. Myiasis, bacterial diarrhoea, and bacterial septicaemia often accompany coccidiosis outbreaks.

Horses The only species occurring in horses, *Eimeria leuckarti*, is not known to cause disease.

Swine At least 12 species of *Eimeria* are thought to occur in swine, and a single species of *Isospora*. *Eimeria* spp are not considered a major cause of disease in pigs and many animals are asymptomatic carriers. *Isospora suis* is the only important species. It causes porcine neonatal coccidiosis, a disease of piglets from about 5–6 days to about 2–3 weeks of age. It is characterized by a yellow, foul smelling diarrhoea, dehydration, occasional vomiting, loss of condition and death, or at the least a temporary check in growth. Morbidity is usually high, mortality low or moderate. Necropsy reveals villous atrophy and a marked, sometimes necrotizing, enteritis of the small intestine. Simultaneous infection of *I. suis* with viruses and *E. coli* results in more severe lesions and clinical disease than with coccidia alone.

Carnivores There are at least 14 species of coccidia in canine faeces: *Isospora canis*, *Isospora (Cystoisospora) ohioensis*, *I. burrowsi*, *I. neorivolta*, *Hammondia heydorni*, *Neospora caninum* and eight species of *Sarcocystis*. A total of at least 15 species exist in cats: *Isospora felis*, *I. rivolta*, five species of *Besnoitia*, *Hammondia hammondi*, *Toxoplasma gondii*, and six species of *Sarcocystis*. In relation to pathogenicity, it is usually the intermediate hosts rather than the dog or cat that are adversely affected. Clinical coccidiosis in dogs or cats is apparently caused by certain species of *Isospora* and by *T. gondii*. For example *I. ohioensis* can cause clinical disease in newborn pups. Diarrhoea is caused by inflammation of the intestinal crypts, with necrosis and massive desquamation of the tips of villi, especially in the lower part of the small intestine.

Immunity and Vaccination Reports from the literature suggest that immunity to coccidia is short-lived in young animals. In contrast, repeated coccidian infections induce a more long-lasting immunity than a single (primary) infection. An increasing infection pressure may cause a deterioration of immunity that is evident by an enhanced number of intracellular developmental stages and so output of oocysts, and possibly the development of clinical signs. The effect of immunity can range from complete (or close on zero) inhibition of oocysts production (premunition) to the passage of smaller numbers of oocysts in the faeces (partial immunity). It is assumed that **premunition** (immunity of the non sterile type) to coccidia depends upon the persistence of some (occult, extraintestinal) development stages in the immune host from initial infection and/or reinfection. Thus, the acquired (often) species- or even life-cycle-stage **specific immunity** of the host to coccidia plays an important role in the control of parasites and may depend on various host factors such as individual immune status, age of the host and its genetic background (breed type: innate immunity). It seems that development of natural immunity to coccidiosis through digestion of sporulated oocysts is rather slow and may take several weeks or even longer. Therefore, using **vaccines** to prevent coccidiosis in the short life span of fattening young animals appears problematic because protective immunity resulting from vaccination may be insufficient. This is true particularly in case of broilers, which life span lasted about 35–40 days only. On the other hand, breeder replacement chicks and commercial layers are able to profit from the immunity protection against coccidiosis. They may be exposed to a controlled number of sporulated coccidia oocysts, i.e., to **live virulent vaccines**. Today, various types of live vaccine for the control of poultry coccidiosis are available (see Table 1). Coccivac® or Immucox® (sporulated oocysts of wild-type) has been used mainly for replacement birds, which represent a relatively small market in comparison with that of the broiler industry. In using such vaccines it appears economically acceptable to have some loss of performance as result of immunization. However, live virulent *Eimeria* vaccines for control of coccidiosis in broiler chicken have rarely been used because loss of body weight, and its effect on feed conversion is not acceptable. **Attenuated parasites** for delivery in the drinking water are 'sub-lines' of sporulated oocysts from chicken derived from the progeny of single oocysts. They may be attenuated after long-term passages through the chorioallantoic membranes of embryonating eggs (Livacox®), or they may be recovered after only a few passages with selection for early development by passages through chickens (Paracox®, Livacox®) or rabbits. **Precocious lines** of coccidia species are drug sensitive and thus any use of anticoccidial drugs should be avoided during the month following vaccination. Precocious lines are characterized by a shorter life cycle, i.e., decrease in prepatent period, number and size of endogenous stages (e.g., deletion of the terminal generation of schizonts of the wild-type parents). They can induce protective immunity against coccidiosis in spite distinct attenuation of their virulence. Such parasites (oocysts) are no longer able to cause infections with severe clinical signs in chickens or rabbits (the production of precocious lines in cattle obviously failed). Problems arising from vaccination in the poultry industry or elsewhere may be the delivery and application practice of such live vaccines. Thus the application management of vaccines seems to be more sophisticated than that of feed-in products in widely used chemoprophylaxis. Another problem may arise through genetic instability in precocious sublines. In addition, cost of production of live **"cocktail"-vaccines** containing all (Paracox®) or only certain *Eimeria* spp. (Livacox®) may be considerable and above that of in-feed delivered anticoccidial drugs.

The chances for developing protective **recombinant vaccines** against coccidiosis in **chickens** appear promising though *E. maxima* shows anti-

Table 1. Live vaccines on the market

Trade name	Pathogenicity	Administration	Producer
Immucox®	virulent	drinking water	Vetech Labs. Inc., Canada
Coccivac®	virulent	drinking water or feed	Sterwin Labs. Inc., USA
Paracox®	attenuated	drinking water	Mallinckrodt Vet., USA
Livacox®	attenuated	drinking water	Biopharm, Czech Republic

Table 1. Anticoccidial drugs used in the poultry industry.

Chemical group (approx. dose, ppm[a]); relevant information	**Non proprietary name,** *brand name (manufacturer, company); other information	**Characteristics**[b] and miscellaneous comments
SULFONAMIDES		
(120–250); recommended dose levels vary and depend on mode of administration, i.e., feed or water, and pharmacokinetic properties of drugs; [c1]1940 [$]not licensed for prevention of coccidiosis in poultry or other animals under European Commission Guidelines (additives in feeding stuffs), and feed regulations of Food and Drug Administration (FDA) in the USA	[c1]**sulfanilamide** sulfaguanidine sulfamethazine sulfaquinoxaline sulfachloropyrazine sulfadimethoxine sulfadimidine sulfamerazine, and others *various	opened up the field for practical use; sulfonamides have a broad spectrum of activity against *Eimeria* spp. of the anterior and lower part of the intestine but only a moderate effect on *E. tenella* (ceca) in chickens; in swine and rabbits, prophylactic feeding of **sulfaguanidine, sulfaquinoxaline, sulfachloropyrazine** (others) may prevent clinical signs and reduce oosyst production thereby allowing development of protective immunity; drugs resistance may limit their use as prophylactics; water solubility renders sulfonamides useful for treatment of coccidiosis in ruminants, swine, (game)birds, and dogs (withdrawal time may exceed 14 days in food-producing animals); drugs are active against first- and second-generation schizonts and probably against sexual stages; their action seems to be coccidiocidal at higher doses and coccidiostatic at lower doses; large doses used for therapeutic applications often cause toxicity (hemorrhagic syndrome, kidney damage, growth depression); sulfonamides interfere with cofactor synthesis, i.e., block dihydropteroate synthetase and hence synthesis of tetrahydrofolate (cofactor is required for cellular methylation reactions); their action can be nullified by excess p-aminobenzoic acid (PABA); in combination with pyrimethanmine and other diaminopyrimidines, long-acting sulfonamides (e.g., sulfadoxine or sulfamethoxin) are highly active antimalarials and antibacterials
Diaminopyrimidines (dose varied depending on combination used) [$](see sulfonamides)	**diaveridine** **ormetoprim** **pyrimethamine** **trimethoprim** * various	usually used in combination with long-acting sulfonamides (see above); as antifolates they synergize (potentiate) anticoccidial action of sulfonamides by blocking the same biosynthetic pathway; they are useful for treating various types of coccidiosis (eimerioses, toxoplasmosis, sarcocystosis, neosporosis), malaria (→ Malariacidal Drugs), and bacterial infections; drugs interfere with synthesis of tetrahydrofolate by inhibition of dehydrofolate reductase reaction
2,4-Substituted aminobenzoic methyl ester (doses: **see** amprolium)	**ethopabate**	has a good innate activity against *E. acervulina* but is less active against *E. maxima* and *E. brunetti* (no activity against *E. tenella*); drug is only used in combination with other anticoccidials (see amprolium); as an analogue of p-aminobenzoic (PABA) it interferes (like sulfonamides) with dihydropteroate synthetase reaction, during which conjugation of a pteridine moiety and PABA occurs; PABA excess neutralizes action of the drug
PYRIDINOLES		
(125) [c2]1968	[c2]**meticlorpindol →clopidol** *Coyden (Dow Agriculture, marketed by Merial) generics (others)	licensed for use in broiler chickens, partridge (guinea fowl, pheasants: restricted use in some countries) and rabbits (125–200 ppm); it arrests sporozoite and trophozoite development, i.e., is coccidiostatic, and withdrawal leads to relapse of infection; it is effective against all *Eimeria* spp. in chickens but considerable problems have been reported with *E. acervulina*; very safe for chickens and safe in mammals; drug resistance problems limit its use to shuttle programs (e.g., last 1–3 weeks of broiler growout); it potentiates the action of 4-hydroxyquinolines since chemical structure of the drug is related to this series of compounds

Table 1. (Continued) Anticoccidial drugs used in the poultry industry.

Chemical group (approx. dose, ppm[a]); relevant information	**Non proprietary name,** *brand name (manufacturer, company); other information	**Characteristics**[b] and miscellaneous comments
4-Hydroxyquinolines *(82) [c3]1967 **(20–40) [c4]1970 ***(110)[c5]1974	[c3]**buquinolate** *Bonaid (Norwich), discontinued [c4]**decoquinate** **Deccox (Rhône Poulenc, Rorer; (methyl benzoquate) *Statyl (earlier ICI), discontinued [c5]**methyl benzoquate** (8.35 parts) + **meticlorpindol** (100 parts) ***Lerbek (now Merial)	*buquinolate* was "commercially dead" within 6 months (sudden and dramatic appearance of drug resistance); hydroxyquinolines are almost entirely coccidiostatic against sporozoites and trophozoites of all *Eimeria* spp. in chickens (see also meticlorpindol); as single compounds they have only limited success (also **methyl benzoquate** at 10ppm and higher doses) as a result of serious and immediate drug resistance in the field; it has been shown that methylbenzoquate-resistant *Eimeria* spp. strains cannot be controlled by the drug at any level. **decoquinate** is licensed for use in broiler chickens and replacement chickens (approval for treatment and prophylaxis of coccidiosis in lambs and calves may be restricted to some countries); **methyl benzoquate** plus **meticlorpindol** exhibit synergistic activity: the combination is licensed for use in broilers, replacement chickens and turkeys; it is used mainly in shuttle (rotation) programs; target of mode of action of this series is energy metabolism: compounds block electron transport down the cytochrome chain in the mitochondria of coccidia and hence inhibit NADH oxidation and ATP synthesis as well
NITROBENZAMIDES		
*(62.5–125) [c6a]1960	[c6a]**dinitolmide** ***Zoalene** (earlier Dow chemical) **Zoamix (Alpharma, USA)	is licensed for use in broiler chickens and poultry; the drug arrests development of first- and second-generation schizonts but would not interfere with development of immunity; it is completely protective against *E. tenella* and *E. necatrix* infections but has limited activity against *E. acervulina*; to extend activity or growth promotion, drug has also been combined with sulfanitran or roxarsone; it may chiefly be used in breeder or replacement chickens (up to 16 weeks of age); mode of action is unknown
(250) [c6b]1958	[c6b]**nitromide** *Unistat (Salsbury)	first nitrobenzamide to be sold was nitromide; to extend its activity spectrum, drug was combined with sulfanitran and roxarsone (= *Unistat); it has the same biological properties as dinitolmide (see above)
(250) c6c1965	[c6c] **aklomide** *Aklomix (Salsbury)	Marketing of aklomide has been discontinued; it had no advantage over nitromide or dinitolmide
Organic arsenicals *(50) [c7a] 1946 **(400) ***(20) [c7b] 1949 [§]not licensed for prevention of coccidiosis in poultry or other animals under European Commission Guidelines (additives in feeding stuffs)	[c7a]**roxarsone** *3 Nitro(Salsbury, Rhône Poulenc) [c7b]arsanilic acid **Pro-Gen (Abbott) ***arsenosobenzene	primary application of arsenicals has been growth promotion; roxarsone should have some activity against *E. tenella* and *E. brunetti* used alone or in combination with nitrobenzamides; arsenicals were almost completely eliminated from the market by environmental problems

Table 1. (Continued) Anticoccidial drugs used in the poultry industry.

Chemical group (approx. dose, ppm[a]); relevant information	**Non proprietary name,** *brand name (manufacturer, company); other information	**Characteristics**[b] and miscellaneous comments
Carbanalide derivatives (100–125) [c8]1955	[c8]**nicarbazin** (Coffolk, Israel)(equimolar complex of dinitrocarbanilide and dimethylpyrimidinol) *Nicrazin *Nicarb (Merck, Sharp, and Dohme → MSD AgVet, now Merial, other distributers)	first drug with "broad-spectrum" activity against *Eimeria* spp. in chickens and the most effective of the older drugs; it shows synergistic effect with polyether antibiotics (see: combination with **narasin**); its action is directed against developing second-generation schizonts; it is licensed for use in broiler chickens only and still has wide application in the poultry market, especially in shuttle programs (starter feed only) in winter or cooler month; for that reason resistance of coccidia to nicarbazin is not yet widespread; there may be problems with side effects – it can cause increased sensitivity to heat stress during summer which results in growth depression and mortality in broilers; it may be used in replacement pullets (125 ppm, up to 6 weeks of age); the drug should not be fed to (laying) hens because of toxic side effects (reduced hatchability, interruption of egg laying); mortality may be caused by cell degeneration processes in liver and kidneys; its mode of action, and thus mechanism of selective toxicity, seems to be unknown in coccidia
NITROFURANS		
recommended dose levels may vary: (50), preventive use; (120), therapeutic use [c9]1948 [$]not licensed for prevention of coccidiosis in poultry or other animals under European Commission Guidelines (additives in feeding stuffs), and feed regulations of Food and Drug Administration (FDA) in the USA	[c9]**nitrofurazone** (**nitrofural**) **furazolidone** **furaltadone** *various (various manufacturers)	possess antimicrobial and limited coccidiostatic activity; **nitrofurazone** affecting second-generation schizonts of *E. tenella* and *E. necatrix* has also been used for control of coccidiosis in lambs and goat kids; **furazolidone**, which has been used for treating established infections and preventive control of coccidiosis, produces neurologic symptoms at higher concentrations (400 ppm) when fed in combination with either dinitolmide (1–5 ppm) or amprolium (125 ppm); it has also been administered to chickens, turkeys, and swine for control (shuttle programs only) and treatment of various digestive tract infections (bacterial enteritis, dysentery, giardiasis); **furaltadone** has been applied successfully for treatment of *Salmonella* and *Mycoplasma* infections in chickens; toxic and chemical properties of nitrofurans have restricted their widespread use; nitrofurans are suspected to be carcinogenic, and in many countries there are activities against this series of compounds
Thiamine analogues (62.5–125 and higher doses) [c10]1960	[c10]**amprolium** (all products MSD Agvet, now Merial)	**amprolium**, and **amprolium+ethopabate** (feed additives) are licensed for use in chickens, guinea fowl, and turkeys for prevention of coccidiosis; active on *E. tenella* and *E. necatrix* (to a lesser extent against *E. maxima*) of chicken and pathogen *Eimeria* spp. of turkeys ; activity is directed against first- and second-generation schizonts (coccidiostatic at lower, coccidiocidal at higher doses). Amprolmix may still be used chiefly in turkey poults because of its good tolerability and allowing development of immunity needed after withdrawal of the product (in replacement layers/poults up to 10–16/8–10 weeks of age); amprolium resistance in replacement pullet farms is a problem and limits its use; **amprolium+ethopabate** have been combined with sulfaquinoxaline, and pyrimethamine to extend their activity spectrum and to improve efficacy against amprolium-resistant Eimeria spp. strains; these combinations have been discontinued in some countries because of residue problems; they have also been used therapeutically; **amprolium** probably acts by inhibiting thiamine uptake by parasites; this vitamin (thiamine pyrophosphate) is a cofactor of several decarboxylase enzymes playing a role in cofactor synthesis; amprolium cannot be pyrophosphorylated (lacks hydroxyethyl group); **ethopabate** (an arylamide containing one phenyl ring, belonging to monocyclic aromatics) is a very safe drug, and a competitor of PABA for absorption by the parasite;
*(66.5–133)	**amprolium** (25 parts)+**ethopabate** (1.6 parts) *Amprolmix Premix	
**(165→100+5+60)	**amprolium** (18 parts)+**ethopabate** (0.9 parts)+**sulfaquinoxaline** (10.8 parts) **Pancoxin (discontinued in some countries)	

Table 1. (Continued) Anticoccidial drugs used in the poultry industry.

Chemical group (approx. dose, ppm[a]); relevant information	**Non proprietary name,** *brand name (manufacturer, company); other information	**Characteristics**[b] and miscellaneous comments
***(170→100+5+60 +5)	**amprolium** (20 parts)+**ethopabate** (1 part)+**sulfaqinoxaline** (12 parts)+**pyrimethamine**(1 part) ***Pancoxin Plus (MSD)	outside the product license, amprolium and Amprolmix (other combinations may be discontinued) has been applied for control and treatment of coccidiosis in various animals such as pheasant (not active against all *Eimeria* spp.), sheep and cattle (medicated feed: amprolium/ethopabate: 250/16 ppm), sows (to control disease in suckling pigs pre and post farrowing: see latter concentrations), or rabbits (medicated feed, amprolium plus ethopabate:125+8 ppm to control intestinal *Eimeria* spp., amprolium is ineffective against hepatic coccidiosis in rabbits); in young ducklings tolerance of Amprolmix appears to be some what erratic and its use is best avoided
(125)	**beclotiamine** *Coccidien (Sankyo)-dimethalium	there are several related thiamine analogues which are similar in their anticoccidial activity and mode of action to amprolium; beclotiamine or dimethalium have never been used extensively since they have no advantage over amprolium and have common drug resistance problems
QUINAZOLINONES		
(2–3) [c11]1976	[c11]**halofuginone** (hydrobromide) *Stenorol (Intervet)	feed additive for prevention of coccidioses is licensed for use in broiler chickens, replacement pullets (up to 16 weeks of age) and turkey poults (up to 12 weeks of age) and usually used in shuttle (or rotation) programs with one of the ionophores; it has broad spectrum of activity against all pathogenic coccidia of chickens and turkeys, and affects asexual stages, particularly during first-generation schizogony maturation; action of drug is coccidiocidal/coccidiostatic but in case of *E. acervulina* not as strong as with other species; drug resistance may occur if drug is used for too long periods in continuous (straight) medication programs; it is well tolerated at 3 ppm in birds licensed; however, in anseriformes (ducks, geese, swans), water fowl, guinea fowls, partridges or other game birds, and rabbits 3 ppm may cause serious side effects and mortality after continuous medication; mode of action of halofuginone in coccidia seems to be unknown; in skin fibroblasts of chickens it interferes with collagen synthesis thereby decreasing skin strength and increasing incidences of skin tears during processing; shuttle programs in which halofuginone is included in grower feed seems to maintain skin integrity in broilers
unpalatable at recommended concentrations for ducks, geese, guinea fowl, partridge, quail and rabbits; this drawback may cause reduction in feed intake and/ or toxic reactions, mortality	originally derived from a plant extract (Dichroa febrifuga Lour); **febrifugine** had an antimalarial and anticoccidial effect but a narrow safety margin; synthetic variations previously made by American Cyanamid led to halofuginone	
Guanidine derivatives (30–36, in birds) (50–66, in rabbit) [c12]1972	[c12]**robenidine** *Cycostat 66 (Europe) *Robenz (USA) (earlier Cyanamid; now Roche)	feed additive licensed for prevention of coccidiosis in broiler chickens, turkeys and rabbits for meat production; has broad-spectrum activity (rabbits intestinal *Eimeria* spp. only); it is most effective against late developing stages of first and second -generation schizonts, and possibly it exhibits some activity against gamonts (sexual stages); its action is first cocidiostatic and then coccidiocidal; unexpected quick development of drug resistance on poultry farms in USA and Canada within a year of its introduction has limited its use; drug is believed to interfere with energy metabolism by inhibition of respiratory chain phosphorylation and ATPase activity in rat liver mitochondria; other guanidine derivatives which lack anticoccidial activity also share this inhibitory activity on oxidative phosphorylation process
if not withdrawn in prescribed time (5 days), or used at higher concentrations drug produces an unpleasant (medical) flavor in edible broiler tissue or eggs of layers		

Table 1. (Continued) Anticoccidial drugs used in the poultry industry.

Chemical group (approx. dose, ppm[a]); relevant information	**Non proprietary name**, *brand name (manufacturer, company); other information	**Characteristics**[b] and miscellaneous comments
Purine analogues (60) [c13]1980	[c13]**arprinocid** *Arpocox (Merial) (discontinued) quick emergence of drug resistance in coccidia	feed additive licensed for prevention of coccidiosis in broiler chicks and replacement pullets (up to 16 weeks of age); it has broad-spectrum activity but *E. tenella* is not as completely affected as other species; it inhibits sporulation of oocysts and may be coccidiostatic (after short medication development of sporozoites and trophozoites is arrested) or coccidiocidal (partial damage to development of first- and second-generation schizonts) after long medication; drug is well tolerated in broilers at 60 ppm; quick emergence of resistant strains in the field has largely limited its use (mainly to shuttle programs); main metabolite is arprinocid-1-N-oxide; is excreted in urine as parent compound and as arprinocid-1-N-oxide; arprinocid acts against coccidia by inhibiting hypoxanthine transport, whereas arprinocid-1-N-oxide causes dilation of rough endoplasmic reticulum in coccidia
POLYETHER IONOPHOROUS ANTIBIOTICS		
additives in-feed licensed either for prevention of coccidiosis in poultry or improving feed conversion efficiency and/or weight gain in cattle or swine; latter products (performance enhancers or promoters) used at lower doses than the analogous anticoccidials may affect coccidia in ruminants and pigs; various ionophores show incompatibility with sulfonamides and/or antibiotics in feed; equines are very sensitive to most of the ionophores;	in vitro studies with monensin, salinomycin, or lasalocid reveal that sporozoites of *E.tenella* are damaged after 24 h incubation, but are not if incubation is shortened (20 min); under laboratory conditions polyethers allow a relatively high oocyst output due obviously to their lack of activity against sporozoites in vivo; it has been shown that these ionophores do not only cause irreversible damage to free merozoites and mature schizonts of first- and second-generation coccidia, and also to erythrocytic stages of various chloroquine-resistant malaria parasites	ionophores are fermentation products of various *Streptomyces* spp. or *Actinomadura* sp. (maduramicin, semduramicin) containing mycelial biomass (maduramicin can be extracted into granular carrier); continuing commercial success is due to their broad-spectrum activity against pathogenic *Eimeria* spp. in poultry and also to a prolonged absence of serious problems with drug resistance; today, resistance in chicken coccidia to ionophores is a problem; they have good efficacy against coccidia in rabbits but their toxicity limits usefulness; in general, ionophores are relatively toxic to animals (particularly horses), i.e., their safety margin may be narrow in birds and mammals; monensin, salinomycin, and narasin are incompatible in broilers with **tiamulin** (macrolide antibiotic); they interact with the drug, (>40 ppm) resulting in toxicity (weight depression, mortality) due to decreased elimination of the ionophores; lasalocid, maduramicin and semduramicin show compatibility with tiamulin; however, most of the ionophores may interact with sulfonamides, chloramphenicol and erythromycin; dose-related toxicity is evident by weakness, leg paralysis, and sharp reduction of water intake; polyethers are able to interact with physiologically important cations (e.g., Rb^+, Na^+, K^+, Cs^+, or Ca^{2+}) destroying their cross-membrane gradients; however, until now the mode of action of ionophores in coccidia has only been speculated on;
(100–125) in broiler chickens; (100–120) in layer replacement chickens, up to 16 weeks of age; (90–100) in turkeys for fattening, up to 12 weeks of ages [c14]1968	[c14]**monensin**-Na *Coban *Elancoban (Elanco; Eli Lilly) other product license holders distributers and brand names	**monensin** was the first ionophore used in poultry; use levels and withdrawal times licensed may vary in various countries (80–125 ppm; 0–5 days; concentrations cited are on basis of EEC directives); has been used extensively in the broiler industry since its introduction, and today monensin tolerance (reduced drug sensitivity) in coccidia is a problem in the field; this is also true for all other ionophores and synthetic drugs if they are used for longer periods in straight programs (see respective chapter); like other ionophores, it exhibits good efficacy against coccidia in cattle, sheep and rabbits when used prophylactically in feed (see also respective entries); monensin may produce poor feathering if diets with low energy and low sulphur-containing amino acids are fed; at recommended doses it may be toxic to guinea fowl and other avian species, or to chickens and turkeys producing eggs; horses are very sensitive to monensin (LD 50 % : horse, 2–3 mg/kg body weight; cattle, 25 mg /kg; chicks, 200mg/kg)

Table 1. (Continued) Anticoccidial drugs used in the poultry industry.

Chemical group (approx. dose, ppm[a]); relevant information	**Non proprietary name,** *brand name (manufacturer, company); other information	**Characteristics**[b] and miscellaneous comments
(75–125) in broiler chickens; (75–125) in layer replacement chickens up to 16 weeks of age; (90–125) in turkeys up to 12 weeks of age; [c15]1974	[c15]**lasalocid** *Avatec (Roche) may be used (90–120) in pheasants, and partridges up to 12 weeks of age;	Chemistry differs from that of other ionophores; it has stronger affinity to divalent cations than monovalent ions, and this may be the reason for differences in effects on birds and coccidia compared to other ionophores; in birds, it possibly alters the water excretion via dietary electrolytes to the extent that wet litter may be a problem at higher drug concentrations; at lower concentrations (75ppm) activity against *E. acervulina* is insufficient and strongest against *E. tenella*; in the field, lasalocid may improve control of coccidiosis where *E. tenella* strains reveal tolerance to other ionophores; like other ionophores, lasalocid shows activity against coccidia in sheep and cattle when given prophylactically in feed; equines are very sensitive to the drug
(50–70) in broiler chickens [c16]1978 **(50–70) in broiler chickens; **(30–50) in replacement chickens, **(20–25) in fattening rabbits	[c16]**salinomycin-Na** *Coxistac (Pfizer) *Usten (Kaken: patentee) *Eustin (Bayer) *Bio-Cox (earlier Robins; now Agri-Bio, American Home Products Corp.) **Sacox 120® micro-Granulate (Invervet) other companies/brand names	Approved levels (44–70), and animal species licensed may differ in various countries; **salinomycin** has broad-spectrum activity (including coccidia of rabbits) and better activity against *E. tenella* and *E. acervulina* than other related ionophores; this is also true in case of drug-tolerant *Eimeria* spp. in the field; it may cause severe toxicity (growth depression, excitement followed by paralysis with head and legs extended, mortality) if feed containing recommended or lower doses is fed to turkeys for longer periods; growing turkeys tolerate the drug better than adult ones, and toxicity of salinomycin to equines is less than that of monensin; **Salinomycin derivatives** with increased lipophilicity were obtained by acylation of the hydroxyl moiety at C-20); the 3-methyl propanoyl ester was about six times more active than the parent compound; many derivatives of the series proved to be toxic very similar to salinomycin in chemistry ("methyl-salinomycin") have affinity to monovalent cations; their anticoccidial effect and performance in field proved not as good as that of salinomycin; to enhance activity, **narasin** has been combined with **nicarbazin**; synergism of
*(60–70 in broiler chickens) [c17a] 1983 **(80–100 in broiler chickens) [c17b]1988	[c17a]**narasin,** *Monteban (Elanco; Eli Lilly [c17b] **narasin +nicarbazin**(1:1 mixture), **Maxiban (Elanco; Eli Lilly)	both drugs improved coccidiosis control; drug combination may be used in starter phase of shuttle programs followed by a different ionophore in grower-finisher phase; broiler feed containing 70ppm or lower (subnormal) concentrations of narasin may cause mortality if fed to adult turkeys and growing turkeys (mortality to a lesser degree); equines are very sensitive to the drug (see: monensin and salinomycin)
(5) in broiler chickens [c18]1984 (3.75+40) in broiler chickens [c18a]1992	[c18] **maduramicin** ammonium, *Cygro (earlier Cyanamid, now Roche) [c18a] **maduramicin +nicarbazin** *Gromax (Roche)	is fermented from *Actinomadura yumaensis* and has a sugar moiety as side chain (monoglycoside polyether); it has affinity to both monovalent and divalent cations and exhibits at extreme low concentrations similar anticoccidial activity like other ionophores in birds; however, soon after introduction into the market cross-resistance between maduramicin and other polyether antibiotics has been found in field strains isolated in the Netherlands and Western areas of Germany; various birds (turkeys, guinea fowl, pheasants, geese, ducks) and mammals (cattle, sheep, horses, pigs, rabbits) tolerate 5 ppm in feed without showing adverse effects; to take the advantage of synergistic effects between drugs, maduramicin has been combined with nicarbacin; the combination may have favorable influence on performance of chickens

Table 1. (Continued) Anticoccidial drugs used in the poultry industry.

Chemical group (approx. dose, ppm[a]); relevant information	**Non proprietary name**, *brand name (manufacturer, company); other information	**Characteristics**[b] and miscellaneous comments
(25 in broiler chickens) [c19]1991 (first registered in countries of South America)	**semduramicin** *Aviax (Pfizer)	latest ionophore (feed additive) developed for prevention of coccidiosis in poultry; is fermented from *Actinomadura roseorufa* and shows broad spectrum anticoccidial activity with best efficacy against *E. maxima* ; like maduramicin, it shows cross- resistance with other ionophorous products in field isolates derived from litter samples collected in Europe, South America and the US between 1991 and 1994; its introduction and licensing in European countries is questionable.
ASYMMETRIC (1,2,4) TRIAZINES		
(1, in broiler chickens and turkeys) [c21] 1988	[c21] **diclazuril** *Clinacox (Janssen: Johnson & Johnson, various distributers, e.g.Hoechst Roussel Vet) **clazuril** *Appertex (Janssen) non-food producing application, limited distribution (e.g. Europe) **HOE 092 V** (Hoechst) discontinued	there have been several 1,2,4-triazine derivatives under development, e.g., clazuril and diclazuril (US patent application to Janssen 1984), and HOE 092 V (DE patent application to Hoechst AG, 1985); like toltrazuril (see there), they have broad-spectrum of activities against various coccidia (other parasites) in birds and animals but at distinctly lower concentrations (0.5-2 ppm, in feed and water, resp.) than the former; **diclazuril** has a strong anticoccidiocidal activity against developing first- and second-generation schizonts, and gamonts of *E. tenella* and other pathogenic *Eimeria* spp. of chickens but developmental stages most affected by diclazuril varies with the *Eimeria* species; it is highly effective against *E. tenella* (all stages) but not so against *E. maxima* (gamonts only); it has been developed as a feed additive and is licensed for the prevention of coccidiosis in broiler chickens and recently turkeys; if used for longer periods in straight programs, coccidia in chickens may develop resistance to the drug; for that reason it is now frequently used in shuttle programs; complete cross resistance with toltrazuril may be evident in the field; Diclazuril has recently been introduced for the treatment of rabbit coccidiosis; it shows high activity against hepatic and intestinal coccidiosis, and outbreaks of the disease can successfully be treated in rabbits; diclazuril appears to be compatible with other feed additives and is also a very safe drug in various animals tested (broiler breeder, turkey, guinea fowl, quail, duck, mouse, rat, rabbit, dog, piglet, horse, cow, sheep , goat); **clazuril**, which has limited action against some chicken coccidia, is highly active against coccidiosis in pigeons; dispensing the drug may need prescription by a licensed veterinarian; **HOE 092 V** proved to be highly active against all developmental stages of pathogenic *Eimeria* spp. in chickens, turkeys and rabbits; it also exhibited a very broad spectrum of activity against various protozoan fish and crustacean parasites as well as a variety of gill- and skin- parasitizing monogeneans; in contrast to toltrazuril, a single application (10 µg/ml, 4h: medical bath) of HOE 092 V was sufficient to eradicate trophozoites of *Ichthyophthirius multifiliis*; development of the drug as a feed additive was discontinued because of adverse effects on endocrine system occurring during toxicological investigations in rodents

Table 1. (Continued) Anticoccidial drugs used in the poultry industry.

Chemical group (approx. dose, ppm[a]); relevant information	**Non proprietary name,** *brand name (manufacturer, company); other information	**Characteristics**[b] and miscellaneous comments
Symmetric (1,3,5) triazines [c20]1987 [$]not licensed for prevention of coccidiosis in poultry or other animals under European Commission Guidelines (additives in feeding stuffs)	[c20] **toltrazuril** *Baycox (Bayer) 2.5% solution for treatment of coccidiosis in chickens (25 ppm, water administered), and other animal species; licensed dosages (10–20 mg/kg body weight) for target animals (chicken, turkey, swine) may differ in countries where drug is registered; for dispensing the drug a prescription by a licensed veterinarian is necessary	has a strong anticoccidiocidal activity against developing first- and second-generation schizonts, and gamonts of pathogenic *Eimeria* spp. of various birds (chicken, geese, ducks, others) and mammals (cattle, sheep, goats, pigs, others); these effects render **toltrazuril** a strong therapeutic drug, and thus outbreaks of coccidiosis as well as subclinical coccidiosis can successfully be treated with the drug; its broad activity is also directed against parasites of fish (microsporidia, myxozoa, monogeneans), and bees (*Nosema* sp.), and other parasites; flexible administration of toltrazuril via drinking water (use of intermittent medication instead of conventional continuous medication) may help to establish efficient metaphylactic programs against coccidiosis in chickens thereby allowing development of protective immunity; the drug is not licensed for the prevention of coccidioses in poultry, because of a too long withdrawal time in broiler chickens under practice conditions; it exhibits a long residual activity and may retard coccidiosis infections in chickens for at least 2 weeks. There is cross resistance between toltrazuril and diclazuril; toltrazuril is well tolerated in birds, mammals and other animals
Experimental drugs	**cationomycin** (Kaken) **D42067** α (Cyanamid) **portmicin**(Kaken) **cytosaminomycins** A, B, C, D (Kitasato Institute, Tokyo) **xylanases, glucanases** and **pectinases** ***Artemisia annua*** (dried leaf supplement) and pure artemisinin	several potent polyethers have previously been under investigation, e.g., **cationomycin** which showed relatively low toxicity in mice and remarkable coccidiocidal activity against all pathogenic *Eimeria* spp. in chickens at 80–100 ppm, or the nonsynthetic broad-spectrum anticoccidial **D42067** α which seemed to act on the early stages of the life cycle and provided excellent control against all important species of chicken coccidia; the polyether antibiotic **portmicin** derived from a *Nocardiopsis* sp. showed activity against gram-positive bacteria, including mycobacteria and coccidia, e.g., *E. tenella* at 6.2–25 ppm; **cytosaminomycins**, novel anticoccidial nucleoside antibiotics derived from *Streptomyces amakusaensis* exhibit activity against *E. tenella* in vitro; intestinal digesta viscosity reducing enzymes as **xylanases, glucanases** and **pectinases** may improve the nutritional value of wheat- or maize-based feed (diets) in poultry but enzyme mechanisms involved in reducing growth depression by coccidial infection is not yet known and need to be elucidated; leaf supplement of *A. annua* and pure artemisinin (see → Malariacidal Drugs, antimalarial activity) prove to be active against *E. tenella* and *E. acervulina* in chickens when given in feed over 3 to 5 weeks

[a] Concentrations (ppm → part per million → mg/kg in-feed
[b] For more details on biochemical action of drugs see ***Drugs***
[c] First practical or commercial application (approx. year)
Data given in the table have no claim to full information.
The primary original manufacturer is indicated.

genic diversity, and live *Eimeria* vaccines may show differences in their virulence and immunogenicity in the individual recipient and different breeds of poultry. The development of effective recombinant vaccines (genetically engineered *Eimeria* antigens) against poultry and farm livestock coccidiosis has become a major goal in modern parasitology. Since coccidiosis involves the intestinal immune system, understanding of the **complex gut-associated immune system** is most important in the development of immunological control strategies to coccidian parasites. **Different types of antigens** (surface-, internal-, secretory antigens of sporozoites, merozoites, gamonts) that induce parasite-specific immunity have been identified by means of monoclonal antibodies against various Coccidia species. Recombinant proteins derived from these antigens have been shown to induce either humoral or cellular response or both, whereas protection against live challenge (sporulated oocysts) proved to be insufficient to weak or partial only. For instance, responses of different poultry breeds vary considerably to such recombinant antigens (epitopes). Sub-unit vaccines developed so far lack epitopes that induce strong protection and this seems to be also true for optimal delivery systems that release these epitopes at the site of infection. Suitable live recombinant vectors derived from bacteria or viruses will be necessary to induce a persistent stimulation of the local immune system by presentation of 'perfect' recombinant target antigens to adequate immune effectors. Such approaches are only possible by using biotechnology that requires great skills of molecular biology as well as a profound knowledge of host immune responses to coccidian infection. Until now, there are large gaps in our existing knowledge concerning precise definition of target antigens and immune effectors. Consequently, the development of protective recombinant coccidiosis vaccines will be in any case a long-term and high-risk research project. Furthermore, such a vaccine must not only confer resistance but also be cheap and fit in with current management practice in the poultry industry and farm livestock. An exiting new approach to vaccine development may be highly attenuated bacterial vectors that have the ability to enter epithelial cells and directing plasmid DNA to the cytoplasm of the host cell for protein synthesis and processing for antigen presentation. **Delivery of DNA-encoded antigens** should permit mucosal immunization against the parasite simultaneously with multiple antigens that can stimulate T helper cells and antibody production, especially the proper folding and conformational epitopes for the immunoglobulins A (IgA) and G (IgG). Aside from the practical oral application of bacterial DNA delivery, this type of vaccines does not need DNA purification and can be produced for the fermentation, lyophilization and packaging.

Therapy

→ Coccidiocidal Drugs.

Codworm Disease

Disease due to infection with anisakid worms (→ Anisakiasis).

Coenurosis, Animals

Coenurosis is a nervous system disease caused by the presence in the cranial cavity of *Coenurus cerebralis*, the larva of *Taenia multiceps* (→ Nervous System Diseases, Carnivores). The infection occurs in sheep and less commonly in other ruminants. It is rare in horses and man. In lambs, an acute meningoencephalitis may develop if a large number of immature stages migrate in the brain. More commonly, the infection follows a chronic course and is associated with the presence of one or two coenuri in the brain, 4–6 months after infection. The coenurus acts as a space-occupying lesion anywhere in the central nervous system. It occurs most often in the cranial cavity, sometimes between the hemispheres and cerebellum, and sometimes underneath, in the region of the mid brain. Pressure signs develop slowly and are usually preceded by dullness, cessation of feeding, loss of weight and the habitual resting of the head against any support. The ultimate signs depend upon the site of the cyst, and include blindness and incoordination, as well as the turning movement which is said to be characteristic of the infection. Other locomotor signs include reacting, stumbling, abnormally high or low carriage of the head, and lunging forwards. These signs occur in attacks of some minutes duration followed by a remission of one or more hours, but the background of dullness and abnormal head carriage is continuous.

Eventually the animal becomes recumbent, and death soon follows. A case of coenurosis has been described in a bullock, which resulted in the development of hydroencephalus.

Therapy
→ Cestodocidal Drugs.

Coenurosis, Man

Coenurosis, is an infection with larval forms of the genus → *Multiceps* (Vol. 1) of canids, which gives rise to small cysts in the brain typically with prominent multiple scolices, but without daughter cysts. These are surrounded by a layer of fibrous tissue. Degenerating cysts give rise to intense inflammatory reaction with symptoms depending on location. Fibrosis and calcification follow.

Therapy
→ Cestodocidal Drugs.

Colorado Tick Fever

Synonym
CTF

Colorado tick fever is caused by the CTF virus (ungrouped → arbovirus (Vol. 1)) and is chiefly transmitted by the tick *D. andersoni*. After an incubation period of 3–6 days there may be encephalitis and bleeding in children. Fatalities are rare.

Congo Floor Maggot Myiasis

Disease of humans in Africa due to the activity of nightly sucking larvae of the fly species *Auchmeromyia luteola* (see page 356 ff).

Conjunctivitis

→ Eye Parasites.

Connatal Infection

(*Lat.* connatus = during birth). Infection of the newborn during birth (e.g. malaria and AIDS may then become transmitted).

Connatal Toxoplasmosis

Disease due to infection with → *Toxoplasma gondii* (Vol. 1) being already present during birth (prenatal infection; *lat.* natus = birth) via the intrauterine way (= congenitally). This infection may only occur in pregnant women who were infected for the first time during pregnancy. See → Prenatal Toxoplasmosis.

Constipation

Clinical symptom in animals due to parasitic infections (→ Alimentary System Diseases, → Clinical Pathology, Animals).

Control

→ Disease Control, Epidemiological Analysis, → Disease Control, Methods, → Disease Control, Planning, → Disease Control, Strategies, → Disease Control, Targets, → Drugs, → Chemotherapy.

Control of Malaria

Factors see → Disease Control, Epidemiological Analysis/Table 2

Cooperiosis

Trichstrongylid trematodes of the genus *Cooperia* (→ Trichostrongylidae (Vol. 1)) live in the upper part of the small intestine of ruminants. The important species include *C. curticei*, which mainly infect sheep and goats, and *C. pectinata*, *C. punctata*, and *C. oncophora*, which mainly infect cattle. *C. oncophora* is less pathogenic for cattle than either *C. pectinata* or *C. punctata*. *C. curticei* is of little pathological significance in sheep. *C. oncophora* only produces pathological lesions in animals experimentally infected with very large numbers of infective larvae (250,000 or more). The lack of pathogenicity is probably explained by the rapid acquisition of resistance by the host, and the superficial character of the intestinal lesions induced. They rather brace or coil themselves among villi to maintain their place. *C. punctata* worms have been reported as penetrating into the mucosa or submucosa. Villous atrophy with reduction in the brush-border enzymes only occurs after heavy infections. Signs of the infection include soft faeces, intermittent or continued diarrhoea, progressive emaciation, reduced feed consumption, weight loss, and listlessness. There is no anaemia. The lack of any significant effect on serum calcium or phosphorus concentrations, or on the skeleton suggest little interference with intestinal absorption. The slight hypoalbuminaemia may, however, indicate macro-molecular leakage into the gastrointestinal tract. Mixed infections with *Ostertagia ostertagi* produce severe adverse alterations of the metabolism in calves, which exceed those produced by infection with either of the species alone. The presence of *C. oncophora* in the small intestine seems to limit the extent of the compensatory digestive responses to *O. ostertagi* infection of the abomasum (→ Alimentary System Diseases, Ruminants).

Therapy

→ Nematocidal Drugs, Animals.

Copra's Itch

→ Mites (Vol. 1).

Corridor Disease

→ Piroplasms (Vol. 1), → Theileriosis.

Cost Effectiveness

→ Disease Control, Planning.

Cowdria ruminantium

Rickettsial agent of → heartwater disease in cattle, sheep and goats in Africa transmitted by bites of *Amblyomma hebraeum* ticks.

Creeping Eruption

Route of wandering larvae of → hookworms (Vol. 1) in human skin (→ Cutaneous Larva Migrans).

Creutzfeldt-Jacob Disease

→ Prions; see also → BSE.

Crimean-Congo Hemorrhagic Fever

Synonym

CCHF

General Information

Crimean-Congo hemorrhagic fever is caused by the CCHF virus (→ Bunyaviridae (Vol. 1)). It is a disease which shows hemorrhagic symptoms, together with serious acute febrile conditions. The virus has been isolated mainly from *Hyalomma ticks* (*H. marginatum marginatum*, *H. m. rufipes*, and other species) in South Europe and the Southwest of the former USSR, and several areas in Africa. These two foci are linked by migrating birds which can carry ticks. Ticks in other genera, including one-, two-, and three-host ticks, have also been found to harbor the virus. Recently, this disease has been associated with mortality in scientific and medical staff in South Africa after laboratory handling of ticks.

Crust

Clinical and pathological symptoms of infections with skin parasites (→ Skin Diseases, Animals, → Ectoparasites: New Approaches).

Cryptosporidiosis, Animals

Pathology

Cryptosporidiosis is an important infectious disease affecting mucosal surfaces commonly found in snakes, lizards, birds and mammals. *Cryptosporidium* species have been reported in all domestic animals including cats, dogs and pigs. In horses, symptoms have been particularly observed in animals with inherited or acquired immuno-deficiency. Cryptosporidiosis is economically most significant in calves and lambs. Symptoms are: enterocolitis in newborn animals manifested by diarrhoea of variable severity; signs of depression, anorexia and weight loss are also seen. The most prominent lesions occur in the ileum. They include blunting and some fusion of villi, and hypertrophy of the crypts of Lieberkühn. There are great individual variations in clinical reactions to *Cryptosporidium* infection and controversy exists as to whether the parasite occurs as a primary pathogen or whether it causes disease by interacting with other enteropathogens.

Immune Responses

As the replication and development of the parasite is confined to mucosal surfaces, local immune functions of the mucosa-associated lymphoid tissue (MALT) are of special importance. Interest in the nature of these host responses is heightened by a unique feature of parasite's development. Merozoits reside intracellularily in epithelial cells and remain anchored at the lumenal surface segregated from the host cell cytoplasm. Since *C. parvum* infections of animals are generally found in neonates and refractoriness to infection develops 2–3 weeks after birth, most investigations on the immune responses have been performed in suckling or immunocompromised mice. Some studies have been alternatively performed with *C. muris* which is able to infect adult immunocompetent mice.

Innate Immunity As with other coccidian, e.g. *T. gondii*, macrophage activation by IFN-γ leads to partial control of microbial reproduction. In T cell deficient individuals such as AIDS patients or nude or SCID mice a chronic infection characterized by gradual expansion of inflammatory foci develops, which in the case of immunodeficient mice eventually results in morbidity and death. Several studies, however, have demonstrated the existence of partially protective innate immune mechanisms mediated by IFN-γ. Treatment of SCID mice with anti IFN-γ or anti IL-12 antibodies increased susceptibility to infection and reduced the time till death from > 12 to 5–6 weeks. Although cryptosporidium sporocoites or oocyst antigen preparations enhanced IFN-γ production by splenic NK cells in vitro, treatment of mice with anti-ASGM1 antibodies to deplete NK cells had no effect on susceptibility to infection. In other experiments *C. parvum* infection of SCID mice was not influenced by IL-2 treatment which activates NK cells. Thus, the in vivo data on the role of NK cell activation conflict with the in vitro findings. Obviously, the cell type in the mucosa which produces IFN-γ in the absence of T cells still has to be identified. Likewise, the accessory cells involved in the local inflammatory response have also to be identified, since the gut epithelium contains only few macrophages.

B Cells and Antibodies Different subclasses of parasite-specific Igs increase in serum and mucosal secretions during *C. parvum* infections of humans and animals and IgG titers especially may persist for up to several years postinfection. Coproantibody titers of IgA and IgM increase during infection and decrease after resolution of the disease. The antibodies produced recognize a number of immunodominant sporozoite antigens of approximately 11, 15, 20/23, 44,100, 180 and > 200 kDa. Different antibodies induced by immunizations possess protective capacity as shown by (1) inhibiting the *C. parvum* development in vitro or in suckling mice by administering antibodies orally or by (2) effectively attenuating the clinical symptoms of human cryptosporidiosis by treating immunocompromised patients with hyperimmune bovine colostrum containing high concentrations of anti-cryptosporidia antibodies. Some of the antigens recognized by protective antibodies have been molecularily cloned, but the exact localization and characterization of the function of these proteins awaits further studies.

Although antibodies induced by immunization with antigens could inhibit parasite development, there is considerable doubt about the protective role of antibodies during natural infection. AIDS patients with severe cryptosporidiosis have high titers of *C. parvum*-specific secretory IgA. Furthermore, depletion of B cells in neonatal mice by anti-μ chain treatment, did not alter the ability to control crytosporidiosis. On the other hand, breast-fed babies are less likely to experience cryptosporidiosis while patients with congenital hypogammaglobulinemia sometimes develop chronic cryptosporidiosis.

T Cells *C. parvum* induces inflammatory infiltrates in the lamina propria of the gut containing lymphocytes, macrophages and plasma cells. In the Peyer's patches of mice increased numbers of $CD4^+$ and $CD8^+$ cells were found and purified T cells from spleens of infected animals showed a proliferative response towards parasite antigens. A parasite-specific delayed type hypersensitivity (DTH) can be elicited in *C. parvum*-infected rats by oocyst antigen injection. The functional importance of a specific T cell immune response was demonstrated by the increased pathology villus atrophy, crypt hyperplasia, erosions of gut epithelium found in T cell deficient hosts such as nude rats and mice, SCID mice as well as in normal mice depleted of T cells by administration of anti-T cell antibodies. In studies with *C. muris*, chronic infections in T cell-deficient mice developed similar pathology to those found in immunocompromised patients with cryptosporidiosis. In neonatal mice $\gamma\delta$ as well as $\alpha\beta$ TCR-expressing T cells are both involved in immunity against *C. parvum* while in adult animals the infection appears mainly to be controlled by $\alpha\beta$ T cells. Clinical studies of human cryptosporidiosis in HIV patients and experiments with mice both indicated that $CD4^+$ T cells are the major players in host protection. Susceptibility to and severity of *C. parvum* infection increases with the decrease of $CD4^+$ cell counts in AIDS patients. In mice the continuos administration of anti CD4 antibodies allowed chronic *C. parvum* infection to develop. The protective effect of lymphocyte transfer to SCID mice is abrogated by the depletion of $CD4^+$ cells. In contrast to the dominant role of $CD4^+$ T cells, $CD8^+$ cells play a negligible role as shown by cell depletion experiments in vivo or by using mice deficient for MHC class I expression. Only some investigators reported a small increase in oocyst production and/or a prolongation of patent infection as a result of treatment of mice with anti CD8 antibodies. Using intraepithelial lymphocytes (IELs) from immune donor mice to adoptively transfer protection it was found that also in this cell population the $CD4^+$ subpopulation is the most effective.

One of the most important effector molecules produced by protective T cells is IFN-γ as shown by enhanced susceptibility of mice as a consequence of treatment with IFN-γ-neutralizing antibodies. Production of this cytokine by IELs was found to be upregulated during the course of infection. However, IFN-γ independent mechanisms of immunity may also exist, since mice continuously treated with anti-IFN-γ antibodies nevertheless eventually overcome the infection with Crytosporidia. In line with this, mouse strain-dependent differences in susceptibility to infection (BALB/b versus BALB/c mice) were not linked to differences in IFN-γ production.

The mechanisms by which IFN-γ mediates control of the parasite remain to be determined. Treatment of mice with inhibitors of nitric oxide production did not influence reproduction of *C. parvum*, unlike the effects found in other experimental infections with parasites such as leishmania. Alternatively, documented IFN-γ effects such as enhancement of the production of the secretory component of IgA or the enhancement of the respiratory burst may be operative in the IFN-γ-mediated control of Cryprosporidium infection. IL-12 appears to be critically involved in the upregulation of IFN-γ production as shown by treatment of newborn mice with anti IL-12 antibodies resulting in enhanced disease susceptibility. Another cytokine involved in the protective immune mechanisms may be IL-2 since treatment of mice with anti IL-2 increased oocyst production and in a human study IL-2 therapy resulted in less severe diarrhoea and oocyst shedding in AIDS patients.

Recently, a protective role of Th2 cytokines in cryptoporidial infection has been reported. Although the amounts of IL-4 or IL-5 produced during the infection appear to be low, treatment of mice with anti IL-4 or anti-IL-5 antibodies, especially when both were combined, increased oocyst shedding. In addition, adult IL-4-deficient mice excreted oocysts in feces approximately 23 days longer than control mice. Mast cells or eosinophils as effector cells stimulated by these cytokines are unlikely since a) these cells were not found in significant numbers by histopathological observation in the epithelial infiltrates and b) mast cell-deficient mice were not significantly more susceptible to infection than control mice. Other studies, however, suggest that overproduction of IL-4 might correlate with increased susceptibility to infection. BALB/b mice produced significantly more IL-4 than the less susceptible BALB/c mice and onset of recovery in BALB/b mice coincided with a reduction of IL-4 production.

Related Entry

→ Alimentary System Diseases.

Therapy

→ Treatment of Opportunistic Agents.

Cryptosporidiosis, Man

The coccidium → *Cryptosporidium parvum* (Vol. 1), the oocysts of which are 4–5 μm in diameter, containing 4 sporozoites, parasitizes the microvillar border of the intestinal epithelial cells projecting into the lumen. With heavy infections the bile ducts, trachea, and possibly conjunctive may also be involved. In addition to the regular coccidian cycle, thin-walled oocysts are formed that sporulate in the intestine and are a source of superinfection; the thick-walled oocysts pass to the outside and develop four sporozoites. In healthy volunteers, acute infection can be produced by ingestion of 30 –300 or more oocysts accompanied by abdominal pain, cramps, diarrhea and fever lasting for 3 - 10 days.

C. parvum causes debilitating gastrointestinal illness in humans and other mammals and is a frequent opportunistic pathogen in AIDS patients. Chronic infection in immunosuppressed patients is associated with flattened villi resulting from loss of epithelial cells and lack of regeneration (→ Pathology/Fig. 6A). This is accompanied by often copious and frequent diarrhea, → malabsorption, and weight loss. Both the light and evanescent infections in the immunocompetent and the heavy infections in the immunosuppressed show little inflammatory reaction. As studied in experimental animals, developing immunity is accompanied by the infiltration of lymphocytes into the lamina propria.

Main clinical symptoms: Abdominal pain, heavy diarrhoea (in immunocompromised persons)
Incubation period: 1–2 days
Prepatent period: 2–4 days
Patent period: 12–14 days
Diagnosis: Microscopic determination of oocysts in fecal samples
Prophylaxis: Avoid contact with human or animal feces
Therapy: Treatment see → Coccidiocidal Drugs and → Treatment of Opportunistic Agents.

Culicidosis

Disease (e.g. Urticaria) due to bites of Culicid mosquitoes (Compare Vol. 1).

Cutaneous Larva Migrans

Cutaneous larva migrans is caused by → hookworms (Vol. 1) mainly of dogs and cats. The itching dermatitis often takes the form of tracks or "creeping eruption" indicating the route of migration of the larvae in the epidermis of the skin. These papular and serpiginous lesions sometimes become vesicular and hemorrhagic, and are often secondarily infected with bacteria after scratching. These larvae never mature into adults, to live in the intestine, but will die within 3 months at the latest.

Therapy

Local application of a 15% tiabendazole-ointment.

Cutaneous Leishmaniasis

→ Leishmania (Vol. 1).

Cyclosporiasis

In diarrhoeic stools of man especially in AIDS-patients the 8–10 μm sized, unsporulated oocysts of → *Cyclospora cayetanensis* (Vol. 1) have been described. At first they were misdiagnosed as CLB (cyanobacteria-like bodies). They apparently induce watery intermittent diarrhoea (3–4 times per day), which last for 2–9 weeks and may disappear without treatment. Infection apparently occurs by inoculation (with contaminated food) of sporulated oocysts containing 2 sporocysts with two sporozoites. **Treatment** may be carried out with Cotrimoxazol (2 x 800 mg Sulfamethoxazol/160 mg Trimethoprim). The source of the oocysts is not yet clear, since *Cyclospora* exists in many animals without clinical symptoms.

Cysticercosis

Pathology

Cysticercosis results from the development of larval tapeworms in humans harboring adult → *Taenia solium* (Vol. 1) (autoinfection) or from ingesting soil containing eggs shed in the feces of humans, in areas where there are no latrines, or where they are so filthy that they are not used. Humans are accidental intermediate hosts and pigs are the normal intermediate hosts; their meat being measly pork. Oncospheres released from eggs penetrate the intestinal mucosa and develop into bladderlike, cysticercus larvae of 1–2 cm which develop in many tissues, mostly in skeletal muscle and subcutaneous tissue. Clinically they are most serious when located in the central nervous system or in the eye where they persist for months to years. The intact cysticerci are surrounded by a fibrous capsule and rarely give rise to symptoms, unless they involve special areas of the brain such as the aqueduct or are present in large numbers. However, degenerating cysticerci give rise to fever with an intense eosinophilic inflammatory reaction (→ eosinophilic reaction) accompanied by tissue swelling, being especially serious in the brain. The dead parasites often calcify and become demonstrable by radiography; the living cysticerci can be diagnosed by computerized axial tomography and magnetic resonance imaging and should correlate with positive serological findings.

Main clinical symptoms: Dysfunction of the organs, within which the cysticerci are located
Incubation period: 8–10 weeks
Prepatent period: 8 weeks
Patent period: 2 years
Diagnosis: Serodiagnostic methods, computer tomography, → serology (Vol. 1)
Prophylaxis: Avoid contact with human feces
Therapy: Treatment with praziquantel, see → Cestodocidal Drugs

Immune Responses

The cysts have developed mechanisms to avoid the host inflammatory and immune response. The analysis of experimental infections of rodents with *T. crassiceps* or *T. taeniaeformis* has significantly contributed to our understanding of the host-parasite relationship.

Complement and Granulocytes Destruction of oncospheres in sera of patients or immune animals is mediated by the classical complement pathway. However, also the primary resistance to *T. taeniaeformis* infection in naive mice is complement-dependent. C3b is deposited on the surface of oncospheres and protooncosphere larvae but only in resistant and not in susceptible murine hosts is C5a produced. In contrast, complement has little effect on viable metacestodes.

In experimental infections, the developing metacestode is surrounded by neutrophils and eosinophils, which appear to have no detrimental effect on the parasite. The formation of infiltrates may be directly initiated by the parasite via the production of several chemotactic factors.

B Cells and Antibodies Antibodies are thought to play a decisive role in the immune response to *Taenia* oncospheres. Passive immunization studies have shown that protection can be transferred with antibody alone, while T cells are involved in the generation of protective immunity. In contrast, antibodies have little effect on metacestodes. However, most viable parasites contain immunoglobulin on their surface, of which the majority is not specific for the parasite. It has been suggested that the parasites may have Fc receptors for host immunoglobulins, which could be taken up, digested and thus serve as a primary protein source for the parasite.

T Cells Acute infections with taeniid oncospheres as well as viable cysts are associated with suppression of the host immune response. For example, spleen cells from acutely infected rats displayed decreased mitogenic responses, and in infected pigs a decreased number of $CD4^+$ T cells in the peripheral blood has been reported. The immune-suppressive effects seem to be mediated, at least in part, by modulation of the role of macrophages as antigen-presenting cells and appear to be dependent on the presence of viable parasites.

In mice infected with *T. crassiceps*, elevated levels of IgG1 and IgE suggested a dominant Th2 response. Increased production of IL-4, IL-6, and IL-10 has been detected, and active infection was also associated with a suppression of proinflammatory and Th1 cytokines.

The death of the parasite is associated not only with a granulomatous response but also with a switch to IgG2a production, a pattern associated with IFN-γ production. In humans, parasite death is accompanied by elevations of neopterin within the spinal fluid, a marker for macrophage activation. In addition IL-12 and IFN-γ expression has been detected in granulomas surrounding dying cyst, consistent with the idea, that death of the metacestodes results in a shift to a Th1 response.

Evasion Mechanisms Molecules able to detoxify reactive oxygen intermediates, such as superoxide dismutase and glutathion-S-transferase, were purified from *T. taeniformis*. Paramyosin, excreted by *T. solium*, is able to bind to C1 thereby inhibiting the classical complement activation. The protease inhibitor taeniastatin, a glycoprotein secreted by *T. taeniaeformis* metacestodes, not only inhibited the classical and alternative pathway of complement activation but also suppressed mitogen- or IL-1-induced proliferation of rat spleen or mouse thymus cells, respectively. Furthermore, taeniaestatin interfered with neutrophil chemotaxis and aggregation in vitro. Prostaglandin E2 isolated from the excretory fluid of *Taenia* metacestodes may contribute to the inhibition of Th1 responses.

Cytauxzoonosis

Disease of domestic cats and bobcats (reservoir) in the USA due to infection with *Cytauxzoon felis*, a 2 µm sized piroplasmean parasite of lymphocytes and erythrocytes being transmitted by ticks (e.g. *Dermacentor variabilis*). Some authors believe that *Cytauxzoon* is synonymous to *Theileria*. The disease shows rapidly progressing clinical signs such as high fever, icterus, lethargy, shortly followed by death.

Therapy
Unknown

D

DCL

Diffuse Cutaneous Leishmaniasis

Death

Some parasites have severe lethal effects on some of their hosts-especially in combination with an existing immune deficiency (see → Opportunistic Agents, Man). Parasites with a high mortality rate in a given population (e.g. children) are among others *Plasmodium falciparum*, hookworms, schistosomes etc.

Dehydration

Clinical symptom in animals due to parasitic infections (→ Alimentary System Diseases, → Clinical Pathology, Animals).

Demodicosis, Animals

→ Mange, Animals/Demodicosis, → Skin Diseases, Animals.

Demodicosis, Man

Demodicosis occurs in the hair follicles and sebaceous glands usually of the face (→ *Demodex* (Vol. 1)). The elongate mites feed on the contents of the sebaceous glands and also penetrate the follicular epithelium with their mouth parts. A mild dermatitis may be produced with inflammation and fibrosis. Immunosuppression may accentuate the severity of infection as observed in certain breeds of dogs infected with another species of *Demodex* (→ Demodicosis, Animals).

Main clinical symptoms: Dermatitis, alopecia, pyodermia
Incubation period: 2 weeks
Prepatent period: 2 weeks
Patent period: Years
Diagnosis: Microscopic inspection of hair bulbi and sebum from skin
Prophylaxis: General hygiene; avoid body contact with heavily infected people
Therapy: Treatment see → Acarizides (Vol. 1); use of ivermectin.

Dengue

Four types of virus diseases (hemorrhagic fever) transmitted by bite of mosquitoes (→ Arboviruses (Vol. 1)).

Dermatitis

Clinical skin symptom due to infections with skin penetrating parasites (e.g. cercariae, hookworm larvae) or blood sucking arthropods (e.g. ticks, fleas, bugs, lice, *Culicoides* sp.) (→ Skin Diseases, Animals).

Derrengadera

Disease of horses due to *T. brucei evansi*: transmitted by → vampire bats (Vol. 1) in Venezuela.

Diarrhoea

Clinical symptom in animals and humans due to parasitic infections (→ Alimentary System Diseases, → Clinical Pathology, Animals).

Dicrocoeliasis, Man

Disease due to infections wit the digenetic trematode → *Dicrocoelium dendriticum* (Vol. 1) by oral uptake of infected ants being attached e.g. to salad.

Main clinical symptoms: Abdominal pain, liver enlargement
Incubation period: 2–4 weeks
Prepatent period: 7–8 weeks
Patent period: Years
Diagnosis: Microscopic determination of eggs in fecal samples
Prophylaxis: Clean thoroughly salad, plants etc. from attached ants
Therapy: Treatment see → Trematodocidal Drugs

Dientamoeba fragilis

The trophozoites of this species reach a size of 3–12 µm, possess mostly two nuclei, include many food vacuoles and live in the colon of humans and monkeys (often in Zoological Gardens). They may cause as facultative parasites abdominal pain and diarrhoea. The transmission occurs by oral uptake of cysts from feces. It is discussed that trophozoites may become included in the eggs of the nematode *Enterobius vermicularis*.

Therapy
→ Antidiarrhoeal and Antitrichomoniasis Drugs.

Dinobdella ferox

Leeches of this species enter the nasal cavities of domestic animals in Southern Asia to suck blood (→ Leeches (Vol. 1)).

Dipetalonemiasis, Man

Zoonotic dipetalonemiasis is usually an infection of porcupines, beavers, or other mammals which is transmitted accidentally to man by mosquitoes. The adults live subcutaneously or in body cavities of the natural host, and are found subcutaneously or in the eyes of humans. Very rarely microfilariae are found, but their specific affinities usually remain undetermined. Living worms give rise to little inflammation, but dead worms cause → hypersensitivity necrosis with eosinophils, followed by granulomatous reaction and fibrosis (compare → Onchocerciasis, Man).

Therapy
Nematocidal Drugs, Man.

Diphyllobothriasis, Man

Disease due to infection with the tapeworm → *Diphyllobothrium latum* (Vol. 1) (→ Pseudophyllidea (Vol. 1)) by oral uptake of larvae (plerocercoids) in undercooked fish.

Main clinical symptoms: Abdominal pain, anaemia due to deprivation of vitamin B12
Incubation period: 3 weeks, if symptoms occur
Prepatent period: 21–24 days
Patent period: 10 years
Diagnosis: Microscopic observation of eggs in fecal samples
Prophylaxis: Avoid eating raw fish
Therapy: Treatment with praziquantel, see → Cestodocidal Drugs

Dipylidiasis, Man

Disease due to the infection with the tapeworm → *Dipylidium caninum* (Vol. 1) which is common in cats and dogs. Infection occurs via oral uptake of tapeworm larvae in crushed fleas.

Main clinical symptoms: Diarrhoea, urticaria, loss of weight and anal pruritus
Incubation period: 10–25 days
Prepatent period: 19–25 days
Patent period: 1 year

Diagnosis: Occurrence of typical proglottids in the feces
Prophylaxis: Deworming of dogs, cats and treatment against fleas
Therapy: Treatment with praziquantel, see → Cestodocidal Drugs

Dirofilariasis, Man

Synonym
Zoonotic filariasis

Pathology
Dirofilariasis is an infection of dogs, raccoons, bears, etc. caused by → *Dirofilaria immitis* (Vol. 1). It is occasionally transmitted by mosquitoes to humans. The young adult worms wander to the right heart and are usually propelled into a pulmonary artery branch, which they thrombus (→ Pathology/Fig. 28B) giving rise to a localized pulmonary infarct. The worm is usually dead in the thrombus. The release of worm antigens provokes an intense hypersensitive reaction, with central necrosis accompanied by eosinophils and a granulomatous or fibrotic reaction peripherally. Some of the older lesions calcify and become visible on radiological examination of the chest.

Main clinical symptoms: Undifferentiated heart pain, hypertrophy of heart, ascites
Incubation period: 3–9 months
Prepatent period: 7–9 months
Patent period: 6–7 years (at least in dogs)
Diagnosis: Difficult, since no microfilariae are formed in humans; serological tests
Prophylaxis: Avoid bites of mosquito vectors
Therapy: Treatment see → Nematocidal Drugs, Man

Disease Control, Epidemiological Analysis

Parasitic diseases originate from an interplay between parasite, vector or carrier, and principal host in an environment that is suitable for the parasite's maintenance or propagation. It is customary to differentiate broadly between food-borne, water-borne, arthropode-borne and directly invasive parasitoses. However, this rather coarse classification will be of little use in identifying targets of control intervention.

Some of the principal modes of infection are given in Table 1, but this grouping does not provide more than a summary orientation. With the individual parasitoses it will be indispensable to analyse, in detail, all stages of the parasite's life cycle and the biological and environmental factors governing its transmission, i.e. to consider the epidemiology of the disease. As an example, the main epidemiological components of the malaria situa-

Table 1. Principal modes of infection in parasitic diseases of man and other animals (according to Wernsdorfer)

Mode of infection (principal host)	Examples of parasite species
No alternation of host species	
Ingestion of infective stage, followed by self-replication within principal host	*Hymenolepis nana*
Ingestion of infective stage, often followed by self-contamination (immediate infectiousness)	*Enterobius vermicularis*
Ingestion of infective stage the development of which requires a period of non-parasitic existence in a suitable environment	*Ascaris lumbricoides*
Active invasion of infective stage the development of which requires a period of non-parasitic existence in a suitable environment	*Ancylostoma duodenale*
One or several alternations of host species	
Ingestion of free infectious stage (often on carrier material)	*Fasciola hepatica*
Ingestion of infectious stage within alternate host	*Gnathostoma spinigerum*
Active invasion of free infective stage	*Schistosoma* spp.
Active invasion, through bite wound inflicted by alternate host	*Wuchereria bancrofti*
Introduction of infective stage with insect bite at blood meal	*Plasmodium* spp.
Deposition of infective stage with feces of vector at time of blood meal	*Trypanosoma cruzi*

Table 2. Major factors related to the transmission and control of malaria (according to Wernsdorfer)

Parasite	Human host	Anopheline vector	Environment
Species	Susceptibility to infection	Species	Topography (plains, hills, valleys)
Pathogenicity	Relative immunity (age)	Susceptibility to infection	Surface water (types, extension,
EE and E schizogony	Gametocytaemia	Feeding habits (hosts)	depth, seasonality)
Hypnozoite infection	Exposure to vector contact	Feeding frequency	Vegetation
Recrudescence pattern	Occupation	(gonotrophic cycle)	Agricultural utilization
Relapse pattern	Age	Resting habits	Meteorology
Duration of infection	Migration	Flight span	Rainfall
Gametocytogony	Economic status and	Life span in relation to	Periodicity
Infectivity to	literacy	relative humidity	Abundance
anophelines	Housing conditions	Breeding habitat	Temperature (seasons)
Temperature	Settlement pattern,	Type of water collection	Wind
dependence of	population density	Stagnant, slow, or fast	Relative humidity
sporogony	General morbidity and	flowing	Man-made
Sporozoite yield	mortality	Shade/sun	malariogenic environments
Sporozoite infectivity	Rural and periurban	Vegetation	Water impoundments
(Drug sensitivity)	economy	Temperature/time	Irrigation systems
	Malariogenic habits	correlation of breeding cycle	Borrow and construction pits
	Awareness of malaria	(insecticide susceptibility)	Intra- and peridomestic artificial
	(Drug tolerance and		breeding places
	compliance)		Predators of anophelines

tion are summarized in Table 2. The resulting picture will be specific for a given area and important inter-area differences are commonly seen. Since the epidemiological situation may change over time, resulting from modifications of environment, host-vector relationships and other factors, the analysis requires regular updating. The approach to analyzing the factors governing the transmission of other parasitic diseases is similar, and should yield the elements required for the identification of targets of intervention.

Mathematical models can be quite useful in epidemiological analysis inasmuch as they facilitate a quantitative appreciation of particular features in the parasite's life cycle and permit projections of the expected efficacy of specific interventions. Such models exist for malaria and some other diseases of major public health importance. The quality of results obtained from such models depends primarily on reliability and exhaustiveness of the data input (→ Mathematical Models of Vector-Borne Diseases).

Disease Control, Methods

General Information

The approaches to the control of parasitic diseases are crucial components of the control strategy. The realization of the various approaches is dependent upon the use of individual measures the selection of which will be determined inter alia by expected efficacy, convenience, economy and acceptability. In most cases a variety of measures will be required simultaneously. This applies particularly to → anthropozoonoses and → zooanthroponoses with highly adaptable biological systems. Such diseases are the most difficult to control, especially if nondomestic animals are involved as reservoirs of infection. Another general aspect is man's awareness of parasitic diseases affecting humans and livestock, and the motivation for taking remedial action. If such broad motivation is lacking among the afflicted population, it is likely that imposed control programmes will have only limited and ephemeral success. → Health education in the widest sense, encompassing both the health of humans and of domestic animals, should therefore prepare the ground for a systematic effort against the diseases affecting the community.

The simplest measures for achieving a set purpose are usually the best, but in their planning and execution due attention should be paid to acceptability, compatibility with cultural and religious background, and technical feasibility. For instance, it would be expecting too much if the dietary patterns of large populations were to be changed abruptly. Here the practical solution will consist of rendering the incriminated food safe rather than banning it. The adoption of particular individual protective measures will depend on the person's economic status. Wearing shoes will generally protect against ancylostomiasis, and the use of impregnated bed nets supports protection against malaria. However, shoes and bed nets have their price and not everybody may be able to afford them or even be willing to use them. Selection of the most appropriate measures for disease control requires therefore a sound appreciation of advantages, limitations and disadvantages of the methods.

Parasitic diseases encompass a wide range of biological systems. Hence, the control of these diseases has many facets, implying a host of different measures the most important of which are detailed in the following sections.

Water Supplies

Water is one of the most important vehicles of parasitic diseases. It harbours a number of pathogens which can reach the human or animal host through transdermal penetration (e.g. → Schistosomiasis, Man) or through ingestion (e.g. → dracunculiasis). It is also the medium through which the larvae of many parasitic species reach molluscan hosts, ultimately to be transmitted to man or livestock as a food-borne pathogen. Safe water is therefore an important means of controlling numerous parasitic diseases, especially helminth infections (in addition to controlling the transmission of nonparasitic water-borne pathogens). Safe water should be available for consumption, bathing, washing and leisure activities. Ideally, piped, treated water should be available for household use. However, this will not be feasible as yet in most of the vast rural areas in the tropics and subtropics. Well-maintained deep wells with elevated rims made out of masonry or concrete for the prevention of contamination will be an acceptable and feasible alternative in many places. The use of traditional step wells or ponds should be discouraged, but, if there is no other source, boiling or sieving water through a fine mesh may render it largely innnocuous. The installation of a supply of piped, treated water may permit the abolition of unhealthy water collections. There may be public objection to this if the water collections are used for producing food (e.g. fish, crabs) or for the irrigation of crops. However, larger pools can be constructed on a community basis and maintained in such a way that they do not permit the transmission of pathogens while fulfilling the purpose of pisciculture and serving as a source of water for agricultural and household needs.

Excreta Disposal

Most helminth eggs or larvae have to reach water or humid ground for further development. They achieve this as a result of urination or defecation into water or onto wet soil. This may be part of a deliberate pattern, e.g. for the fertilization of family fish ponds in some parts of eastern Asia. In other instances it is due to an ingrained behavioural pattern or due to the lack of appropriate facilities for the safe disposal of excreta, or a lack of incentive for using available facilities. Health education is probably the most important remedial factor in such situations. It is not advisable to embark on a major programme of building latrines before the population is willing to use them. This applies especially to rural areas in which the population has easy access to various types of surface water. The acceptability of latrines or of even better facilities for excreta disposal is usually higher in urban areas where the installation of sewage treatment will also often prove to be feasible and cost effective. If the right type of sewage treatment plant is chosen, the resulting sludge will be biologically safe and usable as fertilizer.

Agricultural Hygiene

The agricultural, pastoral and piscicultural environment is often intimately associated with the transmission of parasitic diseases. Agricultural labourers may serve as a source of infective material, especially if they do not dispose of their excreta in a safe way whilst in the fields. They are also exposed to a variety of pathogens, particularly in irrigated areas. In addition to these occupational aspects, the use of unsafe biological fertilizers (fecal matter) on vegetables will promote the spread of some parasitoses, e.g. amoebiasis and ascariasis. Another important feature is the grazing of livestock in wetland areas. Again, health education and community efforts towards the development of safe grazing areas, e.g. through drainage, will be required to remedy the situation. Particular precautions should be taken when using wastewater and excreta in agriculture and aquaculture. Improperly managed water resource development entails the risk of the propagation of parasitic diseases. This should be avoided through appropriate water management.

Personal Hygiene

Apart from the obvious impact of unsafe excreta disposal, the lack of personal hygiene is a leading cause of infection with a variety of parasitic pathogens such as *Giardia lamblia*, *Entamoeba histolytica* and *Enterobius vermicularis*. Washing hands after defecation and before eating would largely reduce the transmission of these pathogens. The use of water and soap would also impede the transition from reversible lymphoedema to irreversible elephantiasis in lymphatic filariasis. However, personal hygiene must go further than water and soap. It should include the seeking of treatment if there are symptoms of disease, and the avoidance of dangerous foodstuffs and of situations conducive to the contraction of infections. Health education will be an important vehicle for imparting the necessary knowledge, awareness, and habits. This process should start at as early an age as possible and schools will have to play a major role in this endeavour.

Housing

Siting and type of human habitations are closely related to the risk of contracting certain parasitic diseases. Houses with cracked masonry, mud walls and/or earth floors were found to be a particularly suitable environment for reduviid bugs responsible for the transmission of Chagas disease. Simple housing improvement was found to reduce or even remove the risk of infection. Siting of settlements away from mosquito breeding grounds was an empirical yet highly effective means of protection against malaria. The siting of settlements at a long distance from irrigation canals (accompanied by the provision of safe household water) is an effective preventive measure against schistosomiasis since it will reduce the frequentation of the canals for washing, bathing and swimming.

Type and standard of housing play a major role in allowing the entrance and exit of disease-carrying mosquitos. It also determines the feasibility of mosquito and fly proofing, and the efficacy of ancillary vector control measures such as mosquito coils and knock-down sprays.

Environmental Management

Some measures of → environmental management as a means of disease control have been known since ancient times. Environmental management was the mainstay of malaria control before the advent of residual insecticides and synthetic antimalarials. It is making a comeback due to the limitations of other methods. The applicability of such measures in the control of parasitic diseases is very wide. The most important methods belong to environmental sanitation and water management. Water collections of various types are known to be breeding grounds for arthropod vectors of disease and the homestead of intermediate hosts of many helminthic organisms. Unless local economic (piscicultural and agricultural) and ecological reasons militate against them, filling, levelling and draining operations (drains, canals, and use of trees) will be appropriate measures. The same applies to the sanitation of wetlands to be converted into safe land for agriculture and livestock.

Environmental sanitation, including peridomestic areas and the safe disposal of waste, is a field in which individual and community initiative can be used to great advantage, the more so when the necessary equipment is easily available and cheap (e.g. pick-axes and shovels) or obtainable on loan from various government departments (e.g. earth-moving machinery).

Water management applied to water-storage reservoirs (level management) and irrigation systems (watering and drying cycles) will facilitate disease control by rendering the areas unsuitable as a habitat of intermediate hosts or vectors of parasitic diseases. Water management should be an integral part of design and operation of water impoundments and irrigation schemes.

Control of Vectors and Intermediate Hosts

The control of vectors and intermediate hosts of parasitic diseases may, to a large extent, be achieved through environmental sanitation. However, in some situations the applicability of such measures will be severely limited, as for example, in the control of *Simulium* spp., the vectors of onchocerciasis. Similarly, widespread temporary breeding places occurring during the rainy seasons in the tropics may pose insurmountable obstacles to environmental management. Alternative control approaches are therefore necessary.

At the beginning of the twentieth century mosquito control was improved by the use of light oils and chemicals such as Paris Green. In spite of their efficacy the application of these measures has remained quite limited due to the need for repetitive use and high cost. The introduction of chemical insecticides has not fundamentally changed the situation. Moreover, ecological considerations and non-target effects against the aquatic fauna and flora restrict the widespread repetitive use of insecticides. Chemical larvicides, rapidly biodegradable insecticides with low non-target toxicity, are still useful in the rapid control of epidemics of some vector-borne diseases. The same applies to the control of *Cyclops* spp., the intermediate hosts of *Dracunculus medinensis* and various other helminths.

Biological methods for larval control have a long tradition inasmuch as larvivorous fish, e.g. *Gambusia affinis*, have been used since the beginning of the twentieth century. Although appealing as a natural solution to a natural problem, larvivorous fish have a limited usefulness since seasonal and shallow breeding places are not suitable for their maintenance. It may also be difficult to find a local species of larvivorous fish. The introduction of non-local species may have a disastrous impact on the local aquatic fauna and interfere seriously with the production of food fish species.

Bacterial toxins from → *Bacillus thuringiensis* and *Bacillus sphaericus* are selectively directed against mosquito larvae and being used in the control of *Culex* spp. and *Aedes* spp. As the microorganisms sink to the ground they are not suitable for controlling *Anopheles* spp. (surface feeders). *B.thuringiensis* does not reproduce in the breeding places, but *B.sphaericus* does to some extent but not sufficiently to relinquish the need for regular retreatment of the breeding places.

The intradomiciliary application of residual insecticides such as chlorinated hydrocarbons (DDT), organophosphorus compounds (malathion), carbamides (propoxur), fenitrothion and synthetic pyrethroids is suitable and often quite cost-effective for the control of adult endophilic mosquitoes. However, the occurrence of specific resistance, aided and abetted by the agricultural use of insecticides of the same chemical groups, the presence of exophilic mosquitoes, increasing cost of insecticides and labour, and rising ecopolitical constraints have reduced their usefulness or applicability. Their use is still important, though, in the control of threatening or manifest epidemics where they are generally applied on a focal basis. These are situations where the ultra-low-volume (ULV) dispersal of suitable insecticides, may also show rapid effect. On the whole, the use of integrated vector control opens better prospects for an environmentally acceptable control of arthropod-borne parasitic diseases.

Pyrethroid-impregnated bed nets (deltamethrin or permethrin) or impregnated curtains and screens bar or reduce the contact between man and vector. They gained a firm place in malaria control,. especially in areas with moderate or intensive transmission. Their efficacy is due to a repellent effect rather than specific insecticidal action, promoting also epidemiologically desirable vector deviation to animals.

The control of aquatic snails continues to present a serious problem. Environmental management is the only effective and widely acceptable procedure for snail control. The available molluscicides are either not sufficiently effective (e.g. copper sulfate) or they are too toxic for the non-target fauna, including fish.

Diagnosis

The ability to diagnose the presence of infections is an important factor in guiding the treatment of individuals, and forms the basis of epidemiological assessment which should enable the health authorities to determine the dimensions of the specific human and/or animal health problem. It is also an essential tool for monitoring the impact of disease control activities. Macroscopic and/or microscopic diagnosis of parasitic diseases may be relatively simple and reliable with some parasite species, especially intestinal helminths, but exceedingly difficult with others, mostly tissue-dwelling parasites, e.g. certain types of nematodes. Serological methods based on the detection of specific antibodies usually reflect past or present host-parasite contact and therefore do not provide proof of current infection. Similarly, relatively fresh infections may not have given rise to detectable antibodies as yet and show seronegativity in spite of the living pathogen's presence. Demonstration of circulating antigens is a more reliable and specific basis for diagnosing current infections. Rather simple and reliable antigen detection tests have been developed for several human parasitoses, e.g. infections with *P. falciparum* or *W. bancrofti*. They are based on the detection of highly specific parasite antigens. However, relatively high costs continue to restrict their use in the framework of control programmes. The same still applies to tests for the detection of lactate dehydrogenase from malaria parasites.

Identification of infections by polymerase chain reaction (PCR) has been developed for numerous parasite species. However, the routine use of PCR is currently limited to research institutions and to diagnostic laboratories in prosperous countries. Its cost and operational requirements are too high to be affordable by most of the tropical countries.

In order to be widely practicable, diagnostic techniques for the most important human and animal parasitoses must be simple, cheap, and undemanding in terms of sophisticated equipment, electricity supply and operator skill. Human and veterinary health services in many parts of the world still lack the infrastructure required for establishing reliable data on prevalence and incidence of major parasitic diseases and associated mortality. This accounts for serious deficiencies in national and international disease statistics.

Treatment

Effective agents for the treatment of numerous parasitic diseases are available (see chapters on disease control). Some are reasonably cheap, such as those for the treatment of intestinal nematode infections of humans and domestic animals. Other medicaments are expensive, such as third-line drugs for the treatment of falciparum malaria. High costs may limit their use or encourage suboptimal medication, and consequently allow a parasite reservoir to be maintained that will be an obstacle to the effective control of the disease concerned. There are, however, a large number of parasitoses for which the therapeutic armamentarium is grossly deficient, e.g. Chagas disease, kala-azar and liver fluke infections.

With regard to malaria the situation was relatively satisfactory after the wider introduction of the 4-aminoquinolines in the late 1940s. However, the advent of chloroquine resistance in *P. falciparum* has compromised the efficacy of this group of drugs in wide parts of tropical Asia and South America. In the hyper- and holoendemic areas of tropical Africa juveniles and adults continue to derive therapeutic benefit from chloroquine, but young children whose immunity is not yet sufficiently developed generally require treatment with alternative drugs. → Resistance (Vol. 1) to the first-line alternative drugs, namely combinations of sulfonamides with pyrimethamine, already affects large areas in southeastern Asia and South America, and is rising in parts of tropical Africa, necessitating the use of second-line alternative drugs which are considerably more expensive. Resistance to mefloquine and structurally related quinine occurs in Cambodia, parts of southern Viet Nam and eastern Myanmar and in some parts of Thailand bordering on Myanmar and Cambodia. Here combined treatment with artesunate or artemether with mefloquine still yields satisfactory results.

On the whole the veterinary health sector has had greater success in the development of antiparasitic drugs than the human health sector. This is largely due to a better financial endowment of agricultural and livestock development and to the more stringent toxicological requirements governing the registration of medicaments for use in man. The situation is compounded by the fact that the development of medicaments against human parasitoses holds little attraction for the pharmaceutical industry since the main market for such drugs is in poor tropical countries.

Immunization

In many parasitic diseases there is evidence of the natural development of immunity to the specific pathogen. Such immunity rarely induces total refractoriness to reinfection, but it will restrict parasite reproduction or acceptance and induce tolerance to the pathogen. The development of immunity is quite slow. Considerable efforts have been made in the field of immunization against parasitic diseases, especially against those caused by protozoa. In bovine babesiosis, attenuated live *Babesia bigemina* is being used for inoculation of livestock. The resulting immunity is satisfactory, but there is still significant mortality associated with vaccination which is, on balance, economically acceptable. Such approaches are not feasible in human parasitic diseases except for agents with low virulence, e.g. *Leishmania major.*

Although immunization holds substantial promise in the control of many parasitic diseases and the progress in gene technology and polypeptide synthesis is likely to pave the way to economically acceptable products, there is still a long and arduous way to go before well-tolerated and reliable vaccination will become a reality in the control of parasitic diseases.

Clinical Relevance

Diagnostic and therapeutic measures have direct clinical relevance. Other disease control measures have indirect clinical relevance inasmuch as they are geared to the reduction of the community's disease burden, and thus a lessening of the pressure on the health services.

Disease Control, Planning

An almost universal shortage of resources, particularly marked in tropical developing countries, renders the simultaneous control of all major parasitic diseases difficult in most areas of the world. The available resources must therefore be used to address important issues with a reasonable prospect of success. Parasitic infections may be important and if the disease causes severe symptoms and kills humans or livestock people will be aware of it. Other parasitoses, however, though less spectacular, may cause even more damage on account of their wide distribution, but public awareness may be low due to the disease's unobtrusive manifestations.

The shortage of resources and competition for those available make it necessary to set priorities on the basis of human suffering and death, or on economic loss caused by particular diseases of man or domestic animals. Many such diseases are important obstacles to development but their impact needs to be quantified in terms of disease-associated loss in order to provide a solid basis for priority ranking by governments or other interested groups. Public awareness of the disease in question may influence priority selection via political pressure and may promote community participation, but there is a risk that priorities so selected may not represent the most appropriate choice.

Once the control of a particular disease or group of diseases has been tentatively allocated high priority, the time has come to review the existing approaches that seem to be feasible in the local situation. Knowledge of the local epidemiological features and earlier experience in control within or near the area will be an asset. The feasible approaches and the specific measures required to implement them need to be projected, on a provisional basis, in terms of requirements for resources, skills, infrastructure, and expected results. At this stage, or earlier, it may be necessary to strengthen epidemiological information.

After the feasibility of possible approaches has been scrutinized the preliminary objectives should be determined. These may range from elementary forms of control for the prevention of death from parasitic disease(s) to the complete elimination of the parasitosis from a particular area, country, or geographical region. The setting of the objectives should be realistic and take into account the available resources. A staged procedure may be envisaged that permits the future upgrading of objectives in keeping with the growth of resources and general development.

Provisional objectives should then be put to the test in realistic feasibility studies to be undertaken under qualified technical guidance through the service structure that is expected to be ultimately responsible for the implementation of the control activities. There is little room for so-called pilot projects since they usually operate with relatively greater resources, more qualified and more motivated personnel, and much higher staffing levels than those ultimately available to the large-scale control programme. Results from pilot projects should therefore not be used for the extrapolation of the probable impact of more extensive routine operations. Feasibility studies of single or combined approaches will permit the validation of approaches as well as technical adjustments to improve their efficacy. Such studies will also clarify requirements in terms of financial and manpower resources, training, equipment and supplies, logistics, mechanisms of evaluation and remedial action, and will provide the elements for determining cost effectiveness.

In the light of feasibility studies, which need to be done in each major epidemiological and operational stratum, it will be possible to conclude whether the provisionally set objective can be reached with the available resources, whether the objective should be upgraded or downgraded or, in the worst of cases, whether attempts at any form of control would be futile under the present circumstances. Some thought should also be given to the capability of sustaining the control effort in the future when prevalence and/or incidence of the disease have been reduced since potentially infective reservoirs will in most situations still be present and cause serious repercussions with any slackening of the control effort. For this reason it may be more appropriate to use an existing general service structure (with technical backup and guidance from a specialized group) rather than a vertical, disease-specific service structure which is subject to political and budgetary vagaries and often lacks popular support. Control of parasitic diseases of man can be incorporated in the development and delivery of general health care, while livestock health can become part of community development activities. This approach would promote community understanding and involvement which are recognized as essential prerequisites for sustained effort and success.

Preliminary and intermediate feasibility assessment and the scrutiny of the inventory of requirements are followed by the preparation of a master plan for the control of a specific disease or group of diseases. Its successful implementation will largely depend on continuous and well-qualified technical guidance and evaluation, timely recognition of technical and administrative problems, and rapid remedial action.

Current strategies for the control of widespread and economically important human diseases such as → malaria, → schistosomiasis and → filariasis are described under the respective headwords.

Remote sensing by meteorological satellite technology is currently being developed as a tool for forecasting epidemics of mosquito-borne diseases. The application holds particularly high promise for areas where epidemics are known to occur as a result of cyclical changes of rainfall (e.g. the Sahel zone of Africa).

Although providing promising leads, efforts directed to the development of vaccines against malaria and schistosomiasis have so far not been successful enough to make projections for the availability of operationally deployable products in the near future.

Disease Control, Strategies

General Information

Parasitoses are widespread in the animal kingdom and in plants, causing a wide spectrum of pathologic effects ranging from little more than commensalism to severe, even fatal disease. Many parasitic diseases cause considerable suffering and have an important impact on human health, and on livestock and crop production. Some parasitic diseases have a relatively low incidence but show high severity and a potential for producing epidemic outbreaks. Others are widespread, rarely causing gross pathology, but sapping the strength and health of those affected. The mode of transmission also shows considerable variety, ranging from direct contact to obligatory passage through several host species. The distribution of such hosts and environmental factors may therefore limit the occurrence of specific parasitoses. Thus some parasitic diseases are cosmopolitan while others may be restricted to very small ecological niches, with many nuances between these extremes. Some parasites have a rather wide choice of vertebrate hosts, others are highly stenoxenic. In addition, naturally acquired immunity may modify the manifestations of many parasitoses and lead to age-specific patterns of disease.

The detrimental impact of many parasitic diseases makes their control desirable if not indispensable in the interests of health and economy. The great variety of parasitic diseases and of the factors involved in their transmission requires widely different control approaches and strategies tailored to the specific epidemiological conditions and the goals of intervention.

In planning control strategies, it is helpful to classify the various parasitic diseases of man and other animals, irrespective of the causative species' taxonomic standing, into those affecting only humans (→ anthroponoses), those affecting only other animals (→ zoonoses), and those affecting both man and other animals (→ anthropozoonoses or zooanthroponoses). (Major representatives of the three groups are listed under the respective headwords in order to facilitate an appreciation of the wide range of causative agents and the modes of transmission involved.) In the majority of parasitic diseases, the causative organism's life cycle provides the essential elements for developing a control strategy. Such strategies invariably aim at the interruption or reduction of transmission at vital points.

Approaches to the control of parasitic diseases are based, essentially, on an analysis of the biological system characteristic of the given parasite species, the determination of targets of intervention, the critical consideration of measures suitable for such intervention, and an inventory of available experience in the control of parasitoses.

Some measures directed against parasitic diseases may exert a non-target environmental impact that needs to be considered in the planning of control activities. This may require environmental compatibility studies before clearing specific methods for wider application.

In all human parasitoses with gross pathology, e.g. malaria, visceral leishmaniasis (Kala azar), African and American trypanosomiasis, onchocerciasis and lymphatic filariasis, clinical relevance is the motor for developing and enacting disease control strategies. These aim at improving human health and reducing disease incidence or prevalence. Similar considerations apply to diseases affecting livestock, where also economic aspects play a substantial role.

Disease Control, Targets for Intervention

Knowledge of the parasite's life cycle is a prerequisite for determining feasible targets of control intervention. However, before such knowledge became available it was not rare for communities afflicted or threatened by certain parasitic diseases to take, empirically or intuitively, appropriate measures such as the prohibition of pork as a means of avoiding trichinosis, the siting of habitations to evade malaria, or the use of *Cinchona* bark as a febrifuge.

A rational approach to the control of parasitic diseases requires a full knowledge of potential targets. Measures directed against such targets may be practical, i.e. feasible and effective with currently available means and technology. Other targets may be too elusive for basing a strategy on their successful control. These are, nevertheless, worthy of exploration since a multifaceted approach to disease control usually offers better prospects for success than the deployment of a single measure. Practical and potential targets of intervention and appropriate control approaches are identified in the following examples.

Disease control is, by definition, clinically relevant. Community participation in disease control programmes depends largely on the visibility of results. Effective treatment is usually the best advertisement for a control programme and promotes the understanding for and acceptance of other programme activities. Programmes without a therapeutic component are generally unpopular. In order to be viable, programmes for the control of acutely life-threatening diseases, e.g. malaria in areas with *P. falciparum*, require an efficient mechanism for the rapid and competent management of severe and complicated cases.

Related Entries
Detailed information on known targets for intervention against specific diseases is given under the respective headwords.

Diseases of the Eye

→ Eye Parasites.

DNA-Synthesis-Affecting Drugs I: Alkylation Reactions

Table 1

Structures
Fig. 1

Nitroimidazoles

Important Compounds Metronidazole, Nimorazole, Ornidazole, Tinidazole.

Synonyms Metronidazole: SC-32642, Artesan, Flagyl I.V., Clont, Arilin, Cont, Danizol, Deflamon, Fossyol, Gineflavir, Klion, Orvagil, Sanatrichom, Trichazol, Trichocide, Tricho Cordes, Torgyl Forte, Tricho-Gynaedron, Tricocet, Trivazol, Vagilen, Vagimid.

Nimorazole: N-2-morpholinoethyl-5-nitroimidazole, Nitrimidazine, K 1900, Acterol, Esclama, Naxofen, Naxogin, Nulogyl, Sirledi, Radanil, Rochagan

Ornidazole: Tiberal

Tinidazole: CP 12574, Fasigin, Fasigyn, Pletil, Simplotan, Sorquetan, Tricolam

Clinical Relevance The 5-nitroimidazoles exert a wide variety of activities against different pathogens. Their antibacterial activity is useful in different indications. Thus, they are useful against obligatory anaerobic intestinal bacteria, intraabdominal gynaecological infections, aspiration pneumonia, superinfected bronchial carcinomas, brain abscesses, *Bacteroides fragilis* infections, gut wall necrosis as well as polymicrobial pelveoperitonitis. The antibacterial activity of metronidazole is furthermore useful in the treatment of *Dracunculus medinensis* infections. Moreover, 5-nitroimidazoles are applied in chronic non-bacterial diseases of the intestine like Morbus Crohn. In addition, metronidazole or tinidazole are effective in the treatment of peptic ulcera in combination with omeprazole (lansoprazole or pantoprazole) and clarithromycin (with or without amoxicillin).

5-nitroimidazoles are in addition useful in protozoal infections caused by *Giardia lamblia*, *Trichomonas vaginalis*, *Entamoeba histolytica* (Table 1), in urogenital infections caused by *Tritrichomonas foetus* in cattle, as well as in intestinal trichomoniasis, giardiasis and amoebiasis in dogs, cats, monkeys. The drug has additional activities

Table 1. Degree of efficacy of giardicidal, trichomonacidal and amoebacidal drugs.

		Mastigophora				Sarcodina	Ciliata
Year on the market	Drugs	*Leishmania* spp.	*Trypanosoma cruzi*	*Giardia lamblia*	*Trichomonas vaginalis*	*Entamoeba histolytica*	*Balantidium coli*
	Iodoqionol					xx	
1962	Metronidazole a)	xxx	xE	xxx	xxx	xxx	xx
	Furazolidone b)			xx			
1956	Diloxanide c)					xxx	
(1937)	Chloroquine c)					xx	
1912	Emetine (Dehydroemetine) c)	x	xE			xxx	
	Erythromycin, Paromomycin d)			xx			
	Tetracycline			xx			
1979	Albendazole and Mebendazole e)			xxx			

xxx = high efficacy at least against some developmental stages and diverse species; xx = partially effective (regarding developmental stages and diversity of species); x= slightly effective; E = active experimentally; a) other 5-nitro-imidazoles : Ornidazole, Tinidazole, Nimorazole; b) nitrofuran, as active as metronidazole, not as widely used, unavailable in Australia; c) alone or in combination with each other or with metronidazole; d) recommended during pregnancy; e) benzimidazole, suitable alternatives to 5-nitroimidazoles

against *Blastocystis hominis* in-vitro and in-vivo and *Balantidium coli* infections of pigs.

The antiprotozoal action of 5-nitroimidazoles is directed against *G.lamblia* trophozoites and cysts in the duodenum, against *T. vaginalis* trophozoites in vagina, cervix, urethra, epididymis, prostate and in *E. histolytica* infections against trophozoites (minuta forms) in the phase of binary fission and cysts in the intestinal mucosa and trophozoites (magna forms) in liver and other extraintestinal organs. The activity against *B. coli* is directed against the intestinal trophozoites which divide by binary fission and cysts in the feces.

Molecular Interactions The antibacterial action of metronidazole relies on its activation in anaerobic bacteria (*Bacteroides, Clostridium*) mediated by pyruvate-ferredoxin-oxidoreductase (PFOR). Metronidazole is generally not toxic to mammalian cells, because they lack electron transport proteins like PFOR with sufficiently negative redox potential for drug activation. The biochemical target of 5-nitroimidazoles is also the enzyme PFOR in *Giardia*, *Trichomonas* and *Entamoeba* (Fig. 2). In *T. vaginalis* PFOR is located in the hydrogenosomes. Comparison of the genes from *E. histolytica*, *T. vaginalis* and *G. lamblia* encoding PFOR show 35–45% sequence identity. The PFOR from these parasites are dimeric or tetrameric proteins of 240 kDa subunits. 5-nitroimidazoles exert their activity only after the reduction of the nitro group by PFOR which occurs in single electron steps (in total 4 electrons) and results in the formation of hydroxylamine. The formation and disappearance of the nitro-free anion radical could be detected in trichomonads (Fig. 2). Until now there is only indirect evidence for cytotoxicity of the intermediates nitroso-free radical and hydroxylamine. It is presumed that the interaction of toxic intermediates with various cellular macromolecules (DNA, proteins, membranes) leads to an irreversible cellular damage by DNA-alkylation. Indeed, there is a correlation between the reduction of the nitro group of metronidazole and DNA damage in vitro and in vivo.

Resistance PFOR is in the centre of the resistance mechanism in *Giardia duodenalis* resulting in reduced production of toxic radicals by decreased PFOR activities. Indeed, in *Trichomonas vaginalis* there is a correlation between increased metronidazole-resistance and decreased activity of the PFOR and hydrogenase. Besides PFOR in *T. vaginalis* ferredoxin (Fd) also seems to play an important role in the resistance mechanism. Thus, in

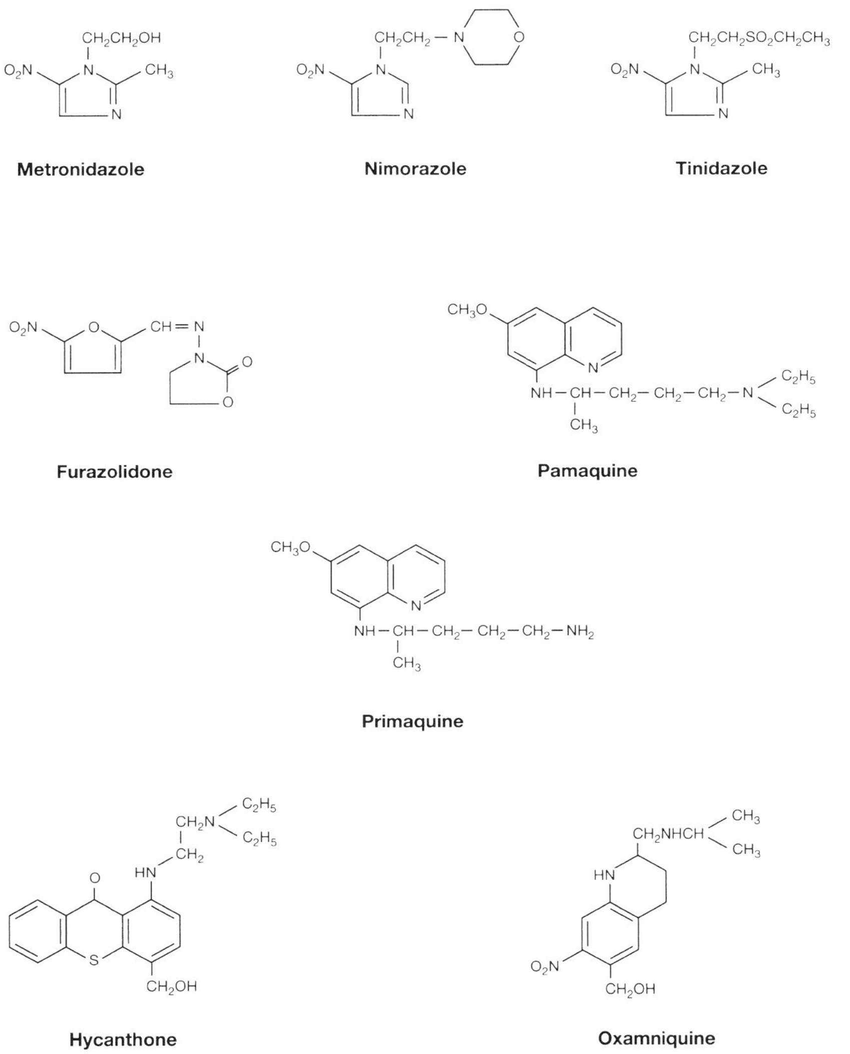

Metronidazole

Nimorazole

Tinidazole

Furazolidone

Pamaquine

Primaquine

Hycanthone

Oxamniquine

Fig 1. Structures of drugs affecting DNA-synthesis by alkylation reactions.

metronidazole-resistant *T. vaginalis* a decrease of intracellular Fd levels by > 50%, a decrease of Fd mRNA levels by 50–65% and a reduced transcription of Fd gene can be observed. Moreover, there is a correlation between resistance and the appearance of point mutations in the 5' flanking sequences of the gene. Two mutations could be identified with a reduced binding affinity of a 30kDa protein to a 28 bp region within the mutated region upstream of the Fd gene. Thus, it seems that metronidazole resistance strongly correlates with an altered regulation of the Fd gene transcription. The limitation of the ability of the cell to activate metronidazole by reduced gene transcription finally results in decreased intracellular levels of Fd, so that metronidazole is less efficiently reduced to its cytotoxic form.

There is an alternative hypothesis of the resistance mechanism in *T. vaginalis* in which a half-type P-glycoprotein should be overexpressed by a mechanism other than gene amplification. However, there is no clear correlation between levels of expression of the gene for this putative transporter protein and levels of resistance. Thus, the role of this gene in resistance remains doubtful.

Furazolidone

Synonyms 3-(5-nitrofurfurylideneamino)-2-oxazolidinone, NF180, Furovag, Furoxane, Furoxone, Giarlam, Giardil, Medaron, Neftin, Nicolen, Nifulidone, Ortazol, Roptazol, Tikofuran, Topazone.

Clinical Relevance This drug is active against *Giardia lamblia* (Table 1). The action is directed

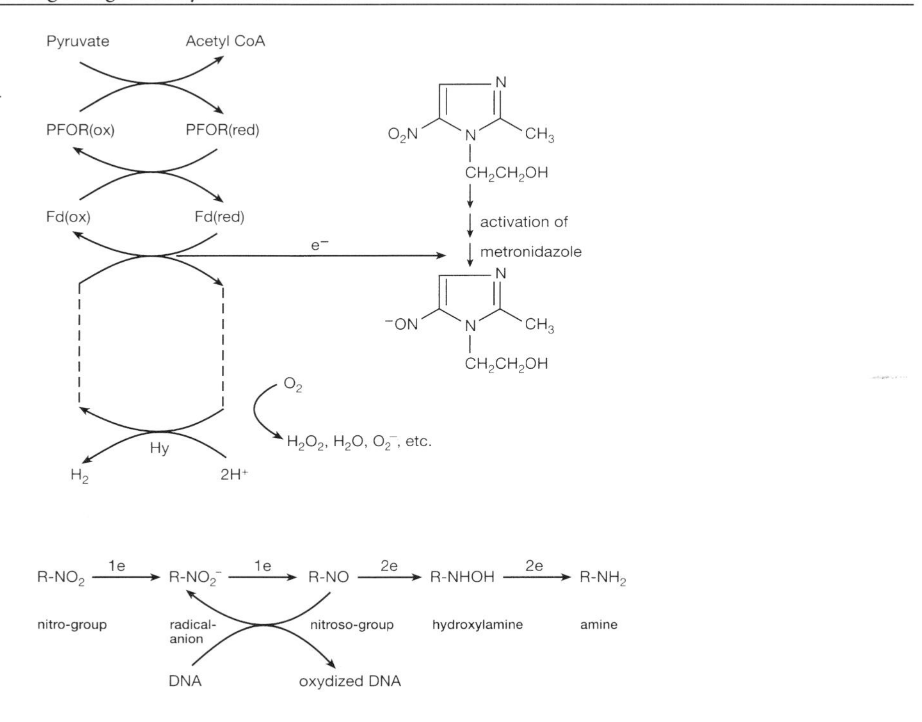

Fig. 2. Model of the mechanism of action of Metronidazole and other 5-nitroimidazoles.

against *G. lamblia* trophozoites in the small intestine which divide by binary fission.

Molecular Interactions The enzyme PFOR seems to be of great importance in the furazolidone action. The reduction of this nitro compound in-vivo to cytotoxic products is assumed to be similar to that of 5-nitroimidazoles (Fig. 2). The reduction potential of furazolidone is regarded as being even greater than that of metronidazole. An additional reduction mechanism of furazolidone via an NADPH/NADH oxidase to its nitroanion radical is also in discussion.

Resistance The molecular mechanism of furazolidone resistance is unclear to date. An involvement of thiol detoxification pathways is discussed. But any correlation between a decrease in furazolidone sensitivity and an increase in thiol cycling in *Giardia lamblia* is very doubtful since these parasites lack the enzymes of glutathione metabolism.

Primaquine/Pamaquine

Synonyms Primaquine: 8-(4-amino-1-methylbutylamino)-6-methoxyquinoline, SN 13272.

Pamaquine: Aminoquin, Beprochine, Gamefar, Plasmochin, Plasmoquine, Praequine, Quipenyl.

Clinical Relevance Both drugs belong to the chemical class of 8-aminoquinolines. Pamaquine was discovered in 1924, primaquine was introduced in 1950. Primaquine is a very effective prophylactic antimalarial agent and very effective in preventing relapses of malaria so that it can be used for a radical cure (→ Membrane-Function-Disturbing Drugs/Table 1). Exoerythrocytic stages of malarial parasites in the liver (sporozoites, hypnozoites, schizonts) and the erythrocytic gamonts get severely damaged (→ Hem(oglobin) Interaction/Fig. 2). Primaquine has an additional influence on the sporogony in the mosquito vector, however, it has no effect on erythrocytic schizonts (parasitic stages responsible for fever). Besides the antimalarial activity primaquine shows an additional activity against *Theileria sergenti* infections.

Molecular Interactions Primaquine is presumably metabolised in-vivo to products including 5,6-quinoline diquinone which structurally resembles hydroxynaphthoquinones. It is therefore assumed that the respiratory chain of the parasites is disrupted and the pyrimidine nucleotide synthesis is inhibited. In an alternative theory the generation of free radicals during the primaquine interaction with the respiratory chain is believed to be of great importance.

Oxamniquine

Synonyms Mansil, Vansil, Vancil.

Clinical Relevance Oxamniquine was introduced in 1973. The antitrematodal activity is directed only against *Schistosoma mansoni*. Oxamniquine has no activity against *S. haematobium* or *S. japonicum* (→ Membrane-Function-Disturbing Drugs/ Table 3).

Molecular Interactions The in-vitro activity of oxamniquine is very similar to that of the structurally related hycanthone. Oxamniquine leads to a delayed death of schistosomes until day 14. Thereby, an in-vitro exposure of only 1h is sufficient for the delayed death of the worms. Because the worm motility is increased at oxamniquine concentrations comparable to hycanthone, an anticholinergic action of oxamniquine was formerly proposed. However, this hypothesis is disproved in the meantime. Indeed, it could be shown that nucleic acid synthesis becomes irreversibly inhibited in drug-sensitive worms, in drug-resistant worms, in *S. japonicum* and in immature worms. The inhibition is more pronounced in male than in female schistosomes. The mode of action of hycanthone and oxamniquine is summarised in Fig. 3 according to the model of Cioli et al. 1995. At first hycanthone (and also oxamniquine) is converted to an ester (sulphate, phosphate or acetate) by a specific schistosomal enzyme. Thereafter, the ester spontaneously dissociates resulting in the formation of an electrophilic reactant which is capable of alkylating schistosomal DNA. Thus, the initial drug esterification is the only enzymatic step in the whole pathway (Fig. 3). The validity of this model is supported by experiments using the N-methylcarbamate esters of hycanthone. These hycanthone esters have been shown to be equally active against sensitive and resistant worms, since the first enzymatic esterification step can be surconvented by these hycanthone esters. As a result

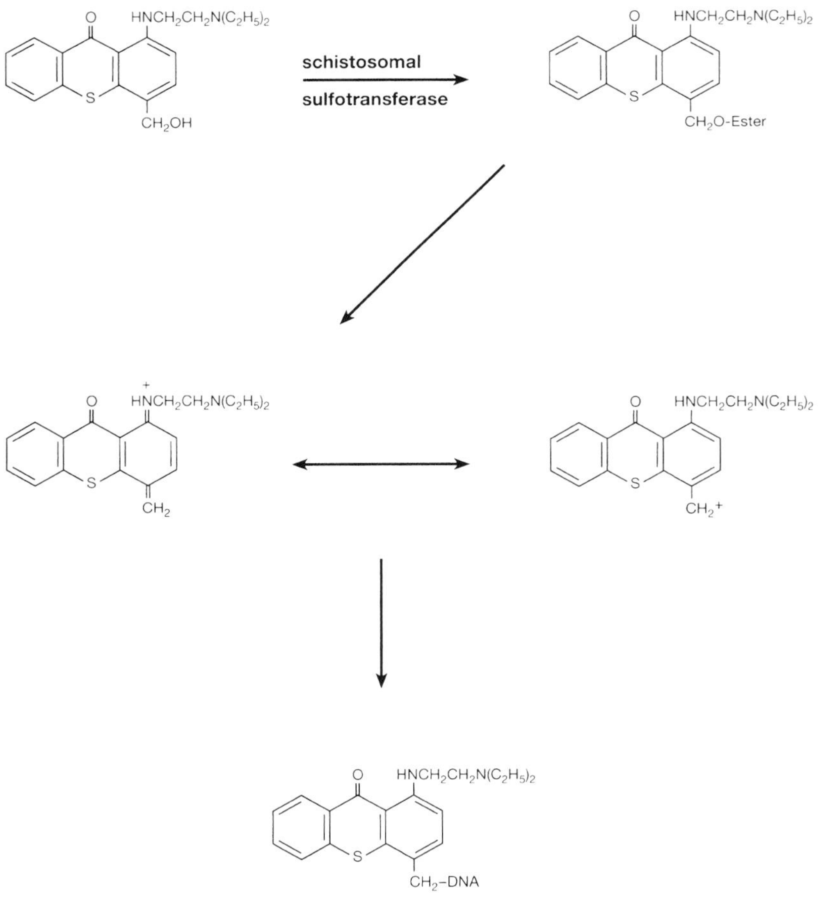

Fig. 3. Proposed mechanism of action of hycanthone and oxamniquine.

covalent binding of hycanthone and oxamniquine to macromolecules including DNA occurs in-vitro. In female and in immature schistosomes binding of hycanthone and oxamniquine to DNA is diminished compared to males. Adducts of hycanthone with guanosine residues of schistosomal DNA are formed.

ATP, Mg^{++} and another unknown small molecule are cofactors during the esterification step by the schistosomal enzyme. The activity of this enzyme can be restored by sulfate ions. Thus, this enzyme may function as a sulfotransferase with a molecular weight ranging between 30 and 35 kDa. The real function of this sulfotransferase is still unknown. A possible detoxifying function is discussed as well as an involvement in modifying male and female steroid hormones. Such a sulfotransferase is presumably also present in *S. haematobium* and *S. japonicum*. However, obviously structural differences of the sulfotransferases between the different *Schistosoma* spp. may be responsible for different binding of hycanthone and oxamniquine to this enzyme. While there is a strong binding of both drugs to *S. mansoni* sulfotransferase, only hycanthone can be bound by *S. haematobium* sulfotransferase. The sulfotransferase of *S. japonicum* is even unable to bind hycanthone or oxamniquine.

Additional Features There are also differences in the mutagenicity between hycanthone and oxamniquine. Oxamniquine has very low mutagenic activity compared to hycanthone. The mutagenicity of hycanthone is due to production of frameshift mutations as a consequence of its ability to intercalate between DNA base pairs resulting in an unwinding and distortion of the double helix. Therefore, hycanthone has been withdrawn as an antischistosomal drug. Oxamniquine also has some minor intercalative properties which are not enough for strong mutagenic effects.

Resistance Resistance against hycanthone and oxamniquine is controlled by a single, autosomal, recessive gene. Resistant schistosomes lack the activity essential for antischistosomal effects of oxamniquine in sensitive worms, because the enzymatic esterification step is missing in resistant parasites similar to the susceptible *S. japonicum*.

DNA-Synthesis-Affecting Drugs II: Interference with Purine Salvage

Structures

Fig. 1

Diloxanide

Synonyms 2,2-dichloro-4'-hydroxy-N-methylacetanilide, Furamide, Entamide, Ame-Boots.

Clinical Relevance Diloxanide is exclusively used as fuorate ester. The antiprotozoal activity is directed against trophozoites and lumen cysts of *Entamoeba histolytica* (→ DNA-Synthesis-Affecting Drugs I/Table 1). Diloxanide has only minor activity in acute ulcerative intestinal amoebiasis. Combinations of diloxanide with 5-nitroimidazoles, emetine or chloroquine for the treatment of amoebic dysentery and liver abscesses can be very useful.

Molecular Interactions It is suggested that it interferes with the purine salvage system by impairing

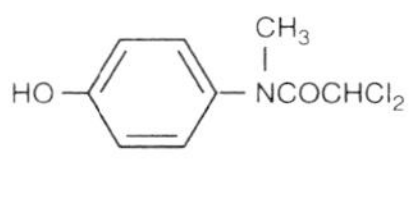

HO–C₆H₄–N(CH_3)$COCHCl_2$

Diloxanide

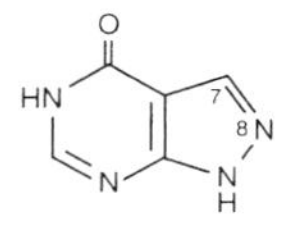

Allopurinol

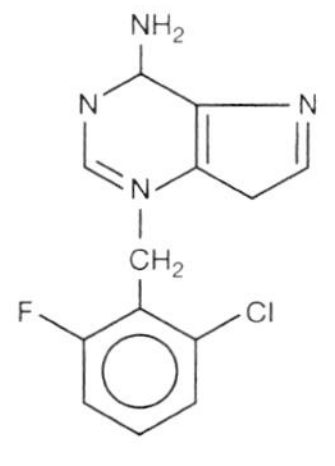

Arprinocid

Fig. 1. Structure of drugs affecting DNA-synthesis by interfering with purine salvage.

Table 1. Degree of efficacy of important drugs against kinetoplastid protozoa.

Year on the market	Drugs	*Trypanosoma brucei* group	*Leishmania* spp.	*Trypanosoma cruzi*
1920	Suramin	xxx 1)		x
1949	Melarsoprol	xxx 2)		
About 1958	Diminazene aceturate	xxx 2)	xE	
	Quinapyramine	xxx 1)		
1982	Eflornithine	xxx 2)	xx	
	Glucantime (Meglumine-antimonate)		xxx	
	Sodium-Stibogluconate		xxx	
About 1938/1942	Pentamidine/(Hydroxy)Stilbamidine	xxx 1)	xx	
	Allopurinol	xE	xx	
	Amphotericin B		xxx	
1972	Nifurtimox	xxx		xxx
1978	Benznidazole			xxx

xxx = high efficacy at least against some developmental stages, and diverse species; xx = partially effective (regarding developmental stages and diversity of species); x = slightly effective; E = experimentally effective ; 1) in blood (acute phase); 2) in liquor (late phase).

adenine incorporation during the RNA-synthesis in *E. histolytica*.

Allopurinol

Synonyms Allopurinol, Zyloric

Clinical Relevance Allopurinol is in medical use against *Leishmania* spp. In addition, it has experimental activity against *Trypanosoma* spp. (Table 1). Furthermore, allopurinol is clinically used in the treatment of gout as urikostatic drug.

Molecular Interactions The antiprotozoal mode of action against *Trypanosoma cruzi* and *Leishmania* spp. relies on the metabolization of allopurinol to adenosine nucleotide analogues. These are then incorporated into RNA with the result that the growth rate of sensitive parasites is significantly reduced.

The mode of action in gout is quite different from the antiprotozoal action presumably by the inhibition of xanthine oxidase.

Arprinocid

Synonyms 9-(2-chloro-6-fluorobenzyl)adenine, MK-302, Aprocox.

Clinical Relevance Arprinocid exhibits good anticoccidial and anticoccidiostatic activity (→ DNA-Synthesis-Affecting Drugs IV/Table 1). Its activity against *Eimeria tenella* is weaker compared to that against other *Eimeria* spp.

Molecular Interactions Arprinocid is simultaneously a purine and pyrimidine analogue. The activity is directed against sporozoites, merozoites and first generation schizonts. The mechanism of action remains unclear. Many pyrimidine nucleotide-requiring enzymes are inhibited. Also, the uptake of hypoxanthine and guanine in infected eukaryotic cells is inhibited. The anticoccidial activity of arprinocid is mediated by the N-1 oxide metabolite. Interestingly, the N-1 oxide metabolite itself is not a potent inhibitor of the biochemical assays in spite of the very potent in-vitro and in-vivo activity. Electronmicroscopically, a vacuolization and degeneration of intracellular membrane systems of coccidia can be observed.

DNA-Synthesis-Affecting Drugs III: Interference with Polyamine Metabolism and/or Trypanothione Reductase

Mode of Action (Fig. 1)

Structures (Fig. 2)

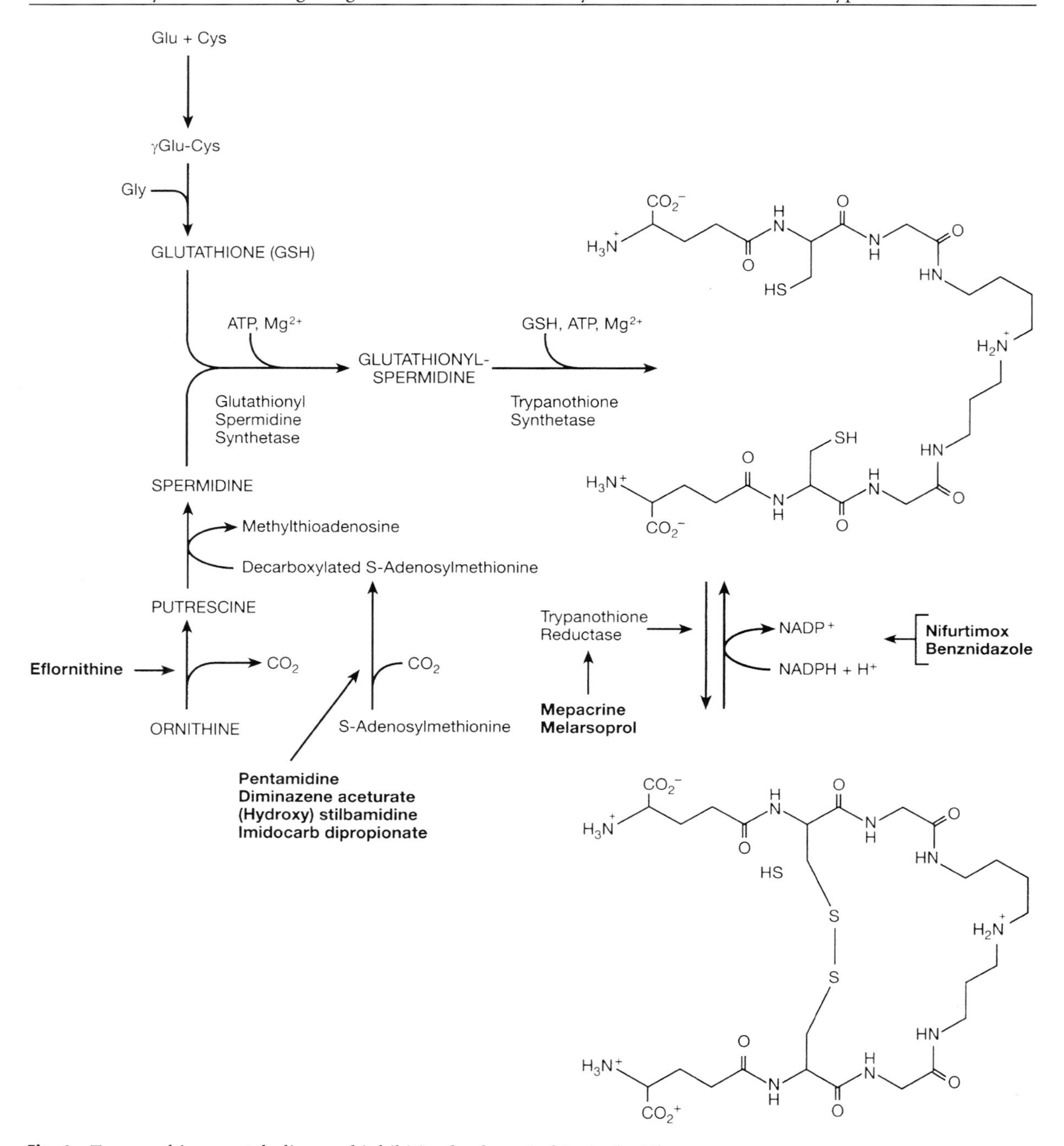

Fig. 1. Trypanothione metabolism and inhibition by drugs in kinetoplastid protozoa.

Melarsoprol

Synonyms Melarsen oxide, Mel B, Arsobal.

Clinical Relevance Melarsoprol was explored in 1949. It was the drug of choice for late phase of infections with *Trypanosoma b. gambiense* and *T. b. rhodesiense* until 1990 (Table 1). An intravenous application is necessary with 3.6 mg/kg b.w. in 3 to 4 series of 4 injections separated by at least one week. Melarsoprol possesses serious toxic side effects such as reactive encephalopathy in 5%–10% of the cases with a mortality rate of 1%–5%. In veterinary medicine melarsoprol is exceptionally used against *T. equinum* in horse. It has only low efficacy against *T. simiae* in pigs.

Molecular Interactions Melarsoprol is a trivalent organic arsenical. The activity is directed against trypomastigotes in the liquor. An inhibition of trypanosomal pyruvate kinase (PK) as the mode of action was proposed very early on. There is a loss of motility of drug-treated trypanosomes, and cell lysis occurs within minutes. However, there is no correlation between melarsoprol-induced lytic effects and inhibition of pyruvate kinase. In the meantime trypanosomal phosphofructokinase (PFK) ($K_i < 1\ \mu M$) and fructose-2,6-biphosphatase ($K_i = 2\ \mu M$) are found to be better melarsoprol targets than PK ($K_i = 100\ \mu M$), resulting in a complete inhibition of the formation of fructose-2,6-bisphosphate.

As an alternative hypothesis the inhibition of trypanothione reductase (TR) by complexation of

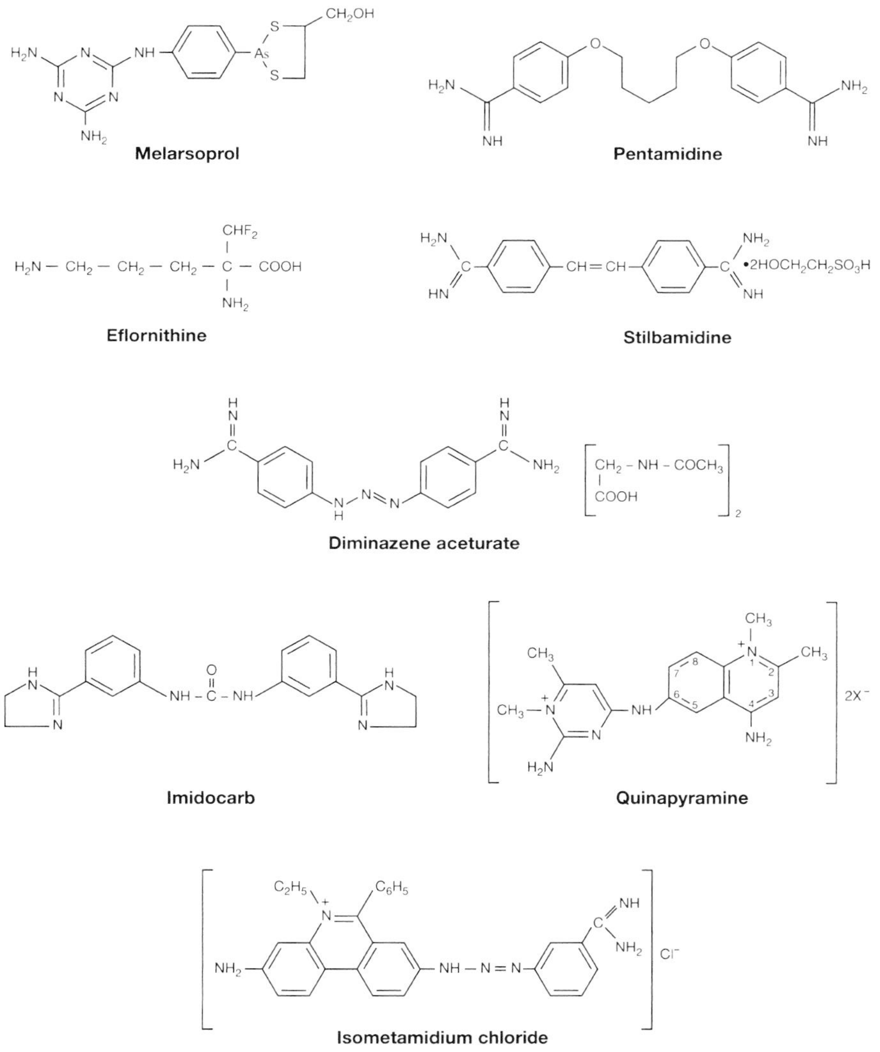

Melarsoprol

Pentamidine

Eflornithine

Stilbamidine

Diminazene aceturate

Imidocarb

Quinapyramine

Isometamidium chloride

Fig. 2. Structures of drugs affecting DNA-Synthesis by interference with Polyamine Metabolism and/or Trypanothione Reductase

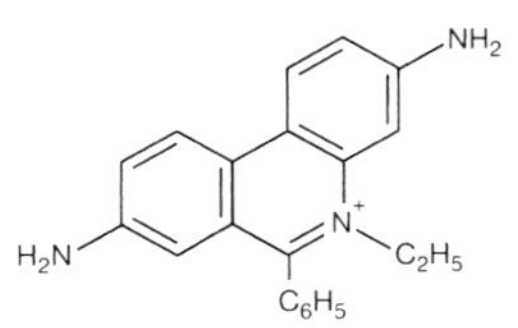

Homidium

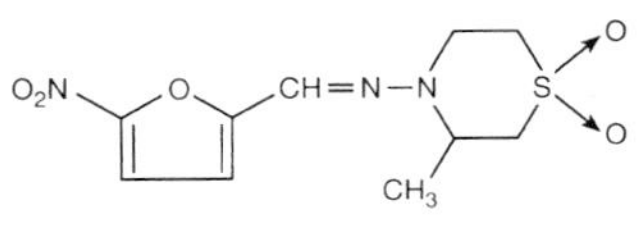

Nifurtimox

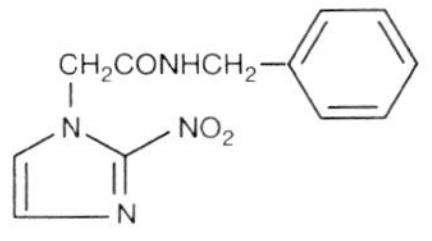

Benznidazole

Fig. 2. (continued) Structures of drugs affecting DNA-Synthesis by interference with Polyamine Metabolism and/or Trypanothione Reductase

trypanothione with melarsoprol or melarsen oxide is discussed (Fig. 1). This complexation would lead to a complete disturbance of the redox balance within the trypanosomal cell. An inhibition of TR would have lethal effects on trypanosomes. The melarsen-trypanothione adduct Mel T has a stability constant of 1.05 x 10^7 M^{-1}. The inhibition of glutathione reductase and the *T. b. brucei* TR is characterised by K_i values of 9.6 and 17.2 µM, respectively.

The main argument against TR as target for melarsoprol is the 18-fold-higher K_i value for the inhibition of TR compared to PFK. Furthermore an association of trypanothione with trivalent arsenicals is much weaker than that of 2,3-dimercaptopropanol or lipoic acid. Thus, the selective advantage of the presence of trypanothione in kinetoplastids has remained speculative until now. Melarsoprol is very efficient in forming adducts with a variety of dithiols (coenzyme dihydrolipoate, some proteins with cysteine residues). A non-specific inhibition of many different enzymes may explain many severe toxic side effects of melarsoprol.

Resistance The resistance against melarsoprol is a general serious problem in the treatment of sleeping sickness. Resistant trypanosomes are not lysed by melarsoprol or by the chemically related melarsen oxide at concentrations even higher than 100 µM. A significant decrease in free trypanothione levels could not be detected in resistant strains, whereas a rapid decrease in trypanothione levels was described in sensitive strains just before lysis. Isolated TR from resistant or sensitive strains was shown to be equally inhibited by the melarsoprol-trypanothione adduct Mel T, indicating that TR may not be a validated target for melarsoprol.

A new hypothesis for the resistance mechanism of melarsoprol relies on alterations of melarsoprol transport in resistant trypanosomes with a possible participation of drug efflux mechanisms similar to multi-drug resistant cancer cells. Indeed, there is no lysis of melarsoprol resistant strains in the presence of melarsoprol plus Ca^{++} channel-blockers (verapamil, diltiazem or nifedipine). The mechanism of this so-called melarsen-based drug resistance in trypanosomes is due to their absolute purine requirement. Two non-identical purine transporters P1 and P2 have been identified in trypanosomes. The transporter P2 is responsible for the uptake of melarsen or melarsoprol, adenine and adenosine. The melarsen oxide-induced trypanosomal lysis can be inhibited by adenine, adenosine and dipyridamol, an inhibitor of nucleoside transport in mammalian cells. Adenine, adenosine and melarsoprol thus compete for the transporter P2 in *T. b. brucei*. There is a reduction of the rate of adenosine transport by 80% in melarsen oxide resistant *T. b. brucei* compared to sensitive strains indicating a possible lack of transporter P2 in melarsen oxide-resistant *T. b. brucei*.

There is high cross-resistance between melarsoprol and the diamidine berenil in strains of *T. b. rhodesiense* clinical isolates, *T. b. brucei* veterinary isolates, in laboratory strains of *T. evansi* and others, but low cross-resistance between melarsoprol and pentamidine. The cross-resistance between melarsoprol and pentamidine can be correlated with differences in their uptake rates in *T. b. brucei*.

Resistance against the so called phenyl-based tryparsamide has been known since the 1930s. Interestingly, melarsen-resistant strains are sensitive to phenylarsenoxide and there is no cross-resistance between melarsoprol and phenylarsenoxide. This fact may be explained by the existence of another transporter (P1) for adenosine and inosine, which, however, does not transport melarsen oxide. Thus, phenylarsenoxide-induced lysis of trypanosomes cannot be inhibited by adenine or adenosine in sensitive trypanosomes which supports the idea of different uptake mechanisms of melarsoprol and phenylarsenoxide into trypanosomes.

Eflornithine

Synonyms DFMO, DL-a-difluoromethylornithine.

Clinical Relevance Eflornithine was introduced in 1982 for the treatment of Westafrican sleeping sickness (*Trypanosoma b. gambiense*). It is active against early and late stages of *T. b. gambiense* (Table 1). It was registered by the US Food and Drug administration in 1990 and the European Committee for Proprietary Medicinal Products in 1991. Eflornithine is characterised by a remarkably great safety index but possesses a relatively weak overall efficacy and a short duration of activity. An intravenous application is necessary with a dosage of 400 mg/kg b.w. per day in 4 equal doses every 6 h for 14 days. Eflornithine also has activity against *T. b. brucei*, *T. b. rhodesiense* and *T. congolense* and against multiresistant strains of *T. b. gambiense* (bloodstream forms and liquor forms). For the activity of eflornithine an intact immune system is necessary. A disadvantage of eflornithine is its ineffectivity against *T. b. rhodesiense* infections.

In addition eflornithine exerts activity against different opportunistic parasites such as *Pneumocystis carinii*, *Cryptosporidium* in AIDS patients as well as in-vitro or in-vivo activity against exoerythrocytic schizonts of *Plasmodium berghei*. Moreover, there is a report about some antitumor activity.

Molecular Interactions Eflornithine is a fluorinated amino acid derivative with zwitterionic properties. Under physiological conditions, it is poorly absorbed and rapidly excreted in the urine. The mechanism of action is well established and the *T. brucei* ornithine decarboxylase (ODC) is a fully validated therapeutic target. In trypanosomes, which are fully dependent on their own polyamine biosynthetic machinery, DFMO acts as an irreversible specific suicide inhibitor of ornithine decarboxylase ODC (Fig. 1) by formation of a covalent adduct between the decarboxylated and defluorinated DFMO and the residue 360 in ODC. As a result the trypanothione biosynthesis is inhibited (Fig. 1) and also the biosynthesis of the polyamines putrescine, spermidine and spermine. Thus, an in-vitro and in-vivo depletion of the putrescine and spermidine from dividing (binary fission) *T. brucei* trypomastigotes in the blood and liquor occurs. As a result of the inhibition of polyamine metabolism many different cell functions are impaired, e.g., the differentiation into non-dividing short-stumpy-like forms. In addition, DFMO-treated *T. brucei* are kept in the dormant G1 phase by loss of ODC activity. Moreover, the synthesis of variant surface glycoprotein is inhibited. The selective toxicity of eflornithine must be seen in direct connection with the slow turnover of ODC of *T. brucei* compared to the mammalian enzyme. Mouse ODC possesses an extra 36 amino acid peptide at the C-terminus (PEST sequence) triggering in-vivo degradation of mammalian ODC. This is responsible for the short half-life of ODC of about 20 min in mammalian cells. By contrast the in-vivo half-life of *T. brucei* ODC is longer than a day because of the lack of the PEST sequence. A further consequence of the DFMO-induced increased levels of adenosylmethionine may be an inappropriate methylation of proteins, nucleic acids or lipids. The lack of polyamines which are essential for the trypanothione synthesis may itself be sufficient for the explanation of the death of trypanosomes caused by DFMO.

Resistance Until now there are no reports about DFMO-resistance because of the short time of its clinical use. Experimental resistance was examined in-vitro using either procyclic forms of trypanosomes or naturally resistant strains. However, the mechanism of resistance on the molecular level remains unclear. In some resistant strains a reduced DFMO-uptake could be observed accompanied by an increase of intracellular concentrations of ornithine, whereas in other resistant strains no such reduced DFMO-uptake was detectable. There is no increased ODC activity in *T. brucei rhodesiense* field strains. However it could be shown that the ODC in DFMO-resistant *T. rhodesiense* possesses a rather short half-life compared to ODC in DFMO-susceptible *T.b. gambiense* which has an extraordinarily long half-life. As a result of the rapid in-vivo turnover rates and synthesis of new active ODC molecules in resistant strains DFMO-inhibited ODC molecules are rapidly replaced. Thus, cells with a rapid ODC turnover are much less affected by the inhibition of ODC. A further difference between DFMO-resistant and -susceptible trypanosomes seems to be the different increase in the adenosine methionine content by DFMO. Thus, in resistant strains an only 7-fold increase is observable compared to the up to 100-fold increase in sensitive strains. There also seems to be a correlation between DFMO-resistance and decrease in adenosine methionine synthetase.

Pentamidine/(Hydroxy)stilbamidine

Synonyms Pentamidine isethionate, 4,4'-diamidinodiphenoxypentane, M&B800, RP2512, Lomidine, Pentacarinat.

Clinical Relevance The diamidines are in clinical use since 1937. Pentamidine is a therapeutic and prophylactic drug against blood-stream forms in sleeping sickness (Table 1). It is, however, active only against early stages of *Trypanosoma b. gambiense* infections, but not against the liquor forms. 7 to 10 intramuscular injections are necessary with 4 mg/kg b.w. daily or on alternative days. Pentamidine also exhibits activity against *Leishmania donovani* and *L. chagasi* in spleen, liver and skin, but it has only minor activity against the American mucocutaneous leishmaniasis (*L. brasiliensis*). Furthermore pentamidine possesses antibabesial activity and is becoming increasingly important in replacing chloroquine against infections with *Babesia* spp. Pentamidine has no effect against *Trypanosoma cruzi*. (Hydroxy)-stilbamidine has a similar antiprotozoal spectrum. It is useful in antimonial-resistant Kala-Azar (*L. donovani*).

The main indication for pentamidine is its activity against opportunistic parasites. It is the drug of choice for the *Pneumocystis carinii* pneumonia in AIDS patients. Moreover, pentamidine has some antifungal activity against north American blastomycosis (*Blastomyces dermatitidis*).

Molecular Interactions The mechanism of action of pentamidine is directed against trypomastigotes in the blood which divide by binary fission. Following the uptake into the bloodstream forms of *T.b. brucei* via a carrier-mediated process the binding of pentamidine to nucleic acids is believed to be of great importance. Recently a pentamidine-dodecanucleotide-complex could be identified by cocrystallisation. Drug binding occurs in the 5'-AATT minor groove region of the duplex, preferentially to the minor grooves of the kinetoplast DNA in *T. brucei*. Thereby, the amidinium groups of pentamidine become H-bonded to adenine N_3 atoms. As a result the kinetoplast DNA is disrupted so that dyskinetoplastic cells are generated with intact mitochondrial membranes but lacking detectable kinetoplast DNA. Pentamidine has no effects on the trypanosomal nuclear DNA.

Furthermore, a 13-fold increase in lysine and 2.5-fold increase in arginine content is induced in the trypanosomes at the therapeutic dose. Electronmicroscopically, an intercalation of diamidines into the kinetoplast DNA (kDNA) could be detected resulting in lampbrush chromosomes. This observation supports the proposed inhibition of DNA synthesis by diamidines. Additional nuclear aggregation in diamidine-treated trypanosomes may also explain the inhibition of ribosomal RNA synthesis. The disintegration of kDNA begins at the periphery of the kinetoplast, where DNA replication starts. Because of their close structural similarity to pentamidine, (hydroxy)stilbamidine presumably has the same mechanism of action (Fig. 1).

Resistance Resistance to diamidines is well established under field conditions. Resistant strains are characterised by a diminished ability to import pentamidine into the cells. However, it is unclear to date whether an impaired pentamidine uptake, drug efflux or drug metabolism is responsible for the mechanism of pentamidine-resistance.

Diminazene

Synonyms Berenil, Diminazene aceturate, Diminazene diaceturate, Azidin, Ganasag, Trypan, Veriben.

Clinical Relevance Berenil was originally introduced in 1955 as a trypanocide and babesiacide. It is active against *Trypanosoma b. gambiense* and *T. b. rhodesiense* (Table 1). Of especial interest is the activity against liquor forms of *T. b. rhodesiense*. Berenil is often used in chronic human infections, although it is a veterinary product. The treatment of human trypanosomiasis with berenil is recommended in cases of arsen resistance or before starting treatment with melarsoprol. In veterinary medicine berenil is used against *T. brucei brucei*, *T. congolense*, *T. vivax* and *Babesia* spp. in cattle, sheep and goats in Africa. Higher dosages are usually necessary for curative effects against *T. equiperdum* in horses. Berenil has only minor activities against *T. simiae* in pigs, *T. evansi* in camels and cattle and *T. equinum* in horses. It should be mentioned that berenil also has experimental effectivity against *Leishmania* spp. (Table 1).

Furthermore, berenil has activity against piroplasms of domestic animals (*Babesia* spp. with large erythrocytic parasitic stages). It has good efficacy against *Babesia bigemina*/cattle, *B. ovis* and *B. motasi*/sheep, *B. caballi*/horse, *B. canis*/dog, *B. hepailuri*/cat. However, the drug has far less activity against *Babesia* spp. with small erythrocytic stages (*B. bovis* and *B. divergens*/cattle, *Theileria* (formerly *Babesia*) *equi*/horse, *B. gibsoni*/dog) and apparently no effect against *B. felis*/cat.

Molecular Interactions Berenil as an analogue of pentamidine exerts a similar mode of action. The activity of berenil is directed against trypomastigotes in the blood and liquor. In berenil-treated *Leishmania tarentolae* the kDNA content is greatly reduced. It is reported that berenil binds to the minor groove of DNA with a higher affinity to 5'-AATT-3' than to 5'-TTAA-3'. The attachment to specific sites in DNA occurs via electrostatic and H-bond forces. In addition, it is discussed that berenil may inhibit the kinetoplast topoisomerase II in trypanosomes, resulting in the cleavage of 2% of the minicircle DNAs in the presence of 1 μM drug. Also a possible interference of berenil with the trypanothione metabolism by inhibiting the decarboxylation of S-Adenosylmethionine is worth mentioning (Fig. 1).

Resistance Interestingly, there is no widespread development of berenil resistance in the field in spite of long-term use. There are reports on cross-resistance between quinapyramine, melarsomine and berenil in laboratory and field strains. The mechanism of resistance to berenil is possibly due to a diminished drug uptake by resistant trypanosomes.

Imidocarb Dipropionate

Synonyms Carbesia, Imixol, Imizol, Imizocarb, 4A65.

Clinical Relevance The antiprotozoal activity of this drug is directed against *Theileria* (formerly *Babesia*) *equi* and *Babesia caballi* in donkeys and mules. In addition, babesiosis of cattle may be controlled relatively easily by imidocarb.

Molecular Interactions Imidocarb is a diamidine derivative. Thus, its action may be similar to that of berenil (→ DNA-Synthesis-Affecting Drugs I/Fig. 1).

Quinapyramine, Homidium, Isometamidium

Synonyms Quinapyramine : M7555, Antrycide, Triguin, Trypacide

Homidium chloride : RD1572, Novidium, Babidium; Homidium bromide : Ethidium, Dromilac

Isometamidium : Metamidium, M&B4180, Samorin, Trypamidium.

Clinical Relevance Quinapyramine (Table 1) has activity against *Trypanosoma equiperdum* in horses and donkeys. Homidium and Isometamidium are used in chemoprophylaxis against *T. brucei evansi* in cattle, sheep and goats in Africa, and they have also activity against *T. vivax* and *T. congolense*.

Molecular Interactions Quinapyramine probably acts indirectly by inhibition of protein synthesis by displacement of magnesium ions and polyamines from the ribosomes. Homidium bromide belongs to the phenanthridinium derivatives. Its antitrypanosomal activity has been known for about 50 years. It is routinely used for staining nucleic acids in research laboratories because it intercalates into nucleic acids. It possesses mutagenic properties. The mechanism of antitrypanosomal action of homidium is unclear. There are reports on an interference with glycosomal functions, interference with the function of an unusual AMP binding protein, on impaired trypanothione metabolism and impaired replication of kinetoplast minicircle (2% of total minicircle become linearized by 1 μM homidium) in trypanosomes.

As isometamidium is structurally related to homidium and berenil, the properties and activities may be similar. Isometamidium has a great acute toxicity to mammals which is not observed with homidium or berenil. The acute toxic effects of isometamidium in mice can be reversed with atropine indicating probable inhibitory effects on acetylcholinesterase. However, the mechanism of antitrypanosomal action of isometamidium is not yet fully understood. A linearization of 6% of the total minicircle DNA from *T. equiperdum* at 1 μM may also contribute to the drug's action. In-vitro an intercalation between the base pairs of the DNA can be observed which may explain the interruption of DNA functioning observed in-vivo.

Resistance The extensive use of homidium in the 1960s and 1970s has greatly reduced its usefulness by widespread trypanosomal resistance. The mechanism of resistance is so far unknown. The mechanism of resistance against isometamidium is presumably associated with reduced accumulation of the drug in trypanosomes. There are reports on cross-resistance between isometamidium and homidium, which supports the idea of a similar mode of action of both drugs.

Nifurtimox

Synonyms Lampit, Bay2502.

Clinical Relevance Nifurtimox was introduced in 1972 as a causal therapeutic drug for American trypanosomiasis (= Chagas disease) caused by *Trypanosoma cruzi* (Table 1). It has curative effects in acute, subchronic and chronic disease. Infection-induced damages of organs, however, are not improved by this drug.

Molecular Interactions Nifurtimox is a nitrofurfurilidene derivative. It induces a destruction of non-dividing trypomastigote bloodstream forms and intracellular amastigote tissue forms in the muscles of heart, skeleton, oesophagus and intestine, in the lymph nodes and in the nervous system. One possible action may be the inhibition of trypanothione reductase (Fig. 1). As *T. cruzi* has only low detoxification capacity, it is completely dependent on the trypanothione metabolism. As another action the generation of reactive oxygen derivatives (superoxide, H_2O_2 and hydroxyl radicals) is discussed, which cause peroxidation of lipids and damage of the nucleic acids.

Resistance At present there are no great problems concerning clinical resistance against nifurtimox.

Benznidazole

Synonyms Ro7-1051, Radanil, Rochagan.

Clinical Relevance Benznidazole was introduced in 1978. It damages trypomastigote bloodstream forms and amastigote tissue forms of *Trypanosoma cruzi* (Table 1). There are reports on a low efficacy against the Brazilian cutaneous leishmaniasis.

Molecular Interactions The activity is directed against the same stages as by nifurtimox. Probably the trypanothione metabolism becomes disturbed and an involvement of generation of free radicals similar to nifurtimox is discussed (Fig. 1). There are additional reports on an inhibition of protein- and RNA-synthesis and a damage of DNA.

DNA-Synthesis-Affecting Drugs IV: Interference with Cofactor Synthesis

Table 1, Table 2, Fig. 1

Structures
Fig. 2

Sulfonamides

Important Compounds Sulfachloropyrazine, Sulfadiazine, Sulfadimethoxine, Sulfadimidine, Sulfadoxine, Sulfaguanidine, Sulfalene, Sulfamethazine, Sulfamethoxazole, Sulfametoxypyridazine, Sulfanitran, Sulfaquinoxaline/Pyrimethamine, Sulfaisoxazole, Sulfathiazole.

Synonyms Sulfachloropyrazine: Cosulfa, Cosulid, Nefrosul, Prinzone, Sorilyn, Vetisulid

Sulfadiazine: Adiazine, Debenal, Diazyl, Eskaiazine, Flamazine, Flammazine, Pyrimal, Silvadene, Sterazine, Sulfolex

Sulfadimethoxine: Agribon, Albon, Ancosul, Bactrover, Diasulfa, Diasulfyl, Dimetazina, Dinosol, Madribon, Maxulvet, Memcozine, Metoxidon, Neostreptal, Radonina, Retardon-N drops, Roscosulf, SDM

Sulfadimidine: Sulfadimidine 33% Forte, Sulfadimidine powder, Unidim

Sulfadoxine: Fanasil, Fanzil, in combination with pyrimethamine : Fansidar

Sulfaguanidine: Abiguanil, Aterian, Diacta, Ganidan, Guamide, Guanicil, Resulfon, Ruocid, Shigatox, Suganyl, Sulfaguine, Sulfoguenil, Enterosediv

Sulfalene: Farmitalia, Dalysep, Kelfizina, Longum, Polycidal

Sulfamethazine: Azolmetazin, Diazil, Dimezathine, Dimidin-R, Mefenal, Neazina, Pirmazin, S-Dimidine, S-Mez, Sulfa 25% powder, Sulfadine, Vesadin, Vertolan

Sulfamethoxazole: Abacin, Apo-Sulfatrim, Bactramin, Bactrim, Bactromin, Baktar, Drylin, Eltranyl, Eusaprim, Fectrim, Gantanol, Gantaprim,

Table 1. Degree of efficacy of important anticoccidial drugs on various protozoan parasites.

Year on the market	Drugs	*Eimeria* spp. (chicken)	*Toxoplasma gondii*	*Babesia* spp.	*Theileria* spp.	*Plasmodium* spp.
1968	Quinolones	xxxR	x a)			xE
1968	Clopidol	xxx				
1972	Robenidine	xxx				
1984	Clopidol/Methylbenzoquate	xxx				
1945	Sulfonamides	xx	x			xx
	Sulfaquinoxaline/Diaveridine	xxx				
	Sulfonamide/ Pyrimethamine	xx	xxx		xx	xxx
1956	Nicarbazine	xxx				
1960	Amprolium	xx	x			
about 1963	Amprolium/Sulfonamide/Ethopabate	xxx				
1980	Arprinocid	xxx				
1987	Toltrazuril	xxx	xxx		x	
1971	Polyethers b)	xxx	x a)			
1986	Narasin/Nicarbazine	xxx				
	Dinitolmide (DOT)	xxx				
1960	Zoalene	xx				
1976	Halofuginone	xxx			xxx	xx
1993	Diclazuril and Clazuril	xxx				

xxx = high efficacy at least against some developmental stages, and diverse species; xx = partially effective (regarding developmental stages and diversity of species); x = slightly effective; E = active experimentally; R = resistances arose quickly; a) prophylactically only; b) e.g. Monensin, Lasalocide, Maduramicin, Salinomycin, Semduramycin, Narasin.

Table 2. Drugs used against *Toxoplasma gondii*, *Neospora caninum*, *Sarcocystis* spp. and *Cryptosporidium* spp.

	Toxoplasmosis		Neosporosis	Sarcocystosis		Cryptosporidiosis	
	Human medicine	Veterinary medicine	Veterinary medicine	Human medicine	Veterinary medicine	Human medicine	Veterinary medicine
Decoquinate		xxx					
Pyrimethamine/Sulfonamide	xxx	xxx					
Sulfonamides				xx			
Epiroprim	xx E						
Epiroprim/Dapsone	xxx E						
Trimethoprim/Sulfamethoxazole/ Clindamycin	xxx						
Clarithromycin/Sulfonamide	xx a)	xx					xE
Clindamycin/Sulfonamide			xxx				
Pyrimethamine/Trimethoprim			xx				
Pirithrexim, Clindamycin, Diclazuril, Robenidine, Pyrimethamine			x E				
Toltrazuril		xxx			xxx		
Letrazuril						xx	
Spiramycin	xxx						
Paromomycin						xx	xxx
Monensin		xxx					

xxx = highly effective, xx = good effective, x = low activity; E = active experimentally; a) low tolerability

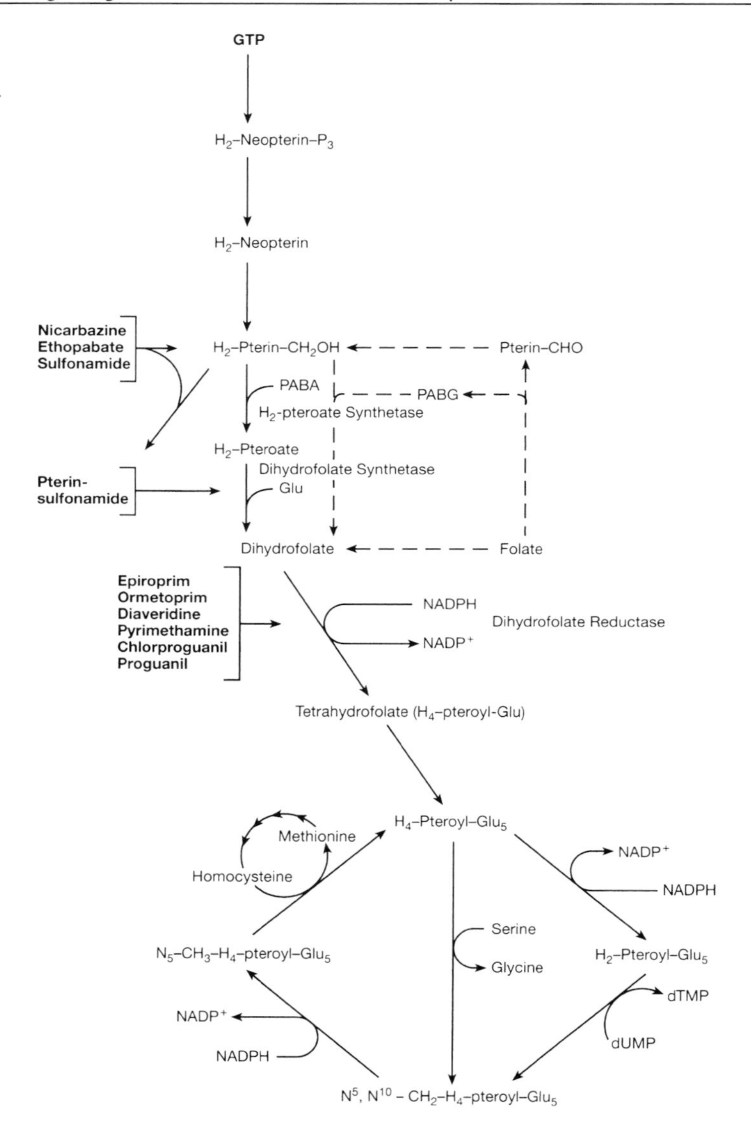

Fig. 1. Model of the de novo synthesis of pyrimidines and folate in apicomplexa

Gantrim, Kepinol, Linaris, Micotrim, Momentol, Nopil, Omsar, Septra, Septrim, Sigaprim, Sinomin, Sulfotrim, Sulfotrimin, Sulprim, Sumetrolim, Suprim, Tacumil, Teleprim, TMS480, Trigonyl, Trimesulf, Trimforte, Uro-Septra

Sulfametoxypyridazine: Davosin, Depovernil, Durox, Kynex, Lederkyn, Lentac, Midicel, Midikel, Myasul, Mylo-Sulfdurazin, Sultirene, Vinces

Sulfanitran: Novastat, Polystat, Unistat

Sulfaquinoxaline/Pyrimethamine: Sulka TAD, Coccex solution, single drug : Aviochina, Embazin, Dr. Hess SQX, Quinatrol, Quinel, Solquin, Sol-Quinel, S.Q., Sulfa-Nox, Sulfa-Q, Sul-Q-Nox, Sulfaquinoxaline 100% powder , Vineland Liquid Sulfaquinoxaline

Sulfaisoxazole: Entusil, Entusul, Gantrisin, Gantrosan, Neazolin, Renosulfan, Sosol, Soxisol, Soxo, Soxomide, Suladrin, Sulfalar, Sulfazin, Sulfium, Sulfoxol, Sulsoxin, V-Sul

Sulfathiazole: Eleudron solution

Clinical Relevance Sulfonamides exert activities against a variety of pathogens. They possess a broad-spectrum activity against gram positive and gram negative bacteria such as *Nocardia* spp., *Chlamydia* spp., *Yersinia* spp. and atypical mycobacteria (*Mycobacterium scrofulaceum*).

The first anticoccidial sulfonamides were introduced in 1946. Sulfonamides currently used in veterinary medicine are sulfaquinoxaline, sulfadimi-

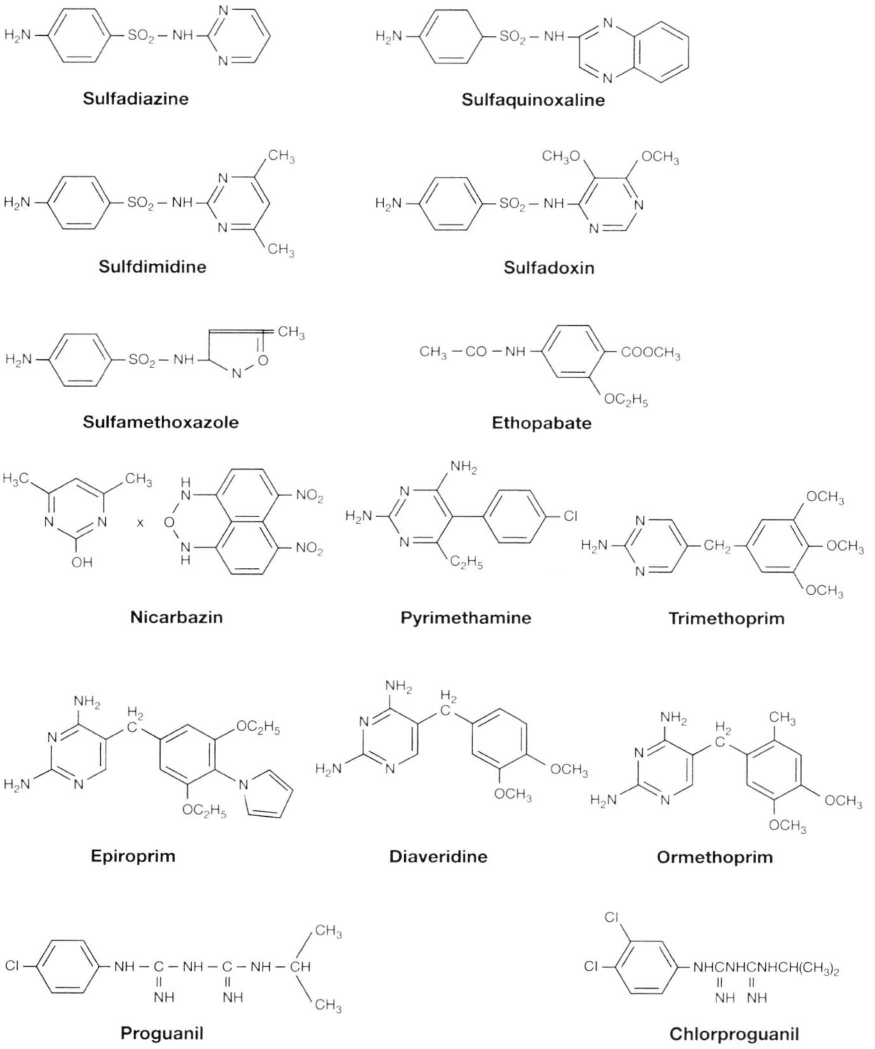

Sulfadiazine

Sulfaquinoxaline

Sulfdimidine

Sulfadoxin

Sulfamethoxazole

Ethopabate

Nicarbazin

Pyrimethamine

Trimethoprim

Epiroprim

Diaveridine

Ormethoprim

Proguanil

Chlorproguanil

Fig. 2. Structures of drugs affecting DNA-Synthesis by interfering with Cofactor Synthesis

dine, sulfamethazine, sulfadimethoxine, sulfachlorpyrazine and sulfathiazole. They are characterised by a narrow anticoccidial spectrum against *Eimeria* spp. residing in the small intestine (*E. acervulina*), and they have only minor activity against *E. necatrix* and *E. tenella*. Sulfaquinoxaline is used in combination with amprolium or in a multidrug combination together with ethopabate and pyrimethamine (Table 1). Other combinations of veterinary importance are those comprised of sulfonamide/ ormetoprim or sulfonamide/diaveridine. Sulfonamides have been very useful against *Plasmodium falciparum* as single drug or in combination with pyrimethamine (sulfadoxin/pyrimethamine), because of their schizonticidal effects against both exoerythrocytic and erythrocytic stages (→ Hem(oglobin) Interaction/Fig. 2). In general, they have greater efficacy against *P. falciparum* compared to *P. malariae*, *P. ovale* or *P. vivax*. Other indications in which sulfonamides are used are infections with *Toxoplasma gondii*, *Pneumocystis carinii*, *Sarcocystis* spp., *Cystisospora* spp., *Isospora* spp. and others. A sulfadiazine/clindamycin-combination is active against *Neospora caninum* in young dogs, if started very early in the disease (Table 2).

Molecular Interactions The antisporozoal activity of sulfonamides is primarily directed against second schizont generations. There is also activity against first schizont generation and sexual stages (→ Hem(oglobin) Interaction/Fig. 2). The activity of antimalarial sulfonamides such as sulfadoxin is directed against exoerythrocytic liver schizonts, erythrocytic schizonts and oocysts in the gut of mosquitoes. Sulfadoxin belongs to the long-acting sulfonamides. The first hint about the mode of action of sulfonamides came from the observation that the anticoccidial activities of sulfonamides can be reversed by paraminobenzoic acid (PABA), an intermediate in folate biosynthesis. Now it has been known for a long time that the action of sulfonamides like sulfathiazole, sulfaguanidine, sulfanileamide, sulfadoxin and of sulfones like dapsone relies on the inhibition of dihydropteroate synthase in intracellular sporozoa *Eimeria*, *Toxoplasma* and *Plasmodium*. Thus, the coccidial folic acid biosynthesis is inhibited by sulfonamides which is lethal to the coccidia because they do not utilise exogenous folate but synthesise folate as cofactor of DNA synthesis de-novo (Fig. 1).

Resistance Dihydropteroate synthetase (DHPS) is a validated target enzyme of sulfadoxin. This could be shown for sulfadoxin-resistant *Plasmodium falciparum* isolates. DHPS from resistant and sensitive strains differ in their amino acid sequences. Indeed, there is a correlation between point mutations in the bifunctional DHPS and sulfadoxin resistance. Interestingly, DHPS of *P. falciparum* is a bifunctional enzyme which includes the dihydro-6-hydroxymethylpterin pyrophosphokinase at the amino terminus.

Ethopabate

Synonyms Methyl 4-acetamido-2-ethoxybenzoate, Ethyl pabate, in combination with amprolium: Amprol Plus.

Clinical Relevance Ethopabate is only a narrow anticoccidial spectrum drug against *Eimeria acervulina*. It has no or only slight activity against *E. maxima*, *E. necatrix*, *E. tenella* or *E. brunetti*. Today, ethopabate is applied only in combination with amprolium (Table 1).

Molecular Interactions Ethopabate is a 2-substituted PABA derivative (= 4-acetamido-2-ethoxybenzoic acid methylester) and functions as a prodrug. Its activity becomes potentiated by pyrimethamine and antagonised by the simultaneous administration of PABA. Thus, the mode of action of ethopabate is similar to that of sulfonamides or sulfones (Fig. 1).

Nicarbazine

Synonyms 4,4'-dinitrocarbanilide, Altek, Elancocin, Nicarb, Nicoxin, Nicrazin.

Clinical Relevance Nicarbazine is used against coccidiosis in poultry. It has a coccidiostatic action by impairing the oocyst formation of the late life cycle stages. The numbers of oocysts are reduced and a latent infection up to the last life cycle stages is always detectable. Thus, nicarbazine-treated animals can develop immunity against coccidia.

Molecular Interactions The activity is directed against schizonts of the second generation of *Eimeria* spp. An inhibition of folate biosynthesis is proposed (Fig. 1).

2,4-Diaminopyrimidines

Important Compounds Pyrimethamine, Trimethoprim, Diaveridine, Ormetoprim, Epiroprim, Pirithrexim.

Synonyms Pyrimethamine: Daraprim, RP4753, Chloridin, Darapram, Malocide, Tindurin; in: Fansidar, Suldox, Malocide, Maloprim, Metakelfin

Trimethoprim: 2,4-diamino-5-(3,4,5-trimethoxybenzyl)pyrimidine, Monotrim, Proloprim, Syraprim, Tiempe, Trimanyl, Trimopan, Trimpex, Wellcoprim; in: Borgal, Cosumix, Leotrox, Protox, Tribissen, Trafigal, Vetoprim, Bactrim, Eusaprim, Septrin, Sultrim

Diaveridine in: Darvisul, Rofenon

Ormetoprim in: Rofenaid-40, Ektecin

Epiroprim: none

Pirithrexim: none

Clinical Relevance 2,4-diaminopyrimidines (pyrimethamine, trimethoprim) are active against a wide variety of human pathogens. Thus, they possess a broad antibacterial spectrum. Antiprotozoal active 2,4-diaminopyrimidines used in human medicine are pyrimethamine and trimethoprim. They are used for prophylaxis in malaria, but the onset of antimalarial activity is very slow. They have no general effect on gamonts and no effect on hypnozoites of *Plasmodium vivax* (→ Hem(oglobin) Interaction/Fig. 2). Because of the frequently occurring resistance against pyrimethamine they are applied in combinations with sulfonamides (sulfadoxin) (Fansidar). In general, combinations of pyrimethamine and trimethoprim with sulfa drugs are of great medical importance for treatment of toxoplasmosis and *Isospora belli* infections in AIDS-patients (Table 1, Table 2) and *Cyclospora cayetanensis* infections in AIDS. Thus, combinations of pyrimethamine with sulfonamides such as sulfadiazine, sulfadimidine or sulfadoxin belong to the standard treatment for human toxoplasmosis (Table 1). However, these combinations are restricted for the treatment of prenatal toxoplasmosis in the time after the 20th week. In addition, a pyrimethamine-sulfonamide combination is the drug of choice for postnatal infections with *Toxoplasma gondii* in children and in AIDS patients. For tolerability reasons and especially in prophylactic treatment a combination of trimethoprim and sulfamethoxazole sometimes in a combination with clindamycin is recommended. There are only anecdotal reports of treatment of microsporidiosis in AIDS with trimethoprim/sulfamethoxazole.

2,4-diaminopyrimidines used in veterinary medicine are pyrimethamine, ormetoprim, epiroprim, pirithrexim and diaveridine. They are routinely used against coccidiosis in combinations such as ormetoprim/sulfadimethoxine or sulfaquinoxaline/amprolium/ ethopabate/pyrimethamine (=Supracox). A combination of clarithromycin and pyrimethamine has been shown to be effective against toxoplasmosis in animals, but is not well tolerated. In addition, a pyrimethamine/sulfamethazine-combination is reported to be effective against *T. gondii* in sheep and goats (Table 2). Trimethoprim or pyrimethamine as single drugs or together in combination show curative effects in *Neospora caninum* infections, when motor nerve disturbances have already occurred (Table 2). Epiroprim is a new 2,4-diamino-pyrimidine for the treatment of toxoplasmosis. It possesses an in-vitro activity against *T. gondii*. There are promising results in animal experiments with a dapsone/epiroprim-combination. Pirithrexim, another 2,4-diaminopyrimidine, and pyrimethamine have been shown to be active against *N. caninum* tachyzoites in cell cultures.

Besides the antibacterial and the antiprotozoal activities 2,4-diaminopyrimidines are useful as anticancer drugs and in therapy of rheumatic diseases.

Molecular Interactions Pyrimethamin is a selective inhibitor of the dihydrofolate dehydrogenase (DHFR) of exoerythrocytic schizonts (in the liver) and erythrocytic schizonts of malarial parasites (→ Hem(oglobin) Interaction/Fig. 2) and other sporozoa such as *Toxoplasma* and *Eimeria*. Schizonts become damaged by pyrimethamine only during nuclear division. There is no effect on the parasitic ring forms of *P. falciparum*. Pyrimethamine has an additional activity against oocysts in mosquito gut. Besides pyrimethamine the parasitic DHFR is also inhibited by the other 2,4-diaminopyrimidines trimethoprim, epiroprim, diaveridine and ormethroprim (Fig. 1). It has been well established for a long time that there are synergistic effects between the 2,4-diaminopyrimidines as DHFR-inhibitors and sulfonamides as inhibitors of dihydropteroate synthetase, when given in combination (Fig. 1).

Resistance Most knowledge about the resistance mechanism against pyrimethamine at the molecular level comes from experiments with plasmodia. A modification of the target receptor DHFR in the folic acid pathway in resistant strains very likely results in a decrease of DHFR sensitivity to inhibition. There are several reports that the molecular basis of pyrimethamine resistance in naturally resistant isolates of *Plasmodium falciparum* are single point mutations (e.g., Ser-108 ⇒) in the DHFR active site (Fig. 3). There is a good correlation between mutations with natural pyrimethamine resistance in a variety of geographically distant isolates. In addition, the two ancillary mutations Asn-51 ⇒ Ile-51 and Cys-59 ⇒ Arg-59 are associated with increased pyrimethamine resistance in the presence of Asn-108.

Another mechanism of pyrimethamine-resistance in *P. falciparum* has been proposed. Thereby, an overproduction of DHFR achieved either by gene duplication or by other mechanisms resulting in increased expression is discussed as being responsible for resistance on the molecular level.

Chlorproguanil

Synonyms Chlorguanil, 1-(3,4-dichlorophenyl)-5-isopropylbiguanide, M5943, hydrochloride: Lapudrine.

Clinical Relevance Chlorproguanil serves as a causal prophylactic, but not as a therapeutic drug for malaria. Its effectivity is not different from proguanil.

Molecular Interactions The target of chlorproguanil is the dihydrofolate reductase of the parasitic stages. It has an influence on sporozoites and the exoerythrocytic stages of all four *Plasmodium* spp. (→ Hem(oglobin) Interaction/Fig. 2). The onset of the effect on erythrocytic stages is very slow. The activity is directed against exoerythrocytic schizonts, erythrocytic schizonts and oocysts in mosquito gut (→ Hem(oglobin) Interaction/Fig. 2).

Resistance There is cross-resistance between chlorproguanil and pyrimethamine indicating the same mode of action by inhibition of the DHFR (Fig. 1, Fig. 3).

Proguanil

Synonyms Chlorguanide, 1-(p-chlorophenyl)-5-isopropylbiguanide, Chloroguanide, M4888, RP3359, SN12837, Diguanyl, Drinupal, Guanatol, Palusil, Tirian.

Clinical Relevance Proguanil was explored in 1945. It was developed for prophylaxis against malaria, but the prophylactic activity is not complete. It leads to a damage of exoerythrocytic schizonts. It has no effect on the hypnozoites in the liver.

Molecular Interactions Proguanil belongs chemically to the biguanids. It is a prodrug, which is converted to cycloguanil, a cyclic triazine with antimalarial activity. The activity is directed against exoerythrocytic schizonts in the liver, erythrocytic schizonts and oocysts in mosquito gut (→ Hem(oglobin) Interaction/Fig. 2). Recently proguanil was introduced in combination with atovaquone for the treatment of acute uncomplicated Malaria tropica. The mode of action is the selective inhibition of the DHFR of malarial parasites (Fig. 1, Fig. 3).

Resistance There are specific DHFR point mutations responsible for resistance to cycloguanil which could be shown in a variety of independent *Plasmodium falciparum* clones and isolates. From examination of the point mutations of DHFR it becomes clear that there is a different molecular basis for resistance to cycloguanil and pyrimethamine, depending on the positions of the mutations and on the residues involved. Point mutations result in an inhibition of pyrimethamine binding at the active site of the reductase (Fig. 3). But there are presumably different effects of the DHFR point mutations on pyrimethamine and cycloguanil, since proguanil is sometimes effective against pyrimethamine-resistant *P. falciparum*, indicating that there may be different binding sites for both drugs. In other cases there is cross-resistance between proguanil and pyrimethamine observable.

DNA-Synthesis-Affecting Drugs V: Interference with Dihydroorotate-Dehydrogenase

Structures
Fig. 1

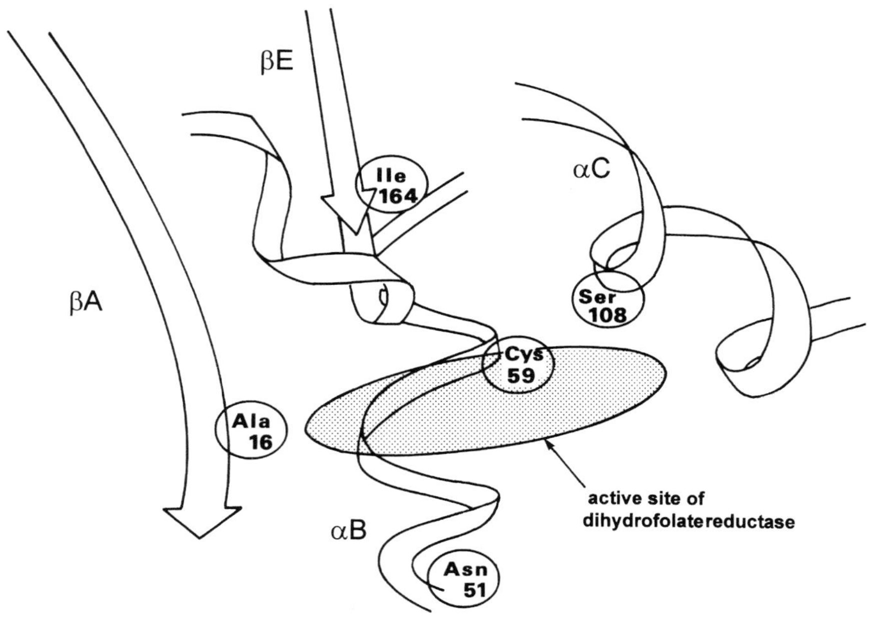

Fig. 3. Schematic model of the point mutations in the active site of dihydrofolate reductase of *Plasmodium falciparum* (Gutteridge (1993) In : Cox FEG (ed) Modern Parasitology, Second Edition, Blackwell Science, pp. 219–242)

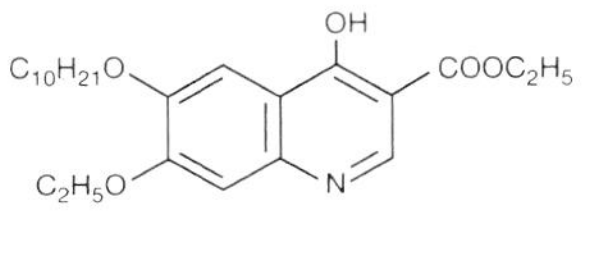
Decoquinate

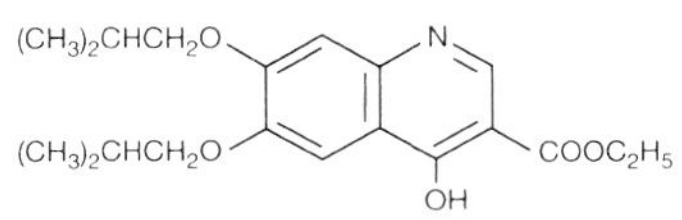
Buquinolate

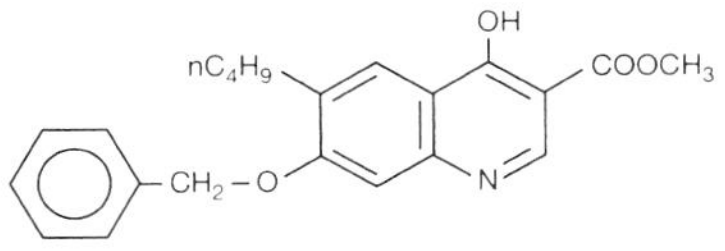
Methylbenzoquate

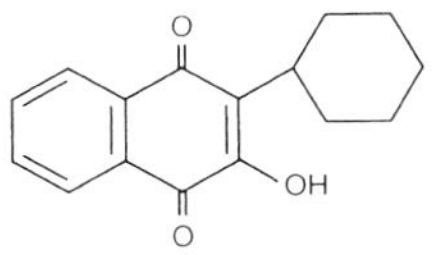
Paravaquone

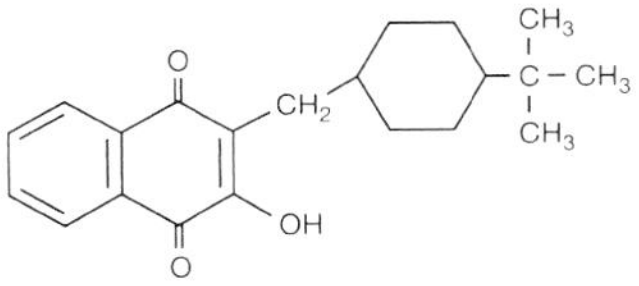
Buparvaquone

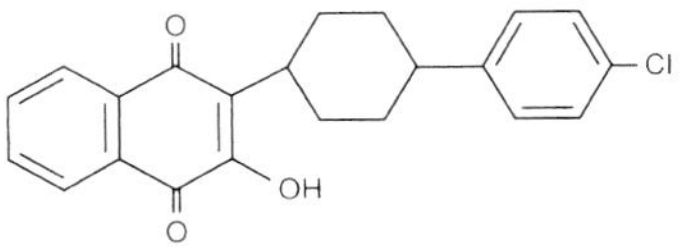
Atovaquone

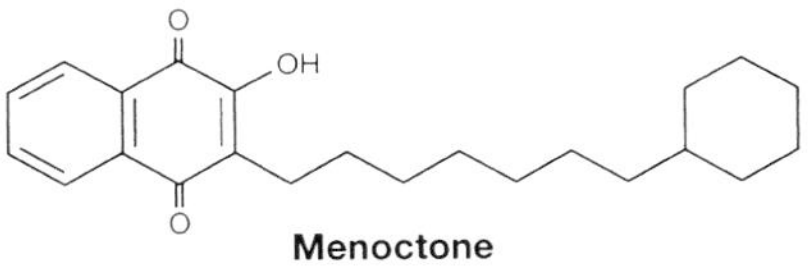
Menoctone

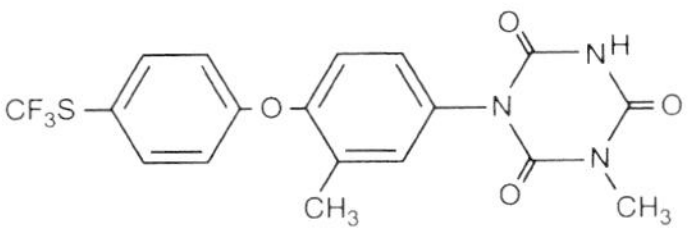
Toltrazuril

Fig. 1. Structures of antiparasitic drugs affecting DNA-Synthesis by interference with Dihydroorotate Dehydrogenase

Hydroxyquinolines

Important Compounds Amquinate, Buquinolate, Decoquinate, Methylbenzoquate.

Synonyms Amquinate: none

Buquinolate: 4-Hydroxy-6,7-diisobutoxy-3-quinolinecarboxylic acid ethyl ester, Ethyl 6,7-diisobutoxy-4-hydroxyquinoline-3-carboxylate, Bonaid

Decoquinate: Ethyl 6-(n-decyloxy)-7-ethoxy-4-hydroxyquinoline-3-carboxylate, M&B15497, Deccox

Clinical Relevance 4-hydroxyquinolines have antiparasitic activities against *Toxoplasma* spp., *Pneumocystis carinii* and *Cryptosporidium parvum*, and they furthermore possess anticoccidial, antimalarial and antitheilerial activity. Decoquinate is active against *T. gondii* in cats (→ DNA-Synthesis-Affecting Drugs IV/Table 2) and is still used in shuttle programs against coccidiosis. For such shuttle programs it is of great advantage when the single components exert synergistic activity. This is the case for methylbenzoquate and meticlorpindol, which are used in combination in such shuttle (rotation) programs. However, as single compounds they have only limited success as anticoccidials.

Hydroxynaphthoquinones

Important Compounds Parvaquone, Buparvaquone, Atovaquone, Menoctone.

Synonyms Buparvaquone: Butalex; Menoctone: Menocton; Parvaquone: Clexon.

Clinical Relevance The potential of this class of 2-hydroxynaphthoquinones was realised over 50 years ago. Parvaquone and buparvaquone have antibabesial and antitheilerial activity (*Theileria parva* in cattle). Against theileriosis parvaquone acts against the schizonts in lymphoid cells. Thereby, it selectively destroys infected lymphocyte cells thus setting free the schizonts which are not protected against the host defence system in contrast to merozoites with their surface coated pellicle. Menoctone (= 2-hydroxy-3-(8-cyclohexyl-octyl)-1,4-naphthoquinone) is active in-vitro in infected bovine lymphoid cell cultures and against theileriosis in cattle *in vivo*.

Atovaquone exerts antimalarial and anticoccidial activity. It is in clinical development for treatment of malaria and opportunistic infections in AIDS (*Toxoplasma gondii* cysts in the brain of mice). It has high efficacy in humans suffering from malaria after oral applications. Very recently a new drug combination atovaquone/proguanil (Malarone) was introduced for treatment of acute uncomplicated Malaria tropica. It is also effective for prophylaxis. There are no more clinical trials of atovaquone against cryptosporidiosis.

Molecular Interactions The action of **atovaquone** against multidrug-resistant strains of *Plasmodium falciparum* may be explained by its new mode of action being different from that of the other antimalarial drugs. It is assumed that the antimalarial actions of 2-hydroxynaphthoquinones and 4-hydroxyquinolines are identical. Their activity is directed against schizonts in lymphocytes or against erythrocytic schizonts. There is additional activity against liver and mosquito stages of *P. berghei*. Besides, the formation of ookinets from mature gametocytes of *P. falciparum* is inhibited. As analogues of ubiquinones these compounds have structural similarity to reduced coenzyme Q, and they act through an inhibition of the electron transfer at complex III of the mitochondrial respiratory chain of parasites. Moreover, there are several reports of an inhibition of cellular respiration by 4-hydroxyquinolines and an inhibition of mitochondrial succinate dehydrogenase and NADH-dehydrogenase activities from different sources. In isolated mitochondria from *P. falciparum* atovaquone is bound strongly and selectively to the ubiquinol-cytochrome c reductase site of the respiratory chain (= complex III). The point of block is located between the ubiquinone and cytochrome b. The inhibition can be reversed by addition of coenzyme Q. The high therapeutic index and thus the high selectivity of 4-hydroxyquinolines correlates with the lack of inhibition of cell respiration in the chicken host. The reason for this is presumably the great difference of the electron transport chains between coccidia and their vertebrate hosts. Indeed, atovaquone is 2000 fold more active against the plasmodial respiratory chain compared to the corresponding rat liver mitochondria.

There is another hypothesis for the action of hydroxyquinolines, which relies on the inhibition of pyrimidine nucleotide synthesis at the level of dihydroorotate dehydrogenase (Fig. 2). This then would result in an inhibition of sporozoite and trophozoite development in the intestinal epithelium. Indeed, a strong inhibition of pyrimidine nucleotide synthesis could be shown for amquinate, buquinolate, decoquinate and meticlorpindol probably due to an inhibition of dihydroorotate dehydrogenase (DHOD) (Fig. 2). In *Plasmodium*, where ubiquinone plays an important role as an electron acceptor for dihydroorotate dehydrogenase (DHOD), atovaquone is believed to inhibit the pyrimidine synthesis at this level.

Resistance There are high frequencies of serious drug resistance in the field against 4-hydroxyquinolines. Thus, resistance against buquinolate already appeared within 6 months. There is cross-resistance between different quinines as indication for a similar mode of action. Mitochondria from 4-hydroxyquinoline-resistant *Eimeria tenella* are insensitive to drug inhibition. The mitochondrial respiration of amquinate resistant cells is nearly 100-fold less sensitive to 4-hydroxyquinolines, but the real mechanism of resistance on the molecular level is still unknown. Until now resistance of malaria parasites against the 2-hydroxynaphthoquinone atovaquone does not play any role.

Toltrazuril

Synonym Baycox

Clinical Relevance Unlike other anticoccidials toltrazuril acts on all intracellular developmental stages of all known *Eimeria* and *Isospora* spp. In addition, toltrazuril exerts an activity against the schizogonous and gametogonic stages of *Toxoplasma gondii* in the cat (→ DNA-Synthesis-Affecting Drugs IV/Table 1). For extraintestinal infections of *T. gondii* in cats longer treatment periods with toltrazuril are necessary. Toltrazuril has additional activity against sarcocystosis (→ DNA-Synthesis-Affecting Drugs IV/Table 2).

Molecular Interactions Toltrazuril is a symmetrical triazinetrione. Its action is directed against first and second generation schizonts, microgamonts and macrogamonts. It has no activity against free sporozoites and merozoites. Toltrazuril probably acts by inhibiting the mitochondrial respiration and nuclear pyrimidine synthesis in the parasite. A destruction of the wall-forming bodies II can be observed in the macrogamonts. Histochemical and biochemical studies reveal that dihydroorotate dehydrogenase may act as a further target of toltrazuril (Fig. 2). There is ultrastructural evidence that plastid-like organelles are present in *T. gondii*, *Sarcocystis muris*, *Babesia ovis*, and *Plasmodium falciparum* containing protochlorophyllidae a as well as traces of chlorophyll bound to the photosynthetic reaction centers PS I and PS II. These plastid-like structures have been described as membranous cytoplasmic structures containing a circular DNA molecule of 35-kb length having a plastid ancestry. It is assumed that the sensitivity of apicomplexans to toltrazuril depends on the interaction of this drug with the D1 protein, a vital constituent of the photosynthetic reaction center II.

Until now there are only reports of isolated cases of resistance against toltrazuril. Under laboratory conditions resistance is relatively difficult to achieve.

Dourine

Chronic veneral disease of horses caused by → *Trypanosoma equiperdum* (Vol. 1) (→ Genital Sys-

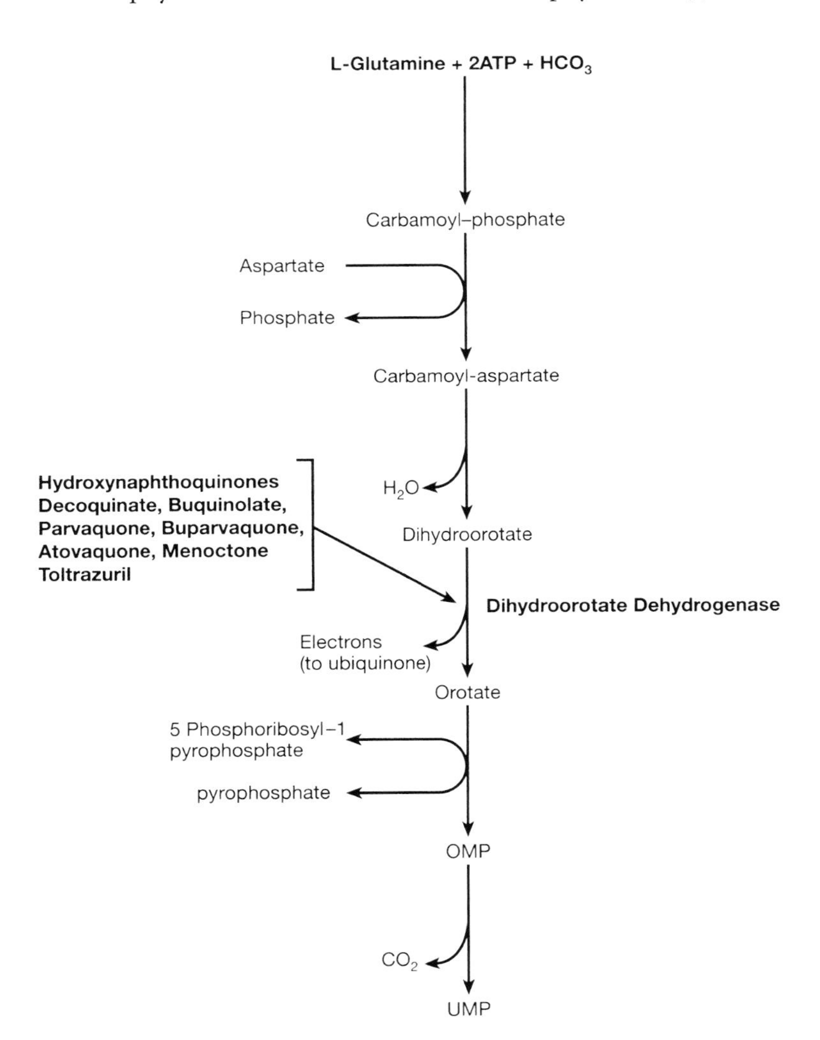

Fig. 2. Model of dihydroorotate dehydrogenase as target for anticoccidial drugs.

tem Diseases, Animals, → Nervous System Diseases, Horses).

Dracontiasis

Synonym

→ Dracunculiasis.
→ Dracunculus medinensis (Vol. 1).

Dracunculiasis

Synonym

Guinea worm infection, Dracontiasis, Medina worm disease, Fil d'Avicenne

General Information

Dracunculiasis is acquired by the swallowing of copepods infected with larvae of *Dracunculus medinensis* with drinking water. The larva molts and may leave the copepod to be ingested by another one. Or, the ingested L3 larva matures in the deep tissues of humans, and as mature female migrates to a subcutaneous site (usually arm or leg). Larvae are deposited into the tissue and because of hypersensitivity a cutaneous vesicle is formed which may ulcerate. Upon immersion of the extremity into water, the adult projects into the ulcer or vesicle and releases the larvae which wait to be ingested again by copepods. Two types of lesions are produced in man, vesicles which ulcerate and subcutaneous or deep abscesses around dead adult worms. Inflammatory reaction in the microabscess includes epithelial and giant cells, lymphocytes, and plasma cells; neutrophils and eosinophils are also present close to degenerating worms. These lesions eventually calcify.

Targets for Intervention

The infective larvae of *Dracunculus medinensis* are ingested by the human host with water while they are still contained within their copepod host (*Cyclops* spp.). Fig. 1 illustrates that dracunculiasis is a disease that lends itself to simple and highly effective control. The major targets of intervention are the cycle of infection and the environment of the copepod host. The approaches to control include detection of infected persons, extraction of the adult worm, prevention of contamination of water used for human consumption and control of *Cyclops* spp. in such water collections, as well as the use of safe drinking water.

Main clinical symptoms: Allergic skin reactions, skin necrosis
Incubation period: 3–4 months
Prepatent period: 10–14 months until emergence of the female
Patent period: Female die usually within 2–6 weeks after emerging from the skin
Diagnosis: Macroscopic inspection of appearing females

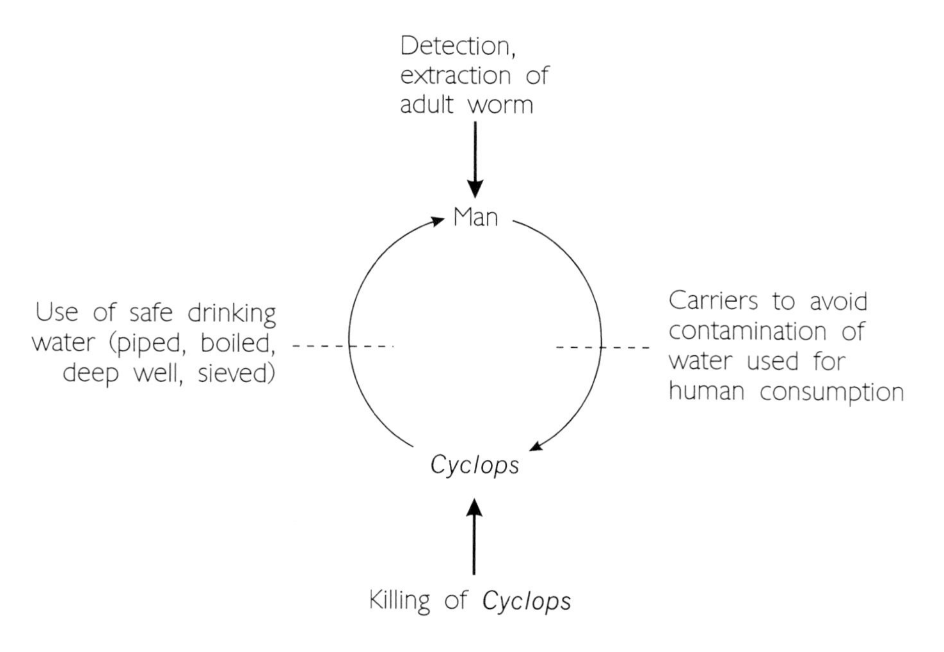

Fig. 1. Targets and approaches for the control of dracunculiasis

Prophylaxis: Avoid drinking uncooked water in endemic regions
Therapy: Surgical withdrawal of adult females, treatment see Nematocidal Drugs

Drug

Synonyms

Compound, substance, agent, medicament, product.

Definition

Drug can be broadly defined as any chemical compound that affects living processes and is used in the diagnosis, prevention, treatment (cure) of disease(s), or for controlling or improving any physiological or pathological disorder or for relief of pain in animals and humans.

General Information

→ Chemotherapy/Drugs.

Related Entries

→ Acanthocephalacidal Drugs, → Antidiarrhoeal and Antitrichomoniasis Drugs, → Arthropodicidal Drugs (→ Blocker of Chloride Channels, → Blocker of Cholinergic Neurotransmission, → Blocker of Sodium Channels, → Blocker of Aminergic Development, → Energy-Metabolism-Disturbing Drugs, → Repellents (Vol. 1)), → Babesiacidal Drugs, → Cestodocidal Drugs, → Chemotherapy, → Coccidiocidal Drugs, → Control, → DNA-Synthesis-Affecting Drugs I, → DNA-Synthesis-Affecting Drugs II, → DNA-Synthesis-Affecting Drugs III, → DNA-Synthesis-Affecting Drugs IV, → DNA-Synthesis-Affecting Drugs V, → Ectoparasitocidal Drugs (→ Arthropodicidal Drugs, → Blocker of Aminergic Transmission, → Blocker of Chloride Channels, → Blocker of Cholinergic Neurotransmission, → Blocker of Sodium Channels, → Blocker of Arthropod Development, → Energy-Metabolism-Disturbing Drugs, → Repellents (Vol. 1)), → Energy-Metabolism-Disturbing Drugs, → Hem(oglobin) Interaction.

Insecticides see → Ectoparasitocidal Drugs.

→ Leishmaniacidal Drugs, → Malariacidal Drugs, → Membrane-Function-Disturbing Drugs, → Microsporidiosis, → Microtubule-Function-Affecting Drugs, → Myxosporidiacidal Drugs, → Mode of Action, → Nematocidal Drugs, Animals, → Nematocidal Drugs, Man, → Protein-Synthesis-Disturbing Drugs, → Repellents (Vol. 1), → Sarcocystosis, → Theileriacidal Drugs, → Treatment of Opportunistic Agents, → Trematodocidal Drugs, → Trypanocidal Drugs, Animals, → Trypanocidal Drugs, Man.

Mode of Action and Resistance of Drugs

→ Blocker of Cholinergic Neurotransmission, → DNA-Synthesis-Affecting Drugs I, → DNA-Synthesis-Affecting Drugs II, → DNA-Synthesis-Affecting Drugs III, → DNA-Synthesis-Affecting Drugs IV, → DNA-Synthesis-Affecting Drugs V, → Energy-Metabolism-Disturbing Drugs, → Hem(oglobin) Interaction, → Membrane-Function-Disturbing Drugs, → Microtubule-Function-Affecting Drugs, → Protein-Synthesis-Disturbing Drugs.

Drugs Against Microsporidiosis

General Information

→ Microsporidians (Vol. 1) are eukaryotes of ancient origin (lack of mitochondria) and obligate intracellular parasites entering host cells via a polar tube within a spore. The organisms are ubiquitous and may occur in humans and a wide range of animals (wild, fish, and arthropods) but also dogs and other domestic animals commonly associated with humans. Today, there are no efficient control measures for prevention of microsporidiosis in humans and animal reservoirs. Spores seem to be resistant to various physical effects such as sonification, freezing or thawing, ultraviolet or gamma radiation.

Humans

Serious microsporidiosis may develop in immunocompromised individuals (e.g. AIDS patients, travellers going to tropic areas). Ocular infection is manifest by conjunctival, corneal and/or stromal invasion, predominantly by *Encephalitozoon hellem*, and *Vittaforma cornea* (syn. *Nosema corneum*). Intestinal infections are due to *Enterocytozoon bieneusi*, *Encephalitozoon intestinalis*, and *E. cuniculi*. Clinical signs may be severe diarrhea, malabsorption, and wasting. Disseminated infections may be produced by *E. hellem*, *E. cuniculi*, and *Pleistophora* spp. and are often accompanied by concurrent HIV infection (see → Treatment of Opportunistic Agents/Drugs Acting on Cryptosporidiosis of Mammals and → Antidiarrhoeal and Antitrichomoniasis Drugs/Drugs Acting on Giardiasis, respectively).

Pathology Immunosuppressed humans have been sentinels of microsporidial infection, with enteric, neurologic, ocular and pulmonary manifestation being recognized. Their symptomatology, pathology and differential diagnosis, based on ultrastructure and the polymerase chain reaction, has been widely established. Microsporidiosis appears to be a common asymptomatic infection, that is not clinically recognized; about 10% of animal handlers were reported to have antibody to *Encephalitozoon* sp. The organism grows intracellularly, destroying the infected cells. There is little inflammation. The Brown-Brenn stain (Gram stain for tissues), basic fuchsin, toluidin blue, Azur II-eosin, the Warthin-Starry silver impregnation and polarization facilitate recognition of microsporidia in tissue sections. Tissue imprints (smears), dried, fixed and stained as for blood smears, are useful for diagnosis of corneal and conjunctival lesions. Stool and sputum smears can be stained with a modified trichrome stain employing chromotrope 2R or with a fluorochrome chitin stain, such as Calcofluor.

A fatal disseminated infection of a 4-months-old thymic alymphoplastic baby with *Nosema connori* involved the smooth musculature, skeletal muscles, the myocardium, parenchyma! cells of the liver, lung and adrenals. *Encephalitozoon* sp. was isolated from the cerebrospinal fluid of a 9-year-old Japanese boy with meningoencephalitis who recovered (Matsubayashi). Intestinal microsporidiosis has been described in several patients with AIDS due to *Enterocytozoon bieneusi* also with cholangitis and due to *Encephalitozoon* (*Septata*) *intestinalis*. The patients had diarrhoea, with weight loss from malabsorption. Inflammation was minimal and the diagnosis was made ultrastructurally. Disseminated *Encephalitozoon cuniculi* infection was noted and *E. hellem* had been isolated from AIDS patients with nephritis and prostatitis and from others with keratoconjunctivitis, bronchitis and sinusitis . Microsporidial myositis due to *Pleistophora* and *Trachypleistophora hominis* was reported in patients with AIDS. Intraocular microsporidiosis was diagnosed from the cornea next to Descemet's membrane with a subacute to granulomatous inflammatory reaction; other HIV-negative cases with corneal stromal infection were linked to *Nosema ocularum* (possibly *Vittaforma corneum*).

Treatment Treatment is problematic since the gram-positive staining spores (most familiar stage of microsporidia) are resistant to most drugs (→ Treatment of Opportunistic Agents/Table 1). Several topical antimicrobial and antiinflammatory drugs have been used for treating ocular disease such as keratoconjunctivitis. **Topical drugs** such as propamidine isethionate or the water-soluble fumagillin derivative Fumidil-B (eyedrops) may resolve ocular symptoms caused by *E. hellem*, which reoccur, however, when treatment is stopped indicating a static rather than a cidal action of drugs. For lesions due to *V. corneae*, topical therapy is generally not effective and keratoplasty may be required. Oral administration of **albendazole** may improve ocular, nasal and enteric symptoms; the drug has a marked static effect on intestinal microsporidians like *Ent. bieneusi* and *E. intestinalis*. The latter species proves more susceptible to treatment with albendazole. The drug seems to inhibit polymerization of microtubules within intranuclear spindles in dividing nuclei only, and for that reason growth of parasites will continue in the absence of nuclear division. In the current situation almost nothing is known about epidemiology of human microsporidiosis; an open question is whether microsporidians in man are solely human infections or are some episodes of zoonotic origin.

Fish

With respect to chemotherapy of microsporidiosis in fish, it was shown that the antibiotic **fumagillin** acts against *Glugea plecoglossi* in the Japanese ayu (*Plecoglossus altivelis*), but in some cases the mortality of treated fish was higher than that of untreated fish. The mode of action of fumagillin is thought to inhibit DNA or RNA synthesis. More recent investigations have shown that a symmetric triazine **toltrazuril** and an asymmetric triazine (HOE 092V) kill the merogonic and the sporogonic stages of the microsporidian *G. anomala*. The drugs are also highly active against other fish and crayfish parasites. As revealed by ultrastructural investigations, the effects of the triazine derivatives on *G. anomala* comprised a decrease in the number of ribosomes, enlargement of the smooth ER, depletion of the nuclear membrane, and destruction of the nuclear structures. In a further study it was demonstrated that different **benzimidazole** derivatives (albendazole, mebendazole and fenbendazole) disturbed the intracellular development of the microsporidian *G. anomala* by damaging its merogonic, sporogonic and prespore stages as well as the mature spores.

Drugs Against Sarcocystosis

Sarcocystis spp. are obligate, heteroxenous coccidian parasites. Definitive host may be carnivores, felidae, humans and primates. Their asexual reproduction may occur in herbivores (horse, cattle, sheep, goat, camel), omnivores (pigs, humans) and also birds (chicken, ducks). Consumption of feed contaminated with sporocysts (oocysts) from the feces of the definitive host leads to infection of herbivores.

Pathology and Treatment

Experimental sarcocystosis in cattle *(S. bovicanis)* or that of horses *(S. equicanis)* has been associated with acute and chronic myositis caused by intramuscular sarcocysts. Long-term treatment with orally administered pyrimethamine, and trimethoprim + sulfamethoxazole is necessary to achieve remission of clinical signs. This can be applied also to cases of naturally occurring equine protozoal myeloencephalitic (EPM) caused by *S. neurona*. The treatment lasted between 45 and 211 days; adverse effects in horses were mild to severe, including abortion. **Halofuginone** (→ Coccidiocidal Drugs/Table 1) appears to be effective against acute sarcosporidiosis in goats and sheep at 0,67 mg/kg on two successive days. In humans, who may serve as accidental intermediate host for several unidentified *Sarcocystis* spp. sarcocysts have been found in striated muscles. Their clinical significance is unknown in naturally occurring life cycles. For **control** of sarcocystosis, carnivores should be excluded from animal houses, and from feed, water and raring facilities for livestock or dead livestock. Cooking or heating (60°C for at least 20 min.) of contaminated material will kill sarcocysts.

Vaccination

There is no protective vaccine against sarcocystosis in animals or humans.

Drugs with Unknown Antiparasitic Mechanism of Action

Structures

Fig. 1

DL-Propranolol

Clinical Relevance The antiparasitic activity of propranolol is directed against *Giardia lamblia*.

Nitrobenzamides

Important Compounds Dinitolamide, Zoalene, Nitromide.

Clinical Relevance Dinitolmide, zoalene have activity mainly against first and second generation schizonts of *Eimeria tenella* and *E. necatrix*. They have only limited effects against *E. acervulina*. They do not interfere with immunity of the chicken hosts. Nitrobenzamides are assumed to act as a nicotinamide antagonist.

Diclazuril/Clazuril

Clinical Relevance Diclazuril and clazuril belong to 1,2,4-triazine-derivatives. They have broad spectrum activity in *Eimeria* infections. They exert additional activity against *Neospora caninum* tachyzoites in cell cultures. The mode of action is unknown. The activity of diclazuril is directed only against specific endogen stages of *Eimeria* species. Thus, diclazuril is active against second generation schizonts of *E. acervulina*, *E. mitis* and *E. necatrix*, against gamonts, late schizonts of *E. brunetti*, zygotes of *E. maxima* and first and second generation schizonts and sexual stages of *E. tenella*. The sporulation becomes delayed by diclazuril.

Febrifugine

Clinical Relevance Febrifugine is an antimalarial drug from traditional Chinese medicine.

Pyronaridine

Clinical Relevance Pyronaridine has high efficacy in clinical studies in China against chloroquine-resistant *Plasmodium falciparum*. There is cross-resistance to 4-aminoquinolines and quinoline-methanol antimalarials. Morphological effects on mitochondria, endoplasmic reticulum and ribosomes can be observed. Food vacuoles are probably a primary target of pyronaridine.

Halofuginone

Clinical Relevance Halofuginone is a quinazolinone derivative. It is an alkaloid originally isolated from the plant *Dichroa febrifuga*. It is effective against the six pathogenic *Eimeria* spp. of chicken. Halofuginone has an influence on sexual

Propranolol

Diclazuril

Zoalen

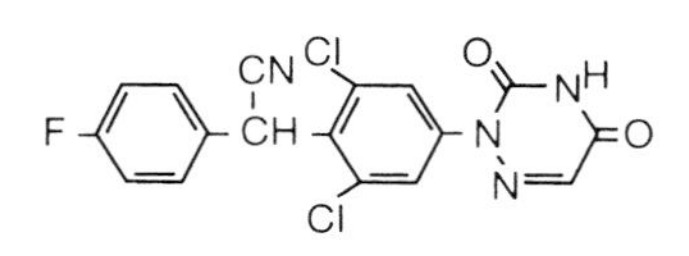

Letrazuril

Febrifugine

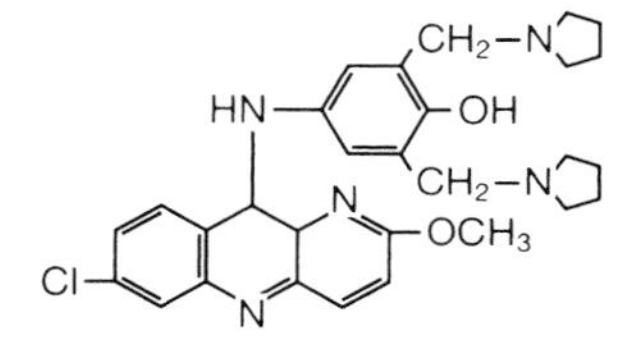

Pyronaridine

Halofuginone

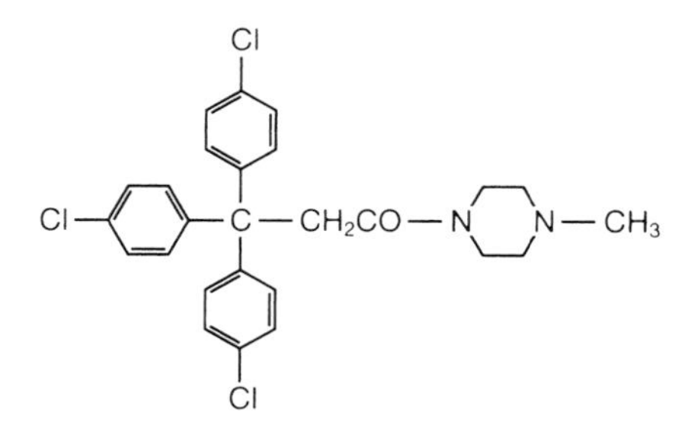

Hetolin

Fig. 1. Structures of antiparasitic drugs with unknown mechanism of action

Cyclosporin A

NHCONH

Quinuronium Sulfate

NHCONH

$H_2N-C=NH$ $H_2N-C=NH$

Amicarbalide

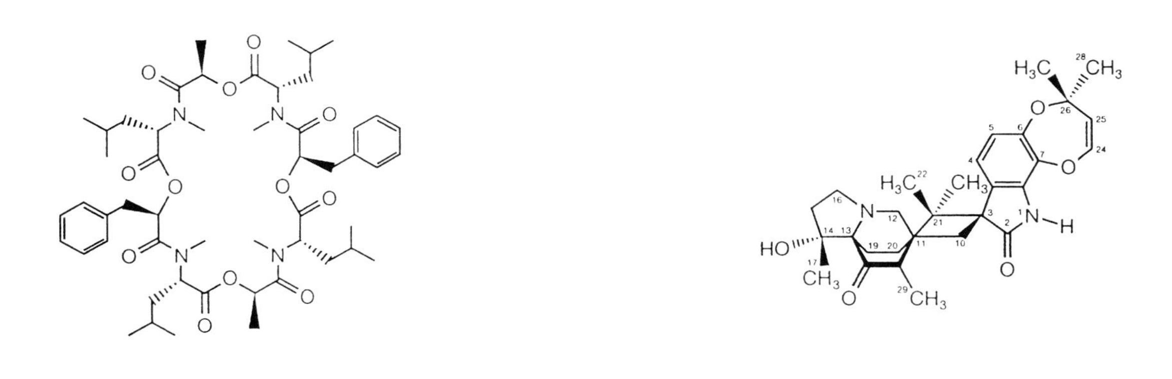

PF 1022 A **Paraherquamide**

Fig. 1. (continued)Structures of antiparasitic drugs with unknown mechanism of action

first-generation schizogony. It specifically suppresses the skin collagen synthesis in mammalian cells in vivo resulting in an increase in skin fragility. In-vitro the incorporation of radiolabelled proline into collagen by avian skin fibroblasts is impaired. Thereby, the expression of a1 gene of collagen type I is specifically depressed but not that of collagen type II in skin fibroblasts and growth-plate chondrocytes. This results in a decrease in synthesis of collagen. The mechanism of action against *Eimeria* spp. is unknown at present. Furthermore, halofuginone has been used in *Theileria parva* and *T. annulata* infections in cattle. The activity is directed against schizonts in lymphocytes. Thereby, infected lymphocytic cells become destroyed thus setting free the schizonts, which in contrast to merozoites with their surface coated pellicle are not protected against the defense system of the host. Halofuginone has additional activity against *Cryptosporidium parvum* in calves.

Halogenated Hydrocarbons

Clinical Relevance The halogenated hydrocarbons carbon tetrachloride and hexachloroethane were used until very recently in veterinary medicine. Now they are no longer used because of mutagenic properties. Hetol (= hexachloroparaxylene) was used to a limited extent in veterinary medicine against *Fasciola hepatica* infections and for human *Clonorchis sinensis* infections (→ Membrane-Function-Disturbing Drugs/Table 4). However, it is characterized by serious side effects.

Hetolin

Clinical Relevance Hetolin is a specific compound against *Dicrocoelium dendriticum* (→ Energy-Metabolism-Disturbing Drugs/Table 1) with erratic efficacies.

Cyclosporin A

Clinical Relevance The antiparasitic activity of cyclosporin A was discovered in 1981. The antiprotozoal activity is directed against *Leishmania* spp., *Toxoplasma* and *Plasmodium*. In addition, cyclosporin A has anticestodal and antischistosomal activity. The latter is directed against schistosomes. Here the compound has a long-lasting prophylactic effect up to 100 days before infection. After treatment the number of worms is reduced. Immature worms and female worms are more affected than male worms. The use of cyclosporin A would mean important qualitative progress in the control of schistosomiasis since there would be protection by a single dose over a good part of the transmission season. A correlation between the antiparasitic and the immunosuppressive effects of cyclosporin A seems very unlikely, since cyclosporin analogs with antischistosomal activity show only minor immunosuppressive activity. Cyclosporin A has lethal effects in-vitro against schistosomes. Cell-mediated immunity does not play a role in the antischistosomal action of cyclosporin A. Furthermore, the compound has antifilarial activity and is used in human medicine as an immunosuppressing drug.

Quinuronium sulfate (= Acaprine), Amicarbalide

Clinical Relevance Babesiosis of cattle may be controlled easily by these two drugs. A single parenteral treatment results in the disappearance of clinical symptoms and premunity against e.g. *Babesia divergens*. In general, however, improvement of parasit-aemia is obtained only after several treatments with higher dosages.

PF1022A

Clinical Relevance This compound has potent antinematodal properties against *Toxocara canis*, *Toxocara cati*, and hookworms in dogs and cats and *Trichuris vulpis* in dogs. In addition, it has high efficacies against gastrointestinal nematodes in horses, sheep, chicken and rodents.

Molecular Interactions PF1022A is a 24-membered cyclic depsipeptide isolated from *Mycelia sterilia*, a fungus which belongs to the microflora of the plant *Camellia japonica*. Recently the chemical synthesis of PF1022A and also the radiolabelled compound have been reported. The anthelmintic action on the molecular level remains obscure to date. It appears that it exerts its activity by interfering with the neuromuscular transmission of nematodes. At low concentrations, the motility of *Angiostrongylus cantonensis* is depressed, and picrotoxin, bicucullin and Ca^{++} can antagonize the action. Thus, the action is explained by an antagonism of acetylcholine receptors and/or gabaergic mechanisms. It could also be shown that PF1022A has a high affinity binding to the $GABA_A$ receptor in *Ascaris suum*.

Paraherquamide

Clinical Relevance Paraherquamide is an oxindol alkaloid metabolite of *Penicillium paraherquei*. It has antiparasitic activity and was first reported in rodent models. It is highly efficaceous as a single oral treatment at dosages above 0.5 mg/kg against adult trichostrongylides in sheep and in addition against L4 larvae of *Cooperia* spp. Moreover, it is effective against ivermectin- and benzimidazole resistant *Haemonchus* and *Trichostrongylus* strains. In calves paraherquamide is effective against adult stages of nine common gastrointestinal and lung nematodes at single oral dosages of 0.5, 1.0, 2.0 or 4.0 mg/kg. However, *Cooperia punctata*, the dose-limiting species, was affected only at the highest dosage of 4.0 mg/kg to 89%. The compound is not toxic in ruminants up to a dosage of 10 mg/kg, but three dogs given an oral dose of 10 mg/kg had severe intoxication within 30 minutes of dosing and two died within 2 hours. On the basis of available data, for a broad-spectrum therapeutic paraherquamide has a therapeutic index of less than 3 at the dosage of 4 mg/kg and is thus inferior to ivermectin. As a narrow spectrum therapeutic used at a dosage of 0.3 mg/kg paraherquamide has a therapeutic index of 33 and could be thus useful as an antiparasitic agent if used in combination with an integrated approach to control resistant parasites.

Molecular Interactions Paraherquamide is an extremely potent competitor at the α-bungarotoxin binding site at the nicotinic acetylcholine receptor in insects. It is also a competitor at the phenothiazine binding site. However, the real mode of action in nematodes has to be elucidated.

Duffy Blood Group

→ Natural Resistance.

Dum-Dum Fever

→ Leishmania (Vol. 1), → Visceral Leishmaniasis.

E

Earcanker

Trivial name of the mite *Psoroptes cuniculi* entering the ear of goats and causing otitis.

East Coast Fever

→ Theileria (Vol. 1), → Theileriosis.

Eastern Equine Encephalitis

→ Arboviruses (Vol. 1).

Echinococciasis

→ Echinococcus (Vol. 1); syn. Echinococcosis.

Echinococcosis

Pathology

→ *Echinococcus* (Vol. 1) *granulosus* is the causative agent of cystic hydatid disease or cystic echinoccocosis whereas infection with *E. multilocularis* in man leads to the more aggressive form of alveolar echinococcosis. In both cases, the primary site of contact with the parasite is the mucosal surface of the host's gastrointestinal tract. Echinococcosis is contracted by the inadvertent ingestion of eggs from the feces of dogs or other carnivores. Thus humans serve as intermediate hosts instead of normally, sheep, mice, or other herbivores. Larval cysts or hydatids can be found in many tissues, most often in the liver, lung, mediastinum, and peritoneum, giving rise only to pericystic fibrosis when intact. Hydatid cysts attain a large size because of asexual reproduction and proliferation of the innermost cyst layer, the germinal epithelium. In *E.granulosus* small protoscolices develop in the hydatid cysts with suckers, hooklets, and calcareous bodies. Detached → brood capsules (Vol. 1) and sometimes daughter cysts develop, again with internal multiplication; the gross appearance may be one of a bunch of grapes of various sizes. In *E. multilocularis,* the external laminated membrane of the cyst is incomplete and the inner germinal epithelium proliferates diffusely in an alveolar pattern, spreading like a neoplasm through the human liver, which is destroyed (→ Pathology/Fig. 16, → Pathology/Fig. 17, → Pathology/Fig. 29E,F). Folded laminated membranes are usually present. Protoscolices are rarely found in humans, although they are common in natural hosts. The cysts of *E. vogeli* usually contain primitive scolices. Production of lesions by all species depends on the location, number, and state of the cysts. Slow seepage of cyst fluid may lead to seeding of new areas, and cystic metastases to brain and lung can occur. Sudden hydatid cyst rupture can produce → anaphylactic shock. Eosinophilia accompanies the infection. Molecular immunopathology is reviewed in detail by Gottstein and Hemphill.

Immune Responses

The cellular and humoral → immune response in humans in response to migrating oncospheres and established metacestodes varies enormously as evidenced for example by different antigens recognized by individual patients with different courses of disease. Hydatide growth despite the presence of humoral and cellular anti-metacestode immune responses may result from the variation or downregulation of the parasite's antigens or from active immunomodulation by these cestodes.

B Cells and Antibodies Most serological studies in humans were performed to monitor patients after long-term chemotherapy or post-operatively. In cystic echinococcosis strain variations of the parasite as well as genetic differences between host populations may significantly affect the antibody response. For example, conventional serological tests were often negative in patients from Kenya or in patients with lung localisations of the cysts. However, rapid seroconversion was observed in most of these patients early after treatment arguing in favour of immunosuppressive mechanisms which are reverted after surgical removal of the cyst. Much emphasis was given to the determination of parasite-specific antibody isotypes. However, only in the case of IgE has a significant (negative) correlation with the response to chemotherapy been reported. IgE bound to basophils may be involved in anaphylactic reactions upon pre-operative or intra-operative cyst rupture.

In most patients with alveolar echinococcosis parasite-specific immunoglobulins of all isotypes can be measured at diagnosis and an association with hyperglobulinemia has been reported. Although protoscoleces and oncospheres of *E. multilocularis* can be lysed by antibody-mediated complement interaction in vitro, antibodies appear unable to control parasite proliferation in vivo. This inability may be due in part to complement-neutralizing factors released by the metacestode or to the inactivation of C3 as it enters the metacestode tissue. Antibodies are produced against many different proteins of the parasite, and some of these antigens were postulated to critically participate at the host-parasite interplay. For example, antibodies against the protein Em2, localized in the laminated layer of the metacestode, were found to be associated with disease susceptibility in experimental murine alveolar echinococcosis. While in resistant C57BL/10 mice anti-Em2 antibodies of the IgG 3 and IgG1 isotype were synthesized, only low levels of anti-Em2 IgG2a were detected in susceptible AKR and C57BL/6 mice.

T Cells In cystic echinococcosis patients there was no correlation between the detection of specific antibodies and the lymphoproliferative response to *E. granulosus* antigens. The fact that seronegative patients showed a positive proliferation assay and vice versa was taken as an argument for the existence of different pathways initiating humoral or cell-mediated responses. In murine cystic echinococcosis a marked reduction of the mean T cell percentage combined with an increase in suppressor activity was reported. An impairment of the host response by the formation of anti-MHC-antibodies or by parasite-derived immune-suppressive or -modulatory substances may account for the enhanced susceptibility to mycobacterial infections close to the parasite lesions. In long-term infected BALB/c mice a higher percentage of $CD4^+$ T cells in peripheral blood and a relative increase of $CD8^+$ T cells in the spleen was observed. These cells appeared to be activated, because they showed a high level of interleukin-2 receptor expression. The finding that PBMCs from patients responding well to chemotherapy produced significantly more IFN-γ and less IL-4 and IL-10 than cells from partial and low responders, indicated an implementation of Th1 cells in protective immunity.

In patients with alveolar echinococcosis the lymphoproliferative response to *E. multilocularis* antigens was highest in cured patients who had undergone radical surgery and significantly lower in patients with partial or no resections. A protective role of T cells has been found in murine models of alveolar echinococcosis. Depletion of T cells or the infection of nude or SCID mice resulted in enhanced metastatic formation and development of *E. multilocularis* accompanied by a drastically reduced host tissue reaction. *E. multilocularis* infection of permissive mouse strains resulted in the depletion of T-dependent zones of lymphoid organs and thymic involution during rapid growth of the metacestode. Although the responsible mechanism has not yet been defined, activated macrophages adhering to the metacestodes in vivo and/or immunomodulatory products of the parasite may account for it. Immunosuppression in murine alveolar echinococcosis appears to be a more general phenomena since there was an enhanced frequency of malignant sarcomas in A/J mice infected with *E. multilocularis* when compared to non-infected controls.

Analysis of cytokine mRNA expression revealed the enhanced production of Th2-type factors such as IL-3, IL-4, IL-10 and predominantly IL-5 in stimulated cells of patients with alveolar echinococcosis. In murine alveolar echinoccocisis there was an enhanced production of IFN-γ, IL-2, IL-5 and IL-10 by stimulated spleen cells in vitro over the first weeks of infection which was almost completely suppressed at 21 weeks of infection. Analysis of the local intrahepatic periparasitic cytokine expression in tissues of patients showed an enhanced expression of IL-1, IL-6 and TNF by activated macrophages. These findings together with the observation that treatment of SCID mice with TNF promoted the formation of granulomatous changes around larval cysts argue for an involvement of the locally secreted proinflammatory cytokines in the development of periparasitic granulomas and fibrogenesis. In resistant hosts activated macrophages appear to kill protoscoleces by arginine-dependent generation of reactive nitrogen intermediates and other, ill-defined destructive effects on the parasite-protective laminated layer surrounding the oncospheres and vesicular cysts.

The genetic basis for either host resistance (no immunosuppresssion?) or susceptibility (→ immune suppression?) is still unclear. Preliminary investigations showed that the frequency of certain HLA allelels (HLA-DR13) was increased in patients with a regressive course of disease after therapy compared to controls or patients with progressive alveolar echinococcosis. However, since inbred mice of the same H-2 haplotype differ significantly in their susceptibility to *E. multilocularis* there are obviously other, non-MHC-linked genes contributing to the disease susceptibility.

Main clinical symptoms in humans: Liver dysfunction, lung problems, ascites, abdominal pain
Incubation period: Years
Prepatent period: Years
Patent period: Years
Diagnosis: Serologic tests, computer tomographic analysis of liver and other organs, → serology (Vol. 1)
Prophylaxis: Avoid contact with infected final hosts (dog, fox, cat)
Therapy: In *E. granulosus* infections hydatids may be removed by surgery, while it is not recommended in alveolar cysts of *E. multilocularis* since the eventual setting free of undifferentiated cells initiates metastasis-like formation of new cysts (thus biopsies are strongly forbidden). Chemotherapy see → Cestodocidal Drugs, → Nematocidal Drugs, Animals and → Nematocidal Drugs, Man (Albendazole)

Echinostomiasis, Man

Disease due to infections of → *Echinostoma* (Vol. 1) species via oral uptake of infected, uncooked snails and clams.

Main clinical symptoms: Diarrhoea, abdominal pain, anaemia, eosinophilia
Incubation period: 1–3 weeks
Prepatent period: 2–3 weeks
Patent period: 6–12 months
Diagnosis: Microscopic determination of eggs in fecal samples
Prophylaxis: Avoid eating raw snails and clams
Therapy: Treatment see Trematodocical Drugs

Ectoinsecticides

Drugs acting against ectoparasites such as ticks, fleas (→ Insecticides).

Ectoparasites: New Approaches

Biological Control

The control of parasitic flies by biological methods has become a viable method in some specific cases.

The release of large numbers of genetically manipulated (either by irritation of genetic transformation, sterile or "less fit") flies into the environment has led to the elimination of the screw worm fly (*Chrysomya*) in the US. This control method has also been applied to blow flies (*Lucilia cuprina*) and is part of a long term strategy to control sheep blow fly. In Australia the use of natural pesticides as another "environmentally friendly" biological pesticide is an area where innovative applications are expected to investigate these opportunities for future products. First biopesticides such as *Bacillus thuringiensis* and *Metarhizium anisopliae* based products reached the crop protection market in the late 1980s. But both approaches are still in their infancy in the animal health sector, since due to their high specificity they were suitable for combating only few fly species and never reached broad regional application.

Vaccination

The growing problems of resistance and persistence of residues in meat and milk have created a renewed interest in antiparasitic vaccines.

It has been known for a long time that certain arthropods stimulate an immune response in infested animals. Apart from two x-irradiated worm vaccines there were no significant parasite immunologicals available until the late 1980s. The discovery and improvement of recombinant DNA technology has created hope that the capability of expressing specific, protective antigens not only against endoparasites, but also parasitic arthropods will lead to new innovative vaccines.

This field is potentially very interesting, but both the life cycle of the arthropods targeted and the limited range of cross reactivity between related strains as well as the technology involved is highly complex.

Table 1 shows where certain activities have been focused, but considering market success as final proof, none of these vaccines achieved a signifi-

Table 1. Vaccine approaches against ectoparasites.

Target parasite	Approach/Status
Warble fly/Cattle grub (*Hypoderma bovis, Hypoderma lineatum*)	The most advanced immunization trials against myiasis have been performed against *Hypoderma*. Some protection against later infections was observed in the field. More detailed investigation revealed, that particularly first instar larvae were sensitive. Studies on the protein composition of *H. lineatum* larvae which might induce protection showed that so called hypodermins were potentially useful antigens. Hypodermins (type A,B,C) are serine proteinases which are of importance for larval tissue migration. Vaccination with hypodermin antigens resulted in up to 90 % protection (hypodermin A) and some degree of cross protection between *H. lineatum* and *H. bovis*. So far, as with almost all insect ectoparasite vaccination approaches no field vaccine is currently available.
One host ticks (*Boophilus microplus*)	Tickgard is a vaccine based on a recombinant hidden antigen, membrane-bound glycoprotein Bm86. The vaccine has achieved some success in Australia against *B. microplus*, particularly in dairy herds. Vaccination results in some mortality in engorging ticks and to significant tick mortality between full engorgement and egg laying. The discovery of these specific tick antigens has opened the possibility of identifying similar proteins in other *B. microplus* strains and probably in other tick species as well. Other interesting antigens (Bm91, Qu13) have been identified, but so far none appear to be as effective as Bm86. Since the current vaccine is particularly effective in reducing the reproductive capacity in engorging female ticks, the continual introduction of ticks or tick-infested animals into a vaccinated herd significantly interferes with a successful strategic vaccination. Highest performance of the vaccination was observed when vaccinated animals were isolated from continual reinfection.
Blowfly strike (*Lucilia cuprina*)	For the time being it is believed that binding of antibodies to the respective antigens leads to a layer covering the peritrophic membrane (PM) which results in restricted permeability of *L. cuprina* larvae fed on vaccinated sheep. Peritrophins of different molecular weight have been identified, which led to starvation through the binding of antibodies to these PM-associated antigens. Despite substantial progress in comparison to the past a useful vaccine is not available so far.
Fleas (*Ctenocephalides felis*)	Hyperimmunized rabbit antisera against concealed antigens of flea midgut, the major digestive organ, revealed when fed in an artificial feeding system, significant decrease of survival rate and egg production of *C. felis* fleas. These preliminary studies demonstrated the feasibility of vaccination against cat fleas. Similar results were obtained following the vaccination of dogs with subsequent challenge. In these preliminary studies statistically significant reduction of about 25 % regarding flea numbers remaining on the animal in comparison to control animals was observed. Despite the fact that at least partially protective antigens have been identified, so far no recombinant vaccine displaying substantial field success is available.
Salmon louse (*Lepeophtheirus salmonis*)	Vaccination might be a valuable method to control copepod infestation which are major parasites of farmed salmonids. Antibodies were raised against various *L. salmonis* antigens. When assessed with immunohistochemical methods gut and ovarial hidden antigens were recognized by respective Mabs*. Then MAbs were used to screen *L. salmonis* DNA libraries in order to identify DNA fragment coding, for proteins which might be useful as a basis for recombinant vaccines.

* Monoclonal Antibodies

cant share either in treatment numbers or in sales, when compared even to "weak" chemicals.

One fundamental problem with vaccine research is the identification of the protective antigens. For complex organisms, such as arthropods, this makes the isolation of useful antigens very difficult, but in the future due to improvements in DNA technology, this technique may start to have a significant impact on treatment against arthropods.

Ectoparasitocidal Drugs

→ Arthropodicidal Drugs.

Ehrlichiosis

Anaemian disease in sheep, cattle, cats and dogs due to the rickettsian agents (e.g. *Ehrlichia canis*, *Haemobartonella felis* and *Anaplasma* species) transmitted by ticks (→ Tick Bites: Effects in Animals). These rickettsial stages are spherical (genera *Anaplasma*, *Haemobartonella*) and are found on the surface of erythrocytes (*Haemobartonella*) or appear ovoid (genus *Ehrlichia*) and are situated inside monocytes. In particular the intraerythrocytic stages of *Anaplasma marginale* of cattle may become misdiagnosed as piroplasms.

Therapy

Haemobartonella: Tetracyclines; *Anaplasma*, *Ehrlichia*: Imidocarb (Imizol*). In humans *Ehrlichia sennetsu* (Japan, South-East Africa) and *E. chaffeensis* (USA) may occur and are apparently also tick transmitted. They cause fever, vomiting, nausea, which may, however, be self-limiting.

Eimeriosis

→ Eimeria (Vol. 1), → Coccidiosis, Animals.

Elaeophoriasis, Elaeophorosis

Elaeophora (→ Filariidae (Vol. 1)) infects deer, elk and sheep. The adult parasites live in the common carotid and internal maxillary arteries and produce microfilariae which are found in the capillaries of the face and the head. In sheep the microfilaria induce a granulomatous reaction in the capillaries of the skin, and a dermatitis. The lesions on the head, feet and abdomen are characterized by intense pruritus resulting in erythrema, alopecia, excoriations, ulcerations, crusts, and haemorrhage. Stomatitis, rhinitis and keratitis also occur.

Therapy

→ Nematocidal Drugs, Animals.

Elephantiasis, Tropical

→ Filariasis, Lymphatic, Tropical; → *Wuchereria* (Vol. 1), → *Brugia* (Vol. 1).

Main clinical symptoms: Lymphangitis, unfeelingness of skin portions; later: chylurie, elephantiasis, i.e. giant swelling of organs
Incubation period: 3–16 months
Prepatent period: 7–24 months
Patent period: 8–10 years (adults live for 18–20 years)
Diagnosis: Microscopic analysis of smear preparations or of membrane filtered material; microfilariae are found mainly at 10 o'clock p.m. in the peripheral blood
Prophylaxis: Avoid bites of vector mosquitoes in endemic regions
Therapy: Treatment see → Nematocidal Drugs, Man

Encephalitis

Inflammation of the brain. Often caused by the viral genus → Flavivirus (Vol. 1) (→ Tick-borne Encephalitis, → Russian Spring-Summer Encephalitis, → Powassan Encephalitis, → Colorado Tick Fever).

Encephalitozoonosis

Synonym

Microsporidiosis, Encephalitozooniasis

General Information

Encephalitozoonosis is a nervous system disease caused by the obligate intracellular microsporidian → *Encephalitozoon cuniculi* (Vol. 1). The disease has been described in rodents, lagomorphs, primates and several species of carnivores. Asymptomatic infection usually occurs in rodents and lagomorphs. In carnivores the neurological signs include repeated turning and circling movements, especially after disturbance, dysmetria, dysergia, blindness and a terminal semi-comatose state. Lesions described are encephalitis and segmental vasculitis. The course of the illness is usually 5–12 days. This parasite is also found in immunodeficient people as pathogen with a generalized spreading over many organs (→ Opportunistic Agents (Vol. 1)).

Therapy

→ Treatment of Opportunistic Agents, see also → Drugs against microsporidiosis.

Endemy

Persistent occurrence of parasites/agents of disease in a defined region.

Energy-Metabolism-Disturbing Drugs

Table 1, Table 2

Structures

Fig. 1

Rotenoids

General Information Rotenone is a natural product isolated from the roots of *Derris* spp. and *Lonchocarpus* spp. The derris root has long been used as a fish poison among the Malaysian natives. The insecticidal properties of derris extracts were known in China long before isolation of the active principle in 1895 by Geoffrey. Rotenone represents the most toxic and abundant member of about 13 rotenoids. Three other members also have been reported to show insecticidal activity but are less potent than rotenone: deguelin, tephrosine, toxicarol with 10%, 2.5% and 0.25% of the rotenone activity, respectively (Fig. 2). In the past forty years rotenone has been intensively used. The world production is in the range of 10 to 20 metric tons.

Molecular Interactions Rotenone acts as an inhibitor of the electron transport system (site one) of the mitochondrial respiratory chain. It inhibits the oxidation of NADH to NAD via co-enzyme Q (reducing quinone to hydroquinone), thus also blocking the oxidation by NAD of substrates such as glutamate, a-ketoglutarate, and pyruvate.

Rotenone is a non-systemic selective insecticide with secondary acaricidal activity with contact and stomach action. It is used for the control of lice, ticks and warble flies and against fire ants in premises as well as mosquito larvae (ponds). Today, its primary use as an ectoparasiticide is against ear mites and demodectic mange and in combination with other ectoparasiticides like organophosphates and pyrethroids against sheep ectoparasites. Inactivation by photo-oxidation limits the use of rotenone as a single product when long residual activity is required. The LD_{50} of acute oral toxicity for rats is 132–1500 mg/kg. The estimated lethal dose for oral application of rotenone for humans is 300–500 mg/kg. Rotenone is highly toxic for mammals upon injection (LD_{50} 0.1–4 mg/kg). Rotenone is highly toxic to pigs but shows no toxicity to bees. The compound is also used as fish toxicant in fish management. Rotenone is metabolised in the rat liver or insect by enzymatic opening and cleaving of the furan moiety or, alternatively, by oxidation of the methyl group of the isopropenyl residue.

Resistance Resistance of ectoparasites against rotenone has not yet been described.

Iodoquinol

Synonyms 5,7-diiodo-8-hydroxyquinoline, SS578, Diodoquin, Di-Quinol, Disoquin, Floraquin, Dyodin, Dinoleine, Searlequin, Diodoxylin, Moebiquin, Rafembin, Ioquin, Direxiode, Stanquinate, Quinadome, Yodoxin, Zoaquin, Enterosept, Embequin.

Cells and Cellular Interactions Iodoquinol is active against *Entamoeba histolytica* and against facultative pathogenic *Entamoeba* spp. such as *Dientamoeba fragilis*. The activity of iodoquinol is directed against trophozoites (minuta forms) in the intestinal mucosa and cysts in the feces (→ DNA-Synthesis-Affecting Drugs I/Table 1).

Molecular Interactions The mechanism of action is unknown. It is tempting to speculate from some structural similarities with hydroxyquinolines that iodoquinol may interfere with the energy metabolism (oxidative phosphorylation, inhibition of ATPase) in amoebes.

Table 1. Degree of efficacy of drugs against *Fasciola hepatica*, *Dicrocoelium dendriticum* and *Paramphistomum* spp.

Year on the market	Drugs	*Fasciola hepatica*	*Dicrocoelium dendriticum*	*Paramphistomum* spp.	Additional Parasites
Halogenated Phenols and Bisphenols					
	Disophenol	x			blood-ingesting nematodes
1968	Nitroxynil	x	x		
1957	Hexachlorophene	x	x	x (only matures)	cestodes
1933	Bithionol	x		x	cestodes, lung and intestinal flukes
	Bithionol-sulfoxid	x		x (only matures)	cestodes
1959	Niclofolan	x		x (only immature)	*Metagonimus*
Salicylanilides					
1960	Niclosamide			x (only immature)	cestodes, intestinal flukes
1963	Tribromsalane	x	x		
1968	Oxyclozanide	x	x	xx	cestodes, intestinal flukes
1966	Clioxanide	x	x		
1969	Rafoxanide	xx		x (only immature)	cestodes, intestinal flukes, blood-ingesting nematodes
1970	Brotianide	x	x	x	
	Bromoxanide	x	x		
	Closantel	xx	x		blood-ingesting nematodes
1969	Resorantel			xxx	cestodes
Benzene sulfonamides					
	Clorsulon	x			
Piperazine derivatives					
1964	Hetolin		x		
Phenoxyalkane-derivatives					
1973	Diamphenethide	xxx	x		
Benzimidazoles					
	Cambendazole		x		nematodes
1961	Thiabendazole		x		nematodes
1971	Mebendazole	x	x		*Giardia*, cestodes, nematodes
1971	Fenbendazole		x		*Giardia*, nematodes
1978	Febantel		x		nematodes
1979	Albendazole	x	x		*Giardia*, cestodes, nematodes
1983	Triclabendazole	xxx			
Isoquinoline derivatives					
1975	Praziquantel		xx		*Giardia*, *Entamoeba*, trematodes, cestodes

xxx = high efficacy at least against some developmental stages and diverse species; xx = partially effective (regarding developmental stages and diversity of species); x = slightly effective

Table 2. Activity of drugs against other trematodes of human importance.

Drugs	Intestinal flukes	Liver flukes	Lung flukes
Chloroquine		*Clonorchis* (x)	
Tetrachloro-ethylene	*Fasciolopsis, Metagonimus, Heterophyes, Echinostoma, Gastrodiscoides, Watsonius, Nanophyetes* (x)		
Niclofolan	*Metagonimus* (x)		
Bithionol			*Paragonimus* (xxx)
Dichlorophene	*Fasciolopsis* (x)		
Hexachloro-paraxylene		*Clonorchis* (x)	
Hexachlorethane	*Fasciolopsis* (x)		
Hexylresorcinol	*Fasciolopsis* (x)		
Resorantel	*Gastrodiscoides* (xxx)		
Niclosamide	*Fasciolopsis, Metagonimus, Heterophyes, Echinostoma, Gastrodiscoides, Watsonius, Nanophyetes* (xxx)		
Praziquantel	*Fasciolopsis, Metagonimus, Heterophyes, Echinostoma, Gastrodiscoides, Watsonius, Nanophyetes* (xxx)	*Clonorchis, Opisthorchis, Metorchis*	*Paragonimus* (xxx)

xxx = high efficacy at least against some developmental stages and diverse species; xx = partially effective (regarding developmental stages and diversity of species); x = slightly effective

Suramin

Synonyms Bayer 205, 309F, Antrypol, Germanin, Moranyl, Naganol, Naganin, Naphuride Sodium

General Information Suramin was explored in 1916 and originally introduced in 1922 as a trypanosomicidal drug in cattle.

Molecular Interactions Suramin is chemically a sulfonic acid and structurally related to dyes such as trypanred, eboliblue or trypanviolet. The six negative charges at physiological pH may be of great importance for the proposed mode of action. The action of suramin is directed against trypomastigotes of *Trypanosoma brucei gambiense* and *T.b. rhodesiense* in the blood dividing by binary fission. The drug is taken up by trypanosomes via receptor-mediated endocytosis in presence of serum proteins 18-fold higher compared to the normal fluid endocytosis alone. Intracellular suramin concentrations (about 100 μM) are equivalent to exogenous drug concentrations. Suramin is bound to albumin and low density lipoprotein (LDL) resulting on the one hand in reduced host toxicity but on the other hand leading to an inhibition of intralysosomal proteolysis in the host cell. The complexation of suramin to albumin is also responsible for a reduced total amount of free drug in the cytoplasm of host cells.

The antitrypanosomal action of suramin on the molecular level is relatively unclear. There are reports on the inhibition of a variety of kinases and dehydrogenases from mammalian, bacterial, and fungal cells. In trypanosomes glycerol-3-phosphate oxidase and glycerol-3-phosphate dehydrogenase become inhibited, resulting in a disturbed redox balance within the trypanosomal cell and a decreased ATP-synthesis rate (Fig. 3). In addition, other enzymes of *T. brucei* such as DHFR, thymidine kinase and a number of glycolytic enzymes (hexokinase, phosphoglucoisomerase, phosphofructokinase, triosephosphate isomerase, aldolase, glyceraldehyde-3-phosphate dehydrogenase, phosphoglycerate kinase, glycerol-3-phosphate dehydrogenase and glycerol kinase) become inhibited. The IC_{50} values of suramin on trypanosomal glycolytic enzymes (between 10 and 100 μM) are much lower compared to the corresponding enzymes from mammalian cells. The reason for the selective toxicity may be explained by the unusually high isoelectric points (between 9 and 10) for most of the glycolytic enzymes from *T. brucei*. The basic properties of the trypanosomal glycolytic enzymes resulting in additional positive charges on their surfaces may fascilitate the binding of the highly negatively charged suramin. Thus, there are likely to be potentiated inhibitory effects by electrostatic interactions between posi-

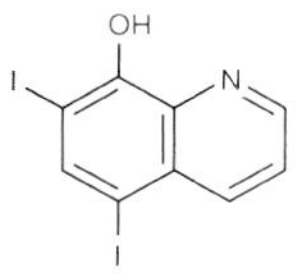

Iodoquinol

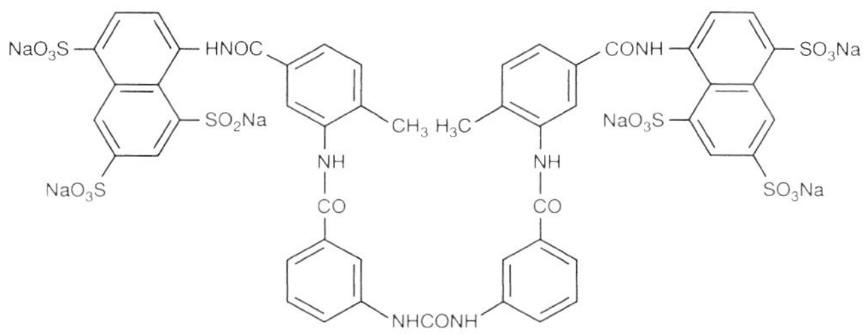

Suramin

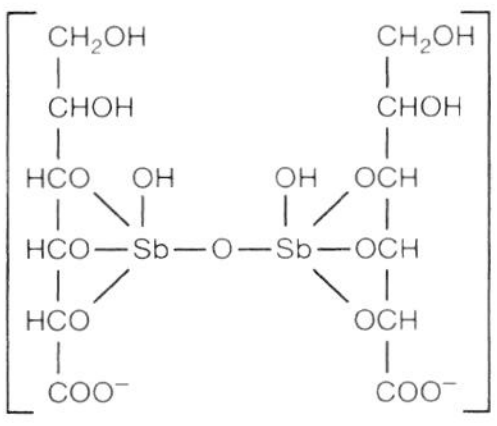

Sodium stibogluconate

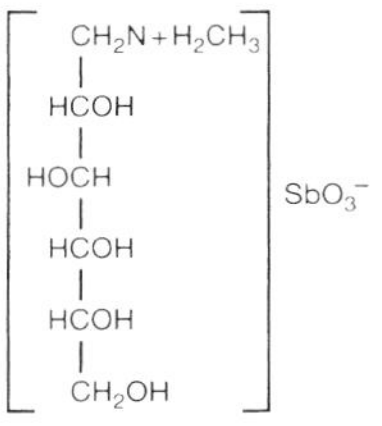

Meglumine antimonate

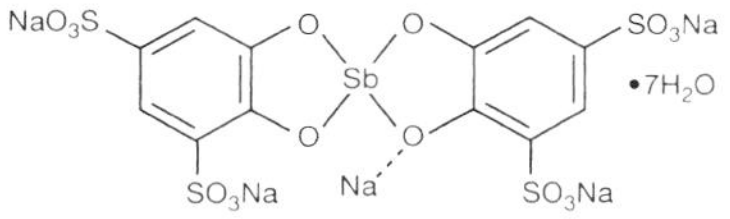

Stibophen

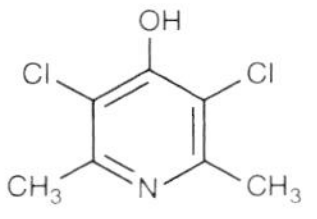

Clopidol

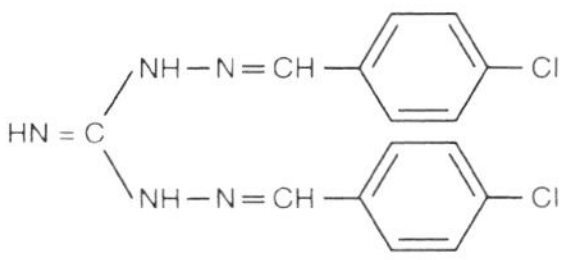

Robenidine

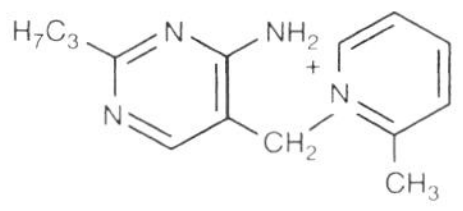

Amprolium

Fig. 1. Structures of drugs affecting energy metabolism.

tively charged trypanosomal enzymes and negatively charged suramin, and there is no direct inhibition of trypanosomal glycolytic enzymes in glycosomes.

A further hypothesis for the suramin action on the molecular level is as follows. The nine glycolytic enzymes are synthesized on free polysomes in the parasite's cytoplasm. Then these enzymes are imported into the glycosomes posttranslationally without any proteolytic modification within 3 to 5 min and become thus protected from suramin by compartimentalization in the glycosomes. Now, suramin possibly binds to the glycolytic enzymes in the cytoplasm on their way to the glycosome and/or interferes with their import into the glycosomes (Fig. 3). The inhibition of glycosomal protein import is followed by a gradual decrease of enzyme concentrations in the glycosomes. The average half-life of glycolytic enzymes is about 48h inside the glycosomes. By this action suramin induces a slowing down of energy metabolism in suramin-treated trypanosomes. The inhibition of the import of glycosomal protein is either partially or totally with disruption of glycolysis in the trypanosomes.

An additional hypothetical action of suramin is the indirect interaction with the DNA/RNA-replication resulting in retarded trypanocidal activity after 24 to 36 h corresponding to 4 to 7 divisions.

The antifilarial activity of suramin is directed against macrofilariae and to a minor degree against microfilariae of *Onchocerca volvulus* (river blindness). For a long time it was the only drug against adult *O. volvulus* in man, but today the use of suramin is only limited. The antifilarial mechanism against *O. volvulus* is delayed (4 to 7 weeks after treatment). An inhibition of the

Fig. 1.(continued) Structures of drugs affecting energy metabolism.

AsO(OH)2
NHCONH2
Carbasone

NHCOCH2OH
HO – As – O – Bi = O
O
Glycobiarsol

As(SCH2COOH)2
CONH2
Thiacetarsamide

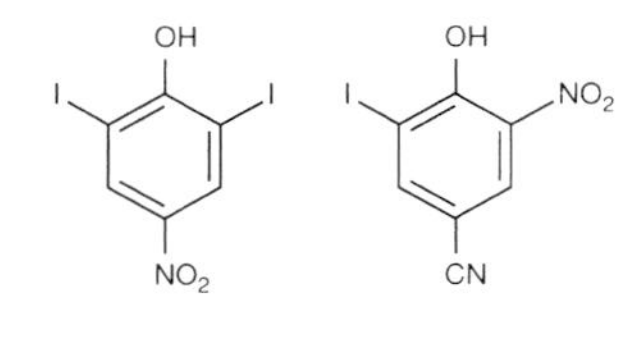

Clorsulon

Bitoscanate

O2N
O
NCS
Nitroscanate

Monophenoles:

OH
I
I
NO2
Disophenol

OH
I
NO2
CN
Nitroxynil

Bisphenoles:

OH
OH
R3
X
R3
R6 R6
R5
R5

	R_3	R_5	R_6	X
Hexachlorophene	Cl	Cl	Cl	CH_2
Bithionol	Cl	Cl	H	S
Bithionol sulfoxide	Cl	Cl	H	SO
Menichlopholan	NO_2	Cl	H	Direct bond

cAMP-independent protein kinase I in *O. volvulus* is discussed. This protein is also the target of suramin and stibophen against *Ascaridia galli* in-vitro. Suramin further exerts curative effects against *Wuchereria bancrofti* filariasis and in addition shows effectivity against all stages of *Litomosoides carinii*. There is a requirement for intravenous injection which is accompanied by severe side-effects. Suramin leads to a sterilisization of adult filariae. *L. carinii* in *Mastomys coucha* are killed within 6 weeks. Despite great chemical-synthetical efforts there is no possibility for improvement of better tolerability without loss of filaricidal activity. In general there is no correlation between filaricidal effects of suramin derivatives and trypanosomicidal activity.

Additional Features Suramin has additional activities against a wide variety of pathogens. Thus it is a potential anti-AIDS drug, due to a potent inhibitory effect on the reverse transcriptase from a variety of retroviruses. In addition, an inhibitory effect on DNA polymerase activity is discussed.

Clinical Relevance The main indication of suramin relies on the activity against early stages (bloodstream forms) of human African trypanosomes (*Trypanosoma brucei gambiense* and *T. b. rhodesiense*) (→ DNA-Synthesis-Affecting Drugs III/Table 1). Suramin is usually given with five intravenous injections at a dosage of 20 mg/kg b.w. once every 5 to 7 days. In veterinary medicine the drug exerts efficacies against *T. b. brucei* (Naga-

Salicylanilides:

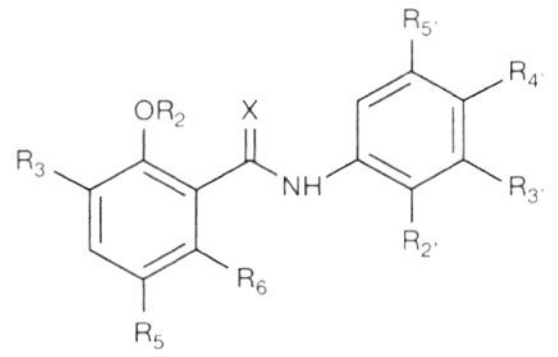

	R_2	R_3	R_5	R_6	X	$R_{2'}$	$R_{3'}$	$R_{4'}$	$R_{5'}$
Tribromsalan	H	Br	Br	H	O	H	H	Br	H
Oxyclozanide	H	Cl	Cl	Cl	O	OH	Cl	H	Cl
Clioxanide	$COCH_3$	I	I	H	O	H	H	Cl	H
Rafoxanide	H	I	I	H	O	H	Cl	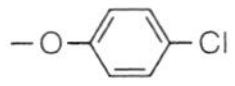 $-O-C_6H_4-Cl$	H
Brotianide	$COCH_3$	Br	Cl	H	S	H	H	Br	H
Bromoxanide	H	$C(CH_3)_3$	NO_2	CH_3	O	CF_3	H	Br	H
Closantel	H	I	I	H	O	CH_3	H	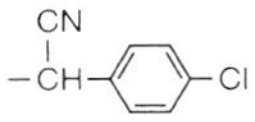 $-CH(CN)-C_6H_4-Cl$	Cl
Resorantel	H	H	H	OH	O	H	H	Br	H
Niclosamide	H	H	Cl	H	O	H	H	NO_2	Cl

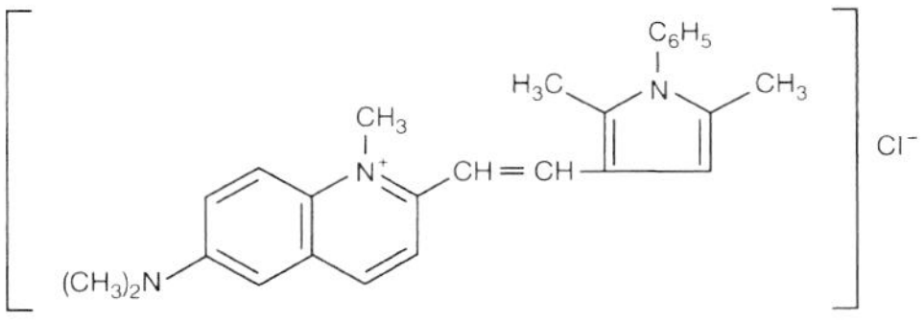

Pyrvinium chloride

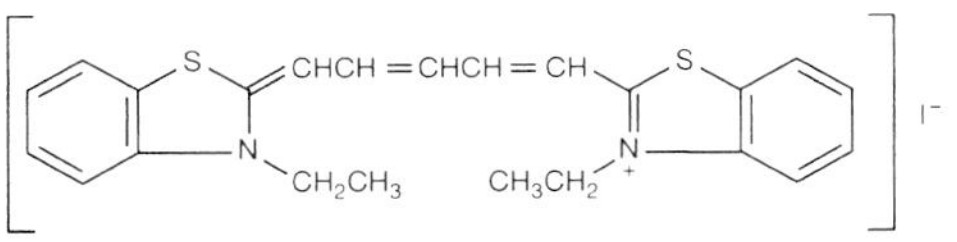

Dithiazanine iodide

Fig. 1.(continued) Structures of drugs affecting energy metabolism.

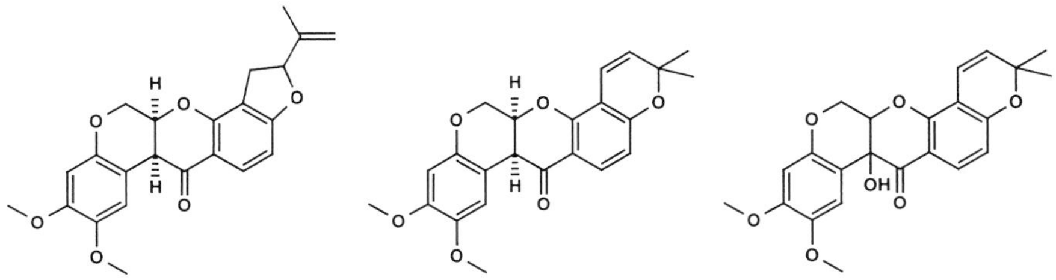

rotenone

deguelin

tephrosine

Fig. 2. Structure of important rotenoids with insecticidal efficacy.

na), *T. b. evansi* (Surra) and *T. equiperdum* (Dourine). In addition, suramin has experimental activity against *Entamoeba histolytica*, *Eimeria* spp. and avian malaria.

Resistance The mechanism of resistance of suramin on the molecular level is unclear. Generally the resistance to suramin is rare even after 70 years of application in trypanosomiasis field clinics. This might be an indirect support for the hypothetical action on multiple targets in trypanosomes.

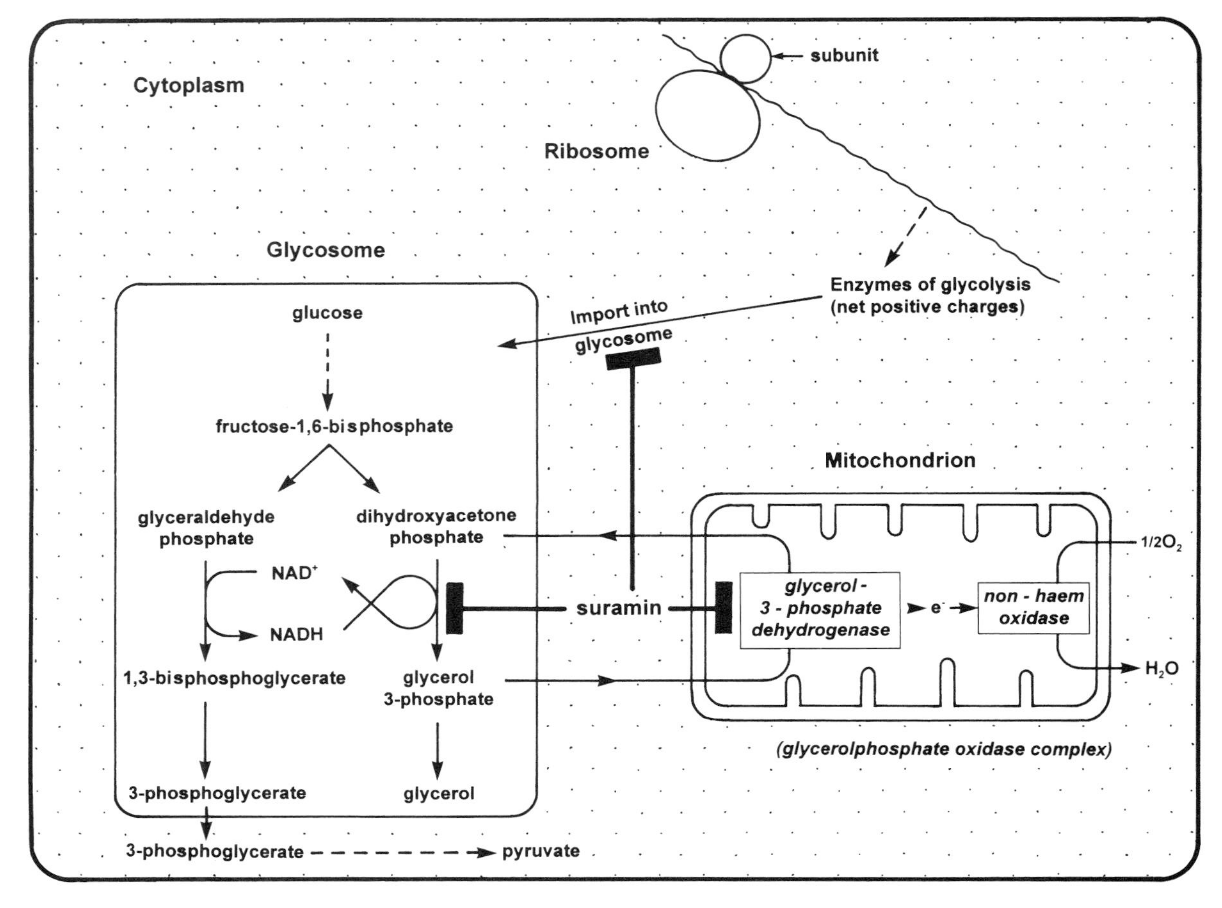

Fig. 3. Target enzymes of suramin; metabolic processes impaired by suramin are indicated by the symbol ⊣; highly schematic representation of trypanosomal cell organelles and their metabolic processes; relationships of the magnitudes of organelles are not correct

Antimonials

Important Compounds Sodium-stibogluconate, Meglumine-antimonate, Stibophen.

Synonyms Sodium-stibogluconate: Antimony sodium gluconate, Pentostam, Myostibin, Solustibosan, Solyusurmin, Stibanate, Stibanose, Stibatin, Stibinol

Meglumine-antimonate: N-methylglucamine antimonate, glucantime

Stibophen: Sodium antimony bis(pyrocatechol-2,4-disulfonate), Sdt 91, Fuadin, Fouadin, Pyrostib, Corystibin, Trimon, Fantorin, Repodral, Neoantimosan, Sodium Antimosan

Cells and Cellular Interactions The antiprotozoal activity of antimonials is directed against cutaneous, visceral and mucocutaneous leishmaniasis (→ DNA-Synthesis-Affecting Drugs III/Table 1). They have no activity against cultured leishmanial promastigotes. Stibophen shows besides the activity against *Leishmania tropica* and *L. mexicana* also activity against *Schistosoma haematobium.*

Molecular Interactions The mechanisms of antiprotozoal action of the pentavalent antimonials sodium-stibogluconate and meglumine-antimonate rely on their reduction to the corresponding trivalent antimonials by host metabolism. The activity is directed against those stages dividing by repeated binary fissions such as *L. donovani* and *L. chagasi* in spleen, liver and skin stages, intracellular amastigotes of *L. tropica*, *L. major* in the skin *and L. brasiliensis* and *L. mexicana* in mucous tissues of nose and mouth. Thereby, presumably those enzymes with sulfhydryl groups become inactivated by trivalent antimonials. Thus, the inhibition of cytoplasmic pyruvate kinase and other kinases with reactive sulfhydryl groups at their active sites results in a decreased flow of glucose into the citrate cycle in *L. tropica* promastigotes in presence of trivalent antimonials, in addition an accumulation of glycolytic metabolites and a disturbance of energy production within the parasites. As an alternative hypothesis a disruption of trypanothione reductase (TR) by antimonials is discussed. The real mode of antileishmanial action of the antimonials is yet unknown.

The antitrematodal action of stibophen may be explained by the inhibition of phosphofructokinase of *Schistosoma mansoni.*

Resistance Resistance against pentavalent antimonials is an increasing problem varying between 5%–70% of patients in some endemic areas. Resistant *Leishmania* isolates tolerate concentrations of antimonials, which are 100-fold higher than the maximal achievable serum levels of drugs in humans.

The real mechanism of resistance against antileishmanial antimonials is still unknown. Biochemical studies indicate that there is a decreased accumulation of sodium-stibogluconate in resistant cell lines. It remains, however, unclear whether there is a decreased uptake or an increased efflux of drug. The resistance of *Leishmania* spp. against pentavalent antimonials can be reversed in-vitro and in-vivo by verapamil indicating a P-glycoprotein-mediated resistance mechanism.

In addition, it was proposed that the resistance mechanism in *Leishmania* spp. against antimonials is based on gene amplifications. Such a mechanism is proposed for the resistance against different compounds such as methotrexate, arsenite, tunicamycine, DFMO, mycophenolic acid and vinblastine.

Clopidol

Synonyms Meticlorpindol, Clopindol, Coyden; in combinations: Lerbek (= Meticlorpindol + Methylbenzoquate).

Cells and Cellular Interactions The antiprotozoal activity of clopidol is directed against *Eimeria* spp. in chicken (→ DNA-Synthesis-Affecting Drugs IV/Table 1). Clopidol leads to an inhibition of the development of sporozoites and trophozoites.

Molecular Interactions Clopidol is a pyridone-derivative structurally related to the quinolones. Its real mode of action is unclear to date. There is no interference with mitochondrial respiration of *E. tenella*. Synergistic effects between clopidol and methylbenzoquate indicate that the action of clopidol may be similar but not identical to that of quinolones. The differences are probably due to an alternative pathway of electron transport in coccidial mitochondria with specific sensitivity to clopidol.

Resistance There is cross-resistance between methylbenzoquate and clopidol in strains of *E. maxima*. The resistance against clopidol in a methylbenzoquate-resistant strain appears after many passages in chicken.

The mechanism of resistance against clopidol is unclear.

Robenidine

Synonyms Robenzidene, Cycostat, Robenz.

Cells and Cellular Interactions The antiprotozoal activity of robenidine is directed against all five economically important *Eimeria* spp. in poultry (*E. acervulina*, *E. maxima*, *E. necatrix*, *E. tenella* and *E. brunetti*) (→ DNA-Synthesis-Affecting Drugs IV/Table 1). Furthermore Robenidine exerts activity against *Neospora caninum* tachyzoites in cell cultures.

Molecular Interactions Robenidine acts against developing first generation schizonts by possible interference with the oxidative phosphorylation and ATPase in mitochondria of *Eimeria* spp. and rat liver mitochondria. In chicken erythrocytes there is an induction of efflux of K^{+}-ions observable.

Amprolium

Synonyms 1-((4-amino-2-propyl-5-pyrimidinyl)-methyl)-2-picolinium chloride, Corid, Amprol, Amprovine; in combinations: Amprol Plus, Amprol Hi-E, Amprolmix, Pancoxin, Supracox.

Cells and Cellular Interactions Amprolium has been used for the therapy of coccidiosis in poultry, zoo birds and mammals since 1960. It has activity against *Eimeria tenella, E. maxima and E. necatrix* (→ DNA-Synthesis-Affecting Drugs IV/Table 1). The action of amprolium is directed against wall-forming bodies II of schizonts of the first and second generation.

Molecular Interactions Amprolium is structurally related to thiamine, but lacks the hydroxymethyl group of this vitamine. Thus, the phosphorylation to the corresponding thiamine pyrophosphate analogue is abrogated. Thiamine is an essential cofactor for pyruvate dehydrogenase activity, which is interestingly also the site of inhibition by antiprotozoal arsenicals. Amprolium competitively inhibits the transport of thiamine across the cell membranes of second generation schizonts. Amprolium possesses a high therapeutic index which may be explained by differences of thiamine transport rates between chicken epithelial cells and *Eimeria* spp..

Resistance The mechanism of resistance against amprolium is explained by a modification of the target receptor resulting in a decreased sensitivity to inhibition. The K_i-value for amprolium is increased nearly 15-fold to 115 μM in the resistant *E. tenella* strain.

Arsenicals

Important Compounds Carbasone, Glycobiarsol, Melarsoprol, Thiacetarsamide, Mel PH, R7/45.

Synonyms Carbasone: N-carbamoylarsanilic acid, Amebevan, Ameban, Amibiarson, Arsambide, Carb-O-Sep, Histocarb, Fenarsone, Leucarsone, Aminarsone, Amebarsone

Glycobiarsol: Bismuth glycoloylarsanilate, Broxolin, Dysentulin, Milibis, Viasept, Wintodon

Thiacetarsamide: Arsenamide, Thioarsenite, Caparsolate, Caparside, Arsphenamide, Filicide, Filaramide

Clinical Relevance The great interest in arsenicals at the beginning of the 20th century relied on their antibacterial activities. Thus, 4-arsanilic acid sodium (**Atoxyl**) and Salvarsan were the first drugs to be active against syphilis. **Carbasone** as one of the antiprotozoal arsenicals was introduced in 1931 against infections with *Trichomonas vaginalis* and *Entamoeba histolytica*. **Glycobiarsol** introduced in 1938 exerts activities against *T. vaginalis, E. histolytica* and *Giardia lamblia*. **Melarsoprol** is still one of the drugs of choice for the treatment of late stage sleeping sickness caused by *Trypanosoma brucei gambiense* or *T. b. rhodesiense*. Arsanilic acid and roxarsone exhibit activity against *E. tenella* sporozoites. **Acetarsone** in combination with arecoline was used as an anticestodal drug in small animals. In addition, sodium arsanilic acid in combination with copper sulfate has antinematodal activities against nematodes in ruminants. The macrofilaricidal efficacy of arsenicals is also known for a long time. Interestingly, all arsenicals obtain almost exclusively adulticidal effects which is clinically proven in man. **Thiacetarsamide** potassium is a drug routinely used against *Dirofilaria immitis* in the dog (→ Inhibitory-Neurotransmission-Affecting Drugs/Table 1). Another arsenical is **melarsomine** used as adulticide against *D. immitis* in dogs. The main disadvantage of the arsenicals are their severe side effects, such as the arsenical encephalopathy. In the meantime new and less toxic organic arsenicals such as **Mel PH** and **R7/45** have been developed. In general, there are no marked reductions of microfilariaemia levels until day 7 p.i. and the adulticidal efficacy of drugs varies with the parasite species. Thus, thiacetarsamide is much more active against *Brugia spp.* than other arsenicals; *Acanthocheilonema viteae* is more resistant to Mel PH and **R7/45** than *Litomosoides carinii* and *B. malayi*. Arsenicals exert better activity against female *L. carinii* and *B. malayi* than against males whereas both worm sexes of *A. viteae* are equally sensitive to arsenicals.

Molecular Interactions The mechanism of action of antiprotozoal arsenicals is presumably due to an inhibition of glycolytic enzymes and/or protein kinases with SH-groups. Also trypanothione reductase may be a target of arsenical compounds (→ DNA-Synthesis-Affecting Drugs III/Fig. 1).

The antifilarial arsenicals lead to an in-vitro and in-vivo inhibition of glutathione reductase as shown in *L. carinii* adults. Thereby, the parasitic enzyme is more susceptible to inhibition than the corresponding mammalian enzyme.

Clorsulon

Synonyms L-631,529, MK-401, Curatrem, in combinations: Ivomec Plus.

Clinical Relevance Clorsulon belongs to the fasciolicidal drugs (Table 1) with good activity against liver flukes (*Fasciola hepatica*) from the age of four weeks. It has low toxicity and is excreted rapidly. Clorsulon is suitable for the use in meat-producing animals. Now clorsulon is used in combination with ivermectin (Ivomec F).

Molecular Interactions It is the only one of the most commonly used fasciolicides whose action is directed against glycolysis on the level of 3-phosphoglycerate kinase and phosphoglyceromutase. There is no great disruption of glycolysis in-vivo by clorsulon. At a concentration of 500 μg/ml for 1 hour the glucose utilisation is decreased by 60%, formation of acetate and propionate is inhibited by 54% and 85%, respectively, and ATP levels are reduced by 67%. The motility of the flukes is gradually suppressed ending in a flaccid paralysis which may be explained by the slow depletion of energy reserves and cessation of feeding. Until now there are no reports about clorsulon resistance.

Isothiocyanates

Important Compounds Bitoscanate, Nitroscanate.

Synonyms Bitoscanate: Jonit, Bitovermol, Sicur.
Nitroscanate: Lopatol, Cantrodifene, Canverm.

Clinical Relevance Nitroscanate was introduced in 1973. Its anticestodal activity is directed against *Taenia* spp., *Dipylidium caninum* and *Echinococcus granulosus* in dogs (→ Membrane-Function-Disturbing Drugs/Table 2). Furthermore, nitroscanate is used in dogs against hookworms and *Toxocara* spp. (→ Membrane-Function-Disturbing Drugs/Table 2). Bitoscanate exerts activities against *Ancylostoma* and *Necator*.

Molecular Interactions There are no data about the mode of action of nitroscanate in cestodes. There is one report showing that the ATP synthesis in the trematode *Fasciola hepatica* is inhibited (Fig. 4).

Halogenated Monophenols

Synonyms Disophenol: Ancylol, DNP, Iodophene, Syngamix.
Nitroxynil: Fasciolid, Dovenix, Trodax.

Clinical Relevance Disophenol and nitroxynil are two members of monophenols with antitrematodal activity against *Fasciola hepatica* (Table 1). Moreover, disophenol exerts antinematodal activity against *Haemonchus contortus*.

Halogenated Bisphenols

Important Compounds Bithionol, bithionol sulfoxide, Meniclopholan.

Synonyms Bithionol: Actamer, Bitin, Lorothidol.
Bithionol sulfoxide: Bitin-S, Disto-5.
Meniclopholan: Niclofolan, Bayer 9015, Me 3625, Bilevon-M, Dertil, Distolon.

Clinical Relevance Bithionol and bithionol sulfoxide (so-called thiobisphenols) were the first members of the class of halogenated bisphenols discovered in the 1930's. Dichlorophene was introduced in 1946 and hexachlorophene in the late 1950's. The anticestodal activity of these drugs is directed against *Taenia saginata*, *T. solium*, *Diphyllobothrium latum*. They have some effects against *Hymenolepis nana*. These compounds are now replaced by more active drugs.

Hexachlorophene has some additional antitrematodal activity against mature *Fasciola hepatica*, *Dicrocoelium dendriticum*, adult paramphistomes (Table 1). Dichlorophene shows activity against *Fasciolopsis buski* (Table 2) and bithionol against *Paragonimus* spp. (Table 2). Furthermore, bithionol is active against immature and adult paramphistomes (Table 1). However, it is toxic at the effective antitrematodal dose rate. Another member of the bisphenols, menichlopholan (= niclofolan) has fasciolocidal activity and high effectivity against immature paramphistomes in sheep (Table 1), but not against immature and adult flukes in cattle. Niclofolan exhibits additional activity against *Metagonimus* spp. (Table 2).

Characteristics The structure-activity relationships for the fasciolocidal activity of bisphenols are similar to that for monophenols. The safety index of monophenols and bisphenols is generally rather low between 1 and 4. Disophenol and nitroxynil obtain as electron-withdrawing substituents halogen, nitro, or cyano groups, which are necessary in at least the ortho- and/or para-positions of the phenol for the fasciolocidal activity. Nitroxynil, niclofolan, bithionol and hexachlorophene have structural similarity to 2,4-dinitrophenol, a known uncoupler of oxidative phosphorylation in mammalian systems (Fig. 4).

Molecular Interactions The mode of action studies with mono- and bisphenols have been carried out in isolated cestodal or mammalian mitochondria but not in liver flukes. Nevertheless, it is proposed that the action of monophenols and bisphenols relies on the decrease of ATP synthesis.

Salicylanilides

Important Compounds Niclosamide, Oxyclozanide, Clioxanide, Rafoxanide, Brotianide, Bromoxanide, Closantel, Resorantel.

Synonyms Niclosamide: Mansonil-P, Lintex-M, Mansonil-M, Yomesan, Bayluscid, Cestocid, Devermin, Fenasal, Radeverm, Sagimid, Tredemine, Vermitin.

Oxyclozanide: 3,3',5,5',6-Pentachloro-2'-hydroxysalicylanilide, Zanil, Diplin, Metiljin.

Clioxanide: 2-Acetoxy-4'-chloro-3,5-diiodobenzanilide, Tremerad.

Rafoxanide: 3'-Chloro-4'-(p-chlorophenoxy)-3,5-diiodosalicylanilide, MK-990, Bovanide, Duofas, Flukanide, Ranide, Ursovermid.

Brotianide: 3,4'-Dibromo-5-chlorothiosalicylanilide acetate, Bay 4059, Dirian.

Bromoxanide: none

Closantel: Flukiver, Seponver.

Resorantel: Resorcylan, Terenol.

Clinical Relevance The first salicylanilide niclosamide was discovered in 1958 and introduced in the early 60's as anticestodal drug. The fasciolicidal activity of diaphene was discovered in 1963.

Niclosamide exerts anticestodal activity. It was the drug of choice in cestodiasis before the discovery of praziquantel. It has high curative efficacy against *Taenia saginata*, *T. solium*, *Diphyllobothrium latum*, *Hymenolepis nana*, *Mesocestoides* spp., and *Dipylidium* spp., but only low activity against *Echinococcus granulosus*. Its absorption from the intestinal tract is poor (only 2%–25% absorption in the first four days), and it is not accumulated in any organ. More than 70% of the drug are excreted via feces and the excretion is completed within 1–2 days. Resorantel, another salicylanilide, exhibits anticestodal activity against *Moniezia expansa* in ruminants.

In general, the antitrematodal activities are the main actions of salicylanilides. Niclosamide exerts activity against immature paramphistomes. It is regarded as the most effective and safe compound for the control of an outbreak of paramphistomiasis (Table 1). However, it has no effectivity against adult paramphistomes in ruminants. In general, drugs with high efficacy against the adult flukes are not active enough for elimination of pasture contamination. Furthermore Niclosamide has activities against intestinal flukes (*Fasciolopsis*, *Metagonimus*, *Heterophyes*, *Echinostoma*, *Gastrodiscoides*, *Watsonius*, *Nanophyetes*) (Table 2), but it is not active against *Fasciola hepatica*.

Diaphene is a mixture of 3,4,5'-tribromosalicylanilide (**tribromsalane**), the main component and fasciolicidal compound, and 4',5-dibromosalicylanilide used as a germicide. Tribromsalane was a new lead structure in the 1960's for a variety of salicylanilides such as **oxyclozanide**, **clioxanide**, **rafoxanide**, **brotianide**, **bromoxanide** and **closantel**. Salicylanilides show an increased potency against adult and particularly also against immature flukes. They possess a greater therapeutic index between 4 and 6 compared to the halogenated phenols with a therapeutic index between 1 and 4. With salicylanilides mass treatment of sheep and cattle is possible for the first time, because they have good activity against 4–6 week old immature flukes.

Resorantel, a 4'-bromo-γ-resorcylanilide-derivative, is a specific and the most effective drug against immature and adult paramphistomes in sheep, goats, and cattle with slightly erratic, but good efficacy. In addition it has some activity against *Gastrodiscoides* (Table 2).

Oxyclozanide is probably the most suitable drug for the control of an outbreak of acute intestinal paramphistomes in calves, especially in concurrent *Fasciola*-infections.

Besides their anticestodal and antitrematodal activity, salicylanilides are also active against some nematodes. Thus, rafoxanide and closantel are active against the blood-ingesting nematode *Haemonchus contortus* (Table 1).

Molecular Interactions The mechanism of the anticestodal action of niclosamide is the uncoupling of oxidative phosphorylation from electron transport in cestodes (Fig. 4). Thereby, protons are translocated through the inner mitochondrial membrane. There is a measurable decrease of ATP synthesis in *Ascaris* muscle mitochondria. The selective toxicity of niclosamide is explained by its poor absorption from the host intestine resulting in a protection of host cells against the uncoupling properties of this drug.

The fasciolicidal salicylanilides possess highly lipophilic groups like iodine, chlorophenoxy, tert-butyl-substituents which are responsible for prolonged plasma half-lifes between 2–4 days. Rafoxanide and bromoxanide have even longer half-lifes of 5–6 days. Moreover, there is slow excretion resulting in persistent drug residues. Therefore, several weeks of withdrawal periods before slaughter are necessary and most of the phenol-type fasciolicides are not used for treatment of milk-producing ruminants.

For rafoxanide, oxyclozanide and closantel there is more direct evidence for an uncoupling action within the fluke (Fig. 4). An increased end-product formation by 32% and decreased ATP-synthesis by 29% by rafoxanide can be detected.

Furthermore, there is an increased glucose uptake, decreased glycogen content, enhanced end-product formation (succinate), increased mitochondrial ATPase activity, reduced ATP levels by closantel. A probable correlation between death of the flukes and reduced ATP levels is discussed. A deformation of mitochondria in many fluke tissues is observable. The Golgi apparatus in the tegumental and gastrodermal cells is reduced in size and contains vacuolated cisternae. The basal infolds of the tegument are swollen and ion pumps associated with the tegumental membranes are inhibited. There is as a result a general induction of rapid spastic paralysis of the adult flukes by the action of the presumptive uncoupler-type fasciolicides such as rafoxanide, oxyclozanide, nitroxynil.

Resistance A selection of resistant strains of *Fasciola hepatica* in Australia has occurred by prolonged use of rafoxanide and closantel. There is cross-resistance between salicylanilides and the halogenated phenol nitroxynil. The resistance is manifest against immature but rarely against adult flukes. There is no side-resistance in rafoxanide- and closantel-resistant liver flukes to oxyclozanide. It is assumed that selection for resistance in the case of rafoxanide and closantel is favoured by possible differences in the mode of action and pharmacokinetic properties between oxyclozanide on the one hand and rafoxanide/closantel on the other hand. Thus, quick peak concentrations in the blood and quick elimination are seen with oxyclozanide in contrast to the strong binding to plasma proteins with rafoxanide and closantel resulting in long-lasting persistance in the blood at therapeutic concentrations for more than 90 days.

Cyanine Dyes

Important Compounds Pyrvinium, Dithiazanine iodide.

Synonyms Pyrvinium pamoate: Pyrvinium embonate, Viprynium embonate, Alnoxin, Molevac, Neo-Oxypaat, Pamovin, Poquil, Povan, Povanyl, Pyrcon, Altolat, Tolapin, Tru, Vanquil, Vanquin, Vermitiber.

Dithiazanine iodide: Abminthic, Anelmid, Anguifugan, Delvex, Dejo, Deselmine, Dilombrin, Dizan, Nectocyd, Partel, Telmicid, Telmid.

Clinical Relevance Pyrvinium exerts antinematodal activity against *Enterobius*, *Trichuris vulpis*, dithiazanine iodide against *T. vulpis*. In addition,

bunamidine
disophenol, nitroxynil,
dichlorophene, hexachlorophene, bithionol(sulfoxide)
niclofolan
salicylanilides (niclosamide, diaphene, tribromsalane, oxyclozanide, clioxanide, rafoxanide, brotianide, bromoxanide, closantel, resorantel)
bitoscanate, nitroscanate

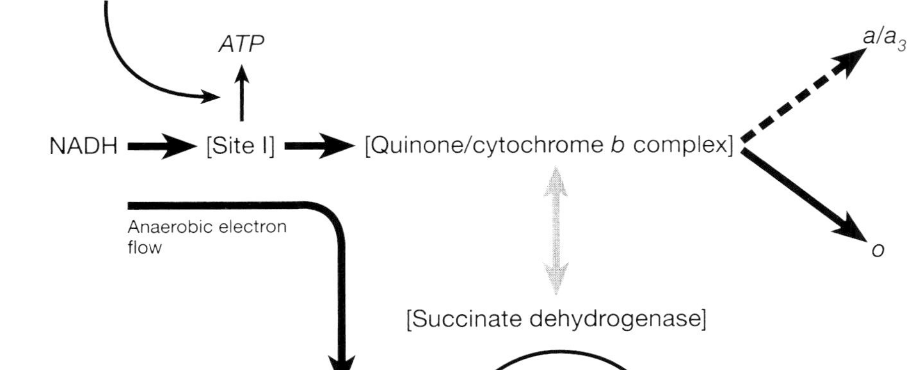

Fig. 4. Action of anthelmintics by uncoupling of oxidative phosphorylation.

dithiazanine iodide has experimentally antifilarial activity against *Litomosoides carinii*.

Molecular Interactions The antinematodal action against *T. vulpis*, a nematode residing in a more anaerobic environment, is probably the inhibition of glucose uptake by dithiazanine and also by pyrvinium. Responsible for the antifilarial action of dithiazanine is presumably the irreversible inhibition of oxygen uptake of adult *L. carinii*, an oxygen requiring nematode.

Entamoebiasis

Disease due to the protozoan
→ *Entamoeba histolytica* (Vol. 1), → Alimentary System Diseases, Carnivores.

Main clinical symptoms in humans: Abdominal pain, bloody-slimy diarrhoea, liver dysfunction in case of liver abscess.
Incubation period: 2–21 days
Prepatent period: 2–7 days
Patent period: Years
Diagnosis: Microscopic determination of cysts in fecal samples
Prophylaxis: Avoid uncooked food/water in endemic regions
Therapy: Treatment see → Antidiarrhoeal and Antitrichomoniasis Drugs

Enterobiasis

Pathology

Enterobiasis is an infection with the ubiquitous pinworm, → *Enterobius vermicularis* (Vol. 1). This small nematode has a simple life cycle in the intestinal lumen (→ Pathology/Fig. 22A). The adult female deposits eggs in the anal canal and on the perianal skin, causing irritation leading to itching. The adults are sometimes found in the lumen and even the mucosa of the vermiform appendix, but their role in causing appendicitis is in doubt. Ectopic worms may be found in the vagina, uterus, migrating up the fallopian tubes into the peritoneum, and occasionally elsewhere, where they tend to die and become surrounded by small granulomas containing eosinophils (→ Pathology/Fig. 27B-D). It has been speculated that eggs of *Enterobius* spp. transmit *Dientamoeba fragilis*, which may cause diarrhea, sometimes with blood and mucus.

Targets for Intervention

Infections with the nematode *Enterobius vermicularis* tend to be transmitted directly from man to

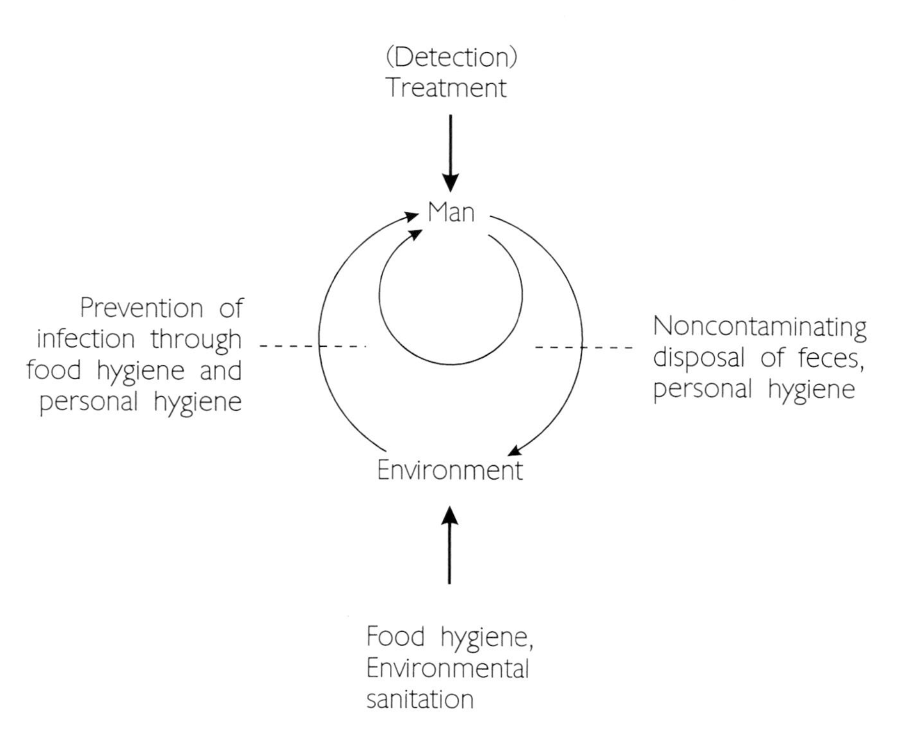

Fig. 1. Targets and approaches for the control of enterobiasis

man (anus-hand-hand-mouth), and self-reinfection is frequent since the worm eggs are immediately infective. However, some infections are also indirectly transmitted through contaminated material. Fig. 1 shows potential targets and approaches of intervention.

The main targets will be infective reservoir and transmission from man to man. The approaches to control consist of detection and treatment of infections, and improving personal hygiene and food hygiene. Blanket presumptive treatment campaigns without prior diagnosis have been successfully conducted, especially in populations of children with high infection rates. The best results were obtained with repeated treatment at an interval of two to four weeks.

Main clinical symptoms: Pruritus analis, diarrhoea, disturbances of sleep
Incubation period: 1–4 weeks
Prepatent period: 5–10 weeks
Patent period: Years due to repeated autoinfections
Diagnosis: Microscopic determination of eggs attached at the skin of the outer anal region
Prophylaxis: Repeated cleaning of toilets and daily cleaning of perianal skin; treatment of the whole family
Therapy: Treatment see → Nematocidal Drugs, Man

Environmental Management

→ Disease Control, Methods.

Eosinophilia

As effect of the occurrence of some parasites in peculiar organs (e.g. *Ascaris* during lung passage, schistosomal eggs in liver etc.) the general number of eosinophilic cells may become considerably increased. This increase is often used as help in diagnosis.

Eosinophilic Reaction

→ Pathology.

Epidemic Spotted Typhus

Disease of humans due to infection with the spherical, 0.3–0.5 µm sized *Rickettsia prowazekii* stages transmitted by the feces of lice (*Pediculus humanus corporis*) (via inhalation or skin scratching). After an incubation period of 10–14 days high fever occurs leading to death in 20% of (untreated) cases.

Therapy
Tetracyclines.

Epidemiology

Epidemiology (expression in medicine) or epizootiology (expression in veterinary medicine) is a science dealing with occurrence, distribution, prevention, and control of disease, injury and other health-related events (e.g. influence of climatic conditions) in a defined animal or human population.

Epizootiology

Epizootiology (expression in veterinary medicine) or epidemiology (expression in medicine) is a science dealing with occurrence, distribution, prevention, and control of disease, injury and other health-related events (e.g. influence of climatic conditions) in a defined animal or human population.

EPM

Equine protozoal myeloencephalitis described as result of infections with *Sarcocystis neurona* in horses (→ Sarcocystis (Vol. 1)). The pathway of transmission remains unclear.

Erythema

Clinical and pathological symptoms of infections with skin parasites (→ Skin Diseases, Animals, → Tick Bites: Effects in Animals, → Tick Bites: Effects in Man, → Ticks as Vectors, → Lyme Disease).

Erythema chronicum migrans

→ Lyme Disease.

Evasion Mechanisms

→ Amoebiasis, → Cysticercosis, → Giardiasis, Man, → Sleeping Sickness.

Excoriation

Clinical and pathological symptoms (loss of surface layers) of infections with skin (*lat.* corium) parasites (→ Skin Diseases, Animals, → Demodicosis, Man).

Eye Parasites

Parasites may enter each organ of the body. Several parasitic stages, however, have developed a special favor for this organ, which of course is especially sensible with respect to human welfare. Table 1 summarizes the parasites found in the different regions of the human eye. The life cycles, morphology, reproduction modes and the pathological effects are described in the entries on the respective organisms). Figures 1–4 show some of the most common aspects of parasites in eyes (Fig. 1, Fig. 2, Fig. 3, Fig. 4). For eye parasites of animals see page 451.

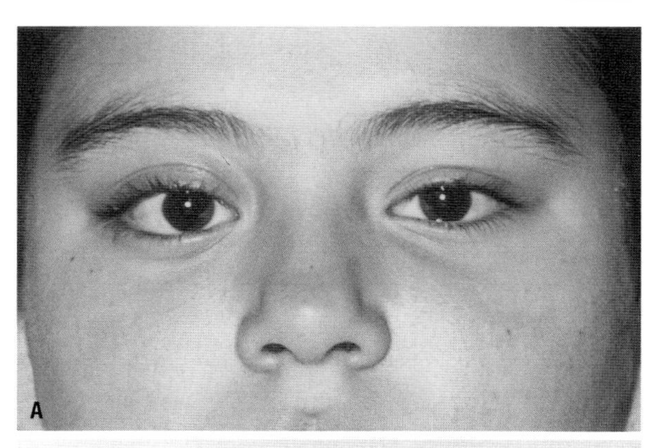

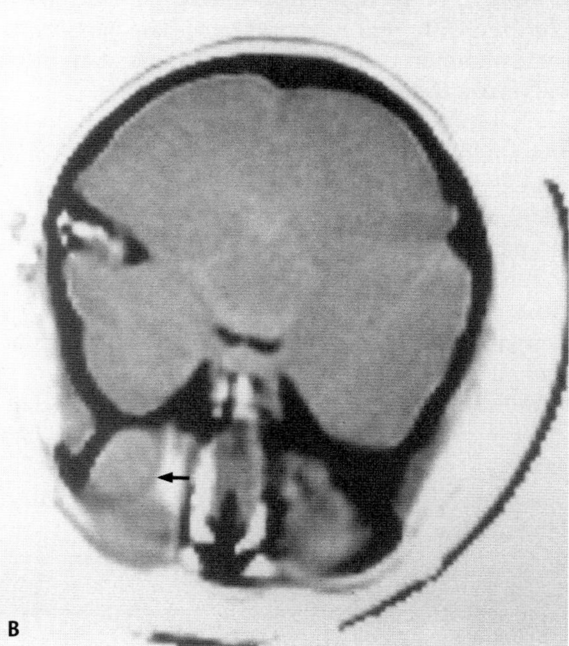

Fig. 1A,B. Effects of hydatids of *Echinococcus granulosus*. A Exophthalmos (right eye) in the orbit of a 10-years-old girl. B Computer tomogram (CT) of a hydatid cyst in the right orbit (by courtesy of Prof. J. Grüntzig, Düsseldorf).

Eye Worm

→ *Filariidae* (Vol. 1), → Loa loa (Vol. 1).

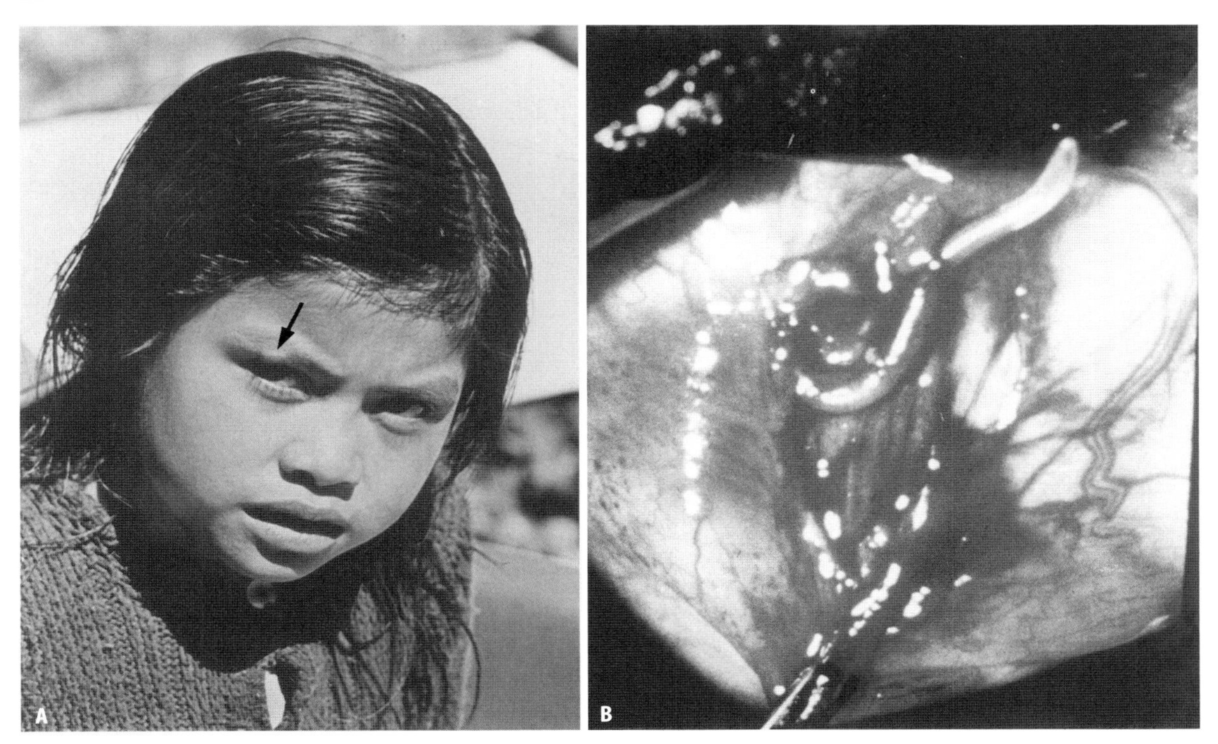

Fig. 2A,B. Adult worms and the eye. **A** Fibroma-like onchocercal nodule (arrow) containing several adult female *Onchocerca volvulus* worms in the right eyebrow of a Mexican child. **B** Adult *Loa loa* worm being surgically removed from the subconjuctival space (by courtesy of Prof. J. Grüntzig, Düsseldorf).

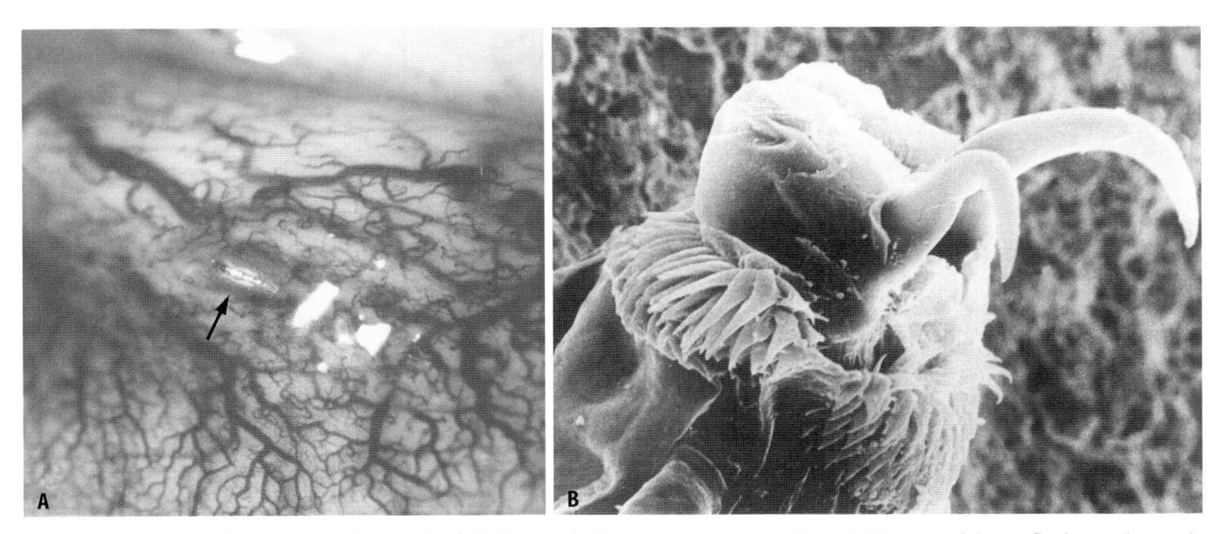

Fig. 3A,B. Effect of fly larva in the eye (ophthalmomyiasis). **A** Acute → conjunctivitis caused by a fly larva (arrow). **B** SEM-micrograph of the anterior pole of a larva of the fly *Oestrus ovis* showing its long mouth hooks (by courtesy of Prof. J. Grüntzig, Düsseldorf).

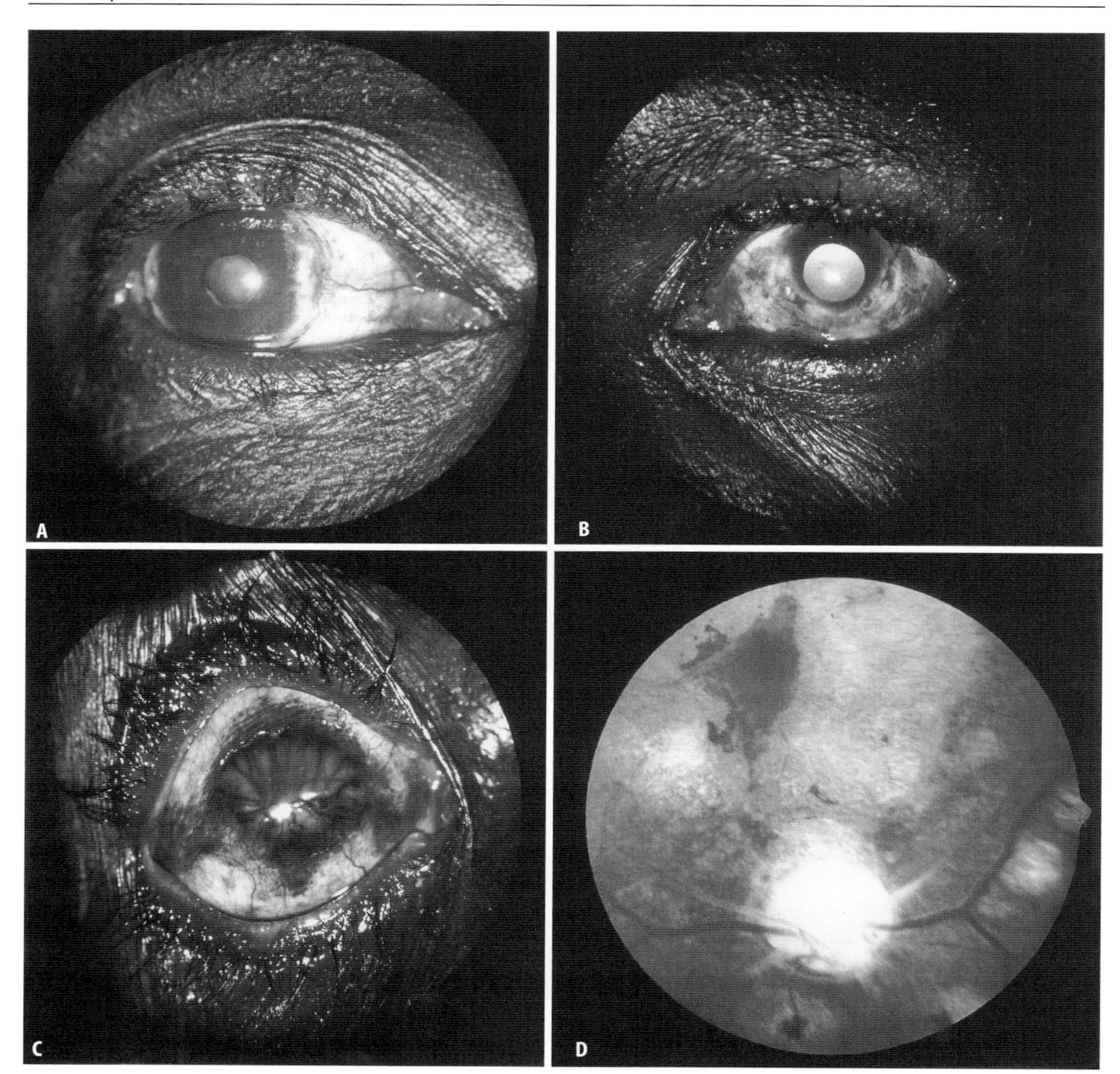

Fig. 4A-D. Effects of → onchocerciasis. A Early sclerosing keratitis in the 2–4 and 8–10 o'clock positions. B Confluent opacification in sclerosing keratitis (keratitis semilunaris). C Advanced sclerosing keratitis. D Optic atrophy with extensive choroidoretinal lesions (by courtesy of Prof. J. Grüntzig, Düsseldorf).

Table 1. Ophthalmologic manifestation of parasitic infections

Ocular regions and ophthalmologic signs	Genera of parasites
Eyebrows and eyelids	*Pediculus, Phthirus, Ixodes*
Eye lid edema	Ixodids, *Pulex*, fly larvae, *Giardia, Trypanosoma, Plasmodium, Schistosoma, Paragonimus, Taenia*, spargana, *Ancyclostoma, Gnathostoma, Toxocara, Trichinella*, filariae
Chalazion and pseudochalazion	*Demodex*, fly larvae, *Trypanosoma, Schistosoma, Onchocerca, Leishmania*
Ptosis	*Taenia solium* (cysticercus)
Inflammations of the eyelid margin (blepharitis)	*Demodex, Pediculus, Phthirus, Plasmodium*
Ophthalmomyiasis	Fly larvae
Lacrimal duets and glands	
Dacryocanaliculitis and dacryocystitis	Fly larvae, *Plasmodium, Ascaris, Mammomonogamus, Trypanosoma, Thelazia*
Dacryoadenitis	*Schistosoma, Plasmodium*
Orbit	
Exophthalmos	*Echinococcus*, coenurus, *Taenia, Schistosoma*, spargana, *Ascaris, Trichinella, Trypanosoma, Plasmodium, Entamoeba, Loa, Dracunculus, Dirofilaria, Gnathostoma*
Ocular muscles	
Diplopia	*Trichinella, Parastrongylus, Ancyclostoma*, spargana, *Taenia, Plasmodium*
Conjunctiva	
Parasites in the cul-de-sac	Fly larvae, *Enterobius, Thelazia*
Subconjunctival parasites	Fly larvae, *Thelazia, Loa, Wuchereria, Brugia, Dracunculus, Porocephalus*
Subconjunctival cysts	*Schistosoma, Taenia, Dirofilaria, Dipetalonema, Habronema, Mansonella*, spargana, *Philophthalmus*
Chemosis	*Ascaris, Trichinella, Giardia, Onchocerca, Trypanosoma*
Hemorrhages	*Schistosoma, Trichinella*
Conjunctivitis	Fly larvae, *Loa, Schistosoma, Entamoeba, Leishmania*
Cornea	
Parasites in the cornea	*Onchocerca, Toxocara, Trypanosoma, Ascaris*
Keratitis, scleritis	*Onchocerca, Toxocara, Trypanosoma, Acanthamoeba, Entamoeba, Leishmania, Ancylostoma*
Scierosing keratitis	*Onchocerca*
Corneal ulcers	*Trypanosoma, Acanthamoeba*
Anterior chamber	
Parasites in the anterior chamber	*Onchocerca, Loa, Wuchereria, Brugia, Entamoeba, Acanthamoeba, Trypanosoma, Schistosoma, Paragonimus, Taenia*, spargana, *Parastrongylus, Ascaris, Gnathostoma, Toxocara, Dirofilaria, Thelazia, Linguatula, Porocephalus, Dipetalonema*, fly larvae
Cysts in the anterior chamber	*Taenia*
Hypopyon	*Entamoeba, Acanthamoeba, Taenia, Gnathostoma, Toxocara*
Secondary glaucoma	*Entamoeba, Giardia, Leishmania, Toxoplasma, Trypanosoma, Plasmodium, Schistosoma, Paragonimus, Taenia, Echinococcus, Parastrongylus, Ascaris, Dirofilaria, Onchocerca, Brugia, Wuchereria, Gnathostoma, Toxocara, Porocephalus, Linguatula*, fly larvae

Table 1. Continued

Ocular regions and ophthalmologic signs	Genera of parasites
Iris	
Mydriasis	*Enterobius, Trichinella, Ascaris*
Miosis	*Enterobius, Ascaris*
Reflectory pupilloplegia (Argyll Robertson)	*Plasmodium*
Distortion of the pupil	*Onchocerca*
Hemorrhages	*Schistosoma, Paragonimus, Loa*
Iritis and iridocyclitis	*Entamoeba, Giardia, Leishmania, Toxoplasma, Trypanosoma, Plasmodium, Paragonimus, Schistosoma, Taenia, Parastrongylus, Ancyclostoma, Ascaris, Trichinella, Toxocara, Onchocerca, Brugia, Wuchereria, Loa*
Lens	
Cataract	*Leishmania*, cysticercus, *Ancylostoma*
Subluxatio	*Linguatula*, fly larvae
Vitreous body	
Hemorrhages	*Ascaris, Schistosoma, Trichinella, cysticercus, Gnathostoma*, fly larvae
Cysts	*Cysticercus, Echinococcus*, coenurus
Parasites in the vitreous	*Parastrongylus, Ascaris, spargana, Dipetalonema, Dirofilaria, Linguatula, Onchocerca, Wuchereria*, fly larvae
Cyclitis	*Schistosoma, Cysticercus, Gnathostoma, Onchocerca, Toxocara, Trichinella*
Optic nerve	
Papilledema	*Entamoeba, Leishmania, Taenia, Parastrongylus, Ancylostoma, Ascaris, Trichinella*
Papillitis	*Entamoeba, Giardia, Trypanosoma, Plasmodium, Paragonimus, Taenia, Ancylostoma, Ascaris, Toxocara, Trichinella, Onchocerca*
Optic atrophy	*Entamoeba, Giardia, Leishmania, Toxoplasma, Trypanosoma, Plasmodium, Paragonimus, Parastrongylus, Taenia, Ancylostoma, Ascaris, Toxocara, Trichinella, Onchocerca*
Retina and chorioidea	
Hemorrhages	*Entamoeba, Giardia, Leishmania, Trypanosoma, Plasmodium, Schistosoma, Ancylostoma, Gnathostoma, Toxocara, Trichinella, Wuchereria, Loa*, fly larvae
Retinal detachment	*Taenia, Porocephalus*, fly larvae
Cysts	*Entamoeba, Echinococcus*
Retinitis and choroiditis	*Entamoeba, Giardia, Leishmania, Toxoplasma, Schistosoma, Taenia, Parastrongylus, Ascaris, Toxocara, Trichinella, Wuchereria, Loa, Onchocerca*, fly larvae

F

Falciparum Malaria

→ Malaria tropica, → *Plasmodium falciparum* (Vol. 1), → Mathematical Models of Vector-Borne Diseases, → Insulin (Vol. 1).

Fascioliasis, Man

Fascioliasis describes an infection of the bile ducts with *Fasciola hepatica,* the liver fluke of sheep, cattle, and man. After the ingestion of the metacercaria encysted on water plants (water cress salad), the larvae wander through the wall of the gut, into the liver parenchyma, and into the bile ducts. The migration tracts are accompanied by an intense inflammatory reaction with prominent eosinophils and Charcot-Leyden crystals, resolving ultimately by fibrosis (→ Pathology/Fig. 3A,B, → Pathology/Fig. 21A,C). The liver may be enlarged and show abnormal function. Blood leukocytosis with eosinophilia, and fever are prominent. After long-standing infection with flukes, bile duct hyperplasia, pericholangitis, periportal fibrosis, and obstruction of the bile duct may develop. F. *hepatica* eggs are shed in the stools.

Targets for Intervention

Among the food-borne zooanthroponotic parasites, *Fasciola hepatica* deserves attention as its control, seemingly simple, may pose major practical problems, foremost the relatively poor response to treatment and the high probability of reinfection. Upon reaching water the eggs release miracidia which infect aquatic snails. After completing development in the snail host, the infective metacercariae leave the snail and attach themselves to plants and so reach their vertebrate hosts, especially sheep, cattle and humans. Fig. 1 (page 224) shows the targets of intervention which consist of the elimination of infection and the interruption of the infection cycle. The practical approaches to control include the detection and treatment of vertebrate carriers, the safe disposal of human feces, the avoidance of raw aquatic vegetables and of those grown in wetlands and, as far as livestock is concerned, the avoidance of grazing in wetlands. The control of aquatic snails is a hypothetical possibility rather than a practical proposition.

Main clinical symptoms: Liver infection, fever, dyspepsy, ascites, eosinophilia
Incubation period: 3–12 weeks
Prepatent period: 3–4 months
Patent period: 1–20 years
Diagnosis: Microscopic determination of eggs in fecal samples, → serology (Vol. 1)
Prophylaxis: Avoid eating raw freshwater plants
Therapy: Treatment see → Trematodocidal Drugs

Fasciolopsiasis

Fasciolopsiasis is an infection of the duodenum and jejunum of humans with adults of → *Fasciolopsis buski* (Vol. 1). It is usually asymptomatic when small numbers of worms are present. However, the multiple attachment sites that become ulcerated, can lead to appreciable blood loss and → abscess (Vol. 1) formation. Intestinal obstruction by large numbers of worms has been reported.

Main clinical symptoms: Diarrhoea, nausea, vomiting, oedema, anaemia, ascites
Incubation period: 1–2 months
Prepatent period: 2–3 months
Patent period: 1 year
Diagnosis: Microscopic determination of eggs in fecal samples

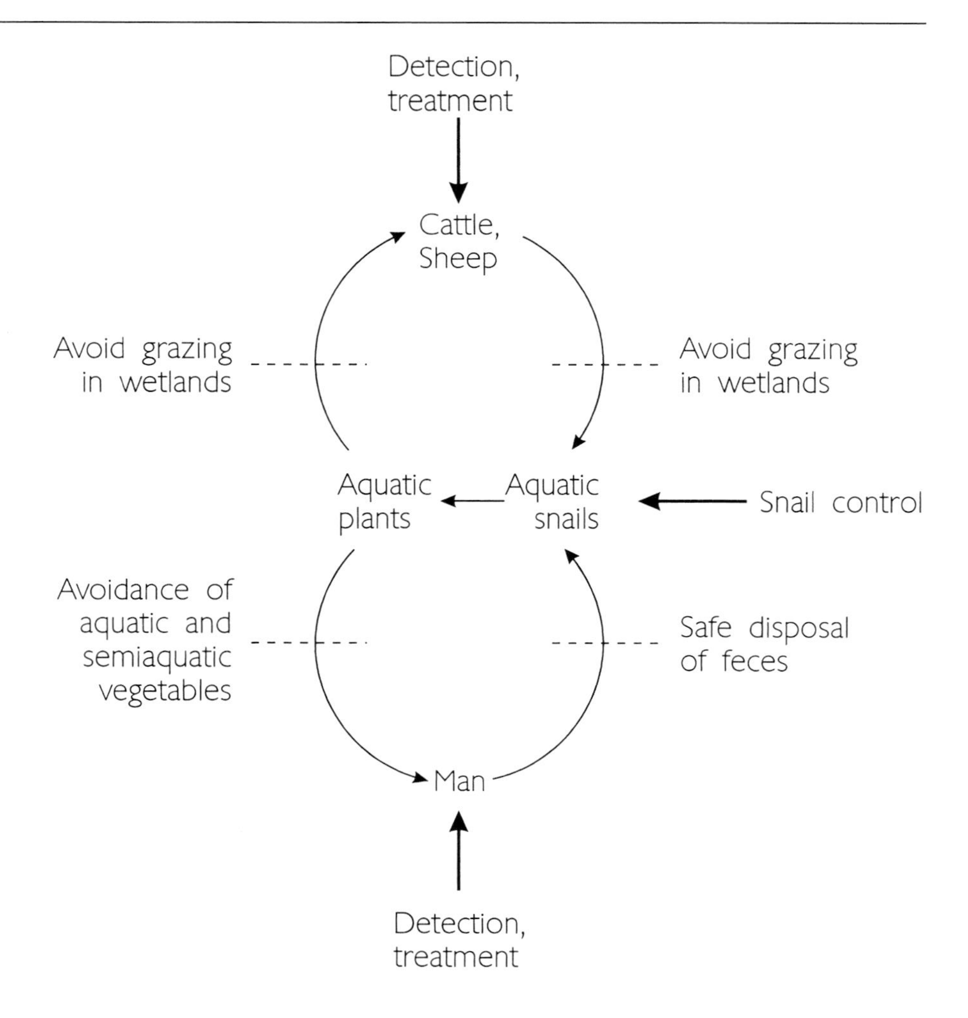

Fig. 1 Targets and approaches for the control of fascioliasis

Prophylaxis: Avoid eating uncooked tropical vegetables or fruit
Therapy: Treatment with praziquantel, see → Trematodocidal Drugs

Fasciolosis, Animals

General Information

Disease caused by the → liver fluke (Vol. 1) genus *Fasciola*. *F. gigantica* occurs in the tropics. *Fasciola hepatica*, the common liver fluke, is the most widespread and important of the group. It is found mainly in sheep and cattle but a patent infection can develop in horses, pigs, wild animals, and in humans (→ Fascioliasis, Man). The pathogenesis of fasciolosis is attributable in part to the invasive stages in the liver and in part to the blood feeding by the adults in the bile ducts. The process in all hosts shows close similarities, but considerable variation in severity occurs.

Pathology

Ruminants The pathological manifestations depend on the number of metacercaria ingested. The disease may follow acute or chronic courses. Acute fasciolosis is less common than the chronic entity and is almost invariably seen in sheep. It is essentially a traumatic hepatitis produced by the simultaneous migration of large numbers of adolescaria. It is towards the end of this development phase, about six weeks after infection, that the signs are apparent, with the major losses occurring 7–8 weeks after infection. Death may occur rapidly or after several days. Animals are disinclined to move, are anorexic and show a distended abdomen which is painful to the touch. This is also the stage of parasitism in which "black disease" occurs (*Clostridium novi* type B).

Fasciolosis is most commonly a chronic disease, with no characteristic clinical signs. Loss of appetite and paleness of the mucous membranes appear to be constant features, and submandibular and udder oedema are occasionally seen. Jaundice is hardly ever a sign in the living animal. Chronic debility with vague digestive disturbances are common There is a substantial effect on milk production, and a reduction in food conversion efficiency with reduced weight gain. A reduction in wool production may occur in sheep, without symptoms of fasciolosis being apparent. Fasciolosis in sheep also has an adverse effect on conception and/or establishment of the foetus.

Changes in serum protein generally take the form of a depression in albumin compared with the globulins. They develop in two stages. The first stage coincides roughly with the period of fluke migration and is characterized by a progressive but usually mild hypoalbuminaemia, with a more pronounced hyperglobulinaemia of variable severity. The second stage, which is associated with the presence of adult parasites in the bile ducts, is attended by further deterioration of albumin as well as a progressive reduction in globulin concentrations. There is little disagreement on the cause of the hyperglobulinaemia, which is generally considered to reflect increase synthesis of immunoglobulins in response to parasitic antigens. On the contrary a more complex nature of the different processes and inter-relationships are involved in the pathogenesis of hypoalbuminaemia. During the migratory stage hypoalbuminaemia is brought on by a combination of reduced albumin synthesis and plasma volume expansion. During the biliary stage of the disease the severity of hypoalbuminaemia is related to the loss of albumin into the intestine and to the rate of albumin synthesis and the fractional and total rates of albumin catabolism. These, in turn, are related to the levels of nutrition, appetite and fluke burden. The increased synthesis of albumin probably diverts available amino acids away from other protein metabolism (muscle, milk, wool), thus accounting for the lowered levels of productivity seen in animals infected with *F. hepatica*.

The anaemia is of the normocytic normochromic type, though some macrocytosis has been reported. The anaemia is well recognized but its etiology is controversial.

Several factors may account for the anaemia :
- anaemia in the migratory phase is caused by accidental damage to hepatic vessels and haemorrhages
- intrabiliary haemorrhage and consequent loss of red blood cells occur when adults arrive in the bile ducts. The ultimate degree of anaemia is not related to the severity of biliary haemorrhage, but rather to the animal's erythropoietic capacity which is influenced by levels of dietary protein and iron.

Good correlation between the bromsulphtalein excretion test, serum glutamate oxalo-acetate transaminase and gamma glutamyl transferase determinations for the assessment of liver damage are found in infected sheep (liver function test).

Horses *Fasciola hepatica* is occasionally found in the equine liver. Heavy infections are rare, and are usually only discovered during postmortem examination.

Swine *Fasciola hepatica* has been found in pigs, but infections are very rare.

Targets for Intervention

→ Fascioliasis, Man/Targets for Intervention.

Therapy

→ Trematodocidal Drugs.

Feasibility Studies

→ Disease Control, Planning.

Fiboblastic Proliferation

→ Pathology.

Filariasis, Lymphatic, Tropical

Synonyms

Filariosis, Brugiasis, Wuchereriasis

General Information

Lymphatic filariasis is an infection with one of several mosquito-borne filarial worms of the species → *Wuchereria bancrofti* (Vol. 1), *Brugia malayi* or *Brugia timori*, which live in the subcutaneous lymphatics or lymph nodes, with larvae circulating in the bloodstream. About one-fifth of the world's population live in areas where lymphatic filariasis is endemic. The disease is world-wide, 110 million people are estimated to harbour such infections. *W. bancrofti* is widely distributed throughout the tropics. *B. malayi* is restricted to parts of Southeast Asia and *B. timori* shows an even more restricted distribution in the Malay Archipelago. Adult *W. bancrofti* is restricted to man, while domestic and wild animals may serve as alternative hosts of *B. malayi* and *B. timori*. *W. bancrofti* is transmitted by *Culex*, *Anopheles* and *Aedes* spp., *B. malayi* and *B. timori* predominantly by *Mansonia* spp.

Clinical manifestations of filariasis are almost exclusively due to the microfilariae shed by the adult female worms. The symptoms include initial filarial fever and lymphangitis which later gives rise to recurring lymphoedema. High adult wormload, and consequently high microfilarial density, favours the development of lymphangitis and elephantiasis.

Pathology

The larvae are injected intradermally with a mosquito bite and find their way to the large lymphatics, where they mature and mate. Swelling of lymph nodes containing adults is a common feature. However, when an adult worm dies severe lymphadenitis with chronic inflammatory to granulomatous reaction results, including eosinophils which ultimately leads to fibrosis. In some multiply infected individuals this may lead gradually to chronic lymphatic obstruction, which in a small percentage of cases progresses to the lymphedematous complication of elephantiasis, usually in an extremity. The newborn larvae circulate in the bloodstream within the internal organs, such as the spleen, and sometimes they migrate cyclically to the peripheral circulation, coincident with the biting/feeding habits of the prevalent transmitting mosquito. Tropical eosinophilic with fever with pulmonary infiltration is often attributed to this infection.

Immune Responses

Because of the very long periods of survival of the macrofilariae (= adult worms) in their hosts (5–15 years), it is obvious, that these parasites must have developed complex mechanisms to evade killing by the host immune defenses. In addition, the host's immune response significantly contributes to the different pathological manifestations of the disease.

There is a broad range of immune reactivity with considerable individual variation. In lymphatic filariasis, microfilaremic individuals (MF) who are clinically asymptomatic have high parasite burdens and little or no parasite antigen-specific cell mediated responses. In contrast, patients with chronic lymphatic disease, e.g. elephantiasis, typically are amicrofilaremic and vigorous T cell responses against the parasite can be detected.

As for most of the other parasitic diseases, models of filarial infection in inbred mice significantly contributed to the understanding of the disease-influencing immunoregulatory events. Since laboratory mice are not permissive for filarial species found in infected humans, immunity to different stages of these filariae (third stage larvae, adults and microfilariae) has been analyzed separately as a surrogate approach. On the other hand, in the mouse model of *Litomosoides sigmodontis* infection, the full developmental cycle can be established in inbred mice, allowing to study immunity during maturation of infective larvae into adult worms.

Innate Immunity Information on the role of components of innate immunity in the early control of filariae is very limited. In a recent study by Babu et al. an unexpected role of NK cells was described. Comparisons of *B. malayi* worm survival in strains of mice with different levels of NK cell activity showed, that host NK cells are required for the growth of this human filarial parasite. While NOD/LtSz-SCID mice with diminished or absent NK cell activity were nonpermissive to worm growth, C.B17 SCID mice with normal NK cell activity were highly permissive. Furthermore, transfer of NK cells into NK-deficient mice rendered these animals permissive. Although the mechanisms by which NK cell allow the growth of filariae is enigmatic so far, these findings clearly point to an interesting role of the innate immune system in the establishment of this parasitic disease. The most compelling evidence for a role of eosinophils in immunpathology of filarial infections comes from analysis of onchocercal dermatitis and keratitis. There is a consistent presence of eosinophils and eosinophil granule proteins at the site of tissue damage, either after parasite death or direct injection of parasite antigens. However, the role of IL-5 and other chemoattractant mediators for the recruitment and activation of eosinophils has yet to be established.

B Cells and Antibodies High levels of parasite-specific IgE and IgG4 are produced in filariasis patients, generally accompanied by eosinophilia. A reciprocal expression of these two isotypes has been found in lymphatic filariasis patients, with asymptomatic patients having much higher ratios of IgG4:IgE than found in elephantiasis patients, suggesting either that IgE is an antifilarial antibody, and/or that high IgE is involved in the pathogenetic pathway of the disease. High quantities of IgG4 can be frequently found in sera of microfilaremic patients, where sometimes up to 95 % of the filarial-specific antibodies are of this subclass. In contrast, in elephantiasis patients IgG1, 2, and 3 dominate the filaria-specific antibody response. It seems likely that these antibodies may contribute to the pathology through ADCC or immune complex formation.

In mice infected with *B. malayi* the clearance of microfilariae was found to be clearly antibody dependent. CBA/N mice which carry the Xid defect have a pronounced impairment of the B1 cell subset and are therefore unable to develop certain T-cell-independent IgM antibodies. The findings that microfilaremia can not be controlled in these animals after i.v. injection of *B. malayi* or implantation of *A. viteae* gravid females strongly suggest that T cell-independent IgM antibodies to the microfilariae's surface are involved in the clearance of microfilariae.

T Cells Immunity to most filarial infections is clearly T cell dependent. Nude mice and rats are susceptible to infection with a number of species (*A. viteae, B. pahangi, B. malayi*) to which their normal, immune competent littermates are resistant, and infection against *B. pahangi* can be established in normally resistant CBA mice when these are deprived of T cells. The resistance of mice to *B. malayi* can be probably mediated by either $CD4^+$ T cells or $CD8^+$ T cells: Resistance to the maturation of infective larvae (L3) was not abrogated in either β2-microglobulin knockout mice, which lack MHC class I molecules and class-I-restricted $CD8^+$ T cells, or in anti-CD4-treated or CD4-deficient mice.

Important clues for the contribution of Th1 or Th2 cells to pathology or control of the parasite came from analysis of onchocerciasis patients. In individuals with generalized infections characterized by high microfilarial loads, low proliferative T cell responses to parasite antigens is accompanied by a production of Th2 cytokines. On the other hand, a minority of patients able to prevent maturation of L3 displays an immune response characterized by the predominance of IFN-γ producing Th1 cells. Several experimental studies have been performed to analyze the relative contributions of Th1 and Th2 cells and their products (cytokines) to the control of filarial parasites in mice. Following infection with L3 or immunization with L3 there is a strong expansion of parasite-specific Th2 cells and of associated immune responses such as IgE production and eosinophilia. In infections with both *Brugia* and *O. volvulus* the Th2 responses appear to be protective, since antibodies against IL-4 or IL-5 resulted in longer survival of larvae. However, resistance to maturation of L3 into adults was not abrogated in IL-4 knockout mice, arguing for compensatory mechanisms in these gene-deficient mice. Infection of BALB/c mice with adult *B. malayi* worms, especially females, also induced strong IL-4 production by splenocytes.

The Th cell response to microfilariae has been most extensively investigated using *A. viteae, B. malayi, O. volvulus* and *O. lienalis*. Interestingly, a dominant Th1 response has been observed during the first 2 weeks of infection, which is followed by an enhanced induction of IL-4 and IL-5 in addition to IFN-γ during the subsequent weeks. Thus, the time of exposure to microfilarial antigens seems to drastically influence the type of Th cell response. Both Th cell subsets may significantly contribute to the control of microfilariae, since both IFN-γ-stimulated macrophages as well as IL-5-dependent eosinophils are operative against microfilariae. Activated macrophages are able to damage microfilariea by releasing NO. In addition, microfilariae of some species(e.g. *O. lienalis*) but not of others (*B. malayi*) are sensitive to H_2O_2 which may be produced by eosinophils.

An important mediator likely to be involved in the crossregulation of the two Th subsets is IL-10. The mechanisms by which Th1 responses are inhibited in microfilaremic patients have not yet been elucidated in detail, but some studies suggest that down-regulatory cytokines such as IL-10 may be involved in this process. Mononuclear cells from microfilaremic hyporesponsive individuals have been found to secrete large quantities of IL-10 spontaneously and in response to parasite anti-

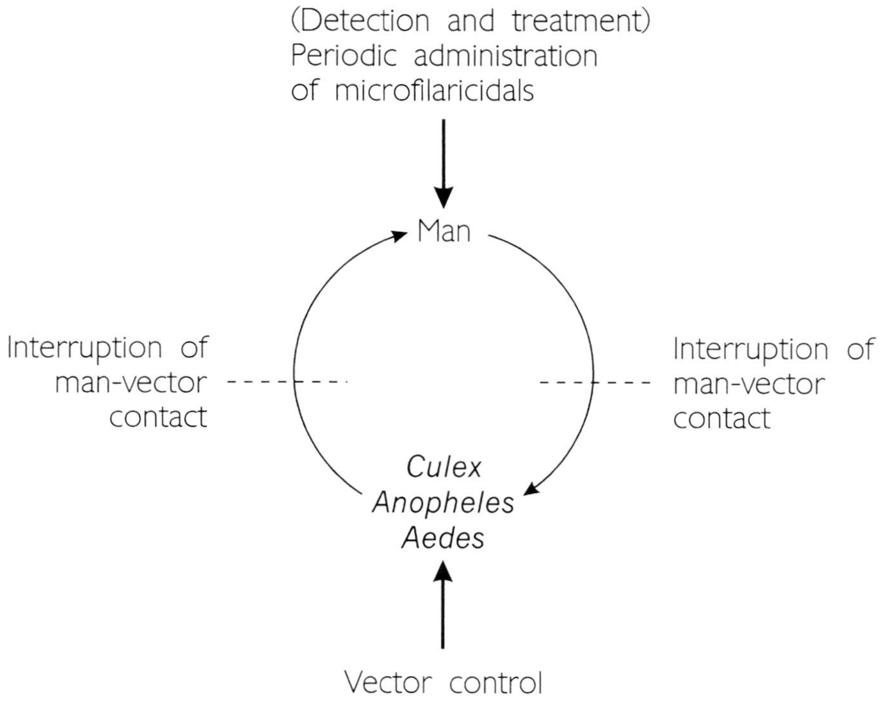

(Detection and treatment)
Periodic administration
of microfilaricidals

Man

Interruption of
man-vector
contact

Interruption of
man-vector
contact

Culex
Anopheles
Aedes

Vector control

Fig. 1 Targets and approaches for the control of bancroftian filariasis

gens, and in a majority of these individuals anti-IL-10 significantly augmented IFN-γ secretion in vitro.

Shaping of the Immune Response by the Parasite
Recent reports suggest that filarial parasites have the capacity to actively shape their immunological environments in their host. For example, secreted products of the nematodes have been found to differentially modulate the expression and activation of protein kinase C isoforms in B lymphocytes. Furthermore, the expression of CD23 on human splenic B and T cells and Th2 responses are enhanced by soluble products of the parasite. Pastrana et al. cloned homologues of the mammalian migration inhibitory factor (MIF) from *B. malayi*, *W. bancrofti*, and *O. volvulus*. The effects of recombinant forms of the parasite MIF and human MIF on human monocytes/macrophages were similar. The inhibition of macrophage migration and the alteration of inflammatory and T cell responses by filarial MIF could provide the parasite with a survival advantage.

Planning of Control
The approaches to the control of lymphatic filariasis were formerly based on the elimination of infection by treatment (diethylcarbamazine = DEC) and vector control for the prevention of infection. Both approaches were only marginally effective due to the poor macrofilaricidal activity of DEC and constraints of controlling *Culex* spp. in urban areas and *Anopheles* spp in the rural environment. Moreover, the first dose of DEC may cause severe and even fatal adverse reactions in persons infected with *Loa loa*, a tissue-dwelling filaria occurring in tropical Africa.

The control of lymphatic filariasis has been revolutionized by the finding that a single dose of ivermectin or DEC or both will eliminate microfilaraemia for several months due to an action against microfilariae and embryonic stages. Although recurring, microfilaraemia will stay below 1 % of the initial level for a year or more. With repeated dosing, once a year, microfilaraemia will not reach the level at which it could cause lymphoedema or elephantiasis. Essentially the same applies to onchocerciasis with regard to the prevention of irreversible ocular lesions. The reduction of microfilarial rates and densities will also lead to a rapid reduction of disease transmission, more effective than it could ever be expected from vector control.

Based on these findings, the WHO has embarked on the global elimination of lymphatic filariasis through the annual single dose administration of a combination of ivermectin and DEC to all persons (approximately 1100 million) residing in areas where the disease is endemic. There is still the caveat of adverse reactions to DEC in people infected with *Loa loa* (→ Loiasis), but this only applies to tropical Africa and can be overcome by using ivermectin alone. The other, yet unknown, factor is the potential role of animal reservoir as an obstacle to the ultimate elimination of human brugian filariasis.

Targets for Intervention

The infective larvae of *Wuchereria bancrofti* leave the arthropod vector at the time of a blood meal, slide along the outer surface of the proboscis stem and actively enter the bite wound. After reaching the lymphatic target organ, they will become adults and mate. The females will produce microfilariae which will periodically or subperiodically enter the blood stream from where they can be taken up by the vector. Fig. 1 shows that the detection and treatment of infected persons, suppression of microfilaraemia, vector control and the interruption of contact between man and vector are potential approaches to the control of bancroftian filariasis.

Main clinical symptoms: Lymphangitis, unfeelingness of skin portions; later: chylurie, elephantiasis, i.e. giant swelling of organs
Incubation period: 3–16 months
Prepatent period: 7–24 months
Patent period: 8–10 years (adults live until 18–20 years)
Diagnosis: Microscopic analysis of smear preparations or of membrane filtered material; microfilariae are found at 10 p.m. in the peripheral blood, → serology (Vol. 1)
Prophylaxis: Avoid bites of vector mosquitoes in endemic regions
Therapy: Treatment see → Nematocidal Drugs, Animals, → Nematocidal Drugs, Man

Related Entry

→ Filariidae (Vol. 1).

Francisella tularensis

Agent of → tularemia being transmitted by contaminated mouth parts of various ectoparasites.

Therapy

Streptomycine, Gentamycine

G

GAE

Granulomatous amoebic encephalitis of man due to infection with → opportunistic agents (Vol. 1): amoeba of the genera → Acanthamoeba (Vol. 1) and → Balamuthia (Vol. 1), which also may penetrate and thus destroy the cornea of the eyes.

GALT

Gut associated lymphoid tissue

Galton-Watson Process

→ Mathematical Models of Vector-Borne Diseases.

Gametocyte

Other word for gamont, i.e. stage proceeding gametes.

Gasterophilosis

Disease due to infestation with *Gasterophilus* species, see Table 1

Table 1. *Gasterophilus* species and Control Measurements

Parasite	Host	Symptoms	Country	Therapy		
				Products	Application	Compounds
Gasterophilus intestinalis (Common bot)	Horse	Most of the time asymptomatic; stomatitis, eating and chewing problems, erosions, ulcera and dilatation of the stomach, chronic duodenitis, proctitis	Worldwide	Eqvalan™ Paste 1.87% (MerckAgVet)	Oral	Ivermectin
Gasterophilus haemorrhoidalis (Nose or lip bot)	Horse		Worldwide	Eqvalan™ Paste 1.87% (MerckAgVet)	Oral	Ivermectin
Gasterophilus inermis	Horse		Europe, Africa, Asia			
Gasterophilus nasalis (Throat bot)	Horse		Worldwide	Eqvalan™ Paste 1.87% (MerckAgVet)	Oral	Ivermectin
Gasterophilus nigricornis	Horse		Europe, Africa, Asia			
Gasterophilus pecorum	Horse		Europe, Africa, Asia			

Gastritis chronica hyperplastica

→ Trichostrongylosis, Animals.

Gedoelstia *Species*

Flies of the family Oestridae, the larvae 1 of which penetrate occasionally into the conjunctival sac of cattle leading to a conjunctivitis. In addition the larvae 2 and 3 are found in the sinus frontalis, in brain and in heart muscles (→ Nervous System Diseases, Ruminants, → Uitpeuloog).

Genital System Diseases, Animals

Parasitic infections of the genital tract are of major economic importance (Table 1), particularly those caused by the widely distributed protozoan parasites *Toxoplasma gondii*, *Neospora caninum* (syn. *Hammondia heydorni*?) and *Tritrichomonas foetus* which are major causes of reproductive failure in ruminants.

T. gondii infections are ubiquitous throughout the world and are an important cause of abortion in small ruminants. In fact, the parasite may invade many different organs in nearly all warm-blooded animals, and the clinical picture in a particular host species depends on the particular involvement of any one or more of these organs. However, in the vast majority of cases, infections remain asymptomatic. During initial exposure of pregnant ewes and does, the parasite primarily invades the placenta and may, at any gestational age, lead to foetal infection, with or without foetal death. Therefore, exposure of susceptible animals may lead to a wide range of clinical manifestations, including early embryonic death, mummification, abortion, stillbirth, neonatal death, or birth of weak offspring. Abortion is generally caused by necrosis of the cotyledons of the placenta. Congenital infection mainly affects the brain of the foetus, and lambs or kids born alive may show signs of encephalitis. In pigs and dogs, cases of congenital toxoplasmosis have been reported, and clinically affected animals often show respiratory signs. *T. gondii* is apparently not a significant foetal pathogen in cattle and horses.

Sarcocystis spp. are common, ubiquitous, sporozoan parasites of herbivores, but they rarely cause clinical signs or abortions in infected animals. Acute forms of sarcocystosis are associated with the massive development of schizonts in endothelial cells of blood vessels. The clinical signs include high fever, anorexia, anemia, ataxia and loss of weight, sometimes with high mortality rates. *Sarcocystis* spp. have occasionally been reported as causing abortion in cattle, sheep, goats and pigs. In some cases, the absence of schizonts from the foetus suggest that the abortion is caused

Table 1. Parasites affecting the Genital System of Domestic Animals (according to Vercruysse and De Bont)

Parasite	Host	Location	Clinical presentation	Principal lesions
Protozoa				
Neospora caninum	Cattle, sheep, goat, horse, dog, cat	Neural tissues	Abortion, neuromuscular signs in congenitally infected calves	Encephalitis in transplacentary infected foeti
Sarcocystis spp.	Cattle, sheep, goat, pig	Vascular endothelium of placenta, foetus	Abortion	Placentitis
Toxoplasma gondii	Sheep, goat	Placental cotyledons, foetus	Abortion	Necrosis in villous part of cotyledons, leukoencephalomalacia in congenitally infected lambs or kids
Tritrichomonas foetus	Cattle	Vagina, cervix, uterus, oviduct, preputium	Early abortion, sterility	Vaginitis, endometritis, balanoposthitis
Trypanosoma equiperdum	Horses	Genitalia, skin, nerves	Genital phase: mucoid, vaginal or urethral discharge, nymphomania, micturition, rarely abortion	Oedema and inflammation of genitalia with ulcera, pigmentation of vulva or penis

by maternal failure. In other cases, where the aborting mother may be otherwise normal, the extensive development of schizonts in the endothelial cells throughout the foetal tissues and within the placenta suggest that abortion is directly caused by the parasitic infection. It is uncertain whether some reported cases of *Sarcocystis* abortion in cattle, in fact, are *Sarcocystis* or possibly, *Neospora* infections.

N. caninum, which has long been misidentified as *T. gondii*, is now known to occur in a wide range of host species, including dogs, cats, horses, cattle, sheep and goats. Dogs have been identified as definitive hosts. Clinical manifestations of neosporosis have mainly been observed in cattle and dogs. Transmission occurs orally, and via the transplacental route. *N. caninum* has a predilection for the central nervous system and skeletal muscles, generally leading to encephalomyelitis and polymyositis. In cattle, neosporosis is regarded as a major cause of abortion, particularly among dairy cattle. Although foetal death probably occurs throughout the gestation period, cows only abort from 3 months of gestation to term. It is likely that 1 to 2-month-old fetuses are killed in utero, resorbed, and the cow returns to heat again. Foetuses which die *in utero* may be resorbed, mummified, aborted or stillborn. *N. caninum* is most often demonstrable in the brain and heart of the foetus, and rarely in other organs, including the placenta. Some congenitally infected calves may be born with signs of neuromuscular dysfunctions, while others may be born clinically normal but chronically infected. In aborted foetuses, the major lesion is a multifocal, necrotizing, non-suppurative encephalomyelitis.

Trichomonosis is an important venereal disease of cattle caused by the flagellated protozoan *T. foetus*. In bulls infections usually go unrecognized. However, they remain chronically infected and transmission of the parasite to heifers or cows occurs at coitus. In the female, a primary infection invariably causes a vaginitis of varying intensity, with swelling of the vulva and vaginal discharge. Thereafter the parasites migrate upward through the cervix and invade the uterus. The inflammatory changes in the endometrium and cervix are relatively mild and nonspecific, although the exudate, mucopurulent in character, may be rather copious. *T. foetus* does not prevent conception as such, but endometritis and uterine catarrh prevent proper fertilization, and result in aberrant oestrus cycles and repeat breeding. Trichomonal abortion may occur at any time, but usually takes place early in gestation. Embryonal or foetal death may be followed by retention and sometimes pyometra, which occasionally result in permanent infertility. Unlike bulls, infections are self-limiting in females, with clearence of parasites after approximately 95 days.

Dourine is a chronic venereal disease of horses caused by *Trypanosoma equiperdum*. The infection occurs in all species and breeds of Equidae, but not with the same intensity. Donkeys and mules are more tolerant to the infection than horses, while among the latter thoroughbreds and imported horses tend to be more susceptible to the disease than native horses. Unlike other species of trypanosome, *T. equiperdum* is transmitted by coitus. It may take several months before clinical signs appear, by which time the infection may have spread significantly over the horse population. *T. equiperdum* is a tissue parasite and infects primarily the mucosa of the genital organs, the skin and nervous tissues. The disease which is insidious in nature, affects both stallions and mares. It usually progresses through three distinct phases. A first phase marked by swelling of the external genitalia is followed by an urticarial phase visible on the skin. These two phases may last several months to several years, during which time the animal gradually becomes cachectic. A third, invariably fatal nervous phase characterized by incoordination and paralysis occurs in some horses. The genital phase is initiated by a mucoid vaginal or urethral discharge, nymphomania and a mild fever with oedema of the genitalia.

In horses, larvae of the nematodes *Strongylus vulgaris* and *S. edentatus* have occasionally been found migrating into the testes, leaving such lesions as inflamed, haemorrhagic migratory tracts. In warm climates, larvae of *Draschia megastoma*, *Habronema muscae*, and *H. majus* deposited near the prepuce by the intermediate host flies (housefly *Musca domestica* and the stable fly *Stomoxys calcitrans*) burrow into the dermis and cause extensive granulomatous inflammation. The infestation known as 'summer sores' is more common in geldings (accumulation of smegma) than in stallions.

Therapy

See treatment of the various species, → Chemotherapy, → Drugs.

Germ Theory of Disease

→ Mathematical Models of Vector-Borne Diseases.

Giardiasis, Animals

Pathology

Giardia species occur in dogs and cats, and giardiosis may represent an important problem in these hosts. It causes anorexia, depression, and a mild recurring diarrhoea consisting of soft, light-coloured stools with a characteristic "oatmeal" texture, frequently containing mucus. Growth retardation and cachexia may occur. The mechanism by which giardial malabsorption and diarrhoea occurs is unclear. Epithelial damage, increased turnover of epithelial cells, villous shortening and disaccharidase deficiency have all been reported as manifestations of giardiosis.

Giardia spp. are usually non-pathogenic inhabitants of the intestine of horses, cattle, sheep and goats. However, they may cause diseases under certain circumstances, such as in animals which are immunocompromised, malnourished, or very young.

Immune Responses

→ Giardiasis, Man/Immune Responses.

Related Entries

→ Alimentary System Diseases, → Giardiasis, Man.

Therapy

→ Antidiarrhoeal and Antitrichomoniasis Drugs.

Giardiasis, Man

Pathology

→ *Giardia lamblia* (Vol. 1) is a binuclear, pear-shaped flagellate which lives in the duodenum and upper small intestine, where it is closely applied or attached to the epithelium by means of a suction disk. Ultrastructural examination of a biopsy of the small intestine often shows the epithelial microvilli to be destroyed at the attachment sites of the flagellates. Acute infections are usually of short duration with diarrhoea and epigastric distress which subsides with the development of immunity. Chronic infections occur in patients with low or absent IgA, IgG, or IgM. In these patients the intestinal villi are often blunted from loss of epithelium which regenerates inadequately. The lamina propria is heavily infiltrated with lymphocytes and granulocytes. Diarrhoea, weight loss, and intestinal → malabsorption with flatulence accompany chronic infections. The trophozoites can be found in diarrheic stools, by duodenal aspiration or biopsy. The four-nucleate → cysts (Vol. 1) are found in stools.

Immune Responses

The establishment of animal models and the production of the whole life cycle of the parasite *in vitro* have greatly facilitated the characterization of stage specific antigens as well of analyzing the contribution of humoral and cell-mediated immune responses in the control of the infection. Most of the current knowledge comes from four different sources: (1) in vitro studies with axenically grown *G. lamblia* trophozoites and immune cells from different hosts; (2) studies with *G. lamblia*-infected mice or gerbils; (3) experiments with *G. muris*-infected mice; and (4) the analysis of immune responses in *Giardia*-infected humans.

Innate Immunity Since neutrophils are circulating blood cells, they are generally considered to play only a minimal role in intestinal infections. However, these cells are able to migrate like amoebae through small cracks in vessel linings to the exterior, where they eventually infiltrate tissues such as epithelial layers of the intestinal tract. It has been shown that certain products of neutrophils, cryptidins and cationic neutrophil peptides, possess anti-giardial activity. The content of granules together with antibodies reduces parasite infectivity and antibody-dependent cytotoxicity (→ ADCC) was demonstrated with human peripheral blood neutrophils in vitro. Since all these anti-giardial effects of neutrophils have been analyzed only *in vitro* so far, the functional role of these cells in vivo remains to be elucidated. Although in mice infected with *G. muris* there was only a small rise in mucosal mast cell numbers, inhibition of mast cell products by cyproheptadine enhanced the infection.

B cells and Antibodies An important function of antibodies in the control of giardiasis is suggested by more severe infections in patients with hypogammaglobulinemia. In B-cell deficient mice, unable to mount an anti Giardia Ig response, the infection with *G. lamblia* could not be resolved and antigenic diversification within the parasite population occurred in an unusually slow manner. Several studies have additionally demonstrated antibody-mediated killing of *Giardia* trophozoites in vitro, which is not in all cases complement dependent. The most important Ig for the control of giardiasis is IgA. The appearance of secretory IgA in the intestine correlates with the elimination of the parasite from the small intestine, but the effect of Ig subclass is complement independent since IgA lacks C1q binding sites in its Fc region required for the activation of the classical complement pathway. IgA may thus mediate its function in a complement-independent manner, e.g. by binding to trophozoite surface proteins thereby causing detachment and aggregation of the parasite. However, a functional role of complement should not be ruled out since (1) proteins of the complement cascade are synthesized by epithelial cells of the intestinum and (2) the alternative pathway of complement activation may be operative. Certain surface molecules, e.g. in the sucking disc of the parasite as well as metabolites released by trophozoites are able to activate the alternative pathway.

T cells There are few studies on the role of specific T cells in the defense against *Giardia*. The latent and acute phases of *G. lamblia* infection are accompanied by an increase of $CD8^+$ cells among IELs while in the elimination phase this population decreases and $CD4^+$ cells increase significantly. Since trophozoites are killed by oxidative burst mechanisms stimulated in macrophages via IFN-γ, it was tempting to speculate that T cells producing this cytokine might be involved in the control of parasite replication.

In addition, IFN-γ leads to enhanced phagocytosis of *Giardia* trophozoites by macrophages. Comparing the antibody and cytokine response in relatively resistant B10 mice and more susceptible BALB/c mice it was found that B10 mice produced IgG2a while BALB/c mice produced IgG1 after *G. muris* infection, suggesting differential involvement of Th1 and Th2 cells. When lymphocytes from mesenteric lymph nodes were stimulated in vitro, only those of B10 mice produced measurable amounts of IFN-γ. The application of neutralizing antibodies against IFN-γ to B10 mice resulted in an enhanced intensity of infection, arguing for a protective role of Th1 cells in this parasitic infection .

Immunopathology The killing of the parasite can lead to injury as shown for example in co-culture experiments with enterocytes and activated macrophages from the gut of mice infected with *G. lamblia*. It is not clear, however, if villus atrophy and crypt hyperplasia observed in response to infection with *Giardia* as well as the phenomena of maldigestion and malabsorption are direct consequences of the parasite load or caused by the host's immune response. Antigenic extracts of *G. lamblia* containing proteins of 32–200 kDa transiently suppressed the activity of dissaccharidases when gerbils were challenged with this fraction. It has been speculated that this short-lived effect could be mediated by lymphokines and/or mediators released by mast cell, neutrophils or macrophages.

Evasion Mechanisms Only recently has the possible involvement of antigen-variation in establishing a *Giardia* infection been noticed. The antigens involved belong to a family of variant-specific surface proteins (VSGs), which are unique, cysteine-rich zinc-finger proteins. After inoculation of a single *G. lamblia* clone expressing one → VSG (Vol. 1) into mice or humans, the original VSP is gradually replaced by many others beginning 2 weeks post infection. Selection by immune-mechanisms is suggested because (1) switching occurs at the same time that antibodies are first detected and (2) the antigenic switching does not occur in SCID mice.

The → antigenic variation (Vol. 1) of *Giardia* parasites may increase the chance of successful initial infection or reinfection.

Main clinical symptoms: Abdominal pain, diarrhoea, malabsorption
Incubation period: 3–21 days
Prepatent period: 3–4 weeks
Patent period: Years
Diagnosis: Microscopic determination of trophozoites and cysts in fecal samples
Prophylaxis: Avoid contact with human or animal feces
Therapy: Treatment see → Antidiarrhoeal and Antitrichomoniasis Drugs

Glomerulonephritis

→ Pathology.

GLURP

Glutamate rich antigen (→ Malaria/Vaccination)

Gnathostomiasis

Gnathostomiasis is an aberrant infection of man with larvae of → *Gnathostoma spinigerum* (Vol. 1) of felines and dogs. It is acquired from contact with meat of infected intermediate hosts (fish, amphibians, reptiles, birds). The larvae from the intermediate hosts enter human tissue and may migrate slowly through many tissues, giving rise to the intermittent subcutaneous swellings. The worm is surrounded by an inflammatory reaction with many eosinophils. The larvae are especially destructive when they die in the brain or eye.

Main clinical symptoms: Eosinophilic encephalomyelitis due to wandering larvae in brain, leucocytosis, blood eosinophilia
Incubation period: 3–7 days
Prepatent period: There is no reproduction in man
Patent period: Months
Diagnosis: Serodiagnostic methods
Prophylaxis: Avoid eating raw meat or undercooked fish and crabs
Therapy: Treatment see → Nematocidal Drugs, Man

Gomori's Silver Impregnation

→ Pneumocystosis.

Granuloma

→ Pathology.

Grocer's Itch

→ Mites (Vol. 1).

Grocott's Modification

→ Pneumocystosis.

H

Habronemiasis, Habronemosis

The larvae of *Habronema* spp. and *Draschia* (→ Filariidae (Vol. 1)) cannot penetrate normal skin. However, cutaneous invasion and lesions occur when larvae are deposited near open wounds by their vectors. Areas frequently involved include the withers, the lower limbs, the medial canthus of the eye, and the urethral process and prepuce of the male. The gross lesions rapidly become progressive and proliferative in nature, comprising ulcerated tumorous masses of red-brown granulation tissue. The sore is already painful before the development of granuloma. This causes an intense pruritus, and infected animals exacerbate the condition by biting and rubbing the lesion.

Therapy

→ Nematocidal Drugs, Animals.

Haemoglobinuria

Clinical symptom of infection with *Babesia* species (→ Babesiosis, Animals).

Haemonchosis

Haemonchosis is a common and severe disease of the ruminant abomasum in many parts of the world. *Haemonchus contortus* infects mainly sheep and goats, while *H. placei* occurs mainly in cattle. The pathogenesis of *Haemonchus* infection is the results of anaemia and hypoproteinaemia caused by the bloodsucking activity of the parasite. Large numbers of *Haemonchus* administered to sheep cause changes resembling those occurring in ostertagiosis, including rises in abomasal pH and increased plasma pepsinogen. However, the latter two effects do not contribute to the spontaneous disease. The following description is mostly based on studies on ovine haemonchosis. The hyperacute form occurs in animals exposed over a short period of time to thousands of parasites, and is rare. The animal bleeds to death within a week, loosing 200–600 ml of blood/day. In many cases there are no preliminary signs and death is sudden. In others there is an extreme anaemia and black faeces. Death occurs before compensatory erythropoiesis can take place, within 7 days. In the acute disease animals of all ages show anaemia, bottle jaw, and dark faeces. There are conflicting reports as to whether anorexia occurs and it has even been reported that sheep eat more than they normally do. Animals lose weight, are weak and lethargic and lose wool. Ewes suffer agalactia, so that their lambs may become emaciated and die from malnutrition. The anaemia of acute haemonchosis develops in three phases. The first phase which occurs during the pre-patent period is characterized by a fairly dramatic fall in packed cell volume, although serum iron at this stage is normal. This is considered to be the result of blood loss caused by immature worms, at a time when the haemopoietic system of the host is not fully mobilised to compensate it. In the second stage (from about 1–2 months) the packed cell volume does not decrease any further, because of the mobilisation of the haemopoietic system and the high serum iron concentrations. However, since the capacity of the sheep infected with *H. contortus* to reabsorb haemoglobin iron is limited, the iron reserves rapidly become depleted, which progressively leads to the third stage of the anaemia i.e. a low serum iron accompanied by a marked drop in packed cell volume. This indicates a dyshaemopoiesis due to iron deficiency and possibly to reduction in the availability of amino acids. Hypoalbuminaemia occurs. Chronic haemonchosis may last for 2 to 6 months. Only a few adults worms (100–1000) can cause a seepage of blood into the abomasum accounting at most for 50 ml/day. Anaemia is not present as compensatory erythropoiesis takes place, which depletes serum iron. The animal looks malnourished, with progressive loss of weight and wool-peeling in adult animals, and stunting of growth in lambs. The condition is aggravated by poor quality grazing as occurs in Africa and other tropical regions. A marked anaemia develops in terminal cases when iron and protein for erythropoiesis are depleted.

Therapy

→ Nematocidal Drugs, Animals.

Halzoun Syndrome

Disease in humans due to infection with the dog pentastomid worm → *Linguatula serrata* (Vol. 1), which block the nasal pathways and may thus introduce oedema and unfeelingness of head regions.

Main clinical symptoms: Oedema, disturbances of organs
Incubation period: 7 days to months
Prepatent period: 6–7 months
Patent period: 15 months
Diagnosis: Microscopic determination of worm eggs in nasal mucus.
Prophylaxis: Avoid contact with dogs in tropical regions
Therapy: Provocation of sneezing, mechanical withdrawal of the worms

Hammondiosis

→ Neosporosis.

Chemotherapy

→ Coccidiocidal Drugs (see → Toxoplasma gondii (Vol. 1))

Happening

→ Mathematical Models of Vector-Borne Diseases.

Health Education

→ Disease Control, Methods.

Heartwater

The significance of some rickettsial diseases is difficult to assess. This applies particularly to heartwater, a disease of domestic ruminants caused by → *Cowdria ruminantium* (Vol. 1), which is transmitted by all African species of *Amblyomma* found on susceptible hosts. Experimentally, it is also transmitted by the American *Amblyomma maculatum*, too, and the disease's recently discovered presence on Guadeloupe and other Caribbean islands (where *Amblyomma variegatum* has been introduced from Africa) poses a severe threat to the cattle and sheep industries in tropical and subtropical mainland America. Heartwater is an acute disease in susceptible animals such as introduced *Bos taurus* cattle but is less apparent in endemic situations which exist in most of subsahelian Africa. After an incubation period of 1–5 weeks following the bite of an infected tick, the first clinical signs are a rise in temperature to over 40 °C; in many cases, few other signs appear until shortly before death. Hypersensitivity and other nervous symptoms appear during the latter stages of the disease, culminating in central nervous effects. Many animals recover without showing any symptoms beyond fever. Definitive diagnosis is usually only possible at post mortem by demonstrating the causative organism in Giemsa-stained smears of brain tissues. Gross lesions commonly seen are pulmonary edema, hydropericardium (from which the disease has its name), hydrothorax, and ascites. The mode of transmission may be through regurgitation of the midgut contents by the tick, but this has not been demonstrated. A crude and fairly inefficient form of immunization is available.

Therapy

Treatment is possible through tetracyclines, but frequently not practically possible because of inadequate diagnostic resources.

Heartworm Disease

→ *Dirofilaria immitis* (Vol. 1), → Cardiovascular System Diseases, Animals.

Helminthic Infections, Pathologic Reactions

Lesions and inflammatory reactions accompanying helminth infections are particularly complex and variable. Immunity against these large organisms is generally less effective than against protozoans and part of a worm's life cycle may be spent in an immunologically privileged state, as with the adult schistosomes, or in a privileged site, as with the helminths living in the lumen of the intestine. Because of the more prolonged infections with schistosomes, and the release of eggs over a long period of time, the histologic reaction to a recently arrived egg can often be seen side by side with reactions to eggs that have been present for a long time as shown by granulomas destroyed eggs and fibrosing (→ Pathology/Fig. 1A-D). One of the hallmarks of the defense against helminths is the eosinophilic granulocyte which is toxic to many worms. This is accompanied by an acidic lysophospholipase of a molecular weight of 13,000 with a free sulfylhydril group which may crystallize as Charcot-Leyden crystals (→ Pathology/Fig. 3B). These crystals have been obtained not only from eosinophils but also from basophils. The role of the latter cells has been studied less frequently than that of the eosinophils. Because basophils and tissue mast cells require special staining for their demonstration, such as Alcian blue at pH 0.5, and are not shown on routine histologic sections, little is known of the role of basophils in the inflammatory and possibly defensive reaction against helminths. More information is available about their participation in the inflammatory reaction evoked by ticks.

Hem(oglobin) Interaction

Mode of Action

Fig. 1, Fig. 2

Structures

Fig. 3

Artemisinin and -derivatives

Important Compounds Dihydroartemisinin, Artemether, Arteether, Artesunate, Bicyclic trioxanes, Tetraoxanes, Tricyclic trioxanes, 11-alkyl,12-deoxy artemisinins, Arteflene.

Synonyms Artemisinin: Qinghaosu.

Clinical Relevance Artemisinin has been used in the meantime for the treatment of malaria in at least 3 million people. The advantage of this drug is its rapid action against cerebral malaria (→ Malaria).

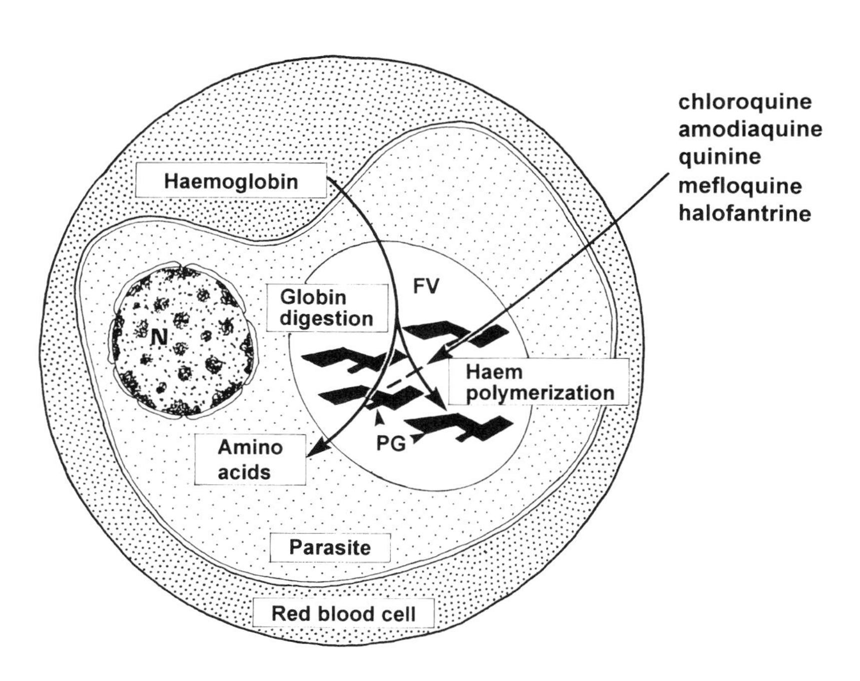

Fig. 1 Model of the formation of the non-covalent heme-chloroquine complex (Gutteridge (1993) In : Cox FEG (ed) Modern Parasitology, Second Edition, Blackwell Science, pp. 219–242)

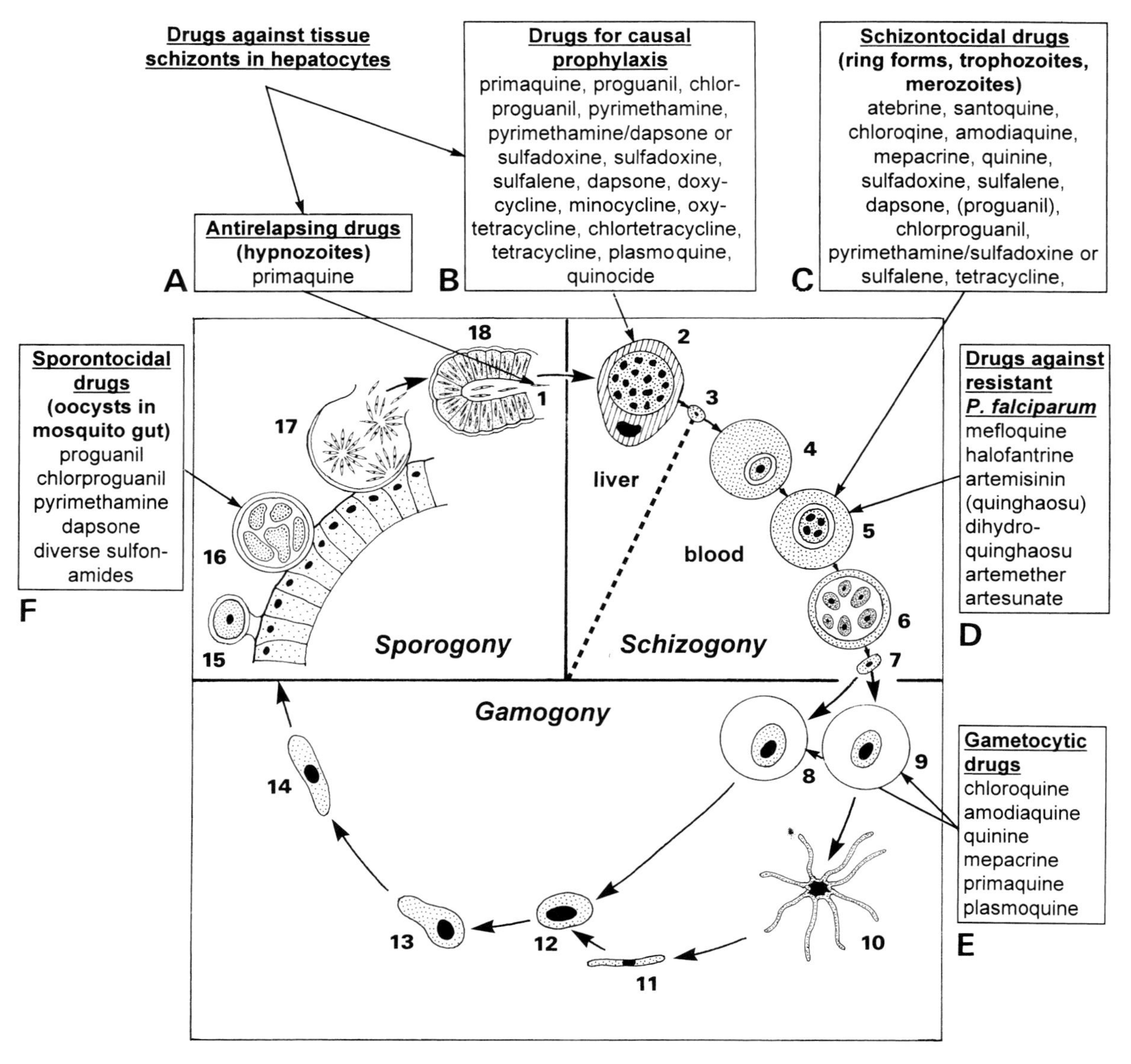

Fig. 2 Action of antimalarial drugs on life cycle stages.

Artemisinin is isolated from the plant *Artemisia annua* and was originally developed in China. It is chemically related to 1,2,4-trioxanes. Among artemisinin and the so-called first generation compounds are esters or ethers obtained from the lactol, dihydroartemisinin.

Molecular Interactions

Artemisinin is characterized by its new structure and new mode of action unrelated to any other known antimalarial drug. Indeed, there is no cross-resistance against any of the known antimalarials. The activity of artemisinin and its derivatives is directed only against erythrocytic schizonts (→ Inhibitory-Neurotransmission-Affecting Drugs/Fig. 2). High drug concentrations are detectable in the region of membranes of intraerythrocytic trophozoites. There are also other measurable artemisinin-induced metabolic alterations: an interference with the energy production, a reduction of DNA synthesis, an inhibition of mRNA polymerase activity and an inhibition of purine synthesis in *Plasmodium berghei* at the level of inosine monophosphate dehydrogenase.

The mode of action procedes in two different steps. In the first step the endoperoxide bridge is

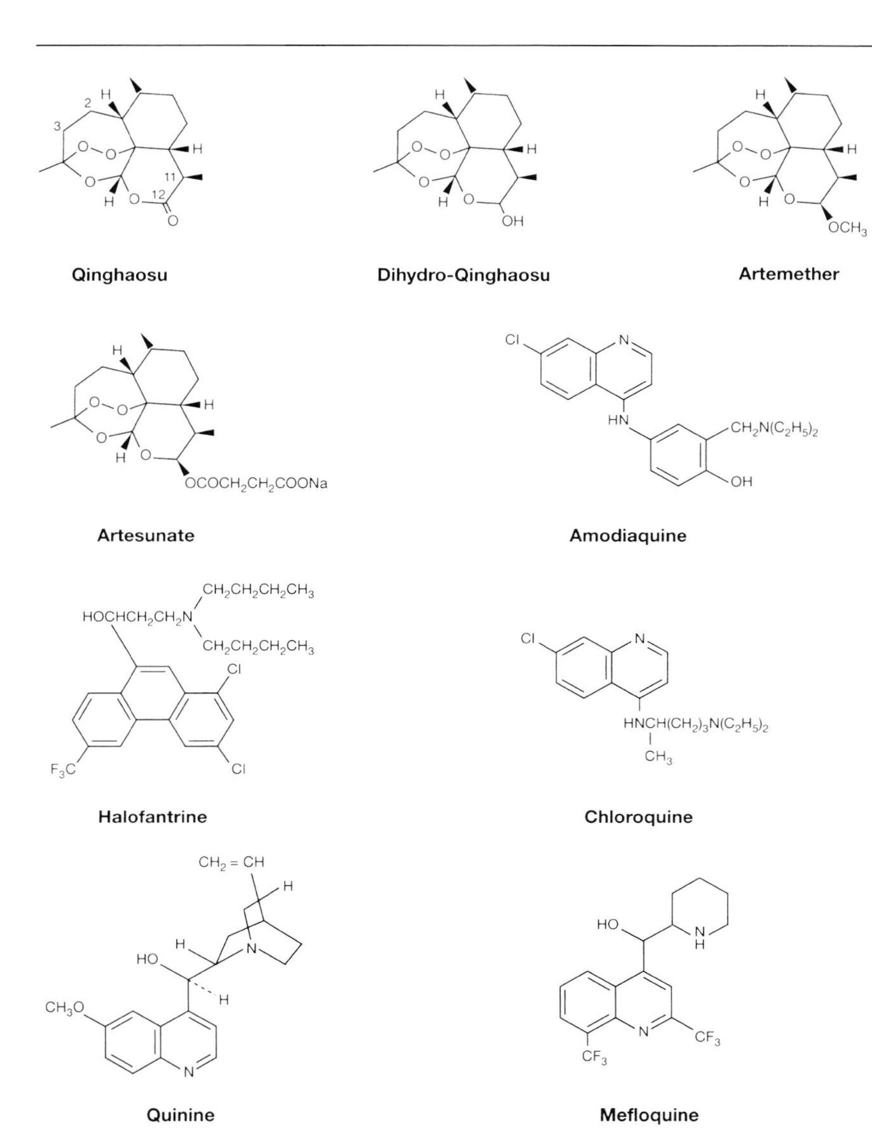

Fig. 3. Structures of antiparasitic drugs affecting hem degradation.

cleaved (Fig. 4). This reaction is catalyzed by intraparasitic iron and heme, and leads to the generation of unstable free radical intermediates. The selective toxicity of artemisinin against malaria parasites is presumably due to this iron catalyzed generation of free radicals, and this is favoured by the fact that just these intraerythrocytic parasitic stages are rich in iron and heme. In the second step of the reaction, specific malaria proteins with molecular masses of 25, 32, 42, 50, 65 and >200 kDa become alkylated. These alkylation reactions take place in parasitic ring forms and trophozoites. Proteins of uninfected red blood cells or of infected red blood cells pretreated with the inactive derivative desoxyarteether are not alkylated. Structure activity relationships of various tricyclic trioxanes reveal that certain rings in artemisinin are redundant. Thus, the high artemisinin-like activity is due to the structurally minimal bicyclic trioxane. A rapid rearrangement of active bicyclic trioxanes and spirocylane ring is induced by ferrous chloride.

Resistance So far there are no reports about resistance against artemisinin or its derivatives. However, the clinical trials of artemether and artemether-benflumetol have led to the observation that artemisinin is able to induce drug metabolizing enzyme and, thus, may contribute to its own clearance.

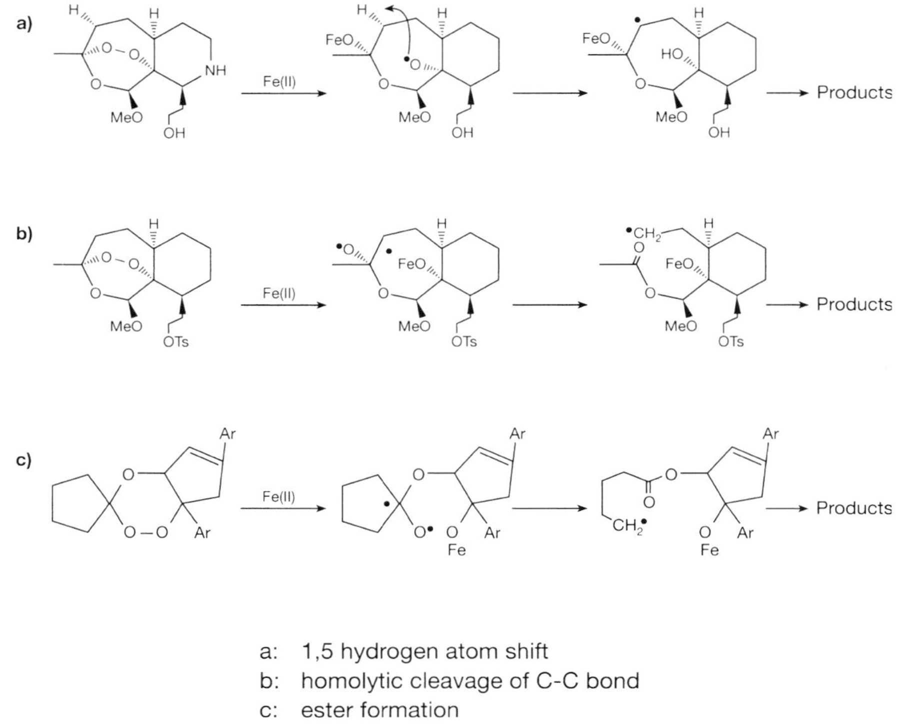

Fig. 4 Cleavage of the Endoperoxide Bridge of Artemisinin by Hem-Iron (Meshnick SR, Jefford CW, Posner GH, Avery MA, Peters W (1996) Parasitol Today 12 : 79–82).

a: 1,5 hydrogen atom shift
b: homolytic cleavage of C-C bond
c: ester formation

Amodiaquine

Synonyms 4-((7-chloro-4-quinolinyl)amino)-a-(diethylamino)-o-cresol, SN10751.

Clinical Relevance Amodiaquine was developed between 1941 and 1945 as an antimalarial drug. Its antimalarial activity is comparable to that of chloroquine.

Molecular Interactions Amodiaquine is a member of the 4-aminoquinolines with an activity, which is directed against erythrocytic schizonts and gametocyts of plasmodia (Fig. 2). Amodiaquine serves as a prodrug, which is converted to desethylamodiaquine responsible for the antimalarial activity. The mechanism of action is presumably identical to that of chloroquine (Fig. 1).

Resistance The mechanism of resistance against amodiaquine is also presumably identical to that against chloroquine which is assumed to be due to an impaired uptake mechanism. This is supported by the appearance of a general cross-resistance between amodiaquine and chloroquine. There are also some few hints of amdodiaquine activity against chloroquine resistant *Plasmodium* strains.

Halofantrine

Clinical Relevance Halofantrine was developed within the Walter Reed Army Institute for Research (WRAIR) antimalarial drug development programme in 1984. It had been in clinical trial since 1974. However, because of severe cardiovascular side effects, halofantrine is no longer used. The antimalarial activity against *Plasmodium falciparum* and *P. vivax* is much better documented than that against *P. malariae* and *P. ovale*.

Molecular Interactions Halofantrine is a phenantrene methanol, and its activity is directed against erythrocytic schizonts (Fig. 2). The action of halofantrine on the molecular level is unknown. It is presumably different from that of quinine and mefloquine, but it may be otherwise similar to that of mefloquine (Fig. 1).

Resistance In Thailand there are mefloquine-resistant strains of *Plasmodium falciparum* which show simultaneously reduced sensitivity to halofantrine, thus indicating cross-resistance between mefloquine and halofantrine. On the molecular level, the resistance mechanism against halofantrine is suggested to be of the MDR phenotype similar to that against mefloquine and quinine, while it may be different from that against chloroquine.

Chloroquine

Synonyms 7-chloro-4-(4-diethylamino1-methylbutylamino)quinoline, SN7618, RP3377, Aralen, Nivaquine B, Sanoquin, Artrichin, Bipiquin, Reumachlor, Bemaphate, Resoquin, Resochin, Chlorochin.

Clinical Relevance Chloroquine has been used in the treatment of malaria for some 60 years since its discovery in the 1930's. It exerts a wide variety of activities against different parasites.

The activity of chloroquine against *Entamoeba histolytica* is directed only against liver abscesses. It has no effect against intestinal stages. Chloroquine especially is used in combination with emetine to improve the curative effects of the latter. Other combinations with significant amoebicidal activity are chloroquine/dehydroemetine or chloroquine/diloxanide furamide.

As antimalarial drug chloroquine is mostly used as a diphosphate salt. It is effective against all four human malaria parasites. Thus, it has high activity against blood schizonts (asexual intraerythrocytic stages of *Plasmodium vivax* and *P. falciparum*) and against gamonts of *P. vivax* (Fig. 2). It has no significant effects against extraerythrocytic stages in the liver. Therefore, chloroquine cannot be used as a causal-prophylactic drug.

Molecular Interactions The amoebicidal action of chloroquine is directed against trophozoites (magna forms) in the liver and other extraintestinal organs. The precise mechanism of action in amoebiasis is unknown. Presumably it is different from the antimalarial action by intercalation into DNA or by inhibition of protein-synthesis.

There are different hypotheses for the mode of action of chloroquine. The uptake of host-derived hemoglobin and its digestion in the food vacuole by developing parasites plays a vital role in survival of these parasites, and this is therefore the Achilles heel, which makes the plasmodia vulnerable to quinoline compounds. During the degradation of globin ferrous heme is released and oxidized to a ferric form, which is toxic for the parasites by damaging parasitic membranes and inhibiting various parasitic enzymes including proteases. To circumvent these toxic effects of ferrous heme plasmodia have evolved specific detoxification mechanisms. These rely on the conversion of the ferric hem into an insoluble, unreactive crystalline material called hemozoin (= malaria pigment). Hemozoin is a polymer of the iron porphyrin ferriprotoporphyrin IX (hemin) residues linked by iron-carboxylate bonding β-haematin. It is at present controversial, whether the formation of this polymer is catalysed by a heme polymerase or proceeds spontaneously without protein. Despite these differences there is no doubt that just this important polymerisation step is susceptible to inhibition by quinoline -containing antimalarial drugs (chloroquine, quinine, amodiaquine) (Fig. 1). This is supported by a strong correlation among a series of quinoline compounds between inhibition of parasite growth and inhibition of the heme polymerization. In addition, the subcellular localization of the inhibition of heme polymerization correlates with the site of drug accumulation. Thereby, the heme substrate is probably converted into a non-covalent complex with the quinoline. This hemin-chloroquine-complex is assumed to be toxic to the parasite by lysing membranes or by inhibiting further aspartic protease-mediated hemoglobin degradation.

In the alternative model of chloroquine action by Warhurst 1995, in which the presence of a heme polymerase is not necessary for β-hematin formation, the formation of β-hematin can be inhibited by chloroquine, amodiaquine and quinine just as they are reported to inhibit heme polymerase. The action of blood schizontocides is thus, according to this model, simply by binding to hemin monomers and preventing their polymerization and detoxification by sequestration with apohemozoin.

For all the known models of the chloroquine action it is proposed that the drug is taken up by the parasites by passive diffusion. However, recently a chloroquine-transporter in the plasma membrane of *P. falciparum* could be characterized. This transport protein is responsible for the accumulation of drug inside the parasite. The uptake mechanism is temperature-dependent, saturable, can be inhibited and follows the Michaelis-Menten kinetics in contrast to the chloroquine uptake into non-infected erythrocytes, which is by simple diffusion. It is proposed that the chloroquine-transporter works as a Na^{+}/H^{+}-exchange protein and is able to regulate the cytoplasmic pH. In contrast to other eukaryonts *P. falciparum* appears not to possess other pH-regulatory systems beside this exchange protein. This is obviously an adaptation to the intracellular parasitism and may explain why this system is an Achilles heel by which the plasmodia become vulnerable to the action of quinolones. Corresponding exchange proteins from human cells do not bind or transport chloroquine The energy for this process is gained via the Na^+-gradient which is built by the Na^+/K^+-ATPase. Chloroquine is able to stimulate the exchange protein thus leading to an enhanced exchange of protons and Na^+-ions. As a result, pH and Na^+-ions are enhanced intracellularly and chloroquine is taken up and accumulated during the initial activation phase. The chloroquine uptake proceeds as long as there is a Na^+-gradient across the plasma membrane. The molecular mechanism of the activatory effect of chloroquine is unclear at present.

Besides the amoebicidal and antimalarial activity, chloroquine has also antibabesial and anticestodal activities. The mechanism of action of chloroquine in these indications is not yet known. In other non-parasitic indications chloroquine is used in rheumatoid arthritis, chronic polyarthritis and Lupus erythematodes. Here the membrane-stabilizing effects of chloroquine may contribute to the antiinflammatory effects. Thereby, the actions of lysosomal proteases become inhibited by this drug.

Resistance Chloroquine resistance in *Plasmodium falciparum* was first observed in the late 1950's. The greatest hindrance for the elucidation of the resistance mechanism was lack of knowledge of the mode of action of the quinoline-based antimalarials on *Plasmodium* spp., which is still discussed controversially. In chloroquine-resistant *P. falciparum* the release of chloroquine is 40–50 times more rapid compared to wild type strains. The efflux rates of chloroquine are different between sensitive and resistant strains. The accelerated chloroquine efflux in resistant strains can be inhibited by calcium-channel blocking agents such as verapamil. Thus, the chloroquine-resistance mechanism resembles that of multidrug-resistant cancer cells. Two genes in *P. falciparum*, pfmdr1 and pfmdr2, could be identified, which are homologous to MDR genes. The gene pfmdr1 encodes a 162 kDa homologue of the P-glycoprotein (Pgh1), located mainly at the membrane of the digestive vacuole of the parasite. Indeed, there is evidence for the involvement of Pgh1 in nucleotide-dependent transport across membranes. However, a correlation between amplification of pfmdr1 and chloroquine resistance remains doubtful.

In an alternative hypothesis, chloroquine resistance may be linked to an impaired chloroquine uptake mechanism. Indeed, it could be shown recently that resistant *P. falciparum* strains possess an altered Na^+/H^+-exchange protein which are continuously in an activated state, so that chloroquine cannot activate it furthermore. As a result, chloroquine cannot be taken up and accumulated by these cells. The molecular biological background for the constitutive activation of the exchange protein in resistant cells is, however, unclear to date.

Quinine

Synonyms 6-methoxy-a-(5-vinyl-2-quinuclidinyl)-4-quinolinemethanol, Aristochin, Aristoquin, Aristoquinine, Biquinate, Coco-Quinine, Dentojel, Diquine carbonate, Quinamin, Quinamm, Quinate, Quinbisan, Quine, Quinoform, Quinsan, Quiphile, Tasteless Quinine.

Clinical Relevance Quinine is the main alkaloid of the Cinchona bark in Peru known as effective febrifuge against intermittent fever since the early 17th century. It has some activity against *Trypanosoma brucei*, but the main activity relies on the antimalarial activity. It is used especially for the therapy of chloroquine- and multidrug-resistant *Plasmodium falciparum* infections. The combination with tetracyclin is used in cases of severe resistance. Furthermore, quinine is used in combination with clindamycin against babesiosis. Another, non-parasitic indication is the usage as antiarrhythmic drug.

Molecular Interactions The activity of quinine is directed against erythrocytic schizonts and gametocyts of all human *Plasmodium* spp., but gamonts of *P. malariae*, *P. ovale* and *P. vivax* are also damaged (Fig. 2). The action of quinine on the molecular level is suggested to be identical to that of chloroquine (Fig. 1).

Resistance Quinine has been used in the meantime for more than 350 years without loosing its general effectiveness. Formerly quinine was, however, probably never used at a high enough frequency against *P. falciparum* to induce resistance. The existence of cases of quinine-resistance was reported for the first time in 1910 by Nocht and Werner in Brasil. The molecular basis of the mechanism of quinine resistance is assumed to be identical to that of mefloquine, but different from that of chloroquine.

Mefloquine

Synonyms Larian, Laricur.

Clinical Relevance Mefloquine was synthesized in 1971 and marketed at the end of the 1980's. The activity is directed against chloroquine-resistant and mostly against multidrug-resistant *Plasmodium falciparum*.

Mefloquine acts against the erythrocytic schizonts of all four *Plasmodium* spp. (Fig. 2). It has additional activity against young gamonts of *P. malariae*, *P. ovale* and *P. vivax*.

Molecular Interactions The mechanism of action of mefloquine may be similar to that of chloroquine or quinine, since mefloquine and chloroquine are believed to share the same drug receptor within the food vacuole of infected erythrocyte (Fig. 1). However, in contrast to chloroquine which has been shown to inhibit the activity of the hem polymerisation by more than 80%, mefloquine does not cause such an inhibitory effect. There is another hypothesis that the antimalarial activity of mefloquine may be due to an interaction with parasite proteins by hydrogen-bond formation.

Resistance There is occasional resistance against mefloquine. The first case of mefloquine-resistance was reported in Thailand in 1982. In the meantime cure rates from eastern Thailand are reported to have dropped to only 41%. The molecular basis of resistance against mefloquine is presumably different from chloroquine resistance since mefloquine still exerts activity against chloroquine-resistant *Plasmodium falciparum*. Enhanced drug efflux as seen in chloroquine-resistant *P. falciparum* has not yet been shown with the class of the structurally related compounds mefloquine, halofantrine and quinine. The neuroleptic agent and calcium-channel-blocker penfluridol increases the susceptibility of mefloquine-resistant strains to mefloquine, and interestingly, there is no increase of mefloquine-activity in mefloquine-resistant strains by verapamil. By contrast penfluridol cannot increase the susceptibility against chloroquine in chloroquine-resistant strains of *P. falciparum*. Mefloquine resistance exhibits patterns more indicative of the MDR phenotype and is often associated with halofantrine and quinine resistance. The EC_{50} values for mefloquine, halofantrine and artemisinin are decreased by penfluridol, but not by agents modulating chloroquine resistance. There are several reports on an association between mefloquine resistance and amplification of pfmdr1. Thus, a mefloquine-resistant clone selected by drug pressure exhibited a two- to four-fold increase in the copy number of pfmdr1 and a significantly higher level of corresponding mRNA compared to the sensitive clone. In ten mefloquine-resistant strains with concurrently reduced susceptibility to halofantrine, pfmdr1 is present in multiple copies, whereas this amplification is not found in the only sensitive strain. A selection for mefloquine-resistance in two clones generated amplification and overexpression of pfmdr1 and overexpression of its product, Pgh1. The overexpression of Pgh1 correlates with increased resistance to halofantrine and quinine and decreased resistance to chloroquine. Thus, there seems to be an inverse correlation concerning Pgh1 expression in chloroquine resistance and a MDR phenotype involving mefloquine, halofantrine and quinine.

Heterophyiasis, Animals

→ *Heterophyes heterophyes* (Vol. 1), → Alimentary System Diseases.

Heterophyiasis, Man

Disease due to infections with → *Heterophyes* (Vol. 1) species via oral uptake of uncooked infected fish.

Main clinical symptoms: Diarrhoea, abdominal pain
Incubation period: 1–3 weeks
Prepatent period: 1–2 weeks
Patent period: 2–6 months
Diagnosis: Microscopic determination of eggs in fecal samples
Prophylaxis: Avoid eating raw fish
Therapy: Treatment with praziquantel, see → Trematodocidal Drugs

Hippoboscidosis

Disease due to infestation with hippoboscids, see Table 1

Histiocytic Reaction

→ Pathology.

Histologic Reactions

→ Pathology.

Hookworm Disease

Pathology

Hookworm disease is produced by *Ancylostoma duodenale, Necator americanus* and several other species (→ Hookworms (Vol. 1)). The larvae live in moist soil and enter the skin, giving rise to allergic dermatitis with a papular and sometimes vesicular focal rash that is intensely pruritic and is sometimes referred to as ground itch. The larvae enter the blood stream, get to the lungs where they give may rise to focal hemorrhages and to allergic pneumonia. The larvae migrate up the bronchial tree to the pharynx and are swallowed. They reach adulthood in the upper small intestine where they attach to the mucosa with their powerful buccal capsule enclosing a tag of mucosa from which they draw blood. This may be accompanied by abdominal pain after 35–40 days. From time to time the worm moves to another site, sometimes leaving the microscopic ulcer to bleed until the vessels thrombose and the site heals. With heavy infection anemia develops. Because the worms take in more blood than they assimilate; black stools, or melena, are common, but some of the iron lost is resorbed in the intestine.

Immune Responses

Data indicate that a Th2-dominated immune response acts to reduce the weight and fecundity of hookworms like *Ancylostoma duodenale* and *Necator americanus* in the human gut. Parasite specific IgE and eosinophilia may be involved in host protection. In elder patients there is significant negative correlation between anti-larval IgG anti-

Table 1. Hippoboscids and Control Measurements

Parasite	Host	Vector for	Symptoms	Country	Therapy		
					Products	Application	Compounds
Melophagus ovinus (Sheep ked, sheep "tick")	Sheep	*Trypanosoma melophagium* (nonpathogenic)	Blood loss, itching, wool loss	World-wide	Butox (Intervet)	Pouron	Delta-methrin
Lipoptena capreoli	Goat						
Lipoptena cervi	Deer, Cattle, Goat		Ear, irritation through movement and blood sucking, itching, wool loss or alopecia through rubbing	Europe			
Hippobosca variegata	Cattle		Tail, irritation through movement and blood sucking, itching, wool loss or alopecia through rubbing	Europe	Ivomec 1% injection	subcutan	Ivermectin
Hippobosca equina (Horse louse-fly)	Horse, Cattle		Anal-/pubic area, irritation through movement and blood sucking, itching, wool loss or alopecia through rubbing	Europe			

bodies and parasite burden. The fact that human hookworms nevertheless survive in an immunological hostile host for many years may be best explained by efficient escape mechanisms preventing the immune system from exerting its full effects. Several molecules secreted by hookworms have a potential to inhibit or antagonize the immune system of the host. In *Ancylostoma* species, the neutrophil inhibitory factor (NIF), a glycoprotein of 41 kDa potently inhibits CD11b/CD18 dependent neutrophil activation and adherence to vascular endothelium. Since *Necator americanus* lacks NIF it may rely on acetylcholinesterase, glutathion-S-transferase and superoxide dismutase to suppress inflammation. Acetylcholinesterase not only inhibits gut peristalsis but may also suppress the release of inflammatory cytokines by lymphocytes stimulated via their muscarinic receptors. Secreted glutathion-S-transferase and superoxide dismutase may represent protective mechanisms of the hookworms against reactive oxygen intermediates produced by activated leukocytes in the gut. An IgA protease present in *Necator americanus* has the potential to produce IgA Fab fragments capable of blocking complement or phagocytic attack mediated by intact IgG or IgM. Since secretory IgA provides a potent signal for eosinophil degranulation, the secretion of an IgA protease would be beneficial to the parasite also on this level.

Targets for Intervention

Both *Ancylostoma duodenale* and *Necator americanus* require specific environmental conditions for their larvae to hatch and become ensheathed, rhabditiform, infective organisms that will actively enter a new host through the skin, usually that of the feet or lower part of the legs. Most of the infections are contracted in the peridomestic environment. Fig. 1 indicates the infected host, the contaminated environment, and the infection cycle as targets of intervention. Suitable approaches to control include the detection and treatment of infected persons, the safe disposal of feces, rigorous environmental sanitation, and wearing shoes as a means of avoiding skin contact with contaminated soil.

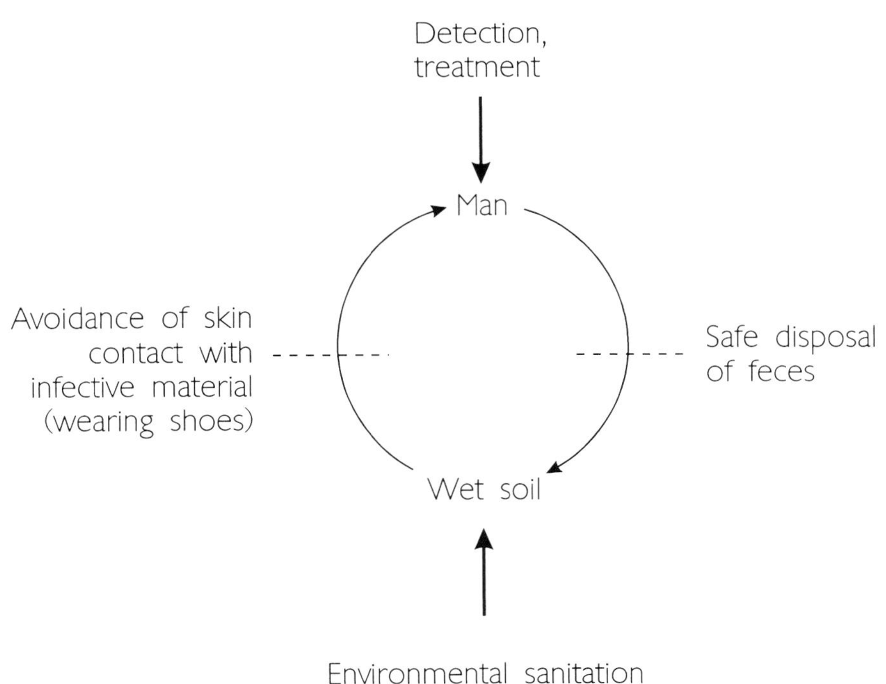

Fig 1. Targets and approaches for the control of hookworm disease

Main clinical symptoms: Pruritus, bronchitis, eosinophilia, red-diarrhoea, fever, anaemia, cachexia, breakdown of circulation
Incubation period: Dermatitis: 4 hours, intestinal symptoms: 2 weeks
Prepatent period: 5–6 weeks
Patent period: 20 years
Diagnosis: Microscopic determination of eggs in fecal samples
Prophylaxis: Use solid shoes in endemic regions and avoid human feces
Therapy: Treatment see Nematocidal Drugs, Man

Horn Fly

Haematobia irritans (→ Diptera (Vol. 1))

Host Response

→ Pathology.

House Fly

Musca domestica (→ Diptera (Vol. 1))

Hydatid Disease

Synonyms

Hydatidosis, cystic → echinococcosis

Hydatidosis

→ Echinococcus (Vol. 1), → Echinococciasis

Hydrocephalus

→ Toxoplasmosis, Man/Pathology.

Hygiene

→ Disease Control, Methods.

Hymenolepiasis

General Information

The cestode *Hymenolepis nana* (syn. → *Rodentolepis nana* (Vol. 1)) is primarily a parasite of rodents, but is also frequently found in man where it is autoreproductive, but in general not very pathogenic. Fig. 1 (page 249) shows the presence of man-man, rodent-rodent, man-rodent and rodent-man cycles.

Targets for Intervention

Possible targets are the rodent reservoir, the human cases, transmission both from and to man, and the environment where infective material may be encountered (Fig. 1). It is evident that one approach on its own will probably do very little to improve the situation.

Main clinical symptoms: Malnutrition, diarrhoea, abdominal pain
Incubation period: 1–4 weeks
Prepatent period: 2 months
Patent period: 2–4 weeks
Diagnosis: Microscopic diagnosis of eggs in fecal samples.
Prophylaxis: Avoid contact with rodent and human feces
Therapy: Treatment with praziquantel, → Cestodocidal Drugs

Hyperplasia

→ Pathology.

Hypersensitivity

→ Pathology.

Hypoalbuminaemia

Clinical symptom in animals due to parasitic infections (→ Alimentary System Diseases, → Clinical Pathology, Animals).

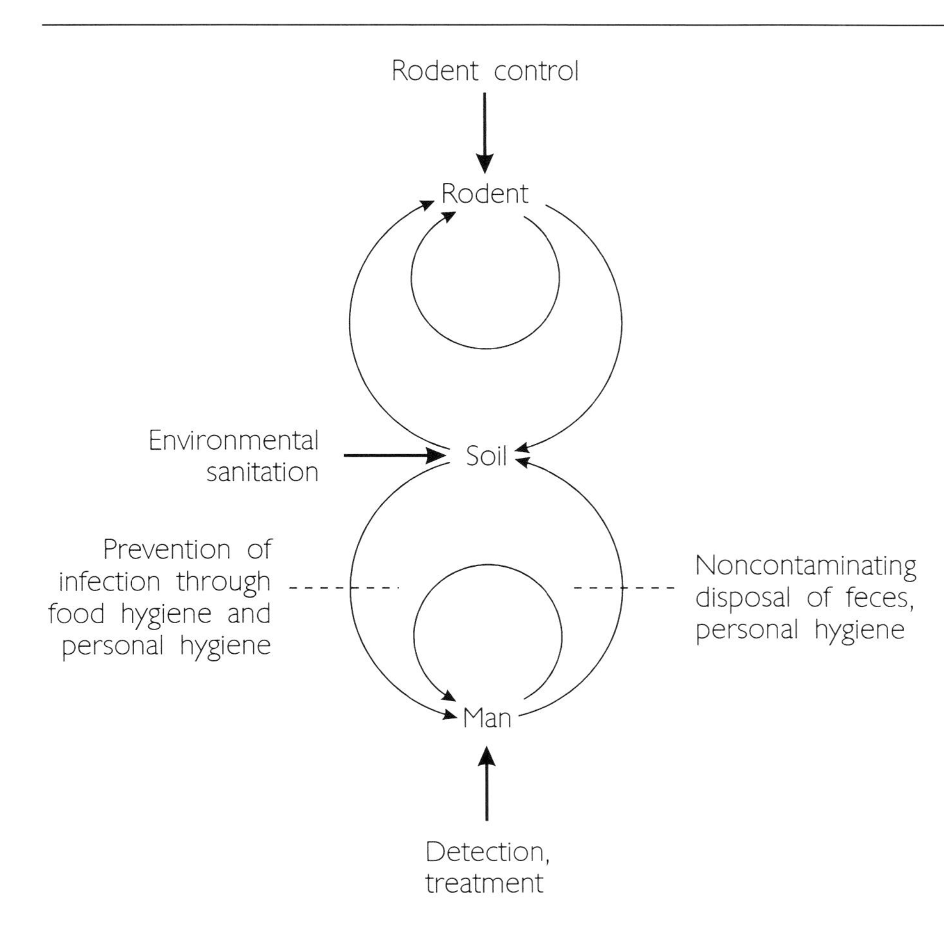

Fig. 1. Targets and approaches for the control of hymenolepiasis

Hypodermosis

Disease du to infestation with grub flies, see Table 1, pages 250 and 251.

Table 1. Grub and Warble flies and Control Measurements

Parasite	Host	Symptoms	Country	Therapy		
				Products	Application	Compounds
Hypoderma bovis (Northern cattle grub)	Cattle, (Horse)	Cattle have a panic fear of flying *Hypoderma* and try to escape; inflammation of skin with strong exsudation, then encapsulation, fistula; high economic loss through skin damages, loss of fattening performance (meat loss), milk loss	Europe, Former Soviet Union, North America, Africa, Asia	Neguvon™ (Bayer): No treatment during migration	Wash or Spray	Trichlorfon/ Metrifonate
				Warbex Famphur Pour-on for Cattle™ (Mallinckrodt)	Pour-on	Famphur
				Co-Ral™ 25% Wettable Powder (Bayer Corp.)	Dip or Spray	Coumaphos
				Ivomec™ 1% Injection for Cattle (Merial)	Injection	Ivermectin
				Dectomax™ (Pfizer Animal Health)	Injection	Doramectin
				Cydectin™ (Bayer)	Injection	Moxidectin
Hypoderma lineatum (Common cattle grub)	Cattle, (Horse)	Creep on the cattle, no fear	In areas with an early spring, early drive to pasture	Co-Ral™ 25% Wettable Powder (Bayer Corp.)	Dip or Spray	Coumaphos
				Warbex Famphur Pour-on for Cattle™ (Mallinckrodt)	Pour-on	Famphur
				Ivomec™ 1% Injection for Cattle (Merial)	Injection	Ivermectin
				Dectomax™ (Pfizer)	Injection	Doramectin
				Cydectin™ (Bayer)	Injection	Moxidectin
Przhevalskiana silenus	Goat, rare Sheep	Inflammation of skin with strong exsudation, then encapsulation, fistula; high economic loss through skin damages, loss of fattening performance (meat loss), milk loss	Mediterranean area			

Table 1. continued

Parasite	Host	Symptoms	Country	Therapy		
				Products	Application	Compounds
Dermatobia hominis (Tropical warble fly or torsalo)	Cattle, Dog, Humans, etc.	Inflammed skin pustules	Latin America	Co-Ral™ 25% Wettable Powder (Bayer Corp.)	Dip or Spray	Coumaphos
				Warbex Famphur Pour-on for Cattle™ (Mallinckrodt)	Pour-on	Famphur
				Ivomec™ 1% Injection for Cattle (Merial)	Injection	Ivermectin
				Dectomax™(Pfizer)	Injection	Doramectin
				Topline™ (Merial)	Pour on	Fipronil

I

Ichthyophthiriasis

→ *Ichthyophthirius multifiliis* (Vol. 1).

Immune Complex

Synonym
Antibody-Antigen Interaction

General Information
The formation of immune complexes with subsequent clearance is a physiological process during the acute phase of an immune reaction. When antibody-antigen aggregates develop in rising quantities they become detectable in the peripheral blood as so called circulating immune complexes (→ CIC (Vol. 1)). The formation of CIC is a common event during chronic parasitic infection like schistosomiasis, filariasis, leishmaniasis, trypanosomiasis, toxoplasmosis and quartan malaria. Only the deposition of immune complexes may cause severe pathological changes in the host.

During parasitic infection various circulating antigens appear at different developmental stages. Due to a selective immune response only few of them are complexed by antibodies. The antibody-antigen interaction may lead to structural changes in the antigen. The clearance of immune complexes depends on the structure of the antigen and the quality (affinity) and quantity of the antibody.

Immune complexes are able to bind complement compounds and to activate the cellular immune response via Fc-receptors. Resulting pathological changes are either membrane proliferative glomerulonephritis or, when the antigen is tissue bound, an → Arthur's Phenomenon.

CICs are a common phenomenon in the sera of *Wuchereria bancrofti*-infected patients. Renal abnormalities like chronic progressive glomerulonephritis are reported for patients with malaria (*Plasmodium malariae*) and bancroftian filariasis. Also, a chronic *Leishmania infantum*-infection in dogs may cause glomerulonephritis.

Immune Evasion

Mechanisms developed by parasites to survive the attack of the host immune system (→ Immune Responses).

Immune Response

See also related entries: → Amoebiasis, → Babesiosis, Animals, → Babesiosis, Man, → Chagas' Disease, Man, → Cryptosporidiosis, Animals, → Cryptosporidiosis, Man, → Cysticercosis, → Echinococcosis, → Filariasis, Lymphatic, Tropical, → Giardiasis, Animals, → Giardiasis, Man, → Hookworm Disease, → Leishmaniasis, Animals, → Leishmaniasis, Man, → Malaria, → Nematode Infections, Man, → Pneumocystosis, → Schistosomiasis, Man, → Sleeping Sickness, → Theileriosis, → Toxoplasmosis, Animals, → Toxoplasmosis, Man, → Trichinelliasis, Man, → Visceral larva migrans, Man.

Immune Responses

A general feature of many parasitic infections by protozoa or helminths is their chronicity and several reasons contribute to this, e.g. weak innate immunity and the capacity of parasites to withstand or to evade destruction by specific immune responses of the vertebrate host. General aspects of the various host immune responses are described in the following. Peculiarities of the immunological response to specific parasites are described in detail under the headwords of the respective diseases.

Innate Immunity

The skin and the linings of the respiratory, gastrointestinal and urogenital tract present formidable physical and chemical barriers to infective organisms and represent a first line of defence. These barriers provide a natural resistance, also called innate immunity, to infection, but they are not perfect. Protozoan and helminthic parasites, have evolved in such a way that they either are able to penetrate the body's barriers directly or are transmitted by insect bites.

In the body the alternative pathway of complement activation provides a first line of defense against many parasites. The complement component C3 is cleaved spontaneously in plasma to produce C3b; once bound to the parasite surface and stabilized to form a C3 convertase, activation of the terminal complement components C5-C9 takes place and the parasites are lysed by the major attack complex MAC. A second line of defense is provided by macrophages and neutrophilic leukocytes, which play a major role in all stages of host defense. These cells by means of their CR3-receptors are able to recognize microbial substances and thus ingest the parasites. As a result of the parasite uptake, the secretion of cytokines by phagocytes is initiated which include interleukin-1 (IL-1), IL-6, IL-8, IL-12 and tumor necrosis factor (TNF). These factors recruit more phagocytes to the site of the infection and an increase in circulating neutrophils. Phagocytes also release other proteins with significant local effects, such as oxygen radicals, peroxides, nitric oxide, prostaglandines, leukotriens, complement components etc. Infection of cells with viruses, but also with parasites induced the production of interferon (IFN)-α and β. These interferons contribute to the inhibition of natural killer cells (NK-cells), cells which are known to function in the initial phase of infection with intracellular pathogens, including parasites such as leishmania. Activated NK-cells secrete large amounts of IFN-γ. This IFN-γ is critical for the control of some parasitic infection before T cells have been triggered to liberate this cytokine. A further effect of IFN-α and -β is to augment the expression of MHC class I molecules, which favors the ability of host cells to present parasite antigenic peptides to $CD8^+$ cells (see below). Once parasites have survived the innate immune response, the acquired immunity comes into effect.

Acquired Immunity

Acquired immunity is mediated by the humoral and cellular immune system, in which the B-lymphocytes are mediators of the humoral responses. Upon direct recognition of the parasites, i.e. antigen, they produce antibodies of different isotypes, that are specific for the antigen. A remarkable difference between bone marrow-derived B and thymus-derived T cells is the inability of T cells to recognize antigens directly as B cells do. T cells, on the other hand, need adequate presentation of antigens mostly by major histocompatibility complex (MHC) molecules expressed on antigen-presenting cells (APCs), such as dendritic cells, Langerhans cells, macrophages, B cells and vascular endothelial cells. The T-cells are distinguished according to their T cell-receptor (TCR) and accessory molecules. The TCR is composed of an α and β chain or an γ and δ chain. Accessory molecules are the CD4 or the CD8 marker.

Antigen Presentation

Before further discussing the functions of the T cells, interest will be focused on the MHC and antigen presentation. MHC genes are organized in a gene complex of about $3 \cdot 5$ mb on chromosome 6 in man. Several classes of molecules are encoded in this gene complex, of which the MHC class I and class II molecules are central to antigen presentation. MHC class I molecules are formed by a variable a polypeptide chain and a constant β-2 microglobulin. The MHC class II molecules represent heterodimers composed of variable α and β chains. Association of antigen with MHC molecules occurs inside the antigen-presenting cell and processing of foreign antigen to peptide fragments is an essential prerequisite for successful association and presentation. The complex of the MHC molecule and antigen peptide is transported to the cell surface and presented to T cells. MHC class I and II molecules can be subdivided into classical and non-classical MHC molecules. Classical class I molecules are encoded in man by the HLA-A, -B and C genes. These highly polymorphic molecules present processed peptides to $CD8^+$ T cells. Non-classical class I molecules are much less polymorphic and are encoded by HLA-E, -F and G. Their function in man is still ill defined. In the mouse, the related Qa-1 and 2 molecules present a restricted set of peptide antigens, for instance a fragment of lysteriolysin, to $CD8^+$ T cells. Classical MHC class II molecules in man are encoded by HLA-DP, -DQ and DR. These molecules present processed peptides to $CD4^+$ T cells. Non-classical class II molecules, such as DMA and DMB, appear to support classical class II molecules in antigen presentation.

CD1 molecules form a non-MHC-encoded family of molecules involved in antigen presentation. As class I molecules, they are composed of a polymorphic α chain and β2-microglobulin. So far, the isoforms CD1 a-e have been described in humans. Recently, it became apparent that CD1 molecules act as restriction elements in the presentation of several mycobacterial lipids, such as mycolic acid, and glycolipid antigens, such as lipoarabinomannan to T cells. Interestingly, CD1 molecules seem to present non-peptide antigens not only to classical T cells but also to the recently described subsets of $CD4^+$ NKT 1.1^+ and $CD4^-$ $CD8^-$ NK 1.1^+ T cells, that are also designated as natural T-cells or NT-cells. These NKT-cells, because of their capacity to produce interleukin 4, can potentially influence the phenotype of the immune response to class II-restricted antigens. Their role in parasite infection, however, is still unclear.

Whether an antigen will be processed and presented with class I or class II MHC molecules appears to be determined by the route that the antigen takes to enter a cell. Exogeneous antigen is produced outside of the host cell and enters the antigen-presenting cells, which degrade the exogenous protein within the phagosome into peptides of 12–15 amino acids length. The peptides are loaded into the cleft within the MHC class II molecules. The MHC class II peptides complex is then exported to the cell surface, where it is recognized by T-cells displaying CD4. $CD4^+$ T cells recognize their antigen MHC class II restricted. Endogenous antigen is produced within the host cell itself. It is either of host cell or of parasite origin. In the cytosol, proteasomes degrade endogenously synthesized proteins to peptides, which are then transported by particular transport-associated proteins so called TAPs to the endoplasmic reticulum. Here the peptides bind to MHC class I molecules. Thereafter the complex is exported to the cell surface. T cells displaying CD8 recognize the complex and are thereby stimulated. Therefore, they are said to be MHC class I restricted.

The great polymorphism of genes coding for MHC molecules allows man to bind and present a vast diversity of different peptides produced by the many parasite pathogens, to a large T cell repertoire, resulting in highly specific immune response.

T-Cell Mediated Responses

Next, the various T cells, to which antigen is presented will be discussed. In man, the T cell population in the periphery amounts to more than 10^{12} cells, of which more than 90% carry the α/β T-cell receptor (TCR) and less than 10% the g/d receptor. The TCR associates with the variable region of the MHC-molecule which carries the antigenic peptide and the CD4 or CD8 molecules bind to the constant regions of the MHC molecules. Binding of the antigen/MHC-complex to the TCR results in the engagement of the CD3-complex, with subsequent signal transduction and T cell activation. This signal transduction is modulated by costimulatory effects induced by the accessory receptor/ligand pairs CD40/CD40 ligand or B7/CD28. CD40-mediated signals seem to affect primariliy $CD4^+$ T cells of the Th1 subtype, whereas the B7 costimulis trigger T cells of the Th2 subtype. The $CD4^+$ T-cells according to their cytokine secretion pattern have been first subdivided by Tim Mossmann into the Th1 and Th2 family, Th1 cells producing IL-2 and IFN-γ, cytokines that serve as stimuli of T-cells or macrophages respectively, and Th2 cells producing IL-4, IL-5 and IL-10, that is T-cells that act as true helper cells for B-cells. The Th1 cells, via IFN-γ, seem to act as classical cells of delayed type of hypersensitivity, i.e. of cell-mediated immunity for instance in the tuberculin reaction. Thus, they are of prime importance in the control of intracellular microorganisms such as toxoplasma, leishmania, mycobacteria, and others. By virtue of their IFN-γ production Th1 cells also downregulate Th2 cells. The Th2 cells, on the other hand represent the classical T helper cells, which are involved in the allergic reaction and in humoral immune responses. IL-4 is central to IgE production and IL-5 to IgA production and IL-10 downregulates Th1 cells. The subdivision of $CD4^+$ T cells in Th1 and Th2 cells should however not be taken too strictly because during a developing immune response, T cells producing both Th1 and Th2 cytokines are found, and Th1 and Th2 cells might well coexist in the tissue.

Differentiation of Th0 cells into Th1 or Th2 is driven by cytokines produced by different cells of the immune system. IL-12 is the major player in the Th1 pathway and IL-4 driving the Th2 cells. Undoubtedly, the signals which in leishmaniasis are decisive for the initiation of a protective Th1 response versus a disease-promoting Th2 response, are by far not clear.

Besides $CD4^+$ T cells, $CD8^+$ cells are involved in immune reactions to most if not all intracellular parasites including plasmodia, toxoplasma and leishmania. $CD8^+$ cells function by their capacity to act as cytolytic killer cells as well as producers of cytokines such as IFN-γ etc. $CD8^+$ T cells lyse infected target cells by cell-to-cell contact subsequent to the recognition of the target peptide that is presented by class I molecules. Target cell lysis is mediated by two separate mechanisms. The first mechanism involves the secretion of perforins and granzymes by the killer cells, which both lead to osmotic lysis of the target cells. The second mechanism requires the cross-linking of the Fas-ligand and the cytolytic $CD8^+$ cells with the Fas-antigen on the target cells with subsequent chromatin condension, DNA fragmentation and cell rupture. This mechanism induces apoptosis, also known as programmed cell death, in the target cells. Cytolytic activity is most prominent in $CD8^+$ T cells, however also some $CD4^+$ T-cells have been shown to act as killer cells as in the case of *Toxoplasma gondii* infected macrophages. $CD8^+$ cells may act in synergy with Th1 cells, when Th1 cells produce the cytokines IL-2 and IFN-γ and $CD8^+$ cells contribute their cytolytic activity to the pathogen-eliminating process.

Specific Aspects of Immune Responses to Helminths

Worms such as nippostrongylus, filaria, ascaris and schistosomes induce high levels of specific IgE antibodies and eosinophilia.This characteristic response pattern is caused by the particular ability of helminths to preferentially stimulate the Th2 subset of $CD4^+$ cells, which secrete the cytokines IL-4 and IL-5, IL-4 involved in the production of IgE antibodies and IL-5 acting as growth factor for eosinophils. In vitro studies suggest that helminths opsonized with specific IgE antibodies are lysed by eosinophils that carry Fc receptors specific for IgE, the toxic product contained in the major basic protein of the eosinophils granules. This effector mechanism however does not relate to cell helminths infections as immunity to schistosoma induces Th1 cells and the production of IFN-γ, which results in activation of macrophages that directly delete the schistosome larvae by means of → nitric oxide (Vol. 1) (NO).

Strategies of Evasion of Immune Mechanisms by Parasites

The capacity of parasites to survive and to persist in their hosts is a result of coevolutionary events that enable the parasites to evade immune effector mechanisms. Most parasites have developed multiple evasion strategies to circumvent both innate and acquired defense mechanisms of the host, which will be discussed separately for each parasite.

Immune Suppression

→ Opportunistic Agents (Vol. 1), → Chagas' Disease, Man, → Echinococcosis, → Sleeping Sickness, → Toxoplasmosis, Animals, → Toxoplasmosis, Man

Immunization

See Vaccination Chapters

Immunoassay

General Information

Immunological assays are available for measuring the humoral immune response after parasitic infections. For the detection and quantitation of antibodies or antigens an immunoassay has to fulfill three basic requirements: highly specific, highly sensitive, and reproducible. The field of application and the quality of antigen/antibody available are the main variables which determine the choice for one specific assay. A test system may be adequate for population screening or for individual diagnosis. Ideally, it is a single definitive test which can equally detect antibodies in low-responders, in people with recently contracted infection or with high titers and clinical signs, and be suitable for a post-treatment follow-up.

The stability of the antigen-antibody complex (avidity) determines the success of all immunological assays.

IFA, EIA, WB

The indirect immunofluorescence (fluorescent antibody) assay (test) (IIFA, IFA, IIFAT, IFT) has long been the best choice in parasitology. By using cryosections from different developmental stages of helminths or free protozoa as antigens the test provides a satisfactory discrimination between specific and non-specific reactions. The differential binding of antibodies to different structures of the parasite or cell is used to distinguish between specific and nonspecific reactivity. Its disadvantage is subjective reading and labour intensiveness. Screening of a large population is practically impossible. The indirect haemagglutination test (IHAT, IHA) is applied for fast screening at low costs of several parasitic infections. A broad application of the indirect enzyme-linked immunosorbent (enzyme immunosorbent) assay (ELISA, EIA) has long been delayed by the use of non-defined, crude parasite extract antigens, resulting in a high degree of cross-reactivity. This disadvantage is now partly overcome by the availability of secretory/excretory (E/S) antigens, synthetic peptides and recombinant antigens. Still, the sensitivity and specificity of the ELISA vary considerably but it is an easy and fast alternative especially for population and herd screening and epidemiological studies. The western blot (WB, enzyme immunoblot, EIB), which provides a species or stage specific diagnosis is a confirmation test and applied as the final diagnostic step when screening results are positive.

Other Test Systems

Today, only few laboratories still use immunodiffusion (ID), gel precipitation, complement fixation test (CFT), Sabin Feldman dye test (DT), counterimmunoelectrophoresis (CIE) or radioimmunoassay (RIA) in routine serodiagnosis. These test systems consume much antigen in comparison to the ELISAs, are less sensitive or use radiolabelled conjugates. Other test systems like card agglutination test (CATT), direct agglutination test (DAT), latex agglutination test (LA), carbon immunoassay (CIA), dot blot, and dip stick antigen assay were developed for a rapid, low cost screening under field conditions.

The rapidly increasing knowledge on immunology and molecular biology may result in a large scale production of characterized parasitic antigens applicable in different test systems. However, the suitability of each product for diagnostic purposes depends on the dynamics of the hosts immune response to that particular antigen.

Characteristics

IIFAT results depend not only on the kind of antigen used but also on its processing. Fixation is one of the most important steps (unfixed, formalin, aceton, methanol, heat, and others) to enhance or decrease the sensitivity or specificity of the test. The reading of the IIFAT requires a specific microscope. Reading of the slides by inexperienced staff may lead to different end-point titers due to subjective impressions. A new generation of IHATs using selected red blood cells of the 0 Rh- type for antigen cross-linking reduces the non-specific agglutination of erythrocytes and improves the specificity of this fast test system. ELISA is a suitable test system for the detection of both antigen and antibody. It is very sensitive and economic for testing a large number of serum samples within a relatively short time. The test principle allows many variations in the performance of the assay and the use of detection systems. Direct and indirect, capture and dip stick and many other ELISA systems are described as efficient in different diagnostic fields. Reading of the test is normally by use of a photometer. In highly specific assays a discrimination between positive and negative results is also possible visually. Indirect ELISAs, which are predominately used in parasitology for antibody detection, measure antibody binding and not antibody levels and therefore provide only a limited quantitative information. The WB analysis demonstrates serum reaction patterns with antigen fractions of either a complex mixture or predefined extracts of parasites. It is used to demonstrate either an individual stage specific antibody response in follow-up studies or species specific reaction for the discrimination of infection with closely related parasites. The CATT is an easily and fast to perform low-cost system which uses the surface antigen of different *Trypanosoma*-species for antibody detection. It is applied as a screening tool for the (West) African sleeping sickness and animal trypanosomiasis caused by *T. brucei/congolense* and *T. evansi.*

Standardisation

Standardisation of immunological assays is a crucial problem in parasitology. The evaluation of the specificity, sensitivity, and negative/positive predictive value of an assay requires either a "gold standard" or a population based evaluation of numerous defined serum samples. Only the DT is an accepted "gold standard" for toxoplasmosis serology. For other parasitic infections either parasite detection, disease, or some other immunological assay with proven diagnostic value are used as "gold standard". Many home made or commercial test systems still work with undefined antigens. Also, reference sera are not available for all parasitic infections. Hopefully, standardisation will improve with the availability of defined antigens.

Clinical Relevance

The clinical value of an immunoassay depends on its ability to allow the identification of an infected individual on the results of one single test. Ideally, the producer provides information on the positive and negative predictive value for his immunoassay. Due to a comparatively low demand for parasitological immunoassays in the low incidence "developed" countries test systems are often produced in parasitological laboratories. At present, only the serodiagnosis of toxoplasmosis is to a high degree commercialized and, in Germany, submitted to an official, external quality control. The choice of a suitable test system is therefore essential for a good diagnosis. For individual diagnosis a quantitative and specific measurement of the serum antibody concentration is required. The approximate antibody concentration is given either by end-point titers (IIFAT, IHAT), units or absorbances (ELISA). The WB pattern can provide only semiquantitative, but highly specific results. In any case, an interlaboratory comparison of test results is only possible when identical assays are performed. The application of intra-and interassay quality controls by use of defined reference material is obligatory. For screening purposes a sensitive assay is preferred for primary diagnosis, followed by a more specific test for definitive diagnosis. Qualitative test results (positive/negative) will be sufficient for population screening or herd control. When interpreting test results it has to be considered that there is not only a population based variability of the antibody responses to parasitic infections but also a great variability of the individual response depending on the parasite load and stage of infection. Finally, antibodies show variation among different techniques due to differences in multivalent binding.

The use of different test systems for seroepidemiological studies may result in different rates of seroprevalence achieved. These differences may reflect basic differences in the sensitivity or specificity between test systems. However, when direct parasite detection as "gold standard" for test evaluation is used the resulting sensitivity depends not only on the experience of the investigator but also the frequency of sampling and the parasite load of the individuals tested. When disease is used as "gold standard" for test evaluation a comparatively high seropositivity will result when the infective parasite exhibits a long incubation period and/or low morbidity rate. The clinical outcome of parasitic infections depends also on the genetic predisposition of the infected individual. A positive test result in one immunoassay may therefore provide evidence for a yet unknown immunological recognition of the parasite, which is undetectable in other assays. Discrepant rates in the prevalence of *Toxocara canis*-antibodies in human serum were reported when either CIEP and IIFAT or E/S-ELISA was applied. The interpretation of unresolved positive values is carried out by considering the incubation period and possible pathological effect of each parasite.

Incidence

Number of new cases of disease with respect to a defined population.

Infestation

Settlement of a parasite in or on a host without local reproduction (e.g. sucking of a tick/flea).

Inflammation

→ Pathology.

Inflammatory Responses

→ Pathology.

Inhibitory-Neurotransmission-Affecting Drugs

Overview see Table 1

Table 1. Drugs active against micro- and macrofilariae.

Year of introduction or discovery	Drugs	Effects on filariae	Effects on other parasites
Drugs with predominantly microfilaricidal effects			
1947/48	Diethylcarbamazine	*W. bancrofti, Brugia* spp., *Onchocerca volvulus, Loa loa*	
1965	Levamisole	*W. bancrofi, B. malayi*	intestinal nematodes
1980	Ivermectin	*Onchocerca volvulus, W. bancrofti, Loa loa, Dirofilaria immitis*	nematodes, arthropods
1993	Doramectin	microfilariae in rodent models*	nematodes; arthropods
1990	Milbemycin A4 oxime	*Dirofilaria immitis*, microfilariae in rodent models*	nematodes, arthropods
1980's	Milbemycin D	*Dirofilaria immitis***	nematodes, arthropods
1990	Moxidectin	*Dirofilaria immitis*, microfilariae in rodent models*	nematodes; arthropods
1955	Metrifonate	*Onchocerca volvulus* microfilariae	*Ascaris, Schistosoma haematobium*, insects
1999	Selamectin	*Dirofilaria immitis*	nematodes, arthropods
Drugs with predominantly macrofilaricidal effects			
1916	Suramin	*Wuchereria, Brugia* (macrofilariae); adult *O. volvulus*	trypanosomes
1971	Flubendazole, Mebendazole, Albendazole	*Wuchereria, Brugia, O. volvulus*, also embryostatic effects	nematodes, cestodes, trematodes, *Giardia*
1984	Arsenamide (Thiacetarsamide)	*Dirofilaria immitis* (dog)	
Drug combinations			
	Ivermectin/DEC Ivermectin/Albendazole	lymphatic filariasis	mites/microsporidia
Drugs with macro- and microfilaricidal effects			
1980	Benzothiazoles	*Onchocerca, Brugia* spp., *Dipetalonema* spp.	cestodes

*, Schares G, Hofmann B & Zahner H (1994) Trop. Med. Parasitol. 45 : 97–106
**, McKellar QA & Benchaoui HA (1996) J.Vet. Pharmacol. Therap. 19 : 331–351

Structures
Fig. 1

Piperazine

Synonyms 75 different synonyms (Chemotherapy of parasitic diseases (Campbell wc, Rew RS (eds), Plenum Press, New York and London, p. 629).

Clinical Relevance The antinematodal activity of piperazine is directed against *Ascaris lumbricoides*, *Enterobius*, ascarids in dogs and cats, adult *Oesophagostomum* in pigs, adult horse nematodes and *Ascaridia* in chicken. Piperazine is only effective against large intestinal nematodes.

Molecular Interactions Piperazine induces a reversible paralysis of *Ascaris suum* in-vitro by exerting hyperpolarizing effects, thus acting as a selective GABA agonist (Fig. 2). The average duration of piperazine induced channel openings is 14 msec, and is thus shorter than the GABA produced openings with 32 msec. Piperazine is about 100 times less potent than GABA in *A. suum*. The difference in potency correlates with the need for higher piperazine concentrations to achieve a similar opening rate to GABA. The *Ascaris* GABA receptor is pharmacologically distinguished from the vertebrate $GABA_a$ receptors. The action becomes potentiated by the presence of a high pCO_2 probable by interaction of CO_2 with the heterocyclic ring of piperazine which partially substitutes for the carboxylgroup of GABA.

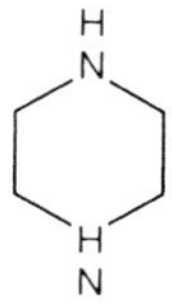

Piperazine

Avermectins

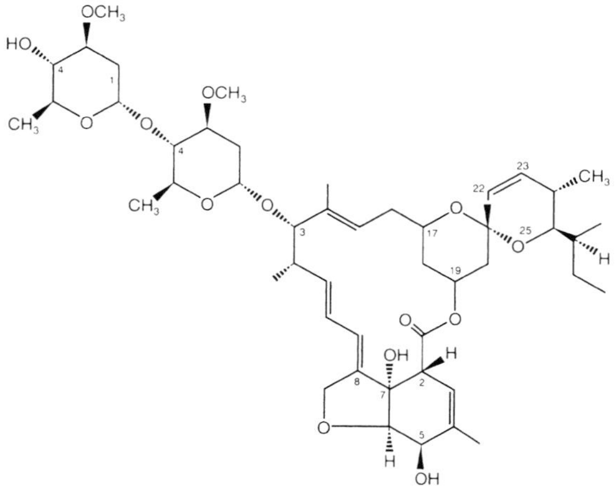

Avermectin B_{1a} (Abamectin)

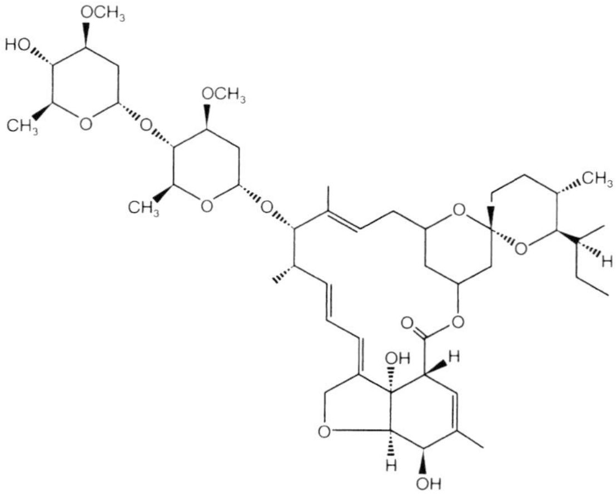

22, 23-dihydroavermectin B_{1a} (Ivermectin)

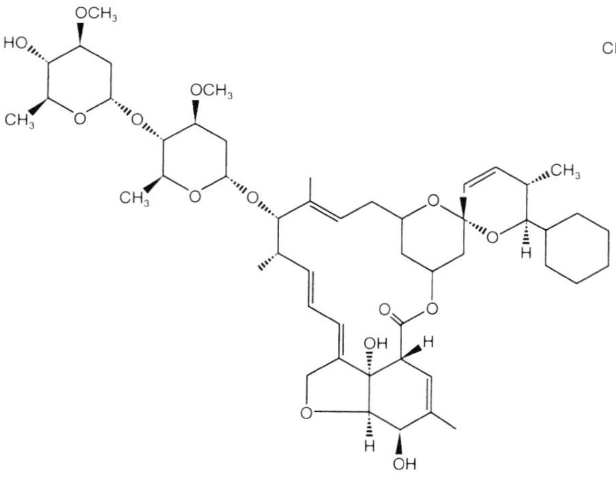

25-cyclohexylavermectin B_1 (Doramectin)

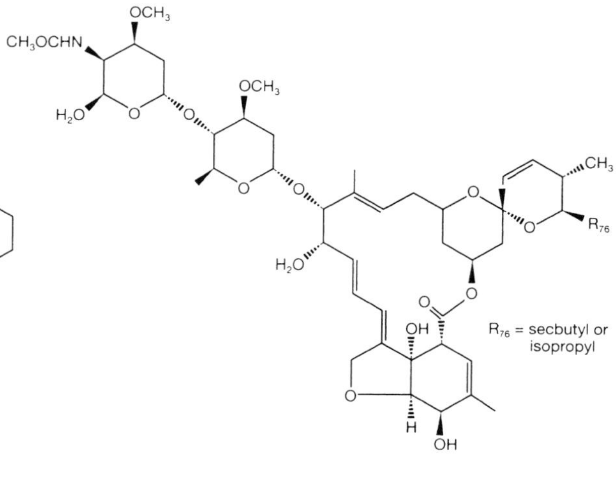

4''-epi-Acetylamino-4''-deoxy-avermectin B_1 (Eprinomectin)

R_{76} = secbutyl or isopropyl

Fig. 1. Structures of antiparasitic drugs affecting GABA- or glutamate-gated chloride channels.

Milbemycins

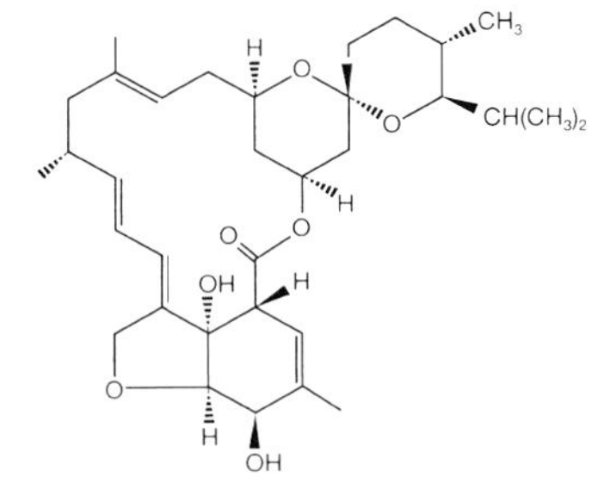

Milbemycin D

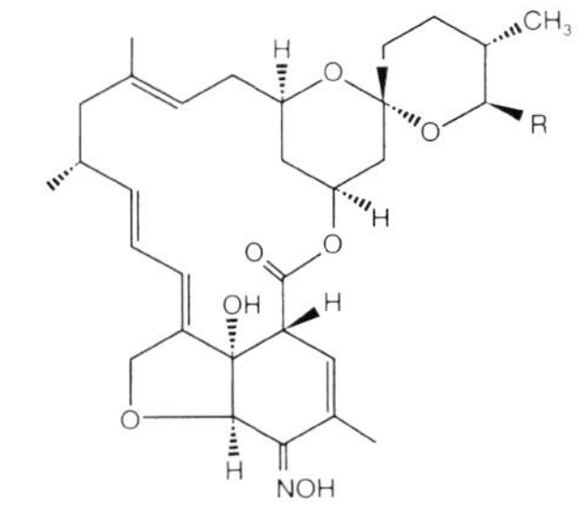

Milbemycin 5-oxime
A3 (R = CH_3)
A4 (R = C_2H_5)

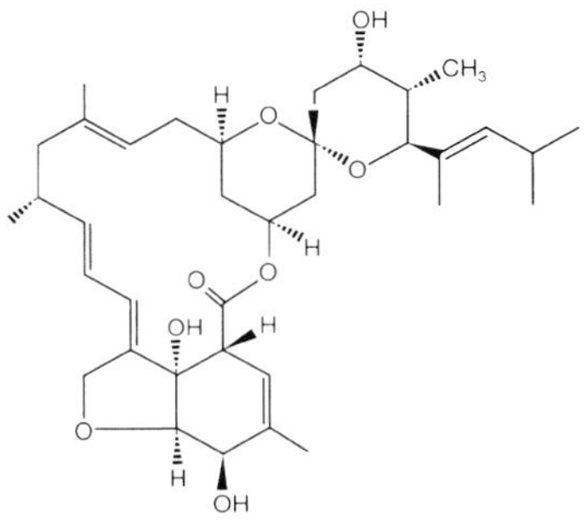

Nemadectin

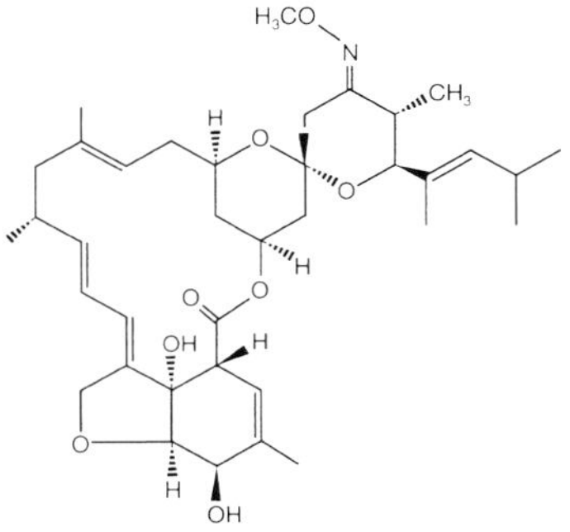

Moxidectin

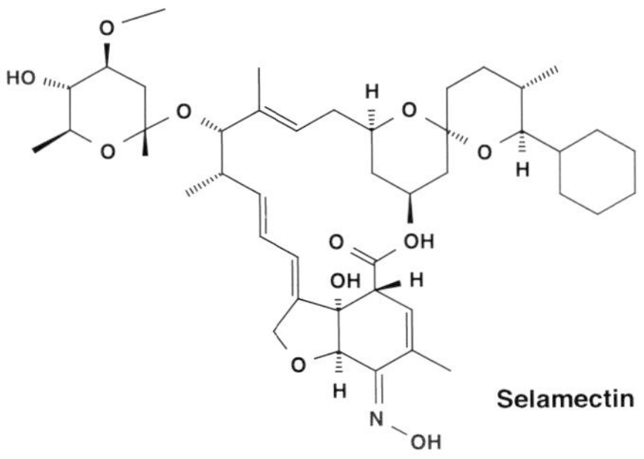

Selamectin

Fig. 1. (continued) Structures of antiparasitic drugs affecting GABA- or glutamate-gated chloride channels.

Macrocyclic Lactones

Important Compounds Ivermectin, Abamectin, Doramectin, Eprinomectin, Milbemycin oxime, Moxidectin, Selamectin.

Synonyms Ivermectin: Baymec, Ivomec, Ivomec-Premix Ivomec-S, Cardomec, Equell, Eqvalan, Furexel, Heartguard 30, Heartgard Chewable, Mectizan, Oramec, Rotectin, Strongid, Zymectrin; in: Ivomec-P, Heartgard Plus

Abamectin: Avomec, Duotin, Enzec; in : Equimax

Doramectin: Dectomax

Eprinomectin: Eprinex

Milbemycin oxime: Interceptor, Interceptor Flavor Tabs; in: Sentinel

Moxidectin: Cydectin, Equest, Pro Heart, Quest

Selamectin: Revolution, Stronghold

Clinical Relevance The group of these macrocyclic lactones is subdivided into 2 subgroups, the **avermectins** and **milbemycins**. To the avermectin anthelmintics belong avermectin (explored 1975), ivermectin (marketed 1980), abamectin (1980), doramectin (1993), eprinomectin (1996) and Selamectin (1999). To the milbemycin anthelmintics belong milbemycin (explored 1973), milbemycin oxime and moxidectin (1993). Avermectin is produced by *Streptomyces avermitilis*, an actinomycete strain. Avermectin was isolated in 1975 and its antiparasitic activity discovered in mice infected with *Nematospiroides dubius*. The components avermectin B_{1a} and B_{1b} exert the highest anthelmintic activities. The chemical reduction leads to dihydro-derivatives with low toxicity. The mixture of 80% dihydro avermectin B_{1a} and 20% B_{1b} is named **ivermectin**.

Macrocyclic lactones are applied as broad-spectrum anthelmintics and ectoparasiticides in horses, cattle, sheep, pigs (Table 2, → Microtubule-Function-Affecting Drugs/Table 1). Ivermectin is the drug of choice against *Strongyloides stercor-*

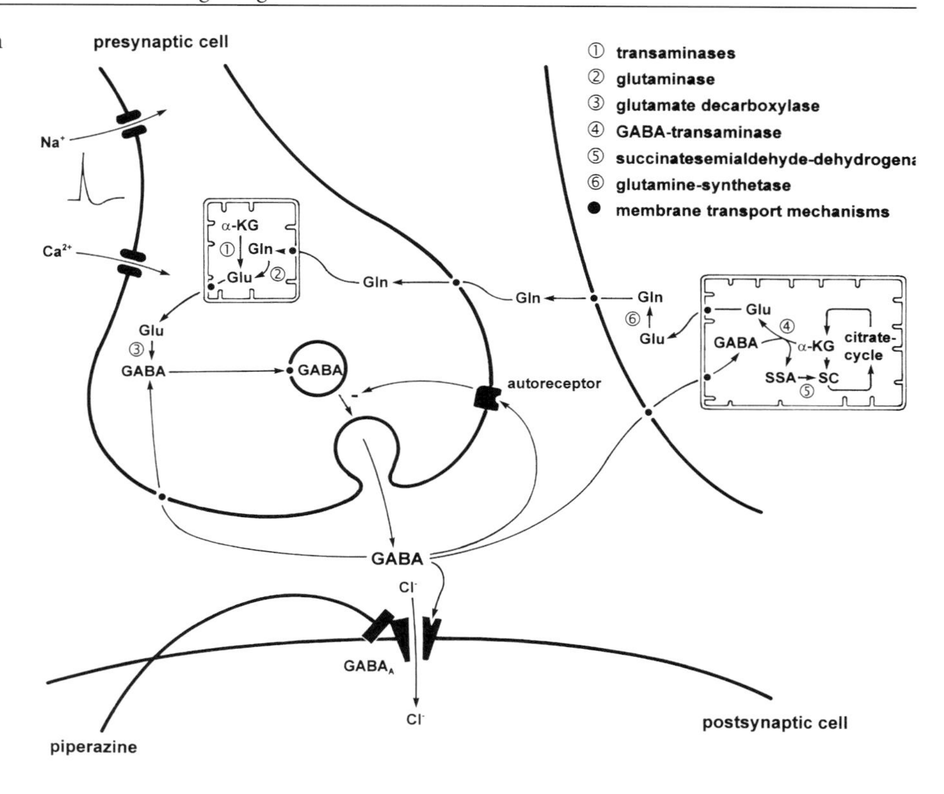

Fig. 2. Model of the action of piperazine on the GABAergic neurotransmission.

alis infections in immunosuppressed patients. In addition, this drug is used for heartworm prophylaxis, in human onchocerciasis it is the drug of choice and it is of growing importance in lymphatic filariasis (e.g., *Wuchereria bancrofti* infections) as single drug or in combination with DEC or albendazole (Table 1). It has microfilaricidal effects and leads to a suppression of embryogenesis in human onchocerciasis. It was introduced to Onchocerciasis Control Program of WHO in 1987 as Mectizan. In addition, ivermectin has some activity against *male Onchocerca volvulus* in man as well as against microfilariae of *W. bancrofti*, *Loa loa*. Ivermectin-activity can be observed against *Litomosoides carinii* microfilariae in the circulating blood but not against microfilariae in the pleural cavity. Ivermectin has some in-vitro activity against filarial parasites, e.g. *Onchocerca* spp.. There is generally a great discrepancy between the good in-vivo efficacy and the minor in-vitro effects. In general, treatment with ivermectin is not accompanied by severe side-effects.

Molecular Interactions Macrocyclic lactones act at the junction of ventral cord interneurons and motorneurons resulting in the immobilization of nematodes and at the neuromuscular junction of arthropods causing paralysis. Macrocyclic lactones are taken up by many gastrointestinal and filarial nematodes via the cuticula and presumably with equal importance by oral ingestion (→ Acetylcholine-Neurotransmission-Affecting/Fig. 3), while in blood-sucking parasites (*Haemonchus contortus*, arthropod ectoparasites) the oral absorption is by far more important. This view is supported by the observation that macrocyclic lactones exert greater activities against sucking lice (*Haematopinus eurysternus*, *Linognathus vituli*) than biting lice (*Damalina bovis*). They also have high efficacy against mites (*Sarcoptes scabiei var bovis*), which are known to be blood consumers.

The mode of action of avermectins and milbemycins relies on the opening of the chloride ion channels in the neuronal membranes of nematodes and the muscle membranes of arthropods (Fig. 3). Thereby, cells become hyperpolarized and can no longer respond to incoming stimuli. All the physiological effects of avermectins and milbemycins can be reversed by picrotoxin, a specific blocker of the chloride ion channels. There is an ivermectin-induced increase of chloride permeability of nerve and muscles membranes of invertebrates observable. The structure and regulation of chloride ion channels of nematodes on the

Table 2. Antiparasitic spectrum of ivermectin, doramectin, eprinomectin and moxidectin in cattle (according the technical manuals of the suppliers).

Ivermectin	Doramectin	Eprinomectin	Moxidectin
1. Gastroinestinal nematodes (adults and L4 larvae)			
Ostertagia ostertagi and arrested larvae	*Ostertagia ostertagi* and arrested larvae	*Ostertagia ostertagi* and arrested larvae	*Ostertagia ostertagi* and arrested larvae
O. lyrata			*O. lyrata*
Haemonchus placei	*Haemonchus* spp.	*Haemonchus placei*	*Haemonchus* spp.
	H. similis		*H. similis*
			H. contortus
Trichostrongylus colubriformis	*Trichostrongylus colubriformis*	*Trichostrongylus colubriformis*	*Trichostrongylus colubriformis*
Trichostrongylus axei		*Trichostrongylus axei*	*Trichostrongylus axei*
Cooperia oncophora	*Cooperia spp.*	*Cooperia oncophora*	*Cooperia oncophora*
C. punctata	*C. punctata*	*C. punctata*	*C. punctata*
C. pectinata	*C. pectinata*		*C. pectinata*
			C. spatulata
		C. surnabada	
Bunostomum phlebotomum	*Bunostomum phlebotomum*	*Bunostomum phlebotomum*	*Bunostomum phlebotomum*
Oesophagostomum radiatum	*Oesophagostomum radiatum*	*Oesophagostomum radiatum*	*Oesophagostomum radiatum*
*Nematodirus helvetianus**	*Nematodirus helvetianus**	*Nematodirus helvetianus*	*Nematodirus helvetianus*
*N. spathiger**	*N. spathiger**		*N. spathiger*
*Strongyloides papillosus**	*Strongyloides papillosus**		
	Trichuris spp.	*Trichuris* spp.	*Trichuris discolor*
2. Lungworms (Adults and L4 larvae)			
Dictyocaulus viviparus	*Dictyocaulus viviparus*		
	Dictyocaulus viviparus	*Dictyocaulus viviparus*	
3. Grubs/Myiasis			
Dermatobia hominis larvae	*Dermatobia hominis* larvae		–
Cochliomyia-hominivorax+ *Hypoderma bovis* *Hypoderma lineatum*	*Cochliomyia-hominivorax*	*Hypoderma bovis* *Hypoderma lineatum*	–
4. Lice			
Linognathus vituli	*Linognathus vituli*	*Linognathus vituli*	*Linognathus vituli*
Haematopinus eurysternus	*Haematopinus eurysternus*	*Haematopinus eurysternus*	
Solenopotes capillatus	*Solenopotes capillatus*	*Solenoptes capillatus*	*Solenopotes capillatus*
Damalinia bovis++	*Damalina bovis*++	*Damalina bovis*	
5. Mites			
Psoroptes ovis	*Psoroptes ovis*		*Psoroptes ovis*
Sarcoptes scabiei var. bovis		*Sarcoptes scabiei*	
Chorioptes bovis		*Chorioptes bovis*	
6. Ticks			
Boophilus microplus *Boophilus decoloratus* *Ornithodorus savignyi*	*Boophilus microplus*		*Boophilus microplus*
7. Flies			
–	*Haematobia irritans*	*Haematobia irritans*	–

*+ prophylactic (injection) or curative (topical) treatment; ++ parasitic control; * only adult stages*

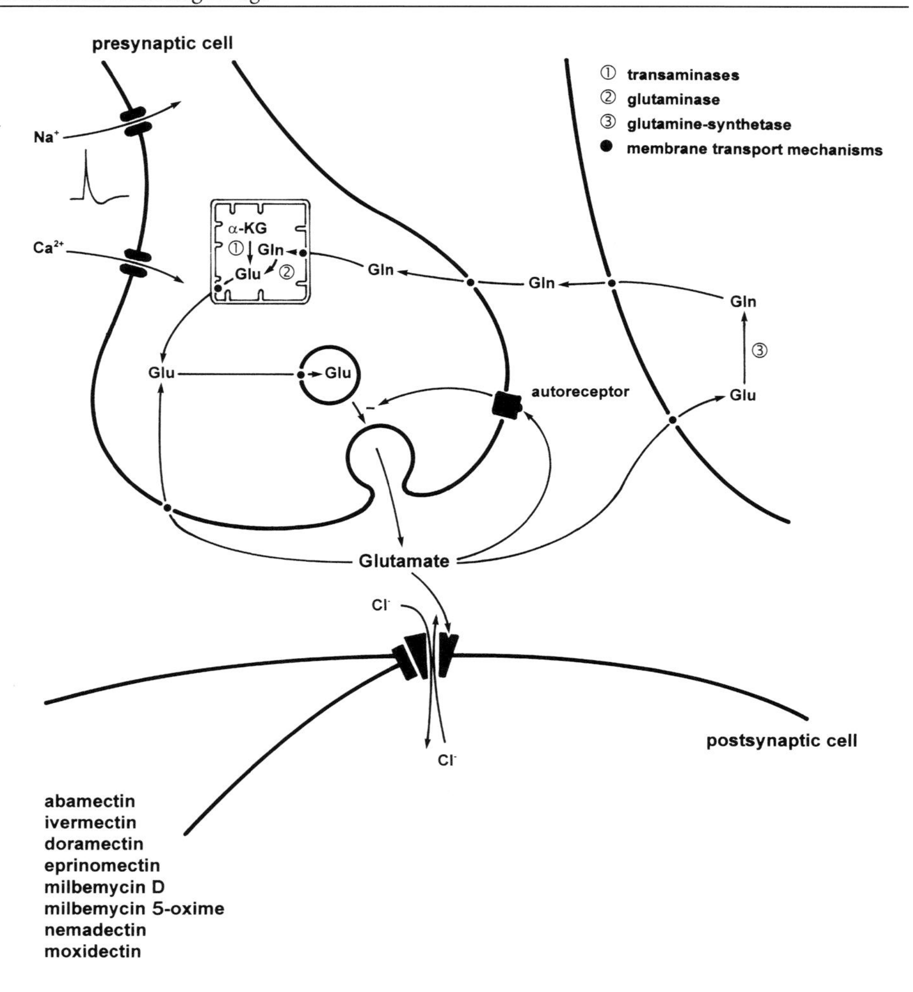

Fig. 3. Model of drugs affecting neuromuscular transmission by inhibitory neurotransmitters in protostomes.

molecular level is unclear at present and also the identity of the target ion-channel is controversial. Results from experiments with crayfish and *Ascaris suum* reveal that there is an interaction with receptors at chloride channels. The action is obviously not mediated by GABA-gated chloride channels. Expression experiments with *Xenopus* oocytes lead to a proposed action of avermectins on a glutamate-gated chloride channel (GluCl). Genes encoding ivermectin-sensitive glutamate-gated chloride channel subunits could be isolated from *Caenorhabditis elegans*. Moreover, the avermectin binding site could be purified from *C. elegans*. A 1.8–2.0 kbp mRNA of *C. elegans* encoding a chloride channel with sensitivity to both ivermectin and glutamate could be identified. The channel is presumably a pentamer similar to the nicotinic receptor and is selectively permeable for anions (chloride). The pentamer consists presumably of GluCl-α subunits with a glutamate binding site and a GluCl-β subunit which contains the ivermectin binding site. Molecularbiological experiments reveal that the GluCl-β subunit of the glutamate-gated channel is expressed in the pharyngeal muscle of *C. elegans*. There are considerable identities of α- and β-subunits of the glutamate-gated ion channels with the α- and β-subunits of mammalian GABA and glycine receptors.

According to a hypothesis the pharyngeal pumping of nematodes is inhibited by ivermectin. In the pharyngeal muscle of *A. suum* a potentiation of the action of glutamate on glutamate receptors that gate chloride channels by ivermectin analogues could be observed. The hyperpolarization of the nerve and muscle membranes leads to a flaccid paralysis of the parasites. The CNS-side effects of ivermectin are explained by the strong action on the receptors in rat brain through potentiation of GABA- and benzodiazepine-binding to open the channel. Nevertheless the relatively safe use of avermectins is due to the inability to cross the blood brain barriers into the CNS in mammals with the exception of Collies. Moreover, the selectivity of macrocyclic lactones is due to different neurotransmitter functioning of glutamate in invertebrates (protostomes) where it acts as an inhibitory neurotransmitter in contrast to vertebrates where glutamate acts as an excitatory neurotransmitter.

Macrocyclic lactones like milbemycin oxime have only slight inhibitory and stimulatory concentration-dependent effects on the motility of filariae such as *Dirofilaria immitis*. It is assumed that definite host factors are required, which are independent of a specific immune status of the host. In *Acanthocheilonema viteae* an ivermectin-mediated cell adherence to the living microfilariae is observable and also cellular cytotoxicity by complement activation via the alternate pathway and/or antibodies. *L. carinii* microfilariae are killed without direct contact between cells and larvae. Here a very short-living mediator, which can be inhibited by the arginin-analogs such as N^G-monomethyl-L-arginine and L-canavanine, seems to be involved in the drug's action. Leucocytes are also required for the in-vitro efficacy of ivermectin against microfilariae of *D. immitis*. The immobilization of the larvae is necessary for cell adherence or cellular toxicity similar to ivermectin-induced paralysis of *A. suum*. In addition, the paralysis of microfilariae of *O. volvulus* presumably fascilitates the phagocytic cell-trapping. The effects of ivermectin on adult worms are not fully understood to date. Electron microscopic studies on *A. viteae* reveal a vacuolization and an increased electron density in all organs beginning 8 days after treatment. In *L. carinii* a degeneration of intrauterine stages and an extreme folding of the uterine wall can be observed accompanied by a generally increased electron density.

Resistance Resistance against macrocyclic lactones is probably inherited by a single dominant allele in *Haemonchus contortus*. The mechanism of ivermectin resistance on the molecular level is still unknown. Membranes of ivermectin-resistant and -susceptible larvae of *H. contortus* contain similar numbers of ivermectin binding-sites with the same affinity characteristics. It could recently be shown that ivermectin-resistance may be caused by an altered P-glycoprotein homolog in *H. contortus*. Thereby the expression of P-glycoprotein mRNA is higher in the ivermectin-resistant *H. contortus* strain than in the susceptible strain. This multidrug resistance mechanism can be reversed by verapamil and there is an increased efficacy of ivermectin and moxidectin against moxidectin-resistant *Haemonchus* in jirds (*Meriones unguiculatus*) in the presence of verapamil. The disruption of the mdr1a gene, which encodes P-glycoprotein in mice, results in hypersensitivity against ivermectin. In another report an involvement of P-glycoprotein in ivermectin resistance in *H. contortus* was excluded. Now there exists a *Caenorhabditis elegans* mutant with a high level of ivermectin resistance, which can serve as a tool for further investigations of alterations in the glutamate/ ivermectin chloride channel receptor and mechanism of ivermectin resistance in future.

Innate Immunity

See related entries: → Amoebiasis, → Chagas' Disease, Man, → Filariasis, Lymphatic, Tropical, → Giardiasis, Animals, → Giardiasis, Man, → Leishmaniasis, Man, → Malaria.

Insecticides

→ Insects (Vol. 1), → Pediculosis, Animals

Isosporiasis, Animals

→ Isospora (Vol. 1), → Coccidiasis, Man, → Coccidiosis, Animals, → Alimentary System Diseases, Animals

Isosporosis, Man

→ *Isospora* (Vol. 1) *belli* undergoes a classical coccidian cycle with schizogony and gametogony mainly in the small intestinal epithelium. Unsporulated oocysts 20–32 μm in size and containing 2 sporoblasts are shed in the stools. In addition, individual encysted zoites are found in the lamina propria and mesenteric lymph nodes (→ Pathology/Fig. 5C). These are similar to those that occur in cats and in rodents, which can serve as the intermediate hosts of feline *Isospora*, which has been reclassified as *Cystoisospora felis and C.*

rivolta. The presence of unizoic → cysts (Vol. 1) suggests that the human *I. belli* may also be heteroxenous, and may better be classified as *Cystoisospora belli*. There is an intense inflammatory reaction in the lamina propria involving plasma cells, lymphocytes, neutrophils, and eosinophilic granulocytes. With chronic infection there is villar atrophy. Intermittent diarrhea, malabsorption, and sometimes fever.

Main clinical symptoms: Diarrhoea, vomiting, loss of weight
Incubation period: 2–13 days
Prepatent period: 7–9 days
Patent period: 2 weeks to 1–2 months (in case of AIDS patients)
Diagnosis: Microscopic determination of oocysts in fecal samples
Prophylaxis: Avoid contact with human feces
Therapy: Treatment see → Coccidiocidal Drugs.

Ixodidiosis

Disease due to infestation with ixodid ticks, see Table 1, Table 2, Table 3

Table 1. One Host Ticks and Control Measurements

Parasite	Host	Vector for	Symptoms	Country	Therapy		
					Products	Application	Compounds
Boophilus microplus (Cattle fever tick)	Cattle	*Babesia bigemina* (Red Water Disease), Anaplasmosis	Blood loss, local dermatitis	Africa, Australia, Asia, Central- and South America	Taktic™ E.C. (Intervet) Topline™ (Merial)	Spray or Dip Pour on	Amitraz Fipronil
Boophilus decoloratus	Cattle	*Babesia bigemina* (Red Water Disease), Anaplasmosis, Spirochaetosis			Bayticol™ (Bayer)	Pour on	Flumethrin
Boophilus annulatus (Southern cattle fever tick)	Cattle	*Babesia bigemina* (Red Water Disease), Anaplasmosis, Spirochaetosis			Taktic™ E.C. (Intervet)	Spray or Dip	Amitraz
Boophilus calceratus	Cattle	*Babesia bigemina* (Red Water Disease), Theileriosis					

Table 2. Two Host Ticks and Control Measurements

Parasite	Host	Vector for	Symptoms	Country	Therapy		
					Products	Application	Compounds
Hyalomma aegypticum	Ruminants	*Theileria annulata* (tropic theileriosis), Rickettsiosis	Blood loss, local dermatitis	Africa, Asia, South Europe	Bayticol™ (Bayer)	Pour on	Flumethrin
Hyalomma marginatum	Ruminants	*Theileria annulata* (tropic theileriosis), Rickettsiosis			Taktic™ E.C.	Spray or Dip	Amitraz
Hyalomma transiens	Ruminants	Sweating sickness		Africa			
Rhipicephalus evertsi	Ruminants	*Babesia bigemina* (Red Water Disease), Anaplasmosis		Africa, Florida			

Table 3. Three Host Ticks and Control Measurements

Parasite	Host	Vector for	Symptoms	Country	Therapy		
					Products	Application	Compounds
Ixodes ricinus	Dog, Cat, Man	Tick borne encephalitis, Babesiosis, Lyme disease	Blood loss, local dermatitis	Central Europe, Northern Africa	Dura Dip™ (Davis)	Spray, Sponge-on or Wash	Rotenone
					Performer™ Flea and Tick Collar (Performer)	Collar	Naled
					Adams™ Flea and Tick Dip (Pfizer Animal Health)	Dip	Chlorpyrifos
					Duocide Flea and Tick Collar (Allerderm/Virbac)	Collar	Chlorpyrifos
					Bayticol™ (Bayer)	Spray	Flumethrin
					Mycodex™ Pet Shampoo, Carbaryl (Pfizer)	Shampoo	Carbaryl
					Kiltix™ (Bayer)	Collar	Flumethrin + Propoxur
					Zodiac Duo-Op™ (Exil)	Spray	Pyrethrin + Piperonylbutoxid + N-octyl bicycloheptene dicarboximide + S-Methoprene
					Defend™Just-For-Dogs Insecticide (Mallinckrodt)	Spray	Pyrethrin + Permethrin + Piperonylbutoxid + N-octyl bicycloheptene dicarboximide
					Exspot™ (Mallinckrodt)	Spot on	Permethrin
Ixodes ricinus continuation	Cattle, Horse, Pig	*Trypanosoma theileri, Babesia bovis, Babesia divergens, Babesia motasi* (northern countries), Louping-ill-virus, tick borne encephalitis, *Borrelia burgdorferi*, Tettnang-virus, Swiss rickettsia, *Dipetalonema* larvae, *Ehrlichia phagocytophila* (Rickettsiosis)	Blood loss, local dermatitis	Central Europe, Northern Africa	Bayticol™ (Bayer)	Pour on	Flumethrin

Table 3. Continued

Parasite	Host	Vector for	Symptoms	Country	Therapy		
					Products	Application	Compounds
Ixodes hexagonus	Dog, (Cat)		Blood loss, local dermatitis	Central Europe	see page 267		
Ixodes canisuga	Fox, Dog						
Ixodes pilosus	Ruminants		Blood loss, local dermatitis, tick paralysis	South Africa			
Ixodes rubicundus	Ruminants			South Africa			
Ixodes scapularis (Black-legged tick)	Ruminants, Dog, (Cat), Man	Anaplasmosis, Ehrlichiosis (human granulocytic ehrlichiosis), *Borrelia burgdorferi, Babesia microti*		North America	Taktic™ E.C. (Intervet)	Spray or Dip	Amitraz
Ixodes holocyclus	Dog, Cattle, Man		tick paralysis	Australia			
Haemaphysalis concinna	Dog, (Cat)		Blood loss, local dermatitis	Central Europe			
Haemaphysalis punctata	Cattle, Horse	*Babesia major, Babesia motasi,* Q-fever		Asia, Africa, Europe			
Haemaphysalis cinnabarina	Ruminants	Babesiosis, Anaplasmosis, Q-fever (*Coxiella burnetii*)	Blood loss, local dermatitis	Asia, Africa, Europe			
Haemaphysalis otophila	Ruminants	*Babesia motasi*		Asia, Africa (not Egypt), Europe			
Dermacentor marginatus	Dog, Cattle, Horse, Man	Babesia canis, *Babesia ovis, Babesia bovis, Babesia caballi, Babesia equi, Theileria ovis, Anaplasma ovis,* Q-fever (*Coxiella burnetii*), Tularemia (*Francisella tularensis*)		Central Europe, China, Iran, Northern Afghanistan			
Dermacentor reticulatus	Cattle, Sheep, Horse	*Babesia divergens*		Europe, China			
Dermacentor pictus	Dog			Central Europe			
Dermacentor albipictus (Winter or Moose tick)	Ruminants	Anaplasmosis, Phantom moose disease (Canada), Colorado tick fever virus		North America, Mexico	Taktic™ E.C. (Intervet)	Spray or Dip	Amitraz

Table 3. Continued

Parasite	Host	Vector for	Symptoms	Country	Therapy		
					Products	Application	Compounds
Dermacentor andersoni (Rocky Mountain wood tick)	Ruminants, Man	Colorado tick fever virus, Powassan virus, *Anaplasma marginale, Anaplasma ovis*, Tularemia (*Francisella tularensis*), Q-fever (*Coxiella burnetii*), Rocky Mountain spotted fever (*Rickettsia rickettsii*)	Blood loss, local dermatitis, tick paralysis	North- and Central America			
Dermacentor variabilis (American dog tick)	Ruminants, Dog, (Cat), Man	*Rickettsia rickettsii* (Rocky Mountain spotted fever), Sawgrass virus, *Anaplasma marginale*		North- and Central America	Taktic™ E.C. (Intervet)	Spray or Dip	Amitraz
Hyalomma mauritanicum	Ruminants, Man	Theileriosis, Crimean-Congo haemorrhagic fever virus	Blood loss, local dermatitis	Africa			
Rhipicephalus sanguineus (Brown dog tick or Kennel tick)	(Import-) Dog, Cat	*Babesia canis, Ehrlichia canis, Rickettsia rhipicephali, Rickettsia conorii*, Crimean-Congo haemorrhagic fever virus, Thogotovirus	Blood loss, local dermatitis	Mediterranean (area) (Africa, South Europe), can be established in buildings worldwide	Dura Dip™ (Davis)	Spray, Sponge-on or Wash	Rotenone
					Performer™ Flea and Tick Collar (Performer)	Collar	Naled
					Adams™ Flea and Tick Dip (Pfizer)	Dip	Chlorpyrifos
					Duocide Flea and Tick Collar (Allerderm/Virbac)	Collar	Chlorpyrifos
					Escort™ (Schering-Plough)	Collar	Diazinon
					Mycodex™ Pet Shampoo, Carbaryl (Pfizer)	Shampoo	Carbaryl
					Frontline™ Top Spot (Merial)	Spot on	Fipronil
					Bayticol™ (Bayer)	Spray	Flumethrin

Table 3. Continued

Parasite	Host	Vector for	Symptoms	Country	Therapy		
					Products	Application	Compounds
					KiltixTM (Bayer)	Collar	Flumethrin + Propoxur
					Zodiac Duo-Op™ (Exil)	Spray	Pyrethrin + Piperonylbutoxid + N-octyl bicycloheptene dicarboximide + S-Methoprene
					Defend™Just-For-Dogs Insecticide (Mallinckrodt)	Spray	Pyrethrin + Permethrin + Piperonylbutoxid + N-octyl bicycloheptene dicarboximide
					Exspot™ (Mallinckrodt)	Spot on	Permethrin
					Taktic™ E.C. (Intervet)	Spray or Dip	Amitraz
Rhipicephalus bursa	Cattle, Sheep, Goat	*Babesia ovis, Babesia motasi, Theileria ovis*, Crimean-Congo haemorrhagic fever virus	Blood loss, local dermatitis, ovine paralysis	Africa, South Europe	Taktic™ E.C. (Intervet)	Spray or Dip	Amitraz
Rhipicephalus appendiculatus (Brown ear tick)	Ruminants	*Theileria parva parva* (East coast fever), *Theileria parva lawrencei* (Corridor disease), Nairobi sheep disease virus, *Theileria taurotragi, Ehrlichia bovis, Rickettsia conori*, Thogotovirus	Blood loss, local dermatitis, fatal toxemia (in susceptible cattle, severe ear damage	Africa, South Europe	Taktic™ E.C. (Intervet)	Spray or Dip	Amitraz
Rhipicephalus capensis	Ruminants		Blood loss, local dermatitis	Africa, South Europe	Taktic™ E.C. (Intervet)	Spray or Dip	Amitraz
Rhipicephalus zambeziensis	Ruminants	*Theileria parva lawrencei* (Corridor disease)		Africa	Taktic™ E.C. (Intervet)	Spray or Dip	Amitraz
Amblyomma americanum (Lone star tick)	Ruminants, Dog, Man	*Ehrlichia chaffeensis* (Human monocytic ehrlichiosis), Tularemia (*Francisella tularensis*) Q-fever (*Coxiella burnetii*), *Rickettsia rickettsii* (Rocky Mountain spotted fever), Lyme disease (*Borrelia burgdorferi*)	Blood loss, local dermatitis, tick paralysis	America	Commando™ Insecticide Cattle Ear Tag (Fermenta)	Ear tag	Ethion

Table 3. Continued

Parasite	Host	Vector for	Symptoms	Country	Therapy		
					Products	Application	Compounds
Amblyomma variegatum (Tropical African bont tick)	Ruminants	Q-fever (*Coxiella burnetii*)	Blood loss, local dermatitis, tick paralysis	Africa	Taktic™ E.C. (Intervet)	Spray or Dip	Amitraz
Amblyomma maculatum (Gulf Coast tick)	Ruminants, Dog, (Cat)	Q-fever (*Coxiella burnetii*),		America	Commando™ Insecticide Cattle Ear Tag (Fermenta)	Ear tag	Ethion
Amblyomma hebraeum (Southern African bont tick)	Ruminants, Man	*Rickettsia conorii* (tick typhus), *Cowdria ruminantium*		Africa			
Amblyomma cajennense (Cayenne tick)	Ruminants, Man (larvae: seed ticks)	*Rickettsia rickettsii* (Rocky Mountain spotted fever)		America			

For further species see page 572.

J

Japanese B Encephalitis

→ *Arbovirus* (Vol. 1) disease in humans being transmitted during the bite of ceratopogonid Diptera (e.g. *Forcipomyia taiwana*).

K

Kala Azar

Synonym
Visceral leishmaniasis (→ Leishmaniasis, Man/ Visceral Leishmaniasis)

Kalabar Swelling

→ *Filariidae* (Vol. 1), clinical symptom due to infection with *Loa loa*.

Katayama Syndrome

Clinical symptoms in humans due to rather fresh infections with *Schistosoma japonicum*.

Kidney Worm

Synonym
→ *Stephanurus dentatus* (Vol. 1), → *Dioctophyme renale* (Vol. 1)

Knobs

→ Malaria, → *Plasmodium falciparum* (Vol. 1)

Kyasanur Forest Disease

Synonym
KFD

General Information
The Kyasanur forest disease in India is associated with the tick *Haemaphysalis* spp. and is related to → Russian spring-summer encephalitis. It is caused by the KFD virus (→ *Flavivirus* (Vol. 1), group B). The mortality rate is approximately 5%.

L

Lambliasis

Synonym
→ Giardiasis, Man.

Larvicides

→ Disease Control, Methods.

Latency

Period of concealment of potential agents of disease inside a host.

Leishmaniacidal Drugs

Table 1

General Information
Leishmania spp. are intracellular parasites (amastigote stages) that affect mainly humans, dogs and rodents. The parasites are transmitted to various hosts by bites of sandflies (*Phlebotomus* spp. and *Lutzomyia* spp., small in size). *Leishmania* invades resting macrophages and reaches cells of the reticuloendothelial system in various organs causing inflammatory processes and immune-mediated lesions. *Leishmania* can cause various disease patterns. Leishmaniasis comprises a variety of syndromes ranging from asymptomatic and self-healing infections (e.g. single cutaneous lesions caused by *L. major* or *L tropica*) to those with a significant morbidity and mortality. The lesions may be confined to skin or disseminated to various tissues as in the case of the potentially fatal visceral leishmaniasis (**VL**). This zoonotic form (Kala-azar, Dum Dum Fever, or Black Sickness) is produced by *L. donovani* in China, India, the Middle East and Africa, by *L. infantum* in North Africa and the Mediterranean region, and by *L. chagasi* in Latin America. Various clinical signs referring to the Old World cutaneous leishmaniasis (**CL**) are due to *L. major*, *L. tropica*, *L. aethiopica* and certain zymodemes of the *L. infantum* complex. *L. mexicana* complex and *L. braziliensis* complex cause the New World cutaneous leishmaniasis and mucocutaneous leishmaniasis (**MCL**) ('Espundia'); they focally occur from Texas (USA) and Mexico southwards throughout Central America and South America as far south as São Paulo state of Brazil. All species except *L. tropica* are essentially zoonoses that occur in scattered foci primarily rural and suburban, but there is a trend towards urbanization. Annually, about 500,000 clinical cases of VL occur worldwide and more than 200 million people are exposed to infection.

Zoonotic VL in **dogs** is a progressive systemic disease characterized by chronic wasting. Initial clinical signs are vague and may be weight loss, fever anorexia, and exercise intolerance. Clinical signs indicative of systemic involvement include nonpruritic skin lesions, peripheral lymphadenopathy, lameness and epistaxis. However, there may be different clinical features depending on individual variations, species of *Leishmania* and phase of the disease. CL is a localized skin disease, which can show cutaneous, or mucocutaneous nodules and ulcerations but does involve other organs. Canine leishmaniasis in the 'Old World' is mainly due to *L. infantum* endemic in parts of Spain and throughout the Mediterranean basin where its incidence may be up to 40 percent. In the 'New World', *L. chagasi* (reservoir crab eating 'fox', *Cerdocyon*, possibly others, dogs serve as domestic reservoir) is the causal agent for American VL in dogs.

Drugs acting on Leishmaniases of Humans and Animals
For 50 years, pentavalent antimony compounds (sodium stibogluconate, identical to sodium anti-

Table 1. Drugs used against *Leishmania* in humans and dogs.

Parasite, disease distribution, pathology	**Stages** affected (location), morphology of eggs	**Chemical class** other information	**Nonproprietary name** adult/pediatric dosage (various routes), comments	**Miscellaneous comments**
Old World and New World Leishmaniasis: are transmitted by sandflies and caused by various *Leishmania* occurring in the Old World and New World; in humans and animals, *Leishmania* produce numerous clinical manifestations attributed to them; thus leishmaniasis comprises a variety of syndromes ranging from asymptomatic and self-healing infections (e.g. single cutaneous lesions caused by *L. major* or *L. tropica*) to those with a significant morbidity and mortality. The lesions may be confined to skin or disseminated to various tissues as in the case of the potentially fatal visceral leishmaniasis (VL); post kala-azar dermal leishmaniases (PKDL) is a relatively common consequence of therapeutic cure from VL caused by *L. (L.) donovani*; amastigote (intracellular) stages can be diagnosed clinical, parasitologic, serologic, and by isoenzyme electrophoresis, and DNA based detection; however, it may be difficult to detect amastigotes in impression smears or in biopsy material; these very small spherical to ovoid stages characterized by large nucleus and a prominent ovoid or rod-shaped kinetoplast may be differentiated from organisms such as *Histoplasma* or *Toxoplasma*; *Leishmania* has recently been divided into two subgenera, *Leishmania* (*Leishmania*) (most species) and *Leishmania* (*Viannia*), e.g. *L. (V) braziliensis* and related species, taking into consideration many factors, including morphology, biochemical and genetic characteristics of the organisms as well as their geographic distribution, clinical manifestations, and epidemiological factors				
DRUGS MAY BE USED FOR TREATMENT OF ALL LEISHMANIASES IN HUMANS AND ANIMALS				
SPECIES OCCURRING IN THE OLD WORLD:				
Cutaneous leishmaniasis (=CL) (oriental score) L. (*L.*) *tropica,* *L* (*L.*) *major* *L* (*L.*) *aethiopica*	**amastigote** (parasites invade resting macrophages of the skin; infection is often confined to the dermis and subcutaneous tissue	**pentavalent antimonials**	**sodium stibogluconate** (drug of choice, 20 mg Sb/kg/d i.v. or i.m. x 20–28d (may be repeated or continued, e.g. for some forms of VL) **meglumine antimonate** (drug of choice, 20mg Sb/kg/d i.v. or i.m. x 20–28d (may be repeated or continued, e.g. for some forms of VL); Sb5+ alone, and sb5+ plus allopurinol (see below) show good activity in drug-sensitive *L. infantum* infections in *dogs*	Sb compounds are generally toxic; frequent fatigue , nausea, muscle and joint pain, increased transaminase, changes in ECG (T wave inversion), occasionally hepatic and renal dysfunction; shock, sudden death (rare)
Visceral leishmaniasis (=VL) (kala-azar) L. (*L.*) *donovani* L. (*L.*) *infantum* complex (certain strains may also cause CL; PKDL is not associated with this species)	**amastigote** (parasites initially invade resting macrophages of the skin and subsequently cells of RHS in liver, spleen lymph nodes and bone marrow)	**polyene macrolide antibiotics**	**amphotericin** B (alternative drug, 0.5 to 1 mg/kg by slow infusion daily or every 2d for up to 8 wks) lipid-encapsulated **amphotericin** B (alternative drug, 15–20 mg/kg total dose over 5 or more days	antibiotic with extreme toxicity; generalized pain, convulsions, anaphylaxis, flushing chills, fever, phlebitis, anemia, thrombocytopenia, nephrotoxicity
SPECIES OCCURRING IN THE NEW WORLD:				
American visceral leishmaniasis (= AVL) *L.* (*L.*) *chagasi* (on rare occasions it may cause CL	**amastigote** (location in host see above) clinical features in children closely resemble those of 'infantile VL due to *L.* (*L.*) *infantum* of Old World	**alkyl phospholipid**	**miltefosine** being developed as oral agent (50 mg capsules); long-term response seems impressive	topical antineoplastic agent (breast carcinoma) (ASTA Medica)
		aromatic diamidines	**pentamidine** isethionate (alternative drug, 2–4 mg/kg daily or every 2d, i.m. for up to15 doses); *L. donovani*: (4mg/kg qod = every other day, x 15 doses); CL: (2 mg/kg qod x 7 doses, or 3mg/kg qod x 4 doses)	frequent hypotension, hypoglycemia often followed by diabetes mellitus, renal damage , pain at injection site, GI disturbance, vomiting

Table 1. (Continued) Drugs used against *Leishmania* in humans and dogs.

Mucocutaneous leishmaniasis (=MCL) *L. (V.). braziliensis* *L. (V.) guyanensis* (and others) may also cause CL *L. (L.) amazonensis* (and others)	**amastigote** (parasites invade resting macrophages of the skin and then they may spread to mucocutaneous junctions to cells of RHS; extensive destruction of dermis and other tissues	**aminoglycoside antibiotics** **paromomycin** Gabbroral (Pharm.Ital), Humatin (Parke Davis)(topical formulation: ointment containing 15% paromomycin and transdermal enhancing 12% methyl benzethonium chloride in soft white paraffin): alternative drug , topically twice daily x 15d, for use in CL caused by *L. major.*	chemical identical to aminosidine (oral , parenteral and topical formulations); the latter is well tolerated; oral use can cause GI disturbance; the drug acts directly on amebae (cf. → Antidiarrhoeal and Antitrichomoniasis Drugs) and cestodes (cf. → Cestodocidal Drugs)
Cutaneous leishmaniasis (CL) *L.(L.) mexicana* *L. (V) lainsoni* *L. (V.) guyanensis* (and others)	**amastigote** (location in host see above; some species may spread to mucocutaneous junctions)	**pyrazolopyrimidine: allopurinol** (purine analogue) **azoles** (**ketoconazole, itraconazole**) **8-aminoquinoline** (WR-6026) there seem to be variations in species sensitivity to these drugs and clinical trials performed in humans have been disappointing; however, **allopurinol** has good activity in *L. infantum* infections in **dogs** after long-term treatment	experimental studies (biochemical, in vitro, animal models) indicated that these drugs should be effective against *Leishmania*)

The letter stands for day (days).
Dosages listed in the table refer to information from manufacturer or literature, and → Further Reading →Control Measurements: Medical Letter, "Drugs for parasitic infections"
More information on adverse effects, manufacturers of drugs and brand names are given in The Medical Letter and partially in → Trypanocidal Drugs, Animals.
Data given in this Table have no claim to full information. The primary or original manufacturer is indicated.

mony gluconate and meglumine antimoniate) have been the first-line drugs for the treatment of leishmaniasis in humans. The precise chemical structure of these drugs is difficult to identify. Thus quality control relies on chemical analysis for **pentavalent antimony** (Sb^{5+}) rather than sugar component, and other physicochemical analyses. Drug tolerance to antimonials in human and canine leishmaniasis is known and there may be considerable rates of treatment failure and relapsing patients; drug tolerance may be due in part to long-term treatment too. Besides unresponsiveness, these drugs may show marked toxic effects such as arthralgia, nephrotoxicity, and cardiotoxicity leading rarely to sudden death. Antimonials are administered either by intralesional infiltration in simple single cutaneous lesions or by intramuscular injection in all cases with systemic involvement. The parenteral administration may be associated with unpleasant side effects. However these drugs seem to be safe if administered in the correct doses. Antimony is excreted quickly from the body so that daily treatment is necessary throughout each course for patients with VL (regimen see Table 1). The polyene antibiotic, **Amphotericine B**, is known to be effective in the treatment of VL, MCL (South America) and systemic mycoses but because of its toxicity it has so far been used only as a second-line drug (regimen see Table 1). There are now lipid formulations of amphotericine B with lower toxicity on the market and all have been on clinical trial for leishmaniasis. Thus the unilamellar liposome formulation, AmBisome, proved highly active against VL in Europe Africa and India. *L. donovani* resistant to pentavalent antimony compounds may respond to lipid-encapsulated amphotericin B (NexStar is partner of TDR, WHO). In the search for non-toxic antileishmanials attention has been directed towards currently used oral antifungal drugs such as the allylamine, terbinafine, N-substituted azoles, ketoconazole and itraconazole. This is also true for the oral purine (hypoxanthine) analogue, **allopurinol** (see Table 1, also → Trypanocidal Drugs, Man/Drug Acting on American Trypanosomiasis (Chagas Disease) of Humans) or parenteral and topical formulations of the aminoglycoside **paromomycin** (= monomycin = aminosidine, Table 1). Most of these drugs and the 8-aminoquinoline WR6026, including synergistic combinations of antimonials either with paromomycin, allopurinol or interferon-γ, which are or were on clinical trial for VL and CL have proved variably effective so far and well tolerated. VL/HIV co-infections present special problems. Indirect methods of diagnosis (serology) frequently fail in treated and relapsing patients and direct invasive methods and skilled microscopy are then required. DNA-based identification of parasites by means of PCR method appears to provide a solution to diagnosis of persistent infections. Standard treatment of VL with conventional drugs gives poor results with HIV patients (about 40% relapsed or showed persistent chronic infections) demonstrating the importance of the immune response during chemotherapy.

Current treatment of **leishmaniasis in dogs** with pentavalent antimony derivatives and/or allopurinol does not always provide complete elimination of parasites and in most cases clinical remission. If treatment period is too short clinical relapses are common. Oral long-term treatment with **allopurinol** for 4 weeks or longer (up to several months) may lead to clinical remission after intermittent administration of the drug. Dose used is usually 10–20 mg/kg b.w. twice daily or higher (up to 30mg/kg b.w. /day) and well tolerated (sometimes vomitus). Simultaneous administration of **meglumine antimoniate** and allopurinol resulted in maintaining clinical remission in dogs. Dose regimen for Sb was 100mg/kg b.w. s.c. for 20 days, followed by discontinuation of treatment for 15 days and repetition of the same regimen for 10 days, and that for allopurinol 30mg/kg b.w. p.o. for 3 months, followed by 20 mg/kg b.w. for seven days each month. Another intermittent regimen, which has successfully been used in treating canine leishmaniasis, is the intravenous (iv.) administration of meglumine antimoniate alone or in combination with oral allopurinol. The intermittent regimen with meglumine antimoniate was 50 mg/kg b.w. (diluted with 0.9% NaCl solution) for 2 days, followed by 100 mg/kg b.w. for 8 days. After discontinuation of treatment for 14 days, the same dosage regimen was repeated. The overall maintaining clinical remission was satisfactory in most patients but bone marrow continued to be PCR positive in the majority (11 of 16) of treated dogs.

Leishmaniasis, Animals

Synonym

Leishmaniosis

Pathology

Leishmaniosis is caused by protozoa of the genus → *Leishmania* (Vol. 1) that affect various mammalian hosts, but disease occurs most commonly in humans and dogs. The disease in dog is caused by *L. infantum*. The parasite is obligatory intracellular. It multiplies within macrophages and other cells of the mononuclear phagocytic system and causes chronic inflammatory processes. Clinically, the disease in dogs is characterized by a chronic loss of weight, non-regenerative anaemia, intermittent pyrexia, and generalized or symmetrical lymphadenopathy. Cutaneous lesions are very common, and include dry exfoliative dermatitis, nodules, ulcers, onychogryphosis (clawlike curvature of the nails), and diffuse, symmetrical or periorbital alopecia. Ocular lesions such as keratoconjunctivitis, uveitis and panophthalmitis may be present. Other signs include intermittent lameness, epistaxis, arthropaties, ascitis and intercurrent diarrhoea. During postmortem examination, generalized lymphadenopathy, and hepato- and splenomegaly are also observed.

Immune Responses

→ Leishmaniasis, Man/Immune Responses.

Therapy

→ Leishmaniacidal Drugs.

Leishmaniasis, Man

Synonyms

Skin Form: Oriental Sore, Aleppo Boil, Delhi Boil, Chiclero's Ulcer, French Bouton d'Orient; *Visceral Form*: Kala azar; *Mucocutaneous Form*: Espundia

General Information

Leishmaniasis is a disease corresponding to a large spectrum of clinical symptoms, including visceral (VL), cutaneous(CL), diffuse cutaneous (DCL) and mucocutaneous (MCL) forms. Recent estimations indicate that more than 400 million people are at risk of catching VL and CL and the annual number of cases of VL or CL has turned into hundreds of thousands. The different species are responsible for various clinical manifestations and exhibit peculiarities of their natural cycle such as animal reservoirs or species of vectors as well as epidemiological features. Therefore a universal control strategy is not possible. Species such as *L. major, L. tropica, L. braziliensis, L. mexicana*, and *L. aethopica* cause mostly single, self-healing cutaneous → ulcers in humans while chronic diffuse cutaneous forms or progressively destructive muco-cutaneous forms occur after infection with *L. mexicana* and *L. amazonensis* or *L. braziliensis*, respectively. The most severe, visceral form (Kala azar), which is fatal, if left untreated, and affects spleen, liver and bone marrow, is caused by *L. donovani* and *L. infantum* (→ *Leishmania (Vol. 1)*).

The infections usually start in the skin after the bite of a phlebotomid sand fly which inoculates promastigotes. In all clinical forms of leishmaniasis (see below) amastigotes multiply in monocytes. Parenchymal cells appear rarely to be involved, suggesting that organisms are phagocytized. Although the organisms are capable of multiplying extracellularly, such as in the gut of the sandfly or in culture, there appears either to be little or no extracellular multiplication in the mammalian host, where such organisms are destroyed by the processes of immunity. Traditionally *L. major, L. tropica, L. aethiopica, L. mexicanum, L. peruviana, L. brasiliensis,* and *L. pifanoi* have been recognized, with several subspecies. The classification of the various leishmanial groups by zymodemes and serodemes and the correlation with clinical forms is in progress.

Pathology

The primarily cutaneous forms of leishmaniasis may be limited to the skin and adjacent tissues, possibly because the temperature optimum of the causative organisms is 33° - 35°C, i.e. as in the skin; also the expression of cellular immunity is impaired at lower skin temperatures. A similar situation appears to account for the superficial localization of leprosy. In contrast to this, the organisms causing visceral leishmaniasis infect the deep tissues even though they are inoculated by *Phlebotomus* spp. bite into the skin.

Cutaneous Leishmaniasis (CL) Cutaneous leishmaniasis occurs in both the Old and New Worlds, produced by *Leishmania tropica, and L. mexicanum* . Lesions start as papules composed of proliferating histiocytes (macrophages) which contain numerous amastigotes (→ Pathology/Fig. 1). The lesions are usually found on the exposed areas of the face or extremities, at the presumed inoculation site. Satellite lesions develop sometimes on skin surfaces with intact epidermis. Diagnosis is easily accomplished in histologic sections or impression smears; however organisms may be sparse. The lesions become infiltrated by varying numbers of lymphocytes and plasma cells and eventually become granulomatous, containing fewer amastigotes after several weeks or months. With the development of delayed hypersensitivity the lesions ulcerate. They become secondarily infected with bacteria and the base of the ulcer contains neutrophils. The amastigotes remain in the epidermally covered areas peripheral to the ulcer and can best be isolated by aspiration from there for diagnosis in culture.

With developing immunity, the ulcers heal with granulation tissue and fibrosis, often leaving a slightly depressed scar. However, chiclero ulcers, typically on the earlobes, do not heal readily in Mexico and Central America, a fact believed to result from the lower body temperature which impairs the expression of cellular immunity.

Main clinical symptoms: Skin nodules, papulae, ulceration, necrosis
Incubation period: 2–4 weeks up to 1 year
Prepatent period: 1–3 weeks
Patent period: Months
Diagnosis: Microscopic determination of amastigotes in skin biopsies, serodiagnostic methods, → serology (Vol. 1)
Prophylaxis: Avoid the bite of the vector
Therapy: See → Leishmaniacidal Drugs.

Mucocutaneous Leishmaniasis (MCL) Mucocutaneous leishmaniasis, or espundia, caused by *Leishmania brasiliensis* complex is also transmitted by sandfly bite in South America. However, skin lesions often metastasize from the site of inoculation to other areas of skin and the mucous membranes, especially of the oro- and nasopharynx. Histologically the lesions are granulomatous with relatively few amastigotes and numerous lymphocytes and plasma cells. The lesions ulcerate, become bacterially infected, and often persist for months or years, at times destroying the cartilaginous nasal septum.

Diffuse Cutaneous Leishmaniasis (DCL) Diffuse cutaneous leishmaniasis may be produced by a distinct species of *Leishmania,* or it may be an individual reaction, as occurs in lepromatous leprosy. It occurs in the Caribbean, Brazil, and Ethiopia (Fig. 1). Huge numbers of macrophages filled with amastigotes accumulate and develop into nodular cutaneous lesions without necrosis, ulceration, or the formation of granulomas, and accompanied by only few lymphocytes and plasma cells (→ Pathology/Fig. 14).

Visceral Leishmaniasis (VL) Visceral leishmaniasis, or Kala-azar, occurs in South Europe, the Middle East, India, Africa, and focally in Central and South America (Fig. 2). It is produced by several forms of *Leishmania* which can be arranged into several groups according to results of isoenzyme analysis, antibody tests, and nucleic acid analysis; these groups may include species other than the classical species, *L. donovani, L. infantum, in the* Mid-east, and *L. chagasi* in Latin America. The reticuloendothelial cells of the viscera are parasitized by amastigotes and multiply greatly, resulting in hepatomegaly and splenomegaly (up to 3000 g) which is palpable through the abdominal wall. Splenomegaly leads to hypersplenism with erythrophagocytosis, anemia, and is accompanied by hyperglobulinemia and hypoalbuminemia. The lymph nodes and bone marrow are usually also involved. Impaired hematopoiesis, leucopenia and thrombocytopenia are commonly found. Histological examination shows that the Kupffer cells of the liver and the histiocytic cells of the spleen are filled with large numbers of amastigotes; the hepatic parenchymal cells often show steatosis and atrophy and the splenic follicles are also atrophic. Immunoglobulins (IgA, IgM, and IgG) are deposited in the glomerular mesangia and around the tubules in the kidney. A long febrile course with progressive cachexia and secondary infection often precedes death. An unknown number of patients recover spontaneously and many do after timely chemotherapy with a regression of the reticuloendothelial hyperplasia. Some of these patients develop post-kala-azar dermal leishmaniasis with amastigote-laden histiocytes accumulating in the skin and producing nodules covered by thin epidermis similar to an anergic cutaneous leishmaniasis (→ Pathology/Fig. 14). Apparently, effector mechanisms of cellular immunity, which operate in the viscera, were not effective in the cooler skin.

Main clinical symptoms: Fever of 39–40 °C, with two peaks in 24 h, anaemia, leucopenia, pale skin, kachexia, bacterial superinfections

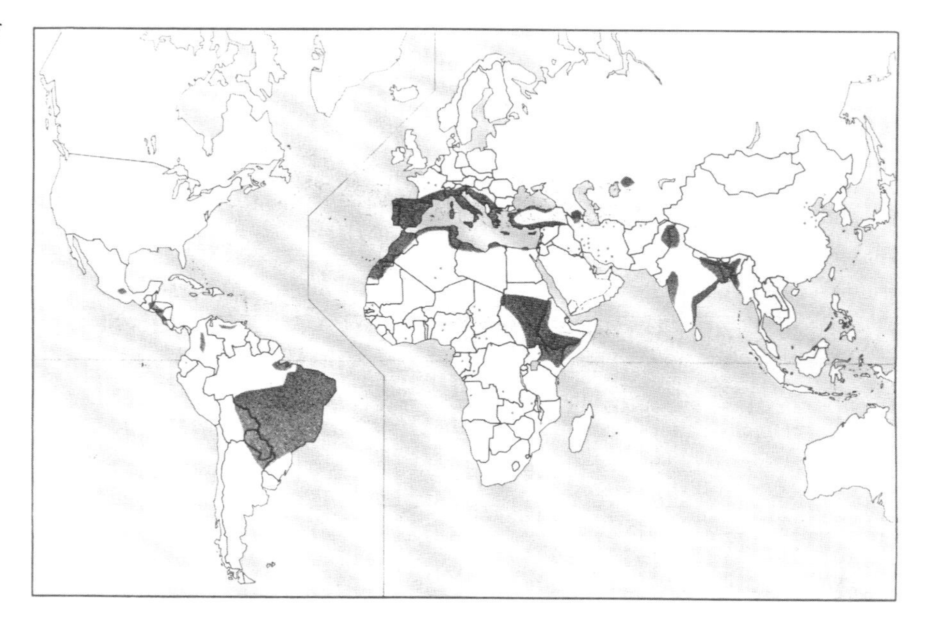

Fig. 1. Distribution map of skin leishmaniasis (according to WHO).

Incubation period: 10 days to 1 year
Prepatent period: 1–3 weeks
Patent period: Months to years
Diagnosis: Serologic tests and microscopic determination of smear preparations of bone marrow, → serology (Vol. 1)
Prophylaxis: Avoid the bite of phlebotomids in endemic regions
Therapy: Treatment see → Trypanocidal Drugs, Man and → Leishmaniacidal Drugs

Immune Responses

In their mammalian host, Leishmania typically reside within macrophages, dendritic cells and fibroblasts which not only serve as potentially safe habitats for the parasite, but may also possess antigen-presenting and/or antimicrobial functions. Experimental infections of mice with either *L. major* or *L. donovani* have greatly attracted many immunologists over the last decades to study the role of innate or acquired immune mechanisms to control an intracellular microorganism. In particular, the existence of inbred mice, which either cure or succumb to the infections has helped to define protective and non-protective functions of the immune system.

Innate Defense Mechanisms While non-infective, procyclic *Leishmania* promastigotes are sensitive to complement-mediated lysis, the infective stages transmitted by sand flies (metacyclic promastigotes) are relatively resistant to direct serum killing. As shown recently by Dominguez and Torano, *Leishmania* promastigotes bind natural anti-Leishmania IgM antibodies within 30 sec, which then activate the classical complement pathway resulting in opsonization by the third component of complement. The opsonized promastigotes then bind quantitatively to erythrocyte CR1 receptors. Progression of infection implies promastigote transfer from erythrocytes to monocytes / macrophages where the parasite uptake is predominantly mediated by CR3. Since crosslinking of the CR3 does not elicit an oxidative burst in monocytes, complement components of the host are used by Leishmania for silent invasion of host macrophages. Macrophages harbor Leishmania and allow the parasite to replicate or when activated by appropriate stimuli such as IFN-γ, kill and destroy the parasites. Toxic nitrogen products, predominantly nitric oxide, which are synthesized by iNOS, are the main parasitocidal molecules produced by activated macrophages. Mice genetically deficient for iNOS or treated with iNOS inhibitors are unable to restrict parasite replication and reactivation of latent leishmaniasis occurred after treatment of long-term-infected C57BL/6 mice with the specific iNOS-inhibitor L-iminoethyl-lysine (L-NIL). In addition, iNOS appeared to have important immunoregulatory functions during the early phase of an Leishmania infection. At day 1 of infection genetic deletion or functional inactivation of iNOS abolished the IFN-

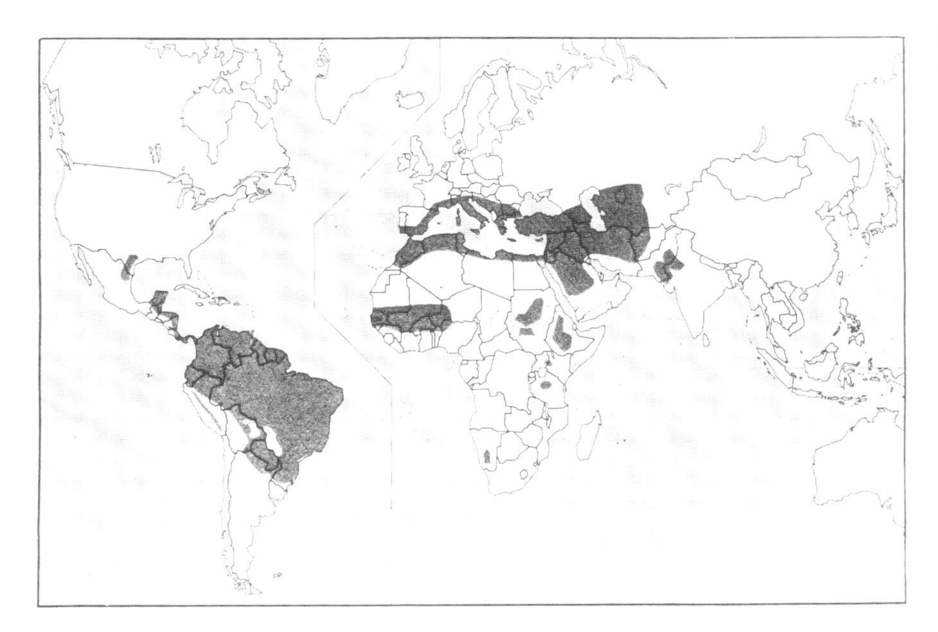

Fig. 2. Distribution map of visceral leishmaniasis (according to WHO).

γ and NK cell response, increased the expression of TGF-β, and caused systemic parasite spreading. Since neutralization of IFN-α/β in vivo inhibited iNOS-expression and mimicked the phenotype of iNOS-deficient mice, type I interferons and iNOS are critical regulators of the innate immune response to *L. major.*

More than macrophages, dendritic cells are extraordinarily efficient in presenting antigen to naive T cells. Langerhans cells of the skin ingest Leishmania parasites, process native antigen, and express relevant epitopes in context with MHC molecules on their surface. The Langerhans cells move to the draining lymph nodes where they activate parasite-specific naive T cells. Evidence suggesting that this takes place not only in experimentally infected mice but also in humans comes from immunohistochemical investigations of biopsy material from patients with cutaneous leishmaniasis: Langerhans cells containing Leishmania antigens have been found in the epidermis and dermis at the site of an oriental sore.

Homogenous populations of mouse mast cells released preformed mediators such as b-hexosamidase or TNF in response to living Leishmania promastigotes. By local cutaneous reconstitution of mast cell-deficient mice, it was found, that the presence of mast cells augmented the lesion size caused by *L. major.* However, there was no influence of mast cells on the cytokine response in the draining lymph nodes or the ultimate outcome of the infection.

Studies with *L. major*-infected mice pointed at NK cells as an important source of IFN-γ during the early course of infection. Genetically resistant mice had a higher NK-activity after infection than susceptible mice and the depletion of NK cells resulted in less IFN-γ production and a transient increase in lesion size. Vice versa, activation of NK cells in susceptible mice by injection of poly I-C enhanced IFN-γ synthesis and led to lower parasite burdens. However, the effects of NK cells appeared to be transient and did not influence the ultimate outcome of experimental *L. major* infections. C57BL/6 mice deficient in NK cell activity due to the beige mutation were less able to control *L. donovani* infection compared to normal control mice and reconstitution with NK cells restored this defect. While in lesions of patients with cutaneous leishmaniasis high numbers of NK cells have been detected, impaired NK activity has been found in the blood of patients with visceral leishmaniasis, which could be restored in vitro by incubation with IL-2.

B Cells and Antibodies In vivo, B cells respond to Leishmania infections by production of parasite specific antibodies, which are generally considered to be not protective against the intracellular Leishmania. The levels of Leishmania-specific antibodies may be very high and in most severe infections an unspecific polyclonal B cell activation leading to hypergammaglobulinemia occurs additionally.

Although B cells can not be infected by Leishmania parasites, activated B cells are able to process and present leishmanial antigens to T cells. It has been proposed, that antigen presentation by B cells is involved in the generation of a Th-2 response. In fact, BALB/c mice treated neonatally with anti IgM were resistant to *L. major,* and BALB/c X-linked immunodeficient (Xid) mice, which lacked the B1 subset of B cells, displayed enhanced resistance to *L. major.* In line with these findings, the co-transfer of B cells converted resistance into susceptibility in T cell-reconstituted, *L. major*-resistant scid mice. However, more recent experiments with mice harboring a targeted disruption of the IgM locus (μMT mice) and therefore lacking B cells, showed no influence of B cells on the polarization of T helper cells: μMT mice on the BALB/c background were susceptible to *L. major* infection and developed a Th2 response.

T Cells and Cytokines A significant increase of γ/δ T cells was found in skin lesions of patients with cutaneous leishmaniasis. Similarly, expansion of gd T cells has been observed in genetically resistant mice following *L. major* infection, indicating that γ/δ T cells may be involved in host defense against this parasite. However, C57BL/6 knockout mice lacking γ/δ T cells (TCR δ -/-) effectively controlled the infection and produced similar levels of IFN-γ when compared with control mice, strongly arguing against an essential protective role of this T cell subset in *leishmania* infection. In contrast, mice depleted of or genetically deficient for conventional α/β T cells were unable to control leishmania parasites. While ample evidence has demonstrated the central role of $CD4^+$ Th cells in the control of a *L. major* infection, the role of $CD8^+$ T cells in cutaneous leishmaniasis is less well defined. Although $CD8^+$ T cells appear to be important for resistance to a secondary challenge with *L. major,* their appears to be no essential function of $CD8^+$ T cells in primary infection. Both, mice genetically deficient for β2-microglobulin and CD8 thus lacking $CD8^+$ T cells were able to mount an effective and long-lasting immune response against *L. major*.

Unlike the *L. major* model, resolution of primary *L. donovani* infection requires not only $CD4^+$ T cells but also $CD8^+$ T cells. Acquisition of resistance involves the secretion of IL-2, IFN-γ and TNF. Similar to cutaneous leishmaniasis, resistance of *L. donovani*-immune mice to rechallenge was strongly dependent on $CD8^+$ T cells.

Mice from the majority of inbred strains (C3H/He, B10.D2, C57BL/6, Sv129/Ev etc.) are resistant to infection with *L. major*, while only mice of a few strains such as BALB/c, develop progressive lesions and succumb to the infection. Healing of lesions induced by *L. major* requires the induction and expansion of specific $CD4^+$ Th1 cells that are restricted by MHC class II and produce IFN-γ, while susceptibility was found to be associated with the development of a predominant Th2 cell immune response. However it has been shown that susceptibility is not an absolute trait but one conditional on parasite dose, since infection with low numbers of parasites (about 1000-fold lower than the number employed (10^5) to define the susceptible phenotype of BALB/c mice) induced long-term protective Th1 immunity in BALB/c mice. Thus in addition to host factors the parasite dose determines the Th1/Th2 nature of the response to *L. major*, and this was found to occur independently of the infection route and parasite strain. A large number of studies has focussed on the immunregulatory mechanisms determining the Th1/ Th2 decision and the role of these different Th cells in *L. major*-infected mice.

The role of IFN-γ in the control of infection with *L. major* was firmly established by experiments showing that genetically resistant mice with disrupted genes for IFN-γ or its receptor failed to resolve their lesions. More recently, the additional importance of the Fas-Fas-L pathway in the elimination of parasites has been demonstrated. In contrast to wild-type C57BL/6 mice gld or lpr mice lacking either a functional Fas or Fas-L were unable to resolve *L. major*-induced lesions although they mounted a normal Th1 response and their macrophages produced normal levels of NO in response to IFN-γ in vitro. Since furthermore IFN-γ upregulated the expression of Fas on *L. major*-infected macrophages, thereby rendering these cells susceptible to apoptotic death by Th1 cells IFN-γ might contribute by at least two mechanisms to the defense against intracellular Leishmania.

Th1 and Th2 cells develop from a common naive precursor. Both accessory molecules and cytokines are known to influence the differentiation of $CD4^+$ T cell precursors in vivo. CD80 (B7-1) and CD86 (B7-2) as well as the CD40 molecule and its ligand have been shown to influence the Th cell differentiation and thus the clinical outcome after infection *with L. major*. Deficiency in either CD40 or its ligand resulted in the inability of mice to generate a Th1 response and to control *L. major* or *L. amazonensis* infections. While the blockade of CD86 by mAb treatment ameliorated the infection and inhibited Th2 development in BALB/c mice, BALB/c mice deficient for CD28, a ligand of CD80 and CD86, remained susceptible to infection. In contrast, the interaction of the CD4 molecule with MHC class II appeared to be of importance for the development of a Th2 cell immune response. There is ample evidence that IL-4 is essential for the development of Th2 cells after infection with *L. major*. The neutralization of IL-4 by mAb or recombinant soluble forms of the IL-4 receptor resulted in Th1 development in BALB/c mice which thereby controlled primary *L. major* infection and became resistant against secondary challenge infections. Confirmatory evidence came from experiments with mice deficient for either IL-4 or STAT-6 (one of the major IL-4 signal-transducing molecules) which were more resistant against *L. major* or *L. mexicana*, respectively, when compared with their control littermates. Only in susceptible BALB/c mice was there a very early IL-4 production by activated $CD4^+$ T cells during the first day after infection with *L. major*. A highly restricted subpopulation of $CD4^+$ T cells expressing the TCR Vβ4 and Vα8 chains specific for a single immunodominant antigen called LACK (Leishmania-activated C kinase) was identified as source for the early IL-4. Interestingly, mice deficient in Vβ-4 mounted a polarized Th1 response and were fully resistant to infection, suggesting that a single epitope of the LACK antigen drives the early IL-4 response and instructs subsequent Th2 differentiation and susceptibility to infection in BALB/c mice. In agreement with this concept, transgenic BALB/c mice expressing the LACK antigen in the thymus were tolerant to this antigen and resistant to infection with *L. major*.

However, since LACK appears not to be the dominant antigen in MHC haplotypes other than H-2^d (N. Glaichenhaus, personal communication) it remains to be determined, which antigen(s) or mechanism(s) are responsible for the susceptible phenotype of BALB congenic mice.

The essential role of IL-12 for the development of a protective Th1 cell response against *L. major* has been demonstrated by several experimental approaches. Neutralization with antibodies or disruption of the IL-12 gene in resistant mice resulted in susceptibility, while treatment of BALB/c mice with recombinant IL-12 during the first week of infection enabled these mice to develop a Th1 response and allowed the resolution of lesions. In line with this, mice deficient for the transcription factor IRF-1 (Interferon regulatory factor 1) were susceptible to *L. major*, most likely due to the impaired ability of their macrophages to produce IL-12. While in resistant C3H mice an enhanced expression of the IL-12 receptor subunits β1 and β1 was detected after *L. major* infection this was not the case in lymph nodes of BALB/c mice unless these mice were rendered resistant by neutralization of IL-4 or treatment with IL-12. Thus, the upregulation and maintenance of IL-12 receptor molecules or its counterregulation by IL-4 on $CD4^+$ T cells may be critically involved in the generation of a protective Th1 cell response.

Several other cytokines are additionally involved in the regulation of immunity against Leishmania. It has been shown that leishmanial infection induced the production of active TGF-β, both in vitro and in vivo. Since application of recombinant TGF-β markedly exacerbated the disease while treatment with anti TGF-β resulted in protection of BALB/c mice after infection with *L. amazonensis*, induction of TGF-β has been regarded as a parasite escape mechanism. TNF, which had no direct toxic effects on leishmania, was found to activate in combination with other cytokines such as IFN-γ the leishmanicidal activity of macrophages in vitro. In vivo, there were no differences in the expression levels of TNF, lymphotoxin (LT) or the TNF receptors I and II (p55 and p75) when susceptible BALB/c and resistant CBA mice were compared. Using knockout mice deficient for either TNFRp55, TNFRp75 or both receptors it was reported that the TNFRp75 plays no essential role in *L. major* infection while the TNFRp55 might be required for optimal macrophage activation. TNFRp55 deficient mice developed larger lesions than control mice and failed to resolve these lesions. However, they were able to eliminate parasites within these lesions. Migration inhibitory factor (MIF), granulocyte-macrophage colony stimulating factor (GM-CSF) and IL-7 were found to enhance leishmania killing by macrophages in vitro. While MIF delivered via a Salmonella-based expression system in vivo enhanced resistance of mice, application of GM-CSF or IL-7 surprisingly caused aggravation of lesions in *L. major*-infected mice. Cytokines such as IL-10, TGF-β (see also above) and IL-13 have been found to deactivate macrophages and to enhance intracellular survival of leishmania. With the exception of TGF-β, the role of these proteins during an immune response against leishmania in vivo remains to be determined.

The expression of chemokines has been analyzed in lesions of patients with localized cutaneous leishmaniasis and diffuse cutaneous leishmaniasis. While high levels of macrophage chemoattractant protein 1 (MCP-1) and moderate levels of macrophage inflammatory protein 1 α (MIP-1a) were detected in the localized forms of leishmaniasis, the pattern was reversed in diffuse cutaneous leishmaniasis, suggesting a functional role of these chemokines in the differential recruitment and activation of macrophages in the different forms of cutaneous leishmaniasis .

It is important to emphasize that susceptibility or resistance to *L. major* most likely involves several mechanisms since it appears to be controlled by several genes. Six loci located on the mouse chromosomes 6, 7, 10,11, 15 and 16 were found to be associated with resistance to *L. major* in BALB/c X B10.D2 backcross mice. Another study analyzing (BALB/c x C57BL/6) F2 mice showed a linkage to the h2 region on chromosome 17 and to chromosome 9.

Although *L. donovani* and *L. chagasi* also readily parasitize and cause noncuring visceral infection in inbred mice, these leishmania species do not regularly provoke an active, functional Th2 response in experimental infection as they seem to induce in human disease. The one reported exception was in mutant C57BL/6 ep/ep (pale ear) mice in which noncuring *L. donovani* infection was related to multiple host defense defects including an active Th2 response. In most other cases noncuring *L. donovani* infection has been ascribed to the failure to properly express a Th1-associated cytokine response rather than to dominant activity of Th2 cells.

In mice the susceptibility to infection with intracellular parasites such as *Salmonella*, *Mycobacteria* and *L. donovani* is controlled by the Nramp1 locus (also known as *Bcg*, *Ity* or Lsh) on chromosome 1. The integral membrane protein Nramp1 is expressed exclusively on professional phagocytes in the late endocytic compartments. Since a single nonconservative amino acid exchange at position 169 of this protein resulted in enhanced susceptibility of mice to *L. donovani*, the Nramp1 protein may alter the intravacuolar environment of the parasite-containing phagosome.

Evasion Mechanisms of Leishmania The complement resistance of metacyclic leishmania promastigotes was explained by the spontaneous shedding of the lytic membrane attack complex from the parasite surface, which might be causally linked to the elongation of the phosphoglycan chain of the surface lipophosphoglycan (LPG). In addition, leishmanial protein kinases have been reported to phosphorylate components of the complement system, thereby inhibiting the classical and alternative complement pathway. The 63 kDa surface metalloprotease (gp63) accelerated the conversion of C3b to a C3bi-like molecule, which acts as an opsonin and facilitates the uptake of leishmania into macrophages. *Leishmania* parasites are able to invade not only macrophages but also host cells devoid of important defense mechanisms such as iNOS. Langerhans cells of the skin as well as cells negative for all classical macrophage and dendritic cell markers, presumably reticular fibroblasts, might function as safe habitats for the parasite enabling its long term persistence. *Leishmania* parasites are able to survive in the phagosome and phagolysosome. LPG is able to inhibit phagosome-endosome fusion and scavenges hydroxyl radicals and superoxide anions which are rapidly produced during phagocytosis. In addition, the protease activity of gp63 has been shown to protect the parasites from intraphagolysosomal degradation and is required for virulence of leishmania. Leishmania parasites are able to interfere with both main antimicrobial effector mechanisms, the release of superoxide and the synthesis of NO. LPG, gp63 and GIPLs, a group of glycolipids related to LPG have been shown to mediate these suppressive effects, at least in part by inhibiting the translocation and activation of the protein kinase C (PKC) of the host cell.

A further, important mechanism by which leishmania influence the host immune response is the modulation of cytokine production. As discussed above, different leishmania species induce the production of TGF-β which has been found to inhibit anti-leishmanial defense mechanisms of macrophages and to aggravate the disease in vivo. The selective suppression of IL-12 p40 synthesis by macrophages mediated by phosphoglycans of the parasite occurring on the transcriptional level appears to be an important mechanism by which leishmania avoid or delay the development of a host protective Th1 response. Interestingly, this effect appears to be cell type-specific, since uptake of *L. major* amastigotes by skin-derived dendritic cells results in activation of these cells and IL-12 release.

The processing and presentation of antigen is also targeted by the parasite. It has been demonstrated that *L. donovani* amastigotes interfered with upregulation of MHC class II molecules on the transcriptional level. Downmodulation of MHC class II occurs additionally on the posttranslational level, most likely by an enhanced internalization and degradation of these molecules. In contrast to other intracellular microorganisms, *L. donovani* prevents the upregulation of costimulatory molecules like CD80 on macrophages. The recently reported finding that gp63 selectively cleaves CD4 molecules from T cells is intriguing, as CD4 via binding to MHC class II, stabilizes the interaction between antigen-presenting cells and T helper cells. A phenomenon called antigen-sequestration not allowing the transport of sufficient numbers of MHC-peptide complexes to the cell surface may let infected macrophages go unnoticed by T helper cells. The molecular mechanism of this phenomenon is not yet defined, and prevention of intracellular protein degradation appears only partially responsible.

In addition to the numerous evasion mechanisms of *Leishmania* species summarized above, the saliva of the parasite-transmitting sandfly exerts various immunmodulatory functions. Saliva components, such as the peptide maxadilan, inhibited killing of *Leishmania* by suppressing the production of NO. Furthermore, sand fly salivary gland lysates were found to down-regulate a Th1, but to upregulate a Th2 response in mice infected with *L. major*. Interestingly, the salivary gland lysates directly upregulated expression of IL-4 mRNA also in the absence of infection with *L. major*.

Vaccination

Although a treatment for leishmaniasis exists, it is costly and difficult to apply because it requires daily injections for weeks. Moreover, resistance against the classical antimonial treatment has been increasing and increased doses, prolonged hospitalization and needs for second treatment can be necessary. Thus, an effective and affordable vaccine remains the only realistic hope of controlling such a parasitic disease. This has been the goal for many years of the tropical disease research program of the WHO, which plays a major role in *Leishmania* vaccine development and several formulations are currently under trial with encouraging results.

The use *L. major* mouse model of *Leishmania* infection has been helpful for the understanding of mechanisms involved in immunity to leishmania. It was first observed that genetically different mice presented a different degree of susceptibility to the *L. major* infection, with some strains such as C57BL/6 being resistant (spontaneous healing of controlled cutaneous lesion) while others, such as Balb/c are susceptible and present progressive disease. It was further demonstrated that the difference in the susceptibility was linked to the expansion of $CD4^+$ T cells secreting different patterns of cytokines. Th1 cytokines such as IFN-γ, in conjunction with the IL-12 and TNF-α secretion by macrophages/dendritic cells are essential for the induction of the inducible nitric oxide synthase (iNOS) leading to large amounts of NO which play a major role in the killing of intracellular *Leishmania*. In contrast, expansion of $CD4^+$ T cells secreting Th2 cytokines such as IL-4, IL-5 and IL-10 but no IFN-γ in conjunction of macrophage-deactivating factors such as TGFβ and/or PGE2, is commonly associated to non-healing infection.

Recent studies have shown that the activation/differentiation of one of the Th type inhibits the induction or expansion of the reciprocal subset, via reciprocal feed-back inhibition by Type 1 or Type 2 cytokines. For instance, IFN-γ inhibits the induction/expansion of TH2, and, IL4 and IL10 inhibit the induction TH1 cells. The mechanisms leading to the expansion of Th1 or Th2 depend on early events, IL-12 secretion in the first case and IL-4 secretion in the last case. IL-12 injection has been shown to induce protection against cutaneous leishmaniasis in susceptible Balb/c mice leading to expansion of a Th1 cell response. From these experiments, the use of IL-12 has been proposed as an adjuvant eliciting a Type 1 response for the delivery of a *Leishmania*-vaccine in particular, but also in general for vaccine against other pathogens.

If these new concepts emerging from the mouse model have allowed important progress in the comprehension of the induction of anti-*Leishmania major* immune response in humans, there is still much to do in understanding the reasons why *L. major* induces generally cutaneous lesions whereas *L. donovani* leads to visceral leishmaniasis and *L. braziliensis* to muco-cutaneous disease. Therefore vaccine development against leishmaniasis has proceeded, so far, entirely within empirical approach. The observation that *Leishmania major* induce usually benign infections with spontaneous healing after 6 to 9 months protecting from pathogenic re-infections was the starting point of vaccine strategies known as **leishmanization**. It consists in injecting viable parasites to produce a controlled lesion in a non visible area of the skin. This induces a significant protection against re-infection. This immunity is essentially T-cell-mediated. Leishmanization was used for a long time and was until recently used in the former USSR, in Israel and in Iran. However, the use of live organisms can induce persistence of parasites in the immune host able to cause serious diffuse or mucocutaneous lesions in cases of change of the immune status. This program has now been abandoned. With the development of *Leishmania* transfection techniques, the production of an avirulent strain lacking the dihydrofolate reductase/thymidilate synthetase gene infecting and persisting in the macrophage is an interesting alternative for attenuated vaccine. This vaccine does not induce side effects but, so far, very little is known about the long-term consequences of such vaccine in particular in HIV infected individuals. Killed parasites have thus renewed interest. Several trials using whole killed parasites with BCG as adjuvant are under evaluation in South America (*L. braziliensis*, *L. guyanensis* and *L. amazonensis*) and in Iran (*L. tropica* and *L. major*) and are encouraging even if they are inferior to live vaccine.

Recently significant progress have been obtained using subunit vaccines. Molecules such as gp63, gp46, PSA-2 and LACK have given interesting results in mouse models using adjuvant not appropriate to humans. Other delivery systems using recombinant bacteria such as *Salmonella typhimurium* or BCG or recombinant vaccinia virus are under study. Beside proteins, the lipophosphoglycan (LPG) seems to be an interesting candidate. It can protect mice from infection with *L. major.* Despite the prevailing dogma that only peptide can induce T cell responses, LPG is presented by Langerhans cells to the T cells, not in the context of classical MHC molecules but by the newly CD1 pathway. Because of possible genetic restriction as well as their partial protective effect, such vaccine candidates have to be mixed in a cocktail vaccine and tested as one vaccine. Subunit vaccine also has the disadvantage of inducing a usually short-lived immune response. One possible solution is to use the vaccine candidates not as proteins or peptides but as their encoding DNA. Indeed DNA vaccine is particularly attractive because it can induce a long-lived immune response. The antigen is constantly produced at low doses inducing an immune response similar to the situation of natural infection. gp63, PSA-2 and LACK delivered as plasmid DNA have already demonstrated efficient protection in mice.

The first generation of *Leishmania* vaccine against CL have already shown a relative efficacy (killed parasites), that needs to be improved by use of appropriate adjuvant. However, because the preparation of such a vaccine is difficult to standardize, research on a second generation against the different forms of leishmaniasis using defined molecules is more than ever necessary.

Leishmanization

The observation that *Leishmania major* induce a usually benign infection with → spontaneous healing after 6 to 9 months protecting from pathogenic re-infections was the starting point of vaccine strategies known as leishmanization. It consists of injecting viable parasites to produce a controlled lesion in a non visible area of the skin. This immunity is essentially T-cell-mediated. Leishmanization was used for a long time and was until recently used in the former USSR, in Israel and in Iran (→ Leishmaniasis, Man/Vaccination).

Lethality

Number of dead individuals in relation to sick people.

Leucocytozoonosis

Disease due to → *Leucocytozoon simondi* (Vol. 1) in domestic and wild ducks and geese transmitted by bite of *Simulium* species.

Lichenification

Clinical and pathological symptom (dry scrub, small papule exanthem) of infections with skin parasites (→ Skin Diseases, Animals, → Lice (Vol. 1)).

Löffler Syndrome

Hemorrhages and inflammation foci within lungs of humans during the migration phase of larval ascarids (→ Ascariasis, Man) being accompanied by dyspnoe, slight fever, blood eosinophilia plus coughing.

Loiasis

Synonym

Eye Worm Disease

Loiasis results from an infection of the subcutaneous and deep tissues with adult → *Loa loa* (Vol. 1), transmitted by a biting fly (*Chrysops*; → *Filariidae* (Vol. 1)). Larvae enter at the site of the fly bite and slowly develop into adults. Adult worms make their appearance after a year or more, when they give rise to symptoms during their subcutaneous or subconjunctival migration (→ Eye parasites). The living worm is not inflammatory, but dead worms give rise to microabscesses (→ abscess (Vol. 1)) with eosinophils. The released microfilariae circulate in the blood and when they die elicit small → granulomas with epithelioid and giant cells; these may give rise to symptoms referable to many organs, including the brain.

Main clinical symptoms: Swellings (sog. calabar-swellings) of the skin = oedema of skin, passage of worms through the eye
Incubation period: 2–12 months
Prepatent period: 6 months 4 years
Patent period: 4–17 years

Diagnosis: Microscopic analysis of blood smear, microfilariae are found at 1–5 o'clock p.m. in the peripheral blood, → serology (Vol. 1)
Prophylaxis: Avoidance of *Chrysops*-bites in West Africa
Therapy: Treatment see Nematocidal Drugs, Man; surgical removement of the worm, when passing the eye.

Loss in Performance

The parasitic load often introduces considerable reduction of the fitness of hosts, i.e. they look tired, their skin is pale, their hairs appear dull, their movements are slow and they grow slowly if at all. Thus several female birds clearly prefer bright shining and highly active male mating partners that indicate health.

Louping ill

Louping ill is a sheep disease found in North Britain which is caused by the LI virus (→ *Flavivirus* (Vol. 1), group B). It is transmissible to human beings in close contact with sheep (laboratory workers, sheep farmers, veterinarians, and butchers), or those exposed to tick bites; at least one human death has been proven.

Louse-Borne Spotted Fever

Human disease due to infection with *Rickettsia prowazeki* being transmitted by the feces of → lice (Vol. 1).

Lucilia sericata

The larvae of this fly causing → Myiasis, Man may be used to clean skin abscesses from the bacterial coat, since they feed only from necrotic tissues and not from healthy ones.

Lyme Disease

Lyme disease is caused by *Borrelia burgdorferi*, a spirochaete-like bacterium transmitted by the ticks *Ixodes dammini*, *I. pacificus* and *I. ricinus*. Transovarial transmission occurs so that the next generation of ticks may be infected. The disease begins with a slowly enlarging focal rash, → *erythema chronicum migrans*, followed by migratory or relapsing arthritis of one or several joints. → Encephalitis or myocarditis also occurs as third phase of the disease. Treatment is done by high dosage of antibiotics (e.g. doxycyclin). In the USA a vaccination has been available since 1998 (see → Ticks as Vectors of Agents of Diseases, Man).

Lymphadenitis

→ Pathology.

Lymphatic Filariasis

→ Filariasis, Lymphatic, Tropical, → Serology (Vol. 1), → *Filariidae* (Vol. 1).

Lymphocytic Meningoradiculitis Bannworth

Symptom of phase two of → Lyme disease.

M

MacDonald Model

→ Mathematical Models of Vector-Borne Diseases.

Major Surface Glycoprotein

→ MSP, → Pneumocystosis, → Surface Coat (Vol. 1).

Mal de Caderas

→ Vampire Bats (Vol. 1).

Malabsorption

→ Hookworm Disease, → Alimentary System Diseases, Animals.

Malaria

Pathology

→ *Plasmodium* (Vol. 1) spp. parasitize the red blood cells, which are metabolized during the schizogonic cycle, leaving pigment granules. Reticular and endothelial cells phagocytize red blood cell fragments and accumulate malarial pigment. *P. vivax* and *P. ovale* predominantly infect the relatively scarce young red blood cells, thus restricting the level of parasitemia. *P. falciparum* and *P. malariae* infect mature cells a few or many of which may be infected, often resulting in anemia. Red blood cells parasitized with *P. falciparum* are sequestrated in capillaries of internal organs by → knobs on their surface reacting with receptors on the vascular endothelium, thereby causing tissue anoxia. This is particularly serious in the brain → Pathology/Fig. 15), where endothelial cells die and capillaries break, giving rise to multiple petechial hemorrhages. Brain anoxia leads to edema and coma, which may be fatal in a few hours. Occasionally glial reactions are seen in response to microinfarcts.

Hepatic damage leads to deep jaundice. Phagocytosis of destroyed red cells imparts a brownish, and after fixation, slate-gray color to the enlarged liver and spleen. However, malarial pigment must be differentiated from formalin pigment or acid hematin. Marked renal tubular necrosis is sometimes seen with hemoglobinuric nephrosis in the presence of anoxia and acidosis, the so-called black water fever. Deposition of immunoglobulin in the glomeruli may lead to mesangial thickening. The preerythrocytic cycle of *P. falciparum* takes place in the liver and large schizonts are occasionally seen in hepatic parenchymal cells. With *P. falciparum*, schizogony proceeds easily in the maternal sinuses of the placenta leading to fetal anoxia, placental and fetal edema and abortion after the third month of pregnancy.

Immune Responses

One of the underlying difficulties still hindering the successful malaria vaccine design at the end of the 20th century is our incomplete knowledge of the precise type(s) of immune responses involved in the control of the different stages of the parasite on the one hand and malaria immunpathology on the other hand. Understanding the nature of the effector mechanisms to blood-borne plasmodia in humans so far have proven intractable since only blood samples on just one or a few time points have been analyzed and in most cases little is known on the detailed parasitological and immune status of the subject studied. The pliability of rodent systems for investigating immunoregulation has provided valuable insight into the balance between protection and pathology in human malaria. However, there are important differences in the antiparasitic responses between humans and rodents. First, immunity in mice to various malaria parasites can develop within weeks following infection, but rapid acquisition of immunity to malaria is not observed in humans. Children in endemic areas can take several years to develop effective immunity, while this occurs much faster in adults. A second obvious difference between the course of infection in both species is that parasitemias are typically much greater in mice than in humans, which might be explained by the fact that rodent plasmodia are not natural parasites for mice.

Nevertheless, various murine models have been developed with parasites isolated from African wild rodents. There are four rodent malaria species (*P. berghei, P. chabaudi, P. vinckei* and *P. yoelii*) which allow to investigation of diverse aspects of the host immune response to malaria. While *P. berghei*, and some *P. vinckei, P. chabaudi* and *P. yoelii* strains are lethal to mice, other strains such as *P. chabaudi adami, P. chabaudi chabaudi, P. vinckei petteri* cause infections in mice which resolve after initial parasitemia and are either eliminated completely (*P. yoelii* strains) or have smaller patent recrudescences. Although no single model reflects infections exactly in humans, the different models together provide valuable information on the mechanisms of immunity and immunopathogenesis.

One of the most studied models is that of *P. chabaudi chabaudi*-infected mice most closely resembling *P. falciparum* infection of humans since (1) it usually infects normocytes (2) undergoes partial → sequestration (although in the liver and not in the brain) and (3) in resistant strains of mice there are recrudescenses by parasites of variant antigenicity.

Protective immunity to the asexual blood stages of malaria parasites, the pathogenic stage of the life cycle, involves both cellular and antibody mediated mechanisms.

B Cells and Antibodies In both humans and mice, passive transfer of antibodies from immune individuals to those suffering from acute malaria resulted in quick and marked reduction of parasitemia. Most recently it has been reported that anti-adhesion antibodies, which limit the accumulation of parasites in the placenta, appear in women from Africa and Asia who have been pregnant on previous occasions (multigravidas). These antibodies were found to be associated with greatly reduced prevalence and density of infection. In addition, infections with *P. berghei* and *P. yoelii* can not be controlled in mice from which B cells are removed by neonatal anti-μ-treatment and the elimination of parasites is also impaired in mice with a targeted deletion of the JH-gene segment of the Ig gene locus. While IgG2a is essential in the mouse model, IgG1 and IgG3 appears to be most effective in humans. In addition several epidemiologic studies have shown a strong association of IgG1 and IgG3 antibodies with immunity to *P. falciparum*, and the same antibody subclasses might also account for the resistance of new-borns delivered by malaria-immune mothers in malaria endemic areas.

IgG1 and 3 in humans and IgG2a in mice are cytophilic isotypes able to promote activation of monocytes and macrophages via Fc receptors. The antibody-dependent killing of parasites in vitro is either dependent on the presence of mouse neutrophils or human monocytes. One mechanism possibly involved in this antibody activity is the induction of TNF by monocytes leading to growth inhibition of intracellular parasites in neighboring cells. In fact, anti-TNF antibodies prevented asexual parasite growth inhibition and parasite inhibitory activity is present in cell free supernatants. Selective agglutination of infected erythrocytes is consistently associated with reduced parasite density. There is growing evidence that both opsonizing and agglutinating antibodies recognize PfEMP1 (*P. falciparum*-infected erythrocyte membrane protein 1), a group of large(200–350 kDa) proteins inserted into the red cell membrane by mature asexual blood stage parasite. PfEMP1 proteins play a critical role in the retention of infected erythrocytes in the blood vessels avoiding sequestration of the parasite in the spleen, since PfEMP1 interacts with a variety of endothelial receptors such as CD36, E-selectin, thrombospondin, vascular cell adhesion molecule 1 and intercellular adhesion molecule 1. Thus, an important function of anti-malaria antibodies might be the inhibition of endothelial-blood cell interaction but facilitating the interaction of opsonized infected erythrocytes with phagocytes in the spleen. Indeed, it has been shown that the spleen is essentially involved in the IgG2a-dependent clearance of *P. berghei* and *P. yoelii* in mice.

T Cells Two stages of the malaria parasite are truly intracellular, that which infects the liver and the asexual stage which resides in red blood cells. Since intracellular parasitism is a strategy for evading antibody-dependent immune responses, T cells most likely are involved in the defense against malaria. Infections with plasmodia stimulate $CD4^+$ as well as $CD8^+$ $\alpha\beta$ T cell receptor expressing T cells and $\gamma\delta$ TCR^+ T cells. While mice genetically deficient for $\alpha\beta$ TCR T cells were very susceptible to *P. chabaudi* infection and died rapidly after infection, there was no difference between $\gamma\delta$ TCR-deficient mice and control mice. However, there is a differential expansion of $\gamma\delta$ T cell subset in the peripheral blood and spleens of mice and humans and it has been shown in the *P. chabaudi adami* mouse model that the $\gamma\delta$ T cell blast response coincides with the remission of parasitemia. Since $\gamma\delta$ T cells proliferate in response to malaria antigens, e.g. heat shock proteins, *in vitro*, the systemic expansion of this T cell subset in vivo might reflect the systemic release of malaria exoantigens liberated from parasitized erythrocytes upon schizont rupture.

$CD8^+$ T cells mediate killing of the liver stage of plasmodia, possibly by producing cytokines (IFN-γ, TNF) which induce the production of NO by infected hepatocytes

The central role of $CD4^+$ T cells for the protective immunity against the asexual blood stages of experimental malaria have been shown by in vivo cell depletion analysis and by cell transfer studies. Since transfer of purified $CD4^+$ T cells or of $CD4^+$ T cell lines to SCID mice or lethally irradiated mice cleared the infection only in the presence of B cells, there is the need for T-B cell interaction in the establishment of a fully protective immune response to malaria parasites.

It has been demonstrated that the host-protective response in *P. chabaudi chabaudi*-infected mice involve both Th1-type and Th2-type $CD4^+$ T cells. The relative contribution of these subsets changes during the course of infection: While Th1 cells predominate during the acute phase, Th2 cells are primarily found during later phases of infection. However, among the non-lethal murine malarias *P. chabaudi chabaudi* can be seen in an intermediate position between two poles of protective immunity. Whereas a strong Th1 response is involved in the response against *P. chabaudi adami* a dominant Th2 cell activation is found after infection with *P. yoelii* and *P. berghei*. The severeness and lethality of an infection appears to be linked to the early and strong induction of a Th2 response. This is best demonstrated by the fact that infection with a non-lethal strain of *P. yoelii* leads to both Th1 and Th2 activation while in the case of an infection with a lethal strain of *P. yoelii* only a Th2 response was detectable. It is unknown, however, why more severe infections fail to trigger an adequate Th1 response.

A protective function of both Th1 and Th2 cells has been indicated by cell transfer experiments to *P. chabaudi*-infected mice. The protective effect of transferred Th1 cells could be blocked by inhibitors of iNOS (L-NMMA), while in contrast resistance conferred by Th2 cells was not influenced. Even in the case of Th1 cells there are clearly NO-independent mechanisms of protection involved, since Th1-mediated protection against *P. yoelii* is not dependent on NO. Protective Th2 cells clones specific for *P. chabaudi chabaudi* drive a strong protective malaria-specific IgG1 response in vivo (see above) which is promoted by IL-4.

An interesting phenomenon observed is the temporal shift from Th1- to Th2-regulated immunity during *P. chabaudi chabaudi* infection of mice. It has been suggested that the type of antigen-presenting cell is critically involved in the Th cell differentiation process. While during the first days of infection professional antigen-presenting cells such as dendritic cells or macrophages might favor the development of Th1 cells, the subsequent activation of malaria-specific B cells appears to be responsible for the observed Th2 cell differentiation. The latter is strongly suggested by the fact that mice rendered B-cell deficient by lifelong treatment with anti-IgM antibodies or by targeted disruption of the Ig-μ-chain gene are unable to generate a malaria specific Th2 response, while the ability to develop Th1 cells is not altered.

In addition to the type of antigen-presenting cell or the cytokine milieu the antigen-concentration is also involved in the regulation of the different Th-subsets. Whereas high-dose challenge of resistant mice led to an enhanced Th1-mediated immune response, low-dose challenge causes more pronounced Th2 cell development.

Given the central protective role of $CD4^+$ T cells in murine models of malaria it is still puzzling that there is no major effect of the HIV pandemic on the incidence or severity of human malaria, as has been observed for tuberculosis, toxoplasmosis or leishmaniasis. It has been speculated, that progression to AIDS involves a preferential depletion of Th1-type $CD4^+$ T cells and not Th2-type cells, and that the latter might be sufficient to allow a protective antibody-mediated immunity to malaria to be maintained.

The early burst of IFN-γ production by $CD4^+$ T cells (Th1) as found in *P. tabai chabaudi*-infected mice appears to be important for the control of primary parasitemia, since neutralization of IFN-γ or injection of rIFN-γ, respectively, exacerbates or inhibits the infection. Part of the protective IFN-γ-mediated effects might be due to the enhancement of TNF production by macrophages. It has been reported recently, that the Th1-associated increase in endogenous TNF in the spleen during early infection correlates with resistance to *P. chabaudi chabaudi*, whereas high TNF levels in the circulation and liver late after infection have a deleterious effect for the host. Because TNF and IFN-γ are not directly toxic for plasmodia, the main effect of these two cytokines contributing to the control of malaria is most likely the enhancement of macrophage cytotoxic activity.

Control of Parasites by Activated Macrophages TNF and IFN-γ together with parasite exoantigens activate macrophages to secrete several products into the local environment amongst which are reactive oxygen and nitrogen derivates. Although reactive oxygen derivates are toxic for plasmodia in vitro, the in vivo studies yielded conflicting results. Injection of the ROI scavenger hydroxyanisole resulted in enhanced parasitemia in *P. chabaudi adami*-infected mice. In contrast, P/J mice with an intrinsic defect for the generation of ROI resolve acute *P. chabaudi chabaudi* infection.

It has been argued that NO might be an improbable defense mechanism against blood stage malaria since the hemoglobin in intimacy with intraerythocytic parasite has a high affinity to NO and may thereby scavenge this molecule. Nevertheless, NO and related compounds are able to inhibit the growth of several plasmodium species in vitro at concentrations in the range generated locally by activated macrophages (40–100 μM). In the presence of hemoglobin, this activity of NO is dependent on the local oxygen tension. While at high O_2 tension the formation of S-nitroso-hemoglobin is favored, at low oxygen tension the S-nitroso-hemoglobin functions as NO donor.

A protective role of NO in vivo has been demonstrated by studies with iNOS inhibitors. Injection of L-NMMA during the ascending parasitemia caused elevated parasitemia of extended duration, and, as mentioned above, the protective effect of Th1 cell transfer was significantly reduced by L-NMMA-treatment. A correlative support for the involvement of NO in the defense against malaria comes from the observation that iNOS expression was inversely proportional to the disease severity in African children with *P. falciparum* infection.

In addition to direct antiparasitic effects of NO, other host-protective functions of this molecule have been proposed: NO is able to inhibit leukocyte adhesion to the endothelium and to increase vasodilation thereby possibly preventing hypoxic tissue damage.

Immunopathology Besides its protective function, the cellular immune response to malaria is also involved in the pathogenesis of the disease. Already 14 years ago it was reported that $CD4^+$ lymphocytes play a major role in the pathogenesis of murine cerebral malaria. The transfer of malaria-specific Th1 cells to SCID mice reduced parasitemia, but the animals died early at low parasitemia, indicating T cell-mediated immunopathology.

There is evidence that the systemic pathology is due to an exuberant inflammatory response to the parasite resulting in manifestations such as diarrhoea, nausea, fever, anemia and cerebral malaria. TNF, produced either by activated macrophages or T cells, appears to be a central molecule in this scenario. In humans, high serum levels of TNF are associated with a poor outcome of cerebral malaria and homozygosity of the TNF2 allele, causing enhanced TNF transcription by a different TNF-promotor, predisposes children to cerebral malaria. This important role of TNF in the pathogenesis of malaria has focussed most interest on malaria toxins as TNF-inductors. Several molecules, such as ring-infected erythrocyte surface antigen (RESA), the merozoite surface proteins MSP1 and MSP2, a soluble protein complex known as AG7 complex as well as lipid moieties have been shown to induce the production of TNF.

Overproduction of IFN-γ is also of relevance to the development of cerebral malaria. In synergy with TNF, IFN-γ stimulates the upregulation of adhesion molecules like ICAM-1 on endothelial cells in the brain, implicated in the pathogenesis of the disease. The critical balance between protective and immunopathologic effects of cytokines like

IFN-γ and TNF is tightly controlled by anti-inflammatory cytokines such as IL-10. In line with this, the susceptibility of IL-10-deficient mice to an otherwise non-lethal infection is not simply due to a dominant parasitemia, but is associated with an enhanced IFN-γ production.

Most recently, it has been shown that mice depleted of γδ T cells by mAb treatment did not develop cerebral malaria after infection with *P. berghei*, suggesting an important function of this T cell subset in the pathogenesis of at least some manifestations of malaria.

Evasion Mechanisms Malaria parasites are obviously able to avoid a protective immune response, since people subject to repeated infections in malaria endemic areas rarely develop complete or sterile immunity to the parasite. Repeats in the structure of parasite surface proteins may help the parasite to evade host immunity by exhibiting sequence polymorphism and preventing the normal affinity and isotype-maturation of an immune response. Furthermore, some of these proteins may act as B cell superantigens and, when expressed in large quantities, capture protective antibodies. Sequence diversity and antigenic variation in non-repetitive parasite molecules located on the surface of infected erythrocytes, for example the existence of 5 different variable antigen types (VAT) in the case of *P. chabaudi*, have also been described as potential mechanisms of immune evasion.

Vaccination

Human malaria is, among animal and human parasite protozoan diseases, the one for which, the most intense effort of research has been accumulated in the last decades in view of the development of vaccines. Scientific literature on this topic accounts for thousands of references every year, particularly concerning *falciparum* malaria. This is comprehensible because of the importance of malaria as a leading cause of morbidity and mortality in the tropical areas of the world, with an estimated 300–500 million cases each year and more than 1 million deaths, mainly among children below five years of age in Africa.

A naturally acquired immunity against malaria is observed in endemic areas where people are exposed to frequent infections. This immunity develops slowly and is characterized, in a first step, by the acquisition of clinical resistance to symptoms, clinical immunity, and later by the ability to control parasitemia at a low level, anti-parasite immunity, usually fully expressed only in adults. Classical experiments of British immunologists working in Africa showed in the 60s that the natural immunity was antibody-dependent, directed against the asexual blood stages of the parasite. More recently, sero-epidemiological surveys in endemic areas have shown the existence of anti-sporozoite specific antibodies as well as antibodies and CTL cells directed against antigens of the hepatic stage. Finally, naturally and artificially raised antibodies against the gametocytes and latter forms of the sexual stage have been described as able to block the development of the parasite in the mosquito. These observations indicate that immunity in malaria is stage specific and this was indeed proved in laboratory experiments with rodent and primate models. Thus, efforts in the construction of vaccines have been directed towards different target alternatives. The starting point was the impossibility of raising vaccines from parasite materials since no culture systems are available for pre-erythrocytic stages of the parasites while culture of blood stages (*P. falciparum*) require growth in human red cells. These constraints made malaria vaccines one of the first domains of medical sciences in which nascent genetic engineering technology was actively introduced with the aim of preparing sub-unit vaccines. In principle the ideal target would be the pre-erythrocyte stages antigens (sporozoites and hepatic forms) since an effective vaccine against these stages would block transmission. However, an inconvenience of such a vaccine is that it would need to induce sterile immunity, because a surviving → sporozoite (Vol. 1) or hepatic schizont would be sufficient to produce erythrocyte → invasion (Vol. 1) and multiplication of the parasite in the blood. Sterile immunity against blood stage, however, is not naturally observed in humans of endemic areas and is usually not experimentally obtained in animal models. In contrast, non sterilizing, partially active vaccines against asexual blood stages would be favorable to avoid the development of high parasitemia and presumably reduce severe malaria outcome responsible for mortality. However, it would poorly interfere at the level of sources of infection in an endemic area and, therefore, in the level of transmission. Anti sexual stage vaccines transmission blocking vaccines would abolish or reduce transmission but would not protect the vaccinated individual from infection (altruistic vaccine). In conclusion, these considerations point to the interest in developping multi-gene, multi-stage vaccination approach like the CDC/NIIMALVAC-1, for which preliminary assays are now in course.

Pre-Erythrocyte Stages Vaccines (Sporozoite and Hepatic Stage Vaccines) After the successful vaccination of human volunteers in the 70s with irradiated sporozoites the search of the antigen(s) responsible for protection was started. With the development of the monoclonal antibody technology it was possible to identify the protein responsible for the circumsporozoite reaction named CS protein. As the anti CSP Fab' monoclonal antibody was able to confer protection in rodent malaria infection by *P. berghei*, CSP was identified as protective antigen. The corresponding antigen of the primate parasite *P. knowlesi* was the first parasite antigen successfully cloned and sequenced by the Nussenzweig, revealing the surprising presence of amino acid repeats in tandem in the central region of the molecule, which turned out to represent the immune dominant B epitope, target of the protective antibodies. Very quickly, the corresponding proteins of *P. falciparum* (containing repeats of NANP peptides) and *P vivax* parasites were cloned and sequenced. Synthetic peptides and recombinant fusion proteins were built containing the repeat tandem units of the corresponding CSPs . Vaccine trials using these molecules were performed in human volunteers in 1986/7 which produced disappointing results with poor protection. However, efforts for improving the immunogenicity of the → CSP (Vol. 1) protein have been recently developed. At the same time, new important knowledge about the structure and functional interaction of the protein with the mosquito and hepatocytes of the human host have been obtained. The region of the CSP molecule responsible for the interaction with hepatocytes (and consequent penetration of the sporozoite) was identified and depends on the presence of an amino acid sequence containing RGD motif (arginine-glycine-aspatic) with heparam sulfate (a protein polysaccharid conjugate) present in the surface of the hepatocytes. The same sequence RGD is present in a second protein of the sporozoite more recently described by the name of TRAP (→ Thrombospondine-Related Anonymous Protein). The name → TRAP was given by the presence of RGD sequence that is found in thrombospondine. These sequences (or associations of CSP and TRAP polypetides) have now been included in new sporozoite vaccines' preparations under study. In other studies, genetic analysis using gene disruption techniques has been performed showing the functional role of CSP and TRAP in sporozoite movement, interaction with the salivary glands of mosquitoes and with hepatocytes. Finally, new adjuvants have been used in association with CSP derived antigens which considerably increased the immunogenicity of the protein. A series of trials (primate and human volunteers) are now in course of development, and sporozoite vaccines are still important hopes for the future of malaria vaccines, particularly in association with other antigens in multi-stage multi-genes vaccines.

In another line of research related to the role of CTL in the cellular immune response, it was shown in animal models that CSP epitopes were expressed in the infected hepatocyte and present in the cell membrane, representing possible target of specific cytotoxic T- cells. In addition to CSP, other liver stage antigen of *P. falciparum* like LSA-1 and LSA-3 (liver stage antigen 1 and 3) and antigens expressed both in sporozoites and liver stages (SALSA sporozoite and liver stage antigen) have been used to vaccinate primates with induction of partial protection against sporozoite challenge.

Asexual Blood Stage Vaccines Asexual blood stage antigens account for the larger number of molecules, essentially polypeptides, described as potential vaccine candidates. Corresponding genes have been isolated and completely and/or partially sequenced. An abundant literature has been accumulated in the last decades about these antigens concerning molecular biology, immunological assays, experimental vaccination of primates, seroepidemiological surveys etc. This is comprehensible in view of the demonstration by the British immunologists in the 60s, confirmed and developed by others, on the ability of antibodies from immune adults in Africa to control blood parasites and symptoms of acutely infected patients. In consequence the search for identification of protective antigens of the asexual blood stage followed two strategies: (1) Differential recognition by antibodies from sera of immune and acutely infected Africans using immuno precipitation and western blot techniques; (2) Recognition by monoclonal or/and mono specific polyclonal antibodies able to inhibit the parasite in ïn vitro assays. In the present article, we will summarize recent progress concerning the main antigens defined as those that have been used with success in experimental vaccination of primates promoting at least partial protection.

→ Merozoite surface protein 1 → MSP1 (Vol. 1) is the most studied malaria antigen for *falciparum* and *vivax* parasites. The original protein of 185–200 kDa undergoes a double processing at the merozoite surface: in a first step three products of 83 kDa (N terminal), 36 kDa and C-terminal 41kDa originate; in a secondary event the 41 kDa fragment is proteolitically cleaved in a 33 kDa and a 19 kDa products. The C-terminal 19 kDa polypeptide is cysteine rich, contains two epidermal growth factor domains and is the only part of the MSP-1 molecule that penetrates the red blood cell with the merozoite. Numerous vaccination experiments have been performed in primates using the whole molecule or recombinant corresponding to the 83, 41 and 19 kDa fragments, or synthetic peptides corresponding to sequences of the N or C terminal regions. The better protection in monkey trials has been obtained with the whole natural molecule but positive results have also been regularly referred with the sub-unit molecules for *falciparum* and *vivax* malaria, as well as in rodent models using the equivalent molecules. Both antibodies (inhibiting red blood cell invasion) and cellular immune responses have been described in the protective mechanisms.

MSP-3 antigen and GLURP (glutamate rich antigen) were described as involved in an antibody dependent cellular inhibition (ADCI) in vitro assay that correlates with natural immunity and has induced partial protection effect in primate vaccination experiments.

Other merozoite antigens that have shown protective effect in monkey trial areas are RAP1/2 (rhoptry antigen), AMA-1 (apical membrane antigen) and EBA-175 (erythrocyte binding protein of 175 kDa) which are discharged on the red blood cell by the merozoite during the invasive process. It might also be referred to as the → SPf66 of Patarroyo et al., a polymerized chimera peptide containing sequences of the MSP-1 N terminal region, two unknown antigen and the sporozoite NANP sequence. SPf66 has been extensively used in monkey and human trials but final evaluations indicate a disappointing poor or absent protective effect.

A second family of asexual blood antigens is represented by proteins secreted by the parasite in the plasma, namely: → PfHRP-2 (histidine rich protein 2), Ag2 and → SERA (serine rich protein). SERA antigen, also described as Pf140 and Pf126 codes for a protease of unknown function and is localized in chromosome 2 where none less than eight copies of closely homologous sequences are present in tandem. Very good protection was obtained in monkey trials using recombinant proteins of the original SERA.

Finally, parasite antigens associated to the red cell membranes have been shown to represent targets of opsonizing antibodies that can mediate ADCC or phagocytosis by macrophages. The Pf332 and PfHRP-2, in this family, have been described as inducing partial protection in primate trials. The PfEMP-1 (erythrocyte membrane protein) has now been identified as the antigen derived from the polygenic family of the variant antigen var gene family with 50–100 copies showing extensive polymorphism. Some of the variant antigens could be associated to severe malaria by promoting sequestration of schizonts in capilary veinules of brain and other tissues. Intensive search for common structures of var genes products, to be used as vaccine targets against cerebral malaria, is now in progress in many laboratories. This goal was recently complicated with the description of other families of variant antigens, under the names of STEVOR and RIF that seems exposed at the infected erythrocyte membrane.

In the slow development of immunity in high endemic areas of falciparum malaria, the first step, already observed in children over 5–7 years is the reduction in disease severity which evolves to complete asymptomatic infections, eventually with high parasitemias. The similarity between the malaria attack and the toxemic shock produced by bacterial LPS induced the search of malaria toxins which became a preferential area of research. Such toxins would be released at the schizont rupture and be responsible for the secretion of TNF-α by immunocompetent cells. TNF-α is found at very high levels in the blood of severe malaria patients and has been shown to correlate with the severity of the illness. The contamination with *Mycoplasma* of *Plasmodium* lines maintained *in vitro* raised numerous questions about potential artefactual nature of some of the work describing the so-called malaria toxin published over the past ten years. Among the many parasite derived molecules proposed to fulfill the function of a toxin in terms of TNF-α production, the GPI anchor of membrane proteins of the merozoite remains the only still accepted candidate. Other molecules such as phosphorylated non-peptidic antigens (termed as phosphoantigens) are also molecules inducing TNF-α production by $\gamma\delta$ T cells, which could contribute to physiopathology of the illness. None of these potential toxins have yet been prepared in conditions allowing use in vaccination trials.

Sexual Stage Vaccines (Transmission Blocking Vaccines) The description of parasite antigens specific of the latter stages of the sexual development of the parasite provide interesting candidates for an altruistic vaccine. Natural antibodies from infected humans and raised monoclonal antibodies have shown, indeed, their ability to block parasite development in the mosquito. These antigens which are not seen by the immune system of infected humans have thus inspired the search for candidates for transmission blocking vaccines. The rationale of such altruistic vaccine is that immunized people would produce antibodies able to inhibit the parasite development in the mosquito and thus interfere with the natural life cycle of the parasite. Among the various candidate antigens, the most promising is the Pfs25 (and equivalent in *P. vivax*) for which a phase I and IIa trial have been completed.

Planning of Control

Attempts to control malaria date back to ancient times, but a rational fight against the disease only became possible after the discovery of the parasite's life cycle, early in the twentieth century. Control then consisted of measures against the anopheline vectors, mainly in their larval forms, and the use of quinine for treatment. Conditions improved with the introduction of potent residual insecticides and highly effective drugs in the late 1940s. At that time three-quarters of the world's population were living in malarious areas. The new tools were considered effective enough for attempting malaria eradication in wide areas of the globe, with the exception of tropical Africa. Malaria eradication was achieved in more than 30 countries, freeing more than one-third of the formerly affected areas from the disease. In other countries the goal of eradication was not attained, but the disease's impact was greatly reduced. Since the late 1960s there has been a stagnation, and in some areas a deterioration, of the malaria situation and the problem of the disease's hard core in tropical Africa is still unresolved.

Today malaria is still endemic in 100 countries and approximately 40 % of the world's population live in malarious areas. The annual number of clinically manifest cases is estimated at 300–500 million, and the annual number of deaths due to malaria at 1.5–2.7 million. There is a large reservoir of chronically infected persons, especially in Africa. Approximately 90 % of malaria cases occur in tropical Africa, nearly 10 % in Asia and western Oceania, and less than 1 % in the Americas. *P. falciparum*, the most dangerous and most widely distributed species of malaria parasites, accounts for approximately 90 % of all infections worldwide. It is the lead-species in tropical Africa, where it is often accompanied by *P. malariae*. In subtropical areas outside Africa *P. vivax* prevails over *P. falciparum*. The occurrence of *P. ovale* is practically restricted to tropical Africa.

In a policy statement on the implementation of the global malaria control strategy the World Health Organization stated that the goal of malaria control is to prevent mortality and reduce morbidity and social and economic loss, through the progressive improvement and strengthening of local and national capabilities. The four basic technical elements of the global strategy are:

- to provide early diagnosis and prompt treatment;
- to plan and implement selective and sustainable preventive measures, including vector control;
- to detect early, contain or prevent epidemics;
- to strengthen local capacities in basic and applied research to permit and promote the regular assessment of a country's malaria situation, in particular the ecological, social and economic determinants of the disease.

In view of the vastly different resources available for malaria control in various countries and of the substantial differences in the intensity of malaria transmission and the capability of conducting health programmes it will be useful to differentiate the following levels of achievement as objectives of antimalaria action:

1. Elimination of mortality and reduction of suffering from malaria;
2. Reduction of the prevalence of malaria;
3. Elimination of malaria.

Level 1 can be achieved by the timely detection and effective treatment of malaria cases and the protection of specific vulnerable groups. This requires the wide availability of health-care facilities throughout the malarious areas, and a well-developed, rapid and efficient referral system that can cope with severe and complicated malaria. Primary health care, backed up by efficient secondary and tertiary health care structure, is the vehicle through which this most elementary objective can be reached. Wherever they were established, malaria clinics proved to be a very useful component of the system. The reliance on drugs as the primary tool necessitates continuous monitoring of drug response and adherence to strict policies for rational drug use. A major constraint is the lack of sufficiently simple and cheap diagnostic techniques which do not require oil-immersion microscopy. In areas with intensive malaria transmission it has been shown that the detrimental impact of malaria (mortality and clinical incidence) can be substantially reduced by the use of pyrethroid-impregnated bed nets.

Level 2 requires measures directed against the transmission of malaria in addition to those needed for achieving level 1. This will be more demanding in terms of resources and skills. Based on sound epidemiological knowledge, proper operational stratification, and the results of feasibility studies, the appropriate approaches are to be selected for each operational area in accordance with the degree of control that is to be achieved. Here the focal application of vector control measures may be required and therapeutic intervention based on the microscopic diagnosis of malaria. Level 2 is therefore more demanding in technical skills and guidance, and specialized manpower may also be required in the periphery. Continuous evaluation is indispensable in order to detect changes in the response to certain measures and to adapt operational procedures accordingly.

Level 3 is still realistic for some countries and may come into the reach of others if and when more effective antimalaria tools become available. Under this objective malaria is regarded as a parasitosis rather than a disease. This implies that every malaria infection, whether symptomatic or not, is of importance and requires radical treatment in order to eliminate infective reservoir. The traditional malaria eradication campaigns relied on a very limited choice of attack measures and on a vertical service structure. This will not be appropriate in the future since the selection of operational approaches will require more flexibility and the fast flow and utilization of epidemiological information. Level 3 should be understood as a logical extension of level 2.

See also → Disease Control, Epidemiological Analysis.

Targets for Intervention

Targets of intervention (Fig. 1) are infected humans, the vector, and the infection cycle. Approaches are numerous and their selection depends on the given epidemiological situation, the available resources and the envisaged level of control. Treatment of infected persons may be suppressive or radical and gametocytocidal. Vector control may be directed against the aquatic stages of *Anopheles*, the adult mosquitoes, or both. The interruption or reduction of man-vector contact is a valuable ancillary measure.

Main clinical symptoms:

a) *Plasmodium vivax* (Malaria tertiana): fever of 40–41 °C for several hours, (after 1 h of shivers) is repeated within 48 h (Fig. 2)

b) *P. ovale* (M. tertiana): as in *P. vivax* infections (Fig. 2)

c) *P. malariae* (M. quartana): rhythmic fevers of 40–41 °C (after shivers) reappear within 72 h (Fig. 3)

d) *P. falciparum* (Malaria tropica): Irregular high fevers of 39–41 °C appear continously after a phase of headache and general abdominal symptoms; fevers may be rhythmic (48 h) or even absent (Fig. 4); eventually followed by coma and death.

Incubation period:

a) *P. vivax*: 12–18 days, occasionally longer

b) *P. ovale*: 10–17 days

c) *P. malariae*: 18–42 days

d) *P. falciparum*: 8–24 days

Prepatent period:

a) *P. vivax*: 8–17 days, occasionally longer

b) *P. ovale*: 8–17 days

c) *P. malariae*: 13–37 days

d) *P. falciparum*: 5–12 days

Patent period:

a) *P. vivax*: up to 5 years

b) *P. ovale*: up to 7 years

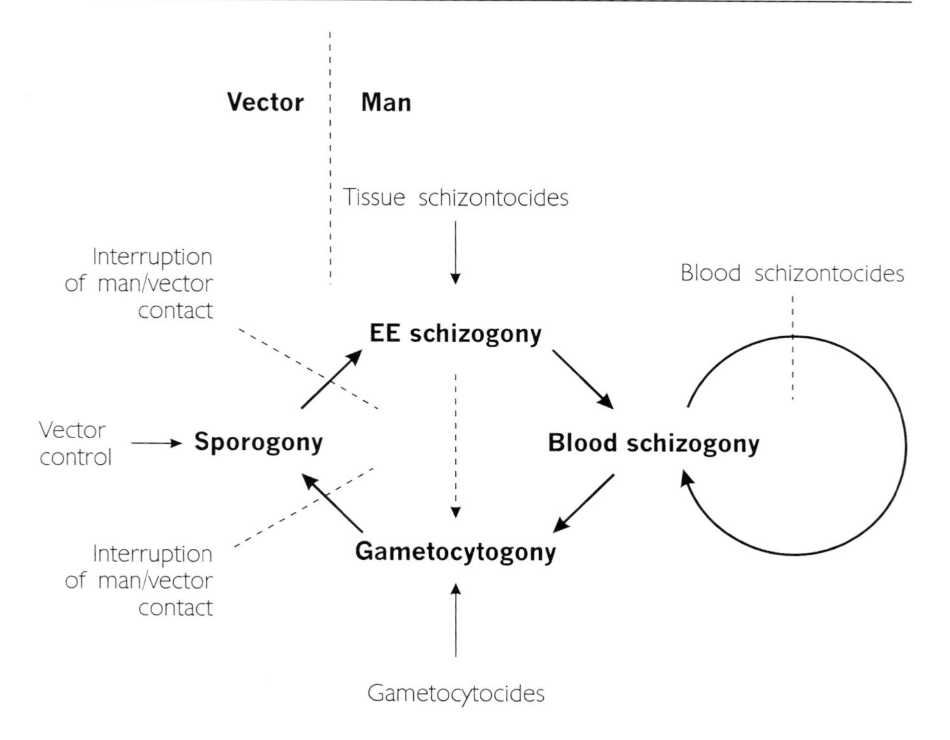

Fig. 1. Targets and approaches for the control of *P. falciparum* malaria.

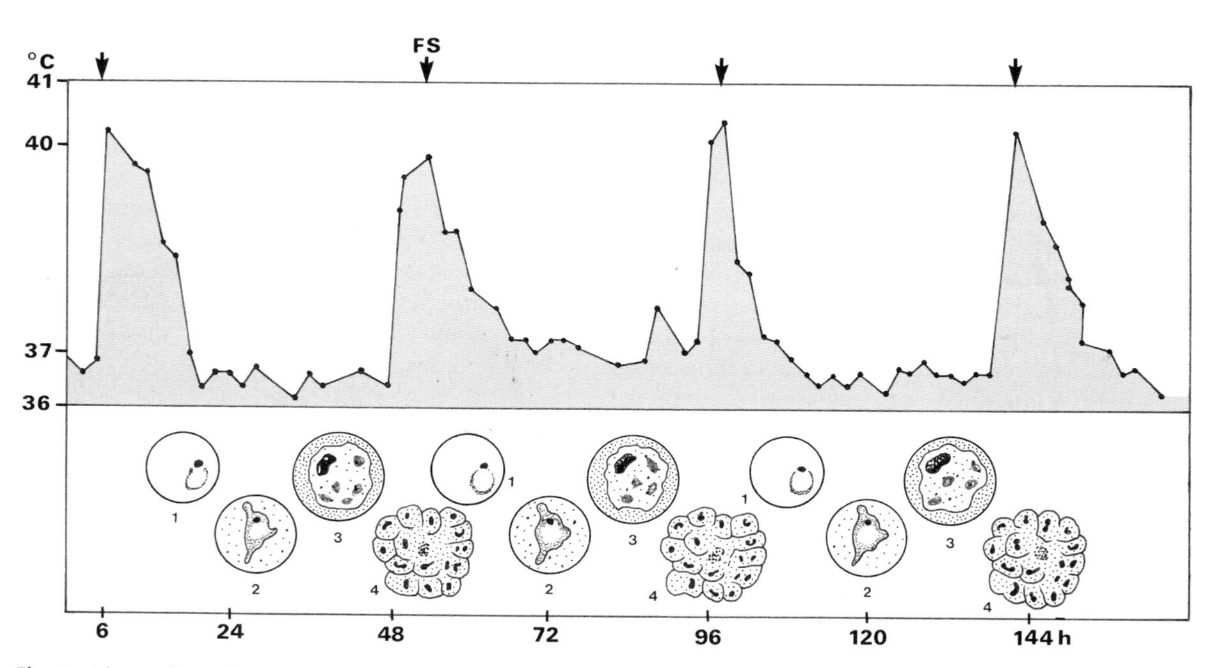

Fig. 2. *Plasmodium vivax*. Diagrammatic representation of the relationships between development of parasites in blood and occurrence of fever in the case of Malaria tertiana (*P. ovale* is similar). **1**, Signet ring-stage; **2**, Polymorphous trophozoite; **3**, Immature schizont; **4**, Mature schizont before formation of merozoites

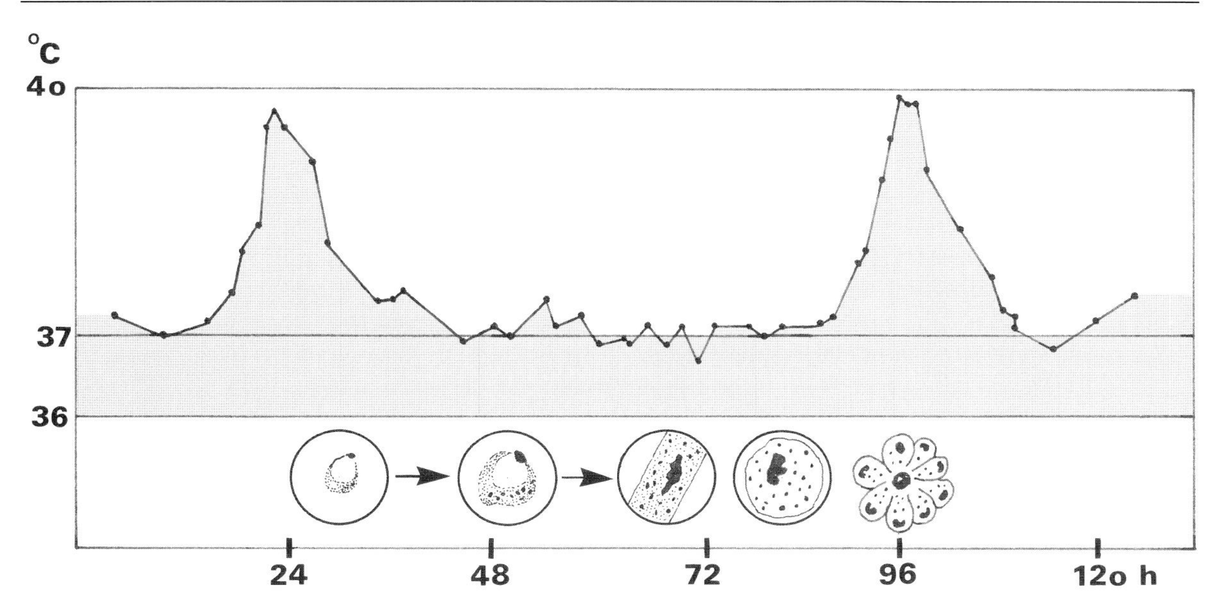

Fig. 3. *Plasmodium malariae*. Diagrammatical representation of the relationships between fever and development of parasites in blood cells during the Malaria quartana.

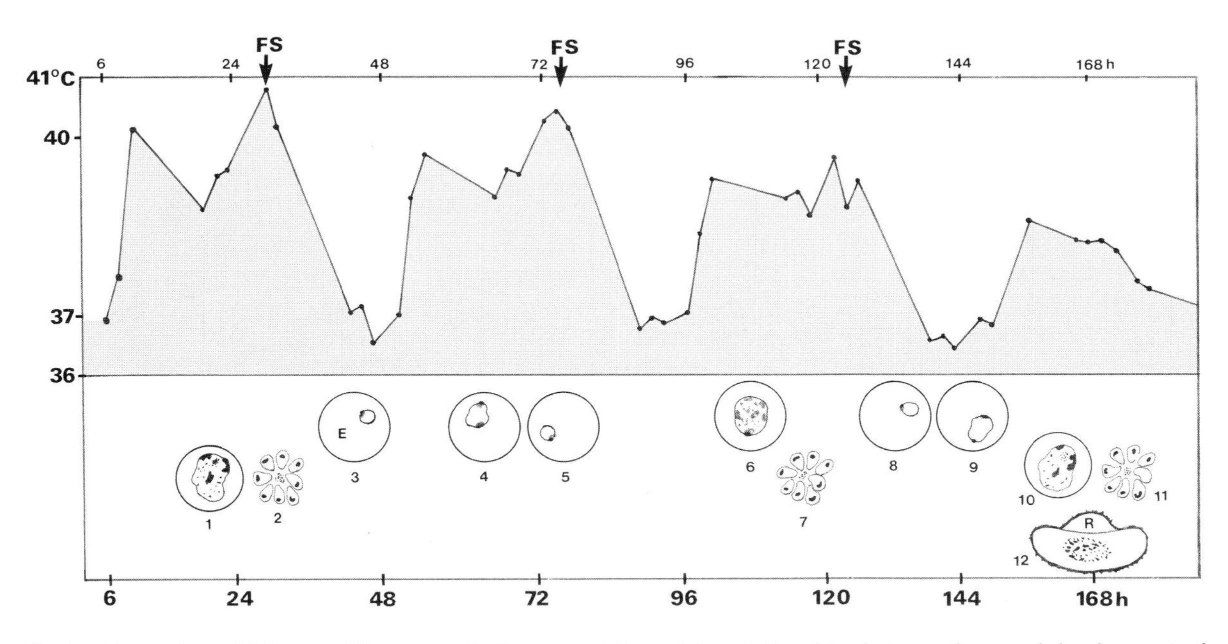

Fig. 4. *Plasmodium falciparum*. Diagrammatical representation of the relationships between fever and development of parasites in blood cells during the Malaria tropica. **1, 6, 10**: Immature schizont; **2, 7, 11**: Mature schizont; **3, 8**: Uninucleate signet ring-stage; **4, 5, 9**: Binucleate signet ring-stage; **12**: Gamont; **E**: Erythrocyte; **FS**: peaks of fever; **R**: residuals of the erythrocyte

c) *P. malariae*: 30 years and more
d) *P. falciparum*: under treatment 4–6 weeks, without treatment 18 months

Diagnosis: Microscopic determination of stages in blood smears and thick droplets; malaria quick test in *P. falciparum* commercially available.

Prophylaxis: Avoid the bite of *Anopheles* mosquitoes, see → Repellents (Vol. 1), → Insectizides (Vol. 1) and use → Chemoprophylaxis in endemic regions

Therapy: Treatment see → Malariacidal Drugs

Malaria Containment

Attempts to eliminate outbreaks.

Malaria Suppression

Attempts to lower prevalence.

Malaria tropica

→ *Plasmodium falciparum* (Vol. 1).

Malariacidal Drugs

Animal Diseases

Hepatozoonosis of Dogs The protozoan *Hepatozoon canis* (→ *Hepatozoon* (Vol. 1)) has been diagnosed in dogs throughout the world and is transmitted by the brown dog tick, *Rhipicephalus sanguineus* (→ Ticks (Vol. 1)). Clinical hepatozoonosis may be accompanied by concurrent **diseases** caused by other hematropic parasites such as ehrlichiosis (*Ehrlichia canis* belonging to intracellular bacteria of fever-group rickettsiae), babesiosis (*Babesia canis, B. gibsoni*), leishmaniasis (*Leishmania infantum, L. chagasi*), canine distemper (a viral infection) or dirofilariasis (*Dirofilaria immitis*). A distinct clinical syndrome involves fever, chronic myositis, debilitation and death. Treatment of hepatozoonosis is problematic; **toltrazuril** (→ Coccidiocidal Drugs), at 5 mg/kg body weight, orally, every 12 h for 5 days may reduce signs of pain, stiffness, and fever so that the initial response to the drug seems excellent. However, the drug failed to prevent relapses, i.e. intracellular schizonts, and cysts are still present during clinical remission. The action of **imidocarb** dipropprionate (→ Babesiacidal Drugs) seems to be inferior to toltrazuril; the drug exhibits moderate effect on *H. canis* and cholinergic side effects in dogs. A combined administration of **clindamycin** and the DHFR/TS inhibitors **trimethoprim** sulfate (antifolates: dihydrofolate reductase/thymidylate synthase inhibitors), and **pyrimethamine** hydrochloride (Table 1), given orally for 14 days, may result in remission of clinical signs; relapses were evident within 3 to 4 months. The antimalarial drug **primaquine** (Table 1) also appears to be effective against *H. ca*nis infection. Palliative treatment with non-steroidal, anti-inflammatory drugs will relieve fever and signs of pain in affected dogs.

Leucocytozoonosis of Poultry *Leucocytozoon* occurs in cells of various organs and the blood, e.g., erythrocytes and leukocytes; *Simulium* flies and other arthropods transmit them. *L. smithi* in turkeys, *L. caulleryi* in chickens or *L. simondi* in geese or ducks may cause death and thus economic loss in the poultry industry in Japan and other countries of Southeast Asia. Though antimalarial drugs (Table 1) have only a limited curative effect they can be used prophylactically (e.g. pyrimethamine plus sulfonamides). A few anticoccidial drugs as meticlorpindol or halofuginone in combination with furazolidone (→ Coccidiocidal Drugs) may exhibit some activity against the parasites when used prophylactically.

Malaria of Birds, Rodents and Monkeys *Plasmodium* of birds occurs in erythrocytes and cells of the reticulohistiocytary system (RHS). Culicine → mosquitoes (Vol. 1) transmit the parasites. Though avian plasmodia infect birds all over the world, clinical signs are infrequently seen. In Europe, *Plasmodium relictum* may occur in songbirds and water birds. Malaria of birds caused by *P. gallinaceum* or *P. juxtanucleare* has only limited veterinary importance. The disease may occasionally produce anemia and high mortality in the domestic fowl or turkeys and occur in subtropical and tropical areas, particularly in Southeast Asia or southern parts of the USA. Antimalarial drugs, which can be used against avian malaria parasites are listed in Table 1.

Table 1. Classification of antimalarial drugs according to the stages of *Plasmodium* affected.

DISEASE *Plasmodium* species (other information)	STAGE AFFECTED (location), other information	CHEMICAL CLASS (other information)	INN, *BRAND-NAME (preferable oral route)	COMMENTS (toxic reactions and other comments)
MALIGNANT TERTIAN MALARIA	*Plasmodium falciparum* infection can be fatal during initial attack; repeated attacks are due to recrudescence (= renewed manifestation of infection due to the survival of erythrocytic forms); the infection seldom exceeds 1 year			
BENIGN TERTIAN MALARIA	*P. vivax* / *P. ovale* infection; repeated attacks are due to recrudescence (see above) or relapses, i.e. renewed manifestations of an infection originating from exoerythrocytic stages of the parasite; the infections die out within 3-4 years			
QUARTAN MALARIA	*P. malariae* infection; recrudescence originate from chronic undetectable erythrocytic infection; the latter tends to persist for many years			
there are no drugs acting directly against sporozoite of *P. falciparum, P. vivax, P. ovale P. malariae*; sporozoites inoculated by female mosquitoes during blood meal temporarily circulate in blood, and then enter liver cell to undergo schizogony				
DRUGS ACTING ON PRIMARY TISSUE STAGES IN THE LIVER (TISSUE SCHIZONTOCISES) current use of these agents (with the exception of primaquine) is principally in conjunction with an appropriate blood schizontocidal drug for radical cure of chloroquine-resistant *P. falciparum* strains and other malarias; primaquine may be used for short-term prophylaxis of malaria , e.g. in G-6-PD normal, non-pregnant, visitors to malarious areas; there are new primaquine analogues of interest				
P. falciparum P. vivax, P. ovale, P. malariae (effect unknown) (failed when used against some chloroquine-resistant strains)	**primary tissue schizonts, hypnozoites** (liver parenchymal cells = hepatocytes)	**8-aminoquinolines** causal prophylactic action prevents vivax and falciparum malaria **quinocide,** similar in structure to primaquine may be used in some parts of the world	**primaquine** (chiefly used for radical cure of vivax and ovale malaria (see antirelapse drugs below)	much less active against erythrocytic stages than tissue stages; adverse effects may be methemoglobinemia (NADH methemoglobinemia reductase deficiency), risk of intravascular hemolysis in G6PD-deficient patients (African or Caucasian type); for interaction with quinacrine see → Antidiarrhoeal and Antitrichomoniasis Drugs/Drugs Acting on Giardiasis)
P. cynomolgi (rhesus monkey) *P. berghei* and *P. yoelii* spp. (rodent malaria, mice)	**primary tissue schizonts hypnozoites** (liver parenchymal cells = hepatocytes)	**8-aminoquinolines** causal prophylactic and radical curative drug	**WR 238 605** being developed by Walter Reed Army Institute for Research, Washington DC	activity compared to primaquine: about >13-times as hypnozoitocidal drug (*P. cynomolgi*), >10-90-times as blood schizontocidal drug (*P. berghei*); it may have utility for the treatment of falciparum malaria
P. vivax	**primary tissue schizonts hypnozoites** (liver parenchymal cells = hepatocytes)	**8-aminoquinolines** causal prophylactic and radical curative drug	**CDRI-80/53** being developed by Central Drug Research Institute, Lucknow, India	may have utility for the treatment of vivax malaria; though not as potent as WR 238 605, it is claimed to be significantly less toxic than primaquine
P. falciparum (*P. vivax:* fleeting inhibitory action on exoerythrocytic (EE) forms only)	**primary tissue schizonts** (hepatocytes)	**biguanides** (antifolate type 2) has been used for causal prophylaxis; drug resistant plasmodia limited its use	**proguanil** (*Paludrine) (= chloroguanide) **chlorproguanil** (*Lapudrine)	currently used only in combination with sulphas (see below) or other antimalarials (e.g. chloroquine) for chemoprophylaxis of malaria; very well tolerated drugs, show tendency to provoke resistance which was widely reported

Table 1. (Continued) Classification of antimalarial drugs according to the stages of *Plasmodium* affected.

DISEASE *Plasmodium* species (other information)	STAGE AFFECTED (location), other information	CHEMICAL CLASS (other information)	INN, *BRAND-NAME (preferable oral route)	COMMENTS (toxic reactions and other comments)
EE forms of human malaria (parasites are believed to be affected by the drug)	**primary tissue schizonts** (hepatocytes) high rates of antifolate resistance in East Africa, SE Asia and now also S-America (DHFR 164 mutation)	**diaminopyrimidines** (antifolate type 2) interfere with dihydrofolate reductase; drugs show tendency to provoke resistance	**pyrimethamine** (*Daraprim) used only in combination with sulphas (see next below)	prophylactic use of combinations may be obsolete because of widespread drug resistance and potential severe adverse effects; pyrimethamine may cause skin rashes, megaloblastic anemia (at higher doses), in rats, evidence of teratogenicity
EE forms of human malaria (parasites possibly affected by sulphas)	**primary tissue schizonts** (hepatocytes)	**sulfonamides sulfones**	**sulfadoxine, sulfalene dapsone** (used in combinations only, see next above)	in rodent plasmodia sulphas exhibit definite causal prophylactic activity at somewhat higher doses than for blood schizontocidal activity
P. falciparum action on EE forms of other species of human malaria has been inadequately investigated; tetracyclines are active against EE stages of *P. vivax* in chimpanzees	**primary tissue schizonts** (hepatocytes) **doxycycline** has been recommended for chemoprophylaxis	**tetracyclines** (causal prophylaxis is not advised on general principles) (may be used for nonimmune patients in areas with high prevalence of multidrug-resistant *P. falciparum*	**doxycycline** (minocycline, oxytetracycline, chlortetracycline, tetracycline) (*various)	use is strictly limited to *treatment* of multiresistant *P. falciparum* infections in conjunction with **quinine** only; hypersensitivity reactions: erythema multiforme, antibiotic-associated colitis; tetracyclines discolour teeth in growing children
DRUGS ACTING ON LATENT TISSUE STAGES OR HYPNOZOITES IN THE LIVER (ANTIRELAPSE DRUGS): are used in conjunction with an appropriate blood schizontocidal drug to achieve a radical cure of *P. vivax* and *P. ovale* infection; primaquine is the prototypical drug to prevent relapse caused by hypnozoites; pyrimethamine may also reveal some of this type of activity against *P. vivax*				
P. vivax, P. ovale highly active during relapse or during latency against latent tissue stages	**hypnozoites** (hepatocytes) **gametocytes** see below (erythrocytes)	**8-aminoquinolones** WR 238605 (=tafenoquine is being developed; it has improved therapeutic index and less elimination compared to primaquine	**primaquine** (*Primaquine) radical cure of benign tertian malaria	tissue schizontozide preventing relapse with poor effect on blood schizonts; for toxicity see under drugs with causal prophylactic action; its interaction with quinacrine see → Antidiarrhoeal and Antitrichomoniasis Drugs/Drugs Acting on Giardiasis
DRUGS ACTING ON ASEXUAL BLOOD STAGES (BLOOD SCHIZONTOCIDES): used for clinical or suppressive cure; these agents interrupt erythrocytic schizogony and terminate clinical attacks (clinical cure)				
P. falciparum, P. malariae, P. vivax, P. ovale: clinical cure of all types of human malaria *P. falciparum, P. malariae*: cure of infections by suppressive action of 4-aminoquinolines	**asexual blood stages:** ring stage, trophozoite, schizont containing merozoites (erythrocyte) 4-aminoquinolines have no effect on primary EE forms or latent EE forms	**4-aminoquinolines** (rapidly acting) had replaced older schizontocides as treatment of choice for of all types of human malaria susceptible to the drug	**chloroquine**; (*Resochin, various others) **chloroquine resistance** seems to have selective advantage and stability, and is consistently high in East Africa, Western Pacific and Southeast Asia	provides simple treatment and effective safe suppressive prophylaxis generally well tolerated after the oral route; following long-term administration there may be skin lesions and ocular damage as reversible neuroretinitis; side effects are common but moderate at curative doses; some individuals may show a high degree of intolerance

Table 1. (Continued) Classification of antimalarial drugs according to the stages of *Plasmodium* affected.

DISEASE *Plasmodium* species (other information)	STAGE AFFECTED (location), other information	CHEMICAL CLASS (other information)	INN, *BRAND-NAME (preferable oral route)	COMMENTS (toxic reactions and other comments)
P. falciparum, P. vivax, P. ovale, P. malariae *P. falciparum, P. malariae*: cure of infections by suppressive action of 4-aminoquinolines	**asexual blood stages:** ring stage, trophozoite, schizont containing merozoites (erythrocyte)	**4-aminoquinolines** (rapidly acting; close analogue of chloroquine) may be effective against some chloroquine-resistant *P. falciparum* strains	**amodiaquine** there are reports of resistant *P. falciparum* in Brazil, Pakistan and elsewhere and cross-resistance between chloroquine and amodiaquine	showing superior antimalarial activity over chloroquine; it can cause frequent neutropenia which may be associated with severe agranulocytosis and toxic hepatitis in one of every 220 – 1700 users
amodiaquine is not recommended as first-line treatment of uncomplicated falciparum malaria; its potent toxicity renders it unsuitable for routine chemoprophylaxis (see column 4)				
P. falciparum, P. vivax, P. ovale, P. malariae *P. falciparum, P. malariae*: cure of infections by suppressive action of 4-aminoquinolines	**asexual blood stages:** ring stage, trophozoite, schizont containing merozoites (erythrocyte)	**9-aminoacridines** (rapidly acting) there is cross-resistance with 4-aminoquinolines	**mepacrine** (syn. quinacrine) obsolete no longer used because it is considered to be too toxic	can be employed in an emergency if no other compound is on hand; yellow skin pigmentation and various skin lesions (lichen planus, exfoliative dermatitis), aplastic anemia; transient psychotic reactions, gastrointestinal disturbances (see → Antidiarrhoeal and Antitrichomoniasis Drugs/Drugs Acting on Giardiasis)
P. falciparum (chloroquine or mefloquine resistant strains) in the 50s, quinine was widely replaced by chloroquine for the radical cure of *P. falciparum* or *P. malariae* infections, and for treatment of acute *P. vivax* or *P. ovale* infections	**asexual blood stages:** ring stage, trophozoite, schizont containing merozoites (erythrocyte) quinine is structurally similar to the other quinolines, especially to mefloquine, see below	**chinchona alkaloids** (rapidly acting; not suitable for causal prophylaxis) **quinidine** (D-stereoisomer of quinine; may be used for treatment of severe *P. falciparum* malaria because of its greater antimalarial action than quinine)	**quinine** (* various) has its origins in Peru in the early 17th century; drug has remained an effective antimalarial for 350 years	oldtimer, now used increasingly in the therapy of falciparum malaria resistant to chloroquine, mefloquine, and other drugs (i.m., i.v.: infusion, p.o.); "general protoplasmic poison", relatively toxic in therapeutic doses: hypersensitivity reactions, intravascular hemolysis; hemoglobinuria, anuria (blackwater fever); agranulocytosis; abortion (overdose), asthma, tachycardia, CNS symptoms, ocular toxicity, tinnitus
P. falciparum may produce clinical cure of falciparum malaria; asexual blood forms of other human malaria parasites seem to be less affected by sulphas; sulphas should not be used alone because of rapid development of drug resistance in plasmodia	**asexual blood stages:** ring stage, trophozoite, schizont containing merozoites (erythrocyte)	**sulfonamides** (slow acting) **sulfones** (slow acting); because of their slow action sulphas should be administered with other synergistic acting antimalarials only	**sulfadoxine,** sulfalene **dapsone** may be used in therapy of uncomplicated *P. falciparum* malaria resistant to chloroquine in combination with pyrimethamine however, widespread use of PS induces also resistance (DHFR 164 mutation)	antifolate Type 1, which blocks incorporation of PABA to form dihydrofolic acid; potentiating effect with pyrimethamine, a Type 2 inhibitor, see causal prophylactics, above; sulfonamides can cause hypersensitivity reactions (Stevens-Johnson type), agranulocytosis; sulphas may rarely cause hemolysis and methemoglobinemia in G6PD-deficient patients (see primaquine); teratogenic risk

Table 1. (Continued) Classification of antimalarial drugs according to the stages of *Plasmodium* affected.

DISEASE *Plasmodium* species (other information)	STAGE AFFECTED (location), other information	CHEMICAL CLASS (other information)	INN, *BRAND-NAME (preferable oral route)	COMMENTS (toxic reactions and other comments)
P. falciparum, P. vivax, P. ovale, P. malariae drugs should only be combined with chloroquine or amodiaquine because their clinical response is slow; cross-resistance with pyrimethamine may occur	**asexual blood stages:** ring stage, trophozoite, schizont containing merozoites (erythrocyte) primary tissue schizonts (hepatocytes)	**biguanides** (slow acting) antifolate Type 2 (see above) treatment of acute malarial attack is not recommended	**proguanil** (*Paludrine) (= chloroguanide) **chlorproguanil** (*Lapudrine); drugs may be used as a partner with other drugs in treating chloroquine-resistant falciparum malaria (e.g. Lapudrine-dapsone)	biguanides are usually used as causal prophylactics (see this table) combined with rapidly acting drugs to retard occurrence of drug resistance; proguanil may quickly induce resistance in *P. falciparum* (dreaded DHFR 164 mutation); *P. vivax* has also been reported to be resistant to biguanides (see atovaquone below)
P. falciparum (Aotus monkeys)	**asexual blood stages:** ring stage, trophozoite, schizont containing merozoites (erythrocyte)	**biguanides** a new class of folic acid antagonists	**WR 250 417** (PS-15) (proguanil analogue) was shown to be highly active against a pyrimethamine-resistant strain of *P. falciparum* and against treatment failures with proguanil	short-acting antifolate developed as prodrug that would be transformed by hepatic cytochrome P-450 to the triazine WR 99210 to overcome its gastric intolerance in humans; it may be used as alternative antifolate in combination with a short-acting sulpha
P. falciparum, P. vivax, P. ovale, P. malariae drug may cause suppressive (radical) cure in *P. falciparum* infection; it is used as a 'plus drug' with another antimalarial in treating uncomplicated chloroquine resistant falciparum malaria	**asexual blood stages:** ring stage, trophozoite, schizont containing merozoites (erythrocyte) primary tissue schizonts (hepatocytes)	**diaminopyrimidines** (slow acting) widespread occurrence of plasmodia resistant to pyrimethamine limited its use as a prophylactic drug though synergy with sulphas should reduce rate of appearance of drug resistance	**pyrimethamine combinations:** plus sulfadoxine (PS), (*Fansidar); plus sulfalene (*Metakelfin); plus dapsone (*Maloprim) *Fansimef see mefloquine below	PS was previously used extensively for prevention of malaria but is no longer effective in Southeast Asia and South America; in Africa are there also *P. falciparum* strains resistant to PS (DHFR 164 mutation); effective treatment for non-severe malaria is the most important malaria control strategy in Africa; combinations can cause neutropenia and agranulocytosis;
Fansidar is no longer recommended for chemoprophylaxis because of associated severe cutaneous reactions				
P. falciparum (other human malarial parasites have not been adequately documented)	**asexual blood stages:** ring stage, trophozoite, schizont containing merozoites (erythrocyte)	**tetracyclines** (slow acting) potent antibacterial agents; use should be restricted	**doxycycline** (concurrent use with quinine for treatment of multiresistant falciparum malaria)	tetracyclines discolour teeth in growing children; skin toxicity; photosensitivity; hepatotoxicity in high doses; is contraindicated in pregnancy and children <8-year old

Table 1. (Continued) Classification of antimalarial drugs according to the stages of *Plasmodium* affected.

DISEASE *Plasmodium* species (other information)	STAGE AFFECTED (location), other information	CHEMICAL CLASS (other information)	INN, *BRAND-NAME (preferable oral route)	COMMENTS (toxic reactions and other comments)
P. falciparum (other human malarial parasites have not been adequately documented)	**asexual blood stages:** ring stage, trophozoite, schizont containing merozoites (erythrocyte)	**macrolide antibiotic** (slow acting) potent antibacterial agents; use should be restricted	**clindamycin** (concurrent use with quinine for treatment of multiresistant falciparum malaria)	lincomycin derivative, side effects may be allergic reactions, diarrhoea (enterocolitis, ulcerous colitis), hepatotoxicity, occasionally hypotension, ECG changes
BLOOD SCHIZONTOCIDES ACTING ON CHLOROQUINE (DRUG)-RESISTANT FALCIPARUM MALARIA *P. falciparum* has developed resistance to chloroquine (**CHQ**), sulpha/pyrimethamine combinations and, to some extent, quinine effective in the treatment of severe and complicated disease; chloroquine resistance of various levels is now common in all endemic countries of Africa, and in many of them, particularly in eastern Africa, high levels of resistance pose increasing problems for the provision of adequate treatment; among the countries with endemic falciparum malaria , only those in Central America and the Caribbean appear to have no serious problems with CHQ resistance; mefloquine (**MEQ**) and halofantrine introduced in the 1980s are effective against multidrug-resistant strains of *P. falciparum* and are being used increasingly for the treatment and prevention (especially MEQ) of falciparum malaria in many parts of the world, particular south-east Asia ; since then there have been reports of decreasing sensitivity and resistance to both these drugs and to the structurally related quinine; in some areas, such as on the Thai/Cambodian and Thai/Myanmar borders, high levels of resistance to MEQ led to the introduction of artemisinin derivatives in 1993; treatment failure rates in children with acute falciparum malaria after administration of high-dose MEQ (25 mg/kg) had exceeded 50% ; recent experiments have suggested that CHQ resistant parasites are more infectious to mosquitoes than are sensitive ones, resulting in enhanced transmission. This fact poses the question, is CHQ resistance in *P. falciparum* associated with a cost or with a benefit? The answer is not yet clear; perhaps the best evidence that resistance is associated with a cost is that resistance was maintained at a low level before the widespread use of drugs; examples that parasites may lose their resistance after discontinuing use of CHQ have been observed in Thailand and Hainan, China				
P. falciparum resistant to chloroquine and antifolates *P. vivax* (other human malarial parasites) as with prophylaxis women reported more side-effects than men; patients with recrudescent infection following initial Lariam treatment were at >7-fold increased risk of severe neuropsychiatric reactions when treated again with high-dose mefloquine 25 mg/kg	**asexual blood stages:** ring stage, trophozoite, schizont containing merozoites (erythrocyte) *Fansimef (pyrimethamine-sulfadoxine-mefloquine) introduced in Thailand in 1984 led to fast development of resistance (dreaded DHFR 164 mutation)	**4-quinolinemethanols** (rapidly acting) (ineffective against EE forms in liver) a prospective study of non-serious adverse effects in 3673 patient with acute falciparum malaria revealed that high-dose mefloquine is well tolerated but should be given as a split dose	**mefloquine** (WR 142, 490) (*Lariam); combinations with pyrimethamine and sulfadoxine (*Fansimef) *Lariam as prophylaxis (250 mg/week, starting 1 week before departure) should not be recommended for short-term travellers (up to 3 weeks) because of possible subsuppressive drug concentration	an outstanding therapeutic and suppressive prophylactic drug that should be reserved for treatment of multiple drug-resistant falciparum malaria; treatment is usually well tolerated and most adverse events are mild and restricted to GI side-effects, fatigue, feelings of dissociation, and dizziness; occasionally frequent vertigo, light-headedness, nausea, GI and visual disturbances, nightmares, headache, insomnia, occasional confusion, and rare psychosis, convulsion, paresthesias, hypotension and coma (for explanation see columns 1+3)
P. falciparum (other human malarial parasites)	**asexual blood stages:** ring stage, trophozoite, schizont containing merozoites (erythrocyte) (ineffective against EE forms in liver)	**9-phenanthrene-methanols** (rapidly acting) have potent blood schizontocidal activity	**halofantrine** (WR 171, 699) (*Halfan) there may be partial cross-resistance with mefloquine	for treatment (or stand-by medication) of acute malaria of children and adults; is generally well tolerated, occasionally GI disturbances, skin rash (pruritus), cardiac arrhythmia as serious prolongation of QTc and PR interval; interaction with mefloquine leads to further prolongation of QT interval

Table 1. (Continued) Classification of antimalarial drugs according to the stages of *Plasmodium* affected.

DISEASE *Plasmodium* species (other information)	STAGE AFFECTED (location), other information	CHEMICAL CLASS (other information)	INN, *BRAND-NAME (preferable oral route)	COMMENTS (toxic reactions and other comments)
[a]the herb *Artemisia annua L.* (sweet wormwood, annual wormwood) has been used for many centuries (over 2000 years) in Chinese traditional medicine as treatment for fever and malaria. In 1971, Chinese chemists isolated from the leafy portions of the plant the substance (crude ether extract) responsible for medicinal action; qinghaosu's poor solubility stimulated Chinese scientists to synthesise more soluble derivatives by the formulation of dihydroqinghaosu (DHQHS); its secondary hydroxy group (-O-H) provides the only site that has been used for derivatization; etherification or esterification of DHQHS led to artemether and artesunate, respectively, and other derivatives; all these derivatives proved to be more effective against plasmodia than the parent compound and seem to be the most rapidly acting of all antimalarial compounds developed so far; the spectrum of activity of all derivatives is similar to that of the parent compound qinghaosu or artemisinin; the efficacy of artemisinin and its derivatives against multiple-drug-resistant *P. falciparum* has been shown in South East Asia and sub-Saharan Africa; attempts to synthesise more soluble, stable and effective derivatives led to artelinic acid. It is likely that a triple combination (artemisinin derivative-chlorproguanil-dapsone) will be available to malaria endemic countries; however, dreaded DHFR 164 mutation, now prevalent in SE Asia, South-America and East Africa, would render chlorproguanil and dapsone useless though the rationale behind the combination is that the artemisinin will protect the partners from resistance				
P. vivax and *P. falciparum* the most striking results achieved with QHS and its derivatives are seen in treatment of cerebral falciparum malaria	**asexual blood stages:** ring stage, trophozoite, schizont containing merozoites (erythrocyte EE forms are not affected	**sesquiterpene lactones** (very rapidly acting) (QHS bears a peroxide grouping appearing to be essential for the antimalarial activity)	[a]**qinghaosu** (QHS) =**artemisinin (=O)** (micronized; sparingly soluble in oil and water) Chinese formulations: suppositories, tablets, capsules, and solution containing groundnut oil for i.m. injection	in 1987 approved for marketing in China; recrudescence rate in *P. vivax* and *P. falciparum* patients may be frequent though multiresistant asexual blood stages of *P. falciparum* are highly sensitive to the drug; QHS is remarkably well tolerated; it appears to be safe in cases complicated by heart, liver and renal diseases of pregnancy
DERIVATIVES OF QINGHAOSU (DHQ) OR ARTEMISININ are now also being produced by pharmaceutical companies outside China				
P. vivax and *P. falciparum* the most striking results achieved with DHQHS and its derivatives are seen in treatment of cerebral falciparum malaria	**asexual blood stages** ring stage, trophozoite, schizont containing merozoites; (erythrocyte)	**sesquiterpene lactol** (retains function of peroxide bridge linkage and is more potent than QHS)	**dihydroqinghaosu** (=dihydroartemisinin) (DHQHS) (-O-H) Chinese oral formulations	semisynthetic compounds of QHS synthesised are more potent than QHS: listed in order of overall antimalarial activity QHS < ethers < esters < carbonates
DERIVATIVES OF DIHYDROQINGHAOSU (DHQHS) retain the potent bloodschizontocidal activity of parent compound with rapid clearance of fever but have a greater solubility than DHQHS				
P. vivax and *P. falciparum* artemether and arteether recent research in Vietnam and other countries of Southeast Asia	**asexual blood stages:** ring stage, trophozoite, schizont containing merozoites (erythrocyte)	**ethers** (some 32 ether derivatives) (very rapidly acting)	**artemether** (-O-CH_3) (*Artenam) oral and parenteral (i.m.) formulations **arteether** (-O-CH_2-CH_3) (short half-life) (oily solution for injection)	limited clinical studies in China; acute toxicity is considerably greater than that of QHS; a decision to develop arteether was made, to avoid toxic problems which might arise with artemether (metabolic formation of formaldehyde and formic acid)

Table 1. (Continued) Classification of antimalarial drugs according to the stages of *Plasmodium* affected.

DISEASE *Plasmodium* species (other information)	STAGE AFFECTED (location), other information	CHEMICAL CLASS (other information)	INN, *BRAND-NAME (preferable oral route)	COMMENTS (toxic reactions and other comments)
P. vivax and *P. falciparum* clinical studies in China; compound is more toxic than QHS but less toxic than artemether; acts rapidly in restoring to consciousness comatose patients with cerebral malaria; recrudescence rate is relatively high	**asexual blood stages:** ring stage, trophozoite, schizont containing merozoites (erythrocyte)	**sodium hydrogen succinate monoester** (very rapidly acting)	**sodium artesunate** [-O-COCH$_2$CH$_2$CO$_2$Na] tablets, capsules for local use; suppositories, parenteral formulations (i.v.; water soluble powder; dual-pack dosage form)	in the pipeline; in Africa, TDR/WHO examines rectal artesunate in children with 'non-severe' falciparum malaria, but whose condition prevents use of oral medication; this might reduce the proportion of children whose condition deteriorates to severe disease
P. berghei	**asexual blood stages:** ring stage, trophozoite, schizont containing merozoites (erythrocyte)	**carbonates** (very rapidly acting)	[-O-C(=O)-O-alkyl or aryl] (oil solubility is similar to that of esters; α-epimer predominates in the products)	most potent derivatives against *P. berghei* in mice; no detailed study of their therapeutic properties has been published
P. falciparum	**asexual blood stages:** ring stage, trophozoite, schizont containing merozoites (erythrocyte)	**artelinic acid** (very rapidly acting)	-OCH$_2$-(phenyl) COOH i.v. formulation	being developed by Walter Reed Army Institute for Research (WRAIR) as a more water-soluble i.v. formulations for the treatment of severe and complicated malaria; currently in clinical trials
stimulus for the synthesis of several novel **trioxane** (1,2,4-trioxanes and cis-fused cyclopento-1,2,4,trioxanes or fenozans) and **tetraoxane** compounds showing promising activity in experimental in vitro and in vivo antimalarial models was the demonstration of antimalarial activity related to both the trioxane ring of qinghaosu and the peroxide group of yingzhaosu (antimalarial moiety of *Atrobotrys unicinatus*); from this series, only Ro 42-1611 (**arteflene**) has been selected for human Phase I and II clinical trials; high recrudescence rates in patients with mild falciparum malaria led to termination of arteflene development				
P. falciparum Malarone appears to be a safe drug during pregnancy and in children; evaluation and international registration is in progress	**asexual blood stages:** ring stage, trophozoite, schizont containing merozoites (erythrocyte)	**hydroxynaphthoquinone**	**atovaquone** (*Mepron) plus **proguanil** (*Malarone)	used as a monotherapy, 30% of patients showed recrudescence; co-administration of atovaquone and proguanil revealed synergestic anti-plasmodial activity and a dramatic effect on cure rates of patients with falciparum malaria (100% cure rate)
P. falciparum	**asexual blood stages:** ring stage, trophozoite, schizont containing merozoites (erythrocyte)	**fluoromethanol** synthesised by Institute of Military Medical Sciences (IMMS), Beijing in the 1970s, registered as antimalarial drug in China in 1987	**benflumetol** or **lumefantrine** plus **artemether** (an oral formulation of the combination is being developed	poorly soluble in water and oils but soluble in unsaturated fatty acid (oleic or linoleic acid); the latter has been used for oral formulation (capsules) in clinical studies in China since 1979 and co-administered orally with artemether; preclinical trials showed synergy between the two drugs

Table 1. (Continued) Classification of antimalarial drugs according to the stages of *Plasmodium* affected.

DISEASE *Plasmodium* species (other information)	STAGE AFFECTED (location), other information	CHEMICAL CLASS (other information)	INN, *BRAND-NAME (preferable oral route)	COMMENTS (toxic reactions and other comments)
P. falciparum	**asexual blood stages:** ring stage, tropho-zoite, schizont containing merozoites (erythrocyte)	**azacrine** synthesised in China in 1970	**pyronaridine** oral and injectable formulations	has been used clinically in China since the 1970s and is now marketed in that country, it may have potential as replacement for oral formulations of chloroquine in many areas; most data are published in Chinese only
DRUGS ACTING ON GAMETOCYTES (GAMETOCYTOCIDES) may inhibit gametocytogenesis or kill mature gametocytes (sexual erythrocytic stages) of plasmodia, thereby preventing transmission of malaria to mosquitoes; damaging effects on gametocytes include severe alterations of morphology, and a marked decrease in numbers of gametocytes; the only drugs that have the potential to interrupt transmission of falciparum malaria are the 8-aminoquinolines (primaquine, pamaquine, and WR-238,605); gametocytogenesis takes place when critical parasite density has been achieved in the blood of host; hence, malaria may be transmitted during the recovery phase of the acute falciparum malaria despite successful eliminating of the asexual stages of the infection by bloodschizontocial drugs having little or no activity against mature gametocytes; patients whose infection recrudesced were nearly 5-times more likely to become gametocyte carrier than those who were treated successfully				
P. falciparum, *P. ovale*, *P. malariae*, *P. vivax*	gametocytes (immature and mature stages) (erythrocyte)	**8-aminoquinolines** (direct and fast action)	**primaquine** **pamaquine** WR-238,605 tafenoquine being developed (more details → above)	only drugs with fast and direct action on gametocytes of *P. falciparum*; high gametocytocidal action on all species of human malarial parasites, rendering the gametocytes incapable of development in mosquitoes
P. vivax, *P. ovale*, *P. malariae*	gametocytes (immature and mature stages) (erythrocyte)	**4-aminoquinolines**	**chloroquine** **amodiaquine**	effective against immature but ineffective against mature gametocytes of *P. falciparum*
P. vivax, *P. ovale*, *P. malariae*	gametocytes (immature and mature stages) (erythrocyte)	**chinchona alkaloids 9-aminoacridines**	**quinine** (p.o., i.m.) **mepacrine**	quinine is a poor drug for general gametocytocidal prophylaxis; mepacrine is no longer used because of its general toxicity
P. falciparum	precursors of sexual stages and early (I-II) gametocytes (erythrocyte)	**artemisinin derivatives**	**artemether** **artesunate**	does not kill mature gametocytes but reduces transmission of falciparum malaria
DRUGS AFFECTING FORMATION OF MALARIAL OOCYSTS AND SPOROZOITES IN INFECTED MOSQUITOES have little or no apparent effect on gametocytes but cause inhibition of subsequent development of sporogonic forms in the mosquito; agents with sporontocidal action can reduce transmission of malaria				
P. falciparum, *P. vivax* sporogony is inhibited for varying periods (dose dependent)	oocysts (mid-gut wall of mosquito)	**biguanides** (highly active) may be valuable for sporontocidal prophylaxis	**proguanil** (*Paludrine) (= chloroguanide) **chlorproguanil** (*Lapudrine)	mosquitoes fed on gametocyte carriers receiving therapeutic doses do not develop intact oocysts, i.e., drugs ablate transmission of malaria by preventing or inhibiting formation of oocyst and sporozoites in infected mosquitoes; infection to man may be interrupted or decreased

Table 1. (Continued) Classification of antimalarial drugs according to the stages of *Plasmodium* affected.

DISEASE *Plasmodium* species (other information)	STAGE AFFECTED (location), other information	CHEMICAL CLASS (other information)	INN, *BRAND-NAME (preferable oral route)	COMMENTS (toxic reactions and other comments)
P. falciparum, *P. vivax* (drug appears to inhibit sporogony)	oocysts (mid-gut wall of mosquito)	**diaminopyrimidines**	**pyrimethamine** (*Daraprim)	no apparent effect on the production, number or morphology of gametocytes
P. berghei (experimental rodent malaria)	oocysts (mid-gut wall of mosquito)	**sulfonamides, sulfones**	**various dapsone**	dapsone may cause increased gametocyte production in falciparum malaria; these gametocytes may not be infectious to mosquitoes

INN = International nonproprietary name
Data given in this table have no claim to full information

For many years avian malaria has played an important role in malaria research as the model of choice for screening of antimalarial drugs prior to the discovery of rodent *Plasmodium* spp. Because the biology of avian malaria differs considerably from that of mammals, various rodent models (e.g. P. *berghei*, *P. yoelii*, or *P. chabaudi* infecting mice, rats and other rodents) met with great interest. These malaria parasites are uncomplicated and easy in handling and available all over the world. The biology of rodent malaria is very similar to that of human malaria parasites, particularly *P. falciparum*, so that results obtained by drug screening in these models have a certain predictive value for the antimalarial activity of drugs against human parasites. Recently, the in vitro cultivation of *P. berghei* has been further optimized now offering new possibilities in molecular screening but also more insight into the conventional screening for chemotherapeutic agents . Today, a variety of in vitro and in vivo models provide a broad basis to investigate drug's mode of action and potential mechanisms leading to drug resistance in rodent and human malaria. Targets may be malarial haemozoin/β-haematin supporting haem polymerization or sequence variations in the *Plasmodium vivax* dihydrofolate reductase-thymidylate synthase gene and their relationship with pyrimethamine resistance. Owing to improved in vitro cultivation techniques, large numbers of different *Plasmodium* developmental stages may be helpful to find out new chemotherapeutic targets. One of these targets is the cytoplasmic ribosomal RNA of *Plasmodium* spp., as it seems to be quite different from the mechanisms in other eukariotic cells. Macaque monkeys are now used extensively not only for AIDS research but also malaria research, involving rhesus monkeys (*Macaca mulatta*) compared to other macaque species (*M. fascicularis*). Rhesus monkeys are highly susceptible to most of the malarias (especially *P. knowlesi*, *P. coatneyi* and *P. fragile*, distinctive less *P. cynomolgi*, *P. fieldi* and others) whereas *M. fascicularis* is not. So a comparison of response to malaria in susceptible rhesus monkey and resistant *M. fascicularis* might be a good starting point. Molecular-genetic studies on their hemoglobins and innate red blood cell polymorphisms would be probably of more value than research on immunity to *P. falciparum* in such artificial hosts as owl monkeys, *Aotus* and *Saimiri*.

Malaria of Humans

Clinical Forms Malaria is a mosquito-borne infection caused by four species of obligate intracellular protozoan parasites of the genus *Plasmodium* of which *P. vivax* is the most common and *P. falciparum* the most pathogenic. Each species has distinguishing morphological characteristics and the disease caused by each is also distinctive. *P. vivax* occurs north and south of the Equator within the 15°–16°C summer isotherms whereas *P. falciparum* is limited to, but widely distributed in, the tropics and subtropics, particularly Africa and Asia. ***P. falciparum*** causes 'Falciparum or malignant tertian malaria' (incubation time 7–14 days), the most dangerous form of human malaria. It can produce a foudroyant infection in nonimmune individuals that if not treated, may result in rapid death. If treated early, the infection usually responds to appropriate antimalarial drugs in chloroquine-sensitive or chloroquine-resistant areas, and recrudescence will not occur. If treatment is inadequate, however, recrudescence of infection may result from multiplication of parasites in the blood. Delay in treatment, especially in patients already having parasites in the blood for a week or so, may lead to irreversible state of shock, and death may occur though the peripheral blood is free of parasites. ***P. vivax*** causes 'Vivax or benign tertian malaria' (incubation time 12–17 days, sometimes several months or >1 year); the disease produces milder clinical attacks than those seen in falciparum malaria. It has a low mortality rate in untreated adults and is characterized by relapses, which may occur as long as 2 years after primary infection. ***P. ovale*** causing 'Ovale or benign tertian malaria' (incubation time 16–18 days or longer) occurs primarily in tropical Africa (especially in West Africa) and in some endemic areas of New Guinea and the Philippines and Southeast Asia. Clinical manifestations are similar to that of *P. vivax* infections (including periodicity and relapses), but are more readily cured. ***P. malariae*** causing 'Quartan malaria' (incubation time 18–40 days) is not found below the 16°C summer isotherms and has a variable and spotty distribution in the tropics and subtropics. Clinical signs are similar to that of 'benign tertian malaria', the febrile paroxysm occurring every 72 hours. Symptomatic recrudescence can occur several years after primary infection and is due to persistent undetectable parasitemia, and not to hypnozoites as in case of *P. vivax* infections.

Biology Though malaria can be transmitted by transfusion of infected blood, humans are naturally infected by **sporozoites** inoculated by the bite of female anopheline mosquitoes. Sporozoites rapidly leave the circulation and initiate **schizogony** in the parenchymal cells of the **liver** (so-called

Table 2. Treatment and prevention of malaria in humans.

Nonproprietary name	Brand name other information	Adult dosage/*pediatric dosage (mg/kg body weight, or total dose/individual, oral route), miscellaneous comments (for adverse effects of drugs see also Table 1)
Malaria parasites: *Plasmodium falciparum* (malignant tertian malaria), *P. vivax, P. ovale* (benign tertian malaria), *P. malariae* (quartan malaria)		
TREATMENT OF CHLOROQUINE-RESISTANT FALCIPARUM MALARIA		
Chloroquine-resistant *P. falciparum* occur in all malarious areas except Central America west of the Panama Canal Zone, Mexico, Haiti, the Dominican Republic, and most of the Middle East (chloroquine resistance has been reported in Yemen, Oman, and Iran)		
quinine sulfate **plus doxycycline**	**drugs of choice**	**quinine**: 650 mg q8h x 3-7d; *25 mg/kg/d in 3 doses x 3-7d; in Southeast Asia, relative resistance to quinine has increased and the treatment should be continued for seven days
or	(many manufacturers)	**doxycycline:** 100 mg bid x 7d; *2mg/kg/d x 7d (slow acting drug); drug is contraindicated in pregnancy and in children less than 8 years old; doxycycline can cause GI disturbances, vaginal moniliasis (candidiasis) and photosensitivity reactions
quinine sulfate plus **pyrimethamine-sulfadoxine**	**drugs of choice** Fansidar (Roche)	**quinine**: 650 mg q8h x 3-7d; *25 mg/kg/d in 3 doses x 3-7d; in Southeast Asia, relative resistance to quinine has increased and the treatment should be continued for seven days **Fansidar:** 3 tablets at once on the last day of quinine; *<1 year: $^1/_4$ tablet; 1-3 yrs: $^1/_2$ tablet; 4-8 yrs: 1 tablet; 9-14 years: 2 tablets; Fansidar tablets contain 25 mg of pyrimethamine and 500 mg of sulfadoxine; resistance to pyrimethamine-sulfadoxine has been reported from Southeast Asia, the Amazon basin, sub-Saharan Africa, Bangladesh and Oceania
or		
quinine sulfate plus **clindamycin**	**drugs of choice** (many manufacturers)	**quinine:** 650 mg q8h x 3-7d; *25 mg/kg/d in 3 doses x 3-7d; in Southeast Asia, relative resistance to quinine has increased and the treatment should be continued for seven days **clindamycin:** 900 mg tid x5d; *20-40 mg/kg/d in 3 doses x 5d; side effects may be allergic reactions, diarrhoea (enterocolitis, ulcerous colitis), hepatotoxicity, occasionally hypotension, ECG changes

Table 2. (continued)

ALTERNATIVE DRUG REGIMENS FOR TREATING MULTIDRUG-RESISTANT FALCIPARUM MALARIA
for treatment of multidrug-resistant *P. falciparum* in Southeast Asia, especially Thailand, where resistance to mefloquine and halofantrine is frequent , a 7-day course of quinine and tetracycline is recommended, artesunate plus mefloquine, artemether plus mefloquine or mefloquine plus doxycycline are also used to treat multiple-drug-resistant *P. falciparum*; at a single high dose (1250 mg) of **mefloquine** for adults (see below), adverse effects including nausea, vomiting, diarrhoea, dizziness, disturbed sense of balance, toxic psychosis and seizures can occur; it is teratogenic in animals and has not been approved for use in pregnancy; mefloquine prophylaxis appears to be safe when used during the second half of pregnancy and possibly during early pregnancy as well; it should not be given together with quinine or quinidine, and caution is required in using quinine or quinidine to treat patients with malaria taken mefloquine for prophylaxis; in the USA and elsewhere, the pediatric dosage has not been approved by government regulatory agencies; this may be also true for other alternative drugs such as doxycycline, clindamycin or atovaquone, including the adult dosage

mefloquine (MEF)	Lariam (Roche)	**mefloquine**: 1250 mg once (750 mg followed 12 hours later by 500 mg); *25 mg/kg once (15mg/kg followed 8–12 hours later by 10 mg/kg (<45kg b.w.); **tablets:** in the USA, = 250 mg = 228 MEF base, other countries 275 mg = 250 mg MEF base; resistance to MEF has been reported in some areas, such as the Thailand-Myanmar border, where 25 mg/kg should be used (adverse effects see Table 1)
halofantrine	Halfan (Smithkline Beecham)	**halofantrine**: 500 mg q6h x 3 doses; repeat in 1 week; *8 mg/kg q6h x 3 doses (<40 kg); repeat in 1 week; halofantrine may be effective in multiple-drug-resistant *P. falciparum* malaria, but treatment failures and resistance have been reported; in adults and children, a single 250-mg dose can be used for repeat treatment in mild to moderate infections; there is variability in absorption; thus it should not be taken 1h before and 2h after meals because food increases its absorption; cardiac adverse reactions may be lengthening of PR and QTc intervals and fatal cardiac arrhythmias and is contraindicated in patients with cardiac conduction defects (cardiac monitoring is recommended). Halfantrene (is possibly being developed for parenteral treatment of severe malaria cases.
atovaquone plus	Mepron (Glaxo-Wellcome)	**atovaquone**: 1000 mg qd x 3d; *11–20 kg: 250 mg; 21–30 kg: 500 mg; 31–40 kg: 750 mg; adverse effects may be frequent rash and nausea, occasionally diarrhoea
proguanil	Paludrine (Wyeth Ayerst, Canada; Zeneca, Europe)	**proguanil**: 400 mg qd x 3d; *11–20 kg b.w.: 100 mg; 21–30 kg: 200 mg; 31–40 kg: 300 mg; occasional adverse effects may be oral ulceration, hair loss, scaling of palms and soles, urticaria, rare: hematuria: (large doses), vomiting, abdominal pain, diarrhoea (large doses), thrombocytopenia
atovaquone-proguanil or	Malarone (Glaxo-Wellcome; Cascan)	Malarone: adult tablet (250 mg atovaquone /100 mg proguanil): 4 adult tablets once/d x 3d; *all adult tablets once/d x 3d: 11–20 kg b.w.: 1; 21–30 kg; 2; 31–40 kg: 3; >40 kg: 4
atovaquone plus	Mepron	**atovaquone**: 1000 mg qd x 3d; *11–20 kg b.w.: 250 mg; 21–30 kg: 500 mg; 31–40 kg: 750 mg; adverse effects may be frequent rash and nausea, occasionally diarrhoea
doxycycline	(many manufacturers)	**doxycycline**: 100 mg bid x 3d; *2mg/kg/d x 3d (slow acting drug); drug is contraindicated in pregnancy and in children less than 8 years old; doxycycline can cause GI disturbances, vaginal moniliasis (candidiasis) and photosensitivity reactions
artesunate plus	(various manufacturers)	4mg/kg/ x 3d; very rapidly acting drug, which may occasionally produce ataxia, slurred speech, neurological toxicity, possible increase in length of coma in severe falciparum malaria, increased convulsions, prolongation of QTc interval; for artesunate formulations see Table 1; the artemisinin derivative is produced in People's Republic of China, Guilin No. 1 Factory, and by Mepha A.G. (Aesch-Basle, Switzerland) in Vietnam

Table 2. (continued)

TREATMENT OF CHLOROQUINE-RESISTANT VIVAX MALARIA

P. vivax with decreased susceptibility to chloroquine has been reported in Papua-New Guinea, Indonesia, Myanmar; India, Irian Jaya, and the Solomon Islands

quinine sulfate plus	**drugs of choice**	**quinine**: 650 mg q8h x 3–7d; *25 mg /kg/d in 3 doses x 3–7d; in Southeast Asia, relative resistance to quinine has increased and the treatment should be continued for seven days
doxycycline or	(many manufacturers)	**doxycycline**: 100 mg bid x 7d; *2mg/kg/d x 7d (slow acting drug); drug is contraindicated in pregnancy and in children less than 8 years old; doxycycline can cause GI disturbances, vaginal moniliasis (candidiasis) and photosensitivity reactions
quinine sulfate plus	**drugs of choice**	**quinine**: 650 mg q8h x 3–7d; *25 mg /kg/d in 3 doses x 3–7d; in Southeast Asia, relative resistance to quinine has increased and the treatment should be continued for seven days
pyrimethamine-sulfadoxine or	Fansidar (Roche)	**Fansidar**: 3 tablets at once on the last day of quinine; *<1 year: $^1/_4$ tablet; 1–3 yrs: $^1/_2$ tablet; 4–8 yrs: 1 tablet; 9–14 years: 2 tablets; **Fansidar tablets** contain 25 mg of pyrimethamine(PY) and 500 mg of sulfadoxine (SD); resistance to PY-SD has been reported from Southeast Asia, the Amazon basin, sub-Saharan Africa, Bangladesh and Oceania
mefloquine	**drug of choice** Lariam (Roche)	**mefloquine**: 1250 mg once (750 mg followed 12 hours later by 500 mg); *25 mg/kg once (15mg/kg followed 8–12 hours later by 10 mg/kg (<45kg b.w.); tablets: in the USA, = 250 mg = 228 mefloquine (MEF) base, other countries 275 mg = 250 mg MEF base; resistance to MEF has been reported in some areas, such as the Thailand-Myanmar border, where 25 mg/kg should be used (adverse effects see Table 1)

TREATMENT OF ALL PLASMODIA (except chloroquine resistant *P. falciparum* and chloroquine resistant *P. vivax*, see above)

CHLOROQUINE-SENSITIVE AREAS

(chloroquine-resistant *P. falciparum* occur in all malarious areas except Central America west of the Panama Canal Zone, Mexico, Haiti, the Dominican Republic, and most of the Middle East (chloroquine resistance has been reported in Yemen, Oman, and Iran; *P. vivax* with decreased susceptibility to chloroquine has been reported in Papua-New Guinea, Indonesia, Myanmar; India, Irian Jaya, and the Solomon Islands)

chloroquine (phosphate)	**drug of choice** Aralen, Resochin, many others (many manufacturers)	1 gram (600 mg base), then 500 mg (300 mg base) 6 hours later, then 500 mg (300 mg base) at 24 and 48 hours; *10 mg base/kg (max. 600 mg base), then 5 mg base/kg 6 hrs later, then 5 mg base/kg at 24 and 48 hrs; if chloroquine is not available, hydroxychloroquine sulfate is as effective; 400 mg hydroxychloroquine sulfate is equivalent to 500 mg chloroquine phosphate; occasional **adverse effects** may be pruritus, vomiting, headache, confusion, depigmentation of hair, skin eruption, corneal opacity, exfoliative dermatoses, eczema, myalgias, photophobia; rare side effects are irreversible retinal injury (may occur when total dosage exceeds 100 grams), nerve-type deafness, peripheral neuropathy and myopathy, heart block, blood dyscrasias, hematemesis (vomiting of blood)

ALL PLASMODIA (ESPECIALLY SEVERE FALCIPARUM MALARIA):

exchange transfusion may be helpful for some patients with high-density (>10%) parasitemia, altered mental status, pulmonary edema or renal complications

quinidine gluconate PARENTERAL ROUTE or	**drug of choice** (many manufacturers)	10 mg/kg loading dose (max. 600 mg) in normal saline slowly over 1 to 2 hours, followed by continuous infusion of 0.02 mg/kg/min until oral therapy can be started; *same as adult dose; **precautions**: continuous EKG, blood pressure and glucose monitoring are recommended, especially in pregnant women and young children; quinidine may have greater antimalarial activity than quinine; thus the loading dose should be decreased or omitted in those patients who have received quinine or mefloquine, if >48 hrs of parenteral treatment is required, the quinine or quinidine dose should be reduced by 1/3 to $^1/_2$

Table 2. (continued)

quinine dihydro-chloride PARENTERAL ROUTE	**drug of choice**	20 mg/kg loading dose i.v. in 5% dextrose over 4 hours, followed by 10 mg/kg over 2–4 hours q8h (max 1800 mg/d) until oral therapy can be started; *same as adult dose; for **precautions**, which should be taken see quinidine above
artemether PARENTERAL ROUTE	**alternative drug** Artenam (Arenco, Belgium, other manufacturers, see column 3)	3.2 mg/kg i.m. then 1.6 mg/kg qd; *same as adult dose; occasional adverse effects may be neurological toxicity, possible increase in length of coma in severe falciparum malaria, increased convulsions, prolongation of QTc interval; for artemether formulations see Table 1; the artemisinin derivative has been produced by Rhône-Poulenc Rorer and Kunming Pharmaceutical Factory in China, and the UNDP/World Bank/WHO Special Program for Research and Training in Tropical Diseases (TDR)
PREVENTION OF RELAPSES: *P. vivax* and *P. ovale* (radical cure of benign tertian malaria) (cf. text below and Table 1: **hypnozoites in liver**)		
primaquine phosphate (it can be also used for prophylaxis, see below)	**drug of choice** various manufacturers (in some countries not directly available in pharmacies)	26.3 mg (15 mg base)/d x 14d or 79 mg (45 mg base)/wk x 8 wks; *0.3 mg base/kg/d x 14d; some relapses have been reported with this regimen, especially in strains from Southeast Asia; relapses should be treated with a second 14-day course of 30 mg base/day; the drug can cause **hemolytic anemia**, especially in patients whose red cells are deficient in glucose-6-phosphate dehydrogenase; this deficiency is most common in African, Asian, and Mediterranean peoples; patients should be screened for **G-6-PD deficiency** before treatment; it should not be used during pregnancy and in lactating women; occasional adverse effects may be neutropenia, methemoglobinemia, GI disturbances, rare are CNS symptoms, arrhythmias and hypertension
MALARIA PREVENTION: No drug regimen guarantees protection against malaria; if fever develops within a year, especially within the first 2 months after travel to malarious areas, travellers should be advised to consult a specialist for tropical medicine; insect repellents, insecticide-impregnated bed nets and proper clothing are important adjuncts for malaria prophylaxis; for prevention of attack after departure from areas where *P. vivax* and *P. ovale* are endemic, which includes almost all areas where malaria is found (except Haiti), some experts prescribe in addition **primaquine** phosphate 26.3 mg (15 mg base) or, for children, 0.3mg base/kg/d during the last 2 weeks of prophylaxis; others prefer to avoid the toxicity of primaquine (see above) and rely on surveillance to detect malaria cases when they occur, particularly when exposure was limited or doubtful		
PREVENTION IN CHLOROQUINE-SENSITIVE AREAS: (chloroquine-resistant *P. falciparum* occur in all malarious areas except Central America west of the Panama Canal Zone, Mexico, Haiti, the Dominican Republic, and most of the Middle East (chloroquine resistance has been reported in Yemen, Oman, and Iran; *P. vivax* with decreased susceptibility to chloroquine has been reported in Papua-New Guinea, Indonesia, Myanmar, India, Irian Jaya, and the Solomon Islands)		
chloroquine (phosphate)	**drug of choice** Aralen, Resochin, many others (many manufacturers)	500 mg (300 mg base) once/week; *5 mg/kg base once/week, up to adult dose of 300 mg base; prophylaxis should be started one week prior to travel and continued weekly for the duration of stay and for 4 weeks after leaving; in pregnancy chloroquine prophylaxis has been used extensively and safely; the safety of other prophylactic antimalarials in pregnancy is less clear; for this reason, travel during pregnancy to chloroquine-resistant areas (see next below) should be considered carefully
PREVENTION IN CHLOROQUINE-RESISTANT AREAS (chloroquine-resistant *P. falciparum* occur in all malarious areas except Central America west of the Panama Canal Zone, Mexico, Haiti, the Dominican Republic, and most of the Middle East (chloroquine resistance has been reported in Yemen, Oman, and Iran; *P. vivax* with decreased susceptibility to chloroquine has been reported in Papua-New Guinea, Indonesia, Myanmar, India, Irian Jaya, and the Solomon Islands)		

Table 2. (continued)

mefloquine is not recommended for patients with cardiac conduction abnormalities, and patients with a history of seizure or psychiatric disorders should probably avoid mefloquine; resistance to mefloquine has been reported in some areas, such as Thailand; in these areas, **doxycycline** should be used for prophylaxis; in children <8 years old, **proguanil** plus sulfafurazole has been used; several studies have shown that daily **primaquine** provides effective prophylaxis against chloroquine-resistant *P. falciparum*; malaria prophylaxis with primaquine was also evaluated in Irian Jaya using 0.5 mg/kg primaquine base daily for 1 year by 126 Javanese men with normal G-6-PD activity; primaquine was well tolerated and effective for prevention of falciparum malaria (94.5%) and vivax malaria (90.4); for prevention of attack after departure from areas where *P. vivax* and *P. ovale* are endemic, which includes almost all areas where malaria is found (except Haiti), some experts prescribe in addition **primaquine** phosphate 26.3 mg (15 mg base) or, for children, 0.3mg base/kg/d during the last 2 weeks of prophylaxis; others prefer to avoid the toxicity of primaquine (see above) and rely on surveillance to detect malaria cases when they occur, particularly when exposure was limited or doubtful; **atovaquone-proguanil** (Malarone: available in the USA, Europe and elsewhere) can be used for oral prophylaxis and treatment of malaria due to blood forms of all human plasmodia, exoerythrocytic forms of *P. falciparum*, and *P. falciparum* resistant to pyrimethamine-sulfadoxine = Fansidar (SE Asia, the Amazon basin, sub-Saharan Africa, Bangladesh and Oceania) or mefloquine (significant problem in Thailand along the borders with Myanmar and Cambodia); co-administration of tetracyclines reduces plasma concentrations of atovaquone by 40 to 50%, proguanil has no known drug interactions		
mefloquine (MEF) or	**drug of choice** Lariam (Roche)	250 mg once/week; *< 5 kg b.w.: no data; 5–9 kg: 1/8 tablet; 10–19 kg: 1/4 tablet; 20–30 kg: 1/2 tablet; 31–45 kg: 3/4 tablet; >45 kg: 1 tablet; **tablets:** in the USA = 250 mg = 228 MEF base, other countries 275 mg = 250 mg MEF base; prophylaxis should be started one week prior to travel and continued weekly for the duration of stay and for 4 weeks after leaving; the pediatric dosage has not been approved in the USA and elsewhere, and the drug has not been approved for use during pregnancy; it has been reported to be safe for prophylactic use during the second half of pregnancy and possibly during early pregnancy as well; women should take contraceptive precautions while taking mefloquine and for 2 months after the last dose
doxycycline or	**drug of choice** (many manufacturers)	100 mg daily; *2 mg/kg/d, up to 100 mg/d; drug is contraindicated in pregnancy and in children less than 8 years old; prophylaxis should be started one day prior to travel and continuing for the duration of stay and for 4 weeks after leaving; contraindications see above; it can cause GI disturbances, vaginal moniliasis (candidiasis) and photosensitivity reactions
primaquine or	**drug of choice** (various manufacturers)	0.5 mg/kg base daily; *same as adult; the drug can cause **hemolytic anemia**, especially in patients whose red cells are deficient in glucose-6-phosphate dehydrogenase (most common in African, Asian, and Mediterranean peoples); patients should be screened for **G-6-PD deficiency** before treatment; it should not be used during pregnancy and in lactating women; occasional adverse effects may be neutropenia, methemoglobinemia, GI disturbances, rare are CNS symptoms, arrhythmias and hypertension
atovaquone-proguanil	**drug of choice** Malarone (Glaxo Wellcome)	250 mg/100 mg (1 tablet) daily; *11–20 kg b.w.: 62.5 mg/25 mg (1 tablet) daily; 21–30 kg: 125 mg/50 mg daily; 31–40 kg: 187,5/75 mg daily; >40 kg: 250/100 mg daily beginning 1 to 2 days before travel and continuing daily for 1 week after leaving malaria zone; it is generally well tolerated

Table 2. (continued)

ALTERNATIVES		
chloroquine phosphate **plus** **pyrimethamine-sulfadoxine** (PY-SD) (for presumptive treatment)	Fansidar (Roche)	**chloroquine** phosphate: 500 mg once/week; *5 mg/kg base once/week, up to adult dose of 300 mg base **Fansidar**: carry a single dose (3 tablets) for self treatment of febrile illness when medical care is not immediately available; * <1 yr: $^1/_4$ tablet; 1–3 yrs: $^1/_2$ tablet; 4–8 yrs: 1 tablet; 9–14 yrs: 2 tablets; tablets contain 25 mg of PY and 500 mg of SD; for prophylaxis of PY-SD resistant *P. falciparum* with **malarone** or **doxycycline** see dosages above
or		
chloroquine phosphate **plus**		**chloroquine** phosphate: 250 mg once/week; *< 5 kg: no data; 5–9 kg: $^1/_8$ tablet; 10–19 kg: $^1/_4$ tablet; 20–30 kg: $^1/_2$ tablet; 31–45 kg: $^3/_4$ tablet; >45 kg: 1 tablet
proguanil (widely available in Canada, Europe and elsewhere)	Paludrine (Zeneca and others)	**proguanil**: 200 mg daily; *< 2 years: 50 mg/daily; 2–6 years: 100 mg/d; 7–10 years: 150 mg/d; >10 years: 200 mg; proguanil is recommended mainly for use in Africa south of the Sahara; prophylaxis is recommended during exposure and for 4 weeks afterwards; proguanil has been used in pregnancy without evidence of toxicity (now also available in the USA)

Abbreviations: the letter <d> stands for day (days); qd = daily (quaque die); qh = each hour every hour ; qd= each day, every day; bid = twice daily; tid = three times per day; qid = four times per day (quarter in die); p.c. (post cibum) = after meals
Dosages and data on toxicity of antimalarial drugs listed in the table refer to information from manufacturer or literature, and → Further Reading, Control Measurements: Medical Letter, 'Drugs for parasitic infections'
Data given in this Table have no claim to full information

preerythrocytic or exoerythrocytic = EE stage of infection), which is asymptomatic and lasts for 5–16 days depending on the *Plasmodium* species. In *P. falciparum* and *P. malariae* infections primary tissue schizonts burst simultaneously within a certain period, leaving no parasite stages in the liver. In *P. vivax* and *P. ovale* infections, some tissue parasites remain 'dormant' (latent forms or hypnozoites) before they proliferate and produce relapses of erythrocytic infection months to years after primary infection.

Mature tissue schizonts rupture in the liver thereby releasing thousands of merozoites; thence, they enter the circulation, and invade erythrocytes (erythrocytic stage or cycle of infection). In red blood cells, parasites undergo asexual development from young ring forms to trophozoites and then to mature schizonts that release several merozoites after erythrocytes being ruptured more or less synchronically. This process produces the febrile clinical attack. The released merozoites invade other naive erythrocytes to continue the cycle, which may proceeds until death of the host or interruption by antimalarials or modulation by acquired immunity will occur. Some erythrocytic parasites differentiated into sexual forms, male microgametocytes and female macrogametocytes.

During the blood meal, the female mosquito ingests gametocytes, and syngamy (fertilization of the macrogamont by a microgamont) occurs in the mosquito's gut. The resulting sedentary zygote transforms to a motile ookinete. Ookinetes are specialized cells able to actively leave the packed blood bolus and invade the mosquito midgut epithelial tissue to reach the hemolymph side and develop as oocyst. In the oocyst, the parasites multiply to sporozoites, which later invade the salivary gland and are subsequently inoculated into another human host during blood meal. Ookinete motility, secretion of chitinase, resistance to the digestive enzymes, and recognition/invasion of the midgut epithelium all may play crucial roles in the transformation to oocyst. A number of target ookinete-stage antigens are currently on the list of malaria transmission-blocking vaccines, and monoclonal antibodies to both Pfs25 and Pfs28 block oocyst development. Pfs25 is at the initial stage of human trials.

Global Control Programs The **prevalence** of malaria is increasing, and in 1998, more patients suffered from malaria than in 1958 . According to the World Health Organization (WHO), more than 500 million people are infected with malaria parasites each year and more then 2 million – mostly

children living in sub-Saharan Africa – die of it. These often quoted figures of malaria-related deaths per year among African children under the age of 5 years originated from analyses of malaria transmission intensities in sub-Saharan Africa, which are typically one or two orders of magnitude greater than those that occur in most other malaria-endemic regions of the world. With the attainment of age- and exposure-acquired protective immunity, there is a rapid decline in the incidence of malaria infection associated with high clinical tolerance and virtually no case fatalities after the ages of 10 to 15 years. In south Asian regions malaria **transmission intensities** are not only typically much lower than those in much of tropical Africa, leading to a very different age distribution of disease, but the health systems for managing the malaria problems are also substantially different. Rapid case treatment appears to be the most suitable measure to save lives at risk under virtually all circumstances of malaria transmission. Where effective early treatment is the main tool in reducing malaria mortality, the emergence of **resistance to antimalarial drugs** is a major concern. High levels of resistance (RIII), especially to drugs such as chloroquine, pyrimethamine-sulfadoxine and mefloquine (Fansidar, Fansimef, Table 1), as for example in Vietnam, have, indeed, been associated with increases in malaria-related deaths . Meanwhile, drug resistant strains of *P. falciparum* are spreading to new territories, including India, South America and the Far East; also the *Anopheles* mosquito vector is gaining greater resistance to insecticides. Antimalarial drugs such as chloroquine, antifolates and mefloquine (Table 2) are becoming frequently less effective against *P. falciparum* and *P. vivax*, the principal infectious agents of malaria. To successfully tackle these serious problems, **malaria research** appears to be drastically underfunded compared to other diseases, such as HIV and asthma, when its relative incidence and its global death toll are taken into account. Total global expenditure on malaria research in 1993 was US$84 million – equivalent to $42 for every death. Calculations for HIV/AIDS gave $3,274 for each death, and $789 for asthma. It is suggested that the results of research have not been sufficiently exploited. On the other hand, obstacles to better exploitation, according to the survey, include poor orientation of research programs to practical problems and public needs. **Topics** having the best prospects for advancing understanding over the next 5 years or so were the genetics and biology of *Plasmodium* and disease epidemiology. Thus many **vaccine projects** have failed after promising starts and malaria researchers say prospects for a workable vaccine are still a long way off. An experimental vaccine devised by the U.S. Army and the SmithKline Beecham company worked well in a preliminary test at Walter Reed Army Institute of Research, Washington D.C. protecting 6 out of 7 people against *P. falciparum* infection after they had been bitten repeatedly by mosquitoes carrying live parasites. The vaccine consists of a synthetic concoction based on proteins that appears on the surface of dead falciparum parasite. Monitor phase I/II trials with RTS, S pre-erythrocytic vaccine (WRAIR/SKB) should be completed in 1998.

The **new global strategy** for malaria control moves away from several outmoded concepts inherited from the times when eradication of malaria still seemed feasible. Particular attention is given to the need for disease-oriented programs, with a reduction of mortality and morbidity. Technical elements of these programs include the provision of **early diagnosis and prompt treatment**, the selective use of sustainable preventive measures, the prevention and control of epidemics, and the strengthening of local capacities in basic and applied research. Some of these issues are development of drug packaging systems at a district level to improve dosing and compliance, better collaboration between public and private sectors to improve case management of malaria particularly childhood illness and improved supervision of drug vendors by district pharmacies. One issue is to replace chloroquine by inexpensive drugs such as pyronaridine and short half-life antifolate drugs (Table 1). Evaluation and development of pyronaridine up to registration including technology transfer to Malaysia is planned (target year 2000).

Vector control is an essential component of the 'Global Malaria Control Strategy', adopted in 1992. Existing tools for vector control, if appropriately used, can help to prevent or to reduce the transmission of malaria and other mosquito-borne diseases (e.g. → lymphatic filariasis). There are detailed guidelines for the use of four main options: indoor residual spraying; personal protection, including **insecticide-impregnated bednets** and other materials (Table 2/Malaria prevention), larviciding and biological control, and environmental management. Thus, efficacy trials with in-

Table 1. Mallophaga and Control Measurements.

Parasite	Host	Vector for	Symptoms	Country	Therapy		
					Products	Application	Compounds
Trichodectes canis	Dog	*Dipylidium caninum?*	Alopecia, eczema, sec. bacterial infections	World wide	Advantage™ (Bayer)	Spot on	Imidacloprid
					Defend™Just-For-Dogs Insecticide (Mallinckrodt)	Spray	Pyrethrin + Permethrin + Piperonylbutoxid + N-octyl bicyclohep-tene dicarboximide
					Bolfo™ Flohschutz-Puder (Bayer)	Dermal powder	Propoxur
Felicola subrostratus	Cat	*Dipylidium caninum?*	Alopecia, eczema, sec. bacterial infections	World-wide	Advantage™ (Bayer)	Spot on	Imidacloprid
					Defend™Just-For-Dogs Insecticide (Mallinckrodt)	Spray	Pyrethrin + Permethrin + Piperonylbutoxid + N-octyl bicyclohep-tene dicarboximide
					Bolfo™ Flohschutz-Puder (Bayer)	Dermal powder	Propoxur
Bovicola bovis	Cattle		Skin secretion; host specific; strong reproduction only possible in ill or weak cattle; constant irritation	World-wide	Tiguvon™ Cattle Insecti-cide Pour on (Bayer Corp.)	Pour on	Fenthion
					Asuntol™-Puder 1% (Bayer)	Dermal powder	Coumaphos
Bovicola caprae	Goat		Itching, constant irritation	World-wide	Sebacil™ Lösung	Wash, Spray	Phoxim
Bovicola limbatus	Goat			Rare in Europe			
Lepikentron ovis	Sheep		Itching, wool loss through rubbing	World-wide	Sebacil™ Lösung (Bayer)	Wash, Spray	Phoxim
Werneckiella equi	Horse	Infectious anaemia?	Itching, strong concern, bite and rubbing wounds	World-wide			

secticide-treated bednets for preventing childhood mortality are on the track, thereby checking promotion of technology, implementation, definition of areas for bednets use, and cost-effectiveness consideration.

Antimalarial Drug Development New antimalarial agents (targets) with novel mode of actions are urgently needed because of increasing drug resistance of malaria parasites in endemic areas, including cities, and islands. In recognition of this need the WHO/TDR Steering Committee on Drugs for Malaria (CHEMAL, NIH/NIAID and MMV = 'medicines for malaria ventures') support studies from the identification of new biochemical targets for drug development to the registration of a drug, usually in partnership with a commercial company. **New drug leads** include selection for development of one lead from 2nd generation peroxidic drugs, phospholipid and antiplasmodial protease (proteinase) inhibitors or protein prenyl transferase inhibitors. The funding agencies now screen a series of compounds (10mg quantities) or compound libraries against three malaria pro-

Table 1. *Cheyletiella* species.

Parasite	Host	Symptoms	Country	Therapy		
				Products	Application	Compounds
Cheyletiella yasguri	(young) Dog	Blood loss, dermatitis	Worldwide	Many	Bathing	Pyrethroids
Cheyletiella blakei	(young) Cat			Stronghold™	Spot on	Selamectin

Table 1. *Demodex* species of animals.

Parasite	Host	Symptoms	Country	Therapy		
				Products	Application	Compounds
Demodex canis	almost only young Dogs	Local - generalised Dermatitis / Alopecia	Worldwide	Ectodex™ (Intervet)	Bathing	Amitraz
				Mitaban™ (Pharmacia & Upjohn)	Bathing	Amitraz
Demodex bovis	Cattle	Often subclinic, no itching, pea-sized nodules, leather damages (economic loss)	Worldwide			
Demodex ovis	Sheep	Often subclinic, hardly itching, often round the eyes, vulva and prepuce	Worldwide			
Demodex caprae	Goat		Worldwide (Switzerland, France)			
Demodex equi	Horse	Often subclinic, hardly itching; transfer only from mother to foal; starts at head, then possibly generalization, bact. sec. Inf.	Worldwide			
Demodex caballi	Horse	Primarily eye (Meibom gland)	Worldwide			
Demodex suis	Pig	Rare; hardly itching, transfer via contact	Worldwide	Point-Guard™ Miticide/ Insecticide (Intervet)	Pour-on	Amitraz

teinases, including two aspartic proteinases (plasmepsin I and II, similar to human cathepsin D, and one cysteine proteinnase, falcipain, analogous to human cathepsin L). For combinatorial libraries, enzymatically active recombinant enzyme can be provided by CHEMAL. In a study, cDNA coding for *P. falciparum* hypoxanthine-guanine-xanthine phosphoribosyltransferase has been cloned in *Escherichia coli* and purified to homogeneity to allow detailed kinetic and structural studies. Significant differences between the human and parasitic enzymes indicate that parasite-specific inhibitors are feasible. Tubulin as a potential drug target is not yet sufficiently 'validated'. Several microtubule inhibitors are potent blockers of various stages of development of *Plasmodium*. Most compounds have been derived from anticancer screening programs as cholchicine-site binders (e.g. cholchicine, colcemid, and anthelmintic benzimidazole carbamates, cf. → Nematocidal Drugs, Man), vinblastine-site binders (vinblastine and vincristine), taxoids (taxol, taxotere) and others (*cis*- and *trans*-tubulozole, and trifluralin). Though trifluralin proved much less toxic than

Table 1. Trombiculid species.

Parasite	Host	Symptoms	Country	Therapy		
				Products	Application	Compounds
Neotrombicula autumnalis	Dog, Cat, **Man**, Horse, Ruminants	Itching, bacterial secondary infections	Worldwide	Kiltix™ (Bayer)	Collar	Flumethrin/ Propoxur
				Bayticol™ (Bayer)	Spray	Flumethrin
Neotrombicula desaleri	Ruminants	No transfer from contact, transfer from plant to animal; often under the tail (cattle), nose (sheep), ears (goat); red stains	Worldwide			
Neoschöngastia xerothermobia	Ruminants					
Trombicula akamushi	Dog, Cat, Cattle, **Man**	Rare, itching, bacterial secondary infections	India, Japan, China, Pacific Islands, North Australia			

the other microtubule inhibitors indications of carcinogenicity and modest tolerability preclude its development as an antiprotozoal drug.

Currently used drugs (Table 2), particular successors to chloroquine have not always met the expectations left by this remarkable compound characterized by low price, low toxicity, optimal pharmacokinetic properties providing safe prophylactic and therapeutics antimalarial activity against all four species of *Plasmodium* that infect humans. Derivatives of artemisinin and new formulations of them (artemether, arteether, and sodium artesunate, Table 1) with potent activity against erythrocytic stages of *P. falciparum* and *P. vivax*, which is at least as effective as the parent compound, are on the target list for further clinical development. Applied field research with artemether has been performed in developing countries (and France) for use of the drug in childhood cerebral malaria, and clinical development of injectable (i.m.) arteether is going on with promising results concerning efficacy tolerability and pharmacokinetics in patients with severe malaria. Some artemisinin-type compounds (artelinic acid, trioxane and tetraoxane analogues) and combinations of benflumetol (a fluoromethanol synthesized in China) plus artemether or mefloquine plus artesunate are at an advanced stage of preclinical or clinical development or already in use for chloroquine resistant falciparum malaria. The development of arteflene was discontinued because of a disappointing bloodschizontocidal activity against falciparum malaria (Table 1). Alternative antifolate combinations with sulfonamides and sulfones exhibiting shorter half-lives than pyrimethamine-sulfadoxine have been studied (WR 250417) (Table 1). Atovaquone, a hydroxynaphthoquinone (Table 1, Table 2), which has novel mode of action, can be used in co-administration with either tetracycline or proguanil to cure multiresistant falciparum malaria. A new 8-aminoquinoline (WR 238605) proves to be more active but less toxic than primaquine. Several plant (root) extracts isolates in China protected mice against *Plasmodium berghei* and *Plasmodium yoelli* infections , and most recently the revealed shikimate pathway in *P. falciparum* may give some hope to new selective therapeutic options and possibly targeted drug development.

In addition, existing validated targets of antifolates such as pyrimethamine or proguanil (e.g. resistance of *P. falciparum* resulting from point mutations of the DHRF domain of the bifunctional thymidylate synthetase with mutations at residues 51, 59, 108, and 164), and still unknown targets of quinoline antimalarials (e.g. precise mode of action and mechanism of parasite resistance to these drugs are still not completely understood), or several other identified enzymes from a number of biochemical pathway in *P. falciparum* have been proposed to be potential drug targets, though few of them have been validated. Possibly **new tools and**

Table 1. *Otodectes* species and Control Measurements.

Parasite	Host	Symptoms	Country	Therapy		
				Products	Application	Compounds
Otodectes cynotis	Dog, Cat	Ear mange, itching, inflammation, "Otitis externa parasitaria" (ear cancer), sometimes generalized	Worldwide	Ultra Ear Miticide™ (A.H.A.)	otic	Rotenone

Table 1. *Notoedres* species and Control Measurements.

Parasite	Host	Symptoms	Country	Therapy		
				Products	Application	Compounds
Notoedres cati	Cat	Head, all ages, mange symptoms	Worldwide	Stronphold™, Revolution™ (Pfizer)	Spot on	Selamectin

Table 1. *Saroptes* species and Control Measurements.

Parasite	Host	Symptoms	Country	Therapy		
				Products	Application	Compounds
Sarcoptes canis	Dog, Man	Head, ear (peripheral), ridge of the nose, eye, lower abdomen, area inside the thigh, itching, later hyper- and parakeratosis, sec. bact. infections	World-wide	Rotenone ShampooTM (Goodwinol)	Shampoo	Rotenone
Sarcoptes bovis	Cattle	Starts often at head, then generalization; alopecia, hyperkeratosis, wrinkling, bact. sec. inf., economic loss	World-wide	Ivomec™ 1% Injection For Cattle (Merial)	Injection	Ivermectin
				Dectomax™ (Pfizer)	Injection	Doramectin
				Sebacil™ Lösung (Bayer)	Wash, Spray	Phoxim
Sarcoptes ovis	Sheep	Often only head, bact. sec. inf.	World-wide	Cydectine™ injectable (Fort Dodge)	Injection	Moxidectin
Sarcoptes rupicaprae (=S. caprae)	Goat	Head mange	World-wide			
Sarcoptes equi	Horse, Man (pseudos cabies)	Starts often at head, then generalization; alopecia, hyperkeratosis, wrinkling, bact. sec. infections	World-wide			
Sarcoptes suis	Pig, Man (pseudos cabies)		World-wide	Ivomec™ 0.27% Sterile Solution (Merial)	Injection	Ivermectin
				Point-Guard™ Miticide/Insecticide (Intervet)	Pour-on	Amitraz
				Sebacil™ Pour-on (Bayer)	Pour-on	Phoxim

Table 1. *Psoroptes* species and Control Measurements.

Parasite	Host	Symptoms	Country	Therapy		
				Products	Application	Compounds
Psoroptes ovis	Ruminants	Cattle-sheep transfer possible, itching, symptoms see *Sarcoptes bovis*, large economic significance in sheep	Europe	Co-Ral™ 25% Wettable Powder (Bayer)	Dip or Spray	Coumaphos
				Ivomec™ 1% Injection For Cattle (Merial)	Injection	Ivermectin
				Dectomax™ (Pfizer)	Injection	Doramectin
				Sebacil™ Lösung (Bayer)	Wash, Spray	Phoxim
				Cydectin™ (Bayer)	Injection	Moxidectin
Psoroptes cuniculi	Goat	Ear Mange, often young (>3 weeks) lambs	Worldwide			
	Horse	Ear				
Psoroptes equi	Horse	Primarily in thick hair, protected areas (e.g. beginning of tail, under the mane), itching, symptoms see *Sarcoptes bovis*, large economic significance in sheep	Worldwide			

Table 1. *Chorioptes* species and Control Measurements.

Parasite	Host	Symptoms	Country	Therapy		
				Products	Application	Compounds
Chorioptes bovis (Leg mange, tail mange)	Ruminants	Starts often at tail, eat flakes; foot mange in sheep	Worldwide	Sebacil™ Lösung (Bayer)	Wash, Spray	Phoxim
	Horse	Clinical symptoms rare (foot mange)				

Table 1. *Psoergates* species and Control Measurements.

Parasite	Host	Symptoms	Country	Therapy		
				Products	Application	Compounds
Psoergates ovis	Sheep	Loss of wool (economic problem)	Australia, New Zealand, South Africa, South America	Sebacil™ Lösung (Bayer)	Wash, Spray	Phoxim

Table 1. *Pneumonyssinus* species and Control Measurements.

Parasite	Host	Symptoms	Country	Therapy		
				Products	Application	Compounds
Pneumonyssoides caninum (Nasal mite)	Dog	Chronic sneezing, epitaxis	Worldwide	Interceptor™ (Novartis)	Oral	Milbemycin, Oxime

technologies (e.g. transfection, DNA microassays and proteomic analysis) and the availability of DNA sequences generated by the Malaria Genome project along with more classic approaches (in vitro and in vivo screening of compounds) will facilitate the development of new antimalarials as well as the generation of a deeper understanding of the molecular mechanism(s) of drug resistance in malaria.

Mallophagidosis

Disease due to infestation with species of Mallophaga, see Table 1 (page 319).

MALT

Mucosa associated lymphoid tissue (→ Immune Responses).

Mange, Animals

Skin disease in animals caused by digging mites such as *Sarcoptes* spp. which make funnels in the skin that become inflamed due to secondary bacterial invasion (→ Acariosis, Animals) of other mites (see below).

Cheyletiellosis
See Table 1 (page 320).

Demodicosis
See Table 1 (page 320).

Trombiculidiasis
See Table 1 (page 321).

Otodectic Mange
See Table 1 (page 322).

Notoedric Mange
See Table 1 (page 322).

Sarcoptic Mange
See Table 1 (page 322).

Psoroptic Mange
See Table 1 (page 323).

Chorioptic Mange
See Table 1 (page 323).

Psoergatic Mange
See Table 1 (page 323).

Pneumonyssoidic Mange
See Table 1 (page 323).

Mange, Man

Skin disease in animals caused by digging → mites (Vol. 1) such as *Sarcoptes* spp. which make funnels in the epidermis that becomes inflamed due to secondary bacterial invasion (→ Acariosis, Animals, → Scabies).

Master Plan

→ Disease Control, Planning.

Mathematical Models of Vector-Borne Diseases

Using simple mathematical transmission models of infectious diseases, one can create and investigate dozens of epidemics in an afternoon, and nobody becomes ill and nobody dies, a feature that makes this an informative and rewarding line of epidemiologic research. Simulation programs on personal computers quickly draw pictures of epidemics and allow rapid explorations of the interactions of populations of hosts and parasites. Even simple host-parasite systems have complex dynamic behavior which initially may appear counterintuitive, but with mathematical models it is possible to educate the intuition and learn about the general behavior of an infectious agent in a particular population. Using these systems one can explore the dynamic behavior of hosts and parasites that is an inherent characteristic of the system.

In particular one can learn to anticipate particularly good or particularly unfortunate behavior of the system for human health. How might the system respond to changes in nature or acts of man? What might be the short and long term effects of interventions of various types at different times? Initially, one needs to learn to avoid an action that inadvertently may cause a perverse outcome, such as provoking an epidemic. Then one can explore the possible beneficial effects of different interventions, and compare their applicability, acceptability, costs and possible adverse effects.

There has been a curious dichotomy in the acceptability of mathematical models in sciences such as physics and engineering, where the use of such models is universal, and infectious disease epidemiology, where mathematical models have only recently been used. Newton's laws of motion are simple differential equation models that are easily tested, and every student in an introductory course in physics in secondary school verifies one of Newton's laws as a first laboratory exercise. Similarly, these mathematical models, expressed as Newton's three body problem, were essential in planning our explorations of the moon.

Scientists make predictions on the basis of theories expressed in mathematical models, and as quickly as possible seek to verify these predictions with experiments in the real world. Mathematical methods were first applied to infectious disease epidemiology by the great Daniel Bernoulli, who was also the author of the → Bernoulli trial in probability theory and of the Bernoulli principle in physics. The transmission of infectious agents (parasites) in populations of hosts was modeled beginning much more recently, after the development of the → germ theory of disease. Until the book of Anderson and May that became an instant classic, there was little effort to gather the vast body of observational data on the occurrence of infectious diseases and epidemics and the mathematical models that might help explain them. Indeed, there is no explanation why, in most areas of science, theories expressed in mathematical models are tested against real data as soon as possible, while in the infectious disease arena such empirical testing has only recently been conducted. In this chapter we will provide readily available modern references that contain the citations to a number of the older original papers.

It is the purpose of this chapter to illustrate the use of mathematical models in understanding and controlling vector borne diseases. This chapter is intended for biologists and field practitioners who have no special training in mathematics. We will use malaria and the main example and we will begin with the simplest models, and add more realistic features in a stepwise fashion so the reader can understand how these models evolved over time, and begin to understand the literature.

Vocabulary of Mathematical Modeling

Microbiologists have not used the words "microparasite" and "macroparasite" as they are used in modeling (in the Anderson and May sense), so these terms will be described here. A microparasite is not a type of creature. Rather, microparasites are whole categories of organisms, usually a bacteria or viruses, that have direct reproduction in the host, usually at high rates. Hosts are either infected or not, but a parasite burden usually has no meaning for microparasites. Microparasites generally are small, have short generation times, and usually produce long lasting immunity against reinfection, as in measles. The duration of infection with microparasites is usually short compared with the expected life span of the host, so the host sees the infection as transient.

Macroparasites such as worms and one celled organisms like malaria have no direct reproduction in the definitive host. They are larger, and have longer generation times, which may be a substantial fraction of the life expectancy of the host. When an immune response is elicited by a macroparasite, it is usually transient, and will rapidly disappear when the parasite is removed, as with chemotherapy. These infections are usually persistent, with hosts being continually reinfected, as in malaria.

By direct transmission modelers mean that the infection moves from person to person directly, with no environmental source, intermediate vector or host. To a modeler direct transmission may take place by contact between mucous membranes as for sexually transmitted diseases, or by droplets aerosolized by a cough or sneeze, as for colds or measles. This may seem very imprecise to a biologist, but what is implied is that for directly transmitted infections one need only model the behavior of the parasites in people.

In contrast, transmission of malaria by mosquitoes would be an example of indirect transmission. The fundamental difference is that if one is modeling malaria transmission, one has to include equations for the behavior of parasites in populations of mosquitoes and also in populations of humans. As a result, modeling indirect transmission is fundamentally more complex.

Modelers use one other term that seems odd to an epidemiologist. In epidemiology we commonly use the word density in a particular way, as in probability density or incidence density. When a modeler uses the concept of density-dependent functions it means number dependent. If a modeler says that the occurrence of an epidemic is density-dependent, that means it depends, for example, on the actual number of susceptibles present in the population.

A concept from ecology that is central to thinking about the transmission of infectious diseases is the basic reproductive number, R_0. R_0 represents the average number of secondary infections produced when a single infectious individual is introduced into a host population in which every individual is susceptible. The time implied is the entire period of infectiousness for the infected case.

For directly transmitted microparasites one is considering a system that includes infectious and susceptible humans, but for indirectly transmitted macroparasites such as malaria, one must consider a system that includes infectious and susceptible mosquitoes as well as infectious and susceptible human populations. For indirectly transmitted infections like malaria, the value of R_0 is for human to human transmission via mosquitoes.

If this reproductive number, R_0, is less than unity (one) then the infection will eventually die out and not persist in that community. There may be some secondary cases, but these will decrease with time, and eventually the infection will become extinct. If this reproductive number is exactly unity, then the infection just barely succeeds in reproducing itself, and there will be a similar number of cases at any later time. If this reproductive number is larger than unity, then the number of cases will increase with time, at least initially, and there may be an epidemic. The nature of the parasite, the nature of the host(s), and the behavior of host(s) all help determine the value for R_0 for a particular infectious disease and community.

When there are some already infected or immune or resistant individuals in the population, then not everybody is susceptible. At that point there is a value of R, the reproductive number for the system at that point, but it is not R_0. R_0 is the upper bound for the value of R, which is usually less than R_0, and the value of R may vary widely

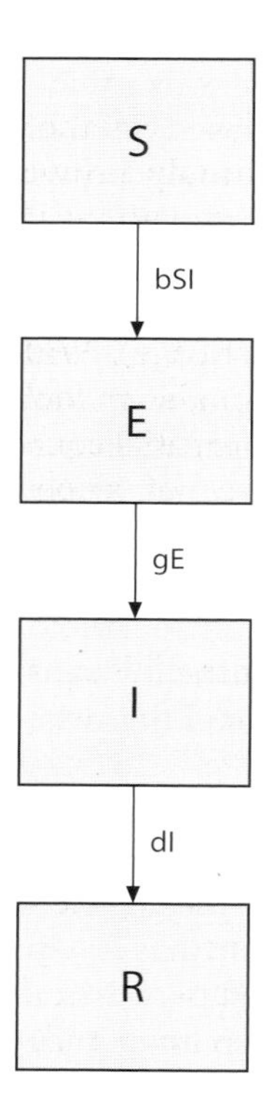

Fig. 1A. The compartment diagram for the four state SEIR model. All individuals are born susceptible into the S compartment. As they become latently infected they progress to the E compartment, and after they have passed through the latent stage they move into the infectious or I compartment. After the infectious period is over and they have developed immunity, they enter the R stage.

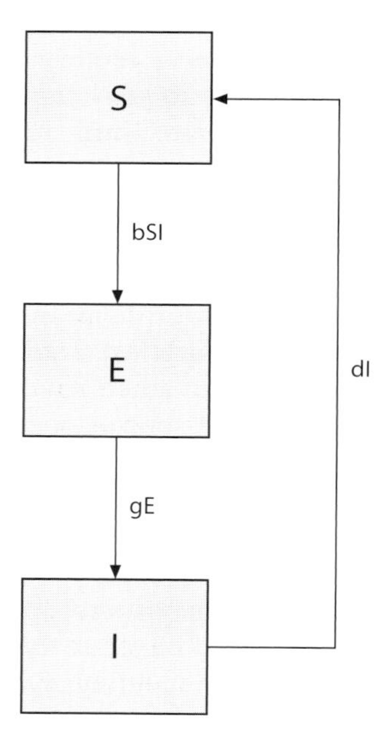

Fig. 1B. The compartment diagram for the three state SEIS model. All individuals are born susceptible into the S compartment. As they become latently infected they progress to the E compartment, and after they have passed through the latent stage they move into the infectious or I compartment. After the infectious period is over there is no long lasting immunity and they revert back to the susceptible or S stage again.

during the course of an infectious disease through a population. Remember, R_0 is a characteristic of the system assuming that everybody is susceptible, while R is the value of the quantity at a particular moment, when some or possibly even most individuals are already infected, immune or resistant.

Timing is a crucial aspect of the study of the epidemiology of infectious diseases, but symptomatology is less so. Transmission can only take place during the period when a host is infectious. There may be no symptoms associated with infection, and when symptoms do occur, they may be apparent in no particular relation to the period of transmissibility. Individuals infected with Human Immunodeficiency Virus (HIV) are asymptomatic but infectious for an average of about ten years before they become clinically ill. In contrast, most infected with tuberculosis (TB) organisms may remain non-infectious for their entire lifetimes, while a few will develop clinical pulmonary TB after a period of months or years. Individuals infected with TB only become infectious to others when they begin to cough. These extreme differences in the behavior of HIV and TB underline the need to separate the latent and infectious periods of a disease from the incubation period and the period of clinical disease. Infectiousness may have little to do with symptoms; an individual newly infected with falciparum malaria will become symptomatic after 7 – 10 days, but will not become infectious to vector mosquitoes for three weeks. Timing determines if transmission of an infectious disease will take place at all, and if it does, timing determines the nature of transmission. While the clinician treats the symptomatic patient, the epidemiologist seeks the infectious individual, who may not be symptomatic, or may be symptomatic at a time when he is not particularly infectious.

The incubation period is the interval from the time a person is infected until he develops clinical disease. The period of clinical disease is the period of symptoms. An infected person may never develop clinical disease, and the period of infectiousness may not correspond very well with the period of symptoms. For many childhood infections, for example, the period of greatest infectiousness is just prior to the appearance of clinical symptoms. This has important implications for control.

The latent period is the interval from the time a person is infected until he becomes infectious to others. The infectious period is the interval during which an individual can transmit an infection. As was noted above, the latent and infectious periods are variably related to clinical symptomatology, but are crucial in the study of the epidemiology and transmissibility of infectious diseases.

In this chapter we will explore mathematical models of disease transmission, as a model can represent aspects of human behavior as well as measurable demographics. As must be evident, mathematical transmission models are totally dependent on knowledge of the latent and infectious period for an infectious disease. The first model we will investigate is the classic SEIR (Susceptible, Latently infected, Infectious, Resistant or immune) model made up of four differential equations. There are different systems of notation used in modeling but in this chapter we will use the most common and point out confusing and conflicting notation (3 – 5). (FIRST CONFUSING NOTATION WARNING: the R and R_0 used to represent reproductive numbers are distinct from the R used to indicate the immune or recovered state for a host.)

Systems of Differential Equations

A differential equation is an algebraic equation that includes a derivative, which is simply a slope. A slope can do one of three things: it can go up, in which case it is positive; it can go down, in which case it is negative, or it can do neither or stay the same, in which case it is zero (no change). With modern simulation programs we can always look at pictures of the performance of differential equations, which translates into pictures of slopes, which in turn means one repeatedly has to answer the question, is this going up, down, or straight sideways?

As those who have studied differential equations know, writing a differential equation is easy; it is the solution that is difficult. Most interesting differential equations remain analytically insoluble for the amateur. What has made mathematical modeling readily accessible in the last decade is the existence of the personal computer with a simulation program for numerical solutions to differential equations. The program does not actually solve the equations, it just presents a picture of what they do, which is what we wanted to know anyway.

One solves an algebraic equation by solving for x in terms of y and z, and then one can draw the picture or graph. One solves a differential equation by integration in order to produce an algebraic equation which one then solves for x in terms of y and z, and then one can draw the picture or graph. The simulation program goes from the differential equations directly to the picture without any intermediate stops. We will present the four differential equations of the SEIR model below to describe the movement of individuals from the Susceptible state to the Latent state to the Infectious state to the Resistant or immune state. We have used this model and the SEIR notation because it is the oldest in the literature and the most commonly used (5), although Anderson and May have used X Y Z instead of S I R throughout their book. In parallel we will also consider the SEIS model as well, a model for an infection with no long lasting immunity, that will become part of our malaria model later. In the SEIS model individuals that have recovered from the infectious stage do not become immune, and revert back to being susceptible again. That is, after recovering from infection they move back into the S compartment again.

The SEIR Model The classic SEIR model uses four derivatives or slopes with respect to time (t);

dS/dt = the change in the numbers of Susceptibles (S) over time,

dE/dt = the change in the numbers of Latents (E) over time,

dI/dt = the change in the numbers of Infectious (I) over time, and

dR/dt = the change in the numbers of Resistant (R) or immune over time.

Other quantities are used as well;

b = the probability of transmission of infection per unit time,

G = the duration of the latent state, in units of time,

g = 1/G or the rate of leaving the latent state per unit time

D = the duration of infectiousness of this disease, in units of time,

d = 1/D or the rate of leaving the infectious state per unit time,

m = birth rate or death rate or rate of entering or leaving life per unit of time.

The movement of individuals from state to state is illustrated in the accompanying compartmental diagram (Fig. 1A). Those who leave one compartment must progress into the next. Entries are (+), departures are (–), and the total number N = S + E + I + R remains constant. In the first model there are no births, no deaths and there is no migration. The susceptibles become latently infected, the latently infected become infectious, the infectious recover and become immune. In an epidemic the number of susceptibles will decrease as they become infected, the number of latents will increase (initially) and the infectious will follow, and the number of immune will increase as the infected recover.

dS/dt = – b x S x I

dE/dt = + b x S x I – E x g

dI/dt = + E x g – I x d

dR/dt = + I x d

In simple algebra this set of differential equations for the SEIR model can be written as,

dS/dt = – bSI
dE/dt = bSI – gE
dI/dt = gE – dI
dR/dt = dI

If this were a disease that produced no long term immunity, then there would be no R state, and those who recovered from the infectious or R state would reenter the S or susceptible state. The SEIS model is given below. Here those recovering leave the infectious state as –dI, and reenter the susceptible state as +dI.

dS/dt = – bSI + dI
dE/dt = bSI – gE
dI/dt = gE – dI

In order to add births and deaths (vital dynamics) to the SEIR model, one would add all of the births to the susceptible state, as mN (the total population), and then subtract deaths from each state. Those dying in the S state would be – mS, those dying in the E state would be – mE, those dying in the I state would be – mI, and those dying in the R state would be – mR.

dS/dt = mN – bSI – mS
dE/dt = bSI – gE – mE
dI/dt = gE – dI – mI
dR/dt = dI – mR

Similarly one could add vital dynamics to the SEIS model. When exact timing is not important it is common to drop the equation relating to the latent period, and to describe epidemics in terms of SIR and SIS models. In these models without latent periods, the newly infected proceed directly from the S state to the I state. The initial malaria model created by Ross did not use latent periods, but as the need for realism increased the latent period for malaria in mosquitoes was added by MacDonald and the latent period in people was added by Anderson and May.

An Intuitive Explanation of Rates of Entering or Leaving States An essential concept for modeling is the rates at which subjects enter and leave various compartments or states. If D is the duration of infection or the duration in the state I for example, and D is 7.0 days, then there is one complete turnover in the I compartment every seven days. On average, one seventh of those in the compartment must come out each day. That is, if the average stay in the compartment is 7.0 days, then the daily rate of recovery from infection must be 1/7.0 In general the parameter for leaving that state is 1/the average duration in that state. For a state with a duration D, on average 1/D individuals leave per unit time.

The last change in convention is that we will call the rate of leaving, d = 1/D. This leads to the mortality rate m = 1/average age at death, or 1/age at leaving life. In a stationary population when births equal deaths them m is also the birth rate. Also we will have g = 1/average duration of latency, and d = 1/average duration of infectiousness. All must be in the same units of time. If we model in units of years, then all parameters must be in years. For example, 7 days is 7/365 or 0.02 years, and the transition parameter is 1/0.02 or 50 per year.

If the average age at death is 74 years, then 1/74 of the living will leave the living state and die in one year.

If the average age at infection for a disease like measles is 5 years, then 1/5 of the uninfected will leave the uninfected state and become infected in one year.

If the average latent period for falciparum malaria in people is 21 days, then 1/21 of people in the latent state will leave this state (and become infectious) in one day.

If the average duration of infectiousness for untreated TB is 5 years (60 months) then 1/5 of the untreated will leave the infectious state by recovering or dying each year, and 1/60 will do so in one month. With appropriate treatment the duration of infectiousness can be reduced to 2 months, so that 1/2 will leave the infectious state in one month. If the average duration of infectiousness for untreated falciparum malaria is nine months, 1/9 will leave the infectious state in one month. With appropriate treatment the duration of infectiousness can be reduced to one month, and all will leave the infectious state in a month. For infectious diseases chemotherapy can shorten the duration of infectiousness. This is also an example of how treatment may also be prevention for an infectious disease.

The Law of Mass Action and Thresholds The above models are based on the law of mass action from chemistry, in that we assume that any individual in a population is equally likely to bump into (and infect) any other individual, like gas molecules moving about in a balloon. Also, we have used S and E and I and R to represent absolute numbers of individuals rather than proportions, because this is most common in the literature, and leads to the evaluation of thresholds, or the minimum number of individuals in a population that could support an epidemic, a topic that is beyond the scope of this chapter.

Deterministic Models versus Random Variation All of the models in this chapter are deterministic models, meaning that they will do the same thing every time they run. Models that include random variation are called stochastic models, but stochastic models are more complex and are beyond the scope of this chapter. Stochastic models are important when populations are small.

Expressions for R_0 Using algebra it is possible to show that for the SIR or SIS models without births and deaths, R_0 is,

$$R_0 = bN/d,$$

from the steady-state SIR or SIS model with vital dynamics we have;

$$R_0 = bN/(m + d),$$

and from the steady-state SEIR or SEIS model we have;

$$\mathbf{R_0 = bgN : (d + m)(g + m)}$$

When both the duration of the latent state, 1/g, and the duration of the infectious state, 1/d, are small (a few days) compared to the length of life or 1/m (50 or 70 years), then all three expressions for R_0 can be approximated as;

$$R_0 \simeq bN/d.$$

Short Term Observation of Populations

Two Kinds of Epidemics in Closed Populations Epidemiologists who deal with acute or short term outbreaks tend to think of these epidemics as occurring in closed populations, because few individuals are born or die, or move into or out of a community in a matter of weeks. We are faced with differentiating two fundamentally different types of epidemics in closed populations; propagated epidemics, and, point source epidemics.

Propagated epidemics must always result from some self-reproducing agent such as an infectious agent, while point source epidemics may be either of infectious or non-infectious etiology. Epidemics of measles are propagated epidemics as each infected individual acquires the measles virus from a person in the infectious stage who was infected in the previous generation of infection. An outbreak of salmonellosis from eating contaminated turkey at a hospital party would produce a point source epidemic with an infectious agent that all of the exposed acquired within a period of a few minutes (eating the main course). In fact, parents with salmonellosis from a point source epidemic may then go home and begin a propagated epidemic among the children in their own families.

In contrast, poisonings must all be point source epidemics, as a toxin cannot reproduce itself. This is true of bacterial toxins (staphylococcal food poisoning) as well as chemical toxins (pesticides) not of microbial origin. Staphylococcal food poisoning often takes place in the absence of living organisms because the toxin is heat stable while the bacteria are not. Cooking may well kill the bacteria and effectively sterilize the food while leaving the toxin unchanged.

The Classical Theory of Happenings This distinction between propagated and point source epidemics was first formulated by Ross (of malaria fame), who described propagated epidemics as "dependent happenings" because the number affected per unit time depended on the number already affected. In contrast, the number affected per unit time during an episode of poisoning was independent of the number of individuals already affected, so these Ross termed "independent happenings."

Indirectly Transmitted Diseases – Vectors Consider malaria as an example of an indirectly transmitted disease, an infection transmitted to humans by a mosquito vector. Both humans and mosquitoes are considered to be born uninfected. An uninfected female mosquito has a blood meal from an infected human and becomes infected with malaria herself. After a suitable latent period she becomes infectious and has another blood meal, this time on an uninfected human, and can transmit malaria to the previously uninfected human. The human infects the mosquito, then the mosquito infects the human. Humans do not infect other humans (except by blood transfusion), and mosquitoes do not infect mosquitoes.

The Malaria Parasite's Guide to the Mosquito-Human Cycle The mosquito is the definitive host for the malaria parasite. That is, sexual reproduction takes place in the mosquito. Only asexual reproduction takes place in humans. Humans can be thought of as warm, friendly, wet reservoirs in which the malaria parasite can survive during hard times for adult mosquitoes. Tropical climates have a wet season and a dry season, and during the dry season adult mosquito populations are greatly reduced, and may disappear altogether. In more temperate climates there is also substantial temperature variation and the cold season is similarly hard on adult mosquitoes. Human reservoirs are essential to tide malaria parasites over until the next season of abundance for adult mosquitoes.

Malaria parasites persist in humans waiting for those wonderful mosquitoes to return, so that the parasites can get back to sexual reproduction. In an evolutionary sense, malaria parasites are just treading water with asexual reproduction in the human. However, natural selection pressure is exerted on humans, so the gene frequency that is sampled by mosquitoes is that in surviving humans. Meanwhile, mosquitoes survive in the form of fertilized eggs, waiting to hatch into larvae when water and warmth return to their part of the earth. Both the human and the mosquito part of the cycle are essential, but for different reasons. Both offer opportunity for intervention. Mosquitoes are seasonal in most places, so that there is usually a six month dry season when there are few mosquitoes and little or no malaria transmis-

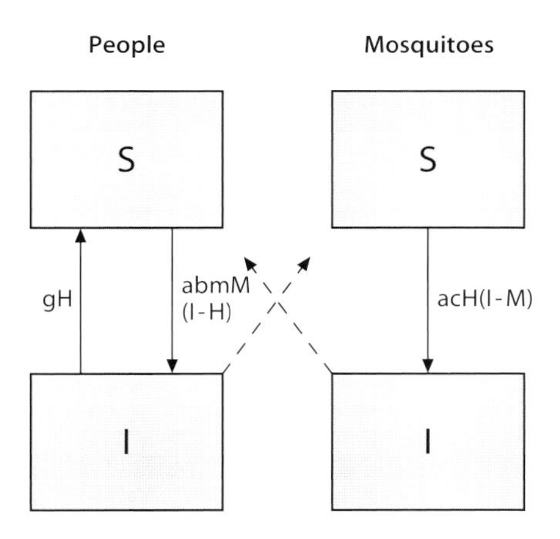

Fig. 2A. A compartment diagram for the Ross malaria model which is an SIS model for humans and an SI model for mosquitoes. The dotted lines indicate that transmission is from infectious mosquito to susceptible human and from infectious human to susceptible mosquito.

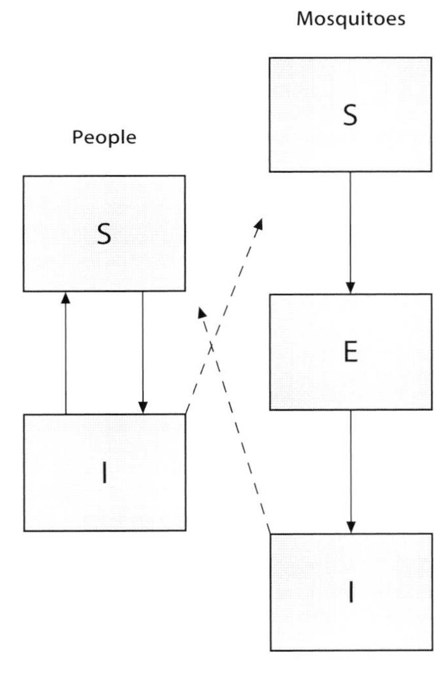

Fig. 2B. A compartment diagram for the MacDonald malaria model which is an SIS model for humans and an SEI model for mosquitoes. The dotted lines indicate that transmission is from infectious mosquito to susceptible human and from infectious human to susceptible mosquito.

sion. There are, however, some places where malaria transmission occurs throughout the year without respite.

The form of the malaria parasite that is infectious for the mosquito is the gametocyte in man.

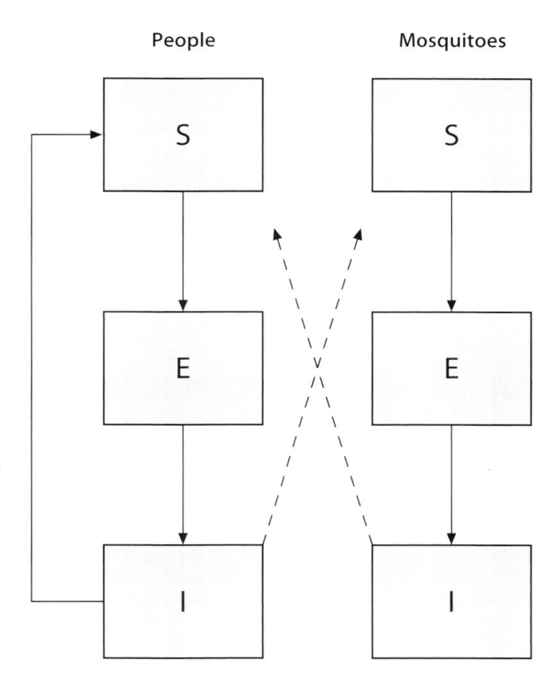

Fig. 2C. A compartment diagram for the Anderson and May malaria model which is an SEIS model for humans and an SEI model for mosquitoes. The dotted lines indicate that transmission is from infectious mosquito to susceptible human and from infectious human to susceptible mosquito.

For → falciparum malaria, gametocytes only appear in the human bloodstream about 21 days after infection, so there is a relatively long latent period from infection to infectiousness in the human. The incubation period, or time until clinical disease in a naive human is about seven to ten days for *P. falciparum* malaria, so the infected human may become severely clinically ill and die a week or more before becoming infectious to mosquitoes.

The form of the malaria parasite that is infectious for man is the sporozoite, which is delivered from the salivary gland of the female mosquito. After a female mosquito has a blood meal on an infectious human, sexual reproduction between ingested gametocytes takes place in the gut of the mosquito. When a human has been infected by multiple mosquitoes (a frequent happening in endemic areas) then multiple different broods of parasites are circulating in a single human and gametocytes from different broods are taken by a mosquito in a single blood meal. After recombination, sexual reproduction then results in sporozoites with different combinations of genes from those in either of the parent broods. There is an approximate 10 day latent period in the mosquito between the time the mosquito has an infectious blood meal from a human, and the time (after sexual reproduction) when infectious sporozoites appear in the salivary glands of the mosquito. Since this latent period may be longer than the average length of life for a mosquito, it is a relatively old and rare (lucky) mosquito that is able to infect a human.

Most malaria mosquitoes are shy, night biting creatures that go relatively unnoticed by their human prey. While feeding, a female mosquito loads up like a tank truck, and can barely fly to the nearest vertical surface, where she spends some hours diuresing most of the fluid she has taken in so that she can move more freely and go about her business and lay eggs. It is during this period that a mosquito is most vulnerable to control measures. This is the time when residual insecticides, such as DDT, are most effective.

To model malaria, Ross used two differential equations: one showing the infection of the human, and the other showing the infection of the mosquito. The simple Ross two differential equation model appears in textbooks, and illustrates some of the important features of malaria. At this point we have to change our conventions and our constant names to make them consistent with the malaria literature.

For SEIR, and SEIS processes described above we have used the absolute numbers of individuals in models so that we could consider eradication and numerical thresholds. With growing or shrinking populations, however, proportions are easier to manage, and for vector borne diseases there are usually so many vectors that human thresholds do not assume much importance. Furthermore, the mosquito vector has elaborate mechanisms to locate prey, thus the name vector, so that the law of mass action is not operative for mosquito-human interactions. As you will recall, if you are enclosed in a tent or a bedroom at night with a hungry female mosquito, she will definitely find you before morning. Chance and the law of mass action are not operating here.

In the Ross model the proportion of infected humans is H, so that the proportion of uninfected humans is 1 – H. The proportion of infected mosquitoes is M, so the proportion of uninfected mosquitoes is 1 – M.

Absolute numbers of people and mosquitoes do not enter this equation, only the ratio, m, between the numbers of female mosquitoes and the numbers of people (CONFUSING NOTATION WARNING: unfortunately, this is another use for the letter m, which was used for the death rate before.) The initial value of m = 40, which would be reasonable for the rainy season.

The man biting rate, a, is the number of bites by a female mosquito delivered on humans per day. This is determined by how often the mosquito needs to feed (the gonotrophic cycle), and whether the mosquito feeds on a human or on another mammal, like a cow or a pig. A common value is 0.25 human bites/day, or one meal on a person every four days.

The proportion of bites by infected mosquitoes on susceptible humans that produce infection in the human is b, that has been measured at 0.09.

The human recovery time is the duration of disease in a human or the time during which an infected human can infect a susceptible mosquito. For falciparum malaria this is in the range of 9.5 months or 285 days. The recovery rate is thus 1/285 or 0.0035/day.

Not all bites on infected humans produce infection in the mosquito. The fraction that do is c, or 0.47.

Although humans spontaneously recover from malaria, mosquitoes do not, and the only way mosquitoes leave the infected pool is by dying. Since an average mosquito lifetime is about eight days, the rate of leaving life for a mosquito is p_m (daily probability of dying for a mosquito) so $p_m = 0.12$/day.

The Ross model consists of two equations, one for humans and one for mosquitoes. Each equation is conceptually similar to the dI/dt equations for SIR models in that each describes entries and departures from the infectious stage. dH/dt is for infection in humans and dM/dt is for infections in mosquitoes. Also in direct analogy to the dI/dt equations, the first part of each equation describes those humans or mosquitoes becoming infectious, and the last part of each equation describes those humans or mosquitoes becoming uninfectious because they recover (humans) or die (mosquitoes). The Ross model does not include latent periods so the equation for human infections represents an SIS model as humans who have recovered return to the susceptible pool, while the equation for mosquito infections represents an SI model as all infected mosquitoes die in that state. As you look at the models you will see that these are models of

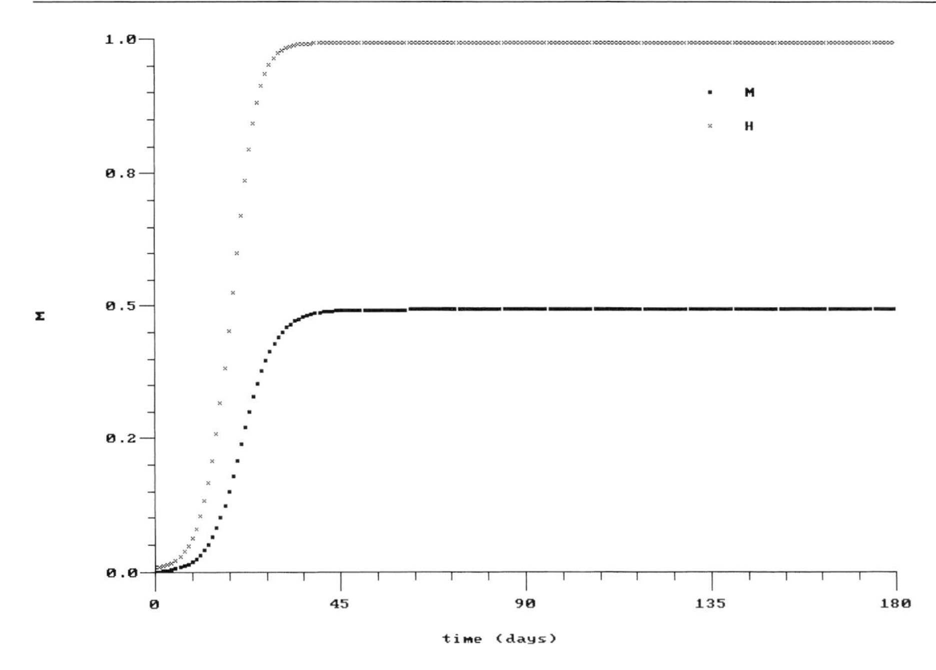

Fig. 3A-C. Results of the Ross Model, the MacDonald Model, and the Anderson and May Model using the same set of parameters listed in the text. In each setting one person infectious with *P. falciparum* malaria was introduced into a community of 100 susceptible people, and the results in the community followed for a six month rainy season. In each setting eventually virtually the entire human population becomes infected in six months, at which point the mosquito density would decrease in the following dry season. **A** The Ross Model without latent periods. Here the progress of the epidemic is far too rapid and the final prevalence of infectious mosquitoes too high.

mosquito-human interaction (Fig. 2). The key difference between malaria models of indirect (vector) transmission and our previous models of direct transmission is that with malaria humans infect mosquitoes and mosquitoes infect humans, but humans do not infect humans nor mosquitoes infect mosquitoes.

The Ross Malaria Model

$$dH/dt = a \times b \times m \times M \times (1 - H) - g \times H$$

$$dM/dt = a \times c \times H \times (1 - M) - p_m \times M$$

The Ross model is presented below in simple algebra,

$$dH/dt = abmM(1 - H) - gH$$

(This is an SIS model for people)

$$dM/dt = acH(1 - M) - p_m M$$

(This is an SI model for mosquitoes).

For the Ross model,

$$R_0 = mao^2bc/gp_m$$

Notice that the female mosquito has to bite twice to complete the cycle, so that the a term is squared. The simple Ross model outlines the basic features of malaria, but does not consider the approximate 10 day latent period in mosquitoes nor the exponential survival of mosquitoes. As a result, the Ross model predicts a too rapid progress for a malaria epidemic in people, and much too high an equilibrium prevalence of infectious mosquitoes. The results of the Ross Model of the progress of *P. falciparum* infection when one infectious person is introduced into a community of 100 susceptible individuals is presented in Fig. 3A.

Mosquitoes that can transmit malaria have a survival pattern that is represented almost perfectly by the exponential distribution in continuous terms, or the geometric distribution in discrete terms. Models built on both distributions are common in the literature, and will be explained in parallel below. In general terms, a fixed proportion of

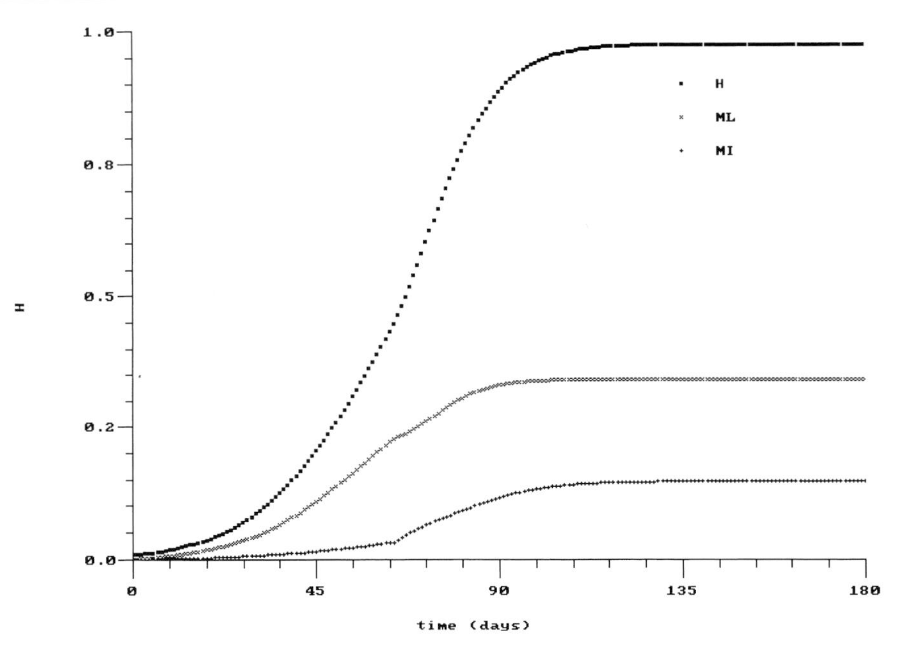

Fig. 3B. The MacDonald Model with a 10 day latent period for mosquitoes. The progress of the epidemic is slower and the final prevalence of infectious mosquitoes lower.

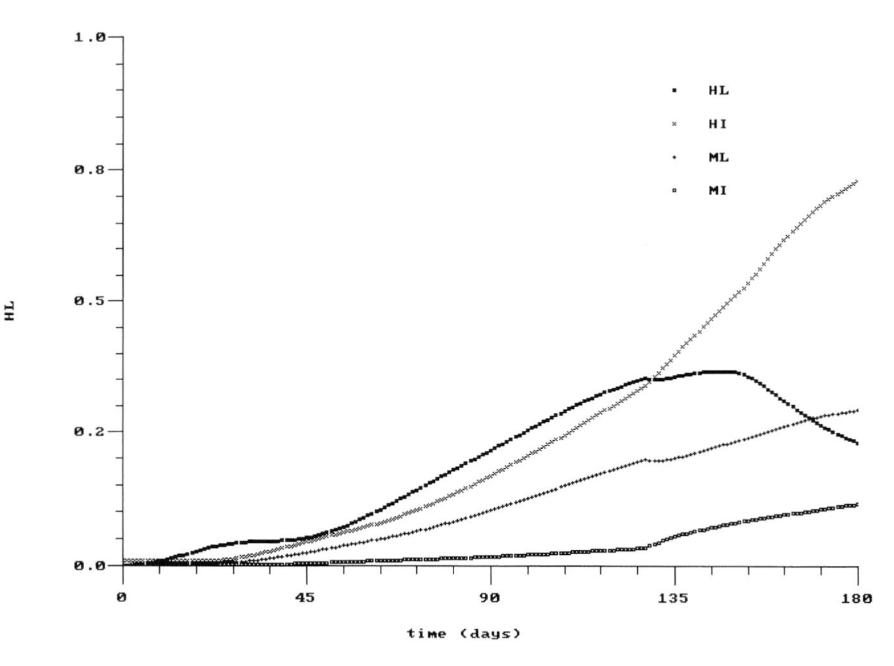

Fig. 3C. The Anderson and May Model with a 10 day latent period for mosquitoes and a 21 day latent period for people. The progress of the epidemic is slower still. Note that the total number of people infected is the sum of those in the latent state and the infectious state, which together add up to virtually 100% infected in six months. This model appears to be reasonably realistic for the short term with no immunity in the population.

mosquitoes survive each day (or die each day), so that a minority, even a tiny minority, survive as long as the latent period and have the potential to become infectious for humans. To transmit malaria a mosquito must have a first blood meal on an infected human, become infected herself, and then survive as long as the latent period to become infectious, and then have a second blood meal on an uninfected human. The Anderson and May formulations of the malaria models include the latent period for mosquitoes and the death rate for mosquitoes using continuous distribution.

In continuous terms, if p_m is the constant mosquito death rate per day and tau_m (greek letter tau, m for mosquito) is the latent period for mosquitoes, then the proportion of mosquitoes that is infectious follows the exponential distribution and is approximately,

$\exp(-p_m tau_m)$,

and this is the multiplier for the expression for the number of infectious mosquitoes.

The discrete counterpart to the exponential distribution is the geometric distribution, where p would be the probability of dying per day and q = 1 – p, is the probability of surviving, which would usually be described in probability terms as the distribution of

q^{tau_m}, where q is the probability of surviving per day.

In the discrete model terminology MacDonald has used p for q and n for tau_m, and –ln(p) is the daily mortality for mosquitoes, similar to p_m in the continuous model. Thus the proportion of infectious mosquitoes in the discrete model as formulated by MacDonald becomes

p^n.

The equivalent continuous and discrete expressions for R_0 for malaria models are,

$R_0 = \{ma^2bc[\exp(-p_m tau_m)]\}/gp_m$, and
$R_0 = ma^2bcp^n/-g\ln(p)$.

They are comparable, with the daily mosquito mortality rate, p_m or –ln(p) in the denominator, and the proportion of mosquitoes surviving the latent period, $\exp(-p_m tau_m)$ or p^n, in the numerator.

In the MacDonald model the lags are for the latent period in mosquitoes, as it is the mosquitoes which have survived the latent period which are infectious now, and they, in turn were infected one latent period ago. For the description below, the first line over an equation is the usual word model, and the second line is an attempted analogy to the familiar dE/dt and dI/dt equations of the SEIR model. dH/dt is analogous to dI/dt for humans, dML/dt is analogous to dE/dt for mosquitoes, and dMI/dt is analogous to dI/dt for mosquitoes. This is an SIS model for humans and an SEI model for mosquitoes. The results of the MacDonald Model of the progress of *P. falciparum* infection when one infectious person is introduced into a community of 100 susceptible individuals is presented in Fig. 3B.

dH/dt = + a*b*m*MI*(1 – H) – g*H

dML/dt = + a*c*H*(1 – ML – MI)

dMI/dt = + a*c*lag_H_tau$_m$*(1 – lag_ML_tau$_m$ – lag_MI_tau$_m$)*exp(–p$_m$*tau$_m$) – p$_m$*MI

To review, the MacDonald model does include the latent period in mosquitoes and the known exponential survival of mosquitoes during the latent period. MacDonald's original form of the model was based on the discrete form as the geometric distribution, p^n, where p was the daily probability of survival for the female mosquito, n was the latent period or time before infectious sporozoites appear in the salivary glands of the infected mosquito, and –ln(p) was the daily mortality for mosquitoes, similar to p_m in the Ross model. Measured values of p range from 0.76 to 0.95, and measured values of –ln(p) range from 0.05 to 0.28. For falciparum malaria, n is about 10.

In contrast, the continuous form of the MacDonald model as presented in Anderson and May is based on the exponential distribution. The latent period in mosquitoes of 10 days is represented by the Greek letter spelled out as tau, and the fact that it is for mosquitoes is indicated by the suffix m. Thus, tau_m is the 10 day latent period in mosquitoes. One way to insert the 10 day difference in time is to lag a variable, so that, for example, lag_H_tau$_m$ is the proportion, H, from 10 days ago. However, the discrete and continuous forms give similar results for the basic reproductive number. The results of the Anderson and May Model of the progress of *P. falciparum* infection when one infectious person is introduced into a community of 100 susceptible individuals is presented in Fig. 3C. Note that in Fig. 3 each successive model adds a latent period, and the apparent progress of the infection through the community is slower. The Ross model (Fig. 3A) contains no latent periods, the MacDonald Model includes a 10 day latent period for mosquitoes (Fig. 3B, and the Anderson and May Model includes a 10 day latent period for mosquitoes and a 21 day latent period for humans.

These systems come to equilibrium because there is a continuous supply of susceptible humans and susceptible mosquitoes. Infected humans recover and reenter the susceptible pool, and new broods of uninfected adult female mosquitoes continue to hatch.

Setting the derivatives equal to zero, in the steady state it is possible to find the equilibrium proportions for infected humans (H*) and infected mosquitoes (M*). For the Ross model;

$H^* = (R_0 - 1)/[R_0 + (ac/p_m)]$.

The equilibrium proportion of infected mosquitoes predicted by the Ross model is much too high, so the relation from the MacDonald model is given below;

$$M^* = [(R_0 - 1)/R_0][(ac/p_m)/(1 + ac/p_m)] \exp(-p_m tau_m)$$

MacDonald defined the quantity ac/p_m, the number of bites on humans per day that produced infection in the mosquito, as the stability index. High levels of ac/p_m, in the range of 2 – 4, indicate that mosquitoes bite man often and have relatively long lifespans, and produce continuous endemic malaria. Macdonald called this stable malaria. Where ac/p_m is low, in the range of 0.5, malaria tends to occur in repeated outbreaks. MacDonald called this unstable malaria (3, 6). One needs also to appreciate that vector density changes orders of magnitude with the seasons, so that malaria occurs in annual epidemics in the rainy season when the mosquito density is high.

Details of a Complex Malaria Model Including Latent Periods for Humans and Mosquitoes Below is a model of malaria modified from that published by Anderson and May that deals with the complexities of humans as well as mosquitoes. In addition to the 10 day latent period in mosquitoes and mosquito mortality, the complex model includes the 21 day latent period in humans (until the appearance of infectious gametocytes), the recovery of humans from both latent and infectious stages, and the death of humans in both latent and infectious stages. This appears to be a reasonably realistic model. The modification of the Anderson and May model was to allow infected humans to recover from the latent stage before they became infectious to mosquitoes. This happens if medical treatment is readily available and individuals who become newly symptomatic at the end of the incubation period (7 – 10 days) are treated before they pass through the latent period (21 days) and become infectious to mosquitoes. This is an SEIS model for human infection and an SEI model for mosquitoes.

$$\mathbf{dHL/dt = a*b*m*MI*(1 - HL - HI) - a*b*m*lag_MI_tau_h*(1 - lag_HL_tau_h - lag_HI_tau_h) *exp((-p_h - p_g)*tau_h)) - p_g*HL - p_h*HL}$$

$$\mathbf{dHI/dt = a*b*m*lag_MI_tau_h*(1 - lag_HL_tau_h - lag_HI_tau_h) *exp((-p_h - p_g)*tau_h)) - p_g*HI - p_h*HI}$$

$$\mathbf{dML/dt = a*c*HI*(1 - ML - MI) - a*c*lag_HI_tau_m*(1 - lag_ML_tau_m - lag_MI_tau_m)*exp(-p_m*tau_m) - p_m*ML}$$

$$\mathbf{dMI/dt = a*c*lag_HI_tau_m*(1 - lag_ML_tau_m - lag_MI_tau_m) *exp(-p_m*tau_m) - p_m*MI}$$

For this malaria model the basic reproductive number is

$$R_0 = \{ma^2bc[\exp(-p_h tau_h - p_m tau_m)]\}/p_g p_m$$

(continuous)

Malaria does produce some immunity, and malaria models including immunity have been developed, but are beyond the scope of this chapter.

Interventions in the Transmission of Malaria

A number of interventions have been developed to limit the transmission of malaria, and it is useful to review how these will appear in the Anderson and May model. A reduction in the number of larvae that will hatch into adult mosquitoes will reduce m, the number of adult female mosquitoes per person. This can be accomplished by eliminating standing water, killing larvae, or putting larva eating fish into ponds that breed mosquitoes. A reduction in the human biting rate, a, can be effected by screening windows and doors, using a plain bednet, using insect repellents, or introducing alternative animals like cows or pigs on which hungry mosquitoes will feed (zooprophylaxis). The duration of life of an adult mosquito can be shortened by increasing the daily mortality of adult mosquitoes (p_m) through the use of residual insecticides on vertical indoor walls. Use of an insecticide-impregnated bednet will combine the last two and both lower a and increase p_m. Chemotherapy, or treating infectious people, will shorten the duration of the infectious period and increase the rate of human recovery, or p_g. Chemoprophylaxis, the taking of drugs (by short stay residents like tourists) to prevent malaria infection reduces b, the probability of infection in a susceptible human from an infectious bite. Note that the dangerous mosquito is the female mosquito who has fed at least once on an infectious human, survived for the 10 day latent period, and has now become infectious herself when she feeds on a susceptible human. Preventing second bites by these infectious mosquitoes would interrupt transmission.

Observing Infection Dynamics and the Effects of Interventions One reason to run this complex model is to observe the dynamics of the two host parasite relationships and appreciate rapidity with which a vector borne disease can run through a population, and the level of infection in humans at which it is stable. Another reason is to try out the effects of various interventions and observe the results on the progress of the disease through the community. In Fig. 4 we present the results of four different interventions, one at a time, so that the effects of each can be observed independently, and the relative efficacy observed.

In Fig. 4A we present the baseline Anderson and May Model, identical to Fig. 3C, but here the results of infections in mosquitoes are not visible, and a third line for human infection is added, the sum of both latent and infectious people. In this baseline it is evident that in a six month rainy season almost the entire population will become infected.

In Fig. 4B we present the result of an intervention on the model in Fig. 4A that reduced m, the number of adult mosquitoes per human by half, from 40 to 20. Reducing m by half resulted in about a 20% decrease in the final prevalence of human infection at the end of the rainy season in that community.

In Fig. 4C we present the result of an intervention on the model in Fig. 4A that reduced a, the human biting rate of adult mosquitoes by half, from 0.25 to 0.125. Reducing m by half resulted in about an 80% decrease in the final prevalence of human infection at the end of the rainy season in that community.

In Fig. 4D we present the result of an intervention on the model in Fig. 4A that doubled p_m, the mosquito mortality per day from 0.12 to 0.24. This reduces the survival of adult at the end of the latent period. Doubling p_m resulted in about a 90% decrease in the final prevalence of human infection at the end of the rainy season in that community.

Comparing the relative effects of different interventions aimed at reducing one or another factor by a similar amount did not reduce the prevalence of malaria equally. Reducing m was least effective, reducing a was substantially more effective, and reducing the length of life of adult mosquitoes (increasing daily mortality) was most effective. Note that this model is non-linear, and that the relative effectiveness of these different interventions is not necessarily intuitively apparent without examination of model results.

Remember that the dangerous mosquito is the female mosquito who has fed at least once on an infectious human, survived for the 10 day latent period, and has now become infectious herself when she feeds on a susceptible human. Preventing second bites by these infectious mosquitoes interrupts transmission. Insecticide-impregnated bednets work well because they affect both a and p_m, but, ironically, kill the fewest mosquitoes. Impregnated bednets simply kill or exclude the most dangerous mosquitoes. This is the sort of insight that can be gained from looking at mathematical models of malaria.

Effects of Interventions and R The relative effects of various interventions can also be appreciated by examination of the expression for the basic reproductive number, or R_0.

$$R_0 = \{ma^2bc[\exp(-p_h tau_h - p_m tau_m)]\}/p_g p_m.$$

Values for R_0 for malaria in endemic areas where transmission is intense are commonly in the range of 100 but may be lower where malaria is unstable and occurs in the form of periodic epidemics.

From the above, it is evident that R_0 is linear in m, but varies as a^2, so any decrease in a will have a larger effect than a similar sized decrease in m. Further, it is clear that p_m appears as an exponent, so that a change in p_m of similar magnitude will be more effective still. These observations have focused attention on spraying indoor vertical walls with insecticide and dipping bed nets in insecticide in order to increase p_m and to the use of screens, bednets and repellents to decrease a.

Also, note that R_0 is linear in p_g, the human recovery rate, so that chemotherapy for individual infected and sick people, while important and life-saving, is not as effective as a community intervention as some of the entomological actions mentioned above. Because malaria transmission can only take place when there are adult mosqui-

Fig. 4A

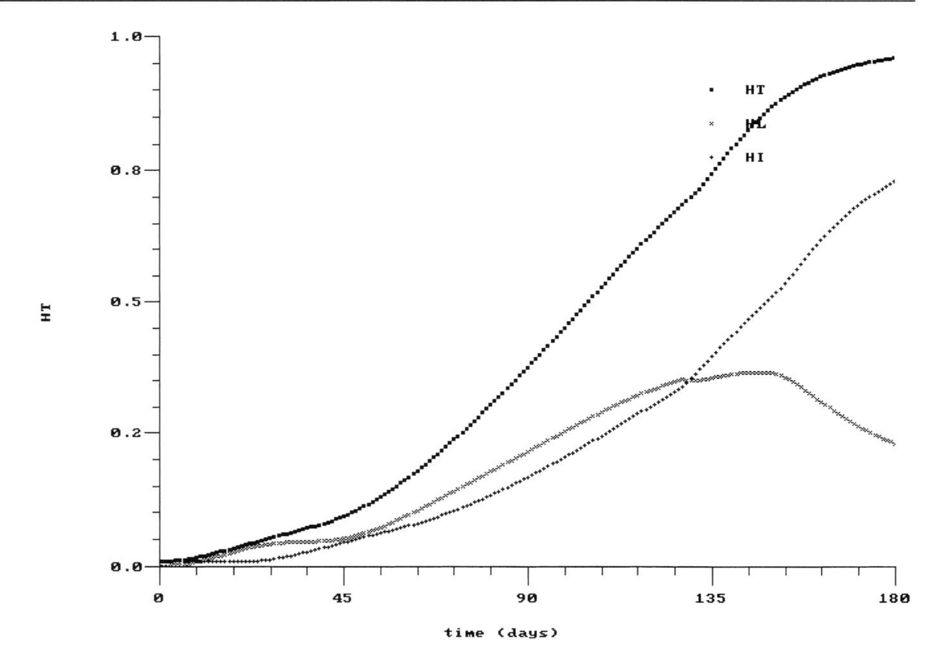

Fig. 4B

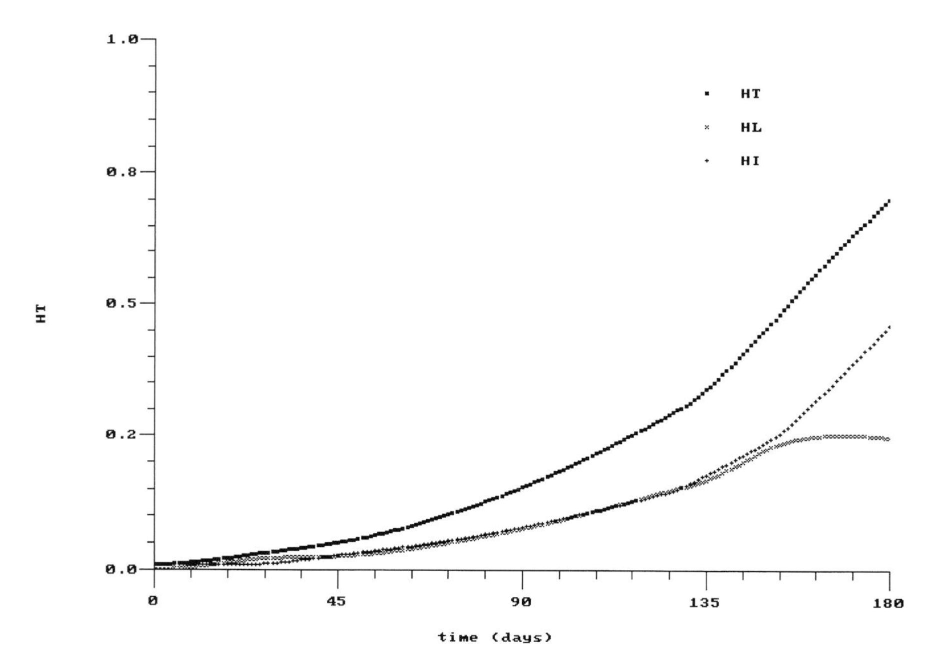

toes feeding, there is a six month hiatus in transmission in locations where a dry season intervenes. This temporary halt in transmission can serve to allow a health care system to catch up and use chemotherapy as a community intervention during the dry season. Seasonality in transmission thus allows chemotherapy to be used as an intervention in some communities where transmission is not continuous.

Historical Use of Malaria Interventions Eliminating water sources where the larvae of malaria mosquitoes hatch has been effective in eradicating malaria in locations as diverse as Italy, Greece, Spain and parts of the US. Impregnated bednets have reduced mortality in areas like the Gambia, where transmission is most intense. Human chemotherapy saves lives and has been effective in interrupting the spread of malaria where there is a dry sea-

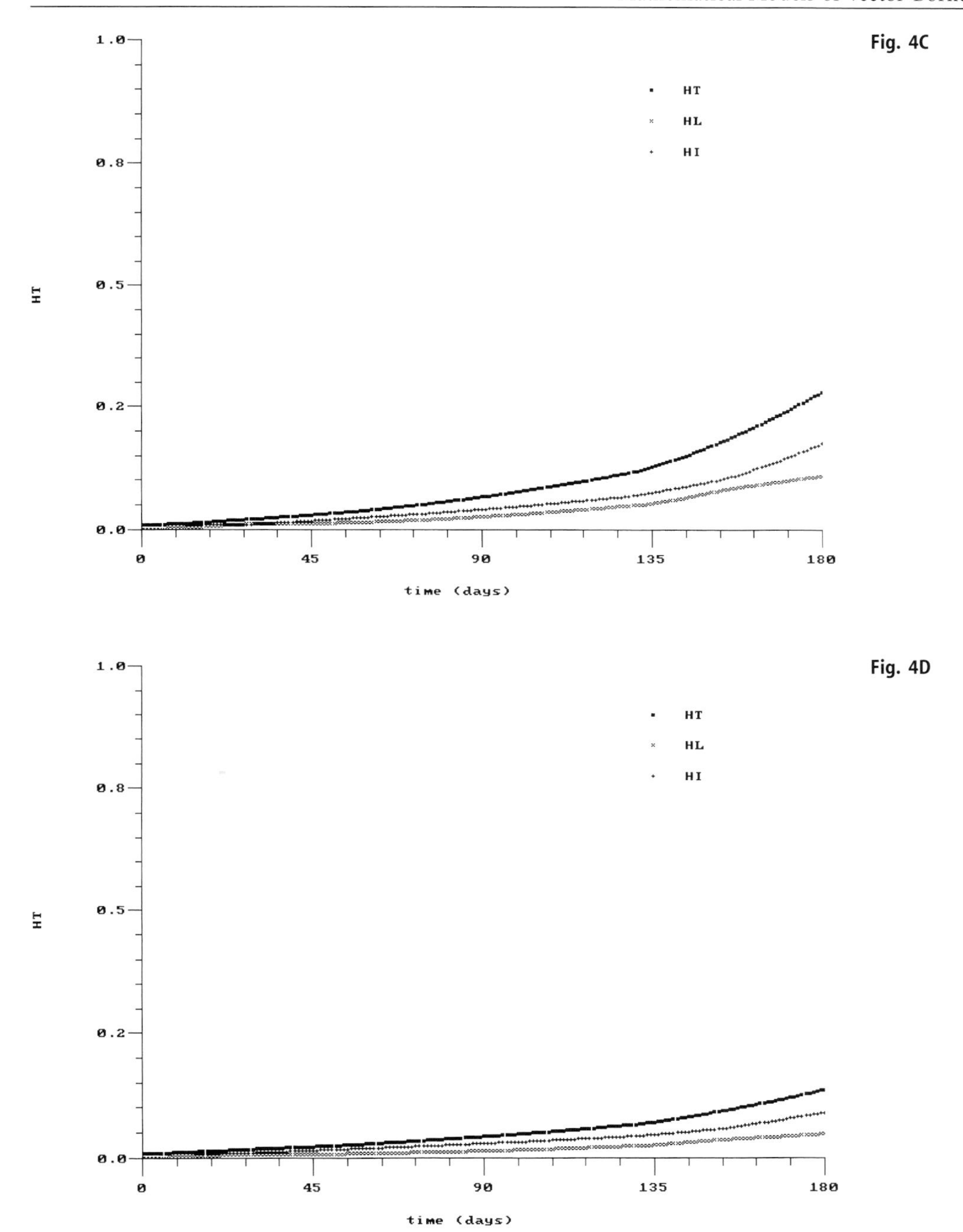

Fig. 4C

Fig. 4D

son during which there are no mosquitoes and transmission naturally stops.

Naive meddling with malaria can be dangerous, however. In endemic areas infants are born with maternal antibodies against disease (not infection) and they begin being bitten and infected immediately. Continuous infection in the individual produces continuous protection from severe clinical disease. If that cycle of continuous infection is broken by an attempt at eradication that drives the prevalence of infection in humans to low levels (but not to zero), a cohort of individuals is created that is susceptible to severe clinical disease. If the eradication program is abandoned and malaria again becomes endemic, the susceptible individuals (now adults) experience severe clinical disease with much excess avoidable mortality. Boom and bust cycles must be avoided in malaria control. The key concept here is that interventions must be sustainable, and once implemented, must never be stopped if they will leave populations of older individuals susceptible to severe clinical disease.

Vectorial Capacity We have spent a considerable amount of time looking at the basic reproductive number for the propagation of infectious diseases in humans, or R_0. It is fitting to end with the same idea from the vector's point of view, vectorial capacity, or V_c. The vectorial capacity is the daily rate at which new infections will occur in humans from a single currently infected human. This depends entirely upon the vector and is simply the basic reproductive number without the duration of infection in the human. For the discrete form of the MacDonald model below, R_0 is the number of humans that will become infected from a single infected human during the entire infectious period for the human, and V_c is the number that will become infected by the vectors in a single day.

Remember here we have switched back to the discrete system where p is the probability of survival and $-\ln(p)$ is mortality;

$R_0 = ma^2bcp^n/-g\ln(p)$ and
$V_c = ma^2bcp^n/-\ln(p)$

If the average case of falciparum malaria is infectious for 285 days, then the recovery rate or g = 1/285 days = 0.0035/day. Thus the magnitudes of R_0 and V_c for this disease differ by a factor of 285, and the vectorial capacity may be smaller than unity while the basic reproductive number is high and virtually all persons in the community are continually infected. In the above example, if R_0 was 100 then V_c would be 0.35, and R_0/V_c = 100/0.35 = 285.

Incompletely Thought Out Interventions Antibodies induced by malaria infection have relatively short term effects and primarily protect infected persons from becoming physically ill. In an area where malaria is endemic neonates are born with antibodies from their mothers that keep them from becoming ill with malaria. These neonates begin being bitten and infected as soon as they are born, and they develop their own antibodies from their own infections as the antibodies from their mothers decline. As these individuals age they are repeatedly infected, remain perpetually parasitemic, but rarely become seriously ill with malaria. An adult that has been infected all of her life is at little risk for illness in an endemic area, while a naive adult would become severely ill very quickly. Malaria mortality is highest among naive young children and women during their first pregnancy.

In 1957 Senators John F. Kennedy and Hubert Humphrey attached worldwide malaria eradication to the Mutual Security Act. This provided substantial funds to produce zero prevalence of malaria in five years, that is, by 1961. In Sri Lanka the prevalence went from approximately one million cases in 1957 to 100 by 1961 and 18 in 1963. Then DDT resistance began to appear, and DDT use in Sri Lanka was dramatically reduced from the initial rate of two million pounds per year. Malaria rates began to creep up and in 1968 there was a huge outbreak involving about half a million cases, and the prevalence kept climbing until it was back to a million cases in 1994.

Before 1957 when malaria was endemic children were infected at birth and most of the population was perpetually infected and relatively asymptomatic. By 1961 there was virtually no malaria, and all of the previously infected adults had lost their protective antibodies. As was mentioned, malaria infection in a naive adult produces severe illness, so when the huge outbreak began in 1968, most of the half million cases were seriously symptomatic. This produced a devastating effect on the economy, and there were hundreds of excess deaths among the adults who had lost their protective antibodies.

Malaria interventions must be carefully thought out so that one does not cause epidemics. Devastating epidemics have occurred as the result of well-intentioned failures. Minimalist interventions cause no epidemics. → Malaria containment involves ignoring the mean prevalence but eliminating outbreaks. Reduce the variance about the mean and prevent new cases in naive adults. → Malaria suppression involves attempts to lower the prevalence. These interventions must be sustainable in the community without outside help, otherwise, when the aid expires, malaria epidemics will occur again in naive adults. Creation of boom and bust behavior introduces chaos.

Other Indirectly Transmitted Infections

African trypanosomiasis or sleeping sickness is another protozoan parasite transmitted by an insect vector, the Glossina or tsetse fly. There are two features of this disease that make control difficult. There are non-human animal reservoirs for the parasite. Also, the parasite populations with an individual human undergo cyclic antigenic variation. The parasite is able to express something on the order of 100 different variable surface antigens (VATs). Every time the host develops antibodies to control one VAT, a different VAT emerges.

Leishmania species are protozoans transmitted by sand flies, and are thought of as New World trypanosomes. There are important non human animal → reservoirs such as dogs.

→ Arboviruses (Vol. 1) are viruses transmitted by insect vectors, and almost 100 arboviruses are known to infect man. Two of the most important are → yellow fever and → dengue.

Yellow fever is transmitted either person to person by mosquitoes, or primate to person, and tends to occur in epidemics. It is still a considerable problem in parts of Asia, tropical Africa and South America. Immunity in humans is lifelong, suggesting control by vaccination, although herd immunity is defeated by primate reservoirs and vertical transmission within mosquito vectors.

Transmission of dengue is similar to yellow fever, except that there are four major types of dengue that all occur together so that a vaccine must be effective against all four types. Infection with one type of dengue is uncomfortable, but infection with a second type after recovery from the first is twenty times as likely to produce the syndrome of Dengue Hemorrhagic Fever that may be fatal.

A Model for Dengue A Dengue model (or a yellow fever model) would be structurally similar to a malaria model, but since humans infected with Dengue develop long lasting immunity to that strain of virus there would be an extra equation for infectious humans who recover and enter the immune or R state. In addition to the latent period in mosquitoes and mosquito mortality, the complex model includes the latent period in humans (until the appearance of viremia), the recovery of humans from both latent and infectious stages, and the death of humans in both latent and infectious stages.

A Model for Eastern Equine Encephalitis The basic dengue model can also be used for eastern equine encephalitis with birds replacing people in the dengue model. The behavior of the system depends on whether the birds die with the infection or develop long term immunity, whether immune birds return to the same roost repeatedly, and the numbers of birds in a roost.

Conclusion

We have shown how basic reasoning was developed beginning with the SIS, SIR SEIS and SEIR compartment models, how the reasoning evolved to include vectors in the malaria models without immunity, and ultimately how the same logic can be extended to other vector borne diseases. The value of mathematical models is that they can educate the public health practitioner about quantitative aspects of host parasite interactions in populations and help guide the choice and application of effective interventions.

MCL

Mucocutaneous leishmaniasis

Meadow Red

→ Babesiosis, Animals.

Mediterranean Coast Fever

→ Theileria (Vol. 1), → Theileriosis of cattle due to infection with *T. annulata* (→ Tick Bites: Effects in Animals).

Membrane-Function-Disturbing Drugs

For overview see Table 1–4

Structures

(Fig. 1)

Amphotericin B

Synonyms Ampho-Moronal, Amphozone, Fungillin, Miaquin.

Clinical Relevance Amphothericin B is used as an antifungal drug against human systemic mycoses, deep organ mycoses, *Candida* spp., *Torulopsis* spp., *Cryptococcus* spp., *Aspergillus fumigatus*, *Mucor* spp., *Coccidioides immitis*, *Histoplasma capsulatum*, *Sporothrix schenkii* and *Blastomyces* spp..

Table 1. Degree of efficacy of antimalarial drugs on various protozoan parasites.

		Mastigophora					Sarcodina	Sporozoa					
Year on the market	Drugs	*Leishmania* spp.	*Trypanosoma cruzi*	*Trypanosoma brucei*	*Giardia lamblia*	*Trichomonas vaginalis*	*Entamoeba histolytica*	*Eimeria* spp. in chicken	*Toxoplasma gondii*	*Babesia* spp.	*Theileria* spp.	*Plasmodium falciparum*	*P. vivax*
1945	Sulfonamides							xx	x			x	xx
1951	Pyrimethamine							x	x			x R4)	x R4)
1945	Biguanide/ Cycloguanide						x E				xx R4)	xxx R4)	
	Atovaquone											xxx	Xxx
1930	Mepacrine	x			xxx		xx E					xxx	xxx
1931	Acaprine			x						xxx			
1951	Primaquine		xx							x	x	x 4)	x 4)
1924	Pamaquine											x 4)	xx 4)
	Spiramycin								xxx				
	Clindamycin						x		xxx			x 4)	
	Sesquiterpene											xxx	xxx
1937	Chloroquine c)				xx		xxx	xx		xx		xxx	xxx
About 1988	Mefloquine				x E							xxx	xxx
1700	Quinine			x								xxx	xxx
1989	Halofantrine											xxx	xxx
About													
2700 B.C.	Halofuginone							xxx			xxx	xx	xx
	Febrifugine											xxx	xxx

xxx = high efficacy at least against some developmental stages, and diverse species; xx = partially effective (regarding developmental stages and diversity of species); x = slightly effective; E = active experimentally; R : resistance arose quickly; 4): predominantly extraerythrocytic stages and gamonts; c) also Amodiaquine

Table 2. Degree of efficacy of important cestodocidal drugs in current use.

Year on the market	Drugs	Cestodes	Trematodes	Nematodes	Protozoans
1. Naphthamidines					
1965	Bunamidine	xx			
2. Isothiocyanates					
1973	Nitroscanate	xx		xxx	
3. Salicylanilides					
1960	Niclosamide	xx	immature *Paramphistomum*		
4. Benzimidazole carbamates					
1979	Albendazole	xx	*F. hepatica, D. dendriticum*	xxx	Giardia
1971	Fenbendazole	xx	D. dendriticum	xxx	
	Flubendazole	xx		xxx	
1972	Mebendazole	xx	*F. hepatica, D. dendriticum*	xxx	Giardia
	Oxfendazole	xx		xxx	
5. Isoquinoline derivatives					
1975	Praziquantel	xxx	flukes (blood, lung, liver, intestine)		
1989	Epsiprantel	xxx	?		

xxx = high efficacy at least against some developmental stages and diverse species; xx = partially effective (regarding developmental stages and diversity of species); x = slightly effective

Table 3. Degree of important antischistosomal compounds in current use.

Year on the market/ or discovery	Drugs	*S. mansoni*	*S. haematobium*	*S. japonicum*	other parasites
1. Antimony compounds					
1973	oxamniquine	xxx	–	–	
2. Isoquinoline derivatives					
1972	praziquantel	xxx	xxx	xxx	protozoans, cestodes
3. Micellaneous substances					
1955	metrifonate (= trichlorfon)	x	xxx	x	cestodes, nematodes (microfilariae)
1981	cyclosporin A	xxx			protozoans, cestodes, filariae

xxx = high efficacy at least against some developmental stages and diverse species; xx = partially effective (regarding developmental stages and diversity of species); x = slightly effective

Molecular Interactions The mode of action relies on a complete abolishment of the barrier function of the plasma membrane in fungi. Amphothericin B has no antibacterial activity, because bacteria lack membrane sterines. The antiprotozoal activity of Amphothericin B is directed against antimony-resistant *Leishmania* spp., and there is some activity against *Trypanosoma* spp. and *Entamoeba histolytica*. The antiprotozoal mechanism of action does not rely on the inhibition of ergosterol biosynthetic activity. The action of Amphothericin B is directed against intracellular amastigotes of *L. brasiliensis* and *L. mexicana* in mucous tissues of nose and mouth probably by an interaction with leishmanial ergosterol. Ergosterol is a membrane component of leishmanial promastigotes. The Amphothericin B-ergosterol interaction results in an enhanced leakiness and permeability of the plasma membrane for ions and small molecules (amino acids and thiourea).

Table 4. Activity of drugs against other trematodes of human importance.

Drugs	Intestinal flukes	Liver flukes	Lung flukes
Chloroquine		*Clonorchis* (x)	
Tetrachloroethylene	*Fasciolopsis, Metagonimus, Heterophyes, Echinostoma, Gastrodiscoides, Watsonius, Nanophyetes* (x)		
Niclofolan	*Metagonimus* (x)		
Bithionol			*Paragonimus* (xxx)
Dichlorophene	*Fasciolopsis* (x)		
Hexachloroparaxylene		*Clonorchis* (x)	
Hexachlorethane	*Fasciolopsis* (x)		
Hexylresorcinol	*Fasciolopsis* (x)		
Resorantel	*Gastrodiscoides* (xxx)		
Niclosamide	*Fasciolopsis, Metagonimus, Heterophyes, Echinostoma, Gastrodiscoides, Watsonius, Nanophyetes* (xxx)		
Praziquantel	*Fasciolopsis, Metagonimus, Heterophyes, Echinostoma, Gastrodiscoides, Watsonius, Nanophyetes* (xxx)	*Clonorchis, Opisthorchis, Metorchis*	*Paragonimus* (xxx)

xxx = high efficacy at least against some developmental stages and diverse species; xx = partially effective (regarding developmental stages and diversity of species); x = slightly effective

Polyether Antibiotics

Important Compounds Monensin, Lasalocid, Salinomycin, Narasin, Maduramicin, Semduramycin.

Synonyms Monensin: Coban, Elancoban, Romensin, Rumensin

Lasalocid: Avatec, Bovatec (Fig. 1)
Salinomycin: Coxistac, Sacox
Narasin: Monteban
Maduramicin: Cygro, Prinocin
Semduramycin: Aviax

Clinical Relevance **Monensin** is the first member of the so-called polyether antibiotics. It is a fermentation product of the fungus *Streptomyces cinnamonensis* introduced in 1971 for the control of *Eimeria*-infections in poultry. Other polyether antibiotics of veterinary importance are **lasalocid** (produced by *Steptomyces lasaliensis*), **salinomycin** (by *S. albus*), **narasin** (by *S. aureofaciens*), **maduramicin** (by *Actinomadura yumaense*) and **semduramycin** (by *Actinomadura roseorufa*).

Monensin is especially effective against the asexual stages of the parasites, the extracellular sporozoites and free merozoites. It has an additional activity against schizogenous and gametogenic stages of *Toxoplasma gondii* in the cat. Maduramicin and alborixin are reported to cause also a reduction of oocyst excretion in human cryptosporidiosis by 71–96%.

Molecular Interactions Polyether antibiotics form complexes with Na^+ and to a lesser extent with K^+. The complexes have a lipophilic surface and move within the lipid regions of membranes, which results in an exchange of sodium ions by H^+ ions. In sporozoites of *Eimeria tenella* 5-fold increased Na^+-concentrations can be measured after exposure to monensin, followed by an increase of the activity of Na^+/K^+-ATPase to restore the physiological electrochemical Na^+-gradient. In addition, a depletion of the amylopectin stores in sporozoites, enhanced lactate formation and decrease of ATP-levels can be observed. The increase of intracellular Ca^{++}-concentrations is due to an enhanced Na^+/Ca^{++}-exchange and an enhanced liberation of these cations from mitochondria. The increased intracellular cation levels are followed by a quick entry of H_2O, alteration of intracellular pH, swelling of cells, vacuolization and damage of intracellular structures. The other anticoccidial polyether antibiotics are suggested to act in the same way.

Resistance In *Eimeria tenella* strains the resistance against monensin is due to reduced drug uptake. There is cross-resistance between monensin

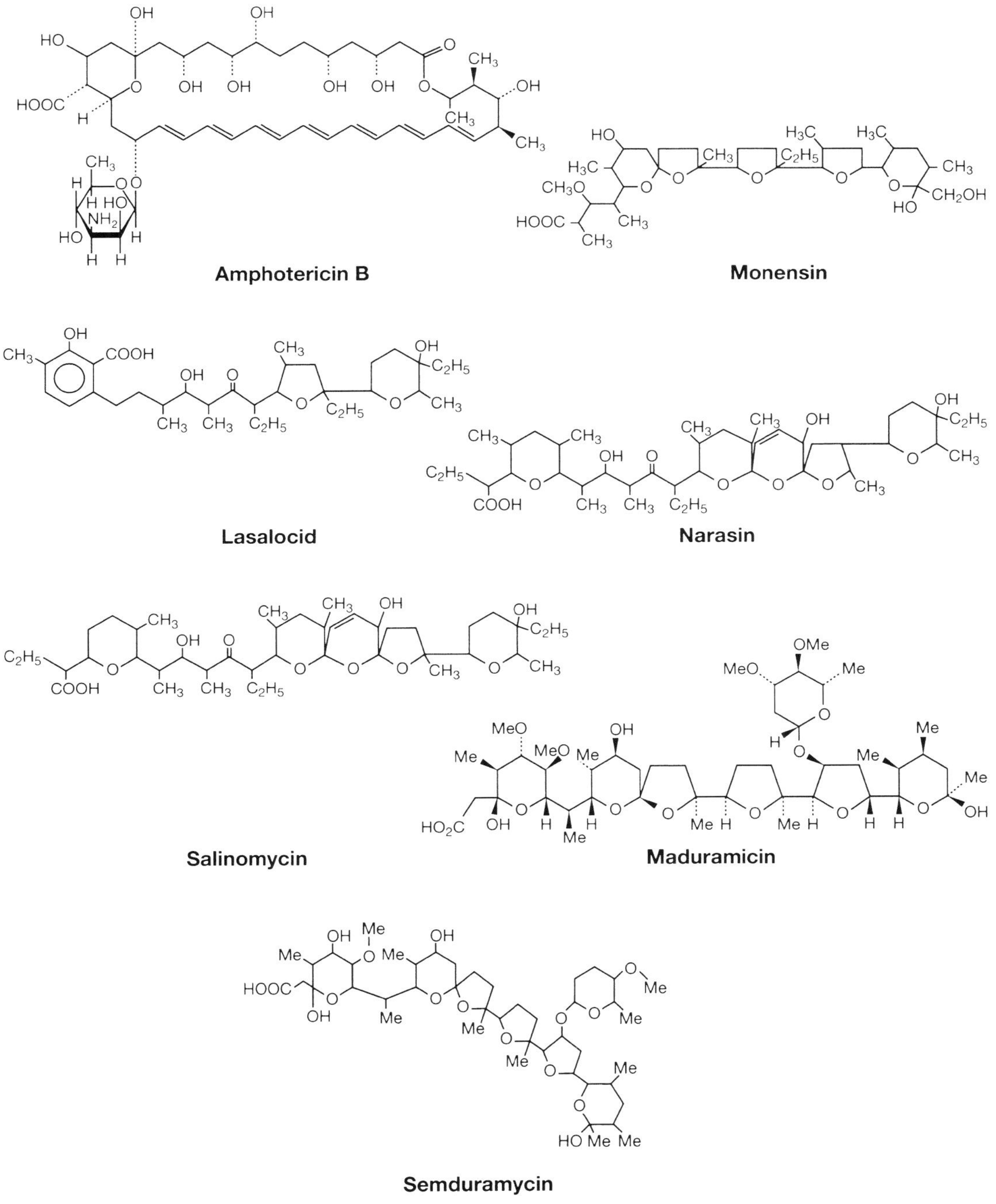

Amphotericin B

Monensin

Lasalocid

Narasin

Salinomycin

Maduramicin

Semduramycin

Fig. 1. Structures of drugs acting by disturbation of membrane function.

Fig. 1. (continued) Structures of drugs acting by disturbation of membrane function.

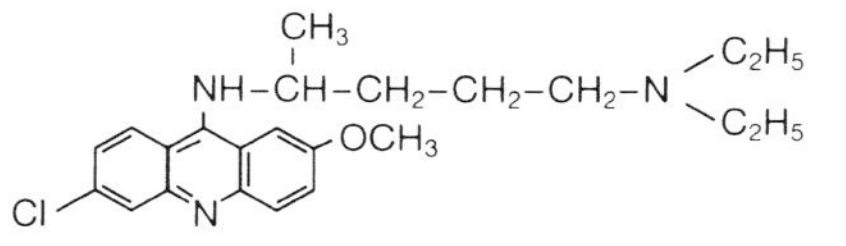

Mepacrine

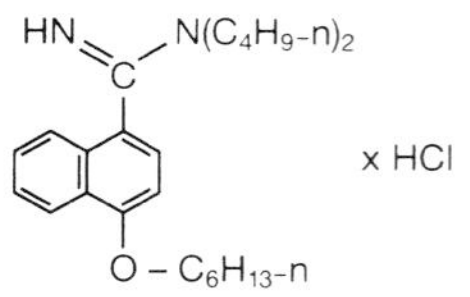

Bunamidine hydrochloride

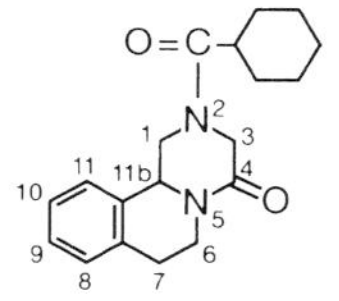

Praziquantel

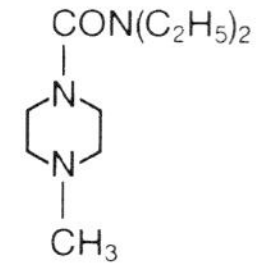

Diethylcarbamazine

and ionophoric polyethers such as narasin, salinomycin and occasionly lasalocid. The resistance is characterized by a great stability throughout many parasite generations. Maduramicin is effective also against monensin-resistant *Eimeria* strains, which is indicative for another mode of action of this drug. The mechanism of resistance is explained by membrane alterations interfering with the penetration of the ionophores into the membranes of the parasites. In resistant strains 20–40-fold higher drug concentrations are necessary to achieve effectivity. There are reports that four overexpressed peptides in sporozoites are associated with ionophore resistance in *E. tenella*. Multidrug-resistance genes coding for an energy-driven efflux mediated by P-glycoprotein are presumably responsible for cross-resistance. Polymerase chain reaction (PCR) methods are now used for the detection of resistance of different *Eimeria spp.* against polyethers and a variety of synthetic anticoccidial drugs.

Mepacrine

Synonyms Acrichine, Acriquine, Atabrine.diHCl, Atebrin.HCl, Chinacrin.HCl, Erion, Italchin, Metoquine, Palacrin, Quinacrine.HCl, Acranil (Fig. 1).

Clinical Relevance Mepacrine was introduced in 1930 as the first synthetic antimalarial compound. It has additional activity against *Giardia lamblia*, *Entamoeba histolytica* and *Leishmania* spp. (Table 1).

Molecular Interactions Mepacrine is an acridine derivative, and acts as an inhibitor of the respiration in *E. histolytica* and *G. lamblia*. A competitive inhibition of trypanothione reductase but not of glutathione reductase (→ DNA-Synthesis-Affecting Drugs III/Fig. 1) may be responsible for the antileishmanial action. Mepacrine specifically interacts with protein side chains. Recently it could be shown that acridine derivatives may have an influence on oxido-reduction-mechanisms involving disulfide reductases. The formation of complexes between DNA and quinacrine, which was believed to be the action of this drug for many years, is presumably not the true mechanism of action. Indeed, there is no accumulation of quinacrine in the nuclei or any structure in trophozoites of *Giardia lamblia*. Instead, blebs of concentrated drug appear prior to the disintegration of the membrane in drug-sensitive trophozoites. The membrane is now believed to be the site of the quinacrine action. In addition, DNA- and RNA-synthesis are inhibited probably by the induction of lesions in the macromolecules.

The antimalarial activity of mepacrine is directed against erythrocytic schizonts and gametocytes (→ Hem(oglobin) Interaction/Fig. 2). It has an influence on erythrocytic schizonts of all four *Plasmodium* species. An intercalation of mepacrine into the DNA of the parasites is discussed. Indeed, a binding of quinacrine to DNA in-vitro could be demonstrated, but the real mechanism in plasmodia remains unknown. There are other mepacrine-induced biochemical changes such as inhibition of replication, damage of ribosomes, inhibition of protein synthesis and also inhibition of respiration.

Besides its antiprotozoal activities, mepacrine exerts anticestodal activity against *Taenia saginata*, *T. solium* and *Diphyllobothrium latum*. The anticestodal action of mepacrine on the molecular level against tapeworms is unknown.

Resistance In *Giardia lamblia* strains resistant to furazolidone are more readiliy resistant to quinacrine indicating a multiple drug resistant phenotype resulting in an active exclusion of quinacrine by resistant trophozoites. Mepacrine resistance in malaria is not as actively researched as that of the newer antimalarials. There is a report on mepacrine treatment failures and the appearance of atebrine-insusceptible or atebrine-resistant *Plasmodium* strains from New Guinea. There is not always cross-resistance between quinacrine and other closely related quinoline containing antimalarials indicating a different mode of action.

Bunamidine

Synonyms Buban, Scolaban.

Clinical Relevance Bunamidine was discovered in 1965. It possesses anticestodal activity against *Taenia* spp., *Dipylidium caninum*, *Mesocestoides*, *Diphyllobothrium* and *Echinococcus granulosus* (Table 2).

Molecular Interactions Bunamidine is a naphthamidine-derivative acting by a disruption of tegumental outer layers of *Hymenolepis nana*. This results in a decrease in the rate of glucose uptake, increase in the rate of glucose efflux and interference with the absorptive surface of the cestodes. The drug also inhibits the fumarate reductase in the mitochondria of *H. diminuta*, thereby interrupting ATP-synthesis (→ Energy-Metabolism-Disturbing Drugs/Fig. 4).

Praziquantel

Synonyms Biltricide, Caniquantel, Cesol, Cestocur, Droncit; in combinations: Caniquantel Plus, Drontal, Drontal Plus.

Clinical Relevance Praziquantel was discovered in 1972. It was first developed as a veterinary cestocide (Droncit). It is the drug of choice for human and veterinary cestode and trematode infections (Table 2, Table 3, Table 4). It possesses broad-spectrum activity against cestodes and trematodes including *Taenia solium* neurocysticercosis, but has low efficacy against larval *Echinococcus* spp. (hydatidosis) and *Fasciola hepatica*.

Importantly, praziquantel is the drug of choice for the treatment of all forms of schistosomiasis (Table 3) in a single oral dose. Praziquantel is now achievable at almost the same price or even cheaper than oxamniquine in South America and the rest of the world. It is applied in extensive control programs in many endemic countries in Africa, South America and Asia, and there are now clinical experiences over the last 20 years. A great advantage is the lack of serious short- or long-term side effects. The drug is safe, effective, and easy-to-handle for treatment of schistosomiasis. Multicenter clinical trials were performed by WHO and Bayer in Africa, Japan, the Philippines, and Brazil. Praziquantel gained a quick dominant role in antischistosomal therapy until today.

Praziquantel has also activity against intestinal, liver and lung flukes (Table 4), but has only low efficacy against infections with *F. hepatica* and *Paramphistomum*. Against *Dicrocoelium* praziquantel is only effective at very high dosages.

Molecular Interactions The tegument and the musculature of the parasites are at least two targets of the action of praziquantel (chemically a pyrazino-isoquinoline derivative), and the action is furthermore partially (or fully?) mediated or supported by the host immune system.

(1) The action of praziquantel against the tegument of cestodes and trematodes: Praziquantel induces an almost instantaneous vacuolization of the tegument of *Schistosoma mansoni* and cestodes (*Hymenolepis nana* and *T. taeniaeformis*). The vacuolization occurs at the base of the syncytial layer. Vacuoles increase in size, protrude above the surface resulting in a final bursting of the blebs. Furthermore changes in ion- and small molecule-permeability across

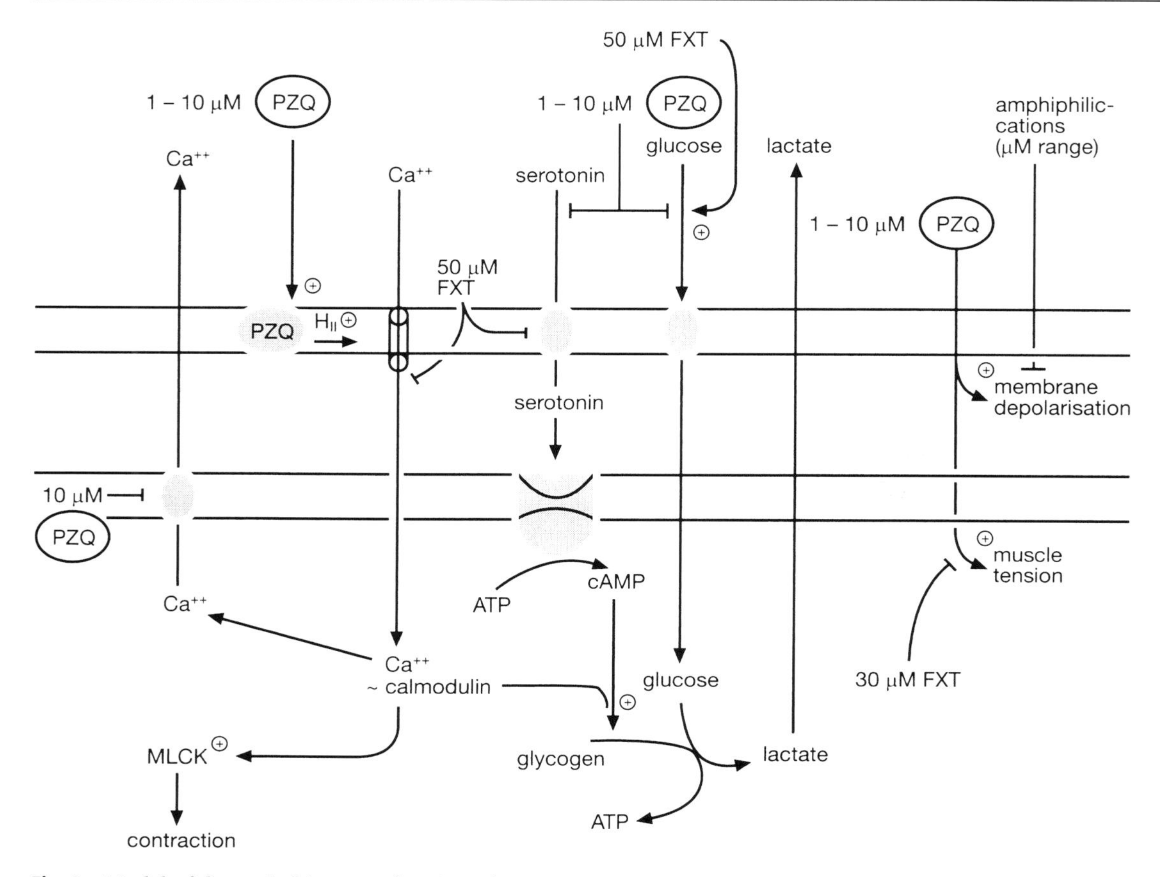

Fig. 2. Model of the antischistosomal action of praziquantel.

the tegumental membrane can be observed. Indeed, damaged parasites loose glucose, lactate, and amino acids into the surrounding medium. In addition, there is an indirect inhibition of some membrane-associated transport proteins (serotonin- or glucose- uptake protein or Ca^{++}-ATPase) by disruption of tegumental environment. The Ca^{++}-ATPase is unable to control the rise of Ca^{++} ion concentration by praziquantel. It was also postulated that different calcium channels might be involved in the impairment of the calcium homeostasis within the parasites. In-vitro a disruption of the bilayer structure of synthetic phospholipid vesicles (phosphatidylserine, phosphatidylethanolamine) can be induced in presence of calcium by praziquantel resulting in the formation of hexagonal structures (Fig. 2). The appearance of such hexagonal structures in the tegumental membranes may thereafter fascilitate Ca^{++} entry into the worms leading to a disturbance of Ca^{++}-homeostasis and overall changes in membrane integrity. Responsible for the perturbations of membranes are presumably interactions between negatively charged phospholipids, Ca^{++} ions and the electrically neutral praziquantel. However, membrane actions of praziquantel are obviously not alone responsible for the drug's action, since the effects on the phospholipids are presumably mediated by a receptor protein. This idea is supported by the finding that there are differences in the anthelmintic activity between two stereo-isomers of praziquantel. A candidate for such a receptor may be a 200 kDa surface glycoprotein.

(2) The action of praziquantel against the parasite musculature:Praziquantel also induces alterations of parasite's muscle physiology/biochemistry. At lower concentrations (below 1 µg/ml) a stimulation of motility of *H. diminuta*, *H. microstoma*, *H. nana* and preadult *Echinococcus multilocularis* is induced, followed by a contraction of the parasite musculature within 10–30 sec. The threshhold concentration of

praziquantel-induced contraction is between 1–10 μg/ml and is the same as that of drug-induced tegumental alterations. In addition, these changes are strongly dependent on the presence of Ca^{++}. Ca^{++} ions as second messengers are responsible for the further contraction of worms and glycogen breakdown.

(3) Involvement of the host immune system in the praziquantel action: The observed vacuolization of the parasite's tegument alone is not lethal. Responsible for the lethal effects may be immune mechanisms of the host. Indeed, an invasion of phagocytic cells into parasite occurs within 17 hours after treatment of the host and a lysis of parasite tissues can be observed within a few days. Therefore, it is assumed that praziquantel induces a fascilitated host immune attack following the tegumental damage and a loss of the ability to repair the tegumental surface lesions. This is caused by the disrupture of tegumental membranes resulting in an exposure of proteins or enzymes, e.g. alkaline phosphatases, in female worms, and direct reactions of parasite antigens at the surface of adult male *S. mansoni* with host antibodies. In this context, it may be important that there are stage-specific capacities of repair mechanisms in schistosomes. Thus, young schistosomules and adult schistosomes have the lowest phospholipid synthesis rate, whereas 11-day-old juveniles possess the highest phospholipid synthesis rate. Praziquantel is most active against 7-day-old schistosomules and schistosomes older than 5 weeks. By contrast, 2- to 4-week-old juveniles are less susceptible to praziquantel. The repair of drug-induced lesions is presumably prevented by the interaction of the immune system with the parasite. Thus, in this model the tegumental damage as the first event of praziquantel's action favours the immune attack by the host. This view is supported by the observation that in immunosuppressed animals praziquantel treatment is less effective than in immune competent animals.

Resistance There are some reports on low cure rates after praziquantel treatment of schistosome-infected humans in Brazil. There is a most alarming case of low cure rate of only 18% after praziquantel treatment from Senegal patients. The mechanism for praziquantel failure is completely unclear and also the molecular mechanism of praziquantel resistance. There are two reports on laboratory selection of praziquantel resistant *S. mansoni*. Praziquantel treatment of mice with bisexual *S. mansoni* infections and successive passage of eggs from worms lead to a strain that had survived treatment.

Diethylcarbamacine (DEC)

Synonyms Tenac, Banocide, Caricide, Carbam, Caritol, Cypip, Decanine, Dicacid, Dicarocide, Difil, Digacid, Dirocide, Diro-form, Ethodryl, Filaribits, Filariosan, Franocide, Hetrazan, Loxwran, Luwucit, Nemacide, Neo Paulvermin, Notezine, Pet-Dec, Pulmocid, Supatonin, Unicarbazan.

Clinical Relevance The microfilaricidal efficacy of DEC had been detected in 1947. DEC has antifilaricidal activity against *Onchocerca volvulus*, *Loa loa*, *Wuchereria bancrofti* and *Brugia malayi* in man (→ Inhibitory-Neurotransmission-Affecting Drugs/Table 1), *Mansonella ozzardi* is not affected. DEC can also interfere with embryogenesis in female worms. The microfilaricidal effects are generally accompanied in man by severe, in dogs often lethal side effects (Mazzotti reaction). In the past few years a combination of DEC with ivermectin has been approved for the control of lymphatic filariasis.

Molecular Interactions The mode of action of DEC is unclear. There are reports on inhibitory effects of DEC on the motility of *Dirofilaria immitis*. Interestingly, there is a general lack of in-vitro activity of DEC indicating that the efficacy in-vivo is mediated by host immune factors. An early trapping of life *Litomosoides carinii* microfilariae by polymorphnuclear cells, macrophages and lymphocytes is observable, also a phagocytosis of microfilariae. The microfilarial sheath becomes destructed by lysosomal enzymes resulting in a final cell-mediated degradation of larvae. This cell-mediated mechanism, however, is not clear on the molecular level. In-vitro an activation of complement on the sheath or the microfilarial surface via the alternate pathway by DEC can be observed. There is also an enhancement of antibody-mediated adherence of cells to larvae, but this seems not to be important for DEC action. Microfilariae become eliminated by a mechanism of DEC which is independent of immune status of the host. Antibodies directed against DEC potentiate the microfilaricidal activity of subcurative

DEC doses in *Seteria digitata* infected *Mastomys coucha*. There is also a participation of other non-antibody mediated mechanisms for DEC action. Thus, there are reports on an inhibition of arachidonic acid pathway in-vitro in DEC treated endothelial cells, an enhancement of macrophage, eosinophil and neutrophil adherence to microfilariae after DEC treatment, a platelet mediated cytotoxicity to microfilariae of *L. carinii* probably by free oxygen radicals and clearance of microfilariae in nude mice.

Meningoencephalitis

→ Pathology, → Tick Bites: Effects in Man.

Merozoite Surface Protein 1

→ Malaria, → MSP-1, → Vaccination.

Metaphylaxis

Method and term are used in veterinary medicine: killing of parasites after an outbreak of disease before serious damage may occur.

Metastasis-like Infiltration

Undifferentiated cells from → *Echinococcus multilocularis* (Vol. 1) cysts (= alveococcus) may give rise to new cysts in many organs, when disseminated during surgery or biopsy.

Micronemiasis, Man

Micronemiasis is an infection by a free-living microscopic nematode, *Micronema* sp., which can give rise to disseminated infection after contamination of wounds with soil or horse manure. This oviparous worm multiplies in the body, building up huge numbers, with the larvae found in many tissues. Only a few cases have been described in humans, in all of whom → meningoencephalitis was present. Similar lesions and granulomatous masses with many worms have been observed in horses.

Therapy

→ Nematocidal Drugs, Man.

Microsporidiosis

Immunosuppressed humans have been sentinels of microsporidial infection, with enteric, neurologic, ocular and pulmonary manifestation being recognized (→ Microsporidia (Vol. 1)). Their symptomatology, pathology and differential diagnosis, based on ultrastructure and the polymerase chain reaction, has been stated by many authors. Microsporidiosis appears to be a common asymptomatic infection, that is not clinically recognized; about 10% of animal handlers were reported to have antibody to *Encephalitozoon* sp. The organism grows intracellularly, destroying the infected cells. There is little inflammation. The → Brown-Brenn stain (Gram stain for tissues), basic fuchsin, toluidin blue, Azur II-eosin, the → Warthin-Starry silver impregnation and polarization facilitate recognition of microsporidia in tissue sections. Tissue imprints (smears), dried, fixed and stained as for blood smears, are useful for diagnosis of corneal and conjunctival lesions. Stool and sputum smears can be stained with a modified → trichrome stain employing chromotrope 2R or with a fluorochrome chitin stain, such as Calcofluor.

A fatal disseminated infection of a 4-months-old thymic alymphoplastic baby with *Nosema connori* involved the smooth musculature, skeletal muscles, the myocardium, parenchyma! cells of the liver, lung and adrenals *Encephalitozoon* sp. was isolated from the cerebrospinal fluid of a 9-year-old Japanese boy with meningoencephalitis who recovered. Intestinal microsporidiosis has been described in a high percentage of patients with AIDS due to *Enterocytozoon bieneusi* also with cholangitis and due to *Encephalitozoon (Septata) intestinalis*. The patients had diarrhoea, with weight loss from malabsorption. Inflammation was minimal and the diagnosis was made ultrastructurally. Disseminated *Encephalitozoon cuniculi* infection was described and *E. hellem* has been isolated from AIDS patients with nephritis and prostatitis and from others with keratoconjunctivitis, bronchitis and sinusitis. Microsporidial myositis due to *Pleistophora* and *Trachipleistophora hominis* was reported in patients with AIDS. Intraocular microsporidiosis was diagnosed from the cornea next to Descemet's membrane with a subacute to granulomatous inflammatory reaction. Other HIV-negative cases with corneal stromal infection were linked to *Nosema ocularum* (possibly *Vittaforma corneum*). For further information see → Microsporidia (Vol. 1).

Main clinical symptoms: Abdominal pain, diarrhoea, loss of weight
Incubation period: 1 week
Prepatent period: 1 week
Patent period: More than 5 months
Diagnosis: Microscopic determination of spores in fecal samples
Prophylaxis: Avoid contact with human/animal feces

Therapy: Curative treatment unknown; see → Treatment of Opportunistic Agents

Microtubule-Function-Affecting Drugs

Table 1

Structures

(Fig. 1).

Benzimidazoles

Important Compounds Phenothiazine, Tiabendazole, Cambendazole, Oxibendazole, Albendazole Albendazole Sulphoxide, Fenbendazole, Oxfendazole, Mebendazole, Flubendazole, Parbendazole,

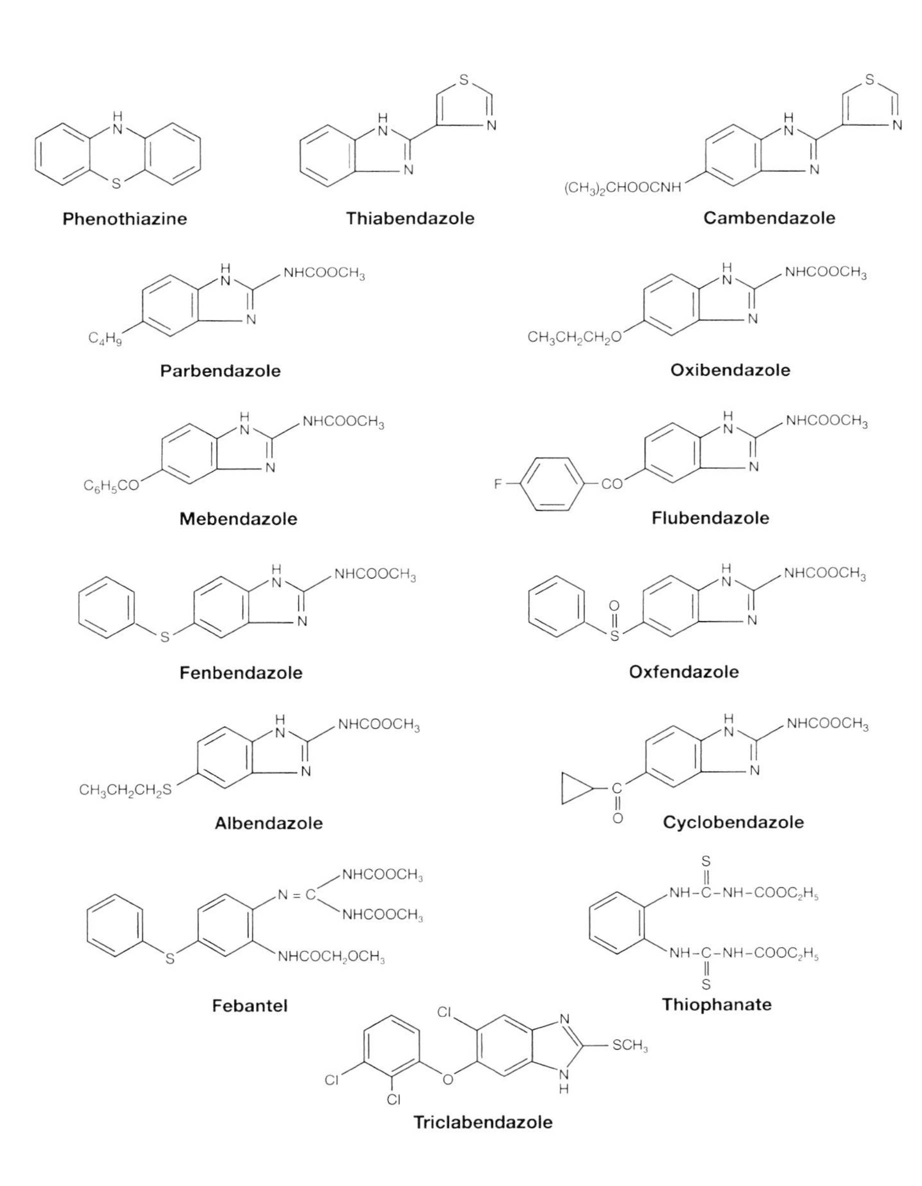

Fig. 1. Structures of drugs against parasites affecting microtubuline integrity.

Table 1. Antiparasitic spectrum of modern antinematodal drugs.

Year on the market	Drugs	Antinematodal activity	Additional antiparasitic activity
1. Tetrahydropyrimidines			
1966	Pyrantel	*Ascaris, Enterobius, Necator, Ancylostoma, Trichinella, Trichostrongylus,* ruminant, pig and horse nematodes	horse cestodes
	Oxantel	*Trichuris*	
	Combination Oxantel/Pyrantel	*Ascaris,* hookworms, *Trichuris, Enterobius*	
	Morantel	ruminant nematodes	
2. Imidazothiazoles			
	Tetramisole	pig nematodes	
1965	Levamisole	*Ascaris,* hookworms, *Strongylus,* pig, ruminant and poultry nematodes	microfilariae
	Butamisole		
3. Benzimidazoles			
1961	Tiabendazole	*Strongyloides, Capillaria,* ruminant, pig and horse nematodes	*Angiostrongylus cantonensis,* cutaneous larva migrans, *Dracunculus medinensis*
	Cambendazole	*Strongyloides,* trichostrongylides, ruminant, pig and horse nematodes	
1966	Parbendazole	Pig and ruminant nematodes	
1973	Oxibendazole	horse and ruminant nematodes	
1971	Mebendazole	*Ascaris, Enterobius, Necator, Ancylostoma, Trichuris, Trichinella, Capillaria,* ruminant and horse nematodes	*Dracunculus medinensis, Giardia,* trematodes, cestodes, macrofilariae
	Flubendazole	*Enterobius,* pig nematodes	cestodes, macrofilariae
1971	Fenbendazole	ruminant, pig and horse nematodes	*Toxocara canis* larvae, cestodes, trematodes
1975	Oxfendazole	ruminant and horse nematodes	*Toxocara canis* larvae, cestodes
1979	Albendazole	*Enterobius, Necator, Ancylostoma, Trichuris, Trichinella,* trichostrongylides	Cutaneous larva migrans, *Giardia,* cestodes, trematodes, macrofilariae
	Cyclobendazole		
4. Benzimidazole prodrugs			
1978	Febantel	ruminant, pig and horse nematodes	
1970	Thiophanate	ruminant and pig nematodes	
5. Avermectins and milbemycins			
1973	Milbemycin	dog nematodes	microfilariae, ectoparasites
1985	Abamectin	ruminant nematodes	microfilariae, ectoparasites
1980	Ivermectin	ruminant, pig and horse nematodes	microfilariae, ectoparasites
1990	Milbemycinoxim	ruminant and dog nematodes	microfilariae, ectoparasites
1992	Moxidectin	ruminant nematodes	microfilariae, ectoparasites
1993	Doramectin	ruminant nematodes	ectoparasites
1996/97	Eprinomectin	ruminant nematodes	ectoparasites
1999	Selamectin	dog nematodes	fleas, ticks, microfilariae

Febantel, Netobimin, Thiophanate, Triclabendazole.

Synonyms Phenothiazine: Contraverm, Coopazine, Fenopur, Helmetina, Neoavilep, Phenovis, Phenoxur, Radiol

Tiabendazole: Bovizole, Coglazol, Equizole, Helmintazole, Hyozole, Mintezole, Nemapan, Omnizole, Polival, Soldrin, TBZ, Thibenzole, in: Equizole A, Equizole B, Ranizole, Suiverm, Thiprazole, Tresaderm, Tricocefal

Cambendazole: Ascapilla, Bonlam, Camvet, Equiben, Equicam, Novazole, Noviben, Porcam

Oxibendazole: Anthelcide, Anthelworm, Equipar, Equitac, Loditac, Verzine, Widespec

Albendazole: Albazine, Valbazen, Zentel

Albendazole sulphoxide: Rycoben

Fenbendazole: Panacur, Safe-Guard

Oxfendazole: Benzelmin, Synanthic, Systamex

Mebendazole: Equivurm, Fugacar, Mebenvet, Mebutar, Multispec, Nemasole, Ovitelmin, Pantelmin, Parmeben, Rumatel, Sirben, Telmin, Telmintic, Vermirax, Vermox

Flubendazole: Flubenol, Flumoxal, Fluvermal

Parbendazole: Helmatac, Topclip, Triban, Verminum, Worm Guard

Febantel: Amatron, Bayverm, Combotel, Provet, Rintal

Netobimin: Hepadex

Thiophanate: Helminate, Wormalac, Nemafax; in: Flukembin, Vermadax

Triclabendazole: Fasinex

Clinical Relevance **Phenothiazine**, an oldtimer, has been used since the 1930s as antinematodal drug in ruminants. In the 1960s it was replaced by the broad-spectrum benzimidazoles for several reasons: resistance had appeared against phenothiazine, benzimidazoles can be applied at much lower dosages, and the latter have a much broader anthelmintic spectrum.

The anthelmintic benzimidazoles can be divided in 4 different subgroups : (1) the **benzimidazole-thiazolyls** (cambendazole, thiabendazole (explored 1961)), (2) the **benzimidazole-methylcarbamates** (albendazole (1979), cyclobendazole, fenbendazole (1971), flubendazole, luxabendazole, mebendazole, oxfendazole (1975), oxibendazole (1973), parbendazole (1966), ricobendazole), (3) the **halogenated benzimidazole-thiole** triclabendazole and (4) the **prebenzimidazoles** febantel, netobimin and thiophanate. Febantel (1978) is converted to the active forms fenbendazole and oxfendazole, netobimin is converted to the active form albendazole, and thiophanate is metabolized to the active form lobendazole.

The benzimidazoles can be used against a wide variety of parasitic pathogens. The antiprotozoal activity of albendazole and mebendazole can be used in the treatment of infections with *Giardia lamblia*. The mechanism of action against *Giardia* is presumably directed against the ventral disc microtubules (→ DNA-Synthesis-Affecting Drugs I/Table 1).

Albendazole has activity against the microsporidia *Encephalitozoon intestinalis*. There is, however, only symptomatic improvement achievable in *Enterocytozoon bieneusi* infections in AIDS patients.

The benzimidazole carbamates (mebendazole, flubendazole, albendazole, fenbendazole) have anticestodal activity against larval stages of *Echinococcus* spp. (hydatidosis). Albendazole and mebendazole are first line drugs for medical treatment of hydatidosis. Flubendazole exerts activity in *Taenia solium*-neurocysticercosis. Fenbendazole, flubendazole and mebendazole are effective against *Taenia* spp. infections in dogs and cats. Albendazole, mebendazole, fenbendazole, oxfendazole and prebenzimidazoles (febantel, netobimin) show activities against cestode infections of ruminants. However, in general high dosages of benzimidazole carbamates are necessary and they have no activity against adult *Dipylidium caninum*, *E. granulosus* or *Mesocestoides* spp. or *Diphyllobothrium*.

Thiabendazole, albendazole, mebendazole and triclabendazole also exert antitrematodal activities. Thiabendazole is the first broad-spectrum anthelmintic benzimidazole with some activity at high dosages against *Dicrocoelium dendriticum*, but no activity against *Fasciola hepatica*. Albendazole and mebendazole possess an anthelmintic spectrum inclusive mature liver flukes in sheep and cattle. Higher dosages are required compared to their nematocidal activity.

Triclabendazole is a benzimidazole-derivative with an unusual chemical structure because of the chlorinated benzene ring. Its efficacy is restricted to *F. hepatica* (chronic and acute fasciolosis) (→ Energy-Metabolism-Disturbing Drugs/Table 1) and paragonimiasis, it has minor activity against other trematodes such as *D. dendriticum*, *Schistosoma mansoni* and *Paramphistomum* spp., but it has no activity against nematodes and cestodes. Triclabendazole is an important fasciolicidal drug with high efficacy against adult and juvenile flukes. Furthermore, it is the drug of choice for human fasciolosis. It is very safe and used at a single dose of 12 mg/kg to be repeated 12 hours later.

The main indication for benzimidazoles relies on their broad-spectrum activity against nematodes in human and veterinary medicine (Table 1). With the exception of triclabendazole, all other (pre)benzimidazoles broad-spectrum anthelmintics have a main action against gastrointestinal and tissue nematodes. Benzimidazoles are mainly orally ingested by nematodes. Of special importance is the efficacy of thiabendazole and albendazole against *Strongyloides stercoralis* in AIDS.

In addition, benzimidazoles have antifilarial activities exerting adulticidal effects e.g. against *Litomosoides carinii* and *Brugia pahangi*. Several compounds have been introduced into clinical trials for human onchocerciasis. They have higher effects against adult and developing parasites than against microfilariae. The need for parenteral application, however, prevented the broad usage of benzimidazoles in human filariasis (→ Inhibitory-Neurotransmission-Affecting Drugs/Table 1). There are severe local intolerabilities after subcutaneous application of flubendazole with intolerable pains. Moreover, after oral administration in man embryotoxic effects have been reported. Flubendazole is the most active filaricidal benzimidazole. There is the following ranking with decreasing activity : flubendazole > mebendazole > oxfendazole, cyclobendazole > albendazole > cambendazole > fenbendazole. All benzimidazoles exert the same type of efficacy with only minor variations. There are additional microfilaricidal effects after the first and second week against *L. carinii* and *Acanthocheilonema viteae* or after the third week against *Brugia spp.*. Such a delayed effect on microfilariae is a common phenomenon with all benzimidazoles. Last but not least, tiabendazole and mebendazole are also tried against *Dracunculus medinensis*.

Molecular Interactions The mode of action of benzimidazoles relies on the impairment of microtubular function. For evaluation of the mode of action most experiments have been performed with nematodes, and there is only few data for cestodes. Very early a disturbance of microtubule shape and function could be observed in *Ascaris suum* intestinal cells by mebendazole as a result of the inhibition of microtubuline polymerization. The mebendazole-induced damage of intestinal cells in *Ascaris* and the damage of tegumental cells are caused by a loss of cytoplasmic tubules. The loss of cytoplasmic tubules is associated with a loss of transport of secretory vesicles and impairment of glucose uptake in intestinal cells. This is, in addition, an indication for an oral ingestion of benzimidazoles by nematodes (→ Acetylcholine-Neurotransmission-Affecting Drugs/Fig. 3).

The great selectivity of benzimidazoles is due to differences in the binding affinity between helminth and mammalian tubulins. There is a correlation between LD_{50} values in developing *Haemonchus contortus* L3-larvae and inhibition of binding of radioactively labelled mebendazole to tubulin. Cestodes are generally less susceptible to benzimidazoles compared to nematodes, and dose rates against *Taenia pisiformis* and *T. hydatigena* are more than 10–15 times higher compared to those necessary for nematodes. This is in line with the binding affinities of mebendazole to the nematode tubulin, which is 2–7 times higher than that to cestode tubulin, and even 10–35 times higher compared to sheep brain tubulin.

On the molecular level benzimidazoles are bound to β-tubulin (Fig. 2). Normally dimers of β- together and α-tubulin polymerize to form microtubule structures inside the cells of nematodes and the hosts. Benzimidazoles compete for the binding site on β-tubulin with colchicine, an inhibitor of cell division in the metaphase. Thereby, the formation of the microtubules by polymerization of tubulin at one end (= positive pole) is inhibited by benzimidazoles. The result is a starvation of the nematodes by intestinal disruption and inhibition of their egg production. The onset of the anthelmintic action of benzimidazoles is in general slower than that of the anthelmintics interfering directly on ion-channels. Embryotoxic effects of benzimidazoles can also be explained by interference with the formation of microtubuli, since rapidly dividing tissues like intrauterine developmental stages are primary targets of benzimidazoles.

In cestodes additional mechanisms besides the inhibition of microtubuli formation are probably responsible for the action of benzimidazoles. There is a reduction in glucose uptake and a decrease in glycogen content of parasites observable. In *Moniezia expansa* a diminished in-vitro and in-vivo ATP synthesis and/or turnover of adenine

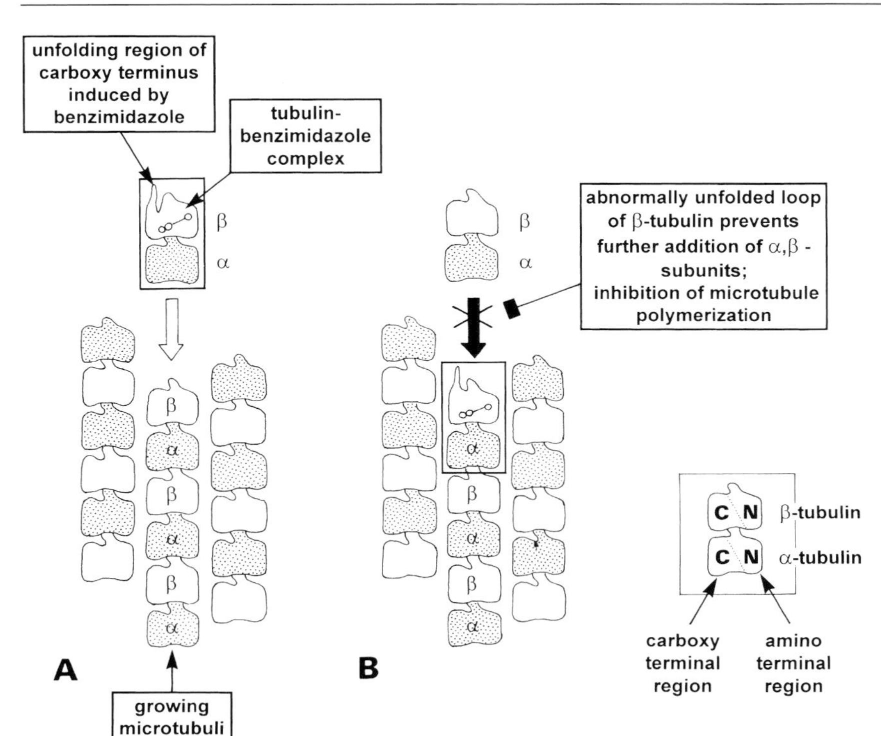

Fig. 2 Model of the mechanism of action of benzimidazoles (Roos MH (1997) Parasitol. 114 : S 137–S144).

nucleotides by mebendazole can be measured. The effects are observed 30 minutes after exposure to mebendazole.

The action of benzimidazoles against flukes are characterized by long-term effects with a gradual decrease of activity of these parasites. Immature flukes are more sensitive to triclabendazole than adult flukes. A gradual hyperpolarisation of the tegumental membrane potential is induced without the involvement of ATP-driven ion pumps. There is a binding of triclabendazole to cytoplasmic microtubules and induction of depolymerization, which is similar to that of the other benzimidazoles by interruption of microtubule-dependent processes in helminths. There is also progressively severe damage of the surface resulting in a total loss of the tegument within 24 hours in the adult flukes. Furthermore, an inhibition of mitotic division of spermatogenic cells, an inhibition of protein synthesis in the tegumental cells, a decline in the number of secretory bodies in the tegument and disappearance of the Golgi complex can be observed.

Recently a model for the mechanism of benzimidazole action on the molecular level has been published (Fig. 2). Thereby, β-tubulin is regarded as a GTP-binding protein. GTP is needed for assembly of the microtubules. Benzimidazoles as nucleotide analogues are bound in neighbourhood of the nucleotide binding domain II near the codon 200. It is now suggested that the binding results in a slight conformational change and induces an alteration of the properties of GTP binding. Thereby, an unfolding region appears at the β-tubulin carboxy terminus while rest of the β-tubulin remains unaltered. Once added to the microtubule the abnormally unfolded loop of β-tubulin prevents further addition of subunits and causes an inhibition of further microtubule polymerization (Fig. 2). Interestingly in *Cryptosporidium parvum* the lack of activity of benzimidazoles correlates with the absence of Glu-198 and Phe-200. This may explain why benzimidazoles have no activity against these sporozoa.

Resistance Resistance of a variety of nematodes in different host animals (sheep, goats, cattle, horse, swine) against benzimidazoles has appeared worldwide. The control of benzimidazole-resistance in *Haemonchus contortus* is recessive. The β-tubulin gene and the gene products of β-tubulin isotype 1 and isotype 2 are involved in benzimidazole resistance in *H. contortus*. At lower resistance levels the specific isotype 1 gene becomes selected, and at higher resistance levels there is a selection of worms with isotype 2 genes. The β-tubulin isotype1 and 2 are encoded by separate genes and numerous alleles. Up to six alleles encode for isotype 1 and up to 12 alleles for isotype 2. In benzimidazole resistant nematodes a reduction in the number of isotype alleles for β-tubulin can be observed resulting in a progressive loss of alleles for isotype 1 and a total loss of alleles for isotype 2. Thus, benzimidazole resistance is presumably characterized by a loss of susceptible phenotypes of β-tubulin and simultaneous survival of resistance phenotypes.

Benzimidazole resistance in fungi is due to the appearance of a different form of β-tubulin. Phenylalanine, present in position 200 on β-tubulin in benzimidazole-suceptible fungi, is replaced in benzimidazole-resistant fungi and in normal mammalian β-tubulin by tyrosin. In *H. contortus* there is a correlation between benzimidazole resistance and a conserved mutation at amino acid 200 in β-tubulin isotype 1. In an interesting experiment a benzimidazole-resistant *Caenorhabditis elegans* strain (ben-1) could be transformed with a β-tubulin isotype 1 gene isolated from a benzimidazole susceptible *H. contortus* population. The expression of this *H. contortus* gene in this formerly resistant *C. elegans* strain switched the phenotype from resistant to susceptible. Thus, the substitution at position 200 in the β-tubulin plays a crucial role in determining benzimidazole susceptibility.

Miner's Disease

Disease due to infection with Old or New World hookworms.

Mode of Action

See → Chemotherapy, → Drug

Models

→ Mathematical Models of Vector-Borne Diseases.

Modes of Infection

→ Disease Control, Epidemiological Analysis.

Morbidity

Number of sick people within a given population (mostly 100,000 people/year).

Mortality

This term (*lat.* mortalitas = dying) describes the reduction of individuals in a given population due to death (with respect to different reasons).

MSP

(1) Merozoite surface proteins (→ Plasmodium (Vol. 1), → Malaria/Vaccination) and (2) major sperm protein (→ Vaccination Against Nematodes).

Murine Spotted Fever

Disease due to *Rickettsia typhi*-bacteria transmitted by rat fleas.

Murine Typhus

Disease in humans due to infection with *Rickettsia typhi* transmitted by bite of → fleas (Vol. 1) and lice.

Murrina

Disease of horses due to *T. brucei evansi*: transmitted by → vampire bats (Vol. 1) in Panama.

Muscidosis

Disease due to infestation with muscid flies, see Table 1 (page 357).

Myiasis, Animals

Disease due to skin infestation with fly larvae, see Table 1 (page 359).

Table 1. Muscid Flies and Control Measurements.

Parasite	Host	Vector for	Symptoms	Country	Therapy		
					Products	Application	Compounds
Musca autumnalis (Face fly)	Ruminants, Horse, Pig	*Corynebacterium pyogenes* (Summer mastitis); Horse: Infectious anaemia, Infectious bovine keratokonjunctivitis (*Moraxella bovis*)	Bothering	Worldwide	Rabon™ 3% Dust (Agri Labs)	Self Treating Dust Bags	Tetrachlor-vinphos
					Vigilante™ Insecticide (Intervet)	Bolus	Diflubenzuron
					Bayofly™ Pour-on (Bayer)	Pour on	Cyfluthrin
					Neporex™ (Novartis)	Spray	Cyromazine
Musca domestica (House fly)	Ruminants, Horse	Horse: Infectious anaemia and *Habronema muscae* and *Draschia megastoma*	Bothering	Worldwide			
Stomoxys calcitrans (Stable fly)	Ruminants, Horse, Pig	Horse: Infectious anaemia and *Habronema majus*	Blood loss, irritation	Worldwide			
Haematobia irritans (Horn fly)	Ruminants, Horse	Infectious anaemia (Horse), *Stenofilaria stilesi* (Cattle)	Blood loss, irritation	Worldwide	RabonTM 3% Dust (Agri Labs)	Self Treating Dust Bags	Tetrachlor-vinphos
					Co-Ral™ 25% Wettable Powder (Bayer)	Dip or Spray	Coumaphos
					Commando™ Insecticide Cattle Ear Tag (Fermenta)	Ear tag	Ethion
					Vigilante™ Insecticide (Intervet)	Bolus	Diflubenzuron
					Moorman's™ IGR Cattle Concentrate	Feed additive	Methoprene
					Bayofly™ Pour-on (Bayer)	Pour on	Cyfluthrin
					Topline™ (Merial)	Pour on	Fipronil
Haematobia irritans exigua (Buffalo fly)	Ruminants, Horse		Blood loss, irritation	Northern Australia, New Guinea, Asia			
Haematobia stimulans (Big Meadow fly)	Ruminants, Horse	Horse: Infectious anaemia	Blood loss, irritation	Europe, Asia, North America	Bayofly™ Pour-on (Bayer)	Pour on	Cyfluthrin

Table 1. (continued) Muscid Flies and Control Measurements.

Parasite	Host	Vector for	Symptoms	Country	Therapy		
					Products	Application	Compounds
Hydrotaea irritans (Head or Plantation fly)	Ruminants, Horse	*Corynebacterium pyogenes* (Summer mastitis); Horse: Infectious anaemia	Blood loss, irritation	Northern Europe (Denmark, Great Britain)	Bayofly™ Pour-on (Bayer)	Pour on	Cyfluthrin
Hydrotaea albipuncta	Horse	Horse: Infectious anaemia	Blood loss, irritation				
Glossina spp. (Tsetse fly)	Animals, Man	*Trypanosoma* spp., "Sleeping sickness"; Nagana (*Trypanosoma vivax vivax* und *Trypanosoma congolense congolense*) in cattle	Blood loss, irritation	Africa			

Table 1. Flies causing myiasis in animals.

Parasite	Host	Symptoms	Country	Therapy		
				Products	Application	Compounds
Lucilia sericata (Blow fly)	Sheep, (Pig)	Blowfly-strike; eggs in wounds, larvae move around, destroy skin; skin inflammation, strong secretion, bact. sec. inf.; large economic loss	Great sheep reproduction countries: Great Britain, Australia, New Zealand, South Africa	Clik™ (Novartis) Zapp™ (Bayer)	Spray on Pour on	Dicyclanil Triflumuron
Lucilia cuprina (Blow fly)	Sheep, (Pig)		Australia, South Africa	(mechanical remove, wound desinfection)		
Chrysomya chloropyga	Sheep		South Africa			
Chrysomya bezziana (Old World screw worm, Oriental fly or Bezzi's blow fly)	Cattle, Sheep	Screw worm disease; blowfly-strike; eggs in wounds, larvae move around, destroy skin; skin inflammation, strong secretion, bact. sec. inf.; big economic loss	Tropic and subtropic areas; screw worm disease (in Africa and Southeast Asia)			
Sarcophaga spp. (Flesh flies)	Ruminants	Blowfly-strike; eggs in wounds, larvae move around, destroy skin; skin inflammation, strong secretion, bact. sec. inf.; big economic loss	Temperate areas			
Wohlfahrtia spp. (Flesh flies)	Cattle		Africa, Asia			

Related Entries

→ Insects (Vol. 1), → Skin Diseases, Animals.

Myiasis, Man

Myiasis is an infection with various fly larvae (→ Diptera (Vol. 1)). Some of these are of species with an obligatory life cycle stage in man or animals (e.g. warble fly *Hypoderma*; human botfly → *Dermatobia hominis* (Vol. 1)). The eggs are deposited either into open wounds, the nose, the ear, scalp or on normal skin. The larvae burrow into the skin and become surrounded by a microabscess 2–3 cm in diameter with acute and chronic inflammatory cells, including eosinophils and granulation tissue surrounded by fibrosis. The mature larvae escapes from the abscess to pupate in the soil, with the lesion healing slowly. A second type of myiasis is produced by opportunistic fly species giving rise to similar, but less persistent lesions. Aseptically reared fly maggots used to be employed to clean necrotic debris in chronic osteomyelitis during the preantibiotic era (e.g. *Lucilia serricata*). Microscopically the fly larvae can be distinguished by the presence of segmentation, a striated musculature, a tracheal system composed of rings, leading to two species-specific posterior stigmata.

The specimens of 80 species of fly larvae are able to enter the body of living and dead humans. According to the place of parasitism the following types of myiasis are differentiated:

- **Intestinal myiasis:** 15 families of Diptera have been found in human intestine apparently on passage, however some parasitize in the region of the rectum.
- **Urogenital myiasis:** larvae of the fly families Muscidae, Sarcophagidae and Calliphoridae as well as mosquitoes of the families Anisopodidae and Scenobinidae are found here.
- **Nasal-pharyngeal myiasis:** 8 families of flies are described which are able to enter the eyes, too (e.g. *Oestrus ovis* into the nose and within the eye).
- **Dermal and subdermal myiasis:** This type is most common, since the eggs or larvae are placed onto wounds or may even enter healthy skin regions. Some larvae (e.g. Hypodermatidae) are able to wander around (→ creeping eruptions (Vol. 1))
  - **Africa:** *Cordylobia anthropophaga* (= Tumbu fly); eggs are laid on sand.
  - **America:** *Cochliomyia* (syn. *Callitroga hominivorax*) (Screw worm); eggs are laid in wounds, *Wohlfahrtia* species attack wounds and healthy skin.
  - *Dermatobia hominis* (only America); eggs are placed on bloodsucking insects which transmit them.
  - *Auchmeromyia luteola* (Africa); larvae suck at night on humans (Congo floor maggot).

These larvae introduce furuncle-like skin swellings which mostly are superinfected by bacteria.

Therapy

Surgical withdrawal of the larvae and antiseptic treatment of the regions. *Lucilia serricata* larvae are used for wound cleaning, since they feed exclusively on necrotic tissues and do not touch healthy ones.

Myocarditis

Clinical symptoms in case of infection with e.g. *Entamoeba histolytica*, *Trypanosoma cruzi*, *Toxoplasma gondii*, *Echinococcus granulosus* or *Dirofilaria immitis*.

Myxosporidiacidal Drugs

The efficacy of **fumagillin** against different species of fish-parasitizing myxosporidians (i.e. *Sphaerospora oenieola*, *Myxidium giardi*, and *Hoferellus carassii*) has been known for a number of years. The deleterious effects of **toltrazuril** (a symmetric triazine) and of an asymmetric triazine (HOE 092V) on developmental stages of gill parasitic *Myxobolus* sp., *Henneguya* sp., and *H. laterocapsulata* have been clearly demonstrated in ultrastructural investigations.

In laboratory trials, **quinine** was found to act on *Myxobolus cerebralis* in rainbow trouts (*Oncorhynchus mykiss*) and in addition, against a gill parasitic *Henneguya* sp. in the tapir fish, *Gnathonemus petersii*. Actually, there is no information on the specific mode of action of the actinomyxosporean chemotherapeutics mentioned above.

:ntries
ts (Vol. 1), → Skin Diseases, Animals.

N

Naegleriasis

Naegleria fowleri, a free-living amoeba is found in lakes, especially warm ones, and in swimming pools (→ amoebae (Vol. 1)). It infects healthy young people. In a small percentage of those exposed it invades the nasopharynx and reaches the brain where it gives rise to acute meningoencephalitis with trophozoites but without cysts. Because of the fast multiplication of the amoebae, clinical infection usually leads to death in a few days. Large numbers of amoebae are present in the subarachnoid space, penetrating into the underlying cortex, but there is little (neutrophilic or monocytic) or no inflammatory reaction (→ Pathology/Fig. 4). Mobile amoebae are often found in the cerebrospinal fluid. Uncal herniation is the usual cause of death. A thick "exudate," mostly amoebae, covers the brain and spinal cord and is most apparent over the sulci, major fissures, and basal cisterns.

Main clinical symptoms: Meningoencephalitis (= primary amoebic meningoencephalitis = PAME), often leading to death within days.
Incubation period: 1–3 days
Prepatent period: 12–14 days
Patent period: 3 weeks (if infection is survived)
Diagnosis: Culture techniques, immunohistological methods
Prophylaxis: Avoid bathing in eutrophic lakes
Therapy: see → Treatment of Opportunistic Agents

Nagana

Synonym

African Trypanosomiasis, Sleeping Sickness of Animals

General Information

Nagana is a very important disease of domestic livestock. According to the Food and Agriculture Organization of the United Nations (FAO), it is probably the only disease which has profoundly affected the settlement and economic development of a major part of a continent. Today, it is still endemic in more than 35 African countries and causes huge economic losses (→ *Trypanosoma* (Vol. 1)).

Ruminants

In cattle the pathogenesis is dominated by three features: anemia, tissue lesions and immunosuppression. The cause of anemia is complex and involves a variety of mechanisms. Although hemolysins are released by trypanosomes, intravascular haemolysis is not a prominent feature, and anemia is rather attributed to erythrophagocytosis by cells of the mononuclear phagocytic system in the spleen, bone marrow, lungs and lymph nodes. These cells are stimulated by the formation of complexes between immunoglobulin specific for trypanosomes and antigen or complements attached to red cells. Other possible contributing factors include increased haemodilution and fragility of the red cells, and a depression of erythropoiesis. Although *T. congolense* and *T. vivax* are mainly intravascular parasites they cause significant tissue lesions, notably myocarditis and myositis. The aetiology of these lesions is unknown but is probably related to the damage induced by parasite products, immune complexes and vasoactive amines to capillary endothelial cells. Finally, chronically infected animals show immunosuppression which, in association with other factors of stress such as malnutrition, pregnancy or lactation, leads to a higher susceptibility to other diseases. African trypanosomosis may follow an acute course, mainly in exotic breeds of cattle which tend to be more susceptible than local breeds. Animals suffer from intermittent fever, quickly lose weight, and may die within 3 to 4 weeks. However, the disease more frequently follows a relatively chronic course characterized by intermittent fever, anemia, lymphadenopathy and progressive emaciation. Animals which have been infected for many months or even years become cachectic, their precrural and prescapular lymph nodes being visible from a distance. African trypanosomosis is usually a herd problem. It reduces the general herd productivity and affects fertility.

T. brucei has more affinity for tissues than for blood, and may cause severe lesions in the tissues it invades. The myocardium is more commonly affected, with degenerative changes and focal necrosis of myocytes, and fibrosis. Lesions due to longstanding infection have also been observed in the pituitary, adrenals, kidneys and gonades. *T. brucei* is generally considered as being of little clinical importance in cattle, but may be responsible for acute and chronic infections in goats and sheep. Mixed trypanosome infections are very common in endemic areas.

Horses

Horses are very susceptible to trypanosomosis. *T. brucei* is certainly the most pathogenic, while *T. congolense* and *T. vivax* produce diseases similar to those seen in cattle. The earliest signs of infection are a stumbling gait, a harsh hair coat and relapsing fever. As the disease progresses, subcutaneous oedema of the limbs, thorax, abdomen and genitalia appears. Anemia is a constant feature and lymphadenitis is usually present. Keratitis and corneal opacity may develop in horses affected by *T. brucei.*

Pigs

Pigs are refractory to infection with *T. vivax*, and are only mildly affected by *T. congolense*, *T. brucei* and *T. suis*. In contrast, they are highly susceptible to infection with *T. simiae*. The latter is highly virulent and may cause death in a few days.

Dogs

African dog breeds are very resistant to most species of trypanosomes. Only *T. brucei* appears to be highly pathogenic and often produces acute disease. Clinical signs include anemia, weakness, loss of weight and development of subcutaneous oedema. Parasitic invasion of the eyes causes inflammatory reactions with reactions of pain and lacrimation. Invasion of the central nervous system with ataxia and paralysis has been reported.

Therapy

→ Trypanocidal Drugs, Animals, → Leishmaniacidal Drugs.

Nairobi Sheep Disease

The Nairobi sheep disease (NSD) virus is transmitted by the tick species *Rhipicephalus appendiculatus* and causes severe losses in sheep in East Africa. The causative virus is passed transovarially by the female tick to the larvae, where it can survive for over 3 months.

Natural Resistance

Some diseases or even variations of the normal physiologic status of some people may have benefits during some parasitic infections, since these persons may become naturally resistant. E.g. *Plasmodium vivax* and *P. knowlesi* merozoites are unable to enter red blood cells lacking the Duffy blood group antigens or *P. falciparum* may not develop in red blood cells of persons suffering from → sickle cell anaemia, from alpha- or beta-thallasaemias or from G6PD deficiency. Thus these negative haemoglobin variations have apparently been maintained in endemic malaria regions due to the selection pressure of the parasitic disease.

Necatoriasis

→ Hookworms (Vol. 1).

Necrosis

→ Pathology.

Nematocidal Drugs, Animals

Chemical Classes of Compounds

Phenothiazine and Piperazines Phenothiazine was the first 'broad-spectrum' anthelmintic agent brought into general use at the end of the 1930s. Structure-activity studies created no useful analogues. In the following years it was extensively used in livestock against a fairly wide range of gastrointestinal nematodes. However, toxicity limits its use to ruminants, horses, and chickens and prevented its use in pigs, dogs, cats, and humans.

For 50 years (discovery of its anthelmintic action in 1949), **piperazine (PPZ)** chemically, diethylenediamine) has been in use as an inexpensive and popular anthelmintic in particular for the treatment of *Ascaris* and *Enterobius* (*Oxyuris*) infections in humans (→ Nematocidal Drugs, Man/Table 1) and animals (Table 3, Table 4, Table 5). Numerous substituted PPZ derivatives have been synthesized and exhibit anthelmintic activity, but apart from diethylcarbamazine none has found a place in animal and human therapeutics. The instability of the PPZ base in the presence of moisture (PPZ hexahydrate: very unstable) is absent in other salts (PPZ adipate, chloride, dihydrochloride, citrate, phosphate and sulfate, all soluble in water). The amount of PPZ base and that of salt moiety differs among the compounds, and hence, doses of compounds to be effective on a same level too. In veterinary practice, the anthelmintic spectrum of PPZ is good for ascarid and nodular worm infections of all species of domestic animals, moderate for pinworm, and variable to zero for other helminths. PPZ compounds has a wide safety index in all animals.

Diethylcarbamazine (DEC), chemical: N, N-Diethyl-4-methyl-1-piperazinecarboxamide, has a high action on microfilariae of *Wuchereria bancrofti*, *Brugia* spp. (lymphatic filariasis of humans), and microfilariae causing onchocerciasis in humans (→ Nematocidal Drugs, Man/Table 1) as well as against microfilariae of *Dirofilaria immitis* producing heartworm disease in dogs (Table 5). DEC has been used also for treatment of lungworm infections caused by *Dictyocaulus viviparus* in cattle (Table 6).

DEC produces alterations in the microfilarial surface membranes, thereby rendering them more susceptible to damage by host immune mechanisms. Massive destruction of the parasites can result directly or indirectly in severe adverse reactions if dose regimen scheme is inadequate. DEC is very useful as a prophylactic treatment for heartworm disease of dogs. It acts not only on infective larvae from the vector mosquito but also against microfilariae residing in the blood of host. This is also true for the prophylactic action of DEC to control → lymphatic filariasis in humans; severe adverse reaction being effectively reduced by a standard treatment scheme (WHO 1992). However, in dogs that are microfilariae-positive at the time of drug administration shock type reaction may occur infrequently and erraticly (sometimes fatal). Therefore, use of DEC is contraindicated in microfilariae-positive dogs.

The predominant effect of PPZ on *Ascaris* is to produce a flaccid paralysis, which results in expulsion of the worm by peristalsis. As in other cholinergic compounds the anthelmintic action of PPZ and DEC may depend upon activation of a GABA-gated Cl^- channel on muscle membrane and/or upon nonspecific blockage of ACh receptors.

Benzimidazole Compounds Subsequent modification to the benzimidazole (BZ) molecular structure in the 1960s and 1970s created improved compounds that were safe and had a wide spectrum of activity (Table 1, Table 3, Table 4, Table 5, Table 6 and → Nematocidal Drugs, Man/Table 1). After the discovery of **thiabendazole** in 1961 (still used in animals and humans), several thousand of BZs for screening for anthelmintic activity have been synthesized by pharmaceutical companies (work is documented in patent literature only) but less than twenty of them have been used commercially (Table 3, Table 4, Table 5, Table 6 and → Nematocidal Drugs, Man/Table 1). BZ compounds in general, and **BZ carbamates** in particular, are crystalline materials with relatively high melting points and are almost insoluble in water. BZ prodrugs include several compounds (e.g., netobimin febantel, thiophanate) that possess little or no anthelmintic activity by themselves, but are designed to undergo either relatively simple (benomyl → carbendazim) or a complex series (netobimin → albendazole) of enzymatic and/or nonenzymatic reactions in the organism to form the active drug. Prodrugs increase the water solubility and therefore the absorption, which renders them suitable for use against systemic infections. In contrast to prodrugs, BZs are more frequently used for intestinal and gastrointestinal nematodes and particularly in veterinary practice because of their broad anthelmintic spectrum and low toxicity. In human practice, only three BZ compounds, albendazole, flubendazole and mebendazole, are currently in use (cf. → Nematocidal Drugs, Man/Table 1 and human hydatid disease: → Cestodocidal Drugs). The low aqueous solubility of BZs requires their formulations as oral suspensions or other oral formulations that deposit the drug directly and wholly within the intestinal tract of humans, or within the rumen of cattle sheep, goats or other ruminants. In the latter animals, the residence time of the drug-digesta complex is shortened if the dose should bypass the rumen due to esophageal groove closure and a proportion of the dose being directed to the abomasum. This physiological phenomenon contributes to treatment failure. Thus drug must be entirely administered over the tongue to reduce esophageal-groove effects and maximize the reservoir action of the drug in the rumen. Time is a crucial element of BZ action and is dependent on the kinetics of the tubulin BZ interaction and parasite expulsion. If the mechanism(s) of removal of the parasite by the host requires a longer period than the residence time of the anthelmintic drug, then selection for drug-resistant nematodes may emerge. Two major enzyme systems of the liver, the cytochrome P 450 family and the microsomal flavin monooxygenases are primarily responsible for the biotransformation of BZs. These processes transform the lipophilic xenobiotic compounds into more polar hydrophilic products that can be easily eliminated. The mode of action of BZs can be directly linked to various interactions of BZs with tubulin. The various aspects of the drug-parasite interaction include structure-activity relationships, species selectivity, drug resistance on a basis of chemical/pharmacological studies and studies on a genetic basis. Benzimidazoles not considered, include **cambendazole** (Merck: Bovicam, Camvet no longer marketed on a wide scale), **cyclobendazole, dribendazole,** and **epibendazole** (do not appear to have been launched, and even though not in major markets such as **luxabendazole;** its high anthelmintic activity is shown in Table 1). Possibly, these drugs had shown side-resistance to BZs on the market, and eventually embryotoxicity, which might have limited their use in livestock, especially in ruminants.

Levamisole, Pyrantel, Morantel Levamisole (**LEV**), an imidazothiazole, is the **S (–) isomer** of **tetramisole;** the latter drug was introduced as an anthelmintic in 1966. Following marketing of racemic tetramisole, it was found that antinematodal action of the racemate based almost solely on the S (–) isomer; as a result of the separation of the enantiomers, the dose could be halved for the S (–) isomer. Levamisole is a highly accepted and widely used antinematodal drug in veterinary practice (Table 1, Table 3, Table 4, Table 5, Table 6). It is also a good drug for the treatment and control of *Ascaris* infections in humans. LEV has besides its antinematodal effect, **immunomodulatory actions,** which have been demonstrated in animals and humans (e.g. cancer patients). It has been shown to enhance immune responsiveness by stimulating the activity of T-lymphocytes and 'correcting' immunological imbalance. Thus the drug may potentiate the rate of T-lymphocyte differentiation, and hence, the promotion and maturation of precursor T-cells into fully functional lymphocytes, which increase the response to antigens and mitogens. The drug induces spastic contraction of worms and then paralysis of nematodes. Several nematode ion channels regulated by neurotransmitters are targets for anthelmintics. A nicotinic acetylcholine receptor (= ACh = primary excitatory transmitter in nematodes) on nematode muscle cells is associated with a cation channel sensitive to LEV.

Like LEV, the related tetrahydropyrimidines **pyrantel** (PYR) and **morantel** (MOR) are cholinergic agonists with a selective pharmacology for nematode receptors. **PYR** (salts: tartrate and pamoate) was introduced as a broad-spectrum anthelmintic in 1966 for use in sheep and has subsequently come to be used in cattle, swine, horse, dogs and also in humans (Table 1, Table 3, Table 4, Table 5, Table 6 and → Nematocidal Drugs, Man/Table 1). **MOR**, the methyl ester analogue of pyrantel, has been developed for anthelmintic use in sheep and cattle. The salts of MOR (tartrate, fumarate and citrate) have a greater activity against gastrointestinal nematodes than the parent compound while their pharmacologic effects are similar. Like other anthelmintics (diethylcarbamazine, a piperazine derivative, cf. Phenothiazine and Piperazines and → Nematocidal Drugs, Man/Table 1), LEV, PYR and MOR can produce nicotine-like paralytic actions in animals that are shared with acetylcholine (ACh) and act by mimicking effects of excessive amounts of this natural neurotransmitter. However, excess amounts of ACh may result in inhibition of autonomic ganglia, chemoreceptors of the carotid and aortic bodies as well as adrenal medullas and the neuromuscular junction. In severely debilitated animals these pharmacologic effects appear to be enhanced (contraindication for use of LEV, PYR and MOR). LEV resistance in trichostrongylids is complex and partly polygenic and in *T. colubriformis* it is mainly ascribed to a single recessive gene, or closely linked group of genes, located on the X-chromosome but not so in case of *Haemonchus*.

Avermectins and Milbemycins Avermectin and milbemycin macrocyclic lactones (Table 1, Table 3, Table 4, Table 5, Table 6 and → Nematocidal Drugs, Man/Table 1), introduced into the antinematodal market in the 1980s, are structurally related and exhibit endectocide activities for prolonged periods at extremely low doses when administered parenterally. Macrolide 'endectocides', as their name implies, may kill both internal (nematodes) and external (arthropods) parasites by opening chloride channels. **Ivermectin** (a mixture of 80% 22,23-dihydroavermectin B_{1a} and 20% B_{1b}) is a semi-synthetic derivative widely used in veterinary medicine as a broad-spectrum endectocide and in human practice for controlling onchocerciasis and lymphatic filariasis (→ Nematocidal Drugs, Man/Table 1). **Abamectin,** avermectin B_1, (natural precursor of ivermectin having a double bond at C22–23 position) is used as antinematodal drug in cattle and as a foliar spray on various plants against diverse agricultural pests (arthropods). **Doramectin** (25-cyclohexyl-avermectin B_1) has a close structural similarity to avermectin B_1. Presumably it is the lipophilic cyclohexyl moiety that causes a fairly long tissue half-life of the drug and thus a high nematocidal and broad spectrum of activity against cattle nematodes. **Eprinomectin** (MK-397) consisting of a (90:10) mixture of two homologues, 4"-epi-acetylamino-4"-deoxyavermectin B_{1a} and B_{1b}, and **selamectin**, a semi-synthetic monosaccharide oxime derivative of doramectin, are the latest members of the avermectin subfamily selected for development as topical endectocides for use in cattle and cats and dogs, respectively. Modifications, especially on the 4" position of avermectin B_1 and ivermectin, that is the introduction of amino groups at this position, revealed optimal compatibility with high bio-activities whereas variations at most other chemically accessible positions often result in less active compounds. A singular feature of eprinomectin among the macrocyclic lactones appears to be a zero withdrawal time for milk in lactating cows whereas selamectin has a wide margin of safety for all dog breeds.

The related milbemycins produced by *Streptomyces* sp. share a common carbon backbone with the avermectins but lack the glycones (structurally equivalent to 13-deoxy-ivermectin aglycones). Semi-synthetic **moxidectin** (23-methoxime LL-F28249α milbemycin), a derivative of naturally occurring nemadectin (LL-F28249α fermentation product from *Streptomyces cyaneogriseus noncyanogenus*) and structurally similar to the milbemycins, differs from them in having an unsaturated C-25 chain. Moxidectin has the same wide spectrum of activity as avermectins but it should be less toxic than abamectin against the dung beetle *Onthophagus gazella*. Thus concerns about possible adverse environmental impact of the avermectins might be less applicable in the case of moxidectin.

Unlike the avermectins, the nematocidal activity of milbemycins is more potent against intestinal nematodes than against heartworm. **Milbemycin oxime** (semi-synthetic derivative of milbemycin A_3/A_4, narrow-spectrum compound) has been developed for strategic control of both *Dirofilaria immitis* (heartworm) and *Ancylostoma caninum* in dogs (Table 5).

Both classes of macrocyclic lactone compounds obviously have a similar mode of action, i.e., they mediate their nematocidal effect via interaction with a common receptor molecule. They open Cl ion channels though to be associated with glutamate-gated ion channels of muscles of the pharynx and probably the somatic musculature. As a result of this interaction worms become paralyzed and starve to death. Several dose (titration), response studies with low dose rates of ivermectin and moxidectin against several species of sheep trichostrongylids point to the fact that side resistance between these drugs is present. It is suggested that the two drugs act at the same site. However, the extent of lipophilicity may influence the relative efficacy of macrolactones in female and male worms. Males with a larger body size than females were more susceptible in an isolate of *Haemonchus contortus* resistant to ivermectin. Their larger body size and the potential for sequestering a lipophilic compound like ivermectin were believed to promote anthelmintic activity.

Narrow-Spectrum Drugs for Use against Drug Resistant Nematodes Substituted salicylanilides and phenols can be used effectively to control some nematode species of ruminants, which have developed resistance to the broad-spectrum anthelmintics. Narrow-spectrum anthelmintics as substituted **salicylanilides** and phenols are anticestodal (→ Cestodocidal Drugs/Table 1) or antitrematodal compounds (→ Trematodocidal Drugs/Table 1). However, some of these compounds, e.g. **closantel** and **rafoxanide** having marked activity against liver flukes, can be used against multiple resistant strains of bloodsucking *H. contortus*, a highly pathogenic nematode of small ruminants (Table 1). If used at appropriate times, taking epizootiology into account, these relatively long-acting drugs can reduce the selection pressure for resistance to the broad-spectrum compounds in *H. contortus* and *Trichostrongylus* spp., thus reducing the contamination of pastures with these species for the rest of the season. For this reason, closantel may be used in strategic treatment programs for sheep and lambs in which the number of treatments with broad-spectrum anthelmintics is kept to a minimum. For example, a broad-spectrum anthelmintic and closantel can be co-administered in sheep in the first two treatments in the new grazing season, followed by closantel. Thus the 'Wormkill' program in Australia demonstrated local eradication of *H. contortus*. However, due to emergence of resistance to closantel and related compounds, this and other programs are increasingly jeopardized. Rafoxanide, closantel, **nitroxynil** and **disophenol** (Table 1) are uncouplers of oxidative phosphorylation in mammalian mitochondria. Some of these compounds are fairly toxic and are detoxified in the host by binding the absorbed drug to plasma proteins. Consequently, drugs at therapeutic dose do not affect host's mitochondria. Therefore, although they may uncouple roundworm mitochondria at low concentration in-vitro, these drugs are initially inactive against the parasite until blood enriched with the drug is sucked in and digested by the parasite, thereby separating drug from plasma albumin. The other possibility is that the bound drug is separated from the plasma in the liver and is excreted in the bile where it contacts and affects parasites residing in the bile ducts or elsewhere.

Organophosphorus compounds had their origin as pesticides and have subsequently been introduced as anthelmintics into veterinary practice. While haloxon, coumaphos, or naphthalophos are preferably used against parasitic infections in ruminants (Table 1), dichlorvos and trichlorfon are mainly used against parasites of horses (Table 3), pigs (Table 4) or dogs (Table 5). In ruminants the anthelmintic efficacy of organophosphates is somewhat restricted, i.e. only parasites of the abomasum (especially *Haemonchus*) are satisfactorily affected whereas nematodes of the bowel are somewhat refractory to a single treatment. To prevent development of drug resistance it is advisable to alternate these compounds with anthelmintics of other chemical classes. When *Trichostrongylus* spp. in sheep are likely to cause resistance problems to both the benzimidazoles and levamisole/morantel, an organophosphorus compound can be used to provide sufficient parasite control. Dichlorvos and haloxon exhibit satisfactory efficacy against small and large strongyles of horses, while trichlorfon (= trichlorphene or metrifonate is converted to dichlorvos at physiologic pH) in combination with any benzimidazole, BZ prodrugs (e.g. febantel) or morantel may provide high activity against *Gasterophilus* (Table 3).

The main effect of organophosphates on worms is ascribed to inhibition of nematodal acetylcholinesterase (AChE). The degree of safety of these compounds for the host is probably related to the host's AChE specificity for a certain drug, and hence, to the 'stability' of the formed drug/AChE complex, which may be reversible or not. The 'affinity' or susceptibility of the host AChE to the organophosphorus compound should be weak, i.e. the formed drug/AChE complex should be limited in time. Conversely it is desirable if nematodal AChE forms an irreversible complex with the organophosphate, as in the case of *Haemonchus* AChE with haloxon. However, the selective toxicity of AChE inhibiting drug to various species of nematode AChE may vary and thus a degree of parasitic drug action resulted. The absence of AChE leads to accumulation of acetylcholine of the parasite and produces disorders of parasite neuromuscular system resulting in paralysis and expulsion of the worm by peristalsis from the host gut. Therapeutic indices of organophosphates are generally smaller than those indices of the broad-spectrum drugs. This should be considered if a higher than the recommended dose is given.

Higher doses can produce illegal residues in milk and edible tissues as well as toxicity in animals (frequent defecation and urination, vomiting, salivation and muscular weakness). To be on the safe side, withdrawal time should be at least 7 days before slaughter. Due to possible cumulating effects, concurrent administration of organophosphates and other AChE-inhibiting drugs such as pesticides (organophosphorus and carbamate insecticides) or muscle relaxants should be avoided; the same is true for the use of organophosphates within 4 weeks of parturition, especially in the equine. Use of insecticides should be restricted to pyrethroids and related compounds or to rotenone and chlorinated hydrocarbons.

Experimental Compounds During the last decade some fermentation products derived from fungus-like microorganisms such as aerobic bacteria *Streptomyces* sp. and fungi have shown promising anthelmintic activities, which may be useful as a tool for identifying a novel molecular target for anthelmintic discovery. Some of these compounds like paraherquamide and PF1022A causing flaccid paralysis may have a specific target within the neuromusculature.

Dioxapyrrolomycin derived from *Streptomyces* sp. has been shown to be effective against *Haemonchus contortus* in sheep, (cleared >99% at 3,1 mg/kg orally); it proved moderately active against *Trichostrongylus colubriformis*. In an in vitro migration assay it was ~6 times less active against closantel-resistant *H. contortus* than susceptible worms. Therefore dioxapyrrolomycin appears to be a narrow-spectrum anthelmintic with a closantel-like mode of action.

Paraherquamide, an oxindole alkaloid of fungal origin (*Penicillium paraherquei* and *Penicillium charlesii*) has been shown to exhibit promising activity against important pathogen nematodes of ruminants. In sheep, a single oral dose of 0.5 mg/kg cleared 98% of *H. contortus* (resistant to ivermectin), *Ostertagia circumcincta*, *T. axei*, *T. colubriformis* (resistant to ivermectin and benzimidazoles) and *Cooperia curticei*. However, this dose proved to be not significantly active against adult *Oesophagostomum* sp. and only variably active against inhibited 4th stage of *O. circumcincta*. In cattle, paraherquamide exhibited action (at least 95%) on roundworms in the abomasum, small intestine, large intestine and lungs at 1 mg/kg orally (*C. punctata* at 4 mg/kg only). In sheep (10 mg/kg) and calves (4 mg/kg) the compound was well tolerated. In dogs, there was an insignificant effect against *Ancylostoma caninum*, *Uncinaria stenocephala*, *Toxascaris leonina* and *Trichuris vulpis* and a significant one against *Strongyloides stercoralis* at 2 mg/kg. However, a dose of 0.5 mg/kg already proved slightly toxic to dogs and a dose of 10mg/kg was quickly lethal.

PF1022A, a cyclodepsipeptide derived from a pool of Mycelia Sterilia PF1022, and insoluble in water, was found to be effective against *Ascaridia galli* in chickens (cleared 91% of the worms at 2 mg/kg orally). It affected *Toxocara canis* and *T. cati* at 0.5 mg/kg, and in cattle *H. contortus* and *O. ostertagia* (dose not specified), and in sheep *H. contortus* at 5 or 10 mg/kg. It induced a flaccid paralysis in *H. contortus*. In mice, PF1022A appears to be safe (no acute signs of toxicity at 1g/kg intraperitoneally or 2 g/kg orally). In vitro and in vivo studies (mice) have shown synergistic action of a new cyclic depsipeptide (BAY 44-4400, a derivative of PF1022A) and piperazine on nematodes.

A novel macrodiolide, **Clonostachydiol,** derived from *Clonostachys cylindrospora*, strain FH-A 6607, and insoluble in water, appears to have a significant activity against *H. contortus* in sheep following a single dose of 2.5 mg/kg subcutaneously (80–90% reduction from pre-treatment egg count).

Chemoprophylaxis and Effects on Protective Immunity against Nematodes and Lungworms Intraruminal boluses (Table 1) are designed to release nematocidal concentrations of an anthelmintic in the reticulo-rumen of cattle in order to kill ingested infective larvae of GI nematodes and those of the lungworm *Dictyocaulus viviparus* for prolonged periods. Some intraruminal devices may release nematocidal concentrations for up to 20 weeks and also other formulations (e.g. for parenteral injection or pour-on) of macrocyclic lactones (Table 1) may protect animals against infections of gut roundworms and lungworms for several weeks postdosing. Thus, a single treatment with an intraruminal device at turnout may ensure protection against parasitic gastroenteritis for the whole grazing season in temperate regions. The suppression of the output of eggs in the early part of grazing season ensures safe pastures for the remainder of the year. There are two strategic regimens, which have been particularly successful in achieving this: repeated treatments with avermectins given 3, 8 and 13 weeks after turnout or the use of intraruminal anthelmintic devices given at turnout.

The control of parasitic bronchitis is more difficult because the epizootiology of *Dictyocaulus viviparus* is complex. Therefore the strategic control is not fully effective against bovine lungworms. During the time of drug release, lungworm infections are prevented; however, thereafter (a rough formula considered approximately 50 days after 'burnout' of intraruminal bolus as critical period) when drug release is exhausted, infections with clinical signs and even cases of fatal parasitic bronchitis may occur. Reinfections are principally due to external sources, e.g. imported infections from other pastures via vectors or game animals. Under natural conditions, immunity to *D. viviparus* is generated much more rapidly than to GI nematodes. An attenuated vaccine (Dictol) has been available for many years but since benzimidazoles (e.g. oxfendazole and fenbendazole) and various avermectins have potent activity against *D. viviparus*, their strategic use for lungworm control has constantly been explored. The occurrence of hypobiotic lungworm infections during the housing period will effectively stimulate immunity build-up. To what extent sporadic lungworm challenges to animals during chemoprophylaxis are capable of contributing to the development of protective immunity without producing hypobiotic infections is still unknown.

The strategic use of anthelmintics such as intraruminal anthelmintic boluses and other long-acting formulations, especially those of the avermectins, raised the question as to whether drug-protected animals are exposed to sufficient antigenic challenge to develop acquired immunity. In general, only first-season grazing calves have to be treated against GI nematodes. During this period substantial reductions in the exposure of calves to infection is normally evident, and it has been shown that not all animals may acquire satisfactory ('functional') immunity to prevent clinical disease in their second grazing season. Thus, elevated pepsinogen levels and heavy worm burdens (e.g. inhibited *Ostertagia ostertagi* L_4 larvae) have been observed during the second grazing season in yearling heifers treated with **ivermectin** at 3, 8 and 13 weeks after turnout in their first grazing period. An 'overprotection' of first-season grazing animals by too strong chemoprophylaxis may result in impaired immunity and therefore to production losses in the second grazing season. It has been shown that ivermectin (possibly other anthelmintics too) may have some direct or indirect immunosuppressive effects in sheep. Lymphocytes from ivermectin drenched lambs had decreased blastogenic activity compared with lymphocytes from control lambs.

As a consequence, prevention of nematode infections in first year grazing calves should be a careful balance between prevention of production loss and support of immunity build up through mild infections sufficient enough to ensure protection against heavy infections during the second grazing season. Current control of GI nematodes in cattle is based on preventive methods (strategic dosing schedules or intraruminal boluses) that provide excellent results in preventing production loss in first-season grazing cattle but have evolved towards abolishing parasite contact with these animals and so induction of a solid immunity. Less use of anthelmintics by reducing the number of treatments and, hence, prolonged drug-free periods may result in moderate pasture infection and build-up of protective immunity against GI nematodes in yearling heifers during the second grazing season.

Gastrointestinal Nematode Infections of Cattle, Sheep and Goats, Timing of Strategic Drug Treatments and Biological Control of Nematode Parasites in Livestock

Ruminants (cattle, sheep, goats) generally become infected by free-living, infective third-stage larvae (L_3) entering the host by oral ingestion (e.g., *Ostertagia* or (syn.) *Teladorsagia* spp. and other trichostrongyle nematodes) and/or the skin (e.g. *Bunostomum* spp.). A variety of other species and their stages (adults, developing and/or inhibited larvae) may reside in the abomasum (e.g., *Haemonchus* spp., *Trichostrongylus* spp. *Ostertagia* spp.), or in the intestine (e.g., *Trichostrongylus* spp., *Cooperia* spp., *Nematodirus* spp., *Bunostomum* spp. = hookworms, *Strongyloides papillosus* in the small intestine; *Oesophagostomum* spp., *Chabertia ovina* in the large intestine, colon) of sheep, goats, cattle and a number of other ruminants throughout the world. Gastrointestinal nematodes may disturb the normal functions of the gastrointestinal tract. Gut disorders lead to a disease syndrome involving diarrhoea, weight loss, anaemia (loss of blood and plasma proteins), mucoid hyperplasia, disorder of pepsinogen produc-

Table 1. Drugs used against gastrointestinal (GI) nematode infections in ruminants

CHEMICAL GROUP, **Nonproprietary name** (approx. dose, mg/kg body weight, oral route) other information	* **Brand name** (manufacturer, company), other information	Characteristics (chemotherapeutic and adverse effects, miscellaneous comments)
BROAD SPECTRUM ANTHELMINTICS: life cycle of nematodes is direct; eggs hatching with adequate warmth and moisture can accumulate in dry weather leading to heavy infections after rain; young larvae feed on bacteria in the feces, molting to give infective third stage larvae; infective larvae can survive for many months on pasture; ingested larvae normally develop into adults in about 3 weeks while larvae, which were inhibited in autumn, resume development in spring or at time of parturition; a number of measures have been recommended to delay the development of resistance to broad-spectrum anthelmintics in major gastrointestinal (GI) nematodes particularly in small ruminants; the sparing use of appropriate drugs and doses (avoidance of underdosing) will help to kill parasites completely thus preventing the escape of resistance survivors; the use of integrated control systems coordinating anthelmintic treatment with appropriate management strategies and the use of drugs of different chemical classes in a slow, 12 monthly rotation, could be useful measures to reduce parasite numbers on pasture and the frequency of the strategic treatment; some programs, in which sheep and cattle graze in rotation, have been shown to provide more effective parasite control than a continuous grazing program for sheep; these programs may all effectively reduce the need for anthelmintic treatment; the timing of strategic treatment and weather conditions would also play an important role in the integrated control; during dry and hot periods there may be high mortality of free-living infective stages, whereas rainfall in the spring or autumn favors the survival and transmission of free-living larvae; thus, a good knowledge of the epizootiology of the parasites may lead to a sensible approach to the strategic use of narrow-spectrum anthelmintics against multiple resistant GI nematodes especially in sheep and goats.		
phenothiazine (400-600) first broad-spectrum anthelmintic agent used for several years that has disappeared virtually from market	powder drug resistance was found in *H. contortus* and other trichostrongylid nematodes	newer drugs are more effective, safer, and easier to handle than the oldtimer; it has only low efficacy against *Ostertagia, Cooperia, Nematodirus, Bunostomum* spp. and may be toxic at therapeutic dose (photosensitization: keratitis, eczema solare); mode of action covers a wide range, such as antipsychotic effects (interaction with dopamine receptors, calmodulin, microtubules), anthelmintic effects (partial inhibition of acetylcholinesterase, interruption of microtubular functions is likely as with benzimidazoles), and antiprotozoal effects (disruption of cytoskeletal microtubules in *T. brucei* in-vitro)
BENZIMIDAZOLES (BZs): exhibit high activity against important GI nematodes; their efficacy against ruminant whipworms, filarial worms (*Onchocerca, Setaria*), tapeworms, and flukes is limited, however; pharmacokinetics: except thiabendazole, albendazole, oxfendazole, only limited amounts of a dose of any of the BZs are absorbed from the GI tract of the host; thus BZs usually are more effective at low dosage regimen for several days (multiple dosing) than at a singly high dosage; in most tissues of treated animals, residues of BZs approach low levels only; however, residues quantities of [^{14}C] labeled parent compounds and their metabolites are detectable in the liver and other organs (<0.3 µg/g tissue) at 2 weeks following a single dose; as a result of detected radioactivity, withdrawal times of BZs before slaughter are necessary for edible tissues and milk intended for human consumption; metabolism of BZs occurs in the liver; phase I reactions involve hem-associated cytochrome P450 and microsomal flavin monooxygenase (MFMO) system catalyzing reactive groups into organic substrate (hydroxy, carboxy, amino and sulphhydril groups); phase II reactions often occur at site of the new functional groups and enable conjugation of the deactivated molecule to amino acids, carbohydrate, sulfate, bile salts and/or glutathione; species of conjugate may relate to route of elimination; **mode of action and resistance** of BZs may be due to interruption of microtubular function, i.e. tubulin polymerization; drug resistance can be attributed to mutations in the tubulin molecule, and possibly to enhanced active cellular efflux of the drugs.		
thiabendazole (= tiabendazole) (cattle, 66-110) (sheep, 44-66; 88 for arrested larvae)	* Equizole * Thibenzole (MSD Agvet) * Omnizole (Merial) paste, suspension (drench), bolus, premix, powder (in feed)	first drug with broad-spectrum activity against adult gastrointestinal (GI) nematodes; developing trichostrongylid nematodes are not so greatly affected (75%-90%); it has no activity against arrested (inhibited) larval stages in cattle at recommended dose; thiabendazole-resistant nematodes are known and frequent; the drug is well tolerated at higher doses, e.g. 100 mg/kg (88 mg/kg for treatment of nematodiriasis in sheep); the upper dose(cattle 110 mg/kg, and sheep 66 mg/kg) is required for treating successfully lungworm infections

Table 1. (continued)

CHEMICAL GROUP, **Nonproprietary name** (approx. dose, mg/kg body weight, oral route) other information	* **Brand name** (manufacturer, company), other information	Characteristics (chemotherapeutic and adverse effects, miscellaneous comments)
BENZIMIDAZOLECARBAMATES		
parbendazole (cattle, 20-30) (sheep, 20)	* Neminil * Helmatac (Hydro, Smith Kline/Novartis) suspension (drench), bolus; premix, powder (in feed) * Verminum * Worm Guard	broad-spectrum drug with high efficacy against adult GI nematodes (more than 90%); immature stages are moderately affected (60%-90%); the drug has no activity against arrested larvae in cattle and sheep; parbendazole-resistant trichostrongyles (sensu lata) are known; it has also teratogenic effects in lambs (anophthalmia; atresia ani; skeletal deformation); treatment of ewes should not be in the first month of pregnancy; drug is not recommended during laying period of birds (fully active against both ascarids and heterakids: 30mg/kg single dose or 0.05% in-feed x2d)
mebendazole (sheep, 15) the drug is embryotoxic in rats at 10 mg/kg; there is no embryotoxic effect in sheep at recommended dose	* Ovitelmin (Janssen) * Mebendan (Mabo) bolus, suspension (drench) * Telmin * Telmintic * Vermox (Janssen) * Chanazole (Chanelle) * Supaverm(Janssen), combination product containing a fasciolicide	broad-spectrum drug highly effective against adult GI (trichostrongylid) nematodes in sheep and goats with somewhat erratic effect against developing stages (50%-90%); it has only negligible activity against arrested larval stages; the drug is used frequently in treatment of nematode and cestode infections of **zoo animals** such as equines (zebra, tapir), and ruminants (giraffe, antelope, gazelle, elk, deer, camel) and given in feed (630 ppm) for 14 consecutive days (equivalent dose → 1mg/kg/d in equines, → 5mg/kg/d in ruminants); treatment of cestode infections: Pinnipedia and Proboscidae 10mg/kg/d x 2-3d, primates (with fruit) 5-10 mg/kg/d x5d, in Rodentia and Marsupialia 15 mg/kg single dose, carnivores (with fish) and Artiodactyla 15mg/kg/d x2d, *Strongyloides stercoralis*, is sometimes fatal in primates: long-term treatment, total 21 days: (25), (50), (25)mg/kg twice daily each dose regimen for 7d, alternating with 7d rest between each dose regimen; zoo birds: 60 ppm in-feed x7d in chicken, turkeys, guinea fowl, and 120 ppm in-feed x14d in pheasants, partridges, geese, and ducks against *Syngamus trachea* ascarids, heterakids, *Capillaria* sp. (dose limiting parasite)and cestodes (about 100% removal of worms)

Table 1. (continued)

CHEMICAL GROUP, **Nonproprietary name** (approx. dose, mg/kg body weight, oral route) other information	* **Brand name** (manufacturer, company), other information	Characteristics (chemotherapeutic and adverse effects, miscellaneous comments)
fenbendazole (cattle 7.5 or 1 SR bolus / animal) (sheep, 5) *Panacur SR Bolus (contains 12 g FBZ) consisting of 10 flat-faced tablets in two magnesium alloy joined and enclosed by plastic rings; bolus releases FBZ continuously in reticulo-rumen of cattle (body weight 100-300 kg) for up to 140days (20 weeks), thereby controlling GI round worms and lungworm (bronchitis) effectively (dose:1 bolus/animal)	* Panacur (Intervet) granules, powder (in feed), paste, pellets, suspensions (drench), boluses (sheep, goats), intraruminal device (SR bolus,cattle), feed blocks *Axilur (Intervet) * Safegard * Orystor (Serum Werk Bernburg) * Fendazole (Osmond & son) * Wormaway (Deosan) * Zerofen (Chanelle)	first broad-spectrum drug with high efficacy against lungworms (Table 5), *Dictyocaulus viviparus*, and GI nematodes adult (95%-100%), developing (95%-100%), and arrested stages (80%-90%) of gastrointestinal nematodes of cattle and sheep; the drug is not embryotoxic or teratogenic in rats, sheep, and cattle and is well tolerated at recommended and higher doses (safety index is more than 500); administration of SR bolus at the beginning of the grazing season will prevent establishment of patent infections throughout the grazing season; reduced pasture contamination in autumn will lower the risk of inhibited ostertagia larvae accumulating in abomasum to cause winter ostertagiasis (inhibited Ostertagia larvae, L_4, are not killed); the period of bolus administration (140 d) should be reduced if cattle are moved to heavily infected pasture; mode of action see thiabendazole; simultaneous use of fenbendazole (FBZ) with bromsalans within 7 days of each other may cause severe (fatal) adverse effects in cattle; in **zoo animals** (lion tiger, cheetah, panther, puma, leopard, jaguar and bear (black, polar, grizzly), infections with *Toxocara cati, Toxascaris leonina, Ancylostoma* spp. and *Taenia* spp. can effectively be treated with FBZ (10 mg/kg/d for 3 days, granules); wild ruminants infected with *Haemonchus* spp. *Nematodirus* spp. *Trichostrongylus* spp. may be treated with medicated feed (2.5 mg/kg/d for 3 days), also so feral swine infected with *Ascaris suum, Oesophagostomum, Stephanurus* (3 mg/kg/d for 3 days), and wild sheep infected with *Protostrongylus* spp. (10mg/kg/d for 3 days); other zoo and park animals seem to tolerate overdoses of the drug equally well (primates: 20mg/kg/d x5d against acanthocephalan *Prostenorchis* and *Physaloptera*; 30-50 mg/kg/d x2 d or 50-100 mg/kg single dose, suspension or granules in food against GI parasites, spirurids, oxyurids, *Capillaria* of reptiles and amphibians); **zoo birds** (chickens, turkeys, guinea fowl, pheasants, partridges, geese, and ducks), 60 ppm FBZ in-feed x7d result in 100% removal of worms (*Syngamus trachea*, ascarids, heterakids, *Capillaria* sp. (dose limiting parasite) and cestodes
oxfendazole (fenbendazole sulfoxide) (cattle, 4.5) (sheep, 5) intraruminal device *Autoworm is a pulse-release bolus that releases 5 pulses of oxfendazole at 3-weekly intervals	* Synanthic (Norden) * Systamex, * Autoworm 5 Pulse Release Bolus with Systamex * Systamex Plus Fluke (Mallinckrodt) * Bovex (Chanelle) * Parafend (Norbrook) * suspension (drench), granules, powder, paste, others	like fenbendazole (FBZ), a broad-spectrum drug with high efficacy against lungworms (Table 5, Table 6), *Dictyocaulus viviparus*, and gastrointestinal nematodes of cattle and sheep, including adults, developing, and inhibited larvae (for more details see FBZ); a controlled dissolution release device (cylindrical polypropylene capsule containing a core of carrier matrix against an orifice by a tensioned spring) is designed to remain in the rumen and release about 0.5 mg/kg per day, which is highly active against established worm burdens of *Ostertagia* spp. in sheep and cattle; doses of 20 mg/kg may cause embryotoxic effects in sheep; oxfendazole (OFZ)-resistant *H. contortus* are known; simultaneous use of OFZ with bromsalans within 7 days of each other may cause severe (fatal) adverse effects in cattle; FBZ metabolism: initial sulfur oxidation step of FBZ to the sulfoxide (= OFZ) is reversible in sheep and resulted in a 4:1 ratio of sulfoxide to sulfide (FBZ) in plasma; establishment of equilibrium is rapid relative to rates of absorption, excretion or further oxidation to the sulfone; however, total bioavailability of both BZs has been found to be ~40% less when the same sheep were dosed with FBZ compared to OFZ; thus administration of FBZ relative to OFZ should not be considered as equivalent

Table 1. (continued)

CHEMICAL GROUP, **Nonproprietary name** (approx. dose, mg/kg body weight, oral route) other information	* **Brand name** (manufacturer, company), other information	Characteristics (chemotherapeutic and adverse effects, miscellaneous comments)
oxibendazole (cattle, 10) (sheep, 10)	* Neplon 100 * Loditac (Smith Kline) paste, suspension (drench); premix, powder (in feed)	broad-spectrum anthelmintic against GI nematodes; its potency is not as pronounced as that of fenbendazole or oxfendazole; it exhibits high efficacy against adults (more than 90%) but less efficacy against developing stages of trichostrongyles in cattle and sheep; the drug is inactive against arrested larvae; an embryotoxic effect may occur in sheep at 4.5 times the dose recommended
albendazole (cattle, 7.5-10) (sheep, 5) **SRC** contains 3.85 g of albendazole delivering 36.7mg/d for ~105 days max. body weight 70 kg	* Valbazen (Pfizer, Smith Kline Beecham) paste, suspension (drench); granules (in feed) * Proftril Captec (SR capsule=**SRC** for sheep, goats)	broad-spectrum drug with activity against lungworms (Table 5, Table 6), *Dictyocaulus viviparus*, and GI nematodes similar to that of fenbendazole and oxfendazole (see above); some varying results concerning efficacy against inhibited larvae of *O. ostertagi* have been reported (*O. circumcincta* is the dose limiting species, i.e. 0.5 mg/kg when using **SRC**); albendazole-resistant *H. contortus* are known to occur frequently; there is an embryotoxic effect (skeletal abnormalities in lambs) at 2 times the recommended dose that may limit use in pregnant animals; activity against **cestode** and **trematode** parasites see relevant TABLES; albendazole (ABZ) is metabolized to the active sulfoxide (= ricobendazole, see below); sulfoxide-sulfide interconversion is evident (cf. oxfendazole) and equilibrium (ratio of sulfoxide: sulfide) heavily favors the sulfoxide
ricobendazole (albendazole sulfoxide) (cattle, 7.5) (sheep, 5)	* Allverm (Crown) * Bental (C-Vet) * Rycoben (Young's Animal Health	its antinematodal and anticestodal spectrum of activity is similar to that of albendazole.(ABZ, see above); its action on trematodes is uncertain; ABZ sulfoxide and sulfone metabolites dominate plasma profile and are the major metabolites in the urine; ABZ sulfoxide is the pharmacological and embryotoxic active agent whereas the sulfone is inactive and non-toxic
luxabendazole (sheep, 7.5-10) *Luxacur (Intervet) suspension (drench) there are two metabolites which proved inactive against helminths; the majority of the dose (83%) is excreted in the feces	broad-spectrum anthelmintic developed for use in sheep in 1987; introduction of sulfoester arrangements -SO_2-O- into a series of benzimidazole carbamates led to the 4-fluoro derivative HOE 216V; **GI nematodes**: controlled trials carried out in both artificially and naturally infected sheep in Europe, Africa, and Australia demonstrated its broad-spectrum activity; it has proved highly active (95%-100%) against adult and immature stages of the major gastrointestinal nematodes of sheep, e.g. *H. contortus, Ostertagia* spp., *Trichostrongylus* spp.; efficacy against nematodes of lesser importance such as *Trichuris ovis, Oesophagostomum* spp., *Chabertia ovina, Gaigeria pachyscelis*, and *Bunostomum* spp. is 80%-100% and against protostrongylids 99%-100%; **trematodes**: the drug is highly active (95%-100%) against adult stages of *F. hepatica* and *D. dendriticum* but less active (70%-90%) against immature stages of *F. hepatica*; it has shown activity (>90%) against *Moniezia* spp. (tapeworms of ruminants); the drug has proved to be ovicidal (flukes and nematodes); at 10 mg/kg it shows a worm reduction rate of 93% against benzimidazole-resistant *H. contortus*; there is a slight side-resistance to other benzimidazoles; it is well tolerated in sheep (> 75 times the therapeutic dose), and there is no indication of mutagenicity or teratogenicity in rats and sheep at doses 250 times and 10 times higher, respectively, than the recommended (therapeutic) doses; in contrast to other benzimidazole carbamates no oxidation, reduction, or hydroxylation in the lipophilic 5-substituent can occur; predominantly as parent compound (71% of total); withdrawal period for meat is approx. 7 days	

Table 1. (continued)

CHEMICAL GROUP, **Nonproprietary name** (approx. dose, mg/kg body weight, oral route) other information	* **Brand name** (manufacturer, company), other information	Characteristics (chemotherapeutic and adverse effects, miscellaneous comments)
PROBENZIMIDAZOLES Prodrugs of some BZs become active after metabolic processing; they undergo a complex series of enzymatic and/or non-enzymatic reactions in the gastrointestinal tract, particularly in the liver; these include complex reactions such as nitro group reduction and cyclization in **netobimin** and **febantel** to form **albendazole** (ABZ), and **fenbendazol** (FBZ), respectively, and their subsequent sulfoxide (active) and sulfone (inactive) metabolites; prodrugs have better water solubility than BZs and may overcome absorption problems seen with the 'directly' active drugs; in all cases, the antiparasitic activity lies with the BZs itself, and in particular with one of the primary metabolites (e.g. ABZ→ABZ sulfoxide or FBZ→FBZ sulfoxide), rather than the parent prodrug; this may be also true for the toxicity, with exception of some reports of embryotoxicity with febantel		
thiophanate (thiourea derivatives) (cattle, 66-132; divided doses for arrested larvae) (sheep, 50)	* Nemafax (Merial) * Provitblock; * Wormalic and others (Dallas Keith) suspension (drench), bolus; granules, powder (in feed), feed blocks	benzimidazole precursor that may be converted by cyclization into the 2-benzimidazole carbamic acid ethyl ester showing marked activity against GI nematodes adult (more than 90%), developing (75%-95%), and arrested stages (75%-90%, and higher) of common gastrointestinal nematodes in cattle and sheep; no embryotoxic effect was seen after the treatment of ewes with 150 mg/kg; the methyl analogue of thiophanate was introduced in 1970 for the control of fungal diseases in plants
febantel (guanidine derivatives) (prodrug of FBZ) (cattle, 7.5; increased for arrested larvae) (sheep, 5)	* Rintal (Bayer) * Bayverm and others suspension (drench), bolus; granules (in feed)	benzimidazole precursor which is rapidly metabolized in the liver to FBZ and FBZ oxide (=oxfendazole, cf. probenzimidazoles above); drug with broad-spectrum activity against lungworms (Table 5, Table 6), *Dictyocaulus viviparus*, and GI nematodes; it is effective against adult stages (more than 90%), developing stages (more than 90% in cattle; 75%-90% in sheep), and arrested (inhibited) larval stages (50%-90%); there is cross-resistance with thiabendazole- and other benzimidazole-resistant *H. contortus* or *T. colubriformis* strains; the drug is well tolerated; at high doses (250 mg/kg), it appears to be teratogenic and embryotoxic in rats; simultaneous use of febantel with **bromsalans** (→Trematodocidal Drugs, Table 1) within 7 days of each other may cause severe (fatal) adverse effects in cattle
netobimin (cattle, 7.5-20) (sheep, 7.5-20) in sheep, it is rapidly metabolized to ABZ and ABZ sulfoxide following intraruminal administration (cf. probenzimidazoles discussion above)	* Hapadex (Schering Plough) suspension (drench), powder, granules (in feed) large doses given to pregnant rats exhibit an embryotoxic effect, but a dose of 18 mg/kg appears to be safe	guanidine derivatives, and prodrug of ABZ with broad-spectrum of activity against **GI** nematodes of ruminants, and horses (Table 2), lungworms (Table 6), *Dictyocaulus viviparus*, cestodes, and trematodes (*Fasciola hepatica*) of ruminants; its activity is largely similar to that of albendazole (see above); at 7.5., 15, and 20 mg/kg it is highly effective (more than 95%) in cattle against adult and immature *T. axei* and adults of *O. ostertagi*, *Haemonchus* spp., and *Cooperia* spp.; 20 mg/kg is active (90%) against developing larvae and inhibited L_4 larvae of *O. ostertagi*; in cattle, there is also high efficacy against *Oesophagostomum* spp., *Toxocara* (syn. *Neoascaris*) *vitulorum*, and *Moniezia benedeni*, and developing larvae and adults of *Dictyocaulus viviparus*; in sheep, the drug is highly effective against *A. suum*, *Oesophagostomum* spp., however, shows reduced activity against benzimidazole-resistant nematodes (*H. contortus*); there is high efficacy against 12-week-old *F. hepatica* and *D. dendriticum* at 15-20 mg/kg,

Table 1. (continued)

CHEMICAL GROUP, **Nonproprietary name** (approx. dose, mg/kg body weight, oral route) other information	* **Brand name** (manufacturer, company), other information	Characteristics (chemotherapeutic and adverse effects, miscellaneous comments)
TETRAHYDROPYRIMIDINES		
pyrantel tartrate (cattle, 12.5-25) (sheep, 25) the mode of action on the worm's neuromuscular system is thought to be similar to that of levamisole	* Banminth (Pfizer) * Banminth Preßlinge N * Exhelm and others suspension (drench), bolus; powder, premix (in feed) use of drug is not recommended in severely debilitated animals probably because of its nicotine-like action	usually administered as a drench via a dosing gun or in-feed; it is effective in cattle, sheep and goats against adult GI nematodes *Haemonchus* (including thiabendazole resistant strains), *Ostertagia* spp.(only 42% effective against 1 week old histotropic stages in sheep and distinctly less effective against these same stages in cattle), *Trichostrongylus axei, T. colubriformis, Nematodirus battus* (100% at 25 mg/kg regardless of stage of parasite), *N. spathiger, Cooperia* and *Bunostomum* but less active against *Oesophagostomum* and *Chabertia*; the drug is not ovicidal and its efficacy against immature and larval stages is not consistently good or unknown; there is only minor activity against arrested larvae; activity in sheep is similar to that in cattle although greater activity has been found against developing larvae of *Haemonchus* spp., *Trichostrongylus* spp., and *Nematodirus* spp.; the drug can be used prophylactically in sheep (3 mg/kg/day may reduce >95% of GI worms compared to untreated animals); it is well tolerated and can be used in pregnant and young animals; toxicity may result from nicotine-like effects on ganglia; pyrantel together with other cholinergic drugs like levamisole may increase toxicity although simultaneous use with other drugs affecting neuromusculature like cholinergic organophosphates appears to be safe
morantel tartrate (cattle, 8.8), (sheep, 10) **SR bolus** for cattle contains 11.8 g morantel base (dose:1 bolus/animal); has no activity against *Dictyocaulus viviparus* in lungs but it reduces risk of parasitic bronchitis by preventing ingested larvae from becoming established	* Banminth II * Paratect Flex Diffuser (**SR bolus** for cattle) (Pfizer Animal Health) suspension (drench), powder, premix (in feed), intraruminal device (bolus) (additional morantel salts: fumarate, citrate)	methyl ester analogue of pyrantel with good efficacy against GI nematodes, e.g. adult trichostrongyles (*Haemonchus, Trichostrongylus, Cooperia*, and *Nematodirus*), which is somewhat greater than that of pyrantel; its effect against developing stages of trichostrongyles is less pronounced (75%-90%) in cattle and sheep; the activity against arrested larvae is less than 50% (cattle); the drug is not ovicidal; administration of SR bolus (continuous, sustained release of **morantel** tartrate for at least 90 days, i.e. ~ 150 mg/days) at the beginning of the grazing season will prevent establishment of patent infections throughout the grazing season; reduced pasture contamination in autumn will lower the risk of inhibited *Ostertagia* larvae accumulating in abomasum to cause winter ostertagiasis (inhibited *Ostertagia* larvae, L4, are not killed); cross-resistance was found in levamisole-resistant *Ostertagia* spp. strains and other trichostrongyle nematodes; drug is well tolerated (better than pyrantel) and can be used in pregnant and young animals; pharmacological properties, and hence, adverse reactions as well as mode of action are similar to those of pyrantel (levamisole).

Table 1. (continued)

CHEMICAL GROUP, **Nonproprietary name** (approx. dose, mg/kg body weight, oral route) other information	* **Brand name** (manufacturer, company), other information	Characteristics (chemotherapeutic and adverse effects, miscellaneous comments)
IMIDAZOTHIAZOLES		
levamisole (8, s.c. or 8, oral: cattle, sheep, and goats) (10, pour-on, cattle) *Cronomintic-Bolus for cattle (Virbac) (dose: 1 bolus/animal) pour-on may be slightly less effective than other preparations when applied to skin in cold weather	* Citarin-L solution * Concurat-L powder (Bayer) *Levacur (Intervet) * Nilverm (Mallinckrodt) * Novaminth (Smith Kline Beecham) * Ripercol 5, 10 solution * Ripercol pour-on solution (Janssen), and many other suppliers and brand names drench, feed additive, in water, parenteral injection, and pour-on	broad-spectrum drug with high efficacy against lungworms (Table 5, Table 6), *Dictyocaulus viviparus*, and good efficacy against GI nematodes (more than 90%) in cattle and sheep (abomasum: *Haemonchus, Ostertagia*, small intestine: *Trichostrongylus, Cooperia, Bunostomum* and *Nematodirus*, large intestine: *Oesophagostomum*, and lung: *Dictyocaulus*); the effect against arrested stages (L_4) is minimal (less than 50%) in cattle, although the drug has activity against many arrested larvae in sheep; there is no ovicidal effect; benzimidazole (BZ)-resistant *H. contortus* may be sensitive to the drug; levamisole (LEV) resistant nematode strains appear to be susceptible to BZs; LEV does not affect fertility in male and female rats and shows no embryotoxic or teratogenic effect; it appears to be a safe drug in pregnant animals; toxicity is rarely seen, however, dosage should be carefully calculated in lambs, particularly with subcutaneous (s.c.) injection (injectable LEV phosphate causes higher blood concentrations than does drenching with LEV hydrochloride); therapeutic index is lower than that of BZs; LEV as cholinergic drug, may produce typical side effects such as salivation, bradycardia, muscular tremor, or death from respiratory failure; it appears to act on the neuromuscular junction; first it may act as a ganglion-stimulating compound (type of nicotinic ganglionic receptor) and then it may induce a neuromuscular inhibition of the depolarizing type causing spastic contraction and then paralysis of muscle of several nematode species.

MACROCYCLIC LACTONES
endectocides with activity against both internal and external parasites such as nematodes and arthropods; they exhibit no activity against cestodes, trematodes or protozoans:

AVERMECTINS
are fermentation products of an actinomycete *Streptomyces avermitilis* first isolated in Japan at the Kitasato Institute from a soil sample in 1975; avermectin itself consists of a mixture of four major components (avermectin A_{1a}, A_{2a}, B_{1a}, and B_{2a}) and f our minor components (A_{1b}, A_{2b}, B_{1b} and B_{2b}) produced by *Streptomyces avermitilis*; there are several major avermectins which differ slightly in their structure; first avermectins available were a mixture of B_1 avermectins; **ivermectin**, derived from this B1 mixture of avermectins, is a highly potent α-L- oleandrosyl-α-L-oleandroside macrocyclic lactone (dose rates are micrograms) consisting of two components, the 22,23-dihydroavermectin B_{1a} (at least 80%) and 22,23-dihydroavermectin B_{1b} (not more than 20%); other avermectins on the market are **abamectin**, a naturally occurring fermentation product of *Streptomyces avermitilis*, and **doramectin**, and **eprinomectin** prepared by

Table 1. (continued)

CHEMICAL GROUP, **Nonproprietary name** (approx. dose, mg/kg body weight, oral route) other information	* **Brand name** (manufacturer, company), other information	Characteristics (chemotherapeutic and adverse effects, miscellaneous comments)

mutational biosynthesis. MILBEMYCINS are fermentation products of *Streptomyces hygroscopicus aureolacrimosus*; they are similar in their structure to avermectins (lack C-13 disaccharide substituent); **milbemycin oxime** consisting of a mixture of two components (80% A_4 and 20% A_3 milbemycin oxime) is the only milbemycin currently marketed for use in dogs (Table 5, Table 6); macrocyclic NEMADECTINS, fermentation products of *Streptomyces cyaneogriseus noncyanogenus*, are classified as milbemycins in that they also lack the disaccharide moiety at C-13; however, owing to a trisubstituted double bond at C-26 in their side chains, nemadectins differ in their structure from milbemycins proper; **moxidectin** (principal component of LL-F28249 antibiotic complex produced by *Streptomyces cyaneogriseus noncyanogenus*), is a chemically modified derivative of nemadectin; it is more lipophilic, and hydrophobic than ivermectin (results in longer effective tissue levels); pharmacokinetics and toxicity of macrocyclic lactones are influenced by various physiochemical and biologic factors such as the specific formulation used, the route of administration, and animal species to which the product is administered; parenteral administration (s.c. injection) may principally result in a greater bioavailability than oral administration; often the parent compound is the major liver residue for up to 7 and 14 days after dosing in sheep and cattle, respectively (e.g. in case of ivermectin); fecal excretion is the main route of elimination of most macrocyclic lactones (e.g. ivermectin, up to 98% in feces, remainder in urine); however in lactating animals up to 5% of the dose may be excreted in the milk; an exception makes the topically applied **eprinomectin** (quite recently introduced avermectin onto the market), which has a zero withdrawal time for milk intended for human consumption; macrolide endectocides have a substantial margin of safety in ruminants, swine, horses and dogs, although in Collie dogs about one-fourth of the animals appear to be extremely susceptible to these compounds. **Selamectin**, a novel avermectin B_1 derivative, which is structurally related to doramectin (it is a monosaccharide oxime derivative of doramectin), has a unique combination of safety in all dog breeds (including Collies) and potency against both external and internal parasites of dogs and cats (biological spectrum see Table 5); its major route of elimination is via the feces (48-68 % in cats, 18-20 % in dogs, 1-3% in the urine: cats and dogs, and partially via sebaceous glands in the skin); tight binding of selamectin to organic matter (main degradation processes will occur in feces and in soil to reduce residue levels) will limit bioavailability and prevent entry into ground water or into surface waters by run-off.

Details see next pages.

Table 1. (continued)

CHEMICAL GROUP, **Nonproprietary name** (approx. dose, mg/kg body weight, oral route) other information	* **Brand name** (manufacturer, company), other information	Characteristics (chemotherapeutic and adverse effects, miscellaneous comments)
ivermectin (0.2, s.c. or oral, cattle) (0.5, pour-on, cattle) (0.2, sheep) **Ivomec SR Bolus** for cattle: contains 1,72 g ivermectin (wax vehicle extruded by an osmotic pump mechanism as a rate of 12.5 mg/day of ivermectin for 135 days (bolus has high action on various arthropods; for prophylactic action on GI nematodes cf. fenbendazole SR bolus, above) (dose: 1 bolus/animal)	* Ivomec (Merial): sterile solution for injection containing 1% ivermectin (w/v), oral liquids (drench), oral paste, pour-on liquid for cattle, intraruminal device (SR bolus for cattle), feed additives * Baymec (Bayer) drench, pour-on (Merial, Farnam Co., Virbac, Intervet, and many others companies and brand names)	has high efficacy (>98%) against GI nematodes adult, developing, and arrested stages of almost all important GI nematodes of cattle and sheep as *Ostertagia* (including inhibited larvae in cattle), *Trichostrongylus, Haemonchus, Cooperia, Bunostomum, Nematodirus* (only adults, *N. helvetianus* is dose-limiting species for both ivermectin and doramectin), *Trichuris, Oesophagostomum*, ovine *Chabertia ovina*, and lungworms (Table 6), *Dictyocaulus viviparus*; benzimidazole-resistant strains of *H. contortus* and *T. colubriformis* in sheep are highly susceptible to ivermectin and are still unaffected by cross-resistance; because the drug is highly persistent in tissues it may protect against the development of infective larvae of several nematode genera for a period of about 2 weeks; ivermectin is highly active against arthropods such as biting and sucking lice (*Linognathus vituli, L. pedalis, Haematopinus eurysternus*), mange mites (*Psoroptes ovis, Sarcoptes bovis* and others), and grubs (*Hypoderma bovis, H. lineatum, Oestrus ovis*), and less active against chewing (biting) lice (*Damalinia* spp.), and *Melophagus ovinus*; adverse reactions have been seen in cattle treated when large numbers of larvae of *Hypoderma* spp. have been present in the esophageal wall or the spinal canal (escape of cytotoxic material from dying grubs can produce anaphylactic reactions and severe paraplegia); the drug has activity against dung-breeding flies (e.g. face fly, *Musca autumnalis*, and hornfly, *Haematobia irritans*) and ticks (interrupts feeding, molting, and egg production); ivermectin is safe in breeding and pregnant animals; the withdrawal period for meat is 38 days; ivermectin should not be injected intramuscularly or intravenously; it is contraindicated in lactating animals
ivermectin + clorsulon (0.2 + 2, s.c. cattle,)	* IvomecPlus * Ivomec F (Merial) sterile solution for injection	injection solution consists of 1% ivermectin and 10% closulon; clorsulon will control trematodes as adult *F. hepatica*; for ivermectin activity against nematodes and arthropods in ruminants see above; combination is well tolerated (safety index of ivermectin is ~ 30, and safety index of closulon ~ 80 (contraindicated in lactating animals)

Table 1. (continued)

CHEMICAL GROUP, **Nonproprietary name** (approx. dose, mg/kg body weight, oral route) other information	* **Brand name** (manufacturer, company), other information	Characteristics (chemotherapeutic and adverse effects, miscellaneous comments)
abamectin (cattle, s.c., 0.2) (also sold as pesticide: *Agri-Mek, *Vertimec, *Avid and others with a broad spectrum of activity against insect and mite pests of agronomic crops), in ruminants abamectin is excreted to a large degree unchanged via the feces	* Avomec (Merial) * Enzec (Janssen) sterile solution (1% w/v abamectin), cattle abamectin is effective against larvae of some **dung-breeding insects** (diptera and beetles) via feces; in environment, it is rapidly degraded by O_2 and light ($t^1/_2$: 4-20h); it is slowly degraded if bound to soil particles and protected from light ($t^1/_2$: 20-48d)	introduced in 1985 (Australia); it is used in cattle to control adults and larval stages of GI nematodes; abamectin is highly effective (up to 99%) against adult and immature stages (also developing L_4 stages) of *Ostertagia ostertagi* (including inhibited larvae), *O. lyrata, Haemonchus. placei, Trichostrongylus axei, Cooperia oncophora,* (also inhibited larvae), *C. punctata* (also inhibited larvae), *C. surnabada, C. pectinata* (also inhibited larvae), *Nematodirus. helvetianus, Bunostomum phlebotomum, Oesophagostomum radiatum* (adults, developing L_4-stage),adult *Trichuris* spp., and lung worms (Table 6), *D. viviparus* (adults, developing L_4-stage); the drug can prevent reinfections with *Ostertagia* spp., and *Cooperia* spp. for at least 7 days after dosing, and *Dictyocaulus viviparus* for at least 14 days after treatment; it has high activity against arthropods, such as cattle grubs (*Hypoderma bovis, H. lineatum*), sucking lice (*Haematopinus eurysternus, Linognathus vituli, Solenoptes capillatus,* prevention for at least 56 days after treatment), and biting lice (*Damalinia* spp. less effective), mange mites (*Psoroptes ovis, Sarcoptes scabiei var. bovis*) engorged female Boophilus *microplus* ticks; abamectin should not be administered by intramuscular or intravenous injection; it should not be used in calves aged 1week to 4 months because of possible adverse reactions; infrequent side effects at site of injection may be swelling, and pain; ocular contact of drug should be avoided; preslaughter withdrawal time for edible tissues may vary (Australia 30 days, UK 42 days, Germany 35 days); abamectin is contraindicated in lactating animals
doramectin (cattle, s.c., 0.2) (1% solution of doramectin in sesame oil-ethyl oleate (90:10 v/v) vehicle) (cattle, pour-on, 0.5) after treatment, infective nematode larvae may fail to establish infection in host for prolonged periods (see prophylactic effect)	* Dectomax (Pfizer) sterile solution for injection in cattle; pour-on for cattle **Prophylactic effect:** doramectin may prevent establishment of infections with *C. oncophora* for 2-4 weeks, *O. ostertagia* for 3-4 weeks and *D. viviparus* for at least 4 weeks (these periods are longer than those seen with ivermectin	introduced in 1993; there is broad spectrum of activity (up to 99%) against GI nematodes (adults and L_4 larvae, including inhibited larvae of *Ostertagia* spp., *Haemonchus placei, Trichostrongylus* spp., *Cooperia oncophora, Oesophagostomum radiatum, Nematodirus helvetianus*) and lungworms (Table 6), *Dictyocaulus viviparus*; there is high activity (up to 98%) against adult stages of *Ostertagia lyrata, Cooperia pectinata, C. punctata, C. surnabada, Trichuris* spp., *Nematodirus spathiger* (*N. helvetianus* is dose-limiting species for both ivermectin and doramectin), *Bunostomum phlebotomum, Strongyloides papillosus* and eyeworms (*Thelazia* spp.); it is highly effective (up to 100%) against arthropods such as cattle grubs (*Hypoderma bovis, H. lineatum*), sucking lice (*Haematopinus eurysternus, Linognathus vituli, Solenoptes capillatus,* less effective against biting lice, *Damalinia bovis*), tropical warble fly (*Dermatobia hominis*), New World screwworm (*Cochliomyia hominivorax*), mange mites (*Psoroptes communis* var. *bovis, Sarcoptes scabiei*), and *Boophilus microplus*; the pharmacokinetic property of doramectin is characterized by a larger area under plasma concentration vs. time curve (AUC) and thus a longer mean residual time compared to that of ivermectin allowing prolonged contact between drug and parasites; preslaughter withdrawal time for edible tissues may vary (RSA 35 days, UK 42 days, Germany 50 days); doramectin is contraindicated in lactating animals

Table 1. (continued)

CHEMICAL GROUP, **Nonproprietary name** (approx. dose, mg/kg body weight, oral route) other information	* **Brand name** (manufacturer, company), other information	Characteristics (chemotherapeutic and adverse effects, miscellaneous comments)
eprinomectin (0.5 pour-on, cattle) compound selection was based on toxicological and efficacy data in sheep and on a milk residue profile in cattle; it has a unique plasma/milk partitioning coefficient, rendering the drug useful for treating beef cattle, and dairy cattle	* Eprinex (Merial, others) pour on formulation: (eprinomectin 0.5% w/v butylated, hydroxytoluene 0.01% w/v, fractionated oils of natural sources quantum satis ad 100% v/v)	introduced in 1996 as topical formulation (pour-on: non-aqueous, non-flammable liquid, which appears to be unaffected by weather, and haircoat, easily to handle and safe for animal and user) for use in beef cattle and dairy cattle; it has a broad-range of activity against GI nematodes: there is high activity (>98%) against immature (L_4) and mature stages of *Ostertagia* spp. *Cooperia* spp. *Bunostomum phlebotomum*, arrested larvae of *O. ostertagi* and *Cooperia* spp., *Haemonchus* spp., *Oesophagostomum* spp., *Nematodirus helvetianus*, *Strongylus papillosus* (only adults), *Trichostrongylus* spp. (adults: appears to be less active, 85%), *Trichuris* spp. (only adults), lungworms (*Dictyocaulus viviparus,* Table 6), and arthropods such as biting lice (*Damalinia (Bovicola) bovis*), sucking lice (*Haematopinus eurysternus, Linognathus vituli, Solenoptes capillatus*), mange mites (*Chorioptes bovis, Sarcoptes scabiei*), grubs (*Dermatobia hominis, Hypoderma bovis, H. lineatum*) and horn flies (*Haematobia irritans*); most drug absorption is within 7 to 10 days post-dose; peak plasma concentrations of 22.5ng/ml are reached 2-5 days post-dose, and then decline to ~1ng/ml by 21days post-dose (mean residence time 165h); drug is not extensively metabolized (parent compound >90% in liver, kidney, fat, muscle, plasma and 85% in feces); safety and pharmacokinetic profile of eprinomectin allows for zero preslaughter withdrawal times for consumption of milk and meat; it is well tolerated and safe in cattle
MILBEMYCINS/NEMADECTINS In contrast to avermectins, they lack the oleandrosyl moiety (C-13 disaccharide substituent) while the macrocyclic lactone ring system is similar to that of avermectins (for differences between milbemycins and nemadectins see avermectins, above); the broad range of biological activities of the milbemycin/nemadectins is generally equal to that of the avermectins; to date, **milbemycin oxime** is the only substitute of the milbemycins proper currently marketed for use in dogs (Table 5); **moxidectin**, a chemically modified derivative of nemadectin (contains at C-23 a N-oxime methyl ether and at C-5 a hydroxyl group) is more lipophilic, and hydrophobic than ivermectin; moxidectin is the only substitute of endectocide nemadectins marketed for use in cattle, and sheep		

Table 1. (continued)

CHEMICAL GROUP, **Nonproprietary name** (approx. dose, mg/kg body weight, oral route) other information	* **Brand name** (manufacturer, company), other information	Characteristics (chemotherapeutic and adverse effects, miscellaneous comments)
moxidectin (0.2, s.c., cattle) (0.5, pour-on, cattle) (0.2, sheep) drug is safe in breeding animals (no adverse effects on reproductive performance of bulls and pregnant cows at 3 times the recommended dose; calves under 100 kg body weight may be susceptible to overdosing (use of correct dose is absolute)	* Cydectin * Vetdectin * Moxidec (American Cyanamid, Fort Dodge, Bayer) (sterile, aqueous solution for injection, cattle, containing 1% moxidectin, propylene glycol, and solubilizers) (pour-on for cattle containing 0.5% moxidectin and oil vehicles) (oral drench for sheep containing 0.1% or 0.2% moxidectin)	introduced in 1990 (Argentina); it exhibits in cattle and sheep a broad-spectrum of activity against GI nematodes; it is highly effective (>99%) against adult and larval stages of *H. placei, H. contortus, O. ostertagi* (including inhibited larvae), *T. axei, T. colubriformis, N. helvetianus* (only adults, >95%), *N. spathiger, C. surnabada, C. pectinata, C. punctata* (cooperids all 92-100%), *Oesophagostomum radiatum, Strongyloides papillosus, Trichuris* spp. (only adults), *Chabertia ovina* (only adults), *Bunostomum phlebotomum* (only adults) and lungworms (Table 6), *Dictyocaulus viviparus* (adults); reinfection with *Ostertagia, Haemonchus, Trichostrongylus Oesophagostomum* is prevented for up to 28 days following 1x s.c. or pour-on, *D. viviparus* (> 28days after 1x s.c. or pour-on); a single s.c. infection is highly effective against arthropods such as cattle grubs (99%: *Hypoderma bovis, H. lineatum*), sucking lice (99-100%): *Linognathus vituli, Haematopinus eurysternus, Solenoptes capillatus*), biting lice, *Damalina bovis* (markedly suppressed after pour-on), mange mites (100%: *Sarcoptes scabiei, Psoroptes ovis*, and *Chorioptes bovis* markedly suppressed); moxidectin should not be administered to lactating animals and to calves younger than 8 weeks; ocular contact, and contact of moxidectin with skin should be avoided; preslaughter withdrawal time for edible tissues may vary (Australia 14 days, UK 45 days, RSA 28 days, Germany 35 days); moxidectin is contraindicated in lactating animals
NARROW-SPECTRUM ANTHELMINTICS Substituted salicylanilides and phenols (cf. anticestodal drugs) can be used for control of highly pathogen GI nematodes such as *Haemonchus contortus* and other trichostrongylid nematodes of ruminants that have developed resistance to the broad-spectrum anthelmintics, particularly benzimidazoles and macrocyclic lactones.		
PHENOL DERIVATIVES		
disophenol (2,6-diiodo-4-nitrophenol) (sheep, 10, s.c.)	* Ancylol (Werfft) * Disophenol (Cyanamid) injectable solution	has been used for treatment of dogs infected with hookworms (Table 5); it is also effective against mature *H. contortus* and may be used in sheep for treatment of **benzimidazole-resistant** *H. contortus* infections at doses similar to those used in dogs and cats; it may prevent the establishment of mature worm burdens for a period of 2 months or longer; pharmacokinetics (strong plasma protein binding) and mode of action (uncoupler of oxidative phosphorylation, and inhibition of fumarate reductase)
nitroxinil (nitroxynil) (sheep, cattle, 10, s.c.) antitrematodal drug effective against the liver fluke *Fasciola hepatica*	* Dovenix * Trodax (May & Baker; Specia) injectable solution (in some countries discontinued or at no time launched)	is more than 99% effective (adults) against **ivermectin**- and **benzimidazole-resistant** *Haemonchus contortus* of sheep; it also exhibits activity against adult stages of *Bunostomum*, and *Oesophagostomum* spp. in cattle and sheep; drug may cause some yellow staining of fleece in sheep, and local reactions at injection site; maximum tolerated dose in sheep is approx. 40 mg/kg; drug is slowly reduced to an inactive metabolite in the rumen and is therefore preferably given by s.c. injection; it is slowly eliminated from body into urine and feces (for 31 days), and milk as well (contraindicated in lactating cows); withdrawal time may vary and may be about 2 months

Table 1. (continued)

CHEMICAL GROUP, **Nonproprietary name** (approx. dose, mg/kg body weight, oral route) other information	* **Brand name** (manufacturer, company), other information	Characteristics (chemotherapeutic and adverse effects, miscellaneous comments)
SALICYLANILIDES		
rafoxanide (cattle, sheep, 7.5) (cattle 3, s.c.)	*Raniden (Merial) * Flukanide (Merial) * Ranide (MSD) * Ursovermit (Serum Werk Bernburg) injectable solution, drench, bolus	antitrematodal agent that is highly active against adult stages of **ivermectin**- and **benzimidazole-resistant** *Haemonchus contortus* in sheep; it may reduce adult stages of other trichostrongyle nematodes in cattle and sheep; the drug can be used in a similar fashion to all other phenols or closantel, to provide specific control of *H. contortus* in sheep, thus reducing the selection pressure for resistance by broad-spectrum compounds; at recommended dose, the drug is well tolerated in sheep and cattle of all ages; drug is extensively bound (> 99%) to plasma proteins and has a long terminal half-life (~17 days); withdrawal time for edible tissues is 28 days (contraindicated in lactating animals); mode of action is uncoupling of oxidative phosphorylation; binds strongly to plasma protein.
+thiabendazole (tiabendazole)	* Ranizol	* Ranizol can be used for simultaneous treatment of liver flukes and gastrointestinal nematodes in sheep an cattle
closantel (sheep, cattle 10) is extensively bound to plasma proteins (>99%) (esp. to albumin); toxic only to worms that ingest and digest blood or reside in the bile duct	* Flukiver * Seponver (Janssen) suspension (drench)	an effective flukicide with high activity also against bloodsucking nematodes such as *H. contortus* in sheep; it is highly effective against adult stages of *Haemonchus contortus* **resistant to benzimidazole, levamisole, morantel,** and **ivermectin** in sheep; it protects against infections of *H. contortus* for up to 1-2 months after treatment; the drug is well tolerated in sheep and cattle (also in reproduction studies in rams, ewes and bulls); safety index in sheep and cattle is ~4 at recommended dose; drug is primarily excreted via feces (80%, urine <1%); withdrawal times for edible tissues are 28 days (cattle) and 42 days (sheep); drug is contraindicated in lactating animals); mode of action is uncoupling of oxidative phosphorylation.
ORGANOPHOSPHATES		
Many organophosphates have been tested for anthelmintic activity and several have been marketed for use in ruminants, pigs, and horses; these compounds are available in a limited number of countries only; their spectrum of activity is not as wide as that of broad-spectrum drugs (particularly in ruminants) and they have a relative narrow range of safety, especially in sheep; toxicity and mode of action are by the same mechanisms, and are attributable to acetylcholinesterase inhibition; typical toxic signs are salivation, diarrhoea, and exaggerated symptoms; death may result from respiratory failure; worms show spastic paralysis and are removed by normal peristaltic action of the bowel; **coumafos, haloxon, naphthalophos** (= naftalofos) and **crufomate** are preferably used against GI nematodes and other parasites of ruminants; their anthelmintic spectrum in cattle is similar to its efficacy in sheep; organophosphates affect principally adult stages (>90-75 %), and less developing larvae (50-75%) of GI nematodes in the abomasum and small intestine; they exhibit no activity against arrested larvae, and only variable effects (20-90%) on adult stages of GI nematodes parasitizing in large intestine; however organophosphates principally control infections due to *H. contortus, Ostertagia* spp., and *Trichostrongylus* spp., and they may be used either alone or in combination with a broad-spectrum anthelmintic where **resistance of trichostrongyles** to both benzimidazoles and levamisole occurs (for more details see drug combination of naphthalophos and fenbendazole below); in general, therapeutic indices of organophosphates are significantly narrower than those of broad-spectrum anthelmintics (benzimidazoles, pyrimidines and macrocyclic endectocides); thus in ruminants and other animal species (horses Table 3, dogs Table 5, and **birds**, see below) exact dosing of organophosphates is necessary		

Table 1. (continued)

CHEMICAL GROUP, **Nonproprietary name** (approx. dose, mg/kg body weight, oral route) other information	* **Brand name** (manufacturer, company), other information	Characteristics (chemotherapeutic and adverse effects, miscellaneous comments)
coumaphos (15, beef and dairy cattle) (2/day for 6 consecutive days, feed supplement or premix, cattle) (8, sheep) (30ppm, laying hens for 14 d; 40 ppm = 0.004% medicated feed for 10-14 d, chicken)	* Asuntol (Bayer) * Meldane * Medane 2 drench (cattle, sheep) * Baymix (Bayer) in-feed supplement, premix, cattle, **birds**; effective against *Capillaria*, ascarids and cecal worms in **chicken**	cholinesterase inhibitor, which is widely used as an ectoparasitic in livestock; it exhibits a cumulative effect on trichostrongyle nematodes if given in-feed (2 mg/kg p.o. daily for 1 week); there is a good activity against *Haemonchus* spp. and *Cooperia* spp. in cattle and sheep; it is less effective against *Trichostrongylus, Ostertagia* spp. and *Oesophagostomum* spp.; the effect on immature stages is minimal; anthelmintic activity can be enhanced if drench passes via the closed esophageal groove directly to the abomasum; this is achieved by premedication of the animal either with sodium bicarbonate in cattle or copper sulfate in sheep; single drench of 30 mg/kg may cause mortality in cattle, and mild signs of toxicity may already occur at 20 mg/kg; in sheep, adverse reactions may be expected in some cases when recommended dose level for drenching is administered (antidote may be atropine); the drug is rapidly metabolized after oral dosing and has a zero withdrawal time for milk and edible tissues, respectively; colored breeds of egg-laying **hens** are more susceptible to the drug than white breeds (should not be treated while they are in production)
haloxon (35-50 sheep) (~44, cattle) (50-100, domestic fowl, turkey, quail, pigeons)	* Halox (Cooper) * Loxon and others drench, (suspension)	still used in many countries for treatment of GI nematodes in cattle, sheep, and goats; it is highly active against adult *Haemonchus*, and *Cooperia* spp. in sheep, and *Neoascaris* in cattle; there is moderate effect against *Ostertagia, Bunostomum, Trichostrongylus,* and *Oesophagostomum* and a rather zero effect against *Nematodirus, Trichuris* and *Chabertia*; toxic effects may occur at therapeutic dose (cholinesterase complex : posterior weakness and paralysis); drug is effective against *Capillaria* infections of **birds** (chicken, turkey, quail, and pigeons) but is ineffective against *Heterakis*; recommended dose range for birds (50-100 mg/kg) is lethal for geese and possibly for other waterfowl
naphthalophos (=naftalofos) (50, cattle, sheep) (10/d for 6 consecutive days in-feed, cattle) (**naphthalophos** 34.5, **+fenbendazole** 5, sheep) was >95 % effective against multiple resistant trichostrongyles on 10 of 13 farms	* Rametin (Bayer) * Maretin drench, in-feed drench (fresh prepared liquid mixture of naphthalophos and fenbendazole)	anthelmintic spectrum in cattle is similar to that in sheep and includes high efficacy (>95%) against all adult *Haemonchus* (99-100% in sheep), *Ostertagia circumcincta*, *Cooperia* and *Trichostrongylus colubriformis*; it is less active against *Ostertagia ostertagi, T. axei,* and *Strongyloides papillosus* (80-85%); its action is erratic (*Nematodirus*), variable or zero (*Oesophagostomum, Chabertia*); when administered in-feed (cattle), the drug shows efficacy against *Haemonchus, Trichuris* and against *Cooperia* (latter species at 20 mg/kg/d x6d); in **birds** (quail) it eliminates infections of *Capillaria obsignata* and *C. contorta* (in-feed 200 ppm x2d,); however, this regimen is insufficient in eliminating *Heterakis gallinarum*; safety margin of drug is variable with respect to animal species: chickens are very sensitive to the drug (single dose of 50 mg/kg is fatal, while a dose of 25 mg/kg is required for satisfactory elimination of *Ascaridia* and *Heterakis*); LD50 in sheep is 300 mg/kg; in cattle, doses above 150 mg/kg may cause poisoning signs (e.g. increased salivation, diarrhoea, loss of appetite)
crufomate (40, cattle) (17/d for 3 consecutive days, cattle)	* Ruelene suspension	used for control of warble fly (pour-on), and as an anthelmintic for control of infections in cattle due to *Haemonchus, Cooperia*, and *Bunostomum* (all >90% efficacy), *Ostertagia, Trichostrongylus axei, Strongyloides papillosus* and *Trichuris* (30-90%); there is insufficient efficacy (0-30) against *Nematodirus, Oesophagostomum*, and intestinal *Trichostrongylus*

Doses listed in this table refer to information from manufacturer and literature
Data given in this table have no claim to full information.
The primary or original manufacturer is indicated if there is lack of information on the current one(s)

tion, and other functional and intestinal disorders such as depressed levels of minerals and depressed activity of some other intestinal enzymes. The effects of nematode parasitism on production are well known and will remain one of the major factors limiting animal productivity on farms that rely on grazing animals on pasture. Economic loss may be due to the reduced skeletal growth (mineral deficiencies), to reduced weight gains (reduced incorporation of amino acids into muscle protein), or to suppressed wool production and reduced wool quality (e.g., break in wool growth due to reduced incorporation of amino acids into protein in hair follicles). Clinical parasitism markedly affects milk production in dairy cows, and subclinical parasitism appears to be of economic importance, as it will also reduce animal productivity. The importance of beef meat production to the animal health market is reflected in some figures. Thus annual beef production (approx. weight in tons) in the USA, China and Germany was 11 million, 2.3 million and 1.5 million, respectively, and that of sheep meat in China, Germany and the USA 1.4 million, 640.000, and 154.000, respectively.

Knowledge of the epizootiology of the parasites permits **strategic timing of drug treatments**. A number of strategies have been suggested to limit the development of drug resistant nematodes but any strategy must fit the particular characteristics (parasite and host biology, epizootiology, etc.) of the target population within a certain region, and hence, strategy will vary considerably between different regions. Therefore, efficient control programs for grazing ruminants (cattle and sheep) are based on seasonal fluctuations of L_3 on pasture. In temperate countries (e.g., Europe) sufficient L_3 may overwinter on pasture to infect susceptible animals next spring. However, the numbers of larvae, which are acquired in spring are seldom sufficient to produce clinical signs. Non-ingested overwintered L_3 may die off in early summer and L_3 that are found in midsummer stem primarily from hatching of eggs deposited the same year. Thus, young cattle should be treated with anthelmintics after they start spring grazing. Removal of cattle to clean pastures may allow treatment to be delayed until early summer and in general there will be no requirement for further treatments during the grazing season. L_3 stages of *Ostertagia* spp. ingested during the fall are arrested at the L_4 stage, primarily in the abomasum. To prevent type II ostertagiasis, susceptible cattle should be treated with an effective drug against these stages at housing. Lack of treatment leads larvae to resume their development next spring, causing severe damage of abomasal mucosa.

During the periparturient period ewes are the major source of pasture contamination for **lambs** since their fecal egg output may rise enormously within this period. Ewes may be treated several weeks prior to lambing and up to 8 weeks thereafter. Lambs are particularly susceptible to nematode infections. At weaning, the pasture may be heavily contaminated even if ewes were drenched before lambing. At this time lambs are usually drenched, followed by a move to a safe pasture. Although the treatment may be effective subsequent contamination with eggs from selected worms on the clean pasture cannot be prevented. Subsequent anthelmintic treatment will continue to select **resistant parasites**. However, lambs are marketed within a relatively short period, thus allowing sufficient control. Pasture should then be grazed by cattle or destocked for conservation. Consequently, resistant subpopulations of worms will die off. This practice may be continued each year without causing a major resistance problem. Ewes should not be grazed on pastures after lambs have been moved since resistant larval worms would be carried over, thus infecting lambs in the next year. In this case, resistance problems would inevitably increase, resulting in drug failure.

Attention should be given to movements of small ruminants from farm to farm and to imports of sheep and goats from countries known for multiple resistant nematodes. Such animals should be examined for anthelmintic resistant nematodes and/or treated with a fully effective drug before being introduced to a farm. Although such preventive measures (quarantining, monitoring, and treating all new replacement stocks) will help to prevent the spread of anthelmintic drug resistance, they are critical and expensive management practices too. Commingling of sheep and goats should be avoided and where practical, alternate grazing of livestock of various species or immune status should be supported.

The aim of any helminth control measure is to reduce parasites to levels that have little impact on animal production and to limit resistance development. A corollary strategic use of anthelmintics appears to be the use of only a single class of drugs within a treatment period or parasite generation. The rotation of anthelmintic classes with a different mode of action on a yearly basis may limit transfer of resistance genes tolerant to the previous class but sensitive to the alternative class early in the selection process. Then heterozygous helminths for the trait may reverse to susceptibility again. Periodic assessment of resistance status should be performed at regular intervals and should be part of any nematode control strategy using antinematodal compounds. The strategies utilizing anthelmintics may vary according to whether meat, milk, or wool production is desired. Furthermore, the choice of a favorable moment for an integrated control measure, the prevailing weather conditions of the region (rainfall and low temperature favor the development of larvae), the anthelmintic formulation, and the anthelmintic itself (narrow or broad spectrum activity), may all exert a great influence upon the result of the control measure. Consequently, the possibility of drug-resistance developing in *Haemonchus contortus*, *Ostertagia* spp. and *Trichostrongylus* spp. of sheep will change the treatment practices. Suppressive drenching schemes may favor the development of drug-resistance, whereas strategic anthelmintic treatment considering the interrelationships between anthelmintic treatment, parasite population dynamics, and weather conditions may prevent or slow down the selection of drug-resistant nematode strains. Therefore, integrated control coordinating anthelmintic treatment and management strategies (e.g., rotational sheep and cattle grazing programs) is an effective tool in preventing development of drug-resistance in trichostrongyle nematodes of sheep.

There may be a number of **causes of treatment failure**, which are unrelated to resistance Thus, in sheep flocks parasitic gastroenteritis may falsely be associated with anthelmintic resistance. However, inquiries often reveal management deficiencies such as flock has been returned to the same pasture immediately after deworming, or periods between treatments have been too prolonged. Other common mistakes are underdosing due to inaccurate estimation of bodyweight and so inaccurate dispensing of the appropriate dose of anthelmintic. Group dosing should be based on the heaviest animal and on label directions that should be followed explicitly concerning dose, route, target parasite, target host, and expiration date. Treatment failures very often result from misdiagnosis (e.g. disease syndromes due to mineral deficiency or plant toxicosis, infections by other parasites), or from use of drugs lacking persistent activity; treated animals may then become rapidly reinfected as a result of local epizootiological conditions (repeated and massive exposure of animals to parasites) and/or poor management. Differences in drug pharmacokinetics may occur between individual animals and between species (e.g. in the order sheep, goat, red deer, higher dosage is needed) or with altered environmental conditions (e.g. reduced drug availibility in pastured calves in contrast to housed animals). Another cause of treatment failure may be the use of inappropriate drugs, so in situations or areas where inhibited larvae are the major cause of the disease and animals suffering from the disease are treated with a drug without claim for these larvae.

For the **biological control** of nematode parasites of livestock hundreds of various antagonistic organisms have been described. Most of these nematode-destroying organisms are found within different groups of microorganisms as viruses, bacteria (*Bacillus* genera) and fungi, or in invertebrates as nematodes, turbellarians, Oligochaeta (earthworm), insects (dung beetles, springtails) or arthropods (tardigrades, mites). In particular certain fungi have shown great potential as biological agent against nematodes pathogenic to livestock. Isolates of **nematode-destroying fungi** can survive ruminant gut passage, germinate and spread in fresh dung and capture large numbers of infective larvae prior to their migration to pasture. However, infective larvae must be small enough to be trapped by the fungus. *Haemonchus*, *Ostertagia*, *Trichostrongylus* and *Cooperia* are readily trapped, others such as the slow-moving and not very active *Dictyocaulus* and nematodes producing persistent egg stages (*Nematodirus*, *Trichuris* and *Ascaris*) are likely to require different antagonistic organisms to act as control agents. Strategic feeding of first-season calves with the fungus *Duddingtonia flagrans* through initial 3 months of the grazing season could prevent severe clinical trichostrongylidosis in the late summer. Larval populations of *Ostertagia* and *Cooperia* were significantly reduced on the pasture grazed by the fungus-treated calves. However, lack of consistent success of relative expensive biological control (=BC) products (mass production of the antagonists is too time-consuming, shelf life should be long enough, BC product must be safe to users, consumers and environment) has left the industry skeptical. Only few companies have shown some interest in developing BC products. Ideally, commercial BC products should be in price and effect similar to that of anthelmintics or consumers will have to accept a higher price of BC products, and hence, higher prices for agricultural products.

Incidence of Drug Tolerant Nematodes in the Field

Since the discovery of the 2-(4'-thiazolyl) benzimidazole, thiabendazole, in 1961, the search for new antinematodal compounds within various chemical groups (Table 1, Table 3, Table 4, Table 5, Table 6 and → Nematocidal Drugs, Man/Table 1) has resulted from sequential improvement in drug properties (spectrum of activity, tolerability, convenience) but chiefly from the development of drug resistance in the field.

Serious problems with anthelmintic resistance may occur in those countries, which have regions where *H. contortus* is endemic and the cause of major losses in productivity. Although *Trichostrongylus* spp. and *Ostertagia* spp. are not as pathogenic to ruminants as *H. contortus* (known as stomach worm or wireworm) subclinical losses in production attributable to ineffective treatment of resistant *Trichostrongylus* spp. and *Ostertagia* spp. are likely to be substantial. Resistance problems are common in Australia, South Africa, the humid semi-tropical regions of South America and other parts of the world where some 300 million sheep are raised. In general, anthelmintic resistance is clearly linked to the frequency of anthelmintic treatment, to the relative importance of the nematode species (being of greatest importance in the regions endemically infected with *H. contortus*) and the prevailing type of grazing management (set-stocked on permanent pastures).

Resistance was first associated with the benzimidazoles and *H. contortus*, but **benzimidazole** resistance in *Ostertagia* spp. and *Trichostrongylus* spp. is also widespread and common. Also **Levamisole** and **morantel** resistance, which at first was slow to be developed, and recently **macrocyclic lactone** resistance (avermectins and milbemycins, the most widely used anthelmintic compounds) are well documented in *H. contortus*, *Ostertagia* spp. and *Trichostrongylus* spp.. Goat farmers are also confronted worldwide with problems of drug failure against *H. contortus*, *Ostertagia* spp. and *Trichostrongylus* spp., which show multiple resistance to benzimidazoles, levamisole/morantel and macrocyclic lactones. As a rule nematodes of **small ruminants** (e.g., sheep, goats) develop anthelmintic resistance much quicker than those of large ruminants (e.g. cattle). A few reports on genuine drug resistance in nematode parasites of cattle (*O. ostertagi*, *Haemonchus* spp., *T. axei*, *C. oncophora*, *Dictyocaulus viviparus* and *Cooperia* spp.) to levamisole/morantel or benzimidazoles and sporadically to ivermectin may be particularly attributable to the less frequent use of anthelmintics, which may minimize selection pressure. Another reason could be that differences in pharmacokinetics of benzimidazoles, as high and long-persisting plasma drug levels in small ruminants, which do not occur in cattle to that extent, may create conditions for a greater selection pressure on nematodes, and hence, more rapid development of resistance. Drug resistance is almost solely confined to the strongylid parasites of the gastrointestinal tract of sheep and goats and appeared in a chronological sequence to phenothiazine, the benzimidazoles, levamisole, morantel and macrocyclic lactones (avermectins, milbemycins). Also cyathostominae (small strongyle parasites of horses) show throughout the world resistance to phenothiazine and the benzimidazoles to a high degree. Although resistance in nematodes to anthelmintics has been slow to develop, it is likely that it will spread to involve other species of parasites and host in areas considered as having fewer problems at present.

Assays for Detection of Resistance to Anthelmintics

Numerous assays (in vitro and in vivo) are available to monitor resistance. However, most results obtained with these assays are either poorly predictive and/or none of them is conductive to field use, i.e. suitable as 'cow-side' test. Often, anthelmintic resistance is first suspected when a farmer reports a poor clinical response in his livestock to anthelmintic treatment. Since clinical signs associated with gastrointestinal parasites such as diarrhoea, weakness, anaemia, and even death are nonspecific, it is important to confirm that parasites are the cause. Information on the control program, the anthelmintic used, the frequency of treatment, group dosing management, stock introductions, grazing management, and nutrition is required prior to the use of an assay. There are three main groups of in vitro tests. In addition to physiological based in vitro tests (egg hatch, larval development, larval paralysis, motility, and larval migration assays), biochemical- (colorimetric tests include tubulin-binding and tubulin-polymerization: benzimidazoles) and PCR based in vitro assays (cloned β-tubulin probes with restriction mapping: benzimidazoles) or isoenzyme analysis using isoelectric focussing (ivermectin) have been, or could be used. The latter tests (exception of simplified colorimetric assays applied for organophosphate and carbamate insecticides in aphids) are not applicable to field use, due to expense and technical requirements.

Drug Formulations, Routes of Administration The administration of drugs by injection, dermal application (pour on) or by the oral route (using granules, drenching, or paste and other formulations) may ensure that each individual receives an adequate dose. If a group of animals is to be treated, doses should be calculated for the heaviest animal in the group. This is important to prevent resistance problems. High doses should leave only few, if any, survivors and may retard genetic variation in parasite populations possessing alleles, which may confer resistance to anthelmintics.

A relative simple application technique, which is widely used in **small ruminants** (sheep and goats, also other animal species), is oral drenching of liquid drug formulations (e.g. oral drench of albendazole, mebendazole, oxfendazole, closantel, fenbendazole, febantel and others).

There may also be the possibility of treating (under controlled conditions) sheep and other animals with in-feed pellets or powdered premix formulations each containing an anthelminthic in a fixed concentration (e.g. pyrantel tartrate and others).

Horses may be treated with pastes (e.g., fenbendazole, febantel, ivermectin, pyrantel and others), or granules and powder (e.g. in-feed medication with fenbendazole, thiabendazole or pyrantel pamoate) or oral drench or by nasogastric tube (e.g. ivermectin). In cattle oral drenching is more difficult, and therefore other application techniques and formulations are used. Treatment can be carried out by intraruminal injection (e.g. oxfendazole), in-feed pellets (e.g. pyrantel tartrate, albendazole, febantel, and fenbendazole), by oral suspensions (levamisole, albendazole, febantel, fenbendazole, oxfendazole), or by subcutaneous injection (levamisole, ivermectin, ivermectin/clorsulon=Ivomec F, and all other avermectin/milbemycin/nemadectin compounds). In cattle pour-on formulations are widely used as dermal applications (e.g. levamisole, avermectins/milbemycins) in cattle. Minor changes on the composition of a formulation can have a major impact on the pharmacokinetic and efficacy profiles of anthelmintic compounds. When using pour-on, it should be considered that as function of the temperature gradient, there might be higher drug absorption during summer than winter, thus favoring the selection of resistant parasites in winter. Pour-on formulations should be weatherproof (not affected by rain, user friendly in extreme temperatures, cold and warm, consistent efficacy if exposed to the sun/light), effective in long haired and short haired cattle, easy to apply, adhesive (no runoff, i.e. dose should stay on animal), odorless, nonflammable, and highly effective (should provide high levels of efficacy against endo- and ectoparasites). For user safety there should be no special requirements for use of protective masks, gloves and aprons.

The use of intraruminal anthelmintic devices as slow (sustained) release boluses (SR boluses) or pulse-release boluses should provide effective control for a period long enough to kill the majority of free-living stages being ingested during the grazing season. The drug release rate must be maintained at a constant level that is high enough to produce a high parasite kill. Furthermore, the release must decline rapidly to zero when the device becomes exhausted. If a significant proportion of worms can survive and reproduce in the presence of the controlled release device there may be the danger of the low drug concentrations allowing resistant worms to emerge. Intraruminal devices are administered via special balling guns in the reticulo-rumen. Slow release cattle dewormer contain antinematodal drugs of different classes, such as morantel tartrate (Paratect Flex-Bolus), levamisole (Chronomintic Bolus), fenbendazole (Panacur SR Bolus), oxfendazole (Systamex 5 Pulse Bolus) or ivermectin (Ivomec SR Bolus). (For efficacy and other characteristics of SR boluses cf. Table 1). Composition, and thus design for drug release (sustained or pulse) of intraruminal boluses may differ, but all devices are designed to release nematocidal concentrations of an anthelmintic in the reticulo-rumen of cattle and so to ensure sufficient control of ingested infective larvae of GI nematodes and those of the lungworm *Dictyocaulus viviparus* for prolonged periods. Some intraruminal devices may release nematocidal concentrations for up to 20 weeks (for effects of chemoprophylaxis with intraruminal boluses on the immune response to GI nematodes in first-season grazing calves).

Efforts have been made to provide simpler treatment techniques that will allow self-medication and mass medication of animals. Self-medication formulations of anthelmintics are available in blocks, licks or as water additives. However, self-medication generally gives a variable intake of a drug (e.g., if there is depressed appetite in ill animals), and using this technique there is a danger that the low variable doses may favor the development of resistant parasites. For this reason self-medication is not so widely used in livestock, but it appears to be a suitable measure for controlling parasite populations in game-reserves and under modern systems of management. Water medication should only be used if there is a guarantee that nonmedicated water is beyond reach. In intensive pig and poultry systems water or feed medication are the most practical means of administration. For example, prophylactic self-medication aims to control subclinical infections, thus enhancing productivity (→ Coccidiocidal Drugs/Table 1).

Preslaughter Withdrawal Periods of Antinematodal Drugs Preslaughter withdrawal periods of anthelmintics designated for the human consumption of edible tissues and milk may vary considerably and depend on the animal species, characteristics of the drug itself (base/salt, formulation, pharmacokinetics), route of administration, dose regimen, and regulatory authorities of various countries. Table 2 shows some examples for the wide range of withdrawal times of some anthelmintics on the market (withdrawal times for other drugs cf. → Cestodocidal Drugs, → Trematodocidal Drugs, and Table 1, Table 4); data given for edible tissues (e.g. meat, liver, kidney and fat) and milk (lactat-

Table 2. Withdrawal time of some *antinematodal and •antitrematodal drugs.

DRUG (various formulations)	EDIBLE TISSUES days	MILK (dairy cattle) days
* piperazine	18 (ruminants, horse, swine)	2
* diethylcarbamazine	4 (ruminants)	4
* trichlorfon	1 (cattle)	0
* coumaphos	0 (cattle)	0
* thiabendazole	0 (cattle, sheep)	0
* parbendazole	14 (cattle, sheep)	5
* oxfendazole	14 (cattle, sheep)	5
* albendazole	28 (cattle, sheep)	5
* fenbendazole (granules)	7 (cattle, sheep, goat)	3
* fenbendazole (SR bolus)	200 cattle	(contraindicated)
* febantel	14 (cattle, sheep)	2
* levamisole	8 (cattle, sheep)	3
* pyrantel	14 (ruminants), 0 (swine)	0
* morantel (SR bolus)	0 (cattle, sheep)	0
* morantel (powder)	7 (cattle, sheep)	0
* ivermectin s.c./pour-on	38/35 (cattle), 28 (swine s.c.)	(contraindicated)
* abamectin, moxidectin	35 (cattle)	(contraindicated)
* doramectin	50 (cattle)	(contraindicated)
* eprinomectin (pour-on)	0 (cattle)	0
• clorsulon	8 (sheep)	4
• oxyclozanide	14 (sheep)	4
• rafoxanide	28 (sheep)	12
• closantel	28 (cattle), 42 (sheep)	(contraindicated)
• nitroxynil	up to 2 months (sheep)	(contraindicated)

ing animals) are approximate figures. Some antitrematodal drugs may be used for the treatment of benzimidazole-resistant trichostrongyle infections in sheep and goats. These narrow-spectrum drugs are strongly bound to plasma proteins of the host. For this reason some of them have fairly long withdrawal periods comparable with those of the avermectins. The low to absent plasma and milk levels of some compounds following single or sustained administration (e.g. slow or sustained-release bolus) allow use of these compounds in lactating dairy animals without a milk-withdrawal restriction.

Gastrointestinal Nematode Infections of Horses

Epizootiology and Characteristics of Various Nematode Infections The most pathogen gastrointestinal nematodes of horses and therefore of economic importance are two species of large strongyles (*Strongylus vulgaris* and *S.edentatus*) and the cyathostomes (small strongyles). In addition, the ascarids (*Parascaris equorum*) can cause major problems in foals before they develop immunity.

Occasional parasites include, lungworms (*Dictyocaulus arnfieldi*), pinworms (*Oxyuris equi*), and stomach worms (*Habronema* spp., *Trichostrongylus axei*), threadworms (*Strongyloides westeri*) and tapeworms (*Anoplocephala* spp. cf. → Cestodocidal Drugs). There are also nematodes known to be nonpathogenic, e.g., *Probstmayria vivipara*, a minute nematode living in the colon. The females are viviparous and produce enormous numbers of larvae without affecting the host. The larvae of several flies (*Gasterophilus* spp., hairy and reduced mouthparts) are parasites of equines and are known as 'bots'. They are attached to the stomach mucosa and other sites, remaining there for several months. *Gasterophilus* larvae may cause pathogenic effects like ulceration of the esophageal region of the stomach though no firm evidence exists that they produce clinical signs.

The life cycle of most GI nematodes is direct though their mode of transmission may differ. Thus, horses ingested infective larvae (L_3) of small and large strongyles while grazing the pasture. Infections with *Strongyloides westeri* (threadworm residing in the small intestine of horses) may be lactogenic, and/or due to larvae penetrating the skin. Another possibility is that horses may become infected by ingestion of eggs containing infective L_2 or L_3 larvae (e.g., *Parascaris equorum* inhabiting the small intestine and *Oxyuris equi* inhabiting the colon and cecum). There are other nematodes belonging to the family of Spiruridae, which as a rule include an intermediate host (arthropods). Infections with *Draschia megastoma* or *Habronema* spp. (residing in the stomach of horses or other sites) are acquired from ingesting the intermediate host, e.g. maggots of flies or by chance of adult flies. However, most often horses become infected when the adult fly feeds and infective larvae pass forwards into the proboscis and are deposited on the lips, nostrils and wound of horses producing cutaneous habronemiasis. In temperate countries strongyle egg output in horse feces usually shows seasonal variation with a gradual increase of egg output in spring followed by a maximum in summer and a remarkable fall in autumn.

The rise in egg counts in spring/summer usually produces peaks of infective larvae in summer/autumn on horse pastures in Europe and northern parts of the USA. Thus pasture in northern latitudes becomes heavily contaminated with strongyle larvae at this time. Many infective larvae of large and small strongyles will survive sub-zero temperatures in winter but are likely to be killed by alternate freezing and thawing ($> 7.5\,^{\circ}C$) before they reach the more resistant (sheathed) infective stage (L_3). Numerous third larvae will additionally die off by early spring when rising temperatures lead to increased activity of the larvae thereby exhausting food reserves. On the other hand, high temperatures and desiccation in summer are lethal to both eggs and larvae, and harrowing the pasture supports this killing effect and thus decreases larval numbers. Seasonal variation is a characteristic of small strongyles (Cyathostominae) living in the cecum and colon of equids throughout the world. Infective larvae of large strongyles (e.g., *Strongylus edentatus*, *S. vulgaris*, *S. equinus*) have a long prepatent period and need more than 6 months to grow to maturity inside the host, which is mostly in the spring of the following year. Buildup of L_3 in pasture usually occurs in midsummer and is attributable to the hatching of eggs of large strongyles deposited in spring of the same year. Because of their relatively short prepatent period of approximately 2 months, small strongyles (Cyathostominae), including various species of different families, may be capable of establishing a second generation of infective L_3 larvae in the same year.

The pathology and clinical signs caused by GI parasites are varied. Larvae of small strongyles (cyathostomes) enter the wall of the large intestine and migrate in the mucosa and submucosa (inhibited or arrested larvae). They injure the gut wall thereby causing numerous nodules containing hypobiotic or encysted larvae that may persist in the tissues for as long as 2.5 years after horses have been removed from infective larvae. This is of practical importance since they can complete their development and cause larval cyathostomiasis and contamination of pastures even after treatment with anthelmintics (including macrocyclic lactones) and movement to clean pastures. The young adult worms (L_5) then return to the lumen of the colon and/or cecum and mature. Cyasthostomes are often found in very large numbers in the colon and cecum, and heavy infections produce a disquamative catarrhal enteritis.

Infective larvae of large strongyles ingested by the host enter the wall of the intestine and migrate inside the blood vessels. Larvae of *S. vulgaris* migrate toward the cranial mesenteric artery causing arteriitis and thrombosis (with all pathologic consequences) pass back as L_4 about 6 weeks after infection via the arterial system to the submucosa of the cecum and colon where they mature to L_5 about 3 months after infection. They then enter the lumen to grow to maturity; egg production may occur 6–7 months after infection. Larvae of *S. edentatus* pass to the liver via the portal system and L_4 migrate in the liver for several weeks; they then pass via hepatic ligaments to reach the parietal peritoneum, thereby causing hemorrhagic nodules (up to several centimeters in diameter) in the right abdominal flank. Three to 5 months after infection they then migrate via the hepatorenal ligament to the submucosa of the cecum and colon causing additional hemorrhagic nodules. Some young adult worms reach the lumen and become mature. Eggs are produced about 10 months after infection. Larvae of *S. equinus* have a similar route compared to *S. edentatus*. Larvae pass through the peritoneal cavity to the liver where they wander for several weeks causing hepatopathy. About 3 months after infection larvae leave the liver via the hepatic ligaments and pass via the pancreas to the peritoneal cavity; they may reach the cecum and colon about 4 months after infection (route is unknown). The prepatent period is about 8–9 months. When clinical signs become apparent, therapeutic treatment with suitable drugs (e.g. macrocyclic lactones, cf. Table 3) should be directed chiefly against migrating larval stages of large strongyles although in heavy infections adult worms at aberant sites should be considered.

A clinical sign produced by *S. westeri* may be a severe acute diarrhoea in foals and in the donkey. In heavy infections migrating larvae of *P. equorum* produce coughing and circulating eosinophilia. Adult worms of *P. equorum* may be the cause of severe catarrhal enteritis and so diarrhoea, general malaise, and debility. Owing to migration of adult worms to abberant sites, complications may occur in foals suffering from heavy *P. equorum* infections. Thus, adults can migrate into the bile duct or penetrate the bowel wall thereby causing peritonitis, or they can obstruct the lumen of intestine by balling up and thus produce colic-like pain.

Epizootiological-based Control Programs against Development of Anthelmintic Resistance Grazing horses are infected with nematodes to a greater or lesser degree throughout their lives. Cyathostomes are now the major threat to equine welfare.The main purpose of prevention is to reduce the intake of infective larvae or larvated (embryonated) eggs, which are responsible for severe pathological damage. **Rotation** of horse and cattle grazing programs provide more effective parasite control than a continuous horse-grazing program. Horses should first graze heavily contaminated pastures about 8 weeks after commencing spring grazing. Meanwhile, ruminants can safely graze pasture; this procedure should be used whenever parasite populations increase too much. In addition to these rotation-grazing programs, hygienic measures may effectively reduce the risk of parasite infection. Stables should be cleaned frequently, and only clean water and food should be supplied. Another control measure is the strategic use of anthelmintics with the aim of preventing clinical disease and building up protective immunity against GI nematodes. Anthelmintics used strategically mean that parasitic development, epizootiology, local management conditions and status of drug resistance to cyathostomes must be considered in the integrated control program. This includes the timing of treatments to obtain the maximum benefit from treatments and management of pasture so that horses are grazing pastures with reduced larval contamination. The treatment frequency of horses should be reduced to a minimum. Foals need to be treated only when their fecal egg counts exceed 100 EPG, because it is important that they are exposed to sufficient immune stimulation. An overprotective treatment schedule will select strongly for drug resistance and increase the damage to the environment with avermectins. However, if drug-free intervals are too long pasture contamination will not be controlled. Thus the simplest way to determine intervals of treatment is to do fecal egg counts at one or two week intervals after treatment until the egg counts reach levels of pre-treatment values. A more prolonged suppression of egg counts can generally be expected with avermectins/milbemycins (e.g. ivermectin/moxidectin) and shorter egg reappearance periods in yearlings than in adult horses when using the same drug.

Highly susceptible **young horses** present a major problem in parasite control; they show only weak response to anthelmintic treatment under intensive grazing conditions even when treated with non-benzimidazole anthelmintics. In spite of treatment, high fecal egg counts may be evident in weanlings and yearlings all the year round. This is at least partly due to lack of immunity in the yearlings to cyathostomes resulting in a greater accumulation of hypobiotic or encysted cyathostome larvae that may emerge and lay eggs after becoming adults soon after treatment. However, good parasite control in these animals will be obtained with simple pasture management strategies such as pasture sweeping or vacuuming twice a week, alternate grazing with ruminants, or prolonged destocking of the pasture, thereby reducing anthelmintic treatments to one treatment per year, still providing satisfactory worm and colic control.

To achieve lasting reduction of infective larvae on pasture in northern latitudes, anthelmintics should be administered at intervals that comply with epizootiological-based strategies (and according to fecal egg counts). Thus, few strategic treatments, i.e. at least during the first months (April to August) of the grazing season, will effectively reduce the spring/summer rise in fecal egg output in adult horses. The strategy of spring/summer treatment has been used successfully in the northern latitudes (Europe, Canada and the northern USA) for more than 15 years with adult horses under intensive grazing conditions. Using this strategic regimen, autumn and winter treatments were found to be unnecessary against nematodes.

Strategic anthelmintic treatment in autumn is directed against inhibited larval stages. However, it is doubtful whether benzimidazoles are capable of affecting inhibited or arrested larval stages of *Cyathostomum* spp., although they may be effective against lumen-dwelling cyathostome adults and larvae.

Anthelmintic resistance in small strongyles is now widespread and common in horses. This phenomenon generally reveals a close association between the prevalence of cyathostome resistance and the frequent use of benzimidazoles. If these drugs had been used on epizootiology-based control programs (minimum number of treatments, epizootiological principles of nematode control, avoidance of introduction of resistant worms in 'clean' environment) there would be fewer problems with anthelmintic (benzimidazole) resistance. Besides epizootiological-based control programs, there may be other possibilities of retarding or even avoiding the development of anthelmintic resistance: (1) For treatment, only effective anthelmintics at their full-recommended dose should be used. (2) One should rotate anthelmintic classes on an annual basis. However, use of drug mixture, that is the concurrent administration of two or more chemically different anthelmintics rather than rotating drugs of different anthelmintic classes, are suggested to be the most effective way to retard selection for resistance.

Horses cannot be completely dewormed before they are placed in a clean environment. None of the available anthelmintics is capable of removing satisfactory hypobiotic or encysted cyathostomes. Nevertheless, to avoid introduction of resistant worms, new arrivals or returning mares should be treated with a non-benzimidazole anthelmintic and kept off pasture for at least two days. This policy (if possible) would reduce the risk of contaminating pastures with the progeny of newly introduced resistant worms from another farm.

Drugs in Current Use against Nematode Infections in Horses The spectrum of anthelmintic activity of some oldtimers such as phenothiazine, piperazine, and organophosphates (trichlorfon, dichlorvos, haloxon) is rather narrow and their innate effect on target parasites may be variable (Table 3) though organophosphates are highly effective against bots (larvae of *Gasterophilus*).

Most **benzimidazoles (BZs)** are highly active against drug-susceptible adults and lumen-dwelling larvae of large and small strongyles of horses (Table 3). Their current use is, however, limited worldwide by anthelmintic resistance to this chemical class. At their recommended dose, BZs are not sufficiently effective against migratory stages of large strongyles. Because of the large safety margin, administration of a single high dose (should be significantly higher than the recommended dose), or administration of several low doses (as total of the high dose) within a day or for consecutive days can raise efficacy in the majority of BZs. The efficacy of BZs against *Trichostrongylus axei* (residing in the stomach), *P. equorum*, and *S. westeri* is varied (Table 3). Common tolerability of BZs is basically good, with the exception of albendazole, which may be toxic if high doses are given repeatedly.

Small strongyles first developed drug resistance against phenothiazine and later to thiabendazole, the first BZ on the market. Today, there is side-resistance among BZs although **oxibendazole (OBZ)** may still be active against strongyles resistant to other BZs. Fortunately, BZ-resistant strongyles can be affected by a variety of older drugs belonging to different chemical classes, such as levamisole, pyrantel pamoate, various drug mixtures, and organophosphates (e.g., dichlorvos) (Table 3). Anthelmintics still considered effective against small strongyles (cyathostomes) include macrocyclic lactones, and with certain reservations OBZ and **pyrantel pamoate (PP)** (Table 3). Resistance to OBZ has developed in situations where the frequency of use has been high. There may be also a dual resistance of parasites to PP and OBZ.

Nowadays, benzimidazoles become increasingly replaced by endectocide **macrocyclic lactones** (avermectins and milbemycins) exhibiting broad-spectrum anthelmintic activity against nematodes, including arthropods like bots. For the current use in equines against GI nematodes and arthropods (e.g. bots) **ivermectin, moxidectin**, and an **abamectin/praziquantel** combination exhibiting additional efficacy against equine tapeworms (*Anoplocephala* spp.) are available (Table 3).

Ivermectin and moxidectin show high activity against the lumen-dwelling cyathostome adults and larvae whereas their activity against hypobiotic or encysted larvae appears to be basically poor in naturally infected ponies. However, differences between these observations and others related to experimentally infected foals were such that a therapeutic dose of ivermectin proved to be active (76.8 %) against developing mucosal stages (inhibited or arrested in development 35 days postdose). Hypobiotic or encysted cyathostome larvae were not enumerated in this study. Moxidectin demonstrated a trend towards greater efficacy than ivermectin (at recommended dose) against encysted cyathostome larvae. It may be less effective than ivermectin against bots (*Gasterophilus* spp.) and is equally ineffective as ivermectin against the ileocecal tapeworm *Anoplocephala perfoliata*. Ivermectin shows activity against immature and adult stages of the horse → lungworm (Vol. 1) *Dictyocaulus arnfieldi* and moxidectin does not (cf. Table 3). Besides the combination abamectin/praziquantel, additional drug combinations are in preparation for the market, like ivermectin/praziquantel.

Biological Control In future, the use of nematopathogenic or nematode-destroying microfungi as biological control agents against free-living stages of horse strongyles might be an alternative or an adjunct to existing control methods. The potential of the nematode-destroying fungus *Duddingtonia flagrans* and other fungi (e.g. *Arthrobotrys oligospora*, and *Dactylella bembicodes*) to reduce the free-living populations of parasitic nematodes of ruminants, horses, and pigs has been demonstrated. Among the nematopathogenic fungi *D. flagrans* belongs to the group of nematode-trapping fungi producing trapping organs such as constricting (active) or non-constricting (passive) rings, sticky hyphae, sticky knobs, sticky branches or sticky networks. Anchoring of the nematode to the traps is followed by hyphal penetration of the nematode cuticle and once inside trophic hyphae grow out and fill the body of the nematode and digest it. There are also so-called endopathogenic fungi having no extensive hyphal development outside the host's body except fertile hyphae (conidiosphores) that release the spores thereby infecting nematodes by spores. Numerous other antagonistic organisms of nematodes such as earthworm (consume nematodes present in soil and feces) or dung beetles (reduce infective larvae of strongyles chiefly by indirect effects such as eroding cow pads, burying fresh dung in the soil, and partially dispersing the remainder).

Gastrointestinal Nematode Infections of Swine

Economic Importance and Disease Patterns of Nematode Infections in Swine Clinical parasitism markedly affects meat production and meat quality in swine, and also subclinical parasitism appears to be of economic importance, as it will reduce ani-

Table 3. Drugs used against gastrointestinal nematode infections in horses.

<table>
<tr><th>CHEMICAL GROUP, Nonproprietary name (approx. dose, mg/kg body weight, oral route) other information</th><th>*Brand name (manufacturer, company), other information</th><th>Characteristics (chemotherapeutic and adverse effects, miscellaneous comments)</th></tr>
<tr><td colspan="3">Basic information on antinematodal anthelmintics are given below; in equines, anthelmintics should be used strategically; this means that the epizootiology of nematode infection, and local management conditions have to be considered in the treatment program; drug tolerance of small strongyles is now widespread and common; anthelmintic resistance in horse parasites generally reveals a close association between the prevalence of resistance in small strongyles (cyathostomes) and the frequent use of benzimidazoles; if these drugs had been used less frequently in the past or in combination with other anthelmintics showing a different mode of action there would be fewer problems with benzimidazole resistance.
Life cycles among GI nematodes may differ considerably in their generation length and thus generation time and can be influenced through local weather conditions and pasture conditions in certain climate zones; it has been suggested that the rotation of anthelmintics belonging to chemically different groups can retard or even avoid selection of resistance; however, rotating drugs may cause selection of multiple drug resistance in strongyles if chemically different anthelmintics are used against parasite populations of the same generation; therefore anthelmintic classes should rotate on an annual basis; an alternative to rotation might be the simultaneous use of two or more chemically different drugs; the use of effective drugs against target parasites and the correct dose should be strictly observed to safeguard the efficacy of drugs; fecal egg counts should be performed twice weekly postdosing to get information on egg reappearance period and thus best intervals between treatments; particular attention should be paid to large strongyle eggs because of the high pathogenicity of large strongyles (Strongylus vulgaris and S. edentatus) to horses.</td></tr>
<tr><td>Phenothiazine</td><td>powder</td><td>oldtimer which is active against adult stages of small strongyles (more than 90%); it has little or no effect on large strongyles, immature stages of small strongyles, and P. equorum; at therapeutic doses there may be side effects, such as anorexia, muscular weakness, icterus, anemia, but seldom mortality; the drug is still available in combination with trichlorfon and piperazine and can be used for treating benzimidazole-resistant strains of small strongyles; phenothiazine-resistant strains of small strongyles were reported as soon as the early 1960s and markedly reduced use of this drug in subsequent years</td></tr>
<tr><td colspan="3">AMINES</td></tr>
<tr><td>piperazine
(90, base; available as adipate, citrate, phosphate, hexahydrate and others)
(all salts have similar activity; as base it easily absorbs water)</td><td>many suppliers powder (water-soluble formulation from different salts may be given in bran mashes or as aqueous solution via stomach-tube)</td><td>oldtimer, anthelmintic activity was recognized in the 1950s; piperazine adipate has been widely used in horses; is effective (>90) against adult stages of small strongyles and P. equorum (adults, developmental stages: because of the 12-week patency period, repeated doses at 10-week intervals are needed in young animals); there is only moderate effect (70–80%) against adult O. equi and adult S. vulgaris (zero-effect against stomach worms (Habronema); it is well tolerated; at higher doses (13 times the therapeutic dose) transient softening of the feces occurs; mode of action is anticholinergic action at myoneural junction in worms.</td></tr>
<tr><td colspan="3">ORGANOPHOSPHATES
had their origin as pesticides and their main effect on animal parasites is inhibition of nematodal acetylcholinesterase; they have activity against some benzimidazole resistant nematodes and arthropods (e.g. bots-larvae of Gasterophilus spp.).</td></tr>
</table>

Table 3. (Continued) Drugs used against gastrointestinal nematode infections in horses.

CHEMICAL GROUP, **Nonproprietary name** (approx. dose, mg/kg body weight, oral route) other information	*__Brand name__ (manufacturer, company), other information	**Characteristics** (chemotherapeutic and adverse effects, miscellaneous comments)
trichlorfon (=**metrifonate**) (35–40) used as a pesticide for plants (Dylox), as an insecticide, and anthelmintic in livestock (Neguvon)	*Neguvon (Bayer) powder *Combot (Haver-Lock-hart) paste, bolus, or powder mixed in a single-day ration	spectrum is rather narrow; more than 90% efficacy against adult and immature *P. equorum* (adult and immature stages), adults pinworms (*O. equi*) and against bots (larvae of *Gasterophilus nasalis* and *G. intestinalis*); at higher doses (60 mg/kg) the drug is active against *S. vulgaris* and small strongyles; trichlorfon is frequently used in combination with mebendazole (Telmin Plus), or febantel (Rintal Plus), or simultaneously with other BZs, or pyrantel pamoate, or piperazine/phenothiazine (drench) for removal ascarids, pinworms, small strongyles (cyathostomes) and all three species of large strongyles (there is insignificant effect against migratory larvae of *S. vulgaris* in walls of mesenteric arteries); at therapeutic dose there may be mild adverse effects (transient softening of feces and mild colic for several hours though horses may tolerate 80 mg/kg in-feed); the conversion of trichlorfon at physiological pH to dichlorvos is believed to contribute to its activity; (its action against *Schistosoma haematobium* see relevant tables)
dichlorvos (35) (20)	*Equigard (Parke-Davis) pellets, slow release resin *Equigel, paste (Shell) *Aquagard *Nuvan (Ciba Geigy)	spectrum of activity is basically similar to that of trichlorfon; however, slow release resin increases activity against large and small strongyles; the drug is also highly active (more than 90%) against mature and immature *P. equorum*, *O. equi* and bots (*Gasterophilus* spp.); pellets show no effect against stomach worms (*T. axei*, *Draschia megastoma* and *Habronema muscae*), which cause summer sores, and cutaneous habronemiasis; at therapeutic doses there may be side effects such as soft feces, salivation, muscle tremors, and incoordination with the paste; treatment causes no ill-effects in mares and foals
haloxon (60)	*Equivurm (Crown Chemical) paste *Loxon (Wellcome) powder *Eustidil,*Halox	is highly effective (more than 90%) against adult stages of *S. vulgaris*, most small strongyles (also benzimidazole-resistant strains), *P. equorum*, and *O. equi*; at 3 times the recommended dose there are no ill-effects; at recommended dose it is a safe drug for pregnant mares; it also exhibits moderate activity against *Schistosoma mattheei* as does trichlorfon; repeated high doses of haloxon (300 mg/kg x 2) are necessary to affect this trematode
BENZIMIDAZOLES (BZs) At regular therapeutic (recommended dose), BZs have little or no activity against migrating larvae of *Strongylus vulgaris* in adventitia of arteries or against stomach worms (*Habronema muscae*, *Draschia megastoma* and others) and bots (*Gasterophilus intestinalis*, *G. nasalis*); efficacy of BZs against lungworms (*Dictyocaulus arnfieldi*) is evident after repeated and enhanced doses; widespread resistance of small strongyles (cyathostomes) against BZs has limited their use in horses in time (action on tapeworms see → Cestodocidal Drugs)		
thiabendazole (50) **combinations:** tiabendazole/piperazine tiabendazole/trichlorfon	*Thibenzole (Merial) suspension, paste, powder; *Equizole A *Equizole B	is highly effective (more than 90%) against adult stages of large and small strongyles (to a lesser extent immature stages), *O. equi*, *P. vivipara* and *S. westeri*; at recommended dose, efficacy against *P. equorum*, *T. axei*, and *O. equi* (L4) is insufficient and dose must be given twice for activity against these parasites; at extremely high dose levels (440 mg/kg x 2) the drug is effective against 14-day-old larvae of *S. vulgaris* and *S. edentatus* (reduced appetite, depression, and mild colic may occur at 24 times the recommended dose); it appears to be a safe drug for mares (also during pregnancy) and foals; tiabendazole-resistant small strongyles show side-resistance to related benzimidazoles

Table 3. (Continued) Drugs used against gastrointestinal nematode infections in horses.

CHEMICAL GROUP, **Nonproprietary name** (approx. dose, mg/kg body weight, oral route) other information	*__Brand name__ (manufacturer, company), other information	**Characteristics** (chemotherapeutic and adverse effects, miscellaneous comments)
cambendazole (20)	*Equiben (Merial) *Ascapilla (Chevita) paste, no more in use	has a broad spectrum of activity; it is highly effective (more than 90%) against adult stages of *P. equorum*, *P. vivipara*, *S. vulgaris*, *S. edentatus*, small strongyles, *S. westeri*, and *O. equi* (the drug is not licensed for use in horses in many countries); it is particularly effective against stomach worm *T. axei* but ineffective against *Draschia megastoma*; at therapeutic dose it is less active (75%–90%) against immature stages of small strongyles; the drug appears to be well tolerated at 8 times the recommended dose (30 times may cause transient depression and softening of feces); there is side-resistance to related benzimidazoles; cambendazole has been found to be teratogenic, limiting its use in pregnant animals
BENZIMIDAZOLE CARBAMATES differ in their structure from thiabendazole (and thiophanate) in having a carbamate substitution on C5 of the benzene ring increasing activity		
mebendazole (8.8) **combination:** mebendazole/ trichlorfon	*Telmin(Janssen) granules, paste; *Telmin plus	has a broad spectrum of activity; efficacy is directed against adult stages of large strongyles (more than 90%) and small strongyles (75%–90%) as well as against *O. equi* (also immature stages) and *P. equorum* (both more than 90%); it is less active (about 75%) against immature stages of small strongyles; migrating larvae of *S. vulgaris* were affected by 13.6 times the recommended dose (120 mg/kg x 2); it is poorly effective against *T. axei*, *S. westeri*, as well as against *Habronema muscae* and *D. megastoma*; a single dose of 20 mg/kg is effective against *Dictyocaulus arnfieldi*; from 5 times the recommended dose upwards there may be slight side effects (fecal softening, diarrhoea); there is side-resistance to the other benzimidazoles; the drug has been found to be teratogenic, limiting its use in pregnant animals; drug combination mebendazole and metrifonate has additional efficacy against *Gasterophilus* spp.
parbendazole (2.5 for 2 days) (10)	*Helmatac *Neminil (SKF) suspension, powder	is highly effective (more than 90%) against adult stages of *S. vulgaris*, *S. edentatus*, small strongyles, and *O. equi*; *P. equorum* is fully affected only at 10 times the recommended dose; the drug is not active against *T. axei*, *S. westeri*, *D. megastoma*, or *Habronema* spp.; there may be transient diarrhoea, anorexia, listlessness; there is side-resistance to the other benzimidazoles; the drug has been found to be teratogenic (Table 1) which may limit its use in pregnant animals
fenbendazole (7.5)	*Panacur, *Axilur (Intervet) suspension, paste, granules	has a broad spectrum of activity; it is highly effective (more than 90%) against adult stages of large and small strongyles and adult and immature stages of *O. equi*, *P. vivipara*, and *P. equorum* (10 mg/kg x 1 or 5 mg/kg x 2); higher doses (e.g. 60 mg/kg) or repeated doses (7.5. mg/kg daily for 5 days) give good control of larval stages of small strongyles in the gut lumen and in the mucosa, and of migrating larvae of *S. vulgaris*, *S. edentatus*, and *S. westeri*; repeated (10 mg/kg x 5) or high doses (30–60 mg/kg) affect *H. muscae*, *D. megastoma*, and *T. axei*; a single dose of 50 mg/kg is effective against *Dictyocaulus arnfieldi*; drug has no teratogenic effects and does not interfere with reproductive function of stallions; drug is well tolerated and shows no adverse effects at 500 mg/kg and higher doses; there is side-resistance to the other benzimidazole compounds

Table 3. (Continued) Drugs used against gastrointestinal nematode infections in horses.

CHEMICAL GROUP, **Nonproprietary name** (approx. dose, mg/kg body weight, oral route) other information	***Brand name** (manufacturer, company), other information	**Characteristics** (chemotherapeutic and adverse effects, miscellaneous comments)
oxfendazole (10)	*Synanthic (Synthex) *Systamex (Wellcome, Essex) suspension, pellets, granules, paste	has broad spectrum of activity; it is highly effective (more than 90%) against adult stages of large strongyles and those of small strongyles (including their immature stages in the gut lumen or in the mucosa) and against *P. equorum* and *O. equi* (including immature stages of the latter species); the drug is particularly effective against *T. axei*; its action against migrating *S. vulgaris* appears to be more variable; oxfendazole shows poor efficacy against *S. westeri*, *H. muscae*, and *D. megastoma*; at 10 times the recommended dose transient softening of feces may occur; oxfendazole has been found to be teratogenic, limiting its use in pregnant animals; there is side-resistance to the other members of benzimidazoles
oxibendazole (10) **combination:** oxibendazole/ dichlorvos (*Equiminthe-plus paste)	*Equiminthe (Laboratories Reading) and many others: *Anthelcide *Equipar, *Loditac *Verzine suspension, paste	has a broad spectrum of activity; it is highly effective (more than 90%) against adult stages of large strongyles and those of small strongyles (including immature stages in the gut lumen) and against *S. westeri*, *P. equorum*, *O. equi*, and *P. vivipara*; the drug shows poor efficacy against *T. axei*, *H. muscae*, *D. megastoma*, and migrating stages of *S. vulgaris*; it may be active against strongyles resistant to other benzimidazoles; oxibendazole appears to be safe for horses (at 3 times the recommended dose there were no side effects, at 4 times the recommended dose; the drug was found to be embryotoxic in rats and sheep)
(**albendazole**)	(brand names and companies cf. Table 1)	although it has broad spectrum of activity, the drug is not used in horses (cause unknown); at 5 mg/kg it is highly effective (more than 90%) against adults of large and small strongyles, *O. equi* (more than 90% against immature stages), and *P. equorum*; its effect against immature larvae (L4) of small strongyles in the gut lumen is moderate (about 70%–90%); at 5 times the recommended dose (three times daily for 5 days) the drug proved highly effective against 30-day-old migrating larvae of *S. vulgaris*; this dosage regimen may, however, cause severe side effects (diarrhoea), and can be fatal; dose regimen of 25 mg/kg 2x/d is effective against lungworm *D. arnfieldi*; albendazole has been found to be teratogenic in lambs, limiting its use in pregnant animals
PROBENZIMIDAZOLES		
febantel (6) **combination:** febantel/ metrifonate (6/30)	*Rintal (Bayer) suspension, paste, granules ***Rintal plus (Bayer)** paste	has a broad spectrum of activity; the drug shows high efficacy (more than 90%) against adult stages of *S. vulgaris*, *S. edentatus*, small strongyles, *O. equi*, and *P. equorum*, and their immature stages; there is only poor activity against migrating larvae of large strongyles, *H. muscae*, *D. megastoma*, *T. axei* (elimination at 20 mg/kg), and *S. westeri* (elimination at 60 mg/kg); as with the other benzimidazole there exists resistance of horse nematodes against febantel; the drug is well tolerated showing no side effects at higher doses; it is quickly metabolized in the liver to fenbendazole and oxfendazole; febantel/ metrifonate combination is highly effective (more than 99%) against adult and immature horse nematodes (*S. vulgaris*, *S. equinus*, *S. edentatus*, small strongyles, *P. equorum*, *O. equi* and *Gasterophilus* spp.; combination is well tolerated; side effects are salivation and restlessness; there are no embryotoxic or teratogenic effects

Table 3. (Continued) Drugs used against gastrointestinal nematode infections in horses.

CHEMICAL GROUP, **Nonproprietary name** (approx. dose, mg/kg body weight, oral route) other information	*__Brand name__ (manufacturer, company), other information	**Characteristics** (chemotherapeutic and adverse effects, miscellaneous comments)
TETRAHYDROPYRIMIDINES		
pyrantel pamoate, insoluble in water (19=6,6 free base) **pyrantel** tartrate, water-soluble (6.6) (in-feed medication on a day-to-day basis appears to render foals more susceptible to parasite challenge than previously untreated animals with greater exposure to parasites)	*Banminth *Strongid *Antiminth *Cobantril *Felex suspension (drench), **paste**, granules, powder *Strongid C *Nemex powder, granules for in-feed on a day-to-day basis in foals (can be started on this drug at 2 or 3 months of age and grain intake is sufficient (Pfizer, other)	whether it is the tartrate or pamoate, both salts of pyrantel (PYR) exhibit similar activity on GI parasites; the two salts are highly active (>90% against adult stages, also early larval stages of *S. vulgaris*, *S. equinus*, small strongyles , *P. equorum* and only moderately (variably: 33–90%) against adult stages of *S. edentatus* and *O. equi* (50–75%; effect against larval mucosal stages of small strongyles (cyathostomes) and other nematodes is minimal; drug is inactive against stomach worms (*T. axei*, *Habronema spp.*, *Draschia megastoma*), *Strongyloides westeri*, and bots (*Gasterophilus* spp.); activity is enhanced by concurrent administration of PYR pamoate and **trichlorfon or dichlorvos**, each at recommended dose; PYR is active against ileocecal tapeworm *Anoplocephala perfoliata* at double the regular dose (13.2 mg/kg); it shows efficacy against BZ resistant strains of cyathostomes; PYR is well tolerated at recommended dose and may be used in pregnant mares, in foals, and in stallions (reproductive performance is not affected); at higher and repeated doses (free base 50 mg/kg) severe toxic reactions (dyspnea, muscular tremor, even death) may occur; the pamoate (=embonate) is used in horses because of its low solubility, thus providing higher concentrations and activity against worms inhabiting the colon and cecum (e.g. cyathostomes)
IMIDAZOTHIAZOLES		
(levamisole)	(brand names and companies cf. Table 1) solution for injection; suspension (drench), powder (in feed)	although highly effective (more than 90%) against adult stages of *S. vulgaris*, *P. equorum*, and *O. equi*, the drug is not marketed for use in horses because of its narrow therapeutic index and spectrum of activity; its efficacy against large stages of *S. edentatus* and small strongyles is limited; the drug is ineffective against migrating larvae of *S. vulgaris*, *T. axei*, *Habronema* spp., and *P. vivipara*; after oral administration toxic effects may occur at 2 times the therapeutic dose (sweating, increased respiration, hyperexcitability, even death); at therapeutic dose (5 mg/kg i.m., 10 mg/kg p.o.) the drug may cause local reactions and signs of a colic after intramuscular injection
MACROCYCLIC LACTONES		

Endectocides with broad spectrum of activity against nematodes and arthropods; they are effective against nematodes resistant to other classes of antinematodal drugs, such as benzimidazoles (BZs); resistance of nematodes to macrocyclic lactones has been observed though degree of side resistance between avermectins and milbemycins is not clear till now; effects of macrocyclic lactones on dung-destroying insects and other environmental impact is discussed below.

Table 3. (Continued) Drugs used against gastrointestinal nematode infections in horses.

CHEMICAL GROUP, **Nonproprietary name** (approx. dose, mg/kg body weight, oral route) other information	***Brand name** (manufacturer, company), other information	**Characteristics** (chemotherapeutic and adverse effects, miscellaneous comments)
AVERMECTINS		
ivermectin (IVR) (0.2) *Eqvalan liquid contains 1% IVR (w/v) (either oral drench or nasogastric tube) Fatal *Clostridium* sepsis has occurred in very few cases with the injectable formulation apparently caused by needle contamination; s.c. formulation has been largely replaced by oral paste or drench formulations	*Eqvalan (Merck) *Ivomec *Ivomec P *Oramec *Cardomec *Zimectrin *Rotectin *Furexel (Merck & Co., Inc., Merial and others) (e.g. *Eqvalan paste *Zimectrin paste with 1.87% IVR (w/v) in a vehicle of 79% propylene glycol plus inert binders)	has a broad-spectrum of activity with a prolonged persistent action on reproductive system of worms (reduction in fecal egg counts may be 2 months or longer); is active (>90%) against adult and most early and late (4th) stage larvae of all pathogenically important small strongyles (cyathostomes including *Triodontophorus* spp., adults) and large strongyles (*S. equinus*, *S. edentatus*, *Strongylus vulgaris*: ~99% against early and late 4th stage larvae reducing markedly acute signs of acute verminous arteriitis within 2days of treatment; resolution of lesions may occur in ~1 month postdosing), ascarids (*Parascaris equorum*,), pinworms (*Oxyuris equi*, adult & immature)), intestinal worms (adult *Trichostrongylus axei*, *Strongyloides westeri*), lungworms (adult and larval stages of *Dictyocaulus arnfieldi*), against migrating or stomach-attached stages of *Gasterophilus* bots (*G. intestinalis* and *G. nasalis*), *Onchocerca* microfilariae (producing skin lesions)and stomach worms (adult and third larval stages of *Habronema* spp and *Draschia* spp. that incite cutaneous 'summer sores' resolving not until after administration of a second therapeutic dose of IVR 1 month after initial treatment); though IVR has high activity against the lumen-dwelling cyathostome adults and larvae their activity against hypobiotic or encysted larvae proved to be basically poor in naturally infected ponies; there is no activity against horse ticks; IVR exhibits full activity against benzimidazole-resistant small strongyles; it is well tolerated at therapeutic dose (substantial margin of safety, 10-fold) and can be used in pregnant mares, foals, and stallions
abamectin/ praziquantel (5.4 g paste/100 kg body weight)	*Equimax (Virbac) (Merial) paste	combination was introduced in New Zealand and Australia in 1997; its anthelmintic spectrum against nematodes may basically be similar to that of ivermectin including all three species of *Gasterophilus* bots (see above); however, additional effect of praziquantel, remove also ileocecal tapeworms (*Anoplocephala* spp.); at recommended dose combination is safe for mares, stallions, and foals; there is a 5-fold margin of safety

Table 3. (Continued) Drugs used against gastrointestinal nematode infections in horses.

CHEMICAL GROUP, **Nonproprietary name** (approx. dose, mg/kg body weight, oral route) other information	***Brand name** (manufacturer, company), other information	**Characteristics** (chemotherapeutic and adverse effects, miscellaneous comments)
MILBEMYCINS (NEMADECTINS)		
moxidectin (0.4) residues excreted in feces of moxidectin treated animals are less toxic to dung beetle larvae than those of animals treated with ivermectin and allow survival of dung beetles or their development to maturity	*Equest (American Home, Fort Dodge) *Quest (Fort Dodge) paste, gel (American Cyanamid, Fort Dodge: *Cydectin for cattle, cf. Table 1)	introduced as horse dewormer in 1996; has a broad-spectrum of activity with a prolonged persistent action on reproductive system of worms (reduction in fecal egg counts may be 3 months or longer); its anthelmintic spectrum is largely similar to that of ivermectin (see there), except lack of action against lungworm *Dictyocaulus arnfieldi* and a trend towards 'greater' efficacy against encysted cyathostome larvae than a therapeutic dosage of ivermectin (difference was not significant); it proved highly efficacious against luminal small strongyle larvae(~100% against L4 , >92% against L3, and some efficacy against encysted or hypobiotic larvae as well); there is 99–100 % efficacy against adults of *Strongylus vulgaris* (larvae >90%) , *S. edentatus* (including larvae), *Triodontophorus* spp. and 22 species of small strongyles (cyathostomes), *Oxyuris equi* (larvae 100%, adults >94%), *Habronema muscae* (adults , larvae, and *Parascaris equorum* (adults, larvae); its activity against bots may be variable (50–100%) and substantially less than that of ivermectin; it exhibits activity against ticks (*Amblyomma cajennense* and *Dermacentor nitens*); at recommended dose, moxidectin is safe for breeding animals (mares and stallions), and foals (at least from 4 months of age: foals below this age, or debilitated animals can infrequently show transient depression, ataxia and recumbency); there may be a 3-fold margin of safety

Doses listed in this table refer to information from manufacturer and literature
Data given in this table have no claim to full information.
The primary or original manufacturer is indicated if there is lack of information on the current one(s)

mal productivity as well. Condemnation of livers, kidneys and the necessary trimming of loins and other valuable parts of carcasses have resulted in important economic losses for the swine industry. The importance of the meat production in the swine industry to the animal health market is reflected in some figures. Thus annual pork meat production (approx. weight in tons) in China, the USA, and Germany was 31 million, 7.6 million and 3.5 million, respectively (recent FAO report). Gradually increasing in the 1970s and 1980s, the latest published figure available concerning liver condemnation index (lb. liver condemned/ numbers of pigs slaughtered) was 0.51 for 1990. Infections of pigs with nematodes can cause reductions in growth rate and efficiency of feed utilization. Thus, parasite infections are continuously found throughout the modern pig industry.

There are several nematode infections which may be economically important. Nodular worms *Oesophagostomum* spp., which reside in the large intestine, the **red stomach worm**, *Hyostrongylus rubidus*, which invades gastric mucosa (gastric glands) and sucks blood often occur in breeding animals. *H. rubidus* can produce mild fever, loss of appetite, diarrhoea, weakness, and reduced weight gain. Weaners and fattening pigs are often infected with *Ascaris suum* (**eelworm**), which inhabits the small intestine, and *Trichuris suis*, which lives in the large intestine.

The intestinal **threadworm**, *Strongyloides ransomi*, in the small intestine may produce severe clinical signs in suckling piglets. Initial anorexia, then diarrhoea, which may become continuous and hemorrhagic characterize strongyloidiasis; pulmonary disorders may also be seen. Since infection with *S. ransomi* is acquired from both milk-borne infective larvae (L_3) entering the host through the mouth (e.g., per os infection with the colostrum) and larvae entering the host through the skin, lesions of the skin, such as erythema and pustular reactions, may also be seen. In heavy infections mortality can reach 50%. Death is mainly caused by a protein-loosing enteropathy.

Acanthocephala (*Macracanthorhynchus hirudinaceus*) occur in the small intestine of the domestic pig and wild boars and are present worldwide. Dung beetles (grubs or adult beetles) of the family Scarabaeidae act as intermediate hosts. Parasites penetrate with their probosces into the intestinal wall, thereby producing inflammation and a granuloma at the site of attachment; perforation of the intestine may cause peritonitis and death. Severe infections lead to reduction in growth or emaciation, while mild infections are not very harmful.

Kidney-worms (*Stephanurus dentatus*), occur in the peritoneal fat, the pelvis of kidneys and in the walls of ureters (aberrant sites are liver or other abdominal organs, sometimes thoracic organs, and spinal canal; for literature and more information on treatment cf. Table 4, doramectin). Infection of pigs with infective larvae occurs per os (earthworms, *Eisenia foetida*, may serve as transport hosts) or through the skin. The kidney-worm is widely distributed in tropical and subtropical areas. The general clinical signs are temporary subcutaneous nodules (early stage of infection), depressed growth rate, loss of appetite and later emaciation; also stiffness of the leg and posterior paralysis may occur.

Ascariasis is extremely common in swine, especially in young animals. Infection usually takes place through ingestion of larvated *A. suum* eggs with food or water or from the soiled skin of the sow in the case of suckling pigs. Eggs hatch in the intestine and the larvae pass through the wall of the gut into the peritoneal cavity and then to the liver where they cause tissue damage and hemorrhage. Larvae then migrate to the lungs, and break out of the alveolar capillary into the alveoli and bronchioles, causing edema and cellular reactions (infiltration of eosinophils). In heavy infections death from severe lung damage may occur, or piglets may remain stunted for a long period. Larvae then migrate from the trachea to the pharynx and are swallowed; the L_3 larvae then may arrive at the intestine 1-week after infection.

Globocephalus spp. (e.g., the **hookworm** *G. urosubulatus*) occurring in the small intestine of wild boars and occasionally in the domestic pig may cause anemia in heavy infection. The life cycle is probably direct. The infection may be due to oral ingestion of L_3 or to infective larvae penetrating the skin.

Trichinella spiralis, the cause of **trichinosis** in almost every country, may lead to serious clinical signs produced by newborn larvae being distributed all over the body via the blood circulation. They grow further, particularly in the voluntary muscles of the tongue, larynx, eye, diaphragm, and the intercostal and masticatory muscles. The larvae then enter striated muscle fibers and become encysted. The capsule is formed from the muscle fiber and the structure of muscle cell is modified (enlargement of nuclei, increase in the number of mitochondria). The so called nurse cell probably plays a role in larval nutrition. Although calcification of the capsule begins after 6–9 months, larvae may live in them and remain infective for several years. Mainly the pig disseminates human trichinosis (→ Nematocidal Drugs, Man/Table 1).

Nodular worms (*Oesophagostomum* spp.) occur in the large intestine of pigs and peccaries throughout the world and have a high incidence of 50%–90% in sows. Nodule formation caused by larval stages (particular *O. dentatum*) is responsible for various clinical signs, such as anorexia and bloodstained feces. In severe infection enteritis may cause death. After ingestion, infective larvae exsheath in the small intestine, causing small nodules (4–5 mm in diameter). The larvae usually reenter the lumen of the large intestine a week after infection, having molted to L_4 larvae. However, some of the L_4 larvae may remain in the nodules for several weeks. Patency is normally reached at about 7 weeks after infection.

The **whipworm** *Trichuris suis* (morphologically identical to *T. trichiura* of man, → Nematocidal Drugs, Man/Table 1) is cosmopolitan in distribution. Pigs become infected by ingestion of larvated eggs, which may reach the infective stage after about 3 weeks under favorable conditions (correct soil moisture and temperature). The eggs may remain viable for several years. After being ingested the larvae hatch in the small intestine and penetrate the small intestine for several days before moving to the cecum where they grow to adults. Infections with *T. suis* occur chiefly in 2- to 4-month-old fattening pigs, and are less common in piglets, sows, and boars. Pathogenicity may be due to the fact that *Trichuris* spp. are blood feeders. Adult worms tunneling into the mucosa cause damage. The mucosa becomes edematous and necrotic thus resulting in catarrhal inflammation of the colon and cecum. In heavy infections colitis and cecitis may lead to watery and bloody diarrhoea. Pigs kept outside or under extensive conditions may occasionally suffer from clinical trichuriasis.

Control Measures and Drugs in Current Use against Nematode Infections in Swine Since production efficiency is of critical importance in the pig industry, precautionary measures such as the all in – all out system, and **strategic herd deworming** must be carried out at regular intervals. Significant worm burdens may be expected when fecal material accumulates and remains accessible to pigs, such as in housings or on pasture in deep litter. Drinking and feeding installations must be kept as clean as possible. Individual treatment has little or no effect on the prevalence of parasites or the degree of infection on the entire stock and most herds are continuously parasitized by several worm species. The sow is thus the most important source of infection for piglets. Reinfection of the entire stock can be markedly reduced by a tactical deworming schedule that should vary with different conditions and the type of farm (e.g., fattening or breeding farm, mixed farm, open or closed, all in – all out, the type of run, and hygienic conditions). This may be achieved by regularly deworming the whole herd simultaneously in a several-day treatment program. New production of large numbers of eggs can be controlled if all animals are treated again as soon as the larvae have grown to maturity. The prepatent period of the worm species should therefore determine the frequency of treatment, i.e., every 2 months in the case of *A. suum*, *Oesophagostomum* spp., and *T. suis*, every 3–4 weeks for *H. rubidus*, and every 8–10 days for *S. ransomi*.

Treatment of **trichinosis** during the muscle phase of infection is unsatisfactory although several **benzimidazoles** show good activity against early stages of *T. spiralis*. **Flubendazole** may eliminate intestinal and migratory stages in experimental infections in pigs when given in feed for longer periods (Table 4). As far as man is concerned, prophylaxis should aim at the thorough cooking of all pork products and the meat of wild animals. The elimination of uncooked garbage such as raw or partly cooked pork and sausages in the feed may prevent infection in domestic pigs.

The oldtimer **piperazine** (most commonly used salts are the citrate in-feed, and the hexahydrate in-water) have been used extensively in swine against adult ascarids and nodular worms. Among the organophosphorus compounds, **dichlorvos** has broad-spectrum of anthelmintic activity, though its effect against migrating and mucosal larval forms of GI nematodes is little. There is no ovicidal effect but a marked action on a portion of freshly hatched and free-living *Oesophagostomum* spp. larvae.

Pyrantel tartrate in-feed is chiefly used for its prophylactic activity against migrating stages of *A. suum*, and *Oesophagostomum* spp and, hence, to prevent establishment of patent infections of these parasites. Most widely used method for deworming pigs with **levamisole** is administering the drug via drinking water or feed; an injectable formulation is also available, which may show higher activity against whipworms (*Trichuris suis*) than oral regimens. The drug is highly active against the majority of other important GI nematodes, including lungworms (Table 6) and kidney worms (*Stephanurus dentatus*) residing in the urinary tract (for more information cf. Table 4, doramectin).

Benzimidazoles (BZs), such as thiabendazole, cambendazole, parbendazole, mebendazole, flubendazole, fenbendazole, and febantel (prodrug of fenbendazole) and another prodrug thiophanate, have a broad spectrum of activity against nematodes of swine (Table 4). BZs, in general, exhibit higher activity at low-level medication for several days than at single dosing. There may be several medicated articles (powders) to make medicated feed for weaners/fatteners or sows for 'long-term' treatment, in that the therapeutic dose (mg/kg body weight) is distributed over 5–15 days. Another dosage regimen may be to divide the therapeutic dose in two and to administer this dose on 3 or 4 consecutive days thereby enhancing the absorption of the drug from the intestinal tract and, hence, increase its anthelmintic efficacy. Most BZs are highly active against adult *A. suum*, *H. rubidus* and *Oesophagostomum* spp. The 'newer' ones show action against *T. suis*, kidney worms and lungworms and a few are effective against immature stages of various GI nematodes (cf. Table 4). BZs appear to be ineffective against spirurid worms (*Macracanthorhynchus hirudinacceus*) occurring in the small intestine of the domestic pig and wild boars (none of BZs claim efficacy for this parasite).

A few avermectins such as **ivermectin** and recently **doramectin** are available as broad-spectrum anthelmintics for use in pigs. The two drugs provide high reduction rates in immature and adult stages of common nematodes, including parasitic arthropods (lice, and mange mite); their action on whipworms (*Trichuris suis*) seems to be variable. While several compounds are effective in treating patent infections of the threadworm *Strongyloides ransomi*, ivermectin appears to be so far the only drug, which exhibits action on somatic third-stages of this parasite in the sow. Thus a premix of ivermectin given to pregnant gilts (daily dose of 100 mcg/kg → equal to 2ppm ivermectin) for 7 days prevented shedding of larvae in sow milk, egg output in feces and the establishment of *S. ransomi* in piglets.

However, as in ruminants and horses, neither the benzimidazoles, including levamisole (or pyrantel) nor the avermectins (ivermectin and doramectin), are uniformly effective against adult and larval stages of all economically important nematodes in pigs (Table 4). In particular, there is a lack of information on the comparative values of broad-spectrum anthelmintics against the larval and immature 5th stages of the GI parasites of swine. In general, these compounds cause marked reduction in both larval and adult stages of GI nematodes and lungworms (Table 6).

Nematode Infections of Dogs and Cats

The veterinary significance of nematode infections of dogs and cats is related to the large number of domestic pet owners in the western industrial countries. In 1998, in the USA estimated numbers for cats and dogs come to about 60 million and 53 million, respectively, and the total number of dogs amounts to about 30 million in some European countries such as France, Italy, Germany and Spain. An example for the economic importance of the food industry in Germany is total annual sales of nearly DM 2.600 million spent on food for around 5 million dogs and 6.2 million cats by pet owners in 1998.

Puppies and kittens are often infected with gastrointestinal nematodes that may cause zoonotic infections such as visceral or **cutaneous larva migrans** in humans. A special hazard may arise for children who have close contact with young puppies. It is the young puppy preferentially infected with the ascarid *Toxocara canis* causing **visceral larva migrans** characterized by severe pathogenic effects causing persistent cough, intermittent fever, loss of weight, and eye lesions (for more detail cf. → Nematocidal Drugs, Man especially → Nematocidal Drugs, Man/Table 1). The main types of gut nematodes found in carnivores live in the small intestine. These may be **ascarids** such as *Toxocara canis* of the dog and fox; *Toxocara cati* of the cat and wild Felidae, *Toxascaris leonina* of the dog, cat, fox, and wild Felidae and Canidae, and blood sucking **hookworms**, which can produce severe anemia in pups and kittens. Common hookworms in the tropics and warm temperate areas are *Ancylostoma caninum* of the dog, fox, wolf, and other wild Canidae, *A. tubaeforme*, the common hookworm of the cat, and *A. braziliense* of the dog, cat, fox, and other wild Canidae. *A. braziliense* can be responsible for **cutaneous larva migrans** or so-called **creeping eruption**, an intensive itching dermatitis in humans (→ Nematocidal Drugs, Man, especially → Nematocidal Drugs, Man/Table 1). Not so common is *A. ceylanicum* of the dog, cat, and wild Felidae occurring in Malaysia and other parts of Asia. *Uncinaria stenocephala* is a hookworm of dogs, cats, and foxes occurring in temperate climates, e.g., the USA or Europe. Canine and feline nematodes living in the large intestine (cecum and colon) are **whipworms** such as *Trichuris vulpis* of the dog and fox; *T. serrata* and *T. campanula* of the cat in South America, Cuba, and the USA. Adult *Trichuris* spp. may cause mucosal damage (necrosis, hemorrhage) by tunneling into the mucosa of the large intestine. *T. vulpis* is a blood feeder and its mouth stylet is used to enter vessels or to injure tissues, giving rise to bleeding; the blood-pools thus created are then ingested by the adults. So-called **heartworms** (*Dirofilaria immitis*) belonging to the superfamily Filarioidea and living in the venous circulation of carnivores are responsible for the debilitating heartworm disease, especially of dogs, which may be enzootic in areas with a tropical or subtropical climate.

Prevention and Treatment of Canine and Feline Nematode Infections Since almost all gastrointestinal nematodes are harmful to dogs and cats and some of them are a hazard to human health (cf. larva migrans in → Nematocidal Drugs, Man or cf. echinococcosis in → Cestodocidal Drugs) effective control measures should be performed to protect cats, dogs, and fur-bearing animals from these parasites. Transmission of nematodes can be reduced by hygienic measures in kennels and cat-

Table 4. Drugs used against nematode infections of swine.

CHEMICAL GROUP, **Nonproprietary name** (approx. dose, mg/kg body weight, oral route) other information	* **Brand name** (manufacturer, company), other information	Characteristics (chemotherapeutic and adverse effects, miscellaneous comments)
AMINES		
piperazine base (110) piperazine citrate, (in-feed) piperazine hexahydrate (in water)	various trade names and suppliers powder (water-soluble) the hexahydrate is very unstable and soluble in water as the citrate	there is 100% elimination of lumen-dwelling (adult) stages of both ascarids (*A. suum*) and nodular worms *Oesophagostomum* spp. after a single treatment; it may still be useful for mass treatment against these GI nematodes; a second treatment 2 months later may be necessary to remove worms that have been in somatic stages during initial infection; withholding of feed or water previous night should be observed to make sure that medicated in-feed or medicated in-water is completely consumed; drug is well tolerated at recommended dose levels; no serious form of intoxication has been seen after 4-10 times the therapeutic dose; preslaughter withdrawal time for edible tissues may be 2 days
ORGANOPHOSPHATES		
trichlorfon =metrifonate (50, for 2 days in-feed)	* Neguvon (Bayer) powder mixed in a single-day ration	shows good efficacy (75%-90%) against adults of *A. suum, T. suis*, and *H. rubidus*; its effect against *S. ransomi* and *T. spiralis* is variable; the cholinesterase inhibitor may cause transient side effects (diarrhoea, muscular tremors) at the therapeutic dose; atropine can be used as an antidote; preslaughter withdrawal time for edible tissues may be 1day
dichlorvos (30-40, single feed) another dose range is recommended	* Atgard * Tenac (Shell) polyvinyl chloride resin pellet; coated drug, powder **atropine** can be used as an antidot against dichlorvos intoxication	since the pure insecticide compound was relatively toxic in pigs, various better tolerated formulations have been developed; the drug has a high efficacy (>90%) against 4th stage larvae, juveniles, and mature adults of *A. suum, T. suis, H. rubidus* (only adults), and *Oesophagostomum* spp.; the drug shows less than 50% activity against migrating and mucosal larvae of *A. suum, H. rubidus*, and nodule worms, but high efficacy against larvae of *T. suis*; its activity against *S. ransomi* is variable (65-100%); coated version in feed (premix) and extruded PVC resin pellet are tolerated; resin pellets appear again in the feces and are toxic to other animals, especially to birds (antidote see atropine); preslaughter withdrawal time for edible tissues may be 2 days
haloxon (30-40, single feed)	* Eustidil * Cavoxon * Loxon (Wellcome, Cooper) powder, premix	is mainly approved as anthelmintic for use in ruminants (GI parasites) and in some countries for use in domestic fowl, turkey, quail and pigeons (*Capillaria*); in swine it is highly active (>90%) against adult *A. suum* and *Oesophagostomum* spp. (also immature stages); its effect against *H. rubidus* and *T. suis* is variable; there may be delayed neurotoxicity (e.g. posterior paralysis)

Table 4. (continued)

CHEMICAL GROUP, **Nonproprietary name** (approx. dose, mg/kg body weight, oral route) other information	* **Brand name** (manufacturer, company), other information	Characteristics (chemotherapeutic and adverse effects, miscellaneous comments)
TETRAHYDROPYRIMI-DINES		
pyrantel tartrate (22, single feed) (maximum: 2g/ animal) (morantel) (oxantel)	* Banminth (Pfizer) powder, granules premix formulation (10.6% pyrantel tartrate) via feed medicated feed is consumed without unwillingness	can be used prophylactically via feed to prevent establishment of ascarid and nodular worm infections (95-99% efficacy against lumen stages *Oesophagostomum* spp); it is active against adults of *A. suum* (some effect against histotropic stages and freshly hatched larvae from ingested eggs), *H. rubidus*, and *Oesophagostomum* spp.; at 5 times the therapeutic dose (side effects) there is activity against immature stages in the intestine and migrating larvae of *A. suum*; it appears to be inactive against whipworms (*T. suis*), *S. ransomi* and lungworms in heavy infections; drug's efficacy may be variable if used therapeutically; **morantel** (see Table 1) is several times more effective against *A. suum* than pyrantel : at 5 mg/kg morantel is highly active (>90%) against adults and immature *A. suum*; **oxantel** is highly effective against *T. suis*, but has no effect on other GI nematodes in pigs; drug is well tolerated at recommended dose (not recommended in severely debilitated animals); concurrent use of pyrantel and levamisole at therapeutic doses enhances toxicity (nicotine like drugs); preslaughter withdrawal time for edible tissues may be zero
IMIDAZOTHIAZOLES		
levamisole (8, single feed/ drinking water) (feed or water should be withheld overnight) (8 s.c.) is widely used in swine via the drinking water or in-feed, less as a subcutaneous injection	many brand names and suppliers (Table 1) granules, powder; feed additive sterile solution for parenteral (s.c.) administration	is highly active (>90%) against adults of *A. suum*, and *H. rubidus*, less so (75%-90%) against adults of *T. suis* and *Oesophagostomum* spp.; there is high efficacy against *A. suum* larvae, less against other larvae in the intestine; adult threadworms (*S. ransomi*) appear to be highly susceptible (90-100%) to the drug at recommended dose; after repeated dosing *S. ransomi* larvae were no longer excreted in the milk; the drug is highly effective (>90%) against hookworms (*G. urosubulatus*) in wild boars, lung worms (*Metastrongylus*, cf. Table 6), and kidney worms (*Stephanurus dentatus*) in the urinary tract (larvae in other parts of body are not affected, cf. doramectin); levamisole is well tolerated at recommended dose and is a safe drug by both the subcutaneous and oral route (in feed or in water); there are no side effects at about 4 times the therapeutic dose; the parenteral route may be used in cases where appetite is markedly depressed as a result of clinically evident parasitism or against whipworm infections because of enhanced activity after s.c. injection of levamisole; preslaughter withdrawal time for edible tissues may be 8 days
BENZIMIDAZOLES		
thiabendazole (50-75, single feed) (500, divided over 5-10 days in feed; 100-500 ppm medicated feed, 'long-term' treatment)	*Thibenzole (Merial) paste, powder, premix	is highly effective (>90%) against adults of *H. rubidus, Oesophagostomum* spp., and *S. ransomi* infections (ineffective against larvae which are passed in colostrum); the drug exhibits minimal or no activity against adult and immature *A. suum* and *T. suis*, and against larvae of *H. rubidus* and *Oesophagostomum* spp.; a combination with **piperazine** increases efficacy against *A. suum*; the drug is well tolerated in pigs and pregnant sows and may be used as an alternative drug; preslaughter withdrawal time for edible tissues may be zero

Table 4. (continued)

CHEMICAL GROUP, **Nonproprietary name** (approx. dose, mg/kg body weight, oral route) other information	* **Brand name** (manufacturer, company), other information	Characteristics (chemotherapeutic and adverse effects, miscellaneous comments)
cambendazole (20, single feed)	* Camdan * Neminil (Merial) * Ascapilla (Chevita) powder, granules	the drug is highly effective (>90%) against adult and immature *A. suum, H. rubidus* (also ovicidal effect), *Oesophagostomum* spp., and *S. ransomi*; there is minimal efficacy against adults/larvae of *T. suis*; since drug has been found to be teratogenic it must not be given during the first 6 weeks of pregnancy; otherwise it is well tolerated in pigs; preslaughter withdrawal time for edible tissues may be 28 days
parbendazole (25-50) (30, single feed)	* Helmatac (SKF) premix, powder	is highly effective (>90%) against adults of *A. suum, S. ransomi, H. rubidus* (somewhat lower), and *Oesophagostomum* spp. (also ovicidal effect); there is variable activity (75%-90%) against *T. suis*; the drug has been found to be teratogenic and embryotoxic in rats (see cambendazole); preslaughter withdrawal time for edible tissues may be 28 days
mebendazole (1.25) (30 ppm medicated feed for 5-10 days)	* Mebenvet (Janssen) powder	is highly effective (>90%) against *A. suum* (also active against migrating larvae in the lungs), *Oesophagostomum* spp. and *T. suis*; efficacy is somewhat lower against *H. rubidus* and *S. ransomi*; there is indication of a narrow safety margin as pigs respond to overdose or inadequate mixing in the feed with softening of feces or even diarrhoea; the drug has been found to be teratogenic, limiting its use in pregnant sows (see cambendazole); preslaughter withdrawal time for edible tissues may be 14 days
flubendazole (5, single feed or divided over 10 days) (30 ppm medicated feed for 5-10 days)	* Flubenol (Janssen) powder may be approved in some countries also for use in domestic fowl	parafluorine analogue of mebendazole with broad spectrum of activity and high potency comparable to those of mebendazole; the drug is active against *T. spiralis*, even against encysted larvae (30-125 ppm for 14 days); it also kills migrating larvae of *A. suum* and is active (75%-90%) against larvae of other gut nematodes; drug is well tolerated in gravid sows or in their piglets at recommended dose (no side effects at 40 times the recommended dose); preslaughter withdrawal time for edible tissues may be 14 days (chicken = zero);
fenbendazole (5, single feed) can be (3, x3-6d or therapeutic dose divided over 5-15 days in feed)	* Panacur (Intervet) powder (4%) premix	is highly effective (>90%) against adult and immature *A. suum* (also active against migrating larvae), *Oesophagostomum* spp., *H. rubidus* and *Stephanurus dentatus* (kidney worm 99%, in all sites: cf. doramectin) and lungworms (*Metastrongylus* spp.); there is high efficacy against adult *T. suis* and good activity against *S. ransomi* when total therapeutic dose is divided over several days(e.g. 3 mg/kg/d x 6 d) or given as medicated feed (17-20 ppm); the drug is well tolerated in gravid sows and their piglets; there is a wide therapeutic index in pigs (>500); preslaughter withdrawal time for edible tissues may be 5 days
oxfendazole (3-4.5)	* Synanthic * Systamex (Synthex/Wellcome, Essex)	may be used in pigs (there is no specific formulation available, cf. Table 1) it is highly effective against adult and immature *H. rubidus* and *Oesophagostomum* spp. and adult *A. suum*; there is a variable effect against *T. suis* and *S. ransomi* despite increase in dose; the drug has been found teratogenic, limiting its use in pregnant animals
oxibendazole (15, single feed) (100 ppm medicated feed for 6 days)	* Loditac (SKF) and others powder, premix	may be used in pigs; it is highly active (>90%) against *A. suum, S. ransomi*, and *Oesophagostomum* spp. (all treatment schedules); efficacy against *T. suis* is somewhat lower although there may be high efficacy at 15 ppm medicated feed for 50 days instead of 100 ppm for 6 days; the drug has been found to be teratogenic, limiting its use in pregnant sows

Table 4. (continued)

CHEMICAL GROUP, **Nonproprietary name** (approx. dose, mg/kg body weight, oral route) other information	* **Brand name** (manufacturer, company), other information	Characteristics (chemotherapeutic and adverse effects, miscellaneous comments)
PROBENZIMIDAZOLES		
febantel (5, **pellets**, single feed) (32 ppm medicated feed for 6 days)	* Rintal (Bayer) powder, granules, premix, pellets	prodrug of fenbendazole and oxfendazole; it is highly active (>90%) against adult *A. suum, H. rubidus, T. suis*, and *Oesophagostomum* spp. when given in medicated feed for 6 days although its efficacy against larvae of common gut nematodes is variable or minimal (immature *T. suis*); a single dose appears to be less effective than medicated feed over 6 days; there is moderate to high efficacy against *T. suis* using doses of 15-20 mg/kg (single feed) and against *S. ransomi* using 40 mg/kg (single feed); the drug is well tolerated at recommended dose in their piglets, also during critical days of gestation, i.e. the first 5 weeks of sows and pregnancy, at 5 times the therapeutic dose; preslaughter withdrawal time for edible tissues may be 14 days
thiophanate (50-100, single feed) (therapeutic dose divided over 14 days)	* Nemafax 14 (May and Baker) premix	is highly active (>90%) against adult *A. suum, Oesophagostomum* spp., *H. rubidus* and *T. suis* using long-term treatment; there is a variable activity against *A. suum* using a single feed; its efficacy against larvae of common gut nematodes is variable as against *A. suum* but somewhat higher against larvae of the other nematodes; the drug is well tolerated at recommended dose
MACROCYCLIC LACTONES		
AVERMECTINS		
ivermectin (0.3 intramuscularly) (premix 2ppm over 7 days to provide 100 mcg/kg body weight)	* Ivomec-S (Merial) parenteral solution, * Ivomec-Premix (MSD) powder, premix	with high efficacy (98-100%) against immature and adult stages of *A. suum, H. rubidus*, and *S. ransomi* (including somatic L3 in pregnant sows), *Metastrongylus, Stephanurus dentatus* (in all sites: cf. doramectin) and intestinal (not muscular) stages of *Trichinella spiralis*; its efficacy against *Oesophagostomum* spp. and *T. suis* (including larvae) is variable; intramuscular (i.m.) treatment of sows 7-14 days before parturition with ivermectin or an in-feed formulation of the drug given for 7 days significantly may prevent the galactogenic transmission of *S. ransomi* to piglets; at regular therapeutic dose it controls lice (*Haematopinus suis*) and the mange mite (*Sarcoptes scabiei* var. *suis*); ivermectin is well tolerated at recommended dose (10-fold safety margin) and is generally safe in breeding and pregnant animals; it should not be fed to sows more than 100 kg body weight; preslaughter withdrawal time for edible tissues may be 28 days for i.m. injection and 7 days for in-feed (premix) medication
doramectin (0.3 intramuscularly) available in the UK and elsewhere since 1998; it provides long-acting control of gastro-intestinal roundworms after i.m. injection; withdrawal time may be 49 d	* Dectomax (Pfizer) injectable solution	efficacies against immature and adult stages of GI parasites appears similar to those of ivermectin: 98-100% against all GI nematodes except whipworms, *Trichuris suis* for which efficacy was variable (54-87% in mixed infections, and in pure infections 95%); it is effective against kidney worm, *Stephanurus dentatus*, in all sites of the sow as fenbendazole or ivermectin; levamisole is effective only against worms in the kidneys (residing sites: majority of worms may be in peritoneal area and kidneys, a few may be scattered in liver, lungs, abdominal muscles and peritoneal cavity); it is highly active (98-99%) against sucking lice *Haematopinus suis*, and the mange mite (*Sarcoptes scabiei* var. *suis*; as other avermectins or milbemycins, not active against eggs of mange mites);doramectin is well tolerated at recommended dose (5-fold safety margin) and it appears safe in breeding and pregnant animals at three times the therapeutic dose

Doses listed in this table refer to information from manufacturer and literature
Data given in this table have no claim to full information.
The primary or original manufacturer is indicated if there is lack of information on the current one(s)

teries; these include regular cleaning of baskets and drinking bowls, and destroying or burning the feces and other waste.

Since rodents and birds may serve as paratenic hosts in the life cycle of ascarids, extermination of rodents and attention to potential infected viscera of birds (e.g. of the domestic fowl) must be included in control programs. (Paratenic host may ingest infective eggs and 2nd stage larvae travel to their tissues where they remain until eaten by a carnivore). With *Toxascaris leonina* periodic deworming of all animals can eliminate this parasite; this parasite lacks a migratory phase in the host and thus infection of uterus by somatic 2nd-stage larvae and mammae by 3rd-stage larvae. With *T. canis*, and *T. cati* controlling parasite stages is more difficult because of the somatic type of 2nd larva migration including the liver, lungs, heart, brain, kidneys and skeletal muscle which is responsible for prenatal and transmammary infection of fetuses or suckling puppies and kittens. Transmammary infection also occurs with *T. cati* but prenatal infection of fetuses is lacking. Long-persisting 2nd-stage larvae of *T. canis* found in various tissues of the body of the bitch and which have undergone no further development are mobilized at each pregnancy, thus transmitting infections to several litters. Puppies should therefore be treated within 2 weeks of birth.

Regular treatment of bitches with effective anthelmintics may prevent prenatal infections (cf. fenbendazole, Table 5). However, the use of anthelmintics for controlling nematode infections in pet animals may be limited because of the lack of suitable formulations and well tolerated (safe) drugs for young pups and kittens, and for enfeebled and pregnant animals. Most anthelmintics have little or no activity against migratory stages of the ascarids. 'Oldtimers', in particular such as plant extracts (→ Cestodocidal Drugs), dithiazanine, toluene and dichlorophen combinations, n-butyl chloride, disophenol or piperazine have a narrow spectrum of activity either against adult hookworm or ascarids only, and their toxicity is rather high (Table 5). Some of these older drugs are still used in the USA and elsewhere. The use of anthelmintics with a narrow spectrum of activity may be indicated if a specific infection is diagnosed regularly. Because labor costs are high and correct diagnosis is too time consuming in certain situations it makes sense to deworm dogs and cats with current products showing activity against all common intestinal nematodes and cestodes. Thus current routine dewormers for dogs and cats are highly effective in eliminating intestinal stages of ascarids (especially *Toxocara canis*), common hookworms, whipworms, and tapeworms (*Taenia* spp. *Dipylidium caninum* and *Echinococcus* spp., for general consideration of tapeworms cf. → Cestodocidal Drugs). Products, which meet all these requirements, are principally drug combinations consisting of benzimidazole carbamates or pro-benzimidazoles and praziquantel (Table 5). Only a few of these products are marketed worldwide and most of them are available in the USA and Europe.

The susceptibility of nematode populations to anthelmintics should be checked regularly on the grounds of the results of parasitological investigations of feces samples. These data may provide the basic information required for preventing the occurrence of nematode resistance and incorporating effective drugs into control programs.

Dirofilariasis of Dogs, Its Epizootiology and Control

Heartworm disease of the dog caused by *Dirofilaria immitis* (family Onchocercidae) is primarily a problem of warm countries where the mosquito intermediate host abounds (>60 mosquito species belonging to different genera, such as *Culex* spp., *Aedes* spp., *Anopheles* spp. and *Psorophora* spp. are susceptible to *D. immitis*). *D. immitis* appears not to be very host specific. Female mosquitoes ingest microfilariae during feeding, and development in the mosquito to infective 3rd stage larvae takes about 2 weeks. Final host is infected by 3rd stage larvae when mosquito takes another blood meal Final hosts are the dog, cat, coyote, dingo, wolf, fox, wild Felidae, sea lions, monkeys, and occasionally humans (→ Nematocidal Drugs, Man, especially → Nematocidal Drugs, Man/Table 1); the domestic cat is not as susceptible to *D. immitis* as the dog. About 6 months following infection of the host, larvae migrate to the subcutaneous or subserosal tissues and undergo 2 molts. Only after the final molt do the young worms pass to the heart via the venous circulation. Ovoviviparous female worms release microfilariae (MF) directly into the bloodstream, and patent infection may be evident by microfilaremia between 6–9 months after infection. Besides the daily periodicity of MF in the bloodstream (highest concentrations from late afternoon to late evening), there is a seasonal periodicity with the highest microfilaremia in spring and summer according to behavior of female bloodsucking mosquitoes. Circulating MF in the host may survive up to 2 years and transplacental transmission may occur with MF being found in various tissues of fetuses. The cardiovascular dirofilariasis is a systemic disease involving the lungs, heart, liver, and kidneys (immune complex glomerulonephritis). The adult filariae reside in the branches of lung artery. In heavy infections they can migrate into the right heart chamber and Vena cava causing severe pathogenic effects, e.g. circular distress (see also Table 5). Large amounts of dying or dead *D. immitis* adult worms (20–30 cm long) as a result of chemotherapy with an adulticidal drug may cause adverse reactions, and pulmonary embolism, which can be fatal in cases of very advanced disease (Table 5). Common clinical signs are cough, blood in the saliva, dyspnoea, and pulmonary hypertension, which may be compensated by right ventricular hypertrophy. In advanced cases permanent pulmonary hypertension may lead to dilatation of the right heart and to congestive heart failure, followed by ultimately chronic passive congestion manifested by liver enlargement (hepatomegaly), ascites and edema accompanying symptoms of ascites. At this stage the dog is weak and listless. The high prevalence of heartworm disease may be due to several vector factors and host vectors. They include ubiquity of the mosquito intermediate hosts (makes control of vectors difficult), their high capacity for rapid reproduction, the short development period from MF to infective 3rd stage larvae in the mosquito, the lack of protective immunity of hosts against *D. immitis*, and the long patency period of the disease of up to 6 years during which time circulating MF are present. For this reason, **heartworm** control is based almost entirely on prophylactic medication of dogs or other animals under risk (Table 5).

Chemotherapy and Chemoprophylaxis of Dirofilariasis of Dogs Surgical removal of heartworms is usually accompanied by mortality rates of about 10 percent. Only if treatment is contraindicated in some severe cases, should heartworms be removed surgically. In mild and moderate heartworm infections, chemotherapy with an arsenical followed by a microfilaricide (Table 5) appears to be a reliable and relative safe method and is recommended with the aim of reducing adult filariae and microfilariae (MF) in time. Prior to start of specific treatment animals should be examined physically, including assessment of heart lung, liver, and kidney function. Pretreatment is indicated in case of cardiac insufficiency. The usual way to treat infected dogs is to administer an adulticidal drug (Table 5) to remove the adult worms. The treatment is often associated with **toxic reactions** resulting from dying worms and thereby resultant embolism; therefore treatment should be performed with extreme care, and the activity of dogs must be restricted for 3–7 weeks. About 6 to 7 weeks later a further treatment with a microfilaricide (Table 5) is given to remove the circulating MF from the bloodstream. For this purpose ivermectin or another macrocyclic lactone may be used, which have in many cases substituted older drugs such as dithiazanine, diethylcarbamazine (DEC) or levamisole, which must be given over several days. With all these drugs, especially with the older ones, there is the risk of adverse reactions to dying MF. Heartworm-free animals are than placed on a prophylactic program and this is considered under control (Table 5).

Lungworm Infections of Domestic Animals

The most pathogenic nematodes in the superfamily Trichostrongyloidea belong to the genus *Dictyocaulus* (family: **Dictyocaulidae**). Members of this genus do not require an intermediate host and thus are 'geohelminths' with a direct life cycle. In contrast to *Dictyocaulus*, nematodes of the superfamily Metastrongyloidea (e.g., families **Metastrongylidae** and **Protostrongylidae**) require intermediate hosts to convey infective larvae to the definite host and thus are 'biohelminths'. Members of both superfamilies are parasites of the respiratory passages and/or blood vessels of the lungs. For example, *Angiostrongylus* spp. mostly occur in the pulmonary artery or cranial mesenteric artery of various species of mammals and occasionally of man (→ Nematocidal Drugs, Man/Table 1). Consequently, lungworms causing parasitic bronchitis especially in young livestock necessitate varying control strategies because of their different life cycles, and hence, epizootiology (for other extraintestinal nematode infections of livestock and wild animals cf. Table 1, Table 3, Table 4, Table 5, Table 6).

***Dictyocaulus* Infections** There are three genera of importance, which may cause parasitic bronchitis in young animals and economic losses during the first grazing season. *Dictyocaulus filaria* occurs in the bronchi of small ruminants (sheep, goats and some wild ruminants and has a worldwide distri-

Table 5. Drugs used against nematode infections of dogs and cats.

CHEMICAL GROUP, **Nonproprietary name** (approx. dose, mg/kg body weight, oral route) other information	*****Brand name** (manufacturer, company), other information	**Characteristics** (chemotherapeutic and adverse effects, miscellaneous comments)

CLINICAL FORMS OF HEARTWORM DISEASE, AND CONSEQUENCES TO USE OF DRUGS

Dirofilaria immitis infection occurring in carnivores primarily in warm countries where the mosquito intermediate host abounds (especially in the southern parts of the USA and Japan, or Australia); the use of the proper drug (drug of choice) in treating heartworm disease depends on both the degree (status) of clinical signs developed in the course of infection and the condition of dog, which may be determined by the amount of adult worms and their location in the venous circulation; thus heartworm disease can be classified in class 1 (defined as asymptomatic to mild heartworm disease sometimes involving occasional listlessness, fatigue on exercise, or occasional cough), class 2 (moderate form of disease, characterized by anemia, mild proteinuria, ventricular enlargement, slight pulmonary artery enlargement, or circumscribed perivascular densities plus mixed alveolar/interstitial lesions) class 3 (advanced form of heartworm disease with cardiac cachexia, wasting, permanent listlessness, persistent cough, dyspnea, right heart failure associated with ascites, jugular pulse, right ventricular and atrial enlargement, signs of thromboembolism, anemia, and proteinuria), and class 4 (severe form with vena cava syndrome, i.e. final stage of congestive right-sided heart failure, *D. immitis* present in vena cava and right atrium of heart; treatment is questionable or not indicated); unsheathed microfilariae (MF) released from female worms into the bloodstream can cause severe adverse effects (anaphylactic-like shock) after being killed by a microfilaricidal drug; as a consequence, dogs with a patent *D. immitis* infection should be cleared from adults and MF prior to start of any prophylactic dosage regimen; this can be done by using a suitable (adulticidal) drug but only in a condition that allows such a treatment, e.g. in animals having mild to significant clinical signs (fall under class 1–3 disease); causal chemoprophylaxis is the most effective measure in preventing establishment of *D. immitis* infection in dogs and other carnivores

OTHER EXTRAINTESTINAL NEMATODES OF VETERINARY IMPORTANCE (cf. also → Nematocidal Drugs, Man/Table 1)

Developing stages (larvae) of these parasites travelling through various tissues of the final host(s) mature to adult worms outside the intestinal tract; adults and migrating larvae are fairly refractory to treatment with anthelmintic drugs as it is also seen with all migrating larvae of gastrointestinal nematodes (see below); *Spirocerca lupi* of Canidae and Felidae frequently occurring in tropical and subtropical areas causes spirocercosis associated with severe damage of esophagus (e.g. granuloma, fibrosarcoma) and aorta (e.g. aneurysm formation); life cycle of this spiruroid includes various intermediate hosts (coprophagous beetles ingesting eggs passed in feces) and paratenic hosts (amphibia, reptiles, domestic and wild birds, and small mammals as hedgehogs, mice and rabbits ingesting beetles or another paratenic host) in which larval worms become encysted; final hosts (e.g. dog, fox, wolf, jackal) become infected by ingesting either infected beetles or infected paratenic hosts; *Filaroides osleri* (*F. hirthi*, and other species) of dog infrequently occurs in the USA, Europe, India, South Africa, New Zealand and elsewhere (a high prevalence may be in dogs kept under kennel conditions); first-stage larva in saliva or feces infects puppies when bitch licks and cleans them (direct life cycle); adult worms living under the mucosa of trachea and bronchi cause development of granuloma and in heavy infections a rasping persistent cough; heavily infected puppies show loss of appetite, emaciation and hyperpnea, and sometimes mortality may occur in infected litters; *Crenosoma vulpis* of dog and (farmed) fox, occurring worldwide, is ovoviviparous and 1st stage larva passes with the feces to be ingested by a land snail containing infective larvae; dogs may eat such snails and, after their digestion, released 3rd stage larvae migrate to the lungs (trachea, bronchi, bronchioles) where they mature to adults thereby producing occlusion of bronchioles or bronchopneumonia; clinical signs are nasal discharge, coughing and tachypnea; outside the host, *Angiostrongylus vasorum* (distribution worldwide, except in the Americas intermediate hosts: land snails, and slugs) has a similar life cycle to *C. vulpis*; in the host, 5th stage larvae enter the pulmonary arterioles and capillaries and may cause chronic endarteritis and periarteritis of the larger vessels or even endocarditis involving tricuspid valve if vascular changes extend to the right ventricle; in longer established and severe infections clinical signs such as tachypnea, cough, painless swellings of lower abdomen and intermandibular space and limbs are present even in resting dogs (for drugs acting on extraintestinal nematodes see diethylcarbamazine, ivermectin or other macrocyclic lactones, disophenol, nitroscanate, pyrantel, levamisole (tetramisole), and benzimidazoles, under 'gastrointestinal (GI) nematodes of veterinary significance in dogs, cats and wild carnivores, below).

Table 5. (Continued) Drugs used against nematode infections of dogs and cats.

CHEMICAL GROUP, **Nonproprietary name** (approx. dose, mg/kg body weight, oral route) other information	***Brand name** (manufacturer, company), other information	**Characteristics** (chemotherapeutic and adverse effects, miscellaneous comments)
DRUGS ACTING ON ADULT HEARTWORMS		
ARSENICALS		
Several arsenicals have been shown to have wide biological activity, including toxicity and to kill adult heartworms (female worms are less susceptible than male worms and there is no action on circulating microfilariae = MF).		
thiacetarsamide sodium (TAS) (synonyms: arsenamide thioarsenite) (dog, 2.2 = 0.44 elemental arsenic, intravenously, twice daily for 2 days) there is no effect against circulating microfilariae (MF); **absolute rest** during first 2 weeks posttreatment is necessary and only limited exercise is allowed during the next two weeks because of risk of embolism)	*Arsphenamide *Caparsolate sodium (Abbott Laboratories) *Caparside, *Filaramide, *Filicide (various companies) **precaution:** function of liver and kidney must be checked before beginning of treatment; severe toxic reactions to TAS can be treated with **dimercaprol** 2.2 mg/kg x 4/day usually gives relief (antidote to poisoning by arsenic, gold mercury, and other metals:	caution must be exercised to avoid perivascular leakage during i.v. injection; TAS is highly irritating to SC tissues and may lead to distinct necrosis of tissues (corticosteroids may reduce aggravating inflammatory reaction); it has been the standard adulticidal drug for the past several decades; its efficacy may vary extremely as a function of worm's age and sex; female worms are less susceptible than male worms, and very young (2 months of age) and very old (2 years of age) were more susceptible than the rest (4/6/12/18 months of age); host-related variations in drug pharmacodynamics may be another explanation for extreme variation in efficacy; adult worms gradually die (within 5–7 days, up to 14 days); dead and dying adult worms washed out of the right heart by blood flow lodge in branches of pulmonary artery and are eliminated by phagocytosis within about 2 months, residues (fragments) of damaged and/or phagocytized worms pose a distinct threat to well-being of animals by embolism occurring principally in the first month following treatment; splitting of daily dose (4.4 mg/kg b.w.) will markedly reduce most of hepatotoxic and nephrotoxic effects of TAS though its recommended dose should not be reduced in very large dogs; tolerability of TAS seems to be best if it is injected 1–2 hours after feeding in the morning or evening; interest in eating may provide some indication of general condition and regimen of treatment is continued if dog does not vomit, is eating well, and there is no indication of hepatic or renal failure; if treatment must be interrupted because of severe toxic reactions, retreatment (entire regimen) is recommended 6 weeks later to prevent liver damage; mortality during or following TAS therapy seems to be related to the degree of clinical manifestation of heartworm disease (asymptomatic = class 1: no loss, mildly symptomatic = class 2: 3–5% mortality, and advanced = class3: up to 50% can be expected

Table 5. (Continued) Drugs used against nematode infections of dogs and cats.

CHEMICAL GROUP, **Nonproprietary name** (approx. dose, mg/kg body weight, oral route) other information	***Brand name** (manufacturer, company), other information	**Characteristics** (chemotherapeutic and adverse effects, miscellaneous comments)
melarsomine dihydrochloride (RM 340, for use in dogs only) **standard regimen for class 1 and 2 class** heartworm disease: (dog, 2.5 = 0.1 ml/kg x 2 intramuscularly, doses given 24 hours apart: first dose right lumbar muscle, second in the left); this 2-dose regimen may be repeated 4 months apart if there is lack of seroconversion or exposure to reinfection **alternative regimen** class 3 heartworm disease: (2.5 i.m. single dose, followed by full 2-dose regimen 1month later) latter schedule reduces risk of complications from pulmonary embolism following treatment	*Immiticide (Merial) (lyophilized sterile powder containing 2 HCl (salt) that is prepared in sterile 0.9% saline solution for deep i.m. injection in the longissimus dorsi superficial injections or leakage should be avoided; repeated administrations should not occur at the same lumbar location **precautions:** after treatment, dogs should be monitored for toxic signs produced by dying worms and residues of dead worms such as aggravating cough, fever, and sudden tachypnea or even orthopnea; dogs should be kept in subdued light and only limited exercise should be allowed (absolute rest post-treatment in the next 2 weeks is a must)	trivalent arsenical of melanonyl thioarsenite group (RM 340) with adulticidal activity against male and female heartworms (*Dirofilaria immitis*) and 4-month-old heartworms in dogs; RM 340 can be used for treatment of stabilized class 1, class 2 and class 3 heartworm disease caused by immature (4-month old L5 larvae) or adult stages of *D. immitis*; it should not be used in dogs suffering from final stage (class 4) of heartworm disease (*D. immitis* present in Vena cava and right atrium of heart causing Vena cava syndrome, i.e. final stage of congestive right-sided heart failure); in class 1 and class 2 of heartworm disease a single 2-dose regimen kills all male worms and about 95% of female worms though complete elimination of all worms resulted in 60–80 % of treated dogs only; elimination of all heartworms results in about 98% of dogs following the sequence of two 2-dose regimens 4 months apart; in severe cases (class 3) initial single dose (see alternative regimen) results in a partial kill of adult heartworms (about 85% of male and 15% of female) leading to some relief of symptoms (e.g. fever, gagging, coughing, tachypnea) thereby reducing the risk (to a certain degree) of embolic shower of whole or partially phagocytized worms in the branches of the pulmonary artery; the 2-dose schedule 1months apart usually kills all worms in about 85% of dogs; pharmacokinetics of RM 340 is characterized by short absorption half-life of 2.6 minutes (peak concentration in blood at 8 min); it has a greater bioavailability than thiacetarsamide thus resulting in a adulticidal effect half the arsenic equivalent of thiacetarsamide and about twice the therapeutic index; unlike thiacetarsamide, which binds to erythrocytes, RM 340 and its metabolites are free in plasma resulting in higher and longer lasting plasma levels than those seen with thiacetarsamide; i.m. injection (1–5% solutions) of RM 340 is well tolerated causing only minor tissue reactions (circumscribed edema); overdosing (e.g. 4.4 mg/kg 3hours apart) results in adverse effects within 30 min after treatment, and most toxic signs last about 1 hour, such as salivation, restlessness, pawing, tachypnea, tachycardia, abdominal pain, hindlimb weakness, recumbency; severe toxicity is characterized by orthopnea, circulatory collapse, coma, and death (**antidote dimercaprol** 3 mg/kg i.m. given within 3 hours after appearance of first toxic signs can reserve toxicity of RM 340 but may reduce its activity against adults); older dogs (> 7 years of age) are more sensitive to RM 340 treatment than younger dogs; safety for use in breeding animals and lactating or pregnant bitches has not been determined and the mode of action of RM 340 in *D. immitis* is unknown
MICROFILARICIDAL DRUGS (for preventive use)		
CYANINE DYES		
dithiazanine (dog, 11–22, for several days depending on target worm species)	*Diazan (Lilly/Pitman-Moore) coated tablets	oldtimer, which exhibits activity (11 mg/kg for 7–14 days) against microfilariae of *D. immitis* and was being used as standard microfilaricidal drug for many years in dogs (now replaced by other and better tolerated and more effective drugs); it may still be used against ascarids (treatment course of 3–5 days), hookworms (treatment course of 7 days), and adult *Strongyloides stercoralis* in the small intestine (treatment course of 12 days); side effects of the cyanine dye derivative may be severe diarrhoea, vomiting and anorexia

Table 5. (Continued) Drugs used against nematode infections of dogs and cats.

CHEMICAL GROUP, **Nonproprietary name** (approx. dose, mg/kg body weight, oral route) other information	*__Brand name__ (manufacturer, company), other information	**Characteristics** (chemotherapeutic and adverse effects, miscellaneous comments)
AMINES (PIPERAZINES)		
diethylcarbamazine citrate (DEC) causal prophylactic use against heartworms (involves treatment prior to infection or within a few weeks after infection): (dog, DEC 6.6 daily) (pet ferrets DEC 2.75–5.5 daily) (DEC 2.75 daily: powder formulation) therapeutic use: (55–110, single dose, tablets, against adult ascarids of dogs and cats; treatment must be repeated in 10–20 days) → strictly **contraindicated** in microfilariae-positive dogs (other animal species)	*Caricide, *Dirocide *Filaricide, *Filaribits *Cypip (powder) (American Cyanamid; Squibb, others) formulations of DEC are given in either dry or wet feed or immediately after feeding tablets (chewable or nonchewable), syrup, powder sold under several brand names) DEC has also been used as a microfilaricide against *Onchocerca volvulus* infections in **humans** (cf. → Nematocidal Drugs, Man/Table 1)	continues administration of the piperazine derivative DEC in low daily doses through the mosquito season and for two months following can be used as a preventive for heartworm disease in dogs, ferrets or sea lions (may frequently occur in amusement parks; occurrence of patent heartworm infections in cats is infrequent); in warmer climates with all year prevalence of mosquito vector transmitting infective larvae of *Dirofilaria immitis*, daily administration of DEC for lifetime should be performed in animals under risk; low daily doses of DEC are also effective in preventing establishment of canine ascarid infection though this dosage regimen is ineffective against adult ascarids; a single high DEC dose may eliminate adult burden of ascarids in both cats and dogs; it has been reported to be effective against the lungworm *Crenosoma vulpis* of dog and farmed foxes; the use of DEC is strictly contraindicated in microfilariae (MF)-positive dogs because of possible but rare shock type (likely non-allergic) reaction (sometimes fatal) produced by liberation of substances from dying or dead MF as a result of DEC treatment; the drug acts on both infective third stage larvae and MF of *Dirofilaria immitis* though its action on circulating microfilariae is not safe; it is more active against preadult developing stages; prophylactic treatment with DEC can be started if a dog is cleared of adult *D. immitis* (and then MF using a microfilaricide) either with arsenicals (see above) or by means of surgery (removal of adult worms in life-threatening conditions such as 'vena cava syndrome' or 'liver failure syndrome' not always amenable to chemotherapeutic treatment); DEC is rapidly absorbed, peak concentration is about 3 hours (h) after oral administration and reaches zero level in 48 h (excretion 70% in urine within 24 h, 10–25% unchanged); it is a relatively nontoxic drug and its side effects are similar to those of piperazine; it may produce vomiting (irritation to gastric mucosa) and has no adverse effect on fertility of male dogs in long term prophylaxis
DEC/styrylpyridinium (3/5) **DEC/oxibendazole** (6.6/5)	*Styrid Caricide *Filaribits Plus (Merial)	DEC/styrylpyridinium chloride or DEC/oxibendazole are effective in preventing the establishment of *Ancylostoma* spp., ascarids and *Dirofilaria immitis* (heartworm) infections in carnivores

Table 5. (Continued) Drugs used against nematode infections of dogs and cats.

CHEMICAL GROUP, **Nonproprietary name** (approx. dose, mg/kg body weight, oral route) other information	***Brand name** (manufacturer, company), other information	**Characteristics** (chemotherapeutic and adverse effects, miscellaneous comments)
MACROCYCLIC LACTONES		
AVERMECTINS		
ivermectin (dog, 0.006 once a month) preventive dose is 100% effective in killing developing 3rd and 4th stage larvae of *D. immitis* (ineffective at any dose against adult worms) experimental studies have shown that the drug has a wide spectrum of activity against various canine GI nematodes (4th-stage and adults) at single s.c. dose of 0.05–0.2mg/kg	*Heartgard-30 chewable tablets (Merial) *Heartgard Chewables (chewable cubes) (Merial) *Toxocara canis*: high reduction rates (approx. 100%) of prenatal and transmammary transmission of 3rd stage larvae from the bitch to her pups can be achieved by treating bitch 10 days prior to and 10 days after whelping with 0.5 mg/kg s.c. each time	endectocide for use in dogs as a preventive for heartworm (*Dirofilaria immitis*); it kills infective 3rd stage larvae (transmitted by mosquitoes) and subsequent developing stages in the subcutaneous or subserosal tissues acquired during the previous approx. 45 days or over the next few months in fresh infections thus preventing establishment of adult heartworm infection in the venous circulation for a prolonged period; elimination of tissue stages of *D. immitis* larvae is usually achieved within 30 days of infection; preventive dose is given at monthly intervals through the mosquito (vector) season and for the two following months thereafter; its administration is prohibited in dogs with an established heartworm infection; ivermectin is also effective against heartworm microfilaria (at 0.05 mg/kg orally) but not approved for this purpose, possibly because of potential (heavy) hypersensitivity reactions produced by dying or dead microfilariae circulating in blood stream; products approved are not recommended for use in dogs under 6 weeks of age causing transient diarrhoea; the drug has a wide safety margin though the Collie is susceptible to ivermectin toxicity at oral doses of 0.1 mg/kg and higher doses; adverse reactions in the Collie are not seen at dose used for heartworm prevention (0.006–0.012 mg/kg) or even 10 times the preventive dose (0.06 mg/kg , monthly for a year) the drug is safe in breeding and pregnant animals; parasitic arthropods such as otodectic, sarcoptic , and notoedric mange and *Pneumonyssus caninum* nasal mites in dogs and cats are affected by ivermectin (0.2 mg/kg x2, s.c., 2 weeks apart), *Cheyletiella* spp. (0.3 mg/kg x2 s.c., 2 weeks apart) or demodectic mange of dogs (0.6 mg/kg x5 at 7-day intervals)
ivermectin/pyrantel pamoate (dog 0.006/5: once a month) (minimum dose during the mosquito season to prevent establishment of heartworm infection)	*Heartgard-30 Plus, or *Heartgard Chewables (beef-based chewable tablets, cubes) (Merial)	marketed in the USA, Australia, Italy, Spain and elsewhere for use in dogs; combination can be used for prevention of canine heartworm *D. immitis* (for more details see ivermectin above); combination provides also monthly control of ascarids (*T. canis*, *Toxascaris leonina*), and hookworms (*A. caninum*, *A braziliense*, and *U. stenocephala*) resulting in nearly 100% reduction of worm load and egg output; the combination is safe for dogs (use is restricted to dogs 8 weeks of age and older); adverse effects may be vomiting, diarrhoea within 24 h postdosing; in puppies, occasionally depression, lethargy, anorexia, mydriasis (anomalous dilation of the pupils), ataxia, staggering, convulsions, and hypersalivation may occur selamectin = 25-cyclohexyl-25-de (1 methyl-propyl)-5-deoxy-22,23-dihydro-5-(hydroxyimino)-avermectin B_1 monosaccharide; for control of heartworms, hookworms and ascarids, fleas and ticks

Table 5. (Continued) Drugs used against nematode infections of dogs and cats.

CHEMICAL GROUP, **Nonproprietary name** (approx. dose, mg/kg body weight, oral route) other information	*Brand name (manufacturer, company), other information	**Characteristics** (chemotherapeutic and adverse effects, miscellaneous comments)
selamectin (cat, dog 6, **topically,** once a month; treatments at a monthly interval only!) coat wetting does not decrease efficacy; there may be transient localized pruritis in treated dogs (0.03%) and cats (0.1%, very rarely alopecia) novel semisynthetic avermectin B_1 derivative related to doramectin	*Revolution, *Stronghold (Pfizer Inc.) isopropanol/dipropylene glycol monomethyl ether based formulation (**spot-on:** single site at base of neck in front of scapulae); ^{3}H selamectin was found in sebaceous glands, hair follicles and on basal layer of epithelium, thus providing a depot for slow release of the drug to skin surface	with high ovicidal/larvicidal/adulticidal activity (99–100%) against fleas of cats and dogs (*Ctenocephalides felis* and *C. canis*); it controls (fast knockdown effect) and prevents (long-term effect) high flea challenges or infestations on dogs and cats living in household environment for a period of 30 days; it is safe and effective in controlling mite and tick infestations due to *Otodectes cynotis* (dog 100%; cat 94%), *Sarcoptes scabiei* (dog 100% after 2 doses), and *Rhipicephalus sanguineus* (dog >95% after 2 doses); **oral application** of selamectin (1x 2mg/kg, gelatin capsules) prevents heartworm disease in dogs (by activity against immature, larval stages of *D. immitis*) or eliminates (100%, or close to 100%) naturally acquired infections of GI nematodes (*A. caninum, U. stenocephala, A. tubaeforme, T. canis, T. leonina, T. cati*); the drug was found to be safe and well tolerated under clinical conditions in the field when administered topically (margin of safety in cats and dogs: 10x, in ivermectin-sensitive strains of Rough-coated Collies 5x the recommended dosage); there is no effect on the health or reproductive status of female or male dogs and cats; for information on environmental safety → AVERMECTINS, Table 1
MILBEMYCINS		
milbemycin oxime (dog, 0.5–0.99 once a month as a preventive for heartworm and intestinal nematodes) a single oral dose (see above) at 30–45 days post-infection with 3rd-stage heartworm larvae prevents establishment of infection completely (incomplete prevention if treatment begins 60–90 days post-infection) cats treated monthly with an oral dose of 0.5–0.9 mg/kg completely prevent establishment of experimental infection with *D. immitis*	*Interceptor (Novartis) chewable tablets (four sizes containing different concentrations of MO) *Interceptor Flavor Tabs (Novartis) *endoVet (Ciba) chewable tablets (can be used for dogs and puppies 4 weeks of age or older and dogs weighing ≤ 2 pounds) dose regimen for dogs having demodicosis (*D. canis*) is 1–4.6 mg/kg/d for at least 2–3 months causing a temporary cure or improvement in most dogs, and non-relapsing (permanent) cure in >50% of treated dogs	milbemycin oxime (MO) consists of the oxime derivatives of 5-dihydromilbemycins in the ratio of approx. 80% A4 and 20% A3; endectocide, having efficacy against both nematodes and *Demodex canis*; it has currently been marketed for prevention of canine dirofilariasis (*D. immitis*) and control of intestinal nematodes *Toxocara canis, Trichuris vulpis*, and *Ancylostoma* (MO does not reliably affect *Uncinaria*) in several countries (Canada, the USA, Japan, Australia, New Zealand, Italy and elsewhere); prior to starting prophylaxis program, dogs should be checked for patent heartworm infections, i.e. for circulating microfilariae in blood and adult heartworms; MO is a potent and fast acting microfilaricide in dogs and a single oral dose of ≤ 0.25 mg/kg results in >98% decline in microfilaremia (blockade of embryogenesis) within a few days, occasionally producing shock-like reactions at the time of treatment; more commonly mild reactions occur such as salivation, coughing, tachypnea, vomiting and depression; concurrent administration of MO and corticosteroids as well as i.v. fluids will markedly reduce these reactions; heartworm-infected dogs given approved dose for *D. immitis* prophylaxis (0.5–0.99 mg/kg monthly) will become free of microfilariae within 6–9 months and the majority of so treated dogs will remain amicrofilaremic following a 4- to 6-month intermittence of prophylaxis (mosquito free winter season); the drug is safe in breeding and pregnant animals; puppies (8-week-old) tolerated several times higher doses than the monthly dose of MO (e.g. 6x 0.5 mg/kg/day for 3consecutive days); three times the monthly dose given to pregnant mother dogs 1 day before whelping, on day of whelping or 1 day thereafter had no adverse effects on the puppies; doses of 1.5 mg/kg MO given to pregnant dogs resulted in measurable drug concentrations in milk, and nursing puppies may show milbemycin related adverse effects; some Collies are more sensitive to MO than other dogs (as with ivermectin) though no adverse reactions are seen at 5 mg/kg (10 times the monthly dose); 12.5 mg/kg (= 25 times the monthly dose) given to rough coated Collies resulted in ataxia, pyrexia, and periodic recumbency in 1 of 14 treated dogs

Table 5. (Continued) Drugs used against nematode infections of dogs and cats.

CHEMICAL GROUP, **Nonproprietary name** (approx. dose, mg/kg body weight, oral route) other information	*__Brand name__ (manufacturer, company), other information	**Characteristics** (chemotherapeutic and adverse effects, miscellaneous comments)
milbemycin oxime/lufenuron (dog, 0.5/10) **lufenuron** acts as an insect development inhibitor by breaking the flea life cycle by inhibiting development of eggs	*Sentinel (Novartis) tablets **lufenuron** interferes with chitin synthesis; it has no effect on adult fleas and prevents most flea eggs from hatching or maturing into adults	approved for prevention of heartworm disease, control of intestinal nematodes (see milbemycin oxime = MO, above) and flea population in dogs and puppies ($\geq$ 4 weeks of age, and dogs weighing $\geq$ 2 pounds) (for toxicity and margin of safety of MO in breeding and pregnant animals or puppies see under MO, above) **lufenuron** is safe in dogs and puppies; given at 20 times the recommended dose it causes only mild adverse effects in 8-week-old puppies (light depression and lack of appetite); in lactating bitches, at 2 times and 6 times the recommended dose the drug is partially eliminated through the milk; excessive overdosing of the drug combination may produce various adverse reactions such as hypersalivation, vomiting, anorexia, lethargy, diarrhoea, pruritus, skin congestion, ataxia, convulsions, and general weakness
moxidectin (0.003 once a month) like ivermectin, the drug shows a prolonged prophylactic effect at a remarkably low dose; it is 100% effective against both 3rd-stage and 4th-stage larvae (1–2-month-old larvae) of *D. immitis*	*Pro Heart (American Home) tablets like milbemycin oxime, moxidectin is highly effective against *Ancylostoma caninum* but less effective against *Uncinaria stenocephala* at a single oral dose of 0.025 mg/kg; whipworms are not affected	approved for prevention of heartworm (*D. immitis*) in dogs ($\geq$ 8 weeks of age) for once-a-month use in dogs; strategic treatment with MOX is as with ivermectin or milbemycin oxime (for details see above); treatment in dogs should begin 1 month after onset of mosquito season and must be continued at monthly intervals, and in the 2 months following termination of mosquito season; moxidectin should only be used in dogs, which proved negative for the presence of heartworms; its strong microfilaricidal action may produce shock-like reactions in infected dogs due to dying or dead microfilariae; thus infected dogs should be treated with an adulticidal drug (see above) for removal of heartworms and microfilariae before initiating a chemoprophylactic regimen; at recommended dose, the drug is safe for a wide variety of dog breeds; MOX tablets were safe at 5 times the recommended (monthly) dose in Collies, and up to 10 times the monthly dose in 8-week-old puppies; adverse effects following overdosing or occasionally recommended dose may be nervousness, vomiting, anorexia, diarrhoea, increased thirst; weakness, lethargy, ataxia, and itching

IMIDAZOTHIAZOLES:
Levamisole was being used as microfilaricide in *D. immitis* infections (11 mg/kg/day orally for 6–10 consecutive days); dogs and cats are much more tolerant of oral than parenteral administration of levamisole; overdose may lead to salivation, vomiting, nausea, muscular tremor, anorexia, depression, and infrequently to ataxia; contraindications are functional disorders of the liver and kidneys, and the simultaneous use of pesticides such as organophosphates or carbamates; boxers and very small "toy" dogs, which seem to be particularly sensitive to chemotherapeutic drugs should not be treated with levamisole (for more details see under GI nematodes, levamisole below)

ORGANOPHOSPHATES:
The use of dichlorvos or trichlorfon in dogs (and cats) infected with heartworms is contraindicated; prior to use of organophosphates for its anthelmintic and pesticidal properties, dogs from endemic heartworm areas should be examined for presence of *D. immitis*; blood levels following therapeutic dose(s) of these drugs may be sufficiently high to produce migration of the adults resulting in occlusion of pulmonary artery (or its branches); they exhibit also a microfilaricidal effect and large numbers of dying or dead microfilariae forming microembolic residues cause disseminated intravascular coagulation; some dog breeds such as greyhounds and whippets appear to be very sensitive to organophosphates and there are many other practical warnings (for more details see organophosphates under GI nematodes, below)

Table 5. (Continued) Drugs used against nematode infections of dogs and cats.

CHEMICAL GROUP, **Nonproprietary name** (approx. dose, mg/kg body weight, oral route) other information	*__Brand name__ (manufacturer, company), other information	**Characteristics** (chemotherapeutic and adverse effects, miscellaneous comments)
GASTROINTESTINAL (GI) NEMATODES OF VETERINARY SIGNIFICANCE IN DOGS, CATS AND WILD CARNIVORES		
There are several important nematodes of carnivores, which may reside in the small and large intestine; small worm loads may be asymptomatic whereas large quantities of migrating larvae and adult worms in the gut of puppies and kittens especially produce severe pathogenic effects and thus clinical signs which are fatal without use of chemotherapy; in general, adult gastrointestinal (GI) nematodes are highly susceptible to various classes of anthelmintic drugs (see below) whereas migrating (developing) larvae of these parasites are fairly tolerant to the majority of anthelmintics even at enhanced and repeated doses; the life cycle of significant GI nematodes is usually direct (without intermediate host), e.g. in **hookworms** occurring endemically in the tropics and warm temperate areas in carnivores, such as *Ancylostoma* caninum of the dog, cat and fox, *A. tubaeforme* of cat, and *A. braziliense* of dog and cat, are responsible for widespread morbidity and mortality, especially in young or debilitated animals due to the bloodsucking activities of these worms in the small intestine (adults are about 1–2 cm long, prepatent period 14–21 days); free-living hookworm 3rd stage larva hatched from egg infects host by skin penetration and undergoes two molts during its migration phase; the oral route of infection by ingestion of infective larva usually occurs with *Uncinaria stenocephala*, a hookworm of dog, cat and fox; highly pathogenic *A. caninum* is characterized by migration of 3rd stage larva via the blood stream through various tissues of host; there is a migratory route through the lungs (3rd stage larva molts in the trachea and bronchi to 4th stage larva) and a transmammary route with galactogenic transmission of 3rd stage larva to nursing pups; this transmammary infection is often responsible for severe anemia in litters of young pups about 3 weeks after whelping; bitches, once infected, can produce transmammary infections in at least 3 consecutive litters; apart from its veterinary importance **ascarids**, especially *Toxocara canis* are responsible for the most widely recognized form of visceral larva migrans in humans (→ Nematocidal Drugs, Man/Table 1); egg (ovoid, yellow-brown, thick sculptural shell) containing 2nd stage larva is infective for dogs and foxes; after hatching of 2nd stage larva in the small intestine it travels via bloodstream to the liver, heart, pulmonary artery, to the lungs (molts to 3rd stage larva) and thence to the bronchi , trachea and via esophagus to the intestine, where 3rd stage larva matures (2 molts) to adults; in the pregnant bitch prenatal infection of the fetus occurs about 3 weeks prior to parturition by 2nd stage larvae migrating to fetal lungs where they molt to 3rd stage larvae; in new born pups the cycle is completed when larvae travel via the trachea to the intestine; a bitch (once infected) harbors enough larvae to infect all her subsequent litters without being reinfected; suckling pups can also ingest infective 3rd stage larva via milk during the first 3 weeks of lactation (transmammary infection); adult worms (females up to 18 cm long, males 10 cm long) may cause pot-belly in pups and occasionally diarrhoea; in heavy infections larval migration can cause pulmonary damage and thus coughing and tachypnea; in pups, which have been heavily infected transplacentally, most mortality may be seen within a few days of birth; rodents or birds may serve as paratenic hosts where L2 travel to their tissues and remain there until eaten by a dog; prepatent period in paratenic hosts is 4–5 weeks, and in prenatal infection 3 weeks; *Toxocara cati* (adults 3–10 cm long, prepatent period about 8 weeks) of cat and wild felines is distributed worldwide; the life cycle is similar to that of *T. canis* but it lacks prenatal infection of the fetus; *Toxascaris leonina* (adult females up to 10 cm long, males 7 cm long, prepatent period about 11 weeks) occurring in the small intestine of dog, cat, fox and wild carnivores in most parts of the world is of less significance because its parasitic phase in the host is non-migratory, i.e., after ingesting the infective larvated egg subsequent development takes place entirely in the wall and lumen of the intestine; **whipworm** (*Trichuris vulpis*, 4–8 cm long prepatent period 11–12 weeks, distribution worldwide) occurs in the cecum and colon (large intestine) of the dog and fox and is characterized by its whip-like body (posterior part being much thicker than the anterior, about three-quarters of the body being made up by the anterior part, which tunnel into the intestinal mucosa); less common is *T. serrata* occurring in cats; infective 1st stage larva within the egg (lemon shaped with a plug at both ends) needs about 1–2 months for development in temperate climate; after ingestion of larvated egg released larva molts four times within the mucosa and emerging adults partly lie on mucosal surface (with their posterior part) while their thin anterior parts are embedded in the mucosa thereby producing marked damage of tissues; pathogenic effects that may result from location and continuous movement of the anterior part of whipworm are lacerate tissues creating pools of blood and fluid, which the adults ingest; in heavy infections this nematode can produce an acute or chronic inflammation, especially in the cecum of the dog		

Table 5. (Continued) Drugs used against nematode infections of dogs and cats.

CHEMICAL GROUP, **Nonproprietary name** (approx. dose, mg/kg body weight, oral route) other information	***Brand name** (manufacturer, company), other information	**Characteristics** (chemotherapeutic and adverse effects, miscellaneous comments)
AMINES (PIPERAZINES)		
piperazine several salts (45–65 base) 2nd dose after 10 days or (45 base for 2 days) piperazine has no effect against *Ancylostoma* spp., whipworms and tapeworms	*Piperazine, and others (Biotec and others) tablet, capsule, paste; liquid formulations (syrups, solutions) and powder for mixing with food (for efficacy of piperazine/thenium closylate see below)	oldtimer that has some erratic efficacy (50%–100%) against adult ascarids; immature stages of ascarids may be affected at increased doses (100 mg/kg piperazine base); there is minimal or no efficacy against intestinal *Toxocara* spp. larvae and no effect against migrating larvae of *T. canis*; it has only a variable effect on *Uncinaria* spp.; the drug is also used in **zoo** canids and felines for removal of ascarids; nursing pups treated at ten-day intervals till 1 month of age (= 3 treatments) show >95% reduction in their worm burden acquired prenatally and lactogenically; the drug appears to be well tolerated in young (nursing) puppies; doses higher than the therapeutic dose may occasionally cause vomiting, nausea and muscular tremor; higher doses should be therefore divided and given on two consecutive days
ETHANOLAMINES (having a quaternary or protonated nitrogen atom at pH 7)		
thenium closylate (500 mg/dog, single dose, dogs weighing > 4.5 kg) (125 mg/dog 2 x, 12h interval, pups weighing 2.5–4.5 kg) **thenium/piperazine** phosphate (2:1 ratio) (2 x 1 tablet, 12 h interval, dog weighing 1–2 kg; 2 x 2 tablets, 12 h interval, >2 kg)	*Canopar (Wellcome/Cooper) tablet (500 mg) *Ancaris (Wellcome/Cooper) tablet: 216 mg thenium + 260 mg piperazine *Thenatol (2:1 ratio) combinations have little effect on *Trichuris vulpis* and none on cestodes	oldtimer, ammonium compound (bephenium analogue cf. → Nematocidal Drugs, Man/Table 1) with good efficacy against hookworms (*A. caninum*, *U. stenocephala*: 90% adults and immature worms and 4th stage larvae, and eliminating 55% 3rd stage larvae of *Uncinaria* from intestine); its activity against canine ascarids is moderate (~ 75%) and it has only weak efficacy against feline ascarids (*Toxocara cati*, 50–75%); due to cholinergic properties of the drug side effects may occur at therapeutic dose (salivation, emesis = vomiting, diarrhoea, and depression); though thenium is poorly absorbed from the intestinal tract of the host, sudden death in dogs (especially in Collies and Airedales) has been reported following routine treatment with the drug; suckling pups or recently weaned pups (< 2.5 kg body weight) must not be treated because there is risk of toxicosis by increased absorption of thenium due to high fat content in the bitch's milk; also felines must not be treated with thenium; combinations act synergistically, thus increasing action on both ascarids and hookworms
HYDROCARBONS		
toluene (dogs, puppies, cats, kittens: 0.22 ml/kg)	*Methacide (Beecham) (others) capsules	oldtimer (derived from coal tars and used as an industrial solvent) has activity against adult ascarids (95%), hookworms (>90%) of dogs and cats, and minimal action on canine whipworm (about 40%); it is fairly well tolerated and vomiting (irritation of digestive tract mucosa) may occur fairly often; puppies and kittens are more sensitive to the drug than older animals (tolerate 5 times the therapeutic dose thereby showing some toxic signs like vomiting, muscular tremor, unsteady gait)
HYDROCARBONS/DIPHENYLMETHANES		
toluene=methylbenzene/**dichlorophen**(e) (dog, cats 264)	*Vermiplex (Pitman-Moore, others) *Tri-plex *Difolin capsules	oldtimer combined with rapidly acting taeniacide dichlorophen is highly active against adult ascarids; it has a variable efficacy against hookworms and a minimal one against whipworms; dichlorophen(e) may cause partial removal (70–82%, destrobilating action only) of common cestodes as *Taenia* and *Dipylidium caninum* in dogs and cats ('ineffective' against *E. granulosus*); overdosing may cause vomiting, CNS involvement (incoordination, unsteady gait)

Table 5. (Continued) Drugs used against nematode infections of dogs and cats.

CHEMICAL GROUP, **Nonproprietary name** (approx. dose, mg/kg body weight, oral route) other information	*__Brand name__ (manufacturer, company), other information	**Characteristics** (chemotherapeutic and adverse effects, miscellaneous comments)
CHLORINATED HYDROCARBONS		
n-**butyl chloride** (dogs, cats, 1 ml: < 2.25 kg b.w., 2 ml: 2.25–4.5 kg, 3 ml: 4.5–9 kg, 4 ml: 9–18 kg, 5 ml: > 18 kg)	can be administered in gelatin capsules (over-the-counter product in the USA and elsewhere)	oldtimer, syn. 1-chlorobutane (C_4H_9Cl) is a colorless highly flammable liquid, which can be used for removal of adult ascarids (90% efficacy) and hookworms (60% efficacy) in young animals; overnight fasting and the use of a laxative may enhance worm expulsion; at recommended dose there is no action on whipworms (50% effect at 3 times the therapeutic dose) or nematodes of other domestic animals; recommended doses are well tolerated in cats and dogs (sometimes vomiting may occur)
PHENOLS		
disophenol (dog, cat 7.5–10, single s.c. injection) subcutaneous injection is easily given and without fasting; i.m. injection often causes pain and local irritation mode of action: uncoupler of oxidative phosphorylation	*DNP (American Cyanamid) injectable 4.5% solution of disophenol (water-polyethylene glycol vehicle) wild felids (leopard, panthers, lions) infected with *Ancylostoma* and *Gnathostoma* have been treated successfully at a single s.c. 6.6 mg/kg	= 2,6-diiodo-4-nitrophenol with a narrow spectrum of activity; at 10 mg/ kg s.c., it is nearly 100% effective in eliminating adult hookworms (*A. caninum*, *A. braziliense*, *A. tubaeforme* and *U. stenocephala*) also in heavily parasitized animals without stress; it does not eliminate histotropic larval forms of hookworms; thus retreatment after 21 days is necessary to remove these worms; disophenol was shown to affect *Spirocerca lupi* of dog, fox, wolf and other wild Canidae and Felidae; *Spirocerca* (intermediate hosts coprophagous beetles, and paratenic hosts) causes severe lesions and scarring of the thoracic aorta and large nodular mass in esophageal wall containing adult worms; cumulative toxicosis (tachycardia,, polypnea, hyperthermia, acidosis, circulatory collapse, lenticular opacity) may occur when drug is administered repeatedly and in intervals less than 21 days; at recommended dose (for accurate dosage each animal should be weighed) it is well tolerated (sometimes vomiting) in young puppies (2 days of age) and pregnant animals; the drug has about a threefold margin of safety in cats and dogs
THIAZOLIDINEDIONE DERIVATIVES		
nitrodan = nidanthel (dog, 100 ppm in-feed for 8 weeks; 230 ppm in-feed for 14 weeks)	*Everfree (Cooper) granules, premix	oldtimer with activity against ascarids and hookworms; continuous feeding of drug (100 ppm) is necessary to achieve variable reduction (20%–90%) of adult hookworm and *T. canis*; the drug appears to be ineffective against *T. leonina*; its efficacy against hookworms is markedly increased (94% reduction of adults) at 230 ppm for 14 weeks; the drug is well tolerated and no contraindications or side effects are known

Table 5. (Continued) Drugs used against nematode infections of dogs and cats.

CHEMICAL GROUP, **Nonproprietary name** (approx. dose, mg/kg body weight, oral route) other information	***Brand name** (manufacturer, company), other information	**Characteristics** (chemotherapeutic and adverse effects, miscellaneous comments)
ORGANOPHOSPHATES		
dichlorvos (puppies, cats 5–11) (dogs 27–33) (dogs 30–35) **trichlorfon** (metrifonate) (dogs 75 x 3 in 3–4-day intervals) there is no activity against migrating larvae of ascarids and hookworms nor any action on cestodes; drug residual passed in feces of treated dogs affects developing infective L_3	*task tablets for cats and puppies *Tenac (Squibb) resin pellets in capsules *Canogard (Shell) capsules *Neguvon (Bayer) *Combot *Dyrex (others, principally for use in horses, cf. Table 3) tablets undigested resin pellets appearing in the feces are toxic for birds	cholinesterase inhibitors that can be used systemically for control of adult GI nematodes and ectoparasites; drugs cause nearly total expulsion of canine and/ or feline ascarids (*Toxocara canis*, *T. cati*, *Toxascaris leonina*), hookworms (*Ancylostoma caninum*, *A. braziliensis*, *A. tubaeforme*, *Uncinaria stenocephala*) and the whipworm (*Trichuris vulpis*, efficacy >90%); formulations for dogs (capsules) should not be used for cats being treated with special (small) tablets after weighing each animal; drug release of capsules is designed to provide maximum activity and minimum side effects; toxic signs result from overdosing, and may involve salivation, vomiting, watery diarrhoea, muscular tremor, and muscular weakness; the margin of safety is narrow in cats and puppies; **atropine** can be used as **antidote**; divided dose (interval 8–24 h) may prevent side effects; the drug is not suitable as a routine anthelmintic; **contraindications** are simultaneous use of pesticides, other anthelmintics, tranquilizers, live vaccines, muscle relaxants, and other cholinesterase-inhibiting drugs; in dogs (and cats) infected with *Dirofilaria immitis*, organophosphates can cause severe clinical signs (see microfilaricidal drugs, above) resulting from dying or dead microfilariae becoming microembolic thereby producing intravascular coagulation, and also from enhanced activity of adult heartworms migrating towards the pulmonary artery, thereby obliterating some of its branches; greyhounds and whippets appear to be very sensitive to organophosphates; there are many other practical warnings (see suppliers' information)
IMIDAZOTHIAZOLES		
levamisole (5 s.c.) *l* isomer of *dl*-tetramisole with almost solely anthelmintic activity of the racemate there is no effect against whipworms or cestodes *dl*-**tetramisole** (mixture of two optical isomers: S (–) levamisole and R (+) dexamisole	*Nemisol (Specia France) *Nemicanisol (Bellon) injectable solution *Levasole *Tramisol *Totalon (others) *Nemicide *Nilwerm *Ripercol	use in dogs and cats is limited because of various potential side effects and contraindications (see below); the drug is highly effective (expelling rate >95%) against ascarids or hookworms; there is some activity against migrating larvae of *A. caninum*; the drug was being used as microfilaricide in *D. immitis* infections (for more details see general discussion on heartworm disease, above); **tetramisole** (10mg/kg/day x 7days, orally, repeated in week-interval x2) has been shown to affect *Spirocerca lupi* infection of dogs (efficacy for levamisole is likely at half the above tetramisole regimen); *Filaroides osleri* of dog may be treated with some success by long-term s.c. injections (7.5 mg/kg/day x 30: numbers of injections may depend on tolerability of drug) to result in regression of symptoms as persistent cough and emaciation (for more information on life cycle, distribution, clinical signs of these extraintestinal living worms, *S. lupi* and *F. osleri*, see general discussion of 'GI nematodes', above) ; the drug is well tolerated at recommended dose; overdose may lead to salivation, vomiting, nausea, muscular tremor, and infrequently to ataxia; contraindications are functional disorders of the liver and kidneys, and simultaneous use of pesticides such as organophosphates or carbamates; boxers and very small "toy" dogs, which seem to be particularly sensitive to chemotherapeutic drugs should not be treated with levamisole

Table 5. (Continued) Drugs used against nematode infections of dogs and cats.

CHEMICAL GROUP, **Nonproprietary name** (approx. dose, mg/kg body weight, oral route) other information	*__Brand name__ (manufacturer, company), other information	**Characteristics** (chemotherapeutic and adverse effects, miscellaneous comments)
levamisole/ niclosamide marketed in France for use in dogs and cats	*Stromiten (Vetoqinol, others)	active against ascarids, hookworms, whipworm, and tapeworms; **niclosamide** shows good activity against *Taenia* spp. but erratic activity against *D. caninum, Mesocestoides corti,* and poor efficacy against *E. granulosus* and *M. lineatus* in dogs
butamisole hydrochloride (dog 2.4 single s.c. dose) intramuscular injection often causes pain	*Styquin injectable solution in small dogs <2 kg b.w. exact dosing is necessary	used in dogs (> 8 weeks of age) for eliminating adult whipworms (*Trichuris vulpis*, expelling rate 99%) and hookworms (*Ancylostoma caninum* expelling rate 92%), several contraindications may limit its use such as severely diseased, debilitated or heartworm-positive dogs, animals with renal or hepatic disorders or concurrent use of cestocidal drug bunamidine which may cause mortality as has seen following treatment of heartworm-infected dogs; concurrent use of butamisole and organophosphate-impregnated flea collars is safe; toxic signs may be vomiting, muscular tremor, unsteady gait, ataxia, convulsions, and lateral recumbency
TETRAHYDROPYRIMIDINES		
pyrantel pamoate (= pyrantel embonate) (dog 15, cat 20–30, paste) (pamoate salt is poorly soluble in water) **pyrantel/oxantel** (14.4/54.3) (combinations with febantel see below)	*Banminth,*Dogminth *Canminth *Nemex-2 suspension *Purina tablets paste, tablets (Pfizer) *Canex plus (Pfizer Animal Health) tablets (100 mg)	at recommended dose highly effective (up to 95%) in eliminating adult hookworms (*Ancylostoma caninum, Uncinaria stenocephala*) and ascarids (*Toxocara canis, Toxascaris leonina*) of dogs (nursing or weaning pups from 2 weeks of age); its efficacy against whipworms is minimal, and there is no activity against cestodes; the therapeutic dose is well tolerated without signs of salivation, vomiting, or diarrhoea; pyrantel also has good efficacy against the common hookworm (*A. tubaeforme*) and ascarid (*Toxocara cati*) of cats at 20–30 mg/kg; cats may refuse to eat the paste after both if injected in the mouth and mixed in their food; therefore, treatment should be repeated 2 weeks later to achieve a safe effect against ascarids; **pyrantel/oxantel** is highly effective against ascarids, hookworms and whipworms of dogs; pyrantel is a safe drug in young puppies, kittens (from 2 weeks of age), pregnant or lactating bitches, or debilitated animals
pyrantel/praziquantel (cat 57.5/5 single dose) for use in cats only (small dogs)	*Drontal cat (Bayer) tablets	effective against common ascarid (*Toxocara cati*), hookworm (*A. tubaeforme*) and cestodes (*Taenia* spp., *Dipylidium*, and *Echinococcus* spp.); it may also be used in small dogs for control of ascarids and cestodes (moderate effect against *A. caninum* only, and ineffective against *Trichuris vulpis, Crenosoma vulpis* (living in trachea, bronchioles) and *Angiostrongylus vasorum* (living in right heart and pulmonary artery)
pyrantel/oxantel/ praziquantel (dog, 14.4/54.3/5)	*Canex Multispectrum (Bomac, Pfizer) *Canex cube (Pfizer) *Popantel (Dover)	marketed in Australia, New Zealand, Brazil and elsewhere; triple combination provides a broad spectrum of anthelmintic activity and is highly effective against common ascarids, hookworms, whipworm, and cestodes of the dog

Table 5. (Continued) Drugs used against nematode infections of dogs and cats.

CHEMICAL GROUP, **Nonproprietary name** (approx. dose, mg/kg body weight, oral route) other information	***Brand name** (manufacturer, company), other information	**Characteristics** (chemotherapeutic and adverse effects, miscellaneous comments)
SUBSTITUTED DIPHENYL ETHERS		
nitroscanate (dog 50: micronized) total elimination of adult *E. granulosus* is not always achieved, even at enhanced doses; therefore it should not be used for this indication	*Lopatol *Cantrodifene (Ciba Geigy) tablets (micronized particles) mode of action in cestodes: drug is believed to act as an uncoupler of oxidative phosphorylation	is used exclusively in dogs for the treatment of gastrointestinal nematodes and common cestodes, including *Echinococcus*; at recommended dose, it has good efficacy against hookworms (*Ancylostoma caninum*, *Uncinaria stenocephala*), and variable activity against ascarids (*Toxocara canis*, *Toxascaris leonina*), *Taenia* spp. and *Dipylidium caninum*; enhanced doses of the drug (100mg/kg x 2, 2day-interval) are necessary to affect *Echinococcus* though 100% elimination of adult tapeworms is not always achieved; its action on *Trichuris vulpis* is poor; good efficacy against *Strongyloides stercoralis* in a Beagle breeding colony has been reported; the drug is poorly absorbed from gastrointestinal tract but irritates gut's mucosa resulting in relatively high incidence of vomiting (10%–20% of treated dogs) within 3–5 h after treatment; fasting, 12–24 h prior to treatment followed by a small quantity of food thereafter will markedly reduce vomiting; nitroscanate should not be used in cats, as it frequently provokes adverse side effects at therapeutic dose
BENZIMIDAZOLES		
(thiabendazole) (experimental drug), may be used in pigs ruminants, and horses	brand names and suppliers see Table 1 various formulations (powder, paste suspension)	active against ascarids and hookworms at 50–60 mg/kg 3 days monthly; in medicated feed (0.025%) given 3 times daily to young dogs it reduces markedly worm burdens of ascarids, and *Trichuris vulpis*; at 50 mg/kg/day for 5 days (repeated after 2 weeks) it affects *Strongyloides stercoralis* (accumulations of this worm are seen particularly in housed dogs); however, this dosage regimen may cause vomiting; in general, thiabendazole is poorly tolerated and can occasionally be fatal in dogs
BENZIMIDAZOLE CARBAMATES		
mebendazole (dog, cat 22 x2–5days) number of doses depends upon worm species to be eliminated it is a safe drug for use in cats and also in dogs infected with heartworm	*Telmin KH (Janssen-Cilag) tablet (100mg) *Telmintic (Mallinckrodt) powder (others) in some countries mebendazole may not be licensed for use in cats	for use in dogs and cats with high efficacy (>90%) against adult ascarids (2-day course), hookworms, whipworms (3–5-day course) but less one against *Taenia* spp. (5-day course), and unsatisfactory one against *Dipylidium caninum*; the drug can also be used in wild carnivores in such a way that drug is mixed into ground meat or put into a meatball (number of doses see above or longer courses, e.g. 3 mg/kg for 10 days depending on target parasite); it has similar activity against feline nematodes and tapeworms as it has in dogs; it is effective against other parasites by reducing level of infection such as *Echinococcus* adults (200 mg twice/day x 5 days), *Strongyloides stercoralis* (22 mg/kg/day x 14–21 days, also used in primate infections with *Strongyloides*), and *Angiostrongylus vasorum* living in right heart and pulmonary artery (22mg/kg/ day x10 days); it appears not to be embryotoxic or teratogenic in dogs (20 mg/kg: 1st day of pregnancy and continuing for 56 days)
mebendazole/ praziquantel (dog, 20/5 x 3 days) (cat >2kg b.w : 100 total/5 x2/d for 2 d; <2kg: 50 total/5 x 3d)	*Duelminth (ATI Pets)	combination available in Italy for use in dogs and cats with activity against ascarids (*Toxocara canis, T. cati, Toxascaris leonina*), hookworms (*Ancylostoma* and *Uncinaria*), whipworm (*Trichuris vulpis, T. serrata*), and common cestodes (*Taenia* spp., *Dipylidium caninum*, and *Echinococcus* spp.)

Table 5. (Continued) Drugs used against nematode infections of dogs and cats.

CHEMICAL GROUP, **Nonproprietary name** (approx. dose, mg/kg body weight, oral route) other information	*__Brand name__ (manufacturer, company), other information	**Characteristics** (chemotherapeutic and adverse effects, miscellaneous comments)
albendazole (not licensed as a single drug for use in the dog or cat, though it show remarkable efficacy against common canine nematodes after repeated daily dosing		its activity (ascarids: 70%, hookworms: 20%, whipworms 10%) is limited against common GI nematodes of dogs and cats when given as a single dose of 15 mg/kg; higher doses (20 or 25 mg/kg) markedly increase its action (>95%) on adult *Toxocara canis* but do not on *Ancylostoma caninum* (60–70% efficacy); nearly 100% efficacy is attained after daily dosing of 15mg/kg for 3 consecutive days against both ascarids and hookworms; some less common parasites such as tapeworm *Mesocestoides corti*, lung nematodes (*Filaroides hirthi*, and *F. osleri*), and urinary bladder worm *Capillaria plica* of foxes and dogs (rarely cats) are also affected by the drug at repeated high doses at 12 hour-intervals (25–50 mg/kg x 5)
albendazole/ praziquantel (dog 50/5)	tablets	combination for use in dogs with activity against ascarids, hookworms, whipworms, *Strongyloides stercoralis*, and cestodes (*Taenia* spp., *D. caninum* and *Echinococcus* spp.); albendazole is teratogenic and may produce weight depression in litters when given to bitch at 100 mg/kg b.w. from day 30 of gestation to day of parturition
fenbendazole (dog, cat 50 x 3 days) (other dose regimens may be used, e.g. 20–50 x 5 days) since overdose presents no risk of adverse effects fenbendazole can be regarded as very safe for Canidae and Felidae; owing to lack of investigations, the drug should not be used in pregnant bitches or she-cats though ill effects are unlikely	*Panacur (Intervet) tablets (tasteless) may be mixed into food drug has **no action** on tapeworms such as *Dipylidium caninum* and *Echinococcus* spp.	for use in dogs and cats with high efficacy (>95%) against adult ascarids, hookworms, whipworms, and common cestodes (e.g. *Taenia* pisiformis) of dogs and cats (e.g. *Taenia hydatigena*); since in carnivores about 50% of fenbendazole is excreted unchanged in the feces within a relatively short period, multiple dosage regimens are more effective than a high single dose (e.g. 125mg/kg); thus daily dosing over several days produces nematocidal drug concentration capable of eliminating adult worms of different species or even migrating larvae of *T. canis* at a daily dose of 50 mg/kg given during the entire gestation period of a bitch resulting in helminth-free and healthy litters; fenbendazole at 20–50 mg/kg/day x 5days may lead to reduction of less common parasites such as the cat lungworm *Aelurostrongylus abstrusus*, and *Crenosoma vulpis*, *Angiostrongylus vasorum* (living in pulmonary arterioles and capillaries of the dog), the stomach worm *Ollulanus tricuspis* (a very small trichostrongyle about 1mm long, living under a layer of mucus in the stomach wall) or *Strongyloides stercoralis* (living in small intestine causing inflammation of mucosa); the drug is also well tolerated with higher dosage regimens (e.g., 500 mg/kg x 1, or 250 mg/kg x 30 in dogs); tasteless drug (10–20 mg/kg x 5days) can be used for treating GI nematodes and extraintestinal living nematodes of wild carnivores (lion, tiger, puma, jaguar, fox, wildcat, and others) by mixing fenbendazole into food (meat or meatballs)
fenbendazole/ praziquantel (dog, cat 50/5 x 3 days)	*Caniquantel Plus *Aniprazol (Animedica), tablets	for use in dogs and cats in Germany, Belgium, the Netherlands and elsewhere with activity against ascarids (*Toxocara*, *Toxascaris*), hookworms (*Ancylostoma*, *Uncinaria*), whipworm (*Trichuris*) and cestodes (*Taenia* spp., *D. caninum*, *Echinococcus* spp.)
flubendazole (cat 22 x 3 days)	*Cat allwurmer, *Flubenol (Janssen)	highly effective against ascarids, hookworms, and *Taenia* spp. in carnivores but like other benzimidazoles, ineffective against *Dipylidium* or *Echinococcus* spp.; it has larvicidal activity against canine heartworm

Table 5. (Continued) Drugs used against nematode infections of dogs and cats.

CHEMICAL GROUP, **Nonproprietary name** (approx. dose, mg/kg body weight, oral route) other information	***Brand name** (manufacturer, company), other information	**Characteristics** (chemotherapeutic and adverse effects, miscellaneous comments)
oxibendazole/ niclosamide (dog, cat 15/200 x 1)	*Vitaminthe (Virbac, and others)	for use in cats and dogs in France and elsewhere; is highly effective against ascarids, hookworms and cestodes (*Taenia* spp., *Dipylidium*) but ineffective against *Echinococcus*
oxibendazole/DEC (dog 5/6.6 daily)	*Filaribit (Merial)	for use in dogs in the USA and elsewhere with activity against hookworms, *Trichuris* and *Dirofilaria immitis* (see also diethylcarbamazine = DEC)
PROBENZIMIDAZOLES		
febantel (dog, <3 months of age: 100 mg/3 kg b.w. x3 in 1 day; > 3 months of age: 100 mg/10 kg/day x 3 days)	*Rintal (Bayer) tablets, pellets	is highly effective in expelling adult and prepatent infections of hookworms (*Ancylostoma caninum*, *A. tubaeforme*, and *Uncinaria stenocephala* >90%), ascarids (*Toxocara canis*, *T. cati*, and *Toxascaris leonina* >95%), whipworm (*T. vulpis* >99%); in pups and kittens; it has a wide range of safety and there are no contraindications for its use in cats or dogs
febantel/praziquantel (dog, cat <6months of age: 15/1.5) (dog, cat >6 months of age: 10/1)	*Vercom Paste paste formulation (available in the USA and elsewhere)	is highly active against adult and prepatent infections of hookworms (*Ancylostoma caninum*, *A. tubaeforme*, and *Uncinaria stenocephala* >90%), ascarids (*Toxocara canis*, *T. cati*, and *Toxascaris leonina* >95%), whipworm (*T. vulpis* >95%) and *Taenia* spp. (*T. taeniaeformis*, and *Dipylidium caninum* >99%); in pups and kittens the combination is well tolerated but it is contraindicated in pregnant dogs and cats because of an increase in frequency of early abortion
febantel/pyrantel embonate (dog 15/14.4 x 1) combination with synergistic effect	*Welpan (Bayer) suspension contraindicated in pregnant dogs and during lactation	for use in pups and young dogs; is effective in expelling adult and prepatent infections of all relevant nematodes of dogs such as *Toxocara canis*, *Toxascaris leonina*, *Ancylostoma caninum*, *Uncinaria stenocephala*, and *T. vulpis*; because of possible prenatal and transmammary infections of litters, treatment of pups should be started 2 weeks after parturition and repeated biweekly; the drug is tolerated at 5 times the therapeutic dose in puppies and young dogs
febantel/pyrantel/ praziquantel (dog 15/14.4/5) (dog 35.8/7/7)	*Drontal plus and others (Bayer and others) tablets	has similar activity to the febantel/praziquantel combination (see above) for canine parasites such as ascarids (*Toxocara canis*, *Toxascaris leonina*), whipworm *Trichuris vulpis*, and tapeworms (including 100% efficacy against *Dipylidium caninum*, *Taenia* spp., *Echinococcus granulosus* and *E. multilocularis*, *Mesocestoides* spp., and *Joyeuxiella pasqualei* (a tapeworm occurring in cats and dogs in Middle East, Africa and other tropical areas); it is more effective (>95%) against hookworms (*Ancylostoma caninum*, *Uncinaria stenocephala*); lungworm *Crenosoma vulpis*, and *Angiostrongylus vasorum*
AVERMECTINS		
selamectin 6 mg/kg body weight (dogs, cats)	*Revolution *Stronghold (Pfizer) Pour-on solutions; Precautions see p. 412	marketed in the rest of the world except for Europe marketed in Europe The drug-kills *Toxocara canis* (dogs), as well as *T. cati* and *Ancylostoma tubaeforme* in cats. Also claimed: litter protection by treating the mother 14 days before and after birth.

bution. Cosmopolitan *D. viviparus* occurs in the bronchi of cattle, buffalo, camel, deer and reindeer and is highly pathogenic to non-immune calves. *D. arnfieldi* occurs in the bronchi of the horse, donkey and other equines and is cosmopolitan in distribution; the donkey appears to be the natural host of the parasite.

In order to survive for longer periods or to overwinter larvae of *Dictyocaulus* spp. need sufficiently high rainfall to prevent them from desiccation. Weather conditions favoring the survival of larvae on pasture are found particularly in temperate areas. L_3 larvae deposited in fall and winter may overwinter on pasture to infect susceptible animals grazing the following spring. Other sources of pasture contamination leading to infection in the following spring may be due to small numbers of lungworm stages residing in the host for longer periods. Thus adult *Dictyocaulus* spp. may survive in the host's lung for several months, and/or inhibited or arrested late L_4 and early L_5 larvae (e.g., of *D. viviparus*) in the local mesenteric lymph nodes or air passages will resume their development in spring. Wind-borne, field-to-field transmission of *D. viviparus* larvae by sporangia of the fungus *Pilobolus* may also play a role in contaminating pastures. The fungus is very common in cattle feces and larvae of *D. viviparus* may accumulate on the surface of the sporangium. When the sporangium explodes it may catapult the larvae several meters (up to 3 m) through the air, moving them from the fecal pats onto the adjacent herbage. In addition, infections with gastrointestinal nematodes leading to loose feces or diarrhoea may also favor the translation of larvae onto the herbage.

Parasitic bronchitis due to *D. viviparus* and *D. filaria* is seen primarily in young calves or spring-born lambs. The outbreaks of disease occur mostly from early summer (July) until early fall (September), or late fall (November) in the Northern Hemisphere, though the heaviest infections in lambs usually occur in the fall because of the increase of larvae on the pasture at that time. Older animals have usually developed strong immunity to *Dictyocaulus* spp. infections by lasting reinfections. However, adult animals may be susceptible to heavy challenge if the rate of acquisition of infection is not sufficient or immunosuppressive agents, other pathogens (diseases) or emaciation hamper immunity build-up.

D. arnfieldi infections in horses may produce coughing, increased respiratory rate and nasal discharge prior to patent infection. Horses are thought to become infected by contact with donkeys though infections in horses infrequently reach patency and thus diagnosis by fecal examination cannot be made.

In ruminants and horses strategic control measures and biological control of nematode are needed. Thus, rotating sheep and cattle grazing programs as well as ruminants and horses grazing programs have been shown to reduce markedly the number of infective larvae on pasture. Annual **vaccination** of all calves before commencing spring grazing in April or May (northern latitudes) with attenuated live larval vaccine (Dictol or Huskvac) may be highly effective in preventing clinical disease, though small numbers of lungworms may develop in the bronchioles of vaccinated calves inducing an additional boost for immunity build-up. Thus post vaccination, pasture commonly remains contaminated with small numbers of infective larvae providing an enduring stimulus to protective immunity in vaccinated calves. However, highly susceptible (naive) or debilitated animals can respond to vaccination with parasitic bronchitis.

The evaluation of drug's efficacy against lungworms (Table 6) is usually based on monitoring the elimination of L_1 larvae in the feces and on the resolution of clinical signs after treatment. Since L_1 larvae may reappear in treated animals after prolonged periods, monitoring of larvae in fresh feces should be carried out for longer periods. The reappearance of larvae after treatment suggests that the drug apparently affects the reproductive organs of the adult worm rather than adult worm itself.

Strategic anthelmintic treatment does not always guarantee survival of sufficient parasite stages in the host to ensure an effective immune response during the grazing season. Animals insufficiently protected should be treated as early as possible after parasitic bronchitis has been diagnosed to prevent severe clinical signs often associated with serious pulmonary tissue damage.

Intraruminal boluses (Table 1, 6) are designed to release nematocidal concentrations of an anthelmintic in the reticulo-rumen of cattle in order to kill ingested infective larvae of GI nematodes and those of the lungworm *Dictyocaulus viviparus* for prolonged periods. Some intraruminal devices may release nematocidal concentrations for up to 5 months and other formulations (e.g. for parenteral injection or pour-on) of macrocyclic lactones (Table 1) may protect animals against infections of gut roundworms and lungworms for several weeks postdosing. However, the sole use of anthelmintics directed towards the prevention of parasitic bronchitis appears not always to be a reliable control measure, since the prophylactic treatment is harmed by the unpredictable occurrence of natural infection and reinfection. Long acting drug formulations may, however, control this challenge to animals. But the action of such drug formulations may interfere with the fairly rapid acquisition of protective immunity and with the maintenance of sufficient levels of immunity following artificial vaccination and natural exposure to infection in endemic areas. On the other hand, there were results that have indicated compatibility of 'concurrent' use of a lungworm vaccine and an **ivermectin** sustained release bolus or an **oxfendazole** pulse release bolus (for more information on SR boluses cf. Table 1). These boluses and other long-term formulations allow development of a protective level of immunity to *D. viviparus*. Thus a **fenbendazole** SR bolus releasing nematocidal concentrations for as long as 5 months or parenteral **doramectin** ('Zero + eight week' treatment program: two vaccinations, each with 1000 attenuated *D. viviparus* larvae 42 and 14 days prior to spring tournout followed by 0.2 mg/kg b.w. doramectin on DO = day of tournout and D56 → the 'zero + eight week' treatment program) had not interfered with the build-up of protective immunity.

There are indications that strategic use of anthelmintics, i.e. in late *Dictyocaulus* spp. infections in autumn, can lead to arrested (inhibited) larvae in the host. Since these larvae grow to maturity in the following spring, vaccination or prophylactic (metaphylactic) treatment before commencing spring grazing will markedly reduce subsequent pasture contamination.

Anthelmintic killing of parasites in extraintestinal tissues often provokes severe systemic reactions and lesions (→ Nematocidal Drugs, Man/Table 1, filariasis). The severity of the resulting pathological lesions appears to be related to the location of the parasitic stages within the respiratory passages. Also, the treatment of parasitic bronchitis may be associated with severe pathological reactions and exacerbation of clinical signs as a result of the rapid killing of adult *Dictyocaulus* spp. and their larvae in the bronchioles and alveoli. The appearance of new severe histopathological lesions (e.g., severe edema of peribronchial tissue, chronic occlusive bronchitis), which sometimes cause fatalities, is due to large numbers of disintegrating worms and larvae in the deeper air passages. Remnants and still 'intact' dead worms and larvae release toxic or antigenic material and cannot be eliminated by the host. As a consequence, these products may elicit severe inflammatory reactions in the host aimed at destroying and eliminating the dead worm material, thereby producing space-occupying lesions occluding vessels, alveoli, and bronchioles.

Only transient reactions (e.g., frequent coughing) are seen following treatment of lungworms and their larvae in the larger bronchioles, the bronchi, and the trachea, e.g., in the case of *Metastrongylus* spp. infections. Severe host reactions are not seen after treating gastrointestinal nematode infections in which remnants or dead parasites are disintegrated by digestion, or in which remnants and intact dead parasites are flushed from the host in the feces.

Protostrongylid Infections of Small Ruminants Protostrongylids are hair-like nematodes living in the alveoli, bronchioles, and parenchyma of the lungs of sheep, goats, wild ruminants (deer), and other species of mammals. *Protostrongylus rufescens* is the most important species, less common are *Cystocaulus* spp., *Muellerius capillaris*, and other genera (Table 6). Eggs released by female worms usually develop in the lungs of the host, and hatched larvae (first stage) are passed via trachea and intestine in the feces. For the further development, the larva requires a snail intermediate host. Prophylaxis against protostrongylid infections is problematic since the extermination of the ubiquitous snail intermediate host is impossible. Thus, pastures remain contaminated for a long period because larvae are protected in the snail, probably for as long as the infected snail lives. Lambs should therefore not be allowed to graze on contaminated pastures. Anthelmintic treatment, so far necessary, with benzimidazoles (fenbendazole 20–80 mg/kg, and albendazole 5 mg/kg, orally) or **levamisole** (20 mg/kg subcutaneously), may lead to a marked decrease (>85%) of larvae in the feces. Often, there is only a transient suppression of egg production despite using relatively high doses of suitable anthelmintics. As a rule, the efficacy of common anthelmintic drugs against protostrongylid worms (especially *Muellerius*) appears to be variable only, particularly with regard to their capability of killing adult worms in sufficient numbers (Table 6).

Metastrongylid Infections of Swine Several members of the family Metastrongylidae are parasites of the respiratory passages and blood vessels of the lungs of especially young pigs and wild boar. Outbreaks of disease do not often occur due to in-house pig husbandry. As far as is known, they require intermediate hosts for their further development. Common lungworms are *Metastrongylus elongatus*, *M. pudendotectus*, *M. madagascariensis* or *M. salmi*, all of which occur in the pig and wild boar (and accidentally in man and ruminants). They dwell in the bronchi and bronchioles of domesticated and wild pigs and may produce pathogenic effects. The life cycle of these nematodes includes various species of **earthworm** (e.g., *Eisenia* spp., *Lumbricus* spp., *and Helodrilus* spp.) as intermediate hosts. Larvated eggs of *M. elongatus* passed in the feces may hatch soon thereafter. Hatched larvae may survive in moist surroundings for several months. To proceed with their development, an earthworm must swallow them. After performing two molts in the intermediate host the larvae are infective and can pass the winter in the earthworm. Pigs usually become in-

Table 6. Drugs used against extraintestinal nematode infections of domestic animals.

CHEMICAL GROUP, **Nonproprietary name** (approx. dose, mg/kg body weight, oral route) other information	***Brand name** (manufacturer, company), other information	**Characteristics** (chemotherapeutic and adverse effects, miscellaneous comments)
CYANOACETIC ACID HYDRAZIDES		
cyacetacide (=2-cyanacetohydrazide and others) (cattle, sheep, 15 s.c. or 17.5 orally)	*Armazal,*Benecid *Cyanazid,*Dictycide *Dictyfuge (many suppliers)	oldtimer with variable activity against adult **lungworms** (*Dictyocaulus* spp.) and with narrow safety margin; results of treatment are erratic despite repeated administration (up to 3 times); drug may cause temporary decrease of larvae in feces, particularly in *Muellerius capillaris*, *Cystocaulus* spp., and *Protostrongylus rufescens* in sheep and goats
PIPERAZINE DERIVATIVES		
diethylcarbamazine (DEC) (cattle: 50, i.m. single dose; 22, i.m. for 3 days; 40, i.m. or orally. for 3 days)	*Franocid (Lederle) more than 20 brand names (many suppliers)	oldtimer with some activity against immature stages of *Dictyocaulus* spp. suppressing early infection but not patent infection; DEC is commonly used as a preventive microfilaricide in humans suffering from lymphatic filariasis (*Wuchereria* spp., *Loa loa*, and *Onchocerca volvulus*, cf. → Nematocidal Drugs, Man/Table 1) or to prevent heartworm disease in dogs (Table 5); it has been used in sheep and cattle for treatment of lungworm infections (*Dictyocaulus* spp.), DEC has shown microfilaricidal action against *Onchocerca* spp. (elongate filariform worms producing microfilariae *in* the skin and connective tissue spaces) in cattle (*O. gibsoni* and *O. gutturosa*)and in horses (*O. cervicalis*)
PYRIDINE DERIVATIVES		
methyridine (cattle, sheep, 200 s.c. or i.p.) (dog 150–200 i.m.)	*Mintic *Promintic (ICI) solution, injectable solution	it has widely been replaced by other drugs showing higher activity and tolerability; oldtimer with efficacy against adult lungworms (*Dictyocaulus* spp.) in cattle and sheep; there is also variable activity against some gastrointestinal trichostrongylids at the recommended dose; it is also active against gastrointestinal nematodes of dogs (*T. vulpis, T. canis, A. caninum*); local reactions (edema, necrosis) may occur at the site of injection, particularly in horses; in severe infections treatment must be repeated; drug has a narrow safety margin; adverse effects in dogs are salivation, vomiting, diarrhoea, and ataxia

Table 6. (Continued) Drugs used against extraintestinal nematode infections of domestic animals.

CHEMICAL GROUP, **Nonproprietary name** (approx. dose, mg/kg body weight, oral route) other information	***Brand name** (manufacturer, company), other information	**Characteristics** (chemotherapeutic and adverse effects, miscellaneous comments)
IMIDAZOTHIAZOLES		
levamisole (cattle, sheep, goats, and pigs: 8 s.c. or orally) (cattle, 10 pour-on) (for more information on anthelmintic effects, tolerability and withdrawal times cf. Table 1) (treatment of parasitic bronchitis caused by *Dictyocaulus* spp. may be associated with exacerbation of clinical signs	*various brand names (Table 1) (Janssen, ICI, Bayer and others) sterile solution for parenteral injection, oral solution, granules, pour-on formulation; bolus for cattle (Table 1)	has for a long time been considered to be the most suitable treatment against verminous bronchitis (*Dictyocaulus* spp.) in cattle; it has a high efficacy (about 99%) against adults and L5 larvae of *D. viviparus*; for efficacy against L4 larvae it is necessary to increase the dose to 10 mg/kg s.c.; however, the double therapeutic dose of the drug may destroy only some of the immature *D. viviparus*, causing impaired pulmonary function and gas exchange; it has variable efficacy against L1 larvae of *Protostrongylus* spp. in sheep and goats; at 3–4 times the therapeutic dose (15 and 20 mg/kg) there is a transient reduction of larvae in the feces; the rapid excretion of the drug may be the cause for the less pronounced activity against L4 larvae; the drug is highly active against adult lung worms (*Metastrongylus* spp.) in pigs (Table 4); the drug shows activity against preadults of the heartworm *D. immitis* in dogs (Table 5) as well as against filariform worms *Setaria equina* (common filariform parasite of equines found in body cavities, sometimes eyes and lungs) and *Stephanofilaria okinawaensis* (filariform parasite in Japan transmitted by flies and causing lesions on muzzle and teat in cattle)→ (7.5 mg/kg x 2); it shows activity against *Crenosoma vulpis* (fairly short worm found in bronchi of fox, dog and wolf causing rhinotracheitis and bronchitis similar to that produced by *Capillaria aerophila*, snails are intermediate hosts)→ (8 mg/kg) and *Capillaria aerophila* (worms of this genus are related to *Trichuris* and found in the trachea and bronchi of dogs, and foxes)→ (5 mg/kg, interval treatment, 3 x at 9-day intervals); there is also efficacy against other extraintestinal nematode infections: *Stephanurus dentatus*, the swine kidney worm (8 mg/kg) or *Thelazia* spp. (spirurids of conjunctival sac or lacrimal duct of mammals and birds) in cattle and other mammals→ (5–12 mg/kg)

Table 6. (Continued) Drugs used against extraintestinal nematode infections of domestic animals.

CHEMICAL GROUP, **Nonproprietary name** (approx. dose, mg/kg body weight, oral route) other information	***Brand name** (manufacturer, company), other information	**Characteristics** (chemotherapeutic and adverse effects, miscellaneous comments)
BENZIMIDAZOLES / [1]PROBENZIMIDAZOLES (=BZs)		
(recommended dose and route of administration see Table 1)	(trade names, suppliers and formulations see Table 1	broad spectrum benzimidazoles (BZs) have in addition to their excellent activity against gastrointestinal nematodes also a high efficacy in the treatment of several extraintestinal nematode infections in ruminants; fenbendazole, oxfendazole, febantel, albendazole (thiabendazole is less active) and luxabendazole (sheep only) are highly effective against adult and immature stages of lungworms *Dictyocaulus viviparus* in cattle and *D. filaria* in sheep (cf. also Table 1); BZs are active (85–90%) at higher dose (2–3-fold the therapeutic dose) against protostrongylid nematodes of sheep and goats such as *Protostrongylus rufescens* (other species), *Muellerius*, *Cystocaulus*, and *Neostrongylus* (intermediate hosts are mollusks); however, BZs action on fecal larvae of these small lungworms appears to be transient and variable, and more information is needed as to whether they kill adult worms; fenbendazole at 1.5–2 mg/kg for 5 days not only affects adult protostrongylid nematodes but also inhibited L4 larvae of *Dictyocaulus* spp. in cattle; similar dosage regimens are highly effective against adult lungworms of swine (*Metastrongylus* spp.): 1.5 mg/kg flubendazole for 5 days, or 5 mg/kg fenbendazole in the feed, divided over 5–15 days (dosage regimens of other benzimidazoles cf. Table 1); some drugs may affect the kidney worm of swine *Stephanurus dentatus* (Table 4); lungworm infections (*D. arnfieldi*) of donkeys and horses (cf. also Table 3) appear to be less susceptible to BZs; *Thelazia lacrimalis* in horses (spirurids of conjunctival sac or lacrimal duct of mammals and birds) may be affected to some degree by fenbendazole 3 mg/kg x 3, flubendazole 1.5 mg/kg x 5; febantel (10 mg/kg); several other extraintestinal nematodes of wild ruminants may respond to BZ-treatment such as *Elaphostrongylus rangiferi* (occurring in lungs of reindeer) → (mebendazole: 6 mg/kg x 10) and *E. cervi* (occurring in connective tissues of breast, thorax, back and CNS of red deer) → (fenbendazole: 7.5. mg/kg x 5); BZs show some activity against lung worm infections in dogs (*Filaroides hirthi*: albendazole: 50 mg/kg twice daily, x 5) or cats (*Aelurostrongylus abstrusus* (adults in terminal bronchioles causing chronic cough with gradual wasting, intermediate hosts are snails and slugs) → (fenbendazole: 50 mg/kg x 3); *Capillaria plica* infections in dogs (worms of this genus are related to *Trichuris* and found in pelvis of the kidney causing cystitis and urinary disorders) may be partially controlled with fenbendazole (50 mg/kg x 3) or albendazole (50 mg/kg x 10); although effects of these drugs proved variably there is an improvement of clinical signs in most cases after treatment
fenbendazole **oxfendazole** **febantel** **albendazole** **luxabendazole** (flubendazole, others cf. Tables 1; 3; 4; 5)	(intraruminal devices e.g. slow or pulse release boluses, for tolerability and withdrawal times cf. Table 1)	
enhancement of anthelmintic efficacy of BZs can be obtained by dividing the therapeutic dose over several days or with in-feed medication (Table 1, Table 3, Table 4)	exacerbation of clinical signs and occasionally mortality may occur in lung worm infected animals a few days following treatment with benzimidazoles	

Table 6. (Continued) Drugs used against extraintestinal nematode infections of domestic animals.

CHEMICAL GROUP, **Nonproprietary name** (approx. dose, mg/kg body weight, oral route) other information	***Brand name** (manufacturer, company), other information	**Characteristics** (chemotherapeutic and adverse effects, miscellaneous comments)
MACROCYCLIC LACTONES		
AVERMECTINS		
ivermectin (cattle, 0.5 pour-on) (cattle, 0.2 s.c. or oral)., (sheep, 0.2 oral) (horse, 0.2 oral) (swine, 0.3 s.c.) (dog, 0.006–0.012 oral). (for more information on anthelmintic effects, tolerability and withdrawal times cf. Table 1)	*Ivomec (Merial) sterile solution for injection, paste, suspension (drench), pour-on, tablet, SR bolus (Table 1) (for strategic use of an **intraruminal slow release device** of ivermectin against *D. viviparus* cf. discussion)	has a very high activity against adults and developing L4 larvae of *Dictyocaulus* spp.; given s.c. the drug produces long persistent plasma and tissue levels (precludes its use in lactating dairy animals); after the s.c. route its prolonged activity exerts an effect on trichostrongylids and lungworms of ruminants (Table 1) for about 3 weeks; residues of the drug persisting in the tissues during this time may provide protection against most gastrointestinal nematode infections for approx. 10 days and against *D. viviparus* infections in cattle for approx. 3 weeks; the drug is effective against lungworms of swine (*Metastrongylus* spp., Table 4), *D. arnfieldi* of equines (Table 3), small lungworms (*Muellerius* spp. and others) infections in sheep, *Capillaria aerophila* (occurring in the trachea and bronchi of canines) and heartworms of dogs (*Dirofilaria immitis*) infections; it was the first macrocyclic lactone, which has been used for the treatment of filarial infections in animals and in man (cf. → Nematocidal Drugs, Man/Table 1); there is microfilaricidal action against larvae of *Onchocerca cervicalis* in horses (0.2 mg/kg p.o.) or microfilariae of *O. gibsoni* and *O. gutturosa* in cattle (there is obviously a prophylactic effect against these species); the drug proves highly effective against *Parafilaria bovicola* of cattle (occurring in tropical areas and transmitted by flies feeding on lacrimal secretions or wounds) at 0.2 mg/kg s.c., reducing totally hemorrhagic nodules produced by this filariform worm in the skin within 2 weeks of treatment; the drug exerts its effect via the reproductive organs of female adult filariform worms (blockade of embryogenesis)
abamectin (cattle, 0.2 s.c.) (horses 5.4 g paste / 100 kg body weight)	*Avomec *Enzec *Duotin *Equimax (Virbac) injectable solution, paste	it is used in cattle as an endectocide and exhibits high efficacy against immature and adult lungworm *D. viviparus* in cattle; in horses it is used in combination with praziquantel (Equimax) and exerts high efficacies against *D. arnfieldi* (Table 3) and *Thelazia spp.* (spirurids of conjunctival sac or lacrimal duct of mammals and birds) and ileocecal tapeworms *Anoplocephala* (for more information on anthelmintic effects, tolerability and withdrawal times in ruminants cf. Table 1 and Table 2)
doramectin (cattle 0.2 s.c.)	*Dectomax (Pfizer) injectable solution	is a very potent endectocide affecting all economically important nematodes and ectoparasites (grubs, lice, mange mites) in cattle (Table 1); it is highly effective against adults and fourth stage larvae of lung worm *D. viviparus* in cattle; it also affects adult *Thelazia* spp. (spirurids of conjunctival sac or lacrimal duct of mammals and birds) (for more information on anthelmintic effects, tolerability and withdrawal times cf. Table 1 and Table 2)
selamectin (dog, cat, 6'spot on'	*Revolution (Pfizer) (for efficacy cf. Table 5)	
MILBEMYCINS		
milbemycin oxime (dog, 0.25 p.o.) (dog 0.5–0.99 for heartworm prophylaxis)	*Interceptor (Novartis) chewable tablet (four sizes)	is a potent microfilaricide for heartworm prophylaxis in dogs (for more information cf. Table 5); besides its strong action on microfilariae of *D. immitis* infections it additionally controls intestinal nematodes of dogs and has a moderate efficacy against *Demodex canis* (mites that may live as commensals in the skin; they can cause squamous or pustular dermatitis = demodicosis especially in young dogs)

Table 6. (Continued) Drugs used against extraintestinal nematode infections of domestic animals.

CHEMICAL GROUP, **Nonproprietary name** (approx. dose, mg/kg body weight, oral route) other information	***Brand name** (manufacturer, company), other information	**Characteristics** (chemotherapeutic and adverse effects, miscellaneous comments)
moxidectin (cattle, 0.2 s.c.) (cattle 0.5 pour-on) (horse 0.4 paste) (dog, 0.003 oral)	*Cydectin (American Home, Fort Dodge) *Equest (American Home) *Quest (American Cyanamid) injectable solution, pour-on	Chemically modified derivative of macrocyclic lactones nemadectin, which has been classed with milbemycins; moxidectin exerts a very high efficacy against early L5 larvae and adults of *D. viviparus* in cattle besides its high endectocide activities; it is highly effective against various nematodes and ectoparasites of cattle (Table 1) and horses (Table 3), including microfilaricidal effects against *Onchocerca* spp. (elongate filariform worms producing microfilariae in the skin and connective tissue spaces) in cattle (*O. gibsoni* and *O gutturosa*) and horses (*O. cervicalis*); it is 100% effective against larvae of *Dirofilaria immitis* (heartworm) of dogs (Table 5)

Doses listed in this table refer to information from manufacturer and literature
Data given in this table have no claim to full information
The primary or original manufacturer is indicated if there is lack of information on the current one(s)

fected by ingesting infected earthworm or by accidentally liberated larvae from an injured or dead earthworm. In the pig, the development of *Metastrongylus* spp. larvae is similar to that of *D. viviparus* and *D. filaria*. The larvae pass through mesenteric lymph glands molting once and travel then to the lungs, where they grow adult after a further molt. Eggs are produced about 3–4 weeks after infection. Common clinical signs caused by *Metastrongylus* spp. may be loss of condition and weight decrease though piglets may have marked bronchitis and sometimes pneumonia associated with secondary bacterial infections. In general, pig lungworms are not as pathogenic as *Dictyocaulus* spp. in ruminants. Anthelmintic treatment is comparable to that practiced in *Dictyocaulus* spp. infections, though higher doses are necessary to affect metastrongylids (cf. Table 1 and Table 6). Anthelmintic killing may lead to moderate and transient bouts of coughing only. To prevent lungworm infection in pigs, the ground should be kept dry and the feces should be disposed of adequately so that the life cycle is interrupted. Since infective larvae can live in the earthworm for an unknown length of time, paddocks and fields may remain contaminated for a considerable period. Young pigs should therefore be run on clean fields only.

Nematocidal Drugs, Man

For Overview see Table 1

Gastrointestinal (GI) Nematodes of Medicinal Importance

Ascaris lumbricoides (roundworm) is a cosmopolitan nematode common particularly in humid tropical climates. According to most recent estimates 1.273 millions of people are infected (prevalence 24%). This large roundworm (20–40 cm long) lives in the small intestine and feeds on gut contents. The eggs are passed with the feces, become infective in about 1 month, and can remain infective in the soil for several years. After uptake of infective eggs, e.g., with vegetables, larvae hatch in the small intestine, penetrate the duodenal wall, migrate through the liver parenchyma and travel via the blood stream to the lungs. Then they break through the alveoli into the bronchioles and bronchi, ascend the trachea, and are swallowed. In the gut developmental stages mature to adults in about 8–10 weeks. Adult females live for about 1 year, and each female may produce about 200.000 eggs per day.

Common → hookworm (Vol. 1) species are *Ancylostoma duodenale* (common in the 'Old World' and occurring from the Mediterranean countries through India to China and Southeast Asia and Brazil) and *Necator americanus* (American or

Table 1. Drugs used against nematode infections of humans.

DISEASE (alphabetical order) parasite, distribution, pathology, **Stages** affected (location)	**Chemical class** other information	**Nonproprietary name**, adult dosage (oral route), comments	Characteristics of compounds
ANCYLOSTOMIASIS (hookworm infection): soil-transmitted helminthic infection; hookworms undergo a cycle of development in the soil (limited by the requirements of developing larvae for warmth and humidity: tropics, subtropics), female worms attached to wall of jejunum by buccal capsule, lay large numbers of eggs passed out with the feces; the hatched larvae become infective after undergoing two molts in the soil to produce infective, filariform larvae, which penetrate the skin of the new host; the classical feature of hookworm disease is severe anemia, and edema and ascites may result from high hookworm loads; each year some 65 000 deaths are directly attributable to hookworm infections; about 44 million pregnant women have hookworm infections, which cause chronic blood loss from intestine and thus predisposition to development of iron deficiency anemia, often of great severity, constituting a major health problem (1998 WHO/CTD homepage, intestinal parasite control); it is estimated that the hookworms currently infect some 800 million people; cases of morbidity may be between 90 and 130 million worldwide, with high prevalence among pre-school and school-age children.			
Ancylostoma duodenale (Old World hookworm) *Necator americanus* (New World hookworm) it is estimated that about a quarter of world's population is infected with hookworms **adults** (small intestine) loss of blood per day and adult worm: *A. duodenale*	**benzimidazole** carbamates (BZs)	**mebendazole** (drug of choice, 100 mg bid x 3d); (dosage for adults and children) **albendazole** (400 mg once or 200 mg/d x 3d) (dosage for adults and children)	BZs may cause infrequent and mild side effects (e.g., epigastric pain, diarrhoea); albendazole and mebendazole have been shown to be embryotoxic and teratogenic in animals; therefore drugs should not be used during pregnancy; it is advisable to refrain from administering **flubendazole** during pregnancy although it failed to demonstrate teratogenicity in rodents; its potency and spectrum of activity is similar to that of mebendazole
0.1–0.2 ml, *N. americanus* 0.02–0.05 ml) **migrating larvae** (lungs) Most serious outcome of hookworm infection is severe hypochromic anemia due to severe blood loss from mucosa	tetrahydropyrimidines	**pyrantel** pamoate (11 mg/kg, max.1-gram/d, x3d) (dosage for adults and children)	pyrantel may produce occasional and mild side effects (e.g., headache, dizziness); although there are no reports of a teratogenic effect, drug should not be given during pregnancy and to children less than 1 year of age; piperazine antagonized depolarization of neuromuscular system caused by pyrantel
damaged by worm attachment; adult worms suck blood directly after destroying villous tissue; treatment of hookworm disease involves individual or mass treatment; blood values should be restored to normal	Imidazothiazoles	levamisole	levamisole has been used principally against *Ascaris* infections; it is less active against hookworms; occasional adverse reactions are nausea, vomiting, abdominal discomfort, headache, dizziness, hypertension
by proper diet and iron treatment before or during treatment; *A. duodenale* is more susceptible to **mebendazole** than to **pyrantel** (however, latter has superior compliance).	quaternary amines	**bephenium** hydroxynaphthoate (contraction of worms can be blocked by piperazine)	bephenium has been used mainly in treatment of *A. duodenale* (results against *N. americanus* are unsatisfactory), it may often cause vomiting, diarrhoea, dizziness, headache; it should not be used in patients with hypertension, and during pregnancy
	halogenated hydrocarbons	**tetrachloroethylene** (TCE): *N. americanus* is more sensitive to TCE than *A. duodenale*	oldtimer TCE has been used specifically in hookworm infections; it shows low intestinal absorption (inhalation causes narcotic effect); major side effects are nausea or vomiting and burning sensation; long-term treatment causes hepatotoxicity

Table 1. (Continued) Drugs used against nematode infections of humans.

DISEASE (alphabetical order) parasite, distribution, pathology, **Stages** affected (location)	**Chemical class** other information	**Nonproprietary name**, adult dosage (oral route), comments	Characteristics of compounds
ASCARIASIS: soil-transmitted helminthic infection; female worms lay large numbers of unembryonated eggs (thick shell) that are passed out with the feces into soil where they undergo development for 2–3 weeks to larvated eggs that contain infective, second stage larvae; these infective eggs can readily contaminate vegetables when night-soil is used as fertilizer; infection of host can be from "hand to mouth" or occurs when larvated eggs are swallowed with contaminated food (e.g. not adequately washed green leaf vegetable) or undercooked food; infective eggs can survive in areas with temperate or cold climate for longer periods; adult worms (male ~15–30 cm long, female ~20–35 cm long) are very active intestinal parasites and in heavy infections overcrowding effects render them aggressive in that they migrate to aberrant extraintestinal sites thereby causing serious pathological damage; a large bolus of round worms expelled from intestine especially of children following anthelmintic treatment may consist of hundreds of worms; distribution of *Ascaris* is global; it is estimated that some 60 000 deaths are directly attributable to *Ascaris lumbricoides* (1998 WHO/CTD homepage, intestinal parasite control) although this mortality rate is rather low compared with the high prevalence of round worm infections worldwide, especially in children; *Ascaris* infects currently about 1000 million people, and cases of morbidity may amount to 120–215 million people.			
Ascaris lumbricoides ('roundworm') **adults** (upper small intestine) **migrating larvae** (liver, lungs); heavy infections, especially in children, may lead to intestinal obstruction and volvulus which can be fatal; adult worms are very motile and have a marked tendency to escape through fistulae or any hole in their vicinity and so may block common bile duct and the appendix itself; in cases of additional hookworm infection care should be taken to avoid unusual activity of *Ascaris* (e.g. perforation of intestine) which can be initiated by therapy; in cases of intestinal obstruction **piperazine** can be used to paralyze and to relax ascarids which are then expelled prior to therapy of ancylostomiasis; **symptoms,** which occur, are associated with migratory larval phase in lungs (pneumonitis) and intestinal phase (worms can form a bolus which obstructs the lumen); adults are known for aberrant migration (e.g., into bile duct); Prevention of ascariasis in rural areas may be supported especially by sanitation and health education measures	**benzimidazole carbamates** (BZs) (albendazole most recent broad-spectrum anthelmintic for use in man; it is used also in veterinary medicine, cf. → Nematocidal Drugs, Animals/Table 1)	**mebendazole** (drug of choice, 100 mg bid x 3d or 500 mg once); (dosage for adults and children) **albendazole** (400 mg once) (dosage for adults and children)	mebendazole and albendazole are suitable for mass treatment; mebendazole is as active as levamisole against ascarids but its overall curative action on soil-transmitted nematodes seems to be superior to that of levamisole; BZs should not be administered during pregnancy (for more information see ancylostomiasis and trichuriasis); in the course of therapy parasites may begin to 'walk'; *Ascaris*-pneumonia caused by migrating larvae is treated with prednisolon (antiphlogistic effect)
	tetrahydropyrimidines (morantel is chemically very similar to pyrantel and thus in its spectrum of activity and potency, → Nematocidal Drugs, Animals/Benzimidazole Compounds)	**pyrantel** pamoate (11 mg/kg once, max. 1 gram) (dosage for adults and children)	pyrantel is well tolerated (side effects, cf. ancylostomiasis); major disadvantage of pyrantel is its lack of activity against widespread whipworms (cf. trichuriasis); it is a reliable and well tolerated drug for individual deworming, also in combination with *oxantel pamoate*, which is closely related to pyrantel and active against *Ascaris* and *Trichuris trichiura*, but inactive against hookworms
	piperazine derivatives (piperazine has been used for mass treatment of ascariasis and enterobiasis)	**piperazine** (syn. diethylenediamine; various salts: hexahydrate, adipate, citrate, phosphate)	piperazine is safe, cheap, and easy to administer; urticarial reactions and fever (obviously due to intoxicated worms) are rare; overdose causes neurological effects; **contraindications** are epileptic seizures and other neurological abnormalities, and pregnancy in the first trimester or hypersensitivity
	imidazothiazoles	**tetramisole:** racemic mixture of isomers, D-(+) dexamisole and L-(-) levamisole; the latter has several times higher anthelmintic activity than dexamisole and same toxic properties	levamisole is highly active against *Ascaris*; a single dose gives parasitological cure; in mixed infections (e.g., hookworm) its curative effect is less satisfactory; it is an immunomodulating agent (cf. → Nematocidal Drugs, Animals/Levamisole, Pyrantel, Morantel) and stimulates parasympathetic and sympathetic ganglia; it is a potent inhibitor of mammalian alkaline phosphatase

Table 1. (Continued) Drugs used against nematode infections of humans.

DISEASE (alphabetical order) parasite, distribution, pathology, **Stages** affected (location)	**Chemical class** other information	**Nonproprietary name**, adult dosage (oral route), comments	Characteristics of compounds
DRACUNCULIASIS (dracontiasis)			
occurs in parts (semi-desert) of Africa , India, the Middle East and Brazil where drinking water is drawn from primitive wells or swallow ponds during the rainy season; copepods (*Cyclops*, water fleas) containing infective larvae are swallowed by humans with the drinking water; the released larvae penetrate the intestinal wall and migrate for about 3 months through connective tissues where male and female worms mate; females then move to the subcutaneous (SC) tissues; preferable location of adult female worms are the foot or lower limbs; about 7 months later adult female(s) (50–80 cm long) emerge from SC tissues to the surface of skin to release thousands of rhabditoid larvae from worm's uterus into the water to be ingested by *Cyclops*; heavy infections may cause clinical signs such as local lesion, intensive burning pain, and secondary infections spreading via an ulcerating papule where the adult female worm reaches the skin surface; bacterial infections of SC tissues may induce a phlegmon of leg and arthritis in the vicinity of joints; prevention measures are filtering or boiling the drinking water, chemical treatment of ponds and preventing infected persons with an emerging worm from entering the water source; causal therapy is not yet established and female worm either becomes extracted (see above) or is removed by means of surgery.			
Dracunculus medinensis (Guinea worm, dragon worm, or Medina worm) **adult female**(s) (subcutaneous tissue, usually of legs; gravid female causes an ulcerated lesion in skin to discharge motile larvae)	**5-nitroimidazoles** benzimidazoles (thiabendazole has also been used because of its antiinflammatory, antipyretic, and analgesic action)	**metronidazole** (drug of choice, 250 tid x 10d) (children: 25 mg/kg; max 750 mg, in 3 doses x 10 d) **mebendazole** (400–800 mg/d x 6d should kill the worm directly)	metronidazole is not curative; its clinical efficacy is probably based on its potency to reduce inflammatory tissue reactions of the host; this may facilitate expulsion and lysis of female worm; it also has antibacterial activity which may control secondary anaerobic infections; action of **niridazole** (5-nitro-thiazole, cf. → Nematocidal Drugs, Animals/Table 3) is probably due to drug's metabolites suppressing enhanced cell-mediated immunity reactions; **traditional treatment** is to remove the worm alive by winding it gradually day by day on a small stick; **surgical excision** of female worm can exaggerate allergic reactions
ENTEROBIASIS (syn. oxyuriasis, pinworm infection)			
larvated eggs (elongated, flattened on one side, thick, colorless shell) are swallowed by humans; transmission may be ano-oral or direct to mouth by hands or caused by dust-born infection; adult worms are small (females: 8–13 mm long, males: 2–3mm long), white, and threadlike; gravid females emerge preferably at night to the perianal surface where they lay some thousands of partially larvated **eggs** (10–15,000 eggs/worm) and then die; fecal examination is unreliable and eggs are best detected by using cellulose tape preparation; eggs on skin are directly infectious on ingestion or larvae that hatch on skin can reenter the anus, or larvae that occasionally move to aberrant sites enter the vagina producing peritonitis and/or ovarian infection; enterobiasis is very common in day nurseries and institutional settings (conditions of familial and group infection); distribution of *Enterobius* is worldwide and its probable prevalence in humans (especially in children) may amount to 300–500 million.			

Table 1. (Continued) Drugs used against nematode infections of humans.

DISEASE (alphabetical order) parasite, distribution, pathology, **Stages** affected (location)	**Chemical class** other information	**Nonproprietary name**, adult dosage (oral route), comments	Characteristics of compounds
Enterobius vermicularis (pinworm) Adults and **larvae** (lumen of descending colon and cecum); Probably still the most common nematode in Europe, the USA and elsewhere because of its ready transmissibility; there is no effective prevention	**tetrahydropyrimidines** (pyrantel dosage for adults and children →) **benzimidazole carbamates** (BZs) (mebendazole or albendazole dosage for adults and children →)	**pyrantel** pamoate (drug of choice, 11 mg/kg once, max.1gram; repeat in 2 weeks) **mebendazole** (100 mg once; repeat in 2 weeks) **albendazole** (400 mg once adults/children or children 10–14 mg/kg; repeat in 2 weeks)	pyrantel and BZs are highly effective against *Enterobius*, and in mass treatment too; individual treatment should always include the whole family or group (side effects see ascariasis/ ancylostomiasis); **piperazine** (various salts) is an inexpensive drug; however, repeated treatment may limit its use in mass treatment (side effects see ascariasis); **pyrvinium** pamoate (red cyanine dye) was shown to be highly active in mass treatment; however, serious side effects such as Stevens-Johnson syndrome and photosensitization may occur and the drug should not be used in patients with renal or hepatic dysfunction; therefore it has been largely replaced by well tolerated anthelmintics as pyrantel and BZs
CAPILLARIASIS Distribution of *C. hepatica* is global but infections of humans are rare; diagnosis is difficult because of the nature of the life cycle of this parasite; adult female worms live in a host-derived capsule within the liver where they feed on cytoplasmic debris and lay unembryonated eggs not passed in the feces (dead end of life cycle in man); unembryonated eggs must be released from the liver by a predator, or by cannibalism or scavenging, and the eggs are passed in the feces of the predator or cannibal; embryonation to infective stage takes 4 weeks at 30 °C; infection takes place by ingestion of larvated eggs; human infection may occur in Zaire, Nigeria and in other parts of West Africa and elsewhere, where people have close contact to numerous definitive hosts such as rat (Gambian rat) and mouse or cricetoma; adult worms of *C. hepatica* can cause severe parenchymal damage of liver involving numerous granulomata consisting of mononuclear cells and eosinophils and finally hepatic fibrosis; diagnosis can only be made by demonstrating the presence of eggs, larvae and adult (very thin 4–12 cm long) in liver biopsy samples *Capillaria philippinensis* occurs preferably in the Philippines, Thailand, Japan , Egypt and Iran; capillariasis is diagnosed on demonstration of characteristic eggs in feces, which are bipolar, small, long-oval and have striated shells; diagnosis is usually made by identifying the characteristic-shaped eggs in feces; man becomes infected by eating raw or undercooked freshwater fish or shrimps (intermediate hosts), which contain infectious, third stage larvae; *C. philippinensis* may occur in epidemic form and probably prevalence in man amounts to thousands; clinical signs are associated with ulcerative enteritis such as uncontrollable diarrhoea and malabsorption syndrome in heavy infections, which is sometimes fatal.			
Capillaria hepatica (common parasite of rodents, rabbits, squirrel, muskrat, opossum, rarely dogs cats, man) **larvae, adults, eggs** (liver parenchyma)	**benzimidazoles**	(thiabendazole) (albendazole)	anthelmintic therapy against hepatic capillariasis is not established; thiabendazole, which is absorbed at relative high concentrations from intestine could be used rather than the BZ carbamate albendazole, which might be effective after long-term treatment
Capillaria philippinensis (parasites of fish-eating birds) **third-stage larvae, adults** (lumen and epithelium/mucosa of posterior small intestine and anterior large intestine)	**benzimidazole carbamates** (mebendazole or albendazole dosage for adults and children →)	**mebendazole** (drug of choice, 200 mg bid x 20d **albendazole** (alternative drug, 400 mg daily x 10d)	follow-up of patient with intestinal capillariasis is necessary; relapses can occur with either compound and then need retreatment; mebendazole appears to be less active against larval stages and requires longer treatment periods; supportive treatment is unsatisfactory in outpatients
		(**thiabendazole**)	thiabendazole is reported to be effective

Table 1. (Continued) Drugs used against nematode infections of humans.

DISEASE (alphabetical order) parasite, distribution, pathology, **Stages** affected (location)	**Chemical class** other information	**Nonproprietary name**, adult dosage (oral route), comments	Characteristics of compounds
FILARIAL INFECTIONS: arthropod-transmitted nematodes of the lymphatic, subcutaneous, and cutaneous tissues; filarial nematodes (some 8 species infect humans) are tissue-dwelling parasite; adult females produce microfilaria larvae, which are taken up by blood-feeding arthropods to produce infective, larval stages, during next blood feed, vectors transmit infective larvae to humans; the most important lymphatic filariae are *Wuchereria bancrofti*, *Brugia malayi* and *B. timori* infecting an estimated 118 million people; in Africa and the Americas about 90 million people are at risk of *Onchocerca volvulus* and ocular and dermatological damage are the main features of onchocerciasis; about 18 million people are infected with *O. volvulus* and at least 0.4 million blinded by the disease; the numbers infected with the less pathogenic filariae *Loa loa* and *Mansonella* spp. are much smaller; though *Mansonella* frequently infects humans it does not always produce clinical signs whereas *Loa loa* does.			
LYMPHATIC FILARIASIS: bloodsucking mosquitoes (e.g., Anopheles) transmit infective third stage larvae of lymphatic filariae; infective larvae penetrate the skin of a new host through the puncture wound made when the mosquito bites and enter the lymphatics where the worms copulate and mature into thread-like adults (adult males of *Wuchereria bancrofti* are about 4 cm long, females 8 to 10 cm); adults can live in lymph glands, e.g. in the groin, for many years thereby producing microfilaria (MF); *W. bancrofti* is widely distributed throughout the tropics (Asia, Africa, Australia, Pacific, and South America), *B. malayi* occurs in South-East Asia, and *B. timori* in Indonesia (islands of the lesser Sunda group and Timor); due to pathological effects of MF and adults acute involvement of lymphatic vessels (lymphangitis) is common, especially in the extremities; local lymphadenitis, acute orchitis in man, associated with hydrocele and fever may be characteristic in the early stage of the disease; lasting lymphatic obstruction and repeated leakage of lymph into tissues produce lymphedema, thickened skin and new adventitious tissue (‘**elephantiasis**’) showing later verrucous growth; injured skin may lead to secondary infections with bacteria and funguses; severe elephantiasis of the scrotum may produce gross and incapacitating deformity, which requires radical surgery to remove the surplus tissue.			
Wuchereria bancrofti, *Brugia malayi*, *B. timori* **third stage larvae (L3)** **thread-like adults** microfilariae (MF sheathed) (L3 enter lumen of lymphatic vessels, i.e., all sites within lymphatic circulation and mature to adults , MF enter bloodstream) clinical features of lymphatic filariasis: lymphangitis, dermatitis, cellulitis associated with fever; later, chronic lymphadenopathy, lymphedema and elephantiasis	**piperazine derivatives** (**antihistamines** or **corticosteroids** may be required to decrease allergic reactions due to disintegration of MF in treatment of filarial infections)	**diethylcarbamazine** (DEC, drug of choice: day 1: 50mg p.c. day 2: 50mg tid day 3:100 mg tid days 4 through 14: 6 mg/kg/d in 3 doses) (full doses may be given from day 1 in patients without MF in blood)	**DEC** has been used for more than 50 years for prevention and mass treatment of lymphatic filariasis; during therapy “allergic” reactions may occur (fever, edema, pruritus); DEC may have a dual effect on MF: (1) direct neuromuscular effect causes immobilization, (2) changes of MF surface coat cf. also → Nematocidal Drugs, Animals); DEC causes MF to leave circulation for the liver where they are entrapped and destroyed by phagocytosis; effect of DEC on adult worms is uncertain; drug itself produces only minor side effects (headache, abdominal pain)
chemoprophylaxis:	**macrocyclic lactones** **benzimidazoles**	COMBINATIONS **ivermectin** (200–400 μg/kg once) plus **albendazole** (400mg once)	additional treatment with **albendazole** has some effect on adult worms (macrofilariae) and may markedly lengthen return of symptoms after remission
DEC: 6 mg/kg once per year or IVER: 400 μg/kg once per year will reduce microfilarial density: DEC to 80–90 %, IVER to 100% (WHO, 1992, Technical Report Series 821, WHO Geneva)		**DEC/albendazole** **DEC/ivermectin** (6 mg/kg/ 400 μg/kg) resulted in high cure rates (99%)	DEC combined with IVER appears to be synergistic; studies in Sri Lanka have shown long-term reductions in microfilaraemia; the combination may be an alternative to DEC in areas with concurrent onchocercosis or loiasis
IVER is extremely effective against MF but does not kill adult worms	**benzimidazoles** **imidazothiazoles**	**mebendazole** **levamisole**	mebendazole and levamisole have shown beneficial effects on lymphatic filariasis when given over 14–25 days

Table 1. (Continued) Drugs used against nematode infections of humans.

DISEASE (alphabetical order) parasite, distribution, pathology, **Stages** affected (location)	**Chemical class** other information	**Nonproprietary name**, adult dosage (oral route), comments	Characteristics of compounds
LOAIASIS large, tabanid flies, *Chrysops* spp., which live in primary rain forests in Africa transmit *Loa loa* to humans (mandrill may be infected with an almost identical parasite); *Chrysops* spp.(mango flies) possess powerful mouthparts by which they injure the skin to form blood pools at the site of wound to feed from blood while infective third stage larvae enter the vertebrate host; blood meals are taken during the daytime; in the host, L3 mature into adults within about one year; adult female worms are about 7cm long and may live for 4 to 12 years; they migrate through the subcutaneous tissues, notably the eye under the conjunctiva; microfilariae (MF) develop from larvae in the female and circulate in the blood with which they are picked up by another fly where they mature to infective L 3 to enter a new host when *Chrysops* takes another blood meal; loiasis is confined to Africa (from Golf of Guinea in the West to the Great Lakes); its probable prevalence in humans may amount to 33 million; "calabar" swellings indicate the tracks of migrating adults and disappear as adults continue their migration; recurrent large swellings are most frequently seen in the hands, wrists and forearms and may be accompanied with itching, erythema and fever; a marked eosinophilia (60–90%) accompanies always this phase of infection; heavy infections may be more common than in other filariases because a single *Chrysops* vector may be the cause for high microfilaremias, which enhance the risk of inducing emboli in capillaries of brain, meninges or retina; degranulation of eosinophils has been reported to be associated with endomyocardial fibrosis; the movement of the adult worm under the conjunctiva may cause considerable irritation and vascular congestion.			
Loa loa (eye worm) **third stage larvae** adults microfilariae (MF sheathed) (MF migrate from subcutaneous tissue, and conjunctiva to blood-stream), adults migrate through subcutaneous tissue or across subconjunctival space inducing local reactions which may cause pain chemoprophylaxis: DEC, 300 mg once weekly, has been recommended for prevention of loiasis	**piperazine derivatives** (**antihistamines** or **corticosteroids** may be required to decrease allergic reactions due to disintegration of MF in treatment of filarial infections, especially those due to *Loa loa*)	**diethylcarbamazine** (DEC) (drug of choice, day 1: 50 mg, oral, p.c. day 2: 50 mg tid day 3: 100 mg tid days 4 through 21: 9 mg/kg/d in 3 doses) (full doses may be given from day 1 in patients without MF in blood)	DEC "destroys" MF and immature stages; more details see *W. bancrofti*; effect of DEC on adult worm is doubtful; killed MF may cause severe "allergic" reactions, which can be reduced by gradually increasing DEC doses; in heavy infections rapid killing of MF can provoke an encephalopathy; apheresis has been reported to be effective in lowering microfilarial counts in patients heavily infected with *Loa loa*; adult worm under conjunctiva of eye must be removed surgically; extraction of adult worms is done by means of fine forceps after anesthetizing the conjunctiva
	benzimidazole carbamates (BZs)	**mebendazole** (300 mg/d x 45d) albendazole	mebendazole or albendazole or ivermectin has been used to reduce microfilaremia; BZs have a more slow onset of microfilaricidal activity and therefore may be better tolerated than DEC or ivermectin; the effect of BZs against adults is erratic
	macrocyclic lactones	ivermectin	
MANSONELLIASIS blood-feeding midges (Culicoides spp.) and/or blackflies (*Simulium* spp.) transmit infective third stage larvae to humans; *M. ozzardi* and *M. perstans* infections may be asymptomatic or associated with allergic reactions such as cutaneous itching, pruritus, arthralgia, inflammation of subcutaneous tissues, inguinal lymphadenitis, moderate abdominal pain and marked eosinophilia; *M. ozzardi* occurs in Central and South America (estimated cases 15 million), *M. perstans* is widely distributed in Africa and South America (estimated cases 65 million.), whereas *M. streptocerca* is confined to Africa (estimated cases : millions); microfilariae of *M. streptocerca* are found in the skin and may cause pruritus and papules, edema, and dermatitis; symptoms may be similar to those of mild onchocerciasis.			

Table 1. (Continued) Drugs used against nematode infections of humans.

DISEASE (alphabetical order) parasite, distribution, pathology, **Stages** affected (location)	**Chemical class** other information	**Nonproprietary name**, adult dosage (oral route), comments	Characteristics of compounds
Mansonella ozzardi *Mansonella perstans* *Mansonella streptocerca* **third stage larvae (L3)** **adults** **microfilariae** (MF unsheathed) (probably in visceral adipose and subcutaneous tissue, abdominal or pericardial cavity, MF enter bloodstream (*M. streptocerca*: MF in skin, subcutaneous tissue)	**macrocyclic lactones**	**ivermectin**	**ivermectin**, 6mg once, has been effective against *M. ozzardi*), DEC has no effect
	benzimidazole carbamates	**mebendazole** (100 mg bid x 30d **ivermectin** (150 µg/kg once)	**mebendazole,** drug of choice against *M. perstans*, levamisole has been reported to have some activity; **ivermectin,** drug of choice against *M. streptocerca* or **DEC**
	piperazine derivatives	**diethylcarbamazine** (DEC) (6 mg/kg/d x14d)	chemotherapy may generally exacerbate hypersensitivity reactions to killed MF (severe pruritus cf. *W. bancrofti*)

ONCHOCERCIASIS (River blindness):
Blackflies (*Simulium* spp.) usually feed on plant juices; only adult females feed on blood, and blood meal is repeated for each ovarial cycle; they transmit infective third stage larvae to humans following the bite of the vector; L3 penetrate the skin through wound and migrate to subcutaneous tissues where they mature into thread-like adult males and females in about one year, adult worms (female 35–70 cm long and male 2–4 cm long) are thin and exhibit sluggish movement; they are found subcutaneously in nodules or free in the tissues of humans; nodules consist of fibrous material, which encloses numerous adults of both sexes; in Africa, fibrous nodules in the subcutaneous tissues are found predominantly in the lower parts of the body, while in South and Central America they are more commonly found in the head region and upper parts of body; larvae produced by females develop to unsheathed microfilariae (MF), which live in skin and eye; MF picked up by another blackfly need about 1 week to become an infective L3; aquatic stages of the vector such as eggs, larvae, pupae are attached to all submerged objects (even crabs) and live in fast flowing oxygen-rich water (streams, rivers, and waterfalls) where larvae and pupae extracted oxygen through head filaments; river blindness has a focal distribution, which is closely associated with the biology of vectors; the disease is endemic in West Africa equatorial and East Africa, Sudan, Central America and in parts of Venezuela and Columbia; about 18 million people are infected with *Onchocerca volvulus*, and about 90 million are under risk; clinical features of the disease vary according to duration and frequency of exposure as well as geographical location; early lesions of the skin are manifest as a papular dermatitis (so-called '**craw craw**' in Africa, i.e. small papules around the MF); in advanced patients a quite common feature of onchocerciasis is thickening and wrinkling of the skin called 'lizard' or 'elephant' skin or dermatitis with lichenification (itching and scratching reactions lead to thickening and hardening of the skin); so-called '**Sowda**' usually of lower limbs is characterized by hyperpigmentation and often involves inguinocrural lymphadenopathy; in late cases (burnt-out onchocerciasis) pretibial atrophy and depigmentation (so-called 'leopard skin') is common; in chronic infections, atrophy of the skin may be evident resulting in 'tissue paper' appearance of the skin; lymphadenopathy of the inguinocrural glands can result in an appearance described as 'hanging groin' and scrotal elephantiasis; lesions of the eye involve early corneal changes due to dead MF such as punctate keratitis, which may clear with time; progressive, sclerosing keratitis commonly producing blindness results from heavy MF infections; chorioretinal lesions (chorioretinitis, iritis, and iridocyclitis) may follow damage by dead MF to anterior segments of the eye; finally optic nerve atrophy may develop; prevention of onchocerciasis may be reduction of man/vector contact (protective clothing, insect repellents), vector control (use of larvicides at blackfly breeding sites), nodulectomy, and chemoprophylaxis with ivermectin (for more details see below).

Table 1. (Continued) Drugs used against nematode infections of humans.

DISEASE (alphabetical order) parasite, distribution, pathology, **Stages** affected (location)	**Chemical class** other information	**Nonproprietary name**, adult dosage (oral route), comments	Characteristics of compounds
Onchocerca volvulus (tissue-dwelling nematode) **third stage larvae** **adults** (subcutaneous tissue forming nodules) **microfilariae** (MF unsheathed) within nodules of adults (subcutaneous tissues), MF migrate to skin, and anterior chamber of eye; there is no evidence that live MF may cause host reaction, only dying	**macrocyclic lactones** cf. → Nematocidal Drugs, Animals/ Avermectins and Milbemycins; initially developed as broad spectrum anthelmintics for use in livestock cf. → Nematocidal Drugs, Animals: Tables 1, 3–6	**ivermectin** (= IVER, drug of choice, 150 µg/kg once, orally, repeated every 6 to 12 months), dosage also for children	well accepted drug (6mg tablets) for mass treatment since 1989 exhibiting sustained microfilaricidal effect; onset of eosinophilia and Mazotti-type reaction are delayed compared to DEC; skin MF density decreases to near zero within a month after treatment and increases to 2%–10% of pretreatment level by 12 months after treatment; it should not be given to children under 5 years or under 15 kg b.w., to pregnant women, breast-feeding mothers within 1 week of delivery, to persons with neurological disorders or severe intercurrent disease
and dead MF induce pathological changes becoming more severe the longer infection has persisted nodulectomy (excising nodules containing adult worms) can prevent serious eye changes thereby eliminating production of MF, the actual pathogenic agents; surgical removal of nodules	**piperazine derivatives**	**diethylcarbamazine** (= DEC as highly microfilaricidal drug is obsolete and no longer in use for controlling MF of *O. volvulus*)	DEC does not kill adult worms; it must be administered under medical supervision; skin lesions were treated with DEC; doses were increased gradually to control "allergic" (systemic) reactions, which proved markedly more severe than those occurring in *L. loa* and *W. bancrofti* infections; destruction of MF greatly increases skin pathology and is the main cause of any inflammatory reaction, so-called Mazotti-type reaction
has been widely used in young patients with' Erisipela della costa' or older patients with 'mal morado' in Central and South America) **Chemoprophylaxis** (150 µg/kg orally once, repeated every year can prevent blindness	**sulfated naphthylamine** (cf. → Trypanocidal Drugs, Animals: suramin was introduced as a trypanocidal drug)	**suramin** (intravenous) (obsolete, no longer in use for controlling adult worms of *O. volvulus*)	first successful drug against adult worms; it produces severe side effects (drug provokes acute allergic reactions in patients with heavy skin MF loads or it causes nephrotoxicity; cf. → Trypanocidal Drugs, Animals/Table 1 and → Trypanocidal Drugs, Man/Table 1; it must therefore be given under medical supervision and was mainly used for treatment of individual cases with recurrent skin lesions and inoperable ocular infections (an initial low dose was given to check for reactivity to the drug)
	benzimidazole carbamates (BZs)	mebendazole **flubendazole** (experimental drugs)	BZs may have a temporary sterilizing effect (up to 1 year) on adult females; use of micronized preparations markedly enhances their absorption from gastrointestinal tract; repeated doses are needed

TROPICAL PULMONARY EOSINOPHILIA (TPE):
cf. lymphatic filariasis; clinical features and pathological aspects of TPE appears to be predominantly associated with lymphatic filariasis and is a result of an atypical hypersensitivity of the host to tissue-dwelling parasites; this abnormal host reaction to the presence of a lymphatic filarial infection (other nematodes?) is most commonly seen in southern India; symptoms are dry coughing resembling asthma attacks and may be due to eosinophilic infiltration of the lungs; characteristic of TPE is the absence of MF in the blood but a positive filarial serology; TPE is also found in areas of the Pacific and East Indies ('Meyers-Kouwenaar syndrome) involving lymphadenitis, enlargement of lymph nodes and spleen (in histologic sections hyperplasia, aggregation of tissue eosinophils and granulomas are evident).

Table 1. (Continued) Drugs used against nematode infections of humans.

DISEASE (alphabetical order) parasite, distribution, pathology, **Stages** affected (location)	**Chemical class** other information	**Nonproprietary name**, adult dosage (oral route), comments	Characteristics of compounds
occult filariasis (parasites of animals or humans ?); absence of MF from blood makes diagnosis difficult microfilariae (MF) in various tissues especially in the lung	**piperazine derivatives**	**diethylcarbamazine** (= DEC, drug of choice, 6 mg/kg/d in 3 doses x 14d) (TPE may resolve rapidly with treatment; DEC can be used also as diagnostic drug)	eosinophilia in blood and tissues is often present at high levels (cf. also *Strongyloides stercoralis* infections below) diagnosis is established by successful response to drug; it may initiate exacerbation of symptoms followed by decrease of eosinophil cell level; X-ray picture of chest may become clear after a few weeks
STRONGYLOIDIASIS: Humans become infected by third stage filariform larvae (L3),which penetrate the skin; infective larvae may arise from free-living rhabditiform larvae outside the body; *Strongyloides* is unique among nematodes because this parasite is capable of undergoing both a parasitic and free living reproductive cycle; parasitic, adult females live threaded into the mucosal epithelium of the small intestine and produce larvated eggs by parthenogenesis, i.e. development from an unfertilized egg; after hatching, larvae may develop through four larval stages into free-living adult male and female saprophytic worms; under certain conditions first stage larvae may develop to L_3 within the intestinal tract initiating internal autoinfection or when larvae emerge to perianal areas and penetrate the skin they can give rise to an external autoinfection; the course of this hyperinfection is often fulminant and sometimes fatal in debilitated or immunosuppressed persons; distribution of strongyloidiasis is global and occurs in subtropical and tropical areas (probably prevalence in humans is about 50–100 million); *S. fülleborni*, which is a common parasite in African and Asian primates is also found in humans in several countries (e.g. in Zambia); clinical signs of *S. fülleborni* infection are anorexia, dullness and characteristic watery mucous diarrhoea causing malabsorption syndrome as result of severe catarrhal (ulcerate) enteritis; passage of larvae through the lungs (as in hookworms) may produce severe coughing, and marked eosinophilia; autoinfection can lead to severe 'creeping eruption', usually on the back, and may occur many years after initial infection; deep migration of the larvae may be associated with 'eosinophilic lung' type syndrome resembling that of tropical pulmonary eosinophilia; incidentally given steroids and immunosuppressive agents (HIV in patients with AIDS) may markedly enhance infection with *S. stercoralis* and can be fatal			
Strongyloides stercoralis (thread or dwarf worm) *S. fülleborni* **parasitic, adult females** larvated eggs (most of them hatch already in mucosa of small intestine, eggs are similar to those of hookworm)	**macrocyclic lactones**	**ivermectin** (= IVER drug of choice, 200 μg/kg/d x 1–2d)	IVER is suitable for therapy of uncomplicated infection; in immunocompromised patients or disseminated disease, it may be necessary to prolong or repeat therapy or use other drugs; if IVER is not approved for disseminated strongyloidiasis, thiabendazole may be preferred thiabendazole is the traditional agent and has significant side effects;
intestinal larvae (L1 → infective L3: autoinfection)	**benzimidazoles** (BZs)	alternative: (**thiabendazole** 50 mg/kg/d in 2 doses x 2d, max. 3 gram/d **albendazole** (400 mg/d x 7d)	all BZs, including **mebendazole,** have varied efficacies only (44–98%);
TERNIDENS INFECTION (false hook-worm infection): Occurs in some areas of East and Central Africa, Mauritius South Africa and Asia; hookworm-like parasites of the large intestine of various monkeys (chimpanzee, gorilla, baboons, macaques) may occasionally infect humans; *Ternidens* has the same size as *Necator*, and eggs may be confused with those of hookworms; infection in man is of minor importance (estimated cases some thousands) and is due to ingestion of vegetables or fruits contaminated with infective third stage larvae; in heavy infections migrating immature stages may cause ulceration (bloodsucking activity of worm?) of the mucosa of the small and large intestine and thus enteritis (and anemia); nodules containing larvae are found in colon only; mild infections are usually asymptomatic.			

Table 1. (Continued) Drugs used against nematode infections of humans.

DISEASE (alphabetical order) parasite, distribution, pathology, **Stages** affected (location)	**Chemical class** other information	**Nonproprietary name**, adult dosage (oral route), comments	Characteristics of compounds
Ternidens deminutus **third stage larvae** fourth stage larvae (in nodules)	**benzimidazoles** (BZs)	**thiabendazole** **albendazole** **mebendazole**	BZs are drugs of choice at recommended dose; cure rates may be more >90%
adults (about 1 cm long) (jejunum, colon)	**tetrahydropyrimidines**	**pyrantel** pamoate	egg reduction rate after pyrantel therapy is low

TRICHINOSIS (Trichinellosis):
Zoonotic infection with global distribution and a probable prevalence in man of about 48 million; *Trichinella spiralis* infection may circulate between rats and other carnivores; a common reservoir of infection may be the wild pig or bear, which initiate isolated outbreaks of human infection following hunting parties; pigs may acquire infection by eating infected rats; cycle of infection also exists in wild Canidae that ingest rodents; humans become infected by ingestion of raw or undercooked muscle (sausages) containing encysted larvae from pig, wild boar, polar bear, walrus, seal and other fur-bearing animals; clinical signs such as diarrhoea, fever, myalgia (stiffness and pain in affected muscles), periorbital edema, eosinophilia, and muscular paralysis may occur when females begin to shed newborn larvae 5–21 days after infection; pathogenic effects are produced by larvae in muscles; crisis is usually reached when larvae become encapsulated; encapsulated larvae may live for several years though their calcification begins already 6–9 months after entering the muscle; they can be detected at biopsy or by serological tests such as fluorescent antibody test, gel-diffusion test, ELISA, and others (e.g. PCR); high antibody titres, which are present in the acute stage of disease are non-protective; essential prevention measures against trichinosis should be thorough meat inspection, eliminating of rats as reservoir host, and regulations to ensure that larvae in pork are killed by cooking or freezing of infected carcasses before marketing; consumer should be instructed that pork or pork products or carcasses of carnivorous game must be cooked sufficiently prior to consumption

Trichinella spiralis **intestinal stages:** encysted larvae, adults, newborn larvae (lumen and mucosa of small intestine) **parenteral stages:** newborn larvae enter the lymph and blood via thoracic duct, and mature in a 'modulated' striated muscle cell (termed 'nurse cell') where they rapidly grow and become encapsulated	**steroids** (adrenal cortex)	**corticosteroids** for severe symptoms	in the acute phase, treatment should begin as early as possible; corticosteroids with high antiphlogistic effects at a high initial dose is recommended to reduce inflammation reactions caused by cell damage (myositis) through migrating larvae; BZs may show some antiinflammatory activity per se and some direct action on adult worms (no proof for removal); the drugs may reduce the intensity of muscle infection; trials in rodents suggest that long-term treatment with some BZs carbamates (also **flubendazole**) will sterilize and kill adult worms
	benzimidazoles (BZs)	plus **mebendazole** (200–400 mg tid x 3d, then 400–500 mg tid x 10d) **albendazole** (400 mg bid x 14d (children 15 mg/kg bid x 14d)	

TRICHOSTRONGYLIASIS:
Several trichostrongylid species are capable of infecting humans; they usually inhabit the digestive tract of herbivores; human infections may occur in many countries (e.g. Iran and Japan) but are rare; transmission of infective larvae is due to close contact with ruminants; humans (like herbivores) become infected by ingestion of vegetables contaminated with night soil containing infective third stage larvae; ingested larvae penetrate the intestinal mucosa forming tunnels beneath the epithelium; infection by the cutaneous route is also possible, especially when persons mold animal dung to 'briquettes' to be dried and burnt as fuel; light worm loads of *T. orientalis* may be asymptomatic but heavy ones may cause enteritis and thus diarrhoea, which may be associated with anemia; adult worms may be confused with human hookworms, however, trichostrongylids are smaller in size and more slender in shape, and bursa is different in form from that of *Ancylostoma duodenale*.

Table 1. (Continued) Drugs used against nematode infections of humans.

DISEASE (alphabetical order) parasite, distribution, pathology, **Stages** affected (location)	**Chemical class** other information	**Nonproprietary name**, adult dosage (oral route), comments	Characteristics of compounds
Trichostrongylus orientalis *T. colubriformis* (at least 8 other species) **third stage larvae,** adults (L3 and head of adults embedded in mucosa of small intestine)	**tetrahydropyrimidines** **benzimidazole carbamates**	pyrantel (drug of choice, 11 mg/kg once, max. 1 gram) **mebendazole** (100 mg bid x 3d) **albendazole** (400 mg once)	*Trichostrongylus* spp. is also affected by other drugs such as levamisole (an imidazothiazoles, cf. → Nematocidal Drugs, Animals/Levamisole, Pyrantel, Morantel), and bephenium hydroxynaphthoate (a quaternary amines, see ancylostomiasis); though levamisole seems to be the most effective drug its use is obsolete because of too many side effects; the same is true for bephenium
TRICHURIASIS: soil-transmitted helminthic infection (cf. ascariasis); female worms lay unembryonated barrel-shaped eggs with thick shells that are passed out with the feces into soil where they undergo development for 2–3 weeks to infective, first stage larvae; larvated eggs (infective L2 in egg shell may survive up to 6 years) can readily contaminate vegetables when night-soil is used as fertilizer; infection of humans with embryonated eggs can be directly from "hand to mouth" or may occur when larvated eggs are swallowed with contaminated uncooked food; the eggs (L2) hatch in the small intestine and the developing larvae pass directly to large intestine where they burrow into the mucosa and mature to adults; maturation (patency) takes about 3 months; adult worms are 3–5 cm long and their whip-like anterior portion becomes entwined in the mucosa of large intestine, the females are slightly larger than the males, which are coiled; distribution of *Trichuris trichiura* is global and the probable prevalence in humans amounts to about 500–1000 million, cases of morbidity may be 60–100 million; chronic inflammation of mucosa of large intestine and lacerations of mucosa caused by feeding activities of worms may lead to secondary bacterial infections; in heavily infected infants and young children, rectal prolapse is often seen followed by chronic bloody diarrhoea associated with rectal bleeding, which contributes to iron deficiency anemia and growth deficits; mixed infections of soil-transmitted nematodes (*Trichuris*, *Ascaris* spp. and hookworms) are very common and in heavily infected patients dysenteric syndrome (severe chronic diarrhoea, colitis, rectal prolapse, anemia) may cause significant growth-stunting in children.			
Trichuris trichiura (whipworm) **infective second stage larvae in egg shell** **hatched L2, larvae in mucosa** **adults** (transverse and descending colon, cecum; anterior portion of worm is embedded within mucosa)	**benzimidazole carbamates** (mebendazole or albendazole dosage for adults and children →) **tetrahydropyrimidines**	**mebendazole** (drug of choice, 100 mg bid x 3d or 500 mg once) **albendazole** (400 mg once; heavy infections: 400 mg/d x 3d) **oxantel** pamoate **oxantel/pyrantel** pamoate	mebendazole is considered to be the safest and most effective drug (also highly active against *Ancylostoma* spp. and *Ascaris* spp.); supportive treatment (e.g. iron substitution) is necessary in anemic patients, **surgical removal** of prolapse and antibiotic treatment in cases with secondary bacterial infections; **oxantel** is active against whipworm but has no action on ascarids; oxantel/pyrantel has similar activity spectrum to that of mebendazole; side effects of oxantel are mild and transitory (headache, abdominal pain, diarrhoea)
ABBERRANT NEMATODE INFECTIONS : Nematodes whose definitive hosts are usually animals can infect humans but are unable to mature to adults in humans			
CUTANEOUS LARVA MIGRANS (CLM): Humans become infected by direct contact with third stage larvae or by ingestion of larvated eggs or intermediate host containing L3 of various animal nematodes; CLM may be removed surgically because there seems to be no specific treatment; CLM is caused by nematodes such as hookworms, *Ancylostoma caninum* and *A. braziliensis* (see below), *Anatrichosoma cutaneum* (common hosts: rhesus monkeys, infective larvated eggs), filariae, *Dirofilaria repens* (common hosts: dog, cat, vector *Aedes* → infective microfilariae causing subcutaneous nodules particularly round the eye of man) or *D. tenuis* (common hosts: racoons, vector mosquitoes → infective microfilariae causing subcutaneous nodules in man); adult worms of *Gnathostoma spinigerum* causing gnathostomiasis is a common parasite of cats, dogs and wild carnivores living in the stomach wall, and *G. hispidum* or *G. doloresi* are common parasites of swine living in the stomach wall; Unlike *G. spinigerum*, *G. hispidum* cannot mature in humans; *Gnathostoma* has two intermediate hosts (=IH):			

Table 1. (Continued) Drugs used against nematode infections of humans.

DISEASE (alphabetical order) parasite, distribution, pathology, **Stages** affected (location)	**Chemical class** other information	**Nonproprietary name**, adult dosage (oral route), comments	Characteristics of compounds
first IH are cyclops ($L_2 \rightarrow$ early L_3), second IH (early $L_3 \rightarrow$ adult L_3) are fresh-water fish) and amphibians (e.g. frogs,) containing encysted infective L_3 (especially in Japan loaches are infected with *G. hispidum*); when infected second IH is eaten by a paratenic host, such as reptiles like snakes, birds, and mammals, larvae (adult L_3) encyst again; accidental host (such as man) becomes infected by ingestion of paratenic hosts or by second IH (raw fish and amphibians) containing infective larvae; early clinical signs produced by migrating larvae may be fever, vomiting, abdominal pain, and weakness, followed by cutaneous symptoms such as creeping eruption associated with erythema, formation of subcutaneous abscesses of the trunk or extremities; disease patterns of visceral larva migrans (VLM) may also develop such as cerebrospinal alterations, migrating tracks in the liver or pulmonary infiltration (cf. *Gnathostoma spinigerum* under VLM, below); ascaridid nematodes of wild felines, carnivores, and rodents such as *Baylisascaris* (parasites of wild felines, carnivores, rodents and didelphoids, opossum) and *Lagochilascaris minor* (latter occur in Surinam and Trinidad) have also been recorded in humans on rare occasions; ingestion of larvated eggs and hatched migrating larvae of these nematodes may cause skin creeping eruption and subcutaneous abscesses; spirurid worms occurring in Far East as *Thelazia callipaeda* is a common parasite of dogs, which lives under the nictitating membrane and in conjunctival sac; intermediate hosts are flies, which may place in the eye region of humans infrequently infective L_3 causing conjunctivitis , pain and excess lacrimation, paralysis of lower eyelid muscles associated with ectropion, and fibrotic scarring most frequent CLM is produced by third-stages larvae of the dog and cat hookworm, *Ancylostoma*; humans become infected by larvae from the soil entering the skin and migrate in it; infective larvae also of other animal hookworm species (e.g. *Uncinaria* or *Bunostomum*) frequently fail to penetrate the human dermis and migrate through the epidermis thereby producing typical serpiginous tracks known as 'creeping eruptions'.			
Ancylostoma caninum (dog) *A. braziliensis* (dog, cat) *A. ceylanicum* (dog, cat, civet) third-stage larvae (epidermis, deeper dermis, subcutaneous tissue)	**benzimidazoles** **macrocyclic lactones**	**thiabendazole** *(topically in DMSO or petroleum jelly) **albendazole** (400 mg/d x 3d) **ivermectin** (150–200 µg/kg once)	other hookworms such as *Uncinaria stenocephala* (dog, cat, fox in temperate climates) or *Bunostomum phlebotomum* (serious pathogen for cattle, and Zebu, worldwide) may cause 'creeping eruption' in skin of humans with intense itching provoked by secretion of proteolytic enzymes of larvae; scratching may be the cause of secondary bacterial infection
VISCERAL LARVA MIGRANS (VLM)): The condition in children is mainly caused by the larval stages of ascarids such as *Toxocara* though larval stages of other nematodes (*Capillaria hepatica* of rodents, see above, and *Lagochilascaris minor* of wild felines, cf. CLM) may also be responsible for causing the VLM syndrome; the entity is characterized by chronic granulomatous, usually eosinophilic lesions; migrating tracks of worms in the inner organs, especially in the liver, lungs, and brain of children leave infiltrates of macrophages, foreign-body giant cells, plasma cells, and eosinophils, sometimes the eye (ocular larva migrans = OLM) and also elsewhere; pathological entity consists of enlargement of the liver (hepatomegaly) and spleen, pulmonary infiltration associated with bronchospasm; CNS symptoms involve seizures, psychiatric manifestations, and encephalopathy; there may be intermittent fever, weight loss, persistent cough, and high (about 50%) persistent circulating eosinophilia; OLM (unilateral vision disorder and strabismus) may primarily occur in older children; invasion of retina produces granuloma formation; retinoblastoma-like masses consisting of cell infiltrates may produce distortion and detachment of retina or diffuse endophthalmitis or papillitis associated with secondary glaucoma leading to blindness, which is common; dying larvae inducing formation of granulomata (of which intensity depends upon numbers of dying larvae and area affected) are responsible for the loss of sight in that eye; the condition is most usually seen in children one to five years of age; children who own a pet (dog or cat) and have unexplained fever and eosinophilia, might be infected with *Toxocara*.			
BAYLISASCARIASIS: *Baylisascaris procyonis* (*Baylisascaris* spp. of wild felines, carnivores, rodents and didelphoids, e.g. opossum) larvated eggs, third-stage larvae (L3 migrate through various organs, CNS, eye and become arrested there to be gradually phagocytosed)		drugs that could be tried include BZs as **mebendazole**, **albendazole**, **thiabendazole** or the **imidazothiazole levamisole** and the macrocyclic lactone **ivermectin**; **steroid therapy** may be helpful, especially in eye and CNS infections to control exaggerated inflammation reactions; ocular baylisascariasis has been treated successfully using laser photocoagulation therapy to destroy the intraretinal larvae	

Table 1. (Continued) Drugs used against nematode infections of humans.

DISEASE (alphabetical order) parasite, distribution, pathology, **Stages** affected (location)	**Chemical class** other information	**Nonproprietary name**, adult dosage (oral route), comments	Characteristics of compounds
TOXOCARIASIS: humans (primarily children) acquire infection by ingestion of embryonated *Toxocara* eggs from soil; children frequently adopt the habit of dirt eating and where soil is heavily contaminated with *Toxocara* eggs (e.g. in soil around doorsteps, garden soil, playgrounds and sidewalks, rural settings such as farms) the ingestion of even moderate amounts of soil may result in intake of large numbers of infective eggs; the custom if giving young puppies to children as playmates, a special hazard may arise since it is the young puppy which is preferentially infected with *T. canis* (clinical signs of VLM and OLM syndrome are described in general discussion above).			
Toxocara canis (common host: dog, cat) *Toxocara cati* (common host: dog, cat) **infective second-stage larvae** in egg shell is ingested **third-stage larvae** (migrate through various organs, CNS, eye and become arrested there to be gradually phagocytosed) **dead larvae** may induce formation of granulomata causing serious clinical signs	**piperazine derivatives** (DEC, mebendazole or albendazole dosage for adults and children →) **benzimidazole carbamates** **diagnosis** of VLM: ELISA (L3 antigen) has sufficient specificity (~92%), and sensitivity (~78%) titer > 1:32 → suspected of VLM; OLM is diagnosed on clinical criteria	**diethylcarbamazine** (DEC) (6 mg/kg/d in 3 doses x 7–10d: dose regimen for adults and children) **albendazole** (400 mg bid x 3–5d: dose regimen for adults and children); **mebendazole** (100–200 mg bid x 5d: dose regimen for adults and children)	there are contradictory reports concerning the use of DEC in treating VLM; thus efficacy of DEC and **thiabendazole** is considered doubtful by some authors; most patients may improve without treatment within 3 months after infection; **mebendazole** (1 gram x 3/d x 21d) at high doses were effective in treating VLM in an adult antiphlogistic **corticosteroids** are useful in suppressing intense inflammation reactions of the eye and may lead to improvement of more serious symptoms, including relief from pain; mistaken diagnosis may result (and on several occasions has resulted) in unnecessary enucleation of the eyeball treatment for OLM include **surgery** (e.g. partial removal of vitreous body = vitrectomy), use of corticosteroids, anthelmintics could be tried
ANGIOSTRONGYLIASIS: Humans acquire infection by ingestion of raw or undercooked intermediate hosts (mollusks: slugs, crustaceans: freshwater prawns) containing infective third-stage larvae; *A. cantonensis* occurring in Australia, Pacific Islands, Taiwan, Malaysia, Far East and India may be the cause of eosinophilic meningitis or meningoencephalitis (thread-like larvae may be found in subarachnoid space); dissemination of the disease in humans is due to one of the best intermediate hosts of *A. cantonensis*, the giant African land snail, *Achatina fulica*, which is a popular item of food in some countries; *A. costaricensis* is widespread in the American continent from the USA to northern Argentina, particularly high in Costa Rica (intermediate hosts are slugs); pathogenesis is attributed to degenerated third stage larvae causing hepatic lesions and thrombus formation destroying arterial walls; adult worms in mesenteric arteries or eggs in the intestinal wall (which fail to hatch in humans) may provoke local inflammatory reactions and necrosis; principal pathological alterations consist of intestinal eosinophilic granulomata resulting from arteriitis, thrombosis and small infarcts; necrotic ulcerations may be found in regional lymph nodes and sometimes peritonitis or involvement of brain with myeloencephalitis may occur; involvement of the eye associated with meningoencephalitis may at times be fatal; other clinical signs may be fever, peripheral eosinophilia, leucocytosis (enhanced when liver is involved), and marked abdominal pain in the right iliac fossa and right flank (palpable tumor-like masses); *A. cantonensis* disease is self-limiting, and recovery usually occurs within 4 weeks following first symptoms.			

Table 1. (Continued) Drugs used against nematode infections of humans.

DISEASE (alphabetical order) parasite, distribution, pathology, **Stages** affected (location)	**Chemical class** other information	**Nonproprietary name**, adult dosage (oral route), comments	Characteristics of compounds
Angiostrongylus cantonensis (rat lung-worm: **third-stage larvae** (capillaries of meninges) *A. costaricensis* (common host: wild rodents) **third-stage larvae, adults, eggs** (cranial mesenteric arterioles and arterioles of cecum) **prevention:** thorough inspection of vegetables for hidden slugs, avoidance of eating raw or not well cooked crustaceans or snails	**benzimidazoles** (mebendazole, albendazole or thiabendazole dosage for adults and children) **imidazothiazoles** (levamisole) and **macrocyclic lactones** (ivermectin) have been used successfully in rats infected with *Angiostrongylus*	**mebendazole** (*A. cantonensis*: drug of choice, 100 mg bid x 5d) **mebendazole** (*A. costaricensis*: drug of choice, 200–400 mg tid x 10d or **thiabendazole** 75mg/kg/d in 3 doses x 3d, max. 3 grams/d: this dose is likely to be toxic and may have to be reduced)	antiparasitic drugs can provoke neurological symptoms in *A. cantonensis* infections and most patients may recover spontaneously without them; **corticosteroids analgesics**, and careful removal of CSF at frequent intervals can relieve symptoms; in *A. costaricensis* infections, **surgical treatment** is generally required for definitive cure; mild disease may resolve spontaneously; it appears therefore doubtful whether killing of larvae with drugs will lengthen or shorten the duration of symptoms; **thiabendazole** with its antiinflammatory properties may be good for reducing the duration of symptoms rather than its anthelmintic efficacy; definitive **diagnosis** is possible by examination of biopsy samples or surgical resection; ELISA and latex agglutination have been employed

ANISAKIASIS: (herring worm disease, codworm)
Common, marine nematodes (distantly related to ascarids) such as adult *Anisakis*, *Contracaecum, Phocanema*, and *Terranova*, live in the lumen of the intestinal tract of sea mammals, e.g., whales, dolphins, seals and sea lions; larvated eggs in ocean water hatch, and larvae are ingested by small crustaceans (krill) where they develop to third-stage larvae; infected krill may be eaten by a wide variety of fish, e.g. salmon, cod, herring, mackerel infected with *Anisakis* spp. (*A. simplex* being the most important species), or cod, pollack, halibut and haddock infected with *Pseudoterranova* spp. (*P. decipiens* being the most important species); humans become infected by ingestion of raw fish (smoked, salted, pickled, poorly cooked) containing infective third-stage larvae of marine nematodes; practice of eating raw seafood (sushi, sashimi, lightly salted 'green' herrings, Tahitian salad and others) in Japan and elsewhere (increasingly in Europe and the USA) has led to increased prevalence of larval anisakid infections; the disease is classified into gastric, intestinal and extra-gastrointestinal (ectopic) anisakiasis; ectopic larvae in abdominal cavity enter various abdominal organs or tissues provoking peritonitis and inflammatory foci; larvae invading gastric mucosa cause acute epigastric pain within a few hours of their being ingested; various symptoms (nausea, vomiting, blood vomitus, ileus, generalized abdominal pain, heart burn, diarrhoea and others) following infection depending on the location of the invasive larvae as they move down the intestine.

third-stage larvae (about 2 cm long, more frequently in stomach wall and/or intestinal tissues; extraintestinal sites: mesenteries and abdominal cavity, esophagus, posterior oropharynx) **ivermectin** being approved for treatment of anisakiasis in some countries;	**treatment of choice:** surgical or endoscopic (by means of biopsy forceps of the endoscope) **removal of anisakid larvae** (anaphylactic reaction is possible when larva to be removed becomes injured); there is no specific treatment; **preventive measures** include the removal of the abdominal viscera of fish as soon as possible after catch (to prevent additional larvae migrating into muscles), freezing of fish (20°C for 3–5 days) or thorough cooking prior to consumption (internal temperature at least 60°C for 10 min); anisakine larvae can survive for some days in soy sauce, Worcester sauce and vinegar	**endoscopy** is a useful tool for diagnosing gastric anisakiasis; X-ray examination (**radiology**) may reveal coiled or thread-like filling defects, inflammatory reactions such as eosinophilic granulomata or ulcer(s); **immunodiagnosis** is essential for chronic infections and extra-gastrointestinal anisakiasis (ELISA assay using a monoclonal antibody recognizing an epitope of anisakine larvae or immunoblot assay using E-S antigens of *A. simplex* detecting IgA, or IgE antibodies specific for E-S antigens of larvae)

Table 1. (Continued) Drugs used against nematode infections of humans.

DISEASE (alphabetical order) parasite, distribution, pathology, **Stages** affected (location)	**Chemical class** other information	**Nonproprietary name**, adult dosage (oral route), comments	Characteristics of compounds
GNATHOSTOMIASIS: Adults of *Gnathostoma spinigerum* live in the stomach of dogs, cats and wild felines or swine (reptiles, birds or mammals serve as paratenic hosts, i.e. encysted third stage larvae ingested with fish encyst in paratenic hosts again); humans (accidental host) become infected either by ingestion of raw or undercooked freshwater fish (also amphibians such as frog) or by ingestion of paratenic hosts (for life cycle see cutaneous larva migrans =CLM: *Gnathostoma*); gnathostomiasis spinigera has several visceral forms and cutaneous forms (see creeping eruption, subcutaneous abscess under CLM, above); the process of larval migration through various organs (wall of small intestine, urogenital tract, striated muscles, liver, lungs, ear, nose, eye and brain) may produce various pathological reactions around invading worms such as acute and chronic inflammatory reactions, local hemorrhage, necrosis, edema, fibrosis, or tumor-like masses; as a result, there are various disease patterns such as gastrointestinal disorder, infiltration of the lungs and liver, ocular disorders associated with visual impairment and often eosinophilic meningoencephalitis or there may be eosinophilic myeloencephalitis associated with obstructive hydrocephalus; death may occur when brain stem with massive hemorrhage in this area is involved; diagnosis can be made by morphological (biopsy) and when recovery of worms is not possible by an ELISA assay using E-S products of advanced third stage larvae for detecting antigen specific IgG or IgE antibodies in sera or spinal fluid.			
Gnathostoma spinigerum (most important species of 12 distinct species) (adult stage can develop in man, but cannot return to stomach) **third-stage larvae** (body of worms is entirely covered with cuticular spines)	**surgical removal** of adults (12–33 mm long), larvae (3–4 mm long) is generally required for definitive cure; supportive, symptomatic and antiphlogistic treatment	**albendazole** (400 mg bid x 21d) has shown some clinical efficacy **corticosteroids, analgesics**, and careful removal of CSF at frequent intervals can relieve symptoms	**prevention** relies on avoidance of raw or inadequately cooked hosts that contain L3 and avoidance of drinking water containing infected cyclops; use of gloves or frequent washing of hands while handling food will prevent larval penetration of skin; larvae are killed by boiling food for 5min or freezing at −20 °C for 3–5 days; increase in world travel and importation of food require greater awareness of potential gnathostomiasis spinigera
DIROFILARIASIS: The common heartworm of dogs and other carnivores, *Dirofilaria immitis* (adults, about 12–30 cm long, live in arteries of the lungs and right ventricle of heart), is primarily a problem of warm countries where the mosquito intermediate host abounds (cf. → Nematocidal Drugs, Animals/Table 5); aberrant infection of humans occur when the female mosquito (various genera) takes a blood meal and transmits infective larvae (about 800–900 μm long); *D. tenuis* (parasite of racoons, southern parts of the USA) and *D. repens* (parasite of dogs and cats in Europe, south-east Asia and Africa) are other filariae which may infect humans; *D. immitis* infection is usually asymptomatic but may be seen as small peripheral lesions ('coin'-size granulomata, each containing a single worm) in lungs on radiography; *D. tenuis* and *D. repens* may occur in subcutaneous nodules on various parts of the body, though those of *D. repens* particularly occur round the eye, in the eyelids and/or retrobulbar tissues; serodiagnostic tests have shown that occurrence of serum antibody to *D. immitis* correlated with prevalence of *D. immitis* infections in dogs (reliable serodiagnostic tests are available)			
Dirofilaria spp. **migrating larvae** infertile young adults	chemotherapy in humans is unknown although several drugs are available for treating heartworm disease of dogs (cf. → Nematocidal Drugs, Animals/Table 5)		**surgical removal** of larvae and infertile adults (microfilariae are not produced) may be indicated if there is harm for patient; possibly, diethylcarbamazine (DEC) or ivermectin may have prophylactic action against *Dirofilaria*
OESOPHAGOSTOMIASIS: *Oesophagostomum* spp. occur worldwide and are common parasites of ruminants, swine and some apes (gorilla, and chimpanzee) and other monkeys (e.g. macaques); human infections are rare and may be most common in Africa (*O. bifurcatum*) with some reported cases in Indonesia, China and South Africa (*O. aculeatum*); humans become infected by oral ingestion (most likely route) of infective third-stage larvae from soil but infection through the skin is also possible; larvae produce nodules in the intestinal wall and occasionally subcutaneous nodules following skin infection of larvae; major pathological consequences of heavy nodule worm infection are solitary, tumor-like inflammatory masses (helminthoma in ileo-cecal region) causing abdominal pain, diarrhoea, intensive rectal bleeding and anemia, though oesophagostomiasis is principally a self-limiting infection.			

Table 1. (Continued) Drugs used against nematode infections of humans.

DISEASE (alphabetical order) parasite, distribution, pathology, **Stages** affected (location)	**Chemical class** other information	**Nonproprietary name**, adult dosage (oral route), comments	Characteristics of compounds
O. aculeatum (syn. *O. apiostomum*) *O. bifurcatum*, *O. stephanostomum* **third-stage larvae** become locked in cysts where they may calcify	**benzimidazole carbamates** **tetrahydropyrimidines**	*albendazole *pyrantel pamoate (cure rate obtained with both drugs may be up to >80%	larvae burrow deeply into the mucosa up to muscularis mucosae of large intestine (and small intestine: *O. bifurcatum*) causing marked inflammatory reactions; rupturing of nodules into lumen can give rise to bleeding and bacterial superinfection, often misdiagnosed as carcinoma, ameboma or appendicitis; diagnosis can be made by barium enema examination, laparotomy or ELISA (worm specific IgG4 antibody, specificity >95%)

Abbreviations: the letter stands for day (days), bid = twice daily; tid = three times per day; p.c. (post cibum) = after meals
Dosages listed in the table refer to information from manufacturer or literature, preferably from Medical Letter (1998) 'Drugs for parasitic infections'. The Medical Letter (publisher), vol 40 (issue 1017): 1–12. New Rochelle New York
More information on antinematodal drugs (biological characteristics, adverse effects, and manufacturers) can be seen in → Nematocidal Drugs, Animals: Tables 1, 36 (nematocidal drugs used in animals) and → Nematocidal Drugs, Animals/General Considerations) or Medical Letter (1998) cited above or WHO (1987) Prevention and control of intestinal parasitic infections. Technical Report Series 749
Data given in this table have no claim to full information.

New World hookworm) occurring in the Americas, Africa and East Asia. *Ancylostoma ceylanicum* is only of local importance. Each female *Necator* may produce 10.000 and each female *Ancylostoma* 20.000 eggs per day. Eggs passed with the feces hatch and develop in soil to infective larvae within 7 days. Infective larvae, which can survive up to 1 month penetrate the skin to infect man (oral route of infection is also possible with *Ancylostoma*). The larvae migrate via the blood stream to the lungs, molt and migrate to trachea, are then swallowed and mature to adults in the small intestine. Adult worms are about 10 mm long and attach to the gut mucosa and suck blood (*Necator* 5–10 times less than *Ancylostoma*). Blood loss caused by hookworms is enormous and especially young children and pregnant women with large worm burdens suffer from severe anemia. It is estimated that approximately 1277 million people are infected with hookworms worldwide and about 60.000 deaths per year occur as a result of heavy hookworm infections associated with iron-deficiency anemia, protein-loss enteropathy and hypoproteinemia.

About 1.200 million people (especially children) are infected with the cosmopolitan **pinworm** *Enterobius vermicularis* with a high prevalence in countries of the Northern Hemisphere. Adult worms live in the lumen of the colon and feed on gut contents. Gravid females leave the anus and deposit eggs around the anus (5.000–10.000 eggs/female). Fully embryonated eggs are infective within a few hours and may infect humans by the oral route or rarely by aberrant routes resulting in peritonitis or adnexitis (Table 1); intestinal larvae mature to adults within about 6 weeks. Reinfections may frequently occur and this should be considered in treatment strategies. A characteristic clinical feature of *Enterobius* infections is the perianal itching occurring preferably at night.

About 902 million people are estimated to be infected with the cosmopolitan **whipworm** *Trichuris trichiura*, which is common in warm and humid climates. Adult worms (3–5 mm long) live attached to the cecal mucosa. The eggs passed with the feces require about 3 weeks to become infective. Eggs are fairly resistant to the environment and can survive in the soil (and thus remain infective) for more than 1 year. The larvae hatch after eggs have been taken up, e.g., with vegetables or dirt, and mature in the intestine within 2–3 months. Adults live in the host for 3–10 years and may be asymptomatic; moderate and heavy infections lead to considerable pathogenic effects of the mucosa of the large intestine accompanied with clinical signs such as abdominal discomfort, diarrhoea, anemia, retardation of growth and mental development in children.

Nearly 100 million people may be infected worldwide with the **threadworm** *Strongyloides stercoralis* occurring mainly in tropical and subtropical climates, extending into areas of southern Europe and those of southern United States. Parasitic female worms live in the small intestine attached to the mucosa on which they feed. They produce larvated eggs by parthenogenesis passed in feces and hatch rapidly either to become free-living adult and female worms or parasitic larvae infecting humans by skin penetration or ingestion. The development of 3rd stage larvae in the host resembles that of hookworms and takes about 17 days to become adult females. Autoinfection is possible if first (filariform) stage hatches in the gut and penetrates mucosa of colon or perianal skin; autoinfection may perpetuate for several years. Strongyloidiasis belongs to the opportunistic infections often being asymptomatic. However, in immunocompromised persons (AIDS-patients) infection can cause serious gastrointestinal symptoms such as catarrhal enteritis associated with severe diarrhoea and central epigastric pain, which may be fatal in heavily infected immunocompromised patients.

Treatment of GI nematodes of humans is given in detail in Table 1. There are highly effective drugs that may remove adult worms from the gut after a single dose, though their action is generally limited against migrating larvae even after repeated dosing. Pathological consequences of GI nematodes due to travelling larvae through tissues of the host (cf. larva migrans below and Table 1) thus remain largely unaffected by chemotherapy.

Cutaneous/Visceral Larva Migrans Syndrome, Trichinosis and Dracunculiasis

Nematode infections of the dog and cat may be responsible for several aberrant larva infections in humans such as echinococcosis (→ Cestodocidal Drugs), cutaneous and/or visceral larva migrans (Table 1). Cutaneous larva migrans is most frequently seen with *Ancylostoma braziliense* larvae, which cause the so-called creeping eruption, an intensive itching dermatitis. Visceral larva migrans is mainly due to migrating larvae of *T. canis* in children but also to a variety of other nematode larvae (Table 1). Larvae can migrate through the inner organs of humans and may produce mechanical damage in the liver, lungs, brain, and in the eye. The disease entity is characterized by chronic granulomatosis, usually eosinophilic lesions, which may lead to hepatomegaly, pulmonary infiltration, and persistent circulating eosinophilia. The clinical signs vary greatly and may be persistent cough, intermittent fever, loss of weight, and loss of appetite. Treatment of larva migrans is problematic and prevention is the most effective measure (Table 1).

Ubiquitous *Trichinella spiralis* (it does not occur in Australia) may infect about. 40–50 million people worldwide. In Europe, annual cost for the control of potential **trichinosis** of the swine may amount to 570 million US$. In the past 20 years, about 2600 cases of trichinosis have been recorded chiefly in France, Italy and other parts of Europe. Humans mainly become infected by consumption of raw or improperly cooked meat (especially pork, also horsemeat) containing encapsulated muscle larvae. After being ingested, larvae become freed in the intestine; they mature to female and male adults, who mate, and females produce larvae (up to 2.000/female) within 2 weeks. Released larvae penetrate the gut wall, enter the lymphatic vessels, disseminate via the circulatory system throughout the body, develop in the muscle, and finally encyst in muscle fibers. Infection is usually asymptomatic during the intestinal phase. The muscle infection may produce fever characteristic eye edema, myositis, eosinophilia, leucocytosis, and muscular pain. Once the diagnosis is made, the majority of larvae has already been produced and become distributed to all tissues, so that removal of adult worms from the small intestine by administering thiabendazole or mebendazole may be too late because of their expulsion by immune mechanism of the host. Migrating larvae rather than larvae after entering the muscle cell remain unaffected by these drugs (Table 1).

During the last 50 years there has been a sustained and rapid reduction in the reported **dracunculiasis** (guinea worm) incidence worldwide. The numbers of humans infected with the Guinea (dragon) or Medina worm, *Dracunculus medinensis* have gradually been decreased in all countries, except in Sudan, from 48 million cases in 1947, 10 million in 1976, 3.3 million in 1986, 892.000 in 1989, to fewer than 35.000 cases in 1996. Today, the disease is distributed focally in West, North, and East Africa, the Middle East from Iran through Pakistan into Indonesia. The ingested larvae develop in the body cavity and deeper connective tissue for about 12–14 months before they become mature. Thence adult females migrate to the subcutaneous tissues, preferable the legs where they emerge thereby causing blisters and ulceration of the skin. When the skin contacts water the emerging female worm releases large numbers of larvae. Copepod crustaceans take them up, and the larvae develop in the body cavity to become infective within 2–3 weeks. Humans are infected by drinking water containing infected copepods. Dogs are known to be reservoir hosts (for more detailed information on traditional treatment Table 1).

Filarial Nematodes of Medicinal Importance

According to new estimates about 120 million people are infected by filariae causing lymphatic → filariasis. *Wuchereria bancrofti* is widely distributed throughout the tropics and subtropics, while *Brugia malayi* occurs focally in India, South East Asia and Japan. *Brugia timori* has only been recognized on islands of Indonesia and Timor. The life cycle of lymphatic filariae involves various mosquito species as intermediate hosts in which infective larvae develop and penetrate the skin of humans when the insect bites. Developing larvae, adult worms and females producing larvae, which transform to microfilariae (MF) live in the lymphatics and may block lymphatic vessels. Thence MF enter the peripheral circulation and are taken up by mosquitoes with a blood meal. Lymphatic obstruction by filariae leads gradually to pathological alterations such as lymphangitis, hydrocele and to moderate to severe elephantiasis in time. Control of lymphatic filariasis relies on chemoprophylaxis with diethylcarbamazine (DEC) or ivermectin (Table 1). Chemotherapy may be problematic if there are already severe chronic pathological changes, e.g. massive elephantiasis, which requires radical surgery to remove new adventitious tissue.

The tissue-dwelling nematode, *Onchocerca volvulus*, has a focal distribution in Africa and South America and causes the so-called river blindness or → onchocerciasis, a disease of public health importance. *O. volvulus* infect about 18 million people in Africa and about 140 000 in South America. The life cycle of *O. volvulus* involves black flies (*Simulium* spp.) as intermediate hosts. Adult worms (macrofilariae) live in subcutaneous nodules and female worms produce MF. These are located in the superficial layers of the skin, where they are taken up with a blood meal by black flies, in which they develop into the infective larvae. Humans are infected when the fly takes another blood meal. Only dying and dead MF induce pathological changes that gradually increase the longer the infection has persisted. There are a variety of acute and chronic lesions of the dermis and the eye. Thus invasion of the eye leads to conjunctivitis, keratitis, iridocyclitis, chorioretinitis, glaucoma, and opacity of the lens resulting in optic atrophy and blindness. Control of onchocerciasis relies on chemoprophylaxis with ivermectin and nodulectomy (for detailed information Table 1).

→ Loiasis is caused by the African eye worm, *Loa loa*, a tissue-dwelling nematode; it infects about 30 million humans in the rain forests of West and Central Africa, where it has a focal distribution. The life cycle involves *Chrysops* flies, in which infective larvae develop. When a fly bites, these larvae enter the vertebrate host where they mature to adults, which migrate through the subcutaneous tissues and under the conjunctiva of the eye. Adult worms (macrofilariae) produce so-called Calabar swelling in hand and arms and considerable irritation and congestion of the eye. MF produced by female worms migrate from the subcutaneous tissues to the peripheral circulation where they are taken up with a blood meal by flies. Control of loiasis may rely on chemoprophylaxis with DEC, which destroys MF and thus interrupts transmission of infection to *Chrysops* flies. However, killed MF may cause severe "allergic" reactions, which can be reduced by gradually increasing DEC doses. The adult worm under the conjunctiva of eye must be removed surgically (for detailed information Table 1).

Other filarial infections such as mansonelliasis or tropical pulmonary eosinophilia, and other nematode infections occurring in humans are listed in Table 1.

Nematode Infections, Man

General Information

The vertebrate intestine is most likely one of the major ancestral sites for parasites. Access to the bodies of vertebrate host as well as to the high concentrations of nutritions available locally may account for the fact that intestinal species are, overall, still the commonest although not the most pathogenic of all parasites. However, the gastrointestinal tract should not be considered as a single homogenous habitat but as series of habitats, each with its own distinct characteristics. Different nematode species prefer certain locations in the intestine to which they are able to actively migrate. In addition, depending on the size and physiology of the worms some species, such as *Ascaris* (*Ascariasis*), live in the lumen, while others e.g. → hookworms (Vol. 1), have an intimate association with the mucosa or live wholly or partially in mucosal tissues like *Trichinella spiralis* (*Trichinelliasis*) or species of *Trichuris* (*Trichuriasis*). In contrast to older views where it has been thought that worms living in the gut lumen were effectively outside the body and could neither initiate nor be affected by immune responses it has now become evident, that intestinal worms clearly are targets for the host's immune response.

Exploitation of habitats other than the intestine of the host is common in the Nematoda. Many species that live as adults in the intestine, e.g. *Ascaris* (*Ascariasis*), hookworms (*hookworm disease*), *Nippostrongylus* and *Trichinella* (*Trichinelliasis*) reside and develop as larval stages in parenteral tissues. Other species, such as filariae are wholly confined to the host tissues and have no contact with gastrointestinal tract (*Filariasis*). These fundamental differences in the developmental cycles of the nematodes must be considered when protection or pathology induced by different immune mechanisms is analyzed.

Important diseases caused by nematodes are listed in Table 3 of chapter Pathology.

Immune Responses

For immune responses induced by other nematodes than those listed below please refer to the respective diseases.

→ Heligmosomoides polygyrus (Vol. 1) **and** Trichuris muris

In mice infected with these nematodes host protective effects of IL-4 have been most prominently demonstrated. While treatment with anti-IL-4 or anti-IL-4 receptor antibodies blocked host immunity to challenge infections and allowed the establishment of chronic infections with *T. muris*, there were apparent differences in IL-4 deficient mice. Such mice failed to control *H. polygyrus* but were still able to expel *T. muris*, suggesting that IL-4-compensating factors, such as IL-13, might be efficient in promoting *T. muris* expulsion, but not *H. polygyrus* expulsion. Treatment of mice with IL-4 complexes displaying a prolonged half-life in vivo cured even established *T. muris* and *H. polygyrus* infections.

Nippostrongylus brasiliensis Infection of mice with *N. brasiliensis* is a well established model to investigate Th2 responses. Infective third stage (L_3) larvae are injected through the skin and migrate to the lungs (day 1–2) where a strong eosinophilic inflammatory response is induced. Larvae are coughed up and swallowed (days 2–4) and mature into egg-laying adults in the jejunum (days 5–8). Adult worms are expelled by day 9–11 after inoculation. Independent of the genetic background of mice, infection is accompanied by blood and lung eosinophilia, intestinal mastocytosis, and high IgE levels.

It has been recently shown that IL-5 is essential for eosinophilic lung inflammation associated with hemorrhage and alveolar wall destruction. Interestingly, the induction of airway hyperresponsiveness was unimpaired in IL-5-deficient mice, demonstrating that eosinophils are not required for the induction of airway constriction following *N. brasiliensis* infection.

IL-4 appears not to be necessary to protect mice from *N. brasiliensis* infections since IL-4-deficient mice expel this nematode normally. However, treatment with exogenous IL-4 completely cured infected SCID mice, also when the animals were additionally treated with anti c-kit antibodies. These findings argue for IL-4-induced mechanisms of *N. brasiliensis*-expulsion which are independent of B-, T- and mast cells. One of the mechanisms involved might be an IL-4-induced increase in intestinal permeability which could result in an inhibition of worm feeding by blocking the nematode contact with the gut mucosa. Expulsion of *N. brasiliensis* from the gut of W/W^v mice, deficient in mast cells, has been described as slow in some, but not all studies. Since restoration of the mast cell compartment by bone marrow transplantation did not correct for slow expulsion of worms, additional defects in W/W^v mice, such as the absence of intraepithelial $\gamma\delta$ T cells, most likely accounts for the defect in worm expulsion in these mutant mice. The IL-4-independent mechanism(s) inducing *N. brasiliensis* expulsion has not yet been identified but is known to be $CD4^+$ Th cell dependent and suppressible by IFN-γ and IFN-α/β. Possible candidates for this are antibody-mediated worm damage, mucus trapping and lipid peroxidation. Antibody-mediated protection has been demonstrated by the fact that serum transfer from immune mice provides protection against *N. brasiliensis*. Mucus trapping preventing adherence or feeding of the worm is suggested by an increase in mucus production and carbohydrate content accompanying *N. brasiliensis* expulsion, although studies in rats have shown that this phenomenon is not essential for parasite expulsion. The possibility that lipid-peroxidation by host-produced oxygen intermediates damages *N. brasiliensis* was suggested by experiments with reactive oxygen scavengers (butylated hydroxyanisole) which suppressed worm expulsion. An interesting hypothesis in this context is that the expression of enzymes such as glutathione reductase, superoxide dismutase or catalase by nematodes offer some protection against reactive oxygen intermediates produced by the host.

Summarizing the studies with the different rodent models of nematode infections four generalizations can be made: (1) $CD4^+$ T cells are essential for host protection, (2) IL-4 production induces either essential or redundant protective mechanisms, (3) IFN-γ and IL-12 inhibit protective immunity, and (4) some cytokines, such as IL-5, that are stereotypically produced in response to gastrointestinal nematode infections appear not to contribute to protective immunity.

Therapy

→ Nematocidal Drugs, Man.

Nematodirosis

Trichstrongylid trematodes of the genus *Nematodirus* species (→ Trichostrongylidae (Vol. 1)) infect the anterior third of the small intestine of ruminants (→ Alimentary System Diseases, Ruminants). The most important species are *N. helvetianus*, which infects cattle; *N. spathiger* and *N. filicollis*, which infect sheep, goats, and cattle; and *N. battus*, which mainly infects sheep. The only species causing disease are *N. battus* in sheep and, to a lesser extent *N. helvetianus* in calves. Nematodirosis, like → cooperiosis, is normally confined to young animals, because of the early development of immunity, partly due to age and partly to experience of infection. Third stage larvae enter the deeper layers of the mucosa, and larvae emerge at the fourth or fifth stage. The presence of large numbers of adult *Nematodirus* worms is associated with the development of villous atrophy. It is not known how adult worms exactly damage the epithelial cells of the host and cause atrophy of the villi, but it may be related to a cell-mediated immune response. This atrophy is associated with the presence of short, sparse microvilli on each individual epithelial cell. Clinical disease usually appears with populations of about 10,000 to 50,000 or more *Nematodirus* worms. Affected animals may lose their appetite and develop a severe dark green diarrhoea. There is very rapid loss of weight and dehydration, as shown by the sunken eyes and the extreme thirst. Lambs may die within 10–14 days of infestation. The clinical signs appear earlier in sheep than in cattle.

Therapy

→ Nematocidal Drugs, Animals.

Neosporosis

Toxoplasma-like disease in dogs, cattle, horses, sheep etc. due to infections with → *Neospora caninum* (Vol. 1) (syn. *Hammondia heydorni*) of the brain of the fetus leading often to abortion. Transmission: Oral uptake of oocysts from the feces of dogs (→ Nervous System Diseases, Animals, → Nervous System Diseases, Carnivores).

Therapy

Toltrazuril, Ponazuril; vaccines are available.

Nervous System Diseases, Animals

Nervous symptoms have been frequently associated with parasitic infections in animals. There is an impressive list of parasites that may be located in the meningeal spaces or may penetrate into the tissues of the brain and spinal cord or eye. Many of these parasites wander in the nervous system aberrantly, especially when they are in an alien host.

The pathological changes are influenced by the route of entry, and the size and mobility of the parasite. These changes fall into three categories, (1) haemorrhagic (2) degenerative and (3) proliferative. Haemorrhagic changes are attributed to parasites in the arterial circulation or to laceration of blood vessels as the parasites move through the tissues. Degenerative changes in neurofilariosis (e.g., *Setaria* and other nematoda) are characterized by disruption of nervous tissues, swelling of axis cylinders and degeneration of neurons. Proliferative changes may be diffuse or focal. Diffuse proliferation includes perivascular hyperplasia of the reticulum, as observed in neurofilariosis and cerebral ascariosis. Focal proliferation usually consists of granulomatous aggregations in the vicinity of the parasite. In some instances the cellular reaction was found to consist mostly of glial proliferation. In contrast, certain nematode infections of the central nervous system show no evidence of cellular reaction in the vicinity of the parasite. Degenerative changes in the vinicity of the parasite probably appear only if the parasite has become quiescent before the host dies. If the parasite is moving at the moment of the host's death, it may be in relatively normal tissue, while extensive damage may be found in other parts of the central nervous system. It is also common to find lesions similar to those produced by migratory parasites without being able to locate the parasite.

Apart from the purely mechanical damage that the parasites may cause, there has been considerable speculation as to whether they may facilitate the entry of virus infections. Nervous symptoms have also been described in parasitic infections where the parasite had not invaded the central nervous system. For instance intestinal parasitism in young puppies may be associated with convulsions that may be produced by a concomitant hypocalcemia or hypoglycemia, or both.

The lesions generated in the nervous tissue are more likely to produce clinical symptoms than aberrant migrations in other tissues. The clinical signs associated with parasites in the nervous system depend on the neuro-anatomic structure affected (Fig. 1). However, most parasites have no specific selectivity for any part of the nervous system and can therefore produce any clinical signs depending on the area they invade. The pathogenesis of infections of the nervous system is too varied to be considered here in detail. However, when known the common symptoms caused by end-disease will be described.

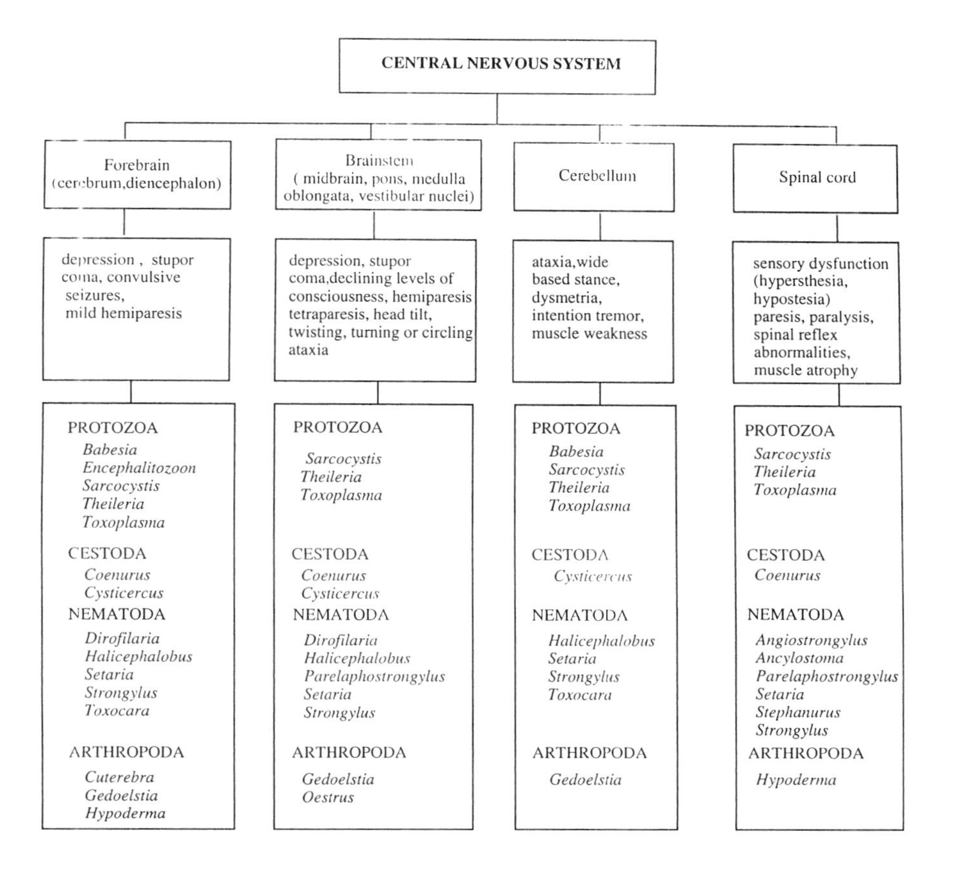

Fig. 1. Localisation and clinical signs of parasites affecting the nervous system

Only few parasites have the eye as predilection site, e.g. *Thelazia* spp. (→ Nervous System Diseases, Ruminants). However, several parasites which normally develop elsewhere in the body have been reported to occasionally penetrate the eye (Table 1) (see also → Eye Parasites).

For detailed information on nervous system diseases in specific host please refer to the following entries:

→ Nervous System Diseases, Carnivores
→ Nervous System Diseases, Horses
→ Nervous System Diseases, Ruminants
→ Nervous System Diseases, Swine

Nervous System Diseases, Carnivores

The common clinical signs and pathology of parasitic infections of the nervous system of carnivores are listed in Table 1.

Protozoa

Several protozoa may cause nervous symptoms e.g. *Babesia canis*, *Encephalitozoon cuniculi*, *Toxoplasma gondii* , *Neospora caninum* and *Trypanosoma* spp. (see species in volume 1).

Infections with some strains of *B. canis* often terminate with signs of cerebral damage such as paddling of limbs, ataxia, mania and coma. This is the result of brain damage caused by obstruction of the brain capillaries by parasitized red blood cells. There is usually no evidence of neuro-

Table 1. Parasites affecting the eyes of domestic animals (according to Vercruysse and De Bont)

Parasite	Host	Clinical signs and lesions
Protozoa		
Encephalitozoon cuniculi	Cat	Keratoconjunctivitis, characterized by multiple superficial corneal opacities arranged in a stellate pattern
Leishmania infantum	Dog	Conjunctivitis, keratouveitis, blindness
Theileria spp.	Cattle	Lacrimation, photophobia, in prolonged cases corneal opacity, blindness
Toxoplasma gondii	Dog, cat, cattle	Focal retinochoroiditis,anterior uveitis, retinal haemorrhage, exudative detachment, blindness
Trypanosoma spp.	Ruminants, horse, dog	Photophobia, lacrimation, conjunctivitis, keratitis, iritis, retinitis, occasionally total blindness
Cestoda		
Coenurus cerebralis	Sheep	Blindness
Nematoda		
Angyostrongylus vasorum	Dog	Impaired vision, retinal haemorrhages
Dirofilaria immitis	Dog	Anterior uveitis, iritis, blindness
Elaeophora schneideri	Sheep	Keratitis
Habronema spp. *Draschia megastoma*	Horse	Granulomatous nodules on the nectitating membrane and conjunctivae near the nasal canthus
Onchocerca cervicalis (microfilariae)	Horse	Keratitis, recurrent anterior uveitis, peripapillary choroidal sclerosis, vitiligo of the bulbar conjunctiva at the lateral limbus
Setaria equina *S. digitata*	Horse	Ocular opacity, photopohobia, lacrimation, corneal leucoma, iridocyclitis and hypopyon
Thelazia spp.	Ruminants, horse, cat, dog	Conjunctivitis, lacrimation, photophobia, oedematous eyelids, mild or ulcerative keratitis
Toxocara canis	Dog	Chorioretinal granuloma
Arthropoda		
Gedoelstia spp.	Ruminants	Lesions vary from a mild conjunctivitis to a destructive ophtalmitis with orbital or periorbital oedema and abscessation

Table 1. Parasites affecting the nervous system (according to Vercruysse and De Bont)

Parasite	Host	Location	Nervous clinical signs	Principal lesions in nervous system
Protozoa				
Babesia canis	Dog	Red blood cells selectively concentrated in brain	Paddling of limbs, ataxia mania and coma	Distention of the capillaria of the gray matter of the cerebrum and cerebellum, dilatation of perivascular spaces and interstitial oedema
Encephalitozoon cuniculi	Carnivores	Brain, kidney and other organs	Desorientation, circling, behavioral changes, convulsions, blindness	Encephalitis and segmental vasculitis
Neospora caninum	Dog	Cranial and spinal nerves	Limb poresis, paralysis	Encephalomyelitis characterised by gliosis, perivascular cuffs and mild necrosis
Toxoplasma gondii	Carnivores	Forebrain, brainstem, spinal cord	Trembling, opisthotonus head tilt, incoordination, paraplegia, blindness	Focal necrosis and vascular damage, glial nodules and scar formation
Cestoda				
Coenurus serialis	Cat, dog	Brain	Alternated state of consciousness, circling, ataxia, vestibular disturbances	Fluid-filled parasitic cyst, 1.5 to 2 cm in diameter compressing brain tissue
Cysticercus cellulosae	Dog	Brain or meninges	No apparent clinical signs in pigs, in dogs neurological disorders	Chronic inflammatory exudate in tissue surrounding the cysticerci
Nematoda				
Angiostrongylus cantonensis	Dog	Larvae in spinal cord and brain	Ascending paralysis, lumbar hyperalgesia	Eosinophilic meningoencephalitis, periradiculoneuritis
Ancylostoma caninum	Dog	Spinal cord	Imbalance, torticollis, tetraparesis and death	Haemorrhagic and necrotic tract in the spinal cord
Dirofilaria immitis	Dog, cat	Meningeal arteries, lateral ventricle	Intermittent convulsion, ataxia, circling	Thrombosis of cerebral artery, ventriculitis
Toxocara canis	Dog	Hypophysis, cerebellum in pigs	Rare	Local eosinophilia, granuloma formation
Arthropoda				
Diptera				
Cuterebra spp.	Dog, cat	Brain	Depression, hysteric convulsions	Acute focal haemorrhagic encephalomalacia

nal degeneration but there is dilatation of the perivascular spaces and interstitial oedema.

Encephalitozoonosis (nosematosis) is caused by the obligate intracellular microsporidian *Encephalitozoon cuniculi*. The disease has been described in rodents, lagomorphs, primates and several species of carnivores. Asymptomatic infection usually occurs in rodents and lagomorphs. In carnivores the neurological signs include repeated turning and circling movements, especially after disturbance, dysmetria, dysergia, blindness and a terminal semi-comatose state. Lesions described are encephalitis and segmental vasculitis. The course of the illness is usually 5–12 days.

The neuropathology associated with canine and feline toxoplasmosis has been described in detail. In these species, toxoplasmosis is characterized by focal necrosis and vascular damage in acute infections, and by glial nodules, repair, and scar formation in chronic infection. Cerebral calcifications, common to chronic toxoplasmosis in children, appear to be rare in animals. In dogs extensive areas of necrosis, gliosis and demyelination are found. Clinical nervous signs include depression, trembling, opisthotonus, head tilt, incoordination, blindness and paraplegia. In puppies, it may resemble distemper, clinical toxoplasmosis occurring sometimes together with this disease. Skeletal muscle atrophy due to damage of lower motor neurons has been associated with a case of clinical canine toxoplasmosis.

Incoordination and spinal paralysis have been reported in dogs infected with *T. brucei brucei.*

Neosporosis (*N. caninum*) is mainly reported in young dogs. Puppies show a hind limb paresis that develops into a progressive paralysis. Neurologic signs are dependent on the site that is parasitized. The hind limbs are more severely affected than the front limbs, and often in rigid hyperextension. The cause of this hyperextension is not known, but is most likely due to a combination of upper motor neuron paralysis and myositis which results in rapidly progressive fibrous contracture of the muscles that may cause fixation of joints.

Cestodes

Cerebral coenurosis, due to *Coenurus serialis* and *Cysticercus cellulosae* has been described in cats and dogs showing neurological disorder.

Nematodes

Various nematode species may invade the central nervous system. *Angiostrongylus cantonensis* is a metastrongylid lungworm of the rat. In unnatural hosts, such as dogs the parasite develops in the spinal cord and to a lesser extent the brain, and usually dies without reaching the lungs. Infection leads to an eosinophilic meningo-encephalitis and a periradiculoneuritis. Clinical signs are slight paresis of the hind legs, uncertain straddle gait and hypersensitivity of the skin.

Cerebrospinal nematodosis caused by *Ancylostoma caninum* has been reported in a dog. A 12-week-old cocker spaniel had signs of imbalance, torticollis and pain on flexion of its neck that eventually progressed to tetraparesis and death. A young adult female *A. caninum* was found in the haemorrhagic cervical spinal cord.

Adult heartworms *Dirofilaria immitis* usually inhabit the right side of the heart or pulmonary arteries of carnivores. Occasionally adult worms have been observed in the brain where they invade the lateral ventricle, or in the meningeal arteries with subsequent occlusion. The clinical course is characterized by intermittent convulsion, blindness, ataxia, behavioral changes and circling. Microfilaria of *D. immitis* have also been reported within the meningeal arteries, deep arteries and capillaries of the brain and extravascularly within the brain.

Larvae of *Toxocara canis* have been recovered from the brains of experimentally infected dogs, but with little clinical illness. A severe granulomatous inflammation of the hypothalamus and adjacent neurohypophysis caused by *T. canis* larvae have been reported in a dog suffering from diabetes insipidus.

Arthropoda

The larvae of *Cuterebra* species (Diptera) normally mature in subcutaneous tissue of Rodentia and Lagomorpha, but occasional infection in dogs and cats may occur. In these abnormal hosts the larvae have been observed in the brain, causing neurologic clinical signs.

Nervous System Diseases, Horses

The common clinical signs and pathology of parasitic infections of the nervous system of horses are listed in Table 1.

Protozoa

Sarcocystis neurona (n.sp.) was proposed as the putative cause of equine protozoal myeloencephalitis (EPM). Lesions occur in the white and grey matter of the brain, and in the spinal cord. They consist of a proliferative inflammation with a variable degree of necrosis of myelin, and axonal degeneration usually referred to as "segmental myelitis". Haemorrhage occurs occasionally. Unlike *T. gondii*, the organisms extensively invade neurons. Clinical signs of EPM are variable in onset and evolution, and depend on the nervous area involved. The onset is either gradual or sudden. Usually, an impairment of action of one limb is the first sign noted. Eventually, ataxia of both limbs or all four limbs will appear, and sometimes circling and depression has been noticed. Recent evidence indicates that → *Neospora caninum* (Vol. 1) may also cause EPM.

The later stages of dourine are characterized by anaemia and nervous disorders such as paralysis of the hind limbs. It is tought that a "toxin" produced by *Trypanosoma equiperdum* causes inflammation and degeneration of the peripheral

Table 1. Parasites affecting the nervous system of horse (according to Vercruysse and De Bont)

Parasite	Host location	Nervous clinical signs	Principal lesions in nervous system
Protozoa			
Neospora caninum	Cranial and spinal nerves	Limb paresis, paralysis	Encephalomyelitis characterised by gliosis, perivascular cuffs and mild necrosis
Sarcocystis neurona	Brain, brainstem and mainly spinal cord, all other organs	Ataxia, circling, incoordination, opisthotonus	Proliferative inflammation of grey and white matter, with a variable degree of necrosis of myelin, axonal degeneration, organisms frequently in neurons
Trypanosoma equiperdum	Lumbar and sacral regions of spinal cord, sciatic and obturator nerves	Paraplegia	Radiculitis and polyneuritis
Nematoda			
Halicephalobus deletrix	Brain	Posterior weakness, ataxia progressing to recumbence, coma	Vasculitis, haemorrhagia, necrosis and malacia
Setaria spp.	Brain and spinal cord	Muscular weakness, incoordination, ataxia to paralysis, death	Focal encephalomyelo-malacia, which in many cases proceeds to liquefaction and cavitation
Strongylus vulgaris	Brain and spinal cord	Chronic incoordination and acute progressive fatal encephalitic disease	Haemorrhagic malacia, tracks in the brain and spinal cord
Arthropoda			
Hypoderma spp.	Aberrant migration in brain and spinal cord	Muscular weakness, localized paralysis, loss of motor control, convulsions	Haemorrhagic tracks in brain or spinal cord

nerves. The motor and sensory disturbances are the direct result of these changes. Incoordination and spinal paralysis have been reported in horses infected with *T. brucei brucei*.

Nematodes

Worms of the genus *Setaria* are commonly found in the peritoneal cavity of ungulates where they are non-pathogenic. The major pathogenic effects of the cattle parasites *S. digitata* and *S. labiato-papillosa* occur when immature forms migrate erratically in the central nervous system of abnormal hosts such as the horse. The lesions are microscopic and may be overlooked. They are usually single tracts left by migrating worms which may be found in any part of the central nervous system. Acute malacia occurs in the track of the worm, with disintegration of all tissues at the centre of the lesion and secondary degeneration of the nerve tracts, with gigantic swellings of the axis cylinders and eosinophilic infiltration. Cavities are occasionally seen. Clinical signs may vary from muscular weakness, incoordination and ataxia, to paralysis and death. The disease in horses is known as kumri.

There are several reports of lesions in the central nervous system, apparently caused by *Strongylus vulgaris*, and resulting in neurological disorders. Lesions are either due to aberrant migration of migrating larvae in the brain or spinal cord, or the result of *S. vulgaris* embolism. The main clinical syndromes are chronic incoordination, paresis and acute progressive encephalitis.

Eye infections due to nematodes of the genus *Thelazia* have been reported in horses in different parts of the world. Whereas some reports have associated the infection with a variety of clinical signs, others were inconclusive as to the role played by this parasite in the production of ophtalmia. Mechanical damage to the conjuctiva and cornea by the serrated cuticula of Thelazia spp. may predispose to bacterial and viral infections (→ Nervous System Diseases, Animals/Table 1).

Halicephalobus (*Micronema*) *delatrix* is a rhabditiform nematode which may accidentally become a parasite. Massive intracranial invasion is reported in horses. The worms are found in the meninges, in the parenchyma of the brain adjacent to blood vessels, in the walls of the vessels themselves and especially in the Virchow-Robin spaces. Horses show early signs of simple lethargy, posterior weakness and mild ataxia, progressing to recumbency, coma, and death.

Arthropoda

The larvae of *Hypoderma bovis* and *H. lineatus* may invade the nervous system of horses and cause central nervous disorders. Larvae are located in the brain and sometimes in the spinal cord, where they leave haemorrhagic tracks, focal areas of haemorrhagic malacia or small abceses. Sudden onset of muscular weakness or localized paralysis is the usual clinical picture, which proceeds to profound loss of motor control, convulsions and death within 1–7 days or so.

Nervous System Diseases, Ruminants

The common clinical signs and pathology of parasitic infections of the nervous system of ruminants are listed in Table 1.

Protozoa

Several protozoa may cause nervous symptoms in ruminants e.g. *Babesia bovis*, *Theileria parva* and *T. mutans*, *Sarcocystis* spp., *Toxoplasma gondii*, *Neospora caninum* and *Eimeria* spp. (see species in volume 1).

Infections with *B. bovis* often terminate with signs of cerebral damage such as paddling of limbs, ataxia, mania and coma. This is the result of brain damage caused by obstruction of the brain capillaries by parasitized red blood cells. There is usually no evidence of neuronal degeneration but there is dilatation of the perivascular spaces and interstitial oedema. A similar pathogenesis involving the central nervous system, known as turning sickness, has been recognized with both *T. parva* and *T. mutans*. Regular and characteristic findings in the central nervous system include intravenous accumulations of lymphoblasts which may be infected with schizonts, venous thrombi, effects of venous alterations, and perivascular lymphocytic infiltrations. Turning sickness occurs in adult cattle in enzootic areas (East Africa) when they are subject to stress, such as calving or a heavy infestation with ticks. The disease was common in the past, but seems to be rare nowadays.

Nervous signs have occasionally been mentioned in cattle suffering from coccidiosis (*Eimeria bovis*, *E. zuernii*). Signs include opisthotonus, medial strabismus, nystagmus, hypersensitivity, tetanic spasms and convulsions. The etiology of the nervous signs remains obscure.

Toxoplasma gondii causes multisystem dysfunction in all domestic animals. The disease is particularly vicious in the new-born infected in uteri and in relatively young animals. The neuropathology associated with ovine and bovine toxoplasmosis has been described in detail. In these species, toxoplasmosis is characterized by focal necrosis and vascular damage in acute infections, and by glial nodules, repair, and scar formation in chronic infection. In sheep and cattle, chronic infections were also associated with vascular mineralisation. Organisms, either free or in cysts, were demonstrated in the majority of cases and, in chronic infections, cysts were most frequently found in the cerebral cortex. Cerebral calcifications, common to chronic toxoplasmosis in children, appear to be rare in animals. In sheep extensive areas of necrosis, gliosis and demyelination are found. Lesions in cattle are mild. Clinical nervous signs include depression, trembling, opisthotonus, head tilt, incoordination, blindness and paraplegia.

Calves infected with *Neospora caninum* may develop neurological signs such as ataxia, decreased patella reflexes and loss of conscious propriocection. Gross lesions consist of malacia, and deviation or narrowing of the vertebral column.

Cestodes

→ Coenurosis is caused by the presence in the cranial cavity of *Coenurus cerebralis*, the larva of *Taenia multiceps*. The infection occurs in sheep and less commonly in other ruminants. It is rare in horses and man.

Arthropoda

The larvae of *Hypoderma bovis*, *H. lineatum*, *Gedoelstia* spp. and *Oestrus ovis* may invade the nervous system of ruminants (see volume 1).

The migration of *Hypoderma* larvae is sometimes known to cause central nervous disorders in cattle (→ Nervous System Diseases, Horses).

The symptoms caused by *Gedoelstia* spp. in cattle and sheep vary considerably but three main forms of disease are clearly distinguishable: (1)

Table 1. Parasites affecting the nervous system of ruminants (according to Vercruysse and De Bont)

Parasite	Host*	Host location	Nervous clinical signs	Principal lesions in nervous system
Protozoa				
Babesia bovis	C	Parasitized red blood cells, selectively concentrated in brain	Paddling of limbs, ataxia, mania and coma	Distention of the capillaria of the grey matter of the cerebrum and cerebellum, dilatation of perivascular spaces and interstitial oedema
Eimeria bovis *E. zuernii*	C	Intestine	Opisthotonus, strabismus hypersensitivity, spasms, convulsions	Vasodilation vessels in brain
Neospora caninum	C	Cranial and spinal nerves	Limb paresis, paralysis	Encephalomyelitis characterised by gliosis, perivascular cuffs and mild necrosis
Sarcocystis spp.	C, S	Brain, brainstem and mainly spinal cord, all other organs	Ataxia, circling, incoordination, opisthotonus	Proliferative inflammation of grey and white matter, with a variable degree of necrosis of myelin, axonal degeneration, organisms frequently in neurons
Theileria parva, T. mutans	C	Parasitized blood cells are selectively concentrated in brain capillaries	Turning and circling, dysmetria, dysergia, blindness	Venous thrombi with haemorrhages, perivascular lymphocytic infiltration, oedema
Toxoplasma gondii	C, S, G	Forebrain, brainstem, spinal cord	Trembling, opisthotonus, head til incoordination, paraplegia, blindness	Focal necrosis and vascular damage, glial nodules and scar formation
Cestoda				
Coenurus cerebralis	S, G	Cranial cavity; rarely spinal cord; mostly on the surface of one of the cerebral hemispheres	Dullness, cessation of feeding, habitual resting of the head against any support, blindness, incoordination, turning and other locomotion abnormalities	Large fluid-containing cyst, 5 cm or more in diameter on surface of brain, compressing brain tissue
Nematoda				
Parelaphostron-gylus tenuis	S, G	Spinal cord and occasionally brain	Tetraparesis, hemiparesis, tetraplegia, spastic gait, scoliosis, vestibular strabismus, blindness	Focal asymmetrical areas of necrosis with minimal inflammation, haemorrhages
Setaria spp.	S, G	Brain and spinal cord	Muscular weakness, incoordination, ataxia to paralysis, death	Focal encephalomyelo-malacia, which in many cases proceeds to liquefaction and cavitation
Arthropoda				
Hypoderma spp.	C	Aberrant migration in brain and spinal cord	Muscular weakness, localized paralysis, loss of motor control, convulsions	Haemorrhagic tracks in brain or spinal cord
Oestrus ovis	S, G	Nasal cavities, sinuses	High stepping gait, incoordination	Erosion of the bones of the skull
Gedoelstia spp.	C, S	Brain	Varies considerably	Encephalomalacia, encephalo-meningitis, haemorrhages and discolorations
Ixodes spp. *Dermacentor* spp.	C	Adult tick produces a “toxin”	Acute ascending flaccid, motor paralysis	Usually no morphological changes in nerves

*Host: C, cattle; S, sheep; G, goats

ophthalmic, (2) encephalitic, and (3) cardiac. The ophthalmic form is characterized by inconspicious conjunctual and intra-ocular haemorrhages, sometimes with a marked protrusion of the eye ball. The nervous symptoms vary considerably as all parts of the brain can be involved.

Oestrus ovis is a common parasite of the nasal cavities and sinuses, especially the frontal sinus in sheep and goats. Erosion of the bones of the skull may occur and even injury to the brain; clinical signs include high-stepping gait and incoordination, which may suggest infection with *Coenurus cerebralis*. For this reason the infection has been called "false gid".

Tick paralysis is a disease of cattle characterized by an acute ascending flaccid motor paralysis. The condition may be fatal unless the tick(s) are removed before respiratory paralysis occurs. Adult ticks, chiefly females, but sometimes nymphs, are responsible. Ticks of the genus *Ixodes* are particulary associated with the condition, although other genera, especially *Dermacentor* may also be concerned. The paralysis-activating substance acts on motor and sensory nerves and on neuromuscular transmission. The nature of the toxin is unknown (see page 572).

Nervous System Diseases, Swine

The common clinical signs and pathology of parasitic infections of the nervous system of swine are listed in Table 1.

Protozoa

The neuropathology associated with porcine toxoplasmosis has been described in detail. It is characterized by focal necrosis and vascular damage in acute infections, and by glial nodules, repair, and scar formation in chronic infection. Lesions in porcine toxoplasmosis are strictly focal and small. Clinical nervous signs include depression, trembling, opisthotonus, head tilt, incoordination, blindness and paraplegia.

Cestodes

Cysticercus cellulosae are commonly found in the brain of pigs. The absence of specific neurologic signs in infected hogs may be related to the absence of hydroencephalus and intracranial hypertension such as is usually observed in affected persons.

Nematodes

Stephanurus dentatus quite frequently invades the spinal canal and may even encyst in the meninges of pigs.

Larvae of *Toxocara canis* have been incrimated as a cause of posterior paralysis in experimentally infected pigs, a host in which the larvae appear to show a special predilection for the cerebellum. Clinical nervous signs were not so much associated with the trauma caused by the numerous migrating *T. canis* larvae, but rather with the development of exuberant tissue reactions around dead or static worms.

Table 1. Parasites affecting the nervous system of swine (according to Vercruysse and De Bont)

Parasite	Location	Nervous clinical signs	Principal lesions in nervous system
Protozoa			
Toxoplasma gondii	Forebrain, brainstem, spinal cord	Trembling, opisthotonus, head tilt, incoordination, paraplegia, blindness	Focal necrosis and vascular damage, glial nodules and scar formation
Cestoda			
Cysticercus cellulosae	Brain or meninges	No apparent clinical signs in pigs, in dogs neurological disorders	Chronic inflammatory exudate in tissue surrounding the cysticerci
Nematoda			
Stephanurus dentatus	Brain, spinal cord	Posterior paralysis	Granulomatous and leucocytic reaction
Toxocara canis	Brain, predilection for cerebellum in pigs	Posterior paralysis in pigs	Local eosinophilia, granuloma formation

Neurocysticercosis

→ *Taenia solium* (Vol. 1).

Neutrophilic Inflammation

→ Pathology.

Nitric Oxide (NO)

One of the principal effector molecules in killing parasites (→ Amoebiasis, → T-Cells).

Nodule

Clinical and pathological symptom of infections with skin parasites (→ Skin Diseases, Animals, → Siphonapteridosis).

Norwegian Scabies

Widespread epidermal scaling in immunosuppressed people due to infections with the mite *Sarcoptes scabiei* (→ Scabies).

Nosematosis

Synonym

→ Bee Dysentery.

Disease of honey bees (leading to diarrhoeal feces, drying and finally to death) due to infection with spores of the microsporidian species → *Nosema apis* (Vol. 1).

Therapy

Fumagillin and similar compounds

Notoedric Mange

→ Mange, Animals/Notoedric Mange.

Nymphomania

Clinical symptom in horses infected with → *Trypanosoma equiperdum* (Vol. 1), see → Genital System Diseases, Animals and e.g. → Filariasis, Lymphatic, Tropical.

O

Oedema

Clinical symptom in animals and men due to parasitic infections (→ Alimentary System Diseases, → Clinical Pathology, Animals).

Oesophagostomosis

Ruminants

Members of the genus *Oesophagostomum* infect cattle, sheep and goats. In sheep and goats two species are present: *O. columbianum* and *O. venulosum*, the former being considerably more pathogenic. Only one species occurs in cattle: *O. radiatum*. The life cycle involves a sojourn in the mucosa of the intestine and it is during this larval histotropic phase that the genus has its most pathogenic effects. *O. columbianum* and *O. radiatum* infections produce lesions principally in the small intestine, while the other species mainly affect the large intestine (caecum, colon). Third-stage larvae penetrate deep into the mucosa and are enclosed into small nodules (1–2 mm) by a fibroblastic reaction. The fourth moult occurs in these nodules. A strong reaction follows superinfection, and larger nodules are produced (1–2 cm) with retention of L4 in the nodules for long periods.

The signs of oesophagostomosis are anorexia, loss of body weight, diarrhoea and sometimes oedema. A moderately severe normocytic, normochromic anaemia appears, together with a decrease in plasma protein, mainly albumin. Considerable exudation of tissue fluids and plasma proteins from the intestinal lesions and haemorrhages caused by larval emergence contributes to the hypoproteinaemia and anaemia. This is exacerbated by impaired coagulation. Reduced growth or loss in condition is mainly the result of the interaction between protein effusion into the gut and loss of appetite. Diarrhoea presumably results from the loss of absorption capacity of the colon. It would appear that secondary complications and bacterial migration play important parts in the disease.

Swine

The two common species found in pigs are *O. quadrispinulatum* and *O. dentatum*. Though the parasites themselves are generally highly prevalent, clinical oesophagostomosis is not common in pigs.

Therapy

→ Nematocidal Drugs, Animals.

Oestrosis

Disease due to infestation with *Oestrus*, see Table 1

Table 1. *Oestrus* and Control Measurements

Parasite	Host	Symptoms	Country	Therapy		
				Products	Application	Compounds
Oestrus ovis (Sheep nose botfly)	Sheep, Goat	Irritation and swelling of nasal mucous membrane, purulent nose discharge; sometimes nervous disorder	World-wide	Neguvon™ (Bayer): No treatment during migration	Wash or Spray	Trichlorfon/ Metrifonate

Omsk Hemorrhagic Fever

Synonym
OHF

General Information
Omsk hemorrhagic fever occurs in Siberia with related diseases possibly occurring in the Ukraine and North Rumania. It is caused by the OHF virus (→ *Flavivirus* (Vol. 1), group B) and is associated with the tick species *Dermacentor pictus*, *D. marginatus*, *Ixodes persulcatus*, *I. apronophorus*, and *I. ricinus*. The incubation period is 3–7 days. Frequently, there is atypical bronchopneumonia, hemorrhagic rash, and extensive internal hemorrhage. Mortality rates are 0.5%–3%.

Onchocerciasis, Man

Synonym
Onchocercosis

Pathology
Onchocerciasis is a tissue infection with → *Onchocerca volvulus* (Vol. 1) transmitted by → blackflies (Vol. 1) (genus *Simulium*), with microfilariae migrating in the connective tissue. The 3rd or 4th stage larvae enter with the bite of the fly. Blackfly bites are often identifiable by a small petechial spot, sometimes with blood seepage, surrounded by marked inflammatory reaction and itching or with a small scab on the surface. The larvae develop subcutaneously and when they become adults they frequently congregate to form subcutaneous nodules. Microscopy shows intertwined worms sometimes surrounded by an eosinophilic coagulate of plasma, suggesting the → Splendore-Hoeppli phenomenon and separated by connective tissue containing lymphocytes, plasma cells, and eosinophils (→ Pathology/Fig. 18E). The → microfilariae (Vol. 1) are born live and migrate through superficial tissues, especially the skin and the cornea. When they are intact they appear innocuous, but when they die they elicit edema, chronic granulomatous inflammation, and fibrosis. With heavy infections there is marked dermatitis with desquamation. Microabscesses (→ abscess (Vol. 1)) or → granulomas with eosinophils are sometimes found around individual dead microfilariae in the skin. There is a relationship between the number of microfillaria dying and the intensity of the proinflammatory response, and an inverse relationship to specific mechanisms suppressing inflammation. The pathogenesis can be duplicated in an accentuated fashion after an injection of diethylcarbamazine, which kills many of the microfilariae and produces intensified inflammation and itching, followed by desquamation. In Africans and Ameridians the lesions are more benign, but contain more filariae; in Europeans and Yemenites lesions are more inflammatory, although with fewer microfilariae. Onchocercal lymphadenitis may give rise to lymphedema with skin swelling and scarring and occasional elephantiasis.

Ocular onchocerciasis results from migrating mirofilariae, some of which die in the eye (→ eye parasites). This can give rise to punctate keratitis around each dying larva in the cornea, progressing to diffuse sclerosing keratitis with numerous microfilariae, and with the inflammatory reaction followed by vascularization and opacification of the cornea (→ Eye Parasites/Fig. 4). Microfilariae in the anterior chamber give rise to iridocyclitis with a possibility of fibrosis, formation of adhesions, or synechiae, and the development of glaucoma. Microfilariae dying in the retina lead to retinochoroiditis with destruction of retinal cells, and depigmentation alternating with proliferation of the pigment epithelium. Inflammation of the optic nerve eventually leads to optic atrophy. All of these lesions, especially those in the cornea and retina, impair vision, and sometimes lead to blindness (e.g., '→ river blindness', → Roble's disease) with heavy or prolonged infections.

Main clinical symptoms: Skin nodules, chronic dermatitis, xerodermic conjunctivitis, keratitis, chorioretinitits, atrophia of the nervus opticus
Incubation period: 3–4 months
Prepatent period: 9–30 months
Patent period:10–16 years
Diagnosis: Microscopic determination of microfilarie from skin-snips, serodiagnostic methods
Prophylaxis: Avoid the bite of the vector
Therapy: Treatment see → Nematocidal Drugs, Man

Onchocercosis, Animals

Several species of *Onchocerca* (→ Filariidae (Vol. 1)) occur in horses, donkeys, cattle, sheep and goats.

The adults live in nodules within the connective tissue of the host. The specific location of these nodules depends on the species of *Onchocerca* involved. The adults produce microfilariae which migrate through the connective tissues to the upper dermis. In cattle *O. gibsoni*, *O. dukei* and *O. ochengi* produces subcutaneous and intradermal nodules in the brisket, and occasionally elsewhere. Infected animals show no other clinical signs. In horses, adult worms live in various ligaments and tendons, and the microfilariae migrate to the dermis. Cutaneous onchocercosis in horses is characterized by pruritis, alopecia, depigmentation, erythrema, and crusting. The lesions occur on the face, neck, tail head and ventral midline. The pathogenicity of microfilariae in horses remains controversial. The lesions may be rather attributed to *Culicoides* hypersensivity (summer dermatitis or sweet itch) which often occurs simultaneously. Ocular lesions such as a periodic ophthalmia are reported.

Therapy

→ Nematocidal Drugs, Animals.

Onicola canis

Acanthocephalan worm in the intestine of wild carnivores and occasionally in dogs and cats (→ Acanthocephalan Infections).

Opisthorchiasis, Man

Clonorchiasis, an infection with the Chinese liver fluke, and Opisthorchiasis, an infection with any of several → Opisthorchis (Vol. 1) spp. are bile duct infections after ingestion of undercooked fish or crustaceans containing the metacercariae. Luminal infections of the bile duct with a small parasite burden may be asymptomatic (→ Pathology/Fig. 22B). The worms are attached to the wall of the bile ducts with their two suckers giving rise to local inflammation. Large numbers of parasites may introduce heavy biliary obstruction, with resultant jaundice and secondary infection leading to cholangiohepatitis, liver abscesses, cholecystitis, and pancreatitis. The worms survive for 20 years or longer, accompanied by adenomatous hyperplasia of the bile ducts, increased mucus production, and sometimes adenocholangiocarcinoma, usually mucin-producing. In the pancreatic ducts both squamous metaplasia and adenomatous hyperplasia may occur. Eggs pass out of the bile duct and are found in the stools.

Main clinical symptoms: Abdominal pain, oedema, diarrhoea, icterus
Incubation period: 2 weeks
Prepatent period: 2–4 weeks
Patent period: 20 years
Diagnosis: Microscopic determination of eggs in fecal samples
Prophylaxis: Avoid eating raw freshwater fish in endemic regions
Therapy: Treatment: praziquantel, see → Trematodocidal Drugs

Opportunistic Agents, Man

Opportunistic infectious agents do not cause obvious symptoms by their presence in immunocompetent individuals, but can proliferate into fulminant infections in immunodeficient or immunocompromised hosts.

→ Immune suppression reactions have increased recently worldwide in humans and animals. This occurs not only due to spreading of virus transmission (inclusive HIV), but also as a result of increasing use of drugs (e.g. cortisone) with immune decreasing side effects. This deficiency of the immune system enables a broad series of agents – including viruses, bacteria, fungi and parasites – to reproduce in a much higher degree than it occurs under normal (i.e. immunocompetent) conditions. Such infections are due to the fact that opportunistic agents – especially the parasites → *Pneumocystis carinii* (Vol. 1), → *Cryptosporidium* species (Vol. 1) and → *Toxoplasma gondii* (Vol. 1) are the three main reasons for deaths in AIDS-patients. The most common parasites from the recently enlarging group of opportunistic agents are listed in Table 1.

Related Entry

→ Opportunistic Agents (Vol. 1).

Table 1. Opportunistic agents in immunocompromized humans

Species	Infected organ	Transmission/ stage	Symptoms of disease	Infected humans/ death per year
→ *Blastocystis hominis* (Vol. 1)	intestine	oral – cysts	diarrhoea	30–40 millions/ thousands
→ *Pneumocystis carinii* (Vol. 1)	lung	inhalation – cysts	pneumonia	400 millions/ 100thousands
→ *Giardia lamblia* (Vol. 1)	intestine	oral – cysts	diarrhoea	450 millions/ thousands
→ *Leishmania* (Vol. 1) species	skin, inner organs, generalizing	bite of sand flies	general destruction of organs	15 millions/ thousands
→ *Entamoeba histolytica* (Vol. 1)	intestine, liver	oral – cysts	diarrhoea, abscess	500 millions/ thousands
→ *Naegleria* spp. (Vol. 1)	liquor, ZNS	via nose – at bathing	encephalitis, PAME	thousands/ few
→ *Acanthamoeba* (Vol. 1) species	ZNS, liquor	via nose – at bathing	encephalitis, destruction of cornea	100thousands/ thousands
→ *Isospora belli* (Vol. 1)	intestine	oral – oocysts	diarrhoea	100thousands/ thousands
→ *Cryptosporidium* (Vol. 1) *parvum*	intestine	oral – oocysts	diarrhoea	40 millions/ 100thousands
→ *Toxoplasma gondii* (Vol. 1)	ZNS, generalizing	oral – oocysts, meat	cerebral destruction	50–60 millions/ 100thousands
→ *Cyclospora* species (Vol. 1)	intestine, generalizing	oral – oocysts	diarrhoea	thousands/ few
→ Microspora (Vol. 1) (e.g. → *Encephalitozoon* (Vol. 1))	ZNS, kidney, intestine, generalizing	oral – spores	destruction of organs	100thousands/ thousands
→ *Balantidium coli* (Vol. 1)	intestine	oral – cysts	diarrhoea	thousands/ few
→ *Hymenolepis nana* (Vol. 1)	intestine	oral – larvae	diarrhoea	thousands/ none
→ *Strongyloides stercoralis* (Vol. 1)	intestine	percutan/ oral – larvae	diarrhoea	100 millions/ thousands
→ *Sarcoptes scabiei* (Vol. 1)	skin	body contact	scabies	20 millions/ thousands
→ *Demodex folliculorum* (Vol. 1)	skin	body contact	rosacea	thousands/ none

Oriental Sore

Disease due to infection with *Leishmania major* and *L. tropica* transmitted by bite of → sand flies (Vol. 1) (see → Leishmaniasis, Man).

Orientia *(nov. gen)* tsutsugamushi

Rickettsial agent of the mite transmitted Tsutsugamushi fever in Asia (syn. *Rickettsia*).

Oroya Fever

Visceral form of disease starting 15–40 days after injection with *Bartonella bacilliformis* bacteria during the bite of sand flies. Clinical symptoms are high fever, lymphadenitis, spleno- and hepatomegaly. The decreasing number of erythrocytes being lysed by the *Bartonella*-stages introduces anaemic symptoms. Months after this acute phase skin symptoms (→ Verruga peruana) may occur. The disease is restricted to the Western and Eastern valleys of the Andes.

Ostertagiosis

Ostertagiosis is probably the most important parasite in grazing sheep and cattle in temperate climatic zones throughout the world. It causes subclinical losses of production and disease. The clinical disease is characterized by diarrhoea, weight loss, decreased production, rough hair coats, partial anorexia, mild anaemia, hypoalbuminaemia, dehydration and in some cases, death. *Ostertagia ostertagi* in cattle and *O. (Teledorsagia) circumcincta* in sheep and goats are the most important species. Related species and genera are *O. leptospicularis*, *Skrjabinagia lyrata* in cattle, and *O. trifurcata*, *Teladorsagia davtiani* and *Marshallagia marshalli* in sheep and goats.

Clinical ostertagiosis occurs under three sets of circumstances called type I, pre-type II and type II diseases. The type I disease is seen in calves at pasture, shortly after a period of high availability of infective larvae. It is due to the direct development of large numbers of L3 larvae to adult worms over a relatively short period of time. In contrast, type II disease is due to the synchronous maturation and emergence of large numbers of hypobiotic larvae from the mucosa, and occurs when intake of larvae is likely to be low or non-existent. It occurs in cattle, mainly yearlings or heifers, during the winter in the northern hemisphere, or during the dry summer period in Mediterranean climates. The almost asymptomatic condition which precedes type II ostertagiosis has been called pretype II. In this phase the abomasum carries a pathogenically adequate burden of inhibited larvae which are still quiescent, but from which disease type II may erupt unpredictably if a sufficiently large number of larvae resume development to maturity. In sheep the same forms occur, but in ewes the course of type II is very rare and more chronic. Clinical signs start at the time the parasites reach maturity; they begin to emerge from the gastric glands (in type I after 18–21 days and in type II after 4–6 months), and marked cellular changes appear. The functional gastric gland mass, particularly the hydrochloric acid (HCl)-producing parietal cells, is replaced by undifferentiated cells. It has also been shown that the secretory activity of parietal cells is blocked. The undifferentiated and hyperplastic mucosa is abnormally permeable to macromolecules following the destruction of the intercellular junctions. This happens not only in the parasitized gastric gland but also in the surrounding glands. These structural changes result in: (1) an elevation of the pH of the abomasal fluid from 2 to 5 or even higher. This leads to a failure to activate pepsinogen to pepsin and to denature proteins. There is also a loss of bacteriostatic activity, which is followed by an increase in the number of bacteria; (2) an enhanced permeability to macromolecules resulting in hypoalbuminaemia, the albumin in the plasma passing into the abomasum. Any loss of protein macromolecules is usually accompanied by loss of electrolytes, mainly $Na+$ and $Cl-$. The onset of diarrhoea increases the loss of electrolytes. Continued loss may lead to increased hypoalbuminaemia, retention of fluid and the development of oedema; (3) elevated plasma pepsinogen concentrations of more than 3 U tyrosine. The mechanism responsible for this increase is not yet completely understood. A multifactorial cause has been postulated, involving direct stimulation of zymogenic cells by factors released from the parasite, indirect stimulation via elevated circulating concentrations of hormones such as gastrin (*vide infra*) and leakage from abomasal fluid between poorly differentiated epithelial cells; (4) raised gastrin levels. The gastrin levels have been found to increase considerably during *Ostertagia* infection in sheep and cattle. The consequences are not clear. Gastrin has a multiplicity of actions e.g. stimulating HCl and pepsinogen secretion, inhibiting reticulo-ruminal motility, trophic effects on the gastric and intestinal mucosa. The cause of hypergastrinaemia has not been established but the presence of the parasite seems to be critical.

Related Entry

→ Hypobiosis (Vol. 1).

Therapy

→ Nematocidal Drugs, Animals, → Drug.

Otodectic Mange

→ Mange, Animals/Otodectic Mange.

Oxyuridosis

→ Oxyurosis.

Oxyurosis

Synonym

→ Enterobiasis.

Disease due to infections with *Oxyuris equi* in horses, *Passalurus ambiguus* in hares and rabbits and *Oxyuris* spp. in reptiles and amphibians.

P

Pale Mucosa

Clinical symptom in animals due to parasitic infections (→ Alimentary System Diseases, → Clinical Pathology, Animals).

Paludisme

French name for → Malaria

PAME

→ Amoebae (Vol. 1).

Pandemy

Number of infections in a given period (not limited by geographic borders).

Papataci Fever

Synonyms

Sand Fly Fever, *Phlebotomus* Fever

This disease, which is found around the Mediterranean Sea, in countries in the Near and Middle East regions, in Central Asia and in East Africa, occurs due to infections with two immunologically different → arboviruses (Vol. 1) being transmitted during the bite of the sand fly *Phlebotomus papatasii*, which may also infect transovarially its next generation. The incubation period is three days then the disease starts with sudden and high fever (41 °C) for 3 further days, during which the following symptoms may occur: headache, vomiting, diarrhoea, myalgia.

Papula

Clinical and pathological symptom (reactions at the biting site) of infections with skin parasites (→ Skin Diseases, Animals, → Mosquitoes (Vol. 1), → Fleas (Vol. 1)).

Parafilariasis, Parafilariosis

Parafilaria bovicola (→ Filariidae (Vol. 1)) occurs in cattle and *P. multipapillosa* in horses. Both have a very similar pathogenesis. Adult worms occur in nodules in the skin and subcutaneous tissue. A bloody exudate appears when female worms rupture the nodules to lay eggs, eventually causing a matting of the hair. As erupted nodules regress, fresh ones appear. Infection is highly seasonal, occurring when the weather is warmer.

Therapy

→ Nematocidal Drugs, Animals.

Paragonimiasis, Man

Paragonimiasis is a lung fluke infection contracted by the consumption of freshwater crabs or crayfish (→ Paragonimus (Vol. 1)). The metacercariae penetrate the intestinal wall, migrate through the peritoneum across the diaphragm, and enter the pleural cavity to reach the lungs. The worms develop in bronchioles and when mature shed their eggs into the bronchi (→ Pathology/Fig. 21D). The worms elicit an exudate with neutrophilic and eosinophilic granulocytes, generally developing a cyst or an abscess which may be surrounded by a fibrous capsule. Hemorrhage into the cyst often occurs and the brownish mucoid exudate containing eggs is coughed up. Degenerating eggs give rise to a granulomatous inflammatory reaction. Aberrant sites of infection include the abdominal cavity, soft tissues, and the brain, where the birefringent *Paragonimus* eggs must be differentiated from the nonbirefringent schistosome eggs. In general the metabolic products of the adults give rise to microabscesses, and the degenerating eggs to → granulomas. These are accompanied by eosinophils and → Charcot-Leyden crystals, and are surrounded by fibrosis or gliosis with scattered plasma cells and lymphocytes. The lesions and symptoms vary by site. In the brain, cavities measuring 10 mm in diameter may be produced, surrounded by a connective tissue capsule.

Main clinical symptoms: Haemoptysis, bronchitis, thoracic and/or abdominal pain
Incubation period: 9–12 weeks
Prepatent period: 10–12 weeks
Patent period: 20 years
Diagnosis: Microscopic determination of eggs in sputum or fecal samples
Prophylaxis: Avoid eating raw crustaceans in endemic regions
Therapy: Treatment with praziquantel, see → Trematodocidal Drugs

Paramphistomosis

Infections by the digenean genus *Paramphistomum* may cause significant intestinal problems in ruminants. The adult worms live in the rumen, but the pathological effects of infection are caused by the immature stages within the small intestine. Many genera and species are involved. Their pathogenic processes are similar. The most pathogenic species are thought to be *Paramphistomum microbothrium*, *P. ichikawai*, *P. cervi*, *Cotylophoron cotylophoron* and various species of *Gastrothylax*, *Fishoederius* and *Calicophoron*.

Paramphistomosis is largely a disease of young animals, because repeated infections of low intensity generally produce an almost complete immunity. The immunity results not only in a marked reduction in the worm burdens from challenge infections but also protects the host against the lethal effects of these infections. The pathological effects of infection are almost entirely caused by the immature stages within the first part of the small intestine. The immature worms penetrate the mucosa of the small intestine as deeply as the muscularis and become attached, with a plug of mucosa drawn into their acetabula. This causes strangulation and the eventual necrosis of the piece of mucosa, leading to the development of erosions and petechiae. These lesions cause intestinal discomfort and a reduction of appetite. At the same time, plasma albumin is lost by seepage and hypoalbuminaemia results. Protein loss into the gut, coupled with loss of appetite, seems to be the most important pathophysiological consequence of paramphistomosis. The low plasma protein concentration causes generalized oedema, seen as hydropericardium, hydrothorax, pulmonary oedema, ascites and submandibular oedema. Clinical signs are profuse and foetid diarrhoea, anorexia, marked weakness, often leading to death. The animals are thirsty and drink frequently. There is no indication of anaemia. After massive infection, migration of the immature worms to the rumen is delayed, and flukes may persist for months in the duodenum prolonging the course of disease.

Therapy

→ Trematodocidal Drugs.

Paramyosin

Protective antigen in → Schistosomiasis, Man.

Parasitaemia

Amount of parasitic stages within the blood in percents of red blood cells.

Parasite Load

The quantity of parasites within a host; this load is mostly responsible for the severity of clinical symptoms (→ Clinical Pathology, Animals).

Parastrongyliasis

→ *Parastrongylus cantonensis* (Vol. 1).

Pathogenicity

Ability of a pathogen to cause a disease.

Pathology

Pathology describes the effects of the parasite on the host, the morphologic and functional changes produced, and the host response. While pathology implies a static description, **pathogenesis** refers to the dynamic events and interactions. An overall successful host-parasite interaction occurs in the life cycle of many parasites and generally depends on development of immunity by the host. The infective agent may be eliminated, or persist as a chronic inactive infection. Some parasites have learned to circumvent some of the immune mechanisms that tend to contain or eliminate them.

Examples of unsuccessful interactions include accidental parasitisms where usually the invader, and occasionally the host, is rapidly killed. In contrast to this, successful parasitism implies the ability to persist and to reproduce in a host without giving rise to lesions that prevent survival as a species of either parasite or host.

In order to develop a tolerable relationship it is likely that both parasite and host have evolved together for at least some time, selecting each other for successful parasitism with host survival. Toleration is most advanced in what we call a mutualistic relationship, which exists for example with flagellates in the gut of termites and ciliates in ruminants. No such relationship with protozoan or metazoan parasites is recognized in humans. Microbes that are not injurious we designate as **commensals**; some parasitic examples are listed in → Protozoan Infections, Man/Table 1.

Individuals who are parasitized, may get sick, or with development of immunity, recover. When the capacity to develop and maintain immunity is impaired, as in AIDS, active infection progresses, and may kill the host.

Many parasite species exert a pathogenic, lesion-producing effect. These lesions may explain clinical symptoms, but more importantly the lesions indicate the pathogenic mechanisms that give rise to disease, and that might be subject to therapeutic intervention. Some species may at times be commensals and at other times parasitic. It is important to differentiate between the biological fact of infection and disease.

→ Host responses to parasite species are varied but a certain "theme and variation" approach can be developed to classify the lesions, with widely varying individual responses. Some of the → inflammatory responses are associated with protection but in → hypersensitivity the inflammatory response contributes to the pathogenic effects regularly produced by the parasite. Occasionally, the responses are so typical that they are helpful in the determination of a parasitizing species.

The relative roles of the parasite and the host in development of lesions are sometimes in doubt. "Opportunistic organisms" are often erroneously cited as giving rise to infections and lesions. However all organisms should be regarded as opportunistic. The concept of "opportunistic infections" is usually invoked when the host is immunosuppressed. The change in host-parasite effect is therefore caused by a compromised capacity to develop or maintain immunity, usually to organisms that are not pathogenic to immunocompetent hosts.

A number of common reactions are listed in Table 1, indicating the limited nature of host response; the same histologic patterns can be elicited by different agents. Also, the character or lesions change with time. In toxoplasmosis the histologic reaction changes from granulocytic to mononuclear (lymphocytic, monocytic, and macrophagic) with the development of immunity, and to hypersensitive inflammation during chronic infections. Destruction of cells by intracellular parasites, or infection necrosis, must be differentiated from hypersensitivity necrosis, mediated by cytolytic T lymphocytes, and from infarction necrosis.

If, during a chronic infection such as schistosomiasis, new eggs are produced over a long period of time, each egg initiates a histologic reaction with a typical course of development, resulting in a pleomorphic pattern in adjacent tissues affected by eggs of different ages (Fig. 1). The courses of a number of inflammatory reactions are outlined in Table 2. Parasitic infections in an immunosuppressed individual are associated with a modified

Table 1. Histologic reactions seen in parasitic infections (according to Frenkel)

Minimal histologic reaction:

Immunologic mimicry – *Schistosoma* adults (Fig. 23A)

Impermeability of living worms – *Angiostrongylus* adults (in rat) (Fig. 23B)

Impermeability of cysts – *Toxoplasma, Sarcocystis, Isospora*, and *Trypanosoma cruzi* (Fig. 5)

Metabolic products drained – *Clonorchis* in bile duct (Fig. 22B); nematodes, trematodes, and cestodes in gut lumen (Fig. 22A)

Atrophy (enhanced attrition of epithelial cells): giardiasis, cryptospordiosis (Fig. 6)

Immunosuppressed host – *Pneumocystis* (Fig. 2A), stronyloidiasis (Fig. 2F), acanthamebiasis (Fig. 4F), toxoplasmosis (Fig. 10), isosporosis (Fig. 5C), cryptosporidiosis (Fig. 6)

Tissue anoxia from obstruction of capillaries: *P. falciparum* malaria (Fig. 15)

Infection necrosis of parasitized cells: toxoplasmosis (Fig. 10), Chagas' disease (Fig. 13D)

Infection necrosis of adjacent cells: amoebiasis (Fig. 4), cyst rupture in toxoplasmosis and sarcocporidiosis

Histiocytic reaction: diffuse cutaneous leishmaniasis (Fig. 14)

Eosinophilic reaction (often with Charcot-Leyden crystals; (Fig. 3B): schistosomiasis (Fig. 1, Fig. 24), migratory *Ascaris* (Fig. 27A), filariasis (Fig. 28A), sarcocystosis, amoebiasis (Fig. 4), cysticercosis, echinococcosis (Fig. 18A, Fig. 29E, F), coenurosis, sparganosis (Fig. 18), fascioliasis (Fig. 3, Fig. 21A–C), paragonimiasis (Fig. 21D), trichuriasis, larva migrans (Fig. 28C), capillariasis, anisakiasis (Fig. 28D), trichinelliasis (Fig. 18C,D), gnathostomiasis, *Angiostrongylus* infection (Fig. 29B–D), dracunculiasis, filariasis (Fig. 28A), onchocerciasis (Fig. 18E), dipetalonemiasis, scabies (Fig. 30B), tick bite (Fig. 18A,B).

Splendore-Hoeppli reaction: schistosome eggs (Fig. 1B), *Wuchereria* (Fig. 28A), *Onchocerca, Anisakis* (Fig. 28D).

Neutrophilic inflammation: *Trichomonas, Balantidium*

Mixed inflammation: *Isospora belli, Toxoplasma* tachyzoites (Fig. 12A) and liberated bradyzoites (Fig. 5A,B, Fig. 13C).

Abscess: dead adult schistosomes in liver, amoebiasis in liver (Fig. 4D)

Ulcerating lesions: myiasis, tick bite (Fig. 13A,B), dracunculiasis, tungiasis, scabies (Fig. 30B), amoebiasis (Fig. 4), dermal leishmaniasis (Fig. 30A)

Granuloma: schistosome eggs (Fig. 1), migrating worms (Fig. 3, Fig. 21A–C), paragonimiasis, filariasis, onchocerciasis (Fig. 18E), larva migrans (Fig. 28C)

Hypersensitivity necrosis: schistosome eggs (Fig. 1) and dead adults disintegrated cysts of *Toxoplasma* (Fig. 18), *Sarcocystis*, and pseudocysts of *T. cruzi* (Fig. 13), tissue migration of *Ascaris* (Fig. 27A), *Anisakis* (Fig. 28D), *Enterobius* (Fig. 27B), *Fasciola* (Fig. 3A,B, Fig. 21C), tick bite (Fig. 13A,B)

Fibroblastic proliferation (organization or encapsulation): schistosomiasis (Fig. 1D, Fig. 24B,C), onchocerciasis (Fig. 18E), filariasis, cysticercosis, enchinococcosis (Fig. 18A), pentastomiasis (Fig. 18F), sparganosis (Fig. 18B), coenurosis

Infarction necrosis, after thrombosis of vessel: toxoplasmosis of central nervous system in immunosuppressed and in newborn (Fig. 12A,B), falciparum malaria, dirofilariasis (Fig. 28B)

Lymphadenitis or lymphoreticular hyperplasia: Postprimary toxoplasmosis (Fig. 11), African trypanosomiasis, Chagas' disease, filariasis, pentastomiasis

Placental involvement: toxoplasmosis, Chagas' disease, malaria

Glomerulonephritis (immune complex): malaria, kala-azar, toxoplasmosis (rare), schistosomiasis, scabies with streptococcosis

Meningoencephalitis: *Acanthameeba* (Fig. 4F), *Naegleria* (Fig. 5E), *Toxoplasma* (Fig. 5A,B, Fig. 10A,B, Fig. 12A,B), microsporidiosis, African trypanosomiasis, Chagas' disease, *P. falciparum* malaria (Fig. 15), cysticercosis, echinococcosis, coenurosis, paragonimiasis, schistosomiasis (Fig. 24C), visceral larva migrans, trichinelliasis, micronemiasis, Angiostrongylus cantonensis, gnathostomiasis

Necrosis followed by calcification: neonatal toxoplasmosis in brain, cysticercosis, coenurosis, schistosomiasis, *Trichinella* in brain, cysticercosis, coenurosis, schistosomiasis, muscular trichinelliasis (Fig. 18D), dracunculiasis, dirofilariasis, pentastomiasis

Hyperplasia and neoplasia: scabies (Fig. 30B); cholangiocarcinoma – opisthorchiasis,, clonorchiasis; carcinoma of bladder – *S. haematobium* schistosomiasis (Fig. 24B)

Accidental parasitism: anisakiasis (Fig. 28D), toxocariasis (Fig. 28B), dirofilariasis (Fig. 28B)

Comparison of histologic reactions in natural and aberrant host: angiostrongyliasis, echinococcosis (Fig. 29)

and usually diminished inflammatory response (Fig. 2) so that the lesion may be atypical and reflect more microbial damage than host defense or hypersensitivity.

While the histologic picture may be useful to indicate one of several etiologic agents it is rarely diagnostic. Although eosinophils may have evolved as a defense against this wormy world, their presence is not pathognomonic (Fig. 1, Fig. 3, Fig. 24). Therefore, identification of an etiologic agent itself, whether protozoan, helminthic, bacterial, fungal, viral, etc., is essential; alternatively, assignation of the process as microbial, and inflammatory, degenerative, or nepotistic, are prerequisite to a definitive diagnosis of the process occurring in the tissue, and together with immunologic and serologic information necessary for an interpretation of its pathogenesis.

Pathologic findings, whether from biopsy or autopsy are useful diagnostically because the etiologic agent is generally found in or near the lesion. However, the etiologic organisms may not be frequent enough to be seen in routinely stained sections, so that special staining for suspected organisms may be necessary, or that cultures or subinoculation to animals must be employed for diagnosis. Histopathologic findings may suggest the mechanisms by which lesions are produced (the pathogenesis), for example, infection necrosis from microbial destruction of cells, hypersensitivity accentuating the lesions and histologic reactions, and immunity which tend to delimit or repair them (the immunopathology). However histopathology is rarely specific and must be complemented by the identification of the organism or its DNA in the lesions, and, if a rare etiologic agent is

Table 2. Sequential progression of parasite-induced inflammatory reactions over time (according to Frenkel)

Spectrum of Initial Inflammation Response	Inflammation modified by immunity and hypersensivity	Repair
1. No reaction →	Worm dies → Tissue necrosis → Fibroblastic prolif → Fibrosis Cyst rupture (hypersensitivity)	
2. Localized inflammation →	Histiocytic reaction → Cyst rupture (hypersensitivity)	
3. Diffuse histiocyte reaction →	Progressive histiocytic reaction	
↓ Lymphoid infiltration		
↓ Granuloma →	Necrosis →	Fibroblastic reaction (capsule formation)
Granuloma →		Fibroblastic reaction (capsule formation)
4. Intracellular infection (no inflammation) →	Destruction of individual infected cells (infection necrosis) → Granulocytic inflammation →	Macrophagic infiltration → Fibroblastic reaction (capsule formation)
	Granulocytic inflammation → Eosinophilic inflammation (Charcot-Leyden crystals) →	Macrophagic infiltration
5. Granulocytic inflammation →	Macrophage Infiltration → Granuloma →	Fibroblastic reaction (capsule formation)
↓ Abscess →	Drainage → Granulation tissue →	Fibroblastic reaction (capsule formation)
Abscess →	Mixed leukocytic inflammation → Lymphocytit inflammation → Granulation tissue	
6. Any of the above →	Involvement of vessel wall with infection or inflammation → Thrombosis of vessel → Infarction (tissue necrosis from ischemia) - - - → Hemorrhage	

Table 3. Main groups of parasite pathogens in humans (according to Frenkel)

1. Intestinal Protozoan Infection
   → Amoebic infections
   → Acanthamoebiasis
   → Naegleriasis
   → Entamoebiasis
2. Lumen-Dwelling Flagellates
   → Giardiasis
   → Trichomoniasis
   → Blastocystosis
3. Intestinal Ciliate Infection
   → Balantidiasis
4. Intestinal Coccidiosis
   → Isosporosis
   → Cyclosporiasis
   → Cryptosporidiosis
   → Intestinal Sarcosporidiosis
5. Tissue Protozoan Infections/Disseminated Tissue Infections
   → Toxoplasmosis
   → Muscle Sarcosporidiosis
   → Microsporidiosis
6. Leishmaniasis
   → Cutaneous Leishmaniasis
   → Mucocutaneous Leishmaniasis
   → Diffuse Cutaneous Leishmaniasis
   → Visceral Leishmaniasis
7. Trypanosomiasis
   → Gambian Sleeping Sickness
   → Rhodesian Trypanosomiasis
   → Chagas' Disease
8. Pulmonary Pneumocystosis
9. Blood Protozoan Infections
   → Malaria
   → Babesiosis
10. Acanthocephalan Infections
11. Cestode Infections
    → Taeniasis
    → Cysticercosis
    → Echinococcosis
    → Coenurosis
    → Sparganosis
12. Trematode Infections
    → Fasciolopsiasis
    → Fascioliasis
    → Clonorchiasis
    → Opisthorchiasis
    → Paragonimiasis
    → Schistosomiasis
13. Intestinal Nematode Infection
    → Enterobiasis
    → Trichuriasis
14. Intestinal and Tissue Nematode Infections
    → Ascariasis
    → Cutaneous Larva Migrans
    → Hookworm Disease
    → Visceral Larva Migrans
    → Strongyloidiasis
    → Capillariasis
    → Anisakiasis
15. Tissue Nematode Infections
    → Trichinelliasis
    → Gnathostomiasis
    → Micronemiasis
    → *Angiostrongylus cantonensis* Infection
    → Abdominal Angiostrongylosis
    → Dracunculiasis
16. Blood and Tissue Nematode Infections
    → Filariasis
    → Loiasis
    → Onchocerciasis
    → Dirofilariasis
    → Zoonotic Dipetalonemiasis
17. Pentastomiasis
18. Arthropod Infections
    → Scabies
    → Tungiasis
    → Pediculosis
    → Myiasis
    → Demodicosis
    → Insect Bites
    → Tick Bites

identified, by if possible fulfilment of Koch's postulates. Serologic reactions are also useful. Parasitologists are focusing on the **pathogenesis** and **immunopathology** of parasitic diseases, because so little is known about them. The experimental analysis of both depends heavily on the use of model infections or, in absence of a good model, by partial analogies. Sometimes spontaneous animal models in domestic, wild or laboratory animals are informative, such as for pneumocystosis of rats. Another approach is to compare infection in a variety of inbred strains of mice differing in innate resistance and ability to acquire immunity, to provide a spectrum of reaction patterns that by analogy may provide information applicable to this infection in the genetically diverse human population.

More detailed information on the lesions and on other pathologic effects are given under the headword of the respective diseases and are summarized in Table3.

Table 4. Survey of the pathology of cestode and trematode infections of man (according to Frenkel)

	Main pathogenic stage	Main lesions, location of the worm	Necrosis	Eosinophilia	Granuloma	Adult site (other hosts)	Larval site (other hosts)	Method of infection
Cestode infections								
Taeniasis	Adult	None, small Intestine	–	–	–	Small intestine (carnivores)	(cattle, pigs)	Ingestion of infected meat
Cysticercosis	Larva	Cysts, many tissues	After death of worm	Common	After death of larva	Small intestine (carnivores)	Many tissues, omnivores	Ingestion of eggs from feces
Echinococcosis	Larva (hydatid)	Cysts, many tissues	Especially *E. multilocularis*	Common	After death of hydatid	Intestine (of dogs,) wolves, foxes)	Many tissues	Ingestion of eggs from feces
Coenurosis	Coenurus cyst	Cyst, brain, eye	After death of cyst	Rare	After death of cyst	Intestine (of dogs)	Brain, eye	Ingestion of eggs from feces
Sparganosis	Larval tapeworm	Nodules, subcutaneous or visceral	After death of worm	Common	After death of worm	Intestine (dogs, cats)	Various tissues	Ingestion of eggs from feces
Trematode infections								
Fasciolopsiasis	Adult	None, duodenum, jejunum	Abscesses	–	Ulcers	Intestine (pigs, dogs)	(Water plants)	Eating, drinking
Fascioliasis	Larva, adults	Liver and bile duct dysfunction	Enlarged liver, dysfunction	Common	–	Bile duct (cattle)	Larvae wander through liver	Eating, water plants, drinking
Clonorchiasis	Adult	Obstruction, bile ducts	–	Slight	–	Bile duct	Snails, fish	Ingestion of fish or crustaceans
Paragonimiasis	Adult	Bronchopneumia, brain abscess	Around adults	Common	Around eggs	Lung (cats, pigs, dogs, etc.)	Snails, fresh watercrabs	Freshwater crustaceans
Schistosomiasis	Eggs	Multiple, gut, liver, bladder, brain	Around dead adults	Usual	Around eggs and dead adults	Portal or mesenteric veins	Snails, water	Drinking or skin contact with snail-infested water

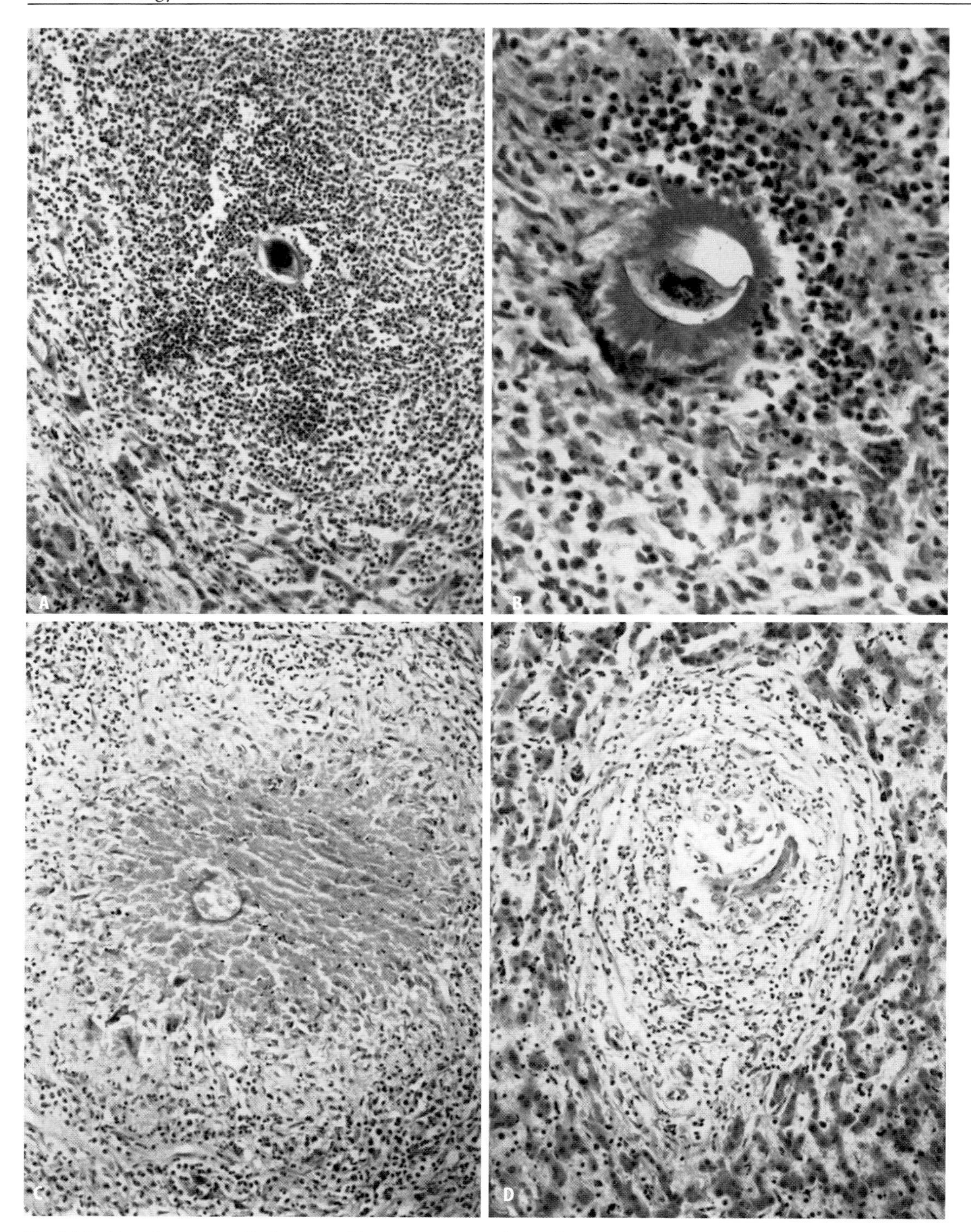

Fig. 1A-D. Development of lesions to schistosome eggs in liver of a patient who died with a 1 month history of *Schistosoma japonicum* infection. A Egg with intense infiltration of eosinophilic leukocytes. B Egg surrounded by stellate eosinophilic matrix, the Splendore-Hoeppli reaction, and predominantly eosinophilic infiltration. C Destroyed egg with Splendore-Hoeppli reaction surrounded by zones of necrosis, epithelioid cells, and eosinophils. D Giant and epithelioid cells occupy the center of the granuloma; the egg has probably been digested. In the periphery a fibroblastic reaction is depositing collagen, loosely infiltrated with eosinophils and a few lymphocytes. The surrounding liver cells are beginning to regenerate as shown by binucleate liver cells. Hematoxylin and eosin (HE): A, C, D x 120, B x 300

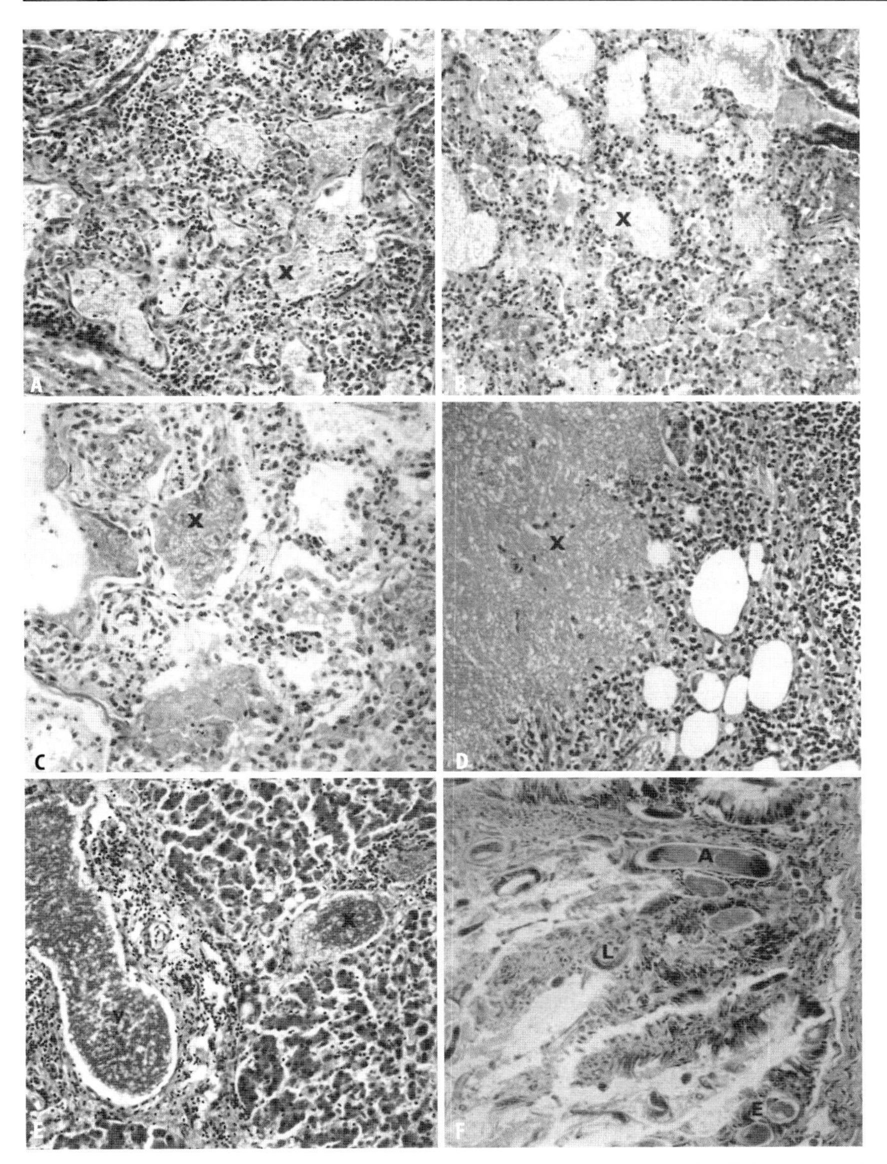

Fig. 2A-F. Lesions with minimal reactions because of suppression of cellular immunity and delayed type hypersensitivity. **A** *Pneumocystis sp.* plasma cell pneumonia in a malnourished infant. Both trophozoites and cysts of *Pneumocysts* sp. are present in the alveoli and appear as a "foamy" colony (X). Plasma cells are the predominant cell in the widened alveolar walls. Type 2 pneumocytes are prominently lining the alveoli. This is interpreted as a progressive primary infection. HE x 120. **B** *Pneumocystis* sp. pneumonia in 2-year-old baby with acute leukemia and also immunosuppressed by virtue of treatment with prednisone, methotrexate, and 6-mercaptopurine for 6 months. Foamy colonies (X) are again visible, but only few lymphocytes are present in the alveolar walls. *Pneumocystis* sp. colonies are not present in the bronchiole (upper right). This could be a primary infection or reinfection. HE x 120. **C** *Pneumocystis* sp. pneumonia in 27-year old male with lymphocytic lymphoma treated with prednisone, methotrexate, and chlorambucil for 3 months. Foamy colonies (X) composed of trophozoites and cysts fill some alveoli. This is either a recrudescent infection or a reinfection. Periodic acid Schiff (PAS)x120. **D** *Pneumocystis* sp. (X) growing free in supraclavicular lymph node extending into fibrofatty tissue. From a 63-year-old woman with chronic lymphocytic leukemia. Slide courtesy of Dr. Michael Coughron. HE x 120. **E** *Pneumocystis* sp. in liver of adult with hypoproteinemia. Groups of organisms are seen both free (X) in the liver and surrounded by fibrosis and lymphocytic infiltration (A). PAS x 120. **F** *Strongyloides* sp. in the colon of a 31-year old woman from Central America who had lived in Los Angeles for 17 years. The patient developed acute lymphoblastic leukemia and was treated with prednisone, vincristine, and 6-mercaptopurine. Adults (A), eggs (E), and larvae (L) can be seen. Although there is some post-mortem autolysis, it is clear that little inflammatory reaction is present in this immunosuppressed patient. This is a recrudescent chronic infection. HE x 120

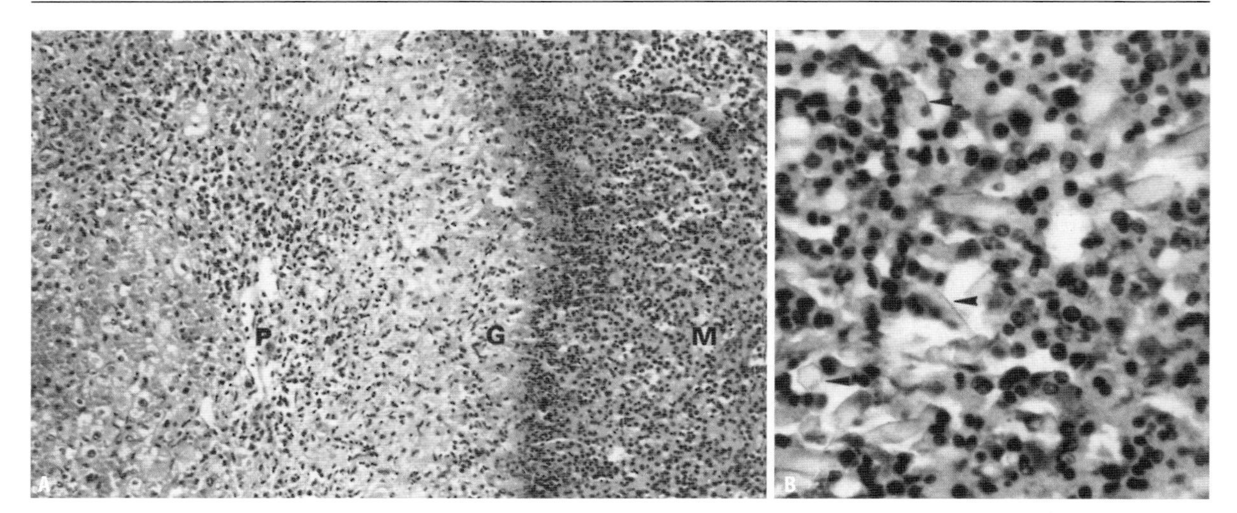

Fig. 3A,B. Eosinophil and basophil infiltration may be accompanied by Charcot-Leyden crystals. Biopsy of liver of a 33-year-old woman from Central America with hepatomegaly, who was passing eggs of *Fasciola hepatica* in the stools, and had a history of the ingestion of watercress. **A** Apparent migration tract (M) in liver, with eosinophilic granulocytes and Charcot-Leyden crystals. This is bordered by an epitheloid cell granuloma (G) and hemorrhagic liver parenchyma (P) infiltrated with lymphocytes and plasma cells. HE x 128. **B** Rhomboid and hexagonal profiles of Charcot-Leyden crystals *(arrowheads)* among eosinophil granulocytes. HE x 510.

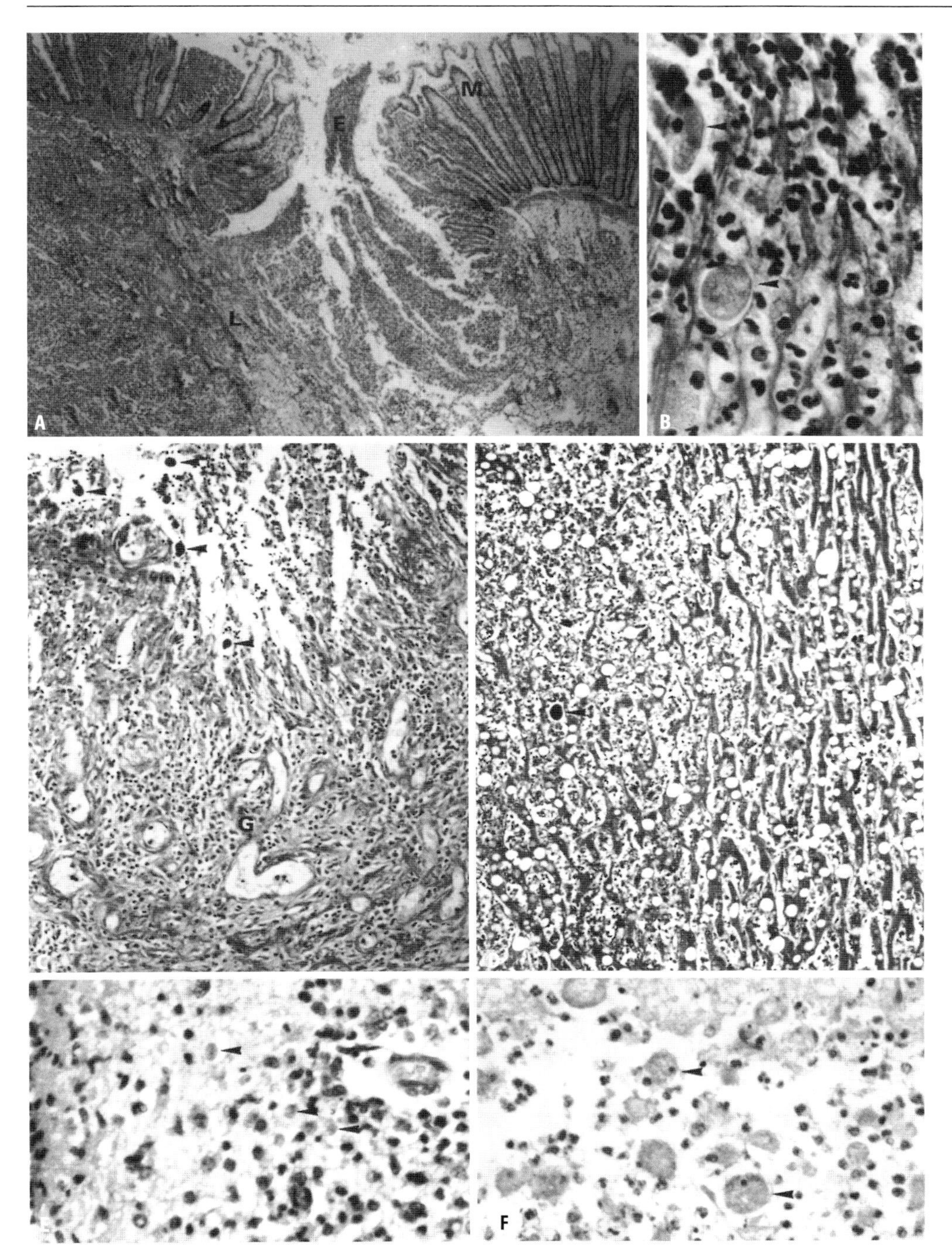

Fig. 4A-F. Infection necrosis of adjacent cells in amoebiasis. **A** Flask-shaped ulcer with narrow neck containing exudate *(E)* extending through mucosa (M), into lamina propria (L) of colon. HE x 36. **B** Fibrinous exudate with granulocytes and two trophozoites of *Entamoeba histolytica* (arrowheads). HE x 480. **C** Loose granulation tissue (G) forms the base of a deep ulcer extending through the muscularis externa of the colon. Trophozoites (arrowheads) stain intensely with the PAS technique. Several ulcers extended into the serosa and/or several perforated into the peritoneum. PAS x 120. **D** Margin of liver abscess with hepatitis and Fatty change. One trophozoite is embedded in necrotic liver accompanied by neutrophile leukocytes and macrophages. PAS x 120. **E** *Naegleria* sp. encephalitis with trophozoites (arrowheads) infiltrating the outer cortex in which neuronal nuclei undergo early pyknosis accompanied by granulocytes. The meninges are densely infiltrated with granulocytes. HE x 300. **F** Acanthamebic encephalitis with trophozoites (arrowheads) in necrotic brain accompanied by few granulocytes and macrphages HE x 300.

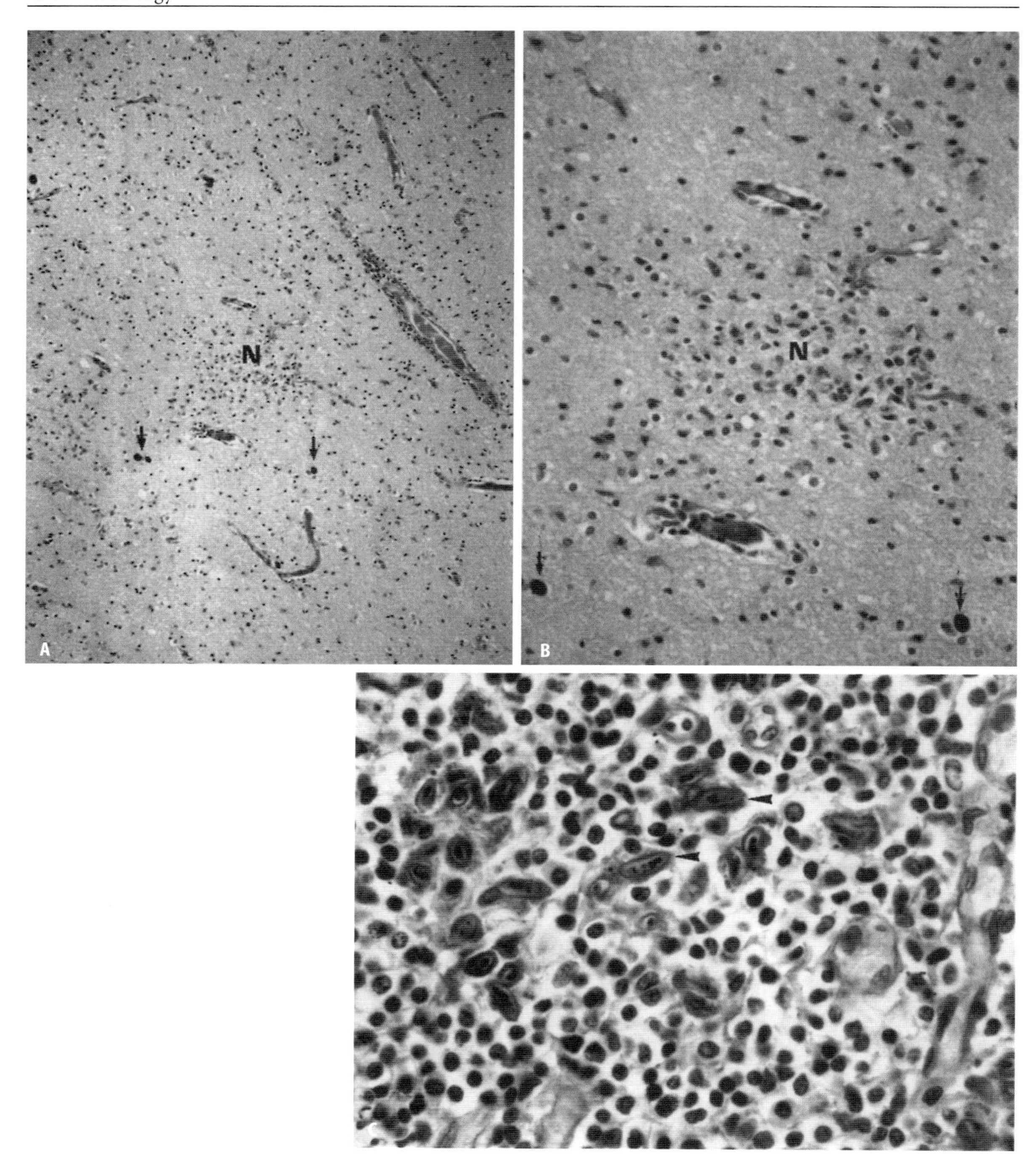

Fig. 5A-C. Lack of inflammatory reaction around protozoan cysts indicates impermeability of cyst or cell membrane to antigen; however, active inflammation follows cyst rupture. **A** Several intact *Toxoplasma gondii* cysts not accompanied by inflammation are indicated by arrows. A glial nodule *(N)* and perivascular infiltration are shown in the center PAS x 72. **B** Toxoplasmic encephalitis (enlargement o] A). Two *Toxoplasma* cysts devoid of inflammation. The glial nodule *(N)*, in the center, is probably formed from the rupture of a cyst. The liberated bradyzoites have been destroyed by the patient's immunity and no secondary infection with tachyzoites or young cysts was found. However, similar glial nodules accompany proliferating tachyzoites. PAS x 180. **C** Hypnozoites of *Cystoisospora belli* in a PAS positive envelope *(arrowheads)* in lymphocyte-depleted lymph node of a patient with AIDS. Slide courtesy of Dr. Carlos Restrepo. PAS x 480

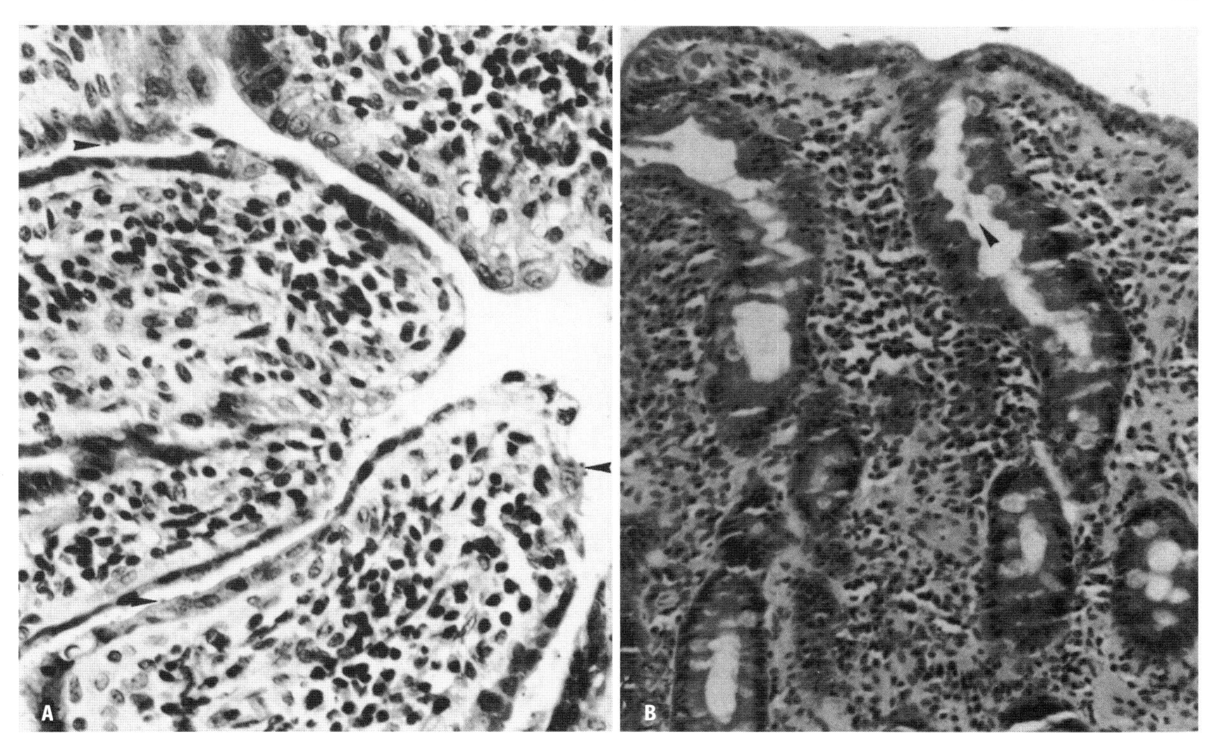

Fig. 6A,B. A Atrophy of epithelium due to attrition of cells. Chronic cryptosporidiosis in patient with AIDS, with flattened epithelial cells covering intestinal villi. Cryptosporidia indicated by *arrowheads*. HE x 375. **B** *Giardia* infection in a patient with AIDS shows shortened villi and plasma cell infiltration. HE. x 450. Slide by courtesy of Dr. Linda Ferrell

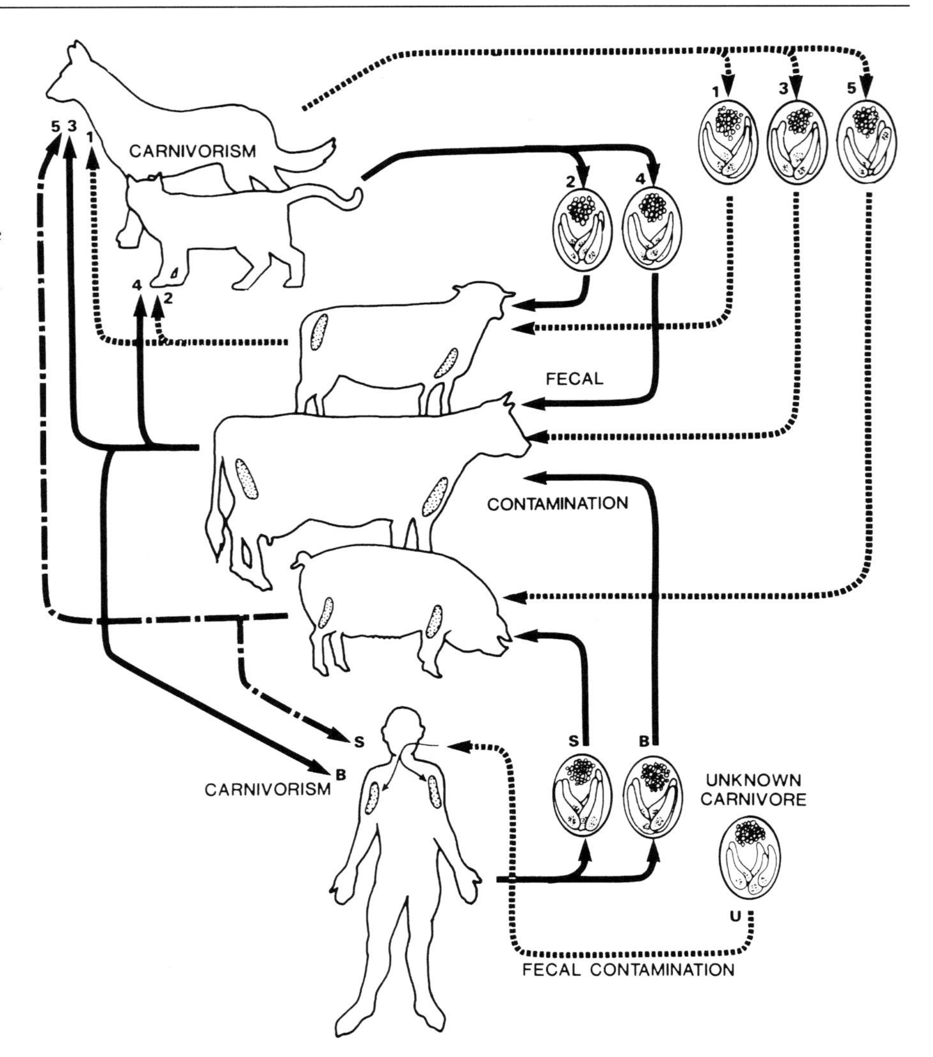

Fig. 7. *Sarcocystis* transmission cycles involves man and some domestic animals. Man is the definitive host of S. *bovihominis* **(B)** and S. *suihominis* **(S)**, the two human intestinal sarcosporidians. Man is the accidental intermediate host with *Sarcocystis* of unknown species **(U)**, where skeletal and heart muscles are parasitized. Also indicated are the cycles of S. *ovicanis (1)*, S. *ovifelis (2)*, S. *bovicanis (3)*, S. *bovifelis (4)*, and S. *suicanis (5)*. Compare *Sarcocystis* in Vol. 1.

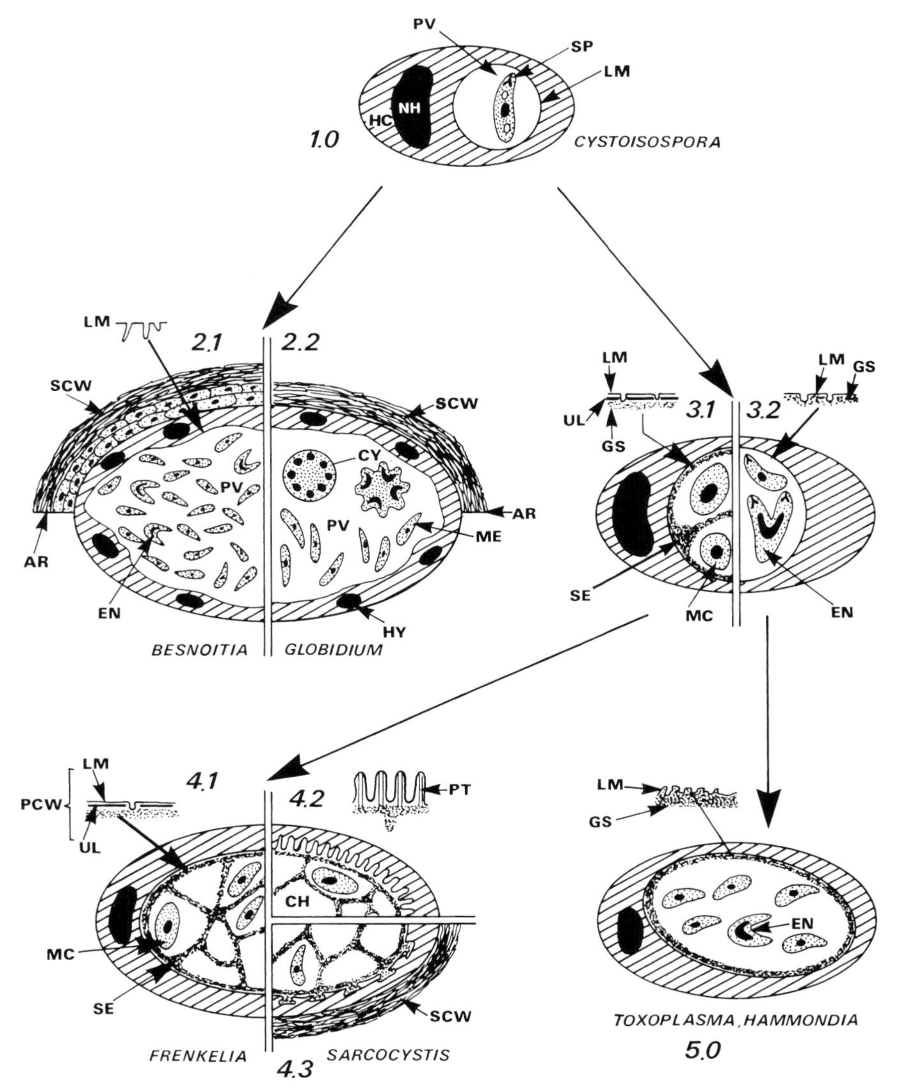

Fig. 8. Diagrammatic representation of cysts in different cyst-forming → coccidia (Vol. 1). 1 The simplest cyst formation. A parasite (sporozoite) is included into a parasitophorous vacuole (PV) which is bounded by a single cell membrane *(LM)*. This is representative of the "monozoic" cysts of *Cystoisospora felis*, C. *rivolta*, and C. *ohioensis* in transport (i. e., *paratenic)* hosts (such as mice). 2 In *Besnoitia* spp. (2.1) and *Globidium* spp. (2.2) cysts the original parasitophorous vacuole (PV) is enlarged and is filled by numerous parasites reproducing by *endodyogeny* (2.1) or *schizogony* (2.2). Even in old cysts the PV is bounded by a single unthickened cell membrane *(LM)*. A *secondary cyst wall (SCW)* consisting of fibrillar material is always present; the host cell nuclei generally undergo hypertrophy and hyperplasia. 3 Young cysts of *Frenkelia* spp. and *Sarcocystis* spp. (3.1), and *Toxoplasma gondii* and *Hammondia* spp. (3.2) show the indicated features. In cysts of *Frenkelia* spp. and *Sarcocystis* spp. (3.1) spherical *metrocytes (MC)* are present (in chamberlike spaces) and divide by endodyogeny, whereas in *Toxoplasma gondii* and *Hammondia* spp. the slender parasites divide by endodyogeny. 4 Mature tissue cysts of *Frenkelia* and *Sarcocystis* are characterized by typical septa *(SE)* formed by the ground substance (GS). In *Frenkelia* spp. and some *Sarcocystis* spp. (4.1) the primary cyst wall *(PCW)* never forms long protrusions, whereas in other *Sarcocysts* spp. typical protrusions occur (4.2; 4.3). With cysts of *S.ovifelis*, a secondary cyst wall *ECU)* surrounds the parasitized muscle fiber (4.3). 5 The primary cyst wall of mature *Toxoplasma gondii* and *Hammondia* spp. cysts remains smooth; the cysts are tightly filled with cyst merozoites (bradyzoites). Typical septa as well as metrocytes never occur. AR, Artificially interrupted SCW; *CH*, chamber-like space filled with parasites; *CY*, cytomere; *EN*, endodyogeny; *CS*, ground substance; *HC*, host cell; *HY*, hypertrophic host cell nuclei; *LM*, limiting single membrane of PV; *MC*, metrocyte; *ME*, merozoite; N, nucleus; *NH*, nucleus of host cell; *PCW*, primary cyst wall; *PT*, protrusion of PCW; *PV*, parasitophorous vacuole; *SCW*, secondary cyst wall; *SE*, septum formed by GS; *SP*, sporozoite; *UL*, underlying dense material. (From Mehlhorn and Frenkel 1980)

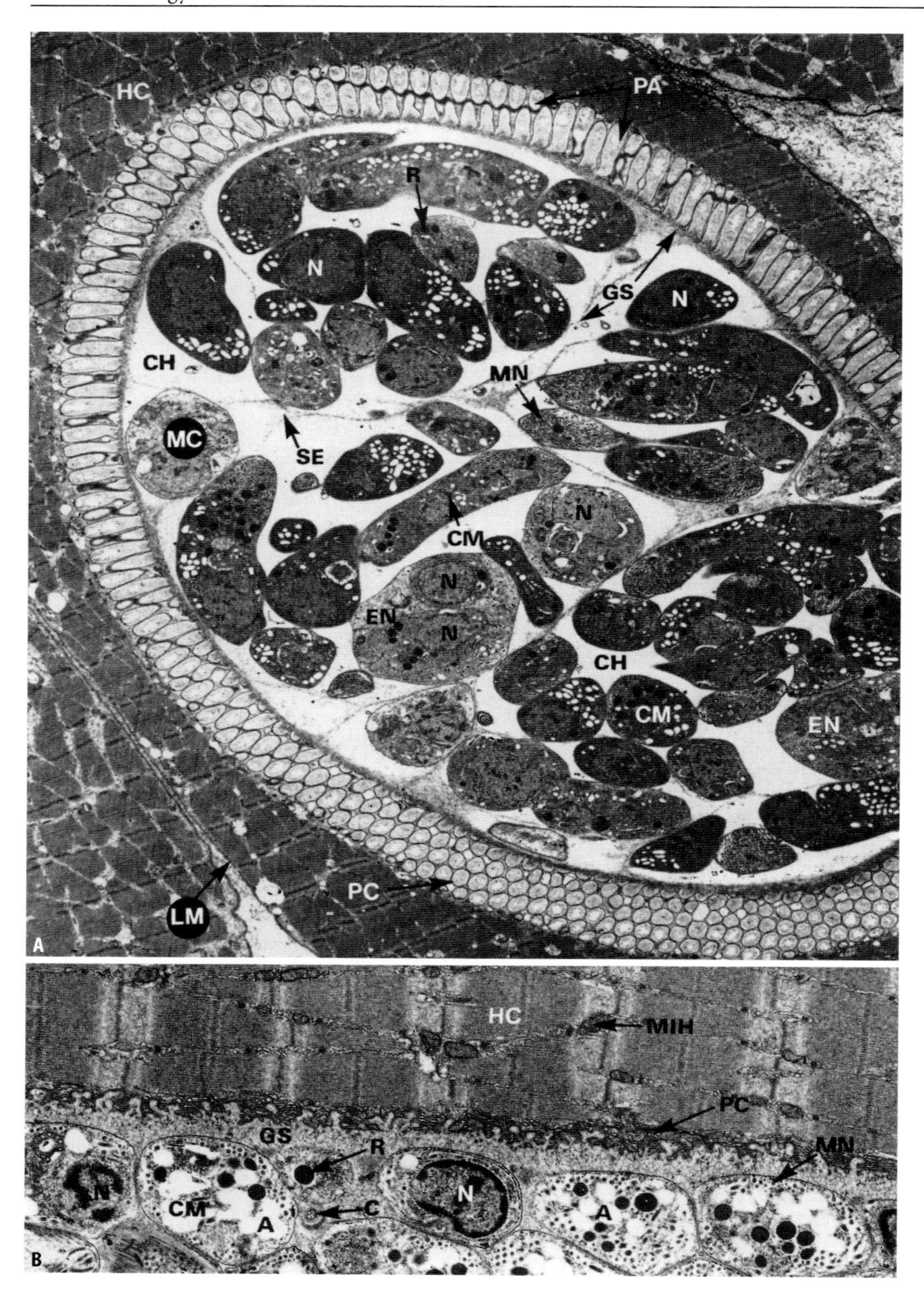

Fig. 9A,B. Transmission electron micrographs of section through tissue cysts of *Sarcocystis ovicanis* (**A**) and *Toxoplasma gondii* (**B**), which are situated within a host cell (for interpretation compare Fig. 8; from Mehlhorn and Frenkel 1980). x4.000, 8.000. *A*, Amylopectin; *C*, conoid *CH*, chamber-like space; *CM*, cyst merozoite (=bradyzoite) *EN*, endodyogeny stage; GS, ground substance; *HC*, host cell; *LM*, limiting membrane; *MC*, metrocyte; *MIH*, mitochondrion of the host cell; *MN*, micronemes; *N*, nucleus; *NH*, nucleus of the host cell; *PA*, palisade-like protrusions of the PC; *PC*, primary cyst wall; *R*, rhoptries; *SE*, septum formed by ground substance

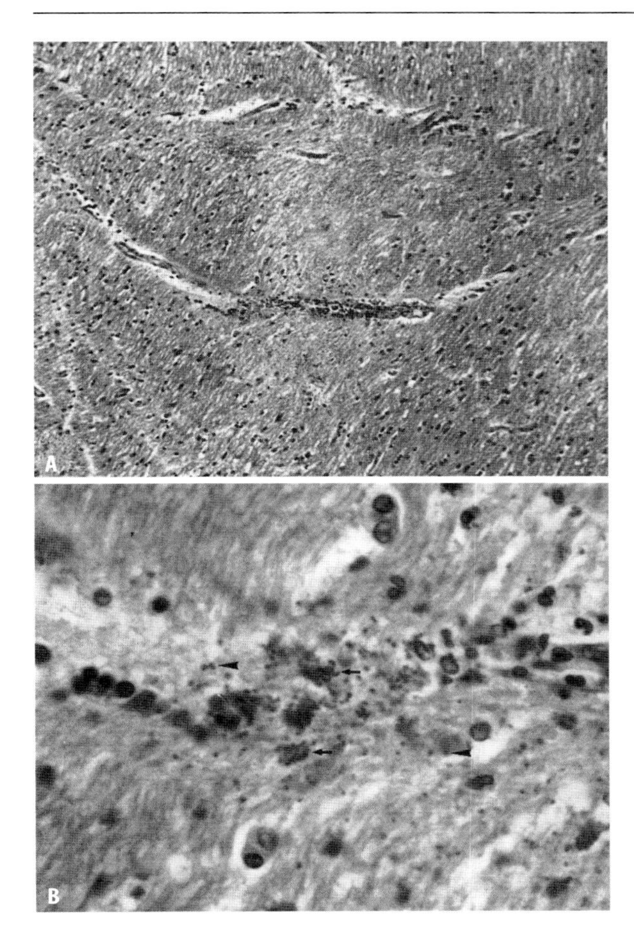

Fig. 10A,B. Necrosis of parasitized cells. **A** Recrudescent toxoplasmic brain in patients with Hodgkin's disease treated with 8–12 replacement doses of prednisone and with cyclophosphamide. The central area devoid of nuclei, above and below the blood vessel, contains numerous *Toxoplasma gondii.* HE x 90. **B** Enlargement of A showing the area of cell necrosis with intracellular (arrows) and scattered *Toxoplasma gondii* (arrowheads). HE x 570. (From Frenkel 1971)

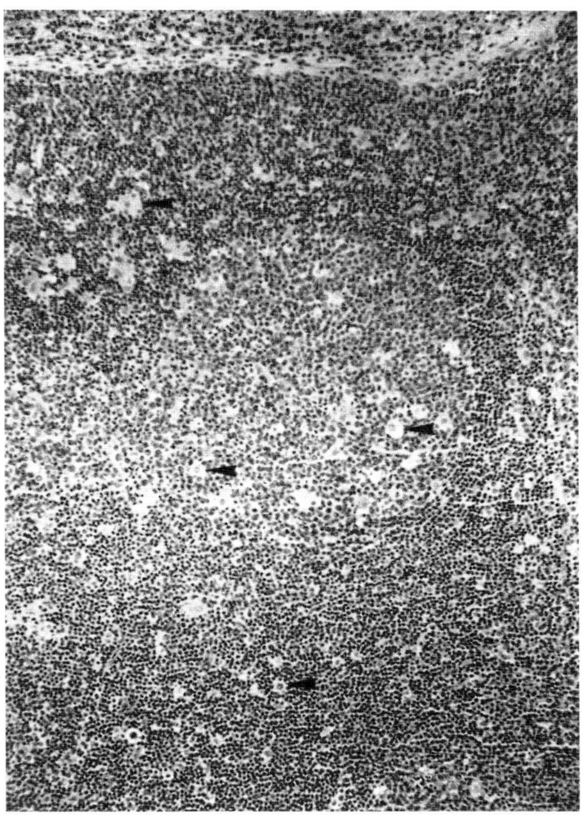

Fig. 11. Lymphoreticular hyperplasia accompanying acquisition of immunity. Cervical lymph node of 21-year-old women who had delivered a toxoplasmic baby 6 weeks earlier. The clear cells are histiocytes (arrowheads). HE x 100. (From Frenkel 1971)

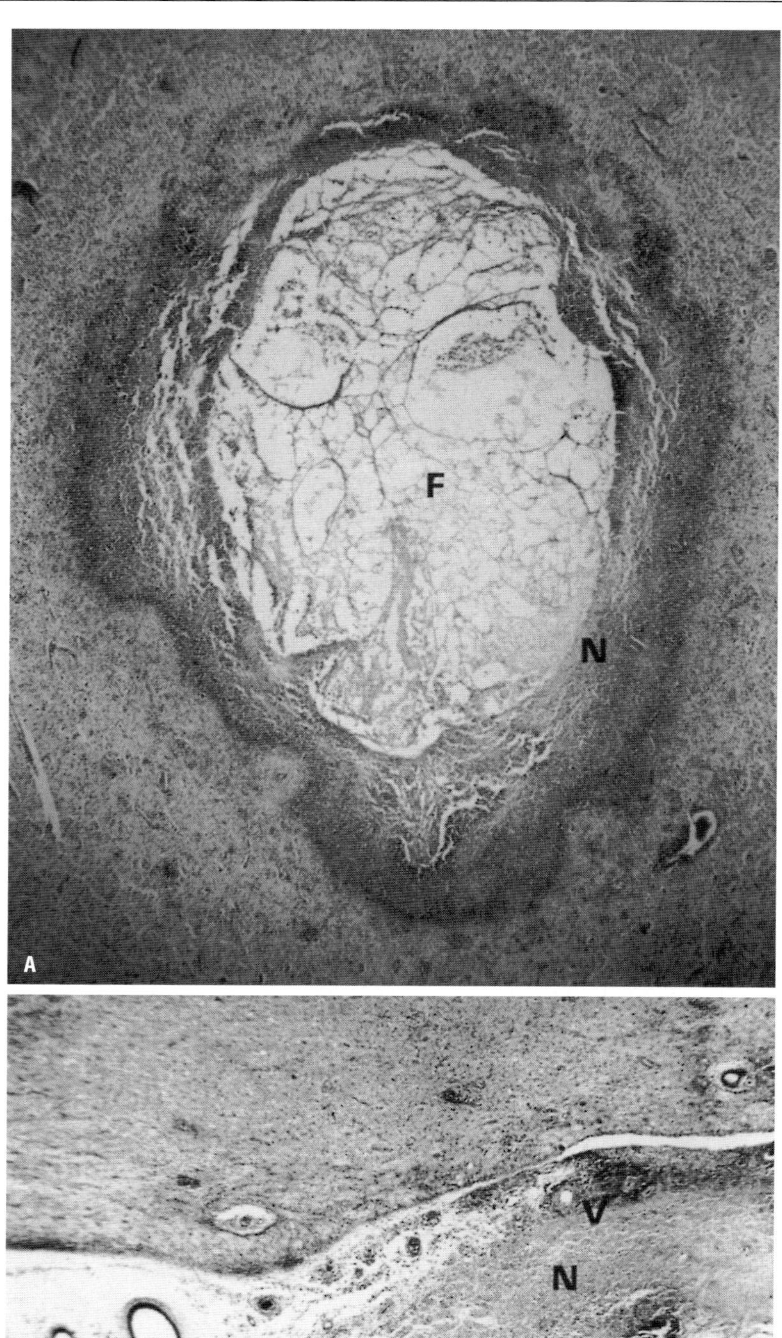

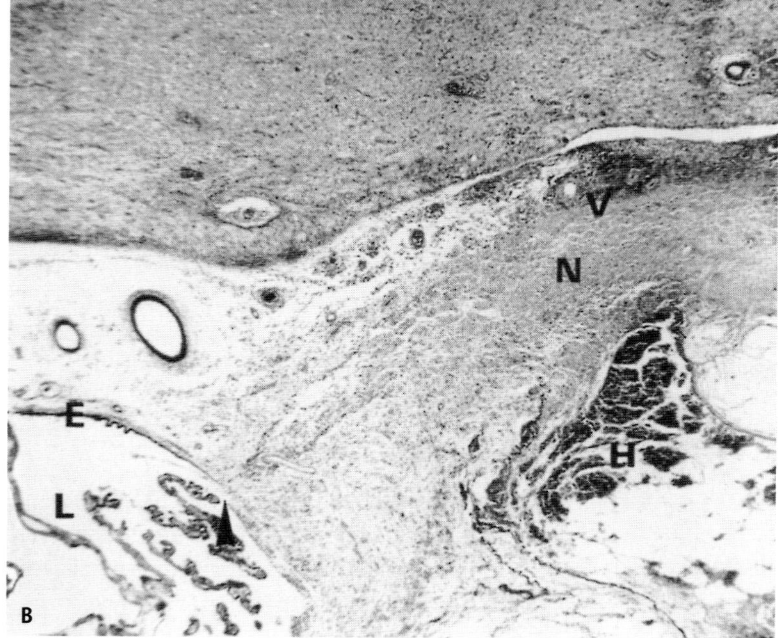

Fig 12A,B. Periventricular necrosis is a unique and pathognomonic lesion of congenital toxoplasmosis. **A** Aqueductal obstruction with fibrinous (E) exudate and periaqueductal necrosis *(N)* at level of pons in 6-week-old infant. HE x 17. **B** Transition from ependyma (E)-lined lateral ventricle (L) to periventricular vasculitis (v) and necrosis *(N)*. From the edge of the ependymal epithelium (arrowhead), the ulcer shows an increasing basophilia and vasculitis with necrosis at right. Hemorrhage *(H)* and necrotic brain *(N)* and vascular leakage account for the yellowish ventricular fluid with high protein content typically found in such babies. Phosphotungstic acid hematoxylin x 17. (From Frenkel 1971)

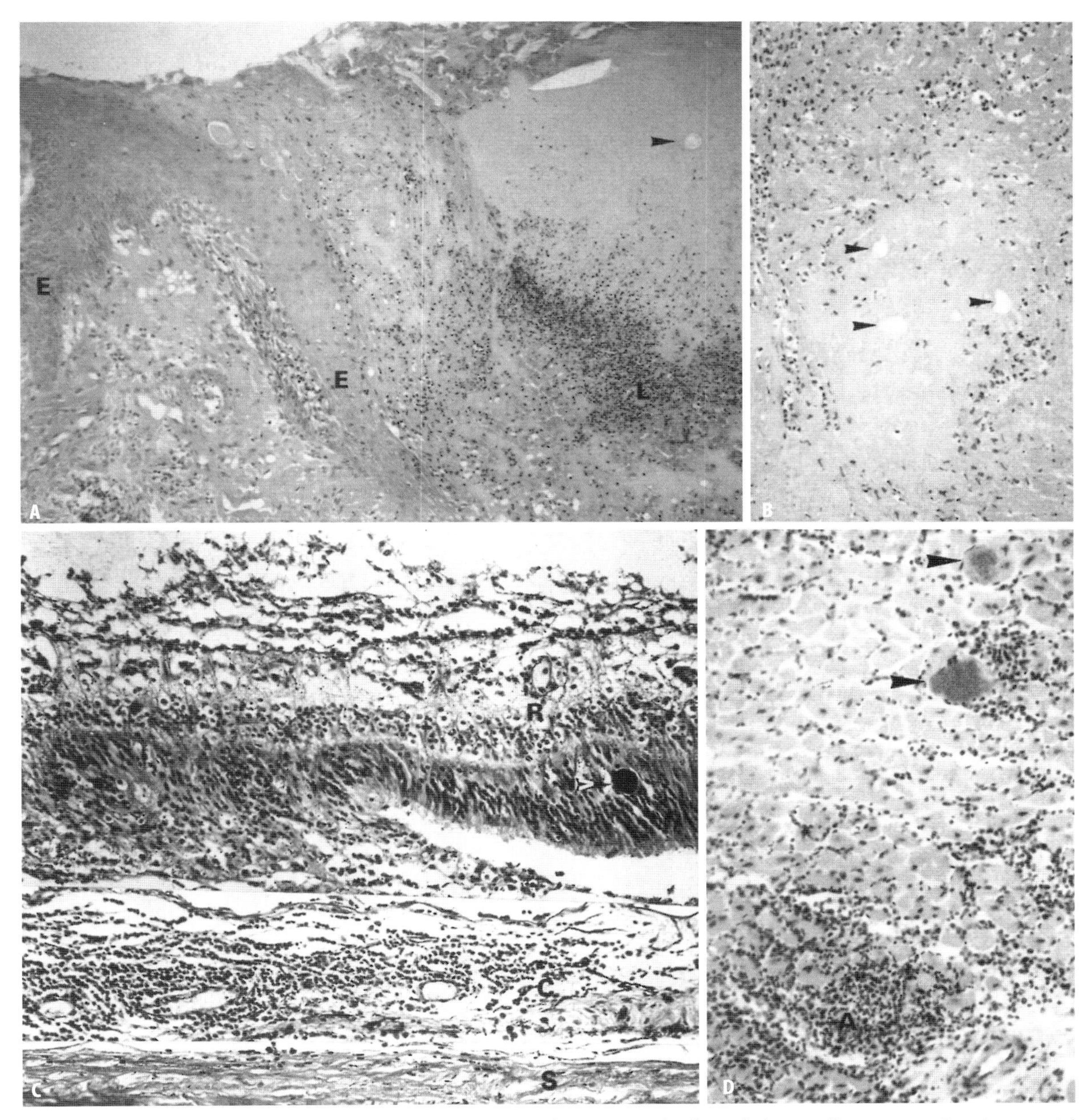

Fig. 13A-D. Delayed-type hypersensitivity inflammation with necrosis of cells and tissue adjacent to antigenic material or parasite. **A** Tick bite lesion 4 months after removal of tick, (probably *Dermacentor variabilis)* from knee of 30-year-old male. The skin ulcer shows a necrotic scab (top right) covering the ulcer which contains a chitinous fragment (arrowhead) from mouth parts of the tick. At the base of the ulcer are leukocytes (L) undergoing necrosis, and at its margin acanthosis of epithelial rete pegs (E), hyperkeratosis, and parakeratosis. Deep to the ulcer chronic inflammation and fibrosis extend through the dermis into the subcutaneous connective tissue. HE x 85. **B** At a depth of 2 mm, three other chitinous fragments surrounded by necrosis (arrowheads) and granulocytic inflammation are found in the same sections. HE x 128. **C** Toxoplasmic retinochoroiditis (left) probably following rupture of *Toxoplasma gondii* cyst, which when intact (arrowhead) was not chemotactic. R, Retina; C, choroid; S, sclera. PASx128 (From Frenkel 1971). **D** Chagas' myocarditis with two intact pseudocysts *arrowheads)* and neutrophilic abscesses (A) and destruction of myocardial fibers probably from rupture of a pseudocyst. There is focal and diffuse lymphohistiocytic infiltration with some plasma cells. HE x128. For additional examples see Fig. 1C, Fig. 27, Fig. 28

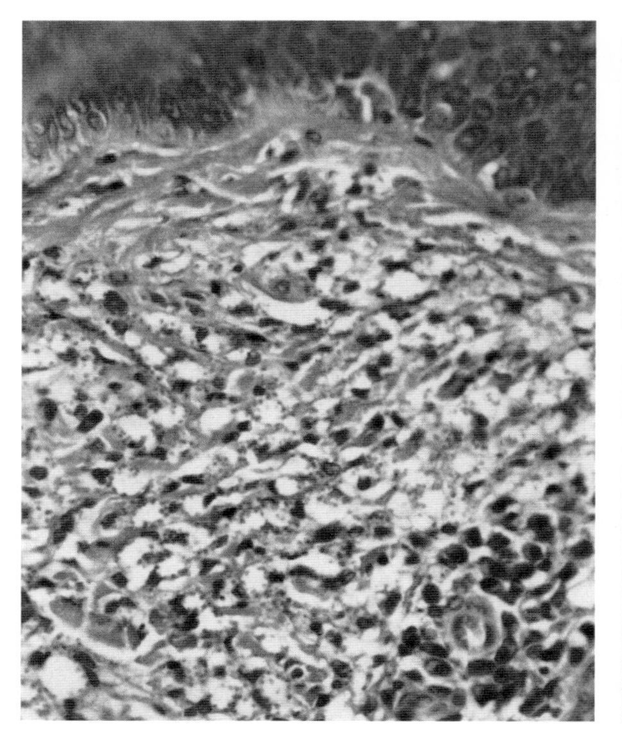

Fig. 14. Histiocytic reaction with relative anergy. Diffuse cutaneous leishmaniasis gives rise to nonulcerating raised nodules composed of histiocytes each of which contain numerous parasites in each of the many vacuoles visible. The epidermis is slightly stretched. A plasma cell infiltration is present around the blood vessels. HE x 350

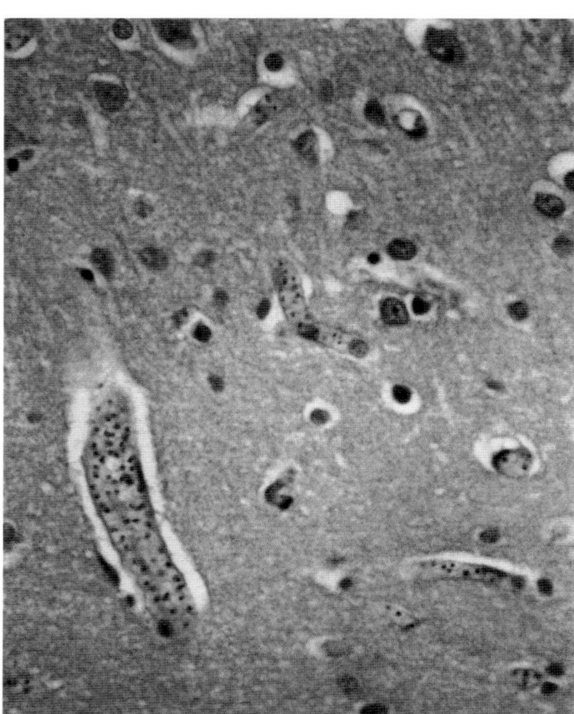

Fig. 15. Tissue anoxia from obstruction of capillaries with erythrocytes infected with *P. falciparum*. The capillaries in the brain of this young male are filled with parasitized erythrocytes each marked by its pigment granule. The flow of parasitized blood has been compared to the flow of sludge. Hence the brain becomes anoxic and edematous, sometimes weighing 1.700 g instead of the normal 1.200 HE x 375

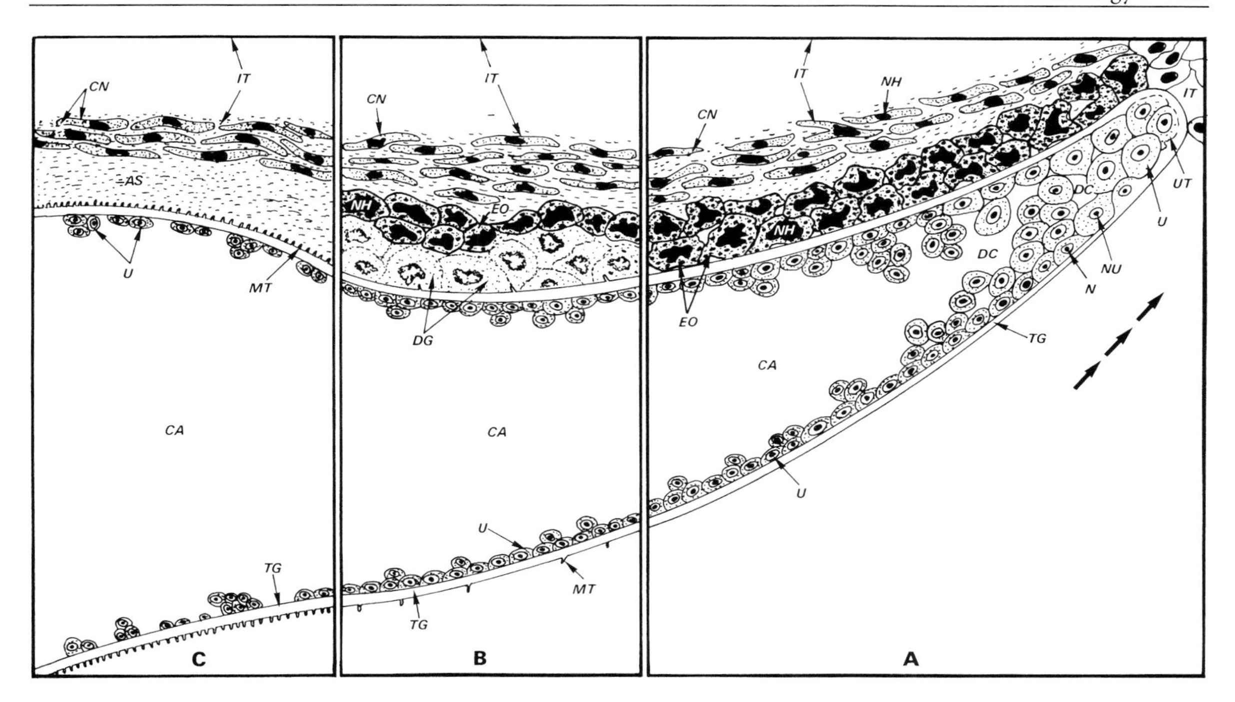

Fig. 16A-C. Diagrammatical presentation of longitudinal section in different regions (**A**-**C**) of parasite protrusions of larval *E.multilocularis*. Such steps of development were observed in all material studied (in experimentally infected rodents and in natural infections of humans). Arrows in **A** indicate the direction of growth. In section **C** formation of brood capsules may start from accumulation of undifferentiated cells (From Mehlhorn et al. 1983). *AS*, Amorphous substance (=laminated layer); *CA*, cavity; *CN*, connective tissue; *CO*, collagen; *DC*, developing cavity; *DG* degenerating host defense cells; *Dl*, division of undifferentiated cells; *KG*, eosinophilic granules; *EO*, eosinophilic granulocvtes: *GR*, granules: *IF*, infiltration zone of host's defense cells; *IT*, intact tissues; *M*, membranes of fusing undifferentiated cells; *Ml*, mitochondrion; *MT*, microtriches of tegument; *N*, nucleus; *NH*, nucleus of host cells; *NU*, nucleolus; *PT*, protrusion of tegumental surface; *TG*, tegument; *U*, undifferentiated cells; *UT*, undifferentiated cells when fusing with tegument; *V*, vacuole

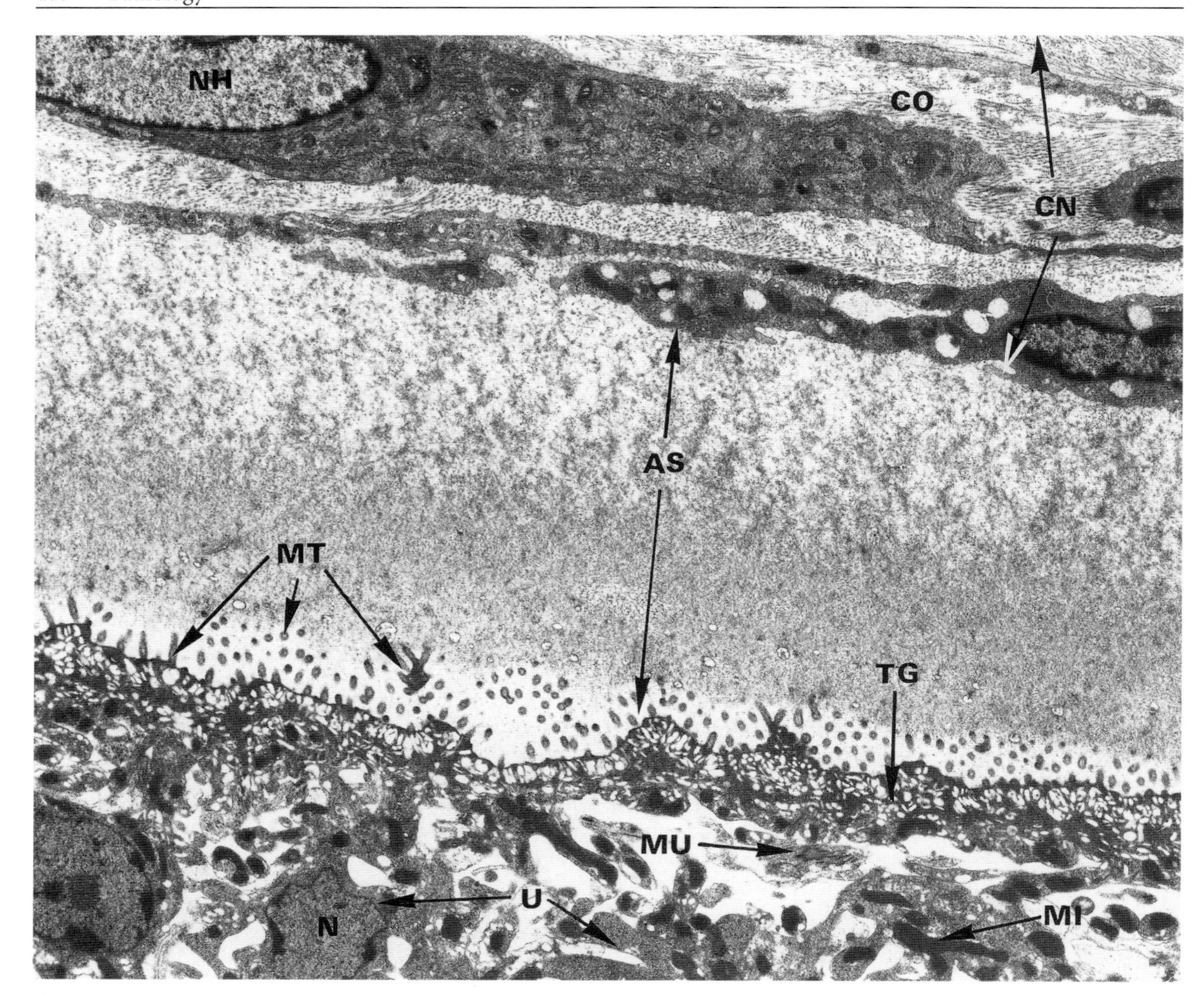

Fig 17. *Echinococcus multilocularis*: transmission electronmicrograph of a section through the periphery of a "cyst from rodents." Note the occurrence of two surrounding layers of host origin (AS, CN) (From Mehlhorn et al. 1983) x2.500. *AS*, Amorphous substance; *CN*, connective tissue; *CO*, collagen; *MI*, mitochondria; *MT*, microtriches; *MU*, muscle cell; *N*, nucleus; *NH*, nucleus of the host cell; *TG*, tegument; *U*, undifferentiated cells

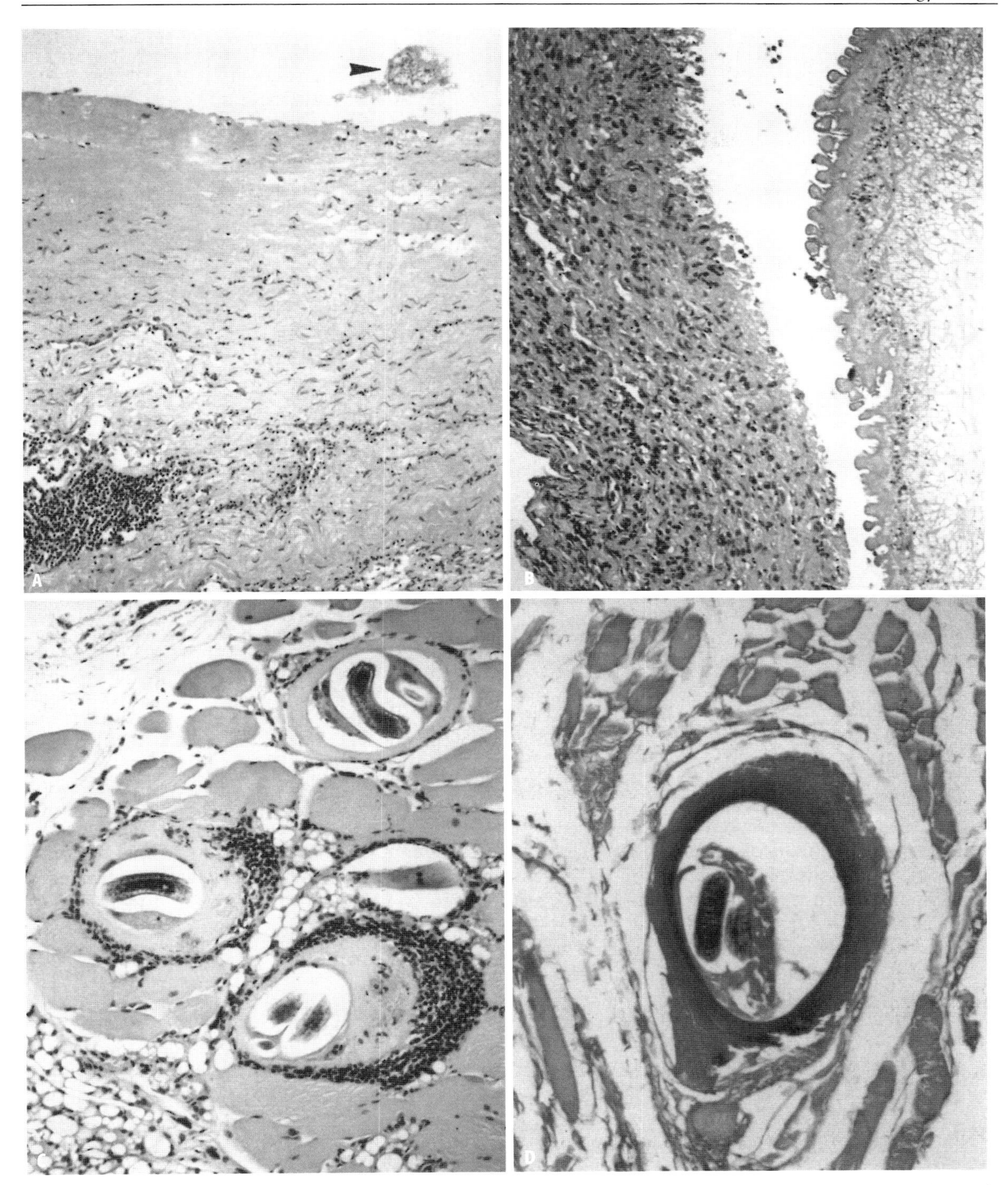

Fig. 18A-D. Encapsulated parasites and the production of fibrosis. **A** The brood capsule of E. *granulosus* in the lung of a 10-year-old Eskimo girl is surrounded by a thick connective tissue capsule. A lymphocytic nodule is present in the outer capsule and a protoscolex in the lumen (arrowhead). HE x 100. **B** Sparganosis. A plerocercoid larva is shown in a subcutaneous nodule from the neck of a Mexican male. The thick capsule is collagenous with eosinophil, lymhocyte, and plasma cell infiltration. HE x 120. **C** Intracellular encapsulation of *Trichinella spiralis* in skeletal muscle of mouse. Some of the larvae are accompanied by slight lymphocytic infiltration. HE x 120. **D** Calcified *Trichinella* sp. and capsule in skeletal muscle of 50-year-old man who died from an unrelated disease. This was found many months or years after a primary infection, of which no history could be obtained. HE x 120

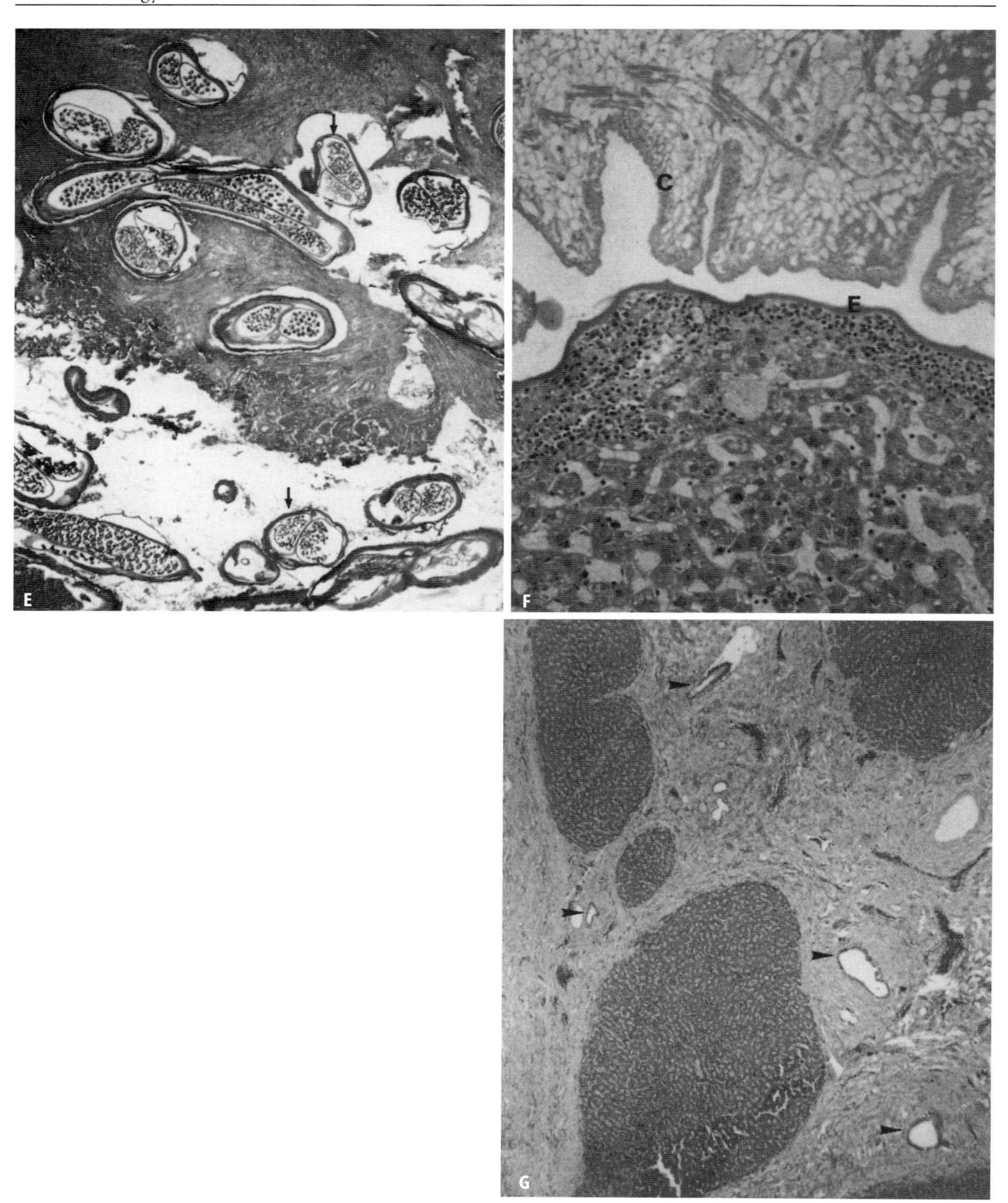

Fig. 18E-G. Encapsulated parasites and the production of fibrosis (continued). **E** A densely fibrotic nodule encloses adult *Onchocerca volvulus* shown in longitudinal and cross sections. Several contain microfilariae (arrows). Such nodules are surgically removed, for example in Guatemala, to reduce the load of microfilariae produced. HE x 35. **F** Nymph of *Porocephalus* sp. thinly encapsulated *(E)* in the liver of a marmoset monkey, accompanied by lymphocytic infiltration. In humans encapsulation in the omentum is more common. Nuclei of cuticular cells (c) of parasite stain more weakly than host cell nuclei. HE x 118. **G** Pipe stem fibrosis in liver of patient with schistosomiasis of 17 years duration. Regenerating liver lobules are separated by dense fibrous bands containing bile ducts (arrowheads). Masson x 35

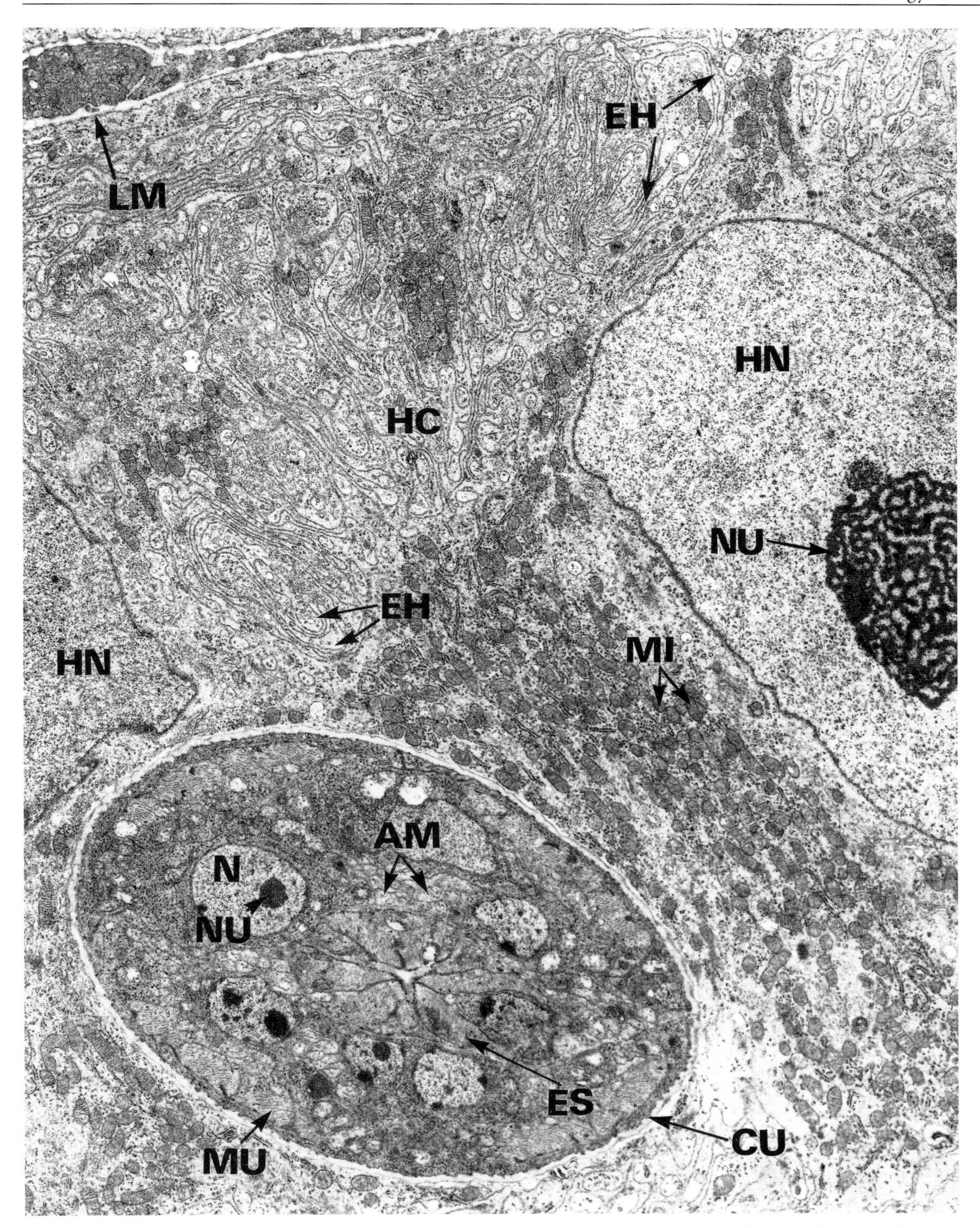

Fig. 19. *Trichinella spiralis:* transmission electron micrograph of a larva within an altered muscle fiber, which is not yet surrounded by a capsule of connective tissue. Note the occurrence of large strands of mitochondria *(Ml)* and endoplasmatic reticulum.x 7.000 (Original: Mehlhorn and Niechoj). *AM,* Amphids; *CU,* cuticle; *EH,* enlarged endoplasmic reticulum; *ES,* esophagus; *HC,* host cell; *HN,* hypertrophied host cell nucleus; *LM,* limiting membrane of the host cell; *Ml,* mitochondria; *MU,* muscle cell containing larva; *N,* nucleus; *NU,* nucleolus

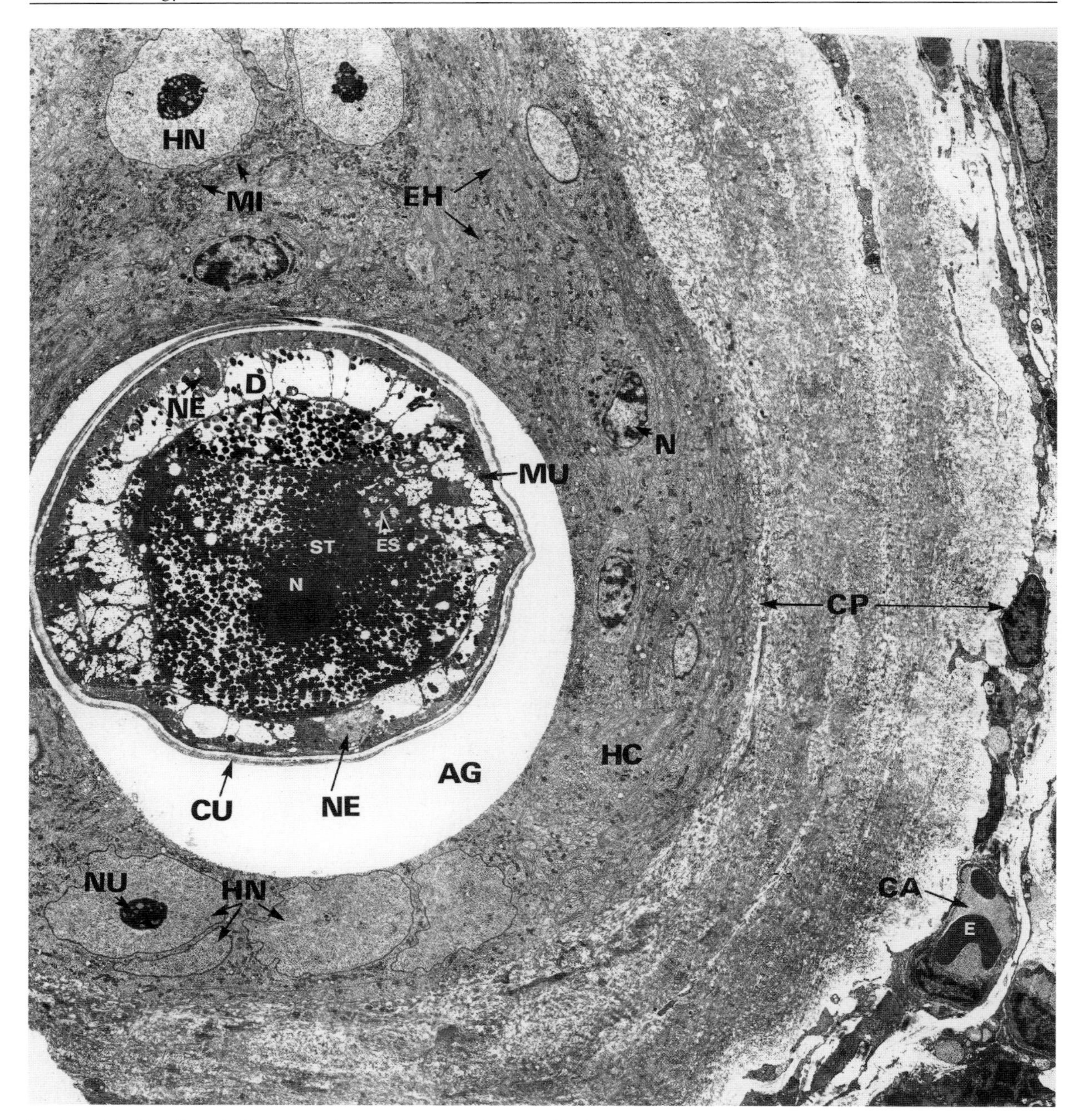

Fig. 20. *Trichinella spiralis:* transmission electron micrograph of an encysted larva. Note the complete alteration of the former muscle fiber and the surrounding capsule formed by collagen and layers of the connective tissue. Some host cell nuclei hypertrophied. For size compare the erythrocyte.x 5.000 (From Mehlhorn and Niechoj). *AG*, gap due to shrinking of larva; *CA*, capillary; *CP*, capsule of host tissue; *CU*, cuticle of the larva; *D*, droplets of stichosome gland; *E*, erythrocyte; *EH*, enlarged endoplasmic reticulum; *ES*, esophagus; *HC*, host cell; *HN*, hypertrophied host cell nucleus; *MI*, mitochondria; *MU*, muscles of the larva; *N*, nucleus; *NE*, nerve cord; *NU*, nucleolus; *ST*, stichosome

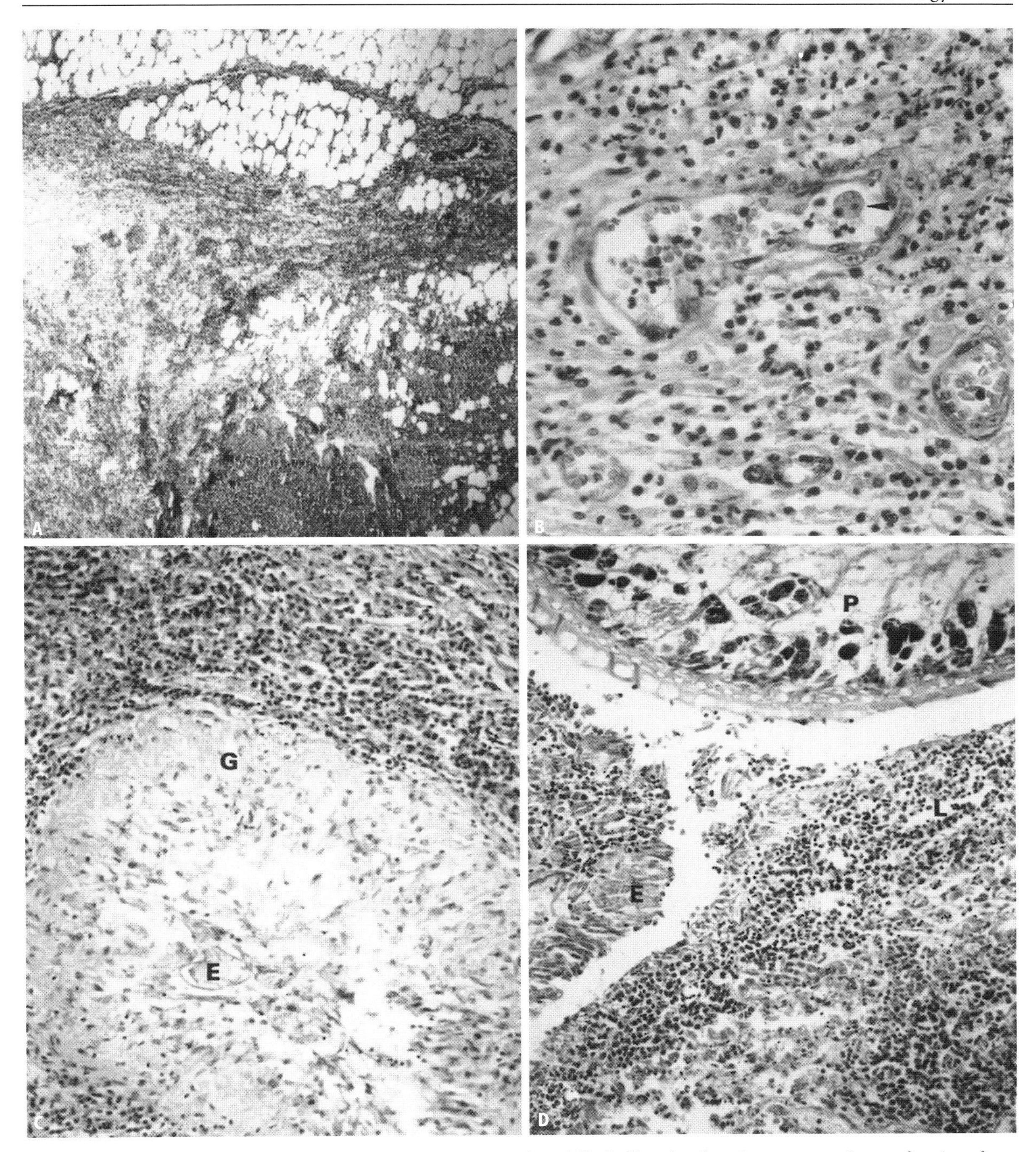

Fig. 21A-D. Parasite migration through tissues. **A** Intense eosinophilic infiltration in subcutaneous tissue of patient from Central America who passed *Fasciola hepatica* eggs, and had a history of eating watercress. HE x 35. **B** *Entamoeba histolytica* in portal vein radicle near the base of the ulcer shown in Fig. 4A. Such amebae are swept into the liver and may initiate amebic hepatitis and abscesses (Fig. 4D). HE x 290. **C** Fibrosing granuloma (G) with egg *(E)* of *Fasciola hepatica* in subcutaneous nodule. HE x 118. **D** *Paragonimus* sp. (P) in bronchus accompanied by lymphocyte (L) and plasma cell infiltration. The bronchial epithelium (E) was desquamated in the main bronchus occupied by the worm but preserved in a branch bronchus. HE x 118

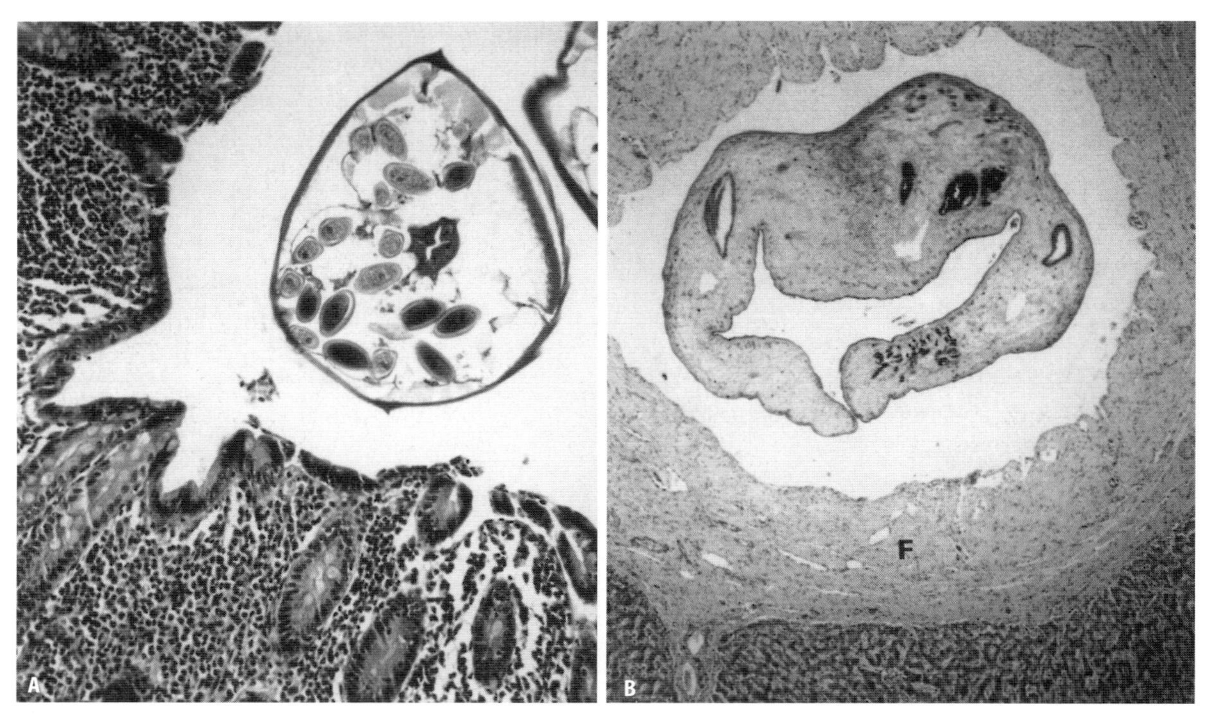

Fig. 22A,B. Absent or minimal inflammatory reaction accompanying luminal worms. A *Enterobius vermicularis* in lumen of appendix. There is no apparent inflammatory reaction in the mucosa. HE x 120. B *Clonorchis sinensis* adult in intrahepatic bile duct showing slight fibrosis (F). Patient had left endemic area 25 years prior to his death. Giemsa x 35

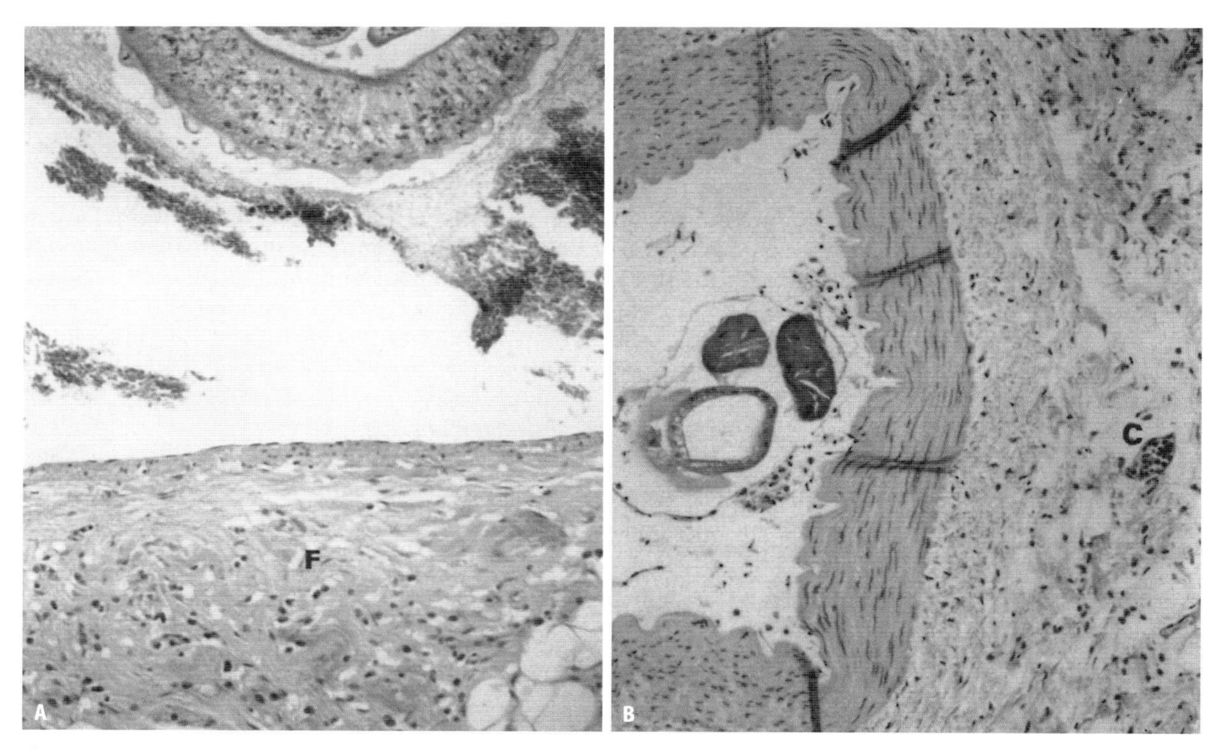

Fig. 23A,B. Intravascular persistence of helminths. A *Schistosoma mansoni* adults in mesenteric vessel showing slight fibrosis (F). HE x 120 .B *Angiostrongylus costaricensis* adult in mesenteric artery branch which appears normal. However, the capillary (C) is filled with eosinophils. HE x 120

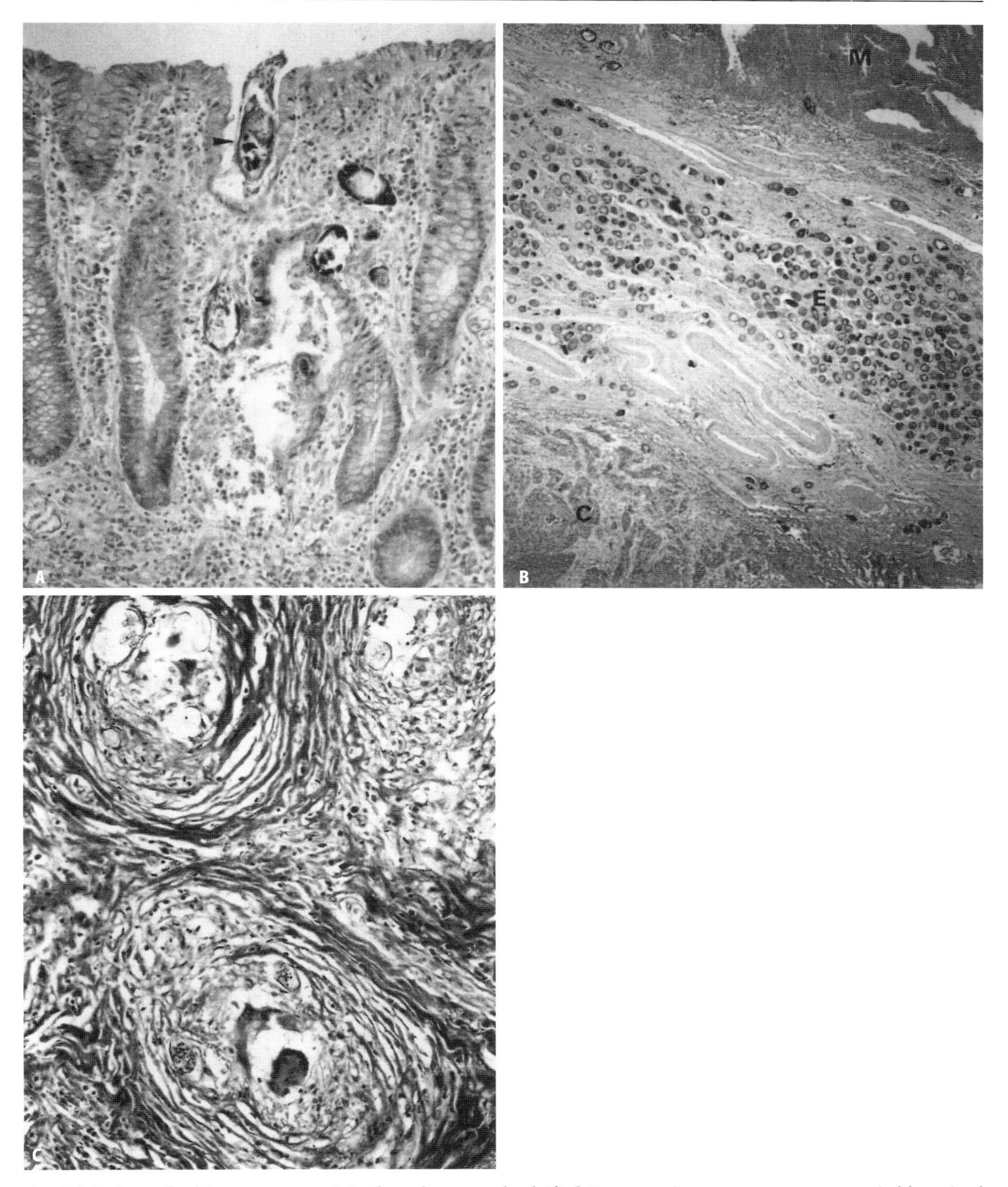

Fig. 24A-C. Fate of schistosome eggs. **A** In the colon several calcified *S. mansoni* eggs are seen accompanied by mixed inflammation. One egg, although nonviable and surrounded by a giant cell, is seen in a mucosal gland and is about to enter the lumen (arrowhead). Other eggs are in the lamina propria accompanied by mononuclear inflammation and in the submucosa surrounded by giant cells. HE x 120. **B** Under the bladder mucosa (M) is a dense plaque composed of masses of calcified *S. haematobium* eggs (E) which are accompanied by fibrosis. A poorly differentiated carcinoma of the bladder (C) is shown below. HE x 35. **C** In the brain are seen three fibrosing granulomas surrounding S. *haematobium* eggs. Masson x 120. For further examples see Fig. 1.

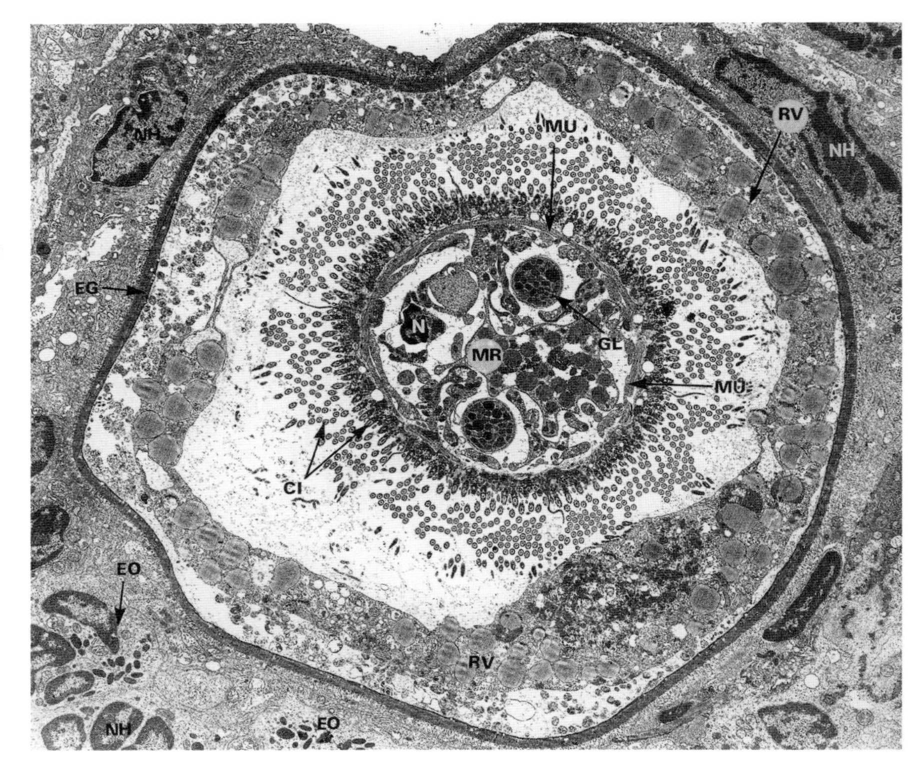

Fig. 25. *Schistosoma mansoni:* transmission electron micrograph of a section through an intact egg containing a fully developed miracidium *(MR)* within the host's liver. Note that the eggshell *(EG) is* closely surrounded by host defense cells *(EO),* the disintegration of which leads to the formation of the granuloma. (Original: Mehlhorn and Bettenhäuser). x 4.000. *Cl,* Cilia of the miracidium; *EG,* eggshell; *EO,* eosinophilic granulocytes; *MR,* miracidium; *MU,* muscle cell layer; *N,* nucleus; *NH,* nucleus of host cell; *RV,* remnants of vitelline cells

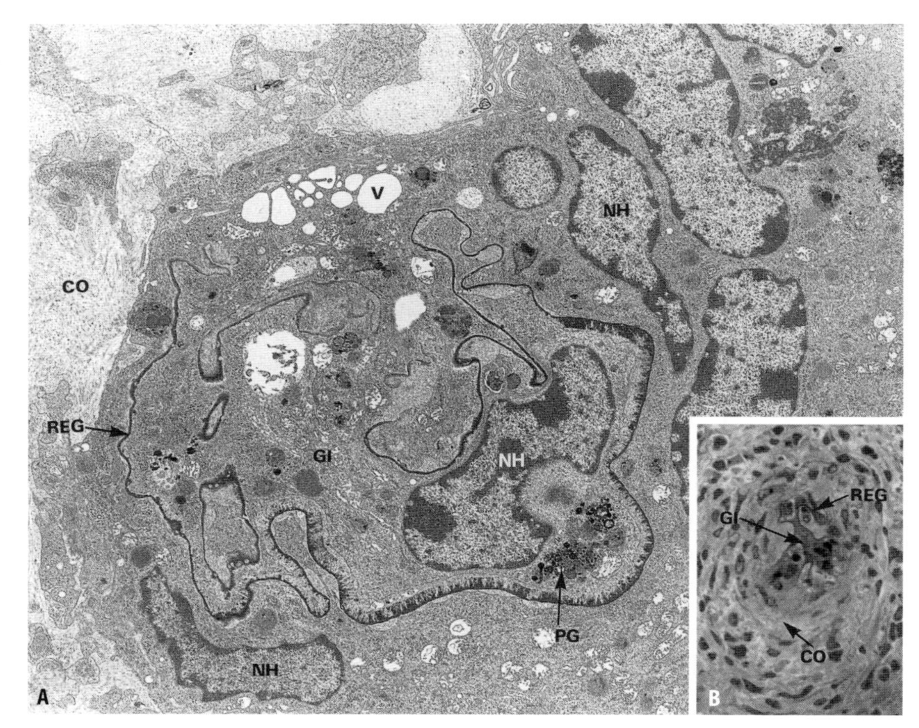

Fig. 26A,B. *Schistosoma mansoni* late stage granulomas. **A** Transmission electron micrograph; **B** light micrograph. The egg is nearly completely dissolved. Only remnants of the former eggshell *(REG)* and cytoplasmic residuals *(PG)* are seen within a multinucleate giant cell *(Gl).* Note that in the egg remnants are closely surrounded by collagen and concentric layers of host cells. (From Mehlhorn and Bettenhäuser). A x 3.500, B x 350. *CO,* Collagen; *Gl,* giant cell; *NH,* nucleus of the host cell; *PG,* pigment (=remnants of egg cytoplasm); *REG,* remnants of the eggshell; *V,* vacuole

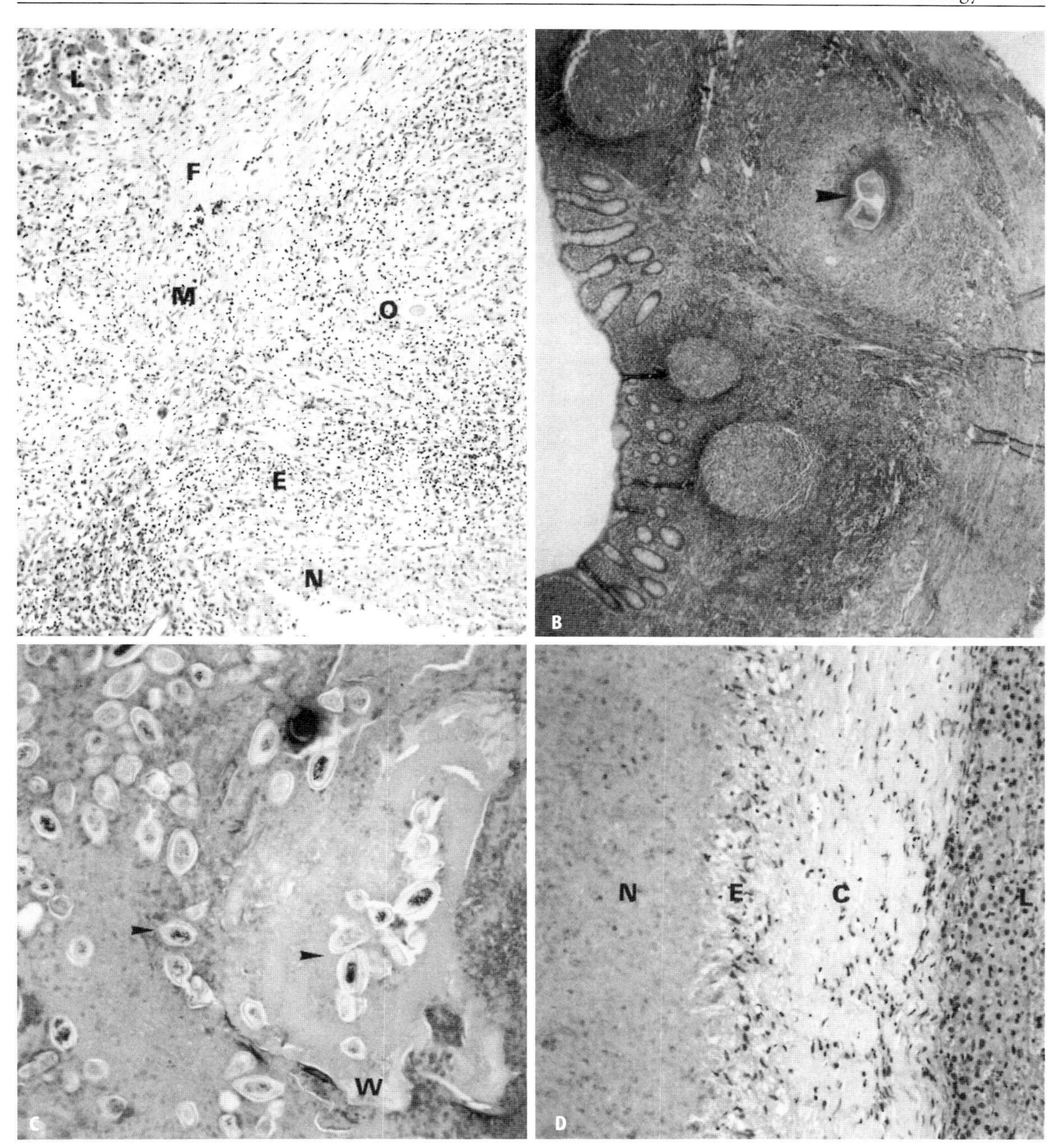

Fig. 27A-D. Aberrant migration of worms. **A** *Ascaris sp.* migration tract through the liver (L) surrounded by necrosis (N), eosinophilic infiltration (E), and mononuclear inflammation (M) with ovum (O) and fibrosis (F) next to liver parenchyma (upper left). HE x 80. **B** Aberrant migration probably of adult *Enterobius* sp. in the wall of appendix. Surrounding the dead worm *(arrowhead)* there is a zone of necrosis, followed by palisading epithelioid reaction and eosinophils. The rest of the appendix appears normal. HE x 35. **C** A later stage of *Enterobius* migrating through liver can be identified by the eggs (arrowheads) deposited. The cuticle of a dead worm (W) can also be seen. This was located in the center of a necrotic nodule 6 mm in diameter, produced by delayed hypersensitivity. See **D** for periphery. HE x120. **D** Periphery of lesions shown in **C**. The central necrosis (N) is surrounded by a thin zone of epithelioid (E) cells and a dense connective tissue capsule (C), separating the dead worm from normal liver (L). HE x 120

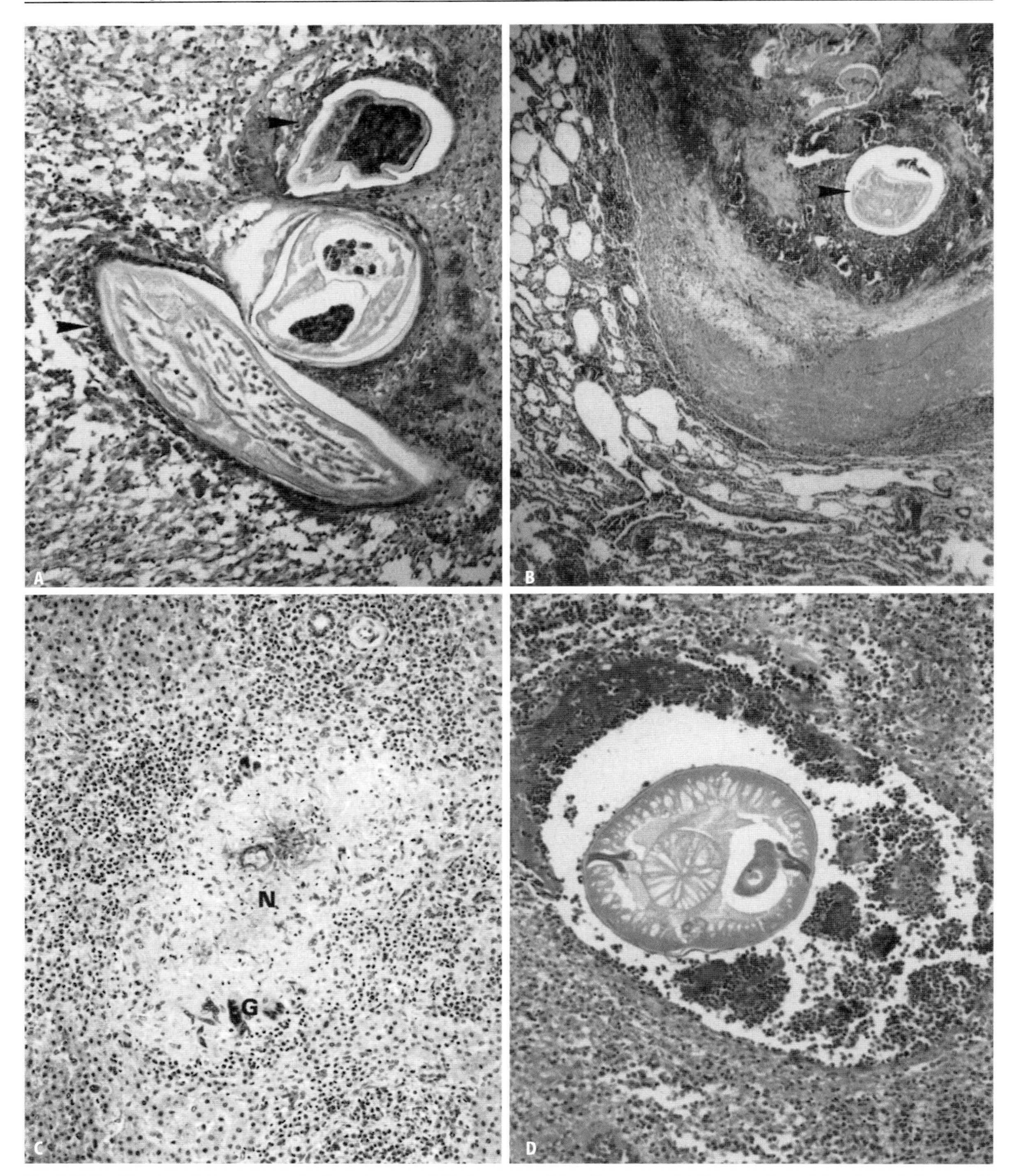

Fig. 28A-D. Dead or dying worms leak antigen and elicit a necrotizing hypersensitivity reaction. A *Wuchereria bancrofti*, two dead (arrowheads) and one viable, in lympnode surrounded by fibrin and necrosis. x 12. B Dead *Dirofilaria immitis* (arrowhead) in pulmonary artery embedded in thrombus (T). HE x 35. C Probable *Toxocara* hepatitis in child with history of eating soil. Focal granuloma with giant cells (G) and central necrosis (N). Eosinophils are in periphery. HE x 120. D Anisakiasis. Dead adult in wall of segment of ileum resected from a 20-year-old man surrounded by eosinophilic abscess. Slide courtesy of Dr. Tomo Oshima. HE x 100

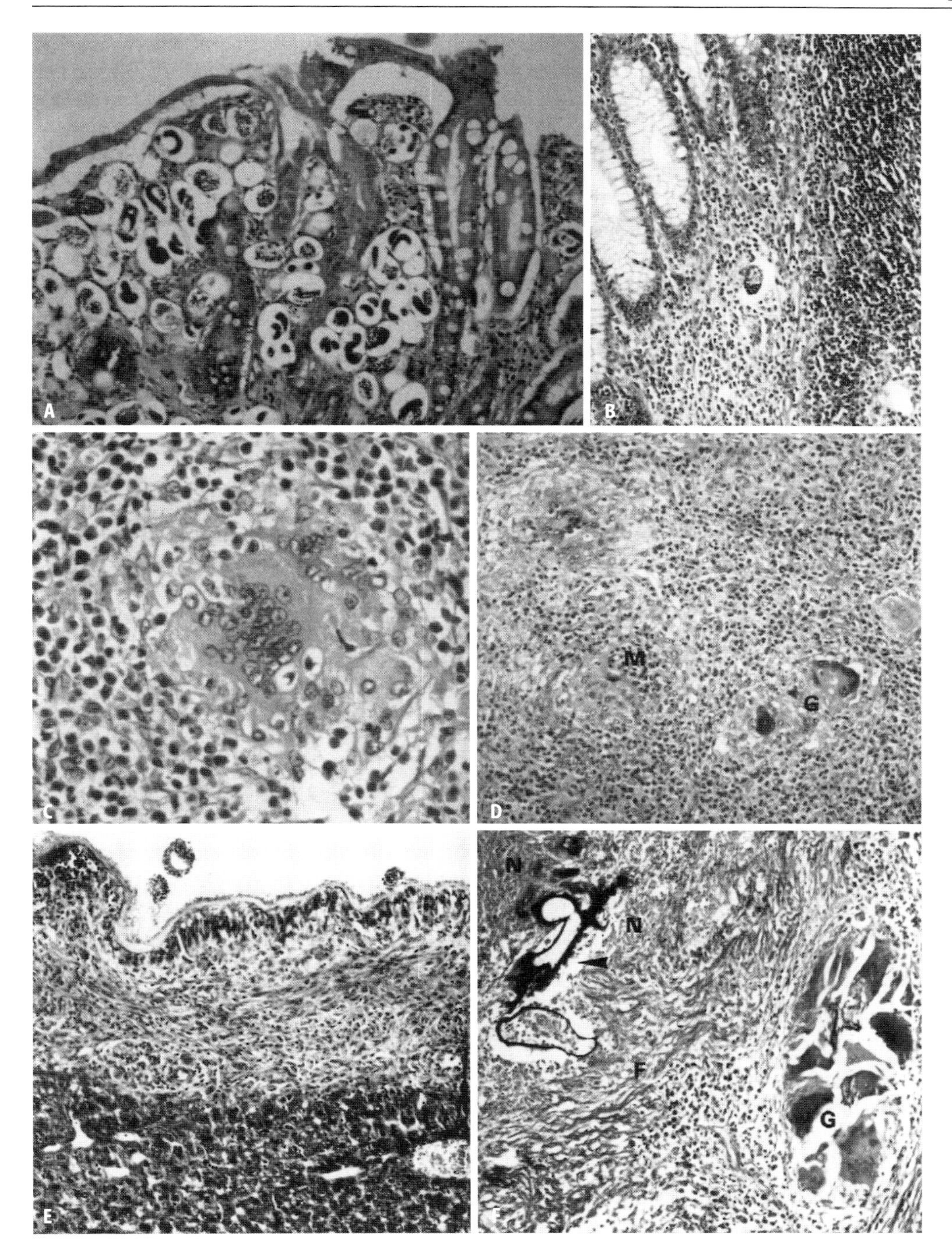

Fig. 29A-F. Contrasting histologic reaction in natural and aberrant host. A-D *Angiostrongylus costaricensis*. **A** Infection in rat showing mostly embryonated eggs in lamina propria without inflammatory reaction. Slide courtesy of Dr. Pedro Morera. HE x 120. **B** Eggs in appendix of child accompanied by eosinophilic infiltration. Human slides courtesy of Dr. Jorge Piza. HE x 120. **C** Two collapsed *Angiostrongylus* egg shells in multinucleate giant cells surrounded by eosinophils, macrophages and fibrosis in appendix of child. HE x 300. **D** Egg granuloma with giant cells (G), and eosinophilic microabscess (M) and fibrosis in appendix of child. HE x 120. **E, F** *Echinococcus multilocularis*. **E** Infection in mouse shows brood capsule with germinal membrane and protoscolices, slight mononuclear infiltration, and thick fibrous capsule. Masson x 120. **F** In human, germinal membrane (arrowhead) is accompanied by necrosis (N), fibrosis (F), giant cell reaction (G) and compression of liver parenchyma (not shown). PA x 120

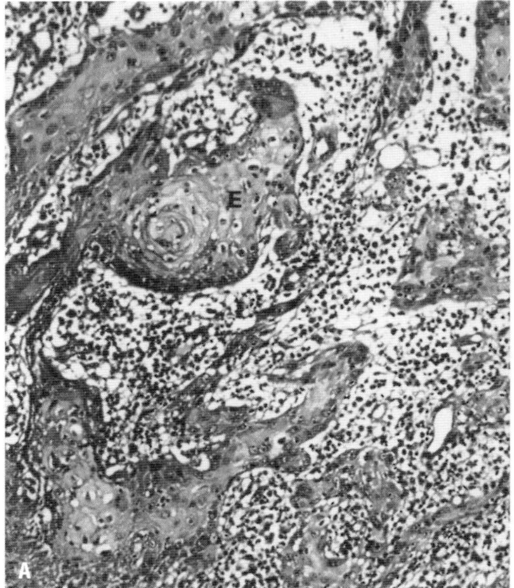

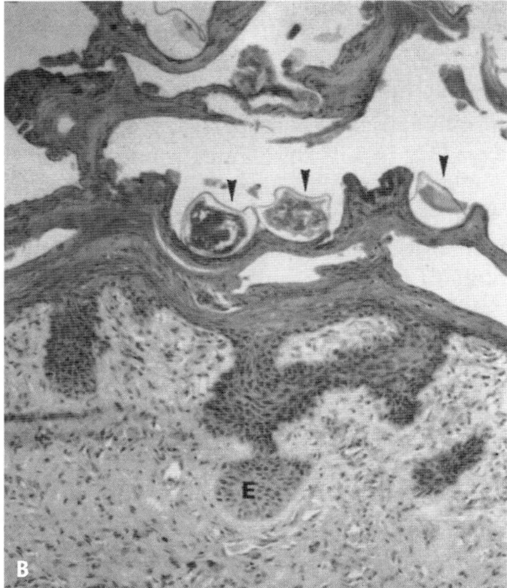

Fig. 30A,B. Hyperplasia. **A** Pseudoepitheliomatous hyperplasia of epidermis (E) adjacent to ulcerating leishmaniasis. A moderate number of *Leishmania peruviana* (not visible at this magnification) are present in histiocytes accompanied by lymphohistiocytic inflammation. HE x 120. **B** Scabies may be accompanied by marked hyperkeratosis, and burrows containing mites and eggs (arrowheads). Acanthosis of epidermis (E). HE x 120. For example of neoplasia, see Fig. 23B

PcP

Pneumonitis caused by → *Pneumocystis carinii* (Vol. 1) (→ Pneumocystosis).

Pediculosis, Animals

Several species of → lice (Vol. 1) infest large and small animals. Domestic animals may suffer from infestations with both biting (Mallophaga) and sucking (Anoplura) lice. Lice are extremely host-specific. Infection is a seasonal problem and the signs associated with pediculosis are extremely variable. Most lesions result from skin irritation and pruritus. They include alopecia alone, papulocrustous dermatitis, and damage to wool or hide caused by rubbing or biting. Sucking lice may induce an anaemia. Constant irritation during lice infestations causes a loss of weight and a decrease in milk production.

Therapy

→ Insectizides (Vol. 1).

Pediculosis, Man

Pediculosis is a superficial skin infection with head or body → lice (Vol. 1), *Pediculus humanus capitis* or *corporis* or crab lice, *Phthirus pubis*. The lice hold on to the body hairs with specialized claws and attach their eggs to the hair. Lice penetrate the epidermis with their mouth parts and suck blood, giving rise to mild dermatitis, often accentuated by scratching.

Therapy

Use of → insectizides (Vol. 1) in wash lotions. Note that repeated washings are needed within 10 days since the drugs have a limited activity on the lice eggs/larvae. Complete removal of the hair is recommended in heavy infestations. In body-louse-infections all bed covers and clothes have to be cleaned (hot-washed, deep-frozen etc.).

Pentastomiasis, Man

Pentastomiasis is an infection with nymphal pentastomes of several genera, the adults which are found in dogs or other carnivorous mammals (→ *Linguatula* (Vol. 1)), or in large snakes (→ *Armil-*

lifer (Vol. 1), → *Porocephalus* (Vol. 1)). Infections are acquired by the ingestion of eggs from the feces of mammals or snakes such as pythons, or the ingestion of undercooked snake filet. The nymphs of *Linguatula* spp. may be found in the throat, giving rise to intense allergic inflammation referred to halzoun in the Middle East, after the ingestion of raw kibbe from mutton containing infected lymph nodes; lesions in cervical lymph nodes of humans, the eye or subcutaneous tissues have also been reported. The larvae of *Armillifer* spp. and *Porocephalus* spp. usually encyst in the abdominal mesentery, liver, and peritoneum. Pentastomids are recognized in sections by the presence of a digestive tract containing ingested blood, by striated muscle, and by a distinctive, often serrated, cuticle and the absence of tracheal rings. The whole parasite shows four claws and a mouth leading to the misinterpreted term pentastomid. These parasites are surrounded by an often hyalinized connective tissue capsule, which may be incomplete, giving way to a segment of granulomatous reaction (→ Pathology/Fig. 18F). The larvae eventually die and may be transformed into an → abscess (Vol. 1), later into a hyalinized scar, and eventually into a calcified nodule.

Pepper Ticks

Trivial name for larval ticks.

Pepsinogen Increase

Clinical symptom in animals due to parasitic infections (→ Alimentary System Diseases, → Clinical Pathology, Animals).

PfHRP-2

Plasmodium falciparum histidine rich protein 2 (→ Malaria/Vaccination).

Phagocytosis

Feeding process in macrophages and amoebae (→ *Entamoeba histolytica* (Vol. 1)).

Phlebotomidosis

Disease due to infestation with phlebotomids, see Table 1

Phobia

Many (mostly elderly) people suffer from the fear of existing pests (such as cockroaches, spiders, mice etc.) or from imaginated items. In the latter case the disease represents a form of schizophrenia and cannot be cured, since the patients have composed a whole story, which they defend against any rational evaluation.

Pigmentation

Clinical and pathological symptoms of infections with skin parasites (→ Skin Diseases, Animals, → Ectoparasite (Vol. 1)).

Table 1. Phlebotomids and Control Measurements

Parasite	Host	Vector for	Symptoms	Country	Therapy		
					Products	Application	Compounds
Phlebotomus spp. (Sand flies)	Dog, Man	*Leishmania donovani, Leishmania infantum, Leishmania tropica, Leishmania braziliensis*	Edema, allergic reactions	Tropic, subtropic areas of Asia, Africa, America, Europe (Greece, Spain, Italy)	1% Vapona insecticide™ (Durvet)	Spray	Diclorvos

Pinworm Disease

Synonym

→ *Enterobius vermicularis* (Vol. 1), → Enterobiasis.

Piroplasmosis

→ Babesiosis, Man, → Babesiosis, Animals, → Theileriosis, Animals.

Placental Involvement

→ Pathology.

Plague

Flea-transmitted bacterial disease (*Yersinia pestis*) (see → Fleas (Vol. 1), → Insects/Fig. 7 (Vol. 1)).

Platyhelminthic Infections, Man, Pathology

Trematodes and Cestodes may lead to severe diseases in man and animals (see separate entries). The clinical symptoms they provoke are mostly correlated to the pathologic effects listed in Table 4 of pathology (page 471).

Treatment

→ Trematodocidal Drugs, → Cestodocidal Drugs.

Pneumocystosis

General Information

Pneumocystis jiroveci Frenkel, 1976 of man and → *Pneumocystis carinii* (Vol. 1) of rats and those from other vertebrates live in the pulmonary alveoli of their respective hosts (Fig. 1); the latter more euphonious name is often applied to the human parasite. Although there is considerable genetic heterogeneity among *Pneumocystis* [*PC1*] *i*solated from different hosts and host specificity is high, *Pneumocystis* from different hosts have usually been distinguished as subspecies of *P. carinii*, and there has been hesitancy to raise these subspecific designations to specific rank. Gene sequence studies of *Pneumocystis* of human and rat origin have indicated a closer relationship to Ascomycete fungi than to Protozoa. However, *Pneumocystis* is neither close to the fungi nor to any other organism. So that one should consider it a phylogenetically old parasite without close relatives. That the organism grows only slowly in axenic cultures hampers many studies, nonetheless a remarkable amount of information on the on *Pneumocystis* spp. has been developed.

Prior to the AIDS epidemic, the organism was only sporadically reported (e. g., malnourished children, and patients undergoing immunosuppressive therapy for cancer or solid organ transplants). In the 1980s as the HIV infection spread, and AIDS became a serious public health problem in the USA and in Europe, *P. carinii* was found to be the most prevalent infectious agent and the most common immediate cause of death among these patients. Thereafter, interest and research on *P. carinii* accelerated. The pneumonitis caused by the organism, *P. carinii* pneumonia (PcP), is a useful indicator of the onset of AIDS among HIV^+ individuals (AIDS-defining illness), since more than half (60% in the mid 1980s; and 43% in the early 1990s) of AIDS patients in the USA were infected with *P. carinii*. About 80% of AIDS patients develop PcP at least once. This organism is now known as the paradigm of pathogens referred to as opportunistic infections. Opportunistic infectious agents do not cause obvious symptoms by their presence in immunocompetent individuals, but can proliferate into fulminant infections in immunodeficient or immunocompromised hosts. Thus, they can be found in low numbers (colonization) in the lungs in domestic and wildlife mammals with low or no virulence.

Dissemination of the infection to other parts of the body is well documented. In non-AIDS patients, the lymph nodes are the main sites for extrapulmonary pneumocystosis. In AIDS patients, the organism has been found in almost all major organ systems, but has been most often detected in the lymph nodes, spleen, liver and bone marrow. It is generally assumed that infection is initiated by the inhalation of the organism into the respiratory system, and the organism escapes upper respiratory defenses. Adhesion to type I epithelial cells is important in the initiation of infection in the lung alveolus. Dissemination is

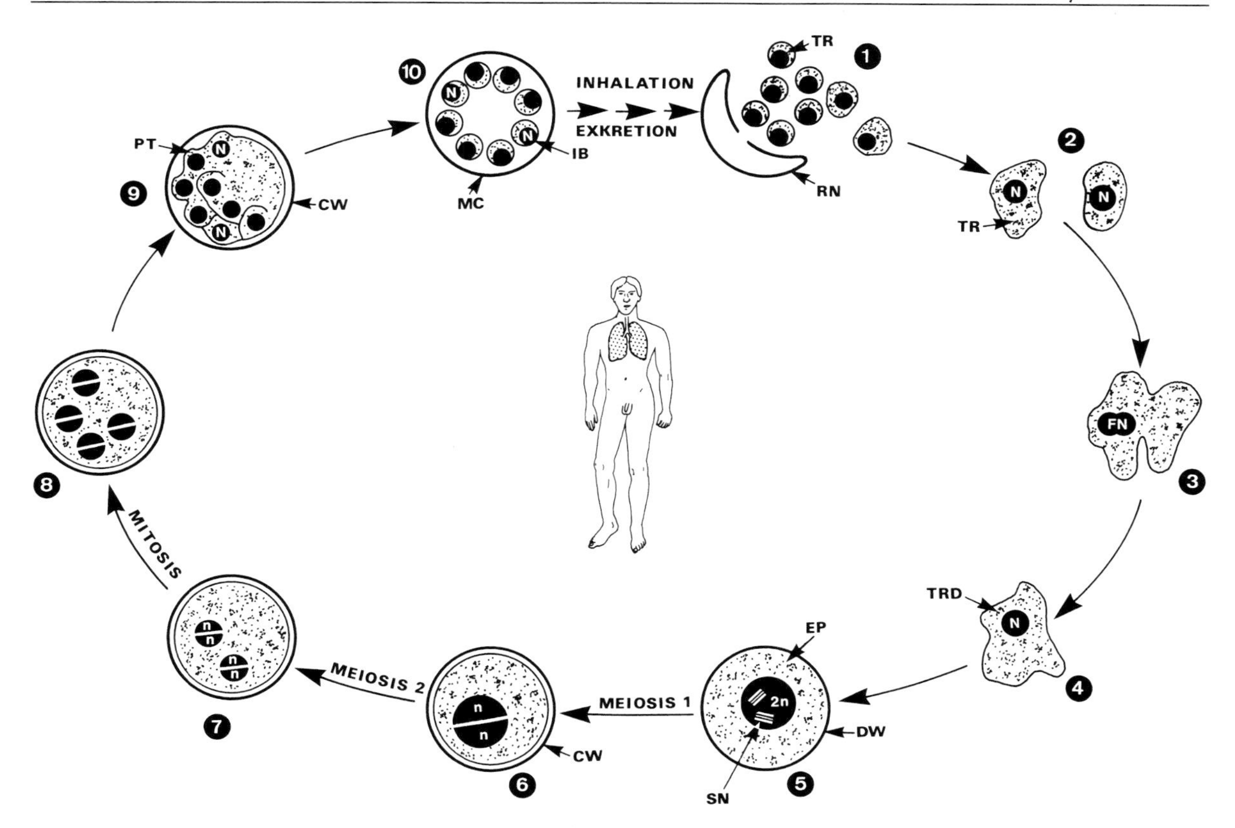

Fig. 1. Life cycle of *Pneumocystis carinii*. Hosts are immunosuppressed humans and animals (e.g. rodents; compare volume 1). 1 Mature cysts are inhaled. After rupturing of cyst wall, eight young haploid trophozoites (= intracystic bodies) are set free along the surface of lung alveoles. 2–4 Two of these trophozoites fuse with each other and give rise to a zygote (= diploid trophozoite; 4). 5 The zygote is surrounded by a developing cyst wall (*CW*) and thus is called an early precyst. 6–8 Formation of eight nuclei via two meiotic divisions and mitosis. 9 Formation of eight intracystic bodies (= young trophozoites) by division of the cytoplasm. This stage is called late precyst. 10 Mature cyst with intracystic bodies. This stage may be released or its wall becomes ruptured in the lung of the same host. In persons with an immunosuppressive disease (e.g. AIDS) these autoinfections lead to an extremely high parasitemia. *CW*, cyst wall; *DW*, developing precyst; *EP*, early precyst; *FN*, fusing nuclei; *IB*, intracystic bodies (= young trophozoites); *MC*, mature cyst; *N*, nucleus (n = set of chromosomes); *PT*, protruding trophozoite; *RN*, remnant of cyst wall; *SN*, synaptonemal complex; *TR*, trophozoite (haploid); *TRD*, trophozoite (diploid) (According to Yoshida and Mackenstedt et al.)

thought to occur after damage to the lung alveolar epithelium has occurred at late stages of the infection. Physical breaches in the alveolar epithelium could give the organism access to the rest of the body.

Distribution, Epidemiology and Transmission

Since over 75% of the population in the USA and Europe are sero-positive to *P. carinii* antigens by age 4, it is believed that the organism exists in high numbers in the environment. Its geographical distribution is worldwide, and is found in humans and other mammals.

Airborne transmission has been demonstrated by controlled experiments using laboratory animals, and *P. carinii* DNA has been detected in indoor and outdoor air samples. Possible modes by which organisms could be acquired include an environmental reservoir, or directly from PcP patients, immunocompetent individual with subclinical infections, or neonates whose immune systems are not fully developed.

Asymptomatic colonization of *P. carinii* in mammalian lungs is well documented. Fulminant infections (PcP) may result from activation by immunosuppression of latent, quiescent organisms residing in the host acquired early in life or organisms growing slowly (colonization). Another mode of infection is by reinfection of the host from an external source of organisms after an infection has been cleared. Thus immunocompetent individuals can harbor transient colonizations and PcP patients treated and cleared of the infection can develop a subsequent episode in which a different genetic population of organisms is detected.

Pathology

The primary infection in man is usually asymptomatic, probably without lesions. Pneumonia occurs only in immunologically compromised individuals. Symptomatic infections develop in children who are malnourished, who suffer from one of the primary immunodeficiency diseases such as X-linked agammaglobulinemia or T-cell defects, or who are treated with anti-neoplastic agents, or, because of organ transplants, with immunosuppressive drugs.

In later life immunosuppressed patients may develop pneumocystosis either from recrudescent latent infection or from reinfection. The trophozoites of *P. jiroveci* grow in the alveoli, giving rise to colonies variously described as foamy, honeycombed, eosinophilic, and PAS-positive masses (→ Pathology/Fig. 2A-C). These stick to the alveolar walls by means of filopodia and are not usually expectorated. The trophozoites are visible mainly by their hematoxylin-staining nuclei and the slight PAS positivity of the ground substance surrounding them The precysts and cysts develop in these colonies and their walls stain intensely with PAS, toluidine blue, and → Grocott's modification of → Gomori's silver impregnation.

The principal effect in the rat and probably also the human infection is disappearance of type I pneumocytes to which the *Pneumocystis* spp. attach; these are replaced by type 2 pneumocytes. The organisms are covered by a liquid alveolar lining layer. The inflammatory reaction is variable but includes neutrophils and macrophages, often accompanied by a transudate containing a variable amount of fibrin. In malnourished children there may be intense plasma cell infiltration of the alveolar walls, giving rise to a form of plasma cell pneumonia (→ Pathology/Fig. 2A). In X-linked agammaglobulinemics the plasma cells are absent and, as in leukemic babies, only a few lymphocytes and monocytes infiltrate (→ Pathology/Fig. 28). Similarly, in patients immunosuppressed by corticosteroids, cytostatic or alkylating agents, or by the human immunodeficiency viruses, little monocytic inflammation is found (→ Pathology/Fig. 2C).

During the active infection organisms are rare in bronchioles. After specific chemotherapy or after restoration of immunity following withdrawal of antitumor therapy, the masses of *Pneumocystis* spp. dislodge from the alveolar walls and can often be found in the bronchi and in expectorated sputum either by tissue imprint or PCR. Macrophages infiltrate the zoogleal *Pneumocystis.* colonies, phagocytose them, and may carry them to the regional lymph nodes. Pulmonary fibrosis from organization of the exudate may follow therapy, especially when complicated by other pathogenic factors (compare page 472).

Rare cases of disseminated pneumocystosis have been observed in severely immunosuppressed patients, with colonies of actively multiplying organisms in the liver, lymph nodes (→ Pathology/Fig. 2D,E), heart, and other organs sometimes giving rise to pressure atrophy of tissues. The luxuriant growth of *P. jiroveci* in extrapulmonary tissues of highly immunosuppressed patients suggests that immunity may suppress growth of the organisms in those that are only moderately immunosuppressed.

The lungs of patients who die from pneumocystosis are diffusely involved, and have a rigid, firm, uniform, pale appearance without necrosis or pleural exudate. After opening the chest cavity and sectioning the lungs, they usually remain expanded, although trapped air can sometimes be seen in alveoli or blebs. The diffuse infiltration extending from the hylum of the lungs presents a fairly characteristic picture radiographically.

Immune Responses

The relationship between *P. carinii* and the host is complex. Part of the difficulty in determining the interaction is that fulminant infections do not occur in normal hosts, whether they be humans or other mammals. Therefore, the abnormalities in immunity reported with PcP may be a result of the infection itself, or could be part of the susceptibility of the host for the infection.

The most common immunologic deficiency associated with PcP is cell-mediated immunity. This form of immunity is characterized by the T-cell, a critical cell in the host defense against infections such as tuberculosis, certain viral infections (such as cytomegalovirus), fungal infections (such as *Cryptococcus neoformans*), and *P. carinii.* All these infections are more common in conditions where T-cells, especially CD4-positive ($CD4^+$) T-lymphocytes, are reduced either by the disease or therapy. These include patients with HIV-infection and solid organ transplants.

The evidence supporting the role of the T-cell as the major host defense against PcP is based on both experimental animal and human epidemiologic studies. In animal studies, the first models of PcP occurred in animals that were treated with corticosteroids and lost weight. This led to a reduction in the number and function of many inflammatory cells, especially lymphocytes. Other investigators have been able to induce PcP in animals with more selective defect, including severe combined immunodeficient (SCID) mice, and mice made more susceptible by CD4 depletion using monoclonal antibodies directed against this cell population . In humans, in both the solid organ transplantation and HIV-infected individuals, the CD4 cell is either depleted by therapy (in transplant patients) or by infection (HIV). In the AIDS population, the risk of PcP is strongly related to the CD4 cell count in the blood. Patients with a CD4 count of >250 cells/ml (normal is >600 lymphocytes/ml) rarely acquire infection and therefore are not considered candidates for prophylaxis treatment. On the other hand, patients with a CD4 count of <200 lymphocytes/ml are at risk for PcP and should be treated by a prophylactic regimen.

Infection with *P. carinii* leads to an antibody response. Most humans have antibodies to the organism by age six, many within the first three years of life . This information has been cited as demonstrating that the organism is ubiquitous. Since antibodies to *P. carinii* are commonly found in the blood of patients infected with the organism, it is obvious that humoral response alone is not sufficient to control PcP. Unfortunately, the antibodies to *P. carinii* are to its → major surface glycoproteins (MSG) which can be varied by the organism itself, perhaps as a method of avoiding neutralization by immunoglobulins. This variation of antigenicity leads to a variable antibody response. Also, immunosuppressed patients with PcP may not be able to mount a vigorous humoral response. Therefore, the use of antibody titers to diagnose clinical infection has not proved useful. It has been used to detect subclinical reinfection (or reexposure) in healthy subjects.

There has been some suggestion that antibody is important in controlling infection. In one study, animals treated with gamma globulin derived from the serum of litter mates infected with *P. carinii* were subsequently exposed to *P. carinii* in the environment. The result was a reduced rate of infection.

The inflammatory cells of the lung are mostly the neutrophils. Although they normally make up a small percentage of the cells in the lower respiratory system, they can be quickly recruited into the lung, e. g., during bacterial pneumonia. The neutrophil response is variable, with some patients having a minimal neutrophil response; this response is nonspecific and perhaps harmful. An increase in neutrophils has been observed in the → bronchoalveolar lavage fluid (→ BALF) of HIV^+ patients. There is a clear relationship between increased neutrophils and worse prognosis for the HIV^+ patient. This inflammatory response is also seen with eosinophils, a cell not normally found in BALF, and an increase in eosinophils is also associated with increased mortality.

Several cytokines have been associated with this inflammatory response . Interleukin-8 and tumor necrosis factor-alpha (TNF-a) are probably the most important mediators. Despite the fact that patients with PcP have an underlying cell-mediated immune defect such as AIDS, they are still able to mount an inflammatory response. These cytokines are mostly released by macrophages, which are not killed by the HIV infection, thus, this inflammatory response remains intact. The use of corticosteroids to treat HIV-infected patients with PcP (see below) is used to reduce this inflammatory response.

Antigens

The most-studied *P. carinii* antigens are those with M_r of 95–140 kDa (major protein component in the organism). The high MW proteins (major surface glycoproteins, glycoprotein A) are encoded by a multi-gene family of about 100 different subtelomeric genes clustered at the ends of chromosomes. Only one is expressed at any given time in an organism; its expression involves transcription at a single site (upstream conserved sequence). The antigen appears to undergo processing by proteolytic action during maturation. These antigens are highly glycosylated, especially with mannose residues, and are involved in adherence to extracellular matrix substrates, host cells and other molecules (e. g., lung surfactant proteins). The MSG are exposed at the cell surface.

Circulating antibodies in mammalian serum most frequently recognize a protein complex with apparent molecular mass of 35–45 kDa in *P. carinii hominis* (*P. jiroveci*) and 45–55 kDa in *P. carinii carinii*. These antigens are not externally exposed, but reside within the cell wall of the organism.

Diagnostics

For any physician, the diagnosis of disease relies on the suspicion of that disease in the patient presenting symptoms. In the case of PcP, some relatively specific features of the history and physical examination can suggest the diagnosis. It is based on these features that the clinician will move onto either more specific tests (sputum induction, bronchoscopy) or an empiric trial of therapy.

In the patient's medical history, the immune status of the patient is crucial. For the HIV^+ patient, the CD4 count has proved predictive of PcP development. In several large studies, the risk for PcP was found to rise dramatically once the patients' CD4 count fell below 200–250 CD4/ml. Current prophylaxis regimens are based on this priniciple. In the transplant patient, the maximal immunosuppression therapy usually occurs immediately following transplant. The effect of immunosuppression treatment is that it takes about six weeks to deplete the memory T-cells. In that time frame, the patient is susceptible to certain infections, e. g., *P. carinii* and cytomegalovirus. As the immunosuppression is reduced (usually by a year), the risk decreases. It is only a problem again if the patient has received another round of intense immunosppression (e. g., for an acute rejection episode).

The clinical history for PcP includes cough, shortness of breath, and fever. Fever is usually seen, but is a nonspecific finding. For HIV^+ patients, these symptoms can be prolonged and subsequent weight loss is common. The duration of symptoms until diagnosis of PcP for an HIV^+ differs from that of a transplant patient; the HIV^+ patient averages more than two weeks of symptoms prior to diagnosis of the infection, whereas the transplant patient becomes rapidly ill.

The physical examination is often unrevealing in a patient wit PcP. Evidence for immunosuppression such as oral thrush is useful to characterize the patient. The only direct physical finding is crackles, which are heard on auscultation in advanced PcP. Often the patient will cough in deep inspiration, making crackles difficult to appreciate.

Of the routine laboratory tests, the chest roentgenogram and level of oxygenation must be assessed in all patients with suspected PcP or any other form of pneumonia. The classic chest roentgenogram of PcP patients show diffuse infiltrates (Fig. 2). Patients on aerosol pentamidine may have a predominance of the upper lobe. A normal chest roentgenogram is seen in about 10% of patients diagnosed with PcP. The presence of a pneumothorax is an unusual roentgenographic finding, ocurring in up to 5% of PcP cases in HIV^+ patients. Since most other opportunistic and routine pneumonias do not cause pneumothorax, the finding of a pneumothorax in an immunosuppressed patient is an indication for further evaluation to rule out PcP. Oxygenation can be assessed during the noninvasive oximetry testing. However, a patient who is hyperventilating may raise his level of oxygen but still have significant lung disease. The arterial blood gas is more precise and the alveolar-arterial gradient (A-a gradient) should be calculated. Normally, this is less than 15 mm Hg. Patients infected with HIV with PcP and an A-a gradient of greater than 35 have a pro-

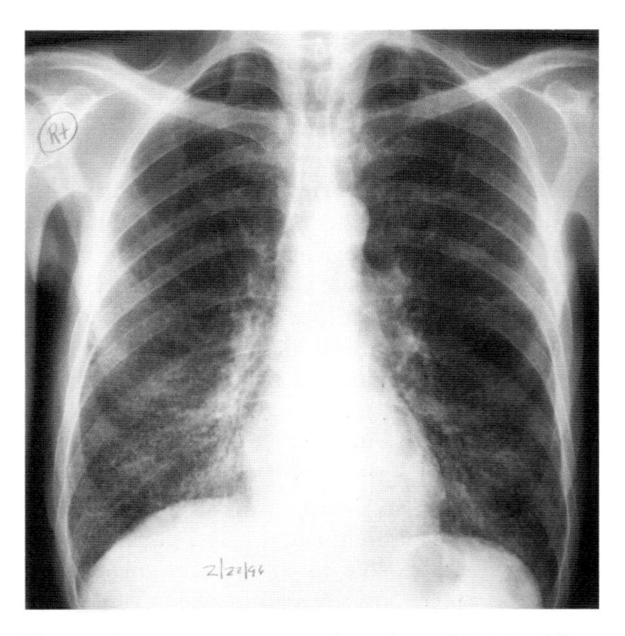

Fig. 2. Chest roentgenogram of a patient with PcP. Diffuse infiltrates are evident, as indicated by the light areas in this x-ray photograph of the lung. (Photo: Baughman and Kaneshiro)

jected mortality of 35% (prior to the use of corticosteroids).

Lactate dehydrogenase (LDH) has been proposed as a diagnostic marker for PcP, since the lung is one source of LDH, and PcP can cause elevation of the enzyme activity. However, this response is fairly nonspecific since other pneumonias as well as lymphomas can lead to a rise in LDH levels. There is some suggestion that the height of the LDH activity level is positively correlated with mortalilty from PcP.

Other proposed techniques for diagnosis have been gallium scan, computer tomography (CT) scan, and exercise oxygenation. The gallium scan and CT are more sensitive than routine chest roentgenogram, but are not very specific. Both tests are costly, and since they are not definitive, they do not seem worth their cost for the routine diagnosis of PcP. The exercise testing is useful for detecting early interstitial lung disease, such as PcP. However, most pulmonary patients (Fig. 2) will already have abnormal oxygenation. The desaturation of oxygenation with exercise is not specific for PcP, but may be useful in evaluating unexplained dyspnea in the otherwise healthy HIV^{+} patient.

The specific diagnosis of PcP depends on the demonstration of the organism. Methods for culturing the organism are limited, therefore routine microbiologic techniques are not useful in diagnosing PcP. Fortunately, the organism is relatively unique in appearance, and visible organisms in respiratory specimens can give a precise diagnosis.

There are several methods of obtaining respiratory specimens (Table 1). These include sputum, bronchoscopy samples including bronchial wash and biopsy, and open lung biopsy. Bronchoalveolar lavage (BAL) is a specific technique in whch the bronchoscope is advanced as far distally in the airways (wedged) and fluid is instilled and withdrawn. The yield is much higher than simple washing samples. It compares favorably with more hazardous biopsy techniques and many clinicians prefer to perform BAL without biopsy to diagnose PcP.

Various methods and stains are used to visualize *P. carinii* (Table 1, Fig. 3). The simpler stains such as modified Wright Giemsa are easy to perform and results can be obtained within minutes. The silver stain is the 'gold standard' of the pathologic diagnosis (Fig. 3), which can take up to a day to process. The use of DNA amplification by polymerase chain reaction (PCR) to identify *P. carinii* has been developed, but there are some technical difficulties since there are distinct genetic populations of organisms infecting humans. Since PCR takes somewhat longer to perform, it may be faster to proceed to a better-understood/defined sample (e.g. bronchoscopy rather than sputum). Overall, the relationship between the samples and the detection method is reciprocol: less sensitive detection methods are needed for samples with more alveolar material (Table 1). For example, PCR may be useful for enhancing the diagnostic yield for PcP if one is looking at induced sputum samples. However, it is probably not a neccesary routine technique for examining → BALF or open lung biopsy samples.

Therapy

There are several issues to consider in the treatment of PcP. The first is choosing the correct antibiotic once a diagnosis has been made. Ancillary treatments include oxygen supplementation and supportive care. The use of corticosteroids as an anti-inflammatory drug has been shown to benefit some patients. Finally, one must look for other infections. Up to 20% of patients with PcP will have other infections.

Several antibiotics have been shown to be useful against PcP. Since the organism cannot be

Table 1. Diagnostic yield of *P. carinii* cysts analyzed by silver staining (according to Baughman and Kaneshiro)

Specimen source	Median percentage (range)
Sputum	50 (15–94)
Bronchial wash	65 (60–70)
Single-area bronchoalveolar lavage	90 (60–100)
Two-area bronchoalveolar lavage	95 (85–100)
Transbronchial biopsy	97 (89–100)

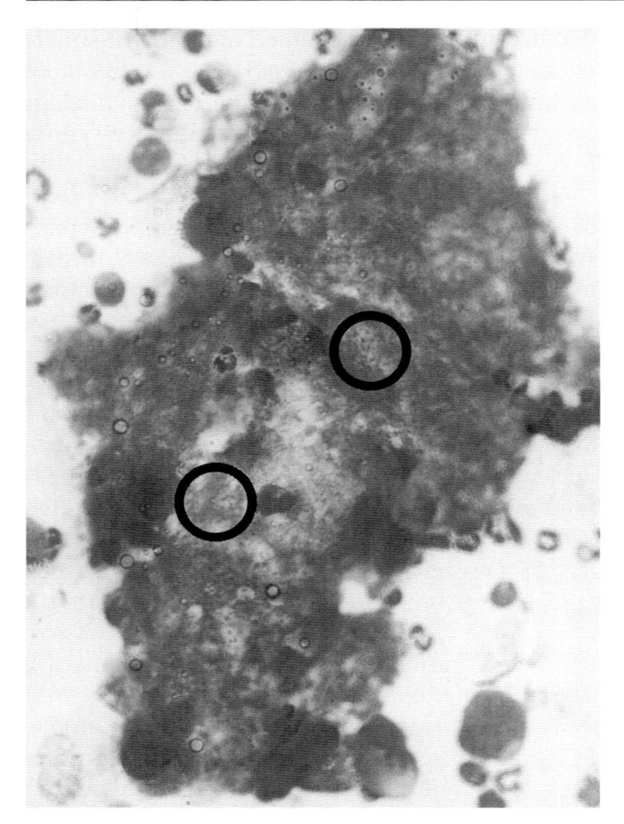

Fig. 3. A modified Wright Giemsa (Diff-Quik) staining of a clump of *P. carinii* in bronchoalveolar lavage fluid (BALF) taken from a PcP patient. Lung surfactant, host cell and cell debris are also present in the clump. The *P. carinii* organisms are abundant and their nuclei are seen as tiny dots (e. g., within the circled areas). (Photo: Baughman and Kaneshiro)

readily grown in culture, in vitro testing of antibiotic effectiveness is problematic. Antibiotic effectiveness has been shown by either treatment of animal models of PcP or the treatment of humans with PcP. Fortunately, the animal models seem to predict which antibiotic will be effective in the clinical treatment of humans.

Trimethoprim with sulfamethoxazole (TMP-SMX, trim/sulfa) has proved to be the most effective antibiotic for PcP. Available for both intravenous and oral use, the drug is currently regarded as the 'gold standard' to which other drugs are compared. The dose recommended is 15 mg/kg trimethoprim. The clinical response rate is greater than 80% . Failures can be due to lack of efficacy of the drug or drug toxicity. Allergic reactions include rash, fever, and myalgias. Most reactions can be surmounted with desensitization and supportive care. However, some cases of reaction are so severe requiring withdrawal of the drug. Patients who are withdrawn from TMP-SMX because of toxicity usually do well with a new agent. However, patients who are changed from TMP-SMX to another agent because of clinical failure have a less than 50% chance of responding to the new agent.

Pentamidine was the first drug shown to have clinical efficacy for PcP. Since there is no oral form of the drug, it is usually administered intravenously. It can also be given by aerosol, which has proved a useful alternative for prophylaxis therapy. For acute disease, it is similar in efficacy toTMP-SMX, however, it is associated with greater toxicity. The major problems with pentamidine include hyopglycemia and arrhythmias. The drug is also irritating to the vein. Aerosol pentamidine is less effective than TMP-SMX as a prophylactic agent, and it is usually reserved for the TMP-SMX-intolerant patient.

Trimetrexate is an analogue of methotrexate, and was originally developed as an antineoplastic chemotherapeutic agent. Since it blocks the dihydrofolate reductase enzyme, it was found to have activity in the treatment of PcP in both animals and humans . In a trial comparing trimetrexate to TMP-SMX, there was no significant difference in the efficacy of the two agents, although trimetrexate was more difficult to use and was associated with more leukopenia. It has to be given with leukovorin to reverse the potential toxicity to humans from the drug. Its major role is its apparent effect in some of the clinical failures to TMP-SMX, who seem to still respond to trimetrexate.

In patients with mild to moderate disease, oral therapy is popular. The definition of mild to moderate disease is based on the A-a gradient. Patients with an A-a gradient of less than 35 mm Hg have a good prognosis overall. In the TMP-SMX-intolerant patient, several agents have been studied. Atovaquone is an oral agent which does not have an intravenous form. It has been useful for the mild form of the disease, especially in the TMP-SMX-allergic patient. Clindamycin and primaquine, as a combination, has also been reported as helpful. Dapsone with trimethoprim is another combination used for PcP. Interestingly, patients who are sulfa-allergic may still be able to take dapsone, although there are some cross-over reactions.

Given the effectiveness of many of the agents, especiallyTMP-SMX, some physicians have proposed empirical therapy in the appropriate patient. In the HIV$^+$ patient who is not receiving effective prophylaxis and presents fever, cough, and diffuse pulmonary infiltrates, at least half the cases, *P. carinii* will be found to be the cause of the pneumonia. Therefore, it has been argued that empirical therapy should be given. However, several lines of evidence have shown that this may not be the best policy. For one, up to 20% of HIV$^+$ patients in that situation will have another infection, either in addition to *P. carinii*, or as a sole agent. This can lead to a delay in diagnosis. Since a common other agent is *Mycobacterium tuberculosis*, this represents not only a potentially treatable infection, but also a public health hazard. Recent analysis of clinical outcome of groups of patients treated empirically versus patients undergoing bronchoscopy, showed that the patients undergoing bronchoscopy had a better clinical outcome.

Supportive care of the patient starts with assessing the need for supplemental oxygen. The patient is usually hypoxic, especially with exercise. The need for oxygen may persist for many days despite therapy for pneumonia. Although arterial blood gas measurement is the most accurate method for assessing the level of hypoxemia, oximetry measurements can be used to follow the patient during therapy. Prior to discontinuation of oxygen, it is useful to be sure the patient is no longer significantly desaturating with ambulation.

At one time, the mortality of PcP in HIV$^+$ patients in the USA was fairly high, e. g., the patient with an A-a gradient of >35 Torr had a 35%mortalilty, despite appropriate antibiotic therapy. In a randomized trial, it was shown that corticosteroids given at the time of diagnosis led to a significant reduction in mortality . It has been proposed that part of the cause of morbidity and mortality from PcP is the inflammatory response. This includes the neutrophil response. Corticosteroids reduce this response by directly blocking cytokines such as IL-8 and TNF-α, and hence improving survival of PcP patients. However, steroids increase the risk for other opportunistic infections; patients on corticosteroids who are worsening may have an alternative diagnosis.

The best medicine is preventive medicine. For *P. carinii*, this is clearly true. There are several different agents which have been shown to prevent *P. carinii* infection. Used in patients at risk for PcP, these can prevent infection. In the HIV$^+$ patient with less than 250 CD4 cells/ml, several forms of prophylaxis have proved effective. TMP-SMX has been shown to be extremely effective, with less than 1% of patients developing PcP. In patients who are TMP-SMX-intolerant, other agents have been used. Aerosol pentamidine as a monthly therapy reduces the chances of PcP, but up to 30% of at-risk patients will develop PcP while on aerosol pentamidine. Dapsone is more effective than aerosol pentamidine, but less effective than TMP-SMX. Among the solid organ transplant patients, the use of low dose TMP-SMX (three times/ week) has essentially eradicated *P. carinii* as a cause of pneumonia (see → Pathology).

Overall, *P. carinii* infection can usually be treated effectively. The major issue is recognizing the disease; improvements in diagnostic testing are needed to make the diagnosis (see pages 592 ff).

Pneumonia

Inflammation of the parenchymal layer of the lungs often as a result of parasitic infections (e.g. *Pneumocystis carinii*, *Paragonimus* spp., *Leishmania* spp. etc.).

Pneumonyssoidic Mange

→ Mange, Animals/Pneumonyssoidic Mange.

Postnatal Toxoplasmosis

Primary infection with *Toxoplasma*-stages (eating tachyzoites, bradyzoites in raw meat or by oral uptake of oocysts from cat feces) after birth. In immuno-competent people: mostly no clinical symptoms except of subacute lymphadenitis in 1% of the cases. However, in immuno-deficient people severe disease may occur (see → Toxoplasmosis, Man).

Powassan Encephalitis

Synonym
POWE

Powassan encephalitis which is caused by the POWE virus (→ *Flavivirus* (Vol. 1), group B) is a North American → RSSE-like disease associated with transmission by bites of *Ixodes* spp. and *Dermacentor andersoni.*

Prenatal Toxoplasmosis

Disease due to transmission of *Toxoplasma gondii* stages from mother to fetus in the case of the mother's first infection during pregnancy. In these cases (0.1–0.7% of the European newborn children), 75% of the infections remain subclinical (with 15% having no damage, but up to 85% with chorioretinitis), 15% have mild symptoms (with 99% chorioretinitis, 1% brain damages) and 10% with severe clinical symptoms (85% brain damages, e.g. hydrocephalus, 15% perinatal death). The children of the first two groups appear mostly healthy after birth, but symptoms may occur later; eye diseases often start after 10 or even 20 years.

Prions

Protein-like infectious organizations (agents). They consist of cellular proteins (PrP^{c}), that are transformed into an infectious, abnormal isoform (PrP^{Sc}). After long incubation periods (years) so-called transmissible encephalopathies (TSE) may occur known under different names (BSE = bovine spongious encephalopathy in cattle, Scrapie in sheep, Creutzfeldt-Jacob Disease in man). Transmission occurs by feeding undercooked (below 141 °C) nerve/brain portions of infected animals or by eating contents of flies, that had fed on such material (experimentally proven in Scrapie-infections).

Productivity Loss

Clinical symptom in animals due to parasitic infections (→ Alimentary System Diseases, → Clinical Pathology, Animals).

Protein-Synthesis-Disturbing Drugs

Structures

See Fig. 1

Emetine/Dehydroemetine

Synonyms

Cephaeline methyl ether/2,3-dehydroemetine, 2-dehydroemetine, Mebadin

Clinical Relevance Emetine and dehydroemetine exert activities against a wide variety of pathogens such as bacteria, protozoans, trematodes and fungi. In addition, it has antiviral activities which are directed against *Herpes zoster* infections and those flaviviridae causing tick-borne encephalitis.

The antiprotozoal activity of both drugs is directed against invasive intestinal and extraintestinal stages of *Entamoeba histolytica* ("Magna forms") resulting in a quick clinical improvement. The drug level in the liver is sufficiently high for damaging *E. histolytica* liver stages. However, both drugs possess severe side effects and therefore have only limited medical use. They induce a direct destruction of tissue trophozoites (intestine, liver), however, they have no effect against "Minuta" forms in the gut lumen or cysts. In general, higher cure rates are achieved by application of combinations of (dehydro)emetine and chloroquine or (dehydro)emetine and 5-nitro-imidazoles. In veterinary medicine there exist positive experiences in the treatment of *E. invadens* infections in reptiles with dehydroemetine. There are reports about some activity of emetine against *Leishmania tropica* and *L. major* infections as well as in-vitro activity against *Blastocystis hominis.*

The antitrematodal activity of emetine is restricted to *Schistosoma japonicum*, *S. haematobium* and *Fasciola hepatica*. In veterinary medicine both drugs exert activity against *F. hepatica* in sheep at higher dosages. Thereby the drug is more active after intravenous than after intramuscular application. However, the antitrematodal efficacy is uncertain. Because of a variety of severe side effects the drugs were quickly replaced by safer drugs.

Further indications of (dehydro)emetine are pulmonal aspergillosis and they may be useful against scorpion stings.

Molecular Interactions Emetine as the active principle was isolated from roots of the south- and

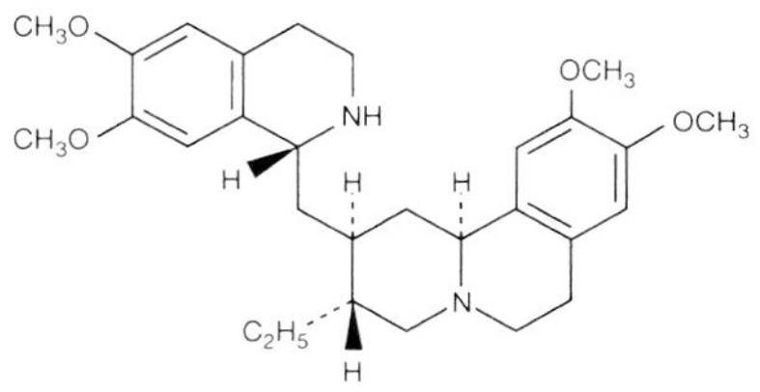

Emetine

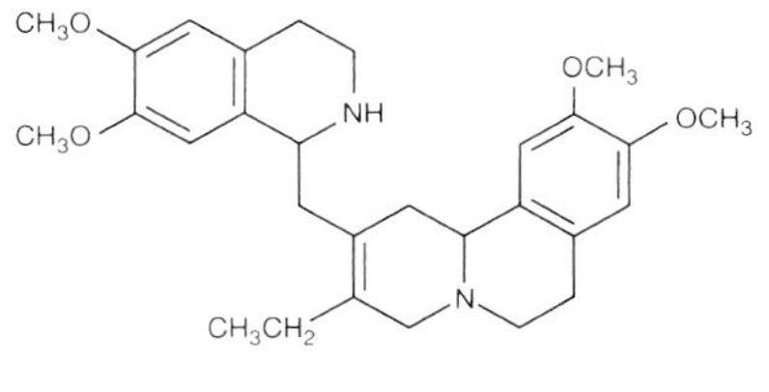

Dehydroemetine

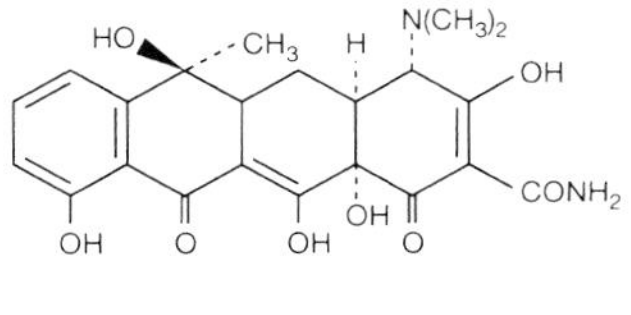

Tetracycline

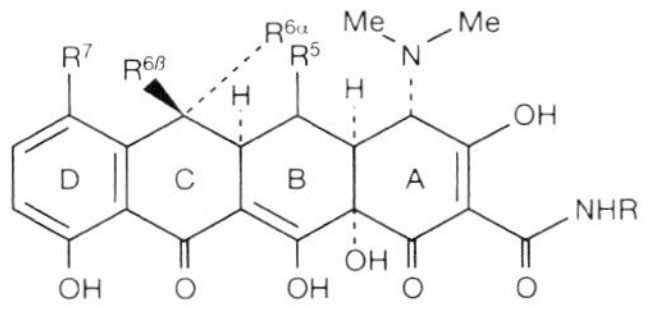

	R^5	$R^{6\alpha}$	$R^{6\beta}$	R^7	R
Tetracycline	H	CH_3	OH	H	H
Chlortetracycline	H	CH_3	OH	Cl	H
Oxytetracycline	OH	CH_3	OH	H	H
Doxycycline	OH	CH_3	H	H	H
Minocycline	H	H	H	$N(CH_3)_2$	H

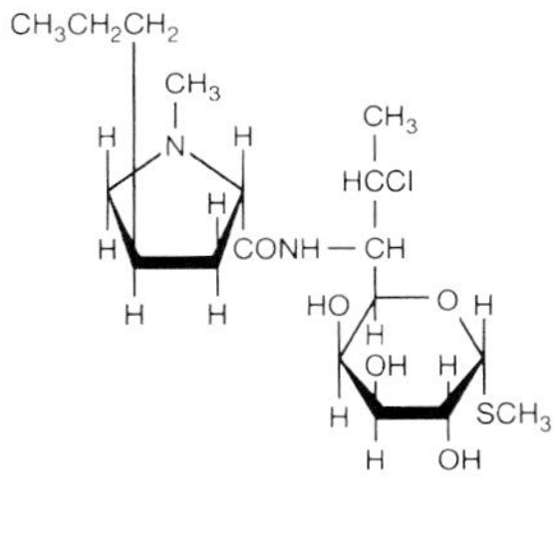

Clindamycin

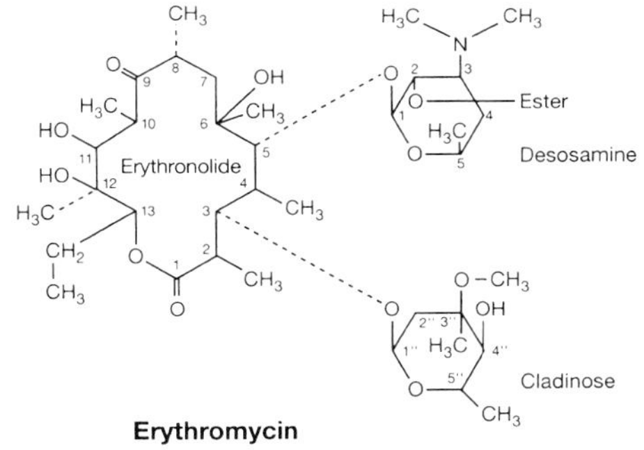

Erythromycin

Fig. 1. Structures of antiparasitic drugs affecting protein synthesis.

centralamerican *Rubiacee* ipecacuanha. The (-)-emetine enantiomer possesses the specific amoebacidal activity whereas the (+)-emetine is much less effective. Emetine is accumulated in the liver. It is proposed that the eukaryotic protein synthesis becomes inhibited. Indeed, there is a correlation between amoebacidal activity and inhibition of translation by various emetine-derivatives. On the molecular level there is an irreversible, but noncovalent binding to the peptide-chain elongation site of the 60S subunit of ribosomes. It is assumed that a unique region within the emetine molecule may be responsible for its irreversible binding. The selective toxicity against *E. histolytica* compared to the vertebrate host is explained by a more slow recovery of the parasites from inhibition of protein synthesis compared to the situation in mammalian cells. In mammalian cells binding of emetine occurs to protein S14 of the 40S ribosomal subunit. Thereby the elongation factor 2-dependent translocation is prevented.

Resistance There is a correlation between resistance against emetine and resistance to cycloheximide, an inhibitor of protein synthesis. It could be shown that the emetine resistance is due to a mutation and not to any post-translational modification. After cloning and sequencing of mRNA coding for emetine-resistance and mutagenization of *E. histolytica* and selection of a clone resistant to emetine it could be shown that overexpression of a P-glycoprotein homologue may be responsible

Fig. 1. (continued) Structures of antiparasitic drugs affecting protein synthesis.

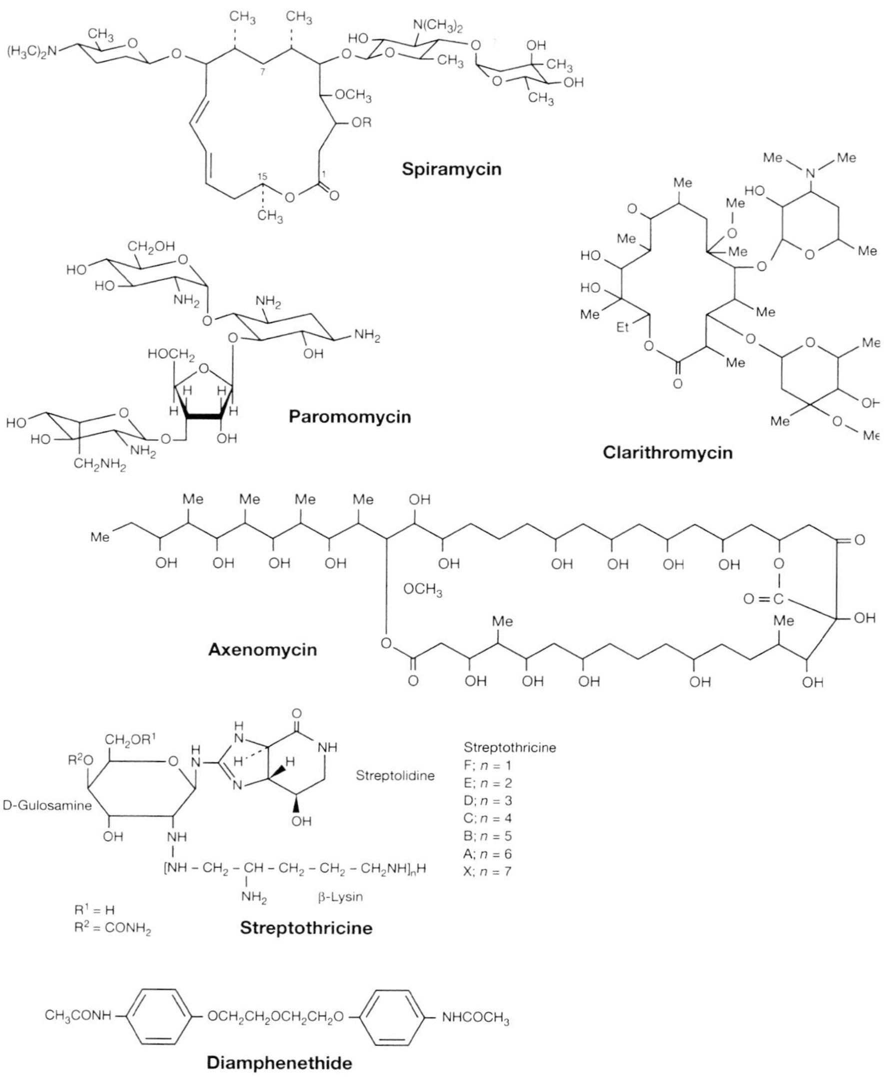

for resistance. This resembles the multi-drug-resistance (MDR) phenotype, as there is cross-resistance to hydrophobic drugs, increased efflux of emetine and reversal of resistance by verapamil. Thus, it is very likely that a protozoan P-glycoprotein is involved in emetine resistance. In the meantime, emetine-resistant mammalian cells could be isolated and an emetine-resistant *Caenorhabditis elegans* strain was detected in which P-glycoprotein genes were overexpressed.

Tetracyclines

Important Compounds Tetracycline, Doxycycline, Minocycline, Oxytetracycline, Chlortetracycline

Synonyms Tetracycline: Deschlorobiomycin, Tsiklomitsin, Abricycline, Achromycin, Agromicina, Ambramicina, Bio-Tetra, Bristaclicina, Cefracycline suspension, Criseoclicina, Cyclomycin, Democracin, Hostacyclin, Omegamycin, Panmycin, Polycycline, Purocyclina, Sanclomycine, Steclin, Tetrabon, Tetracyn, Tetradecin

Doxycycline: α-6-deoxy-5-hydroxytetracycline monohydrate, GS-3065, Doxitard, Liviatin, Vibramycin, Vibravenös

Minocycline: none

Oxytetracycline: Glomycin, Terrafungine, Riomitsin, Hydroxytetracycline, Berkmycin, Biostat, Imperacin (tablets), Oxacyclin, Oxatets, Oxydon, Oxy-Dumocyclin, Oxymycin, Oxypan, Oxytetracid, Ryomycin, Stevacin, terraject, Terramycin, Tetramel, Tetran, Vendarcin, Vendracin

Chlortetracycline: 7-chlorotetracycline, Acronize, Aureocina, Aureomycin, Biomitsin, Biomycin, Chrysomykine; (in combinations): Verman, Salmocarp

Clinical Relevance The main indication of tetracyclines is their general antibacterial activity. This is directed against numerous gram positive, gram negative bacteria, mycoplasma, chlamydia and rickettsia.

Tetracycline is especially used against chronic bronchitis (Haemophilus influencae and others), atypical pneumonia's caused by mycoplasma, *Chlamydia psittaci* (=ornithosis) infections, *Coxiella burnetti* (= Q fever) infections, non-gonorrhoic urethritis caused by *Chlamydia trachomatis*, mycoplasms, ureaplasms, Lymphogranuloma inguinale, prostatitis, Acne vulgaris, Lyme borreliosis with penicillin allergy, Cholera, heavy shigellosis, yersiniosis, pseudotuberculosis; aerobic-anaerobic mixinfections such as Morbus Whipple or actinomycosis with penicillin allergy and septicaemia caused by brucellosis, leptospirosis, tularaemia, rickettsiosis, melioidosis, pest. Tetracyclines have been known of since 1948.

Doxycycline was explored in 1967. It is used against plasmodia in cases of resistance against the usual drugs.

Minocycline was explored in 1967. The antibacterial activity is used against individual forms of acne, nocardiosis and infections with sensitive atypical mycobacteria. In addition, it is used against plasmodia in cases of resistance against the usual drugs.

Oxytetracycline has been known of since 1950. The antibacterial activity is directed against *Vibrio cholerae* and enteritis bacteria. In addition, oxytetracycline is used against plasmodia in cases of resistance against the usual drugs. It can also be used against gut lumen forms of *Entamoeba histolytica* because of its low resorption.

Chlortetracycline has been known of since 1947. The antimalarial activity is directed against exoerythrocytic schizonts in the liver (→ Hem(oglobin) Interaction/Fig. 2). For a long time chlortetracycline was the only drug with prophylactic activity on *T. parva* infections. The activity is directed against schizonts in lymphocytes. There is also additional anticoccidial activity.

Molecular Interactions It has been known for a long time that tetracyclines inhibit bacterial protein synthesis. Thereby, tetracycline is bound preferentially to the small ribosomal subunit, but also binds to the 50S ribosomal subunits. The binding of amino acid charged tRNA is inhibited resulting in the prevention of chain elongation of the nascent peptide chain. Tetracycline interferes with the binding of aminoacyl-t-RNA at the acceptor site of ribosomes at the interphase between the large and small subunit of bacterial 70S ribosomes.

The antiprotozoal activity of tetracycline is directed against *Giardia lamblia*. A combination of quinine and tetracycline is used against chloroquine-resistant plasmodia. Thereby, the action is directed against exoerythrocytic and erythrocytic schizonts (→ Hem(oglobin) Interaction/Fig. 2). In addition, tetracycline is active against *Balantidium coli*. Here the activity is directed against trophozoites in the intestine, which divide by binary fission, and cysts in the feces. The mechanism of antiprotozoal action of tetracycline may be similar to the antibacterial action, but the real target site(s) and mechanism(s) of action remain(s) unclear.

The action of doxycycline, minocycline, oxytetracycline and chlortetracycline is directed against exoerythrocytic schizonts (→ Hem(oglobin) Interaction/Fig. 2). The mode of action is presumably identical with that of tetracycline.

Lincosamides

Important Compounds Clindamycin

Synonyms Clindamycin: n7(S)-chloro-7-deoxylincomycin, U-21251, Cleocin, Dalacin C, Sobelin

Clinical Relevance The main indication of lincosamides relies on their antibacterial activity. It is used especially in patients with penicillin/cephalosporin allergy, after an initial therapy with a penicillin/aminoglycoside combination against intraphagocytic persistent bacteria, in chronic osteomyelitis, against anaerobic bacteria accompanying polymicrobial mixed infections (pelveoperitonitis, aspiration pneumonia).

The antibacterial activity of clindamycin comprises furunculosis, erysipelas, tonsillitis, abscess with penicillin allergy, acute infections with anaerobic bacteria (adnexitis, liver abscess, beginning aspiration pneumonia), curative therapy of osteomyelitis and endocarditis prophylaxis with penicillin allergy.

Clindamycin is the only lincosamide which possesses additional antiprotozoal activity. Thus, it has some activity against *Entamoeba histolytica*, and it is effective against *Neospora caninum* tachyzoites in cell cultures (→ DNA-Synthesis-Affecting Drugs IV/Table 2). The combination clindamycin and sulfonamide is highly effective against neosporosis. Also the multi-drug-combination of pirithrexim, clindamycin, diclazuril, robenidine and pyrimethamine is experimentally active against neosporosis (→ DNA-Synthesis-Affecting Drugs IV/Table 2). In combination with trimethoprim and sulfamethoxazole, clindamycin is highly effective against human toxoplasmosis. Clindamycin can also be used as antimalarial drug, especially in cases of resistance against the common drugs. The combination of clindamycin and quinine exerts antibabesial activity.

Molecular Interactions Clindamycin acts as a peptidyl transferase inhibitor. It binds to the 50S subunit of bacterial ribosomes. This mechanism of action is assumed to be the same in plasmodia and babesia. The activity is directed against erythrocytic schizonts (→ Hem(oglobin) Interaction/Fig. 2). In in-vitro cell cultures and in in-vivo assays clindamycin prevents replication of *Toxoplasma gondii*, but does not inhibit protein labelling as does cycloheximide. However, PCR amplification of total *T. gondii* DNA identified an additional class of prokaryotic-type ribosomal genes, similar to the plastid-like ribosomal genes of the *Plasmodium falciparum*. Ribosomes encoded by these genes are predicted to be sensitive to the lincosamide/macrolide class of antibiotics, and may serve as the functional target for clindamycin, azithromycin, and other protein synthesis inhibitors in *T. gondii* and related parasites.

Makrolide Antibiotics

Important Compounds Erythromycin, Spiramycin, Clarithromycin, Azithromycin

Synonyms Erythromycin: none

Azithromycin: none

Spiramycin: Rovamycin, Selectomycin, Suanovil, Sequamycin, RP5337, Foromacidin, Rovamicina, Provamycin

Clarithromycin: none

Clinical Relevance The main activity of the macrolide antibiotics is their antibacterial activity. They are used as alternatives for penicillin in A-streptococcosis (tonsillitis, erysipelas, prophylaxis of rheumatic fever, scarlet, diphtheria). They are useful against lues and gonorrhoea in cases of penicillin allergy. They are alternatives for aminopenicillin in the treatment of otitis media, sinusitis, tracheobronchitis, beginning pneumonia; pertussis. In addition they are alternatives for ampicillin in the treatment of listeriosis. As alternatives for tetracyclines they can be used in the treatment of interstitial non-viral pneumonia caused by mycoplasms, chlamydia, rickettsia, non-gonorrhoic urethritis, Acne vulgaris, *Mycobacterium*-marinum-infections of the skin; *Legionella pneumonia*; Furthermore they can be used in the treatment of *Campylobacter-jejuni*-enteritis.

Erythromycin has been known of since 1952. The antibacterial activity is directed against gram positive bacteria, small gram negative bacteria (*Neisseria*, *Haemophilus*, *Bordetella*, *Legionella*, *Brucella*, anaerobic bacteria), mycoplasma, chlamydia, rickettsia, *Treponema*, *Borrelia*, and *Campylobacter*.

Spiramycin has a less antibacterial activity compared to erythromycin or azithromycin. As antiprotozoal drug spiramycin has activity against *Toxoplasma gondii* as only indication. It can be used in the treatment of acute prenatal toxoplasmosis up to the 20th week (→ DNA-Synthesis-Affecting Drugs IV/Table 2).

Clarithromycin is used in combination with a sulfonamide with good activity against human toxoplasmosis (→ DNA-Synthesis-Affecting Drugs IV/Table 2), however, the tolerability is low. Clarithromycin leads to significant reductions in *Cryptosporidium parvum* burdens in rodent models. A pre-treatment with this drug for a possible prevention of cryptosporidiosis is under investigation. Clarithromycin is useful against peptic ulcera caused by *Helicobacter pylori* in combination with metronidazole and amoxicillin and/or omeprazole (→ Antidiarrhoeal and Antitrichomoniasis Drugs).

Azithromycin is used against *Giardia lamblia*, and shows potent activities against *Cryptosporidium parvum* in tissue cultures and in rodent models. Furthermore, this drug is active against *Plasmodium* spp.. It concentrates intracellularly and tissues with a serum half-life of 2.4 days. In mice it is more active against the hepatic stage of

P. yoelii and against the erythrocytic stage of *P. berghei*. A pilot study from the Walter Reed Army Institute of Research showed the prophylactic efficacy of azithromycin against *P. falciparum* at an oral dose of 500 mg followed by 250 mg daily for seven more days. The drug seems to be as active as doxycycline.

Molecular Interactions The molecular level of bactericidal activity of erythromycin and azithromycin is the inhibition of protein synthesis in the elongation phase of polypeptides at the 50S subunit of bacterial 70S ribosomes. Thereby, the translocation of peptidyl-t-RNA from the acceptor to the donor site is inhibited.

The antiprotozoal activity of erythromycin is directed against *G. lamblia*. Thereby, the action in giardiasis is presumably indirect. The intestinal bacteria, which serve as a food supply for trophozoites in the small intestine, are eliminated by this antibiotic.

The mechanism of action of spiramycin is the inhibition of bacterial protein synthesis similar to that of erythromycin. Thereby, the polypeptide positioning at the exit channel of ribosomes becomes presumably disturbed. The action is directed against the centre of peptidyltransferase by inhibition of the peptide bond formation.

The mode of action of clarithromycin is presumably similar to that of erythromycin, spiramycin or azithromycin.

The mechanism of action of azithromycin is presumably similar to that of erythromycin. In *T. gondii* there is a distinct class of prokaryotic-type ribosomal genes encoding ribosomal proteins which may be a target of azithromycin, clindamycin and other protein synthesis inhibitors.

Aminoglycoside Antibiotics

Important Compounds Paromomycin

Synonyms Paromomycin: Aminosidina, Humatin

Clinical Relevance Aminoglycoside antibiotics possess a broad-spectrum activity. They have primarily bactericidal activity and are useful in the indications tuberculosis, Endocarditis lenta, gonorrhoea and acute septic infections.

Paromomycin was isolated in 1959. It is a fermentation product of *Streptomyces rimosus* var. *paromomycinus*. The antibacterial activity is directed against gram-positive and gram-negative bacteria. It is active against *Pseudomonas aeruginosa*, and enterobacteriaceae resistant against other antibiotics. It is used to reduce the aerobic bacterial gut flora preoperatively in patients with granulocytopenia or coma hepaticum.

Paromomycin is the only member of the aminoglycoside antibiotics with useful antiparasitic activity. It is active against *Giardia lamblia* and intestinal *Entamoeba histolytica*, and experimentally active in-vitro and in-vivo against antimony-susceptible and -resistant *Leishmania* strains. In the treatment of cryptosporidiosis a combination of letrazuril and paromomycin shows some promising results. Paromomycin is the most consistently effective anti-cryptosporidiosis agent, but its curative effects in AIDS cryptosporidiosis is erratic. An inhalation therapy with paromomycin is reported to be successful in human respiratory tract cryptosporidiosis. Clinical trials with letrazuril have now been stopped by the manufacturer. Cryptosporidiosis in young and older cats, cattle, sheep, pigs and poultry can be successfully treated with paromomycin. The particular localisation of cryptosporidia is in general responsible for difficulties in treatment since they are located intracellularly but extracytoplasmatically in the striated border of intestinal mucosa cells. Paromomycin exerts some activity against plasmodia, and there are only few reports on effects of paromomycin on microsporidiosis (*Enterocytozoon bieneusi*) in AIDS.

Moreover, paromomycin exerts activities against cestodes (e.g., *Taenia saginata* and *Hymenolepis nana*). The medical application is, however, limited because of the availability of alternative highly effective drugs. The main disadvantages of paromomycin are the long duration of treatment and often side effects.

Molecular Interactions The antibacterial action of paromomycin relies on the binding of paromomycin to 30S subunits of ribosomes. This results in misreading during protein synthesis and the appearance of nonsense proteins.

The antiprotozoal action of paromomycin is presumably the inhibition of protein synthesis. The real mechanism of action is, however, completely unknown in cryptosporidiosis. Paromomycin affects intracellular but not extracellular parasites. This can presumably be explained by the entry of paromomycin into the intracellular cryptosporidia via overlaying apical host membranes.

The anticestodal mechanism of action is unknown. It is assumed that the action against *T. saginata* is possibly due to a disruption of the tegumental membrane thus making the parasite susceptible to the host's digestive system.

Glutarimide Antibiotics

Important Compounds Axenomycin

Clinical Relevance Axenomycin is a fermentation product of *Streptomyces lisandri*. It has an anticestodal activity against *Hymenolepis nana*, *Taenia pisiformis*, *Diphyllobothrium* spp. and *Dipylidium caninum*.

Molecular Interactions The mechanism of action is due to an inhibition of EF2- and GTP-dependent translocation of peptidyl-t-RNA from the ribosomal A- to the P-site.

Glycopeptide Antibiotics

Important Compounds Streptothricin

Clinical Relevance Streptothricin has been known of since 1943. It is a fermentation product of *Streptomyces griseocarneus*. The anticestodal activity is directed against *Taenia pisiformis*, *T. taeniaeformis*, *T. hydatigena* and at higher dosages against *Dipylidium caninum*.

Molecular Interactions The mechanism of action is due to an interaction with protein-synthesis.

Diamphenethide

Synonyms Diamphenetide, Coriban

Clinical Relevance Diamphenethide was explored in 1973. It exerts exclusively antitrematodal activities. It is a unique fasciolicidal drug with higher activity against juvenile than against adult *Fasciola hepatica* (→ Energy-Metabolism-Disturbing Drugs/Table 1), being even active against 1-day-old flukes. The drug also has high efficacy against early immature flukes up to 6 weeks after infection. Thus, diamphenethide is probably successful for suppression of fasciolosis by regular treatments during periods with expected high contamination with metacercariae at intervals of 6 weeks. Thereby, much of the liver damage caused by migrating liver flukes can be prevented. Diamphenethide is regarded as an alternative drug with respect to resistance against other fasciolicides with activity against immature flukes. In addition, diamphenethide has some activity against *Dicrocoelium dendriticum*.

Molecular Interactions Diamphenethide is a phenoxyalkane-derivative. It is a prodrug which is converted by deacetylation in the host liver to the active amine compound which exerts flukicidal activity. Locally high concentration in the liver is necessary for activity against juvenile flukes. There is an age-related onset and severity of changes in the tegumental and gut cells of the flukes caused by diamphenethide. Flukes with increasing age become less susceptible to the drug. Paralysis of worms begins within 1.5–2 hours after drug exposure, surface alterations are detectable from 3 hours onwards, internal tegumental changes are seen after 6 hours and tegument flooding begins after 9 hours. An inhibition of protein synthesis is measurable from 6 hours onwards. These changes are observable at a drug concentration of 10 µg/ml near the maximum in-vivo blood level. Indeed, there is a correlation between inhibition of protein synthesis and high activity of diamphenethide against juvenile flukes. The juvenile flukes are characterised by a very active phase of growth and differentiation accompanied by a higher demand of production of tegumental secretory bodies and glycocalix turnover in juvenile flukes compared to adult flukes. The inhibition of protein synthesis is a novel mode of action for modern fasciolicides. The formerly used emetine is the only other drug against liver fluke infections in rodents, sheep and man which acts as inhibitor of protein synthesis. Emetine is also only active against intrahepatic juvenile flukes, but not against adult flukes in the bile duct. Thus there are similarities between the diamphenethide and emetine action.

Furthermore, other metabolic events are induced by the diamphenethide action. Thus, malate levels become elevated after 3 hours, end-product formation (acetate, propionate, lactate) is increased between 6 and 24 hours and ATP levels decrease by 47% after 24 hours. The drug-induced paralysis of the flukes leads to a starvation of the worms. The neuromuscular effect is the quickest event in the diamphenethide action, but the neuromusculatory action on the molecular level itself is still unclear.

Protozoan Infections, Man

Protozoan parasites (Table 2, page 517) may introduce in humans and animals a variety of more or less severe diseases, the clinical symptoms of which are strongly correlated to the pathogenic effects (listed in Tables 1, 2; pages 516, 517) and their location (in tissues, lumens, blood).

Tissues

The lesions produced by tissue protozoan infections generally are not diagnostically distinctive. Morphologic identification of the individual tissue protozoan is necessary, usually by light microscopy or aided by ultrastructural examination. This not only enables the different protozoans to be distinguished, but allows many other pathogenic microbes and non-microbial lesions to be considered histopathologically in the differential diagnosis.

The main distinguishing features of tissue-inhabiting protozoans are reviewed and contrasted in Table 1 and Table 2. *Toxoplasma gondi* (→ Toxoplasmosis, Man) and *Sarcocystis* spp. (→ Sarcosporidiosis, Man) form similar cysts which, when intact, are not accompanied by an inflammatory reaction (→ Pathology/Fig. 5). The ultrastructural characteristics of the cyst wall can be used definitively to separate the two genera (→ Pathology/Fig. 8, → Pathology/Fig. 9). The microsporidia *(Encephalitozoon* and *Nosema)* (→ Microsporidiosis) can be distinguished because their spores are acidfast, Gomori methenamine silver positive (GMS), contain one PAS-positive granule, and a coiled polar filament that can be seen with ultrastructural examination. All these are lacking in *Toxoplasma gondii*, which, however, contains as the *Sarcocystis* species many PAS-positive granules in each cyst organism. Furthermore, they possess an apical complex visible with ultrastructural examination. The leishmanias (→ Leishmaniasis, Man) and trypanosomatids (→ Trypanosomiasis, Man) can be distinguished from all the others by their → kinetoplast (Vol. 1).

RES, Reticuloendothelial system; EM, by electron microscopy; + positive finding; – negative finding; ( ) slight or rare; absence of symbol- variable, non-diagnostic; - 10± variable, diagnostic

Pruritus

Clinical and pathological symptoms (itching, scratching) of infections with skin parasites (→ Skin Diseases, Animals, → Ectoparasite (Vol. 1)).

Psoroptic Mange

→ Mange, Animals/Psoroptic Mange.

Table 1. Intestinal and other lumen-dwelling protozoans of man (according to Frenkel)

	Size (μm)		No. of nuclei in cyst	Usual location	Pathogenicity	
	Trophozoite	Cyst			Effects	Inflammatory reaction
Entamoeba histolytica	8–30	10–20	1–4	Colon	Ulceration, liver abscess	Eosinophils, few PMN blood, Charcot-Leyden crystals; granulation tissue
E. hartmanni	4–12	5–10	1–4	Colon?	None	None
E. coli	15–15	10–35	1–8	Lumen, colon	None	None
E. gingivalis	5–20	None	–	Mouth	None recognized	? Suppuration
Endolimax nana	6–15	5–14	1–4	Cecum, colon	None	None
Iodamoeba buetschlii	8–20	5–20	1	Cecum, colon	None	None
Dientamoeba fragilis	3–18	None	1–4	Cecum, colon	Diarrhea (?), abdominal discomfort	? Mucus
Blastocystis hominis		5–30	Several		None	
Acanthamoeba spp.	10–45	7–25	1	Exogenous, pharynx	Keratitis, necrosis, meningoencephalitis, abscesses	Macrophagic and granulomatous
Naegleria fowleri	7–20	7–10	1	Exogenous invader	Meningo-encephalitis	Hemorrhagic necrosis, little inflammation
Trichomonas tenax	5–12	None	–	Mouth	Commensal, lung abscess?	? None
T. hominis	5–14	None	–	Colon	None	None
T. vaginalis	5–23	None	–	Vagina, prostate gland	Pruritus, discharge, dysuria, cervical erosion, prostatitis, urethritis	Mucus, neutrophils, desquamating epithelia
Giardia lamblia	5–21	8–12	2–4	Duodenum	Epigastric pain, flatulence, diarrhea, steatorrhea, blunting of villi, microvillar border damage	Neutrophils, lymphocytes
Balantidium coli	40–200	45–75	2	Colon	Diarrhoe, ulceration, abscess, necrosis	
Cystoisospora belli		10–19 × 20–23[b]	8[b]		Abdominal discomfort, fever, diarrhea	Neutrophils, lymphocytes
Cryptosporidium spp.	2–5	4–5	4[b]	Entire intestine (bile duct, trachea)	Diarrhea	None, or lymphocytes
Sarcocystis suihominis		9 13[c]	4[c]		Vomiting, diarrhea	Mixed inflammation, with eosinophils
S. bovihominis		9 15[c]	4[c]		? None	?

[a] Trophozoite, [b]oocyst, [c]sporocyst

Table 2. Diagnostic features and characteristics of lesions produced by tissue-inhabiting protozoans (according to Frenkel)

	Toxoplasmosis	Sarcocystosis	Microsporidiosis	Pneumocystosis	Leishmaniasis			Trypanosomiasis		Malaria (*P. falciparum*)
					Visceral	Cutaneous	Mucocutaneous	cruzi	African	
Diagnostic										
Kinetoplast	–	–	–	–	+	+	+	+	+	–
Pigment in RES	–	–	–	–				–		+
Conoid (EM)	+	+	–	–	–	–	–	–	–	+
Polar granule or filament (EM)	–	–	+	–	–	–	–	–	–	–
Silver + cyst wall	+	+	–	+	–	–	–	–	–	–
PAS + bradyzoites	+	+	–	–	–	–	–	–	–	–
Acid fast spores	–	–	+	–	–	–	–	–	–	–
Grocottsilver + spores	–	–	+	+	–	–	–	–	–	–
Location										
Heart	+	+	+	(–)				+		
Lung	+	(+)	+	+						
Liver	+	?	+	(–)	+					Enlarged
RE cells	(+)	–	+	–	+	+	+			Enlarged
Brain	+		+	–				+	+	In vessels
Skeletal muscle	+	+	+	–						
Kidney	+	–	+	–						
Lesion										
With indiv. Organisms	+	±	+	+	+	±	+	+	+	–
Cysts	±	±	+	+	–	–	–	+	–	–
Neutrophils	+	+	–	–	+	+	+	+	+	–
Eosinophils	±	+	–	–	?	+	+	+	+	–
Macrophages	+	+	–	+	+++	+++	+++	+	+	+
Lymphocytes	+	+	+	+	+	+	+	+	+	+
Plasma cells	+	+	–	+	+	+	+	+	+	
Granuloma	(+)	–	+	–	+	+	+	–	+	–
Microabscess	+	+	+	–	–	+	+	+	+	–
Necrosis	+	+	+	–	Rare	Ulcer	Ulcer	+	–	–
Fibrosis	+	+	+	+	Late	Late	Late	+	+	–
Anemia and pigment deposition	–	–	–	–	+	–	–	–	+	+
Nephritis	(+)		+		+				+	+

Q

Q Fever

Synonym Query-fever

Disease of man, cattle, sheep and goats due to infection with *Coxiella burnetii*-stages by inhalation or dust-contaminated mouth parts of argasid ticks.

Therapy
Tetracyclines

R

Rabies

→ Vampire Bats (Vol. 1).

Recombinant Vaccines

→ Vaccination Against Nematodes.

Recrudescence

→ Recrudescence (Vol. 1) (e.g. *Plasmodium malariae*, babesiae) (→ Plasmodium (Vol. 1)).

Redwater Disease

→ Babesiosis, Animals.

Relapse

→ Relapse (Vol. 1), e.g. *Plasmodium vivax* (→ Plasmodium (Vol. 1))

Relapsing Fever

Relapsing fever is caused by spirochaetes (*Borrelia* spp.) and is transmitted by *Ornithodoros* tick species. Mortality in endemic areas (Central and South Africa, Asia, and America) is 2% – 5%, but can reach 50% in epidemics. Treatment is possible with antibiotics (penicillin, tetracycline). The causative organism is transmitted by the bite of the tick as well as through infected coxal fluid and can penetrate unbroken skin.

Repellents

Important Compounds

N,N-diethyl-m-toluamide (DEET), dimethylphthalate, picaridin, piperidin.

General Information

Among methods to protect human beings and particularly companion animals against blood sucking arthropods, repellents play an important role. Different modes of action are discussed for repellents. Electrophysiological observations indicate that inhibitory effects on the arthropods receptors responsible for host odours recognition or an interference with the olfactory input important for a host-characteristic response pattern might cause the repellent effect.

DEET probably was the most frequently used repellent. In a variety of different formulations it displays a broad repellent efficacy against biting arthropods. DEET is superior to another drug, dimethyl phthalate against some ticks and mosquito species. Particularly DEET displays some undesirable properties in that it sometimes causes irritation effects and is incompatible with some synthetic matrices.

Recently a new repellent was discovered and developed, picaridin/hepidamin, an acylated 1,3-aminopropanol, which is even superior with regard to its biological activity against insects and ticks and additionally does not display adverse solvent action on plastic materials.

Most recently a substance (Bayrepel*) was introduced which is a derivate of the 2-(2-hydroxyethyl)-piperidin. This product has an enlarged period of protection (ticks: 4 hours; mosquitoes: 6 hours) compared to DEET and better properties upon contact with skin and clothes.

Related Entries

→ Synergists, → Arthropodicidal Drugs.

Reservoirs

Animals containing identical stages of parasites as found in humans, but the symptoms of disease are mostly less strong, so that these animals are often the source for human infections.

Resistance Against Drugs

Genetically transmitted decreased sensitivity against drugs. See → Resistance (Vol. 1), → Drug and → Chemotherapy.

Respiratory System Diseases, Animals

General Information

A wide variety of clinical signs and pathological lesions may result from invasion of the respiratory system by parasites. The severity of the clinical manifestations depends largely on the number of parasites present and on which part of the tract they normally invade. It is also determined by individual variations between hosts in the anatomical and physiological features of the respiratory tract, or in the nature of the response to infection. As for many other parasitic diseases, the parasites that cause respiratory problems do not all live in the respiratory system as adults; some (e.g., *Ascaris*, *Ancylostoma*, *Strongyloides*) pass through the lungs in the normal course of their migration whereas others may penetrate the system by error (e.g., *Fasciola*). Finally, parasites residing primarily in other systems may cause syndromes in which respiratory difficulties are one of the foremost presenting signs (e.g., *Dirofilaria immitis*).

Manifestations and Pathophysiology

As with other parasitic diseases the clinical signs resulting from infection of the respiratory tract may vary from mild to very severe. The principal clinical manifestations of respiratory dysfunction are hyperpnea, tachypnea, dyspnea, respiratory noises and nasal discharge. Table 1 provides a list of the different parts of the respiratory tract and of the clinical signs which may be observed when they are affected.

Table 1. Clinical signs associated with specific anatomic involvement of the respiratory tract (according to Vercruysse and De Bont)

Anatomic area	Clinical signs
Nasal cavity	Nasal discharge, snorting, sneezing, nasal rubbing
Sinuses	Nasal discharge
Larynx	Dyspnea, coughing
Trachea	Harsh resonant cough, dyspnea
Bronchi or bronchioles	Coughing, dyspnea
Alveoli	Cough when associated with bronchial pathology, hyperpnea, dyspnea

Parasites living in the nasal passages and/or the sinuses induce inflammatory reactions of the mucosa. Rhinitis and sinusitis start as catarrhal inflammations characterized by a nasal discharge which is serous initially, but rapidly becomes mucoid and purulent. Sneezing is also a characteristic (Fig. 1). Wheezing and stertor may be present when both nostrils are partially obstructed. The irritation may cause the animal to shake its head, or rub its nose against the ground or its front legs. Inflammation of the larynx, trachea and bronchi are characterized by cough, noisy inspiration, and some degree of inspiratory dyspnea (Fig. 1, Fig. 2).

Pneumonia is an inflammation of the lung, usually accompanied by inflammation of the bronchioles and sometimes by pleuritis (pleuropneumonia). It is manifested clinically by an increase in respiratory rate, cough, dyspnea and sometimes nasal discharge (Fig. 1, Fig. 2). The most important parasites causing bronchiolitis and alveolitis are helminths belonging to the Dictyocaulidae and Metastrongyloidea. The reaction to the presence of eggs, larvae or adults is comparable to the one induced by foreign bodies, with accumulation of masses of eosinophils, macrophages and giant cells. The lesions may cause the dysfunction of large masses of lung tissue. Emphysema commonly accompanies pneumonia, particularly in cattle. It is presumed to be caused by a combination of blockage of bronchioles and violent coughing so that air retained in the alveoli exerts enough pressure to cause a rupture of the alveolar walls.

The larvae of several intestinal nematodes (e.g., ascarids, hookworms and *Strongyloides*) normally migrate through the lungs. Once in the bronchioles, they are coughed up and swallowed, to even-

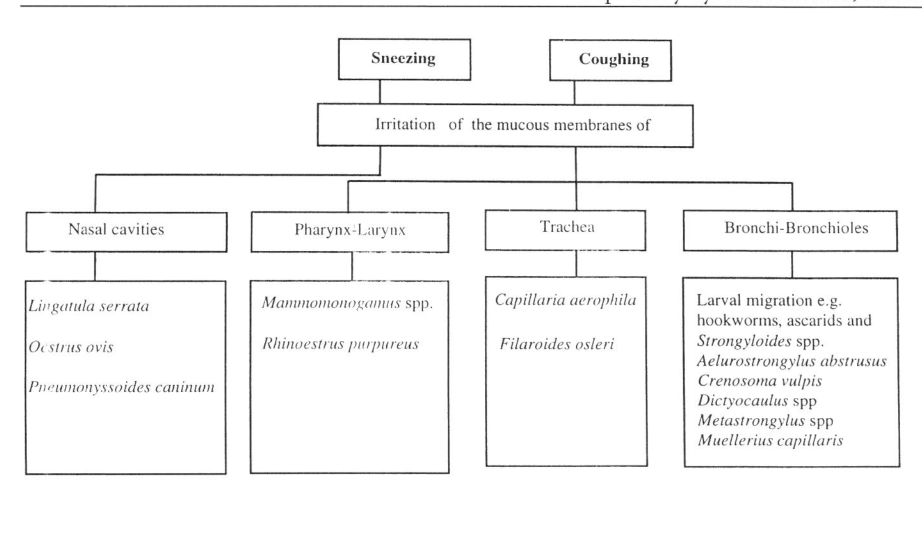

Fig. 1. Parasites causing sneezing and coughing

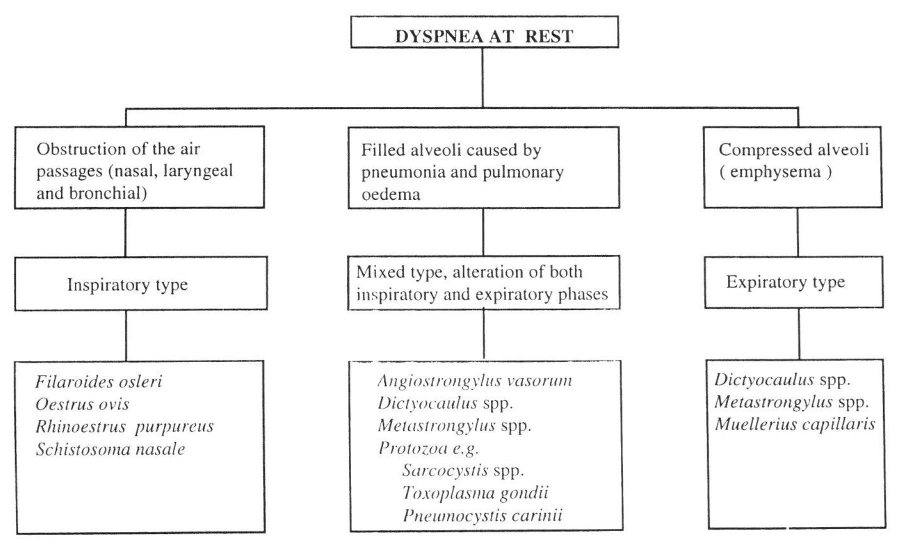

Fig. 2. Parasites causing dyspnea at rest

tually reach the small intestine where they mature. In the lungs, they cause a transitory eosinophilic alveolitis and bronchiolitis. Although the pathogenesis of the lesions is attributable mainly to the physical damage caused by larvae, there is also evidence of hypersensitivity phenomenon in the modulation of infection.

Most parasitic infections of the respiratory tract are associated with peripheral eosinophilia.

Related Entries

→ Respiratory System Diseases, Ruminants, → Respiratory System Diseases, Horses, Swine, Carnivores.

Treatment

→ Chemotherapy, → Drugs.

Respiratory System Diseases, Horses, Swine, Carnivores

The common clinical signs and pathology of parasitic infections of the respiratory system of horses, swine and carnivores are summarized in Table 1.

Nasal Cavity and Sinuses

Linguatula serrata is a pentastomid which infects the nasal passages of dogs and rarely horse, goat and sheep. The adult parasites are up to 12 cm long and lie on the surface of the nasal mucosa. No symptoms of nasal irritation are apparent except for occasional sneezing, and sometimes epistaxis.

Table 1. Parasites affecting the respiratory system of horses, swine, dogs (according to Vercruysse and De Bont)

Parasite	Type	Host	Location	Clinical presentation	Principal lesions
Protozoa					
Pneumocystis carinii	1	Dog	Alveoli	Clinical signs when impairment of host resistance: dyspnea, tachypnea, cough, cyanosis	Interstitial pneumonia with massive mononuclear cell infiltration
Sarcocystis spp.	3	Carnivores	Vascular endothelium	Anorexia, fever, weight loss, anaemia and dyspnea	Lungs: mild interstitial pneumonitis and vasculitis
Toxoplasma gondii	1	Dog	Alveoli	Anorexia, fever and lethargy accompanied by dyspnea	Lungs: focal areas of coagulation necrosis, adjacent to and involving small vessels and bronchioles
Trematoda					
Paragonimus spp.	1	Cat, dog	Lung parenchyma	Mild intermittent coughing, expiratory wheezing and, on occasion, acute dyspnea	Cystic lesions, eosinophilic granulomatous pneumonia
Nematoda					
Rhabditida					
Strongyloïdes spp.	3	Pig, dog	Small intestine, larvae migrate through lungs	Light coughing	Transitory eosinophilic alveolitis and bronchiolitis
Strongylida					
Aelurostrongylus abstrusus	1	Cat	Terminal bronchioles	Occasionally coughing, sneezing, oculo-nasal discharge	Light bronchiolitis, hypertrophy of the arterial and arteriolar smooth muscles
Ancylostoma spp. and other hookworms	3	Carnivores	Small intestine, larvae migrate through lungs	Soft cough	Petechial haemorrhages on the lungs with transistory alveolitis and bronchiolitis
Angiostrongylus vasorum	3	Dog	Pulmonary artery and rarely in right ventricle	Dyspnea and dead of cardiac insufficiency	Pulmonary oedema, granulomatous interstitial pneumonia
Crenosoma vulpis	1	Dog, cat	Bronchi, occasionally trachea	Coughing and dyspnea	Catarrhal eosinophilic bronchitis
Dictyocaulus arnfieldi	1	Donkey, horse	Bronchi	Clinical signs mainly in horses: coughing, hyperpnea and nasal discharge	Bronchitis, bronchiolitis, pneumonitis
Filaroides spp.	1	Dog	Alveoli and bronchioles	Usually no clinical signs, respiratory distress has been noted	Foci of granulomatious interstitial pneumonia
Filaroides osleri	1	Dog	Trachea-bronchial bifurcation	Non-productive cough, dyspnea, exercise intolerance, cyanosis	Submucosal, firm nodules
Metastrongylus spp.	1	Pig	Bronchioles, small bronchi	Husky cough, slight dyspnea	Bronchitis, bronchiolitis, pneumonitis
Spirurida					
Dirofilaria immitis	3	Dog, cat	Right heart ventricle, the pulmonary artery	Deep, soft, cough, haemophtysis, dyspnea	Endarteritis, secondary pulmonary parenchymal lesions

Table 1. (continued)

Parasite	Type	Host	Location	Clinical presentation	Principal lesions
Ascaridida					
Ascaris suum	3	Pig	Small intestine, transit of larvae through lung parenchyma	Coughing	Transitory eosinophilic alveolitis and bronchiolitis, lesions more marked in repeated infections
Parascaris equorum	3	Horse	Small intestine, transit of larvae through lung parenchyma	Cough and nasal discharge	Transitory eosinophilic bronchitis and bronchiolitis
Toxocara canis	3	Dog	Small intestine, transit of larvae through lung parenchyma	Coughing, in young dogs dyspnea, death	Transitory multifocal interstitial pneumonitis
Enoplida					
Capillaria aerophila	1	Dog, cat	Nasal cavity, trachea	Nasal discharge, mild cough, in heavy infestations dyspnea	Mild catarrhal rhinitis and tracheitis
Arthropoda					
Acaridia					
Pneumonyssus caninum	1	Dog	Nasal passages, sinuses	Occasionally sneezing	Mild rhinitis, sinusitis
Diptera					
Rhinoestrus purpureus	3	Horse	Nasal passages, sinuses, pharynx	Nasal discharge,sneezing dead (larval aspiration)	Rhinitis, sinusitis, pharyngitis
Pentastomida					
Linguatula serrata	1	Dog, more rarely in horse	Nasal cavities, occasionally paranasal sinuses	Occasionally sneezing with mucous discharge	Catarrhal rhinitis

1, A primary parasite of the respiratory system. 2, Affects the lungs through normal migration or proliferation. 3, Parasites of another organ system that produces respiratory symptoms

The mite *Pneumonyssus* (*Pneumonyssoides*) *caninum* is occasionally found in the nasal passages and sinuses of dogs. It is usually an incidental finding not associated with clinical signs or the development of lesions. However, there are reports of the mites causing mild rhinitis, sinusitis, and even bronchitis. Clinical signs include chronic sneezing, head shaking, epistaxis, and impaired scenting ability.

A naso-pharyngeal myiasis of horses, with similar clinical symptoms as *Oestrus ovis* is caused by *Rhinoestrus purpureus*, the Russian gadfly. Leeches such as *Dinobdella ferox* commonly enter the nasal cavities of domestic animals, mainly in southern Asia. They suck blood, generally induce inflammation, and may impede breathing.

Larynx and Trachea

Capillaria aerophila (*Eucoleus aerophilus*) is found mainly in the trachea and the nasal cavity of dogs and cats. Mild infestations are asymptomatic and provoke a mild catarrhal inflammation with nasal discharge and a mild cough. In heavy infestations a persistent dry cough and intermittent dyspnea may be observed.

Filaroides (Oslerus) osleri is an ovoviviparous, filiform worm parasitizing the dog and related species. The worms cause submucosal nodules of up to 10 mm in diameter in the region of the tracheal bifurcation. The clinical symptoms are proportional to the severity of infection and the number and size of the tumors. The most frequently recorded symptom is a sporadic but persistent non-productive cough. The most severe clinical cases, which usually occur in pups under 1 year of age, show persistent coughing, respiratory distress and emaciation, with up to 75% mortality rates in affected litters.

Lower Respiratory Tract

As for ruminants, some species such as the lungworms use the lower air passages and more rarely the lung parenchyma as final habitat. Others species just pass through the lungs during their migration, causing various degrees of damage according to the nature and intensity of the host-parasite interaction.

Protozoa → *Toxoplasma gondii* (Vol. 1) can infect a wide variety of cell types in nearly all warm-blooded animals, and the clinical picture in a particular host species depends on the particular involvement of any one or more of these organs. In dogs and cats, involvement of the lungs is a common feature. In dogs, clinical toxoplasmosis is most frequent in puppies, and is often complicated by the simultaneous manifestations of distemper. The clinical syndrome is variable in course but anorexia, fever and lethargy accompanied by dyspnea are the essential features. Proliferation of the organism in the lungs leads to focal areas of coagulation necrosis adjacent to and involving small vessels and bronchioles and exudation of fibrin.

Pneumocystis carinii is a protist of uncertain taxonomy which inhabits the pulmonary alveoli of dogs, horses, goats, pigs and humans, and which may give rise to severe respiratory distress in hosts who suffer from immunodeficiency. The onset of pneumocystosis can be acute or insidious. Signs of dyspnea, tachypnea, cough, periodic cyanosis and weight loss were noted in otherwise alert animals. The disease may take a progressive course and death can occur in a few weeks if the patient is not treated.

Nematodes In contrast with horses, donkeys are only slightly affected by even heavy infections with → *Dictyocaulus arnfieldi* (Vol. 1). It has been concluded that donkeys are the natural host of the parasite. Infections in horses are generally non-patent, but may be associated with clinical signs such as coughing, increased respiratory rate and nasal discharge.

There are three important species of *Metastrongylus*: *M. elongatus* (*M. apri*), *M. pudendotectus* and *M. salmi*. They are all parasitic in the bronchi and bronchioles of pigs. The lesions caused by the parasites resemble those caused by *Dictyocaulus* spp. in ruminants but, generally speaking, with much lighter clinical signs. Infected pigs often develop a husky cough which, if infections are heavy, may get superimposed by often fatal bacterial or viral infections.

Aelurostrongylus abstrusus is a small metastrongyle that develops in the bronchioles of cats and is capable of eliciting an extensive bronchiolitis and interstitial pneumonia. In clinically infected cats abnormal respiratory signs with coughing, sneezing and some degree of oculo-nasal discharge appear 6–12 weeks after infection. Most cases show little clinical disturbance, because the lesions regress spontaneously as immunity develops.

Crenosoma vulpis, a small metastrongyle, occurs in the bronchi and occasionally the trachea of foxes, dogs and cats. The adult worms induce a catarrhal eosinophilic bronchitis with heavy coughing and dyspnea.

Filaroides milski and *F. hirchi* live in the alveoli and bronchioles of dogs causing an interstitial pneumonia. Clinical signs of infection are rare although respiratory distress and even mortality has been reported.

Paragonimus kellicotti is a digenetic fluke that inhabits fibrous cysts in the lungs of wild carnivores and domestic cats and dogs. P. westermani occurs in the lungs and more rarely in the brain and spinal cord of dogs, cats, wild animals and humans. The clinical signs include mild coughing, expiratory wheezing and on occasion, acute dyspnea. Pneumothorax may be a rare complication in both dogs and cats. Lesions are mainly due to the adult flukes which are found, usually in pairs, in inflammatory cysts in the pulmonary parenchyma and occasionally the bronchi of predominantly the right caudal lung lobes.

Angiostrongylus vasorum occurs in the pulmonary artery and rarely in the right ventricle of the dog and the fox. They cause a proliferative endarteritis comparable to that induced by *Dirofilaria immitis*, but the more severe damage is caused by eggs that lodge in arterioles and capillares. Together with the larvae they provoke a chronic inflammation in which fibroplasia predominates. The larvae break into the alveoles and migrate in the respiratory passages. Affected animals suffer from dyspnea and may die of heart failure. The fatal outcome is largely attributable to pulmonary oedema and pneumonia.

Infection with *D. immitis* in dogs generally gives rise to respiratory involvement. The pneumonitis is caused by granuloma formation around microfilariae trapped in the lung. The clinical signs of heart worm disease are discussed in the section on parasitic vasculitis (→ Cardiovascular System Diseases, Animals). The larvae of several intestinal nematodes pass through the lungs during the course of their normal migration. Lesions and respiratory signs are most pronounced with the larvae of the ascarids *Ascaris suum*, *Parascaris equorum* and *Toxocara* spp, in pigs, horses and carnivores, respectively. The migration of *P. equorum* larvae through the lungs provokes an inflammatory reaction which may result in mild to severe respiratory distress. Larvae of *T. canis* are often present in the lungs of newborn puppies (after transplacentary migration), and heavy infections result in substantial haemorrhage into the alveoli during the first day of life. This may prove fatal. Pulmonary lesions are characterized by granulomas and multifocal interstitial pneumonitis.

Acute respiratory signs (coughing) may appear in young animals a few days after exposure to *Strongyloides westeri* (horses). The respiratory signs caused by migration of hookworms (dogs) are less pronounced.

Aberrant migrations through the respiratory system of parasites such as *Fasciola* spp. or *Spirocerca lupi* do not usually cause clinical signs and are only detected post-mortem.

Treatment/Therapy

→ Chemotherapy, → Drugs.

Respiratory System Diseases, Ruminants

The common clinical signs and pathology of parasitic infections of the respiratory system of ruminants are summarized in Table 1.

Nasal Cavity and Sinuses

The larvae of a number of flies of the family Oestridae are parasites of nasal cavities and sinuses of domestic animals. The most ubiquitous is the nasal botfly of sheep and goats, *Oestrus ovis*. The flies cause great stress when they attack the sheep to deposit larvae near the nostrils of the host, a process which significantly interferes with grazing and rumination. The larvae which develop in the nasal cavities may get up to 3 cm in length and cause severe discomfort, partly because of the damage caused by the oral hooks and cuticular spines of the larvae, but also because of hypersensitivity phenomena. Nasal discharge and sneezing are common features in affected sheep, with caked dust obstructing the nostrils. Head shaking and nose rubbing may sometimes be seen. Extension to the cranial cavity via the ethmoid causes nervous symptoms and is usually fatal, but is also very rare.

Schistosoma nasale lives in the nasal veins of a variety of domesticated animals on the Indian subcontinent, including cattle, water buffalo, sheep, goat, and rarely horse. The lesions which are granulomatous in nature are caused by the passage of eggs through the wall of the nasal cavity. The condition is associated with cauliflower-like growths on the nasal mucosa causing partial obstruction of the cavity and snoring sounds when breathing. The lesions tend to get more severe in older animals and may become very spectacular in cattle, leading to "snoring disease".

→ Leeches (Vol. 1) such as *Dinobdella ferox* commonly enter the nasal cavities of domestic animals, mainly in southern Asia. They suck blood, generally induce inflammation, and may impede breathing.

Larynx and Trachea

Mammomonogamus laryngeus occurs in the larynx and the trachea of cattle and humans in South-East Asia and South America, and *M. nasicola* is found in nasal cavities, trachea, larynx and bronchi of sheep, goats and cattle in Africa and South America. In animals there are no apparent symptoms except for a light coughing. There is one record of a fatal infection in sheep believed to be due to a respiratory obstruction initiated by *M. nasicola*.

Lower Respiratory Tract

Parasitic infections of the lower respiratory tract of ruminants are very common and important. Some species such as the lungworms use the lower air passages and more rarely the lung parenchyma as final habitat. Others species just pass through the lungs during their migration, causing various degrees of damage according to the nature and intensity of the host-parasite interaction.

Table 1. Parasites affecting the respiratory system of ruminants (according to Vercruysse and De Bont)

Parasite	Type	Host	Location	Clinical presentation	Principal lesions
Protozoa					
Sarcocystis spp.	3	Ruminants	Vascular endothelium	Anorexia, fever, weight loss, anaemia and dyspnea	Lungs: mild interstitial pneumonitis and vasculitis
Trematoda					
Schistosoma nasale	1	Cattle	Nasal mucosal veins	Muco-purulent discharge, dyspnea, snoring	Granulomas in nasal mucosa
Nematoda					
Rhabditida					
Strongyloïdes spp.	3	Ruminants	Small intestine, larvae migrate through lungs	Light coughing	Transitory eosinophilic alveolitis and bronchiolitis
Strongylida					
Hookworms	3	Ruminants	Small intestine, larvae migrate through lungs	Soft cough	Petechial haemorrhages on the lungs with transistory alveolitis and bronchiolitis
Dictyocaulus filaria	1	Sheep, goat	Small bronchi	Coughing, hyperpnea and dyspnea	Bronchitis, bronchiolitis, pneumonitis
D. viviparus	1	Cattle	Bronchi, bronchioles	Coughing, hyperpnea and dyspnea	Bronchitis, bronchiolitis, pneumonitis, pulmonary oedema, emphysema
Mammomonogamus spp.	1	Ruminants	Nasal cavities, larynx, trachea and bronchi	Light coughing	Chronic inflammation with small ulcerations of the mucosa of upper airways
Muellerius capillaris	1	Sheep, goat	Alveoli, pulmonary parenchyma, subpleural tissue	Usually no clinical evidence, sometimes persistent coughing, dyspnea	Bronchiolitis, pneumonitis with nodular lesions
Protostrongylus spp.	1	Sheep, goat	Bronchioles	Usually no definite clinical signs	Bronchiolitis, lobular pneumonitis
Ascaridida					
Ascaris suum	3	Cattle	Small intestine, transit of larvae through lung parenchyma	Coughing	Transitory eosinophilic alveolitis and bronchiolitis, lesions more marked in repeated infections
Arthropoda					
Acaridia					
Diptera					
Oestrus ovis	3	Sheep, goat	Nasal passages, sinuses	Nasal discharge, frequently sneezing, nasal rubbing	Catarrhal rhinitis and sinusitis

1, A primary parasite of the respiratory system. 2, Affects the lungs through normal migration or proliferation. 3, Parasites of another organ system that produces respiratory symptoms

Protozoa → *Sarcocystis* (Vol. 1) spp are very common ubiquitous parasites of ruminants, but they rarely cause clinical signs in infected animals. Acute forms of sarcocystosis are associated with the massive development of schizonts in endothelial cells of blood vessels. The clinical signs include high fever, anorexia, anemia, ataxia, loss of weight, and dyspnea. A multifocal interstitial pneumonitis and vasculitis are responsible for the respiratory signs.

Lungworms → *Dictyocaulus viviparus* (Vol. 1) is certainly the most important lungworm of cattle in temperate areas. The severity of the clinical signs depends on the susceptibility of the host and on the number of invading larvae. Cattle are most susceptible to infection when they are first exposed to contaminated pastures. Since the occurrence of the primoinfection varies, dictyocaulosis (husk) can be seen in all age classes. In early infections lesions are mainly found in the alveoli, which the larvae penetrate from lymphatics and blood vessels. An eosinophilic exudate accumulates in the alveoli and the terminal bronchioles. Hyperpnea and coughing may become noticeable as soon as 10–14 days after heavy infections. Occasionally, fatal pulmonary oedema and emphysema develop at this stage, probably as a result of hypersensitivity reactions. During the patent period (25–55 days after infection), adults reside and lay eggs in the bronchi where they induce a hyperplasia of the mucosa. Eosinophilic exudate obstructs the lumen of the bronchi, which result in atelectasis of the alveoli distal to the plugs. In addition, eggs aspirated into the alveoli initiate foreign body reactions. The overall consequence of dictyocaulosis is a diffuse consolidation of the lungs. The animals show dyspnea and coughing, with rapid loss of condition. Harsh respiratory sounds with ronchi and emphysemaous crackling sounds can be heard. The post-patent phase of the disease is often one of gradual recovery in that the respiratory rate decreases, weight gain is resumed and the coughing abates.

D. filaria causes outbreaks of pulmonary nematodosis in sheep and goats in most temperate areas of the world, often with high mortality rates. The pathogenesis and clinical signs appear to be similar to those of *D. viviparus* infections in cattle.

The Protostrongylidae are common lungworms of sheep and goats. They include *Muellerius capillaris*, *Protostrongylus rufescens*, *P. brevispiculum*, *P. kochi* and *Cystocaulus ocreatus*. These parasites are of minor pathogenic importance. Infections with *Muellerius* and *Cystocaulus* are generally associated with small, spherical nodular lesions in the lung tissue, whereas *Protostrongylus* causes irritation and local inflammatory reactions in the bronchioles resulting in small foci of lobular pneumonitis. Generally animals show no clear symptoms although in the rare heavy infections, and especially with *Protostrongylus* in sheep and *Muellerius* in goats, there may be severe and even fatal disease.

Other Parasites Larvae of *Ascaris suum* may be responsible for an atypical interstitial pneumonia in grazing cattle. Signs of acute respiratory distress such as severe dyspnea, expiratory grunt, hyperpnea and moist cough appear about 10 days after application of contaminated pig manure as a slurry to the pasture.

Treatment/Therapy

→ Chemotherapy, → Drugs.

Rhinoestrosis

Disease due to infestation with *Rhinoestrus*, see Table 1

Rhinoestrus purpureus

Agent of → Myiasis in horses found in the conjunctival sac and in the choanes.

Rickettsiae

Family of intracellularly reproducing bacteriae transmitted by ticks (e.g. *R. rickettsi*) and lice (e.g. *R. prowazeki*, *R. quintana*). Rickettsial agents in animals see → Ehrlichiosis.

RIF

Variant antigen (→ Malaria/Vaccination)

Table 1. *Rhinoestrus* species

Parasite	Host	Symptoms	Country	Therapy		
				Products	Application	Compounds
Rhinoestrus spp.	Horse	Swelling of nasal and pharyngeal cavity, cough, loss of strength, death	Worldwide	Eqvalan (Merial)	Oralpaste	Ivermectin

River Blindness

→ Onchocerciasis, Man, → *Filariidae* (Vol. 1).

Roble's Disease

→ Onchocerciasis, Man, → *Filariidae* (Vol. 1).

Rochalimea

Lice-transmitted bacteria now belonging to the genus *Bartonella* (→ Trench Fever).

Rocky Mountain Spotted Fever

Rocky Mountain spotted fever is caused by *Rickettsia rickettsi*. It is found from Canada to South America and is associated with various ixodid ticks (*D. andersoni* and *D. variabilis* in North America) which can transmit the pathogen transovarially to the next generation. All tick stages can harbor and transmit agents of the disease. Argasid tick species may also be involved. The disease can be acquired through the tick bite or through contact with tick tissues when the tick is crushed. The incubation period is 2–5 days in severe cases and up to 14 days in mild cases. In untreated cases, death occurs 9–15 days after onset of symptoms. Mortality was formerly high but has been reduced through antibiotic treatments (tetracyclines).

Related Entries
→ Boutonneuse Fever, → Tick Typhus.

Rosacea migrans

→ Lyme Disease.

Ross Malaria Model

→ Mathematical Models of Vector-Borne Diseases.

Roundworm Disease

Synonym
→ Ascariasis due to *Ascaris lumbricoides* (→ Ascaris (Vol. 1)).

RSSE

Synonym
→ Russian Spring-Summer Encephalitis.

Russian Gadfly

Rhinoestrus purpureus leads to nasopharyngeal → Myiasis in horses and donkeys.

Russian Spring-Summer Encephalitis

Synonym
RSSE

The Russian spring-summer encephalitis which is caused by the RSSE virus (→ *Flavivirus* (Vol. 1), group B) has a mortality rate of up to 25%–30%. It is a complex of viruses with a wide geographical range from East Germany to Siberia and the Soviet Far East, and possibly into North China. It is mainly associated with the tick *Ixodes persulcatus*, but also with *Haemaphysalis concinna* and other tick species, and may also have other arthropod reservoirs. In endemic areas (such as in taiga forests), over 50% of residents may have antibodies without showing symptoms, while newcomers to these areas more frequently exhibit clinical symptoms.

S

Sabin's Tetrad

→ Toxoplasmosis, Man.

SAG

Surface antigens (→ Toxoplasmosis, Man)

Salmon Poisoning

Disease in dogs with a high mortality rate due to infections with *Neorickettsia helminthoeca* imported into the dog with infections of the trematode *Nanophyetus salmincola* (→ Alimentary System Diseases, Carnivores).

Sarcocystosis

→ Sarcosporidiosis, Man, → Sarcocystis (Vol. 1).

Sarcoptes Mange

→ Sarcoptes (Vol. 1), → Scabies

Sarcoptic Mange

→ Mange, Animals/Sarcoptic Mange, → Scabies

Sarcosporidiosis, Animals

→ Sarcocystis (Vol. 1).

Sarcosporidiosis, Man

Pathology

Intestine → *Sarcocystis suihominis* (Vol. 1) and S. *bovihominis* are ingested as bradyzoites from cysts in infected pork and beef (→ Pathology/Fig. 7, → Pathology/Fig. 8, → Pathology/Fig. 9). There is no schizogony in the human gut. The ingested bradyzoites develop directly into gametocytes in the lamina propria; fertilization then occurs, followed by development of a zygote and an oocyst. Sporulation takes place in the gut wall, with two sporocysts, each forming four sporozoites. Judging from a few isolated observations, where, however, the possibility of the presence of other pathogen could not be excluded, there may be intense mixed inflammatory reaction that includes eosinophils. Experimental infections suggest the absence of a strong immune reaction, and the possible hypersensitivity reactions, resulting in the greater pathogenicity of *S. suihominis* than of *S. bovihominis.* Sporulated sporocysts are usually shed in the feces and appear to be infectious only to pigs (*S. suihominis)* or cattle *(S. bovihominis).* In older references these two organisms were not distinguished and identified as *Isospora hominis.* The *Sarcocystis* stages found in human muscle result from infection with different species.

Muscle Sarcocysts different from those in the intestine are diagnosed occasionally in the myocardium and skeletal muscle of humans (→ Pathology/Fig. 7) who are accidental intermediate hosts for sarcocysts normally found in monkeys and possibly cattle. Mammalian, avian, or reptilian carnivores may be the definitive hosts ingesting bradyzoites in cysts and shedding oocysts in their feces which are accidentally ingested by humans. Intact cysts with bradyzoites measuring up to 100 µm in diameter and 325 µm in length and unaccompanied by inflammation are the usual findings (→ Protozoan Infections, Man/Table 1). However, cases with young cysts and diffuse lymphocytic and eosinophilic infiltration have been described, probably from recent infections. Whether the inflammation is a consequence of the prior schizogony in blood vessels, or of young cysts that degenerate, or both, has not been determined. Evidence of lesions from cyst rupture has been reported, where an eosinophilic myositis with lymphocytes, and later fibrosis, was diagnosed by biopsy of one of a number of suddenly appearing spontaneous lesions. These events occurring over many years, were accompanied by painful muscle swelling, initially associated with erythema and occasionally with bronchospasm, and lasted for 2 days to 2 weeks.

Intestinal sarcosporidiosis:
Main clinical symptoms: Vomiting, sweating, diarrhoea
Incubation period: 4–8 hours
Prepatent period: 5–10 days
Patent period: 6–8 weeks
Diagnosis: Microscopic determination of sporocysts in fecal samples
Prophylaxis: Avoid eating raw meat of pork or cattle
Therapy: Treatment see → Coccidiocidal Drugs and → Drugs Against Sarcocystosis.

Scabies

Scabies is a human infection of the skin with the → mite (Vol. 1) → *Sarcoptes scabiei* (Vol. 1) or a transitory infection with similar species from animals where they produce mange. The entire life cycle of the mite takes place in the keratinaceous layer of the epidermis which often undergoes hyperplasia with marked hyperkeratosis around papules or burrows created by the mites, often on the hands, feet, genitalia or axilla. Adults create burrows in the keratinaceous layer where they live and lay eggs, which give rise to larvae, nymphs and new adults (→ Pathology/Fig. 30B). Metabolic products of the mites give rise to inflammation with lymphocytes and eosinophils in the dermis. In the absence of a thickened stratum corneum, mites elicit more inflammation, more itching and often ulcers from scratching, sometimes with secondary bacterial infection. Glomerulonephritis may result from secondary streptococcal infection of the lesions.

In immunosuppressed individuals there may be confluent infection with thousands of mites and widespread epidermal scaling, so-called Norwegian scabies, that can be diagnosed by scrapings or skin biopsies.

Main clinical symptoms: Pruritus, exanthemes, skin scaling
Incubation period: 1–2 weeks
Prepatent period: First eggs are laid after 15 days
Patent period: Years, since many generations of females may follow each other
Diagnosis: Microscopic determination of mites in skin scrapings
Prophylaxis: Avoid contact with infested people and exchange of clothes; use fresh bed covers etc.
Therapy: Treatment see → Acarizides (Vol. 1); use of ivermectin.

Related Entry

→ Mange, Man.

Scabies crustosa

Synonym

Scabies norvegica

Severe skin symptoms in persons with immuno-deficiency; highly infectious due to formation of huge numbers of mites.

Scale

Clinical and pathological symptom of infections with skin parasites (→ Skin Diseases, Animals, → Ectoparasite (Vol. 1)).

Schistosomiasis, Animals

Pathology

Different species of the genus → *Schistosoma* (Vol. 1) give rise to infection in several domestic animals. *Schistosoma japonicum* have been found within the mesenteric and hepatic portal veins of pigs and dogs. Although *Schistosoma* infections in ruminants are highly prevalent in certain regions, the general level of infestation is often too low to cause clinical disease or losses in productivity. Levels sufficiently high to cause outbreaks of clinical schistosomosis do occur occasionally and infestation becomes manifest either as an intestinal syndrome which is usually self limiting, or as a chronic hepatic syndrome, which is usually progressive. The intestinal syndrome is caused by the deposition of large numbers of eggs in the intestinal wall and usually follows a heavy infestation in a susceptible animal, i.e. an animal in which the capacity of the host to suppress the egg laying of the parasite has not been stimulated by previous infestations. This has been reported among cattle, sheep and goats infected with either *S. bovis* or by *S. mattheei*. As the faecal egg counts rise sharply with the onset of egg production the animal develops a mucoid and then haemorrhagic diarrhoea, accompanied by anorexia, loss of condition, general weakness and dullness, roughness of coat hypoalbuminaemia and paleness of mucous membranes. Death may occur a month or two after the onset of clinical signs. In most cases, the animal makes a spontaneous but slow recovery. The primary cause of the diarrhoea is the passage of large numbers of eggs through the wall of the intestine. The → anaemia is usually due to an increased rate of red cell removal from the circulation; while haemodilution and the inability to mount a sufficiently effective erythropoietic response are of secondary importance. The underlying cause of the hypoalbuminaemia is hypercatabolism of albumin due to substantial loss of protein in the gastro-intestinal tract (→ Cardiovascular System Diseases, Animals).

Vaccination

→ Schistosomiasis, Man/Vaccination.

Immune Responses

→ Schistosomiasis, Man/Immune Responses.

Therapy

→ Trematodocidal Drugs.

Schistosomiasis, Man

Synonym

Bilharziosis, Bilharziasis

General Information

This complex of diseases is caused by *Schistosoma haematobium*, *S. intercalatum*, *S. japonicum*, *S. mansoni* and *S. mekongi*. Schistosomiasis is acquired from free-swimming fresh-water cercariae that penetrate the skin or are swallowed with fecally contaminated water from snail-infested sources (→ *Schistosoma* (Vol. 1)). As they penetrate the skin cercariae lose their tail and become **schistosomules**. Growing couples of monogamous male and female migrate to intestinal (*S. mansoni* and *S. japonicum*) or urogenital (*S. haematobium*) venules where females lay hundreds to thousands of eggs per day. The great variety of lesions are superbly described and illustrated by McCulley et al. and Lichtenberg. The granulomatous response is subject to multiple immunologic mechanisms.

Distribution

There are an estimated 200 million people, in 74 countries, infected with schistosomes. Intestinal schistosomiasis caused by *S. mansoni* occurs widely in tropical Africa, in some parts of North Africa and Southwest Asia as well as parts of South America and the Caribbean. *S. intercalatum* is limited to parts of tropical Africa, *S. japonicum* occurs in some countries bordering on the western Pacific, *S. mekongi* is restricted to parts of the central Mekong basin, and *S. haematobium*, the cause of urinary schistosomiasis, is widely distributed in Africa and also found in Southwest Asia. The geographical distribution of the various species follows that of the obligatory snail host in association with suitable environmental temperatures for the parasite's extrinsic development.

Pathology

Acute schistosomiasis occurs in immunologically naive, previously uninfected hosts, e.g. tourists and peace corp workers, and is characterized by hyperreactivity to schistosome worm and egg antigens. In contrast, chronic schistosomiasis mainly affects people born and residing in endemic areas. Most individuals chronically infected with Schistosomes have few or no symptoms, but 5–10 % develop severe disease. Portal hypertension, as a consequence of liver fibrosis, is the major cause of morbidity and mortality in *S. mansoni* and *S. japonicum* infection *while S. hematobium* infections often result in mass lesions of the bladder and ureters.

Dermatitis Dermatitis starts as a macular, and later papular, rash covering the areas in contact with contaminated water. This persists for 2–3 days if infection is with species which mature in man. However, schistosomules from bird schistosomes may give rise to urticarial eruption with intense itching, vesicles, and, if secondarily infected, pustules. Depolymerization of the interstitial ground substance may be seen surrounding the schistosomules and protein precipitates may be found at their oral and genital pores. Histologically, a mixed inflammatory exudate is present usually with eosinophilia, especially with non-human schistosomules, most of which probably die in the skin.

The pathology is complex and will be discussed as a central theme, followed by the variations seen in the several organs involved. The variable picture that can be seen in the same patient is explained in part by the chronic active infection; with a continual arrival of new eggs, so that old and recent inflammation may be side by side and intermixed. Differences between patients are believed to depend on usual multifactorial variables of intensity and duration of infection, nutritional state, and the various elements of immunity and hypersensitivity, and the presence of other infections.

Acute Schistosomiasis Acute schistosomiasis is acquired from exposure to large doses of cercariae when drinking and swimming in fecally contaminated snail-infested water. It is seen as acute febrile illness 3–4 weeks after exposure, with abdominal symptoms coincident with the onset of oviposition. The intestinal mucosa is edematous and hyperemic with small hemorrhages, early granulomas, and shallow ulcers with eosinophils.

Adult schistosome pairs are present in the radicles of the portal vein around the colon, or in the pelvic and vesical plexuses around the urinary bladder. The female is usually surrounded by the male in the lumen of a vessel without an apparent lesion or inflammatory reaction (→ Pathology/Fig. 23A). The male attaches to the venous wall and prevents the pair from being swept away by the bloodstream. To lay eggs, the pair migrates into the wall of the viscus and the female wedges its body into the small veins to deposit its eggs there. The adult worms are long lived with reports of persistence for 20–50 years after leaving an endemic area.

Eggs Eggs appear in the stools 40–80 days after primary *S. mansoni* infection. Dysentery with blood, mucus, and necrotic tissue may accompany the eggs. Early in the course there is diffuse eosinophilic hepatitis, which is followed by granuloma formation around individual eggs.

Eggs are found typically in the walls of the gut or the bladder, and atypically in many other organs to which they have usually been swept by the bloodstream (→ Pathology/Fig. 1, → Pathology/Fig. 23). The eggs become surrounded by inflammatory reaction, usually a granuloma, whether the embryo is alive or dead (→ Pathology/Fig. 24). Antigenic material exudes from the eggs, particularly through the spine when present. The inflammatory reaction is quite variable. The eggs may be surrounded by eosinophils, with a mixed eosinophilic and neutrophilic inflammatory reaction peripherally. These inflammatory cells may undergo necrosis (→ Pathology/Fig. 1C, → Pathology/Fig. 26). Other eggs may be surrounded by well-developed epithelioid and occasionally giant cells, followed successively by zones of lymphocytes and fibrosis. The granulomas are larger, more focal, and more structured during the early infections when fewer eggs are present. The granulomas during later infection may be largely necrotic, or may have undergone fibrosis. Some eggs are surrounded by a layer of eosinophilic fibrinoid material, found also around other chronic antigenic sources, the so-called → Splendore-Hoeppli reaction (→ Pathology/Fig. 1B). During chronic infections of long standing the inflammation is varied, with lymphocytes, eosinophils, and fibrosis interlaced with various stages of granuloma associated with eggs or egg masses.

Evidence has been presented which shows that together with the spines of *S. mansoni*, *S. mekongi*, and *S. haematobium*, the inflammatory reaction is instrumental in propelling the eggs towards the lumen of the gut or bladder before the contained miracidium dies. Eggs with dead miracidia gradually lose their inflammatory reaction and their shells often calcify (→ Pathology/Fig. 24A,B).

Intestinal Lesions Intestinal lesions produced by eggs are either granulomas or diffuse or segmental fibrosis of the submucosa, mainly of the colon. Inflammatory polyps may extend into the lumen, with egg masses forming the nidus. Both fibrosis and polyps may lead to obstruction. *S. japonicum* also gives rise to lesions in the small intestine. Localized masses of eggs trapped in the serosa are sometimes referred to as **bilharziomas**; they may form polypoid projections into the peritoneum. The mesenteric lymph nodes may be enlarged from lymphoid hyperplasia during earlier infection, or they may be small from lymphocytic depletion during late chronic infection.

Liver Lesions Liver lesions are produced by eggs and adult worms carried there in the portal vein. The eggs produce small granulomas similar to those in the gut or bladder wall. However, dead adult worms carried to the liver elicit large lesions, with necrosis and around the worm, accompanied either by a mixed granulocytic inflammatory reaction, a **granuloma**, or both. Much liver tissue is destroyed. This is particularly so after chemotherapy when the worms dislodge and large numbers are swept simultaneously into the liver. Scarring is seen principally in the portal areas and, when pronounced, is called Symmers' pipe stem fibrosis (→ Pathology/Fig. 18G) This term alludes to the gross appearance of a cross-section of fixed liver, which looked to Symmers (1904), as if "a number of white clay-pipe stems had been thrust at various angles through the organ." This fibrosis leads to portal hypertension with the eventual formation of dilated venous collaterals, or varices, usually around the lower esophagus, connecting the portal with the general venous circulation. The esophageal varices may bleed when ulcerated. The lobular architecture of the liver is generally preserved and so is liver function, unlike in cirrhosis.

Splenic Enlargement Splenic enlargement secondary to the liver fibrosis is often found with chronic schistosomiasis. It is usually characterized by venous congestion rather than specific egg-related lesions.

Urogenital Schistosomiasis Urogenital schistosomiasis is usually due to *S. haematobium* which deposits its eggs in the venous plexuses around the bladder, ureter, seminal vesicles, prostate, fallopian tubes, etc.. The bladder often contains focal polypoid mucosal lesions or plaques of large masses of eggs (→ Pathology/Fig. 24B) attributed to relatively sessile single pairs of adults. Eggs of *S. haematobium* in the bladder and ureteral wall appear to have a tendency to calcify giving a "sandy" appearance to these focal lesions and making them visible roentgenologically. The microscopic lesions are similar to those described earlier except that diffuse inflammatory reaction and fibrosis are more common than the large granulomas. Ureteral polyps, strictures, and obstruction may lead to pyelonephritis and hydronephrosis. Cystitis with squamous metaplasia and ulceration leading to hematuria are common findings throughout the course of *S. haematobium* infection. Carcinomas of the bladder (→ Pathology/Fig. 24B), of which half are squamous cell carcinomas, and almost half transitional carcinomas, with a few adenocarcinomas, are late complications.

Main clinical symptoms: Unspecific fevers, haematuria, feeling of a burning in the urethra, possibly later development of carcinomas
Incubation period: 4–7 weeks
Prepatent period: 9–10 weeks
Patent period: 25 years
Diagnosis: Microscopic detection of eggs in the urine
Prophylaxis: Avoid entering lakes and rivers in endemic regions
Therapy: Treatment with praziquantel, see → Trematodocidal Drugs

Pulmonary Schistosomiasis Pulmonary schistosomiasis is produced by *S. haematobium* eggs escaping into the general venous circulation and by *S. mansoni* and *S. japonicum* eggs that pass through the portal drainage via the collaterals into the general circulation. These eggs occlude the pulmonary arterioles giving rise to thrombi, granulomas, and arteritis. With heavy infections there is significant obstruction, with pulmonary hypertension leading to cor pulmonale dilatation and hypertrophy of the right heart.

Central Nervous System Granulomas Central nervous system granulomas can be produced by any of the schistosome species, but especially by *S. japonicum* due to its proclivity to reach and grow in the cerebrospinal venules. Because of the calvarial exoskeleton, and space-consuming lesion within it impacts on the brain. So the cerebral or meningeal granulomas surrounding egg; of *S. japonicum* (→ Pathology/Fig. 24C) may give rise to focal epileptic convulsions and those of the cord to transverse myelitis. Because of the association of central nervous system involvement with light, aberrant, or early infections, the expected diagnostic findings of eggs in the stool or urine may be absent.

Miscellaneous Lesions Miscellaneous lesions include ectopic eggs which may be found in many other organs, but because the eggs are usually few in number the lesions produced tend to produce few or no symptoms. Development of dilated anastomoses between portal and systemic veins has been commented upon. They may bleed, usually into the esophagus. → Bilharziomas are granulomatous and fibrotic lesions that develop around egg masses away from the mucosa. Glomerulonephritis associated with the deposition of immune globulin in glomeruli has been described in patients with *S. mansoni* infection. Chronic *Salmonella* infection is sometimes associated with schistosomiasis, with bacteria growing within the adult schistosomes. This may resemble a systemic infection or pyelonephritis. Pigmentary deposits of hematin are often found in the sinusoidal lining cells of the liver which resemble those formed in malaria, or artifactually by acid formalin. The schistosomal pigment is produced by the breakdown of blood ingested by the adult schistosomes.

Schistosomiasis is often associated with malnutrition and other infectious agents, such as hookworm, *Ascaris* spp., malaria, tuberculosis. amebiasis, and bacillary dysentery, and lesions and symptoms may overlap. Infection with *S. mekongi is* often accompanied by *Opisthorchis viverrini*. Incidentally to the investigation of these other clinical entities, subclinical schistosomiasis may be diagnosed. Because of the long survival of adult schistosomes and the worldwide travel of the human host, chronic infections may be found away from the areas of endemicity. There are several species of schistosomes parasitic in animals which are also capable of giving rise to light infections in humans.

Intestinal Schistosomiasis in general:

Main clinical symptoms: Dermatitis due to penetrating cercariae, later intermittent fevers (Katayama syndrome), abdominal pain, swellings of liver and spleen, blood in stool, eosinophilia, liver dysfunctions, liver fibrosis

Incubation period: *S. mansoni*: 2–3 weeks, *S. japonicum*: 1–3 weeks, *S. intercalatum*: 4–7 weeks

Prepatent period: *S. mansoni*: 4–7 weeks, *S. japonicum*: 4–5 weeks, *S. intercalatum*: 6–8 weeks

Patent period: 5–25 years

Diagnosis: Microscopic determination of eggs in fecal samples, → serology (Vol. 1)

Prophylaxis: Avoid entering lakes and rivers in endemic regions

Therapy: Treatment with praziquantel, see → Trematodocidal Drugs

Immune Responses

One of the main histopathological findings in schistosomiasis is the formation of granulomas around schistosome eggs, which is also frequently found in experimental schistosome infections of mice, monkeys and other hosts. Although the circumoval granulomas in experimental infections are massive compared to infections in humans, much has been learned about the responsible immunoregulatory processes especially by analyzing granuloma formation in response to *S. mansoni* eggs in mice.

B Cells and Antibodies A variety of antibodies can be produced against different antigens of adult worms, but protective immunity against reinfection appears to be mainly operative against larval stages. Numerous in vitro studies in experimental and human schistosomiasis have clearly pointed out the essential role played by antibodies in various effector and regulator mechanisms according to their isotypes. However, the course of infection as well as the relative importance of antibodies varies in different experimental hosts. Antibodies of the IgG and IgE isotypes are directly involved in the in vitro killing of schistosome larvae in association with effector cell populations such as eosinophils, macrophages and platelets. These antibodies also induce protection against a schistosome challenge when transferred to naive rats. In rats and rhesus monkeys there is a short self-limiting infection after which persistent low worm burden is controlled by concomitant immunity. In rats the protective immunity involves antibodies of IgE and IgA isotypes. In humans infected with *S. mansoni* a parallelism of the generation of IgA antibodies against the protective recombinant 28 kDa glutathione-S-transferase and acquisition of resistance to reinfection has been observed. Functional analysis revealed that these IgA antibodies not only inhibited the activity of the glutathione-S-transferase but also markedly impaired schistosome fecundity, by suppressing both the egg production by adult female worms and the hatching capacity of schistosome eggs into viable miracidia.

Beside these protective antibodies, several blocking antibody isotypes have been reported. In humans, IgM and IgG_2 antibodies specific for glycanic schistosome antigens prevented the eosinophil-dependent killing by the IgG fraction of the same sera. Furthermore, in *S. haematobium*-infected children immunity to reinfection correlated with increased levels of IgE and decreased levels of IgG_4 antibodies.

In addition to their role in antibody production B cells participate in the modulation of granuloma formation as has been demonstrated in *S. mansoni*-infected B cell-deficient (μMT) mice. Due to mechanisms not defined so far, the B cell-deficient mice displayed an increased hepatic fibrosis and an enhanced Th1-type T cell response.

T Cells The formation of granulomas surrounding eggs of schistosoma is clearly T cell-dependent and immune serum was shown to be not important for the formation or modulation of these lesions. Th2 cytokines are of paramount importance for the granuloma development in experimental infections with *S. japonicum* and *S. mansoni*. Seven to 10 days after injection of eggs or 5–6 weeks after infection with *S. mansoni*, the time when egg laying begins, the cytokine response of mice evolves from a Th0 to a Th2 pattern. While in the lung model using *S. mansoni* eggs treatment with anti-IL-4 antibodies markedly reduced the size of granulomas in mice, injection of anti IFN-γ mAb enhanced both the granuloma size and the parasite-specific Th2 response. However, the transfer of Th0, Th1 and Th2 clones or cell lines all augmented granuloma formation in naive recipient mice.

In *S. mansoni*-infected mice anti-IL-4 treatment had only a moderate to minimal effect on the size of granulomas but hepatic fibrosis was greatly reduced. Surprisingly, both the granuloma-formation as well as the development of hepatic fibrosis was not significantly altered in IL-4-deficient mice when compared to wild type mice, suggesting compensatory mechanisms in these knockout mice. Another Th2 cytokine, IL-5, appears to be of importance for the development of tissue eosinophilia, but anti-IL-5 treatment did not significantly influence the granuloma size. A central, non-redundant role of IL-2 is demonstrated by the findings that anti-IL-2 treatment or infection of mice depleted of IL-2 receptor expressing cells by injection of an IL-2-fusion toxin both resulted in the reduction of granuloma sizes and hepatic fibrosis. Since concomitantly there was a reduced Th2 response, IL-2 is most likely essentially involved in the generation of disease-aggravating Th2 cells.

Cytokine treatments of infected mice generally had the effects expected from the anti-cytokine treatments. While IL-2 and IL-4 administration increased the granuloma sizes, IFN-γ application had the opposite effect and reduced hepatic fibrosis. In line with the latter finding, IL-12 dramatically downregulated the granuloma formation, largely through the stimulation of IFN-γ production from NK cells.

In summary, the formation of egg-induced granulomas in experimentally infected mice may be viewed as Th2-driven process supported by chemotactic factors derived from the eggs. Although the information on the immune response in humans is much more scarce, the immunoregulatory processes might be similar: There were elevated levels of IL-4 in sera of schistosome-infected patients and IL-4 levels after in vitro polyclonal stimulation correlated positively with the intensity of schistosome infection while there was a negative correlation with the amounts of IFN-γ produced. However, enhanced production of IL-4 and IL-5 by cells from Egyptian patients in response to *S. haematobium* adult worm antigens correlated with immunity against reinfection.

One of the key features of schistosomiasis in mice is the immune downregulation of the granulomatous response during chronic infection. In *S. mansoni*-infected mice the size of newly formed granulomas peaks at 8 weeks post infection. The subsequent downregulation of granulomas is accompanied by decreased cutaneous reactions to soluble egg antigens (SEA) and a decreased proliferative response and cytokine production of $CD4^+$ cells. Several findings are consistent with the idea of active suppression of Th responses. Diminution of hepatic granulomas was observed upon transfer of spleen cells from chronically infected (16–24 weeks) mice and this adoptive suppression required the presence of histocompatible $CD8^+$ T cells. Since adult thymectomy results in almost complete lack of the suppressive T cell population, recent thymic emigrants are obviously required for the maintenance of the suppressor cell population. The mechanisms by which $CD8^+$ T cells suppress $CD4^+$ Th populations are still a matter of debate. There is experimental evidence for an involvement of T suppressor cell circuits and idiotypic interactions, e.g. Ia^+ suppressor inducers (termed Ts1) and I-J-restricted $CD8^+$ suppressor effectors (Ts2) have been postulated. These cells have been shown to produce soluble effector molecules which have been described as containing TCR a chains and resembling solubilized T cell receptor heterodimers. It has been recently proposed that antigen-specific $CD4^+$ T cells might shed their TCR against which an anti-idiotypic response is generated, e.g. by $CD8^+$ T cells. These may then shed anti-idiotypic receptors which, by mechanisms unknown so far, might clonally anergize a target T cell population.

An alternative, not necessarily exclusive, possibility for suppressive activity of $CD8^+$ T cells has been proposed involving IFN-γ as key regulatory molecule. In $CD8^+$ T cells from *Schistosoma*-infected mice the frequency of cells expressing the activation phenotype CD44 ^{high} L-selectin ^{low} is increased and these cells produce IFN-γ in the presence of IL-2 after TCR stimulation. Since some of these $CD8^+$ T cells are responsive to schistosome antigens and $CD8^+$ T cells have been found in close proximity with $CD4^+$ T cells within granulomas it is tempting to speculate that the egg-induced lesions could be the site of CD4 / CD8 T cell interaction.

In infected humans, a failure to develop immune-downregulatory mechanisms has been observed which correlates with clinical disease: Antigen-induced proliferative responses of PBLs in vitro were high in the majority of acutely infected patients and amongst ambulatory patients with hepatic or hepatosplenic disease, while asymptomatic chronically infected patients were predominantly low to moderate responders. Interestingly, immune-responsiveness is regained after curative chemotherapy, presumably by removing egg antigens which might be of importance for sustained immunoregulatory constraints.

Fibrosis underlies most of the chronic pathology associated with schistosome infections. While in mice most hepatic fibrosis is associated with granulomas, in humans the relation appears to be less stringent. Type I and III are the predominant collagen isotypes synthezised in both human and murine infections, and a switch from predominant type I to type III collagen synthesis has been reported to occur during the chronic phase of infection. The fibrogenic process is regulated by T cells and macrophages interacting with cells of the mesenchymal/fibroblast lineage. Cytokines are certainly involved in these interactions, since anti-IL-4 treatment and application of IFN-γ decreased fibrosis. At least part of these profibrotic and antifibrotic effects of IL-4 and IFN-γ may be directly on the proliferation and collagen synthesis of fibroblasts.

A novel cytokine, fibrosin, which has been recently cloned, together with TGF-β1 is associated with fibrosis in murine granulomas and downregulation of both cytokines coincided with immune downregulation and reduction of granuloma sizes.

Vaccination

The best characterized vaccine model for plathelminths is the vaccination of mice with γ-irradiated cercariae of *S. mansoni*. Optimally irradiated cercariae stimulate the host's immune system and confer high levels of resistance without causing the severe pathological symptoms of schistosomiasis. They penetrate the host's skin as successfully as non-attenuated larvae do, but their migration is delayed, causing them to spend a prolonged time in the skin, lymph nodes and lungs. Because optimally attenuated schistosomes die immaturely during their passage from the lungs to the liver of the host, pathology in the form of egg granuloma is completely circumvented. First studies testing irradiated cercariae as a possible vaccine against schistosomiasis were performed about 4 decades ago. Although the experimental designs were not comparable among investigations, these early data prompted further research on the vaccine model resulting in a thorough analysis of this method of immunization.

Conditions for Immunization The dose of irradiation used to attenuate the cercariae was recognized early on as a parameter affecting the level of resistance. At first, doses of 2.5 – 10 kilorad were regarded as optimal. Later, presumably due to technical improvements, the positive correlation between irradiation dose and level of resistance was found to continue up to and level at a range of 24 – 56 kilorad, decreasing gradually with doses beyond. More recently, irradiation doses of 15 – 20 kilorad have consistently resulted in higher levels of protection than doses of 50 kilorad or more.

As with other vaccination protocols, the level of resistance depends on the number of boosts with and the dose of antigenic material, in this case the number of exposures to and the load of irradiated cercariae. Studies comparing levels of protection induced by up to eight monthly exposures demonstrated that five immunizations result in an optimal level of resistance. A four-week interval generates best results and this regimen is commonly applied. If mice are vaccinated once, the number of immunizing cercariae does not appear to affect the level of resistance. However, the cercarial number has a significant influence in multiply-vaccinated mice and when higher irradiation doses are used. Mice vaccinated with 500 – 1000 cercariae achieve higher levels of resistance than do those vaccinated with 20 – 100 cercariae.

Employing the parasite's natural behavior, mice are generally exposed percutaneously to irradiated cercariae. The skin of the shaved abdomen, tail or ear pinna serves as entry site. Distant sites are sometimes chosen for penetration of immunizing and challenging cercariae to avoid non-specific local inflammatory responses. Such reactions, however, do not appear to significantly influence the level of resistance and any of the three entry sites may be used.

Protective immune responses stimulated by vaccination with irradiated cercariae are most effective seven to 30 days post-vaccination, with the 30-day period between vaccination and challenge considered optimal. If mice are challenged 15 weeks after exposure to immunizing cercariae, their levels of resistance are somewhat reduced. Eighteen months post-vaccination, resistance is lost in some mouse strains (CBA/Ca, BALB/c), whereas others maintain partial resistance (CF1). Thus, the immunity induced by irradiated cercariae appears to be relatively long-lasting.

The genetic background of the murine host also influences the absolute level of resistance that may be achieved. Data on resistance of C3H/HeJ (C3H) and CBA/J (CBA) mice are variable, ranking them as non- or moderate responders. In contrast, C57BL/6J (C57) mice are consistently regarded as high responders. Differences in the degree of immunity are caused, in part, by variations in the major histocompatibility complex, as was demonstrated by cross and backcross experiments of congenic mice differing in their H-2 haplotypes. However, outbred mice are also effectively immunized by irradiated cercariae and may be more representative of natural host populations.

Interestingly, the resistance stimulated by vaccination with irradiated cercariae seems to be species-specific. Mice vaccinated with irradiated *S. mansoni* cercariae and challenged with other *Schistosoma* spp or vice versa are not protected. However, cercariae of geographically distinct isolates of *S. mansoni* successfully crossprotect mice, indicating that the major antigens relevant to protection are common to these isolates. Similarly, no difference in the ability to stimulate resistance exists in clones of schistosomes, or in parasites that have been passaged selectively for their resistance to the host's immunity. Thus, results obtained using one *S. mansoni* strain may be extrapolated to other strains.

Although percutaneous exposure to irradiated cercariae stimulates the highest levels of resistance, mechanically transformed irradiated schistosomula may be administered instead. Advantageously, schistosomula may be stored by cryopreservation without losing their immunogenicity. The effectiveness of vaccination with irradiated schistosomula varies, however, with the route of injection. Intravenous or subcutaneous injection of irradiated schistosomula is only marginally protective, and intraperitoneal, intratracheal and intramuscular injection rank intermediate, whereas the intradermal injection of attenuated schistosomula induces particularly good protection against challenge infection. Intradermally administered schistosomula persist in the skin and are able to migrate to the draining lymph nodes and the lungs. Only schistosomula administered intradermally share similar migration patterns with penetrating cercariae.

Migratory Pattern of Immunizing and Challenging Schistosomes To identify sites where the protective immunity is stimulated, the migratory pattern and the attrition site of irradiated schistosomes have been compared to those of non-attenuated parasites. Various methods were applied such as histological investigations, counting of parasites upon their exit from minced lung tissue, and detection of radio-labeled parasites by compressed organ autoradiography. Generally, the time of survival and the migratory pattern of attenuated larvae depend on the irradiation dose used (Fig. 1). A dose of one kilorad seems to have little effect on parasite development, although an increased number of dead eggs are found. At two to three kilorad, very few and stunted adult worms are recovered from the liver. Because the occasional eggs they produce are not viable, this dose is considered sterilizing. Doses of four kilorad or higher do not permit survival of parasites to adulthood. Parasites irradiated with a 20-kilorad dose migrate more slowly than do their non-attenuated counterparts. They remain in the skin for up to one week, and thus their passage to the lungs is delayed. There, they are observed at least until day 21, at which time non-attenuated parasites have left the lungs and entered the liver. Lying within alveoli, many schistosomula seem to lose the capacity of onward migration and only a quarter of 20-kilorad irradiated parasites reaches the liver. Irradiation with a 50-kilorad dose results in further retardation of the parasites' migration from the skin to the lungs, and only few worms are found in the liver. As early as two weeks after penetration, the majority of these parasites have died, leaving residual inflammatory foci. Most 90-kilorad irradiated parasites fail to leave the skin, indicating that increasing irradiation doses diminish the parasites' ability to migrate through the tissues. Abnormal constrictions are observed in attenuated larvae as early as six days after 20-kilorad irradiation that are not apparent in non-attenuated parasites. This impairment appears to restrict the motility of irradiated parasites and may explain their prolonged presence in the skin and lung tissue as well as the inhibition of their onward migration.

Excision experiments have been performed to determine for how long, and where within the host, attenuated parasites must be present in order to stimulate resistance. Excising the site of skin penetration within the first four days following exposure completely blocks induction of resistance, possibly because most attenuated parasites are removed before they are able to disseminate in the host. Removal of the penetration site between the fifth and eighth day permits the development of low but significant levels of resistance compared to non-treated mice. Skin excision thereafter fails to affect resistance. If axillary and inguinal lymph nodes draining the abdominal penetration site are surgically removed five days before exposure to irradiated cercariae, the level of resistance is reduced by two thirds. Thus, factors such as the duration of host-parasite contact, maturation of

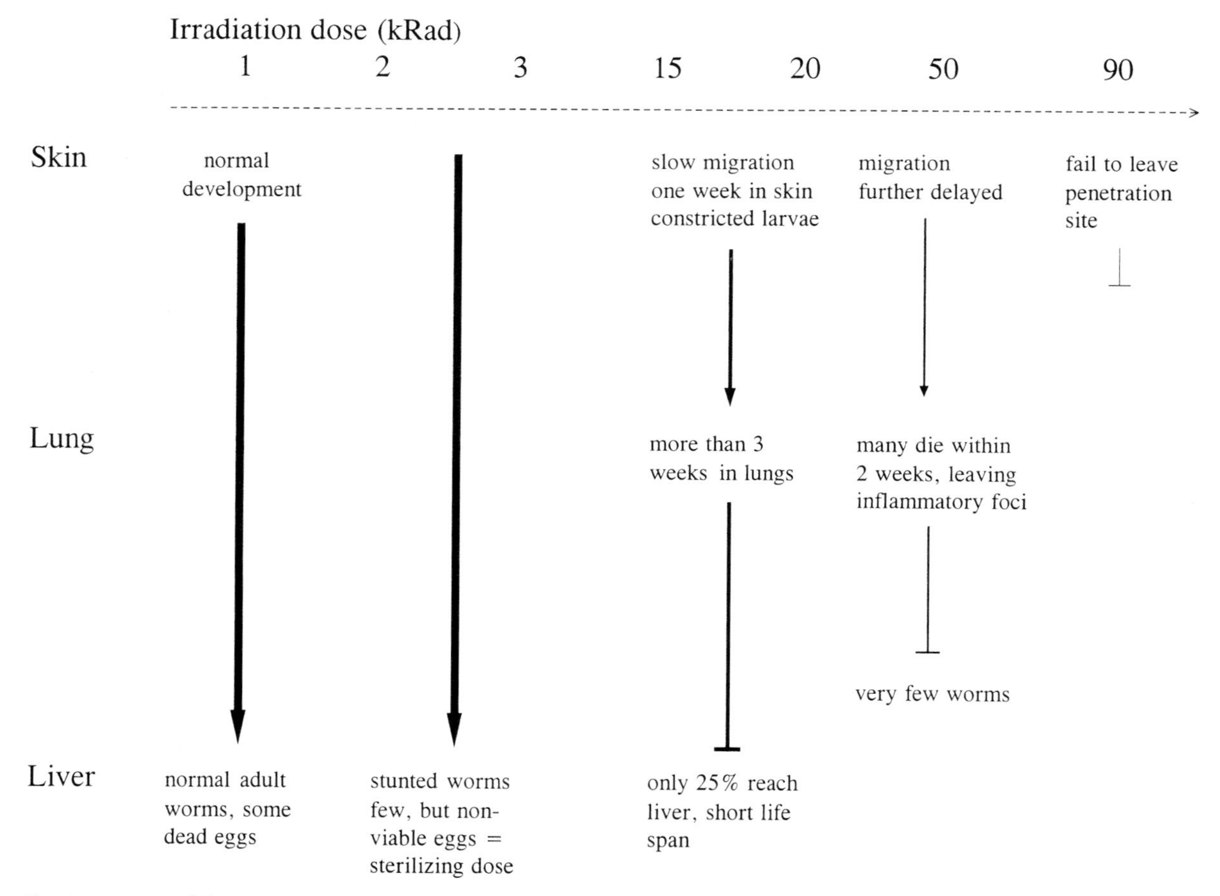

Fig. 1. Impact of the irradiation dose on migration and survival of attenuated *Schistosoma mansoni* in mice.

the immunizing parasite and migration to a post-skin site appear to be relevant to the development of a protective immune response.

In vaccinated mice, the migration pattern and attrition site of non-attenuated parasites delivered as challenge infection have been investigated. In mice vaccinated with 20-kilorad irradiated cercariae, the migratory pattern of challenge parasites is delayed but otherwise similar to that observed in naive mice. Challenge parasites in mice vaccinated with 50- or 56-kilorad irradiated cercariae migrate even more slowly to and from the lungs. Some studies identify the skin as the major site of immune elimination, others the lungs. Differences in methods or mouse strains may cause these variable results. In both sites, inflammatory foci are observed and may be relevant to the immobilization and elimination of challenge parasites. It is generally agreed that in optimally vaccinated mice challenging schistosomes are eliminated before they reach the liver.

Although the morbidity-causing host response in the form of egg granulomas does not result from exposure to irradiated cercariae, attenuated cercariae induce inflammatory responses during their migration through the host. The extent of host tissue response depends on the irradiation dose applied to the immunizing cercariae. Whereas 50-kilorad irradiated parasites induce dermatitis and vasculitis, 5- or 2.5-kilorad irradiated larvae cause granulomatous foci in lungs or liver, respectively. Cercariae irradiated with 24 kilorad induce more lesions than the slightly less protective 48-kilorad irradiated cercariae. Unlike granulomas developing around schistosome eggs, the inflammatory foci formed around irradiated parasites appear not to harm the host, because they are not systemic and disappear with time. In fact, inflammatory foci may even benefit the host by trapping challenge parasites. As a result, migration of challenge parasites is slowed; they leave the skin or lungs later than do schistosomes in naive mice, and eventually die within such foci. Thus, focal inflammatory responses in vaccinated mice may be advantageous rather than detrimental to the induction of resistance.

Humoral Immune Responses The role of antibodies in the protective immunity induced by irradiated cercariae was demonstrated early on by passively transferring resistance to naive mice using serum of vaccinated mice. The protective capacity is restricted to sera obtained from multiply-vaccinated mice. The serum may be administered one hour before or several days after challenge, depending on whether the skin or lung stage, respectively, is to be targeted. Serum administration by intravenous, intraperitoneal or subcutaneous injection is equally effective. The IgG isotypes, particularly IgG_1, appear to be protective and may be enhanced synergistically by the presence of IgM.

The successful transfer of resistance prompted further analysis of the humoral immune response in vaccinated mice. Parasite-specific antibodies are detected as early as two weeks after vaccination. Their titers peak at weeks 5 – 6, then gradually decline, but antibodies are still detectable 15 weeks after vaccination. Antibody titers are enhanced by repeated exposure to irradiated cercariae or after challenge infection with non-attenuated cercariae, showing a typical anamnestic response.

Although the presence of antibodies is essential, as determined by the failure of B-cell depleted mice to generate any resistance, no consistent association between overall antibody titer and level of resistance is apparent. Whereas levels of resistance are not affected by the absence of IgM antibodies in x-linked immunodeficient (xid) mice, non-protected mice of the P/N strain produce only small amounts of schistosome-specific IgM antibodies as compared to highly protected mice of other strains. Similarly, upon comparison of three mouse strains, titers of antibodies binding to crude antigen mixtures seem not to correlate with levels of resistance. In contrast, overall levels of schistosome-specific antibodies are greater in mice vaccinated with 15- or 20-kilorad irradiated cercariae than in less protected mice vaccinated with 40- or 50-kilorad irradiated cercariae. Idiotypic regulation may also be important, because anti-idiotypic immunization suppresses the development of resistance. Therefore, antibody specificity rather than quantity appears to be relevant to protective immunity.

In order to examine antibody specificity of vaccine sera, surface- or metabolically labeled antigens of schistosomes have been immunoprecipitated. Although these studies differ in irradiation doses, mouse strains and parasite stages used, various antigens having molecular mass of 15, 17, 19 – 20, 22 – 23, 32, 38, 43 – 45 and 92 – 94 kDa are consistently detected. A direct comparison of two mouse strains (C57 and C3H) as well as three irradiation doses (5, 25 and 50 kilorad) using immunoprecipitation failed to demonstrate differences in the pattern of antigens recognized.

The antibody specificity in an array of different vaccine sera has been analyzed by immunoblot, probing whole parasite extracts. Antigens of 22 – 23, 28, 31 – 32, 70 and 97 kDa were identified as the integral membrane protein Sm23, glutathione-S transferase = GST, triose-phosphate isomerase = TPI, cathepsin B, hemoglobinase, → heat shock protein (Vol. 1) 70 (HSP70) and paramyosin, respectively. In contrast to other studies, immunoblot analysis demonstrated that both the irradiation dose used to attenuate cercariae as well as the genetic background of the mice influence the titer and the specificity of antibodies to particular antigens (Fig. 3).

Cellular Immune Responses T cells are essential for the induction of protective immunity in this model, because athymic mice fail to develop resistance following exposure to irradiated cercariae. Proliferative responses of lymphocytes to schistosomal antigens peak during the first two weeks after vaccination, waning after the fourth. Lymphocytes derived from draining lymph nodes respond considerably more strongly than those derived from spleen. The relevance of regional rather than systemic stimulation is supported by the observation that attenuated parasites release significant amounts of antigenic material during their passage through skin, lymph nodes and lungs. As a result, the time and site of lymphocyte priming coincide closely with the parasites' migration, i.e. proliferation is observed first in skin- and later in lung-draining lymph nodes. Because greater amounts of antigens are released over an extended period in axillary and inguinal lymph nodes of vaccinated mice than in other lymph organs or as compared to mice infected with non-attenuated cercariae, lymphocytes in these lymph nodes, as well as in mediastinal nodes, proliferate most strongly. In contrast, no proliferation is detected in cells from brachial, periaortic or mesenteric nodes. In primed lymph nodes, the number of T cells increases relative to that of B cells. The irradiation dose used to attenuate the cercariae has a profound effect on their lymphostimulatory capacity, because it affects their migratory pattern. Whereas longer-lived 20-kilorad irradiated cercariae stimulate extensive proliferation in draining lymph nodes, 50-kilorad irradiated cercariae induce modest responses, and non-protective 80-kilorad irradiated parasites that fail to leave the skin penetration site induce only a transient increase in cell number. Therefore, optimally attenuated parasites deliver themselves to sites where antigen processing is intense. While remaining there for a prolonged period, they release antigenic material priming lymphocytes required for successful vaccination.

Removal of draining lymph nodes before vaccination reduces the level of resistance. Because removal a week after vaccination eliminates priming parasites as well as primed lymphocytes, only low levels of immunity develop. Lymphadenectomy at later times does not abrogate resistance, because primed lymphocytes have begun to circulate. The pool of these peripheral lymphocytes expands vigorously, reaching its maximum 3 – 4 weeks after vaccination, and persists at an elevated level. Simultaneously, numerous activated lymphocytes infiltrate the pulmonary parenchyma and airways. Lymphocytes of the draining lymph nodes as well as those recruited to the lungs participate in the induction of immunity. Successful vaccination correlates with percutaneous exposure or intradermal injection of attenuated parasites, because only these routes of administration allow the parasites to migrate through the draining lymph nodes as well as through the lungs.

T-cell subsets have distinct effects on the induction of immunity, as demonstrated by depletion studies. Depletion of $CD4^+$ T cells decreases the level of resistance. In fact, resistance to challenge falls below that observed in athymic mice, if mice are depleted of $CD4^+$ T cells before vaccination. In contrast, depletion of $CD8^+$ T cells reduces morbidity and enhances resistance to a level higher than that observed in non-depleted control mice. The ratio of reactive $CD4^+$ T-cell subsets is shifted by the number of exposures to irradiated cercariae. Cytokines produced by Th1 cells predominate in once-vaccinated mice and dissipate in multiply-vaccinated mice with a concurrent increase in cytokines produced by Th2 cells. Repeated vaccination results in an overall decreased proliferative response. Both events might explain why $CD4^+$ T cells are essential to the induction of resistance in once-vaccinated mice, but appear less critical in twice-vaccinated mice.

The Th subsets participating in the induction of protection were further characterized by cytokine studies. Upon *in vitro* stimulation, lymphocytes of mice vaccinated with the more-protective 15-kilorad irradiated cercariae secrete significantly more of the Th1 cytokine IFN-γ than do mice vaccinated with less protective 50-kilorad irradiated cercariae and lymphocytes of the latter mice produce far more IFN-γ than non-protected P/N mice. Further, in vaccinated IFN-γ-receptor knock-out mice or in vaccinated mice depleted of IFN-γ by antibody treatment, protective immunity is abrogated by about 50–90%. In such mice, mRNA expression of Th2 cytokines, such as IL-4, IL-5, IL-10 and IL-13, is elevated and that of Th1 cytokines diminished. The kinetics of IFN-γ production in vaccinated mice coincides with the migratory pattern of the immunizing parasites. This cytokine is initially produced by lymphocytes obtained from axillary and inguinal lymph nodes five days after vaccination, peaking two weeks later. At this time, lymphocytes from mediastinal lymph nodes only begin secreting IFN-γ, while lymphocytes obtained by bronchoalveolar lavage secrete high titers of IFN-γ, coinciding with macrophage activation. In contrast, depletion of cytokines produced by Th2 cells, such as IL-4 and IL-5, does not affect resistance in vaccinated mice. Neither IgE nor eosinophils appear to be required in vaccinated mice. Therefore, secretion of Th1 cytokines correlates with the degree of vaccine-induced immunity, whereas production of Th2 cytokines shows an inverse relationship.

A potent inducer of IFN-γ production is IL-12. This cytokine affects the differentiation of Th cells by stimulating the expansion of Th1 cells while suppressing the differentiation of Th2 cells. In mice vaccinated once with irradiated cercariae, administration of IL-12 enhances the vaccine-induced protection against a challenge infection by about 20%. At the same time, mRNA levels of IFN-γ and IL-12 increase, while those of the Th2 cytokines, IL-4 and IL-5, eosinophilia and titers of IgE are reduced. Although the responses of the Th2 subset prevail in mice multiply vaccinated with irradiated cercariae, exogenous IL-12 is capable of further augmenting the degree of protection, even achieving complete protection in some individuals. In such multiply vaccinated mice, IL-12 appears to enhance the production of parasite-specific antibodies, particularly those isotypes that are Th1-associated. Studies on IL-12 knock-out (IL-12KO) mice vaccinated with irradiated cercariae confirm the fundamental role this cytokine plays in generating protective immune responses. Vaccinated IL-12KO mice produce cytokines and antibody isotypes that correspond to the Th2 phenotype and, possibly as a result, develop a significantly greater worm burden from a challenge infection than do their wild-type counterparts. Inflammatory foci in the lung of vaccinated IL-12KO mice have a looser appearance and contain more eosinophils than do those in wild-type mice and, as a result, may be less efficient in blocking the migration of challenging schistosomes. In IL-12KO mice, recombinant IL-12 permanently restores the ability to generate protective Th1 responses. If this cytokine is administered during the first week after vaccination, such knock-out mice produce levels of IFN-γ, develop inflammatory foci around challenge larvae and achieve levels of protection comparable to those of wild-type mice. Thus, IL-12 serves an important role in inducing protective Th1 responses in mice vaccinated with irradiated cercariae.

The importance of Th1 responses is further evidenced by studies on the delayed type hypersensitivity (DTH) (type IV). Vaccinated mice exhibit a profound DTH reaction to soluble schistosomal antigens *in vivo*, starting ten days after exposure to irradiated cercariae and peaking one week later. Significantly decreased DTH reactivity coincides with reduced levels of resistance in athymic mice, P/N mice, or in mice depleted of $CD4^+$ or IL-2-receptor bearing cells. On the other hand, mice deficient in mast cells or IgE production develop levels of resistance comparable to that of non-deficient controls. The immediate hypersensitivity (type I) response observed in mice of the P/N and C57 strains does not differ. Delayed but not immediate hypersensitivity appears to be relevant to the immunity induced by irradiated cercariae.

Th1-cell responses appear to participate in the development of inflammatory foci. After vaccination, large numbers of Th cells infiltrate the lungs. Virtually all express high levels of the CD44 molecule, identifying them as effector/memory cells. Because binding of CD44 to its ligand, which is present in the lung, promotes cell aggregation and cytokine release, it may serve to initiate and maintain inflammatory responses. Upon challenge, the cellular composition of these foci is characteristic of a DTH response. The cells are found in tightly compact aggregates. In vaccinated IL-12KO or IFN-γ-receptor-KO mice, however, cell aggregates are much larger and looser, concurrent with a reduction in immunity. In addition, IFN-γ

affects the expression of inducible nitric-oxide synthase, which is associated with inflammatory foci developing around challenge schistosomula in the lungs of vaccinated mice. Thus, challenge parasites appear to be trapped in compact pulmonary inflammatory foci of vaccinated mice.

Candidate Vaccine Antigens The antigens that are recognized by the humoral compartment of mice vaccinated by irradiated cercariae also induce lymphocyte proliferation in these mice (Fig. 2, Fig. 3). The response to this array of antigens, consisting of Sm23, GST, TPI, cathepsin B, Sm32, HSP70, and paramyosin, has been further characterized by quantifying levels of antigen-specific isotypes as well as their lymphostimulatory capacities in different groups of vaccinated mice. Experimental groups of mice achieve various degrees of resistance due to their genetic background, the number of exposures to irradiated cercariae and the irradiation dose with which the immunizing cercariae had been attenuated. Vaccinated C57 mice develop a higher degree of protection than do CBA mice which can be further enhanced by multiple exposures to irradiated cercariae. Those mice that are vaccinated with 15-kilorad irradiated cercariae are better protected against challenge infection than those vaccinated with 50-kilorad irradiated cercariae. GST is recognized by the humoral and the cellular immune compartment of all groups of vaccinated mice. Protective sera, passively transferring resistance, contain particularly high titers of IgM antibodies to GST. These antibodies bind predominantly to carbohydrate epitopes of this antigen. GST stimulates proliferation of Th2 cells in both C57 and CBA mice, whereas proliferation of Th1 cells is restricted to vaccinated CBA mice. Antibodies specific for the integral membrane protein Sm23 are present in all vaccine sera tested. The highest levels of protection in mice multiply vaccinated with 15-kilorad irradiated cercariae coincides with the highest levels of Sm23-specific antibodies. Highly protective sera contain large quantities of IgG_{2b} antibodies binding to Sm23. In contrast, the humoral and cellular responses to recombinant TPI are restricted to once-vaccinated mice in this vaccine model. Thus, protective vaccine sera derived from mice immunized with irradiated cercariae appear not to contain TPI-specific antibodies. The digestive enzymes, Sm32 and cathepsin B, are recognized by mice vaccinated with 15-kilorad, but not 50-kilorad irradiated cercariae. Only worms that are able to survive to the hemoglobin-digesting stage seem to induce antibody production to these developmentally regulated antigens. The humoral and cellular recognition of Sm32 is strain-specific, i.e. limited to vaccinated CBA mice. In contrast, HSP70 is predominantly recognized by vaccinated C57 mice. Strong responses of both immune compartments coincide with the highest level of resistance in this strain. In highly protected mice, the response to HSP70 appears to shift from a mixed Th1/Th2 cell population to an exclusive Th2 cell population upon multiple vaccinations. The highest levels of anti-paramyosin antibodies correlate with the lowest degree of protection in vaccinated CBA mice. Although vaccinated C57 mice fail to produce paramyosin-specific antibodies, a distinct cellular response to paramyosin is associated with a high degree of immunity in these mice. Factors such as the irradiation dose used to attenuate the immunizing cercariae, the genetic background of the host and the number of vaccinations have distinct effects on the recognition of individual antigens in this vaccine model.

The results of extensive studies on the irradiated cercariae vaccine model demonstrate the importance of analyzing diverse aspects of the immune responses that are induced by attenuated parasites. They allow us to begin to understand the complex mechanisms that are involved in generating vaccine-induced protection against plathelminths. Conditions for immunization, such as irradiation dose, number of cercariae, route, site and schedule of application, have been compared and optimized. Further, the migratory pattern and attrition site of immunizing and of challenge parasites have been outlined. The compartments of the host's immune system that participate in the induction of protective immunity have been elucidated. This has initiated an analysis of where, when and how the immune compartments interact. Subsequently, antigens were identified that stimulate humoral as well as cellular immune responses of vaccinated mice. As a result, the complex kinetics of immune responses to irradiated cercariae are being increasingly understood and may facilitate the development of a vaccine against schistosomes. It will likely consist of a cocktail of antigens in order to address the heterogeneous responsiveness to particular antigens among a majority of individuals to be protected. It may need to be administered in a way that mirrors the effi-

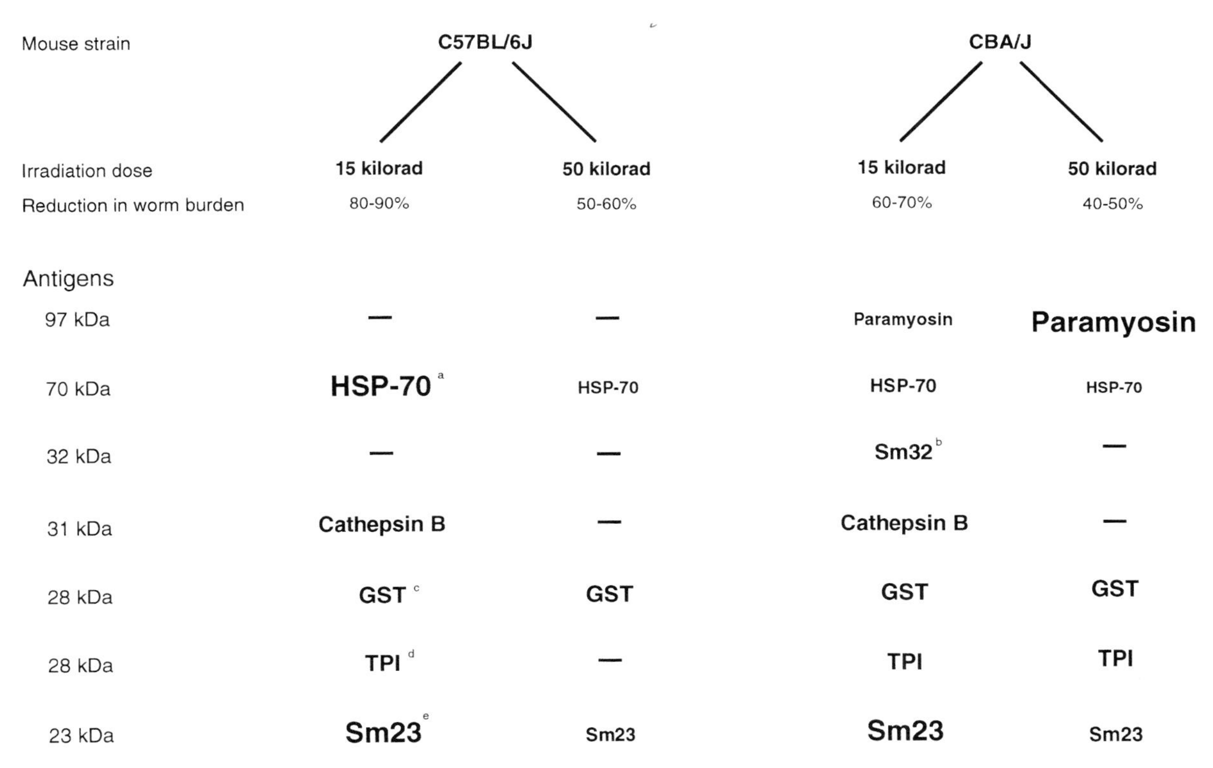

Fig. 2. Antigens recognized by antibodies of mice vaccinated with irradiated cercariae of *Schistosoma mansoni*. The experimental groups of mice differed in the degree of protection achieved against a challenge infection. Specific antigens were identified by immunoblot analysis. Enzyme-linked immunosorbent assay (ELISA) using purified native or recombinant antigens served to determine the intensity of antigen recognition by antibodies. Font size indicates relative intensity. Generally, levels of antibodies specific for these antigens are higher in multiply-vaccinated mice than in once-vaccinated mice. [a]Antibodies of vaccinated CBA/J mice detect heat-shock protein (HSP-70) only in ELISA, not in immunoblot. [b]Sm32 is also known as hemoglobinase. [c]IgM antibodies dominate the response to glutathione S-transferase (GST) and about half recognize carbohydrate epitopes. [d]Triosephosphate isomerase (TPI) is solely recognized by once-vaccinated mice. [e]In response to the integral membrane protein Sm23, C57BL/6J mice vaccinated with 15-kilorad irradiated cercariae produce also antibodies of the IgG_{2b} isotype that are not detected in other experimental groups.

cient mode of antigen presentation by moderately irradiated cercariae and, thus, optimally stimulates protective Th1 responses.

Planning of Control

Intestinal and urinary schistosomiasis may cause gross pathology and shorten the human host's life expectancy. Worm load is recognized as a major determinant of the severity of the disease. For a long time, the measures available for the control of schistosomiasis were water management, especially in connection with water impoundments and irrigation schemes, rather inefficient use of molluscicides, safe excreta disposal, health education, and less than satisfactory treatment. The introduction of oxamniquine has rendered infections with S. *mansoni* treatable, and that of praziquantel has radically improved the treatment of all forms of schistosomiasis. Both drugs are well tolerated and simple to use.

Molluscicides still pose problems: copper salts, though cheap, are not sufficiently effective, while niclosamide, though highly effective, is expensive and associated with strong non-target effects on the aquatic fauna. Much can be achieved with appropriate water management and other forms of environmental management.

Biological methods of snail control, e.g. snail pathogens and the replacement of host species by nonsusceptible snail species, are being explored

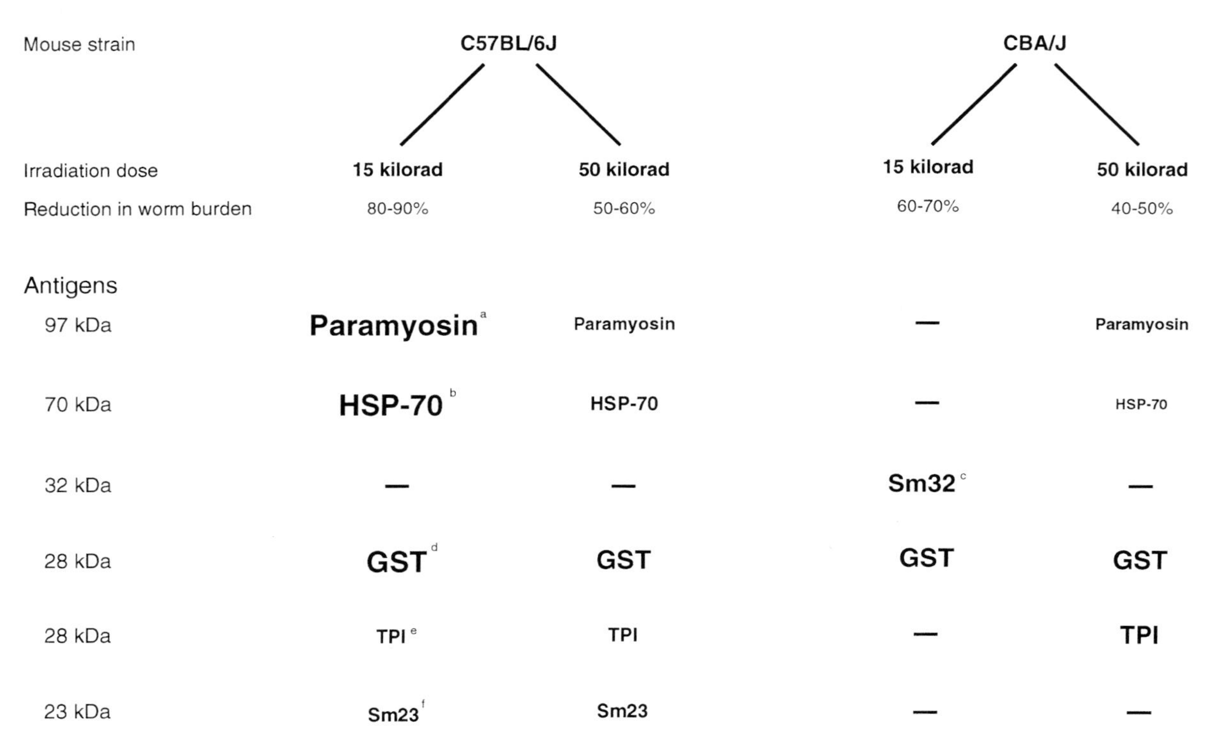

Fig. 3. Antigens recognized by lymphocytes derived from draining lymph nodes of mice vaccinated with irradiated cercariae of *Schistosoma mansoni*. The experimental groups of mice differed in the degree of protection achieved against a challenge infection. Proliferation assays stimulating lymphocytes of axillary lymph nodes with purified native or recombinant antigens served to determine the intensity of antigen recognition by lymphocytes. Font size indicates relative intensity. Lymphocytes of once-vaccinated mice proliferate more strongly in response to these antigens than do those of multiply vaccinated mice. [a]In response to paramyosin, lymphocytes of C57BL/6J mice vaccinated with 15 kilorad irradiated cercariae produce IL-2, but not IL-4. [b]In the presence of heat-shock protein (HSP-70), IL-2 production by lymphocytes of C57BL/6J mice vaccinated with 15 kilorad irradiated cercariae decreases after repeated vaccination, while IL-4 production increases. [c]Sm32 is also known as hemoglobinase. [d]Glutathione S-transferase (GST). [e]Triosephosphate isomerase (TPI). [f]Integral membrane protein Sm23.

but their non-target effects need to be assessed very carefully before a wider use may be contemplated. The provision of safe water supplies is an important factor in reducing the transmission of schistosomiasis.

Currently the primary objective of schistosomiasis control is the reduction or elimination of morbidity. Appropriate operational approaches for the attainment of this objective are available. However, control programs offer a reasonable prospect of success only if they have sufficient and well-qualified man power, and organizational/managerial structures capable of ensuring correct planning, smooth implementation and continuous evaluation of operations and results. A key factor for success is local and national commitment to such a program, expressed in adequate financial resources to maintain it without external assistance. The strategy of schistosomiasis control consists, essentially, of three phases, namely:

1. A planning period devoted to the collection and analysis of epidemiological baseline data, the determination of feasible control approaches, the preparation of a master plan of action, the allocation of resources, and training.
2. An intervention period during which intensive operations are conducted, aimed at reducing as rapidly as possible the reservoir of infection and curbing transmission. Treatment will be in-

strumental during this phase which is usually shorter than the first.

3. A maintenance phase during which the gains of the second phase are to be consolidated and safeguarded and, if possible, a further reduction of disease prevalence and intensity of infection are achieved. Apart from the environmental management activities, this phase will be less demanding in resources than phase 2, and most of the diagnostic and therapeutic maintenance effort will be the responsibility of the general health care services which undertake the bulk of surveillance and monitoring.

Targets for Intervention

Infection results from the active transdermal invasion of the free-swimming cercariae. These originate from the water-borne miracidium which emerges from the egg and develops further in a suitable aquatic snail host. Targets of intervention (Fig. 4) are the infected human host, the snail intermediate host and the infection cycle. Suitable approaches to control consist of the detection and treatment of cases, the safe disposal of excreta, the use of safe water for drinking, washing, bathing and swimming, and snail control through environmental management.

Therapy

→ Trematodocidal Drugs.

Schizodemes

→ Amoebiasis.

Scrapie

Brain disease of sheep due to infections with → prions that introduce spongious degenerations. These prions may become transmitted orally by ingestion of contaminated brain or by ingestion of fly larvae and pupae which have eaten infected brain.

Screw Fly Disease

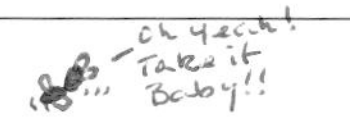

Myiasis due to infection with *Callitroga* (syn. *Cochliomyia*) or *Chrysomya* species (→ Myiasis, Animals, → Myiasis, Man).

Scrub Typhus

→ Tsutsugamushi Fever.

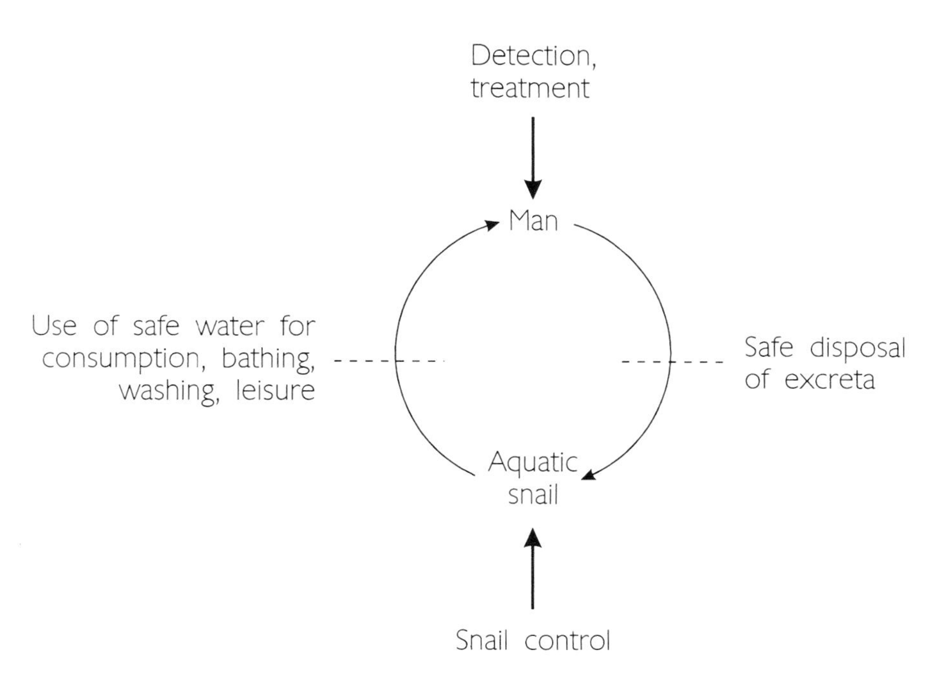

Fig. 4. Targets and approaches for the control of human schistosomiasis

Secondary Antibody

Synonym
Conjugate, Anti-Immunoglobulin antibodies

General Information
In indirect serological methods, which are common for serodiagnosis of parasitic infections, the antigen-specific antibody is detected by use of a secondary antibody. The secondary antibody (anti-immunoglobulin antibody) is available as species, class, subclass, or domain specific. It can be used as a general reagent for screening of all antibodies or as a selective reagent to identify one type of antibody class or molecule. Different degrees of cross-reactions occur at all levels between subclasses, classes, and species and may decrease the specificity of an assay. A reduction of cross-reactions is possible by use of cross-adsorbed preparations. The conjugate provided by suppliers should be specific for the stated isotype. Labeling of the secondary antibody is mainly by enzymes (horse-radish peroxidase, alkaline phosphatase), fluorochromes (fluorescein, rhodamine) or biotin. The choice of label depends on the technique used, the available detection system and the field of application.

Characteristics
Fluorescein-labeling is unstable and does not allow a repeated reading or long-term storage of the preparations. The conjugated enzyme and its substrate are chosen for sensitivity and convenience. In general, alkaline phosphatase is simpler to use but horse-radish peroxidase is probably more sensitive.

Seed Ticks

Trivial name for the small larval ticks.

SEIR Model

→ Mathematical Models of Vector-Borne Diseases.

Sequestration

Parasitized red blood cells are retained in capillaries (→ Malaria).

SERA

Serine rich protein (→ Malaria/Vaccination)

Sibirian Tick Typhus

→ Tick Typhus.

Sickle Cell Anaemia

Blood disease that is fatal in humans being homozyyous carriers of the defect gene. On the other hand individuals being heterozygous for the gene responsible for sickle cell haemoglobin (HbS), in which a substitution of valine for glutamic acid occurs in the beta-chain of the molecule, are strongly (90%) protected against severe Malaria tropica (due to *Plasmodium falciparum*). This effect is based on the fact that *P. falciparum* may not develop into mature schizonts after having entered the red blood cell due to the low oxygen tension and leakage of potassium from host cells during sequestration in capillaries thus considerably reducing the pathologic effects e.g. in the brain.

Simuliidosis

Disease due to bites of simuliids, see Table 1 (page 549)

Siphonapteridosis

Disease due to → flea (Vol. 1) bites, Table 1 (pages 550, 551).

Table 1. Simuliids and Control Measurements

Parasite	Host	Vector for	Symptoms	Country	Therapy		
					Products	Application	Compounds
Simulium spp. (black flies)	All animals, Man	*Onchocerca volvulus* (human filariasis)	Edema, allergic reactions (simulio toxicosis)	World-wide	1% Vapona insecticide™ (Durvet)	Spray	Diclorvos
Simulium reptans	Ruminants, Horse, Pig		Edema, allergic reactions (simulio toxicosis)	Central-Northern Germany, Austria, foothills of the Alps			
Odagmia ornata	Ruminants, Horse, Pig	*Onchocerca gutturosa*		World-wide			
Wilhelmia equina	Ruminants, Horse, Pig			Switzerland, Germany			
Boophthora erythrocephala	Ruminants, Horse, Pig			Germany, Switzerland, Poland, Czechoslovakia, Italy, France			

Skin Diseases, Animals

General Information

Parasitic diseases of the skin are of major economic importance. Discomfort and pruritis interfere with the normal rest and feeding of the animal, and the loss of protective function of the skin facilitates bacterial infection. In addition, the commercial value of the hides is often reduced. Some ectoparasites, such as blood-sucking flies and some species of ticks are of great economic importance because of the diseases they transmit. However, the skin lesions produced by these parasites are of relatively minor significance. Finally, parasitic infections of the skin and the sometimes ugly lesions they cause affect the general appearance of the animal, and upset the owner. It is intended to deal only with those parasites which are of importance owing to the damage they cause to the skin (Table 1). Most of these are arthropods. However, some protozoa and helminths may occasionally be responsible for localized skin lesions and these will be referred to briefly.

Protozoa

Cutaneous lesions occur in several systemic protozoal infections, including besnoitiosis in cattle and horses, dourine (*Trypanosoma equiperdum*) in horses and leishmaniosis in the dog.

Cattle – and rarely horses – serve as intermediate hosts of *Besnoitia besnoiti*. In cattle, the parasite mainly infects cells of the connective tissue and produces characteristic cysts with a very thick wall. Clinical besnoitiosis in cattle is characterized by two sequential stages: an acute febrile stage and a chronic seborrheic stage. Persistent high fever is the first clinical sign. During the febrile stage cattle may develop a photophobia, anasarca, diarrhoea, and swelling of the lymph nodes. This is followed by a progressive thickening and wrinkling of the skin and the development of a marked alopecia. During the seborrheic stages denuded parts are covered by a thick scurfy layer. In chronic cases the skin remains alopecic, lichenified and scaly. The cysts may be visible macroscopically in the scleral conjunctiva or nasal mucosa as small, round, white foci. Death may occur in severe cases.

→ *Trypanosoma equiperdum* (Vol. 1) causes typical oedematous swelling of the external genitalia and ventral abdomen in horses. Raised urticarial plaques, 4–5 cm in diameter, called silver dollar plaques, may also appear, especially on the flanks.

The cutaneous lesions of leishmaniosis commonly observed in dogs include a dry exfoliative

Table 1. Fleas and Control Measurements

Parasite	Host	Vector for	Symptoms	Country	Therapy		
					Products	Application	Compounds
Ctenocephalides canis (Dog flea)	Dog, Cat (fleas in general not very host-specific)	*Dipylidium caninum, Dipetalonema reconditum,* Bartonellosis (Cat scratch fever)	Blood loss, local skin reaction, strong itching, flea allergic dermatitis	World-wide	Advantage™ (Bayer)	Spot on	Imidacloprid
					Performer™ Flea and Tick Collar (Performer)	collar	Naled
					Adams™ Flea and Tick Dip (Pfizer Animal Health)	Dip	Chlorpyrifos
					Duocide Flea and Tick Collar (Allerderm/Virbac)	Collar	Chlorpyrifos
					Cyflee™ (Boehringer Ingelheim)	Oral	Cythioate
					Escort™ (Schering-Plough)	Collar	Diazinon
					Tiguvon™ (Bayer)	Spot on	Fenthion
					Mycodex™ Pet Shampoo, Carbaryl (Pfizer Animal Health)	Shampoo	Carbaryl
					Cap Star™ (Novartis)	Oral	Nitenpyram
					Frontline™ Top Spot (Merial)	Spot on	Fipronil
					Kiltix™ (Bayer)	Collar	Flumethrin + Propoxur
					Zodiac Duo-OpTM (Exil)	Spray	Pyrethrin + Piperonylbutoxid + N-octyl bicycloheptene dicarboximide + S-Methoprene
					Defend™Just-For-Dogs Insecticide (Mallinckrodt)	Spray	Pyrethrin + Permethrin + Piperonylbutoxid + N-octyl bicycloheptene dicarboximide
					Exspot™ (Mallinckrodt)	Spot on	Permethrin
					Program™ Tablets (Novartis)	Oral	Lufenuron
					Ovitrol Flea Egg-Control Collar (Vet Kem)	Collar	Methoprene

Table 1. (continued)

Parasite	Host	Vector for	Symptoms	Country	Therapy		
					Products	Application	Compounds
					Mycodex™ Fast Act IGR Flea and Tick Spray (Pfizer)	Spray	Pyriproxyfen (+Pyrethrins)
					Vapona™ (Pharmacia & Upjohn)	Collar	Diclorvos (DDVP)
					Bolfo™ Flohschutzband (Bayer)	Collar	Propoxur
					Faszin™ (Albrecht)	Collar	Diazinon (Dimpylate)
					Fleegard (Bayer)	Spot on	Pyriproxyfen
Ctenocephalides felis (Cat flea)	Dog, Cat (fleas general not very host-specific)	*Dipylidium caninum*, *Dipetalonema reconditum*, Bartonellosis (Cat scratch fever)	Blood loss, local skin reaction, strong itching, flea allergic dermatitis	World-wide	Advantage™ (Bayer)	Spot on	Imidacloprid
					Revolution™ (Pfizer)	Spot on	Selamectin
					Stronghold ™ (Pfizer)	Spot on	Selamectin
					Performer™ Flea and Tick Collar (Performer)	Collar	Naled
					Adams™ Flea and Tick Dip (Pfizer Animal Health)	Dip	Chlorpyrifos
					Duocide Flea and Tick Collar (Allerderm/Virbac)	Collar	Chlorpyrifos
					Cyflee™ (Boehringer Ingelheim)	Oral	Cythioate
					Escort™ (Schering-Plough)	Collar	Diazinon
					Tiguvon™ (Bayer)	Spot on	Fenthion
					Mycodex™ Pet Shampoo, Carbaryl (Pfizer Animal Health)	Shampoo	Carbaryl
					Cap Star™ (Novartis)	Oral	Nitenpyram
					Frontline™ Top Spot (Merial)	Spot on	Fipronil
					Kiltix™ (Bayer)	Collar	Flumethrin + Propoxur
					Zodiac Duo-OpTM (Exil)	Spray	Pyrethrin + Piperonylbutoxid + N-octyl bicycloheptene dicarboximide + S-Methoprene

Table 1. (continued)

Parasite	Host	Vector for	Symptoms	Country	Therapy		
					Products	Application	Compounds
					Defend™Just-For-Dogs Insecticide (Mallinckrodt)	Spray	Pyrethrin + Permethrin + Piperonylbutoxid + N-octyl bicycloheptene dicarboximide
					Exspot™ (Mallinckrodt)	Spot on	Permethrin
					Program™ Tablets (Novartis)	Oral	Lufenuron
					Ovitrol Flea Egg-Control Collar (Vet Kem)	Collar	Methoprene
					Mycodex™ Fast Act IGR Flea and Tick Spray (Pfizer)	Spray	Pyriproxyfen (+Pyrethrins)
					Vapona™ (Pharmacia & Upjohn)	Collar	Diclorvos (DDVP)
					Bolfo™ Flohschutzband (Bayer)	Collar	Propoxur
					Faszin™ (Albrecht)	Collar	Diazinon (Dimpylate)
					Fleegard (Bayer)	Spot on	Pyriproxyfen
Tunga penetrans (Sand flea, jigger)	All animals, Man		Man: flea penetrates skin (often foot), local skin reaction, strong itching, purulent inflammation; bact. sec. inf. (clostridial inf.)	Tropic areas			
Pulex irritans	Man, (Pig)	*Dipylidium caninum*		World-wide			
Xenopsylla cheopis	Rodents, Man	*Yersinia pestis* (plague)	Blood loss, local skin reaction, strong itching	Flea: world-wide;			
		Plague: North-, South America, Central-, East-, South Africa, Madagascar, Central- and Southeast Asia					

Table 1. Parasites affecting the skin and subcutaneous tissue (according to Vercruysse and De Bont)

Parasite	Clinical aspects													Localisation				
	1	2	3	4	5	6	7	8	9	10	11	12	13	1	2	3	4	5
CATTLE																		
Protozoa																		
Besnoitia besnoiti			+			+				+		+	+	+				
Helminths																		
L3-*Bunostomum* and *Strongyloides, Schistosoma* spp.	+				+											+	+	
Onchocerca spp.			+														+	
Parafilaria bovicola			+					+					+					+
Stephanofilaria spp.	(+)	+	+				+	+		+		+	+			+	+	
Arthropoda																		
Mites																		
Chorioptes bovis	+			+					+		+					+	+	+
Demodex bovis			+			+								(+)	+			
Psoroptes bovis	+	+		+			+		+	+		+	+	(+)			+	+
Sarcoptes scabiei	+	+					+		+	+	+	+	+	(+)	+			
Lice																		
Damalinia bovis, Haematopinus eurysternus, Linognatus vituli	+					+			+			+		+	+			+
Diptera																		
Hypoderma bovis, H. lineatum			+						+									+
Chrysomyia bezziana, Callitroga hominovorax	+						+	+					+	+				+
SHEEP and GOATS																		
Helminths																		
L3-*Strongyloides, Bunostomum, Gaigeria pachicelis*	+				+											+	+	
Elaeophora schneideri	+				+		+	+	+			+	+		+	+	+	
Arthropods																		
Chorioptes ovis	+						+									+		
Psoroptes ovis	+	+		+			+		+			+	+				+	
Psoroptes cuniculi (G)	+						+											
Psorergates ovis	+						+		+				+				+	
Sarcoptes scabiei	+	+													+			
Demodex ovis			+				+									+	+	
Damalinia spp., Linognatus ovillus	+						+							+				
Melophagus ovinus	+						+						+	+				
Blowfly strike (Lucilia spp., *Calliphora* spp., *Phormia* spp.*)*	+						+	+	+				+		+		+	+
HORSE																		
Helminths																		
Habronema spp.	+							+	+						+		+	
Onchocerca spp.	+				+	+	+				+	+			+	+	+	
Parafilaria multipapillosa			+					+										

Table 1. (continued)

Parasite	Clinical aspects													Localisation				
	1	2	3	4	5	6	7	8	9	10	11	12	13	1	2	3	4	5
Arthropods																		
Chorioptes equi	+						+		+			+				+		
Psoroptes equi	+			+		+			+	+		+	+	+				+
Sarcoptes scabiei	+	+					+		+	+	+	+	+	+	+			
Werneckiella equi,	+					+	+		+			+		(+)				+
Haematopinus asini	+																	
Culicoides spp.	+					+	+		+		+	+	+		+		+	+
PIG																		
Arthropods																		
Demodex phylloides															+		+	
Sarcoptes scabiei	+	+				+	+		+				+	+	+			+
Haematopinus suis	+						+							+				
CARNIVORES																		
Protozoa																		
Leishmania spp.	+					+		+				+	+	+				
Helminths																		
Dirofilaria immitis	+		+					+					+		+	+	+	
L3-*Ancylostoma* spp.	+	+			+									+		+	+	
Arthropods																		
Demodex canis	+					+				+	+	+	+	+	+			
Notoedres cati		+					+							+			+	
Sarcoptes scabiei	+	+					+		+						+			
Otodectes cynotis	+						+								+			
Cheyletiella yasguri/blakei	+	+		+		+	+					+					+	+
Trichodectes canis, Felicola subrostratus	+					+								+				
Ctenocephalides canis, C. felis	+	+			+								+	+				
Blow and flesh flies, Cordylobia	+		+					+						+				

Clinical aspects: 1, Pruritus; 2, Papule; 3, Nodule; 4, Vesicle; 5, Erythrema; 6, Scale; 7, Crust; 8, Ulcer; 9, Excoriation; 10, Lechinification; 11, Abnormal pigmentation; 12, Alopecia; 13, Systemic signs;
Localisation: 1, Generalized/not specific; 2, Head, neck; 3, Limbs; 4, Thorax, ventral abdomen; 5, Back, hindquarters

dermatitis, ulcerations, a periorbital alopecia, diffuse alopecia and onycogryphosis.

Helminths

The skin is the natural site of entry for a number of parasites that have their final habitat in the gastrointestinal tract or elsewhere, e.g. the nematodes *Strongyloides*, *Ancylostoma*, *Bunostomum* and *Gnathostosma*, and the trematode *Schistosoma*. The passage of these parasites rarely causes cutaneous lesions in animals, except for *Ancylostoma*. Hookworm dermatitis begins with the appearance of red papules on those parts of the body which are often in contact with the ground; later these areas become uniformly erythematous, and then thickened and alopecic. Pruritis is mild but evident, especially during initial larval penetration. Repeated hookworm penetration of the foot in dogs may result in secondary bacterial invasion producing gross enlargement of the feet and paronychia. As a result of the paronychia the claws may become deformed.

The helminth infestations that remain more or less localized to the dermis are the filariid parasites most commonly seen in cattle, sheep and horses (→ Onchocercosis, Animals, → Stephanofilariosis, → Elaeophoriasis, Elaeophorosis, → Parafilariasis, Parafilariosis, → Habronemiasis, Habronemosis). Some filariid worms have been reported to occasionally cause cutaneous lesions, e.g. *Dirofilaria* (→ Dirofilariasis, Man), *Brugia*, *Dipetalonema*.

Arthropods

Mite Infestation Several mites infest animals and cause significant dermatological diseases. The lesions are the result of mechanical damage to the skin and probably also of hypersensitivity reactions to toxic secretions (→ Acariosis, Animals).

Tick Infestation Ticks, like the other mites, are important arachnid parasites of both large and small animals. They play a major role as vectors of a large number of diseases. Ticks also harm their hosts more directly by causing local injury at the site of attachment. Ticks suck blood and heavy infestations may cause anaemia. Sites of tick bites attract flies and may become the site of development of → myiasis.

Lice Infestation (Pediculosis) Several species of lice infest large and small animals. Domestic animals may suffer from infestations with both biting (Mallophaga) and sucking (Anoplura) lice. Lice are extremely host-specific. Infection is a seasonal problem and the signs associated with pediculosis are extremely variable. Most lesions result from skin irritation and pruritus. They include alopecia alone, papulocrustous dermatitis, and damage to wool or hide caused by rubbing or biting. Sucking lice may induce an anaemia. Constant irritation during lice infestations causes a loss of weight and a decrease in milk production.

Flea Infestation Fleas are the most common ectoparasites of dogs and cats. The clinical manifestations are highly variable. Some animals remain asymptomatic carriers, others develop a flea-bite dermatitis which is a reaction to irritant substances in the flea's saliva. The infection may also cause a mild papulocrustous dermatitis with a mild pruritis. An acute flea-bite allergic dermatitis may develop in dogs, causing intense pruritus and erythrema. Secondary lesions which result from self-excoriation include breaking of hair and local alopecia, and occasional areas of acute dermatitis. Fleas may also induce anaemia in heavily infested animals (see Table 1, page 550).

Flying and Biting Insect (Diptera) Infestation Flying and biting insects are ubiquitous pests for domestic animals. Not only do they cause a loss of productivity by continuously annoying the animals, they may also cause diseases. Direct or indirect pathological effects of these flies include the deposition of larvae on or into the skin (myiasis), the local irritation (→ dermatitis), the injection of antigens inducing hypersensitivity reactions, the blood-feeding activities leading to anaemia; and the inoculation of pathogenic organisms.

The most important flies are those species whose larvae are highly destructive, facultative or obligate parasites. Infestation with such larvae is called myiasis. Warbles, caused by *Hypoderma bovis* and *H. lineatus* occur chiefly in cattle (see volume 1). They form on the back of the animal multiple nodules with breathing pores which may be painful upon palpation. Affected animals may manifest signs referable to the migration path of the individual grub prior to the development of nodules. Economic losses are due to gadding, milk and meat loss, and depreciation of the carcass and hide. Destruction of *Hypoderma* larvae in the infected host may cause severe clinical reactions, which are sometimes fatal. These toxic manifestations appear in the form of local and systemic effects which include, in the instance of *H. bovis*, inflammatory lesions in the spinal canal accompanied by stiffens, and ataxia, paraplegia, and collapse. *H. lineatum* larvae cause inflammation in the oesophageal wall, dysphagia, drooling of saliva, and bloating. Shock-like cardiorespiratory signs may accompany any of these conditions. The exact nature of these adverse signs is not known but it seems possible that, in the living host, adverse reactions to dying larvae may comprise both direct (toxemic) and indirect (→ anaphylactic shock) components.

Myiasis caused by screwworms has been a cause of great financial loss in the livestock industry. There are two important species of "screwflies": → *Callitroga* (*Cochliomyia*) *hominivorax* (Vol. 1) and → *Chrysomyia bezziana* (Vol. 1). These flies are obligatory parasites which may infect all domestic animals, and which only lay eggs on fresh wounds. The infection causes intermittent irritation and pyrexia. A cavernous lesion is formed, characterized by progressive liquifactive necrosis and haemorrhage that oozes a foul-smelling liquid. A gross fibrous involution follows the larval exodus. Significant haematological and biochemical changes include an initial neutrophilia, anaemia, and decreased total serum protein with a progressive rise in serum globulins. A significant loss in body weight occurs in infested animals.

Cutaneous infestation by blowfly maggots causes heavy mortality in sheep and significant losses in wool production in many countries. A large number of species belonging to the genera *Lucilia*, *Calliphora*, *Phormia* and *Chrysomya* are capable of causing the disease. Moisture and warmth are essential for the hatching of eggs and the development of the larvae. The breech is by far the most common site involved because of soiling and excoration by the soft faeces and the urine of the animal. The affected sheep are restless, do not feed, tend to bite or lick at the "struck" area. Examination shows a patch of discoloured, greyish-brown, moist wool with an evil odour. In very early cases the maggots may be found in the wool attached to the skin, while in the latter stages the maggots burrow into the tissues causing an inflamed wound which produces a foul-smelling liquid. There may be fever, and death may follow.

The larvae of the Tumbu-fly *Cordylobia anthropophaga* penetrate the skin of dogs, cats and humans in sub-Saharan Africa and produce painful boil-like swellings.

The housefly (*Musca domestica*), stable fly (*Stomoxys calcitrans*), face fly (*M. autumnalis*) and the horn fly (*Haematobia irritans*) are responsible for considerable annoyance of animals. Wheals, crusts, and cutaneous papules and nodules have mainly been associated with the biting flies *Stomoxys* and *Haematobia*.

Culicoides spp. are small flies which inflict extremely painful bites. In horses they are associated with → hypersensitivity reactions leading to a pruritic dermatitis (referred to as sweet itch or summer dermatitis). Pruritus is most intense along the base of the mane and tail and on the withers. However, the condition can also involve other parts of the body. Lesions consist of self-inflicting hair loss, excoriations with crusting and scaling, and – after a period of incessant rubbing – striking hyperkeratosis and thickening of the skin.

Treatment

→ Arthropodicidal Drugs, → Ectoparasitocidal Drugs.

Sleeping Sickness

Synonym

African Trypanosomiasis

Pathology

The African → trypanosomes (Vol. 1) are parasites of humans and domestic animals causing the disease African trypanosomiasis Trypanosomiasis, Animals, → Trypanosomiasis, Man) or sleeping sickness. *T. brucei gambiense* infection occurs in Central and West Africa and is slowly progressive and generalized. Initially the trypanosomes are present extracellularly in the subcutaneous tissue at the site of the bite of the → tsetse fly (Vol. 1) and give rise to a papular and later ulcerating lesion, often called a chancre, which persists for about 2 weeks. In the second stage trypanosomes enter the bloodstream and multiply there. In the third stage there is fever and lymphoid hyperplasia leading to enlargement of the spleen and especially of the cervical lymph nodes, which contain trypanosomes useful for diagnosis by puncture and smear. The fourth stage is central nervous invasion associated with intermittent fever. Trypanosomes in the neuropil and the cerebrospinal fluid produce diffuse meningoencephalitis with lymphocytes, plasma cells, and histiocytes infiltrating mainly the gray matter, the Virchow-Robin perivascular spaces, and the vessel walls. Notable are the plasma cells with multiple eosinophilic proteinaceous globules, or morula cells, found in the meninges. Trypanosomes are not easily found but can be cultured from the spinal fluid. Neuronal loss and demyelination are not prominent, but the cortical microglial and astrocytic gliosis is impressive, especially in the superficial cortex. This inflammatory reaction extends to the spinal ganglia and cranial and spinal nerve roots. These central nervous system lesions are accompanied by headache, apathy, wasting of musculature, emaciation, tremors, inability to walk, and eventually to somnolence, paralysis, coma, and death, usually after a course of 1–3 years. Often the disease is complicated by other infections such as → malaria, → schistosomiasis, animals, → schistosomiasis, man, → hookworm disease, or pneumonia. Hematologically there is anemia and granulocytopenia, sometimes with lymphocytosis and hyperglobulinemia, especially of IgM.

Immune Responses

These kinetoplastid protozoa live only extracellular and have evolved remarkable → immune evasion mechanisms. Since *T. brucei* infects laboratory rodents readily, the mouse model has been used for almost all immunological studies. Different inbred strains of mice exhibit different resistance to *T. brucei*, but all eventually succumb to the parasite. As in other experimental parasitic diseases resistance is under polygenic control, but the exact nature of the genes involved is not known.

B Cells, Antibodies and Antigenic Variation The metacyclic and blood stream forms of *T. brucei* are uniformly coated with the variant surface glycoprotein (VSG). This GPI-anchored protein has a highly polymorphic N-terminus forming the exposed domain. Up to 1000 different VSG genes distributed throughout the genome are present in the genome of the parasite. At any one time, only a single VSG gene is actively transcribed at a telomeric expression site. The switching from one VSG variant to another by translocation events or telomeric in situ activation occurs in a spontaneous manner at a surprisingly high rate of 10^{-4} to 10^{-5} per cell division. Antigenic variation leads to the characteristic fluctuating parasitemia, as successive VSG variants elicit an antibody response and are then destroyed. The primary antibody response is a T cell-independent IgM response, but T cell-dependent IgG responses against normally buried nonvariant VSG epitopes can be also initiated after phagocytosis of trypanosomes.

Immunosuppression Immunosuppression, which occurs both in infected animals and humans, has been extensively studied in the mouse model of trypanosomiasis. Lymph node enlargement and splenomegaly are accompanied by massive accumulation of B cells and null cells. The massive polyclonal B cell activation manifests in elevated IgM levels and auto-antibody production. The alterations in the cellularity and architecture of the lymphoid system are accompanied by a dramatic suppression of T and B cell responses to antigens and mitogens. These effects are mediated by intermediate cells, in particular suppressor macrophages. The transfer of as few as 40 000 peritoneal or splenic macrophages from trypanosome-infected donor mice were capable of causing a 50 % suppression of recipient mitogen responses. The suppressor macrophages display an activated phenotype and especially two secreted products of these cells, PGE2 and NO, have been shown to mediate suppressive effects. Both, inhibitors of iNOS (L-NMMA and L-NAME) and cycloxygenases (indomethacin) led to a partial abrogation of suppressor macrophage activity, and when used in combination there was a complete restoration of proliferative responses in splenocyte cultures from *T. brucei*-infected mice. The role of NO during murine trypanosomiasis has also been studied in vivo. Treatment of infected mice with L-NMMA resulted in a restoration of mitogen-driven T cell proliferation in the spleen and surprisingly also in a significant reduction of the first parasitemic peak. In contrast to other parasites, NO does not significantly affect trypanosome proliferation. While *T. brucei* is killed by NO in vitro, the presence of red blood cells, as in the natural habitat of the parasite, the bloodstream, abolishes this effect completely. This is a result of hemoglobin acting as high affinity sink for → Nitric Oxide (NO).

While the suppressor macrophage products PGE2 and NO are clearly damaging to the host, TNF has a more ambiguous role. TNF has been originally identified as the cachexia-inducing factor in chronic trypanosomiasis. However, treatment of primed mice with anti-TNF antibodies reduced the duration of survival, and subsequent studies showed a direct trypanolytic activity of TNF.

T Cells Given the central role of activated macrophages in African trypanosomiasis, IFN-γ which has been found to be significantly upregulated in infected humans and mice most likely plays a central role. Recent studies have shown three independent cellular sources of IFN-γ in experimental trypanosomiasis. First, VSG-specific MHC class II restricted $CD4^+$ Th1 cells have been identified. A second source is a $CD8^+$ T cell population which is directly activated by a 42–45 kDa trypanosomal protein designated trypanosome-derived lymphocyte triggering factor (TLTF). A third source for IFN-γ early after *T. brucei* infection (e.g. days 2–4 of infection) are NK cells, as revealed by (1) a significant T-cell-independent production of IFN-γ in infected nude mice and (2) the reduction of IFN-γ production after depleting NK cells with anti-asialo-GM antibodies. A contribution of NK cells to the pathogenesis in murine trypanosomiasis is further suggested by the finding that NK cell-deficient beige mice showed prolonged survival time after *T. brucei* infection. Most interestingly and in addition to its role in the generation of activated suppressor macrophages IFN-γ appears to have also a direct stimulatory effect on the growth of the parasite. The proliferative response of *T. brucei* to IFN-γ has been demonstrated in axenic cultures. In line with the parasite growth-promoting activity of IFN-γ parasitemia in IFN-γ-deficient mice were consistently lower than in wild type mice. In contrast, IFN-γ receptor -/- mice displayed a more rapid parasitemia and shorter survival time, ascribed to higher levels of free plasma IFN-γ in these mice. Thus, IFN-γ represents another host cytokine in addition to epidermal growth factor described earlier which is able to directly influence the growth of *T. brucei.*

In addition to IFN-γ there appear to be factor(s) produced by the parasite acting as co-stimulators for macrophage activation. This activity was named trypanosomal macrophage-activating factor (TMAF) and although there is currently no information on the molecular nature of TMAF it appears to be distinct from VSGs of *T. brucei.* However, since TMAF was able to induce both PGE2 and NO synthesis by host macrophages, it may represent a virulence factor of the parasite contributing to the immunosuppression observed in trypanosomiasis.

Main clinical symptoms: Fever, local oedema, evtl. polyadenitis, neural complications, death
Incubation period: local oedema: 1–21 days, fever: 3 weeks; cerebral disorders: 3 months in *T. b. rhodesiense*, 9–12 months in *T. b. gambiense*
Prepatent period: 1–3 weeks
Patent period: Years in chronic cases
Diagnosis: Microscopic determination of blood stages, serologic methods
Prophylaxis: Avoid bite of tsetse flies in endemic regions
Therapy: Treatment see → Trypanocidal Drugs, Animals, → Trypanocidal Drugs, Man

Snoring Disease

Disease due to infection of cattle with the trematode *Schistosoma nasale* (→ Respiratory System Diseases, Ruminants).

Sowda

See Onchocerciasis, page 436.

Sparganosis, Man

Sparganosis is an infection with a larval tapeworm, usually *Spirometra* of dogs and cats (see also → Pseudophyllidea (Vol. 1)). Humans become infected by drinking water containing infected copepods, eating infected amphibians, or reptiles, the intermediate hosts. The spargana give rise to small, sometimes migratory, nodules or abscesses containing the elongated, segmented worm-like structures without scolex, sucker, or cyst (→ Pathology/Fig. 18B). The larvae in the → abscess (Vol. 1) are surrounded by an intense mixed inflammatory reaction, often containing large numbers of eosinophils and → Charcot-Leyden crystals. The nodule is delimited by granulation tissue. Rarely, the spargana proliferate in humans, giving rise to tracts or to space-occupying cysts in the brain, surrounded by inflammation variable in size and shape over time, that can be observed by tomography and magnetic resonance imaging.

Therapy

→ Cestodocidal Drugs.

SPf66

Polymerized chimera peptide (→ Malaria/Vaccination)

Spinose Ear Tick

Otobius megnini, the bite of which may introduce severe secondary bacterial ear canal infections in cattle (→ Tick Bites: Effects in Animals).

Splendore-Hoeppli-Reaction

→ Pathology, → Onchocerciasis, Man.

Spontaneous Healing

→ Leishmanization, → Leishmania (Vol. 1).

Spotted Fever

Diseases due to infection with tick-transmitted rickettsiae; *R. rickettsii*: Rocky Mountain spotted fever (RMSF); *R. conori*: fièvre boutonneuse; *R. sibirica*: North-Asian spotted fever.

Therapy
Tetracyclines

SREHP

Serine-rich *Entamoeba histolytica* protein; its excretion leads to abscesses, e.g. in liver (→ Amoebiasis).

St. Louis Encephalitis

Disease in the Americas due to → Flaviviridae (Vol. 1) transmitted by *Culex* mosquitoes.

Stable Fly

Stomoxys calcitrans (→ Diptera (Vol. 1))

Stephanofilariosis

Several species of *Stephanofilaria* (→ Filariidae (Vol. 1)) cause cutaneous lesions similar to those of → onchocercosis in cattle, but on different parts of the body. The adult worms live in cystic diverticula at the base of the hair follicles. The lesions develop over several years. Initially they appears as small papules which coalesce to form a larger lesion covered with crusts. Eventually the skin becomes thickened, there is loss of hair, hyperkeratosis, an ulcerating core and haemorrhage. The lesions are mildly pruritic. After healing, the affected areas remain as hairless lichenified plaques.

Therapy
→ Nematocidal Drugs, Animals.

STEVOR

Variant antigen (→ Malaria/Vaccination)

Strategies

→ Disease Control, Strategies.

Streptothricosis

Bovine streptothricosis is associated with the tick species *Amblyomma variegatum*. It is an acute or chronic, local or progressive and sometimes fatal, exudative dermatitis of cattle and other domestic and wild hosts caused by the bacterium *Dermatophilus congolensis*. It is characterized by a serious exudate which dries to mat the hair into paintbrush-like tufts, or to form crusts and thick scabs. It is widespread in tropical areas of vector distribution, where its appearance is mainly seasonal, occurring more during the rainy seasons.

Therapy
→ Antibiotica.

Strongyloidiasis, Man

Strongyloidiasis is produced by → *Strongyloides stercoralis* (Vol. 1), a parasite of humans, dogs, and cats, and occasionally other species , normally from other hosts. Larvae penetrate the skin and give rise to dermatitis at the site of entry. They then migrate either though the lung, where they may cause allergic pneumonia, or by other pathways to the intestine where they become adults. The females most commonly reach adulthood in the duodenum and upper jejunum and enter the mucosa, to lay eggs. Mucosal inflammation, and later atrophy, leading to → malabsorption and emaciation, are the consequence of heavy infections. The larvae, which are soon released from the thin walled eggs, are shed in the stool. However some reenter the mucosa or the perianal skin maintaining a chronic infection which is characterized by fleeting urticarial rashes on the abdomen, buttocks, thighs, and often perianally. In immunosuppressed hosts this autoinflection can lead to a marked increase in tissue invasion by larvae and of adult females in the gut (→ Pathology/Fig. 2F), which may contribute to a fatal outcome.

Main clinical symptoms: Bronchitis, bronchopneumonia, diarrhoea, loss of weight, eosinophilia, anaemia, death
Incubation period: Skin: 12–18 hours, lung: 1 week, intestine: 2 weeks
Prepatent period: 14–21 days
Patent period: 40 years (due to repeated autoinfections)
Diagnosis: Microscopic determination of larvae in feces or duodenal fluid, → serology (Vol. 1)
Prophylaxis: Use solid shoes in endemic regions and avoid human feces
Therapy: Treatment see → Nematocidal Drugs, Man

Strongyloidosis, Animals

Pathology

Ruminants *Strongyloides papillosus* occurs in cattle, sheep and goats. This nematode lives in tunnels within the epithelium of the villi of the anterior part of the small intestine. Severe infections cause villous atrophy, with a loss of plasma proteins and a reduced activity of several enzymes (alkaline phosphatase, lactase, saccharase and maltase). Clinical outbreaks principally affect young suckling animals. Signs include anorexia, loss of weight, diarrhoea (rarely haemorrhagic), dehydration, slight to moderate anaemia. Severe infections may be fatal. Studies in Japan demonstrated that *S. papillosus* could cause sudden death in calves.

Horses The only species in the small intestine of horses is *Strongyloides westeri* (→ Alimentary System Diseases, Ruminants). Clinical outbreaks principally affect young suckling foals. Signs include anorexia, loss of weight, coughing, diarrhoea (rarely haemorrhagic), dehydration, slight to moderate anaemia. Severe infections may be fatal.

Carnivores *Strongyloides stercoralis* occurs in dogs. Though not common, infection in young animals may have severe consequences. There is enteritis with erosion of the mucosa of the small intestine, and haemorrhages. Bloody diarrhoea occurs in heavy infections. Dehydration develops rapidly, and death may occur.

Swine Strongyloidosis caused by *Strongyloides ransomi* occurs in swine. Clinical outbreaks principally affect piglets. Signs include anorexia, loss of weight, diarrhoea (rarely haemorrhagic), dehydration, slight to moderate anaemia. Severe infections may be fatal.

Therapy

→ Nematocidal Drugs, Animals.

Subunit Vaccines

→ Vaccination Against Nematodes.

Sulfamethoxazole

→ Pneumocystosis.

Summer Influenza

Disease (flu) due to the Tahyna virus which is (at least occasionally) transmitted during bites of blood-sucking mosquitoes in the European summer (see also page 461).

Summer Ostertagiosis

Infection due to → Ostertagia (Vol. 1), see also → Nematodes (Vol. 1).

Superinfection

Occurrence of a second infection besides an existing primary one.

Surra

Trypanosoma evansi (syn. *T. equinum*) has a wide range of hosts and is pathogenic to most domestic animals (→ Trypanosomiasis, Animals). Camels, horses, dogs and Asian elephants are highly susceptible. The infection in horses (called surra) and dogs is severe and probably uniformly fatal in the absence of adequate treatment. Cattle are mildly affected and act as reservoir.

Therapy

→ Trypanocidal Drugs, Animals.

Sweating Sickness

Disease of cattle due to toxicosis associated with tick bite (*Hyalomma truncatum*) in Africa (→ Tick Bites: Effects in Animals).

Swimmers' Itch

Synonym

→ Cercarial Dermatitis.

Synergists

Important Compound

Piperonylbutoxide (PBO)

General Information

Synergists usually enhance the efficacy of an → arthropodicidal drug without displaying toxic effects by themselves. In most cases synergists are thought to act by inhibiting metabolism of a given drug and in this context can be used at least, sometimes to prolong the activity of a compound in resistant parasite strains. PBO, one of the most frequently used synergists, inhibits cytochrome P-450 microsomal monooxygenases and is used together with DEET (→ Repellents (Vol. 1)) and pyrethrins (→ Arthropodicidal Drugs/Pyrethroids and DDT) to control lice and biting midges on companion animals.

T

T-Cells

→ Immune Responses.

Tabanidosis

Disease due to infestation with tabanids, see Table 1

Taeniasis, Animals

→ Alimentary System Diseases, Animals, → Taenia (Vol. 1), → Platyhelminthes (Vol. 1).

Taeniasis, Man

Taeniasis with *T. saginata* and *T. solium* tapeworms (→ Taenia (Vol. 1)) in the small intestinal

Table 1. Tabanids and Control Measurements

Parasite	Host	Vector for	Symptoms	Country	Therapy		
					Products	Application	Compounds
Tabanus spp. (Horse flies)	Ruminants, Horse	Bact. infections (e.g. Leptospirosis, Listeriosis)	Females suck blood; irritation, allergic reactions, economic loss	World-wide	Many	Pour on	Pyrethroids
Tabanus bromius	Ruminants			Central Europe			
Tabanus spodopterus	Ruminants						
Tabanus atratus	Ruminants						
Tabanus sudeticus	Horse						
Hybomitra ciurea	Ruminants						
Chrysops caecutiens (Deer flies)	Ruminants, Horse						
Chrysops relictus (Deer flies)	Ruminants						
Haematopota pluvialis	Ruminants, Horse				Bayofly™ Pour-on (Bayer)	Pour on	Cyfluthrin
Haematopota italica	Horse						
Tropic tabanids	Horse	*Trypanosoma brucei evansi* ("Surra"); *Trypanosoma brucei equinum* ("Mal de Caderas")	Females suck blood, bothering	Tropic areas			

lumen is largely asymptomatic. Microscopic lesions have not been described, except for slight → eosinophilia. However, after ingestion of *T. solium eggs* humans can act as intermediate hosts. The larval cysts develop in almost any tissue and can cause serious damage especially when they involve special areas of the brain (→ Cysticercosis).

Main clinical symptoms: Loss of weight, abdominal pain, anal pruritus
Incubation period: 8 weeks
Prepatent period: 8–18 weeks
Patent period: 25 years in man
Diagnosis: Occurrence of typical white proglottids in feces
Prophylaxis: Avoid eating raw meat
Therapy: Treatment with praziquantel, see → Cestodocidal Drugs

Related entry → Taenia (Vol. 1).

Targets for Intervention

→ Disease Control, Targets.

Texas Fever

→ Babesiosis, Animals.

Theileriacidal Drugs

For overview see Table 1

Epizootiology

The piroplasms are tick-transmitted blood cell parasites of vertebrates occurring in lymphocytes, erythrocytes, and other blood system cells (→ *Theileria* (Vol. 1)). The occurrence of vectors (ticks) determines the geographical distribution of *Theileria* spp. in tropical and subtropical areas.

T. parva parva causing East Coast fever (ECF) is enzootic in South, East and Central Africa and may be lethal for *Bos taurus*, *B. indicus*, and *Bubalus bubalis* (water buffalo) as well as for imported cattle. *T. p. lawrencei* causing corridor disease occurs primarily in African buffalo (*Syncerus caffer*) and is endemic in East and Central Africa and Angola as well. It may produce a mild disease in buffalo but a fatal one in cattle and water buffalo. *T. annulata* causing tropical theileriosis occurs in the northern subtropical and Mediterranean regions from Morocco through the Middle East, Southern region of Russia, and neighboring countries to the Indian subcontinent and China. It may be lethal to cattle but may produce only mild disease in buffalo. *T. orientalis* (syn. *T. sergenti*) has a low to moderate pathogenicity, and its distribution coincides over large areas of Asia with that of *T. annulata*. *T. mutans* regarded as mildly pathogenic or non-pathogenic may be severely pathogenic in cattle under various stress situations; this species is ubiquitous. *T. hirci* and *T. ovis* may be common in sheep and goats. *T. hirci* occurs in Asia, Africa, and South Europe and may cause a severe disease (50%–100 % mortality), mainly in newly introduced animals. Its morphology is similar to that of *T. annulata*. *T. ovis* has a low pathogenicity and is morphologically indistinguishable from *T. hirci*. However, *T. ovis* is more widely distributed than *T. hirci* and occurs in Africa, Europe, parts of the former USSR, India and West Asia.

Strategic Control Programs

Strategic control programs used in enzootic areas with theileriasis in livestock are similar to that applied in babesiasis (→ Babesiacidal Drugs). They rely on measures, such as premunization (live, attenuated vaccines, or "infection-treatment"), tick control (regular acaricide dipping, other application techniques), quarantine (especially with regard to importation of cattle from *Theileria*-free areas into enzootic regions where tick vectors exist), chemoprophylaxis, and finally chemotherapy.

To prevent cattle from areas and farms contaminated with infected ticks stock-proof fencing is essential. Therefore, the farm area must be cleaned of infected ticks before susceptible animals are brought in, and at least weekly dipping (in case of *Rhipicephalus appendiculatus* at least two treatments per week) should be carried out to control *T. p. parva* infections. Enzootic stability as a means of controlling theileriasis (as partially practiced in controlling babesiasis) may be achieved by natural challenge in indigenous cattle and thus development of some degree of resistance against theileriosis. Stock recently introduced into infested areas and coming from regions free of *Theileria* spp. or areas with different strains of *Theileria* spp. may need, however, a year

Table 1. Antitheilerial drugs for use in cattle.

Chemical group (approx. dose, mg/kg body weight[a], route)	**Nonproprietary name**, *brand name (company, manufacturer)	**Characteristics**[b] and miscellaneous comments
TETRACYCLINES		
(cattle, 1.5, p.o., daily for 28 days) (cattle 4, i.m., daily for 4 days) (cattle, 20, i.m., 1–2 times) [c1] 1953	[c1] **chlortetracycline** (American Cyanamid) rolitetracycline (Hoechst, others; not approved for use in animals) oxytetracycline * Terramycin LA 200 (long-acting formulation; Pfizer, others)	Already 1953 the prophylactic effect of chlortetracycline was demonstrated against *T. p. parva*; tetracyclines may arrest schizogony and thus reduce parasitemia of *Theileria* spp. infections; however, drugs must be administered just prior to or simultaneously with infection; sufficiently high active drug levels are necessary throughout incubation period; tetracyclines are useful tools in chemoimmunization programs (rolitetracycline; long-acting oxytetracyclines)
HYDROXYNAPHTHOQUINONES		
	Menoctone (discontinued) (Wellcome; Sterling Winthrop)	first drug with high antitheilerial activity causing marked degeneration in appearance of macroschizonts and suppression of parasitemia (piroplasms) in established *T. p. parva* infections in cattle at 5–10 mg/kg b.w.; intramuscular injection of the drug is followed by marked but transient pain, even at 2.5 mg/kg; drug has then been replaced by parvaquone; mode of action in *Theileria* spp. is not yet proved but in malaria drug blocks, as ubiquinone analogue, electron transport at the ubiquinone level; mechanism of selective toxicity might be due to difference between parasite and mammalian ubiquinone (see also mode of action of 4-t-butyl derivative BW 58C;
(20, i.m. in *T. annulata* infections; single dose of 20 mg/kg may be divided into two equal doses of 10 mg/kg (i.m.) with an interval of 48 h between doses: regimen used in treating *T.p.parva*, *T.p.lawrencei* and *T.mutans* infections) [c2] 1985 (Kenya)	[c2] **parvaquone**→BW 993 C * Clexon (Pitman-Moore, Mallinckrodt Vet)	although not as active as menoctone, drug was selected as the most cost-effective compound from a series of naphthoquinones for further development as antitheilerial drug; drug proved highly active against *T. p. parva* and *T. annulata* infections in cattle when treatment was performed in early stage of infection (macroschizonts detectable, fever starts); treatment 8 days after infection allows development of protective immunity without apparent clinical signs; preslaughter withdrawal periods for milk and edible tissues are at least 14 and 28 days, respectively; there should be no apparent discomfort after deep i.m. injection (occasionally, localized "painless" edematous swelling); for mode of action see menoctone
(2 . 5, i.m., single dose: may be repeated in case of heavy infections after 48 or 72 h)	**buparvaquone** (BW 720 C) * Butalex (Pitman-Moore, Mallinckrodt Vet)	a parvaquone analogue in which cyclohexyl moiety is 4-substituted by an alkyl group; latter group obviously slows down metabolic degradation of parent compound, and thus increasing *in vivo* efficacy against *T. p. parva* and *T. annulata*; in vitro it proves 20 times more active than parvaquone; when tested in *T.p.parva* and *T. annulata*-infected cattle drug shows a marked therapeutic effect at 2.5 mg/kg; buparvaquone appears to be the most active (and also most expensive) compound in the hydroxynaphthoquinone series; it proved to be active during incubation period, and after outbreak of bovine theileriosis (also in advanced cases); preslaughter withdrawal periods for milk and edible tissues may be 2 and 42 days, resp.; mode of action see menoctone

Table 1. (Continued) Antitheilerial drugs for use in cattle.

Chemical group (approx. dose, mg/kg body weight[a], route)	**Nonproprietary name**, *brand name (company, manufacturer)	**Characteristics**[b] and miscellaneous comments
QUINAZOLINONES		
(1. 2, p.o.) [c3] 1986 (Kenya)	[c3] **halofuginone** lactate (Hoechst Roussel Vet.) * Lerioxine, * Terit (discontinued) aqueous solution prepared from tablets)	it is active against *T. annulata* and *T. p. parva* infections in cattle; treated animals were immune against high challenge infections with homologous strain; high activity is evident when treatment is started with first clinical signs (body temperature above 40 °C) and macroschizonts are detectable by lymph node biopsies; morphological alterations of the drug on *T. parva*-infected lymphoid cells have been described; it is less active when administered during incubation period; there is only weak activity against piroplasms, and persistent carrier states with *Theileria* spp. may therefore occur or be maintained depending on start of treatment; drug proved active against *T. p. lawrencei* but was less active than in previous trials with *T. p. parva*; drug is well tolerated at recommended dose but safety margin is relatively narrow; 3 mg/kg may cause subnormal temperature, profuse diarrhoea, cachexia, and purulent eye discharge; preslaughter withdrawal periods for milk and edible tissues was 8 and 24 days respectively; for data concerning anticoccidial activity, see → Coccidiocidal Drugs/Table 1; drug seems also to be effective against acute sarcosporidiosis in goats and sheep (*Sarcocystis capracanis* and *S. ovicanis*, respectively at 0.67 mg/kg on two successive day) and Cryptosporidiosis of calves (*Halocur)

[a]Doses listed refer to recommended dose of the manufacturer and/or to literature on the subject
[b]For more details on biochemical action of drugs see entries on → drugs
[c]First practical (commercial) application (approx. year)
Data given in the table have no claim to full information
The primary or original manufacturer is indicated.

longer for partial immunity to be developed by application of various vaccination schemes. Tetracyclines (Table 1) are mainly used in chemoimmunization programs (infection-treatment methods). They may suppress or eliminate infections in areas where cattle have already developed a certain degree of protective immunity against *T. p. parva* and *T. annulata*. Tetracyclines administered simultaneously with infected ticks or vaccines can modify the course of the infection so that proliferation of parasites is limited and allow the development of protective immunity. As a result of infection-treatment methods mild clinical symptoms may occur while immunity is built up in the host. Another successful method of immunization against ECF seems to be the infection of cattle with live sporozoites (derived from standardized stabilates of *Rhipicephalus appendiculatus* followed by the administration of a long-acting oxytetracycline or parvaquone. Although treated animals showed a solid resistance to homologous challenge, they were not protected against a challenge with parasites unrelated to those initiating the primary reaction. To overcome this very specific immune response so-called "cocktail" vaccines derived from different strains have been prepared and used in large-scale field trials of immunization against cattle theileriosis. A satisfactory schizont vaccine against *T. annulata* does not require simultaneous drug treatment, and so saves costs.

Economic Importance and Pathogenesis

There are several *Theileria* spp. with different pathogenic features (→ Theileriosis). Parasites of economic importance in cattle are *T. p. parva* and *T. p. lawrencei* which cause East Coast fever (ECF) and corridor disease, respectively, as well as *T. annulata*, which produces tropical theileriosis. Commercial dairy herds and high-performance beef cattle on pastures must be protected against these *Theileria* spp. since their pathogenicity is generally high. Thus, mortality in fully susceptible cattle infected with *T. p. parva* may reach 90% – 100% although fatal cases in all are lower in endemic areas, and zebu cattle commonly show a high level of natural resistance. Fatal infections caused by *T. annulata* in cattle may vary considerably (10%–90%). Under natural conditions the interaction between host and parasite depends on host susceptibility, age, and variations in virulence of *Theileria* spp. strains, and intensity of challenge, i.e., the number of parasites transmitted by total number of ticks. The course of infection may therefore vary from peracute, acute, or subacute, to chronic, although the acute form is the usual one in susceptible animals (calves are more resistant to infection). Clinical signs in the acute and subacute forms are high to irregular intermittent fever, markedly swollen (enlarged) lymph nodes (hyperplasia), edema, diarrhoea, dyspnea, anemia, leukopenia (below 1000 cells/mm^3 is fatal in most cases) and general weakness. The pathogenesis and clinical symptoms are associated with the multiplication of parasites within transformed lymphoblastoid cells. Damaging effects on lymphoid tissues follows repeated schizogony synchronized with division of host cells. The lymphodestructive processes are characteristic for infections with *T. p. parva*, *T. annulata*, and *T. hirci*, while *T. orientalis* (syn. *T. sergenti*) and *T. mutans* infections are associated with invasion and destruction of erythrocytes resulting in anemia. Animals that recover from lymphoproliferative theileriosis have usually eliminated macroschizont stages and are solidly resistant to homologous challenge; however, challenge with heterologous parasites may lead to a partial or complete breakdown of immunity.

Drugs Acting on Theileriasis and Theileriosis

For a long time the search for an active drug was hampered by the lack of a suitable laboratory model of the infection. Chlortetracycline was shown to be the only drug, which exhibited prophylactic activity on *T. p. parva* infections. Today, long-acting oxytetracyclines are used which may arrest development of parasites in lymphocytes and thus reduce the rate of invasion of the cells (Table 1). To exhibit a reliable action tetracyclines must be given simultaneously with the initial *Theileria* spp. infection. As a result they curtail clinical symptoms in susceptible cattle types and consolidate immunity to virulent *Theileria* spp. strains; tetracyclines cannot be used for curative treatment and are ineffective once clinical signs become evident.

Using infected bovine lymphoid cell cultures, McHardy et al. (1976) demonstrated a high level of antitheilerial activity in vitro of the 2-hydroxy-3-(8-cyclohexyloctyl)-1,4-naphthoquinone, i.e., **menoctone**. Subsequent tests in cattle artificially infected with *T. p. parva* confirmed the activity of menoctone against East Coast fever (ECF) , even in cases with already established infection. Prior to the discovery of the antitheilerial effect hydroxynaphthoquinones had been shown to be active against coccidia and malaria parasites; recently, **atovaquone** (→ Malariacidal Drugs/Tables 1; 2) has also been proved to exhibit a high anti-plasmodial activity. Menoctone was then replaced by the closely related analogue **parvaquone**, which has a lower activity but is better tolerated by cattle than menoctone. The most active antitheilerial compound in the hydroxynaphthoquinone series is **buparvaquone**; it has been shown to be distinctly more active (factor 4) than parvaquone (Table 1). Like the quinazolinone derivative **halofuginone lactate** (anticoccidial effect cf. → Coccidiocidal Drugs/Table 1) both drugs are effective in treating the early stages of *T. p. parva* and *T. annulata* infections. Once established neither halofuginone nor parvaquone or buparvaquone can sterilize *Theileria* spp. infections. Prophylactic administration of hydroxynaphthoquinones obviously prevents infection completely and thus development of any serological/immunological response. Treated cattle were fully susceptible to a later *T. p. parva* challenge. The use of halofuginone lactate in the treatment of ECF in Tanzania had certain limitations. Field studies revealed that losses from theileriosis could be prevented only following diagnosis of early cases of the disease in several herds by repeated oral administration of 1.2 mg/kg body weight. Cattle suffering from the early stages of ECF recovered completely after this dosage regimen whereas two out of six cattle treated in the late stages died. **Imidocarb** diproprionate (→ Antibabesial Drugs/Table 1) and **primaquine** diphosphate (antimalarial, cf. → Malariacidal Drugs/Tables 1; 2) have been shown to be effective in reducing *T. orientalis* (syn. *T. sergenti*) parasitemiae in cattle at doses of 1.2 mg/kg body weight (x1) and 1 mg/kg body weight (x2), respectively.

B. equi has been renamed as *Theileria equi*. Infections are more refractory to therapy, and some cases may be cleared with imidocarb. Thus recommended doses of drugs must be enhanced to obtain elimination of carrier infections. However, marked increase of dose level and repeated injection often lead to undesirable acute side effects and occasionally to mortality. Obviously there are different susceptibilities among *B. (T.) equi* strains to drugs. Antitheilerial compounds parvaquone and buparvaquone (Table 1) have also been shown to have some efficacy against parasitemia of initial *B. (T.) equi* infection indicating an action chiefly directed against schizontal stages.

Theileriasis

Synonym Theileriosis.
→ Tick Bites: Effects in Animals.

Theileriosis

Synonym

→ Theileriasis.

General Information

→ *Theileria* (Vol. 1) species are tick-borne protozoan parasites which cause infections of ruminants characterized by successive developmental stages in leukocytes and erythrocytes. In contrast to other apicomplexan protozoans such as → Plasmodium (Vol. 1) and → Babesia (Vol. 1) which cause disease by destruction or sequestration of erythrocytes, pathology produced by *Theileria* is attributable mainly to the intraleukocyte stage. The severe, often fatal diseases in ruminants cause huge economic losses to the cattle industry, primarily in East and Southern Africa. In the horse, *Babesia equi* which has now been redescribed as *Theileria equi*, is also a major pathogen.

In cattle, the most important species in East and Southern Africa is *T. parva*. Two types are generally recognised: *T. parva parva* which causes East Coast Fever and *T. parva lawrencei* which causes → corridor disease. *T. annulata* causes theileriosis in North Africa, the near Middle East, Southern Europe and Central Asia (e.g. → Mediterranean Coast Fever). *T. mutans* has a very wide distribution and is generally benign, except in some parts of East Africa where it is reported to cause severe disease. *T. lestoquardi* (*T. hirci*) and *T. ovis* are parasites of sheep in North Africa, Southern Europe and Asia.

Pathology

When introduced into a non-endemic region, **East Coast Fever** normally kills 90% or more of the susceptible cattle population. In endemic areas the severity of the disease is greatly reduced, particularly in calves which develop only subclinical infections when exposed to moderate tick challenge. Susceptible animals introduced into enzootic areas rarely survive the infection. The disease is characterized by high fever, lymphadenopathy, severe pulmonary oedema, and wasting. The fever follows a severe panleukopenia and remains high until recovery or death, which commonly occur after a course of 5 to 25 days. Oedema of the eyelids and lacrimation are often present. Dyspnoea and a soft moist cough usually occur in terminal stages, sometimes with a voluminous frothy nasal discharge. At necropsy, the most obvious finding is often a severe pulmonary oedema. Lymph nodes are enlarged, oedematous, and may contain hemorrhages. Multifocal lymphoid hyperplasia may be visible as white spots within the renal cortex. Infection with *T. parva lawrencei* induces clinical signs and pathological changes very similar to those of → East Coast Fever. The differentiation between the two diseases is mainly based on epidemiological grounds: Corridor Disease is transmitted to cattle by ticks of the African buffalo and occurs only in areas within the distribution of *Syncerus caffer*. The clinical signs of Tropical Theileriosis are very similar to those of East Coast Fever, except maybe that in the former disease anaemia becomes more severe and icterus may be present. Leukocytopaenia does not usually develop in tropical theileriosis. However, the geographical ranges of *T. annulata* and *T. parva parva* do not overlap except, perhaps, in some parts of East Africa. The pathogenesis of *T. mutans* is entirely associated with the proliferation of the intraerythrocytic piroplasms. The disease is normally mild in character, but cases of anemia, icterus and haemoglobulinuria have been reported from East Africa. *T. lestoquardi* (*T. hirci*) is pathogenic in sheep and goats, clinical signs resemble those of East Coast Fever in cattle.

The horse parasite *T. equi* has recently been redescribed. It is highly pathogenic and induces the following clinical signs: high fever, listlessness, lacrimation, oedema of the eyelids, severe anaemia, haemoglobinuria and icterus. Pathological findings include emaciation, anaemia and icterus, hepatomegaly and splenomegaly, lymphadenopathy, oedema of the lungs, ascites, hydrothorax and hydropericardium.

Immune Responses

In the following section we will focus on *Theileria parva.*

Development from sporozoite to schizont inside leukocytes induces activation and proliferation of the host cell. This process results in rapid clonal expansion of parasitized cells and thus allows the parasite to remain intracellularily. As a consequence, immunological control of established infections is largely T cell mediated. In vitro studies have demonstrated that *T. parva* can infect $\alpha\beta$ T cells (both $CD4^+$ and $CD8^+$) $\gamma\delta$ T cells and B cells with similar frequencies. However, in vivo the vast majority of infected cells are $CD4^+$ and $CD8^+$ T cells and infection of unfractionated blood mononuclear cells usually gives rise to $\alpha\beta$ T cell lines. Infected cells constantly express high levels of class I and II MHC molecules.

The reservoir host of *T. parva*, the African buffalo, does not develop disease despite the fact that its cells are equally susceptible to infection and transformation by the parasite *in vitro*. Since in contrast experimental infection of susceptible cattle results in an acute fatal disease in the majority of animals, most studies on the immunity to *Theileria* species have been performed in an "infection and treatment model". In this model, the treatment with a slow-release formulation of oxytetracycline at the time of infection allows animals to recover from infection.

B Cells and Antibodies Studies with immune sera and mAbs have failed to demonstrate expression of *Theileria* antigens on the surface of parasitized lymphoblasts. Because of the apparent absence of such antigens on the cell surface and the intracellular localization throughout this stage of development it is not surprising that passive transfer of immune serum to susceptible animals failed to give any protection. However, sporozoite-specific antibodies capable of neutralizing infectivity *in vitro* have been detected in cattle in endemic areas. The major specificity recognized by these sera appeared to be a 67 kDa protein (p67) which has been cloned and used as recombinant immunogen in vaccination studies. Although all immunized animals generated strong antibody responses with similar titres, isotype patterns, avidity and epitope specificity, only 70 % of the cattle were protected against challenge infections. Thus it is not clear yet whether protection induced by p67 immunization relates to antibody function.

T Cells Immunization of cattle by infection and treatment elicits a strong MHC-I restricted specific response of $CD8^+$ cytotoxic T cells (CTL), whereas parasite-specific CTL are not detected at any stage of the primary infection with *T. parva*. In immune animals the frequency of detectable CTL precursors is relatively constant (1:2 000–1:12 000) at 5–6 weeks post infection but reaches levels as high as 1:30 in efferent lymph to 1:600 in PBMC following challenge with sporozoites. The strain specificity of CTL responses varied between animals immunized with the same parasites but only a restricted, immunodominant subset of epitopes appeared to be recognized. Furthermore, in many MHC-heterozygous animals a single class I molecule determined the epitopes that were recognized by CTL. The molecular and cellular events leading to immunodominance are not understood but are likely to involve variation in the concentrations of antigenic peptides bound to the presenting MHC molecules and the composition of the T cell receptor repertoire. Cell transfer experiments utilizing identical twin calves produced by embryo splitting have shown, that $CD8^+$ T cells mediate protection. Whether they exert their effect solely by killing parasitized cells or by cytokine-mediated mechanisms has not been investigated in detail. There is some evidence that TNF as well as other cytokines may inhibit early Theileria development inside lymphocytes, but none of the cytokines tested so far inhibited established infections. In addition, a wide range of cytokines, such as IL-2 and IL-10, appears to be expressed by *T. parva*-infected cells themselves.

Although the majority of studies have focussed on the role of $CD8^+$ T cells, there is no doubt that parasite-specific $CD4^+$ T cells are generated following immunization with *T. parva*. Some cloned $CD4^+$ T cells have been shown to produce IFN-γ indicating that they are Th1-like. However, the functional role of $CD4^+$ T cells in *T. parva* infection, for example as helper cells for the establishment of a $CD8^+$ T cell response or as antiparasitic effector cells, has not been analyzed.

During primary infection of cattle with *T. parva* the lymph node draining the site of infection undergoes a dramatic 3–4-fold increase in size 7–9 days after infection. Proliferation of mainly $CD8^+$ T cells is induced which contain a large subset of $CD2^-$ T cells. The latter cells may be derived from a rare pre-existing population and are polyclonal in origin as shown by analysis of their TCR repertoire. Analyses of the stimulatory requirements and function of these $CD2^-CD8^+$ T cells have been hampered by the inability to culture them *in vitro*. Since only a minor part of the lymphoblastic cells in the lymph nodes is parasitized, the question has been raised, whether or not the T cell response in naive animals contributes to the pathogenesis of the disease by potentiating the growth of parasitized cells. Since parasitized cell lines express cytokine receptors such as CD25 (IL-2 receptor, p55) and their growth could be enhanced by cytokines like IL-2, IL-1 and IL-10 *in vitro*, it is likely that the induction of an early cytokine expression in the infected host could help parasitized cells to multiply. Transfer studies with lymphocytes infected *in vitro* with *T. parva* sporozoites clearly demonstrated that the cell type infected influences the pathogenicity of the parasite. While infected $CD4^+$ or $CD8^+$ T cells induced severe, potentially lethal infections, infected B cells produced only mild self-limiting disease. The factors that determine the difference in the pathogenic potential of parasitized B and T cells have not been identified so far.

Vaccination

A number of attempts to produce commercially available products have not yet been successful. *T. annulata* can be grown in culture using lymphocytes or fibroblasts as host cells. If this cultivation is done over a longer period of time, the virulence of the parasite is attenuated. In Israel, vaccination of cattle against *T. annulata* was introduced using attenuated cultured schizonts which are distributed in frozen form. The use was generalised in wide scale in North Africa and Asia and provides very effective control.

The search for attenuated vaccines against *T. parva* has not yet reached the same stage. However, in spite of the high costs, vaccine using live parasites has been used. The strategy of such a vaccine is based on the injection of live schizonts followed by treatment of the animal with tetracycline. More recently an attenuated *T. parva* isolate Boleni – has been used with success without concurrent tetracycline therapy. The main hindrance is that suitable infective parasite materials, i. e., sporozoites, can only be obtained from the salivary glands of the ticks. Sporozoites have been successfully used to obtain protection in cattle by injecting them frequently together with tetracycline. In vaccination trials, cattle immunized against *T. parva* with this technique developed immunity against the homologous sporozoites as well as those of *T. parva lawrencei* and a number of other *T. parva* strains, indicating that there is a common protective antigenic determinant on the sporozoites used for vaccination. Doherty and Nussenzweig mentioned in 1985 that a number of monoclonal antibodies do not distinguish between sporozoites of the various *T. parva* strains.

A 67 kDa protein on the surface of the sporozoite of *T. parva* was identified as target antigen. Corresponding recombinant molecules have been used in vaccination experiments but producing partial protection. An equivalent recombinant protein SPAG-1 from *T. annulata* was also used, also producing partial protection. Intensive efforts have been focused on the identification of schizont-specific components for incorporation in a second-generation multi-component product, but these efforts remain so far inconclusive. Attenuated vaccines lines of *T. annulata* and *T. parva* are at the present time the only practical solutions for theileriosis control. Thus the molecular biology studies on the mechanisms of parasite virulence remain a priority in the hope of achieving a rational attenuated parasite line for vaccination.

Therapy

→ Theileriacidal Drugs.

Therapy

See chapters on Disease Control, Control and → Chemotherapy, → Drugs, → Biological Control (Vol. 1).

Threadworm Disease

Synonym

Strongyloides stercoralis (→ Strongyloides (Vol. 1)).

Thrombosis

→ Pathology, → Malaria.

Thrombospondine-Related Anonymous Protein

→ Malaria, → TRAP, → Vaccination.

Tick Bites: Effects in Animals

General Information

Ticks play a major role in the human economy by causing significant losses in animal production through their own direct effects, as well as by transmitting diseases to domestic stock. In temperate and tropical countries, ticks surpass all other arthropods in the number and variety of diseases they transmit to animals. Disease agents transmitted to animals by ticks include viruses, rickettsiae, bacteria, fungi, and protozoans. In a few cases, helminths have also been found in ticks. A summary of the most important pathogens transmitted by ticks to animals is shown in Table 1. Direct effects of ticks are manifested as mechanical injury, anemia, paralysis, and toxicosis. These may be extended by secondary infections with bacteria, fungi, or myiasis.

Lesions

In most cases the bite of a tick is not felt at first, but it can be painful in *Amblyomma* spp., which have very long mouthparts. Even if ticks are removed by grooming, persistent lesions can remain.

Mechanical injury to the host is initiated by the penetration of the mouthparts into the skin, during which the chelicerae serve to cut into the epidermis, causing some damage to capillaries and tissues. The recurved denticles of the hypostome serve to anchor the tick within the lesion. The depth to which inserted mouthparts enter the host's skin depends on the morphology and feeding habits of the tick species. Those with long mouthparts, such as members of the genera *Amblyomma* and *Hyalomma*, penetrate much deeper than ticks with short mouthparts, such as members of the genera *Boophilus*, *Dermacentor*, *Haemaphysalis*, and *Rhipicephalus*, which are superficial feeders. Female *A. americanum* ticks insert their mouthparts well into the lower reticular area of the dermis to the layer of adipose tissue. *Hyalomma asiaticum* penetrate to a similar depth in sheep.

The extent of tissue damage caused by tick feeding is not always clearly differentiated from host reactive tissue damage. The involvement of host reactions leading to tissue damage may be dependent upon recruitment and degranulation of mast cells, resulting in the release of heparin and histamine (or 5-hydroxytryptamine in bovines) from the granules, leading to inflammatory responses characterized by dermal cell infiltrates which form the lesion. The type of cell which infiltrates is dependent upon the tick/host systemy involved and can be predominantly neutrophil or mononuclear. The cell populations can change after multiple infestations. At the feeding site of *Hyalomma a. anatolicum* on rabbits, the extent of collagen destruction is parallel to the degree of neutrophil infiltration. Mated females *A. americanum* caused a more substantial inflammatory response than unmated females, which produced, on average, smaller cavities and lesions over the same duration of attachment. Lesions from unmated females were the same size after 240 h as those from mated females after 48 h.

In fast-feeding argasid ticks, the leukocyte response is less marked than in the slow-feeding ixodid species. After primary feeding by *Ornithodoros parkeri* and *O. tartakovskyi*, basophils were found to accumulate after as little as 24 h in guinea pig skin. There was also a marked blood basohphilic response, with weak eosinophilia.

The extent of local injuries can be influenced by the tick species and the site. There is a higher frequency of severe udder damage in cattle infested by *Amblyomma hebraeum* than in those infested by *A. variegatum*. Larvae and nymphs of *Otobius megnini*, the spinose ear tick, cause marked irritation in the ear canal, with secondary bacterial infections extending inwards, sometimes resulting in serious complications.

Theoretical estimates of blood loss and ensuing damage due to ticks do not always apply to the field situation. In many cases the severity of effects is likely to be enhanced by toxic salivary excretions. Acute and even fatal anemia can occur in heavy infestations. *Argas persicus* can completely exsanguinate chickens, particularly young ones, within short periods.

Table 1. Tick-borne bacteria, rickettsiae, and protozoans in domestic animals

Tick vector	Pathogen	Host	Disease
Argas reflexus	*Borrelia anserina*	Poultry	Avian spirochaetosis
	Aegyptianella pullorum	Poultry	
Amblyomma spp.	*Cowdria ruminantium*	Ruminants	Heartwater
	Theileria mutans	Ruminants	Benign theileriosis
	Theileria velifera	Bovines	
Amblyomma variegatum	*Dermatophilus congolensis*		Streptotrichosis
	Ehrlichia bovis	Bovines	Ehrlichiosis
Boophilus spp.	*Anaplasma marginale*	Bovines	Anaplasmosis
	Anaplasma centrale	Bovines	Anaplasmosis
	Babesia bigemina	Bovines	Texas fever
	Babesia bovis	Bovines	Babesiosis
	Borrelia theileri	Bovines	Borreliosis
Dermacentor andersoni	*Anaplasma marginale*	Bovines	Anaplasmosis
Haemaphysalis leachi	*Babesia canis*	Dogs	Babesiosis
	Ehrlichia canis	Dogs	Ehrlichiosis
	Babesia felis	Felines	Babesiosis
Haemaphysalis longicornis	*Theileria orientalis*	Bovines	Theileriosis
Haemaphysalis punctata	*Babesia major*	Bovines	Babesiosis
Hyalomma anatolicum, Hyalomma spp.	*Theileria annulata*	Bovines	Mediterranean Coast Fever
Hyalomma spp.	*Trypanosoma theileri*	Bovines	
Ixodes persulcatus	*Babesia divergens*	Bovines	Babesiosis
Ixodes ricinus	*Babesia divergens*	Bovines	Babesiosis
	Ehrlichia phagocytophila	Sheep	Tick-borne Fever
	Staphylococcus aureus	Sheep	Tick pyemia
	Borrelia burgdorferi	Many	Lyme-Disease
Rhipicephalus appendiculatus	*Theileria parva parva*	Bovines	East Coast Fever
	Theileria parva lawrencei	Bovines	Corridor Disease
	Theileria taurotragi	Elands	Theileriosis
Rhipicephalus bursa	*Theileria ovis*	Sheep	Theileriosis
Rhipicephalus evertsi	*Theileria ovis*	Sheep	Theileriosis
Rhipicephalus pulchrellus	*Theileria taurotragi*	Elands	Theileriosis
Rhipicephalus sanguineus	*Babesia canis*	Dogs	Babesiosis
	Ehrlichia canis	Dogs	Ehrlichiosis
	Haemobartonella canis	Dogs	Bartonellosis

Paralysis

→ Paralysis (Vol. 1) caused by ticks is reversible when the causative ticks are removed, but it is sometimes difficult to clearly separate paralysis from toxicosis. Paralysis is usually associated more with female ticks and can be produced by a single tick. In Australia, paralysis is produced by *Ixodes holocyclus*, particularly in dogs, but sometimes also in other animals. In South Africa, Karoo tick paralysis is caused by *I. rubicundus*, mainly in sheep, as well as in other domestic stock. Also in South Africa, *Rhipicephalus evertsi evertsi* and *R. e. mimeticus* cause spring lamb paralysis while *Argas* ticks in the subgenus *Persicargas* cause paralysis in poultry. Other important paralysis-producing species are *Dermacentor andersoni*, the Rock Mountain wood tick, which affects sheep and cattle, and *D. variabilis*, the American dog tick, which paralyzes dogs. A number of other species in several genera have also been incriminated. The distribution of paralysis due to a certain species of tick does not necessarily coincide with the total area of distribution of the species, and pathogenicity may also change within this area.

A paralysis-inducing toxin has been isolated in *I. holocyclus* and used as an iminunizing agent. *Rhipicephalus evertsi* females are only able to produce toxin while they are within a specific weight range, which is of short duration when they are mated but prolonged in the absence of mating; thus there is increased risk of paralysis when females feed in the absence of males. This explains the higher frequency of some forms of paralysis in the field than in the laboratory, where in the past females have usually been provided with adequate numbers of males.

Toxicosis

Tick toxicosis is also associated with tick bite. A well-known African form is → sweating sickness, a disease of cattle, with profuse serum exudation onto the skin. It is caused by the bite of *Hyalomma truncatum*, a species of tick common throughout subsahelian Africa. Reactions to argasid ticks are more frequently caused by juvenile instars. *Ornithodoros savignyi* toxicosis can cause losses among cattle and *O. lahorensis* among sheep. The toxic component of oral secretions of *O. savignyi* was found to have a molecular weight of 15 kilodaltons.

A variety of different pathogens can use lesions caused by ticks to infect animals. The bite of *Ixodes ricinus* is associated with tick pyemia, a disease of 2 to 6-week-old lambs in the UK caused by *Staphylococcus aureus*. A similar, usually fatal condition, caused by the same bacteria, has been found in rabbits infested with *Rhipicephalus appendiculatus*. It is thought to be related to immunosuppressive components of tick saliva.

Clinical Relevance

Clinical signs due to ticks are found in birds, reptiles, and mammals. There are reports of large tick burdens on some wild animals but the effects of ticks on individual animals are best known from domestic stock. Threshold values for production loss in domestic animals, i.e., changes in average daily weight gains, differ depending on tick species and hosts. For instance, weight gain was shown to be affected by 15 *Amblyomma americanum* females per calf, but far larger numbers of ticks are considered acceptable under other challenge situations.

Nevertheless, systematic tick control nearly always results in improved weight gain and yield in domestic stock. On the other hand, it has also been shown that under extensive husbandry conditions in dry African rangelands, cattle may make better weight gains in the absence of acaricide treatment, despite tick challenge.

Enormous tick challenge can be encountered in some parts of the world. The classic report of heavy infestation was by Theiler, who in 3 days removed half the *Boophilus decoloratus* ticks from a horse which had died of acute anemia and found them to weigh about 7 kg. In addition to blood loss and anemia, host animals are also affected by tick salivary excretion into the host, which may amount to at least the equivalent in weight of the engorged ticks. However, it must be accepted that most reports of massive harmful infestations with ticks involve either naive animals which have had no opportunity to acquire resistance, notoriously tick-susceptible species/breeds, e.g., Friesian (Holstein) cattle, or animals stressed by other causes.

Transmitted Pathogens

Many tick-borne diseases (Table 1) are characterized by a high degree of endemic stability, i.e., in areas with well-established populations of host animals, disease is experienced by most individuals, without more than passing clinical signs, and there is a high degree of immunity or premunity. Stability can be disturbed by the introduction of susceptible, nonadapted hosts, after which serious outbreaks can occur in the new animals, inducing altered epidemiological conditions.

Virus diseases transmitted by ticks have been described from a large number of different hosts and are associated with a large number of different species of ticks (e.g. → African swine fever, → Nairobi sheep disease). The presence of viruses in ticks in an area is not always associated with disease. In addition, ticks transmit protozoans (*Babesia*, *Theileria*), rickettsiae and bacteria which cause severe diseases (→ heartwater, → streptothricosis).

Related Entries

→ Tick Bites: Effects in Humans, → Protozoan Infections, → Ticks as Vectors

Tick Bites: Effects in Humans

General Information

Ticks are a hazard to human health through direct effects as well as through the transmission of viral, rickettsial, bacterial, fungal, and protozoan diseases (→ Ticks/Important Species (Vol. 1)). Their capacity as vectors for the transmission of human diseases is surpassed only by that of → mosquitoes (Vol. 1). A summary of some major diseases transmitted by ticks is shown in Table 1. Direct effects of tick activity include mechanical lesions and paralysis.

Lesions

Some lesions are complicated by the mouth parts of ticks breaking off during manual detachment. This is rare in many tick species, but can occur in up to 50% of *Ixodes* spp. female detachments. This can be followed by the formation of abscesses at the tick feeding site. The bite may not be known to the patient, particularly in cases where the tick is found on the scalp. This is a preferred site of *Rhipicephalus sanguineus,* a tick species which has spread rapidly during the recent past, attaching to children in temperate European countries. Some human cases of ear infestation with lanal and nymphal spinose ear ticks, *Otobius megnini,* have also been reported.

Tick bites from ixodid ticks are sometimes biopsied because they give rise to a persistent ulcer with a necrotic base. In the center of the lesion, up to 2–3 mm into the dermis, one can sometimes find remnants of the chitinous mouth parts (hypostome) of a tick which was removed incompletely several months earlier. These irregular brownish fragments elicit an intense hypersensitivity reaction and are embedded in necrotic connective tissue surrounded by fibrosis and an intense lymphohistiocytic reaction including eosinophils and basophils. The epithelium peripheral to the lesion undergoes acanthosis, hyperkeratosis, and focal parakeratosis.

The attachment of ticks is usually not felt, but may result in persistent irritation long after the tick has been removed. Larval ticks, also called → pepper ticks or → seed ticks, cause a small lesion which may lead to a relatively large inflammatory area remaining for several weeks. Large numbers of larvae are picked up when the offspring of a female tick crowd together on the vegetation. They will frequently attach on the ankles just above the socks and can cause intense and persistent itching.

Paralysis

Tick → paralysis (Vol. 1) in man has been associated with *Dermacentor andersoni* and *D. variabilis* in North America, *Ixodes hexagonus* in Britain, and *I. holocyclus* in Australia, among others. Female ticks are usually involved. It is caused by a toxin and can lead to complete locomotory paralysis and death through respiratory paralysis.

Table 1. Some important tick-borne viruses, bacteria, rickettsiae, and protozoans in humans

Tick	Pathogens	Diseases
Ornithodorus spp.	*Borrelia duttoni*	Relapsing fever
Dermacentor spp. and other ixodid species	*Rickettsia rickettsii*	Rocky Mountain spotted fever
	Francisella tularensis	Tularemia
Ixodes dammini	*Borrelia burgdorferi*	Lyme disease
	Babesia microti	Babesiosis
Ixodes holocyclus	*Rickettsia australis*	Queensland tick typhus
Ixodes ricinus	*Babesia microti*	Babesiosis
	Babesia divergens	Babesiosis
	Borrelia burgdorferi	Lyme disease
	Toga-/Flavi-virus	Meningoencephalitis
Rhipicephalus sanguineus	*Rickettsia conori*	Boutonneuse fever
Several ixodid species	*Rickettsia sibirica*	Siberian tick typhus
Amblyomma spp.	*Francisella tularensis*	Tularemia

Paralysis due to ticks is characterized by an acute ascending flaccid motor paralysis which can be fatal if the causative tick is not removed. In North America tick paralysis, the first symptoms usually appear 4–6 days after exposure. The cause of paralysis may not always be determined correctly, since the tick may be hidden by hair, which can lead to incorrect diagnosis. It is manifest at first as difficulty in walking, followed by inability to walk, limb numbness, and complete locomotory paralysis within 24 h, difficulties in speech, respiratory paralysis, and death. Removal of the culprit tick results in rapid and complete recovery. In cases of extensive paralysis, recovery may be delayed for 1–6 weeks. Paralysis caused by *Ixodes holocyclus* in Australia does not usually peak until about 48 h after the tick involved has been removed. Accompanying symptoms are vomiting and acute illness.

Transmitted Pathogens

Ticks are more important as transmitters of disease (Table 1, → Ticks/Important Species (Vol. 1)) than through their own direct effects. It is wise to regard all live ticks collected in the field for laboratory maintenance as potential sources of infection and they should always be handled with extreme caution. There are, unfortunately, too many instances where neglect of this principle has resulted in loss of life or severe illness. The association of many tick species with migrating birds can allow a rapid movement of virus-infected ticks into new areas. It should be kept in mind that tick tissues or excretory products may contain infective agents which can be transmitted through contact or aspiration alone. It may be necessary to investigate ticks for the presence of unwanted virus infections before general handling can take place. Even apparently very closely related tick species may differ completely in their capacity to transmit pathogens. It can therefore be of great importance to identify tick species correctly.

Many tick-transmitted pathogens are likely to be carried by natural wild hosts (reservoir), to which they are well adapted and in which they therefore cause no or only mild symptoms. At the same time, they can pose a serious threat to human beings or domestic stock, which may be severely affected and respond pathologically to infection. On the other hand, viruses have also been detected in ticks in areas where no disease is reported.

The → arboviruses (Vol. 1) associated with ticks and human disease belong to the → Togaviridae (Vol. 1), Flaviviridae (→ Flavivirus (Vol. 1), group B), to the → Reoviridae (Vol. 1), *Orbivirus*, and several ungrouped viruses.

Tick-transmitted rickettsiae usually do not persist for long in the peripheral circulation and the ticks themselves therefore serve as the main reservoir of disease. The best known is → Rocky Mountain spotted fever caused by *Rickettsia rickettsii*, which is found from Canada to South America and is associated with various ixodid ticks (*D. andersoni and D. variabilis* in North America) which can transmit the pathogen transovarially to the next generation. → Tick typhus is caused by other rickettsiae (e.g. *Rickettsia sibirica*). → Boutonneuse fever is caused by *Rickettsia conori,* which is widespread in Africa, the mediterranean region, and parts of Southeast Asia. Queensland tick typhus is caused by *Rickettsia australis* and is found in coastal Queensland, where *Ixodes holocyclus* appears to be the main vector. The transmission of *R. prowazeki*, the agent of *epidemic typhus*, by ticks has not been confirmed, although at one time they were thought to be vectors (→ Lice/Feeding Behavior and Transmission of Disease (Vol. 1)).

Ticks are also known or suspected of harboring a number of bacterial infections of man. → Relapsing fever is caused by spirochaetes (*Borrelia* spp.) and is transmitted by *Ornithodoros* tick species. → Tularemia, caused by *Francisella tularensis,* is found in North America, reaching as far south as Venezuela, and is also found in parts of Asia and Europe. It is transmitted by *Dermacentor andersoni* (in the USA) and other tick species such as *Rhipicephalus sanguineus.* → Lyme disease is caused by *Borrelia burgdorferi*, a spirochaete-like bacterium transmitted by the ticks *Ixodes dammini, I. pacificus* and *I. ricinus*.

Finally, ticks are vectors of many protozoan infections in man (e.g. → Babesiosis, Man).

Therapy

Treatment and control see → Acarizides (Vol. 1).

Tick Fever

→ *Babesia* (Vol. 1).

Tick Typhus

Tick typhus is caused by different rickettsiae. *Rickettsia sibirica* causes Siberian tick typhus which is transmitted transstadially and transovarially by ticks of the genera *Dermacentor* and *Haemaphysalis* as well as *Rhipicephalus sanguineus* and *Hyalomma asiaticum*. Ticks can serve as long-lived → reservoirs of disease. Other arthropods may also play a role. Queensland tick typhus is caused by *Rickettsia australis* and is found in coastal Queensland, where *Ixodes holocyclus* appears to be the main vector.

Related Entries
→ Rocky Mountain Spotted Fever, → Boutonneuse Fever.

Tick-Borne Encephalitis

Synonym
TBE

General Information
Tick-borne encephalitis is caused by the TBE virus (→ *Flavivirus* (Vol. 1), group B) which is also called hypr virus. It is a disease clinically similar to → Russian spring-summer encephalitis. A frequent route of infection is oral, by drinking the milk of infected domestic animals. The main vector is *Ixodes ricinus*, a tick found throughout Europe, including Turkey, and the Atlas region of North Africa. The TBE virus has a tendency to spread westwards throughout the area of vector distribution. In Austria it is a serious menace and it has more recently occurred in West Germany, where mortalities were initially high. *I. ricinus* is also associated with → louping ill.

Therapy
→ Vaccination.

Ticks as Vectors

→ Tick Bites: Effects in Animals.

Ticks as Vectors of Agents of Diseases, Man

Introduction
Human populations are becoming increasingly vulnerable to infection by tick-borne pathogens (→ Arboviruses (Vol. 1), → *Babesia* (Vol. 1), → Theileria (Vol. 1), → Bacteria (Vol. 1)). These infections are more intense and diverse now than they appeared to be during the mid 1900s. Our recent experience with Lyme disease, human babesiosis and human ehrlichiosis illustrates this pattern of emergent tick-borne infection. This trend is driven by a proliferation of vector ticks in many parts of the world and wide-spread human encroachment into forested sites. The purpose of this review, therefore, is to identify the environmental determinants of tick-borne disease. In particular, we shall review concepts concerning the diverse modes of perpetuation of tick-borne agents and identify conditions leading toward human infection. Although Lyme disease will be emphasized, this review will include other tick-borne pathogens that share similar ecological features.

Biology of Ticks
→ Ticks (Vol. 1) are arthropods, as are insects, but are classified with the mites and spiders. They may readily be distinguished from insects by their characteristic flat, unsegmented bodies, absence of antennae and four pairs of legs in the nymphal and adult stages. All ticks share various structural features employed for finding and feeding on vertebrate hosts. Their mouth parts include retractable chelicerae that penetrate the vertebrate skin and a multispined hypostome that affixes the feeding tick to its host. Olfactory setae, located on the anterior ends of their legs, serve to detect the presence of such animals. Hook like-structures at the ends of their legs enable ticks to attain contact with passing hosts. These features facilitate the parasitic mode of life.

The two major taxa of ticks, the Argasidae (soft ticks) and the Ixodidae (hard ticks), differ radically in structure, life-history and pathogen associations. The cuticle encasing the bodies of soft ticks is leathery while that of hard ticks is rigid. It becomes elastic, however, as a hard tick fills with blood. Soft ticks are endemic to arid regions and are closely associated with the nests or burrows of birds or rodents. Such ticks feed frequently during each trophic stage and become attached to their hosts only briefly, generally for no more than an hour. Each of these feeding episodes provides a tick-borne pathogen with another opportunity to infect susceptible hosts. Hard ticks, in contrast, feed for several days or more and do so only three times during their entire lifespan: once during each of the three trophic stages. Although pathogens of hard ticks have fewer transmission opportunities than do soft ticks, they maintain a diverse array of viral, bacterial, protozoan and metazoan pathogens. Soft-ticks in contrast, transmit only relapsing fever spirochetes, a complex of pathogens that is restricted to a few particularly arid sites. The following discussion, therefore, will focus exclusively on hard ticks and their associated pathogens.

Ticks require vertebrate blood to grow and to reproduce. Molting follows each feeding episode by a subadult tick. The life cycle culminates after adult ticks feed and the female deposits a batch of eggs. Ticks ingest enormous quantities of blood, hundreds of times their prior weight. A complex series of physiological events within the vertebrate host and vector tick facilitates this feeding process. The mouth parts of the tick penetrate the skin and secrete a cement-like substance that affixes it to the host. During blood feeding, anti-hemostatic and anti-inflammatory components of the saliva are secreted into the host to prevent platelet activation and suppress the immune response of the host. The tick ingests a mixture of blood and tissue fluids from the skin of its host in a manner that is poorly understood. The exoskeleton then becomes plastic and unfolds, accordion-like, to accommodate the final phase of blood feeding. Vast quantities of blood are ingested during the final 24 hours of feeding. The bloated tick then withdraws its feeding apparatus and proceeds to digest its meal of blood. Replete ticks either molt to the next developmental stage, if subadult, or, if adult, produce a clutch of eggs. Death follows oviposition.

Ticks may survive for months or even years between feeding episodes. They remain motionless in a dormant state until environmental conditions permit them to resume activity. Day-length and temperature serve as seasonal cues to initiate or suppress questing activity. During their questing season, ticks generally ambush their hosts. They leave their sheltered habitats, ascend on vegetation to a height commensurate with the body-form of their host, and wait for any passing object. Those ticks that hunt more aggressively may actively pursue their hosts. Such ticks will move great distances in response to carbon-dioxide and other host-related stimuli to locate suitable hosts.

Perpetuation of Infection

Virtually all tick-borne microbes that cause human disease are zoonotic, in that they perpetuate mainly as parasites of certain non-human reservoirs. Each human infection, therefore, constitutes a diversion that reduces the force of transmission. Ticks that narrowly focus their feeding on a particular reservoir population most effectively amplify the natural cycle of transmission of the pathogen. Vector ticks that feed most frequently on people would seem to contribute least to the enzootic cycle of transmission because their host-range is broad. Infections tend to perpetuate most readily in those ticks that seem innocuous because they rarely come in contact with "dead-end" human hosts. This paradox, that is common to all zoonoses, is resolved in sites where species diversity is limited. In such sites, vectors that fail to discriminate between hosts may sustain enzootic transmission while allowing for episodes of human infection to occur.

Contribution of Vector Ticks

The intensity of transmission of a tick-borne pathogen is determined, in part, by a series of vector-related physiological and ecological variables. The salient properties of the tick population include (1) competence, (2) abundance, (3) site-fidelity, (4) longevity, (5) seasonality and (6) narrowness of host range. The term "vectorial capacity," which is based on a comprehensive synthesis of these six entomological properties, describes the number of new infections derived from each originally infected reservoir animal per unit of time. The relative contribution of each of these variables to the force of transmission of a tick-borne pathogen remains poorly defined. These variables will be discussed, in turn, in the discussion that follows.

Vector competence describes the physiological suitability of a particular kind of arthropod as host for a microbe. This parameter is measured in the laboratory and estimates the proportion of ticks that acquire, maintain and transmit a pathogen between vertebrate hosts. Vector ticks, therefore, must be able to ingest sufficient infectious organisms for the pathogen to become established, must maintain infection transtadially through the relevant molt and must deliver a sufficiently large innoculum to infect a particular vertebrate host. In addition to this horizontal mode of transmission, certain pathogens are maintained vertically, by inherited infection. The various babesial infections of cattle, for example, illustrate this pattern of transmission by inheritance, while the rodent babesias rely exclusively on horizontal passage between the larval and nymphal stages of development. Although Lyme disease spirochetes (*Borrelia burgdorferi*) mainly perpetuate in a similarly horizontal cycle, occasional episodes of inherited infection seem to occur. Transmission by co-feeding, that is, direct passage of a pathogen from the mouth parts of an infected tick to those of a non-infected tick, has been demonstrated in laboratory experiments but not in the field. These cycles defy generality.

Vector abundance constitutes an important variable in the force of microbial transmission. When vector ticks are sparsely distributed, individual reservoir hosts may not sustain the requisite number of vector contacts; only a few reservoir animals would acquire and subsequently pass-on the pathogen. They might only rarely acquire infection or acquire infection so late in the transmission season that they would infect few ticks. Intensity of transmission correlates directly with vector density relative to the density of reservoir hosts.

Because the environmental requirements of ticks tend to be highly specific, their distribution is discontinuous. Vector ticks, therefore, must be sessile enough to preclude dispersal from their point of origin. Although ticks generally remain close to their point of origin, some may migrate several hundred meters in response to such host-associated stimuli as carbon-dioxide. Those that attach to vagile hosts, such as birds, may readily be carried away from a permissive habitat and be lost to the transmission cycle. In this manner, tick-borne infections simulate the classical Russian concept of the "nidality of disease."

The proportion of the tick population that survives long enough to become infectious also influences the force of transmission. Incidence of infection in the reservoir population, therefore, depends directly on interstadial survival of the vector tick. Indeed, the vast majority of ticks that feed as larvae fail to feed once again as nymphs. In the northeastern U.S., for example, 3.5 times as many larval as nymphal deer ticks (*Ixodes dammini*) attach to white-footed mice (*Peromyscus leucopus*). Almost a third of these ticks appear to survive to feed again. In addition to longevity, however, this estimate of survival assumes that each relevant developmental stage of the tick responds similarly to the array of available hosts. Trans-stadial survival has not been estimated directly, and the magnitude of its contribution to the force of transmission remains unknown.

The seasonality of feeding activity of the vector tick relative to the density of the reservoir population may also affect transmission. In the case of the American vector of the agent of Lyme disease, seasonality is highly punctuated. The nymphal stage of the tick feeds, each season, before the younger larval stage. This inverted pattern of feeding serves to intensify transmission of pathogens because the reservoir population receives its infectious innocula before the larval recipient stage of the tick commences feeding. Because European wood ticks (*Ixodes ricinus*) lack such a precisely punctuated developmental cycle, the cycle in Europe seems less efficient than in eastern North America. The requirement for precision would be exacerbated in the event that reservoir hosts remain infectious only for a brief period of time. The pathogen must then become available to the vector population precisely when the appropriate stage of the vector quests for hosts. Seasonal events may profoundly affect transmission.

Narrowness of host range of the vector tick powerfully affects transmission because hard ticks feed only three times per generation. At least two of these feeding episodes must be directed toward the population of reservoir animals that enables vectors to successfully acquire and ultimately transmit the pathogen. For example, only larval and nymphal stages of the deer tick feed on rodent reservoirs, while adults parasitize larger non-competent hosts such as deer. Diversion of either of the sub-adult stages to non-competent hosts, therefore, negates transmission for the other feeding episode.

Contribution of Reservoir Hosts

Although vector ticks may acquire infection from an array of hosts existing in nature, only one generally serves as the main reservoir of the pathogen. "Reservoir capacity" expresses the relative number of infected ticks derived from each host species. An effective reservoir host must be (1) competent for the pathogen, (2) sufficiently abundant, (3) parasitized by numerous vector ticks, (4) parasitized by at least two developmental stages of the vector tick and (5) continuously resident in the enzootic site. These biological properties, together, define the capacity of reservoir populations to perpetuate tick-borne pathogens.

Reservoir competence is a measure of the physiological ability of a vertebrate host to exchange a pathogen with vector ticks and generally is analyzed experimentally in the laboratory. A competent reservoir must readily acquire infection, sustain its development and ultimately present the pathogen to the vector. This parameter should be measured over a span of time that corresponds to that of the seasonal activity of the vector tick. For the Lyme disease agent in North America, for example, reservoir hosts must become and remain infectious over the two month interval spanning the maximum feeding activity of nymphal and larval deer ticks. The white-footed mouse fulfills this criterion because it attains infectivity within 2 weeks of infection and remains infectious for life. A competent reservoir, therefore, must remain infectious long enough to pass infection to the relevant stage of the tick.

Reservoir hosts should be sufficiently abundant in nature that vector ticks are likely to encounter them before encountering other less suitable but tick-attractive hosts. Although the force of transmission initially increases with reservoir density, greater host density might dilute the vector population such that individual hosts that become infected are unlikely to encounter and infect non-infected ticks. The presence of tick-attractive but pathogen-incompetent hosts would divert vector ticks, a relationship known as "zooprophylaxis." Transmission, therefore, tends to be most intense in ecological island sites where host diversity is restricted and particular reservoir hosts predominate.

A complex set of ecological and physiological properties of reservoir and vector populations determines the frequency of vector-host contact. Effective reservoirs, of course, must occupy the same habitats as do vector ticks. Ticks position themselves on the vegetation at an appropriate height above the ground, thereby insuring a degree of host specificity. The stature of particular hosts influences the probability of encountering a tick in nature. In addition, reservoir hosts must forage at a time of day when ticks actively seek hosts. Ticks quest most effectively at night and during the morning and evening hours when the atmosphere is sufficiently humid. Once a questing tick attains host-contact, it must successfully feed without invoking an inflammatory response. Poorly-adapted hosts develop an inflammatory response against tick bites after repeated exposure. Such resistant animals feed fewer ticks due to the direct effects of host immunity and irritation induced by tick bites which increases host grooming. The physiological and ecological variables that regulate host-tick contact remain poorly understood.

Reservoir hosts must have sufficient contact with pathogen-acquiring and infecting stages of the tick to perpetuate the pathogen. Entomological inoculation rate (EIR) describes the frequency of vector ticks delivering infection to the reservoir population. This variable depends on the frequency with which pathogen-infective stages of the tick feed on a particular reservoir population and the prevalence of infection in these ticks. To complete transmission, reservoir hosts must be abundantly parasitized by the pathogen-receptive stage of the tick. Reservoir inoculation rate (RIR) describes the number of infections generated in the vector population per unit of time. Together, these variables describe the ability of particular kinds of hosts to receive (EIR) and deliver (RIR) infection to and from the vector population.

Reservoir hosts should remain within the enzootic site throughout the transmission season. Mobile hosts such as birds tend to be ineffective reservoirs because they may readily disperse the pathogen to an inappropriate site that lies outside of the focus of transmission. Although migratory hosts may fail to maintain infection locally, they may passively transport ticks into new permissive sites. In this manner, excessively mobile hosts may reduce the force of transmission of a pathogen locally while accelerating the expansion of its range.

Risk of Infection

The potential for tick-borne pathogens to infect human hosts depends on the questing density of infected ticks and the behavior of human hosts. Herein, we explore the conditions that may favor human infection.

The density of ticks largely correlates with that of their main vertebrate host(s). Definitive hosts, in particular, powerfully affect tick abundance because they comprise the main food-source for the reproductive stage of the tick. Successful feeding of the adult stage of the tick results in huge increments of increase: thousands of larvae may result. In contrast, feeding success by subadult ticks merely promotes development. Adult deer ticks, for example, feed mainly on deer and proliferate solely where deer are abundant. This relationship was tested experimentally by depriving deer ticks of access to their cervid hosts. Deer inhabiting an island site were virtually eliminated, which resulted in a diminished tick population. The density of larvae per mouse declined five fold during the year following the intervention, and that of nymphs somewhat more slowly, extending over several years. This relationship, however, appears to be non-linear because incremental decreases in deer density may fail to reduce tick densities. Modest reductions in deer abundance may simply cause more ticks to feed on each remaining host. The quantitative relationships between tick and host density have not been defined precisely.

Seasonality in the questing density of ticks may profoundly affect the shape of the epidemic curve representing any pathogens that they transmit. In North America, human Lyme disease infections tend to occur most frequently during July because fewer people engage in risk-promoting activities during May and June, when deer tick densities are greatest. Nymphal densities decline greatly by July and are virtually nonexistent in August. Fewer human infections occur during the fall and winter months, although adult ticks, which quest at that time of year, are far more frequently infected than are nymphal ticks. Few people, however, are exposed and those that enter forested sites then are fully clothed. Then too, adult ticks are more readily discovered before they can feed long enough for transmission to occur. Risk of human infection is modified by a complex interaction of the stage-specific activity of vector ticks and human behavior.

Pathogen-infected vector ticks may be more abundant in certain sites than in others and infection far more prevalent than in the case of insect-borne disease. The Lyme disease spirochete, for example, infects 20–40% of deer tick nymphs and 40–70% of adults in the northeastern and north-central U.S. but rarely infects ticks south of Maryland. Likewise, human infections cluster in space and time mainly in the upper Midwest and Northeast, but also in several sites in California. Elsewhere, the scattered distribution of human cases suggests that infected ticks may be imported, perhaps carried by south-migrating birds. Enzootic transmission implies that both vector and pathogen populations propagate locally. Larval ticks, for example, would outnumber nymphs where transmission is stable; a preponderance of nymphs would imply that the vector population is sustained by importation from some remote enzootic site. In a stable zoonotic focus, the RIR (prevalence of infection in the reservoir and feeding density of vector ticks on those hosts) would be consistent with the EIR (stage-specific density of infected ticks). Stable transmission requires long-term constancy in the incidence of infection in both vector and reservoir populations.

Outbreaks of tick-borne disease may emerge when people encroach upon previously silent transmission foci. In this manner, focused contacts between reservoir hosts and vector ticks may be altered and redirected toward human hosts. The first outbreak of Rocky Mountain spotted fever erupted, for example, when pioneers cleared land in the Bitterroot Valley of Montana. The result was devastating, nearly preventing this fertile region from developing. Tick-borne encephalitis, likewise, became intensely prevalent when forestry workers and trappers relocated into undisturbed tracts of Siberian forest. Human disruption of enzootic cycles serves to produce sporadic outbreaks of tick-borne disease.

Environmental change may promote epidemics of tick-borne disease when particular hosts and ticks become extraordinary abundant. Reforestation of previously cultivated land in the eastern United States has permitted deer to proliferate, often in close proximity to residential communities. Lyme disease emerged as a significant health problem when deer and their associated tick ectoparasites increased in abundance and expanded in distribution. The first outbreak of Crimean-Congo hemorrhagic fever occurred when population densities of hares and *Hyalomma marginatum* vectors exploded after hunting was prohibited and fields were abandoned. Massive outbreaks of human infection followed the resulting proliferation of these apparent reservoir hosts. Any disruption in the balance of vertebrate hosts may support an overabundance of vector ticks.

Example of Lyme Borreliosis

Although *Borrelia burgdorferi* is the sole causative agent of human Lyme disease in North America, the etiologic agents in Europe and Asia are more diverse. They include the "genospecies" designated as *B. burgdorferi*, *B. afzelii* and *B. garinii*. These spirochetal agents of human disease are transmitted by members of the *Ixodes ricinus* complex, including *I. dammini* in eastern North America, *I. pacificus* in western North America, *I. ricinus* in Europe, *I. persulcatus* in eastern Europe and Asia. People become infected mainly via the bites of nymphal ticks, although some infections may be derived from the adult stage of the tick. In certain communities, Lyme disease transmission may be particularly intense. Lyme disease spirochetes may infect as many as 40% of nymphal ticks and virtually all of the rodents. Human seroprevalence may approach 25%, and 5% of residents may become infected each year. Currently, Lyme disease accounts for more than 90% of all reports of vector-borne disease in the United States.

In the northeastern United States, the Lyme disease spirochete perpetuates in a cycle involving vector deer ticks (*I. dammini*) and white-footed mouse reservoir hosts (*Peromyscus leucpous*). White-tailed deer (*Odocoileus viginianus*) are not directly involved in the transmission cycle, but play a vital role in maintaining tick densities because they serve as the preferred hosts of the adult stage of the vector tick. Although various other vertebrate hosts may inhabit zoonotic sites and come in contact with vector ticks, white-footed mice provide the main source of spirochetal infection to the nymphal stage of the vector tick. These mice serve as effective reservoirs because they are locally abundant in zoonotic sites, are the main hosts for the larval and nymphal stages of the vector, are frequently infected in nature and readily infect vector ticks. Estimates of reservoir capacity suggest that one white-footed mouse infects as many ticks as do 12 chipmunks or 221 meadow voles. Some kinds of passerine birds may also transmit infection to the vector population. The greater mobility of avian hosts diminishes their contribution to local transmission but aids in dispersing vector ticks and Lyme disease spirochetes to new sites.

The apparent diversity of vector ticks, spirochete variants and the vertebrate reservoir fauna of the western United States renders the epizootiology of these microbes more complex than in the Northeast. Although *Ixodes pacificus* serves as the principle vector to people in this region, relatively few harbor Lyme disease spirochetes. The diversity and abundance of non-competent hosts in western United States appear to contribute to low infection rates in *I. pacificus*. *I. neotomae*, in contrast, narrowly focuses its feeding on wood rats and kangaroo rats and may effectively maintain Lyme disease spirochetes in an enzootic cycle involving these hosts. Lyme disease spirochetes may perpetuate in this *I. neotomae*-wood rat cycle and occasionally infect the *I. pacificus* population. It is not clear, however, whether wood rats and kangaroo rats represent the main source of spirochetal infection within the *I. pacificus* population.

The force of transmission of the agent of Lyme disease in western Europe tends to be weaker than in northeastern North America. European *I. ricinus* ticks transmit these microbes less efficiently than do their North America counterpart, *I. dammini*, because each trophic stage feeds most frequently on different kinds of hosts. Larvae tend to parasitize rodents, and nymphs to feed on medium-sized mammals, birds and lizards. Adults feed mainly on deer or sheep. So few of these rodents are parasitized by nymphs that the EIR may be limited. Nevertheless, certain kinds of rodents harbor sufficient infectious nymphs to ensure perpetuation of the pathogen. Edible door mice (*Glis glis*) and black-striped mice (*Apodemus agarius*) serve as particularly efficient reservoirs because they are frequently infested by both the infectious and pathogen-acquiring stages of the wood tick. Norway rats (*Rattus norvegicus*), too, may support transmission of the Lyme disease agent in particular urban sites. Because shrews and voles are far more abundant than mice in Sweden, these small mammals may perpetuate the life cycle in certain Scandinavian sites. The relative importance of each kind of vertebrate host as a reservoir of infection differs according to local conditions.

In eastern Asia, Lyme disease spirochetes appear to circulate in a cycle involving the taiga tick (*I. persulcatus*) and rodents of the genera *Clethrionomys* and *Apodemus*. To date, solely *B. garinii* and *B. afzelii* have been isolated from tick vectors and rodent hosts inhabiting these regions. The bank vole (*Clethrionomys glareolus*) predominates in western Russia, whereas *C. rufocanus*, *C. rutilus* and *Apodemus peninsulae* dominate further to the east. Although numerous *Borrelia* isolates have been derived from these hosts, their relative contribution to infecting the tick population by xenodiagnosis remains uncertain.

Co-Infecting Pathogens

Although public attention has focused on Lyme disease, *Ixodes* ticks may transmit numerous other agents of human disease. Human babesiosis, a malaria-like illness, is caused by the protozoan parasites *Babesia microti* in North America and *Babesia divergens* in Europe. Signs of this illness become evident mainly among elderly or immunocompromised subjects and may be fatal if not treated promptly. Viruses of the tick-borne encephalitis complex induce a potentially fatal form of encephalitis endemic to Europe and Asia. More recently, a new member of this viral complex was discovered in North America; transmission was attributed to deer ticks. Finally, two closely-related pathogens (*Ehrlichia phagocytophila* and *E. equi* in eastern and western North America respectively) were recently implicated as agents of human disease. These rickettsial pathogens infect leukocytes, and human cases may also terminate fatally. The diverse array of pathogens transmitted by *Ixodes* ticks, thereby, severely burdens human health.

In the northeastern United States, the agents of Lyme disease, human babesiosis and human granulocytic ehrlichiosis perpetuate mainly in a cycle involving white-footed mice. Vector ticks, thereby, tend to acquire more than one of these microbes from reservoir rodents. This implies that individual human hosts tend to be vulnerable to co-infection. Serological surveys indicate that 10 to 60% of Lyme disease patients had been co-infected by *B. microti*. Interestingly, these pathogens tend to synergize in human hosts such that the resulting illness is more severe than would be anticipated as the sum of symptoms produced by each pathogen. More symptoms are experienced, and the duration of illness is prolonged. The particularly severe manifestations of *Ixodes*-borne disease that occur in certain enzootic sites may reflect a peculiar combination of coinfecting pathogens.

Anti-Vector Interventions

Once established, local transmission cycles of tick-borne zoonoses tend to persist in the face of public health interventions. Various interventions, however, have been devised, and certain of them appear promising. The following discussion describes selected strategies designed to reduce the public health burden presented by vector ticks.

Individual residents of enzootic sites may practice preventive measures that effectively reduce their risk of infection by tick-borne pathogens. They should: (1) avoid tick infested habitats whenever feasible; (2) wear light-colored trousers with cuffs that are tucked into their socks; (3) apply tick repellent containing DEET to exposed parts of their skin and permethrin to their clothing; (4) periodically examine the surface of their clothing and skin and remove any ticks that have attached using fine tipped forceps. Prompt removal of attached ticks generally aborts transmission because transmission of many of these infections tends to require extended periods of host attachment. The arboviral agents may constitute an exception. Although the efficacy of these measures has not systematically been evaluated, they appear to provide an important degree of protection against tick-borne disease.

Depriving ticks of access to their main vertebrate hosts may effectively reduce the density of ticks. This intervention strategy, however, is practical solely in sites that such hosts would not rapidly re-invade. Deer inhabiting a study site on Great Island, MA were virtually eliminated, which resulted in decreased abundance of deer ticks. Host reduction proved to be effective largely because the relative isolation of the site restricted the movements of deer. Similar efforts on the mainland proved to be impractical and excessively costly. Thousands of small rodents, for example, were destroyed in Montana in order to suppress the density of American wood ticks (*Dermacentor andersoni*), the vectors of the agent of Rocky Mountain spotted fever. Any gains were transient, however, because wood ticks from nearby undisturbed sites rapidly re-invaded the intervention site. Anti-tick measures based on the removal of their vertebrate hosts require that the site be isolated in order to limit immigration from adjacent sites.

Although broad-scale applications of acaricides may destroy numerous ticks, environmental damage tends to result. Acaricidal applications focused around residential sites may alleviate the immediate tick burden. Residual pesticides such as carbaryl, chlorpyrophos and diazinon temporarily render such sites virtually tick-free. Less toxic materials, containing pyrethroids, may also reduce tick density. These products, however, lack long-term residual activity and require at least monthly application to maintain satisfactory freedom from ticks. Regardless of the kind of acaricidal compound that is applied, pesticide resistance should always be anticipated. Intensive and extensive applications of killing chemicals can only be temporary. Loss of acaricide susceptibility renders acaricidal interventions inherently unsustainable.

Innovative strategies have been developed for delivering acaricides directly to the hosts of vector ticks. Various self-medicating devices for destroying ticks on deer or other ungulates are in various stages of development and evaluation. In general, such devices deliver acaricide from a dispenser that the animal contacts when feeding on a bait contained within. Another host-targeted strategy distributes grain impregnated with systemic acaricides, such as ivermectin. When deer are the targets of such interventions, they must be habituated to the bait-station, and this requires delivery of large quantities of grain, frequently a maize-molasses mixture. This has the undesirable side effect of promoting the density of various rodents as well as the targeted deer, themselves. A cotton-baited acaricidal formulation has been implemented to target rodent hosts such as the white-footed mouse in eastern North America. This method is designed to reduce the force of transmission of *Ixodes*-borne pathogens by eliminating those ticks that feed on the rodent reservoirs. Host-targeted acaricidal formulations are attractive because they limit any environmental damage that might be induced by these biologically active chemicals.

Ticks are vulnerable to destruction by various parasitic or predatory organisms. Although certain *Dermacentor* and *Amblyomma* ticks secrete a pheromone that deters attack by ants, *Ixodes* lack such protection against predation. To the extent that fire ants are important predators of these ticks, their presence might benefit public health. A chalcid wasp (*Hunterella hookeri*) frequently parasitizes larval *Ixodes* ticks in northeastern North America and Europe and destroys them in their nymphal stage. Although these wasps infect as many as a third of the nymphal deer ticks in eastern North America where Lyme disease is enzootic, none infect spirochete- or *B. microti*-infected ticks. Efforts to use these wasps to reduce risk of Lyme disease, therefore, would fail. Certainly, tick densities seem unaffected in the face of this natural burden. The applicability of biocontrol efforts against vector ticks remains speculative.

The density of vector ticks may be reduced by removing understory vegetation and leaf litter, either mechanically, chemically or by fire. Where few buildings are present, the undergrowth or ground-cover that shelters ticks is most readily destroyed by burning. In a Massachusetts site, burning and mowing reduced deer tick densities by as much as 80%. Similar efforts in Tennessee greatly reduced the density of Lone Star ticks. Safety considerations, however, limit the wide-scale application of this measure. Limited areas such as along a road, may be mowed, and this would seem to protect people from contact with ticks. The efficacy of this measure, however, remains ill-defined. Routine herbicidal applications are poorly tolerated by many people and may excessively harm the environment. While vegetation management provides effective protection against ticks, it must be reapplied on a yearly basis.

Tissue Anoxia

→ Pathology.

Toxicosis

→ Tick Bites: Effects in Animals, → Tick Bites: Effects in Humans.

Toxocariasis, Man

Synonym

→ Visceral larva migrans (VLM).
See → Toxocara (Vol. 1).

Therapy

→ Nematocidal Drugs, Man.

Toxocarosis, Animals

→ Toxocara (Vol. 1), → Alimentary System Diseases, Animals.

Toxoplasmosis, Animals

→ Nervous System Diseases, Carnivores, → Nervous System Diseases, Ruminants, → Nervous System Diseases, Swine

Toxoplasmosis, Man

Pathology

Ignored for a long time, human toxoplasmosis was universally recognized as a genuine toxoplasmic disease only 50 years ago while the *Toxoplasma* parasite was first identified at the begining of the century. Not being highly virulent, → *Toxoplasma gondii* (Vol. 1) is indeed a typical parasite which is found worldwide and often at very high prevalence. About half of the human population are asymptomatic carriers. The rapid multiplication of the invading → tachyzoite (Vol. 1) stage, leads to a mild to subclinical phase. Recovery is associated with parasite sequestration into → cysts (Vol. 1) containing → bradyzoites (Vol. 1) located particularly within skeletal and heart muscle and in the central nervous system and remaining latent for life.

Infection originates from infected house cats (the definitive = final host) excreting oocysts in their feces or from undercooked meat with bradyzoites in tissue cysts. Indeed, *T. gondii* infection is also widespread in farm animals in particular, goat, sheep and pigs and causes large financial losses in Australia, New Zealand and England (→ Toxoplasmosis, Animals).

In man *T. gondii* disseminates from the site of entry via the bloodstream and the lymphatics to involve many tissues. The sites and character of lesions depend on the vascular supply of the tissue and the regenerative ability of the host cells. Tachyzoites proliferate approximately until immunity develops, at which time more slowly multiplying bradyzoites develop in tissue cysts. These cysts are common in the brain, skeletal and heart muscle, and sometimes the retina. The cysts persist for months or years, in a biological sense waiting to be eaten by a cat; hence a chronic latent infection persists, and we have an infection-immunity rather than a sterile immunity. The intact cysts are not chemotactic (→ Pathology/Fig. 5A,B). However, leaking or ruptured cyst; elicit necrosis and an inflammatory reaction, interpreted as manifestations of hypersensitivity. The inflammatory reaction is mixed, involving neutrophils, lymphocytes, and macrophages, followed by fibrosis, and in the brain gliosis.

Two stages are distinguished in tissues: → tachyzoites (Vol. 1) and → bradyzoites (Vol. 1). **The tachyzoites** multiply rapidly, destroying the cells they parasitize, giving rise to diffuse lesions, as in connective tissue or lungs, or to focal lesions, as in the liver and brain (→ Pathology/Fig. 10). Tachyzoites destroying cells during acute toxoplasmosis lead to interstitial pneumonia, hepatitis, encephalitis, and myocarditis. A maculopapular rash may develop from small foci of *T. gondii* multiplying in the dermis. Many tissues may be only microscopically involved, and clinical symptoms do not draw attention to all of them. Lymphocytes and macrophages are the main inflammatory cells, with an admixture and neutrophils. When blood vessels are involved in the brain infarcts may result.

As already mentioned, individuals develop a non sterilizing immunity with indefinitely latent infection when infected postnatally. This immunity is congenitally transferred to the foetus if acquired before pregnancy. However, at the occasion of an intercurrent infection or modification of the host immunocompetence, slight such as pregnancy or more profound such as malignancies, organ transplants or the acquired immune deficiency syndrome (AIDS), mid syndrome to severe life threatening pathologies can be observed. Notably, when a primary infection is contracted by pregnant women, congenital infection can lead to abortion or neurological sequels and ocular disorders in the fetus. While fetal death and abortion have been attributed to rapidly dividing tachyzoites, the CNS lesions and chorioretinitis in congenital infection or transplant or AIDS patients are caused directly by the cysts containing **bradyzoites** or by reactivation of this so-called dormant stage. The emergence of toxoplasmosis as a major opportunistic infection in AIDS leading to toxoplasmic encephalitis, in up to 48 % of the AIDS patients in areas where *Toxoplasma gondii* is highly prevalent, has considerably stimulated interest in the last decade. Numerous scientists from all the fields of the biology have taken up the challenge and have begun to study various aspects of the host-parasite relationship. They have particularly concentrated their efforts on the understanding of the immune response against the parasite in order to develop vaccines and new immunotherapic approaches able to prevent (1) primary infection during pregnancy that eventually results in congenital infection in the fetus or abortion (i.e. immunity against the tachyzoite stage) (2) reactivated toxoplasmosis in immunocompromised patients (i.e. immunity against the cyst stage).

Lymphoreticular Hyperplasia Lymphoreticular hyperplasia with prominent histiocytes is often present in the posterior cervical lymph nodes following febrile acute, or asymptomatic toxoplasmosis (→ Pathology/Fig. 11). This is not associated with tachyzoites or with cell or tissue necrosis; even bradyzoites in cysts are rare. Serologic tests indicate high antibody titers, and so the lymphoreticular → hyperplasia is interpreted as an immune reaction. The diagnosis can be suspected histologically and confirmed serologically, or vice versa.

Placental Toxoplasmosis Placental toxoplasmosis occurs in 20–40% of primary infections acquired during pregnancy, with microscopic but no gross lesions. This leads to toxoplasmosis in the fetus which most often is asymptomatic, but which in 10%–20% of infected babies is accompanied by clinical manifestation. Initially the lesions are generalized with hepatitis, splenomegaly, pneumonia, rash, anemia, extramedullary hematopoiesis, and failure to gain weight. As partial immunity is developed extraneural lesions subside. Sabin's tetrad, hydrocephalus, retinochoroiditis, intracerebral calcification, and psychomotor retardation characterizes persistent central nervous system infection.

Hydrocephalus The pathogenesis of hydrocephalus in fetal and neonatal toxoplasmosis is unique. *T. gondii* tachyzoites reaching the central nervous system via the bloodstream give rise to microglial nodules throughout the brain (→ Pathology/Fig. 5, → Pathology/Fig. 6). In addition the tachyzoites are disseminated through the ventricular system, resulting in widespread infection and necrosis of ependyma and subjacent tissues. This leads to obstruction in the narrow aqueduct of Sylvius (→ Pathology/Fig. 12A). As a consequence metabolic products of *T. gondii* and liquefied necrotic brain material accumulate in the lateral and third ventricles. *T. gondii* antigen in the ventricles diffuses through ependymal *ulcers* and interacts with antibody in the periventricular blood vessels (→ Pathology/Fig. 12B). These become surrounded with inflammatory cells as they approach the ventricles, further inwards the vessels leak protein and they become thrombosed close to the ventricles, leading to infarction necrosis. This in vivo antigen-antibody reaction and the zone of necrosis surrounding the lateral and third ventricles and the aqueduct are pathognomonic of congenital toxoplasmosis. Because the fourth ventricle fluid is drained through the open foramina of Luschka and Magendi; ependymal ulcers are not accompanied by this reaction. Calcifications, often visible radiologically, develop in the areas of periventricular and aqueductal necrosis, and focally in areas of vasculitis throughout the infant's brain.

Retinochoroiditis Retinochoroiditis, although often found in infected babies, also occurs in children, adolescents, and adults. It rarely results from acute infection, but usually develops during chronic infection, after *T gondii* cysts persisting in the retina, disintegrate. Most of the retinal lesions are believed to have followed infection acquired in utero. The rupture of a cyst results in destruction of the bradyzoites if immunity is intact (→ Pathology/Fig. 13C); if not, proliferation of tachyzoites results. After either event, retinochoroiditis develops. The release of *T. gondii* antigen is inflammatory in the presence of hypersensitivity, the usual state during chronic infection, and reflected by a positive skin test. In the absence of an effective cellular immunity multiplication of tachyzoite results in destruction of retinal cells and inflammation. Because of the concentration of function in the retina, cyst rupture will often be symptomatic, whereas similarly sized lesions would not be noted in muscle or even the brain. Retinal lesions from recrudescent multiplication of tachyzoites are a dangerous complication of immunosuppression from corticosteroid or tumor chemotherapy, or with AIDS, and if untreated may lead to blindness.

Cerebral Toxoplasmosis Cerebral toxoplasmosis is seen in immunosuppressed adults such as those treated for Hodgkin's disease, other lymphomas or carcinomas, in corticosteroid-treated patients after organ transplantation, and in patients with AIDS. Encephalitis usually results during chronic infection probably after an accidental cyst rupture. Normally the bradyzoites would be destroyed with a glial scar remaining (→ Pathology/Fig. 5A,B). However, because of the immunosuppression, the released bradyzoites have time to develop into tachyzoites and to multiply, wandering from cell to cell and producing an ever-expanding focus of necrosis accompanied by little inflammation (→ Pathology/Fig. 10A,B). One or numerous focal lesions have been found in the brains of immunosuppressed patients. They can be identified by computerized tomography and magnetic resonance imaging where they resemble abscesses, although microscopically they are focal necrosis without granulocytes or pus, the hallmark of an abscess.

Immune Responses
Interaction between *T. gondii* and the functional immune system does not result in parasite elimination, but rather to a reduced parasite load and changes in morphology and surface antigen expression. The adapted parasites persist as bradyzoites in cysts located in different tissues for the remaining life span of the host. Aquired immunosuppression leads to reactivation of the parasite resulting in life-threatening toxoplasmic encephalitis. The sexual part of the life cycle of Toxoplasma takes place in the intestine of the definitive host, the cat. Despite the sexual life cycle in cats, only three major clonal lineages have been identified with little of the recombination one would expect if these animals fed on prey infected with different *T. gondii* strains. Thus, under normal conditions, the protective immune response against *T. gondii* is effective and long-lasting thereby preventing infection of animals with multiple strains of *T. gondii*.

Intracellular Survival and Host Cell Activation *T. gondii* is able to replicate in nearly all nucleated cells of mammals. The intracellular fate of the parasite depends on the type and activation state of the host cell. Depending on the ability to restrict replication of intracellular toxoplasma, all cell types analyzed so far can be arranged into three groups. Group A consists of cells that are able to restrict the growth of toxoplasma without prior activation, for example human monocytes. Group B, to which the majority of cell types belong, includes cells that restrict the growth of the parasite only after activation with cytokines, e.g. IFN-γ. Microglia cells and macrophages of mice as well as human fibroblasts belong to this group. Cells of group C, such as murine astrocytes and fibroblasts or human EBV-transformed B cells, are unable to restrict the growth of *T. gondii*, even after activation with IFN-γ. Cells of group C might thus function as safe harbor and transport vehicle for the spread of the parasite throughout the body. However, depending on the species of the host, the precise nature and frequency of these C-type cells differ, possibly influencing the susceptibility of the mammal species to *T. gondii* infection.

For example, human astrocytes and fibroblasts were found to restrict toxoplasma growth after activation with cytokines, while these cell types from mice support growth of the parasite even after IFN-γ activation. Thus the more severe illness in mice as compared to humans may in part be a consequence of the high frequency of these type C cells in rodents.

Among the cytokines able to induce anti-toxoplasma effector mechanisms IFN-γ appears to be the most potent. TNF was found to enhance the activation by IFN-γ in a synergistic manner in vitro and induces protection of mice in vivo. Likewise, IFN-β and IL-1 were described to stimulate toxoplasmacidal effects in human cells in vitro and induced protection in mice infected with *T. gondii*. The protective effects of cytokines such as IL-2, IL-7 and IL-12 in experimental infections of mice are most likely indirect, caused by an enhanced production of IFN-γ. At least three different anti-parasitic effector mechanisms induced by IFN-γ have been defined: (1) production of toxic oxygen radicals, (2) production of nitric oxide, and (3) degradation of L-tryptophan.

The role of the oxidative burst in control of toxoplasma growth is still a matter of debate, since many conflicting results have been published. Free radical scavengers such as catalase and superoxide dismutase have been shown to inhibit toxoplasmastasis in human and murine macrophages. Although some investigators reported that toxoplasma infection induces oxidative burst, others have been unable to detect even traces of oxidative burst products in comparable cells. Moreover, Toxoplasma tachyzoites are at least in part resistant against the damaging effects of toxic oxygen radicals since they possess reactive radical scavengers. The pathway by which toxoplasma enters a cell determines the fate of the parasite. While antibody-coated toxoplasma were rapidly killed in human granulocytes by strong superoxide anion production, parasites without antibodies on their surface were able to replicate and induce only a minor oxidative burst. Thus, toxic oxygen radicals may be operative under certain circumstances, while other mechanisms of defense are definitively also able to inhibit *T. gondii*, even in oxidatively incompetent cells of patients with chronic granulomatous disease.

Nitrogen intermediates such as → nitric oxide (NO) contribute to anti-microbial activity of rodent macrophages. The inducible form of NO synthase (iNOS) is induced by IFN-γ and there is no doubt that NO production is the key defense mechanism against toxoplasma (as well as other intracellular pathogens) in murine macrophages and other rodent non-professional APCs. In addition, human astrocytes stimulated with IL-1 and IFN-γ are also able to inhibit toxoplasma growth by producing NO. However, as shown by experiments with iNOS inhibitors, NO does not contribute to the defense of human monocyte-derived macrophages against *Toxoplasma*. In addition an IL-12-mediated mechanism of protection independent of iNOS has been recently described in IRF-1-deficient mice infected with *T. gondii*. Although macrophages isolated from iNOS-deficient mice (iNOS -/-) displayed defective microbicidal activity against the toxoplasma in vitro, iNOS-deficient mice survived acute infection and controlled parasite growth at the site of infection. By 3–4 weeks post infection, however, iNOS-deficient mice did succumb to *T. gondii* and enhanced parasite expansion and pathology were evident in the CNS. This suggests that the protective effects of NO might be tissue / and or infection phase specific.

In many different human cell types IFN-γ induces the indolamine 2,3-dioxygenase (IDO), an enzyme capable of degrading tryptophan. Since tryptophan is an essential amino acid for *T. gondii*, the depletion of this amino acid results in parasite growth inhibition. This has been confirmed by several different experimental findings, e.g. (1) the fact that tryptophan supplementation partly antagonized anti-parasitic effects induced by IFN-γ in human cells (2) the absence of anti-parasitic effector mechanisms in a mutant cell line lacking the IDO gene and (3) the capability of trypB gene-transfected *T. gondii*, which are no longer tryptophan auxotroph, to replicate in human IFN-γ treated fibroblasts. The induction of IDO appears to be inhibited by NO in rodent cells, since the addition of iNOS inhibitors resulted in detectable IDO activity in stimulated murine macrophages.

The existence of further defense mechanisms in addition to oxidative burst, NO production and tryptopohan degradation is suggested by experiments in which the inhibition of all three pathways in human endothelial cells did not abolish the IFN-γ-induced anti-parasitic effect.

Innate Immunity An increased percentage of neutrophils has been detected in the blood of mice infected with *T. gondii* by gavage and depletion of neutrophils by specific antibodies resulted in increased disease severity and death during acute toxoplasmosis. Another cell type possibly contributing to the very early innate defense against *T. gondii* are platelets. It was observed that tachyzoites of *T. gondii* induced activation of human platelets and that platelet-derived growth factor inhibited intracellular growth of the parasite.

In experimental infection of mice, one of the first events occurring after infection with *T. gondii* is the activation of NK cells. However, in contrast to human IL-2-activated NK cells, murine NK cells are unable to lyse *T. gondii*-infected target cells. Instead, IFN-γ produced by mouse NK cells appears to be of importance for early resistance against *T. gondii*. Although T cell-deficient SCID mice eventually succumb to infection, IFN-γ produced by NK cells leads to control of the parasite soon after infection. In addition, administration of IL-12 to SCID mice known to stimulate NK cells resulted in a remarkable delay in time till death, while treatment with anti IL-12 resulted in early lethality following infection. Antigen preparations of *T gondii* can activate NK cells in vivo and in vitro, presumably via stimulation of the secretion of monokines such as IL-1, IL-12 and TNF by macrophages or as shown recently for IL-12, by dendritic cells. In addition to these NK-stimulatory cytokines, cell-cell contact involving CD28 on NK cells and CD80 or CD86 on macrophages is able to amplify the IL-12-driven IFN-γ production of NK cells. Only in the absence of a functional NK compartment as seen in mice deficient for the common γ-chain of cytokine receptors, are $CD4^+$ cells able to confer early IFN-γ-dependent resistance.

The fall of NK cell activity shortly after the initial peak of activation appears to be mediated by IL-10 and TGF-β. Following infection with *T. gondii*, the expression of both of these cytokines is upregulated and treatment of SCID mice with neutralizing antibodies against IL-10 or TGF-β delays the time period till death.

B Cells and Antibodies An important immune reaction of mammalian hosts is the production of IgM and IgG antibodies directed against *T. gondii*, which eventually activate complement by the classical pathway, resulting in efficient killing of extracellular parasites. In contrast, activation of complement by the alternative pathway, does not result in destruction of *T. gondii*. A protective role of IgA is suggested by the finding that secretory IgA obtained from toxoplasma-infected cats reduced the parasite´s cell penetrating activity. The humoral immune response is mainly involved in the acute phase of *T. gondii* infection, possibly hampering hematogenous spread of extracellular tachyzoites.

T Cells In addition to NK cells, T cells driven by a recently described superantigen of *T. gondii* play an important role after infection with the parasite. This manifests as expansion of $CD8^+$ $V\alpha5^+$ cells producing IFN-γ soon after infection followed by nonresponsiveness of this population during chronic infection. Surprisingly and for reasons unclear at the moment, mice expressing the highest levels of $V\alpha5^+$ cells display also the highest mortality level. IL-12 and IFN-β expressed during the early response play a decisive role in the development of Th1 cells, which in turn produce IL-2 thereby driving the expansion of $CD8^+$ cells. Adoptive transfer and in vivo depletion experiments in various mouse stains confirmed the paramount importance of $CD8^+$ T cells in controlling the acute infection as well as in preventing toxoplasmic encephalitis. $CD8^+$ cells mediate their protective effects through three different mechanisms: (1) production of IFN-γ important for the activation of macrophages (2) MHC class I restricted cytotoxicity for *Toxoplasma*-infected cells and (3) direct tachyzoite cytolytic activity. Perforin-mediated cytotoxicity by T and NK cells, however, plays a limited role in host resistance to *T. gondii*, since *T. gondii*-vaccinated perforin-deficient mice were completely resistant to a challenge infection and only in the chronic stage of toxoplasmosis was there a three- to fourfold enhancement of brain cyst numbers in the mice lacking perforin.

The role of $CD4^+$ T cells during *T. gondii* infection remains controversial. On the one hand depletion of $CD4^+$ cells in vivo exacerbated the course of the disease, increased parasite burden and promoted recrudescence of latent infection. On the other hand, $CD4^+$ cell depletion reduced brain inflammation and Th1 cells were associated with the development of necrotic lesions in the ilea of susceptible mice. In contrast, the coproduction of Th2 cytokines together with IFN-γ by $CD8^+$ cells in the gut-associated lymphoid tissue (GALT) may be essential to limit the pathological inflammatory response.

Given the decisive role of T cell for the immune defense against *T. gondii*, it is important to mention, that infection of murine macrophages with the parasite results in down-regulation of MHC class II molecules and inability to upregulate class I molecules. The interference with antigen presentation may be an evasion strategy of *T. gondii* to facilitate intracellular survival.

In addition to the activation of T cells expressing $\alpha\beta$ TCRs, an activation of $\gamma\delta^+$ T cells has been described in humans as well as in mice. These cells were found to play a role in the induction of hsp 65 expression of macrophages and were mainly involved in the early defense against *T. gondii* in the gut by mechanisms which await clarification.

Toxoplasma Immunity in the Brain In the course of toxoplasmosis the central nervous system is almost always involved. The limited access of cells of the immune system to the brain may, in part, explain the persistence of the cyst stage in the brain which is responsible for the neurological symptoms in congenital toxoplasmosis and reactivation toxoplasma encephalitis in immune-compromised individuals.

The unique immune status of the brain is a consequence of a number of factors. Resting T cells, antibodies and cytokines are unable to cross the blood-brain barrier, and a primary immune response does not usually occur. In addition, glial cells are known to be able to suppress T cell responses. Since furthermore the CNS lacks a proper lymphatic system and only low levels of endogenous MHC class I and II molecules are expressed, it may be easy for *T. gondii* to evade the full consequences of the host immune system at this site.

Although *Toxoplasma*-encephalitis is one of the most common manifestations of clinical toxoplasmosis there clearly is evidence for antiparasitic effector mechanisms operative in the brain. The essential role of T cells is illustrated by the fact that there is no toxoplasmic encephalitis in immune-competent individuals, while the disease occurring in AIDS patients is almost invariably associated with very low $CD4^+$ T cell counts ($< 100/mm^3$) with a corresponding reduction in $CD8^+$ T cells. In addition to the impaired T cell function in AIDS patients, macrophages infected with HIV have a reduced ability to kill *T. gondii*. Furthermore, a direct link between replication of *T. gondii* and HIV is suggested by the fact that *T. gondii* infection of HIV-1-transgenic mice stimulated proviral transcription in macrophages. Thus, infection with *T. gondii* might increase the viral replication, thereby hastening the loss of T cells and allowing toxoplasma encephalitis to develop.

In murine toxoplasma encephalitis adult immunocompetent mice were found to develop cellular infiltrates composed of $CD4^+$ and $CD8^+$ T cells and macrophages surrounding toxoplasma cysts and tachyzoites. Cell transfer experiments showed that especially $CD4^+$ T cells are able to confer protection against cyst reactivation in the brains of infected SCID mice. In addition, it was found by cell depletion experiments that $CD8^+$ T cells participate in the control of cyst numbers.

Analysis of cytokine production in the brains of *T. gondii*-infected mice indicated that the outcome of the encephalitis is dependent on the differential production of Th1 or Th2 cytokines. IFN-γ appears to be the most important protective cytokine since (1) administration of rIFN-γ to chronically infected mice reduced disease severity which (2) was enhanced in mice treated with neutralizing antibodies against this cytokine and (3) IFN-γ receptor deficient mice with a resistant genetic 129 background rapidly died following infection. Studies showing that administration of the NO inhibitor aminoguanidine resulted in increased severity of toxoplasma encephalitis strongly suggest that NO synthesis is an important mechanism of IFN-γ-induced protection. TNF is another cytokine which appears to be centrally involved in the control of toxoplasma encephalitis since treatment with TNF-neutralizing antibodies or the disruption of the TNF receptor p55 gene resulted in increased disease severity. In line with these findings, the enhanced *T. gondii*-resistance of male mice compared to females is associated with more rapid production of IL-12, IFN-γ and TNF.

The role of other cytokines is less clear. IL-6 neutralization using antibodies in mice with established encephalitis led to reduced brain inflammation and decreased parasite burden, while IL-6-deficient mice were found to be more susceptible than their immunocompetent counterparts. The function of the Th_2 cytokines IL-4 and IL-10 is also uncertain. Elevated levels of expression of both these cytokines have been detected in brains of mice with toxoplasma encephalitis. While studies with mice deficient in IL-4 or IL-10 demonstrated an enhanced disease susceptibility others reported that IL-4 deficient mice were resistant to toxoplasma encephalitis. The differences between these studies may be related to the different strains of parasites and genetic backgrounds of the mice.

Immune Pathogenesis *T. gondii* infection during both the acute and the chronic phases of disease is controlled by a delicate balance between different inflammatory and regulatory cytokines. As described above, the production of IL-12, IFN-γ and TNF is a prerequisite for protective immunity, but on the other hand their overproduction can also be deleterious to the host. For example, mouse strains producing the highest levels of IFN-γ also have the highest mortality rates. In addition, administration of rTNF results in earlier mortality in immunocompetent and SCID mice. In addition, mice lacking IL-10, a counter-regulator of cell-mediated immunity, are more susceptible to toxoplasmosis.

Vaccination

T. gondii is one of the most successful parasites able to infect virtually all nucleated cells from a broad host range including birds, farm animals, wild animals and humans. Moreover this parasite is easy to grow, is haploid in most stages and can easily be manipulated by transfection strategies. Although there is no evidence of sophisticated mechanisms such as antigenic variation, the parasite has developed a strategy to infect the host cells allowing to escape to the immune system and/or trigger different immune effectors. It has been shown that the parasitophorus vacuole formed upon penetration of tachyzoites into macrophages, or fibroblastes is fusion incompetent. This would allow the parasite to avoid direct cellular destruction by macrophage machinery. Nevertheless, like a number of other intracellular pathogens, such as *Leishmania*, macrophages activated by cytokines, notably those derived from activated T cells, acquire the capacity to kill or inhibit the development of intracellular pathogens. *In vitro* studies as well as experiments in Toxoplasma-infected mice whose cytokine genes or cytokine receptor genes have been invalided have shown that IFNγ and TNF-α are the most cytokine implicated. The general consensus is that *T. gondii* induces TNF-α and IL-12 production by macrophages . This induces IFN-γ release by NK cells that synergize with TNF-α for the parasite killing by a combination of oxygen-dependent and independent mechanisms. Intracellular macrophage killing can not be sufficient to account for the immune control of the infection since *Toxoplasma* (in contrast with *Leishmania* parasites) is able to develop in practically every nucleated cell. Although the role of T cells in anti-*Toxoplasma* immunity has been recognized for a long time, it is only recently that important progress has been made in understanding the control mechanisms involved, particularly the role of $CD8^{+}$ T cells. $CD8^{+}$ would have not only a direct tachyzoite cytolytic activity but also would prevent high cysts burden and *Toxoplasma* encephalitis through IFN-γ secretion. Although cell-mediated immunity is the major component, antibodies also have been protective in models. While antibodies don't seem to play an essential role in maintaining a steady-state equilibrium between the parasite and the host during the chronic phase of the infection, they could participate in the protection during the primary infection in vaccinated animals.

A vaccine comprising live attenuated tachyzoites (the infecting stage) is already successfully used in sheep to prevent abortion, but such vaccine is inappropriate in humans. Therefore an important research area is the identification of molecules involved in the invasion process by the tachyzoite of the parasite in its host cell. It is a difficult issue because the parasite has evolved a complex family of redundant receptors able to match to virtually all nucleated cells within an enormous number of animals from birds to humans. The new advances of the reverse genetics have made it possible to clarify the respective roles of the tachyzoite surface antigens in the invasion process and therefore has pointed out some of these antigens as interesting targets for vaccine development. Among the five originally described surface antigens three have been extensively studied: SAG1 (P30), SAG2 (P22) and SAG3 (P43). SAG1 and SAG3 correspond to homologous proteins with 24% of amino-acid identity and conserved cysteines leading to similar secondary and tertiary structures. SAG2 shows no apparent similarities with SAG1 and SAG3, other than that these three tachyzoites are anchored in the membrane by glycosyl-phosphatidyl inositol (GPI) moeties. Polymorphism analysis of these antigens has shown that in contrast to what is observed for *Plasmodium* merozoite antigens (MSP1 and MSP2 for instance), the number of allele is extremely limited which is encouraging for vaccine development. For instance, the analysis of the SAG1 locus (encoding the P30 surface antigen) has shown the existence of only three alleles, two of them (CEP and ME49) encoding for identical proteins and two alleles have been described for SAG2. Vaccination experiments using one or the other of these two antigens, SAG1 and SAG2, as recombinant antigens or synthetic peptide have given drastically different results, reducing or increasing the mice or rats mortality, depending on the adjuvant. Indeed recent knowledge on immune responses against *T. gondii* have readily established that induction of $CD8^{+}$ T cells and Th1 cells is pivotal for the induction of a successful control of *T. gondii* infection. Adjuvant inducing a Th1 response such as immunostimulatory complexes (ISCOMs) induce generally protecting immunity whereas adjuvant promoting Th2 responses such as Aluminium hydroxyde (the only adjuvant licenced for use in humans) exacerbate the disease in rodent models. Recombinant cytokines such as IL-12, that is determinant for Th1 differentiation, could provide the adjuvant activity for new generation vaccine as proven by successful experiment for *Leishmania* vaccination. However the surproduction of pro-inflammatory cytokines, such as IFN-γ and TNF-α, may increase morbidity or/and mortality. Indeed, recent elegant experiments have found a correlation between the susceptibility of inbred mouse strains to *Toxoplasma* encephalopathy (TE) and the structure of the gene to TNFα. Thus, differences in TNFα expression could be one of the factors implicated in the TE susceptibility.

In contrast to SAG3 which is expressed on tachy- and bradyzoites surface, SAG1 and SAG2 are abundant tachyzoite specific antigens, highly immunogenic. Mutant parasites lacking either SAG1 or SAG3 clearly show an impaired invasion, suggesting a role in the attachment of the parasite to the host cell. SAG2 would rather play a role in the reorientation of the tachyzoite during the invasion process as indicated by blocking experiments with mabs against SAG2. Invasion and establishment in the parasitophorus vacuole is a key event for the survival of this parasite. The GRA proteins, discharged from the dense granule play a major role in the modification of the parasitophorus vacuole, and may also be relevant as vaccine component. Immunization of mice with GRA2 induces the increase of the survival of mice from 10% to 75%. Recent advance in the *Toxoplasma* genome project has yielded over 10000 expressed sequence tags

(ESTs) from *Toxoplasma gondii*. Sequencing of ESTs has already lead to the discovery of four new SRS genes (SAG1 Related Sequences), increasing the SAG family to eight members. As SRS antigens are less abundant antigens they were missed by classical biochemical techniques. This illustrated the fantastic potential advance coming from the postgenome period in the discovery of new vaccine candidates. Due to the stage specificity of Toxoplasma antigens as well as the differential immunological control of the tachy- and bradyzoite stages, an efficient vaccine preventing fetal damages and abortion as well as reducing cysts formation has to contain multistage antigens. Thus future vaccine research has not only to target antigens from different stages but also to induce the appropriate immune responses. Further progress in the dissection of protective defense mechanisms in experimental models as well as naturally infected individuals will help for vaccine design.

Once established in host cells and disseminated, *T. gondii* differentiates from tachyzoites to the persistent encysted bradyzoite stages. Knowledge about factors which lead to bradyzoite, or oocyste formation is still very poor and almost nothing is known about the genes which are involved in this transition. Concerning cysts wall composition, Sims et al have recently reported the presence of toxoplasma antigens, but the interaction between these antigens and the host immune system has not yet been investigated. The understanding of the mechanisms by which this stage becomes reactivated in immuncompromised host, is a major challenge for researchers, in order to develop strategies preventing this transition. Decreasing levels of IFN-γ observed in AIDS patients would play a role via the decrease of expression of indoleamine 2,3-dioxygenase (IDO) and inducible NO synthase (iNOS), which are important for the stabilization of the cyst stage. New approaches in molecular genetics, cellular microbiology and immunology will clarify the complex interrelationship between the different forms (tachyzoites, bradyzoites and oocysts) as well as the suitable balance in the induction of immune mechanisms, knowledge essential for vaccine development.

Lastly, recent experiments have highlighted the possibility of expressing *P. falciparum* antigens in *T. gondii* opening the possibility of using *T. gondii* as carrier for vaccination.

Main clinical symptoms: In immuno-incompetent people mostly symptomless, but in acute infection the following are common: lymphadenitis, iridozyklitis, chorioretinitis, myocarditis, meningoencephalitis-myelitis; see → Connatal Toxoplasmosis
Incubation period: Hours to 2 days
Prepatent period: 1 day to weeks (strain dependent)
Patent period: Years
Diagnosis: Serodiagnostic tests, → serology (Vol. 1)
Prophylaxis: Avoid eating raw meat and having contact with cats that feed on mice.
Therapy: see → Treatment of Opportunistic Agents, → Coccidiocidal Drugs.

Transmission Models

→ Mathematical Models of Vector-Borne Diseases.

TRAP

Thrombospondine related anonymous protein (→ Malaria)

Treatment

See chapters on Disease Control and Control

Treatment of Opportunistic Agents

Drugs Acting on [*P. carinii*] Pneumonia (PCP) in Humans

Pneumocystis carinii, a unicellular eukaryote, develops extracellularly in the alveoli of lungs of animals and humans thereby undergoing encystment during one phase of its life cycle (→ *Pneumocystis carinii* (Vol. 1)). The **clinical symptoms** of acute pneumonitis in immunosuppressed children and adults both with and without HIV infection are dyspnoea, tachypnoea, cough and fever. In immunocompetent hosts the infection is latent without any clinical signs and widely distributed in a variety of domestic and wild animals. Direct or close contact to carriers, and airborne transmission seems to be the common route of infection in hu-

Table 1. Drugs acting on diseases caused by unicellular opportunistic parasites in humans

DISEASE **nonproprietary name** (chemical group)	**Brand name** other information	**Adult dosage/*pediatric dosage** (mg/kg body weight, or total dose/ individual, oral route), miscellaneous comments
PNEUMOCYSTOSIS *(Pneumocystis carinii* pneumonia = PCP) is an acute, usually a bilateral and diffuse pneumonitis caused by pulmonary infection of *P. carinii*; this unicellular eukaryote that usually develops extracellularly in the lungs of host animals undergoes encystment in one phase of its life cycle; its biological characteristics resemble those of both fungi and the protozoans; *P. carinii* is widely distributed in nature and occurs in many mammals serving as natural hosts; in immunocompetent hosts, the infections are latent and asymptomatic; *P. carinii* is an important pathogen causing opportunistic pneumonia in immunosuppressed hosts and in the 1980s it became the most significant opportunistic pathogen in AIDS patients (> 60% of AIDS patients in the USA and Europe were estimated to develop pneumocystosis); extrapulmonary infected lesions (all tissues) are more common in AIDS patients than in other immunocompromized persons; prevention of PCP and its treatment is entirely dependent on current chemotherapeutic drugs, which all may produce severe side effects; early diagnosis is essential for effective treatment consisting of both specific chemotherapy and supporting therapy; chemoprophylaxis against PCP is necessary for individuals suffering from conditions such as AIDS, aids related complex (ARC), acute lymphoblastic leukemia, and those having received solid organ transplants or bone marrow transmission; aerosolized pentamidine (AP) given by inhalation is extensively used for PCP prophylaxis in children > 6-year old and adults producing moderate to severe cough as adverse effect; undesirable side effects may require termination of AP but this may also be true for all other treatment regimens used to prevent PCP; in PCP, accompanied by moderate or severe hypoxia, adding prednisone (corticosteroid) at the start of treatment has decreased the incidence of respiratory deterioration and death; corticosteroids may also improve tolerance for high-dose trimethoprim-sulfamethoxazole; oral candidiasis and reactivation of herpes simplex infections can occur; at present no protective vaccine is available.		
trimethoprim (TMP)/ **sulfamethoxazole** (SMX) (diaminopyrimidine/ sulfonamide)	Bactrim, and others (drug of choice)	TMP 15 mg/kg/d, SMX 75 mg/kg/d oral or i.v. in 3 or 4 doses x 14–21d; *same as adult dose; adverse effects may be folate deficiency, neuropenia, thrombocytopenia, agranulocytosis, rash, fever, headache, depression, jaundice, diarrhoea (rare) and others; is the treatment of choice for PCP and extrapulmonary *P. carinii* infections; episodes of toxicity may require discontinuation of the drug; subsequent desensitization may lead to renewed drug tolerance
pentamidine isetionate (aromatic diamidine)	Pentam (alternative drug)	3–4 mg/kg i.v. qd x 14–21d; *same as adult dose; side effects may be sharp fall in blood pressure after rapid i.v. injection (orthostatic hypotension); it can induce pancreatitis (hypoglycemia and hyperglycemia), reversible renal dysfunction, abortion, peripheral neuritis (rare), cardiac arrhythmias; drug is contraindicated in diabetes
trimetrexate plus **folinic acid** (diaminopyrimidine)	Neutrexin (alternative drug)	45 mg/m^2 i.v. qd x 21d plus folinic acid 20 mg/m^2 per os or i.v. q6h x 21d; antifolate drug approved for treatment of moderate to severe PCP; is not as effective as TMP/SMX; folinic acid prevents bone-marrow suppression; (Neutrexin is licensed in the USA and elsewhere)
trimethoprim plus **dapsone** (diaminopyrimidine/ sulfone)	Trimpex, and others (alternative drugs)	5 mg/kg per os tid x 21d plus dapsone 100 mg per os qd x 21d; antileprosy sulfone dapsone given concurrently with trimethoprim can be used in treatment of mild to moderate PCP; adverse effects may be nausea, rash and methemoglobinemia and hemolytic anemia in patients with G-6-PD deficiency; (drugs are not licensed for this dosage regimen in the USA but considered investigational for this condition by the FDA)
atovaquone (hydroxynaphthoquinone)	Mepron, others (suspension) (alternative drug)	750 mg bid per os x 21d; can be used for treatment of mild to moderate PCP; it is less effective than TMP/SMX but better tolerated; side effects may be gastrointestinal disorders, hepatitis, and rash; (licensed in the USA and Europe).

Table 1. (continued) Drugs acting on diseases caused by unicellular opportunistic parasites in humans

DISEASE **nonproprietary name** (chemical group)	**Brand name** other information	**Adult dosage/*pediatric dosage** (mg/kg body weight, or total dose/individual, oral route), miscellaneous comments
primaquine phosphate (8-aminoquinoline) plus **clindamycin** (7-chloro-lincomycin)	(alternative drugs) Cleocin, and others	30 mg base per os qd x 21d plus clindamycin 600 mg i.v q6h x 21d, or 300–450 mg per os q6h x 21d; concurrent use of i.v. or oral clindamycin with oral primaquine can be used in patients with mild to moderate PCP; primaquine can frequently cause hemolytic anemia, especially in patients whose red cells are deficient in glucose-6-phosphate dehydrogenase, this deficiency is most common in African, Asian, and Mediterranean peoples; patients should be screened for G-6-PD deficiency before treatment, it should not be used during pregnancy; (not licensed in the USA but considered investigational for this condition by the FDA)
PRIMARY AND SECONDARY PROPHYLAXIS: in HIV-infected patients, *Pneumocystis carinii* pneumonia can be prevented by oral TMP/SMX or other alternative treatments		
trimethoprim/ sulfamethoxazole (diaminopyrimidine/ sulfonamide)	Bactrim, and others (drug of choice)	1 tablet (single or double strength/=DS) per os qd or 1 DS tab 3x/week; *TMP 150 mg/m^2, SMX 750 mg/m^2 in 2 doses per os on 3 three consecutive days per week; oral TMP/SMX is the prophylactic agent of choice; it can prevent PCP in most HIV-infected patients; adverse effects are frequent, particularly nausea, rash and fever; reduction of dosage may reduce toxic episodes or patients have to discontinue the drug (for details see TMP/SMX above) and take an alternative drug
dapsone (sulfone) (not licensed in the USA but considered investigational for this condition by the FDA)	(alternative drug)	50 mg per os bid or 100 mg per os qd; *2 mg/kg (max. 100 mg) per os qd; frequent rash, GI irritation, anorexia, infectious mononucleosis-like syndrome, occasionally methemoglobinemia, hemolytic anemia (G-6-PD deficiency), nephrotic syndrome, liver damage and others, rare optic atrophy, agranulocytosis; (dapsone alone and with pyrimethamine has been used as an alternative to TMP/SMX prophylaxis: see *Toxoplasma*).
dapsone (sulfone) plus **pyrimethamine** (diaminopyrimidine) plus folinic acid	(alternative drugs)	50 mg per os qd or 200 mg each week plus pyrimethamine 50 mg or 75 mg per os each week plus 25 mg folinic acid with each dose of pyrimethamine; (dapsone is not licensed in the USA but considered investigational for this condition by the FDA); pyrimethamine occasionally causes blood dyscrasiasis, folic acid deficiency, rare rash, vomiting, others
atovaquone (hydroxynaphthoquinone)	Mepron, others (alternative drug)	750 mg bid; frequent rash, nausea, occasionally diarrhoea; (licensed in the USA and elsewhere); antimalarial drug in combination with proguanil (Malarone, Glaxo-Wellcome, cf. → Malariacidal Drugs/Table 8b)
pentamidine (aerosolized) (diaminopyrimidine)	Nebupent, others (alternative drug)	300 mg inhaled monthly via Respirgard II nebulizer; *>5 year-old: same as adult dose; (not licensed in the USA but considered investigational for this condition by the FDA, licensed in Europe)
CRYPTOSPORIDIOSIS though *Cryptosporidium parvum* is a coccidian parasite and should be affected therefore by anticoccidial drugs (cf. → Coccidiocidal Drugs/Table 1), it turns out that this monoxenous coccidian parasite proves considerably refractory to any known chemotherapeutic drug; only a very few chemotherapeutic agents seem to have moderate clinical effects on life- threatening diarrhoea in immunocompromized persons (AIDS patients) and in young animals (cf. Drugs Acting on Cryptosporidiosis of Mammals); management of cryptosporidiosis has to include fluid therapy, nutritional support and the use of antidiarreal agents; in a controlled trial, paromomycin was found to reduce diarrhoea in AIDS patients		

Table 1. (continued) Drugs acting on diseases caused by unicellular opportunistic parasites in humans

DISEASE **nonproprietary name** (chemical group)	**Brand name** other information	**Adult dosage/*pediatric dosage** (mg/kg body weight, or total dose/individual, oral route), miscellaneous comments
paromomycin (aminoglycoside antibiotic)	Humatin (drug of choice)	25–35 mg/kg/d in 3 or 4 doses; (not licensed in the USA but considered investigational for this condition by the FDA and others); may frequently cause GI disturbances and occasionally auditory-nerve damage, vertigo, pancreatitis and others

TOXOPLASMOSIS

The definitive host of *Toxoplasma gondii* is the cat, which passes infective oocysts in its feces; there is no satisfactory treatment, which eliminates completely oocyst shedding in cats; a combination of antimalarial drug pyrimethamine and sulfadiazine is effective against tachyzoites, but not so bradyzoites; clindamycin affects murine toxoplasmosis, and like pyrimethamine will reduce but not eliminate oocyst output in cats; infection of humans may be postnatally acquired or congenital; the majority of acquired infections are asymptomatic and widespread among humans though prevalence varies locally (about 500 million humans have antibodies to *T. gondii* or in most countries about 60% of adults are seropositive); in immunosuppressed patients, including AIDS patients rupture of 'dormant' tissue cysts may lead to transformation of bradyzoites into tachyzoites and new multiplication; thus HIV infected patients often develop CNS (central-nervous-system) toxoplasmosis characterized by a focal encephalitis; human chemotherapy and chemoprophylaxis rely on drugs that affect tachyzoites rather than bradyzoites 'encapsulated' in the tissue cyst; atovaquone appears to be the most cyticidal among drugs tested in the mouse model; in ocular toxoplasmosis, corticosteroids should also be used for an anti-inflammatory effect on the eyes; the antifolate pyrimethamine given with sulfadiazine is the treatment of choice for CNS toxoplasmosis; folinic acid is given concurrently to attenuate bone marrow suppression caused by pyrimethamine; length of treatment is determined by clinical response to therapy and lasts for weeks; **feline toxoplasmosis** may be treated with clindamycin (12.5–25 mg/kg b.w. p.o. or. i.m. q12h x 2 weeks) or sulfadiazine (30 mg/kg b.w.) plus pyrimethamine (0.25–0.5 mg/kg b.w.) p.o. q12h x 2 wks plus folinic acid (5 mg/d)

pyrimethamine (diaminopyrimidine) plus **sulfadiazine** (sulfonamide)	Daraprim (standard treatment, drugs of choice)	25–100 mg/kg/d x 3–4 weeks plus sulfadiazine 1–1.5 grams qid x 3–4 weeks plus folinic acid 10 mg with each dose of pyrimethamine; *2mg/kg/d x 3d, then 1mg/kg/d (max. 25mg/d) x 4 weeks plus sulfadiazine 100–200 mg/kg/d x 3–4 wks plus folinic acid 10 mg with each dose of pyrimethamine; congenitally infected newborns should be treated with pyrimethamine every 2 or 3 days and sulfonamide daily for about 1 year;
atovaquone (hydroxynaphthoquinone) plus **pyrimethamine**	(alternative regimen in sulfa-intolerant patients)	atovaquone plus pyrimethamine appears to be an effective alternative in sulfa-intolerant patients
pyrimethamine plus **clindamycin**	alternative treatment	50–100 mg/d x 3–4 wks plus clindamycin 450–600 mg per os or 600–120 mg i.v. qid plus folinic acid, 10mg, with each dose of pyrimethamine
spiramycin (for prophylactic use during pregnancy)	Rovamycine, others (alternative drug)	3–4 grams/d; *50–100 mg/kg/d x 3–4 weeks; if it is determined that transmission has occurred *in utero*, therapy with pyrimethamine and sulfadiazine should be started.

ALTERNATIVE REGIMENS TO TREAT CNS TOXOPLASMOSIS:

in HIV-infected patients with cerebral toxoplasmosis, some clinicians have used pyrimethamine 50–100 mg/d after a loading dose of 200 mg with a sulfonamide and, when sulfonamide sensitivity developed, have given clindamycin 1.8–2.4 g/d in divided doses instead of the sulfonamide; clindamycin with pyrimethamine (see above) has been proved an effective alternative for treatment of cerebral toxoplasmosis; also atovaquone has been effective and well tolerated in some patients

CHRONIC SUPPRESSION OF TOXOPLASMOSIS:

pyrimethamine and sulfadiazine or pyrimethamine and clindamycin are the most commonly used regimens for chronic suppression of toxoplasmosis; daily pyrimethamine and sulfadiazine appears to be more effective than a twice-weekly regimen

Table 1. (continued) Drugs acting on diseases caused by unicellular opportunistic parasites in humans

DISEASE **nonproprietary name** (chemical group)	**Brand name** other information	**Adult dosage/*pediatric dosage** (mg/kg body weight, or total dose/individual, oral route), miscellaneous comments
pyrimethamine plus **sulfadiazine**	standard treatment	25–50 mg per os daily plus sulfadiazine 500 mg–1gram per os q6h plus folinic acid, 10mg, with each dose of pyrimethamine
pyrimethamine plus clindamycin	alternative treatment	50 mg per os daily plus clindamycin 300 mg per os qid plus folinic acid, 10mg, with each dose of pyrimethamine
PRIMARY PROPHYLAXIS OF TOXOPLASMOSIS: in HIV patients: with <100 CD4 cells, either trimethoprim-sulfamethoxazole, pyrimethamine plus dapsone or pyrimethamine plus sulfisoxazole can be used; pyrimethamine plus folinic acid should be considered in HIV patients with < 100 CD4 counts who are intolerant to trimethoprim-sulfamethoxazole; doses of trimethoprim-sulfamethoxazole used to prevent *Pneumocystis carinii* pneumonia (PCP, see above) may also prevent first episodes of toxoplasmosis; daily dapsone and weekly pyrimethamine or both twice weekly may also prevent first episodes of toxoplasmosis		
MICROSPORIDIOSIS current information indicates that immunocompromized patients (as in HIV infected individuals) are at the greatest risk of developing microsporidial disease patterns such as ocular infections involving conjunctival, corneal epithelium and even corneal stroma (keratoconjunctivitis) or enteritic infections associated with enteritis, colangitis and diarrhoea; there may also be multiorgan infection or systemic dissemination of microsporidians, including the liver, lungs and kidneys; treatment of microsporidial infections is problematic because of the intracellular habitat of the parasite stages and the resistant nature of the spores (for more information see → Microsporidiosis)		
OCULAR INFECTIONS due to *Encephalitozoon hellem, Encephalitozoon cuniculi, Vittaforma corneae* (*Nosema corneum*)		
albendazole (benzimidazole carbamate) (licensed for treatment of various helminths animals (cf. Nematocidal Drugs, Animals Benzimidazole Compounds) and humans (cf. → Nematocidal Drugs, man/Table 1)	Albenza, others (drug of choice)	400 mg bid (not licensed in the USA but considered investigational for this condition by the FDA, licensed in Europe and elsewhere); ocular lesions due to *E. hellem* in HIV infected patients have also responded to fumagillin eyedrops prepared from Fumidil-B, a commercial product, used to control a microsporidial disease of honey bees; for lesions due to *V. corneae*, topical therapy is generally not effective and keratoplasty may be required
INTESTINAL INFECTIONS due to *Encephalitozoon bieneusi*, and *Encephalitozoon* (*Septata*) *intestinalis*		
albendazole (benzimidazole carbamate)	Albenza and others (drug of choice)	400 mg bid (not licensed in the USA but considered investigational for this condition by the FDA); octreotide (a somatostatin analogue, Sanostatin) has provided symptomatic relief in some patients with large volume diarrhoea; oral fumagillin has been effective in treating *E. bieneusi* but has been associated with thrombocytopenia
DISSEMINATED INFECTIONS due to *Encephalitozoon hellem, Encephalitozoon cuniculi, Encephalitozoon intestinalis,* and *Pleistophora sp.*		
albendazole (benzimidazole carbamate)	Albenza, others (drug of choice)	400 mg bid (not licensed in the USA but considered investigational for this condition by the FDA); there is no established treatment for *Pleistophora*

Abbreviations: the letter d stands for day (days); qd = daily (quaque die); qh = each hour; bid = twice daily; tid = three times per day; qid = four times per day (quarter in die); p.c. (post cibum) = after meals

Dosages listed in the table refer to information from manufacturer or literature, preferably from Medical Letter (1998) 'Drugs for parasitic infections'

Data given in this table have no claim to full information

mans. The incidence rate is high and may exceed 80 % in children An effective treatment of infection with chemotherapeutic drugs is dependent on early diagnosis. Treatment of choice for PCP and extrapulmonary *P. carinii* infections in AIDS patients is oral (PO) or intravenous (IV) administration of trimethoprim +sulfamethoxazole (Bactrim, and others). The dosage is 15mg/kg body weight/day (based on trimethoprim component) given in 3 or 4 doses x 21days. In PCP accompanied by severe hypoxia, oral prednisone at the start of treatment has decreased incidence of respiratory deterioration and may improve tolerance for high-dose trimethoprim +sulfamethoxazole. However, the corticosteroid may reactivate herpes simplex infections or other opportunistic infections like candidiasis. Alternative treatment (for regimens Table 1) may be used for patients with mild to moderate or even severe PCP who have failed or have shown intolerance to the standard treatment (trimethoprim +sulfamethoxazole). For cases of moderate to severe PCP, alternative drugs may be parenteral pentamidine or trimetrexate (antifolate drug) plus folinic acid (to prevent bone-marrow suppression) or dapsone (sulfone with antileprosy activity) given concurrently with trimethoprim. All these drugs appear to be less active but better tolerated than trimethoprim +sulfamethoxazole. This seems to be true also for the antimalarial atovaquone given as suspension (750 mg PO, bid x 21 days). Simultaneous use of parenteral (i.v.) or oral clindamycin with oral primaquine (→ Malariacidal Drugs) has been successful in patients with mild to moderate PCP. For primary and secondary chemoprophylaxis of PCP standard treatment is oral trimethoprim +sulfamethoxazole (for dosage cf. Table 1). For alternative treatment dapsone may be used alone, or concurrently with pyrimethamine plus folinic acid (dosage cf. Table 1). Aerosolized pentamidine (e.g. NebuPent, 300mg inhaled monthly via a nebulizer) is well tolerated in PCP prophylaxis but less effective than trimethoprim +sulfamethoxazole and, in patients with < 100 CD4 cells, less active than dapsone. Chemoprophylaxis against PCP is also needed in patients with AIDS-related complex diseases or those with bone marrow or solid organ transplants. In absence of prophylaxis most of heart-lung, and lung allograft recipients will probably develop PCP. Supporting treatment as elevation of arterial oxygen pressure, support of lung function, or corticosteroid therapy along with specific causal agents against PCP will reduce the frequency of mortality in immunocompromised patients, particularly in those suffering from AIDS.

Drugs Acting on Cryptosporidiosis of Mammals

Species of the genus *Cryptosporidium* are coccidian parasites that infect epithelial cells (extracytoplasmic) of the intestinal and respiratory tract of vertebrates (see also → Cryptosporidium (Vol. 1) for life cycle and → Coccidiocidal Drugs/Drugs Acting on Cryptosporidiosis in Birds). Although immunocompetent hosts show no or only mild clinical signs after *C. parvum* infections immunosuppressed hosts may suffer from life threatening watery diarrhoea caused by enteritis of the small intestine. Infections are due to infective oocysts (= sporocysts) passed in feces of carriers in the environment. Transmission may occur by food/ feed or water supplies containing sporulated oocysts, or by droplet infections. Outbreaks of the disease are mainly seen in young animals or neonates (calves, lambs, piglets, foals, and zoo, pet or laboratory animals). The disease in livestock is associated with intensive husbandry, seasonal breeding, and mixed grazing practices (feed, water and holding facilities contaminated with oocyst,). Clinical signs are weakness, dullness, rough coat, weight loss and mortality. In **humans** the severity of infection depends on immunocompetence of the patient. Immunosuppressed humans suffering from AIDS can show intractable diarrhoea causing dehydration, weakness, considerable weight loss, and even mortality.

Causal **therapy** and chemoprophylaxis of cryptosporidiosis in animals and humans is problematic (Table 1). Only a few drugs show some activity against *C. parvum* infections and are approved for the use in animals rather than humans for this indication. Many approaches to anticryptosporidial efficacy of commercial drugs have failed in improving symptoms in ruminants suffering from *C. parvum* infections. Several anticoccidials, like sulfonamides, lasalocid sodium, halofuginone lactate, decoquinate, or paromomycin (available as additives in-feed, cf. → Coccidiocidal Drugs, or other dosage forms for oral or parenteral administration) have been found to be insufficiently effective in controlling or even eradicating *C. parvum* infection in calves and kids. They may exhibit positive clinical short-term effects such as improvement of watery diarrhoea and reduction of oocyst output in feces. Thus drugs are static rather than cidal in their action on cryptosporidia. Monoclonal antibodies raised against *C. parvum* may reduce clinical signs in *C. parvum* infected laboratory animals. Therefore, drug treatment should be associated with **strict measures of hygiene** (which should also include the farm personnel) and sanitation, such as disinfection (ammonium hydroxide) and thorough cleaning in contaminated farms. During the calving period, calving cows must be separated from other animals and new-born calves too. Oral and parenteral rehydration therapy is essential in animals with severe diarrhoea to maintain the fluid balance. Also management in AIDS patients has principally included fluid therapy, use of antidiarrhoeal agents and nutritional support. For chemotherapy applied to humans see Table 1.

Drugs Acting on Toxoplasmosis of Mammals

The cyst-forming coccidian parasite *Toxoplasma gondii* is widespread in human beings and many warm-blooded animals, and cats including wild Felidae, are the only definitive hosts excreting *T. gondii* oocysts in their feces. The domestic cat appears to be the major source of contamination with oocyst since a cat can excrete millions of oocyst surviving for long periods under ordinary environmental conditions (e.g. in moist soil). Despite the fact that cats are frequently infected clinical signs are rare. Feline toxoplasmosis can be treated with clindamycin or sulfadiazine plus antifolates (cf. Table 1).

Ovine toxoplasmosis may be associated with abortion in ewes and perinatal mortality in lambs. Toxovax, commercially available in Europe and New Zealand, is a live vaccine containing *T. gondii* tachyzoites of the S 48 "incomplete" strain. The vaccine is used for the control of ovine toxoplasmosis and reduces fetal losses in sheep in endemic areas.

Human toxoplasmosis is most often the result of ingestion of tissue cysts in raw or undercooked meat from pigs, sheep and rabbits (less prevalent in cattle). Exposure to heat (70°C) and cold (-15°C) can kill parasites in meat. Good hygiene and sanitation can help to control infection; hands of people preparing raw meat and all materials (cutting boards, knives, etc.) coming in contact with uncooked meat should be washed with soap and water, and rinsed thereafter thoroughly with tap water. To avoid oral infection with oocycts shed by cats gloves should be worn while gardening, and vegetables should be washed thoroughly before eating because of possible contamination with cat feces. Pet cats should be fed only cooked food. **Pregnant women**, in particular, should not eat raw or uncooked meat and avoid contact with cats and soil because of risk of a congenital infection. The infected fetus may develop full tetrad of signs, i.e., retinochorioditis, hydrocephalus, convulsions and intracerebral calcification.

Patients with AIDS (acquired immune deficiency syndrome) often develop CNS (central-nervous-system) toxoplasmosis (focal encephalitis). In the USA, the human illness losses due to congenital toxoplasmosis have been estimated at US$ 0.4–8.8 billion annually (the wide range reflects the uncertainty about the number of infected babies. **Currently used drugs** and dosages for treating the disease in individuals are shown in Table 1. In AIDS patients, 'dormant' developmental stages of *T. gondii* (tissue cysts containing bradyzoites) are reactivated thereby altering the latent infection to an acute one. The drug of choice for treating CNS toxoplasmic encephalitis is oral pyrimethamine (50–100 mg per os daily), usually given for 3 to 5 weeks depending on clinical response, plus folinic acid (10 mg, with each dose of pyrimethamine) plus oral sulfadiazine (1–1.5 mg per os q6h, Table 1). Concurrently given drugs may cause severe adverse reactions in approx. 40% of the so treated patients, though folinic acid may reduce bone marrow suppression caused by pyrimethamine. Alternative regimens can be used in patients with failure to pyrimethamine and sulfadiazine treatment or intolerance to these drugs. On the other hand these regimens may be used for treatment of mild to moderate cerebral toxoplasmosis as oral pyrimethamine (50–100 mg per os daily) given concurrently with clindamycin (per os or intravenously, dosage cf. Table 1). The hydroxynaphthoquinone atovaquone (Mepron, dosage see under pneumocystosis in Table 1). Among numerous experimental drugs tested in mice atovaquone have shown the most cysticidal activity. For chronic suppression of toxoplasmosis a commonly used regimens is oral pyrimethamine with reduced daily doses given concurrently with oral sulfadiazin (dosage see Table 1), or alternatively with oral clindamycin (dosage see Table 1). Primary prophylaxis of first episodes of toxoplasmosis can be done with the fix drug combination trimethoprim-sulfamethoxazole (Bactrim, Eusaprim, and other products, dosage see under pneumocystosis in Table 1). Also oral dapsone, given daily concurrently with oral pyrimethamine plus folinic acid, weekly can be used (dosage see Table 1). Experimental drugs with good activity against *Toxoplasma gondii* in mice are diclazuril and toltrazuril (→ Coccidiocidal Drugs). Although the action of diclazuril on tachyzoites can be enhanced by combination with pyrimethamine this drug mixture is not able to prevent tissue cyst formation in surviving mice.

Trematode Infections

→ Platyhelminthic Infections, Man, Pathology.

Trematodocidal Drugs

See Table 1

Economic Importance of Fascioliasis and Other Trematode Infections

The most common and pathogenic liver flukes in cattle, sheep, and goats are *Fasciola hepatica*, (common liver fluke) and *F. gigantica*. In humans fascioliasis (caused by *Fasciola hepatica*) is focal in distribution and sporadic while in ruminants the infection is principally endemic and of greatest economic importance. *Fasciola hepatica* is widespread and about 250 million sheep and 350 million cattle are at fascioliasis risk worldwide. Thus in many countries about a quarter of the sheep and cattle population is exposed to the infection causing severe economic loss (approximately US$2 billion worldwide) in domestic livestock; economic damage is due to mortality, liver condemnation, secondary infections, reduced milk and/or meat production, abortions and fertility reduction. During the migration phase through the abdominal cavity, liver parenchyma, and wall of bile ducts young liver flukes may produce serious, acute inflammatory tissue reactions and blood loss owing to mechanical trauma. The young flukes then reach maturity in the bile ducts. Death of the host is mostly due to severe anemia and failure of liver function. Cattle seem to be more resistant to fascioliasis than are sheep, possibly because their stronger tissue reactions form a certain mechanical barrier against reinfection. Young *Dicrocoelium dendriticum* or *D. hospes*, which migrate directly from the small intestine into the biliary system through the common bile duct do not cause mechanical tissue damage to liver parenchyma. These small liver flukes produce mainly *chronic tissue reactions* of the liver, such as fibrosis of small bile ducts, portal veins, and hepatic artery (Glisson's capsule), leading to biliary cirrhosis. Heavy infections may result in loss of weight and emaciation. Another hepatic trematode, *Fascioloides magna*, a common trematode of deer in North America and brought into several European countries and South Africa, may cause severe disease and death in cattle, sheep, and goats by continuous tissue migration. Thus, a single migrating fluke can eventually cause the death of the host.

Infections due to intestinal trematodes (e.g. family Paramphistomatidae) in domestic and wild animals are of minor economic (veterinary) importance although they are common throughout the world in cattle and sheep. The pathogenicity of these flukes is lower than that caused by the hepatic trematodes and intestinal nematodes. Young paramphistomes are deeply embedded into the mucosa of the small intestine causing mechanical damage of the epithelial cells by their large posterior (ventral) sucker. Adult worms living in the forestomachs of ruminants may produce only slight clinical signs. However, large numbers of immature worms in the small intestine and abomasum may cause loss of appetite, diarrhoea (dehydration), anemia (loss of protein), retarded growth, and even mortality, especially in young susceptible animals. *Gigantocotyle explanatum* (a common paramphistome in cattle in India) which occurs in the bile ducts rather than in the intestinal tract may cause similar pathological changes like those caused by *Fasciola hepatica* or *F. gigantica*. There are many other genera and species of intestinal trematodes in ruminants, equines (e.g., *Gastrodiscus aegyptiacus*), and pigs (e.g., *Fasciolopsis buski*, *Gastrodiscus aegyptiacus*). In dogs there are also a large number of intestinal flukes, such as heterophyids, echinostomes, and diplostomes, which show little host specificity (Table 2). Adult worms are nearly without pathological findings and, in general, do not cause clinical disease. Intestinal flukes also frequently occur in birds and rodents, which may form reservoirs for infections in humans and domestic animals. Because there is little host specificity intestinal trematodes are of-

Table 1. Drugs used against trematode infections of domestic animals

CHEMICAL GROUP **Nonproprietary name** (approx. dose, mg/kg body weight, oral route) other information	*__Brand name__ (manufacturer, company), other information	**Characteristics** (chemotherapeutic and adverse effects, miscellaneous comments)
HALOGENATED HYDROCARBONS		
carbon tetrachloride (sheep, 80, or 0.05 ml/kg, p.o.), (sheep 80–160, i.m. in liquid paraffin; cattle 40, i.m., not recommended)	introduced for treatment of *F. hepatica* infection in 1921	first effective drug against *Fasciola* spp., especially *F. hepatica*; is still used in some countries to control subacute and chronic fascioliasis in sheep (routine dose 1 ml/animal); recommended dose is only active against flukes older than 10 weeks; CCL4 poisoning causes liver and kidney dysfunction; i.m. application reduces risk of liver toxicity but can produce necrosis at injection site, particularly in cattle; maximum tolerated dose in sheep is 160–800 mg/kg, p.o.
hexachloroethane (cattle, sheep, 200–300)	first use: against *F. hepatica* in 1928	has been used mainly in cattle; efficacy against mature *Fasciola* spp. is approx. 90%; drug is less toxic than carbon tetrachloride although liver toxicity and mortality may occasionally occur; drug exhibits marked activity against mature paramphistomes in rumen; maximum tolerated dose in sheep is approx. 1200 mg/kg; drug is no longer used in the USA and other countries because of possibly mutagenic properties of polyhalogenated hydrocarbons
hexachloroparaxylene (chemical name:1,4-bis(trichloromethyl)-benzene) (sheep, 150) (cattle 125)	*Hetol (former Hoechst, discontinued), *Bitriben *Hexichol	old-timer which has been used with satisfactory results against adult *F. hepatica*; erratic effect against *Dicrocoelium dendriticum*; the drug exhibits marked effect (80% - 90%) against mature paramphistomes in rumen; drug is well tolerated in cattle and sheep at recommended dose and seems still to be widely used in parts of the former USSR, and China for treating clonorchiasis in man; maximum tolerated dose in sheep is approx. 600 mg/kg
CARBON ACID PIPERAZINE DERIVATIVE		
1-ß,ß,ß,-tris-(p-chloro-phenyl)-propionyl-4-methyl-piperazine hydrochloride)	*Hetolin (former Hoechst, discontinued) *Dicroden	old-timer which has been used with somewhat erratic results (80%–90% efficacy) against adult *D. dendriticum* infections in sheep at dose levels of 20–40 mg/kg; in general, treatment of dicrocoeliasis is uneconomic and doubtful with regard to improvement of liver function
HALOGENATED BISPHENOLS		
hexachlorophene (sheep, 15, p.o. or s.c.) (cattle 20, p.o. or s.c.)	*Distodin *Fasciophene	has antiseptic and anthelmintic activity (trematodes, canine cestodes); approx. 90% activity against mature *F. hepatica* and *F. gigantica* (at least 12 weeks old) in sheep and cattle; drug is moderately active against *Paramphistomum* spp.; liver toxicity may occur in sheep and cattle at maximum tolerated dose of 40 mg/kg, and possibly after s.c. injection; narrow chemotherapeutic index requires exact dosing; metabolite (glucuronide) excreted into bile is highly active against the liver flukes
menichlopholan (= **niclofolan**) (sheep and cattle, 4) (cattle, horse 0.8 s.c., not recommended in sheep, pig 3–5 s.c.),	*Bilevon-R *Bilevon-M (Bayer) *Distolon, *Dertil (tablets, injectable solution; latter are widely used	currently used drug which is highly active against mature *F. hepatica* (12-week-old, at progressively higher doses also juvenile flukes but limited by safety), immature paramphistomes in sheep (75%–95% at a dose of 6 mg/kg, safety margin =2) and adults of *F. gigantica*; drug is also effective against fascioliasis (adult worm) in pigs and horses; maximum tolerated dose per os in sheep is approx. 12 mg/kg; side effects may be fever, tachypnea, sweating rarely mortality; is excreted in milk of cattle (5–8 days posttreatment up to 0.1 ppm, limit for consumable milk is 0.01 ppm); injectable solution: safety index = 2.5–3; drug has also been used in man for treating *Metagonimus yokogawai* infection (Table 2)

Table 1. (continued) Drugs used against trematode infections of domestic animals

CHEMICAL GROUP **Nonproprietary name** (approx. dose, mg/kg body weight, oral route) other information	***Brand name** (manufacturer, company), other information	**Characteristics** (chemotherapeutic and adverse effects, miscellaneous comments)
bithionol (sodium salt) bithionol sulfoxide (sulfene in Russia) (cattle 30–35) (sheep 40–75) +**hexachlorophene** (see text, characteristics)	*Actamer *Bitin-S (Tanabe), *Bitin *BTS *Disto-5 *Lorothidol	used as bithionolate sodium (drug of choice) against *F. hepatica* infection in man (Table 2); it shows moderate efficacy against adult *F. hepatica* in cattle (60%–70%) and may be used in chronic *F. hepatica* infections in sheep (95%–100% efficacy); bithionol (30 mg/kg) combined with **hexachlorophene** (5 mg/kg) may enhance fasciolide action against adult *F. hepatica* (over 12weeks old) in cattle and sheep (up to 100%); drug is effective against immature paramphistomes (at higher dose up to 100%) but less so against adult paramphistomes in sheep and cattle; adverse effects (diarrhoea, skin reactions such as urticaria, also seen in man) may occur at recommended dose; maximum tolerated dose in sheep is 75 mg/kg; safety index =1 at recommended dose in sheep!
ORGANOPHOSPHATE (masked bisphenol derivative)		
bromophenophos (bromofenofos) (cattle, 12) (sheep, 16)	*Acedist (Merial)	is highly active (91%–99%) against mature *F. hepatica* (10- to 14-week-old flukes); its action on immature stages is less pronounced; maximum tolerated dose in sheep is approx. 50 mg/kg; safety index =3 at recommended dose rate in sheep; contraindicated during pregnancy; withdrawal times are 21 days (edible tissues) and 7 days (milk)
MONOPHENOLIC DERIVATIVES		
nitroxinil (*nitroxynil*); (sheep, cattle, 10, s.c.) (in some countries discontinued or at no time launched)	*Dovenix *Trodax (May & Baker)	reasonably good activity against *F. hepatica* (50%–90%, flukes aged 6–8 weeks, activity is erratic; 90%–99%, flukes aged 10–14 weeks), *F. gigantica* but not against paramphistomes; the drug is slightly less active than rafoxanide against immature flukes at recommended dose; in cattle and sheep it also displays activity against some nematodes such as *H. contortus* (cf. rafoxanide, and → Nematocidal Drugs, Animals/Table 1) and *Parafilaria bovicola*; drug may cause some yellow staining of fleece in sheep, and local reactions at injection site; maximum tolerated dose in sheep is approx. 40 mg/kg; drug is slowly reduced to an inactive metabolite in the rumen and is therefore preferably given by s.c. injection; it is slowly eliminated from body into urine and feces (for 31 days), and milk as well (contraindicated in lactating cows); withdrawal time may be about 2 months
dichlorophenol + **bithionol**	*Trematol	was combined with bithionol (96%–97%) = Trematol to increase intrinsic activity of bithionol against adult paramphistomes in sheep, and cattle; dichlorophenol has been shown also to be effective against *Fasciolopsis buski* in man
RESORCYLANILIDES		
resorantel (sheep, cattle, 65)	*Terenol (former Hoechst)	highly effective against *Paramphistomum* spp. in cattle and sheep (immature stages in small intestine 80%–99% efficacy, adults in rumen 85%–100% efficacy); it has also marked activity against adult cestodes (→ Cestodocidal Drugs) and *Gastrodiscoides aegyptiacus* infections in horses; drug is well tolerated at recommended dose

Table 1. (continued) Drugs used against trematode infections of domestic animals

CHEMICAL GROUP **Nonproprietary name** (approx. dose, mg/kg body weight, oral route) other information	***Brand name** (manufacturer, company), other information	**Characteristics** (chemotherapeutic and adverse effects, miscellaneous comments)
SALICYLANILIDES		
bromsalans (various mixtures of *tribromsalan* and *dibromsalan*) (sheep, cattle, 20)	*Fascol *Diaphene (1:3 ratio) *Hilomid (1:1 ratio) *Trionoin (various)	drug mixtures show marked but somewhat erratic activity (40%–98%) against mature *F. hepatica* (12-week-old and older flukes); efficacy against juvenile flukes (8- to 10-week-old is reasonably effective (80%–95%); maximum tolerated dose in sheep is approx. 60 mg/kg (safety index =3); in cattle bromsalans are not compatible with benzimidazole carbamates (→ Nematocidal Drugs, Animals, fenbendazole, oxfendazole should not be given within 7 days because of possible mortality)
niclosamide (sheep, 50–90)	*Mansonil (Bayer) *Yomesan (for human use) *Various others	has little activity against *Fasciola* spp., but is highly effective against tapeworms (see → Cestodocidal Drugs); drug shows nearly 100% activity against immature *Paramphistomum* spp. in sheep and some activity against *Fasciolopsis buski* in man (Table 2); drug is well tolerated and also safe at higher doses
oxyclozanide (cattle, 10) (sheep, 15) **+levamisole** (cf. also → Nematocidal Drugs, Animals/Table 1)	*Zanil *Diplin (ICI, Boehringer Ingelheim Vetmedica) *Diplin Combi discontinued	currently used drug which is highly active against mature *F. hepatica* (90–99% against flukes aged 10–14 weeks); repeated doses (3 x 15 mg/kg) show some activity against immature flukes; therefore drug may be used in acute fascioliasis if highly active drugs are not available; there is marked efficacy against paramphistomes (immature: 60% cattle, 80–92% sheep, and mature: 70%–90% cattle, sheep); drug (15 mg/kg) was found partially effective against immature and mature *Fascioloides magna* in cattle; however, this effect is considered unsatisfactory since a single migrating fluke can kill the host; it exhibits efficacy against *Notocotylus attenuatus* in ducks (15 mg/kg b.w. per os, or 30 mg/kg in-feed, well tolerated); its combination with levamisole makes simultaneous treatment against gastrointestinal nematodes and lungworms of ruminants possible; maximum tolerated dose in sheep is 60 mg/kg safety index <4); following absorption it is excreted as an active glucuronide metabolite into bile (terminal half-life 6.5 days) and shows low residual levels in cattle and sheep; preslaughter withdrawal time for edible tissues is 14 days, that for milk 4 days
clioxanide (sheep, 20); not recommended in cattle	*Tremerad (Parke Davis)	is active against mature *F. hepatica* (50%–90% against 8- to 9-week-old flukes; 91%–99% against 10- to 14-week-old flukes); there is some effect on paramphistomes in cattle, as determined by egg reduction; maximum tolerated dose in sheep is approx. 100 mg/kg

Table 1. (continued) Drugs used against trematode infections of domestic animals

CHEMICAL GROUP **Nonproprietary name** (approx. dose, mg/kg body weight, oral route) other information	*__Brand name__ (manufacturer, company), other information	**Characteristics** (chemotherapeutic and adverse effects, miscellaneous comments)
rafoxanide (cattle, sheep, 7.5) (cattle 3, s.c.) **+thiabendazole** (syn. tiabendazole)	*Ranide *Raniden (Merial) *Flukanide (Merial) *Ursovermit (Serum Werk Bernburg) *Ranizol (simultaneous treatment of liver flukes and gastrointestinal nematodes in sheep an cattle)	currently used drug highly active against mature *F. hepatica*, and *F. gigantica* (91%–99% against flukes aged 7–14 weeks), and immature stages of these flukes at a higher dose (10–15 mg/kg: 50%–90% against flukes aged 4–6 weeks at); drug may be used for strategic treatment and chemoprophylaxis to reduce pasture contamination (cf. also diamfenetide); it seems to be 100% effective (10 and 15 mg/kg) against immature and mature *Fascioloides magna*; there is a reasonably good activity (15 mg/kg) also against immature paramphistomes in sheep but not so against adults; drug is also active against some gastrointestinal nematodes in cattle (e.g., adults of *Haemonchus contortus* and *Bunostomum*, cf. → Nematocidal Drugs, Animals), and *Oestrus ovis* (nasal bot) of sheep; maximum tolerated dose per os in sheep is approx. 45 mg/kg; there may be local reactions after the parenteral route; at recommended dose, the drug is well tolerated in sheep and cattle of all ages; drug is extensively bound (> 99%) to plasma proteins and has a long terminal half-life (~17 days); withdrawal time for edible tissues may be 28 days (contraindicated in lactating animals); mode of action is uncoupling of oxidative phosphorylation
brotianide; (sheep, 7.5) (*F. gigantica*, cattle, 15) **+thiophanate** (see also → Nematocidal Drugs, Animals/Table 1)	*Dirian (Bayer) *Vermadax	drug in current use for treatment of acute, subacute or chronic fascioliasis with high activity against *F. hepatica* in sheep (91%–99% against flukes aged 7–14 weeks, 50%–90% against flukes aged 6 weeks); at 15 mg/kg it has marked efficacy in cattle against immature and mature (also sheep) paramphistomes (85%–90%), and in cattle and sheep against *F. gigantica* (treatment in 60-day intervals throughout the 'snail season'); obvious lack of action against *Dicrocoelium dendriticum*; it has been combined with the probenzimidazole **thiophanate** for treatment against flukes and nematodes; maximum tolerated dose of brotianide in sheep is approx. 25 mg/kg; doses of 15 mg/kg significantly reduce milk production in cattle and presumably in sheep
closantel; (sheep, cattle 10) like rafoxanide, it is extensively bound (> 99%) to plasma proteins (mainly albumin) and has a long terminal half-life (~15days); prolonged residuals may control *Haemonchus* and *Fasciola* infections for longer periods (up to 60 days)	*Flukiver (Janssen) *Seponver (Janssen, Smith Kline)	currently used drug which is highly effective (90%–98%) against adult *F. hepatica* but slightly less effective than **rafoxanide** against immature fluke at recommended dose; it was found nearly 100% effective against 8-week-old *Fascioloides magna* in sheep (15 mg/kg oral, or 7.5 mg/kg i.m.); drug is more toxic after parenteral application (see also rafoxanide); closantel has no action on paramphistomes but is effective against certain nematodes as *Haemonchus contortus* in sheep (particularly against, levamisole-, morantel-, rafoxanide-, and benzimidazole-resistant strains: closantel tolerant *H. contortus* strains are also known), *Strongylus vulgaris* in horses, and *Ancylostoma caninum* in dogs (only adults, ineffective against somatic stages); there is also activity against arthropods (e.g., nasal bot *Oestrus ovis* in sheep, or *Gasterophilus* in horses (→ Nematocidal Drugs, Animals); the drug is well tolerated in sheep and cattle (also in reproduction studies in rams ewes and bulls); safety index in sheep and cattle is ~4 at recommended dose; drug is primarily excreted via feces (80%, urine <1%); withdrawal times for edible tissues may be 28 days (cattle) and 42 days (sheep); drug is contraindicated in lactating animals

Table 1. (continued) Drugs used against trematode infections of domestic animals

CHEMICAL GROUP **Nonproprietary name** (approx. dose, mg/kg body weight, oral route) other information	***Brand name** (manufacturer, company), other information	**Characteristics** (chemotherapeutic and adverse effects, miscellaneous comments)
BISANILINO COMPOUNDS (PHENOXYALKANES)		
diamfenetide (syn. acemidophene, former USSR)); (sheep,100)	*Coriban (Wellcome) (is approved in the USA)	currently used drug (a bisacetamide, which may also be regarded as a bisacetanilide or masked aromatic diamidine), is enzymatically converted in liver cells (deacetylation by deacylases) to the bisanilino compound; this amine metabolite is highly effective (100%- 91%) against early immature *F. hepatica* in sheep aged 1 day to 9 weeks; thus very juvenile flukes passing through the liver parenchyma are rapidly killed by high local concentrations of the bisanilino compound ; older and mature flukes located in bile ducts may survive because of obviously quick catabolism of the very toxic bisanilino compound; thus quantities of the active metabolite reaching mature flukes appear to be small (sub-therapeutic concentrations); thus there is a gradually lower activity with aging of the fluke (efficacy 70–50%); diamphenethide is a very useful drug for treatment of acute fascioliasis; it may be also used in strategic control programs (chemoprophylaxis) against fascioliasis: either as single drug (repeated treatments in 6- to 8-week intervals, twice in spring, twice in autumn), or in combination with a flukicide acting against flukes aged 6–14 weeks; it is also active at high dose rates (200 mg/kg) against adult *D. dendriticum* (efficacy 85–93%); it is well tolerated at recommended dose (no teratogenicity in pregnant ewes, or adverse effects on fertility in ewes or rams, no other contraindications) maximum tolerated dose in sheep is 400 mg/kg (1600 mg/kg cause low incidence of mortality); it is slowly absorbed (greatest concentrations in liver and gallbladder 3 days after dosing, and then declining to negligible values after 7 days); withdrawal time for edible tissues may be 7 days
BENZIMIDAZOLE CARBAMATES		
triclabendazole (cattle, 12; sheep,10; goats 5)	*Fasinex (Novartis)	a thiobenzimidazole derivative with high activity against early immature (aged 1day → 98–100% efficacy at 15 mg/kg) and mature *F. hepatica* (aged 14-weeks →98–100% efficacy at 2.5 mg/kg); in general it seems to be the most useful drug for the treatment of either acute subacute, or chronic fascioliasis in doses of 5 (goats), 10 (sheep) and 12 mg/kg (cattle); it has also been used successfully for chemoprophylaxis in annual programs (Control Measures and Epizootiology of Trematode Infections); it proved effective against *F. hepatica* infections in horses (12mg/kg), *F. gigantica* in cattle (12 mg/kg), and *Fascioloides magna* in sheep or deer (20 mg/kg); triclabendazole is a safe and well tolerated drug at recommended dose (safety index >10), and can be simultaneously used with nematocidal drugs (e.g. fenbendazole); maximum tolerated dose in sheep is 200 mg/kg; drug has no effect on nematodes, including *H. contortus*; the drug is rapidly metabolized to its sulfoxide and sulfone (maximum plasma concentrations 12–38 hours after dosing); metabolites are bound to albumin and persist in plasma for about 7 days; it is chiefly excreted via bile; withdrawal times for edible tissues and milk may be 12 days (dairy cow) and 4 days (dairy sheep)

Table 1. (continued) Drugs used against trematode infections of domestic animals

CHEMICAL GROUP **Nonproprietary name** (approx. dose, mg/kg body weight, oral route) other information	***Brand name** (manufacturer, company), other information	**Characteristics** (chemotherapeutic and adverse effects, miscellaneous comments)
other benzimidazoles: **luxabendazole** **albendazole** **oxfendazole** **cambendazole** **fenbendazole** **thiabendazole** **mebendazole** **netobimin**	for more information see → Nematocidal Drugs, Animals probenzimidazole of albendazole	**luxabendazole** appears to be the most active drug against trematodes (for nematocidal or cestocidal action see → Nematocidal Drugs, Animals and → Cestodocidal Drugs); it is highly active against adult *F. hepatica* and *Dicrocoelium dendriticum* (7.5–10 mg/kg x 1, sheep); **albendazole** and **oxfendazole** show marked activity (approx. 95%) against mature *F. hepatica* in sheep and cattle at dose level of 7.5 mg/kg and 15 mg/kg, respectively; some other benzimidazole proved to be active (up to 90%) against adult *D. dendriticum*, e.g. albendazole (7.5–15 mg/kg x 2, weekly interval), cambendazole (25 mg/kg x 1), fenbendazole (150 mg/kg x 1 or 20 mg/kg x 5, daily), thiabendazole (200 mg/kg x 1), mebendazole (20 mg/kg x 1); however, these high doses proved uneconomical; simultaneous use of **bromsalans** with fenbendazole or oxfendazole within 7 days of each other can cause severe adverse reactions, which can be fatal in cattle; **netobimin** (→ Nematocidal Drugs, Animals) is primarily active against nematodes (7.5 mg/kg); at higher doses (up to 20 mg/kg) it exhibits activity against adult stages of flukes as *F. hepatica*, (90%)and *D. dendriticum* (90–99%)
PYRAZINOISOQUINOLINES		
praziquantel (for details see → Cestodocidal Drugs, and Table 2)	*Droncit (Bayer)	primarily active against various cestodes (→ Cestodocidal Drugs) and schistosomes it shows also marked activity against *D. dendriticum* in sheep with some erratic dose-activity relationship (20 mg/kg, 98%; 40 mg/kg, 76%; 50 mg/kg, 98%; total elimination of flukes cannot be achieved); the drug is highly effective against various intestinal flukes in man (*Heterophyes* spp., *Metagonimus yokogawai*, cf. Table 2) and domestic animals
HALOGENATED BENZENESULFONAMIDES		
clorsulon (cattle, sheep, 7) (approved in the USA for use in cattle) clorsulon / **ivermectin** (cattle, 2 +0.2, s.c.)	*Curatrem (MSD AGVET, now Merial) *Ivomec F (Merial) sterile solution: clorsulon (10%) +ivermectin (1%) for simultaneous treatment of *F. hepatica* and nematode infections in cattle (is well tolerated and safe in cattle; action of ivermectin cf. → Nematocidal Drugs, Animals)	extensive chemical modification of halogenated sulfanilamide derivatives showing fasciolicidal activity led to clorsulon (=MK-401) with 100% activity against adult *F. hepatica* at recommended dose; its activity decreases the younger the flukes are (15 mg/kg: 91–99% efficacy 8–6-week-old stages; 30 mg/kg: 99–100% 3-week-old stages, and 85% 2-week-old stages); there is also reasonable efficacy against other fluke species (*F. gigantica* → 7 mg/kg x5, cattle: 100% adults, 92% immature stages; *Fascioloides magna* → 21 mg/kg x1, cattle, sheep: >92% immature stages 8-week-old, 72% 16-week-old stages); the drug shows low toxicity and is considered safe in breeding and pregnant animals or male fertility studies in rodents and cattle; maximum tolerated doses in sheep (>200 mg/kg x1) and cattle (175 mg/kg x1) have revealed neither gross toxic manifestations nor clinical and histopathologic alterations; rapid absorption (maximum plasma concentration in rats 4 hours postdosing, that in flukes 8–12 hours postdosing) and rapid excretion (short half-life) renders the drug suitable for use in meat producing animals; withdrawal times for edible tissues and milk may be 8 days and 4 days, respectively; it is compatible with other anthelmintics (e.g., fenbendazole or other benzimidazoles); its mode of action involves inhibition of glycolytic enzymes of *F. hepatica*, such that its main source of metabolic energy via Embden-Myerhof glycolytic pathway is blocked

Doses listed in this table refer to information from manufacturer and literature
Data given in this table have no claim to full information
The primary or original manufacturer is indicated if there is lack of information on the current one(s)

Table 2. Drugs used against trematode infections of humans

PARASITE, DISEASE distribution, pathology	**Stages** affected (location), morphology of **eggs**	**Chemical class** other information	**Nonproprietary name** adult/*pediatric dosage (oral route), comments	**Comments**
BLOOD FLUKES				
SCHISTOSOMIASIS (snail-mediated helminthiases): infections are caused by cercariae penetrating human skin during contact with freshwater; prevention on an individual level requires that people must avoid any contact with infested freshwater, or that all individuals defecate and urinate in sanitary facilities; at the community level the following control measures must be considered: public health education, sanitation, eradication of snail vector, and chemotherapy; the aim of the chemotherapy is to suppress egg production which is responsible for pathological damages				
Schistosoma mansoni endemic in Africa Middle East and parts of South America; intestine and liver are particularly affected; damage caused by *Schistosoma* eggs (all species) is related to host response: inflammatory acute phase as reactions to soluble egg antigens is followed by chronic irreversible changes in intestine and liver (fibrosis), other organs	**adults** (venous system of intestine) **eggs** (embryonated, large, oval, with lateral spine) pass into feces; the latter must be deposited in fresh water so that miracidia can hatch and reach appropriate snails	**pyrazinoisoquinolines** (represents a major therapeutic breakthrough)	**praziquantel** (drug of choice, 2x 20 mg/kg x 1d); cure rates 80%-100%) (*dosage as adults)	side effects are common but mild: headache, diarrhoea, rash, fever; single dose treatment results in a very high cure rate
		tetrahydroquinolines (may be used in areas in which praziquantel is less effective; has no useful efficacy against other schistosome species)	**oxamniquine** (15 mg/kg once; cure rates 70%-95%; regional differences in efficacy) (*20 mg/kg/d in 2 doses x 1 d, or 30–40 mg/kg in East/South Africa)	side effects are common but mild; in rare cases convulsions (history of epilepsy); minor increases of transaminase activities; contraindicated in pregnancy
S. japonicum endemic in the Far East, SE Asia, Philippines; acute systemic reactions (Katayama fever); chronic stage with hepatomegalia caused by portal fibrosis, and splenomegaly; liver is the organ most affected (fatal fibrosis)	**adults** (mesenteric venules) **eggs** (embryonated, large but <*S. mansoni*, globular, lack a large spine) pass into feces; latter must be deposited in fresh water so that miracidia can hatch and reach appropriate snails	**pyrazinoisoquinolines**	**praziquantel = PZQ** (drug of choice, 3x 20 mg/kg x 1d); cure rates may reach 80%– 92%) (*dosage as adults)	damage induced by *S. japonicum* is more severe than that of *S. mansoni*; generally a single dose of PZQ has the same efficacy as several smaller doses given at intervals of several hours; thus frequency of side effects is greater with a large single PZQ dose; **diagnostic problems** see *S. mekongi*
		sesquiterpene lactol (→Malariacidal Drugs) in rabbits or dogs infected with cercariae there was 93–98% efficacy against schistosomula at 3x 15 mg/kg p.o. (→d7/14/21 after infection)	**artemether** (field trials (>4500 individuals) conducted in China among high-risk groups have shown that artemether is an effective drug against schistosomula at 6 mg/kg p.o. 1x/15d x4)	
S. mekongi ('minor' species), endemic along Mekong river (Laos, Kam-puchea, S. Thailand)	**adults** (mesenteric venules) **eggs** (resemble closely *S. japonicum* eggs) pass into feces	**pyrazinoisoquinolines**	**praziquantel** (drug of choice, 3x 20 mg/kg x 1d) (*dosage as adults)	**diagnostic problems**: fecal debris adheres to shell of *S. japonicum* and *S. mekongi* eggs; thus eggs may be overlooked in fecal preparations (spine is inapparent and difficult to see)
S. intercalatum ('minor' species), found in West and Central Africa	**adults** (mesenteric venules) **eggs** (resemble closely *S. haematobium* eggs but are larger) pass into feces	**pyrazinoisoquinolines**	**praziquantel** (drug of choice, 3x 20 mg/kg x 1d)	

Table 2. (continued)

PARASITE, DISEASE distribution, pathology	**Stages** affected (location), morphology of **eggs**	**Chemical class** other information	**Nonproprietary name** adult/*pediatric dosage (oral route), comments	**Comments**
S. haematobium endemic in Africa and the Middle East; causes urinary bilharziasis; involvement of urinary tract (hematuria, obstruction of ureters, hydronephrosis); accumulation of eggs around bladder and ureters resulted in inflammation of bladder, formation of granulomas and finally fibrosis; bladder epithelium can transform into squamous cell carcinoma in untreated patients	**adults** (venous plexus of urinary tract mainly bladder)	**pyrazinoisoquinolines**	**praziquantel** (drug of choice, 2x 20 mg/kg x 1d, cure rates >91%)	
	eggs (embryonated, large oval, with terminal spine) pass into urine; latter must be deposited in fresh water so that miracidia can hatch and reach appropriate snails	**organophosphates** (has no useful efficacy against other schistosome species; they are also used as insecticides)	**metrifonate** (5-10 mg/kg x3 at 2 week intervals) (transformed in vivo to dichlorvos; inhibition of plasma cholinesterase)	side effects common but mild (abdominal pain, diarrhoea); seems to cause no overt toxic effects in patients activity
		5-nitrothiazoles	**niridazole** (alternative drug: usually 25 mg/kg, maximum 1.5 g/d, for 7d) urinary metabolites have been reported to be mutagenic); drug is used in *D. medinensis* is infectious too	contraindicated in CNS disorders and/or severe hepatic disturbances; risk of toxic effects on CNS (hallucinations, confusion, convulsions) may be high if drug is used in other schistosome infections
INTESTINAL FLUKES (other species see under lung flukes) infections are acquired by consumption of littorial vegetation, raw or undercooked fish, or mollusks contaminated/infected with encysted metacercariae				
Nanophyetus salmincola (eastern Siberia, north-western area of the USA) in various fish-eating mammals (dog, cat, fox otter, mink, lynx and some piscivorous birds, man	**Adults** (small or large intestine) **eggs** (unembryonated, indistinct operculum, much smaller than those of *P. westermani*) pass into feces	**pyrazinoisoquinolines**	**praziquantel** (drug of choice, 3x 20 mg/kg x 1d	*N. salmincola* transmit to Canidae *rickettsial* agents of 'salmon poisoning' causing high mortality, and also 'Elokomin fluke fever' (wider host range, including man, high morbidity)
HETEROPHYIASIS				
Heterophyes heterophyes (small intestinal fluke); uncommon but widely distributed (Middle East, Turkey, E+SE Asia); occurs in dog, cat, fox, and man	**adults** (attached to wall of small intestine)	**pyrazinoisoquinolines**	**praziquantel** (drug of choice, 3x 25 mg/kg.x1d	many species of fish act as second intermediate host; only heavily infected individuals may exhibit nonspecific diarrhoea
	eggs (small, embryonated inconspicuous operculum, resemble *C. sinensis* eggs) pass into feces	salicylanilides	niclosamide	
		bisphenols	menichlopholan =niclofolan (Table 1)	

Table 2. (continued)

PARASITE, DISEASE distribution, pathology	**Stages** affected (location), morphology of **eggs**	**Chemical class** other information	**Nonproprietary name** adult/*pediatric dosage (oral route), comments	**Comments**
METAGONIMIASIS				
Metagonimus yokogawai (small intestinal fluke); most common heterophyid fluke in the Far East (also found in Mediterranean basin); occurs in dog, cat, pig, and man	**adults** (attached to wall of small intestine **eggs** (small, embryonated, resemble *C. sinensis* and *Heterophyes* eggs but obvious operculum)	**pyrazinoisoquinolines**	**praziquantel** (drug of choice, 3x 25 mg/kg.x1d)	several species of freshwater fish act as second inter-mediate host; only heavily infected individuals may develop nonspecific diarrhoea
		salicylanilides	niclosamide	
		hydrocarbons	tetrachloroethylene	
		bisphenols	menichlopholan = niclofolan (Table 1)	
ECHINOSTOMATIASIS				
Echinostoma ilocanum E. lindoense; they are primarily parasites of birds and mammals (rodents); may occur in Korea and the Philippines	**adults** (attached to wall of small intestine) **eggs** (usually large, oval, unembryonated) pass to feces)	**pyrazinoisoquinolines**	**praziquantel** (drug of choice, 3x 25 mg/kg.x1d)	fish, clams, and tadpoles are second intermediate host; heavy infection can be accompanied with moderate diarrhoea
		salicylanilides	niclosamide	
		hydrocarbons	tetrachloroethylene	
GASTRODISCIASIS				
Gastrodiscoides hominis occurs in India southeast Asia and parts of former USSR; pigs (natural host), man, monkeys, field rats serve as hosts; incorrect **egg** diagnosis may occur	**adults** (attached to wall of colon and cecum) **eggs** (unembryonated large, ovoid; resemble closely *F. hepatica* and *F. buski* eggs) pass to feces	**pyrazinoisoquinolines**	**praziquantel** (drug of choice, 3x 25 mg/kg.x1d)	man acquires infection by eating uncooked aquatic plants; only in massive infections may a mucous diarrhoea develop
		salicylanilides	niclosamide	
		hydrocarbons	tetrachloroethylene	
FASCIOLOPSIASIS				
Fasciolopsis buski (large intestinal fluke) occurs in man and pig in Far East (China, India, other parts of Asia); symptoms are diarrhoea, edema, eosinophilia, in severe infections ascites	**adults** (attached to wall of small intestine) **eggs** (unembryonated, large, broadly ellipsoidal, operculum indistinct, resemble closely *F. hepatica* eggs) pass into feces	**pyrazinoisoquinolines**	**praziquantel** (drug of choice, 3x 25 mg/kg.x1d)	infection is acquired from uncooked aquatic plants (water caltrop, water chestnut, water bamboo, etc.); in massive infection (thousands of worms) obstruction of common bile duct and small intestine can occur
		hydrocarbons	tetrachloroethylene (adverse reactions)	
		salicylanilides	niclosamide (less effective)	
		phenol derivatives	dichlorophenol	

Table 2. (continued)

PARASITE, DISEASE distribution, pathology	**Stages** affected (location), morphology of **eggs**	**Chemical class** other information	**Nonproprietary name** adult/*pediatric dosage (oral route), comments	**Comments**
LIVER FLUKES infections are acquired from consumption of raw or undercooked fish or crustaceans infected with encysted metacercariae, or from ingestion of raw or undercooked plants contaminated with metacercariae				
FASCIOLIASIS				
Fasciola hepatica is cosmopolitan in herbivores grazing on wet pasturage contaminated with metacercariae, occasionally man), dog, cat, pig, horse, kangaroo *F. gigantica* is cosmopolitan in herbivores (the Americas, Hawaii, Middle East, Africa and Asia)	**adults** (bile ducts, liver tissue, and aberrant sites, e.g. lung or subcutaneous tissue) **eggs** (unembryonated, large, broadly ellipsoidal, operculum indistinct, resemble closely *F. buski* eggs) pass into feces; *F. gigantica* > *F. hepatica* (adults), eggs resemble closely)	**bisphenols** **benzimidazoles** bisphenols alkaloids benzimidazoles pyrazinoisoquinolines	**bithionol** (drug of choice 30–50 mg/kg on alternate days x10–15 doses) or **triclabendazole** (10 mg/kg once) menichlopholan = niclofolan (Table 1) and **emetine** show minor activity, albendazole and praziquantel show only poor activity at usual doses	infections are caused by ingestion of raw vegetables contaminated with encysted metacercariae; symptoms are malaise, intermittent fever, pruritus, eosinophilia, abdominal pain, jaundice, enlarged liver, anemia ,aberrant adults (subcutaneous tissue) can be removed surgically
Dicrocoelium dendriticum is cosmopolitan in herbivores, rabbit, pig, dog, deer, rarely in man; not as pathogen as *F. hepatica*	**adults** (fine branches of bile ducts, gallbladder) **eggs** (embryonated, ovoid, small, indistinct operculum, brown shell) pass into feces	**pyrazinoisoquinolines** benzimidazoles	**praziquantel** (3x 25 mg/kg x 1d) albendazole	intermediate (i.) hosts are land snails (1. i. host) and ants (2. i. host); in advanced cases extensive cirrhosis of liver clinical signs are anemia, edema, and emaciation
CLONORCHIASIS				
Clonorchis sinensis (Chinese or oriental liver fluke); common in the Orient; only heavy infections are clinically significant (diarrhoea, abdominal pain, icterus; ascites resulting from cirrhosis of liver); in severe chronic infection cholangiocarcinoma of the liver may develop	**adults** (bile ducts, sometimes pancreatic duct and duodenum) **eggs** (embryonated, ovoid, small, seated operculum) pass into feces; man becomes infected by ingestion of raw freshwater fish (Cyprinidae)	**pyrazinoisoquinolines** **benzimidazoles** bisphenols hydrocarbon	**praziquantel** (3x 25 mg/kg x 1d) or **albendazole** (10 mg/kg x 7 d) (drugs of choice), bithionol (alternative), regimen, see fascioliasis; hexachloroparaxylene, no longer in use because of erratic and serious side effects (Table 1)	occurs in fish-eating mammals (weasel, mink dog, cat, pig, serve as reservoirs for human infections); imported pickled fish containing viable metacercariae may lead to infection in countries where *C. sinensis* is not common; worms may live in host for up to 25 years

Table 2. (continued)

PARASITE, DISEASE distribution, pathology	**Stages** affected (location), morphology of **eggs**	**Chemical class** other information	**Nonproprietary name** adult/*pediatric dosage (oral route), comments	**Comments**
OPISTHORCHIASIS				
Opisthorchis felineus (syn. *O. tenuicollis*); E+S Europe and parts of former USSR) *O. viverrini*, Far East, mainly Thailand); both species occur in dog, cat, fox, pig, and man	**adults** (bile ducts of liver) **eggs** (embryonated, small, seated operculum, difficult to distinguish from those of *C. sinensis*) pass into feces	**pyrazinoisoquinolines** bisphenols benzimidazoles	**praziquantel** (drug of choice, 3x 25 mg/kg x 1d) bithionol **mebendazole** (has been reported to be effective)	man becomes infected by ingestion of raw freshwater fish (see clonorchiasis); clinical signs are comparable to those seen in *C. sinensis* infections
Metorchis conjunctus (North American liver fluke) in dog, cat, fox, mink, racoon, man; pathogenic effects are similar to those of *Opisthorchis* spp.	**adults** (gall bladder, bile ducts of liver) **eggs** (small, operculum) pass into feces	**pyrazinoisoquinolines**	**praziquantel** (drug of choice, 3x 25 mg/kg x 1d	*M. albidus* is a similar species occurring in Europe, former parts of the USSR, N America; cyprinid fish serve as second intermediate host
LUNG FLUKES	*Paragonimus* infections are usually caused by consumption of raw or undercooked freshwater crustaceans (crabs) infected with encysted metacercariae			
(cf. Intestinal flukes)	*Nanophyetus salmincola* (intestinal fluke) and *Paragonimus* spp. belong to the same family (Troglotrematidae). *N. salmincola* is a parasite of the intestine which penetrates deeply into the mucosa of the duodenum or attaches to the mucosa of other parts of the small and large intestine causing superficial or hemorrhagic enteritis; cercariae encyst on fish such as of the family Salmonidae and infect man and animals when raw or undercooked fish is eaten (**for treatment see intestinal flukes**)			
PARAGONIMIASIS				
Paragonimus westermani (Asiatic species) *P. uterobilateralis* ***P. africanus*** (African species) ***P. kellikotti*** (North America species) other species in Central and South America; common in crustacean-eating wild (e.g. bush rat, raccoon, otter, fox, mink) and domestic animals (e.g. dog, cat, pig), man	**adults** (parenchyma of lung, forming lung cysts or capsules; aberrant sites such as brain, liver, intestine, muscles, skin, and testes) **eggs** (unembryonated, prominent operculum, different sizes: *P. westermani* much larger than others, dark shell) pass up from lung into sputum; eggs may be confused with those of *Diphyllobothrium latum* (smaller than those of *P. westermani*)	**pyrazinoisoquinolines** phenol derivatives benzimidazoles	**praziquantel** (drug of choice, 3x 25 mg/kg in 3 doses x 2d) bithionol (alternative drug, 30-50 mg/kg on alternate d, x10-15 doses) triclabendazole (alternative drug 5mg/kg once daily x 3d, or 10 mg/kg, twice x 1d)	infection of man is due to ingestion of raw crabs in form of uncooked paste containing metacercariae; metabolic products of worms may cause inflammatory response in lungs: fibrotic lesions, hyperplasia of bronchioli, pneumonia (bacterial superinfections), pneumothorax; chronic infection causes clubbing of fingers and toes; brain lesions produce seizures

The letter d stands for day (days); Dosages listed in the table refer to information from manufacturer or literature, preferably from Medical Letter (1998) 'Drugs for parasitic infections'. The Medical Letter (publisher), vol 40 (issue 1017): 1-12. New Rochelle New York
Triclabendazole (a veterinary fasciolide) has been found safe and effective in humans. Data given in this table have no claim to full information.

ten found in unusual hosts. For some time schistosomiasis in ruminants (particularly in cattle) has been recognized as a veterinary problem in Africa, the Middle East, and Asia (India and China). (For information on schistosomiasis in humans see Control Measures and Epizootiology of Trematode Infections, Drugs Acting on Fascioliasis and Other Trematode Infections, and Table 2).

Control Measures and Epizootiology of Trematode Infections

The main flukes parasitic in domestic animals and humans belong to the subclass *Digenea* of the class Trematoda (phylum → Platyhelminthes (Vol. 1)). The life-histories of all genera and species belonging to this subclass are indirect, i.e., trematodes require one, two or more intermediate hosts to complete their life-cycle in the definitive host; obligatory intermediate hosts are snails. Human infections with the cosmopolitan liver flukes *Fasciola hepatica*, and *Dicrocoelium dendriticum* may occur by chance in endemic areas depending on people's eating habits. The chronically infected sheep and cattle serve as the definitive host for *Fasciola hepatica* and play an important part in contaminating pastures with million of embryonated eggs (an animal may produce 1–2 million worm eggs per day). Domestic animals other than sheep and cattle, and wild animals may serve as other reservoir hosts for *Fasciola hepatica*. Fascioliasis in ruminants is found throughout the Americas, southeastern United States, Africa, Europe and China. The clinical disease occurs mainly in late autumn and winter. Important prerequisites of successful prevention of the development of acute, subacute or chronic fascioliasis in sheep and cattle include potential prophylactic programs (metaphylaxis). Thus strategic treatments at epizootiological appropriate times (in late summer to avoid outbreaks of disease, and in late winter plus early spring to reduce contamination of pastures with eggs before grazing commences) are designed to minimize subsequent snail infection in the autumn and spring. Annual programs of control for fluke may be three **triclabendazole** treatments combined with other broad-spectrum anthelmintics. Successful chemoprophylaxis may also be performed with **diamfenetide** (~100% efficacy against 1-day-old to 9 week-old flukes) and **rafoxanide** (~86–100% efficacy against 6-week-old to adult flukes, cf. Table 1). Several agents such as bithionol, hexachlorophene, bromsalans, bromophenophos, oxyclozanide, menichlopholan (=niclofolan), albendazole, clorsulon and clorsulon+ivermectin (Ivomec F, cattle) have activity against adult flukes only. At doses approaching toxic levels they may affect immature flukes (8–10 weeks old) and are thus unsuitable for treatment of acute fascioliasis and only relatively suited to strategic treatments (metaphylaxis). Only their frequent use (which is uneconomic) may prevent economic loss. Besides the regular use of antitrematodal drugs, integrated control of fascioliasis should comprise measures modifying environment to the disadvantage of the intermediate host. Thus the number of snails in endemic habitats as well as the overall frequency of flukicide application can be reduced by physical means such as effective drainage systems, proper dams, and fencing off swampy areas. Biological control may also help to reduce molluscan vectors; living antagonists or predators such as ducks and frogs may ingest snails in endemic areas. On the other hand, chemical control by the use of highly active and for the environment toxic **molluscicides** can be applied. However, these agents not only kill snails but also other invertebrates and fish. Currently available molluscicides are **niclosamide ethanolamine** salt (Bayluscid), **N-tritylmorpholine** (Frescon), **sodium pentachlorophenate** or copper sulphate. Strategic applications of molluscicides must be carried out with appropriate techniques and at regular yearly intervals to reduce effectively the snail population including those which are intermediate hosts for trematodes. Because the use of molluscicides has been severely restricted in most countries management of fluke infections relies chiefly on treatment of cattle. A **geographic information system** (GIS) forecast model based on moisture and thermal regime has been developed to assess the risk of *F. hepatica* infections in endemic areas. GIS can be used to complement conventional ecological monitoring and modeling techniques; it permits database management of standard maps, aerial photographs, satellite images, climate zones and ground survey maps. Careful definition of factors affecting dynamics of fascioliasis on a geographic basis may be useful for researchers when making decisions on resource allocation and setting priorities. There are no efficient control measures for preventing **dicrocoeliasis** (caused by the small liver fluke *D. dendriticum*) in sheep and cattle. None of the antitrematodal drugs available (Table 1) is effective enough to remove adult parasites entirely and to obtain permanent improvement of liver function and regeneration of liver parenchyma. Nevertheless, strategic treatment in early spring summer and autumn may prevent production loss in sheep and cattle. However, required doses of fasciolicides (various **benzimidazoles**, **praziquantel**) are too high for an economic chemoprophylaxis. Intermediate hosts (snails, ants) of *Dicrocoelium* spp. are known to be widespread and have high reproductive rates. This is also true for intermediate hosts (aquatic snails, freshwater fish) of other liver flukes as *Opisthorchis* spp. and *Clonorchis sinensis* widely distributed in Japan, Korea, Vietnam, and China. Dogs, cats and pigs serve as reservoir hosts. Man becomes infected by consumption of raw or improperly cooked cyprinoid fish chiefly coming from contaminated aquacultures. Ammonium sulfate added to egg-contaminated feces may interrupt transmission to freshwater snails. Thus, the application of molluscicides (see above) is unlikely to be sufficiently effective in controlling snail vectors and the distribution of ants can be successfully reduced only by destroying their nests in certain districts. However, Formicinae are protected animals and control of ants is prohibited in most countries. Farm management techniques for controlling intestinal **paramphistomiasis** caused by rumen flukes in cattle, sheep and goats are similar to those performed in integrated control of fascioliasis. In general intestinal paramphistomes respond to flukicides used for controlling of fascioliasis in ruminants (Table 1), and some anticestodal agents (e.g. niclosamide, resorantel, cf. → Cestodocidal Drugs/Table 1). However, compounds may reveal different chemotherapeutic activities in ruminants such as cattle, or sheep or their actions on juvenile flukes in the intestinal and abomasal walls and adult flukes in the rumen may differ as well. For instance, bithionol exhibits better efficacy against juvenile than adult flukes whereas oxyclozanide or resorantel are effective against both stages. Large numbers of immature stages of *Paramphistomum* spp. cause clinical signs, such as anorexia and diarrhoea, during the migration phase in the duodenal wall. The young rumen flukes are then highly pathogenic to young, naive (previously uninfected) sheep and cattle. As a rule, susceptible calves, and lambs should not be grazed together with adult animals, as these are chronically infected in most cases. If an outbreak of disease occurs all animals should be treated and removed immediately from pasture.

Cattle schistosomiasis caused by as many as ten different species (e.g. *S. bovis*, *S. mattheei*, *S. japonicum*, *S. spindale*, and others) is known to be highly focal in endemic areas because of aggregated distribution of the intermediate snail hosts and the restricted stock movement from one farm to another. Thus individuals from the same herd are likely to be exposed to similar ranges of cercarial challenge. The per-oral route of infection may be of importance in cattle, particularly when animals drink infrequently and swallow large volumes of water. In China, *S. japonicum* infections are known to be an important zoonosis; millions of cattle may be infected and develop a natural acquired immunity to this schistosome infection. In most parts of endemic regions *chemotherapy* appears not to be a suitable mean for controlling schistosomiasis in domestic livestock though it is the current method of choice in the control of human schistosomiasis (Table 2).

Human schistosomiasis (*Schistosoma mansoni*, *S. haematobium*, *S. japonicum* and other species) is a major medical problem in many tropical and sub-tropical regions. Over 200 million people may be infected worldwide, and effective long-term control has proved difficult. Population chemotherapy and other intervention strategies (application of molluscicides and biological agents for eradication of snail vectors, sanitation) have reduced morbidity. Thus transmission control needs new techniques (assessment of worm burdens, antigen detection assays) to improve the diagnostic and epidemiological toolbox. Thus epidemiology of human schistosomiasis has received particular attention. A major factor restricting epidemiological studies of human infections is that it is not possible to assess worm burdens in vivo. The intensity of infection is currently estimated indirectly, i.e. by counting the number of *S. mansoni* eggs in calibrated feces samples (eggs per gram → EPG). There are reasons to assume that worm numbers in endemic areas are markedly higher than is usually perceived. Meanwhile, epidemiological tools have to rely on conventional techniques as well as statistical and mathematical approaches. In 1994 the International Agency for Research on Cancer (IARC) decided to include, in its monograph series on **carcinogenic risks** to man, schistosomes and liver flukes. The evidence here comes from epidemiological studies (helpful animal cancer models are not available). *Opisthorchis viverrini*, which is endemic in Thailand, Laos, and Indian subcontinent with ~9 million infections, and *S. haematobium* are classified as carcinogenic (category 1, main cancer cholangiocarcinoma and bladder cancer, respectively). *Clonorchis sinensis* (~30–40 million infections in Taiwan, China, Korea, Japan and Philippines) and *S. japonicum* endemic in Southern China, the Philippines and in Japan with ~30 million of infected people are classified as probably or possibly carcinogenic (category 2A/2B, main cancers cholangiocarcinoma/gastrointestinal cancer). *O. felineus* found in the former USSR and in some European regions (~2 million infections) and *S. mansoni* is non-classifiable because of insufficient human evidence (category 3).

The human lung fluke, *Paragonimus westermani*, and related species of *Paragonimus* may cause **paragonimiasis** (~30 million infected people) in certain areas of Southeast, West Africa and the Americas. Numerous reservoir hosts (Table 2) makes control of *P. westermani* nearly impossible; boiling the freshwater crabs (second intermediate host; all organs of the crab can harbor encysted metacercariae) for several minutes until the meat has turned opaque can kill metacercariae.

Intestinal trematodes in humans (Table 2) are of minor medical importance. There are numerous species living attached to the epithelium of the small intestine. They are widely distributed throughout Southeast, the Indian subcontinent, West Africa, and Mediterranean countries (*Heterophyes heterophyes* found especially in Egypt). Symptoms caused by these trematodes are dependent on the number of worms present. In massive infections with thousands of worms occlusion of common bile duct and small intestine can occur. Thus heavily infected individuals may develop nonspecific diarrhoea and experience abdominal pain similar to that due to peptic ulcer. Eosinophilia is a common feature. Dependent on the worm species, humans become infected with encysted metacercariae by consumption of raw/undercooked aquatic plants or raw mollusks and fish. For instance, *Fasciolopsis buski* (giant intestinal fluke) is transmitted by the consumption of raw aquatic plants (e.g., water chestnuts and bamboos) contaminated with metacercariae of this fluke. There may be 15–20 million people infected with *Fasciolopsis buski* in areas of the Far East. Metacercariae of *Echinostoma ilocanum* encyst in freshwater mollusks (snails or clams). Metacercariae of *Heterophyes heterophyes* (~10 million infected people, uncommon but widely distributed) and those of *Metagonimus yokogawai* (most common heterophyid fluke in areas of the Far East and Mediterranean basin) encyst under the scales or in the skin of various fish species. The primary control measure against these infections (transmission of eggs to intermediate hosts) is prevention of contamination of water supplies with fecal material. Reservoir hosts like fish-eating mammals may play a certain role in the maintenance of intestinal trematodes in the environment.

Drugs Acting on Fascioliasis and Other Trematode Infections

Animals Flukicides are marketed in most livestock-producing countries of Europe, Australia, the USA, Africa, and the Americas. In 1921, carbon tetrachloride (CCl_4) was the first agent used for treatment against fascioliasis in ruminants. The drug itself has no effect on the isolated liver fluke; an active metabolite appears to affect the fluke. Today there are mainly six distinct chemical groups of anthelmintics (Table 1), which can be used in controlling fascioliasis. (1) **Halogenated hydrocarbons** (most drugs of this group are no longer used because of adverse effects and variable efficacy against *Fasciola hepatica*). (2) **Halogenated phenols** *and* **bisphenols** as hexachlorophene (used since the late 1950s), bithionol (with a sulfur bridge between the two phenol rings), and the corresponding structural variants bithionol sulfoxide and bithionolate sodium. They were predominantly used as antimicrobials against bacteria and fungi. Other drugs are menichlopholan (two phenol rings are linked directly via a carbon→carbon bond), and the monophenol nitroxynil (nitroxynil) with the electron-withdrawing nitro group, and a cyano group in ortho- and para-position. (3) **Halogenated salicylanilides** that may be regarded as close analogues of the bisphenols with a carboxamide group connecting the two aromatic rings as in case of bromsalans discovered in 1963. Bromsalans consist of a certain mixture of 3,4', 5-tribromosalicylanide (tribromsalan, the active principle), and 4', 5-dibromosalicylanilide (dibromsalan, a germicide). They led to a number of modified salicylanilides such as oxyclozanide, clioxanide, rafoxanide, brotianide, bromoxanide, and closantel. (4) **Benzimidazole carbamates** such as albendazole and luxabendazole have a broad anthelmintic spectrum against nematodes and mature *Fasciola hepatica.* The chlorinated methylthio-benzimidazole derivative triclabendazole has a marked activity against *F. hepatica* in cattle and sheep but only a poor action on nematodes. (5) **Bisanilino compounds** initially synthesized as mono- and bisanilino structures in the late 1960s had caused serious toxic side effects (visual disturbances and blindness) in ruminants. Only diamfenetide (introduced much later in the early 1970s) in form of the bisacetylated prodrug exhibited an excellent activity against immature *F. hepatica* and was tolerated well in sheep. Its rapid deacetylation in the liver leads to the active metabolite whose effectiveness decreases, as the flukes grow old. (6) **Benzene sulfonamides** to be subjected to extensive modification in a series of halogenated sulfanilamide derivatives led to clorsulon in the late 1970s. It shows high action on both immature and mature *F. hepatica* and is safe for use in breeding and pregnant animals. Thus the flukicidal spectrum as well as efficacy of these compounds vary considerably against immature stages of *F. hepatica* and those of other trematodes (for more details see Table 1). Because of its high activity against very young flukes (e.g., 1-day-old to 6-week-old *F. hepatica*) diamfenetide is useful for the treatment of acute fascioliasis in sheep. It can also be used for prophylaxis in combination with an agent active against flukes from 4 weeks of age to adult flukes (Control Measures and Epizootiology of Trematode Infections). Today, chemotherapeutic treatment of *Fascioloides magna* infections in domestic ruminants attracts more attention because of the high pathogenicity of this fluke. Closantel (15 mg/kg >90% efficacy, 8-week-old flukes), triclabendazole (sheep, 20 mg/kg 99% efficacy, 12-week-old flukes), and rafoxanide (cattle, 10–15 mg/kg 100% immature and mature flukes) show high activity against this liver fluke. Because of economic considerations older flukicides such as halogenated hydrocarbons and some phenol derivatives remain in use in various countries. They are chiefly active against adult (mature) flukes and may be used to control subacute and chronic fascioliasis. In general, currently used compounds exhibit higher activity against *Fasciola* spp. than against *Dicrocoelium*, paramphistomid flukes, and schistosomes. Organophosphates such as bromphenophos (a masked bisphenol effective against adult *F. hepatica* in cattle), and metrifonate (=trichlorphon used in horses against various nematodes, cf. → Nematocidal Drugs, Animals) have shown some effect against schistosomiasis in cattle (*Schistosoma bovis* and *S. mattheei*). By contrast, the pyrazinoisoquinoline praziquantel is highly active against schistosomes in cattle. However administration of the drug may produce large numbers of killed worm pairs and thus portal obstruction with more serious consequences than the disease itself (For more information on antischistosomal agents used in human schistosomiasis see Table 2).

Flukicides used in meat and milk producing cattle or sheep may have a wide range of preslaughter **withdrawal periods** depending on their pharmacokinetic properties (terminal half-life, residues in edible tissues). Thus some salicylanilides as closantel and rafoxanide, which show additional activity against *H. contortus* (→ Nematocidal Drugs, Animals), are bound extensively to plasma proteins (mainly to albumin) resulting in long terminal half-lives in sheep and cattle. For instance, animals (cattle, sheep, or goats) for human consumption are permitted to be slaughtered 2 months after the treatment with nitroxinil, and 7 days with diamfenetide. Some flukicides such as oxyclozanide, and bromphenophos has been approved for use in dairy cows because of their relatively rapid elimination from the body (4 and 7 days withdrawal time for milk, respectively, cf. Table 1).

In general, **side effects** due to flukicides are negligible (occasionally loosening of feces); only some older halogenated hydrocarbons may rarely show erratic toxicity including mortality. At recommended dose, the majority of therapeutic indices of current flukicides seem to be safe for ruminants (halogenated phenols 1–4, diamphenethide 3–5, salicylanilides 4–6, clorsulon 5, triclabendazole 20–40).

Drug resistance in *F. hepatica* to various flukicides (rafoxanide, closantel, and triclabendazole) has been identified in endemic areas of Australian sheep farms and may become an increasinglproblem in future. The long and regular use of salicylanilides, particularly rafoxanide and closantel but also the benzimidazoles triclabendazole and luxabendazole support the selection of drug tolerant field strains of this fluke. By contrast, attempts to select clorsulon resistant *F. hepatica* strain in the laboratory were unsuccessful. For this reason several drug combinations (some with synergistic effects) have been used successfully in Australia and elsewhere, e.g. triclabendazole + luxabendazole, or + clorsulon, or + closantel; closantel + luxabendazole, or + clorsulon, or + nitroxinil; clorsulon + nitroxinil, or + luxabendazole. The application of such drug combinations may be indicated if one partner of the combination has developed reduced efficacy and the counterpart is still fully active against *F. hepatica* field strains; this may slow down the development of further spreading of drug-tolerant *F. hepatica*.

Humans Praziquantel (PZQ) is the current drug of choice for the treatment of human schistosomiasis (Table 2). It is effective against all species of schistosomes and liver flukes infecting man. The drug is well tolerated with few, minor side effects and is suitable for mass treatment because of single dose regimen. Today the question is posed whether resistance or tolerance to PZQ in schistosomes is a fact or artifact. Drug tolerance should now be considered a fact in strains selected in the laboratory. There is also evidence of drug tolerance to PZQ in a few endemic human populations; PZQ tolerant strains of *S. mansoni* proved fully susceptible to oxamniquine. This drug is principally active against *S. mansoni* (second drug of choice) whereas metrifonate and niridazole show good activity against *S. haematobium* infections; the latter drug is still used in China. Hycanthone, active against *S. mansoni* and *S. haematobium*, is no longer used because of its potential to cause mutagenic effects in mammals. PZQ is an excellent, and widely used drug against other human trematodes such as *Clonorchis sinensis, Paragonimus westermani*, and numerous intestinal trematodes. Clinical disease of human *F. hepatica* infection may be treated best with bithionol. Unlike infections with other flukes, *Fascioliasis of man* does not respond well to praziquantel. Amoscanate (benzeneamine derivative) and oltripaz (dithiole-3-thionine derivative) that had shown promising antischistosomal efficacy have been dropped from further consideration because of their toxic side effects. Antischistosomal drugs of former importance have been the antimonials (e.g. tartar emetic and astiban), lucanthone and hycanthone. All these compounds were not as safe and effective as the more modern drugs (Table 2). Today their use cannot generally be justified. For action of artemether against *S. japonicum* schistosomula infections see Table 1.

Trench Fever

Disease with 3–8 repeated fever phases (all 5 days = therefore also called five-days fever) due to infection with *Bartonella quintana*-bacteria. These bacteria being previously members of the genus *Rochalimea* are transmitted by → lice (Vol. 1).

Therapy

Tetracyclines

Trichinelliasis, Man

Synonym

Trichinellosis, Trichinosis

Pathology

Infection with → *Trichinella spiralis* (Vol. 1) and several subspecies is acquired by ingestion of undercooked meat from pig, bear, walrus, and certain other omnivorous species (→ Pathology/Fig. 18C,D, → Pathology/Fig. 19, → Pathology/Fig. 20). Encysted larvae in muscle are set free during digestion, and enter the intestinal epithelial cells where they become mature in the first week, generally giving rise to diarrhoea and severe eosinophilic inflammation especially in reinfections. The worms mate and produce larvae which invade the intestinal wall and enter the bloodstream. After some migration, they enter skeletal muscle in the second and third week of infection. With heavy infection myositis, edema, and high fever with eosinophilia make their appearance in the second week when larvae invade the muscle fibers in which they encapsulate (→ Trichinella (Vol. 1), → Pathology/Fig. 18C). Blood eosinophilic is pronounced 3–5 weeks after infection. Myocarditis and encephalitis may result from transitory worm migration. While the larvae grow intracellularly, the muscle fibers form an inner capsule and an outer capsule, the endomysium, which becomes hyalinized. The coiled larvae may persist for many years. Calcium may be deposited in the capsule and muscle and eventually the larvae dies (→ Pathology/Fig. 18D). Eosinophilic inflammatory foci caused by occasional degenerating larvae are found in muscles.

Immune Responses

Within 10–15 days *Trichinella* are completely removed from the intestine of infected rats or mice.

The worm loss is associated with profound inflammatory changes such as infiltration of the mucosa with mast cells, villus atrophy and crypt hyperplasia, net secretion and accumulation of fluid in the gut lumen, and increased peristalsis. These changes in the environment appear to make the intestine inhospitable to the worm, so that it is no longer able to maintain its preferred position in the small intestine. The inflammatory changes are dependent upon the local activation of $CD4^+$ Th2 cells that develop in the lamina propria and draining mesenteric lymph nodes. These cells do not mediate worm expulsion by themselves, but instead promote the differentiation and activation of mast cells. Several experimental findings support this scenario: (1) Nude mice and mast cell deficient mice allow prolonged *Trichinella* persistence and restoration of mast cell responses restores the ability to expel worms (2) worm loss correlates with the release of mucosal mast cell-specific proteases and (3) blocking mast cell development with antibodies against c-kit (stem cell factor receptor) prevents worm expulsion. Th2 cytokines such as IL-3, IL-4 and IL-9 participate in the development of a protective mastocytosis. The accompanying infiltration of the mucosa with eosinophils could be blocked by treatment of mice with anti IL-5 antibodies. The finding that this treatment did not stop worm expulsion suggests, that if eosinophils do have a role it is not essential.

Unexpectedly, it has been recently shown that IL-4 is not only required for worm expulsion but also involved in the development of enteropathy. Moreover, abrogation of severe pathology in TNF-receptor-deficient mice did not prevent parasite expulsion. These findings suggests (1) a novel interplay between IL-4 and TNF and (2) that IL-4-mediated protection operates by mechanisms other than merely the gross degradation of the parasite's environment as a consequence of immune enteropathy.

The role of antibodies in worm expulsion is questionable. Passive transfer experiments suggest that IgA and IgG antibodies may interfere with worm growth and reproduction, but do not directly cause worm loss during primary infections. Experience of a primary infection with *Trichinella* however leads to a dramatically faster expulsion of worms following a secondary infection. This is associated with a number of electrophysiological changes in the epithelial cells of the mucosa which are induced by IgE- and IgG bound to mucosal mast cells leading to an anaphylactic reaction mediated via 5-hydroxytryptamine.

Main clinical symptoms: Abdominal pain, diarrhoea, vomiting, oedema, fever for days to weeks, muscle pain, eosinophilia
Incubation period: 1–28 days
Prepatent period: 5 days
Patent period: 20 years
Diagnosis: Serodiagnostic methods, microscopic determination of larvae in muscle biopsies, → serology (Vol. 1)
Prophylaxis: Avoid eating raw meat
Therapy: Treatment see → Nematocidal Drugs, Man

Trichinosis

→ *Trichinella spiralis* (Vol. 1), → Trichinelliasis, Man.

Trichomoniasis, Man

→ *Trichomonas vaginalis* (Vol. 1) is a flagellate, 10–30 μm in size, a pale-staining nucleus, 4 free flagella, and an undulating membrane (→ Trichomonadida (Vol. 1)). It lives in the vagina and prostate gland. It is best recognized supravitally by its motility or in a smear. Trichomoniasis gives rise to acute and chronic vaginitis accompanied by a neutrophil exudate and a change in bacterial flora. Infection is chronic, but if cured by chemotherapy reinfections can occur giving rise to renewed symptoms. Immunity in the vagina appears to be poor. Cervical dysplasia is occasionally observed but appears to be the result of concomitant infection with one of the papilloma viruses. Males may have infection in the prostate gland which is usually asymptomatic but is accompanied by acute and chronic inflammation. *T. tenax* from the mouth and *T. hominis* from the gut are regarded as commensals.

Main clinical symptoms: Occurrence of whitish mucus (fluor), feeling of burning in vaginal and urethral regions
Incubation period: 4–24 days
Prepatent period: 4–20 days
Patent period: Months – years

Diagnosis: Microscopic detection of trophozoites in mucus samples.
Prophylaxis: Avoid unprotected sexual intercourse
Therapy: Treatment see → Antidiarrhoeal and Antitrichomoniasis Drugs

Trichostrongyliasis

→ Trichostrongylidae (Vol. 1), → Trichostrongylosis, Animals.

Trichostrongylosis, Animals

Stomach

→ *Trichostrongylus axei* (Vol. 1) worms occur in the stomach of horses and rarely in pigs. These nematodes are rarely pathogens on their own, most infections are chronic and mild. However, *T. axei* induces typical lesions in horses. The condition has been described as a *gastritis chronica hyperplastica et erosiva circumscripta* for the main lesion is a pad- or cushion-like thickening in the glandular part of the stomach.

Abomasum

Trichostrongylus axei lives in the abomasum of cattle, sheep and goats. In ruminants, *T. axei* infections are usually part of a mixed abomasal helminthosis and its effects cannot be dissociated from those of other worm species. The worm is rarely a pathogen on its own, as most infections are mild. Animals experimentally infected with large numbers of *T. axei* show a decrease of blood albumin, haemoconcentration and a rise in serum pepsinogen. The clinical signs include diarrhoea, anorexia, progressive emaciation, listlessness and weakness.

Small Intestine

Some members of the genus *Trichostrongylus* parasitize the anterior part of the small intestine of ruminants, and are particularly important in sheep. The most common species in sheep and goats are *T. colubriformis* (also found in cattle), *T. vitrinus* and *T. rugatus*. *T. vitrinus* appears to be more pathogenic than the other two species. They all cause a similar syndrome which may range in intensity from a subclinical but significant loss of production to overt disease. Trichostrongylosis is characterized by anorexia, soft faeces, intermittent or continued diarrhoea, weight loss, listlessness and osteoporosis in growing lambs. Severely affected animals become dehydrated and some may die. Changes in blood constituents include a light anaemia, hypophosphataemia with normocalcaemia, and a characteristic hypoalbuminaemia. A reduction in thyroxine concentrations and an increase in circulating levels of alkaline phosphatase of intestinal origin has been reported in chronic cases.

Although considerable progress has been achieved in our understanding of the physiopathology of *Trichostrongylus* parasites, it is still difficult to explain the signs of trichostrongylosis. Practically all stages of the parasite live in tunnels beneath the epithelial cells of the intestine, causing mucosal and villous atrophy or flattening. Sparse stunted microvilli, epithelial hyperplasia are also present, with infiltration of lymphocytes and neutrophils in the damaged area. This atrophy leads to a reduction of the effective glandular mass and of the levels of brush-border enzymes (notably dipeptidase, alkaline phosphatase and maltase). In addition, there is evidence that the parasite alters the levels of gut hormones (e.g. secretin and cholecystokinin) and induces a progressive inhibition of abomasal, duodenal and cranial jejunal motility, which reduces the passage of digests. Potential causes of diarrhoea, when it occurs, may be an alteration in ruminal and abomasal functions, increased plasma loss into the intestine, or other effects of the worm on water, Na^+ and osmotic loading of the small intestine. The decrease in productivity does not appear to be related to malabsorption, since net absorption of nutrient over the length of the small intestine is not severely affected. It is rather caused by the combination of loss of appetite, enteric losses of protein, and increased protein metabolism in the intestinal tissue, which all together cause a movement of amino acid nitrogen from the muscle, and possibly the skin, to the liver and intestines. This decreases the possibility for growth, and production of milk and wool. The reduction in feed intake is the main factor limiting the availability of energy for maintenance and/or growth. Another reason for the less efficient use of metabolic energy is the marked increase in synthetic rates of blood proteins and proteins in the gastrointestinal tissue, to compensate for the losses of plasma protein into the alimentary tract and for the increased sloughing of epithelial cells. The reduced mineralisation of bones leading to osteoporosis in growing lambs may be attributable to reduced intestinal absorption of calcium and phosphorus.

Related Entry

→ Alimentary System Diseases, Ruminants.

Therapy

→ Nematocidal Drugs, Animals.

Trichrome Stain

→ Microsporidiosis.

Trichuriasis, Animals

→ *Trichuris* (Vol. 1) spp., the whipworms, inhabit the caecum and occasionally the colon of ruminants. The most important species are *T. discolor* and *T. globulosa* in cattle, *T. suis* in pigs, and *T. ovis* and *T. skrjabini* in sheep and goats. *Trichuris* is highly prevalent in all parts of the world but rarely causes clinical signs. Heavy infections associated with severe and often haemorrhagic typhlitis or typhlocolitis has been rarely reported in cattle. Clinical manifestations include anorexia, dysentery, dehydration, weight loss and terminal anaemia. In severe cases the faeces may be markedly haemorrhagic or even all blood. The lesions are caused by the adult worms boring tunnels into the mucosa of the large intestine. Penetration of the mucosa by the parasites produce nodules in the intestinal wall. There is little evidence that *Trichuris* spp. of ruminants ingest measurable quantities of blood. The signs of the disease appear to be primarily related to a reduction of the absorption capacity of the colon, an effusion of protein into the lumen, and a loss of blood through haemorrhages.

Therapy

→ Nematocidal Drugs, Animals.

Trichuriasis, Man

Pathology

Trichuriasis is an infection with a small lumen-dwelling whipworm → *Trichuris trichiura* (Vol. 1), of worldwide distribution. The thin anterior end of the worm is embedded in the epithelium of the colon from which it ingests intercellular fluids. Depending on the degree of infection, the degree of inflammation produced may be severe, with a mixed inflammatory reaction and with bloody mucus, containing eosinophils and → Charcot-Leyden crystals (→ Pathology/Fig. 3). Rectal prolapse from tenesmus has been described in heavily infected children. Although it does not actively suck blood, the daily blood loss was calculated as 0.005 ml per worm, supporting its role in causing anemia in iron-deficient children together with malnutrition.

Immune Responses

→ Nematode Infections, Man/Immune Responses.

Main clinical symptoms: Red-diarrhoea, anaemia, colitis, eosinophilia
Incubation period: 2–3 months
Prepatent period: 3 months
Patent period: 15–18 months
Diagnosis: Microscopic determination of eggs in fecal samples
Prophylaxis: Avoid eating uncooked vegetables and contact with human feces
Therapy: Treatment see → Nematocidal Drugs, Man

Trickle Infections

Application of low doses (given continuously at short intervals) of parasites to produce a persistent parasitic load in a given laboratory host. E.g. it is proven that *Nippostrongylus brasiliensis* given in this way produces large and persistent infections in rats and the normal spontaneous cure response does not occur. Trickle infections may establish worms even in immune hosts.

Trimethoprim

→ Pneumocystosis.

Trombiculidiasis

→ Mange, Animals/Trombiculidiasis, → *Neotrombicula autumnalis* (Vol. 1).

Tropical Elephantiasis

→ *Filariidae* (Vol. 1).

Trypanocidal Drugs, Animals

Table 1

Disease Patterns of African Trypanosomiasis

African trypanosomiasis caused by tsetse-borne heteroxenous trypanosomes (*T. vivax vivax*, T. *congolense congolense*, and *T. brucei brucei*) is known as → **Nagana**. Formerly the term was restricted to infections caused by *T. b. brucei*. Today, the term "trypanosomiasis" is also used as a collective word for all animal trypanosomiasis. The severity of disease may depend on several factors, such as trypanosome species, strain variants, infection dose (low or high tsetse risk), and species of host. Infections can vary from acute (*T. c. simiae* infections in pigs, *T. b. evansi* infections in camels) to usually mild or almost inapparent (*T. b. brucei* infections in cattle). In typical cases, African trypanosomiasis is a wasting disease with clinical signs like anemia, leucopenia, thrombocytopenia, plasma biochemical changes and lesions in some tissues and organs. The disease produces slowly progressive loss of condition accompanied by increasing weakness and extreme emaciation, leading eventually to collapse and death. *T. v. vivax* causes the most important form of trypanosomiasis in cattle in West Africa and elsewhere. The infection may be asymptomatic, subacute, peracute or chronic. Hemorrhagic *T. vivax* outbreaks have been reported from farmers in Kenya and Uganda with considerable deaths of cattle. Symptoms were anemia, bleeding through the skin and ears (prior to death), petechial hemorrhages on the tongue and enlarged spleen (for more information on hemorrhagic *T. vivax* see: Use of Drugs in the Field to Control Cattle Trypanosomiasis). *T. c. congolense* produces the most severe form of animal trypanosomiasis in East and Central Africa. Serious disease and death may occur in cattle, horses, and dogs. *T. b. evansi* also occurs in a dyskinetoplastic form in Central and South America (synonyms include *T. equinum* and *T. venezuelense*) where it is regarded as a separate species. *T. brucei equiperdum* produces a venereal disease (= → **Dourine**) in equids in Northwest Africa, Ethiopia, Central and South America, the Middle East and Asiatic Russia. The disease is usually transmitted by coitus, and infrequently by biting flies or infective discharge. Apart from the typical salivarian or stercorarian pathway of infection any trypanosome can also be transmitted mechanically (e.g., artificially by "syringe passage") without undergoing cyclical development in a vector as *T. vivax* infections of ruminant livestock in South and Central America. Noncyclical transmission can be done in nature by blood-sucking insects, such as *Tabanus* spp. and *Stomoxys* spp. flies (→ Diptera (Vol. 1)). In South America → vampire bats (Vol. 1) should also be a vector transmitting *T. brucei evansi* infections in horses. The disease is known as → **Murrina** (Panama) or → **Derrengadera** (Venezuela). *T. b. equiperdum* infections in horses and donkeys may be transmitted by coitus.

Economic Loss in Livestock

Economic loss due to cattle trypanosomiasis is difficult to assess, but the fact that livestock in Africa are treated with more than 30 million doses of trypanocidal drugs each year may give some indication of the importance of this problem. The impact of disease extends over approximately 9 million km^2 of Africa between the southern border of the Sahara in the north and the Limpopo in the south (sub-Saharan Africa), and threatens more than 50–70 million animals in 37 African countries. Partly a result of this disastrous situation is that Africa produces about 70 times less animal protein per unit area than Europe.

Dissemination of Trypanosomes in the Body of Host and its Influence on Drug Action

There are two groups of tsetse-transmitted organisms, which can be distinguished: (1) the hematic group, including *T. c. congolense* and *T. v. vivax* and confined to the blood and lymphatic systems and (2) the humoral group, including *T. b. brucei*, *T. brucei rhodesiense* and *T. b. gambiense* (→ Trypanocidal Drugs, Man/Drugs Acting on African Trypanosomiasis (Sleeping Sickness) of Humans). In addition to occurring in plasma, species of the humoral group are also present in body cavity fluids and intercellular tissue. Parasites of this group are parasitemic only in the terminal stages of the infection, and the chief pathological changes caused by these trypanosomes are extensive inflammatory, necrotic, and degenerative reactions (tissue damage), probably associated with release of kinins and fibrinogen degradation products. In contrast, parasites of the hematic group produce

Table 1. Drugs used against trypanosome infections of domestic animals

CHEMICAL GROUP, **nonproprietary name** (approx. dose, mg/kg body weight, parenteral route) other information	****Brand name** (manufacturer, company); other information	**Characteristics** (chemotherapeutic effects, adverse effects, miscellaneous comments)
TRIVALENT ANTIMONY COMPLEXES		
potassium antimony tartrate (tartar emetic) (1–1.5 g i.v., repeatedly, 5% aqueous solution) (3–6 g/100 kg i.m. or s.c. repeated doses required at weekly intervals) [c]1908	**Therapeutic use** antimosan, stibophen (sodium salt of antimosan)	the only satisfactory and cheap compound available prior to discovery of phenanthridine derivatives; it was useful for over 40 years in treating *T. c. congolense* and *T. v. vivax* infections in cattle and *T. b. evansi* infections in camels; extravascular injection causes severe necrosis; narrow chemotherapeutic index (about 6% mortality in routine treatment); antimosan and stibophen were found to be effective against *T. c. congolense* and *T. v. vivax* but less against *T. b. brucei*
SULFATED NAPHTHYLAMINES		
suramin (standard dose: camel, 10, slowly i.v., horse, three doses in 1 week; dog, dose may be repeated for several days)[c]1920	Therapeutic use *Germanin, *Bayer 205, *Naganium© and others (Bayer) suramin does not cross blood-brain barrier and is not active against secondary CNS stages of subgenus *Trypanozoon*	developed by Bayer in Germany during the 1914–1918 war (Bayer 205); first report on *T. evansi* activity in1925; shows high efficacy against trypanosomes of subgenus *Trypanozoon* (*T. b. brucei*, *T. b. evansi*, *T. equiperdum*) and onchocerciasis in man (→ Nematocidal Drugs, Man); drug of choice for *T. b. evansi* infections (surra) in camels and horses; it may be toxic in equines (slow i.v. injection) causing edema of sexual organs, lips, eyelids or painful hoofs; i.m. or s.c. administration can cause severe necrosis at injection site (beware paraveneous injection); subdosing (less than 1 g/ 100 kg b. w.) may lead to suramin-resistant strains which are usually sensitive to quinapyramine dimethylsulfate; drug is embryotoxic in mice; drug is bound to almost 100% to plasma proteins and slowly eliminated via kidneys; plasma t50% is about 32h, which may be problematic in animals intended for human consumption
<**suramin** (anhydrous) -*quinapyramine* sulfate complex>	**Prophylactic use** [c] 1966 →pig; [c] 1971 →horse	strongly anionic suramin is able to form a sparingly soluble salt complex with cationic groups of other known trypanocidal drugs; as a result toxicity is reduced and prophylactic effect considerably prolonged; "depot effect" of suramin/ quinapyramine against *T. b. evansi* infection in horses may last up to 6 months; it can be used prophylactically against *T. c. simiae* infections in pigs; trypanosomes resistant to the complex can be treated with isometamidium; mode of action is energy metabolism and hydrogen transport; it blocks NADH oxidation by inhibition of α-glycerophosphate dehydrogenase and oxidase
AMINOQUINALDINES		
quinapyramine dimethosulfate (ruminants, pig, dog, 5, s.c.; equines, camel 3–5, s.c.; dose should be divided, and given at 6 h intervals because animals are more sensitive to drug than bovines [c]1949	**Therapeutic use** *Trypacide sulphate (May and Baker, Rhône-Merieux) *Antrycide (Alkaline Chemical Corp., India; Bella Trading Khartoum, Sudan) *Noroquine (Norbrook) *Quintrycide (Gharda) (usually used as a 10 % aqueous solution)	is highly active against *T. c. congolense*, *T. v. vivax*, *T. b. brucei*, and *T. b. evansi* and reaches therapeutic levels quickly; drug can cause local and systemic reactions (salivation, shaking, trembling, diarrhoea, collapse) in cattle, horse, dogs, and pigs within minutes of treatment; effects resemble those of curare; stress (heat, fatigue, fear, etc.) should be avoided before and after treatment; unexpected acute toxicity and rapid development of drug-resistant strains of *T. c. congolense* have limited its operational area in treating trypanosomiasis in cattle; however, drug seems to be safe and efficient for treating surra (*T. b. evansi*) in camels and horses as well as *T. b. evansi* infections in pigs; quinapyramine-resistant strains are usually controlled by isometamidium; quinapyramine is active against suramin-resistant strains (*T. b. evansi*, *T. b. brucei*)

Table 1. (continued) Drugs used against trypanosome infections of domestic animals

CHEMICAL GROUP, **nonproprietary name** (approx. dose, mg/kg body weight, parenteral route) other information	***Brand name** (manufacturer, company); other information	**Characteristics** (chemotherapeutic effects, adverse effects, miscellaneous comments)
quinapyramine dimethosulfate (water soluble) + chloride (insoluble in water) (3:2, w/w) (7.4, s.c.) [c]1950 **quinapyramine** chloride (unstable thick suspension, must be shaken during use) (pig, s.c., behind the ear) [c]1961	**Prophylactic (Therapeutic) use** *Trypacide Prosalt (May and Baker, Rhône-Merieux) *Antrycide Prosalt (Alkaline Chemical Corp., India; Bella Trading Khartoum, Sudan) * Noroquine Prosal (Norbrook) * Quintrycide Prosalt (Gharda) (usually used as a 16,7 % aqueous solution)	salt mixture has the same spectrum of activity as the dimethosulfate; s.c. injection of the mixture results in formation of a depot from which drug is slowly released; it can also become enclosed in a fibrous capsule or abscess resulting in loss of efficacy (observed chiefly in horses); protective activity may last about 2–3 months depending on severity of tsetse fly challenge; unexpected acute toxicity (see dimethosulfate), and signs of a delayed toxicity can infrequently occur about 14 days after treatment; signs are loss of condition, weakness, collapse, and death as a result of severe kidney and liver damage; drug is selectively localized in these organs quinapyramine chloride is not commercially available (special order necessary); it has been used prophylactically in pigs; *T. c. simiae* infections in growing pigs can be protected by the chloride, given 50 mg/kg at 3-month intervals; pigs and cattle are remarkably tolerant to the drug; its "depot" effect is due to an "egg-like" deposit from which drug is slowly released giving protection for 3 months in low tsetse fly challenge; target of action of quinapyramine is protein synthesis; it seems to act by displaying Mg ions and polyamines from cytoplasmic ribosomes; there is a similar type of kinetoplast DNA condensation as in diminazene, and an extensive loss of ribosomes
PHENANTHRIDINE DERIVATIVES (PHENANTHRIDINIUM COMPOUNDS)		
homidium bromide (soluble in warm water) (1; cattle, deeply i.m.; small ruminants, pigs, horse, i.v.) [c]1952 **homidium** chloride (soluble in cold water) (1; cattle, deeply i.m.; small ruminants, pigs, horse, i.v.) [c]1955	**Therapeutic use** * Ethidium (FBD Ltd.) (2.5% aqueous solution) *Novidium (2.5 % aqueous solution); is as active as the bromide) (May and Baker, Merial) homidium does not cross blood-brain barrier and is not active against secondary CNS stages of *T. b. brucei*	both salts have a somewhat higher selective effect on *T. v. vivax* infections in cattle than they have against *T. c. congolense*; *T. b. brucei* is less susceptible; homidium can be used for treating *T. v. vivax* and *T. c. congolense* infections in horses and dogs; its limited protective activity in cattle depends on severity of challenge and may last 3–5 weeks; mass treatment with homidium resulted in appearance of resistant *T. c. congolense* strains in East and West Africa; homidium-resistant trypanosomes can be controlled by diminazene or isometamidium (enhanced doses); homidium is generally well tolerated at recommended dose and also at higher dose levels (no systemic toxicity); drug may be irritant at site of injection; deep i.m. injection effectively reduces local irritations; severe reactions may occur in horses after i.m. injection whereas i.v. injection seems to be well tolerated (paraveneous injection can lead to severe damage of jugular vein); homidium may be used for prophylactic treatment of slaughter cattle if tsetse fly challenge is moderate and cattle are trekked over not too long distances; it interferes with nucleic acid synthesis by intercalative DNA binding; interaction with DNA depends on length and nature of linking chain causing unwinding and extension; drug binds well to kinetoplast DNA older drugs of this series are **phenidium** chloride and **dimidium** bromide (precursor of homidium), which cause a high incidence of delayed toxicity (marked liver damage) and severe local reaction at injection site; they were replaced by better tolerated homidium

Table 1. (continued) Drugs used against trypanosome infections of domestic animals

CHEMICAL GROUP, **nonproprietary name** (approx. dose, mg/kg body weight, parenteral route) other information	***Brand name** (manufacturer, company); other information	**Characteristics** (chemotherapeutic effects, adverse effects, miscellaneous comments)
pyrithidium bromide [c]1956	**Prophylactic use** * Prothidium (Boots) has been discontinued	activity spectrum was similar to that of homidium (highly active against *T. v. vivax*, *T. c. congolense*, less against *T. b. brucei*); it was widely used in East Africa (2 % – 4 % solution, 2 mg/kg i.m.); its protective effect was 3–5 months depending upon tsetse fly challenge; trypanosomes rapidly developed resistance to this synthetic hybrid, showing cross-resistance to both quinapyramine (its pyrimidyl moiety) and homidium (its phenanthridinium moiety) in mass treatment; resistant strains were usually sensitive to diminazene or isometamidium; chemotherapeutic index was narrow (5 mg/kg led to mortality in cattle); unacceptable local reactions and weight loss occurred in cattle at 2.5 mg/kg
AROMATIC DIAMIDINES		
diminazene aceturate (3.5, cattle, sheep, horses, i.m.) [c]1955 in case of resistant trypanosomes dose can be raised up to 8 mg/kg b.w. but total single dose should not exceed 4g per animal diminazene does not cross blood-brain barrier and is not active against secondary CNS stages of *T. b. evansi* and *T. b. brucei*	**Therapeutic use** [c]* Berenil (Hoechst) * Ganaseg (Squibb) and others (7% aqueous solution) 1g of granules contains 445 mg diminazene aceturate and 555mg phenyldimethyl pyrazolone (*Antipyrin, an analgesic acting as solvent mediator)	is highly effective against *Babesia* spp. (Babesiacidal Drugs/Table 1), *T. c. congolense*, and *T. v. vivax*, but less active against *T. b. brucei* and *T. b. evansi* infections (5–10 mg/kg); drug shows no activity against *T. c. simiae*; it seems to have a wide therapeutic index in cattle: subcutaneous doses up to 21 mg/kg were reported to be tolerated in cattle without serious side effects; as 'sanative' drug it is used alternately with isometamidium, which does not cause mutual cross-resistance; its relative 'rapid' excretion was believed to reduce risk of parasites becoming resistant (Pharmacokinetics of Trypanocides and Chemical Residues in Edible Tissues and Milk); trypanosomes resistant to other drugs (except quinapyramine) are commonly susceptible to diminazene; routine and mass treatment may lead to development of diminazene-resistant *T. v. vivax* and *T. c. congolense* strains; as a rule, diminazene-resistant strains are susceptible to isometamidium; local reactions can occur in cattle (slight swelling after s.c. injection) and in horses (skin may slough off after s.c. injection, abscess formation after i.m. injection); severe systemic reactions may be evident in equines after higher than recommended doses; camels seem to be most sensitive to diminazene; treating *T. b. evansi* infections, severe toxic reactions and death can occur at 3.5 and 7 mg/kg, respectively; unexpected side effects (hypotensive, hypoglycemic, and neurotoxic effects: tremor, nystagmus, ataxia, convulsions, vomiting) have been observed in dogs at the recommended dose; for that reason diminazene should not be used in dogs and camels; besides its trypanocidal and babesiacidal action, diminazene may exert some anti-inflammatory (antihistaminic) effect; diminazene and pentamidine (Trypanocidal Drugs, Man, Pharmacokinetics of Trypanocides and Chemical Residues in Edible Tissues and Milk and Chemotherapy/Withdrawal Time of Drugs in Target Animals); diminazene interferes with nucleic acid synthesis; and bind to DNA *in vitro* (particularly well to kinetoplast DNA) by a non-intercalative mechanism; drug blocks DNA and RNA synthesis (Search for New Drugs)

Table 1. (continued) Drugs used against trypanosome infections of domestic animals

CHEMICAL GROUP, **nonproprietary name** (approx. dose, mg/kg body weight, parenteral route) other information	***Brand name** (manufacturer, company); other information	**Characteristics** (chemotherapeutic effects, adverse effects, miscellaneous comments)
isometamidium chloride (0.25–0.5, cattle, sheep goats deeply i.m.) (1, dog, buffalo) [c]1958, launched in 1961	**Therapeutic use** *Samorin (May and Baker) * Trypamidium (Rhône-Merieux; Specia) (1% aqueous solution) isometamidium does not cross blood-brain barrier and is not active against secondary CNS stages of *T. b. evansi*	metamidium was itself a mixture of two isomers; subsequently the much more soluble, red, highly active isomer was isolated for field trials and named isometamidium; it is a synthetic hybrid like pyrithidium consisting of diazotized *p*-aminobenzamidine moiety of diminazene molecule linked into 7-position with homidium chloride; drug is highly active against *T. v. vivax* infections in ruminants and horses as well as against *T. c. congolense* infections in ruminants, horses, and dogs; it is less active against *T. b. brucei* and *T. b. evansi* infections in horses, ruminants, camels, and dogs; location of the latter *Trypanosoma* spp. in tissues and body cavities makes them less susceptible to drug action, and treatment with suramin or quinapyramin is therefore suggested; trypanosomes resistant to the drug are usually susceptible to diminazene; acceptable daily intake (ADI, cf. Chemotherapy/Withdrawal Time of Drugs in Target Animals) of isometamidium for humans is 6mg (total intake); maximum residue limit (MRCL, cf. Chemotherapy/Withdrawal Time of Drugs in Target Animals) suggested for meat, fat, and milk is 0.1 mg/kg, for liver 0.5 mg/kg , and kidney 1mg/kg resulting in a withdrawal time of at least 30 days (excluded injection site) for recommended dose
isometamidium chloride (0.5–1, deeply i.m., cattle; s.c. injection in dewlap of cattle may avoid muscle necrosis) (0.5, i.v. 1% glucose solution over 30 min horses and camels)	**Prophylactic use** *brand names see under therapeutic use	isometamidium provides extended protection at higher doses, 2–4 months depending on tsetse fly challenge; medium tsetse fly challenge may require 0.5 mg/kg, heavy challenge 1 mg/kg every 2 months; recommended dose is usually well tolerated by cattle; however, i.m. injection can cause severe local reactions like extensive fibrosis at injection site (muscle of neck); i.v. injection in horses and camels may avoid local reaction but may cause systemic toxicity (salivation, tachycardia, profuse diarrhoea, hindleg weakness, collapse due to histamine release); drug can reversibly block neuromuscular transmission and stimulation in cholinergic receptors; extensive accumulation of drug occurs in liver and kidney
MELAMINOPHENYL ARSENICALS		
melarsamine HCl [c]1989	**Therapeutic use** *Cymelarsen (Merial)	effective against trypanosomes of the *T. brucei* group (*T. b. evansi*); it is suggested that trivalent cationic arsenicals interact with trypanothion to form the stable adduct mel T

Doses listed in this table refer to information from manufacturer and literature on the subject.
[c]First practical (commercial) use of drug (approx. year)
Data given in this table have no claim to full information. The primary or original manufacturer is indicated

mainly a severe anemia, which determines the severity of disease. Although the anemias produced by *T. v. vivax* and *T. c. congolense* are equally serious, the mechanism of pathogenicity may be different for each species. *T. c. congolense* can also develop outside the circulatory system. Thus, the different distributions of the trypanosomes in the body of host result in varying susceptibilities to trypanocides depending on their pharmacodynamics (mechanisms of drug action) and pharmacokinetics (disposition and fate of drugs in the body). Relapse of infection, i.e., return of patent parasitemia after its apparent cessation by drug administration, may occur in chronic *T. b. brucei* infections (→ Trypanocidal Drugs, Man/Drugs Acting on African Trypanosomiasis (Sleeping Sickness) of Humans: late stage of trypanosomiasis = sleeping sickness of man). The relapse due to the appearance of trypanosome populations from privileged sites, such as the cerebrospinal fluid and/or intercellular tissue spaces (parasites from the latter site may also be the cause of relapse in *T. c. congolense* infections). Commonly used drugs, such as **diminazene**, **isometamidium** and **homidium** do not have the ability to cross the blood-brain barrier or produce constant trypanocidal concentrations in body cavity fluids and intercellular tissues that kill trypanosomes. Relapse in chronic *T. b. brucei* infections is evident when chemotherapy was started too late. This is of considerable interest because drug sensitivity changes as the infection progresses. Complete cure is usually achieved when drugs are given in the early stage of infection. In late-stage *T. b. brucei* infections with CNS involvement treatment with non-arsenic drugs gives rise to an apparent cure since parasites disappear from the circulation but, after a period of weeks, they reestablish themselves in the circulation. The natural immunity of humans to the cattle pathogen *T. b. brucei*, but not to the morphological indistinguishable human pathogens *T. b. rhodesiense* and *T. b. gambiense*, is probably a result of the selective killing of this species by normal human serum containing trypanolytic factors. Unlike in animal trypanosomiasis, the most prominent symptoms of sleeping sickness may result from the marked damage to the CNS in late-stage *T. b. gambiense* (and *T. b. rhodesiense*) infections. **Melarsoprol** and related arsenicals (known for their high systemic toxicity (→ Trypanocidal Drugs, Man/Drugs Acting on African Trypanosomiasis (Sleeping Sickness) of Humans) are able to cross the blood-brain barrier. A long-term model of African trypanosomiasis in mice producing meningo-encephalitis astrocytosis and neurological disorders can be used to understand the pathogenesis of human African trypanosomiasis from initial infection to advanced stages and to evaluate drug efficacy in the late stage of disease. Trypanocides may also be suitable tools in diagnosis of chronic (subpatent) *T. c. congolense* infections in cattle. For this purpose drugs are applied intravenously before and in combination with the indirect fluorescent antibody test. Rapid flushing of cryptic trypanosomes from the microcirculation may lead to increase of jugular parasite concentrations within 6–10 min of the administration of diminazene, pentamidine, or homidium chloride. However, diamidines given by the intravenous route are liable to give rise to hypotension and other severe, alarming reactions, some of which are due to histamine release.

Current Control Measures

Measures currently used to control trypanosomiasis are diagnosis and treatment, chemoprophylaxis, tsetse fly control or eradication of tsetse flies, and the utilization of so-called trypanotolerant breeds. However, this most challenging task in Africa is complicated and hampered by several specific factors. The number of tsetse flies (and thus the occurrence of disease) fluctuates greatly over periods of several years and makes assessment of the actual risk to which livestock are exposed difficult. In addition, control of trypanosomiasis is hindered considerably by the fact that African trypanosomes are able to establish chronic infections in their mammalian hosts because of their highly developed system of antigenic variation. Individual members of the parasite population change the composition of their surface coat so that variations in the composition of these variant surface glycoproteins (VSGs) allow the parasite to escape the host's immune system. Thus fluctuating parasitemias produced by *T. brucei* are associated not only with the phenomena of variable antigen type (VAT) but certainly also the potential for regulation of trypanosome growth by environmental factors such as epidermal growth factor (EGF), transferrin and low-density lipoprotein. This makes effective immunoprophylaxis unlikely.

In areas with low tsetse fly density the method of choice for controlling African trypanosomiasis seems to be the eradication of the vector. For the time being the spraying of insecticides dominates in tsetse fly eradication. In regions with very low levels of infestation, e.g., by riverine → tsetse (Vol. 1) species (*Glossina palpalis* group, e.g., *G. palpalis, G. fuscipes*), trypanosomiasis can be controlled by surveillance and treatment only. Nevertheless, flies of the savanna (and thicket) group (*G. morsitans* group, e.g., *G. morsitans, G. pallidipes, G. austeni*) may give rise to severe trypanosomiasis in susceptible stock even if their numbers are low. In these areas commercial cattle ranching may be possible under chemoprophylactic protection. However, tsetse fly density and thus contact between cattle and vector must be reduced by additional spraying of insecticides with residual effects (e.g., synthetic pyrethroids) and by setting up impregnated traps and screens. In areas with medium tsetse fly density the further exploration and logical exploitation of trypanotolerant cattle, including crossbreeding trials with European breeds to increase milk and meat productivity of indigenous trypanotolerant cattle, may offer a realistic alternative to not yet available vaccination. At least in areas with high tsetse fly density even trypanotolerant animals may not survive unless they are treated prophylactically against trypanosomiasis. The control of the disease in fully susceptible stock even under chemoprophylaxis seems to be impossible in regions heavily infected with tsetse.

Tsetse flies can detect odors by means of receptors on their antennae. Experience with insect pheromones was used to identify the chemical components of the ox odor, which might attract tsetse flies and led to the discovery of 1-octen-3-ol. It proved highly attractive to flies of the savanna (*G. pallidipes* and *G. m. morsitans*). Thus live bait (e.g. cattle treated with insecticides: spot-on, pour-on), fly traps, and screens impregnated with "essence of ox" and pyrethroid insecticides (e.g., deltamethrin, aplphametrin, or cyfluthrin), and sophisticated ground spraying technology may markedly reduce tsetse infestation in limited areas of riverine woodland or transitional forest-savanna zones. Traps baited with acetone and 1-octen-3-ol have been used in Zimbabwe, Zambia and Malawi to detect the presence and distribution of tsetse flies. It has been shown that **isometamidium** is capable of eliminating the insect vector form of *T. v. vivax*. This experimental finding may be of potential significance in the control of trypanosomiasis in the field, particularly in the operation of the sterile insect technique (e.g. in Nigeria).

Effects of infections on vector survival are of interest for the evolution of parasite-vector interactions since trypanosome transmission depends strongly on vector survival and the frequency of genetic factors controlling vector susceptibility depending on the fitness of infected vectors. In several species of tsetse flies, males from natural populations, and from laboratory-bred colonies, are more likely to develop mature trypanosome infections than females.

Today, there is neither a breakthrough in biological control of tsetse flies nor are there promising solutions for a vaccine against African trypanosomes.

Trypanotolerance of Indigenous African Breeds

The term 'trypanotolerance' means reduced susceptibility to trypanosomiasis and denotes an inherited biological property allowing animals to live, breed, grow and survive in a naturally infected environment without exhibiting clinical signs of trypanosomiasis after harboring pathogenic trypanosomes.

In regions where eradication of the vector is not possible with present methods, genetic improvement of trypanotolerant breeds should be attempted. Attention has recently focused on genetic resistance and various selection programs are being discussed to select trypanotolerant animals. Such programs could involve selection of trypanotolerant animals under natural challenge or selection of marker traits (e.g., aspects of the immune response). Selection could also act on polymorphic loci that may affect trypanotolerance, and may be closely linked to genes acting upon tolerance via marker loci. Trypanotolerance is found not only in cattle (all dwarf semiachondroplastic West and Central African types) but also in sheep, goats, and in some rare pony types, such as the Kotokoli of the Ivory Coast. The N´Dama (Hamitic Longhorn of the *Bos taurus* type as well as those breeds of the West African Shorthorn) is a West African breed (e.g., Gambian cattle) noted for its small size and its trypanotolerance. This humpless breed responds very well to improved management and can attain levels of productivity comparable to that of many African beef breeds of the *Bos indicus* type, such as the West African Zebu, the Orma Boran, the Ankole, or the Afrikander. In addition the N´Dama can maintain reasonable production levels under conditions of poor management, climate, nutrition and high tsetse fly densities. Trypanotolerant breeds of Zebus, sheep and goats may also exist in East Africa. Field studies on two types of large East African Zebu (*Bos indicus*) Boran cattle on a beef ranch in Kenyo have demonstrated that a boran type bred by the Orma tribe had a superior response to tsetse fly challenge compared to an improved Boran when introduced to a new locality. Superior resistance to tsetse fly challenge was evident by lower trypanosome infection rate, and when this was untreated, by lower anemia and decreased mortality.

Drug Interactions Associated with Induction of Immunity

Following the successful feeding of tsetse flies infected with *T. c. congolense*, cattle develop local reactions of delayed onset (commonly called a chancre) that persist for several days. The proliferation of the parasite in the hosts skin prior to its passage into the bloodstream via draining lymphatics plays an important role in the induction of immunity, as it is only after regression of the chancre that cattle are immune to tsetse-transmitted homologous challenge. Attempts to induce skin reactions by intradermal injection of bloodstream forms of *T. c. congolense* have failed. Thus, the induction of immunity to trypanosomes may be adversely affected if trypanocidal drugs are given prior to regression of the chancre. In an area of medium tsetse fly challenge it was found that the degree of immunity was greatest in cattle in which infections were established and clinical disease could develop before treatment. Conversely, no immunity developed in cattle treated immediately trypanosomes were seen in the peripheral blood and prior to any evident clinical signs. Induction of immunity to *T. c. congolense* in rabbits by infection and treatment with **homidium chloride** may also be adversely affected if animals are infected concurrently with antigenically different stocks of trypanosomes. There was no marked cellular proliferation in the skin at the sites of secondary infection bites following feeding of a single *G. morsitans*; a chancre failed to develop. Possibly the impaired response was due to drug action preventing trypanosomes from developing extravascularly.

On the other hand, the apparent duration of drug protection has been thought to be influenced by protective immunity which may develop as a result of interactions between insect vector, host, trypanosome population, and drug. These interactive effects may lead to "non-sterile immunity" or "tolerance" in cattle following drug administration and trypanosome challenge. The role of immune responses has been investigated in **isometamidium** treated Boran cattle under single or repeated challenge with *T. c. congolense* infected tsetse flies. Six months after treatment two-thirds of the cattle were resistant to challenge, irrespective of whether animals had received single or multiple challenge. The animals had no detectable skin reactions at the site of deposition of metacyclic trypanosomes and produced no trypanosome-specific antibodies, indicating that drug residues effectively inhibited trypanosome multiplication in the skin and thus subsequent parasitemia. It was concluded that immunological priming of the host had not occurred, and that the protection achieved was not related to the development of immune responses by the host enhancing the length and potency of protection afforded by isometamidium. These findings indicate that development of immunity is not necessary for successful maintenance of cattle in tsetse fly areas provided close control of drug regimes is maintained. The results may also indicate that it is essential to allow multiplication of parasites prior to drug treatment to induce immunity in the host.

Induction of non-specific host defense (e.g., macrophage functional activity) by immunomodulators was demonstrated in 1979 by Murray et al. *Bacillus Calmette-Guérin* (BCG) and *Corynebacterium parvum* were found to enhance the immune response to *T. c. congolense* infections in susceptible A/J mice and more resistant C57Bl/6J mice, both showing reduced parasitemias and increased survival times. This effect could not be transferred from treated to untreated mice of the identical strain by spleen cells or serum. It is not yet clear by which mechanisms immunomodulators influence the course of infection. The development of effective, immunostimulants may provide attractive, complementary tools for combating trypanosomiasis and should be considered as an additional approach to the complex undertaking of a screening program for new trypanocidal drugs and breeding programs for trypanotolerant livestock.

Search for New Drugs

For animal trypanosomiasis no new drugs of any kind have appeared in the field since the introduction of isometamidium in 1961. Nevertheless, aromatic diamidines continue to provide new compounds of high intrinsic activity. Among these, several compounds are highly active on *T. c. congolense* and *T. v. vivax*, while others show a high activity on trypanosomes of the subgenus *Trypanozoon*. Unfortunately, resistance to one trypanocidal diamidine appears to confer resistance to all diamidines and diminazene-resistant trypanosomes have been shown to be resistant to **DAPI** (4′, 6-diamidino-2-phenyl-indole) and other **diamidines** synthesized by Dann and his colleagues. Aromatic diamidines (e.g. pentamidine, diminazene) not only inhibit the growth of protozoans but also of bacteria, fungi and tumor cells, generally at concentrations below those found to be active on the host. DAPI forms fluorescent complexes with double-stranded DNA and is now used for the fluorescent staining of prokaryotic and eukaryotic cells. The drug seems to interact with A-T-rich regions of DNA and thereby to suppress the DNA-directed RNA and DNA polymerases. Several trypanocides (quinapyramine, pentamidine, diminazene aceturate, and isometamidium) and a babesiacidal drug (imidocarb) have been investigated in an activated DNA-directed DNA synthesis assay system catalyzed by *T. b. brucei* DNA polymerases, murine thymus DNA polymerase alpha, and Rauscher murine leukemia virus reverse transcriptase. From the results obtained it was suggested that trypanosomal DNA polymerases are not the selective target of drugs as they showed a similar dose dependent inhibition to other DNA polymerases of eukaryotic cells. Stimulation of reverse transcriptase activity was observed in the presence of quinapyramine and imidocarb but this could be negated by the presence of spermine in the reaction mixture. As part of studies on N-oxidative biotransformation of amidines, potential metabolites of pentamidine have been synthesized. Although several **amidoximes** of pentamidine and diminazene proved highly active against various African trypanosomes in mice, their potency was inferior to that of the parent compounds. Several compounds of a series of **aryl bisbenzimidazoles** have shown excellent activity against diminazene-resistant *T. c. congolense*, *T. v. vivax*, and *T. b. evansi* strains. Unfortunately this series caused delayed toxicity in calves, including serious liver and kidney damage.

Several antitumor antibiotics have revealed unsuspected high activity against trypanosomes in vitro, particularly DNA and RNA synthesis inhibitors such as 5-chloro-puromycin. **Daunorubicin,** an anthracycline antibiotic intercalating with DNA, which is one of the most potent trypanocidal agents in vitro, has proved totally inactive against *T. b. rhodesiense* in infected mice. Limitations of efficacy and problems with toxicity impose severe limitations on the usefulness of antitumor drugs as potential leads to new trypanocides in humans and animals. The antifungal nucleoside antibiotic **sinefungin**, which strongly inhibits S-adenosyl-methionine dependent transmethylation reactions, has a marked effect on African trypanosomes in mice when administered intraperitoneally. Goats infected with *T. c. congolense* and treated with intramuscular doses of 10 or 20 mg/kg b.w. showed relapse of infection; higher doses (up to 50 mg/kg b.w.) were toxic and caused death. Among a series of novel **purine derivatives** (phosphonylmethoxyalkylpurines and pyrimidines) with antiviral activity against a broad spectrum of DNA viruses some of them showed potential activity in vivo against *T. b. brucei* at dosages that were below those toxic for mice. **Ketoconazole** and related **azole derivatives** with high activity against *T. cruzi* infections in mice have proved ineffective against *T. b. brucei* in mice. Among a series of **phthalanilides** and related compounds, **BW 458 C** was the most effective in curing short-term and long-term *T. b. brucei* infections in mice. Cure rates greater than 90% were achieved with the drug at 10 or 25 mg/kg body weight. None of several compounds of a series of **suramin analogues** was more active than suramin against macrofilariae of *Dipetalonema viteae* and various *Trypanosoma* spp. Inhibition of lipid metabolism in the trypanosomes may be central to the therapeutic effects of the garlic extract containing **diallyl-disulfide** (DAD). DAD is known to have a lipid-regulatory effect and a sulfur-rich compound that readily undergoes ionic interaction with SH being a vital component of coenzyme A. The latter is required in growing cells for the provision of activated acetate molecules, which are then channelled into lipid synthesis and other vital cellular processes.

Salicylhydroxamic acid (SHAM), a substituted aromatic hydroxamic acid, inhibits aerobic energy production (L-glycerol-3-phosphate oxidase system) in trypomastigote stages; it can clear temporarily bloodstream infections of *T. b. brucei* in rats if administered concomitantly with glycerol. In practice only one far from ideal drug, **melarsoprol** (Mel B) is available to treat late-stage sleeping sickness. Calcium (Ca) has a synergistic effect on this trypanocide and has been shown to be more critical in its action than SHAM +glycerol. These data may be important in the clinical management of sleeping sickness. If total Ca is reduced in a patient it is possible that supportive therapy to restore Ca concentrations could improve the therapy, especially in late-stage Gambian infections.

The reason that potent chelators are trypanocidal but not toxic to mice may relate to acute competition for Fe between the host's Fe-binding proteins like transferrins and ferritin and the parasite's Fe requirement. Several chelators such as caffeic acid, cuproine, and other commercially available chelators, which had shown heme sparing or inhibition of growth of *Crithidia fasciculata* in vitro were active against *T. b. rhodesiense* in mice after high doses only. Divalent cation chelators such as ethylenediamine tetraacetate (EDTA) or the calcium-specific chelator ethyleneglycol tetraacetate (EGTA) can abolish the synergistic action of heparinized rat blood with SHAM +glycerol. Transferrin may also function as a drug carrier in African trypanosomes in such a manner that complexes of transferrin with isometamidium (Samorin) are targeted directly with high specificity into the lysosome system of *T. c. congolense*.

DL-α- difluoromethylornithine (DMFO = **eflornithine**) is a selective and irreversible inhibitor of ornithine decarboxylase and a key enzyme in polyamine biosynthesis in *T. b. brucei*. The substituted amino acid was shown to have activity against CNS *T. b. brucei* infections in rodents and is the only 'new' drug to be developed for the treatment of sleeping sickness in humans. It has proved to be an effective treatment for late stage infections of *T. b. gambiense* in humans (→ Trypanocidal Drugs, Man/Drugs Acting on African Trypanosomiasis (Sleeping Sickness) of Humans).

There are various areas considered as leads in research relevant to the development of potential new agents for African trypanosomiasis and targets for chemotherapeutic attacks such as, glycolytic enzymes (non-oxidative branch of pentose phosphate pathway = PPP), antigenic variation, and trypanothione metabolism in trypanosomes. Oxidative branch of PPP might be an alternative lead for new drugs. It maintains a pool of NADPH (reduced form of nicotinamide adenine dinucleotide phosphate required for synthesis of fatty acids via phosphogluconate pathway) that serves to protect against oxidant stress and which generates carbohydrate intermediates used in nucleotide and other biosynthetic pathways. Thus 6-phosphogluconate dehydrogenase (6PGDH) in *T. b. brucei* may be a potential target for chemotherapy because in other eukaryotic organisms the deletion of the gene encoding 6PGDH is lethal. The gene encoding *T. b. brucei* 6PGDH has been cloned, and the enzyme purified and crystallized. Suramin inhibits 6PGDH, and trivalent aromatic arsenoxides inactivate the enzyme with marked potency. Considerable attention has been devoted also to topoisomerases of kinetoplastid organisms. This group of enzymes could be another valuable drug target for new trypanocides. Topoisomerases, which mediate topological changes in DNA, are essential for nucleic acid biosynthesis and for cell survival. Topoisomerase II activity has been purified from *Leishmania donovani*, *T. cruzi* (→ Trypanocidal Drugs, Man) and *T. equiperdum*, and topoisomerase II genes have been cloned also from *T. b. brucei* and *T. cruzi*. Studies with purified topoisomerases indicate that the enzymes from kinetoplastids generally exhibit the expected inhibitor sensitivities. Thus activity is reduced by intercalators acting by deforming the DNA substrate, minor groove binders (compounds that bind in the minor groove of the DNA helix) and compounds that compete for binding at the enzyme's ATP site (e.g., novobiocin, coumermycin). Agents that specifically inhibit type II enzymes by trapping the enzyme on its DNA substrate, forming a 'cleavable complex', are the fluoroquinolines and etoposide. Thus, antibacterial fluoroquinolines were shown to exhibit marked activity in vivo against *Leishmania donovani*. Some classical trypanocides such as DNA-binding agents (diminazene and pentamidine: minor groove binders) and intercalators (e.g., ethidium bromide) are well known for their ability to generate dyskinoplastic trypanosomes, which retain mitochondrial membranes but lack detectable kDNA. Selective inhibition of mitochondrial topoisomerase II may be an explanation for the propensity of these drugs to induce dyskinoplastic cells. Because kDNA is not essential for the survival of bloodstream form of African trypanosomes, nuclear rather than mitochondrial topoisomerases should be the preferred target for drug search. Differences in parasite and mammalian topoisomerases may provide the basis for selective toxicity of new trypanocidal compounds. On the other hand, kinetoplasts can be an obligatory target for antitrypanosomal drug action if these organelles are important for successful subsequent cycling into the insect vectors. There must be some mechanism to assure that an organism does not replicate the nucleus and divide until or unless it has replicated its kinetoplast. Drug targeting of kinetoplasts, then, could interfere with cell replication by preempting this regulatory mechanism. Thus identification of regulatory mechanism generating dyskinetoplastic resistance in trypanosomes could possibly provide a basis for new therapeutic approaches. Also molecular biological investigations, as well as inducible gene expression systems (e.g., the tetracycline-responsive repressor of *Escherichia coli*, TetR) in trypanosomes, could suggest potential targets for chemotherapy and pathogenicity.

Drug Combinations with Synergic Effects

For cattle treatment only a few drugs have been developed, and these are involved in resistance problems today. Under such conditions (such as in cancer chemotherapy) exploration of combinations might be an alternative strategy. Therefore, the possibilities of trypanocidal synergic effects of known drugs have been extensively examined in vitro and in vivo using monomorphic laboratory strains of *T. b. rhodesiense*. Only suramin +tryparsamide, suramin +puromycin, suramin +diminazene, and 9-deazainosine +DL-α-difluoromethylornithine (cf. also Search for New Drugs) have been shown statistically significant synergy. Another example of a successful combination therapy is the suppression of chronic *T. b. brucei* infections in mice (CNS involvement) by diminazene diaceturate or suramin, each combined with a substituted 5-nitroimidazole (e.g., fexinidazole or MK 436). None of these drugs administered singly caused 100% permanent cure. Only fexinidazole (Hoe 239) was able to cure a high percentage of the mice when given repeatedly at relatively high dose levels of 250 mg/kg. In several experimental studies fexinidazole has also been found to exhibit a strong effect against *T. cruzi*, *T. vaginalis* and *E. histolytica in vivo.*

Chemoprophylaxis of Cattle Trypanosomiasis

Although there are different ways of combating cattle trypanosomiasis, each of the control methods in use at present has serious limitations. Because of the major economic importance of trypanosomiasis in cattle the great majority of control measures have been aimed primarily at the protection of these animals by the use of suitable trypanocidal drugs. In the absence of a suitable vaccine, chemotherapy and chemoprophylaxis are the most important tactics, which are available as part of any strategy of trypanosomiasis control. They are still considered to be the most effective measures for trypanosomiasis control.

Drugs used for the treatment and prophylaxis of animal trypanosomiasis center on a small number (Table 1). They can be characterized on the basis of their ionizaton at blood pH as cationic or anionic drugs. Cationic drugs are quaternary ammonium trypanocides (quinapyramine, homidium, pyrithidium, isometamidium), and aromatic diamidines (diminazene and pentamidine). The only anionic drug currently in use is presented by suramin. It is a sulfated naphthylamine derivative that readily binds to plasma proteins; it is still widely used in the treatment of equine trypanosomiasis.

The risk of infection to which cattle are exposed is closely related to the density and the species of tsetse fly present. The incidence of tsetse flies thus chiefly determines the frequency of treatment, which is in most regions regulated by the government. The nomadic habits of the major cattle-owning peoples have given rise to the widespread use of trypanocidal drugs and, in general, treatment of individual animals is not practiced. Ormerod (1979) pointed out that rationale for treating cattle trypanosomiasis is entirely different from that for treating sleeping sickness, for the following reasons. (1) Trypanosomes are very much more common, so that any animal, which becomes infected by tsetse flies, is liable to be infected. Therefore treatment of individuals (or even herds) has no general sanitary significance. (2) Drugs are often given prophylactically to cattle on their way to slaughter in Africa. Cattle are usually moved over long distances to provide meat in urban areas and prophylactic drugs are administered so that animals can pass through the "fly belt". Prophylactic treatment is particularly afflicted with problems concerning variations in the length of protection resulting from varying field situations and the rate of drug elimination from the body (preslaughter withdrawal time). The duration of chemoprophylaxis thus not only depends on the degree of tsetse fly challenge but also on the timing of treatment in relation to occurrence of infection. Insufficient drug protection may result if infection by tsetse flies occurs too early in the trek, and cattle may then succumb to infection before reaching their destination.

The period of effectiveness of prophylactic drugs thus varies with environmental conditions, the tsetse fly challenge and activity of the treated animal as well as actual concentration of the active drug (there is the risk of producing a too long-lasting subcurative concentration in blood and tissues). Thus, the period of protection may be considerably reduced particularly during strenuous activity, especially when trade cattle pass through a fly belt in the course of their journey.

Drug Complexes with Enhanced Prophylactic Activity

The prophylactic action of drugs has been prolonged by preparing complexes of **suramin** (anionic drug) with cationic/basic drugs, thereby reducing systemic toxicity of the drugs. Although a **homidium-suramin complex** gave extended protection (6–12 months), it caused unacceptably severe reactions at injection sites. A **quinapyramin-suramin complex** proved active at a single dose of 50 mg/kg body weight in protecting adult pigs and piglets for at least 6 and 3 months, respectively. However such long acting drugs often cause rapid development of drug resistant trypanosomes. Encapsulation of drugs, either in polymers or in artificial phospholipid membranes (liposomes) has been known for a long time. Preparing a complex of isometamidium with the well-defined polyanion dextran, thereby reducing its toxicity, could enhance the duration of protection produced by the drug. Entrapping homidium bromide in bovine carrier erythrocytes has caused slow release of the drug. However, the various short comings of such preparations, like drug quality problems (standardization), marked drug residues (possibly posing a human health hazard) and severe reactions at sites of injection, have hindered the further preclinical development of such preparations. Recently, more promising results were obtained using different types of subcutaneously implanted slow release devices (SRD) containing polycaprolactone/homidium bromide SRD or more readily biodegradable poly (D, L-lactide) or poly (D, L-lactide-co-glycolide) SRD containing either isometamidium chloride or homidium bromide for intramuscular administration. As a result local toxicity was minimized and prophylactic effects in comparison with the parent compounds were markedly prolonged. When breakthrough isolates derived from SRD-treated animals (rabbits) were compared with the original *T. congolense* strain, such isolates showed some loss of sensitivity to homidium only.

Drug Tolerance in the Field and Assays to Assess Intrinsic Trypanocidal Activity

The conditions under which 'man made drug tolerance' develops in the field are derived basically from under-dosing due to incorrect estimation of body weight; this is difficult to avoid when mass treatment is involved. A high incidence of trypanosomiasis in conjunction with the irregular use of prophylactic and therapeutic drugs also favors the emergence of drug-resistant trypanosomes. Thus, drug-resistant parasites may emerge in any situation where prophylaxis and therapy are inadequate for the degree of tsetse fly challenge. This may be the case particularly in regions of high tsetse fly challenge. The misuse of drugs leads consequently first to "individual" resistance and then to "area" resistance. Generally, prophylactic drugs induce resistance more rapidly in trypanosomes than do 'therapeutic' drugs. The latter drugs reach trypanocidal plasma levels relatively quickly and may be more rapidly metabolized and excreted from organisms than 'prophylactic' drugs. In areas with a high incidence of tsetse flies, subcurative drug levels may already exist towards the end of the protection period, and this is the case particularly after using drugs for prophylaxis. Treatment must therefore be repeated to restore trypanocidal plasma concentrations.

The phenomenon of 'natural (intrinsic) drug tolerance', i.e., variation in drug sensitivity that is not dependent on previous exposure to the drug concerned, has been demonstrated in *T. v. vivax* and *T. c. congolense*. Thus, West African *T. v. vivax* strains seem to be more susceptible to homidium than are East African *T. c. congolense* strains. By contrast, *T. c. congolense* strains appear to be more susceptible to diminazene than are *T. v. vivax* strains. It is likely that the initial appearance of homidium-resistant *T. c. congolense* strains and diminazene-resistant *T. v. vivax* strains can be connected directly with the varying intrinsic sensitivity of these species of a given drug. Some of this variation in drug sensitivity may also be the result of persistent cross-resistance induced by quinapyramine, which was extensively used for therapeutic and chemoprophylactic treatment before homidium became the drug of choice. Differences in drug sensitivity of stocks of the subgenus *Trypanozoon* have also been reported. In an in vivo assay designed to minimize the influence of host-parasite interactions using X-irradiated trypanosomes, it was demonstrated that isoenzymically defined West African *T. b. brucei* stocks were not as sensitive to pentamidine and diminazene as typical East African stocks. This test sought to measure the intrinsic sensitivity of a trypanosome population by reducing the influence of extrinsic determinants of drug sensitivity, in particular trypanosome "penetration" of tissues inaccessible to drugs and host antibody-mediated relapses of parasitemia.

Problems involved in using inappropriate in vivo models for testing drug sensitivity may be overcome by culturing trypanosomes in vitro, allowing precise detection of intrinsic sensitivity of all stages in the life cycle of trypanosomes. The use of simple in vitro assays using feeder layer-free in vitro systems may help to obtain rapid information on the susceptibility of isolated trypanosome strains to the drug concerned. However, based upon the ability of these assays to predict potential drug efficacy in vivo, not all fresh isolates or clones of trypanosomes can be grown in feeder layer free systems. Thus, a combined mammalian feeder layer-trypanosome culture system may make it possible to determine different effects of a compound on host cells (general toxicity) versus parasites (selective toxicity). Calcium antagonists of several chemical classes including verapamil, cyproheptidine, desipramine and chlorpromazine, alone and in combination with various trypanocidal drugs (suramin, diminazene and others), were unable to reverse resistance in *T. evansi* to any of the trypanocides tested in vitro. These results are in contrast with those occurring in *T. cruzi*, *Plasmodium*, *Leishmania* and cancer cells, in which calcium antagonists have successfully reversed resistance.

Use of Drugs in the Field to Control Cattle Trypanosomiasis

Current limitations on drug efficacy are due to the occurrence of trypanosomes showing multiple drug tolerance to several drugs with close chemical relationship. This has been true also for a *T. v. vivax* strains first reported from Kenya in 1985. Thus a 'cocktail' of 11 *T. v. vivax* isolates has proved resistant to all drugs on the market, e.g., to isometamidium chloride (2 mg/kg b.w.), diminazene aceturate (3.5 mg/kg b.w), homidium chloride (2 mg/kg b.w), and quinapyramine sulfate (5 mg/kg b.w.). This finding appears to have implications of considerable importance to East African cattle producers. The ability of *T. v. vivax* to cause a hemorrhagic syndrome has also been discussed. Hemorrhages apparently do not occur in all cases and not in all stages of the disease. This form of trypanosomiasis can be acute or peracute and is responsible for severe losses in unprotected stock (for more information on hemorrhagic *T. v. vivax* infections cf. Use of Drugs in the Field to Control Cattle Trypanosomiasis).

The main problem in chemotherapy and chemoprophylaxis is to control the widespread cross-resistance in trypanosomes to the few drugs on the market (Table 1). Resistance to a drug, which has developed as a result of previous exposure of trypanosomes to a different drug of the same series or to a drug of an unrelated series can only be effectively controlled by using drugs that do not induce resistance to each other. If this is the case, then they can be used alternately when resistance to either drug appears in the field. Already in the early 60s significant knowledge of cross-resistance patterns was obtained from studies of large numbers of cattle maintained under controlled field condition in East Africa. Insufficient response of trypanosomes to certain prophylactic and curative drugs at recommended doses led to the strategic use of 'sanative' pairs of drugs in the field. Such drug pairs include homidium/diminazene and isometamidium/diminazene, which show no cross-resistance although quinapyramine-resistant trypanosomes confer resistance to each of these drugs. Moderate side-resistance may also be present between homidium, pyrithidium and isometamidium, which belong to the same chemical class of phenanthridines. Increased doses of isometamidium (1–2 mg/kg b.w.) may, however, control resistance to homidium and pyrithidium.

In curative field programs homidium may be used until evidence of resistance appears. It should then be replaced by diminazene, which generally controls infections in cattle reinfected with homidium-resistant parasites. Homidium may be used again after a year or so. Isometamidium and diminazene may be used alternately in prophylactic field programs. However, the appearance of drug-tolerant strains is believed to be inevitable in these programs, particularly in high-risk areas where isometamidium chloride is used at the standard dose of 1 mg/kg b.w. every 3 months. This dose may protect cattle against trypanosomiasis for 6–12 weeks if tsetse fly challenge is not too high. Higher dose levels can cause local reactions, a problem, which is common to all prophylactic drugs currently used (for comments see Table 1).

Control of the disease has been maintained when quarterly prophylactic injections with isometamidium were supplemented by block treatment with diminazene at regular intervals, i.e., every 6 months, 1 month prior to routine treatment with isometamidium. 'Sanative' diminazene will not control the situation if the challenge becomes too high as a result of increasing rates of reinfection with resistant trypanosomes. *T. v. vivax* and *T. c. congolense* strains, which survive isometamidium doses of 1 mg/kg b.w. and are cross-resistant to homidium can be controlled, however, by repeated administration of diminazene aceturate at a dose of 7 mg/kg b.w.

Field observations on the **stability of drug-resistance** in trypanosomes undergoing cyclical transmission are contradictory. Some observations suggest that drug resistance is stable and transmissible, while other investigators have assumed that drug-resistance in a trypanosome population is transient in the absence of drug pressure and infected cattle. In a series of experiments drug tolerance to curative doses of trypanocides was shown to be of stable nature, while *T. v. vivax* and *T. c. congolense* were transmitted through tsetse and cattle. However, it was assumed, that in the field competition between resistant and sensitive parasites in the trypanosome population might lead to an advantage for sensitive forms resulting in a gradual disappearance of drug-resistant parasites.

Pharmacokinetics of Trypanocides and Chemical Residues in Edible Tissues and Milk

There has been increasing public health concern about the consumption of trypanocidal drug residues in foods. A survey conducted recently in central Kenya has shown significant quantities of trypanocides in cattle meat from various slaughterhouses. Previously, the phenantridines (Table 1) **isometamidium** (for MRLs see Table 1) and quinapyramine have been believed generally to maintain trypanocidal blood concentrations for longer periods than diminazene. This led to the assumption that storage in and release from deep compartment are due to a process different from that occurring with diminazene. Thus, diminazene has been found to have only a limited prophylactic effect, and patent parasitemia has often been detected 2 weeks after treatment. This indicates that diminazene may be rapidly removed from bloodstream if given as the readily water-soluble aceturate. In contrast, the virtually water-insoluble diminazene dihydrochloride (or embonate) yielded in rats a fairly long protection period of 56–70 days at subcutaneous doses of 1x 16.5 and 1x 33 mg/kg b.w. against a high challenge of *T. b. rhodesiense* and *T. b. gambiense*. It was also demonstrated that diminazene diaceturate was "rapidly" removed from the plasma in mice whereas its tissue concentration remained relatively high for several weeks. In rhesus monkey (*Macaca mulatta*) the elimination of diminazene aceturate (single intramuscular dose of 20 mg/kg b.w.) occurred in two phases with half-lives of 2.1–2.7 h and 15.5–23.3 h; the protection period against a high challenge of *T. b. rhodesiense* was 21 days. Similar biphasic elimination of the drug was observed in rabbits after intramuscular injection of 3.5 mg/kg b.w. Seven days after treatment 40%–50% of the dose had been excreted in the urine and 8%–20% in the feces; the highest diminazene residues were found in the liver and corresponded to 35%–50% of the dose given.

Pharmacokinetic studies in cattle provided further evidence for the validity of a two-compartment model in the case of diminazene; there were a biphasic profile and two phases of distribution. Pharmacokinetic properties of **diminazene diaceturate** [bisphenyl-$U^{14}C$] (3.5 mg/kg b.w. i.m.) were investigated in healthy calves. Levels of radioactivity were determined in the blood, plasma, urine, feces, and edible tissues. There was a rapid onset of absorption, which led to high blood, and plasma levels (4.6 nEq/ml). The decrease in concentration followed a biphasic process with half-lives of 2 and 188 h; 20 days after administration 72.2% and 10.3% of the dose had been excreted in the urine and feces, respectively. The main product in urine was unchanged diminazene. Radioactivity could be detected in blood and plasma for up to 20 days after administration. Distribution studies revealed low concentrations in edible tissues, particularly in skeletal muscle and fat. From these results it was concluded that diminazene is not as rapidly and entirely metabolized (or biotransformed) in the body as suggested previously. Following the results a preslaughter withdrawal time of 21 days for all edible tissues (also liver) was recommended for cattle. A similar long preslaughter withdrawal period (14–20 day) was estimated for sheep after a single intramuscular dose of diminazene 3.5 mg/kg. Drug concentrations were determined in plasma and equilibrium dialysis and high-performance liquid chromatography. As expected, dairy goats that had received two successive intramuscular doses of diminazene aceturate 2 and 3.5 mg/kg b.w. showed somewhat different pharmacokinetics from those that received a single injection. The estimated preslaughter withdrawal period was between 28 and 35 days. Dairy cows repeatedly infected with different strains of *T. congolense* and treated with different dose of radiolabelled diminazene aceturate have been investigated for dependence of drug residue levels in milk. Results of this study indicate that the degree of parasitemia (anemia) affects the distribution, disposition, and elimination of diminazene. At 3.5 mg/kg b.w. 0.4% of the dose was excreted in milk after 21 days, while 0.54% of the 7 mg/kg b.w. dose was excreted during the same time. On the basis of data of half-lives for the second phase (elimination phase) milk from treated animals should not be consumed for at least 3weeks posttreatment. In rabbits treated with a single intramuscular dose of [^{14}C] **homidium** bromide 1 or 10 mg/kg b.w. blood and tissue levels reached a maximum within 1 h then fell rapidly. After 4 days 80%–90% of the radioactivity injected had been excreted, 33% in the urine and 66% in the feces. In view of the rapid rate of drug excretion it was assumed that the time and level of infection relative to the time of drug administration might markedly affect the protective action of ethidium. Some doubt must therefore remain about the value of this drug for the prophylactic treatment of slaughter cattle as recommended previously.

Changes in the Field of Animal Trypanosomiasis over the Past 40 Years

Based on a review of the literature with special reference to control of animal trypanosomiasis in Africa, the conclusion was that each of the control methods in use has serious limitations. This appears to be true also for the present situation. Today and in the past, the effectiveness of chemoprophylaxis and chemotherapy has been reduced markedly by the widespread development of drug resistance. The enormous cost involved in research on and development of new drugs, which industry has to consider, have meant that very little research on potential trypanocides has been done. Thus, 20 years ago some pharmaceutical companies were still involved in research on the chemotherapy of trypanosomiasis, despite the financial considerations of the relatively small market and uncertain financial returns. Economic and ecological constraints on trypanosomiasis control are still evident, and the high cost involved in a continuing program of eradication of tsetse flies, or even of isolated tsetse-belts, is often beyond the reach of individual countries. Furthermore, tsetse fly clearance remains an unreliable control measure when continued surveillance is not guaranteed, and tsetse fly eradication will not necessarily result in the eradication of trypanosomiasis since *T. v. vivax* and *T. b. evansi* can cause infections without cyclical transmission and can be spread mechanically by biting dipterans. On the other hand, attempts to eradicate tsetse flies by chemical control, e.g., by massive aerial insecticide spraying, are always associated with considerable adverse effects on environment.

Biological and genetic control methods are still at an early stage of development (as 30 years ago) and control of trypanosomiasis by immunological tools will only be achievable on a long-term basis. It seems likely that research into the response of trypanotolerant cattle will be of special value although host-parasite relationships and thus, trypanotolerance are still poorly understood. All these limitations would be less important if the present control measures were used in an integrated control program; the importance of international cooperation in combating trypanosomiasis should be stressed. However, the problems encountered in the organization of control programs differ greatly according to whether the method of control is directed against the parasite or the vector tsetse fly. Campaigns directed against the vector are much more a matter of straightforward organization, logistics, and cost (e.g., considerations of the economic return from development after tsetse fly eradication) than are those which involve the attack of infections of the vertebrate host using curative or prophylactic drugs. Perhaps the most valuable use of trypanocidal drugs is in the development of cattle rearing and production in areas where tsetse fly eradication cannot be achieved in the near future. In such areas, conditions can gradually be created under which operations against tsetse flies may be undertaken. Thus, attention must be drawn to one of the chief remaining constraints on the improvement and multiplication of trypanotolerant livestock, i.e., the relatively low reproductive performances of certain cattle breeds under traditional management systems.

Trypanocidal Drugs, Man

Table 1

Drugs Acting on African Trypanosomiasis (Sleeping Sickness) of Humans

In humans, *Trypanosoma brucei gambiense* and *T. b. rhodesiense* cause African trypanosomiasis

Table 1. Drugs used against trypanosome infections of humans

PARASITE, DISEASE distribution	**Stages** affected (location), comments	**Chemical class** other information	**Nonproprietary name** adult/* pediatric dosage (various routes), comments	**Miscellaneous comments**
AFRICAN TRYPANOSOMIASIS (SLEEPING SICKNESS): epidemic/endemic in a belt across central Africa south of the Sahara Desert and transmitted by the bite of infected tsetse flies (*Glossina* spp.); in tsetse, trypomastigotes transform into epimastigotes, which divide during a complicated migration in the fly and then transform into metacyclic trypomastigotes infective for man and **reservoir animals** (*T. rhodesiense*); chemotherapy in patients with CNS involvement (sleeping sickness) is generally problematic because of potential, severe side effects (e. g. fatal encephalopathy) caused by arsenicals				
Early phase of disease *Trypanosoma brucei gambiense* (West and Central Africa, probably only human reservoir) *T. b. rhodesiense* (East Africa, zoonotic infection, reservoir hosts: antelope, hartebeest cattle, lion, hyena, and others	**trypomastigote** (extracellular: early hemolymphatic stages and later stages in interstitial spaces) Gambian disease (chronic with low parasitemias; incubation time: months to years) Rhodesian form (acute with high parasitemias; incubation period: days to weeks)	**sulfated naphthylamines** (anionic urea compound) (does not pass blood-brain barrier; high protein-binding activity) is also used to clear blood of trypanosomes prior to treatment with **melarsoprol** (CNS involvement)	**suramin** sodium (drug of choice, test dose: 100-200 mg slowly i.v., then 1 gram iv. on d1, d3, d7, d14 and d21) very effective in treating early cases of sleeping sickness (Gambian and Rhodesian forms)	side effects may be shock, loss of consciousness (rare), urticaria, colic, heavy proteinuria, severe toxic effects on kidney (degenerative changes, contraindication for use of drug: renal diseases), peripheral neuropathy as paresthesia, hypoesthesia, agranulocytosis and hemolytic anemia (rare)
Early phase of disease *T. b. rhodesiense* *T. b. gambiense*	**trypomastigote** (extracellular: early hemolymphatic stages and later stages in interstitial spaces)	**aromatic diamidines** (does not pass blood-brain barrier) used against suramin resistant strains and to clear blood of parasites prior to treatment with melarsoprol in patient with CNS involvement	**pentamidine** isethionate (alternative drug, 4mg/kg /d i.m. x 10d) (4mg/kg/d i.m.,d1 and d2:) before starting melarsoprol regimen in late cases of disease very effective in treating early cases of *T. b. gambiense* infections);	contraindication diabetes; drug can induce pancreatitis hypoglycemia and hyperglycemia); sharp fall in blood pressure after i.v. administration; renal dysfunction is reversible; abortion and peripheral neuritis (rare)

Table 1. (continued)

PARASITE, DISEASE distribution	**Stages** affected (location), comments	**Chemical class** other information	**Nonproprietary name** adult/* pediatric dosage (various routes), comments	**Miscellaneous comments**
Late phase of disease with CNS involvement *T. b. rhodesiense* *T. b. gambiense*	**trypomastigotes** (late stage patients with CNS involvement and parasites in cerebrospinal fluid; later stages in interstitial spaces hemolymphatic stages)	**melaminophenyl arsenicals** (trivalent cationic compound, highly toxic) (Arsobal, Rhône-Poulenc Rorer) (pass blood-brain barrier) used in late secondary CNS stages of (Gambian) sleeping sickness	**pretreatment** with either suramin or pentamidine on d1 and d2 to clear hemolymphatic stages **melarsoprol** (Mel B) (drug of choice, doses mg/kg/d, i.v. on days: (1.2) d4; (2.4) d5; (3.6) d6; (1.2) →d17; (2.4) d18; (3.6) d19, d20; (1.2) d30; (2.4) d31; (3.6) d32, d33 (for more details see WHO 1986 Technical Report Series No. 739)	most serious side effect is reactive encephalopathy (1–10 % of treated patients with mortality rates of 1–5 %; hospital supervision is necessary; fever, joint pain, renal damage, myocarditis, peripheral neuropathy, gastrointestinal disturbance, hypersensitivity, hypertension
Late phase of disease with CNS involvement *T. b. rhodesiense* *T. b. gambiense*	**trypomastigotes** (late stage patients with CNS involvement and parasites in cerebrospinal fluid; later stages in interstitial spaces hemolymphatic stages)	**neutral aromatic arsenicals** (pentavalent compound) highly toxic but less toxic than **atoxyl; tryparsamide** was used (1924-1950) in patients with late stage sleeping sickness on a wide scale in Belgian Congo and Cameroon	**tryparsamide** (alternative drug, one injection of 30 mg/kg (max.2g), i.v. every 5d to total of 12 injections; may be repeated after 1 month plus **suramin**: one injection of 10mg/kg i.v. every 5d to total of 12 injections; may be repeated after 1 month)	frequent nausea, vomiting; occasionally fever, impaired vision, optic nerve atrophy, tinnitus, allergic reactions, exfoliative dermatitis (in 1932, related arsenical **atoxyl** had caused blindness in 800 sleeping sickness patients after receiving a too high dosage regimen)
Late phase of disease with CNS involvement *T. b. gambiense* (in Rhodesian infections of humans DMFO is variably effective or ineffective (cause is not yet known)	**trypomastigotes** (hemolymphatic stages, preferably later stages in interstitial spaces and cerebrospinal fluid of late stage patients with CNS involvement	**substituted amino acid** (pass blood-brain barrier) (drug is available from WHO only (?), which holds 'license')	**eflornithine** (α-Difluoromethylornithine = DFMO) (Ornidyl, Aventis) (some clinicians have given 400 mg/kg/d i.v. in 4 divided doses for 14d, followed by 300 mg/kg/d p.o. for 3–4 wks)	drug still on clinical trial; good efficacy in late Gambian infections; frequent anemia and leukopenia; occasionally thrombocytopenia, seizure, diarrhoea, hair loss (rare)

Table 1. (continued)

PARASITE, DISEASE distribution	**Stages** affected (location), comments	**Chemical class** other information	**Nonproprietary name** adult/* pediatric dosage (various routes), comments	**Miscellaneous comments**
AMERICAN TRYPANOSOMIASIS (CHAGAS' DISEASE): zoonosis with an extensive mammalian reservoir (armadillos and opossums, some domestic animals and humans) is endemic in Central and South America, being found only in the American Hemisphere. *T. cruzi* may be transmitted to humans in two ways, either by blood-sucking infected reduviid, or directly by transfusion of infected blood (iatrogenic transmission); the vector bugs infest poor housing and thatched roofs; acute phase of disease is seen in children with and without acute clinical manifestations (all patients must be treated with a trypanocidal drug); lesions of chronic phase irreversibly affects internal organs such as heart, esophagus, colon and peripheral nervous system (treatment is indicated in recent chronic infection of children); chronic cases with established pathology appear to be unable to benefit from long-term treatment (approx. 60–90 days: hospitalization or careful monitoring may be needed)				
Trypanosoma cruzi dividing (A) forms produce pseudocysts; daughter (A) transform back to (T); these enter blood and may then reinvade various muscular tissue *T. rangeli* (appears to be non-pathogenic to man; only (T) forms in blood of humans, resemble *T.cruzi*	**trypomastigotes** (=T) **amastigotes** (=A) (T) in blood and (A) in: cardiac muscle, smooth muscle of gut, skeletal muscle *T. rangeli* differs from *T. cruzi* by longer and better developed undulating membrane, and small subterminal kinetoplast	**nitrofuran derivatives** (no longer readily available) addition of γ-interferon for 20 days may shortened acute phase of disease (RE McCabe et al., J Infect Dis 163: 912, 1991) mode of action may be production of free oxygen radicals enhancing oxidative stress on (T) + (A)	**nifurtimox** Lampit (Bayer)(drug of choice, 8–10 mg/kg/d orally in 3–4 doses x 90–120d * 1–10 yrs: 15–20 mg/kg/d orally in 4 doses x 90d * 11–16 yrs: 12.5–15 mg/kg/d orally in 4 divided doses x 90d)	side effects common (50 %): gastrointestinal complaints as anorexia, nausea, vomiting; vertigo insomnia, headache, peripheral neuritis, myalgia, arthralgia, neurological reactions: excitability; rare: convulsion,; rashes, pulmonary infiltrates and pleural effusion; side effects can lead to interruption of treatment
PRECAUTIONS Neither benznidazole nor nifurtimox should be given to pregnant women; in patients with illnesses associated with Chagas disease potential risk of severe adverse effects should be considered carefully	**trypomastigotes** (=T) **amastigotes** (=A) (T) in blood and (A) in: cardiac muscle, smooth muscle of gut, skeletal muscle	**2-nitroimidazoles** (drug of choice in Brazil because of fewer drug-tolerant strains than elsewhere), mode of action may be interactions of its metabolites with DNA	**benznidazole** Radanil (Roche) Rochagan (Roche Brazil) (alternative drug, 5–7 mg/kg/d orally x 30–90d *up to 12yrs: 10mg/kg/d orally in 2 doses x 30–90d	side effects common: immediate and frequent hypersensitivity reactions (rashes in 30% of treated patients), bone marrow suppression, psychic and GI disturbances, peripheral polyneuritis, leucopenia, agranulocytosis (rare); side effects can lead to interruption of treatment

The letter d stands for day (days)
Dosages listed in the table refer to information from manufacturer or literature, preferably from The Medical Letter (1998) 'Drugs for parasitic infections'. The Medical Letter (publisher), vol 40 (issue 1017): 1-12. New Rochelle New York
Data given in this table have no claim to full information. The primary or original manufacturer is indicated.

(sleeping sickness); the disease is transmitted by the bite of infected tsetse flies (*Glossina* spp.). These two subspecies are morphologically indistinguishable but differ in their pathogenicity and thus disease pattern. In animals, *T. b. brucei* and other *Trypanosoma* spp. cause the disease 'nagana' (→ Trypanocidal Drugs, Animals). *T. b. gambiense* infection is widespread in West and Central Africa whereas *T. b. rhodesiense* is restricted to the East and East Central areas with some overlaps between both species. In 37 countries of sub-Saharan Africa, 22 of which are among the least developed countries in the world, more than 55 million people are at risk of African trypanosomiasis.

Differences in host specificity are due to a non-immune killing factor (trypanosome lytic factor = TLF) in human serum causing lysis of *T. b. brucei* in vitro and in vivo. African sleeping sickness trypanosomes are resistant to this factor which is long known and may support chemotherapy. In the bloodstream of infected humans the trypanosomes grow and multiply extracellularly as long slender (LS) forms. After several divisions they transform into first intermediate (I) forms and then non-dividing short stumpy (SS) forms, which are infective for tsetse. The latter possess a functional mitochondrion. **Eflornithine** (DMFO, cf. → Trypanocidal Drugs, Animals/Table 1), interfere with the division process of LS forms and reduce hemolymphatic trypomastigotes and those in the central nervous system (CNS) (for more information see below and Table 1).

West African trypanosomiasis produced by *T. b. gambiense* is chronic in nature lasting up to 4 years. In the absence of chemotherapy, patients with Gambian infection become progressively more wasted and comatose. Involvement of (CNS) disorders, damage of the heart, and other organs generate the classical picture of sleeping sickness. Disease caused by *T. b. rhodesiense* is more acute and may last rarely longer than 9 months. Without treatment, death often occurs from toxic manifestations before CNS changes are evident. Because Rhodesian form is a zoonotic disease (reservoir animals) treatment of infected humans has less effect on incidence of infection in humans. In contrast, control of Gambian form and hence treatment of infected individuals relies mainly on surveillance of human population and direct field diagnosis by mobile teams. In general, cure rates are higher when infected individuals are treated in the early phase of the disease.

Only three drugs are available which may be used for the treatment of African trypanosomiasis (→ Trypanocidal Drugs, Animals/Search for New Drugs). **Suramin** and **pentamidine**, which were discovered in the first two or three decades of this century, are still used for clearing blood and the hemolymphatic system from trypanosomes in the early phase of the disease. The trivalent arsenical **melarsoprol** (Mel B), which crosses the blood-brain barrier, is used for the treatment of later CNS stages of the infection. The compound is derived from melarsen and its phenylarsenoxide with the melaminyl moiety in the *p*-position. It requires parenteral administration as other standard trypanocides. It may be extremely effective but highly toxic in all advanced CNS cases and need hospitalization and considerable care in its use (→ Trypanocidal Drugs, Animals). Melarsoprol exhibits a rapid lethal effect on trypanosomes in the CNS causing the so-called Herxheimer-Jarisch type of reactive encephalopathy in up to 10% of the treated patients with a mortality rate of up to 5%. Recent data on clinical pharmacokinetics led to an alternative regimen of melarsoprol, which could reduce its toxicity. The only 'new' drug developed for the treatment of sleeping sickness is **eflornithine** (DL-α- difluoromethylornithine = DMFO, Ornidyl, cf. → Trypanocidal Drugs, Animals/Search for New Drugs). In *T. b. brucei* polyamines are synthesized from ornithine, and DMFO is a highly inhibitor of ornithine decarboxylase (ODC), which catalyzes the decarboxylation of ornithine to yield putrescine and then spermidine. Polyamines play an important role in cell division and differentiation of eukariotic cells. Depletion of putrescine (and thus spermidine) leads to inhibition of the transformation of the LS form to the SS form and therefore to inhibition of trypanosome growth. When administered in drinking water, DMFO selectively blocks multiplication of the parasites and eliminats the infection. It was shown to have activity against CNS *T. b. brucei* infections in rodents, and has proved to be an effective treatment for late stage infections of *T. b. gambiense* in humans, even in arsenical-refractory CNS patients. Although eflornithine is highly active against Gambian trypanosomiasis its use is very limited (available from WHO). Clinical trials with the drug have been performed and are going on to evaluate its trypanocidal efficacy and systemic toxicity in patients with Gambian sleeping sickness. Dosage regimen, which had successfully been used, was 400 mg/kg/day intravenously in 4 divided doses for 14 days, followed by oral treatment with 300 mg/kg/day for 3–4 weeks. The drug may have some 'minor' drawbacks such as variable activity against *T. b. rhodesiense* infections and the need for high parenteral doses in late cases which makes treatment management costly. A combination of DMFO and suramin is on trial for the treatment of CNS involved Rhodesian infections.

The current regimen for the treatment of early *T. b. rhodesiense* and *T. b. gambiense* infections is **suramin**. In late stage Rhodesian infections, patients are treated first with suramin to clear the blood and lymph from parasites and then with multiple injections of **melarsoprol**. If there is tolerance to suramin **pentamidine** isethionate is an alternative drug for the treatment of early (primary) *T. b. gambiense* infection. It is also used in late (secondary) stage patients with CNS involvement to eliminate hemolymphatic trypanosomes prior to administration of melarsoprol (for route and dosage cf. → Trypanocidal Drugs, Humans/Table1). In addition, it is used in prevention and control projects (FAO/WHO initiative initiated in 1993, now PAAT = program against African trypanosomiasis) in epidemic/endemic regions of Angola, Cameroon, Central African Republic, Congo, Gabon, Equatorial Guinea, Uganda, Sudan, Chad and Zaire.

Retrospective long-term study with the diamidine, **diminazene** aceturate (Berenil), by follow-up of 99 human patients with early-stage disease of sleeping sickness showed that there was satisfactory efficacy after the parenteral route, and side effects were no more serious than those produced by suramin. However, the drug exhibits reduced activity after oral administration because its extensive hydrolysis in stomach results in two metabolites: one is 4-amidinophenyl-diazonium chloride, which exerts distinct trypanocidal activity, the other is 4-aminobenzamidine dihydrochloride, which proves ineffective against *T. b. brucei*.

Drug Acting on American Trypanosomiasis (Chagas Disease) of Humans

Trypanosoma cruzi is the causative agent of Chagas disease (American trypanosomiasis) and occurs only in the Western Hemisphere from the Central American countries in the North to the Andean countries and Southern Cone countries in the South. Uruguay was certified free of vectorial and transfusional transmission of Chagas disease in 1997 (WHO, CTD homepage January 1999). The disease affects 16–18 million people and some 100 million, i.e. about 25% of the population of Latin America is at risk of acquiring Chagas disease. Rural migration to urban areas changed the traditional epidemiological pattern of Chagas disease; it became an urban disease, as unscreened blood transfusion created a second way of transmission. *T. cruzi* is primarily an intracellular parasite (amastigote stages) occurring as pseudocysts in cardiac and smooth muscle cells, glial cells of the brain, and mononuclear phagocytes. However, immediately after the infection of the host by the reduviid bug (various species *Rhodnius, Triatoma, Panstrongylus*) first trypomastigote stages circulate in the bloodstream. During this acute phase of infection, which may last up to 60 days, trypomastigote forms and can be detected by direct examination of peripheral blood (by wet smear or after staining) along with the detection of IgM anti- *T. cruzi* antibodies. All patients suffering from acute Chagas disease must be treated since cure rate (parasitological and serological) in the acute phase of infection may be 50–60% only. Positive serological reactions in children 6 months after birth are indicative of congenital transmission and xenodiagnosis or hemoculture may corroborate the serological finding; specific treatment should be started immediately. Treatment is indicated in recent chronic infections, especially in all children with positive serological reactions whose infection occurred a few years (<10) ago. Patients (selected cases) with the indeterminate form, slight cardiac form and digestive form may be treated as well. To assure the free passage of the chemotherapeutic drug and its absorption, symptomatic treatment of dysphagia is recommended in cases with megaesophagus. However, it is not clear whether etiological intervention will stop the progression of the disease in chronic patients. Operational problems should be envisaged in the majority of chronic infections because long-term treatment demands thorough and proper follow-up of adverse effects caused by chemotherapy. Reactivation of Chagas disease may occur in immunosuppressed patients (e.g., HIV infection, AIDS related infections, or any organ transplantation) but clinical manifestations usually differ from those of the acute phase. For this reason, adequate monitoring of a potential *T. cruzi* infection should be done. In cases where parasitological reactivation is evident, chemotherapy should be envisaged though the risk of severe side effects may increase, or chemoprophylaxis has been suggested for these patients.

Current treatment is based on the nitrofuran, **nifurtimox** (no longer readily available), and the 2-nitroimidazole, **benznidazole**. Both drugs are administered orally for prolonged time (dosage and side effects cf. → Trypanocidal Drugs, Humans/Table 1). Successful treatment requires either hospitalization or careful monitoring of the patient. There may be marked variation in the drug response of different *T. cruzi* strains. Therefore, in the acute phase of infection, cure rates with both drugs are not total and vary regionally. There is limited or no action of both drugs on the chronic phase of the disease. However chemotherapy is essential in immunocompromised patients showing meningo-encephalitis.

Allopurinol (structural analogue of hypoxanthine), a drug used to decrease the excessive amounts of uric acid in the blood caused by gout and other metabolic disorders, may be a low cost, non-toxic alternative for treatment of *T. cruzi* infection. However, results from clinical trials indicate some doubt of its efficacy. The drug is taken up by the purine salvage pathway and prevents formation of ATP and also interferes with nucleic acid synthesis (for its action in animal models cf. → Trypanocidal Drugs, Animals/Table 1 and that on *Leishmania* spp. cf. → Leishmaniacidal Drugs/Table 1).

Sterol biosynthesis inhibitors such as the antifungal azoles, **ketoconazole** and **itraconazole** and the more recently investigated D0870, the *R* (+) enantiomer of ICI 195,735 continue to be potential chemotherapeutic agents against Chagas disease, especially at the chronic stage. However, previous studies in humans have failed to corroborate the action of ketoconazole on *T. cruzi* reported from animal models (for more information on chemotherapy including regimen and adverse effects of current drugs see → Trypanocidal Drugs, Humans/Table 1).

For the control of vector-transmitted infection through triatomid bugs (popular names: 'vinchuca', 'barbeiro', 'chipo' and others) the objective is to interrupt transmission of *T. cruzi* by insecticide spraying, insecticidal paints, fumigant canisters, housing improvement and health education in rural and suburban areas. The basic strategy includes house spraying with modern pyrethroids such as deltamethrin, cyfluthrin or λ-cyhalothrin, followed by long-term, community-based surveillance designed to report any 'residual' infestation that then can be selectively treated. For the control of blood-transmitted infections the goal is to screen all blood donors from endemic countries for *T. cruzi* antibodies (including HIV and hepatitis B). **Gentian violet** (Aksuris, Oxiuran, Viocid) toxic and mutagenic to mammals is still used as a disinfectant against *T. cruzi* in blood samples of potential infected donors. In blood banks of urban areas, transmission of *T. cruzi* by blood transfusion remains a major problem. Between 1960 and 1989 the prevalence of infected blood in blood banks in selected cities ranged from 1.7% in Sao Paulo, Brazil, to 53% in Santa Cruz, Bolivia, a percentage far higher than that of hepatitis or HIV infection (WHO homepage January 1999).

Trypanoplasmosis of Fish

→ Trypanoplasma (Vol. 1).

Trypanosomiasis, Rhodesian

→ *Trypanosoma brucei rhodesiense* (Vol. 1) infection occurs in East and Central Africa. It is characterized by a rapidly progressive clinical course and often leads to death within months, in the third stage of the disease, without pronounced central nervous system involvement. The early course of the disease, chancre and parasitemia, is similar to the Gambian disease. Fever and lymph node enlargement are especially prominent, often with myocarditis, weakness, weight loss, preceding death. Myocarditis is characterized by the presence of lymphocytes and plasma cells, and is accompanied by trypanosomes in myocardial cells and by pericardial effusion.

Therapy

→ Trypanocidal Drugs, Man.

Trypanosomiasis, Animals

Synonym

Trypanosomosis

General Information

Trypanosomosis is a cardiovascular system disease (→ Cardiovascular System Diseases, Animals) which results from infection with protozoan parasites of the genus → *Trypanosoma* (Vol. 1). With the exception of *T. equiperdum*, the cause of dourine, they are all transmitted by haematophagous arthropods, such as → kissing bugs (Vol. 1) for *T. cruzi* (the cause of → Chagas' disease in South America), biting flies for *T. evansi* (the cause of → Surra in Asia, central America and North Africa), or bloodsucking flies of the genus *Glossina*, for *T. congolense*, *T. vivax* and *T. brucei* (the cause of → Nagana in Africa).

Therapy

→ Trypanocidal Drugs, Animals.

Trypanosomiasis, Man

The following → trypanosomes (Vol. 1) commonly have been found in humans: *T. brucei gambiense* and *T. b. rhodesiense* (→ Sleeping Sickness), *T. cruzi* (→ Chagas' Disease, Man), and T. *rangeli*. The first three produce disease; the last gives rise to asymptomatic infection in Latin America and must be distinguished from the less frequent, but pathogenic *T. cruzi*.

Since *Trypanosoma cruzi*, restricted to America, is a mainly intracellular protozoan parasite, while the African trypanosomes (*T. brucei* sp.) live extracellularily, the immune defense mechanisms against these parasites as well as the mechanisms of pathogenesis are distinct (→ Chagas' Disease, Man/Immune Responses, → Sleeping Sickness/Immune Responses).

Therapy

→ Trypanocidal Drugs, Man.

Tsutsugamushi Fever

→ Mite (Vol. 1)-transmitted disease in humans (→ Orientia)

Tularemia

Tularemia, caused by the bacterium *Francisella tularensis*, has a → reservoir (Vol. 1) in lagomorphs and some rodents from which it can be transmitted through direct contact, such as when rabbits are skinned, or via the bite of ticks and tabanids. Mortality is low. Antibiotics such as streptomycin can be used for treatment. *D. andersoni* (in the USA), and other tick species such as *Rhipicephalus sanguineus* can transmit the pathogen. It is found in North America, reaching as far south as Venezuela, and is also found in parts of Asia and Europe.

Tumbu-Fly Disease

Myiasis due to the larvae of the fly *Cordylobia anthropophaga* (→ Myiasis, Man).

Tumor Necrose Factor

Synonym

TNF, α-TNF (Cachectin) and β-TNF (Lymphotoxin α) belong to the cytokines which regulate the immune reactions of the host.

Tungiasis

Tungiasis is an infection of the epidermis, usually the foot, with sand fleas (→ *Tunga penetrans* (Vol. 1)), usually acquired by walking barefoot on moist sand or soil in the tropics. The parasitic stage is a mated female flea which attaches to the skin and feeds on blood, enlarging in size up to about 1 cm in diameter when eggs develop. The flea becomes surrounded by keratin except for the head. which extends into the papillary dermis. The whole tumor is surrounded by chronic inflammatory reaction including eosinophils. The flea sucks blood and has a digestive tract which ends superficially where the keratin layer is incomplete. Large numbers of eggs develop in the flea and are discharged to the outside. The flea dies when it has discharged all of its eggs, its body collapses, and an intraepithelial abscess develops which eventually drains and heals. The flea larvae live in the soil. After pupation, mating occurs, the male flea dies after copulation, the place of which is not yet fully understood, and the female flea attaches to the foot of any of a number of hosts and burrows into the keratinaceous layer.

Therapy

→ Control of Insects, → Insecticides, → Arthropodicidal Drugs, → Ectoparasitocidal Drugs, see also Siphonapteridosis.

Tungidosis

Synonym

→ Tungiasis.

Turning Disease

→ Coenurosis, Animals, → Coenurosis, Man.

Typhus exanthematicus

Synonyms

Louse Borne Typhus, Spotted Typhus

Human disease due to infection with *Rickettsia prowazeki* transmitted via feces of lice.

U

Uitpeuloog

Infection with → *Gedoelstia* larvae (Diptera) leading to thrombo- and endophlebitis in animals, as well as to the bulging eye disease of cattle (→ Cardiovascular System Diseases, Animals).

Ulcer

Clinical and pathological symptom of infections with skin parasites (→ Skin Diseases, Animals, → Ectoparasite (Vol. 1)).

Ulcerating Lesions

→ Pathology, → Skin Diseases, Animals.

Unthriftiness

Clinical symptom in animals due to parasitic infections (→ Alimentary System Diseases, → Clinical Pathology, Animals).

Urinary System Diseases, Animals

Parasitic infections of the urinary system are not of major importance in domestic animals. Clinical signs are generally absent, or not seen, and the presence of the parasites is often only detected incidentally through the observation of sporocysts, eggs or fragmented worms in the urinary sediment. Four nematodes (*Capillaria plica*, *C. feliscati*, *Dioctophyma renale* and *Stephanurus dentatus*) and one protozoan (*Klossiella equi*) are parasitic in the urinary system of domestic animals (Table 1). In addition, *Toxocara canis* in dogs and *Schistosoma mattheei* in cattle, may occasionally invade the urinary system and lead to the formation of small granulomas around the larvae or eggs. Haemoglobinuria (or "→ red water"), a clinical sign which may be caused by → *Babesia* (Vol. 1) multiplying in erythrocytes, is not associated with the presence of parasites in the urinary system.

Klossiella equi is found in the convoluted tubules of the kidney in the horse and its relatives. These sporozoan parasites cause a destruction of renal epithelial cells, and histopathological examination of the kidneys may reveal an interstitial nephritis in some animals. However, the infection produces no clinical manifestation of disease.

The trichurid nematodes *Capillaria plica* and *C. feliscati* live in the urinary bladder and occasionally the ureters and renal pelvis of dogs and cats, respectively (see volume 1). They are about 3–6 cm long and produce little or no pathogenic effect, probably because of the superficial attachment of the worms to the epithelium of the urinary bladder. However, the parasites may occasionally invade the mucosa and cause an inflammatory reaction, leading to haematuria, dysuria and pollakiuria.

The giant kidney worm *Dioctophyma renale* occurs in the kidneys and peritoneal cavity of the mink and dog and other fish-eating mammals, but has also occasionally been found in the pig, cattle, horse and man. In the dog, adult worms have primarily been found in the peritoneal cavity, which suggests that it is not a natural definitive host of the parasite. *D. renale* is the largest known nematode. Female worms may reach up to 100 cm in length and the pathogenesis of the infection is related to the considerable space that the parasite takes inside the renal pelvis. The enlargement of the cavity occurs at the expense of the parenchyma, to such an extent that the kidney may even-

Table 1. Parasites affecting the urinary system of domestic animals (according to Vercruysse and De Bont)

Parasite	Host	Location	Clinical presentation	Principal lesions
Protozoa				
Klossiella equi	Horse	Kidney	None	Interstitial nephritis
Trematoda				
Schistosoma spp.	Ruminants	Kidney, bladder	Haematuria	Granuloma formation around eggs
Nematoda				
Capillaria feliscati	Cat	Bladder, sometimes the pelvis of the kidneys	Mostly asymptomatic, in heavy infections haematuria, dysuria and pollakiuria	Light inflammation of bladder mucosa
C. plica	Dog, fox, wolf			
Dioctophyma renale	Dog, mink, cat, fish-eating mammals, rarely horse, pig, **man**	Free in abdomen, kidney	Asymptomatic when one kidney is involved, otherwise uremia may occur	Kidney parenchyma (usually right) is destroyed, often only the capsule being left
Stephanurus dentatus	Pig	Perirenal fat and adjacent tissues	No typical signs, weight loss	Adults: cysts, filled with greenish pus

tually be reduced to an empty capsule. Partial obstruction of the ureter may also take place, leading to hydronephrosis. The severity of clinical disturbance and outcome of the infection depend on the ability of the host to maintain a normal renal function. In dogs, the right kidney is more frequently invaded than the left. Such unilateral infection generally leads to a compensatory hypertrophy of the opposite kidney, and does not cause clinical disturbance. When both kidneys are parasitized the animal may die of uraemia without sufficient time for extensive pathologic changes to develop in the kidneys. Although the presence of adult worms in the abdominal cavity has occasionally been associated with ascites and haemorrhages, it does not usually cause clinical signs.

Stephanurus dentatus, the kidney-worm of swine, is generally found in the perirenal fat and adjacent tissues. About 2–4 cm in length, these worms form cystic cavities that communicate with the renal pelvis and allow the eggs to be passed in the urine. Most of the pathogenesis of this tropical and subtropical parasite is related to its larval stages. They are particularly aggressive in the liver where their migration causes severe inflammatory reactions and eventually results in extensive portal fibrosis. The adult parasite encysted near the ureters is not markedly pathogenic, although thickening of the ureters and cystitis have occasionally been reported. Aberrant migration of *S. dentatus* to other abdominal or thoracic organs where it causes local purulent tissue reaction, appears to be a common feature. As in many helminthosis of the pig, the major clinical sign in most infections is a failure to gain weight.

The urinary form of schistosomiasis in domestic animals appears to occur only in cattle heavily infected with *Schistosoma mattheei*, a parasite of the mesenteric and hepatic portal veins of ruminants in southern Africa. The lesions in the urinary bladder may range from scattered individual granulomas with petechia to widespread polypoid and granular patches. Urinary manifestations such as haematuria are rare and seen only in super infected animals.

Therapy

See entries of the different species

Urticaria

Skin reaction to previous bites of many bloodsucking insects and mites.

V

Vaccination

Long before the parasites and the causes of parasitic diseases were known, it was observed that if an individual recovered from a disease, rather than succumbing to it, he rarely developed the same illness again. It must have been these types of observations that led people in Asia some hundreds of years ago to deliberately infect infants on the backside with infective material from *Leishmania* lesions (→ Leishmanization), thus inducing an immunity that protected immunized individuals from reinfections with disfiguring lesions (→ Immune Responses). After Edward Jenner in 1796 succeeded in immunizing a young boy with material from cowpox blisters against a challenge infection with smallpox, the first attenuated live vaccine (after "vacca," the Latin word for cow) was born. Up to now the vast majority of successful vaccines against virus, bacteria or parasites are attenuated live vaccines based on the principle that avirulent organisms can confer protection against virulent, disease-causing pathogens (→ Vaccination Against Nematodes, → Vaccination Against Protozoa, → Vaccination Against Platyhelminthes, → Amoebiasis, → Chagas' Disease, Animals, → Chagas' Disease, Man, → Trypanosomiasis, Animals, → Trypanosomiasis, Man, → Leishmaniasis, Animals, → Leishmaniasis, Man, → Malaria, → Toxoplasmosis, Animals, → Toxoplasmosis, Man).

Vaccination Against Nematodes

Vaccine research always has to compete with the development of anthelmintics. Whereas the latter increasingly cover a broad range of endo- and even ectoparasites, vaccines are efficient against only one parasite species. However, with increasing resistance of parasites against anthelmintics and ecological awareness about drug residues, vaccines become more and more important.

Irradiation Attenuated Live Vaccines

The first and undoubtedly most successful anti-nematode vaccine to date is the irradiation attenuated live vaccine against the bovine lungworm, → *Dictyocaulus viviparus* (Vol. 1). Following the observation that cattle who survived natural infections with *D. viviparus* larvae obviously developed a certain degree of immunity that protected them from clinical disease after a natural challenge, first immunization trials were carried out in the mid 50s. It was soon found out that crude preparations of somatic antigens from adult worms or larvae did not stimulate sufficient active immunity to prevent the disease. It was then assumed that the stimulation of protective immunity might depend on substances which were only elaborated by living worms. After intensive search for methods to weaken the infective larvae without killing them, X-ray irradiation proved to be the method of choice. Whereas irradiation below 200 Gy did not produce a sufficient degree of alteration of the larvae, 600 Gy considerably damaged the larvae which failed to stimulate a protective immunity. Eventually best results under both experimental and field conditions were obtained by 2 oral administrations of 1000 infective, 400 Gy attenuated third-stage larvae, each, 4 weeks apart. This vaccine was launched on the market in 1958 and has been commercially available ever since (Dictol®, Bovilis®, Intervet).

Vaccination with irradiated lungworm larvae does not confer a sterile immunity. During subsequent challenge infections animals are protected against clinical disease and worm burdens are reduced by 95–98 %, but low numbers of worms may reach maturity and produce new larvae. Immunity from vaccination protects for up to 12 months (similar to a single natural infection), i.e.

for the following grazing season. If no natural infections booster the immunity, animals are susceptible to disease again thereafter. Therefore vaccination is recommended for endemic *D. viviparus* areas, where natural boosts can be expected and life-long protection may be achieved.

Although calves may be vaccinated already from two weeks after birth the vaccine usually is administered to susceptible calves in early spring. The second immunization has to be given at least 2–3 weeks before turnout to pasture. Within 2 weeks after the last vaccination no anthelmintic treatment should be given.

As freezing kills the larvae and higher temperatures reduce their lifespan, the vaccine must be stored between 2 °C and 8 °C. Under these conditions the shelf-life is 3 months.

The mechanism of X-ray attenuation is unknown. It has been speculated that irradiation induces abnormalities in protein structure producing highly immunogenic molecules, which in their normal configuration are only weak immunogens. However, they still have to be common enough to native antigens to stimulate an immune response that is able to interact with antigens from normal organisms. Besides modified antigen conformation additional modes of action have been discussed, such as prolonged exposure of particular antigens, induction of specific cytokines and dynamics of cell-mediated responses.

The mode of action of the irradiated *D. viviparus* live vaccine and the relevant antigens have not been identified yet. Efforts have been made to identify protective antigens on a molecular level and will be described under "Recombinant vaccines".

The successful vaccination of cattle against *D. viviparus* inspired the initiation of a similar programme and the production of a commercial vaccine against *D. filaria* in sheep in India in 1971 and Iraq shortly thereafter. Infective, third-stage larvae are irradiated with 500 Gy. As with bovine lungworms, irradiated larvae confer high levels of immunity when given orally to young lambs in two doses of 1000 and 2000 larvae, 4 weeks apart. Without reinfection protection lasts between 12 and 24 months and varies between sheep breeds. The efficacy of the vaccine has been demonstrated under experimental and field conditions. Within the first ten years about 360.000 sheep in endemic areas had been successfully vaccinated, however the demand is much higher and by far exceeds the production capacity of the vaccine producers. The vaccine is also used for the immunization of goats. Irradiated larvae have to be maintained at 4 °C and have a shelf-life of only 2 weeks.

After irradiation proved to be a successful method to attenuate infective helminth larvae for live vaccines a number of immunization experiments with all kinds of nematodes were carried out. A second, although only temporarily available commercial vaccine was developed against the canine hookworm, → *Ancylostoma caninum* (Vol. 1) and released in the USA in 1973. After washing and sterilization of eggs collected from faeces, larvae were cultured in sterile medium, harvested by filtration and irradiated (400 Gy). Due to sterile culture conditions, a shelf-life of six months was achieved at 10–15 °C. The vaccine conferred a high degree of protection but no sterile immunity allowing single worms to become adult and produce eggs. Therefore, the veterinarians preferred the simultaneously introduced modern anthelmintics, which immediately eliminated eggs from faeces. This together with increasing production costs eventually led to the withdrawal of the vaccine in 1975. Trials with 5-fluorouracil or UV attenuated hookworm larvae also showed a considerable degree of protection but remained at an experimental stage.

Immunization experiments with attenuated infective larvae under both experimental and field conditions have been performed with a large number of nematodes (Table 1). Most of them never reached beyond an experimental stage and did not lead to a commercial vaccine. Problems arose with vaccination of young animals. Juvenile animals often are not yet immunocompetent at the time of vaccination or still have maternal antibodies that kill and eliminate vaccine larvae before they can stimulate immunity.

Subunit Vaccines

A major advantage of live vaccines is that during invasion, tissue penetration and development a whole range of antigens is presented and generally a solid protective humoral and cellular immune response is stimulated. However a major disadvantage is always that shelf-life is short and insufficient attenuation may lead to pathogenic effects and the spread of the parasite. Compared to attenuated live vaccines, subunit vaccines contain only a small number of defined antigens but are

Table 1. Studies on the protective potential of irradiated vaccines against nematodes (according to Schnieder)

NEMATODE	HOST	ATTENUATION
Rhabditida:		
Strongyloides papillosus	sheep	UV
Strongyloides ratti	rat	gamma
Strongyloides avium	chicken	x-ray
Strongylida:		
Strongylus vulgaris	horse	gamma
Oesophagostomum columbianum	sheep	gamma
Syngamus trachea	chicken, pheasant	x-ray, gamma
Stephanurus dentatus	pig	UV
Ancylostoma caninum	dog	x-ray, UV
Ancylostoma ceylanicum	hamster	UV
Bunostomum trigonocephalum	sheep	x-ray
Gaigeria pachyscelis	sheep	gamma
Amidostomum anseris	goose	x-ray
Haemonchus contortus	sheep	x-ray
Ostertagia circumcincta	sheep	UV
Ostertagia ostertagi	cattle	x-ray
Cooperia punctata	cattle	x-ray
Cooperia oncophora	cattle	x-ray
Trichostrongylus colubriformis	gerbil, guinea pig, sheep	gamma
Trichostrongylus vitrinus	guinea pig	gamma
Trichostrongylus tenuis	grouse	x-ray
Heligmosomoides polygyrus	mouse	x-ray
Nippostrongylus brasiliensis	rat	x-ray
Dictyocaulus viviparus	cattle	x-ray
Dictyocaulus filaria	sheep	x-ray
Metastrongylus apri	guinea pig	x-ray
Ascaridida:		
Ascaris suum	pig	UV
Toxocara canis	mouse	x-ray, UV
Ascaridia galli	chicken	gamma
Spirurida:		
Dirofilaria immitis	dog	x-ray
Onchocerca volvulus	chimpanzee	x-ray
Brugia pahangi	cat	gamma
Brugia malayi	gerbil	gamma
Litomosoides carinii	rat	x-ray, gamma
Enoplida:		
Capillaria obsignata	chicken	x-ray
Trichinella spiralis	in vitro, mouse	x-ray, UV

safe in that no viable parasites are administered and development and reproduction cannot occur.

One of the nematodes, where major research efforts have been undertaken and where a commercial vaccine seems to be not too far away is → *Haemonchus contortus* (Vol. 1), a sheep nematode that causes major economical losses in sheep breeding countries. Strategies for vaccine development against haemonchosis basically take two different approaches: (1) identification of "natural antigens" (or conventional antigens), that are presented to the host's immune system during natural or experimental infections and (2) search for "hidden antigens" (sometimes also called concealed, covert or novel antigens), that are extracted from internal parts of the parasite, mostly the gastrointestinal tract, and that are not "seen" by the host during the course of infection.

Natural antigens usually comprise excretory/secretory or surface antigens and in *Haemonchus* are mostly derived from the early infective larval stage (L3). In early immunization experiments whole L3 extracts and excretory/secretory (ES) antigens gave no significant protection, whereas a high molecular weight (> 30,000) fraction from both antigen preparations reduced wormburdens after challenge by 59 %. Immunization with a purified metabolite of exsheathed in vitro cultured L3 considerably reduced the egg production but not the number of worms. A new approach for the identification of natural protective antigens was reviewed by Newton. It is based on the observation that B-cells are rapidly recruited to the site of infection with a pathogen and to the local draining lymph node before they migrate into the target tissue and differentiate into antibody-secreting cells (ASC). It was therefore assumed that antibodies secreted by parasite activated B-cells are likely to recognize antigens important in rejection of the parasite by the host. Compared to serum antibodies, ASCs from immune animals, harvested during stimulation by the parasite from the lymph node draining the site of infection, recognized a much more limited and different group of antigens. ASCs isolated from sheep 5 days after a challenge infection with large doses of *H. contortus* specifically recognized L3/L4 antigens at approximately 44–48 kDa and a broad band at 70–83 kDa. These antigens did not react with ASCs from older infections (without challenge) whereas serum from both groups showed complex patterns of antigens, emphasizing the advantage of using ASCs compared to serum. The 70–83 kDa band proved to be a glycosylated L3 surface antigen. Immunization trials with this purified antigen showed a significant reduction in total faecal egg counts (FEC) in vaccinates of 54 % and 50 % in adult worm numbers. Although natural antigens described above conferred only about 50 % protection, efficacy could possibly be enhanced with optimal antigen presentation and different adjuvants.

A number of ES antigens from adult worms and L3 with protease activity have been described and some have been evaluated for their protective potential. Three apical gut surface proteins of adult *H. contortus*, p46, p52 and p100, were able to induce protective immunity to challenge infection in goats. All three proteins are encoded by a single gene, GA1, and initially expressed in adult parasites as a polyprotein, p100GA1. p46GA1 and p52GA1 are related proteins with 47 % sequence identity. GA1 proteins occur in the abomasal mucus of infected lambs, suggesting that they are ES antigens and possibly presented to the host immune system during infection.

In recent experiments partly purified low molecular weight antigens obtained by gel filtration of whole worm homogenates or total adult ES antigens were tested for their ability to induce protective immunity against *H. contortus*. Except for one animal low molecular weight fraction vaccinates showed a highly significant reduction of adult worms of 97.6 % and FEC reduction of 99.9 %. Vaccination with ES antigens conferred a lower protection of 63.7 % (adult worms) and 32.2 % (FEC). Further analysis of the ES fraction of adult worms revealed two immunogenic low molecular weight proteins of 15 and 24 kDa, the first not glycosylated the latter containing some glycosylation. Sheep immunized with either fraction showed more than 70 % reduction in worm burdens and FEC, respectively. Corresponding recombinant proteins were isolated from a *H. contortus* L5 cDNA library and expressed in *E. coli*. As mRNA encoding both ES products occurred only in the parasitic stages, expression appears to be developmentally regulated. Both recombinant antigens were recognized by sera from *H. contortus* hyperimmunized sheep, suggesting that antigenic determinants were also present on the recombinant proteins.

Whereas immunity against natural antigens is boostered by natural infections, immunity against hidden antigens is not. The hidden antigen strategy has been simultaneously developed for the cattle tick → *Boophilus microplus* (Vol. 1) and for the nematode *H. contortus*, both hematophageous parasites. The immunization of an animal against "Achilles heels" of the parasite such as surface proteins from the gastrointestinal tract or metabolic or detoxifying enzymes will result in the production of specific circulating antibodies. After ingestion of these antibodies by the parasite they bind to the corresponding antigen and disrupt its structure or function. Although this strategy is most effective in blood-feeders the ingestion of immunoglobulins has also been described for non-blood feeding nematodes such as *Ostertagia* and *Dictyocaulus*. The first hidden antigen described for *H. contortus* was a polymeric helical structure, associated with the surface of the intestinal epithelium, called contortin. Although immunization trials showed protection levels of >90 % purification proved to be difficult. With the discovery of H11 contortin was not further developed. H 11 (or H110D) is a heavily glycosylated 110 kDa integral membrane protein of intestinal microvilli of adult *H. contortus*. DNA and amino acid sequence analysis showed a high identity to mammalian microsomal aminopeptidases. Their function is the cleavage of dipeptide products of digestion to amino acids for transport across the plasma membrane. Therefore the efficacy of the vaccine is most likely based on the inhibition of enzyme activity and subsequent starvation of the worms for essential amino acids. Numerous immunization trials with different breeds of sheep showed that, unlike irradiated vaccines, H11 is effective in very young lambs already and average protection levels exceed 90 % in terms of FEC reduction and 75 % reduction in worm burdens. The higher effect on egg production is at least partly caused by the fact that the vaccine affects female worms to a greater extent than males. Additionally, surviving female worms are smaller than those from control animals and presumably produce fewer eggs. As the parasites have to suck blood before protective antibodies can be effective, PCV values fall after challenge, but less marked than in controls and animals do not show clinical signs of haemonchosis. Although the native H11 is highly protective and recombinant clones have been produced, there are no reports to date about successful immunization trials with the recombinants. It may be assumed that in vitro expression of the heavily glycosylated protein in both procaryotic and eucaryotic cells does not correctly produce the immunogenic epitopes. Aminopeptidases are functional proteins not only present in *Haemonchus*. Homologues have been described in *Ostertagia* spp. and will most likely be present in numerous nematodes.

During the first steps for the extraction of H11 two additional proteins are co-purified, H45 od also termed P1 and a *H. contortus*-galactose binding glycoprotein (H-gal-GP). P1 is a protein complex that comprises 3 protein bands on SDS-PAGE of about 53, 49 and 45 kDa. P1 is also localized at the surface of intestinal epithelium cells and has aspartyl proteinase activity. It is assumed that it contributes to the digestion of blood meals and may therefore be a promising vaccine candidate. Immunization trials showed only about 30 % reduction of adult worms and about 70 % egg counts. H-gal-GP is a complex of intestinal surface proteins that appear on non-reducing SDS gels as 2 main bands at 230 and 170 kDa and 2 faint bands at 47 and 50 kDa. It is not known whether one of these bands or the complex as a whole confers protection. Immunization with the whole protein complex showed an average FEC reduction of about 90 % and again a lower efficacy against adult worms with approximately 60 % reduction after challenge. Antibodies raised against the corresponding recombinant antigen bound to the luminal surface of the gut in adult *H. contortus*, but to date no data about protective potential of the recombinant antigen are available. The examples show that gut associated proteins of *H. contortus* proved to be a rich source for vaccine candidates and will be further evaluated.

Although non-blood feeding nematodes do also take up host IgG, to date only natural antigens have been examined for their protective potential in other trichostrongylid nematodes. A complex ES antigen preparation (molecular weight > 10,000) of adult *Cooperia punctata* showed various degrees of protection ranging from 16 to > 80 % reduction of worm counts in calves. Immunization trials with ES antigens from exsheathed *Trichostrongylus colubriformis* L3 in guinea pigs showed worm reductions between 0 and 74 %. Among different protective components an immunodominant glycoprotein of 94 kDa was identified.

In *Ostertagia circumcincta* a 31 kDa glycoprotein (GP31) was identified in secretory organelles within the cells of the oesophageal glands of L3. After in vitro cultivation GP31 was shown to be one of the major components of the ES complex. The purified GP31 had no detectable proteolytic activity in protein degradation assays, but homologues were found in *T. colubriformis* and *H. contortus* L3. Immunization of lambs conferred insignificant reduction in total worm counts and FEC. In immunized animals humoral and cellular immune responses could be detected.

Tropomyosin is a protein of muscle cells of evertebrates but can be found in different isoforms in non-muscle cells such as fibroblasts as well. A 41 kDa tropomyosin has been identified from a detergent-soluble fraction of *T. colubriformis* L3. Immunization of guinea pigs with the 41 kDa antigen induced 43–51 % protection in terms of reduced worm burden after challenge infection. The same antigen was isolated from *Acanthocheilonema viteae* and conferred significant protection (up to 65 % reduced adult worms, up to 95 % reduced circulating microfilariae) in jirds. Recombinant tropomyosin cDNA clones from *Onchocerca volvulus* have been used to vaccinate BALB/c mice against challenge infection with *O. lienalis*. Significant reductions (48–62 %) in the recovery of micrifilariae from the skin were achieved. Recombinant tropomyosin clones have also been prepared from *T. colubriformis* and *H. contortus*, however protectivity data have not yet been published.

Paramyosin, a filamental protein, is a substantial part of myosin filaments in muscle cells of many nematodes. It blocks the actomyosin binding in the contracted muscle cell to make possible a persistent contraction without energy consumption. → Paramyosin has been identified as a protective antigen and a major vaccine candidate in schistosomes but also in different nematode species. In filariid nematodes such as *Onchocerca volvulus*, *Dirofilaria immitis* and → Brugia malayi (Vol. 1) paramyosin clones have been identified. Immunization of mice with a native 97 kDa paramyosin homologue of *Brugia malayi* reduced microfilariaemia by 40–60 % after i.v. challenge with live *B. malayi* microfilariae. Vaccination of jirds with a recombinant *B. malayi* paramyosin fused with maltose-binding protein (BM5-MBP) stimulated protection against challenge infection. Adult worm recoveries (43 %) and female worm length (10 %) were significant, blood microfilaria counts slightly reduced compared to MBP vaccinated controls.

The early success and the introduction of a commercial live vaccine against *Dictyocaulus viviparus* in cattle seemed to paralyze the research on lungworm immunity and the search for alternative vaccine candidates for the following years. Because vaccination trials with cattle are expensive, most studies on protective antigens were done in a guinea pig model. However, it is well known that the immune response is different from cattle and antigens such as lyophilized worms that proved to be protective in guinea pigs may have no efficacy in cattle. McKeand and colleagues immunized guinea pigs with either somatic extracts of adult lungworms, somatic extracts of L3 or ES products from adults. Only the adult ES fraction conferred more than 80 % protection.

Acetylcholinesterases (AChE) are found mainly in adult stages of many free-living and parasitic nematode species. Secreted AChE are assumed to have an anticoagulant role and affect glycogenesis. Most importantly, however, they seem to have an immune modulatory effect and reduce inflammation in the vicinity of the parasites. The reason why some benzimidazole resistant nematode strains contain elevated amounts of AChE is still unknown. As in many other nematodes AChE activity was also identified in somatic extracts and ES products of adult *D. viviparus*, the latter containing over 200 times more AChE activity than the first. In Western blots AChE was only recognized by serum from naturally or experimentally infected calves and not by serum from irradiated larvae vaccinated animals, suggesting that AChE is secreted only by adult worms. Guinea pigs immunized with the AChE enriched fraction of ES antigens of adult worms showed significant lower worm burdens after challenge. The AChE was successfully cloned but immunization trials with the recombinant antigen in cattle were not successful. A secretory AChE from *T. colubriformis* had already been used to vaccinate against mixed infections of *T. colubriformis*, *H. contortus* and *C. oncophora* with average worm count reductions of 31 % and up to 58 % in individual cases.

The amoeboid motility of nematode sperm is mediated by cytoskeletal filaments composed of major sperm protein (→ MSP). MSP filaments are constructed from 2 subfilament strands, which are themselves formed from a helical arrangement of subunits. Vectorial filament assembly and filament bundling is critical for sperm cell motility and reproduction of nematodes. Although MSP was investigated thoroughly in *Caenorhabditis elegans* and some parasitic nematodes such as *Ascaris suum*, *A. lumbricoides* and *O. volvulus*, it was surprising that the immunodominant antigen complex in adult *D. viviparus* comprised MSP. In Western blots serum from naturally or experimentally *D. viviparus* infected cattle recognizes MSP, whereas serum from live vaccine immunized cattle does not. Being a substantial part of the sperm cell MSP is adult male specific. Except for using the MSP as highly specific and sensitive diagnostic antigen vaccination of cattle with a purified native MSP containing antigen from adult worms conferred 80 % reduction of adult worms after challenge infection. Surviving worms were significantly smaller (40 %) than worms from control animals (Hofmann and Schmid 1997, European patent application EP 0 785 253 A1). The recombinant *D. viviparus* MSP has been cloned in different procaryotic and eucaryotic expression vectors and will soon be tested in vaccination trials.

Because of its zoonotic importance and lacking prophylactic strategies a vaccine against → *Trichinella spiralis* (Vol. 1) would be a great advantage. A preparation of whole newborn *T. spiralis* larvae killed by freezing and thawing induced a high level of protection against challenge. Muscle larval counts were reduced by 78 % compared to 40 % after immunization with ES antigens from muscle larvae. In a comparative study freeze-thaw/sonicated preparations of newborn larvae, adult worms, muscle larvae and a mixture of all three were used to immunize pigs. Eight weeks after challenge infection muscle larval counts were reduced by 82–93 % (newborn larvae), 66 % (muscle larvae) and 98 % (mixture of all fractions). ES antigen preparations from *T. spiralis* and *T. britovi* conferred significant protection in a mouse model, showing that *T. britovi* was more immunogenic with greater host-protective immunity. A 40-mer synthetic peptide was produced from a 43 kDa immunodominant glycoprotein secreted by *T. spiralis* larvae. Immunization of mice with the 40–80 peptide fraction induced an accelerated expulsion of adult worms from the gut. As this is mainly induced by T-cell mediated inflammatory events in the intestine, the 40–80 peptide was assumed to induce a protective T-cell response. This was the first time a synthetic peptide was shown to confer protective immunity in an intestinal nematode.

Recombinant Subunit Vaccines The production of purified native antigens of defined quality for a commercial vaccine would be too expensive. Promising subunit vaccine candidates must be produced as recombinant antigens. Although recombinant subunit vaccines hold great promise, they do present some potential limitations. Recombinant subunit vaccines generally seem to be less immunogenic than their conventional counterparts because they are composed of a single antigen. In contrast, conventional vaccines contain a mixture of antigens that may aid in conferring an immunity to infectious agents that is more solid than could be provided by a monovalent vaccine. This problem can probably be minimized by using a "cocktail" of recombinant antigens from the same pathogen.

Vaccination with recombinant subunit proteins or synthetic peptides is often hampered by a weak immune response due to inappropriate presentation of the antigen to the host's immune system or the loss of critical immunogenic epitopes during in vitro expression. Additionally protein immunization very often stimulates a B-cell response only where a T-cell response is necessary for protective immunity. Such problems may be circumvented by DNA vaccines (also called DNA based immunization, genetic immunization, naked DNA vaccines, etc.) which has recently emerged as an attractive alternative to conventional vaccines. Numerous studies have already shown that immunization of experimental animals with plasmid DNA encoding antigens from a wide spectrum of parasites leads to protective humoral and cell-mediated immunity. cDNAs encoding protective protein epitopes can be cloned into plasmid vectors containing strong mammalian promoters for high expression. Purified plasmid DNA containing the protective parasite antigen is administered to the host via intramuscular or subcutaneous or intracutaneous injection or needle free application by carbon dioxide pressure (Biojector®, Bioject Inc.) or with particle bombardment (Gene gun®, Powderject Inc.). The DNA is incorporated by professional antigen presenting cells and tissue cells and expressed in enough quantity to induce a potent and specific protective immune response. Expression of the protein antigens of interest directly in host cells can provide appropriate tertiary structure for the induction of conformationally specific antibodies, and also facilitates the induction of cellular immunity. Because the vaccine does not contain genetic elements responsible for replication or infectivity, the vaccine itself is safe and cannot cause the disease. DNA vaccine technology has been successfully used to protect against many different viral, bacterial, mycoplasmal, protozoal, and worm infections. Except for successful trials with a cestode (*Taenia ovis*) and trematodes (→ *Schistosoma* (Vol. 1) species) preliminary DNA vaccination trials against nematode infections have been conducted to date with *Ancylostoma caninum*. A paramyosin homologue was used to intramuscularly immunize mice and dogs against challenge infection. Preliminary results showed a consistant B- and T-cell stimulation and reduced wormcounts.

As mentioned before, one of the disadvantages of using single clones as vaccine candidates is often the narrow immunogenic spectrum that may lead to an inappropriate immune response compared to conventional vaccines. The current approach to include a large number of cDNAs encoding putative protective antigens without knowing beforehand the protective epitopes is called 'expression library immunization' and comprises the vaccination with several thousand different cDNAs at the same time. DNA vaccine technology is only beginning to exploit its possibilities in parasite vaccine development and as its advantages in terms of production costs, storage conditions, safety and efficacy compared to other vaccines are convincing, it can be expected that the number of vaccination trials against nematodes will increase and possibly will soon lead to the development of a commercial product.

Vaccination against Platyhelminthes

General Information

Requirements of Vaccines Against Multicellular Organisms The development of vaccines against multicellular parasites, such as platyhelminths, poses particular requirements. Although many vaccines have been developed and have been marketed to protect against viruses and bacteria, vaccines against parasitic Protozoa and Metazoa are limited to a few that are available solely for animal use. None has yet successfully been developed which eliminates or eases the burden that multicellular parasites inflict on people. Not only have platyhelminths a far more complex genetic composition than do viruses and bacteria, they also pass through an often complex sequence of developmental stages that home to specific sites within their host's body. Thus, the target of an anti-platyhelminth-vaccine constantly changes and moves. In coevolution with their hosts, these parasites have acquired sophisticated mechanisms of immune evasion. As a result, little or no protective immunity develops even in hosts that are frequently re- or superinfected. A vaccine against a platyhelminth ought to target a mixture of diverse antigens.

Advantages of Immunotherapy over Chemotherapy Although chemotherapy successfully reduces worm burden resulting from the most common platyhelminth infections, a vaccine would offer added benefits. Whereas a drug removes worms after they have accumulated in the host's body, a vaccine prophylactically limits or even prevents the worm burden. Drug therapy fails to prevent re-infection and necessitates frequent re-treatment, its anti-parasitic effect, however, is almost immediate, often acting within hours or days. Immunity resulting from vaccination develops slowly, requiring several months, but continues to protect the host for years, particularly when frequently boosted by exposure to the parasite. Drug resistance of platyhelminths is a common outcome of overuse or inadequate administration of a drug. No such genetic adaptation to vaccine-induced immunity has yet been demonstrated for any pathogen. To combine the immediate but short-term effect of chemotherapy with the slowly generated but long-lasting effect of immunoprophylaxis seems the most desirable strategy.

Reduction of Morbidity Versus Sterilizing Immunity In contrast to parasitic Protozoa, platyhelminths, such as schistosomes, fasciolids and taeniid cestodes, do not divide within their definitive hosts. The worm burden is accumulative and chronic. It differs sharply from the sudden acute attack by rapidly dividing protozoon parasites. Vaccines against parasitic protozoa, such as *Plasmodium* species, must produce a sterilizing immunity. Because the morbidity resulting from platyhelminths derives largely from the intensity of infection, even a partially effective vaccine may benefit the host by reducing pathology. To evaluate the protective potential of an experimental vaccine against platyhelminths, the worm burden resulting from a challenge of vaccinated animals is compared with that of challenged non-vaccinated animals. The degree of pathology can also be compared. Immunoprophylaxis would not only limit morbidity, it may also reduce transmission. If the fecundity of the challenge worms is affected by the vaccine, egg production is reduced, subsequently affecting the transmission of the parasite. A vaccine directed against platyhelminths may be effective in reducing morbidity and limiting transmission, even if it fails to eliminate all worms.

Concomitant Immunity and Acquired Resistance Infections with platyhelminths rarely induce acquired immunity. A particular form of immunity has been described as concomitant immunity. Here, a persisting primary infection confers resistance of the host against a secondary infection with the same pathogen. In the murine model of schistosomiasis, however, it has become evident that the presumed concomitant immunity is an artefact. Pathology induced by granuloma formation affects the portal vasculature and impedes the establishment of schistosomes that arrive with a secondary infection. Platyhelminths have evolved successful mechanisms of immune evasion. Schistosomes, for example, adsorb host-molecules on their surface to disguise themselves against the host's immune system. Evidence of acquired immunity against human schistosomiasis remains inconclusive. Although prevalence and intensity of infection decreases with age, children before puberty are completely unprotected.

Parasite Antigens as Vaccine Candidates

Antigen Groups **Fatty acid-binding protein** (FABP), a protein of 12–14kDa, has been described for various platyhelminths, including *Schistosoma, Fasciola,* and *Echinococcus.* Similar to its mammalian counterpart, it is involved in intracellular transport of fatty acids. By electron microscopy, it is detected in lipid droplets in the subtegumental area of male schistosomes and in vitelline droplets of the vitelline glands of female schistosomes. Vaccination of cattle with purified FABP achieves 55% protection against *Fasciola hepatica* and 30% protection against *Fasciola gigantica* (Table 1). It reduces the worm burden resulting from a *F. hepatica* challenge in vaccinated mice by more than 80%.Up to 67% reduction in worm burden can be achieved in FABP-vaccinated outbred mice and about 90% protection in vaccinated rabbits challenged with *Schistosoma mansoni.*

Sm23 and **Sj23**, the integral membrane protein of 23kDa of *S. mansoni* and *S. japonicum*, respectively, is present on the surface of all stages of schistosomes in the mammalian host. It is expressed also by the lung stage which presents a favorable target for a vaccine. Mice vaccinated with its recombinant form or with Sm23 as a multiple antigenic peptide (MAP) achieve 40–60% protection against a challenge with *S. mansoni* (Table 1). A similar degree of protection results from vaccination of sheep with recombinant Sj23 against a challenge with *S. japonicum.* Interestingly, Sm23 and Sj23 are members of a superfamily of membrane proteins of unknown function expressed by hemopoietic and/or malignant cells of mammals.

Glutathione-S-transferase (GST) is an ubiquitous enzyme that initiates detoxification of xenobiotics or endogenous toxic compounds. Several isoenzymes between 23 and 28 kDa are found in schistosomes and fasciolides. It is localized in the parenchyma and the tegument of schistosomula and adult schistosomes as well as of juvenile and adult *F. hepatica.* As much as 70% reduction in worm burden can be achieved over an extended period in vaccinated sheep and cattle challenged with *F. hepatica* (Table 1). Protection against a challenge with *S. mansoni* by vaccination with native of recombinant GST reaches levels of 30–60% in experimental rodent and of 40% in non-human primate models. In addition, it reduces the fecundity of female schistosomes as well as egg viability. As a result, GST may also affect transmission. SmGST28 of *S. mansoni* has been tested most extensively and in diverse vaccine formulation, but SjGST26 of *S. japonicum* and SbGST28 of *S. bovis* are similarly promising vaccine candidates in their respective hosts. Although no protection data are yet available for mice vaccinated with purified plasmid DNA encoding SmGST28, preliminary studies on this DNA vaccine indicate that it effectively stimulates antigen-specific humoral and cell-mediated immune responses.

Triosephosphate isomerase (TPI) is an ubiquitous glycolytic enzyme of 28kDa that has been detected in all stages of schistosomes. It appears to be localized on the surface of cercariae and young schistosomula. Monoclonal antibodies specific for TPI confer protection in the murine model and initiated research on this vaccine candidate. Vaccination with synthetic peptides of TPI in form of

Table 1. Candidate vaccine antigens that reduce worm burden resulting from various platyhelminthic infections (according to Richter)

Vaccine candidate	Molecular mass (kDa)	Species	Kind of host	% Reduction in worm burden
Fatty-acid binding protein	12	*Fasciola hepatica*	Mice	80–100
			Cattle	55
		Fasciola gigantica	Cattle	30
	14	*Schistosoma mansoni*	Mice	67
			Rabbits	89
Sm23	23	*Schistosoma mansoni*	Mice	60
Sj23	23	*Schistosoma japonicum*	Sheep	59
Glutathione-S transferase	23–26	*Fasciola hepatica*	Cattle	50–70
			Sheep	57
	28	*Schistosoma mansoni*	Mice	30–60
			Rats	40–60
			Baboons	38
	28	*Schistosoma bovis*	Goats	48
	26	*Schistosoma japonicum*	Pigs	25
			Sheep	30–60
Triose-phosphate isomerase	28	*Schistosoma mansoni*	Mice	30–60
Cathepsin L	27–29	*Fasciola hepatica*	Cattle	50–70
Oncosphere antigen 45W	45	*Taenia ovis*	Sheep	94
Oncosphere antigen EG95	95	*Echinococcus granulosus*	Sheep	96
Paramyosin	97	*Schistosoma mansoni*	Mice	30
	97	*Schistosoma japonicum*	Mice	60–80
			Sheep	48
Hemoprotein	>200	*Fasciola hepatica*	Cattle	43

multiple antigenic peptides (MAP) reduces the worm burden of a challenge infection with *S. mansoni* in mice by 30–60% (Table 1).

Cathepsin L includes two proteolytic enzymes of 27 (CatL1) and 29 (CatL2) kDa that are secreted in the gastrodermis of *F. hepatica* to aid digestion of ingested liver and blood tissue. Regurgitating worms release these enzymes into the host's blood stream. Both cathepsins may also have extracorporeal functions. CatL1 is able to cleave immunoglobulins and may, thus, prevent antibody-mediated attachment of effector cells. CatL2 cleaves fibrinogen furthering clot formation on the parasite's surface. Vaccination of cattle with CatL1 reduces the worm burden of a challenge infection with *F. hepatica* by about 50% (Table 1). A combination of CatL2 and hemoprotein (see below) achieves more than 70% protection against *F. hepatica* in cattle. In addition, the egg output of *F. hepatica* in vaccinated sheep and cattle is reduced by up to 70% and surviving eggs are less viable than those generated in non-vaccinated hosts. In schistosomes, **Cathepsin B** and **Hemoglobinase** with a molecular mass of 31 and 32 kDa, respectively, are secreted in a similar fashion by those stages of schistosomes that digest blood. Although the protective capacity of these digestive enzymes has not been studied in animals challenged by blood flukes, related enzymes appear to protect hosts against liver flukes.

The cestode-specific **45W antigen** has been isolated from oncospheres of *Taenia ovis*. Its recombinant form induces 94% protection in sheep (Table 1). 45W is the first vaccine against a platyhelminth that is registered for commercial use. However, the amount of 45W expressed by different parasite isolates varies, possibly affecting their susceptibility to anti-45W immune responses.

Studies on the oncosphere antigen (EG95) of *Echinococcus granulosus* demonstrate similarly high degrees of protection. Oncosphere antigens of taeniid cestodes are promising candidates for the vaccination of intermediate hosts, thereby limiting the transmission of these tapeworms to their definitive hosts.

Paramyosin, known to be involved in the catch mechanism of mollusks, is a major component of the thick filament of invertebrate muscles. This 97kDa protein is expressed by all stages of schistosomes, localized in membrane-bound elongate bodies of the tegument. Paramyosin does not appear to be a surface protein that is easily accessible by the immune system. However, metacestodes of *Taenia solium* and adult schistosomes secret this protein, as it is detected in culture supernatants. Electron microscopy studies also suggest that schistosomula release paramyosin from their postacetabular glands, while shedding the highly immunogenic glycocalyx of their previous stage. Paramyosin of these platyhelminths inhibits the classical complement cascade by binding to the collagen-like region of C1. As a result, it may modulate the host's immune system. Native and recombinant paramyosin confer protection against challenge infection; at least 30% reduction of worm burden has been measured in paramyosin-vaccinated mice after challenge with *S. mansoni* and as much as 80% protection has been achieved against *S. japonicum* (Table 1).

Although the **hemoglobin-like protein** found in *Fasciola* has a similar adsorption spectrum to that of hemoglobins, its sequence bears no resemblance to this ubiquitous protein. The function of hemoprotein with a molecular mass exceeding 200 kDa is unknown but may involve oxygen transport or storage. Cattle vaccinated with hemoprotein and challenged by *F. hepatica* harbor a worm burden that is reduced by about 40% (Table 1). A combination of hemoprotein and cathepsin L achieves up to 70% protection with an additional 98% decrease in egg production. Thus, a combined vaccine including hemoprotein may offer the added benefit of limiting parasite transmission.

Parasite-specific Antigens and Ubiquitous Proteins The platyhelminth antigens that are currently under investigation as vaccine candidates can be divided into those that are specific to the parasite and those that are similar to host proteins. Parasite proteins of genetically conserved nature bear the risk of stimulating auto-immune responses in the host. Of the vaccine candidates that resemble host proteins, such as GST, TPI, Sm23, cathepsin and FABP, none has yet been shown to elicit auto-immune responses of the host. Not surprisingly, the epitopes that stimulate protective immune responses are located in the non-conserved, parasite-specific regions of these proteins. If selected epitopes, e.g. in the form of synthetic peptides, are used for vaccination rather than the entire protein, the genetic variability of the host may affect the effectiveness of the vaccine in different host individuals. Parasite-specific proteins such as the hemoprotein of *Fasciola* and the 45W oncosphere protein of *Taenia* constitute unique targets. Their functions remain unidentified, but may relate to the unique requirements of these parasites in their host organisms. Paramyosin, a protein common to invertebrates, may have a different function in platyhelminths, as its location in the parasite suggests. To circumvent the genetic variability of the host organism, a cocktail of a variety of vaccine antigens or of selected epitopes seems most appropriate.

Homologous and Crossprotective Antigens Particular vaccine candidates have been detected in several different platyhelminths. GST is a protective antigen in fascioliasis as well as in schistosomiasis, its immunoreactive epitopes, however, differ. Variances in B and T cell repertoires of the respective definitive hosts may have led to this distinct immunoreactivity. Within the genus of *Schistosoma*, cross-protection by vaccination with GST may be feasible. Paramyosin secreted by schistosomes and taeniid cestodes may share a similar immunomodulating function for both groups of worms. Some of these homologous antigens even cross-protect their hosts against diverse worms. A particular protein fraction of *F. hepatica* not only protects mice and cattle against the worms that served as source for the antigen preparation, it also reduces the worm burden of mice challenged by schistosomes. This cross-protective protein fraction of liver flukes contains an FABP that is homologous to its counterpart in blood flukes. Similarly, vaccination with the recombinant FABP of *S. mansoni* reduces the worm burden of mice challenged by cercariae by 67% and completely abrogates the establishment of *F. hepatica* metacercariae in the same animal model. Thus, one antigen may potentially protect against two kinds of worms.

Vaccination Methods

An array of different vaccination methods has been evaluated for potential anti-platyhelminth vaccines. Exposing hosts to irradiated larvae consistently generates high degrees of protection. In schistosomiasis, the worm burden of challenge infection in mice vaccinated with irradiated cercariae is reduced by an unsurpassed 90% compared to non-vaccinated mice. Because it is not practical to vaccinate people or animals with such short-lived attenuated larvae, this approach is considered as a model system to identify vaccine candidates and understand the requirements for optimal vaccination routes (→ Schistosomiasis, Man/Vaccination). Native antigens purified from the parasites are usually obtained in such small quantities that they are not sufficient for more than the early phase of the development of a vaccine. Commonly, vaccine candidates are produced as recombinant proteins. Large quantities can easily be generated. Different expression systems are available and the optimal system has to be determined for each vaccine candidate. However, recombinant proteins expressed in bacteria often lack glycosylation sites that are characteristic for the native protein and may, thus, stimulate altered immune responses. Synthetic peptides that contain protective epitopes may be presented as multiple antigenic peptides (MAP). This structure consisting of multiple peptides combined with a branching lysine core is large enough to eliminate the requirement of a carrier. The protective capacity of MAPs of GST, TPI and Sm23 is currently being evaluated. The most recent development in vaccinology is the application of DNA vaccines. Instead of vaccinating with a protein, its DNA is injected into the organism that is to be immunized. The cell machinery of the vaccinated organism subsequently transcribes and expresses the antigen. Compared to protein vaccines, the advantages of DNA vaccines lie in fast and economic production and relative stability for storage. The requirement of adjuvants can be circumvented. Although protection data using DNA vaccines against platyhelminths are not yet available, specific immune responses in mice vaccinated with a plasmid encoding the oncosphere antigen of *T. ovis* (45W) or encoding GST or paramyosin of schistosomes offer some promise.

To increase the immunogenicity of an antigen, various adjuvants are available. Freund's adjuvant, an oil emulsion of heat-killed *Mycobacterium tuberculosis*, saponins and various other formulations have commonly been used in experimental animal models. Because of possibly severe local and systemic reactions, they are considered unsafe for use in people and their use in animals is now restricted. Solely aluminium hydroxides and aluminium phosphate are registered for use in people. The efficacy of self-replicating live vaccine vehicles is currently being tested; non-virulent strains of *Salmonella typhimurium*, *E. coli*, Bacillus Calmette Guérin (BCG), or vaccinia virus are transformed with gene fragments encoding the antigen in question. Other antigen vehicles in form of liposomes and proteosomes are similarly under development. Immune stimulating complexes (iscoms) are cage-like structures consisting of the saponin Quil A, cholesterol, phospholipids and antigen. Lastly, co-administration of interleukin-12 (IL-12) augments vaccine-induced immune responses to various pathogens. Besides determining the suitable antigen or cocktail of antigens, it is important to determine the optimal carrier for presenting it to the host's immune system.

Vaccination Against Protozoa

General Information

Parasite-induced diseases, both in animals and in man, represent a considerable medical and economical burden in many countries. Despite the availability of a number of effective drugs for treatment of the most important diseases, a pressing need for development of successful vaccines remains and is in fact increasing. The reasons for this are multifold. One of the most important is the increasing problem of resistance of vectors and parasites to both successfully used and newly developed drugs. This is typified, for example, by human → malaria. In animals the main problem is in factory farming, where the presence of parasitic diseases over a long, seasonal time, requires high amounts of antiparasitic drugs, resulting in the associate problems of drug residues in milk, milk derived products and meat. In addition, such chemoprophylaxis and chemotherapy is often difficult to apply over a long period of time.

Many parasites possess sophisticated immune evasion mechanisms making it – at present – difficult to conceive the development of efficient vaccines. However, the rapid development of immunology and genetic manipulations of cells will perhaps change these perspectives in the near future. Already commercially available vaccines against parasites as well as the state of development of other vaccines which might become applicable for animals or humans in the near future are described in detail under the headwords of the respective diseases.

Immunological Aspects

In general parasites induce a strong humoral and cellular → immune response but this does not ensure that the host becomes protected against the disease and also against re-infection. Parasites indeed have developed mechanisms not only to evade immune responses of the host but also to facilitate their survival in the immunocompetent host and their transmission. Therefore, immune responses in most cases do not impair parasites' development and sometimes even favour their development but in contrast often become harmful to the host. Pernicious effect can originate from an excess of activation of the immune system to an exhaustion of the immune system that becomes unable to construct the convenient protective response; also an unbalanced production of cellular and/or humoral effectors can originate that results in immunopathology.

Various evasion/subversion mechanisms have been described among parasitic protozoa to escape the host immunity. Some of them are very sophisticated such as antigenic variation, well studied in African trypanosomes and more recently in *Plasmodium* species. Antigenic variation depends on polygenic families whose members code for membrane proteins at the surface of the parasite (or the infected cells) and are continuously exchanged. Therefore the immune system is constantly confronted with new parasite surface molecules. Intracellular protozoa like *Leishmania*, *Toxoplasma* and *T. cruzi* have elaborate strategies to escape destruction by lysosome effectors: avoiding fusion of the parasitophorous vacuole with lysosomes by introduction of parasite proteins in the membrane vacuole like *Toxoplasma*; digesting the parasitophorous vacuolar membrane and invading the cytosolic compartment like *T. cruzi* surface structures which resist complement and hydrolitic lysosome enzymes like *Leishmania*. Invasive *Entamoeba histolytica* secrete proteases that degrade antibodies and some protozoans inactivate antibodies producing sophisticated papain-like proteases that split the Fab2 fragments in the phenomenon described as fabulation.

Among the various processes elaborated by parasites to escape host immunity, are the production of parasite products able to interfere with the regulation of the immune response: for instance the production of mitogens and or superantigens inducing polyclonal B and/or T lymphocyte activation well described in trypanosomiasis and in toxoplasmosis. In malaria the polyclonal T and B cell activation seems rather to be the result of a flood of soluble and insoluble antigens released during the schizonte rupture. The presence in numerous surface antigens of degenerate repeats of amino acids sequences but quite immunogenic is another sophisticated method elaborated by *Plasmodium* to escape immunity. These more or less degenerated repeats are dominant B cell epitopes, inducing a strong antibody response with large and diffuse affinity avoiding clonal selection and maturation of B cell clones secreting high affinity specific antibodies. Another strategy to avoid the induction of an efficient immune response is the generation of high levels of antigenic polymorphisms. When concurrent presentation to the immune system of different allelic forms occurs, antigenic competition as well as altered peptide ligand antagonism prevent induction of an efficient memory response. This is another evasion strategy utilized by malaria parasites. Another possibility is the production of a dominant antigen subverting the immune system. This is the case of the LACK antigen (*Leishmania* homologous for activated C kinase) which induces an early secretion of IL-4 by a particular T cell population, thus creating a microenvironment propitious for the development of a Th2 T cell response which prevents the development of a healing Th1 response Some parasites produce macrophage and/or T lymphocyte activators that interfere with the cytokine cascade of the immune response, by an excess production of inflammatory cytokines (TNF, IFNγ), creating a cytokine chock involved in different pathogenic mechanisms like in malaria.

Accomplishments

In the early years of the 80s, with the development of new technologies in biological sciences, in particular recombinant DNA technology and monoclonal antibodies, it became possible to identify individual antigens that were the target of immune (humoral or/and cellular) responses. It became equally possible to identify, clone and sequence the corresponding genes and to produce recombinant or synthetic antigens in considerable amounts to study them as possible vaccine candidates against various pathogenic parasites. The field of anti parasite vaccines then attracted a large number of new participants, particularly many brilliant and competent molecular biologists with, however, poor knowledge of parasitology and immunology. Their first contribution was to discover an additional problem for vaccines construction, represented by the extreme variability and plasticity of the parasite genome, concerning sequences coding for potential vaccine targets. At the end of the century, twenty years after the beginning of this golden age, we still do not have any effective sub-unit anti-parasite vaccine. However, the efforts and investments made by the new generation of vaccinologists associated with progress in the understanding of immune mechanisms, has considerably increased our knowledge on the molecular structure of parasites, on the genetic variability, as well as on the immune responses (protective and escape mechanisms) they develop. This will certainly open new alternative pathways for the development of anti-parasite vaccines. In the present mature status of knowledge in these areas it is reasonable to wait for the birth of recombinant and synthetic vaccines with sufficient effects to be introduced in public health practices for control of parasite infections at the beginning of the new century. In the following, some important progress that has been made in the area since the 1st edition of this book justifying this optimistic view will be summarized.

Progress in Understanding Protective Responses The efficiency of antibodies in protection against parasites has been shown to depend not only on their inhibition/ neutralization activity as observed in virus infections but also on their capacity to interact with immunocompetent cells such as , natural killer cells, monocytes/macrophages or granulocytes to induce ADCC (antibody dependent cellular cytotoxicity), ADCI (antibody dependents cellular inhibition), opsonisation/phagocytosis etc. The desired antibody response for such protective mechanisms is, therefore isotype- specific and depend on cellular effectors. New knowledge arises from the role of cytokines, like the role of IL-12 and inflammatory cytokines such as IL-18 and IFN-γ in regulation of the Th1/Th2 pathways of T helper lymphocytes and in the production of specific cytotoxic CD8 T lymphocytes. Immunologists are accumulating new information on the role of IFN-γ, IL-4, IL13 and IL-10 the regulation of the immunoglobulin switch in humans.

Progress in Genetics Considerable progress has been registered in the last decades on protozoan genome structure, particularly in *Leishmania, Trypanosoma, Toxoplasma* and *Plasmodium* parasites. On the one hand various Genome sequencing programs have now been developed, in particular by US and UK institutions with the support of TDR/WHO and private Foundations. In relation to parasites responsible for human diseases, data banks are organised like the TDR/IMMAL Malaria DNA sequence database open to access by Internet (http://www.monash.edu.au/informatics/malaria/who.html) and the *Leishmania/Trypanosoma cruzi* database. On the other hand, genetic studies of some parasitic protozoa are in progress, due to the success in transfection with navette vectors which are able to grow in *E.coli* and/or yeast as well as in protozoa. These vectors have been used for promoting gene disruption, gene mapping, gene complementation and gene replacement by homologous recombination. These genetic manipulations allow characterization of target proteins by their function and progress in the understanding of parasite virulence mechanisms as well as new rationale for the construction of attenuated parasites that will replace the empirical ones used so far.

Progress in the Preparation of Antigen Carriers and Adjuvants Until recently the only allowed adjuvant used in human vaccines was alum (antigens are adsorbed on aluminum hydroxide) which has poor effects. New adjuvants have been investigated and used in human volunteers and in animal trials. ISCOMS (immunostimulating complexes) are large spherical multilamelar structures of the Quil A adjuvant (complex lipid) associated to different immunogens. Other adjuvants in development are DETOX (cell wall skeleton of *Mycobacterium phlei* associated to lipid A and squalen), MPL (lyposome monophosphoril lipid A), mf59 9 (oil-water emmulsion), QS-21 (saponin-based adjuvant) and detoxified toxin of *Pseudomonas aeru-*

gionosa, tetanus and diphteria toxoid. Enhancement of immune responses has also been achieved using cytokines as adjuvants, in particular IL-1 and IL-12, which act by enhancing antigen presentation and/or processing via MHC pathways. In addition to adjuvants, synthetic peptides usually require carrier molecules containing "universal T cell epitopes" able to induce T helper cells in a genetically diverse population. Tetanus and diphteria toxoids have been currently used as carriers but provoked some undesirable side effects such as hypersensitive reaction. More recently synthetic carriers have been constructed such as MAPS (multiple antigen presentation) composed of 4, 8 or 16 peptide-antigen branched on a lysine core.

DNA Vaccines Finally, an important biotechnological development represented by DNA vaccines has been intensively studied recently and seems promising. This approach has been successfully used in the immunization of experimental animals against a range of infectious and parasitic diseases as well as several tumor model diseases. The technique involves insertion of the gene encoding the antigen of choice into a bacterial plasmid and intramuscular or subcutaneous injection into the host. This lead to the constant production at a low level of the antigen inducing long-lived humoral and cellular immune responses. Among the numerous advantages of this revolutionary approach over the conventional vaccines is the induction of antigen specific protective cytotoxic T-cells as well as humoral immunity. The antigenic specificity of the Cytotoxic T cell response depends in this case on antigens presentation via class II MHC molecules. These advances and the emerging voluminous literature on this subject announce a wide use of this approach for animal vaccines development. At the same time, intense experimental vaccination of non human primates and current trials with human volunteers indicate that DNA vaccines will certainly constitute the third generation of vaccines. Experiments to clarify and develop safety aspects for human use of DNA vaccine are, however, still not available for a generalized use in human trials.

Examples for Vaccination

→ Amoebiasis, → Babesiosis, Animals, → Chagas' Disease, Animals, → Chagas' Disease, Man, → Coccidiosis, Animals, → Leishmaniasis, Man, → Malaria, → Theileriosis, → Toxoplasmosis, Man, → Trypanosomiasis, Animals, → Trypanosomiasis, Man.

Vaccines

→ Arboviruses (Vol. 1), → Vaccination Against Protozoa, → Vaccination Against Platyhelminthes, → Vaccination Against Nematodes.

Validation of Approaches

→ Disease Control, Planning.

Varroatosis

Disease due to infection of honey bees with the mite → *Varroa jacobsoni* (Vol. 1) which sucks at larvae and workers. Symptoms: Occurrence of small, degenerated workers, the number of which becomes constantly reduced.

Therapy

Treatment with Amitraz (Miticur®), Coumafos (Perizin®, Ceteafix®) or similar compounds.

Vasculitis

→ Cardiovascular System Diseases, Animals.

Vector Control

→ Disease Control, Methods, → Insecticides.

Vectorial Capacity

→ Mathematical Models of Vector-Borne Diseases.

Velvet Disease

Fish disease due to infection with the protozoon *Oodinium ocellatum*.

Vermes

Lat. worms, see → Helminth (Vol. 1), → Helminthic Infections, Pathologic Reactions

Verruga peruana

Cutaneous disease due to infection with *Bartonella bacilliformis*-bacteria transmitted by sand flies (*Lutzomyia colombiana, L. verrucarum*). The characteristic papulae appear on the skin of extremities and face several months after survival of the initial → Oroya fever.

Therapy
Tetracyclines, Macrolids

Vesiculation

Clinical and pathological symptoms of infections with skin parasites (→ Skin Diseases, Animals, → Ectoparasite (Vol. 1)).

Visceral Larva Migrans, Man

Synonym
VLM, Toxocariasis

Pathology
→ Toxocara (Vol. 1) *canis* (from dogs) and other larval parasites such as *Baylisacaris procyonis* (from racoons) or → Toxocara (Vol. 1) *cati* (from cats) can cause the visceral larva migrans syndrome (VLM) of humans. Since they cannot complete their development in man they are usually not found as adults in the intestine. Instead the larvae undergo a prolonged migration through various tissues of the human host (→ Pathology/ Fig. 28C) with less tropism towards the lung than in the normal host. Most of the nonspecific symptoms of VLM like fever, cough and abdominal pain or signs such as hepatosplenomegaly, lymphadenopathy, granuloma formation, and eosinophilia can be attributed to the migrating larvae and the host response to them.

The larvae ultimately die in various organs, and each one gives rise to a granuloma, with inflammation containing lymphocyte and eosinophils. Blood eosinophilia is common. Heavy infection such as in children with a craving for pica, the eating of dirt, may lead to myocarditis, encephalitis, and granulomas in liver (→ Pathology/Fig. 28) and lung, together with diffuse inflammation, and is sometimes fatal. Even in light infections, if a larva enters the eye the intense allergic inflammation may give rise to retinal necrosis with formation of granulomas and retinochoroiditis, vitritis, and iridocyclitis. The eye may become blind or be enucleated to exclude a malignant neoplasm, retinoblastoma.

Immune Responses
When laboratory mice are infected with *T. canis* eggs, the larvae disseminate throughout the body and become encapsulated within eosinophil-rich granulomas. A significant peripheral eosinophilia which peaks around day 14 post-infection persists for months thereafter. Both the formation of granulomas and the eosinophilia as well as the IgE and IgG1 antibody responses observed in experimental toxocariasis are largely $CD4^+$ T cell-dependent. Granulomatous reactions which can be found in the musculature, the liver, the kidneys and the heart, begin as accumulations of eosinophils around the worms, and within a week or two these cells are replaced by lymphocytes and macrophages. As late as 6 months postinfection the granulomas have contracted in size and the macrophages have become epitheloid, which is suggestive of a response to the secretion of soluble antigens by viable larvae.

The $CD4^+$ T cell response against *Toxocara* in mice is dominated by Th2 cells. In mice treated with antibodies against IL-5 the pulmonary infiltrates were devoid of eosinophils, while treatment with anti IFN-γ had no effects on the extent or cellular compositions of the pulmonary infiltrates. In line with this, human T cell clones with specificity for *T. canis* consistently produced the Th2 cytokines IL-4 and IL-5 in response to *Toxocara* antigens. However, other cells than $CD4^+$ T cells might contribute to the production of IL-5. In T-cell deficient mice the first wave of eosinophilia, occurring at day 11 post infection could still be detected, while only the second wave around day 21, was absent. Interestingly, it has been reported recently that a population of double negative ($CD4^-$ $CD8^-$) cells can also produce IL-5.

The functional role of eosinophils, one of the most striking features of tissue-invasive worm parasites, is still a matter of debate. Several studies suggest that eosinophils with their low affinity receptors for IgE (CD23) are highly efficient antiparasitic killer cells. The mechanism of killing is presumed to involve the attachment of the eosinophil via CD23 to the worm which had been opsonized or coated with parasite-specific IgE. The eosinophils then degranulate and exocytose their toxic proteins onto the parasite's surface. However, some findings also suggest that eosinophils may not directly kill *T. canis*. First, in contrast to human eosinophils no CD23 has been detected on the surface of mouse eosinophils. Second, although eosinophils could attach to *T. canis* larvae in vitro, the larval surface was shed and the worms were not affected by this interaction. Furthermore, in vivo the *T. canis* larvae obviously survive in the paratenic hosts despite a strong infiltration of eosinophils surrounding them in the tissues.

The question as to whether or not a paratenic host can develop resistance to *T. canis* infection was analyzed in the mouse model in 1960. There was an only partial resistance induced by a primary infection which resulted in approximately 20 % fewer worms recovered from multiply infected mice than from mice infected only once. More interestingly, the worms comprising the subsequent infective doses tended to accumulate in eosinophil-rich granulomatous reactions in the liver, a phenomenon which has been termed "liver trapping". It appeared to be antigen-specific as mice immunized with secreted products of *Toxocara* did trap larvae, while mice immunized with the soluble egg antigens derived from *S. mansoni* failed to do so. Anti-IL-5 treatment depleting eosinophils or passive transfer of immune sera from primed mice did not influence the liver trapping. In contrast, the trapping phenomenon was clearly T cell dependent because nude mice failed to trap larvae upon a second exposure. Since depletion of $CD4^+$ T cells did not completely abrogate liver trapping this host response most likely is a multifaceted reaction.

Therapy

→ Nematocidal Drugs, Man.

Visceral Leishmaniasis

→ Leishmania (Vol. 1).

VL

→ Visceral Leishmaniasis.

Vomiting

Clinical symptom in animals due to parasitic infections (→ Alimentary System Diseases, → Clinical Pathology, Animals).

Warthin-Starry Silver Impregnation

→ Microsporidiosis.

Weight Loss

Clinical symptom in animals due to parasitic infections (→ Alimentary System Diseases, → Clinical Pathology, Animals).

West Nile Fever

Virus disease transmitted by → Culicidae (Vol. 1) (→ Togaviridae (Vol. 1)).

Whipworm Disease

Synonym
→ Trichuris (Vol. 1), → Alimentary System Diseases.

White Dot Disease

→ *Ichthyophthirius multifiliis* (Vol. 1).

Winter Ostertagiosis

→ Ostertagia (Vol. 1).

Wolhynic Fever

Disease in humans due to infection with *Rickettsia quintana* transmitted by body lice.

Xerophthalmy

Drying of eye as symptom of lack of vitamin A.

Yawn

Clinical symptom in horses due to infection with → Gasterophilus (Vol. 1) larvae.

Yellow Fever

Virus disease transmitted either from person to person by bite of mosquitoes, or from primate to human by contact (→ Aedes (Vol. 1), → Arboviruses (Vol. 1), → Insects/Fig. 8B (Vol. 1)).

Z

Ziehl-Neelsen Staining

Method of demonstrating *Cryptosporidium* oocysts in fecal smears.

Zooanthroponoses

Synonym
→ Anthropozoonoses.

Zooanthroponosis

→ Anthropozoonoses.

Zoonoses

Diseases due to agents being transmitted between animals and man (→ Anthropozoonoses). However, there are many diseases affecting only animals. The major representatives of such parasitic "animalosis" are listed in Table 1 (page 666).

See also → Anthropozoonoses, → Opportunistic Agents (Vol. 1).

Zymodemes

Genomic patterns of different strains of a parasite, e.g. *Entamoeba histolytica*, Amoebiasis.

Table 1. Economically important parasites which are usually restricted to animals other than man (according to Wernsdorfer)

Parasite species	Principal hosts	Infective stage	Mode of infection	Other obligatory hosts	Disease
Protozoa					
Trypanosoma equiperdum	Equines	Trypanosome	Sexual transmission	None	Dourine
Trypanosoma evansi	Various domestic animals	Metacyclic trypanosome	Bite of horseflies	*Tabanus* spp.	Surra
Trypanosoma brucei	Bovines, equines, camels, porcines, canines	Metacyclic trypanosome	Bite of *Glossina*	*Glossina* spp.	Nagana
Trypanosoma congolense	Bovines and other domestic mammalians	Metacyclic trypanosome	Bite of *Glossina*	*Glossina* spp.	Bovine trypanosomiasis
Trichomonas gallinae	Avians	Trichomonad	Ingestion	None	Avian trichomoniasis
Tritrichomonas foetus	Cattle	Trichomonad	Sexual transmission	None	*Tritrichomonas* abortion
Histomonas meleagridis	Avians	Ameboid and flagellate forms	Ingestion	None	Blackhead enterohepatitis
Nosema bombycis	Silkworms	Spore (trans-ovarian and regular forms)	Ingestion, transovarian	None	Pébrine disease of silkworm
Nosema apis	Bees	Spore	Ingestion	None	*Nosema* disease of bees
Glugea hertwigi *Glugea mulleri*	Various Freshwater and marine fish	Spore	Ingestion	None	Microsporidiosis of fish
Babesia caballi *Babesia equi* (syn. *Theileria*)	Equines	Sporozoite	Tick bite	Various ticks (*Dermacentor, Hyalomma, Rhipicephalus* spp.)	Equine piroplasmosis
Eimeria tenella *Eimeria averculina*	Domestic poultry	Oocyst	Ingestion	None	Avian eimeriosis (coccidiosis)
Eimeria bovis *Eimeria zürnii*	Bovines	Oocyst	Ingestion	None	Bovine eimeriosis (coccidiosis)
Theileria parva	Cattle	Sporozoite	Tick bite	Ticks (*Rhipicephalus* spp.)	East Coast Fever
Trematodes					
Fasciola gigantica	Equines, bovines	Metacercaria	Ingestion	Aquatic snails	Fascioliasis gigantica
Fascioloides magna	Equines, bovines, sheep	Metacercaria	Ingestion	Aquatic snails (*Galba. Pseudosuccinea, Fossaria* spp.)	Fascioloidiasis
Nematodes					
Ascaris suis	Porcines	Egg containing 2nd-stage larva	Ingestion	None	Porcine ascariasis
Parascaris equorum	Equines	Egg containing 2nd-stage larva	Ingestion	None	Equine parascariasis
Trichuris discolor	Cattle	Embryonated egg	Ingestion	None	Bovine trichuriasis
Trichuris suis	Pigs	Embryonated egg	Ingestion	None	Porcine trichuriasis

Table 1. (continued)

Parasite species	Principal hosts	Infective stage	Mode of infection	Other obligatory hosts	Disease
Nematodes					
Trichuris ovis	Cattle, sheep	Embryonated egg	Ingestion	None	Trichuriasis of cattle and sheep
Ancylostoma caninum *Uncinaria stenocephala*	Canines, felines	Strongyliform larva	Transdermal penetration	None	Canine and feline ancylostomiasis
Strongyloides papillosus	Sheep	Filariform larva	Transdermal penetration	None	Strongyloidosis of sheep
Strongyloides ransomi	Pigs	Filariform larva	Transdermal penetration	None	Strongyloidosis of pigs
Dictyocaulus arnfieldi	Equines	3rd-stage larva	Ingestion	None	Equine lungworm disease
Trichostrongylus axei	Cattle, sheep, horses	3rd-stage larva	Ingestion	None	Stomach worm disease
Haemonchus contortus	Sheep, other ruminants	3rd-stage larva	Ingestion	None	"twisted" stomach worm disease
Metastrongylus apri	Mainly porcines	3rd-stage larva	Ingestion with earthworm	Earthworms (*Lumbricus, Eisenia* spp., etc.)	Swine lungworm disease
Strongylus equinus	Equines	Strongyliform larva	Ingestion	None	*Strongylus* – disease of equines
Protostrongylus rutescens	Sheep, goats	3rd-stage larva	Ingestion with infected snail	Aquatic snails (mainly *Helicella* spp.)	Red lungworm disease
Dirofilaria immitis	Canines, felines	3rd-stage larva	Mosquito bite	*Culex, Aedes, Anopheles* spp.	Heartworm disease of dogs and cats

Further Reading

Arboviruses and Transmission

1. Asnis DS, Conetta R, Teixeira AA, Waldman G, Sampson BA (2000) The West Nile Virus outbreak of 1999 in New York: The Flushing hospital experience. CID 30:413–418
2. Aspöck H (1970) Das synökologische Beziehungsgefüge von Arboviren und seine Beeinflußbarkeit durch den Menschen. Zbl Bakt I Orig 213:434–454
3. Aspöck H (1996) Stechmücken als Virusüberträger in Mitteleuropa. Nova Acta Leopoldina NF 71:37–55
4. Calisher C, Francy DB, Smith GC et al (1986) Distribution of Bunyamwera serogroup viruses in North America (1956–1984) Am J Trop Med Hyg 35:429–443
5. Fields BN, Knipe DM, Howley PM (eds) (1996) Fields Virology, 3rd ed vol I: pp 1504, vol II: pp 1445
6. Heinz FX, Collett MS, Purcell RH, Gould EA, Howard CR, Houghton M, Moormann RJM, Rice CM, Thiel H-J(1999) Flaviviridae. In: Morphy FA, Fauquet CM, Bishop DHL, Ghabrial SA, Jarvis AW, Martelli GP, Mayo MA & Summers MD (eds) Virus Taxonomy. Seventh Report of the International Committee on Taxonomy of Viruses. In press
7. Karabatsos N (ed) (1985) International catalogue of arboviruses. 3rd ed. San Antonio. Amer Soc Trop Med Hyg, pp 1147
8. Kunz, CH (1986) Die Frühsommer-Meningoenzephalitis. In: Gsell O, Krech U, Mohr W (eds) Klinische Virologie. Urban & Schwarzenberg München, Wien, Baltimore, pp 275–285
9. Labuda M, Jones LD, Williams I, Danielova V, Nuttall, PA (1993) Efficient transmission of tick-borne encephalitis virus between cofeeding ticks. J Med Entomol 30:295–299
10. Málková D, Danielová V, Holubová J, Marhoul Z (1986) Less known Arboviruses of Central Europe. A new Arbovirus. Lednice, Prague: Academia Publishing House of the Czechoslovak Academy of Sciences, pp 75
11. Murphy FA, Fauquet CM, Bishop DHL, Ghabrial SA, Jarvis AW, Martelli GP, Mayo MA, Summers MD (eds.) (1995) Virus Taxonomy. Classification and Nomenclature of Viruses. Wien Springer, pp 586

Behavioural Aspects of Parasitism

1. Bakker TCM, Mazzi D, Zala S (1997) Parasite-induced changes in behaviour and colour make *Gammarus pulex* more prone to fish predation. Ecology 78:1098–1104
2. Barnard C J, Behnke JM (1990) Parasitism and behaviour. Taylor and Francis, London, pp 332
3. Clark TG, Dickerson HW (1997) Antibody-mediated effects on parasite behaviour: evidence of a novel mechanism of immunity against a parasitic protist. Parasitol Today 13:477–480
4. Dunlap KD, Schall JJ (1995) Hormonal alterations and reproductive inhibition in male fence lizards (*Sceloporus occidentalis*) infected with the malarial parasite *Plasmodium mexicanum*. Physiol Zool 68:608–621
5. Hamilton WD, Zuk M (1982) Heritable true fitness and bright birds: a role for parasites? Science 218:384–387
6. Helluy S, Holmes JC (1990) Serotonin, octopamine and the clinging behaviour induced by the parasite *Polymorphus paradoxus* (Acanthocephala) in *Gammarus lacustris* (Crustacea). Can J Zool 68:1214–1220
7. Kavaliers M, Colwell DD (1995) Discrimination of female mice between the odours of parasitised and non-parasitised males. Proc R Soc Lond Ser B, Biol Sciences 261:31–35
8. Milinski M, Bakker TCM (1990) Female sticklebacks use male coloration in mate choice and hence avoid parasitised males. Nature 344:330–333
9. Poulin R (1995) "Adaptive" changes in the behaviour of parasitised animals: a critical review. Int J Parasitol 25:1371–1383
10. Rosenqvist G, Johansson K (1995) Male avoidance of parasitised females explained by direct benefits in a pipefish. Anim Behav 49:1039–1045

Control Measurements

1. Adams HR (ed) (1995) Veterinary pharmacology and therapeutics (7th edn). Iowa State University Press Ames, USA
2. Andrews P, Harder A (1998) Antihelmintics, toxicology. In: Ullmann's encyclopedia of industrial chemistry (6th edn). electronic release; Wiley-VCH; Weinheim
3. Arrioja-Dechert A (ed) (1997) Compendium of veterinary products 1997–1998 (4th edn). Publisher Bayley AJ, North American compendiums, ltd. distributed by North American compendiums, Inc. Port Huron, MI 48060, USA/Canada
4. Campbell WC, Rew RS (eds) (1986) Chemotherapy of parasitic diseases. Plenum, New York

5. Collier L, Balows A, Sussmann M (eds) (1998) Topley and Wilson's microbiology and microbial infection (9th edn). Vol 5; Parasitology, Cox FEG, Kreier JP, Wakelin D (vol eds). Arnold, London
6. Condor GA, Campbell WC (1995) Chemotherapy of nematode infections of veterinary importance, with special reference to drug resistance. Adv Parasitol 35:1–84
7. Cioli D, Pica-Mattocia L, Archer S (1995) Antischistosomal drugs: past, present and future? Pharmac Ther 68:35–85
8. Croft SL (1997) The current status of antiparasitic chemotherapy. Parasitology 114:S3–S15
9. Goodman-Gilman A, Goodman LS, Rall THW, Murad F (eds) (1985) Godman, Gilman's: The pharmacological basis of therapeutics (7th edn). Macmillan Publishing Company, New York
10. Long PL (ed) (1990) Coccodiosis of man and domestic animals. CRC Press, Boca Raton, Ann Arbor, Boston
11. Mehlhorn H, Düwel D, Raether W (1993) Diagnose und Therapie der Parasitosen von Haus-, Nutz- und Heimtieren (2. Auflage), Gustav Fischer, Stuttgart, Jena, New York
12. Peters W (1987) Chemotherapy and drug resistance in malaria. Vol. 1 Academic Press ltd, London
13. Schroer Ch, Hempel L (eds) (1996) ABDATA Pharma-Daten-Service (ed office): List of pharmaceutical substances (10th edn). Werbe/Vertriebsgesellschaft Deutscher Apotheker mbH, Eschborn/Taunus, Germany
14. Soulsby EJL (1982) Helminths, arthropods and protozoa of domesticated animals (7th edn). Bailliere Tindall, London
15. Stephen LE (ed) (1986) Trypanosomiasis, a veterinary perspective, Pergamon, Oxford
16. Urquhart GM, Armour J, Duncan JL, Dunn AM, Jennings FW (1987) Veterinary parasitology. Longman Scientific Churchill Livingstone Inc, New York
17. WHO (1986) Epidemiology and control of African trypanosomiasis. Technical Report Series No 739, World Health Organization, Geneva
18. WHO (1991) Control of Chagas disease. Technical Report Series 811, World Health Organization, Geneva
19. WHO (1993) A Global strategy for malaria. Order No 1150405, World Health Organization, Geneva
20. WHO (1995) WHO Model prescribing information: Drugs used in parasitic disease (2nd edn). World Health Organization, Geneva
21. WHO (1996) Manual on visceral leishmaniasis control. WHO/LEISH/96,40, World Health Organization, Geneva

Diseases: Animals

1. Abbott EM, Parkins JJ, Holmes PH (1986) Influence of dietary protein on the pathophysiology of acute ovine haemonchosis. Vet Parasitol 20:291–306
2. Berry CI, Dargie JD (1978) Pathophysiology of ovine fascioliasis. The influence of dietary protein and iron on the erythrokinetics of sheep experimentally infected with *Fasciola hepatica*. Vet Parasitol 4:327–339
3. Burridge MJ (1985) Heartwater invades the Caribbean. Parasitol Today 1:175–179
4. Cawdery MJH, Strickland KL, Conway A, Crowe PJ (1977) Production effects of liver fluke in cattle. I. The effects on infection on liveweight gain, feed intake and food conversion efficiency in beef cattle. Br Vet J 133:145–159
5. Dunn AM (1978) Veterinary helminthology. Heinemann, London
6. Hale OM, Marti OG (1984) Influence of an experimental infection of *Strongyloides ransomi* on performance of pigs. J Anim Sci 58:1231–1235
7. Knight DH (1983) Heartworm disease. In: Ettinger SJ (ed) Textbook of veterinary internal medicine. Diseases of the dog and cat, 2nd edn. Saunders, Philadelphia, pp 1097–1124
8. Nesbitt GH, Schmitz JA (1978) Fleabite allergic dermatitis: A review and survey of 330 cases. J Am Vet Med Assoc 173:282–288
9. Ogbourne CP, Duncan JL (1985) *Strongylus vulgaris* in the horse: Its biology and veterinary importance. Commonwealth Agricultural Bureau, Farnham (CPI miscellaneous publication no 9)
10. Uilenberg G (1983) Heartwater (*Cowdria ruminantium* infection): Current status. Adv Vet Sci Comp Med 27:427–480
11. Vercruysse J, Fransen J, Southgate VR, Rollinson D (1985) Pathology of *Schistosoma curassoni* infection in sheep. Parasitology 91:291–306
12. Zaman V (1994) *Balantidium coli*. In: Kreier JP (ed) Parasitic protozoa, vol. II. Academic, New York, pp 633–653
13. Zumpt F (1965) Myiasis in man and animals in the old world. Butterworths, London, p 267

Ecology and Population

1. Anderson RM, May RM (1970) Regulation and stability of host-parasite population interactions. I. Regulatory processes. Journal of Animal Ecology 47:219–247
2. Combes C (1995) Interactions Durables. Ecologie et Evolution du Parasitisme. Masson, Paris (F)
3. Combes C (1991) Ethological aspects of parasite transmission. American Naturalist 138:866–880
4. Combes C (1991) Where do human schistosomes come from? Trends in Ecology and Evolution 5:334–337
5. Combes C (1997) Fitness of Parasites. Pathology and Selection. International Journal for Parasitology 27:1–10
6. Dawkins R (1982) The Extended Phenotype. Oxford University Press, Oxford (UK)
7. Ebert D, Herre EA (1996) The evolution of parasitic diseases. Parasitology Today 12:96–101
8. Ewald PW (1995) The evolution of virulence: a unifying link between parasitology and ecology. Journal of Parasitology 81:659–669
9. Hamilton WD, Zuk M (1982) Heritable true fitness and bright birds: a role for parasites? Science 218:384–386

10. May RM, Anderson RM (1978) Regulation and stability of host-parasite population interactions. II. Destabilizing processes. J Anim Ecol 47:249–267
11. Poulin R (1995) Phylogeny, ecology, and the richness of parasite communities in vertebrates. Ecological Monographs 65:283–302
12. Poulin R (1998) Evolutionary ecology of parasites. Chapman & Hall, London
13. Poulin R (1999) The functional importance of parasites in animal communities: many roles at many levels? Int J Parasitol 29:903–914
14. Poulin R, Combes C (1999) The concept of virulence: interpretations and implications. Parasitol Today 15:474–475
15. Thompson JN (1994) The coevolutionary process. University of Chicago Press, Chicago
16. Toft CA, Aeschlimann A, Bolis L (1991) Parasite-Host Associations: Coexistence or Conflict. Oxford Science Publications, Oxford

Hormones

1. Beckage NE (1993) Receptor 3:233–245
2. Dhadialla TS, Carlson GR, Lee DP (1998) Annu Rev Entomol 43:545–569
3. De Jong-Brink M (1995) Adv Parasitol 35:177–256
4. Ramasamy R (1998) Biochim Biophys Acta 1406:10–27
5. Spindler K-D (1988) In: Mehlhorn H (ed) Parasitology in Focus. Facts and Trends. 1st ed, Springer, Heidelberg, pp 465–476
6. Spindler K-D (1997) Vergleichende Endokrinologie. Regulation und Mechanismen. Thieme, Stuttgart
7. Spindler K-D, Spindler-Barth M (in press) In: Dorn A (ed) Progress in Developmental Endocrinology, chapter: Nematodes
8. Zuk M, McKean KA (1996) Int J Parasitol 26:1009–1023

Host Finding

1. Ashton FT, Schad GA (1996) Amphids in *Strongyloides stercoralis* and other parasitic nematodes. Parasitol Today 12:187–194
2. Clements AN (1992) The biology of mosquitoes. Vol. I. Development, nutrition and reproduction. Chapman and Hall, London
3. Clements AN (1999) The biology of mosquitoes. Vol. II. Sensory reception and behaviour. CABI Publishing
4. Combes, C, Fournier A, Moné H, Théron A (1994) Behaviours in trematode cercariae that enhance parasite transmission: patterns and processes. Parasitology 109:S3–S13
5. Haas W, Haberl B (1997) Host recognition by trematode miracidia and cercariae. In: Advances in trematode biology. Fried B, Graczyk TK (eds) CRC Press, Boca Raton, pp 197–227
6. Haas W, Haberl B, Schmalfuss G, Khayyal MT (1994) *Schistosoma haematobium* cercarial host-finding and host-reconition differs from that of *S. mansoni*. J Parasitol 80:345–353
7. Haberl B, Kalbe M, Fuchs H, Ströbel M, Schmaluss G, Haas W (1995) *Schistosoma mansoni* and *S. haematobium*: miracidial host-finding behavior is stimulated by macromolecules. Internat J Parasitol 25:551–560
8. Hawdon JM, Schad GA (1991) Albumin and a dialyzable serum factor stimulate feeding in vitro by third-stage larvae of the canine hookworm *Ancylostoma caninum*. J Parasitol 77:587–591
9. Knols BGJ, van Loon JJA, Cork A, Robinson RD, Adam W, Meijerink J, De Jong R, Takken W (1997) Behavioral and electrophysiological responses of the female malaria mosquito *Anopheles gambiae* (Diptera: Culicidae) to Limburger cheese volatiles. Bull Entomol Res 87:151–159
10. Osterkamp J, Wahl U, Schmalfuss G, Haas W (1999) Host-odour recognition in two tick species is coded in a blend of vertebrate volatiles. J Comp Physiol A 185:59–67
11. Sengupta P (1997) Cellular and molecular analyses of olfactory behavior in *C. elegans*. Cell Developm Biol 8:153–161
12. Sonenshine DE (1991 and 1993) Biology of Ticks, Vols I and II. Oxford Univ Press, New York
13. Walladde SM, Rice MJ (1982) The sensory basis of tick feeding behaviour. In: Obenchain FD, Galun R (eds) Physiology of ticks. Pergamon Press, Oxford, pp 71–118
14. Walladde SM, Young AS, Morzaria SP (1996) Artificial feeding of ixodid ticks. Parasitol Today 12:272–278
15. Willemse LPM, Takken W (1994) Odor-induced host location in tsetse flies (Diptera: Glossinidae). J Med Entomol 31:775–794

Immunology

1. Alexander J, Hunter CA (1998) Chem Immunol 70:81–102
2. Babu S et al (1998) J Immunol 161:1428–1432
3. Beklaid Y et al (1998) J Exp Med 188:1941–1953
4. Bogdan C et al (1993) Immunobiol 189:356–396
5. Bogdan C, et al (1996) Curr Opinion Immunol 8:517–525
6. Bogdan C, Röllinghoff M (1999) Parasitol Today 15:22–28
7. Brown D, Reiner SL (1999) Infect Immun 67:266–270
8. Campbell D, Chadee K (1997) Parasitol Today 13:184–190
9. Campbell D, Chadee K (1997) J Infect Dis 175:1176–1183
10. Campbell JDM, Spooner RL (1999) Parasitol Today 15:10–16
11. Cheever AW, Yap GS (1998)) Chem Immunol 66:159–176
12. Chumpitazi BF et al (1998) Clin Exp Immunol 111:325–333
13. Coyle AJ et al (1998) Eur J Immunol 28:2640–2647
14. Croese J (1998) Parasitol Today 14:70–72
15. Däubener W, Hadding U (1997) Med Microbiol Immunol 185:195–206
16. Denkers EY et al (1997) J Immunol 159:1903–1908
17. Dominguez M, Torano A (1999) J Exp Med 189:25–35
18. DosReis GA (1997) Parasitol Today 13:335–342
19. Faubert GM (1996) Parasitol Today 12:140–145

20. Finkelman FD et al (1997) Annu Rev Immunol 15:505–533
21. Fried M et al (1998) Nature 395:851–852
22. Gottstein B, Hemphill A (1997) Chem Immunol 66:177–208
23. Hörauf A, Fleischer B (1997) Med Microbiol Immunol 185:207–215
24. Huber M et al (1998) Infect Immun 66:3968–3970
25. Huston CD, Petri WA Jr (1998) Eur J Clin Microbiol Infect Dis 17:601–614
26. Janeway CA, Travers P (1996) Immunobiology. The immune system in health and disease. 2nd ed. Churchill Livingstone, Edinburgh London New York
27. Jones D et al (1998) Infect Immun 66:3818–3824
28. Kayes SG (1997) Chem Immunol 66:99–124
29. Kierzenbaum F (1995) Parasitol Today 11:6–7
30. Kopacz J, Kumar N (1999) Infect Immun 67:57–63
31. Lawrence CE et al (1998) Eur J Immunol 28:2672–2684
32. Lohoff M et al (1998) Int Arch Allergy Immunol 115:191–202
33. Lotter H et al (1997) J Exp Med 185:1793–1801
34. Louis J et al (1998) Curr Opinion Immunol 10:459–464
35. Luder CG et al (1998) Clin Exp Immunol 112:308–316
36. Maizels RM et al (1995) Parasitol Today 50:56
37. Mblow ML et al (1998) J. Immunol 161:5571–5577
38. McDonald V, Bancroft GJ (1998) Chem Immunol 70:103–123
39. Menon JN, Bretscher PA (1998) Eur J Immunol 28:4020–4028
40. Morrison WI, McKeever DJ (1998) Chem Immunol 70:163–185
41. Omata Y et al (1997) Vet Parasitol 73:1–11
42. Pastrana DV et al (1998) Infect Immun 66:5955–5963
43. Pearce EJ, Pedras-Vasconcelos J (1997) Behring Inst. Mitt. 99:79–84
44. Piedrafita D et al (1999) Eur J Immunol 29:235–244
45. Pritchard DI (1995) Parasitol Today 11:255–259
46. Ramasamy R (1998) Biochim et Biophys Acta 1406:10–27
47. Reed SG (1998)) Chem Immunol 70:124–143
48. Roitt I et al (1996) Immunology. 4th ed. Mosby, London
49. Sayles PC, Johnson LL (1996) Nat Immun 15:249–258
50. Scharton-Kersten T et al (1998) J Immunol 160:2565–2569
51. Scharton-Kersten T et al (1997) J Exp Med 185:1261–1273
52. Seydel KB et al (1997) Infect Immun 65:3951–3953
53. Smith NC, Fell A, Good MF (1998) Chem Immunol 70:144–162
54. Sousa CR et al (1997) J Exp Med 186:1819–1829
55. Stager S, Muller N (1997) Infect Immun 65:3944–3946
56. Stamm LM et al (1998) J Immunol 161:6180–6188
57. Von Stebut E et al (1998) J Exp Med 188:1547–1552
58. Sternberg JM (1998)) Chem Immunol 70:186–199
59. Takamoto M et al (1998) Immunol 95:97–104
60. Tarleton RL (1995) Parasitol Today 11:7–9
61. Taylor-Robinson AW (1998) Internat J Parasitol 28:135–148
62. Taylor-Robinson AW, Looker M (1998) Lancet 351:1630
63. Tran VQ et al (1998) J Infect Dis 177:508–511
64. Venkatesan P et al (1996) Infect Immun 64:4525–4533
65. Venkatesan P et al (1997) Parasite Immunol 19:137–143
66. Walderich B et al (1997) Parasite Immunol 19:265–271
67. White AC Jr et al (1997) Chem Immunol 66:209–230
68. Yanez DM et al (1999) Infect Immun 67:446–448

Insectizides

1. Aiello SE (1999) The Merck Veterinary Manual. 8th edition. Publisher Merck & Co, Inc
2. Arrioja-Dechert A (1997–1998) Compendium of the Veterinary Products, 4th edition. A.J. Bayley (publisher), North American compendiums Inc, Port Huron, MI
3. Bloomquist JR (1993) Comp Biochem Physiol 106:301–314
4. Boeckh J, Breer H, Geier M, Hoever FB, Krüger BW, Nentwig G, Sass H (1996) Pestic Sci 48:359–373
5. Casida JE, Quistad GB (1998) Annu Rev Entomol 43:1–16
6. Clark JM (1999) Molecular action of insecticides on ion channels. ACS Symposium Series 591, San Diego
7. Devonshire AL, Field LM (1991) Annu Rev Entomol 36:1–23
8. Ishaaya I, Degheele D (1998) Insecticides with novel modes of action. Mechanisms and application. Springer, Berlin, Heidelberg, pp. 1–6, 10–11, 15, 17, 50–71, 92–106, 152–167
9. Kagabu S (1977) Rev Toxicol 1:1–14
10. Londershausen M (1996) Pestic Sci 269–292
11. Narahashi T (1996) Pharmacol Toxicol 78:1–14
12. Nolan J, Schnitzerling HJ (1986) In: Chemotherapy of Parasitic Diseases (eds. Campbell WC, Rew RS) Plenum Press, New York, pp. 603–620
13. Price NR (1991) Comp Biochem Physiol 100:319–326
14. Tomlin CDS (1977) The pesticide manual (11th edition). British crop protection council, Bracknell, UK
15. Wesley T (1998) Animal health in 2005 and beyond. PJB Publications Ltd, Animal Pharm Reports
16. Willadsen P (1997) Vet Parasitol 71:209–222

Mathematical Models of Vector Borne Diseases

1. Desowitz RS (1991) Malaria capers: tales of parasites and people. WW Norton & Co, New York
2. Dietz K, Schenzle D (1985) Mathematical models for infectious disease statistics. In: Atkinson AC, Fienberg SE (eds) A celebration of statistics. Springer Verlag, New York
3. Anderson RM, May RM (1992) Infectious diseases of humans: dynamics and control. Oxford University Press, Oxford
4. Anderson RM (1982) Population dynamics of infectious diseases: theory and applications. Chapman and Hall, London
5. Scott ME, Smith G (1994) Parasitic infectious diseases: epidemiology and ecology. Academic Press, San Diego
6. MacDonald G (1957) The epidemiology and control of malaria. Oxford University Press, London
7. Gilles HM, Warrell DA (1993) Bruce-Chwatt's essential malariology (3rd ed). Edward Arnold, London

8. Freeman J, Laserson KF, Petralanda I, Spielman A (1999) Effect of chemotherapy on malaria transmission among Yanomami Amerindians: simulated consequences of placebo treatment. Am J Trop Med Hyg (in press)
9. Struchiner CJ, Halloran ME, Spielman A (1989) Modeling malaria vaccines I: new uses for old ideas. Mathematical Biosciences 94:87–113

Metabolism

1. Baker JR, Muller R, Rollinson D (1997) Advances in Parasitology, vol 39. Academic Press, London New York, pp 141–226
2. Boothroyd JC, Komuniecki R (1995) Molecular Approaches to Parasitology, Wiley-Liss, New York
3. Bryant C, Behm C (1989) Biochemical Adaptations in Parasites. Chapman and Hall, London New York
4. Coombs GH, North M (1991) Biochemical Protozoology. Taylor & Francis, London, Washington
5. Coombs HG, Croft SL (1997) Molecular Basis of Drug Design and Resistance. Cambridge University Press
6. Francis SE, Sullivan Jr DJ, Goldberg DE (1997) Ann Rev Microbiol 51:97–123
7. Kulda J, Nohynkova E (1995) In: Kreier JP (ed) Parasitic Protozoa, vol 10. Academic Press, New York, pp 225–422
8. Marr J, Müller M (1995) Biochemistry and Molecular Biology of Parasites. Academic Press, New York
9. Shapiro TA, Englund PT (1995) Ann Rev Microbiol 49:117–143
10. Vanhamme L, Pays E (1995) Microbiol Rev 59:223–240

Mode of Action of Drugs

1. Arena JP, Liu KL, Paress PS, Easter G, Frazier G, Cully DF, Mrozik H, Schaeffer JM (1995) The mechanism of action of avermectins in *Caenorhabditis elegans:* correlation between activation of glutamate-sensitive chloride current, membrane binding, and biological activity. J Parasitol 81:286–294
2. Boray JC (1994) Chemotherapy of infections with fasciolidae. In: Boray JC (ed) Immunology, pathobiology and control of fasciolosis. Round Table Conference, ICOPA VIII, Izmir, MSD AGVET, pp 83–97
3. Borst P, Ouelette M (1995) New mechanisms of drug resistance in parasitic protozoa. Ann Rev Microbiol 49:427–460
4. Cioli D, Pica-Mattoccia L, Archer S (1995) Antischistosomal drugs: past, present and future? Pharmac Ther 68:35–85
5. Cleland TA (1996) Inhibitory glutamate receptor channels. Molec Neurobiology 13:97–136
6. Cox FEG (1996) Modern Parasitology, a textbook of parasitology, second edition, Blackwell Science
7. Eckert J, Kutzer E, Rommel M, Bürger HJ, Körting W (1992) Veterinärmedizinische Parasitologie, 4. Auflage, Verlag Paul Parey
8. Gräfe U (1992) Biochemie der Antibiotika, Struktur – Biosynthese – Wirkmechanismus. Spektrum Akademischer Verlag, Heidelberg, Berlin, New York
9. Haberkorn A (1993) Protozoenmittel, Hager's Handbuch der Pharmazeutischen Praxis, 5. Auflage, Sachgebiet: Stoffe/Medizinischer Teil, Springer-Verlag Berlin, Heidelberg, New York, London, Paris, Tokyo, Hong Kong
10. Haberkorn A (1996) Chemotherapy of human and animal coccidioses: state and perspectives. Parasitol Res 82:193–199
11. Krauth-Siegel RL, Schöneck R (1995) Trypanothion reductase and lipoamide dehydrogenase as targets for a structure-based drug design. FASEB J 9:1138–1146
12. Krogstad DJ, Schlesinger PH, Gluzman IY (1992) The specificity of chloroquine. Parasitol Today 8:183–184
13. Martin RJ, Robertson AP, Bjorn H (1997) Target sites of anthelmintics. Parasitol 114:S111–S124
14. Mehlhorn H, Eichenlaub D, Löscher T, Peters W (1995) Diagnostik und Therapie der Parasitosen des Menschen, 2. Auflage. Gustav Fischer Verlag, Stuttgart, Jena, New York
15. Roos MH (1990) The molecular nature of benzimidazole resistance in helminths. Parasitol Today 6:125–127
16. Wang CC (1995) Molecular mechanisms and therapeutic approaches to the treatment of african trypanosomiasis. Ann Rev Pharmacol Toxicol 35:93–127
17. Wang CC (1997) Validating targets for antiparasitic chemotherapy. Parasitol 114:S31–S44
18. Warhurst DC (1995) Haemozoin and the mode of action of blood schizontocides : more controversy. Parasitol Today 11:204–205
19. Wellems TE (1991) Molecular genetics of drug resistance in *Plasmodium falciparum* malaria. Parasitol Today 7:110–112
20. Zahner H, Schares G (1993) Experimental chemotherapy of filariasis: comparative evaluation of the efficacy of filaricidal compounds in *Mastomys coucha* infected with *Litomosoides carinii, Acanthocheilonema viteae, Brugia malayi* and *B. pahangi*. Acta Trop 52:221–266

Morphology and Reproduction

1. Anderson RC (2000) Nematode parasites of vertebrates. CABI Publ., Watwaterford
2. Boch J, Supperer R (2000) Veterinärmedizinische Parasitologie. 5th ed, Parey, Berlin
3. Canning EU, Lom J (1986) The microsporidia of vertebrates. Academic Press, London
4. Chen TC (1986) General Parasitology. Blackwell, Oxford
5. Cox FEB (1993) Modern Parasitology
6. Crompton DTW (1985) Reproduction. In: Crompton DTW, Nickols BB (eds) Biology of the Acanthocephala. Cambridge University Press, pp 213–272
7. Dettner K, Peters W (1999) Lehrbuch der Entomologie. G. Fischer, Stuttgart
8. Ehlers U (1989) Das phylogenetische System der Plathelminthes. G. Fischer, Stuttgart
9. Eldridge BF, Edman JD (2000) An update on medical entomology. Kluwer Academ Publ., Amsterdam
10. Garcia LS, Bruckner DAC (1988) Diagnostic Medical Parasitology. Elsevier, New York

11. Grell KG (1973) Protozoology. Springer, Heidelberg
12. Hausmann K, Hülsmann N (1996) Protozoology. Thieme, Stuttgart
13. Kakoma J, Mehlhorn H (1994) *Babesia* of domestic animals. In: Kreier JP (ed) Parasitic Protozoa. Academic Press, San Diego
14. Khalil LF, Iones A, Bray RA (1990) Keys to the cestode parasites of vertebrates. CAB International, Wallingford
15. Lane RP, Crosskey RW (1993) Medical insects and arachnids. Chapman & Hall, London
16. Leak SGA (1998) Tsetse biology and ecology: Their role in the epidemiology and control of trypanosomosis. CABI Publ., Watwaterford
17. Lucius R, Loos-Frank B (1997) Parasitologie. Spektrum, Heidelberg
18. Maggenti A (1981) General nematology. Springer, Heidelberg
19. Mehlhorn H (1998) Cellular organization of parasitic protozoa. In: Cox JP (ed) Topley and Wilson's Microbiology and microbial infections, Vol. 5. Arnold, London
20. Mehlhorn H, Düwel D, Raether W (1993) Diagnose und Therapie der Parasitosen der Haus-, Nutz- und Heimtiere. G. Fischer (Enke), Stuttgart
21. Mehlhorn H, Eichenlaub D, Löscher T, Peters W (1995) Diagnose und Therapie der Parasitosen des Menschen. Urban/Fischer, München, Stuttgart
22. Mehlhorn H, Schein E, Ahmed JA (1994) *Theileria*. In: Kreier JP (ed) Parasitic Protozoa. Academic Press, San Diego, pp 216-304
23. Riley J (1986) The Biology of Pentastomida. Adv Parasitol 25:45–128
24. Rogan MT (ed.) Analytical Parasitology. Springer, New York
25. Sonenshine DE (1991) Biology of ticks, Vol. I/II. Oxford University Press, New York
26. Taraschewski H (2000) Host-parasite relationships in the Acanthocephala. A morphological approach. Adv Parasit 40:1–73
27. Yamaguti S (1958) Systema Helminthum. Interscience Publishers, New York

Neurophysiology

1. Gustafsson MKS (1992) The neuroanatomy of parasitic flatworms. Advances in Neuroimmunology 2:267–286
2. Gustafsson MKS, Lindholm AG, Trenina NB, Reuter M (1996) NO nerves in a tapeworm. NADPH-diaphorase histochemistry in adult *Hymenolepis diminuta*. Parasitology 113:559–565
3. Halton DW, Gustafsson MKS (1996) Functional morphology of the platyhelminth nervous system. Parasitology 113:S47–S72
4. Halton DW, Maule AG, Shaw C (1997) Trematode Neurobiology. Adv Trematode Biology 11:345–382
5. Halton DW, Shaw C, Maule AG, Smart D (1994) Regulatory peptides in helminth parasites. Adv Parasitol 34:163–227
6. Reuter M, Gustafsson MKS (1995) The flatworm nervous system: Pattern and phylogeny. In: Breidbach O, Kutsch W (eds) The nervous System of Invertebrates. An evolutionary and comparative approach. Birkhauser Verlag, Basel, pp 52–59
7. Reuter M, Gustafsson MKS (1996) Neural signal substances in asexual multiplication and development in flatworms. Cell Mol Neurobiol 16:591–616
8. Reuter M, Gustafsson MKS (1999) Developmental endocrinology in platyhelminthes. In: Dorn A (ed) Reproductive Biology of Invertebrates. Wiley-Liss

Pathology

1. Ackerman SJ, Weil GJ, Gleich GJ (1982) Formation of Charcot-Leyden crystals in human basophils. J Exp Med 155:1597–1607
2. Beaver PC, Jung RC, Cupp EW (1984) Clinical Parasitology, 9th ed. Lea and Febiger, Philadelphia
3. Frenkel JK (1973) Toxoplasmosis in and around us. Bioscience 23:343–352
4. Frenkel JK (1976) Toxoplasmosis. In: Binford CH, Connor DH (eds) Pathology of tropical and extraordinary diseases: an atlas, vol. 1. Armed Forces Institute of Pathology, Washington, pp 284–300
5. Frenkel JK (1988) Pathology. In: Mehlhorn H (ed) Parasitology in Focus. Springer, New York
6. Frenkel JK, Nelson BM, Arias-Stella J (1975) Immunosuppression and toxoplasmic encephalitis: clinical and experimental aspects. Hum Pathol 6:97–111
7. Moskowitz LB, Hensley GT, Chan JC, Conley FK, Donovan Post MJ, Gonzales-Arias SM (1984) Brian biopsies in patients with aquired immune deficiency syndrome. Arch Pathol Lab Med 108:368–371
8. Von Lichtenberg F, Smith JH, Cheever AW (1966) The Hoeppli phenomenon in schistosmiasis: Comparative pathology and immunopathology. Am J Trop Med Hyg 15:886

Phylogenic Aspects of Taxonomy

1. Andrews RH, Chilton NB (1999) Int J Parasitol 29: 213–253. Blackwell Science Ltd. Oxford
2. Coombs GH, Vickerman K, Sleigh MA, Warren A (1999) Evolutionary relationships among protozoa. Kluwer Academic Publishers, Dordrecht Boston London
3. Felsenstein J (1973a) Amer J Human Gent 25: 471–492
4. Hillis DM, Moritz C (1990) Molecular Systematics. Sinauer Associates Inc. Sunderland
5. Morrison DA (1996) Int J Parasitol 26:589–617
6. Page RDM, Holmes EC (1998) Molecular Evolution: a phylogenetic approach. Blackwell Science, Cambridge
7. Philippe A, Chenuil A, Adouette A (1994) Development, Suppl:15–25
8. Schlegel M (1991) Europ J Protistol 27:207–219
9. Simpson GG (1961) Principles of animal taxonomy. Columbia University Press, New York
10. Sogin ML, Silberman JD, Hinkle G, Morrison HG (1996) In: Roberts DM, Sharp P, Alderson G, Collins MA (eds) Society for general microbiology. Sympo-

sium: Evolution of microbial life. Cambridge University Press, Cambridge, Ma, pp 167–184
11. Ridley M (1996) Evolution. Blackwell Science, Cambridge

Platyhelminthes Vaccination

1. Bergquist NR, Colley DG (1998) Schistosomiasis vaccines: research to development. Parasitol Today 14:99–104
2. Lightowlers MW (1996) Vaccination against cestode parasites. International J Parasitol 26:819–824
3. Spithill TW, Piedrafita D, Smooker PM (1997) Immunological approaches for the control of fasciolosis. International J Parasitol 27:1221–1235
4. Spithill TW, Dalton JP (1998) Progress in development of liver fluke vaccines. Parasitol Today 14:224–228
5. Wilson RA, Coulson PA (1998) Why don't we have a schistosomiasis vaccine? Parasitol Today 14:97–99

Pneumocystis carinii

1. Baughman RP, Liming JD (1998) Frontiers Biosci 3:1–12
2. Cushion MT (1998) In: Collier L, Balows A, Sussman M (eds) Topley and Wilson's Microbiology and Microbial Infections, vol 4. Oxford University Press, New York, pp 645–683
3. Kaneshiro ES, Wyder MA, Zhou LH, Ellis JE, Voelker DR, Langreth SG (1993) J Eukaryot Microbiol 40:805–815
4. Kaneshiro ES (1998) Clin Microbiol Rev 11:27–41
5. Masur H, Shelhamer J (1996) Ann Int Med 124:451453
6. Mazars E, Dei-Cas E (1998) FEMS Immunol Med Microbiol 22:75–80
7. Pareja JG, Garland R, Koziel H (1998) Chest 113:1215–1224
8. Stringer JR (1996) Clin Microbiol Rev 9:489–498
9. Walzer PD (1994) *Pneumocystis carinii* Pneumonia, 2nd edn. Marcel Dekker, New York.
10. Yoshida Y (1989) J Protozool 36:53–260

Protozoan Vaccination

1. Alarcon JB, Waine GW, McManus DP (1999) DNA vaccines: technology and application as anti-parasite and anti-microbial agents. Adv Parasitology 42:343–410
2. Beverley, SM, Turco SJ (1998) Lipophosphoglycan-LPG- and the identification of virulent genes in the protozoan parasite *Leishmania*. Trends in Microbiology 6:35–40
3. Doolan D, Hoffman SL (1995) Multi-gene vaccination against malaria: a multi-stage, multi immune response approach. Parasitol Today 13:171–178
4. Engers HD, Godal T (1998) Malaria vaccine development. Parasitol Today 14:56–60
5. Handmann E (1997) *Leishmania* vaccines: old and new. Parasitol Today 13:236–238
6. Hommel M (1997) Modulation of host cell receptors: mechanism of the survival of malaria parasites. Parsitology 115:S45–S54
7. Louis J, Himmelrich H, Parra-Lopez C, Tacchini-Cottier F, Launois P (1998) Regulation of protective immunity against *Leishmania major* in mice. Current Opinion in Immunology 10:459–464
8. Miller LH, Good MF, Kaslow DC (1998) Vaccines against the blood stages of *falciparum* malaria. Adv Exp Med Biol 452:193–205
9. Murphy KM (1998) T lymphocyte differenciation in the periphery. Current Opinion in Immunol 10:226–32
10. Nussenzweig V, Nussenzweig RS (1989) Rationale for the development of an engineered sporozoite malaria vaccine. Adv Immunol 45:283–334
11. Scott P, Trincheri G (1997) IL-12 as an adjuvant for cell-mediated immunity. Seminars in Immunology 9:285–291
12. McKeever DJ, Morrison WI (1998) Novel vaccines against *Theileria parva*: prospects for sustainability. Int J Parasitol 28:693–706
13. Vermeulen NA (1998) Progress in recombinant vaccine development against coccidiosis. A review and prospects into the next milenium. Int J Parasitol 28:1121–1130
14. Waters AP et al (1996) Transfection of malaria parasites. Parasitol Today 12:129–132

Serology and Immunodiagnostic Methods

1. Austin DJ, Anderson RM (1996) Parasitology 113:157–172
2. Bradley JE, Trenholme KR, Gillespie AJ, Guderian R, Titanji V, Hong Y, McReynolds L (1993) Am J Trop Med Hyg 48:198–204
3. Corral RS, Altcheh J, Alexander SR, Grinstein S, Freilij H, Katzin AM (1996) J Clin Microbiol 34:1957–1962
4. Deelder AM, Qian ZL, Kermsner PG, Acosta L, Rabello ALT, Enyong P, Simarro PP, Van Etten ECM, Krijger FW, Rotmans JP, Fillie YE, de Jonge N, Agnew AM, van Lieshout L (1994) Trop Geogr Med 46:233–238
5. Gottstein B, Jacquier P, Bresson-Hadni S, Eckert J (1993) J Clin Microbiol 31:373–376
6. Stanley SLJr, Jackson TF, Foster L, Singh S (1998) Am J Trop Med Hyg 58:414–416
7. Taverna J, Bradley J E (1998) In: Roitt I, Brostoff J, Male D (eds) Immunology, 5th edition. Mosby, London Philadelphia St. Louis Sydney Tokyo, pp 243–261
8. Voller A (1993) Trans Roy Soc Trop Med Hyg 87:497–498
9. Wincker P, Telleria J, Bosseno MF, Cardoso MA, Marques P, Yaksic N, Aznar C, Liegeard P, Hontebeyrie M, Noireau F, Morel CM, Breniere SF (1997) Parasitology 114:367–373

Strategy

1. Liese B (1986) The organization of schistosomiasis control programmes. Parasitol Today 2:339–345
2. Warren KS, Mahmoud AAF (eds.) (1984) Tropical and geographical medicine. McGraw-Hill, New York
3. World Health Organization (1984) Malaria control as part of primary health care. WHO, Geneva (WHO Technical Report Series no 712)

4. World Health Organization (1986a) Safe water supply and sanitation: Prerequisites for health for all. World Health Stat Q 39 (1):1–117
5. World Health Organization (1986b) Epidemiology and control of African trypanosomiasis. WHO, Geneva (WHO Technical Report Series no 739)

Test systems

1. Dye C, Vidoe E, Dereure J (1993) Epidemiol Infect 110:647–656
2. Kanmogne GD, Asonganyi T, Gibson WC (1996) Ann Trop Med Parasitol 90:475–483
3. Lally NC, Jenkins MC, Dubey JP (1996) J Parasitol 3:275–279
4. Pansaerts R, Van Meirvenne N, Magnus E, Verhelst L (1998) Acta Tropica 70:349–354
5. Petithory JC, Beddock A (1997) Bull Soc Franc Parasitol 15:199–209
6. Sabin AB, Feldman HA (1948) Science 108:660–663
7. Taverna J, Bradley JE (1998) In: Roitt I, Brostoff J, Male D (eds.) Immunology, 5th edn. Mosby London, Philadelphia St. Louis Sydney Tokyo, pp 243–261
8. Truc P, Formenty P, Duvallet G, Komoin Oka C, Diallo PB, Lauginie F (1997) Acta Tropica 67:187–196
9. Tsang VCW, Brand JA, Boyer AE (1989) J Infect Dis 159:50–59
10. Vekatesan P, Wakelin D (1993) Parasitol Today 9:228–232
11. Verhofstede CP, van Gelder S, Rabaey M (1988) Parasitol Res 74:516–520

Ticks as Vectors

1. Krause PJ, Telford SR, Spielman A, Sikand V Ryan R, Christianson D, Burke G, Brassard P, Pollack R, Peck J, Persing DH (1996) Concurrent Lyme disease and babesiosis: evidence for increased severity and duration of illness. JAMA 275:1657–1660
2. Matuschka FR, Fischer P, Heiler M, Richter D, Spielman A (1992) Capacity of European animals as reservoir hosts for the Lyme disease spirochete. J Infect Dis 165:479–83
3. Sonenshine DE (1993) Biology of Ticks, vol 2, Oxford University Press, Oxford
4. Spielman A, Wilson ML, Levine JF, Piesman J (1985) Ecology of *Ixodes dammini*-borne human babesiosis and Lyme disease. Ann Rev Entomol 30:439–460
5. Telford III SR, Armstrong PM, Katavalos P, Foppa I, Olmeda Garcia S, Wilson ML, Spielman A (1997) A new tick-borne encephalitis-like virus infecting New England deer ticks *Ixodes dammini*. Emerg Infect Dis 3:165–170
6. Telford III SR, Dawson JE, Katavolos P, Warner CK, Kolbert CP, Persing DH (1996) Perpetuation of the agent of human granulocytic ehrlichiosis in a deer tick-rodent cycle. Proc Natl Acad Sci USA 93:6209–6214

Printing and Binding: Stürtz AG, Würzburg